T5-CVT-148

Drug Dependence and Alcoholism

Volume 1
Biomedical Issues

DRUG DEPENDENCE AND ALCOHOLISM

Drug Dependence and Alcoholism

Volume 1
Biomedical Issues

Edited by
Arnold J. Schecter
Professor, Department of Preventive Medicine
Upstate Medical Center, Clinical Campus
State University of New York at Binghamton
and
Director of Health, Broome County Health Department
Binghamton, New York

Plenum Press · New York and London

WILLIAM MADISON RANDALL LIBRARY UNC AT WILMINGTON

Library of Congress Cataloging in Publication Data

National Drug Abuse Conference, 5th, Seattle, 1978.
Drug dependence and alcoholism.

Proceedings of the 1978 National Drug Abuse Conference, held in Seattle, Wash., Apr. 3–8, 1978.
Includes index.
CONTENTS: v. 1. Biomedical issues .–v. 2. Social and behavioral issues.
1. Drug abuse–Congresses. 2. Alcoholism–Congresses. I. Schecter, Arnold. II. Title.
[DNLM: 1. Alcoholism. 2. Drug dependence. WM270.3 D794]
RC564.N37 1978 362.2′9 79-24558
ISBN 0-306-40323-4 (v. 1)

These volumes were supported in part by a grant from the Sandoz Foundation designating Dr. Arnold J. Schecter as a Sandoz Fellow. Funds were also received from the National Institute on Drug Abuse to support the National Drug Abuse Conference from which these proceedings were derived.

First half of the proceedings of the 1978 National Drug Abuse Conference, held in Seattle, Washington, April, 3-8, 1978

© 1981 Plenum Press, New York
A Division of Plenum Publishing Corporation
233 Spring Street, New York, N. Y. 10013

All rights reserved

No part of this book may be reproduced, stored in a retrieval system, or transmitted, in any form or by any means, electronic, mechanical, photocopying, microfilming, recording, or otherwise, without written permission from the publisher

Printed in the United States of America

RC564
.N37
1978
v.1

NATIONAL DRUG ABUSE CONFERENCE

NATIONAL CHAIRPERSON

Arthur Simmons

NATIONAL CO-CHAIRPERSONS

Region 1
Susan Smith Resteghini

Region 2
Ron Gaetano

Region 3
Frank Espada

Region 4
Matt Gissen

Region 5
Rev. Walter Jones

Region 6
Ramon Adame

Region 7
Rev. Richard Gilmore

Region 8
George Effman

Region 9
Tommy Chung

Region 10
June Leonard

NASDAPC Co-Chairperson
William J. McCord

International Co-Chairperson
Ted Schramm

1978 SPONSORS

National League of Cities
National Federation of Concerned Drug Abuse Workers
Joint Commission on Accreditation of Hospitals
National Association of State Drug Abuse Program Coordinators
National Association of Puerto Rican Drug Abuse Programs
National Association on Drug Abuse Problems, Inc.
National Institute on Drug Abuse
Drug Abuse Council

228389

National Council on Drug Abuse
Center for Addiction Services
Chicano Alliance of Drug Abuse Programs
Washington State Department of Social and Health Services
Washington State Council on Crime and Delinquency
Therapeutic Communities of America
Mental Health Association of San Francisco
National Council on Alcoholism
National Institute on Alcohol Abuse and Alcoholism
Washington State Medical Association
Alcohol and Drug Problems Association
National Asian-Pacific Substance Abuse Network

SPECIAL ACKNOWLEDGEMENTS

James M. Shulman, program assistance
James W. Leigh, program assistance
James R. Anderson, program assistance
Bradley S. Williamson, program art work
Sandy Corkrum
Dee Johnson
Vern Parrott
Richard Singer, C.A.S.
Shirley Nelson, C.A.S.
Raymond W. Hummel, C.A.S.
Norm Johnson, C.A.S.
Lois Hansen, C.A.S.
Central Breakthrough
Larry Gregory, King County Drug Commission
Del Schooley, Olympic Hotel
Richard Wegscheider, Olympic Hotel
Julie Hale, Seattle Convention Bureau
Jerry Morrier, Printcraft
Kay Henshaw, Quadracolor
King County Drug Commission
Washington State Office of Drug Abuse Prevention
NDAC '77 Staff
Washington State Congressional Delegation
C.A.S. Staff
C.A.S. Board of Directors

Edward S. Singler, President
Geary Simmons, Vice President
Nova G. Jones, Secretary
Billie J. Caldwell, Treasurer
Dr. Albert Black
Sunny Brite, R.N.
Darrell Fregia
William Hall, Jr.
Dr. Virgie L. Harris
Edith Hilliard
Eddie Rye, Jr.
Maryanne Vandervelde
Linda B. Wohlers

NDAC NATIONAL PLANNING COMMITTEE

Carol Addis
Michael Aldrich
Harold Alksne
James Allen, Jr.
Andy Anderson
Mirijean Anderson
Richard D. Atkins
James Bailey
David J. Baird
Dave Bearman
Charles E. Becker
Howard Becker
David Bentel
Karst Besteman
Peter Bjorklund
Margaret Blasinsky
Ken Blum
Arytur Bolter
Philip C. Bourdette
Peter Bourne
Gordon Brownell
Thomas E. Bryant
Terry Buris
Albert S. Carlin
Lawrence F. Carroll
John Chappel
Tommy Chung
Ron Clark
Allan Y. Cohen
Shirley Coletti
Mary Beth Collins
Joseph Corcoran
Harold J. Cornacchia
Robert Corrado
Grace Dammann
Gerry De Angelis
David Deitch
Marlene De Rios
Risa Dickstein
Sara Dowdy-Green
Gerry Dubin
Robert DuPont
Roberta Duron
George Effman
Frank Espada
Richard Esposito
Mathea Falco
James Fausel
Beryl Feinglass
Sanford J. Feinglass
Harvey Feldman
Loretta Finnegan
Rinna Flour
Douglas Foster
Helen Foster
Herbert J. Freudenberger
Ron Gaetano*
Dan Garrett
Frances Rowe Gearing
Richard Gilmore
Matthew Gissen*
Avram Goldstein
Jack Gordon
John Griffin
Maria Elena Guillen
Seymour Halpern
William Harvey
Tom Hastings
Robbie Hayes
Carmen Helisten
Eileen Henriques
Rudy Hernandez*
Rayburn Hesse
Rae Anne Hickling
Rogers Hoffman
John E. Imhoff
Darryl Inaba
Phillip E. Jacobs
Norman Johnson
Walter Jones*
Manny Karell
Richard Katon
Edward Kaufman
Edward Khantzien
Benjamin Kissin
Ron Kletter

---

* Site Selection Committee

Irving Klompus
Gilbert Koda
Milan Korcok
Day Kuntz
John G. Langrod
Herbert Leibowitz
Frank Lemons
June Leonard*
David Lewis
Ondra Lewis
Walter Littrell
Roy Eugene Lokey
Leroy Looper
Sandra Lowe
Joyce Lowinson
Karin Mack
James F. Maddux
Ira Marion
Jill Marsden
Mary Tuma McAdams
William J. McCord
Harvey Milkman
Robert Milkman
Lonnie Mitchell
Josette Mondanaro
Bob Moore
Merlyn Moore
Bill Morris
Ronald Murphy
Frank Nelson
Marshall Newman
John Newmeyer
Stuart Nightingale
Helen Nowlis
Joe Okimoto
James Oss
Donald Ottenberg
Zelba Owens*
Abe Pasadaba*
Vernon Patch
Gary Payer
Thomas Payte
David Petersen
Bianca Podosta
Thomas Powers
Hale Pringle
Mark Quinones
Richard Rappolt
Richard Rawson
Susan Resteghini*
Gerry Richardson
Anne Salsbury
Jordan Scher
Ted Schramm*
Doris Schwarz
Edward Senay
Frank Seixas
Richard Seymour
Arnold Schecter
Vernon Short
Daniel Joseph Silva
Jane Silver
David Smith
Roger Smith
Stuart Snyder
Charles B. Starr
Arthur Stickgold
Shirley Stone
R. Keith Stroup
Alice Swanson
Jose Szapocznik
Shani Taha
Forest Tennat
Harold Trigg
Kay Tucker
Raoul Tuset
Linda Tyon
Kathleen Bell Unger
J. Thomas Ungerleider
April Vandetta
William P. Vanquez*
Joan L. Wade
Ron Wakabayashi*
D. Leong Way
David Wellisch
Oz White
Charles Winick
Bobby Yoshida
Jim L. Young
Norman E. Zinberg

CONFERENCE STAFF

Linda Howell Schodt
Conference Coordinator

P. Jean D. Williamson
Administrator

Jody Bagnariol
Staff Assistant

Ken Saunderson
Public Relations Director

Carol Strena
Administrative Assistant

Don Christensen
Program Designer

Dan Kelsch
Program Assistant

John Gallant
Press Assistant

PREFACE

The 1978 National Drug Abuse Conference held in Seattle marked the beginning of the second decade of these conferences and their predecessor National Methadone Conferences. They began as small conferences devoted to understanding the problems and promises inherent in methadone maintenance treatment of opiate-dependent patients. The first conference was held about a decade ago in New York City at the Rockefeller University. The attendees consisted of a small group of invited clinicians, administrators, and research workers. Over the years the conferences have increased in both breadth and depth of their coverage.

On a national scale this conference alone considered the issues of alcoholism, opiate dependence, polydrug abuse, and all other forms of substance abuse. The thousands attending each of the conferences came from all walks of life within our field. Lawyers, physicians, and basic and applied research scientists met and interacted with counselors, administrators, government officials, ex-addicts, controlled alcoholics, and others with serious interest in this field. Only at this conference was it possible to attend presentations concerning the newest findings of a cellular, molecular, and chemical basis on one day and participate in discussions of problems of disadvantaged minorities, women, and clinicians on the next day. It was uniquely possible to meet with government officials and question them publicly, as well as in individual private conversations at this conference.

Because the number of papers submitted to this conference was quite large in comparison to the number of papers submitted at previous conferences, we elected to divide this book into two volumes. Biologically and medically oriented papers have been placed in Volume I. Volume II consists of papers involving social and cultural issues. It is hoped that by this organization and division of what would otherwise be one giant volume into two large but manageable volumes, persons may elect to purchase one or the other only, if so desired. Thus, exorbitant cost to participants and readers can, hopefully, be avoided.

This publication marks the first time that the proceedings have been published by such a large, active, and important medical and scientific publishing house. Plenum Press has enjoyed a long and noteworthy role in scholarly publishing, especially within the field of drug abuse, alcoholism, and psychopharmacology. It was the feeling of the participants, organizers, planners, chairpersons, as well as the editor, that by publishing with a major scientific and medical publishing firm, the proceedings would be made available to a markedly larger audience at a somewhat lower cost than would otherwise have been the case without aggressive marketing.

Unfortunately, no funds were available this year to subsidize these proceedings by assisting in the monumental retyping efforts as well as other essential administrative tasks. Thus, it was also not possible for this conference to purchase volumes directly for resale at a lower price for participants, although special arrangements for participants have been made by the publishing house. It is of interest to speculate as to the role these proceedings may play in the course of the conferences. It may well be that the 1978 Seattle Conference will mark the largest of the multi-disciplinary conferences. In 1979 there are a number of smaller groups with special interests. The Committee of Problems of Drug Dependence is of a special interest to pharmacologists and research workers in the basic biological and social sciences. There are two major alcoholism conferences being held; one "women's conference" has been held this year and one "minority" conference. It is uncertain as to whether there will ever again be such a large and heterogeneous collection of outstanding authorities in the field of drug abuse and alcoholism assembled at any one meeting for the purpose of discussing such a broad spectrum of topics within the field of substance abuse. In future years it is possible that no one conference proceedings will be able to pull together such enormous breadth and depth of information to the same degree that was possible here.

It is appropriate in closing to thank the hundreds of persons who took part in this Herculean endeavor. The magnitude of the task this year was almost overwhelming. Collecting, organizing, editing, reorganizing, typing, reviewing, and publishing a conference of this size is a task that could not be done without the goodwill and hard work of hundreds, if not thousands, of workers and colleagues. Many are noted here as the authors of various chapters. Others are cited as chairperson, co-chairpersons, or planning committee members. Thanks are also due to the invaluable office coordination and secretarial skills of Ms. Dorothy (Dot) Benziger, whose work has been outstanding. In addition, Mrs. Lorraine English, Mrs. Rose Jefferson, and Ms. Pam Shiwram made significant and valuable secretarial contributions.

Arnold Schecter

Newark, New Jersey
and
Binghamton, New York
Summer, 1979

CONTENTS

VOLUME I
BIOMEDICAL ISSUES IN DRUG & ALCOHOL ABUSE

PART 1
DRUG ABUSE TREATMENT: GENERAL

PART 2
TREATMENT: THERAPEUTIC COMMUNITIES

PART 4
WOMEN AND CHILDREN

PART 5
INNOVATIONS IN CLINICAL PHARMACOLOGY: NARCOTIC ANTAGONISTS IN TREATMENT

PART 6
INNOVATIONS IN CLINICAL PHARMACOLOGY: LAAM IN TREATMENT

PART 7
GENERAL PHARMACOLOGY

PART 8
EVALUATION OF TREATMENT OUTCOME

DRUG DEPENDENCE

AND ALCOHOLISM

Volume 1
Biomedical Issues

COGNITIVE DISSONANCE IN PERCEPTION OF GOAL ACHIEVEMENT: A PILOT STUDY

Elizabeth Serkin, M.A., and Robert D. Gorodetzer, B.A.

Bucks County Drug and Alcohol Commission

The purpose of this paper is to examine the apparent lack of congruence between clients' self reporting and agency records of treatment outcome. Our interest arose out of monitoring and client follow-up activities conducted by us at a suburban out-patient treatment center for young drug and alcohol abusers. We had been struck by the inadequacy of UDCS discharge data to reflect treatment outcome, by counselors' propensity to assess treatment outcome in a way consistent with the designated reason for termination, and by clients' tendency to describe their treatment experience in a way supportive of their decision to terminate.

The problem has both practical and theoretical significance. For purposes of program evaluation, it is useful to consider methods, approaches, and definitions employed in gathering and analyzing treatment outcome data. Problems of reliability and validity, too, are magnified when client and counselor perceptions do not correspond. We were also interested in exploring some of the theoretical implications of our observations, hoping to arrive at a model for assessing treatment outcome that would reconcile some of the many apparent contradictions.

Cognitive dissonance theory, popular in the social-psychology of the 1950's and 1960's (Brehm and Cohen, 1962; Festinger, 1957; Festinger, 1964), refers to people's tendency to define their disparate responses in their various role relationships in a way consistent with each other, maximizing congruence between feed-back and self concept, and supporting self esteem. Our use of this approach was combined with the more contemporary interactionist perspective to enable us to examine treatment outcome in light of negotiated definitions.

In our exploratory study, we compared client discharge data, as recorded by the counselor, with data collected shortly after discharge in an in-depth clinical follow-up interview conducted by one of the authors. Goal achievement, from the agency's point of view, entailed total abstinence, psychosocial well-being, and improved vocational or educational functioning. Whether the counselor believed that the client terminated with goals fully or partially achieved was reflected in UCDS reporting and discharge summaries. "Irregular" discharges were particularly interesting in that they usually represented "splits," i.e., clients simply stopped making or keeping appointments, and the counselor rarely knew why. We anticipated that the discharge summaries in these cases would tend to attribute lack of motivation or other negative traits to clients, and to rate them low in coping and adjustment, predicting problems for the future. Clients, for their part, we predicted would tend either to report that they had in fact completed treatment, or that they felt the program deficient in some way, generally rating themselves high in coping and adjustment. For those clients who terminated by agreement ("completed treatment, no drug use"), we anticipated more agreement between counselor and client reports, this agreement not necessarily supported by objective evidence, however.

To enable us to assign scores with which to compare clients' and counselors' data, we developed a modified global levels of functioning scale. We agreed with Belasco (1971) and others (Ludwig, Levin, and Stark, 1950) who express the need for multifaceted criteria of successful treatment outcome, as there does seem to be as much asynchrony as there is overlap in sobriety, occupational adjustment, and psychosocial stability. We needed a scale which would incorporate perceptions (e.g., cognitive capacities, reality orientation, and knowledge); feelings (e.g., affective levels, emotional stability, and expressiveness); and coping (e.g., skills and abilities in managing oneself and one's ordinary social circumstances, as in meeting familial and vocational role requirements). Levels of function in three general areas--sobriety, interpersonal relationships, and vocational/educational activities--were assigned on the following basis:

1) Severely dysfunctional. Danger to oneself and others. Requires custodial care.

2) Moderately or mildly dysfunctional. Adequate reality orientation. Able to maintain minimal self care, but difficulties in role functioning and in appropriate expression of feelings; isolated; requires regular supportive services.

3) Copes well in one area (sobriety, interpersonal, or vocational, but requires considerable support and/or intervention in the others.

4) Able to function comfortably in two areas under ordinary circumstances, but requires occasional support or intervention.

5) Functions well in all three areas; may occasionally need supportive services.

6) Able to provide support to others and/or to make a socially meaningful contribution.

Each of the clients in this study were assigned two scores, one based on client self reporting and the other on discharge data. Because we were interested in the degree of association between the two sets of scores, we selected a correlation analysis as an appropriate statistical approach. We discovered a significant discrepancy between counselor and client perception, leading us to conclude that outcome data being utilized by the agency needs to be re-evaluated. An analysis of variance supported our hypothesis that scores varied according to the reason for discharge. Counselors consistently underestimated the level of functioning of "irregular" discharges, whereas they tended to rate clients who terminated by agreement in a level that coincided with the clients' assessment of their status.

A few words of caution and qualification:

1) The low completion rate (approximately 40% of the original sample had moved and could not be located within thirty days of discharge) precludes generalization or definitive conclusions about treatment outcome.

2) We had hoped to link the level of functioning with duration of treatment, but ambiguities in determining the length of time in treatment must first be resolved. For example, under the UDCS system, discharge date may be the day of last service if treatment is terminated by agreement, but a month after the date of last service for "splits."

3) Linking outcome to treatment is further hampered by the obvious but often overlooked fact that treatment is not a single, definable entity. How is one to know what aspect of treatment may be credited with effecting the outcome? Neither is it possible to ascertain the extent to which the program as a whole determined the outcome.

4) Without the input of "significant others" as informants, it is not possible to check the reliability of counselor or client data. Some studies (Ball, 1967) have corroborated the reliability of addict self-reports, but we did not find this to be the case in several situations where we serendipitously received outside information about clients. Problems of confidentiality make access to verifying information difficult.

5) Sobriety, in and of itself, may not be synonymous with treatment "success." Yet it seems to be salient. Furthermore, defining sobriety may not be as simple as it may seem.

6) Finally, the researchers themselves are vulnerable to cognitive dissonance. In their attempt to reconcile their stance of scientific objectivity, their need to feel that clients and counselors are being truthful and open (and that they have the insight to distinguish truth from fiction), and their desire to make some sense out of confusing, often contradictory data, they probably engage in dissonance-reducing efforts which compromise their very aims.

In conclusion, the value of this pilot was in demonstrating the need for a new look at discharge data. Clients who "split" reported significantly higher levels of functioning than was attributed to them in discharge summaries, and they tended to attribute much of their improvement to the program. The program was unable to take credit for their improvement, and consistently predicted problems which follow-up interviews did not bear out. We believe, that with clear definitions or terminologies, and a new look at discharge data, we have the beginnings of worthwhile further study of treatment outcome. Feed-back based on follow-up, moreover, might serve to reduce cognitive dissonance in counselors, as it may be used to support program efficacy.

REFERENCES

Ball, J.C. 1967. The Reliability and Validity of Interview Data Obtained from 59 Narcotics Addicts. *AJS* 72: 65-654.

Belasco, J.A. 1971. The Criterion Question Revisited. *BJA* 66: 39-44.

Brehm, J.W. and A.R. Cohen. 1962. *Explorations in Cognitive Dissonance*. New York: Wiley.

Festinger, L. 1964. *Conflict, Decision, and Dissonance*. Stanford, Calif.: Stanford U. Press.

Festinger, L. 1957. *A Theory of Cognitive Dissonance*. Evanston, Ill.: Row, Peterson.

Ludwig, A.M., J. Levin, and L. Stark. 1970. *LSD and Alcoholism: A Clinical Study of Treatment Efficacy*. Springfield, Ill.: Thomas.

THE JOB SKILLS BANK IN DRUG ADDICTION TREATMENT

Joan Randell, M.A.
Beth Israel Medical Center Methadone Maintenance Treatment Program

Lee Koenigsberg, M.B.A.
Community Treatment Foundation

INTRODUCTION

A major objective of many drug treatment programs is the placement of clients in gainful employment. Since this task is difficult and challenging, success can be enhanced by approaching placement in an organized and systematic fashion. A commonly held misconception fosters the myth that there are few employment opportunities for the job-ready ex-addict. This notion implies that more and more emphasis must be placed on seeking out job opportunities for this population. In reality, the experience of a large, hospital-based drug treatment program and a non-profit organization specializing in the job placement of ex-addicts shows the reverse to be true, i.e., it has not been possible to refer suitable applicants to the majority of the job opportunities open to ex-addicts. This experience does not necessarily imply that there are not qualified applicants; rather, it suggests the magnitude of the problem in identifying the clients who have the prerequisite skills for jobs that are available.

Traditionally, the matching of clients with jobs has been dependent upon the counselor's recall of each clients's employment history and skills. Such an approach tends to be haphazard and highly subjective, particularly in large programs and/or programs with a large caseload per counselor.

The use of a job skills bank, employing either a manual or a computerized system, significantly increases the potential for successful job placements by providing a precise and objective method of expeditiously matching client skills with job opportunities.

This paper describes two job skills bank systems conceptualized by the Vocational Rehabilitation Department at the Beth Israel Medical Center Methadone Maintenance Treatment Program in New York City.

MANUAL JOB SKILLS BANK

This system was developed to be utilized by all counselors who carry a caseload. It was designed to give them the ability to categorize *all* their clients' skills, both primary and secondary, in order to respond quickly and systematically to a given job opportunity. Knowledge of secondary skills is very important in placing job-ready ex-addicts. Many clients, especially older ones, have performed many job tasks that are not easily associated with a specific occupational title. For example, skills needed as a building superintendent can often be translated to entry level jobs in carpentry, plumbing, painting, floor waxing, etc.

In this system a counselor conducts an in-depth vocational interview with the client and completes the attached summary card (Appendix 1) with information relating education, skills, training, work experience, licenses, and medical limitations. The card asks for a primary occupation or skill, which requires the client and the counselor to designate the one job category that best suits the client. In addition, a secondary card or cards (Appendix 2) are completed for each of the skills the client possesses. The secondary classification allows the counselor and client to specify all applicable skills of the client, thereby increasing the possibility of the number of referrals to different jobs. Each secondary card is referenced to the respective primary card, providing the counselor with a more complete patient profile without the need for duplicating all of the information on each secondary card.

The primary and secondary skill cards for all clients are filed in a small file on the counselor's desk. The file is divided into 12 categories (Appendix 3): artists, construction and building trades; driving/trucking/transportation; health occupations; professional; service/sales; skilled trades; stock/warehousing; unskilled occupations; and patients not currently job ready. When a job opportunity arises, the counselor goes to the file to ascertain whether any client on his/her caseload has the specific skill required. If a match is found, the client can then be contacted and interviewed to determine if, in fact, a referral is appropriate.

The advantage of this method of vocational retrieval lies in its simplicity. All drug treatment staff, regardless of their level of sophistication in vocational rehabilitation, can utilize this system. It obliges each counselor to be aware of, and responsible for, the vocational abilities of clients. It allows for flexibility since each counselor can categorize a given occupational skill in the categories that make the most sense to him/her. The occupational index simply serves as a guide. In addition, it alerts staff to clients that are not job-ready. They can then focus on these clients' vocational development.

COMPUTERIZED SKILLS BANK

Some drug treatment programs utilize the services of a computer for accounting procedures, a central registry of clients, or for re-research. If such a program has a centralized vocational unit, it would be highly desirable to develop a computerized vocational retrieval system.[1] A computerized data bank of clients skills will provide highly accurate and comprehensive vocational data on all clients in treatment, so that all who possess the required skills will be considered in response to a given job opportunity.

The system works as follows: counselors at each treatment facility interview clients on their caseload and complete a basic questionnaire on each client's vocational skills, experience, training, limitations, and preferences. Using a coding system, the data is entered into a centralized computer file. When a job opening occurs the specifications of the position, i.e., skills, experience, education level, etc., are entered into the computer. The computer will then automatically search through the client file and identify those clients who meet the stated criteria, and a list of the most qualified clients will be generated. Clients appearing on the list are then discussed with the respective counselor and, if appropriate, the client is referred for placement. The matching and retrieval of clients can be accomplished in a matter of minutes if an on-line computer system is used; alternatively, the system could be implemented on a batch system.

Each job referral emanating from this procedure is entered in the client's record on the computer file with a notation as to whether the referral resulted in an actual placement, and, if not, the reason for the client's rejection. When a placement is made, periodic follow-up information is essential for the analysis and evaluation of the program's experience in job referral, placement, and retention.

A computerized system can produce various reports to serve a number of purposes. Specifically, such a system can serve to:

---

1 Such a system has been designed in New York City and is scheduled for funding this year as a pilot project in a large methadone maintenance treatment program.

1. develop an overall understanding of the vocational demographics of the client population (skills, training, work experience, educational level, and interests). This information can prove to be of immeasurable value for a treatment program in planning vocational rehabilitation priorities, objectives, and activities.

2. foster the development of jobs in those occupational areas where the highest concentration of clients possess skills. This eliminates the time which might be spent developing jobs that cannot be adequately filled due to a lack of suitable clients.

3. identify those clients who have no skills and need training.

4. identify those clients who need retraining or job upgrading due to changing labor market trends.

5. identify those clients who fall into different categories (skilled, unskilled, high school graduates, high school dropouts, etc.) who may benefit from different kinds of counseling services and vocational programs.

CONCLUSION

Jobs skills banks can be a useful tool in the difficult task of placing job ready ex-addicts. The two systems described in this paper serve as a model. Each treatment program should attempt to develop job skills banks that best meet their program and client needs. Evaluation of the systems is recommended to empirically aseess their effectiveness in facilitating placement.

Appendix 1

(front)

BETH ISRAEL MEDICAL CENTER METHADONE MAINTENANCE TREATMENT PROGRAM

______________________________ ______________________________

(last name) (first name) Primary occupation or skill

phone number: ________________

EDUCATION
(check or fill in as appropriate)

____Highest Grade Completed ____Years of College

____HSG or HSE ____College or Graduate Degrees (specify)

PREVIOUS TRAINING

Skill	Dates	Training Completed: Yes or No

Licenses (specify class or type)	Medical or other limitations: (specify)
motor vehicle operator: professional: other:	

(back)

WORK EXPERIENCE
(list most recent first)

Job	Duties	Dates

Other Skills:

Appendix 2

BETH ISRAEL MEDICAL CENTER METHADONE MAINTENANCE TREATMENT PROGRAM

____________________________ ____________________________

last name first name Secondary occupation or skill

See primary card, filed as job title:

____

Under occupational category:

____

Appendix 3

BETH ISRAEL MEDICAL CENTER
Methadone Maintenance Treatment Program

OCCUPATIONAL CATEGORY INDEX

I ARTISTS

actors and actresses
artisans
dancers
floral designers
interior decoraters
jewelry makers
landscapers
musicians
photographers
singers

II CONSTRUCTION & BUILDING TRADES

carpenters
cement masons
dry wall installers
electricians
elevator constructors
floor covering installers
glaziers
operating engineers
painters
plumbers
roofers
sheet metal workers
structural workers & riggers
tile setters
welders

III DRIVING/TRUCKING/TRANSPORTATION

bus driver
mover
taxi driver
ticket agent
tractor trailer driver
train or subway worker
truck driver

OCCUPATIONAL CATEGORY INDEX

IV HEALTH OCCUPATIONS

dental assistant
dental hygienist
dental laboratory technician
medical laboratory technician
nurse
therapy & rehabilitation occupations

V MECHANICAL/REPAIR/TECHNICIAN

air conditioning/refrigerating/heating
appliance repair
auto body repair
auto mechanic
business machine repair
communications technician
computer technician
locksmith
TV & radio technician

VI OFFICE OCCUPATIONS

bank teller
bookkeeper
cashier
computer operator
file clerk
keypunch operator
mail clerk
receptionist
secretary & stenographer
telephone/switchboard operator
typists

VII PROFESSIONAL

accountant
actuaries
architects
attorneys
computer programmers
counselors
psychologists
social workers
teachers

OCCUPATIONAL CATEGORY INDEX

VIII SERVICE/SALES

building service worker
cleaning
day care worker
domestic worker
food service
gas station attendants
guards
hotel worker
mail carrier
model
parking attendant
personal service
porter
waiter

IX SKILLED TRADES

apparel cutter
barber
beautician
dog groomer
draftsperson
dressmaker
dry cleanters
exterminators
gardeners
machinists
meat cutters
photo processers
printers
tailors
tool & die makers
upholsterer

X STOCK/WAREHOUSING

fork-lift operator
loader & unloader
shipping & receiving clerk
stock clerk

OCCUPATIONAL CATEGORY INDEX

XI UNSKILLED OCCUPATIONS

construction laborer
dishwasher
factory worker
floor finisher
handyman
launderer
supermarket clerk
window washer

XII Not currently job ready

TRAINING COUNSELORS IN CRISIS INTERVENTION - A WORKSHOP

Mary Ann Pingalore, M.S. and Elena J. Eisman, Ed.D.

Region IV Association for Human Development

Middletown, Massachusetts

This paper describes a model for training counselors in the skills needed to function as crisis intervention workers. This model and its content and presentation format were developed for use with counselors having at least one year of counseling experience in self-help and/or drug treatment services.

This model was further specialized in that it was developed for a specific program, Project RAP in Beverly, Massachusetts which, in partnership with the State Department of Mental Health, developed a program for early response to crisis for clients in their catchment area.

While it is not the intention of this paper to develop and detail modifications of this model for other groups of trainees or for training in various other situations, the authors suggest that modifications can be easily made in the basic design.

The training model and its delivery will be discussed in three parts: preparation, content of training, and techniques utilized.

PREPARATION

In preparing for the training sequence, three factors needed to be addressed: identification of trainees, resource compilation, and assessment of the target population to be served by the trainees.

It is a well established fact that crisis intervention work is enormously taxing both to the individual worker and to the agency employing him/her. Chronic tension, absenteeism, high personnel turnover and the myriad of "burn-out" symptoms plaguing the workers are often exacerbated by the agency's struggle to schedule services, rotate worker responsibilities, respond to workers' support needs while continuing to finance the service on a cost-effective basis.

In the light of these exigencies, this model pre-selected trainee-applicants who had a minimum of one year of counseling experience, who were presently employed in a human service position, and who intended to augment this employment with part-time crisis intervention work.

Once this pool of applicants had been pre-selected, the group was progressively defined by group problem-solving interviews followed by personal interviews. Special consideration was given to ethnic, sexual and racial mix of the final pool of trainees. (While bilingual capacity was not an issue in this training, it does represent an important consideration for modifications of the model.)

One of the basic tools of the crisis worker is up-to-date accurate referral and resource information. An appendix or manual of all types of services available in the community and surrounding areas was prepared prior to training. This manual was organized according to the type of service available, and included the name of the appropriate service, telephone numbers, hours of service, and the names of at least two contact persons in the agency.

While compilation of this data may seem a formidable task, it was rendered more manageable by utilizing the data bank of the local hot-line. Recourse to hospital, police and fire services augmented this listing as well as contact with local clergy.

Perhaps the most crucial effort to be expanded prior to the actual instructional phase, was the determination of the target population for the service. In other words, who were the individuals the trainees would eventually serve?

Assessment profitably began with "first responders" i.e., police, hospital emergency room personnel, fire fighters, drug and hot-line workers, since theirs are the services generally contacted in crisis situations.

A second wave of assessment included the canvassing of human service agencies, churches, schools, and community project groups. At each level, deficits in services became apparent. Assessment was strengthened too, by following hypothetical "sample crisis cases" through the various systems and agencies expected to respond.

The final results of assessment not only yielded a profile of the target population, but also served as validity checks to the accuracy of the resource manual, and to the qualities and skills of the trainee pool.

CONTENT

Now to address the content area of crisis intervention training. It is the opinion of the authors that people involved in crisis intervention work need to be able to deal with all types of crises that occur. The rationale for this is that people in crisis do not spend time looking through the phone book for the most appropriate agency to handle their particular problem. If they reach the stage of reaching out at all, it will probably be to the first agency they come across in the phone book, the phone number on the bumper sticker in front of them, or on the radio, or the place where their friend received help. In addition, there is a responsibility held by the initial responder to either deal with the crisis or to maintain the individual until they get hooked up with an appropriate resource. For these reasons the authors believe that the content of crisis intervention training should include: a familiarity with the common symptoms and patterns presented by people in any type of crisis, a knowledge of basic assessment and intervention strategies, and specific training in issues likely to be addressed by people experiencing some of the more common types of crises.

The basic introduction to the training focused on three questions: what is a crisis, why does it occur, and how do you identify a person in crisis. The definition used for crisis is "a short period of psychological disequilibrium in a person who confronts a hazardous circumstance that for him constitutes an important problem which he can for the time being neither escape nor solve with his customary problem solving resources" (Caplan, 1964).

Crisis is described graphically through what the authors call the "balloon model." Briefly this describes a person in terms of a dynamic energy flow system. Energy resources such as food, shelter, supportive relationships and money all correspond to air flowing through a balloon representing the individual. Energy drains correspond to pin pricks or air losses in the balloon and include: unresolved past emotional issues such as sibling rivalry, personal losses, or ego development deficiencies; present stress such as illness, marital problems, drug use, or underemployment; and other unexpected traumas or crises such as rape, fires, and losses. This model shows that a person can maintain emotional equilibrium even if there are significant stresses in their life as long as they have sufficient emotional energy resources. If, however, a crisis occurs without adding additional supportive resources, there is a negative energy drain. The person is in effect "deflated." This model is continued through the training and raised when intervention strategies are discussed. This allows the trainees to see that intervention with a client in crisis can be addressed on three levels: to increase the client's resources, to deal with the emotional impact of the present crisis, or to deal with the energy drain of the life stresses the client had been dealing with before the crisis.

In terms of identifying people in crisis, the model used is the one described by Garrison (1975). This identifies personal characteristics of people in crisis including attention problems, preservation, memory problems, feelings of powerlessness, paralysis by indecision, decrease in performance, identity problems, change in the client's relationships, and signs of distress.

Once it has been established that the client is in crisis, the counselors are then trained to identify the problem's scope. This phase focuses on determining how widespread the crisis is in terms of family or friends' involvement. This is felt to be a crucial part of interventional work done by Speck and Atteneave (1973) on family networks. The counselor attempts to reach other people in the client's social network who might also be in crisis. Conversely it also allows the counselor to seek out potential friends or relatives in the client's social network who might serve as a support system to the client. The positive effects of a group of people pulling together to deal with a common crisis or to help an individual deal with his or her individual crisis are quickly realized by the trainees.

The next issue addressed after identifying the scope of the problem is that of problem assessment. Here the counselor is trained to determine the severity of the problem and the need for immediate intervention. Basically this is done by assessing the severity of the symptoms presented, and by comparing the level of functioning of the client with their level of functioning prior to the onset of the crisis.

Finally, the counselor is trained to make a determination of the intervention needed by the individual. There are three basic types of intervention discussed in this phase of training: problem solving, referral, and support. These can be used either singly or in combination to help a client in crisis. Specific emphasis is placed on the clinical/philosophical position that drastic personality restructuring should not be a goal during crisis intervention. The goal of the intervener is to help the client regain their equilibrium after a crisis so that they may continue to function in more or less the same way as they had before the crisis. The major work done, is in helping the client gain insight into their maladaptive patterns of dealing with crises and allowing them to try out, with the support and encouragement of the counselor, alternative and hopefully more successful methods of coping with sudden stress. The concept of "buying time" is also introduced at this point in the training. This intervention concept helps address the seemingly contradictory behavior pattern of impulsivity in people who feel powerless and unable to choose a course of action. This has already been mentioned as a symptom of people in crisis. Buying time is a role that the crisis counselor plays in providing an external support for and expectation that the client will do the best that they can until things begin getting easier for them. The counselor in effect supports the dual input of the individual and the environment. Clients who see everything as bleak and themselves as powerless to change things are encouraged to work and to wait. They are taught to assess each change in their environment in terms of its potential toward helping them feel better. This approach is particularly helpful with suicidal clients.

We have reviewed two of the three major areas of crisis training, namely common factors and symptoms present in people experiencing crisis, and methods for assessment and intervention. The third area addressed in the training was a review of some of the more common types of crises in terms of their unique impact on the individual.

The specific types of crises discussed included: alcohol and drug emergencies, loss and grief in the generic sense (including loss of people, health and property), family disputes and child abuse, rape, and psychiatric emergencies. For each of these sessions, training included a basic overview of the stresses, emotions, and cognitions likely to occur for the individual both during and after the trauma. The authors feel that specific focus on several of the more common crisis situations enhances the counselor's empathic understanding of a client coming to them with these problems. It also helps the counselor to plan their initial assessment of the client and his/her situation by suggesting some important questions to ask or answers to listen for.

TECHNIQUES

The last aspect of training we will address is that of the techniques used to enhance the acquisition of crisis intervention skills. As was said before, the population used for the training were people who already held full-time jobs during the day, and were to do their crisis intervention work at night. For that reason, the training was held for three weeks, three nights a week, for three hour sessions except the first introductory session which lasted four hours. This gave a total of 28 hours.

The format of the training was a combination of mini-lecture, discussion, and role-play. Case studies were presented throughout the training to enhance the learning by grounding theory in practical experience. The case studies presented were designed to increase in complexity as the training progressed to tap the increased skills of the trainees. Participants were encouraged to discuss personal experiences with crisis, either their own, those of friends or family, or those of clients they had worked with in the past.

Role-plays were also designed to mirror the natural development of the training group. The first role-plays were done as round-robins, where the entire training group took turns playing the role of the counselor by making several interventions then moving on to the next trainee. The trainers played the role of the client in this exercise. This was done at the beginning of training to minimize performance anxiety and to build a sense of cohesion among the group. The second series of role-plays were done in a "fishbowl" manner with one trainee playing the role of the counselor and one the role of the client. In all instances, client roles were determined in advance by the trainers and presented to the trainee playing the role of the client. Finally, the role-plays were done with pairs of counselors to better model the needs of their future work relationships since the crisis intervention program they were to be working in used intervention teams. This last type of role-play was used to bring up the significant training issues of intervention style and interdependence of counselors.

Time was left at the beginning and end of each training session to bring up questions and to process what had occurred during the training. The goal here was to build a support network within the training group that would be continued as they began work. The stressful nature of crisis intervention work necessitates the establishment of strong staff supports if counselors are expected to continue working efficiently without burning out. Often however, human service providers work diligently to help clients develop support systems but neglect to do the same for themselves. The trainers used several techniques to enhance this process. Three mentioned before were paired role-plays, round-robin role-plays, and direct verbalization of the need to support each other. Other techniques used were the Art Form Process (Burglass, Bremer, and Evans, 1976), encouragement of self-disclosure on the part of the trainers and trainees, group process comments, and the establishment of mandatory, ongoing, weekly meetings devoted to support and supervision.

The model we have discussed is one which was developed to meet the training needs of a particular staff. This staff have been working effectively for eight months with much positive feedback from the community. The training model is however easily modified to meet the training needs of any group dealing with clients in crisis.

REFERENCES

Burglass, M.E., Bremer, D.H., and Evans, R.J. The Art Form Process: A clinical technique for the establishment of affect management in drug dependent individuals. Presented at the 1976 National Drug Abuse Conference, New York.

Caplan, G. 1964. *Principles of Preventative Psychiatry.* New York: Basic Books.

Garrison, J. 1975. The layperson's guide to crisis intervention. *J. of Mental Health Technology*. V.I. No.I: 1-6.

Speck, R. and Atteneave, C. 1973. *Family Networks*. New York: Pantheon.

HEROIN MAINTENANCE VERSUS DILAUDID TRANSITION

Jordan Scher, M.D., Khalid Baig, M.D., Si Bom You, M.D.

National Council on Drug Abuse and the Methadone Maintenance Institute, Chicago, Illinois

The prevalence of narcotic addiction in Britain and Europe and the apparent need to control narcotic distribution provoked the Hague Convention in 1912. By 1920, Parliament enacted the Dangerous Drug Act, following the lead of the Harrison Act of 1914 in this country. Thus, narcotic control was established in England and the United States, and by 1926 the Rolleston Committee had designated that opiate dependency in Britain was a medical condition, whose treatment required physician prescriptions. A somewhat similar program existed in the United States for the decade following 1912, during which time 12,000 addicts received morphine or heroin from 44 clinics. These were suddenly disbanded in 1924.

In England, however, the problem of narcotic addiction had never reached the proportions that it has in the United States. In fact, it has been claimed that heroin dependency effectively declined between 1922 and 1955 in Britain, so that by the latter date perhaps 1500 people were on the rolls of addicts in Great Britain. By 1961, however, the problems had again accelerated, and the Inter-Departmental Committee on Drug Addiction, chaired by Sir Russell Brain, reinvestigated the heroin dependency issue. By 1966, in a second report, Britain further restricted the right of private physicians to prescribe opiates and cocaine.

Subsequent to that time, a limited number of clinics, staffed by specially licensed physicians, were established, and a registry of addicts was established for Great Britain. Contrary to the American situation, the presumption in Britain has long been that heroin addiction was basically a medically induced problem. As a result, it is felt that addiction in Britain is an issue for medical care, and this has been the case throughout the history of heroin treatment in Great Britain.

British medical thinking, unlike the American point of view, has not been that narcotic treatment of addiction is for the purpose of maintenance. Instead, the various Brain Committes have been far more concerned with the issue of what they call "containment" of addiction (Johnson, 1977) than heroin maintenance. The British Home Office and the Department of Health and Social Security maintains statistical information about the problems of heroin and other narcotic addiction in Britain. Addicted patients in Britain may receive legal prescriptions for heroin, methadone and other narcotic drugs which they then fill at particular pharmacies. The pressure in Britain has been to make it difficult, but not impossible, for a heroin addict to receive his heroin prescription. It has also been a part of the policy of containment to attempt to prevent heroin from being sold to other addicts, or non-addicted individuals, as well as to get heroin addicts to switch to methadone, or as it is called in Britain, physeptone. The British have felt that if they could reduce the amount of legal heroin prescribed, as well as curtail the distribution of "Chinese Heroin", its illegal counterpart, then heroin addiction would be on the way to being controlled.

As a matter of fact, very little actual transfer of heroin addicted patients to methadone has occurred in Britain, if one can rely on the statistical information available. Older chronic addicts, receiving their heroin from either private physicians or drug treatment centers, have tended to continue on the same prescription. (Johnson, 1977) Newer addicts, however, or those who have re-registered, may be placed on methadone instead of heroin. It is also important to note that most methadone in Britain is prescribed for injection, rather than oral use.

Between 1968 and 1974, there seemed to have been a continuous decline in the number of notified heroin addicts in Britain from a high of 2240 to a low of 866. Nonetheless, the total number of addicts notified increased from 2782 in 1968 to 3430 in 1975, since the number of those receiving methadone had been on the increase. That is, the number of those receiving methadone rose from 1011 in 1969 to 1543 in 1975. The number of those under 20 years of age receiving either heroin or methadone declined from 221 in 1969 to 7 by 1975, which may have represented less the actual realities of teenage addiction than the more restrictive attitudes toward prescribing narcotics for younger individuals. The conclusion from these statistics by the Department of Health and Social Security in Britain was that heroin addiction was being contained by 1972.

Among the reasons for this conclusion are the removal of notified addicts, either by death or for other reasons. For example, in 1973, 1145 addicts were dropped for the following reasons: 35% were imprisoned, 3% were in other institutions and 68% were no longer seeking treatment. Of 1407 notified addicts dropped from the list in 1975, similar percentages seemed to occur for the same reasons.

To further complicate the picture, addicts in Britain are characterized by the principal drug of addiction received, so that many of them may be on more than one drug at a time. Thus, patients may be on methadone alone; heroin alone; methadone and heroin; methadone, heroin and cocaine; methadone and morphine, etc. For this reason the issue is far more complex in terms of drug usage than the situation here in the United States with the prescription of methadone alone. Nonetheless, heroin utilization for addicts in Britain declined from 34% in 1969 to 16% by 1975.

Johnson (1977) further divides heroin maintenance in Britain according to long-term, over one year, and short-term, under one year, categories. According to his estimate the number of those involved in long-term heroin maintenance had shrunk from 75% in 1971 to under 4% in 1975. Between 1968 and 1975, 8710 addicts were notified in Great Britain. Not all of these, however, were prescribed narcotic medications. In fact, of 3430 (39%) recorded, only 1954 (22%) were prescribed narcotics by the end of 1975. Blumberg et al. (1974) have suggested that each notified addict may have a friend who has not registered in Britain. If this is the case, then there may actually be as many as 18,000 registered and non-registered addicts in Britain, although this number might be pared down to between 10,000 and 13,000, if one considers that at least some of these non-registered addicts may be known by more than one other addict who is registered.

As mentioned above, not all persons seeking treatment for addiction will be prescribed narcotics, and in practice, one-half to two-thirds of those who register annually will receive legal narcotics. This is because two positive drug urines are required, and many of those desiring treatment may not be successful in showing these. Currently, only 2-3% of addicts are maintained on heroin in Britain, and the average dose for those receiving heroin is approximately 100 milligrams, while the average for injectable methadone is about 50 milligrams. Once an addict has dropped from the register, he may find it very difficult to gain a new prescription in Britain, even if he may clearly be addicted. This, of course, is very different from the situation generally speaking in the United States.

As evidence of the fact that an illicit market in other drugs exists in Britain, one study shows that 16% of those enrolled in a clinic were not using the prescribed opiates, while 70% were using other drugs that had not been prescribed (Bewley, 1970). In another study, 40% of clinic members in Britain received their narcotics illicitly, as well as through the clinic itself (James, 1971). In Britain, about 3% of those enrolled in clinics die annually from overdoses. And, despite the fact that legal prescriptions are readily available for most of Britain's addicted population, approximately 60% are not gainfully employed, while a similar percent are engaged in criminal activities (Johnson, 1977). Thus, the legal prescription of drugs seems to have relatively little to do with either permitting rehabilitation from an economic standpoint in Britain, or the reduction of the criminal activity in which addicts become involved.

In the United States, it has generally been assumed that 50% of crimes against property result from narcotic addiction. In one study in New York, property crimes supported 38% of the heroin used, 43% was financed by the sale and distribution of heroin itself, 6% through prostitution and gambling and 13% from legal sources. (Moore, 1977) In another study in Detroit, it was predicted that a 10% increase in the number of addicts in treatment would result in a 2% reduction in crimes against property.

In 1972, Rector estimated that roughly 10% of the addicted population in the United States was in treatment, and that this was only 10% effective. This is probably a harsh judgment. Perhaps a fairer estimate would be that 25-30% of addicts in this country either are or have been in treatment, and that for those in treatment, the success rate is 25-50% on the average, while it may reach as high as 60-80% in certain clinic situations. It has also been estimated that methadone maintenance and/or detoxification is a viable treatment of choice for probably 60-80% of all addicts, while less than 10% avail themselves of drug free approaches. It would also appear that only 2-3% of those who attempt drug free treatment can successfully complete such programs and remain drug free.

Nonetheless, it would appear that perhaps as many as two-thirds of the suspected 700-800,000 U.S. addicts have not been in treatment of any kind, and the majority of them seem to have considerable resistance to entry into any of the known treatment modalities available. This is the reason, of course, for the recurrent suggestion that heroin maintenance, or some form of it, be introduced into the United States. But, if we are modeling on Britain, the British concept is not so much narcotic maintenance, but narcotic containment. And as to heroin maintenance, or containment itself, the actual percentage of addicts so handled is not more than 4-16% of those actually in treatment in Britain. So, if we are desperately striking out for an exemplary program that we can emulate through the use of heroin for the addict currently recalcitrant to any kind of treatment in this country, heroin maintenance in Britain would not seem to be a viable model, since it is very limited in nature.

Furthermore, in Britain most methadone is used by injection. Again this provides a sharp contrast to the way the drug is used in this country. In the United States, methadone, although quite effective in approximately half of those patients who have been enticed into using it, has suffered from a serious failure of confidence among what would appear to be the majority of street addicts who have not experienced it.

The reasons for this failure to appeal to the majority are curious and fascinating indeed. In effect, there are perhaps five or more independent lobbies arrayed against methadone. Intentionally or not, they, in effect, have established a massive and coordinated propaganda front.

The first of these lobbies is what might be called the "pusher/addict lobby". This lobby has very successfully convinced many addicts that methadone is physically dangerous to patients, is detrimental in pregnancies, is harder to kick than heroin, produces two habits, etc. The second important lobby is that of the drug free people who have been very vocal in their assault against the use of methadone. The third lobby might be called the "third world anti-genocidal lobby", which claims that methadone is a form of White and governmental enslavement and stupefaction of Blacks and Spanish people. A fourth lobby may be called the "bureaucratic lobby", including those in elective offices in government who may be under compulsion to either stamp out drug abuse, claim its eradication or containment, justify the vast sums currently expended in its control and treatment, or rail against those expenditures for having failed to accomplish what they were presumably intended to do. The fifth important lobby is that of the visual and print media. They find in drug addiction a marvelous vehicle for the production of a chronic and unending array of multi-million dollar movies and television extravaganzas. On the other hand, the media find an easy victim in addiction treatment and so inveigh against the horrors of drug abuse and the efforts such as methadone to treat it. One final lobby is that of the enforcement and criminal justice people, whose economics are tied closely to the rate of addiction.

The net result of this considerable propaganda effort against methadone has been that these lobbies have effectively supported and promoted the continual use of street heroin, and kept in bed together groups which would be appalled to find that they have inadvertently established a common front against methadone, the most effective treatment method we now have.

In view of the fact that we currently confront such resistance to methadone on the part of a large majority of those who should be in some kind of treatment, many voices have been raised for the establishment of heroin maintenance programs. At least half a dozen state legislatures have proposed heroin maintenance programming, none of which has gone to the point of actual legislation. A number of enforcement officials in various parts of the country have also proposed heroin maintenance programs, and they have even been tentatively suggested in the highest quarters of government.

The Vera Institute of Justice proposed an experimental program in 1972 to be conducted among treatment dropouts from methadone programs having documented three-year-addiction histories. It was the intention of this proposal to determine how effective a heroin maintenance program would be among methadone failures. Another proposal of a similar nature was made by Dr. Avram Goldstein, which he called STEPS, Sequential Treatment Employing Pharmacological Supports. (1976)

Goldstein's proposal suggested that morphine be used intravenously initially, later subcutaneously, in a stepwise fashion for up to a year. He further proposed progressive transfer to methadone, then LAAM, followed by abstinence aided by the use of an antagonist, abstinence without drugs and finally presumably complete social rehabilitation. The Vera program was intended as one which would be limited to a year of heroin maintenance before transfer to another modality such as methadone. The STEPS program was less definite in duration.

Each of these proposals attempts to answer the problem of resistance to methadone as we currently encounter it. The Vera proposal seems to be somewhat more realistic in terms of its appeal to the street addict in that it permits him to shift from an illegal to legal use of heroin. Goldstein's proposal is excellent except that it perhaps has an unrealistic expectation of the most recalcitrant and difficult group of street addicts. If the unreached group is already so resistant to methadone, it would seem unlikely that they would very readily continue in a program which required the progression of stages that Goldstein would like to apply.

Both of these proposals attempt to translate an interpretation of the British experience to the American scene. However, one must understand that the situation in England is quite different from that in the United States on a number of grounds. First of all, the addicted population is infinitesimally small compared to that in the United States, and is thus much more easily managed. As a result, the British have considerably less reluctance than would we have to give patients heroin or other narcotics for injection, along with injection paraphernalia to take with them on an outpatient basis. As another small indication of the differences between the social structures of Britain and the United States, one should remember that the police in Britain, generally speaking, do not carry guns.

Also, there is considerable reason to believe that even if heroin maintenance were to be made available, it would still fail to attract many of the addicts who currently resist methadone. Many of these addicts seem to prefer the criminal and street lifestyles, difficult though they may be. Despite problems of infection, endocarditis, hepatitis and recurrent incarceration, all too many drug addicts do not suffer the discouragement and wearing down that leads to entry into methadone or other programs. Even in Britain, despite the availability of pure drugs and clean injection materials, it has been reported that 60% of the patients frequently suffer from various infections due to the improper use of dirty needles.

If heroin maintenance were to be introduced into this country, there is little chance that addicts would be given injection materials and drugs as is the case in Britain. Quite the contrary, with the stringent control of methadone, it would be very likely that heroin maintained addicts would either have to be live-in situations or would have to return to the clinic five or six times a day to receive their injections. This is in contrast to the current avoidance of such chronic enslavement to the treatment process that methadone provides. At worst, a methadone patient only has to attend the clinic once a day, six days a week, for the first three months of treatment.

The street addict, as everyone knows, has a full-time job of stealing and shooting, so that little time is left for anything other than pursuit of the drug life. Despite all the problems of street addiction, there clearly is an enormous appeal to this type of existence for all too many people. We are, therefore, confronted with a seemingly insurmountable need to find some other way of reaching this largely unreachable population.

If heroin or morphine were to become available, one of the most important concerns would be how to determine who should be permitted to take advantage of these drugs. Treatment failures, long-term addicts, etc., would be the obvious choices. But how would we prevent or dissuade methadone patients from transferring to heroin, and/or how would we block those patients who might otherwise enter a methadone program from opting for heroin, should it become available?

An alternative possibility would perhaps be to launch a direct appeal through some sort of outreach effort to the street addicts themselves, should a more palatable drug of greater familiarity to the addict become available. To explore this possibility, we at the National Council on Drug Abuse have been conducting a street survey of the acceptability of heroin, methadone, or some other narcotic substitute approach for the past few months.

Paraprofessionals with personal addictive backgrounds have been dispatched to known copping corners to assess the realities of recalcitrant street addiction and the amenability of such addicts toward any kind of treatment approach. To date, approximately 485 street addicts have been surveyed.

Of this group, about one-third have been on methadone programs anywhere from one day to two years. Among the reasons for their dropping out has been the fact that many of them felt they were not permitted to receive the amount of methadone they required since, in Illinois, most public programs are not permitted to dispense above 40 milligrams.

As a result, many of these addicts, in their view, had been compelled to engage in "two habits" even while on the program, treatment methadone and street heroin. Another reason for dropping out of methadone treatment was the fact that many of these individuals refused to comply with disciplined clinic attendance and other demands treatment centers made upon them. Others claimed merely to prefer the street lifestyle, and felt that drug treatment had no meaning for them.

Of the two-thirds who had never been on a treatment program, 20-30% felt they were in control of their addiction and that no treatment was required. Most of the rest of the individuals surveyed had been thoroughly indoctrinated with an overriding fear of methadone and/or were afraid of developing "two habits". When this group was asked if they would possibly enter a heroin treatment program, the street addicts interviewed seemed intensely fearful of any kind of involvement with governmental authorities, whether they were of treatment or enforcement varieties, and they could not seem to be convinced that very frequently treatment and enforcement were not one and the same thing.

In view of all of the above, it is clear that the mere offering of heroin, morphine or other narcotic substance is not going to answer the question of how to appeal to those who are currently so reluctant to enter treatment of any kind. It would almost seem more an effort of desperation even to suggest the development of heroin treatment programs. Perhaps what is needed far more than this is some kind of front line social intervention program to reach street addicts on a continual basis.

Toward this end, some time ago we proposed to the Carter administration that the various enforcement modalities become a kind of front line social service reception area. By this we mean that instead of perpetual recirculation of addicts through the criminal justice system, they might be recurrently referrable to treatment modalities even before conviction and possible diversion into treatment programming at a later date a la currently available TASC programs. If enforcement officers could be indoctrinated toward a pre-conviction treatment mentality, then much could be done to break the cycle of repeated apprehension, release and/or prosecution. To date, this suggestion does not seem to have been greeted with much acceptance at any level.

Regardless of this, if we are to interrupt the recurrent cycle of apprehension and release, only reorientation of enforcement personnel will produce any real possibility of changing the street addiction problem. Larger and larger seizures of narcotics alone will not solve the problem, since it has been variously estimated that no more than 1-10% of all narcotics entering this country is subject to seizure, at best.

One other approach may be mentioned, but it, too, depends upon the effective reorientation of enforcement personnel if it is to have any chances of success. This approach is what might be called Dilaudid

Transition. It suffers from all of the problems of patient appeal, patient selection and others inherent in intruding into recalcitrant street addiction. By "Dilaudid Transition" we mean the offering of dilaudid in acceptable doses to heroin addicts to entice them into clinic settings. Dilaudid itself is known to most street addicts, both through certain physicians who prescribe it and as an available alternative drug on the illicit market. The drug, as a heroin substitute, has a high degree of acceptability to street addicts. It would, of course, be required that they return to a clinic three or five times a day to be injected with dilaudid. In order to avoid the issue of needles, the dilaudid could be injected subcutaneously or intramuscularly with a needle-free syringe, one variety of which is made by Schuco.

Under this proposal, dilaudid would be provided for those addicts whose history demonstrated an extreme reluctance to engage in methadone treatment. That is, only addicts with at least a three to five year history, and who have never had any experience with methadone would be acceptable under this program. Furthermore, such addicts would be recruited directly from the street and/or medical or enforcement sources. The Dilaudid Transition program would last no longer than one to three months, during which time the patient would be progressively switched from full dilaudid programming to a full methadone program, at a schedule to be mutually determined with the patient. To evaluate the effectiveness of such a program, the dropout or retention rate would be critical in determining its usefulness.

In summary, a close inspection of the mythical heroin maintenance program of Britain reveals such a program to be almost non-existent in reality. Regardless of that, we must find alternative means of reaching those addicts who are most resistant to treatment. The proposals of the Vera Institute and Dr. Avram Goldstein represent steps in the right direction. Our own suggestion of Dilaudid Transition is another consideration along these lines. But the most important step toward interdicting and/or rescuing recalcitrant street addicts is that of reorienting enforcement and criminal justice personnel to be the first line of defense and referral, rather than instruments for the continual retreading of these tired and reluctant addicts along their same unsuccessful paths.

Methadone maintenance, properly evaluated, has been highly successful with those on whom it has been tried. Drug free programming has been considerably less successful with narcotic addiction, and is perhaps far more appropriate for poly-drug problems. Despite the problems encountered, there is much to be satisfied with about the current status of the treatment of narcotic addiction. There are, however, very compelling questions, currently unanswered, about how to reach the unreached street addicts. Heroin, morphine and dilaudid are approaches that may work on some individuals. But only if we use all the resources available, including enforcement as an effective treatment modality, can we possibly resolve our most difficult concerns regarding treatment resistant addiction.

REFERENCES

Bewley, T.H., James, I.P., Mayhon, T. 1970. *Evaluation of the Effectiveness of Prescribing Clinics by Narcotics Addicts.* United Kingdom: 1968-1970.

Blumberg, H.H. et al. 1974. British Opiate Users: I. People Approaching London Treatment Centers. *Internatl. Journal of Addict.* 9: 1-23.

Goldstein, A. 1976. Heroin Addiction: Sequential Treatment Employing Pharmacological Supports. *Archives of General Psychiatry.* 33: 353-358.

James, I.P. 1971. The Changing Pattern of Narcotic Addiction in Britain, 1959-1969. *Internatl. Journal of Addict.* 6: 119-134.

Johnson, B.I. 1977. How Much Heroin Maintenance (Containment) in Britain. *Internatl. Journal of Addict.* 12: 361-398.

Moore, M.H. 1977. *Buy and Bust: The Effective Regulation of the Illicit Market of Heroin.* Lexington: Lexington Books.

Rector, M.G. 1972. Heroin Maintenance: A Rational Approach. *Crime and Delinquency.* 18: 241-242.

Stolff, P., Levine, D.B., Spruill, N. 1975. *Public Drug Treatment and Addict Crime.* Arlington: Public Research Institute.

Vera Institute of Justice. 1972. *Heroin Research and Treatment Program Summary.* New York: Vera Institute of Justice.

"DROP-INS," "DROPOUTS," AND "TREATMENT FAILURES" IN METHADONE MAINTENANCE

Jordan Scher, M.D., Barry Crown, Ph.D., Helen O'Connor, R.N., Douglas Reginetz, A.B.

National Council on Drug Abuse and the Methadone Maintenance Institute, Chicago, Illinois

The problem of success or failure in methadone maintenance treatment is a very complex one. Many patterns occur within a patient population regarding their stability as patients and the way they relate to the concept of detoxification.

Most clinics have observed that the initial patient is highly unstable in his ability to relate effectively to the clinic. Various estimates have indicated a dropout rate on new patients of 20-50% within the first one to three months. Dole (Dole and Nyswander, 1966) has said that any clinic which does not have a 75% retention rate is probably not doing its job. It is very difficult to precisely determine the proper retention rate at one month, three months, six months, one year or longer. And since retention rates vary markedly from 20-85%, depending upon the clinical setting one is considering, the reasons for this are crucial. Among these are the patient himself, the clinic, the regulatory stringencies under which clinics operate and the interaction of these, and other, elements.

The Methadone Maintenance Institute has seen 3763 patients since March of 1972. It maintains an average census of about 550 patients. In 1975, there were 565 new patient admissions while 785 new admissions occurred in 1976 and 450 new patients were administered in 1977. Whether or not we can take these figures to reflect the relative number of addicts on the street during these years is unlikely, but the above figures clearly reflect changes in clinic circumstances. MMI competes with some 45 public drug clinics and, within the past year, two other fee-for-service clinics have been opened. It is quite probable that the drop in new patients between 1976 and 1977 reflects the presence of the two other proprietary clinics.

Also, in trying to determine the movement of patients among clinics, possible patient oscillation between various facilities must be considered. This means that over the past year, at least, some patients may have detoxified in our clinic, only to show up at one of the other fee-for-services clinics within a few days to a year. The same situation applies in reverse but, at least in these circumstances, the near outcome for such a patient is fairly clear. Other patients who may drop out of our clinic within a few days of admission, may return to MMI in a few days to a year or two. Another group, after dropping out as either short or long-term patients, will return to one of the other clinics, rather than to us.

A close study of the new patients for 1977 was made in an attempt to discover what patterns of clinic attendance occurred within this group. Of our 450 new patients in 1977, 90 (20%) were found to drop out within the first week of treatment, and within the first month 31 (7%) more dropped out. Of all these early dropouts, about 71 (50%) returned for a shorter or longer period of time within the next six months. It is therefore very difficult to determine clearly at what point a patient may be called a "treatment failure" among these early dropouts.

Programs may, therefore, differ markedly in how they view the patient who tends to be a chronic, or recurrent, dropout. Several years ago, it was our policy to inform recurrent droppers in and out that if they repeated the process more than three or four times within a six month period, they would not be reinstated. Over the last year and a half, this policy was relaxed somewhat in order to study the process of recurrent dropping in and out. Once this policy was changed, some patients dropped in and out on frequencies of once to twice per month, continuing this behavior for periods of three and four months at a time. In order to assess these patterns, a group of 60 successive patients with an erratic pattern of attendance was selected out for special observation.

One of the possible causes of recurrent dropping in and out may have been dosage. In our clinic, most of the patients are maintained on an average of 40-50 mm. of methadone. With the cooperation of the physician, patients may increase their dosages to as high as 100 mm. without too much difficulty. At the clinic, decreases of methadone are negotiated with the nurse on an immediate basis and may be changed with the next dose in the dispensary. The process of increasing the dosage is slightly more complex, but not too difficult, in that the patient pays a special visit to the physician whenever he may feel he needs an increase. Increases or decreases are generally stipulated at 5-10 mm. per day, and usually not more than 20 mm. in any three day period.

In an effort to discover the basis for the frequency of such cyclic dropping in and out behavior in this study group, patients with regular patterns of behavior were interviewed on each return to determine what we could learn in order to help interdict this activity. As a result of reviewing these individuals, it was our feeling that they should be called "drop-ins", rather than "dropouts".

Among the reasons given for such cyclic behavior were the following. Patients stated that their jobs interfered with regular attendance at the clinic for six or seven days per week that may have been required. Others stated that they periodically did not have the money to pay the weekly fee. This, however, is not a valid reason, since patients are not refused medication if they fail to meet their service fee requirements. In fact, many patients have been frequently carried as long as several weeks when they were unable to keep up their payments, and have then been found jobs, transferred or detoxified if money problems persisted. Obviously, the cost to an addict is far higher in every way if he chooses to return to the street. Other patients asserted that they preferred to get high by using their heroin, and consequently preferred to drop in and out as the spirit moved them.

In view of the fact that this group of cyclic, recurrent drop-ins frequently admitted that their motivation for help was not great, it is surprising that they did not, as a group, try to achieve the highest dose of methadone possible. In fact, it would appear that this group actually resented increased methadone which might have achieved something close to a blocking dose. An obvious explanation for this behavior is that these patients did not wish to be on a dose of methadone that would impair their ability to simultaneously get high with heroin or other drugs. Their urine patterns tended to confirm these facts.

These cyclic drop-ins seemed to be bent on establishing a pattern of slight to partial commitment to the treatment process. It was as though they wished to keep the line open to the treatment situation until such time as they were able to seriously commit themselves to the treatment process. Among the group of chronic, transient drop-ins, about a third tend to settle down to regular attendance in the clinic after six or eight months, another third continue the process of oscillating attendance indefinitely and the final third tend to drop out altogether over a year's time. In our study, two of these drop-ins continued to appear at least twice monthly for a two year period of duration, while another small group of these same patients succeeded in very rapid, and apparently successful, detoxification.

At the other end of the spectrum, in our clinic about 45% of the regularly attending active patient population is in various stages of attempted detoxification. Of these patients, anywhere from 5 to 20 will complete the process of detoxification in any month. Among these will be those patients who enter the program and wish to detoxify quickly within the first month or so. Another group are those who have been in the clinic from a month to two years, may decide on a very rapid detoxification within a few weeks to a month, or may even drop out of the program to detoxify on their own, with or without any other medication to see them through. (Scher, et al., 1974; Martin, 1973)

Of those who do succeed in detoxifying, it is often difficult to determine their rate of success. In our clinic, all patients are followed up within two weeks after they become inactive for whatever reason and again in six months. Unfortunately, since we are often dealing with a frequently moving population, telephone or letter follow-ups may be unsuccessful in tracking the patient. Indirect assistance may be provided about the subsequent fate of inactive patients in several ways. The patient may show up at a state clinic or another private clinic, in which case records will be sought from us. Or the patient may return to us, and we will therefore have some indication of what has happened to him.

Of the 450 patients admitted in 1977, approximately 75 detoxified within one to six months. About 150 stabilized their methadone and remained in maintenance. Another 81 dropped out within six months to a year. Of the 75 who detoxified, primarily from indirect evidence, it would seem that 50-60% of these have remained drug free, or at least have not appeared on another local program or returned to our own. Among the new patients (under six months) on the program who attempt to detoxify at a much slower rate, about 20% were able to detoxify in 1977. Of these, approximately half (38) have been drug free six months or longer, from our direct and indirect evidence.

Thus, many different patterns of maintenance and detoxification exist among short-term and long-term patients, and there is no indication that either the rate or style of detoxification is any determinant of the success of that detoxification. Patients who suddenly detoxify from a dose of 60-100 mg., without assistance, within a few days to a month, (Scher, et al., 1974) may be just as successful or more so in remaining drug free six months or longer than those who detoxify slowly over a period of six months to a year. In our experience, older patients, those in the program for six months to a year or longer, tend to be more successful in detoxifying than newer patients of from one to six months. And we do not agree, as some have asserted, (Senay, 1977) that it is more difficult to detoxify from a lower dose than a higher dose of methadone.

If approximately 50% of those who do detoxify voluntarily are successful on the first attempt for six months or longer, another group may be successful on a second, third or fourth attempt at one to six month intervals. Return to the clinic after such a successful "detoxification" can be a very emotional and traumatic experience to a patient who feels that he has successfully bid good-by to methadone forever. In fact, it is sometimes only with great reluctance that such detoxified patients can be induced to return to the treatment situation. Once the patient has gotten over this sense of failure, he may often have a far more successful experience of detoxification the second time around.

Many patients who attempt to detoxify frequently, find that they undergo considerable psychological and physiological turmoil in the process. (Scher, et al., 1975) This may either drive them back to methadone or to a higher dose of it. In fact, several years ago, the emotional turmoil some patients seemed to undergo in the process of detoxification caused us to postulate that, for certain patients, methadone could well be a vital mental stabilizer. This conclusion was drawn prior to the isolation of the endorphins. (Marx, 1976; Goldstein, 1976) Strictly on clinical grounds, it was postulated that some physiological disturbance or deficiency, beyond the mere withdrawal of the narcotic, had occurred. (Scher, et al., 1975) Subsequent to that time, various studies regarding the endorphins have specified the nature of the physiological disturbance more accurately.

Thus, many patients who might have been successful detoxifications have been discovered to be far more significantly physiologically wedded to the narcotic than had been previously thought. Withdrawal itself, in some cases at least, is the smallest part of the problem. This physiological deficit must be made up over weeks to months, and it is well known that patients may be re-adapting as long as a year later. Some patients say that, like the "dry alcoholic", the urge, or the memory, of the narcotic experience is always with them, day and night, for the rest of their lives. So even those patients who may successfully detoxify may find themselves, one to twenty years later, recalled to readdict themselves, despite however much they may wish to resist this urge.

What then is a treatment failure or a treatment success? Despite the fact that addiction is centuries, if not eons old, it is only in the last ten years of intensive, more or less controlled study, that we have begun to understand the stages of addiction; the urge to addict oneself; the accident of the earliest, naive experiences with narcotics; the satisfaction of some patients to remain indefinitely in an addicted situation; the desperate demand of other patients to free themselves from addiction; finally, the difficult passage to complete freedom from addiction, and the painful, perpetual resistance to a return to it.

Treatment failures are clearly a matter of how one defines the ends and goals of treatment. If a treatment success is defined as disruption of the life cycle of street addiction, then many patients have been freed from this burden for a longer or shorter period of time. If treatment success is defined as stabilization on a methadone program, then many other patients qualify for this definition of success. If treatment success is defined as detoxification for six months or longer, then obviously the number of successes markedly diminishes.

Until we understood, as we have better within the last few years, that detoxification from methadone is not only difficult in the first instance, but remains exceedingly difficult for the next six months to a year or longer, we had no concept that merely removing the narcotic was not in itself the final stroke in the process. As a matter of fact, as long ago as 1961, one of us (Scher, 1961) suggested that incarceration, hospitalization and other forms of enforced detoxification, were vital elements in the reinforcement of the addictive process itself. By this we mean that, once freed of the tolerance induced by chronic and regular use of a narcotic, the individual would then be almost compelled to seek a return to the addicted state, once he was released from the restraining situation. In the light of the recent work on the endorphins, this may be understood as evidence of the reduced levels of the hormone, enkephlin, in the brains of these post-addicted individuals.

Ben Yehuda (1977), in an extensive study of our patient population for his doctoral thesis, completed in December of 1977, attempted to define two classes of addicts on an existential spectrum. He defined these as "past-oriented", i.e., those who are oriented to what appears to be an unchanging psycho-social way of operating; and "future-oriented", i.e., those who seem to be more involved in meeting the demands of a current and progressively more acceptably oriented way of operating. It was his observation that the past-oriented individuals tended not to detoxify, or, if they did so, that they would probably not be successful in remaining drug free. The future-oriented individuals, if they attempted to detoxify, appeared to be far more successful in their ability to remain drug free once they had done so. It may be that these two differing orientations may actually represent differing physiological hormone levels, or differing rates of recovery of the endorphin levels of the addicted individuals.

As a consequence of all of the above considerations, the whole range of addictive experiences, which result in refusal to enter clinics, indefinite establishment on methadone programs, spontaneous self-detoxifications, arbitrary and unexplained dropouts, cyclical and interminable drop-ins, and finally successful and unsuccessful short and long-term detoxifications, tell us that all of these patterns represent a spectrum of psychological and physiological continuity and discontinuity at the same time. It is almost impossible to separate out any of these variations from any other to say that here we have a success and there we have a failure.

We are looking at a matrix of indefinite variations and are interrupting a process at arbitrary points along its course. Simple statements about treatment failures and treatment successes have no meaning in the context of this wide ranging, multi-faceted, living movement in process. It will only be possible sometime hence, and at a twenty year remove, for us to say with any validity that our experience in evaluating various aspects of the treatment situation can be called a study of success and failure.

Thus, much as we would like to draw firm conclusions and make strong recommendations about the meaning of success and failure in the treatment of narcotic addicts, at this point, regardless of our efforts and our own desires to draw firm conclusions, all too much remains to be done and understood in the future about the nature and treatment of addiction to be assured of what success and failure will eventually come to mean in the existential experience of narcotic addiction.

REFERENCES

Dole, V.D. and Nyswander, M. 1966. Rehabilitation of Heroin Addicts after Blockade with Methadone. *N.Y. State J. of Medicine*. 66: 2011-2017.

Goldstein, A. 1976. Opioid Peptides (Endorphins) in Pituitary and Brain. *Science*. 193: 1081-1086.

Martin, W. et al. 1973. Methadone - A Reevaluation. *Arch. Gen. Psych*. 28: 286.

Marx, J.L. 1976. Neurobiology: Researchers High on Endogenous Opiates. *Science*. 193: 1227-1229.

Scher, J. *Group Structure and Narcotic Addiction: Notes for a Natural History*. Chicago: NCDA.

Scher, et al. 1974. *Self-selected Preferential Autonomous "Cold Turkey" Detoxification in the Course of Methadone Maintenance Treatment*. San Francisco: North American Cong. on Alcohol and Drug Problems.

Scher, et al. 1975. *Methadone Maintenance as a Mental Stabilizer*. New Orleans: NDAC.

Senay, E.C. 1977. Withdrawal from Methadone Maintenance. *Arch. Gen. Psych*. 34: 361-367.

Yehuda, B. 1977. *The Myth of the Junkie - Towards a Natural Typology of Drug Addicts*. Chicago: doctoral thesis.

BREAD & ROSES -- PRESCRIPTION FOR STAFF BURN-OUT SYNDROME

Peter D. Rogers, Ph.D. and Kathlyn T. Spencer, D.T.R.

"Burn out" is defined as: " . . . to fail, to wear out, or become exhausted by making excessive demands on energy, strength or resources." Workers in the "helping professions" are particularly vulnerable to Burn-Out due to conditions of long hours, short pay, and a clientele composed of people expert at ripping off energy and nurturance. Staff members suffering from Burn-Out Syndrome often manifest the following physical signs/symptoms: 1) feeling exhausted or fatigued (run down), 2) inability to shake a cold, 3) frequent headaches or G.I. disturbances, 4) sleeplessness, 5) depression.

Burn-Out Syndrome may be observed directly by looking for changes in body posture and movement patterns. A literal translation of "inertia" can be observed, where the ongoing pattern of movement--plodding, speeding, etc.--is continued until an outside force (a schedule change, for example) intervenes. The pattern is then changed by virtue of bouncing off this "wall" and a new pattern is started which continues until it meets another outside force. This movement style, combined with the frequent loss of one's relationship to one's weight (impact), can give the appearance of bouncing "off the walls," i.e., the person has lost the stabilizing factor of internal process input. A sinking response to weight--heaviness--takes the place of an active relationship to one's weight. These people literally depress their bodies, themselves.

Staff members suffering from Burn-Out Syndrome may manifest some of the following behavioral signs: 1) A person who used to be an active talker now silent at staff meetings (as though resigned to a hopeless situation); 2) People quick to anger (constant irritability and continually feeling frustrated); 3) "Paranoia"--as though other staff members are out to screw him--followed by self-isolation and feelings of omnipotence (I can do it all by myself); 4) Rigidity--people closed to any new idea (all change seen as threatening); 5) A person who is totally negative (is often seen as the "house cynic").

Burn-Out Syndrome is often manifested in a change in ability to take in information. The person's normal feedback systems become dissociated from the needs of the moment, and his movement style tends to become more extreme, polarized. This is particularly noticeable in the tension-flow pattern, i.e., the flow between movement characterized by a "going-with" quality as contrasted with movement which is held, careful, able to stop. One can observe sharper transitions in movement; for example, a staff member with a "bursting" flow style (who communicates in spurts) might move toward a more explosive pattern (held in longer and released more vehemently).

Dissociation between inner and outer awareness may be noticed in a diffusion of attention, inability to focus, or shift in visual perspective. Dissociation is frequently observable in conflicted body postures and gestures; the body is literally trying to move in two directions at once. Cognitive and emotional conflict may be manifested by the head being far forward while the shoulders are pulled back and down, or the legs can be tapping and fidgeting while the whole upper body is still and heavy.

The quality of diffusion can also impart a sense of blankness, dullness; the person may appear "spaced out". Changes in breath pattern also indicate a breakdown in ther person's integrative abilities, particularly the ability to renew vital energy, to be refreshed.

TREATMENT AND RESISTANCE

In making an assessment of staff Burn-Out the "identified patient" is invariably the staff as a whole, not simply an individual staff member. We see all members of the staff as inextricably linked together in a system which is stuck. It is the consultants' job to find the sticking point(s) and through a combination of leverage and lubrication restore vital movement. Individual work directed toward regeneration and focused on activating the individual's resources for renewing inner rhythms is balanced with an analysis of those systems which keep the staff stuck (family dynamics and communication theory).

Our starting place is an exploration of what staff members see as being a problem *right now*. By "talking about" the problem we focus the resistance on the verbal level (where "helping professionals" are most facile), allowing us to proceed unhampered at a non-verbal, movement level. Our work is then directed toward discovering movement impulse, grounding in a satisfying rhythmic pattern, and exploring contact and withdrawal around points of "stuckness". Skills in relaxation, combined with fantasy and movement are taught, and may be incorporated in the ongoing work style of the group. Participants learn to contact their inner sources of weight-activation, so that they can reclaim their ability to affect the environment.

We are committed to the notion that men and women do not live by bread alone--we need roses too. Our multi-modality approach to staff Burn-Out problems encourages playfulness and exploration of new ways of being in the world. The people we work with learn to re-own personal power and develop life skills which enable them to "get it on", and regain on the job effectiveness.

FOSTERING PATIENT AUTONOMY IN A METHADONE MAINTENANCE PROGRAM

John A. Renner, Jr., M.D. and Howard Shaffer, Ph.D.

City of Boston Drug Treatment Program

> *"The goal of psychotherapy is to relieve a patient's emotional distress and help him modify personality characteristics that are preventing him from realizing his human potential or enjoying rewarding interpersonal relationships . . . At times the methods of psychotherapy are confused with its goals . . . "*
>
> (Weiner, 1975)

This point is clear: the goal of psychotherapy is demonstrable personal change. For the addicted client, therapeutic change must focus on replacing dependence, self-doubt, and denial with self-sufficiency, independence and autonomy. Can one justify the use of an addicting drug such as methadone in a treatment program where the goal is to release clients from their dependency and restore their autonomy? Are the goals of methadone maintenance in opposition to the goals of other forms of addiction treatment? Is there not a logical contradiction in a treatment approach that espouses the goal of autonomy, yet continues to provide its clients with an addicting drug? This paper will explore these questions and concerns. The authors contend that the use of methadone is consistent with the therapeutic goal of autonomy, and will illustrate how a program can maximize a client's opportunities to achieve autonomy in a methadone maintenance treatment setting.

THE NATURE OF AUTONOMY

Autonomy is not an absolute state; it is relative. Few men or women experience absolute autonomy and total independence from the demands of their fellows, their environment or their own bodily needs.

Indeed, there is little reason to believe that total autonomy is either desirable or possible. The mutual sharing of responsibilities is one of the great burdens and joys of the human condition.

Realistically, autonomy means the ability to direct one's life towards the fulfillment of the goals and obligations that one *must* accept and the goals that one *elects* to accept. For example, caring for one's bodily needs and caring for one's infant children are clearly obligations one must accept; seeking fulfillment in a particular area of sports or social interaction should ideally be options for us all. In a psychological sense, autonomy means the ability to *maximize* choices without neglecting obligations.

In discussing personal freedom, Carl Rogers (1961) has noted that "The individual is as free to be afraid of a new venture as to be eager for it; free to bear the consequences of his mistakes, as well as of his achievements. It is this type of freedom of responsibility to be oneself which fosters the development of a secure locus of evaluation within oneself, and hence tends to bring about the inner conditions of constructive creativity."

Given the compelling nature of addiction, addicts have little power to exercise personal freedom or the ability to determine their behavior. The addict's life becomes totally focused on drugs to the exclusion of those creative interests (friends, job, hobbies, etc.) that others enjoy. When viewed as a pattern of excessive behaviors (Gilbert, 1976) addiction excludes other responses and consequently leaves the addict feeling deprived and helpless in the pursuit of other goals.

An effective treatment program must therefore deal with both their client's drug use and the effect that this has had on other aspects of his or her life. To focus only on the addict's drug-seeking behavior is to miss a major part of the problem. Once the drug-seeking behavior is controlled or eliminated, then it becomes possible to deal with the addict's ability or *inability* to function in an autonomous manner. This is the goal of most addiction treatment.

Therapeutic communities are able to achieve control over clients' drug use by placing them in a drug-free environment, from which they have few reasonable options for escape. Once in this environment, treatment can begin. Methadone maintenance programs achieve similar results by shifting the addict from a variety of highly intoxicating, euphorogenic drugs to methadone which prevents withdrawal, keeps the addict feeling physically comfortable and alleviates the need to seek and use other drugs. The degree to which the maintained addict continues to experience euphoria is related to the dose level and the duration of treatment. Once stabilized on a moderate dose of methadone (40 mg. to 60 mg.) most addicts experience little or no euphoria, and certainly do not manifest any of the debilitating signs of intoxication seen during intravenous heroin use. Once such stabilization has been achieved, the addict is comfortable physically, but now must face the often frightening experience of coping with the real/straight world. His or her options greatly expand, but many addicts need considerable assistance to deal effectively with these new opportunities. This then becomes the task of *therapy* with the methadone client.

It is important to recognize that both the therapeutic community and the methadone maintenance program provide a safe, comfortable environment in which treatment can occur. However, the control and the physical comfort provided by methadone, like the restraints and interpersonal comforts provided by a therapeutic community, are not, in and of themselves, the essential core of the therapy. These are, however, necessary requirements if therapy is to occur. Vaillant (1975) has clearly described how a secure, controlled environment is necessary for the psychological treatment of the addict. Similarly, Khantzian (1977) has focused on the chemical supports and controls provided by the use of methadone. Such controls can be present both in the drug itself and in the structure of the treatment program; each is necessary if the addict is to develop any autonomy. In describing how a child achieves autonomy, Erik Erikson (1963) noted that " . . . firmness (parental) must protect him against the potential anarchy of his as yet untrained judgment . . . " Similarly, a drug treatment program must provide a consistent structure within which rehabilitation can occur.

Without even addressing the goal of eventual detoxification from methadone and thus complete independence from either drugs or a treatment environment, it is clear that an addict's options for autonomy are greatly expanded simply by being placed on methadone. Although chemically dependent, the methadone client no longer has to engage in the excessive behavior patterns necessary to maintain a drug habit; therefore, he or she is free to engage in alternative, more productive behavior.

Unfortunately, programs which dispense methadone can often be as dependency inducing as the drug itself. This danger is shared by many other forms of treatment (e.g., psychoanalysis, psychotherapy, therapeutic communities, etc.). All psychiatric institutions are aware of the destructive effects of excessive dependency, yet recognize that some degree of dependency is necessary for the treatment process. The dilemma for the therapist/institution is how to sustain enough dependency for adequate treatment without undermining the client's capacity for autonomy. This problem has been discussed at length in the literature dealing with psychotherapy (Rogers, 1961; Pakes, 1975).

Surprisingly little has been written about this issue as it relates to methadone maintenance treatment. This is unfortunate, since the therapist's dilemma is so acute and obvious in this setting. Fortunately, many programs have developed treatment techniques that facilitate client autonomy and thus counterbalance the dependency inherent in (a) the use of an addictive drug and (b) the process of a supportive treatment setting. Such treatment programming can be difficult because federal and state regulations tend to inhibit the originality and spontaneity that other programs, particularly self-help therapeutic communities, have used to encourage the development of autonomy in settings that are usually heavily controlled, authoritarian and likely to produce dependency.

The therapeutic community client must first transfer his or her dependency from the drug to the group (and/or program) before he or she is able to function without the supports of the group. Therapeutic communities often foster extreme dependency as a necessary step in the progression to autonomy. This is not unlike the peer group dependency that adolescents utilize in their transition from childhood dependence on family, to adult independence.

In spite of the client's tendency to remain "frozen" in a state of dependency on both methadone and the treatment program, a similar therapeutic process can occur in methadone maintenance treatment. Even with the constraints of federal and state regulations, there are many techniques that can be utilized to maximize the client's opportunities for personal growth and to counterbalance the dependency inherent in being a *client*, using *drugs* and participating in a *program*. This paper will explore these three areas and describe techniques used by the City of Boston Drug Treatment Program (and other programs) to increase client autonomy.

THE CLIENT

The primary focus of work with clients must be to help them recognize the consequences of their behavior and their responsibility for their progress or lack of progress in therapy. Rewards and privileges must be clearly earned by positive changes in behavior and not by promises. This can best be achieved by the use of contingency management on the basis of contracting. "The treatment contract in psychotherapy is an explicit agreement between patient and therapist to work together toward alleviating the patient's psychological difficulties. To arrive at this contract, patient and therapist need first to agree that treatment is indicated and will be undertaken. Next, they need to agree on the objectives of the treatment and on the procedures they will follow in working toward these objectives. And, finally they need to agree on such specific treatment arrangements as the time, place, frequency, and fee for sessions." (Weiner, 1975)

Contingency Management on the Basis of Contracting

Impulsive clients typically come to treatment as a result of external pressures, i.e., involuntarily. In order to efficiently and effectively treat a substance abuser, treatment should be voluntary. It is this dilemma (paradox) that often confronts the clinician: can an obviously involuntary situation be reframed so that a client adopts specific ends towards which she or he will work voluntarily? We believe the answer is yes.

The concept of contingency management rests upon the development of a treatment contract which is negotiated by client and therapist. The term "negotiated" cannot be overemphasized. Substance abusers frequently indicate that "things are fine", they are not uncomfortable, and that they are entitled to far more of life's rewards than they have received. Also, addicts often deny negative feelings. They are typically unwilling to share their anxieties, since this would be a display of weakness. In order to develop a treatment plan, clinicians can impose these issues on their client by including them in a non-negotiated treatment plan, e.g., you should acknowledge negative affect, build self-esteem, learn to trust others, etc. Nothing is easier for a character disorder to manipulate then an imposed treatment plan. Imposition of such a "treatment plan" on a client is a challenge from which the client will emerge victorious. Therapy sessions soon will fill to the brim with "negative affect", a sense of self-improvement and worth following a cathartic episode, and ultimately statements like, "I can really trust you--you're different, you care", etc. In this manner, clients conform to the requests of an imposed system only to receive its rewards and respond to its challenge, thereby, "beating the system."

An alternative to the system described above presently exists: contingency management. Contingency management is based on a progressive system of rewards, responsibilites, and freedoms based upon the demonstration of consistent patterns of behavior (which the client and therapist cooperatively choose), not upon possibilities or promises of change. One example of this technique is goal attainment sealing (Kiresuk and Sherman, 1968). A variant of this approach, a level system, has been implemented by the City of Boston Drug Treatment Program.

The Level System

The level system was developed to (a) monitor client progress in treatment, (b) provide an equitable mechanism for reinforcing client's progress with appropriate rewards and responsibilities, and (c) as a management tool to evaluate the effectiveness of each treatment unit.

All program clients are assigned to one of five levels corresponding to their length of time in treatment and their satisfactory compliance with program rules and the terms of their own treatment contracts. For example, all new clients begin on *level five*. After three months, clients can advance to *level four* if they have complied with their contract, have clean urines and not presented any disciplinary problems. *Level three* requires six months of clean urines and appropriate compliance with contract and program rules. *Level one* is reserved for those clients who have met all contract and program requirements and have also completed detoxification from methadone.

Depending on their level, clients are eligible for participation in various program activities and are awarded certain privileges. All clients can vote for client council representatives, but only those clients on levels one through four are eligible to run for the council. There is also proportionally greater representation on the client council for those clients on the more advanced levels. Some program activities are available only to clients on specific levels. For example, clients on the more advanced levels are given priority in the assignment of medication dispensing time slots. In addition, only clients on levels two and three are eligible to apply for take-home methadone privileges.

Individual Treatment Contracts

On an individual basis, therapists should cooperatively develop a treatment contract with each client. This contract should revolve around those areas that the client identifies as requiring change and not upon values and goals which the therapist has determined a priori. For example, a client might say, "really, everything's o.k.--there's nothing I want to change or do differently."

The therapist might respond, "Everything?"

Client, "Well, not everything."

Therapist, "What then?"

Client, "I'd like to make more money--yeah, more money."

Therapist, "You're unemployed, how are you going to make this money?"

Client, "I don't know, get a job, hit the number, go to the track."

Therapist, "There are training programs around, vocational programs, maybe *we* can look into them. I can't help with those other ideas, though--they are based on luck and *neither* of us can do anything about that."

This conversation may continue until a firm agreement is made to acquire and examine brochures from vocational training programs during the next week and discuss them in the next session. Although this example is overly simplified, it illustrates some important therapeutic principles: (1) there is always an area that clients will suggest changing--probing will help identify these areas; (2) given the opportunity, clients will generate--in response to proper conditions and verbal probes--areas which are of real concern to them, areas in which they are most willing to exert some effort; (3) progress can be identified in observable, measurable steps, e.g., acquiring and examining brochures, and (4) purposeful behaviors which are directly a function of the task at hand will be performed with little resistance. Thus, the client above might have to go to the library or local school to find out where brochures and information can be obtained, make telephone calls, etc. In sum, contingency management approaches require the negotiation of a treatment contract; clients must plan, delay gratification, cooperate, improve object relations, exercise ego functions, reality test, etc.

Achievement of contracted goals should be followed by interpersonal praise and discussion within the therapeutic relationship, i.e., how did the client cognitively explain his or her success; how does this affect their attempt to deal with a future situation, etc? In addition, major changes in behavior and achievement should be followed by recognition, increased privileges, and responsibilities within the treatment setting (clinic).

THE DRUG

There are several ways that a program can minimize the potential for the dependency that is inherent in methadone. One avenue is to maintain the patient on as low a dose of methadone as possible.

Dose Ceilings

Since Goldstein (1972) demonstrated the efficacy of low methadone doses in comparison to high doses, many programs have gained considerable experience managing clients on lower doses and have found that it works well. In the dose range of 40 to 60 mg., clients have fewer side effects, are more functional and less likely to feel "narcotized", and there is a decreased sense of psychological dependence on the drug. Detoxification is usually seen as an easier process for those individuals on lower doses.

For new clients on methadone, we have tried to establish a dose ceiling of 50 mg. This has worked well, particularly when coupled with self-regulated dose schedules. Most clients keep themselves well below the maximum allowed. For older clients, especially those stabilized for several years on higher doses, we have not *required* that they lower their dose. However, we have encouraged them to try a gradual self-regulated dose reduction to a lower level. Almost without exception, most clients have been willing to try this and have successfully lowered their dose. The knowledge that they can go back up if they feel uncomfortable seems to be a major factor in their willingness to "risk" a dose cut. Once they recognize that they can manage on a lower dose, even long-term maintenance clients are accepting of this approach.

Open Doses

Permitting clients to know their methadone dose (open dose schedules) can also be a useful way to encourage client autonomy. While it may seem easier for staff and clients when doses are blind (unknown to the client)--the open approach is always more realistic and forces clients to deal with the anxiety they invest in their dose schedule.

Self-Regulated Medication Schedules

Originally, the procedures of methadone maintenance and detoxification protocols required that dosage schedules be controlled by attending physicians (Nyswander, 1957; Langrod, Brill, Lowinson, and Jospeh, 1972; Dole and Nyswander, and Kreek, 1966). These protocols were based on the premise that addicts would (a) attempt to obtain as large a quantity of methadone as often as possible; (b) increase their fixation on drugs and avoid dealing with the more fundamental issues of treatment, and (c) compete with other patients to "see who can maintain the highest dosage level." Despite these beliefs, Renault's (1973) noteworthy research demonstrated that there was "no evidence to support the position that patients should be ignorant of their dosage. On the other hand, strong arguments can be made for a patient knowing his dosage. Knowledge of dosage may encourage a more adult attitude in patients by reducing the mysteries surrounding methadone and by diminishing the view of the physician as an all-knowing and all-powerful figure." Similar results were obtained by Angle and Pawatikar (1973) in a detoxification experiment. This study offered surprising support for the concept of self-regulated dosage since strong competition developed between the subjects in this study to reduce their methadone intake to very low levels. Also, Angle and Pawatikar indicated that patient's somatic complaints were reduced when they controlled their dosage levels as compared to physician determined dosage.

These findings point to some important implications for treatment of the "impulsive personality." The self-regulated schedule permits--in fact requires--that patients control and manage one very important component of their lives, i.e., the decision to utilize drugs according to some rational plan. This is not a task without therapeutic impact. Typically, adddicts are poor planners who do not learn from experience, and treat the events of their lives as discontinuous and separate. Utilizing the presenting problem, opiate addiction, self-regulated dosage permits the clinician to institute a plan of constructive treatment which would otherwise elicit enormous resistance. With few exceptions, patients welcome the opportunity to determine their own level of medication--and why not, they have typically been doing this for years.

In this section, we have reviewed evidence which indicates that self-regulated medication schedules will not increase patients conning and manipulating in order to receive higher doses of methadone. Indeed, the studies examined here suggest that there is a reduction in patients' attempts to "beat" the system. Stern, Edwards, and Lerro (1974) have even considered that the adversary relationship traditionally found in physician-controlled treatment settings is sufficient to produce most of the sociopathic behavior which is observed therein. While acknowledging that many patients are not serious about their rehabilitation, self-regulated dosage permits the clinician to (a) facilitate addict responsibility for their own behavior, e.g., detoxification schedules, without permitting abuse of that responsibility, (b) reduce the inherent resistance to treatment and concurrent behavior change, and (c) facilitate a positive shift in staff-patient interaction. "Since the methadone-on-demand technique seemed to avoid pathological confrontations by placing the major onus of treatment on the addict himself and since this inclusion of the addict in the treatment planning has the effect of appealing to his maturity rather than to the regressive and childish aspects of his personality, we believe there is reasonable justification for this approach to detoxification" (Stern, Edwards, and Lerro, 1974). It is the appeal to a patient's strengths and postive ego functions that facilitates the efficacy and positive patient response to treatment.

Lower and open doses fit easily into the self-regulated medication approach. The program we utilize involves more active physician control during the first few days of treatment when dose adjustments are made based on clinical signs of withdrawal or evidence of over-medication or intoxication.

After a client has been stabilized on a dose for three to four weeks, he or she can then be transferred to a self-regulated schedule. This usually occurs at a time when a stabilized client either requests an increase or desires to begin detoxification, or at the time of a general review of treatment progress.

The physician establishes a maximum daily dose--usually the client's current dose, or a dose five to ten mg. above the stabilized dose. Orders are then written so that the client may request a PRN dose change only once a week from the nurse (on a specific designated day) increasing or decreasing his/her dose from one to five mg. per week. The dose adjustment may be spread over several days (e.g., one mg. cut every other day) if the client so wishes, but the client may not reverse the requested dose change until the following week.

This insures that the client will follow through with the change once the decision is made and it forces him or her to think more carefully before requesting a dose change. An additional advantage of the one change per week rule is that after several days clients have usually passed the minimal discomfort associated with a small dose cut and are therefore much more likely to remain on their new dose level rather than impulsively requesting a return to their previous dose.

Since clients can never exceed their maximum dose, nor go up or down more than five mg. per week, it is virtually impossible for them to either become intoxicated on an excessive dose or to experience undue withdrawal symptoms.

LAAM

Another option that can greatly enhance client autonomy is the use of levo-alpha-acetyl-methadol (LAAM). This longer-acting derivative of methadone lasts from 48 to 72 hours and requires that the client attend the clinic for medication only three times per week. This greatly reduces the client's dependency on the clinic and permits a much more normal lifestyle. Since LAAM presents none of the diversion problems associated with take-home methadone, it is clearly a major technical advance in maintenance treatment.

THE PROGRAM

There are many areas of program management where it is possible to engage the client in an active partnership that can avoid or dilute the dependency inherent in the traditional medical treatment model. Self-help programs made the first giant step in this area (Yablonski, 1967), when they developed a treatment system that gave clients a clear opportunity to advance and ultimately share in the control of the program.

Such a system can be abused if it is only designed to reproduce ex-addict staff willing to work for menial wages--a clear example of how program needs can be better served than the needs of clients. Similarly, the authoritarian structure of some therapeutic communities can be as rigid and counterproductive as the traditional medical model. Given the medical model required for methadone maintenance and the need to meet federal regulations and federal funding criteria, how can program structure foster an environment that encourages client autonomy?

A basic requirement is a well-trained staff that respects clients as fellow human beings and does not see them as inferiors in need of rescue. Despite their good intentions, some traditionally trained staff often feel a need to keep clients in a secondary position and are very threatened by programmatic changes that jeopardize their superior position. In-service training is necessary to deal with these concerns, otherwise the staff may subvert efforts to involve the clients more directly in the operations of the program.

To achieve the goal of increased client autonomy, it is necessary that the program operate in such a way as to share real power with the client. This requires that clients or their representatives become active participants in the management of the program. Some form of elected client government is a helpful vehicle to channeling client input; a program "alumni" or graduate association is also a viable arrangement. We have established an elected Client Council in each of our methadone clinics. Such groups can take an active role in reviewing program policy and rules with final policy being the result of deliberations between staff and client representatives. Similarly, client representatives might participate in the process used to determine whether or not a client has violated program rules. Because of confidentiality regulations, this can obviously occur only with the permission of the affected client. Our program has successfully utilized elected client representatives as client advocates in such proceedings. Clients can also play an active role in community relations work and can be very effective advocates for program needs.

All individuals being hired for staff positions can be interviewed by client representatives or a client representative can participate as a voting member of a search committee established to select senior staff. We have used both of these approaches very successfully in hiring new staff.

Similarly, staff positions should be open for ex-addicts and current clients. Our program has been able to use CETA funds to hire clients still in treatment. It has been our policy, however, that clients still in active treatment should not work in the treatment unit where they are also a client. This helps to make a clear separation in their dual role as both client and staff. Our clients have worked successfully as Research Assistants, Administrative Aides, Clerks and Secretaries. It has been our feeling that clients still in active treatment should not be hired for counseling or senior administrative positions. The unique demands of those positions are more appropriately handled by clients who have completed their own rehabilitation. We recognize, however, that some programs have hired clients as counselors while they are still on methadone and have been comfortable with this staffing model.

An additional word about client fees. We strongly believe that all clients should pay some fee for service, even if it is a minimal one. Free service can imply a dependent relationship on a program--

clients who pay fees feel much more comfortable in demanding (as they should) appropriate quality services.

SUMMARY

Methadone maintenance treatment has been criticized as a substitute addiction that only encourages a client's dependency and fails to deal with the addict's basic problems. This paper shows how treatment concepts utilized in self-help programs and contemporary psychotherapy can be adopted for use in a methadone maintenance program. With increased client autonomy as a basic goal, methadone can be a useful treatment tool and maintenance programs can help clients develop the strengths necessary to eventually lead drug-free lives. Techniques recommended to facilitate this process include treatment contracts based on contingency management, open doses, self-regulated medication schedules, client governments and other forms of client participation in program management. The authors suggest that these techniques will foster client autonomy and discourage the excessive dependency that can occur in methadone treatment.

REFERENCES

Angle, H.V. and Parwatikar, S. 1973. Methadone self prescription by heroin addicts in an inpatient detoxification program. *Psychological Record*. 23: 209-214.

Dole, V.P. and Nyswander, M.E. 1965. A medical treatment for diacetylmorphine (heroin) addiction. *JAMA*. 193: 646-650.

Dole, V.P., Nyswander, M.E., and Kreek, M.J. 1966. Narcotic blockade. *Arch Internal Medicine*. 118: 304-309.

Dole, V.P. and Nyswander, M.E. 1967. Rehabilitation of the street addict. *Arch Environmental Health*. 14: 477-480.

Dole, V.P., Nyswander, M.E., and Warner, A. 1968. Successful treatment of 750 criminal addicts. *JAMA*. 206: 2708-2711.

Erikson, E.H. 1963. *Childhood and Society*. New York: W.W. Norton, 2nd ed.

Garbutt, G.D. and Goldstein, A. 1972. Blind comparison of three methadone maintenance dosages in 80 patients in *Proc. 4th National Conference on Methadone Treatment*.

Gilbert, R.M. 1976. Drug abuse as excessive behavior. *Canadian Psychological Review*. 17: 231-240.

Kanfer, F.H. and Phillips, J.S. 1970. *Learning Foundations of Behavior Therapy*. New York: Wiley.

Khantzian, E.J. 1977. The ego, the self, and opiate addiction: theoretical and treatment considerations in *Psychodynamic Aspects of Opiate Dependence: Research Monograph #12*. Rockville: NIDA.

Kiresuk, T.J. and Sherman, R.E. 1968. Goal attainment scaling: A general method for evaluating comprehensive community mental health programs. *Community Mental Health Journal*, 4: 443-453.

Langrod, J. Brill, L. Lowinson, J. and Joseph, H. 1972. Methadone maintenance: from research to treatment. In Brill, L. and Lieberman, L. (Eds.), *Major Modalities in the Treatment of Drug Abuse*. New York: Behavioral Publications.

Nyswander, M.E. 1967. The methadone treatment of heroin addiction. *Hospital Practice*. 2: 27-33.

Pakes, E.H. 1975. Dependency and psychotherapy: developmental considerations. *Am. J. of Psychotherapy*. 29: 128-133.

Renault, P.F. 1973. Methadone maintenance: the effect of knowledge of dosage. *Int. J. Addictions*. 8:41-48.

Rogers, C.R. 1961. *On Becoming A Person*. Boston: Houghton Mifflin.

Stern, R., Edwards, N.B. and Lerro, F.A. 1974. Methadone demand as a heroin detoxification procedure. *Int. J. Addictions*. 9: 863-872.

Vaillant, G.E. 1975. Sociopathy as a human process. *Arch. Gen. Psychiatry*. 32: 178-183.

Weiner, I. 1975. *Principles of Psychotherapy*. New York: Wiley.

STEPS AWAY FROM METHADONE: A BEHAVIORAL APPROACH TO THE DETOXIFICATION OF METHADONE MAINTENANCE CLIENTS

Demosthenes Lorandos, Ph.D.

Central Michigan University

The step program was creative heresy. It was conceived and implemented at the peak of the Nixon administration's "total war on street level pushers". Instead of trying to hold onto addicts while adjusting them to a crazy world, the step program was designed to pull them out of the dull world of maintenance. With psychologists, boatbuilders, artists and a rag taggle array of do-gooders the step program achieved some startling results. Of substantial value today, it must be cast in the proper historical setting to fully appreciate its impact.

When Richard Nixon became president in 1969 he was happy with the federal response to the drug problem. In 1970 an amazing number of "Crime Control" measures slipped through Congress. 1971 brought devastating changes in the Supreme Court. By 1972 the Nixon administration had increased spending for federal drug activities by eleven hundred percent. The "war on drugs" became a training ground for ambitious young Nixonites.

In eary 1973 there were over forty Nixon agencies engaged in drug law enforcement. The Office of National Narcotics Intelligence was hiding stolen files from the Ervin subcommittee on constitutional rights; the Law Enforcement Assistance Administration was funneling millions to state police for surveillance equipment; the National Institute of Mental Health was sponsoring programs to implant electronic devices in the heads of convicted drug traffickers; and the National Security Agency was testing a device that would gain covert access to every home in America.

Three programs are of special interest. The Office of Drug Abuse Law Enforcement DALE was created for and and run by a Nixon favorite:

Special Attorney General Myles Ambrose. Mr. Ambrose rose to the top in the Nixon drug police through years of outfoxing the Bureau of Narcotics and Dangerous Drugs with his Customs Agents. He was so dedicated that he had a sign on his Washington desk which read: "Bust A Junkie". When Ambrose was sworn in as the drug war general he vowed that his first action would be to "drive the street level addicts into treatment programs".

Unfortunately, there were fewer than ten thousand addicts enrolled in methadone programs in 1971 (Dole, 1971), but in less than two years, ten times that many would be "diverted" into maintenance. Credit for this tremendous upsurge in methadone maintenance belongs to DALE's team mate in the Nixon administration, the Special Action Office for Drug Abuse Prevention.

SAODAP's chief officer, Dr. Jerome Jaffee was especially appealing to the Nixonites. It was Jaffee who suggested photographing the feet of every methadone addict in the country. It was Jaffee who advocated civil commitment, mandatory minimum sentences and "quarantine" to "break the back" of heroin addiction in America. It was Dr. Jaffee who called for mass urine screening in the nation's schools to help cope with the "crisis".

The draconian alliance of SAODAP and DALE gave birth to another agency that completed the circle: The Treatment Alternatives to Street Crime project (TASC). Ambrose would bust 'em, Jaffee would fund the maintenance clinics so TASC was created to grease the works between arrest and "diversion". This was all easily done in the early Nixon years. The prevailing research extolled the merits of methadone in reducing crime by long-term maintenance (Pearson, 1969; Perkins and Bloch, 1970, 1971; Dole, 1971; DuPont and Katon, 1971; Cushman, 1971 et cetera). The "heroin epidemic" ballyhooed by Nixon, Ziegler, Ambrose, Jaffee, DuPont and the national media provided sufficient impetus to boost congressional allotments for drug police and treatment activities by tens of millions each year. It was in this climate that we began our heresy.

METHOD

In the fall of 1972 I was hired by Marin Open House in San Rafael, California to create an ancillary services-job training program. Having come from a background in therapeutic communities and other drug free settings I was none too happy about methadone. However, being a refugee from a Ph.D. program in New York with a new baby on the way, I was happy to find work in this, my home town. MOH was run by a group of radicals and bourgeois that only Marin County, California could assemble. Having recently been evicted from their home so a parking lot could be built, they were searching for a new place to locate. The building eventually picked was a huge old industrial laundry that had eight or ten offices blocked off in front. It was vacated by a Johnson administration agency when the great society went sour.

The building was perfect for a large shop. With almost two thousand square feet of empty floor space in back and twelve hundred square feet upstairs, it was a job training supervisor's dream. Several ancillary services staff members and I sold the treatment staff on the idea of creating a huge two part workshop that would be entirely built, run and governed by a cadre of the more enthusiastic clients.

The philosophy was simple. It was my feeling that people become addicted to stupidity more than to alcohol, heroin, food or cards; that in addition to being addicted to stupidity, our clients were dangling near the end of a long string of failure experiences. To turn this around we needed to provide for success experiences in a way that would allow those who were terrified of "success" to work into it slowly. I knew from past experience that our clients were particularly wary of give-away programs fueled on middle-class guilt. This was apparent every day in the groups run by our *very* middle-class treatment staff. Many of the more clever clients absolutely delighted in manipulating the pants off the do-gooders. It took a lot of bruises and smashed expectations to educate those counselors.

I developed a hodge-podge of instrumental techniques and Glasser's "take responsibility for yourself" reality therapy and set to work. In order to kindle the methadone affected interest of the clients, the dispensing window was located directly adjacent to the large open space in which the shop would be built. My first task was to build a strong, waist-high fence separating the front of the building and dispensing area from the work space. It was designed so that there was no room to sit down and watch the shop activity while awaiting a daily dose of methadone. I hustled up a stereo and set to work building the shop.

When clients began to express an interest in what I was doing I told them that I was creating a workshop for candlemaking, leather crafts, ceramics and woodworking. Eventually I would hint, we would begin a small boatbuilding business. It did not take long for the word to spread and clients to begin asking how they could become involved.

With feedback from the treatment staff, a few of the clients who were thought to be more highly motivated were invited to join me when they expressed an interest. Once across the shop threshold they were immediately reinforced for their curiosity and "enrolled" in our new program. Handing them a broom or a mop I told them that we wanted only clients who were really interested in skilled training and jobs. This first stage of broom pushing and mop swinging became "step one" of our soon to become, eight step program.

Within a week or so the shop was starting to take shape and five or six clients who were beginning to become consistent in their desire to be a part of our new program were pronounced to be in "step two". This placed them squarely in an emerging heirarchy that recognized them for their consistency and allowed them to move on to more interesting tasks.

The approach to all building projects from the beginning of the program was oriented to the maximum utilization of creative expression as a self motivator. All treatment staff who pitched in with the ancillary services team were enjoined to pay close attention to facilitating self expression and creativity without allowing our work space to degenerate to the lowest common denominator of graffiti "art." Responsibility and style were emphasized in all projects. So when the treatment staff needed to divide the upstairs into group rooms, I persuaded them to allow the "step programmers" to do the building.

Everyone was surprised to discover that instead of erecting sheetrock and two by four walls, we built two portable geodesic domes from electrical conduit. We covered them in packing insulation for soundproofing and held it all down with tie-dyed army surplus parachutes. The domes were designed to collapse and store easily in boxes that could be carried on top of a VW. In this way, they could be taken down and used as emergency overdose aid tents at local rock concerts. Moreover, they allowed us to leave the open floor space unbroken. This proved extremely helpful when the step program expanded to encompass the entire building six months later.

By the time the work shop and group room domes were completed we had twelve to fifteen regulars in all of four steps. Word was out in the local artists community that we had free workspace to exchange for crafts instruction and a wonderfully talented ceramics teacher from a local junior college began to become a part of our program.

The craft instruction portion of the step program was designed around several important concerns: First, we needed to offer instruction in crafts that had an immediate "hip" appeal as well as an economic benefit. Breaking through a methadone affected attention is not easy. In this, the heyday of the street vendor, we chose to offer candlemaking and leather craft for those in the beginning steps because they were easily learned crafts which fostered good hand-eye coordination and patience. The items produced were given to the clients to keep or sell as they wished. This kindled their economic self interests right from the start. Second, we offered instruction in crafts that were relatively simple to learn so that we could begin to shape clients into success experiences as soon as possible. Third, we needed a means to evaluate each client and to establish a process as well as a product oriented progression from one step to another.

After clients had served a time as clean-up crew in step one, they were sent upstairs to our crafts program for steps two, three and four. Once upstairs they were each assigned a personal work space and helped to learn the crafts in sequence by the artisan-instructors and the "upper-steppers". Progress from one step to another was clearly defined in terms of behavioral and production goals.

Once through with step four, clients would be directed into one of three separate programs. One choice was an industrial sewing shop set up by a local sail-maker and an ancillary services staff person. Here they received a step five stipend of approximately one dollar an hour to learn the art of canvas accessory sewing. Another choice was advanced ceramics with our resident potter. In this option, clients received the same step five stipend and learned hand building, firing and some work on the potter's wheel. The third option brought them downstairs to the wood shop where I taught them wood working and small boat repair skills, giving them the same stipend as the other options.

Clients in step five had a vote in the "steppers council," so called to distinguish them from the rest of the clients on methadone and to indicate the quality of their self reinforcing motivation. The "steppers" concerned themselves with all problems of discipline and movement of individual clients through the steps. Although I retained a distant veto, this drew the steppers closer together into a tight knit cadre of folks taking responsibility for themselves and handling success one step at a time.

Steps six and seven saw the clients beginning to function semi-autonomously and receiving a higher stipend. They acted as the older sisters and brothers of the clients in the lower steps and through the connections of the ancillary services staff in the community, they were introduced to prospective employers. These interested employers very often dropped in to observe the clients in their step program atmosphere and were always impressed.

Step eight of the program was designed for support while moving from the dull world of methadone maintenance into the world of work. Treatment and ancillary services as well as other steppers formed a strong support network for clients who began to detoxify from methadone in step eight. The sense of belonging fostered so strongly in the lower steps was heavily utilized by step eight members in their struggle to begin a new life free from chemicals in the world of work.

RESULTS

The first nine months of the step program was "a happening". Literally forming itself as it grew according to the philosophy of treatment mentioned above, it flourished in a highly charged atmosphere of creativity. It generated a tremendous surge of interest and curiosity in the client population. This interest was shaped into the participation of more than half of the clinic's client load within a few months.

The first six months of operation was a building and defining time for the program. After the shop and crafts programs were completed, we counted thirty clients involved in steps one through four. One of the unfortunate side effects of this highly charged creative atmosphere was that we were all caught up in the process and paid little attention to the accurate recording of events. There was another consequence of this creative fervor which proved to be of much more devastating impact, but more of that later.

While the step program was developing not only clients, but their associates began to frequent our shop. Some of them were allowed to enroll in the program because we felt them to be in imminent jeopardy of drug addiction or some other debilitating personal problem. A few others who were drawn to the shop atmosphere were enrolled because they represented a population whose personal problems were not characterized by drug abuse, but who were nevertheless caught up in an unbroken cycle of failure. We took them in because there was simply no other place for them to receive the kind of treatment-training we offered and because they were very persistent. At every juncture, it was the upper steppers who passed final judgment upon who should be included and who should not.

The sixth through the ninth month of the step program saw the first successful detoxification from methadone. Three clients detoxified voluntarily as a result of their participation in the program. Two of them moved into apprenticeships with the local carpenters union thanks to the strong support of a friendly contractor. The other, relocated to another state to take a job with an older family member.

Approximately nine months into the program, with several additional clients approaching their final doses, things started to go wrong. Members of the treatment staff who were supportive in the early months began to express feelings of resentment towards the perceived "new direction of the clinic". Clients were much more involved in the step program than they had ever been in group and began to skin counseling sessions to spend their time in steppers council. The council had become a highly motivated subculture in the clinic. Factions began to develop among the staff and the executive director chose that moment to resign.

By this time the step program had grown so large that it required the entire building. While looking for an executive director, the treatment and administrative people had to search for office space as well. This served to further divide and factionalize the staff for there were those treatment persons who preferred to remain behind in the changed atmosphere of the step program.

When the assistant director, a charming young pharmacist with very little experience in drug abuse, assumed the role of the executive director the tension and discord among staff had reached a point of irreconcilable differences. The staff was split again over racial issues in the composition of clients and staff. This problem played an increasingly divisive role in the already tension filled staff meetings.

A discussion of the administrative and racial issues which divided the clinic would take me too far afield from this attempt to semi-objectively report the results of the step program. Suffice it to say, converging trends of personnel communication difficulties, racial tension, lack of leadership, and the almost subliminal machinations of SAODAP's local representative, spelled the end of the step program as it was originally envisioned.

The ninth through the twelfth month of the program's operation saw the ancillary services staff divided along many different lines of staff discord. The clients in the step program began to suffer from the lack of staff cohesiveness and the steppers council voted to join the faction of staff and clients with which I was identified, and strike.

Now a strike on the part of clients receiving methadone is very serious business. The local TASC and SAODAP folks grew increasingly nervous. Finally, in what seemed like the result of some very Machiavellian goings on, a do-nothing board of directors named a cooperative staff person to be the new executive director and pink slips were sent out in the mail. With half the staff resigning or being fired, and the more vocal of the clients discreetly taken off to jail, the step program ended.

DISCUSSION

With a blend of operant techniques, principles of reality therapy and a creative milieu the step program at Marin Open House was able to demonstrate that methadone maintenance clients could be drawn into a job and life skills training program. Within six months the program had enticed over half of the clinic's clients into the training classes.

The clients who participated in the steps began as unemployed, unskilled drug addicts with lives characterized by a continual cycle of failure experiences. With clearly defined behavioral and product goals designed to prepare the clients for the world of work, the step program sought to reverse the trend of continual personal failure. The "steppers" in the program became a self governing, mutually supportive subculture in the larger methadone clinic and began supporting each other to detoxify from methadone and graduate into full time drug-free employment.

Personnel and racial problems among staff, together with the resignation of the clinic's executive director at an unfortunate time brought the step program to an end. A post-mortem examination of the events and personnel which brought this phenomenal project into being, helped failure oriented, unskilled drug addicts to flourish and develop self respect as responsible group members, suggests that the same fervor that created the project also brought about its doom.

Projects of this nature, whether erected during a national trend to long term maintenance or in today's world of goal oriented therapeutic approaches, must have strong and somewhat dispassionate administrators or they will collapse through burn-out or their own lack of attention to detail. In the case of this program, a political naiveté on the part of this author and a lack of understanding for the very real concerns of treatment staff who watch their case loads co-opted by a skill training program, spelled out "the end" all too soon.

REFERENCES

Cushman, P. 1971. Methadone maintenance in hardcore criminal addicts: Economic effects. *NYJSM*. 71: 1768-1774.

Dole, V. 1971. Methadone treatment of narcotics addiction. *LJCPTT*. 4: 429-432.

DuPont, R. and Katon, R. 1971. Development of a heroin addiction treatment program: Effect on urban crime. *JAMA*. 216: 1320-1324.

Pearson, B. 1969. Methadone maintenance in heroin addiction: The program at Beth Israel Medical Center. *AJN*. 70: 2571-2574.

Perkins, M. and Bloch, H. 1970. Survey of a methadone maintenance treatment program. *AJP*. 126: 1389-1396.

Perkins, M. and Bloch, H. 1971. A study of some failures in methadone treatment. *AJP*. 128: 47-51.

RETENTION RATES AMONG METHADONE PATIENTS: AN ANALYSIS OF THE NEW YORK EXPERIENCE*

Lee Koenigsberg, M.B.A. and Ronald Bayer, Ph.D.

Community Treatment Foundation, Inc.

INTRODUCTION

In the early 1970's, the methadone maintenance modality was characterized by its capability to attract and retain addicts in treatment. The results were partciularly impressive since methadone programs essentially relied on voluntary treatment as opposed to the coercive, i.e., criminal justice, retention procedures often used by other modalities. Both the professionals in the field--and the critics--have noted, with increasing concern, the declining retention rates among patients admitted to methadone maintenance programs in recent years.

This paper presents an analysis of retention rates for methadone programs in the New York City metropolitan area since the inception of the first treatment program in 1964 through 1976. Because the New York experience is based upon the most comprehensive "historical" experience and upon the largest patient population base in the United States, the findings should be of interest to all who are concerned about current problems in methadone maintenance.

---

* This publication is based upon research activities supported through a grant from the New York State Office of Drug Abuse Services. The conclusions stated herein are not necessarily those of that office.

PROCEDURES

The analysis is based on data maintained by the Methadone Information Center of The Community Treatment Foundation, Inc. (CTF). CTF operates central registry covering all methadone programs in New York State, under a contract with the New York State Office of Drug Abuse Services. All applicants for admission to a methadone program must first be cleared through the central registry to make certain that they are not already in treatment elsewhere. Once the applicant is cleared for admission, CTF assigns a discrete, permanent identification number to each individual. This identification number is used to record all subsequent treatment episodes for that individual.

Terminations of patients' treatment and transfers between clinics are reported to CTF on an on-going basis. This data base enables CTF to track the treatment history of all patients and to acknowledge the concept of continuous treatment irrespective of transfers from one facility to another. Further, it is readily possible to analyze first admissions to the treatment network and to ignore readmissions in compiling data to support a study of this nature.

From 1964 through 1976, there were 78,498 first admissions to methadone maintenance programs reporting to CTF. This paper is based on an analysis of the retention of each annual admission cohort. The retention data are based on the tracking of each patient's treatment history at six month intervals following the actual date of the first admission.

FINDINGS

Table I contains figures representing the percentage of each annual cohort which remained in treatment at six-month intervals following the first admission to treatment. The figures clearly show the decline in retention rates for each cohort, with virtually each successive cohort exhibiting a lower retention rate than the one which precedes it. The deterioration in retention rates takes on dramatic overtones when one compares the decline in the percentage of patients who remained in treatment at least one year for the 1970 cohort--80%--with the 1976 cohort--44%. Comparative data for two-year retention rates reveals a drop from 68% in 1970 to 29% in 1975.

Within the thirteen years covered by this study, it is possible to identify three distinct periods. The first seven years, i.e., 1964-1970, can be defined as the period during which the methadone "experiment" was being implemented. In this time frame, the difference in retention rates between each admission cohort and its predecessor was comparatively small, even though the general downward trend was already obvious.

Table I. Retention rates of annual first admission cohorts at six-month intervals.

Number of Years	Year of Admission (See Table IA for N)												
	1964	1965	1966	1967	1968	1969	1970	1971	1972	1973	1974	1975	1976
.5	100%	91%	93%	93%	89%	90%	89%	84%	77%	67%	64%	62%	62%
1.0	100	81	88	87	82	81	80	74	63	53	50	47	44
1.5	100	75	82	81	76	75	73	65	52	43	40	38	
2.0	100	67	76	76	73	70	68	58	43	36	34	29	
2.5	100	64	73	70	69	66	63	51	37	31	29		
3.0	100	57	69	68	64	61	58	45	33	27	23		
3.5	100	52	65	65	61	56	53	40	29	24			
4.0	100	48	60	63	57	52	48	35	25	20			
4.5	83	47	58	59	54	49	41	32	23				
5.0	83	43	55	56	52	46	36	29	19				
5.5	83	40	53	54	49	42	33	26					
6.0	83	38	52	51	46	38	30	21					
6.5	83	36	51	48	45	34	27						
7.0	83	35	50	47	41	31	20						
7.5	83	33	50	46	40	29							
8.0	83	33	48	44	38	24							
8.5	83	32	43	42	36								
9.0	83	31	42	40	28								
9.5	83	28	40	38									
10.0	83	26	38	32									
10.5	83	23	36										
11.0	50	20	30										
11.5	50	18											
12.0	33	16											
12.5	33												
13.0	33												

Table IA. Number of first admissions by year.

Year Of Admission	First Admissions (N)
1964	6
1965	90
1966	246
1967	427
1968	535
1969	1,294
1970	3,299
1971	14,208
1972	19,687
1973	14,228
1974	10,743
1975	7,294
1976	6,441

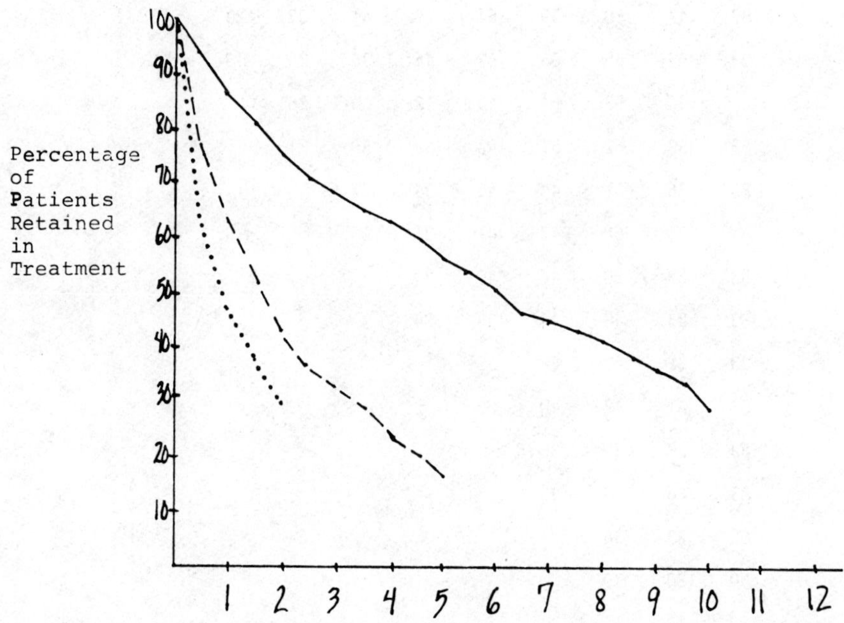

Fig. 1. Retention rates for representative years.

From 1971-1973, the methadone programs in New York experienced a massive and rapid expansion of admissions. (61% of all first admissions in the 1964-1976 period were recorded during these three years). In this period, the retention rate of each succeeding admission cohort dropped markedly.

The final three-year period (1974-1976) represents a time of declining first admissions. During this time frame, the retention rates have begun to level off.

Figure 1 graphically illustrates the successive deterioration of retention rates for representative (mid-point) years in each of the three periods described above.

The most significant data presented in Table I pertains to the serious decline in retention rates during the first six months in treatment, i.e., from 89% for the 1970 cohort to 62% for the 1976 cohort.

The phenomenon of declining retention rates can be best analyzed by a presentation of termination data, as shown in Table II. These figures clearly illustrate that the dropouts from treatment during the first six months severely impact the overall retention rates for the later cohorts. Up until 1970, an average of 9% of all admissions dropped out during the first six months, and approximately 8% more left during the second six months of their treatment. In contrast, 38% of the 1975-1976 cohorts dropped out during the first six months, followed by approximately 16% who discontinued treatment in the second six-month period. On a comparative basis, the dropout rate for

Table II. Comparative termination rates: the first six months versus the second six months of treatment.

Year of Admission	Percent Terminated in First Six Months	Percent Terminated in Second Six Months
1964	0%	0%
1965	9	10
1966	7	5
1967	7	6
1968	11	7
1969	10	9
1970	11	9
1971	16	10
1972	23	14
1973	33	14
1974	36	14
1975	38	15
1976	38	18

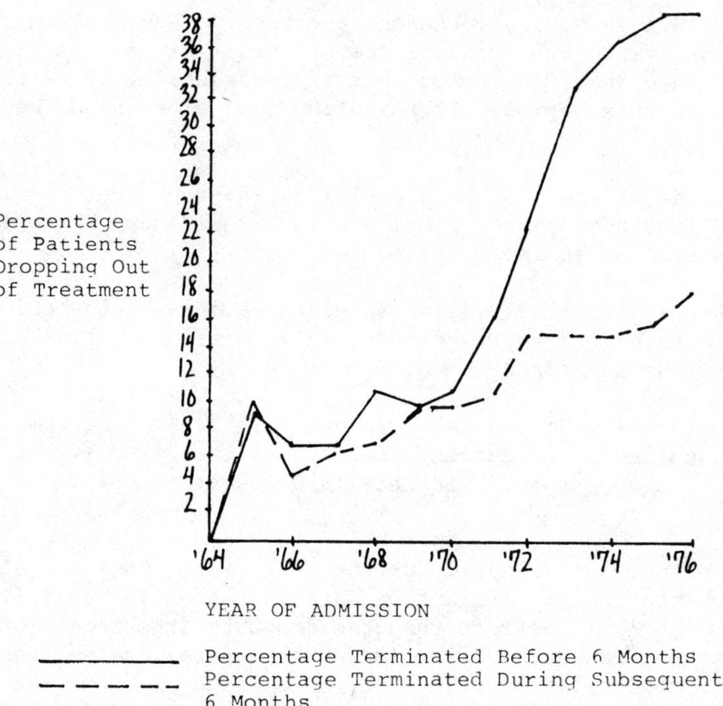

Fig. 2. Comparison of dropout rates during first six months and subsequent six months of treatment.

Table III. Retention rates of annual admission subcohorts (those who have remained in treatment at least six months).

	Year Of Admission						
Number of years	1970	1971	1972	1973	1974	1975	1976
1.0	90%	88%	81%	78%	77%	75%	71%
1.5	82	78	67	63	62	60	
2.0	77	69	56	54	53	47	
2.5	71	61	48	46	45		
3.0	65	53	42	41	36		
3.5	60	47	37	36			
4.0	54	42	33	30			
4.5	46	38	30				
5.0	41	34	25				
5.5	37	31					
6.0	33	25					
6.5	30						
7.0	23						

the second six month period increased only two-fold. The changing pattern of dropouts in each half of the first year of treatment is illustrated in Figure 2.

The importance of the first six months on the overall retention rate for any cohort can best be understood by redefining the cohort to exclude those patients who dropped out of treatment during the first six months. Table III contains figures representing the retention rates of this sub-cohort. While the successive downward trend depicted in Table I is still present, the rate of decline has been tempered considerably; for example, 90% of the 1970 sub-cohort were retained in treatment as compared to 71% for the 1976 sub-cohort. These figures are in sharp contrast to the 80% and 44% retention rates for the entire cohort for these two respective years.

A graphic presentation of the one year retention rates, comparing the total cohort to that of the sub-cohort for each year of admission (Figure 3), clearly illustrates that the ability of the metha-

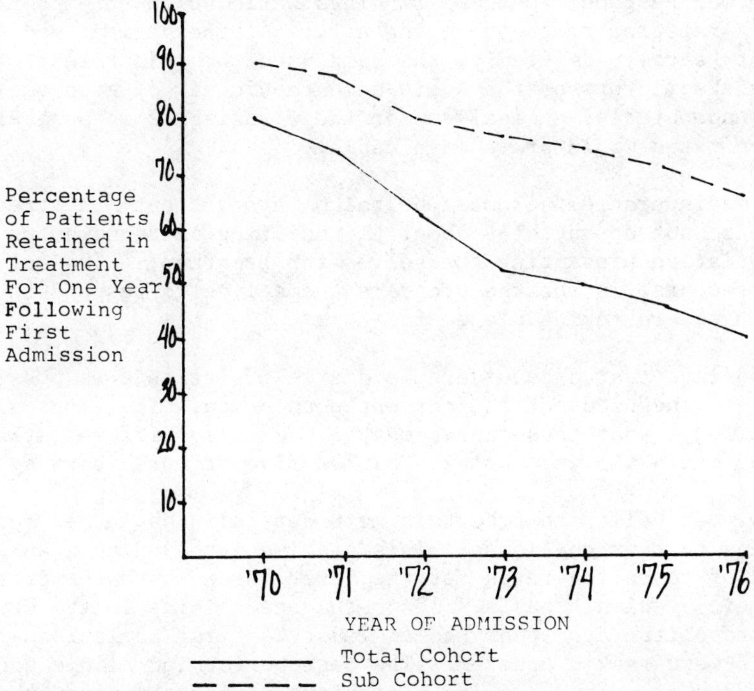

Fig. 3. Comparative one-year retention rates of all first admissions (total cohort) and of those who remained in treatment at least six months (sub-cohort).

done programs to retain a portion of its patient body has not eroded as seriously as the gross data would suggest.

The data which served as the basis for this study did not contain any information regarding the dropout pattern within the first six months of treatment. However, a more limited study of retention data which was recently undertaken for the period of 1974-1976 revealed that 68% of the patients who dropped out of treatment in the first six months actually left during the first three months.

DISCUSSION

The above findings support two conclusions:

1. There is a continuing and serious decline in retention rates of patients admitted to methadone programs in New York and
2. The increasing preponderance of patients who leave treatment during the first few months following admission represents the key factor in explaining this downward trend.

A deemed response to these findings would be an attempt to offer hard data relating to the changing nature of the patient population in recent years. Admittedly, the data used for this retention study excludes the factors of the socio-demographic mix of each admission cohort; undoubtedly, an analysis of these variables as a catalyst of retention rates would prove invaluable.

One indisputable fact must be taken into account: the treatment programs cannot do anything about the shifting patterns among the addict population presenting themselves for treatment. The only productive recourse is for the programs themselves to respond to whatever changes are taking place.

We believe that a considerable degree of latitude is possible in adapting to the needs of the current population. The first step is to acknowledge that these needs may be radically different than those of the patients who were being admitted five or more years ago.

It is our belief that certain methadone clinics in New York are responding to this challenge. This premise is based upon an initial analysis of retention rates for the first three months of treatment on a clinic-by-clinic basis. 104 methadone clinics in the New York City metropolitan area that had at least 25 first admissions during 1974-1976 were ranked ordered. The best performing clinic had a dropout rate of only 3% of its admissions during the first three months of treatment; by contrast, the worst performer was a clinic with a dropout rate of 50%. New York has a number of large programs with multiple clinic facilities. A comparison of three-month retention rates for those clinics within the same administrative program also revealed a wide disparity in the clinic's ability to retain patients for at least three months, e.g., the range ran from 3% to 24% in one such program. Presumably, all clinics within this administrative program operate under the same set of policies and procedures regarding the screening of applicants and terminations of patients. Such wide variations in the ability to retain patients during this crucial initial period of treatment suggest that much remains in the hands of the individual clinic staff.

We believe that more research is needed to identify the factors which account for varying degrees of success in retaining patients. Any failure on the part of methadone programs to take positive action to reverse the serious declining trends in retention rates is a failure of major consequence.

Our original plan was to compare the New York experience with that of other major American cities. We found, however, that comparative data could not be undertaken in the absence of a centralized registry, such as the one maintained by CTF. We would suggest that studies similar to the one presented here be replicated by researchers in other localities. The value of such analyses lies not in their academic interest, but, rather, in the practical implications which may follow for upgrading treatment.

METHADONE DOSAGE AND ILLICIT METHADONE USE: AN INITIAL EXPLORATION*

Lee Koenigsberg, M.B.A. and Ronald Bayer, Ph.D.

Community Treatment Foundation, Inc.

New York, New York

INTRODUCTION

An all too common criticism of methadone programs is that they have become the source of a supply of a new black market narcotic, i.e., methadone diverted for illicit use. Arguments are presented that as long as patients can take home relatively high dosages of methadone, there is a strong likelihood that a portion of the medication will be converted to a commodity for trade.

Two alternatives are generally offered to curtail this alleged abuse: 1) end all take-home privileges, or 2) reduce the dosages of medication so that patients will experience withdrawal discomfort if they do not consume their entire dosage.

The first option, which has been implemented by some clinics in various parts of the country, is considered unsatisfactory by many clinicians in the field. An across-the-board requirement for a seven-day-per-week pick-up schedule constitutes an extraordinary imposition on the patients, particularly those who have been in treatment for a long period of time and are deemed to be responsible and stable individuals. Further, the additional costs involved in keeping a clinic open seven days a week causes a significant financial drain on the

---

* This publication is based upon research activities supported through a grant from the New York State Office of Drug Abuse Services. The conclusions stated herein are not necessarily those of that Office.

operation, which can only be financed at the expense of other supportive services.

As a consequence, the second option has seemed increasingly more appealing as an appropriate response to the charge of methadone diversion. The regimen of reduced dosages also serves to placate to some extent the opponents of methadone who are unduly concerned about the use of this medication on general principles.

This paper presents an examination of the recent experience in New York to ascertain whether the level of methadone dosages has any discernable impact on illicit methadone use.

PROCEDURES

As a means of measuring the extent of illicit methadone use in New York, we examined the standard intake forms of a sample of patients admitted to methadone maintenance programs for the years 1973 through 1977. These forms, based on self-reported data, indicate not only the use of illicit methadone, but the extent of such use. The validity of this element of self-reported data has been repeatedly demonstrated and verified by researchers in the field. We believe that the reliance on these intake forms is wholly appropriate since our primary interest was in the overall trend of illicit methadone use rather than the absolute levels.

The Community Treatment Foundation (CTF) has on file the standard intake form for a number of methadone maintenance programs in New York City, representing approximately 30% of admissions during the period under study (1973-1977). All admission forms covering the first four months of each of these years were examined for this study. From this group, those intake forms relating to first admissions to treatment were selected. The total sample of intake forms used for the study was 1,626.

As part of its routine operation, CTF maintains individual patient dosage information for a variety of methadone programs. In total, such data are available for approximately 30% of the entire active population. The distribution of dosages, by ranges, was based on these files for the years 1974 through 1977. (Data for 1973 were not available.)

FINDINGS

1) The analysis of our sample indicates that a surprisingly high percentage of first admissions to programs during this five-year period claim to be using illicit methadone. Table I presents our findings in this regard. As shown, the incidence of use ranges from 68% in 1976 to 79% in 1975.

Table I. Methadone use among first admissions to treatment (1973-1977).

Year	Sample (N)	Percent of sample using any methadone	Percent of sample using methadone as primary drug	Percent of sample reporting daily use of methadone
1973	797	72%	19%	31%
1974	535	72%	25%	46%
1975	66	79%	25%	30%
1976	144	68%	19%	21%
1977	84	78%	26%	30%

2) Although the proportion of newly-admitted patients who claimed methadone as a *primary* drug of abuse never rose above 26% (1977 admissions), as many as 46% of the 1974 cohort admitted to using methadone daily. This data clearly suggest that a significant number of addicts presenting themselves for treatment use illicit methadone in concert with other substances.

3) The incidence of daily use among the 1976 sample declined to 21%, as compared to 30% of the 1975 sample and 46% of the 1974 sample. However, this downward pattern was reversed in 1977, when the daily use rose to 30% of the sample. In fact, all of the indicators moved upward in 1977 to approximately the same levels as 1975. We are at a loss to explain the deviant nature of the 1976 experience, but it is obvious that the decline is not indicative of an overall downward trend.

4) Table II contains a summary of the distribution of dosage dispensed, by broad ranges, from 1974 through 1977. The figures show a continual decline in dosage levels: in 1977, less than 30% of the patients received 80 or more milligrams daily, whereas in 1974, 44% of the sample population were medicated within this range. Converse-

Table II. Daily methadone dosage distribution for selected programs (1974-1977).

		Percent of patients, by range of daily dosage		
Year	Sample (N)	Less than 40 mg.	40-79 mg.	80 mg. or more
1974	9,314	16.6%	39.4%	44.0%
1975	9,470	20.7%	43.5%	35.8%
1976	10,205	22.0%	46.7%	31.3%
1977	10,004	24.8%	45.8%	29.4%

ly, almost 25% of the 1977 sample were receiving less than 40 milligrams daily, as compared to 16.6% of the 1974 sample.

5) We were unable to find any correlation between the incidence of illicit methadone abuse of addicts entering treatment and the decrease in the amount of methadone being dispensed to patients. While we did not have access to precise data regarding the level of take-home medication for each of these years, we found that there were no major policy changes on take-home doses since 1974. Clinicians associated with the larger programs in New York City confirmed that the average number of take-home doses per patient has not changed materially since that time. Figures 1 and 2 graphically illustrate the patterns of illicit methadone use and dosage distribution.

DISCUSSION

The data we have presented suggests no relationship between the declining trend in methadone dosages and the pattern of illicit methadone use as measured by self-reported data gathered from individuals admitted to methadone programs. We suggest, therefore, that further study is necessary to ascertain the source of the supply of illicit methadone.

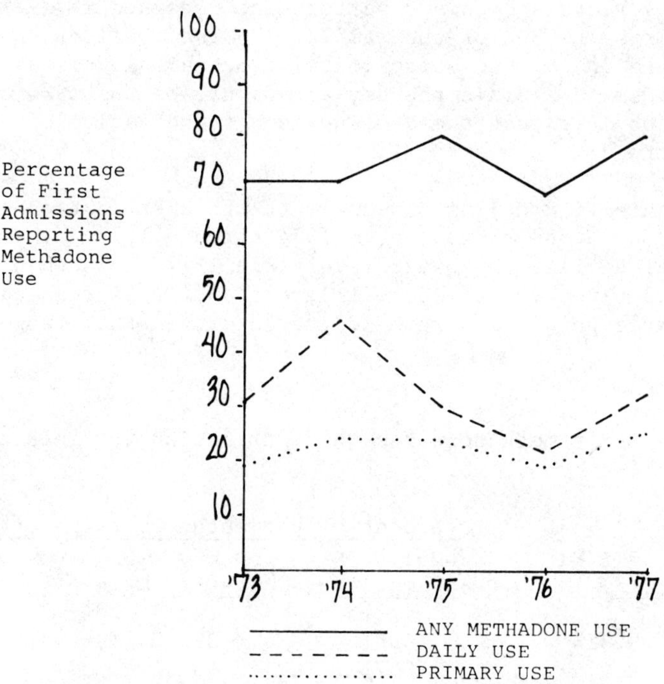

Fig. 1. Methadone use among first admissions to treatment.

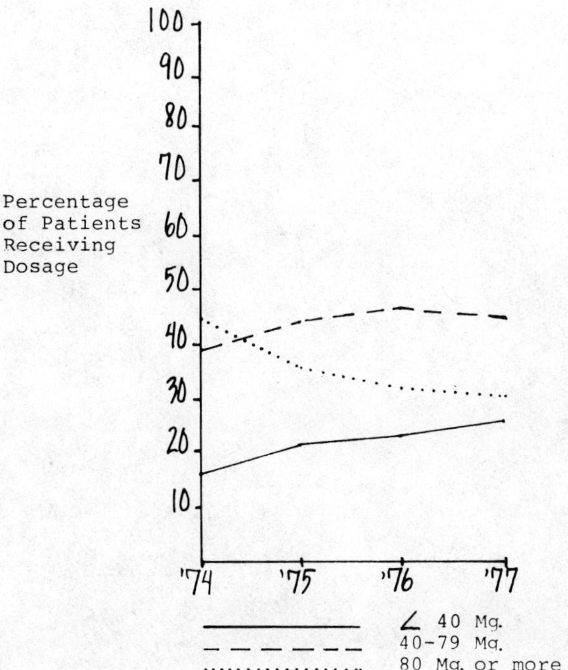

Fig. 2. Daily methadone dosage distribution for selected programs.

We further believe that it is wholly inappropriate to resort to a policy of reducing the dosage levels of patients in treatment as a response to the phenomenon of illicit methadone use. Such course of action may undermine the therapeutic efficacy of methadone treatment without having any positive impact on curbing the supply of illicit methadone.

SUPERVISORY FEEDBACK PROCESS

Robert Klein, Ed.M.,
GKSW, Inc.

Martha Ann Zehner
Private Practice

The Supervisory Feedback Process (S.F.P.) was developed by Robert Klein and Martha Ann Zehner in 1976. It was first used in the Advanced Clinical Training Program in Alcoholism in Rosemont, Pennsylvania. The authors felt after several years experience in doing training in the drug and alcohol and mental health fields that there was a lack of a specific training tool which allowed for the recognition of the inherent sensitivity of the trainee and the opportunity which allowed for specific constructive feedback and teaching by the trainer.

We had seen time and time again the feelings and beliefs of trainees trampled over by a supervisor or trainer who felt that he/she knew what the trainee needed in order to become a competent counselor. There was little or no respect shown toward the trainee's need to grow at his/her own rate. There was no awareness of what the trainee had to offer, the assumption being made and acted upon that the trainer/supervisor knows best. Trainees were either talked down to with little recognition shown concerning useful information the trainees had, or trainees were rushed and pressed ever forward with the belief the more information poured in the more likely progress would be made by the trainee/"receptacle".

However, mistakes have also been made in the opposite direction. Supervisors/trainers have been so reluctant to say anything that the trainee is left with nothing more than he/she came in with except a few approving nods and a lot of meaningless positive feedback. We have understood this reaction as an attempt to not dominate, over power or deluge the trainee. It is also understandable that a trainer/supervisor might react this way especially when using video-tape. The experience of seeing oneself on video-tape can be traumatic enough, let alone hearing feedback that is often heard as doubly threatening because the trainee is also observing him/her self simultaneously.

We felt there were really several basic issues: 1) How do we provide a situation which is sensitive to the trainees' experience of being observed; 2) How do we allow for the trainee to develop at his/her rate; 3) How can we provide on the spot help in a training situation; 4) How can we maximize feedback without allowing those giving feedback to run "rough shod" over the trainee receiving the feedback; 5) How can we allow for the trainer/supervisor to give useful information and support to the trainee; and 6) How can we provide a means of viewing a video-tape which allows for the input of useful information based on the trainees' ability and readiness to hear it and the supervisors'/trainers' awareness of the need to broach new concepts and topic areas.

We obviously spent much time attempting to answer these questions, though clearly none were as well formulated as they are at the writing of this paper. We both knew intuitively and intellectually that the models we had seen, experienced and used in no way answered all of the above articulated questions. As happens in most situations there is a certain amount of serendipity involved. Ours involved a former trainee named Richard Vesper. In August of 1976, he came to visit and teach us a method of training he had been using and found useful. We were all both excited and nervous about learning the system. The system was called Interpersonal Process Recall: A Method of Influencing Human Interaction (Kagan, 1976). We were trained in it and found it exciting, useful and thought provoking. Though it respected the rights and sensitivities of the trainees, allowed them to grow at their own speed, it did not, to our way of thinking, allow for sufficient input from the other trainees and the trainers. We found the trainees not directly involved getting bored, annoyed by only being allowed to ask specific questions and not being permitted to give direct feedback.

Prior to this we had been using a model in doing group therapy which allowed for a very specific type of feedback. A group composed of the trainees was led by one or two of the trainees. The trainees who were clients in the group were asked to be themselves and to discuss a real issue in their lives, not to role-play another, real or fantasized person. This was requested with the understanding that anytime a trainee felt that the material that he/she was talking about got too difficult to handle, they could just say they wanted to stop and that would happen. They were supported in this by the trainers. The trainees also knew that the trainers were available to help them with any unfinished business for that or other experiences.

We specifically ask people not to role-play in that it has been our experience that when people role-play they are actually being themselves no matter how they may try to disguise the material they are talking about in the session. It is when they are role-playing that they "accidently" find themselves talking about very personal issues that they may not have wanted to share with a group. We therefore ask the trainees to be themselves and talk about what is important to them in their lives and support them in stopping when they feel they need to end the session. We thus minimize the likelihood of someone "accidently" revealing more than they want.

This practice group being described lasts about fifteen minutes. When it is finished there are specific rules about feedback being given. First, it must be feedback for the trainees who are acting as counselors, not those who participated as clients. Second, the feedback is "owned" by those giving it. By that we mean that the statements made are prefaced by such words as "It seemed to *me*", or "This is the way *I* saw this happen." Comments are only allowed about what the trainees observed happening, not what they thought the trainee acting counselor should or should not have done. No advice giving is allowed.

From these two basic backgrounds and other personal inputs, we developed what we now call the "Supervisory Feedback Process" (S.F.P.). In training groups we usually work with anywhere from five to fifteen participants. If there are ten or more, we usually divide the group and each of the authors work with half. We then ask who would like to be a counselor and who would like to be a client. The rest of the trainees act as observers, except one who will operate the video-tape camera.

We ask the client to either pick one of the issues that Rick Vesper and people he had worked with at the University of Pittsburgh saw as critical to doing good counseling (trust, competence of the counselor, dependency, hostility, sexuality or separation), other issues that we have found as pertinent in the training (death, racism, sexism, professionalism, etc.) or material the client feels is important for him/her to talk about at the time of the session. We ask the client to relate the issue he/she chooses in terms of their own life. It is not critical for the counselor to guess the issue if he/she does not know it beforehand. Rather it allows the client to have something to focus on in the session. Our real life clients rarely announce, "This is what I am going to deal with today."

The trainer functions as a consultant for the counselor. The trainer sits directly behind the counselor and is available for on the spot consultation should the counselor feel stuck or have some doubts about where to go in the session. The counselor merely signals for a break. The trainee operating the video equipment stops it and the counselor gets his/her on the spot consultation. The session then continues with the camera being restarted. The session can run for any length of time. We have found fifteen minutes to be more than adequate. When the time is completed, the session is stopped and the trainer has a small debriefing session with the counselor and client right in the same spot where they have been sitting. This is done just to make sure they are both feeling okay and are in a sense "back" from the session.

At this time feedback is given by the others observing. Again, specific rules are followed. First, feedback is only for the counselor, not the client. This is so the session does not become one of "open hunting" for the other trainees' theories about the client. Second, the feedback is phrased in terms of what the observer observed happening, not what they thought should or should not have happened. The trainee doing the video-tape talks about how the interaction looked through the camera. This is extremely valuable since this particular trainee rarely hears what is being said because he/she is so busy working with the camera. As a result this person gets a movement or kinesthetic feel for what has been happening in the session. Lastly, the trainer gives his/her feedback about what he/she has seen happening regarding the session. This then becomes a meaningful time to get into specific content or process issues as they specifically relate to the session just completed. The trainer is then in a position to discuss areas of clinical work that are observable to all, in fact, have just been observed. They are not then the purveyors of unrelated and useless material.

Lastly, the video-tape is made available for the counselor and client to view alone if they wish and also definitely with the trainer. The counselor and client may invite any of the other trainees to view the tape if they wish, either separately or with the trainer. When viewing the tape, there are certain rules. Only the counselor and the trainer may stop the tape to discuss what was happening. They may ask each other questions, share options, discuss possible new ways of acting, but not interpret the client's behavior, verbal or non-verbal. The client, if he or she wants to, may also go over the video-tape with the trainer using the same rules.

We have found this process, S.F.P., to be uniquely valuable in training counselors in the drug and alcohol field, mental health field and our assumption is that it can be used in most interactive situations. Video-tape helps, but is not a prerequisite.

Recently we have developed a training delivery system using the S.F.P. as a backbone in teaching clinical supervision to drug and alcohol treatment supervisors in the Commonwealth of Pennsylvania. This is clearly an area of needed help as the drug and alcohol field matures. We discovered that the majority of supervisors in the field were former counselors who not only lacked expertise in clinical supervision, but received little if any support. We found that when confronted with difficult situations, most relied on their therapeutic expertise which may have solved the immediate situation, however, it complicated things further down the line. It intensified dependency issues as just an example, not to mention how it complicated transference problems. Through the use of S.F.P. on the level of clinical supervision we are able to help them to learn to do competent supervision, separate theory vs. training issues, discover their own theories of learning which they practiced daily and look at issues like dependency, sexuality, racism, sexism, anger, and clinical judgment conflicts between supervisor and counselor. We found this to be important because of the lack of material available in clinical supervision. If one does not rely on standard psychiatric texts or attempt to teach through a training analysis, the only material left is in teaching where some help is available (Cogan, 1973 and Goldhammer, 1975).

We have thus found the Supervisory Feedback Process to have value not only as a training device for counselors, but also for those responsible for the training of the counselors, the clinical supervisors.

REFERENCES

Kagan, N. 1976. *Interpersonal Process Recall: A Method of Influencing Human Interaction*. Michigan State University.

Cogan, M. 1973. *Clinical Supervision*. Boston: Houghton, Mifflin Co.

Goldhammer, R. 1975. *Clinical Supervision: Special Methods for the Supervision of Teachers*. New York: Holt, Rinehart and Winston, Inc.

THE WISCONSIN SUBSTANCE USE INVENTORY (WSUI)*

Khalil A. Khavari, Ph.D. and Frazier M. Douglass, Ph.D.

University of Wisconsin - Milwaukee

Considerable concern has appeared during the past few years with problems of "drug abuse" and its complex implications for the individual, his family, and the society at large. The research effort in this field is beset by confusion, methodological and conceptual flaws, absence of needed data, difficulty in obtaining such data, and a multitude of other problems. Research progress has been stymied, on one hand, by lack of standardized and operationalized definitions, terms, and concepts (Elinson and Nurco, 1975) and on the other hand, by the absence of a quantitative and standardized method of measurement (Walizer, 1975).

One stumbling block for the worker in this field is the label of "drug abuse" itself. What constitutes drug abuse? What is legitimate drug use? Who is to determine what constitutes use and what should be viewed as abuse? Many workers object to the term "abuse" for it implies "deviance", "pathology", "weakness", and the like. A recent report (Elinson and Nurco, 1975) reflects the concern of the researchers to operationalize terms, concepts, and possibly even measurement procedures on the field.

Conceptual and semantic difficulties aside, the research literature is replete with a bewildering array of methodologies where one can hardly relate one study to another (Berg, 1970; Glen and Richards,

---

* Supported, in part, by NIDA Grant DA01080 and funds provided by the College of Letters & Science to K.A. Khavari. We thank Mae Humes for her contributions to portions of this work.

1974; Josephson, 1974). Difficulties are encountered with respect to finding any consensus on the methods of assessing frequency of use, the definitions of light, moderate, or heavy use of a given drug or a class of drugs, and the objective and universal classification of drug use categories. Consequently, most studies are designed and carried out within the constraints imposed by the aforementioned deficiences which produce findings of highly limited generalizability.

With respect to use of drugs, measurements and quantification may be achieved by pharmacological assays such as saliva test, blood test and urinalysis, by monitoring autonomic and physiological manifestations such as pupillary dialations and contractions, muscle tone, ataxia, cardiac and respiratory states, by inspecting prescription and medical records, and by simply asking the individual. Clearly, the objective of the particular measurement and the feasibility of each procedure determines the optimal selection in a given case.

Epidemiological research in the area relies heavily on self-report techniques where survey and questionnaire methods are used. Researchers are concerned with: 1) determining factors such as incidence and prevalence of use, 2) systematically quantifying this information for the general population, subsamples, and individuals, 3) exposing parameters which lead to or mitigate against substance use, and 4) devising optimal ways of prevention and treatment. It is in this epidemiological domain that measurement instruments are highly non-uniform and even irreconcilable. Criteria for categories such as heavy, moderate, or light use are often arbitrary and lacking a viable procedure for uniform quantification of measurement.

A number of workers in the field have addressed themselves to one or more of the troublesome issues listed above. Blum and Associates (1970) employed a procedure where the subjects themselves indicated the extent of their own drug use as light, moderate, or heavy. Others have attempted to establish criteria for determining the extent and nature of polydrug use by ascribing weightings to the various types of the psychoactives, in accordance with their judged potencies. Such studies range from arbitrarily deciding what frequency defines heavy use (Eells, 1968) to procedures where average ratings of "experts" is used (Bucky, Edwards, and Thomas, 1974; Nail and Gunderson, 1973). Once a weighting has been defined for each drug, these weightings are multiplied by the frequency of use for each respective drug which yields a measure or "intensity of drug use". As Barter, Minzer, and Werme (1971) point out, in spite of the qualifications of the judges the weightings are still defined subjectively and these subjective weightings are often influenced by many unknown variables. Thus, it is evident that there is a need for an instrument which would provide uniform and quantitative information on drug use. The Wisconsin Substance Use Inventory (WSUI) has been developed to meet this need.

Table I. Means, standard deviations, dispersion units, assigned values (AV) and T scores for each of the 19 substances.

Substance	Mean	S.D.	Dispersion Unit	never tried av=0	but not using av=1	less than monthly av=2	once a month av=3	once a week av=4	several times a week av=5	daily av=6	several times a day av=7
Tobacco	2.9	2.8	3.5	39.7	43.2	46.7	50.2	53.7	57.3	60.7	64.3
Alcohol	3.8	1.4	7.1	23.1	30.2	37.3	44.4	51.5	58.6	65.7	72.8
Marijuana	1.7	1.9	5.3	41.2	46.5	51.7	57.0	62.3	67.5	72.8	78.1
Hashish	.8	1.1	9.3	43.2	52.5	61.7	71.0	80.2	89.5	98.7	108.0
LSD	.2	.6	17.2	45.9	63.1	80.3	97.5	114.8	132.0	149.2	166.4
Other Psychedelics	.3	.6	16.8	45.8	62.6	79.3	96.1	112.8	129.6	146.3	163.1
Opiates	.2	.5	19.4	46.9	66.3	85.7	105.0	124.4	143.8	163.2	182.6
Methadone	.1	.3	30.1	48.9	79.0	109.1	139.8	169.4	199.6	229.7	259.8
Pain Killers	1.0	1.1	9.2	40.6	49.7	58.9	68.1	77.2	86.4	95.6	104.7
Barbiturates	.4	.9	11.5	46.0	57.5	69.0	80.5	92.0	103.5	115.0	126.5
Sedatives	.3	.7	13.7	46.2	59.9	73.6	87.3	101.0	114.7	128.4	142.1
Tranquilizers	.4	1.0	10.1	46.1	56.2	66.3	76.4	86.5	96.6	106.7	116.8
Relaxants	.6	1.3	7.7	45.1	52.7	60.4	68.1	75.8	83.5	91.2	98.9
Amphetamines	.5	1.1	9.1	45.1	54.2	63.3	72.4	81.5	90.6	99.7	108.8
Cocaine	.3	.7	13.9	45.8	59.7	73.6	87.5	101.4	115.3	129.2	143.1
Diet Pills	.4	1.2	8.6	46.2	54.8	63.4	72.1	80.7	89.3	98.0	106.6
Antidepressants	.1	.6	16.7	48.0	64.7	81.3	98.0	114.7	131.4	148.0	164.7
Over-the-Counter	.9	1.2	8.6	42.2	50.8	59.4	68.0	76.6	85.2	93.8	102.4
Antiinfectious	1.2	1.3	7.9	40.9	48.8	56.6	64.5	72.3	80.2	88.0	95.9

DEVELOPMENT OF THE INVENTORY

The inventory is based on responses from 1,121 adult (median age= 23 years) males (517) and females (604). Of the total, 310 are members of the work force and 811 are college students. There were 954 white, 124 black, and 33 other respondents. Mean education was 12.8 years for workers and 14.9 years for students. Although no attempt was made to use probability sampling procedures, stratified quota sampling or other techniques to obtain a representative sample of the general population, the subjects represent a very heterogeneous segment in an urban environment. The procedure for development of the scale employs techniques similar to those used by Hathaway and Mckinley for construction of the Minnesota Multiphasic Personality Inventory (Welsh and Dahlstrom, 1956). Basically, the mean standard deviation for reported use of each drug is calculated. Then, individual values are converted into T scores by the standard formula (Runyon and Haber, 1971). This distribution provides for a population mean of 50 and standard deviation of 10. Thus, $T=50 + [10(x-\bar{x})/SD]$.

Data from the drug questionnaire (a self-administering paper and pencil test requiring about eighth-grade reading ability) were converted into T scores by assigning a value from 0 to 7 to each of the eight response categories (never tried = 0, tried but currently not using = 1, less often than monthly = 2, about once a month = 3, about once a week = 4, several times a week = 5, daily = 6, several times a day = 7). The mean, standard deviation and the dispersion unit for each of the 10 substances of the questionnaire were calculated from these assigned values (see Table I). T values for each level of use, for each drug, were then calculated by treating its respective assigned value as the X in the equation. By this procedure, we constructed the T distribution for the 19 products.

RESULTS AND DISCUSSION

As we pointed out earlier, the WSUI was derived by using widely accepted statistical procedures. But questions still may be raised regarding the decisions breaking down usage into arbitrary intervals (i.e., not at all, less often than monthly, etc.) and arbitrarily assigning incremental equivalent unit numerical values to each of the usage categories which are not equal distances apart (e.g., less often than monthly = 2, . . .several times a day = 7). These two objections can be effectively countered. Some kind of discrete categorization is indispensable. Basically, we selected these categories for two reasons. First, they are widely used by workers in the field. Second, they are more discriminating in the range of more frequent usage (i.e., the main area of concern). Specifically, all information relating to drug use behavior is of interest and concern for one objective or another, but, the WSUI is not designed to be a comprehensive measurment and quantification device for all the component behaviors. The focus is on whether the individual has ever tried the substance, and his current status of use. Furthermore, emphasis is placed on finer segregation at the critical region of high use. With reference to the apparent decision of assigning numerical values to the categories, the following points must be made. First, we do not imply or assume that the usage of a certain drug "about once a week" is mathematically twice the usage when reported as "less often than monthly". In fact, it is evident that usage reported as "several times a day", with a numeral assignment of 7 is hardly only one mathematical unit greater than "daily" usage with assigned value of 6. Second, in spite of this lack of equivalency of steps between and among categories of use, the assignment of numerical values is defensible on the ground that each numeral has an explicit definition of use associated with it; and the derivation of norms, against which the individual or group of individuals are compared, is also based on the same set of conventions.

Validity

The issue of validity is a universal concern of all measurements, but it becomes even more critical in this field. How do we know that we are, in fact, measuring what we want to measure? How do we know that the testee is telling the truth? Is the individual trying to fake the test? What about his memory? Does he accurately recall his drug use behavior? What about his ability to recognize drug names and keep them segregated? What about all the other drugs which are not listed by trade names, generic names, and street names?

None of the above concerns can be fully discounted. However, the extent to which these artifacts can distort the results may be minimized. First, since we are dealing, in main, with "current" usage, the risks associated with poor recall is reduced considerably. Second, our own experience shows that testees display a remarkable degree of candor and cooperation in taking the tests. Over 4,000 drug questionnaires have been administered under a variety of conditions to a highly heterogeneous population. Faking, noncooperation, and "phony" answers were virtually absent. For example, not one person indicated that a parent was the primary source of suggestion to use LSD. In contrast, a good many respondents indicated that parents were indeed the primary source which suggested the use of alcoholic beverages to them. In the latter case, we can see the tradition of "family drinking" on special occasions in a western society. Third, the validity of the self-report technique in studies of drug use is well-documented (Smart, 1975).

General Merits and Utility of the WSUI

The inventory can be used for a variety of research as well as

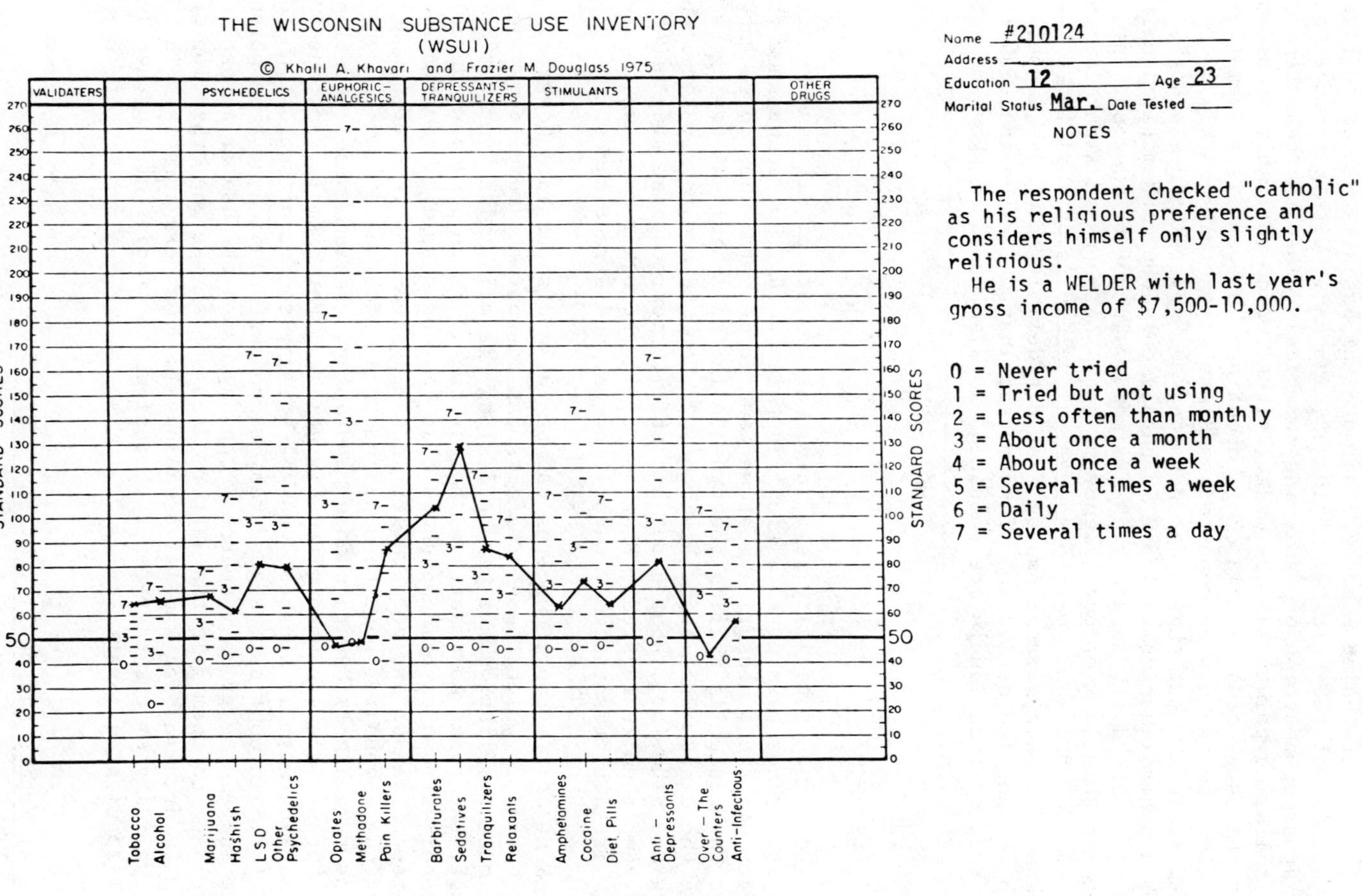
THE WISCONSIN SUBSTANCE USE INVENTORY
(WSUI)
© Khalil A. Khavari and Frazier M. Douglass 1975
Name #210124
Address
Education 12 Age 23
Marital Status Mar. Date Tested
NOTES
The respondent checked "catholic" as his religious preference and considers himself only slightly religious.
He is a WELDER with last year's gross income of $7,500-10,000.
0 = Never tried
1 = Tried but not using
2 = Less often than monthly
3 = About once a month
4 = About once a week
5 = Several times a week
6 = Daily
7 = Several times a day
VALIDATERS
PSYCHEDELICS
EUPHORIC-ANALGESICS
DEPRESSANTS-TRANQUILIZERS
STIMULANTS
OTHER DRUGS
STANDARD SCORES
Tobacco
Alcohol
Marijuana
Hashish
LSD
Other Psychedelics
Opiates
Methadone
Pain Killers
Barbiturates
Sedatives
Tranquilizers
Relaxants
Amphetamines
Cocaine
Diet Pills
Anti-Depressants
Over-The Counters
Anti-Infectious

Fig. 1.

clinical objectives. It is important to note that data provided by the WSUI represent *dependent variables*. That is, they reflect quantitative measures of drug use *behavior*. There are at least nine attractive features about the WSUI: 1) the scale allows for a quantitative assessment of the individual's drug use with respect to each of 19 drugs or drug categories, 2) at a glance, the individual's drug profile can be compared to that of a sub-population or the general population, 3) a great number of arbitrary "authoritative", and judgmental decisions which have plagued the field become unnecessary (for example, notions such as heavy, moderate, and light level of usage for a substance can be defined in terms of SD units), 4) assignment of "weight" to each drug can be made on the basis of difference between any two or more points on the T score distribution, 5) the scale can provide operational definitions of a quantitative nature for concepts such as "drug abuse","polydrug use" and so forth, 6) patterns of drug use for a specific group can be readily compared with other groups or the general population, 7) cross-cultural comparison of usage profiles can be carried out, 8) changes of use patterns, over time for individuals and/or groups can be determined, and 9) a comprehensive quantitative profile of each individual's drug use can be readily constructed for diagnostic and treatment objectives.

Some Profiles and Interpretations

Fig. 1 represents the drug use profile of an individual we tested. A glance at his profile provides the viewer with a set of independent, yet interrelated, information. Starting at the left of the profile, we notice that the respondent is a regular smoker (tobacco=7=several times a day; T=64.29) and a regular drinker (alcohol=7=daily; T=65.71). Further inspection of his profile shows some use of marijuana and hashish, but significant use of the following drugs: LSD, other Psychedelics, Pain Killers, Barbiturates, Sedatives, Tranquilizers, Relaxants, Cocaine, and Anti-Depressants. Since the usage level for all the above drugs is at least two standard deviation units above the population mean, then it can be concluded that his usage of these agents is significantly greater ($p<.01$) than the population average. Clearly, he is a polydrug user. But, depending on treatment or research concerns, his profile can be further categorized along the indicated objectives. For example, the physician, the therapist, or the counselor can employ a quantitative cut-off to derive a more focused label. The individual may be labeled as a polydrug user of *Depressant-Tranquilizers-Pain Killer-Psychedelic type*. The Depresant-Tranquilizers dominate his polydrug use profile. Additionally, we note that he consumes alcoholic beverages (depressants) daily, which further substantiates the notion that the individual's dominant drug consumption behavior evolves around "DOWNERS". In all instances, the scale provides us with exact quantitative data about the individual's "standing" vis-a-vis the population. For research purposes, on the other hand, a priori cut-offs can be established to mark off the

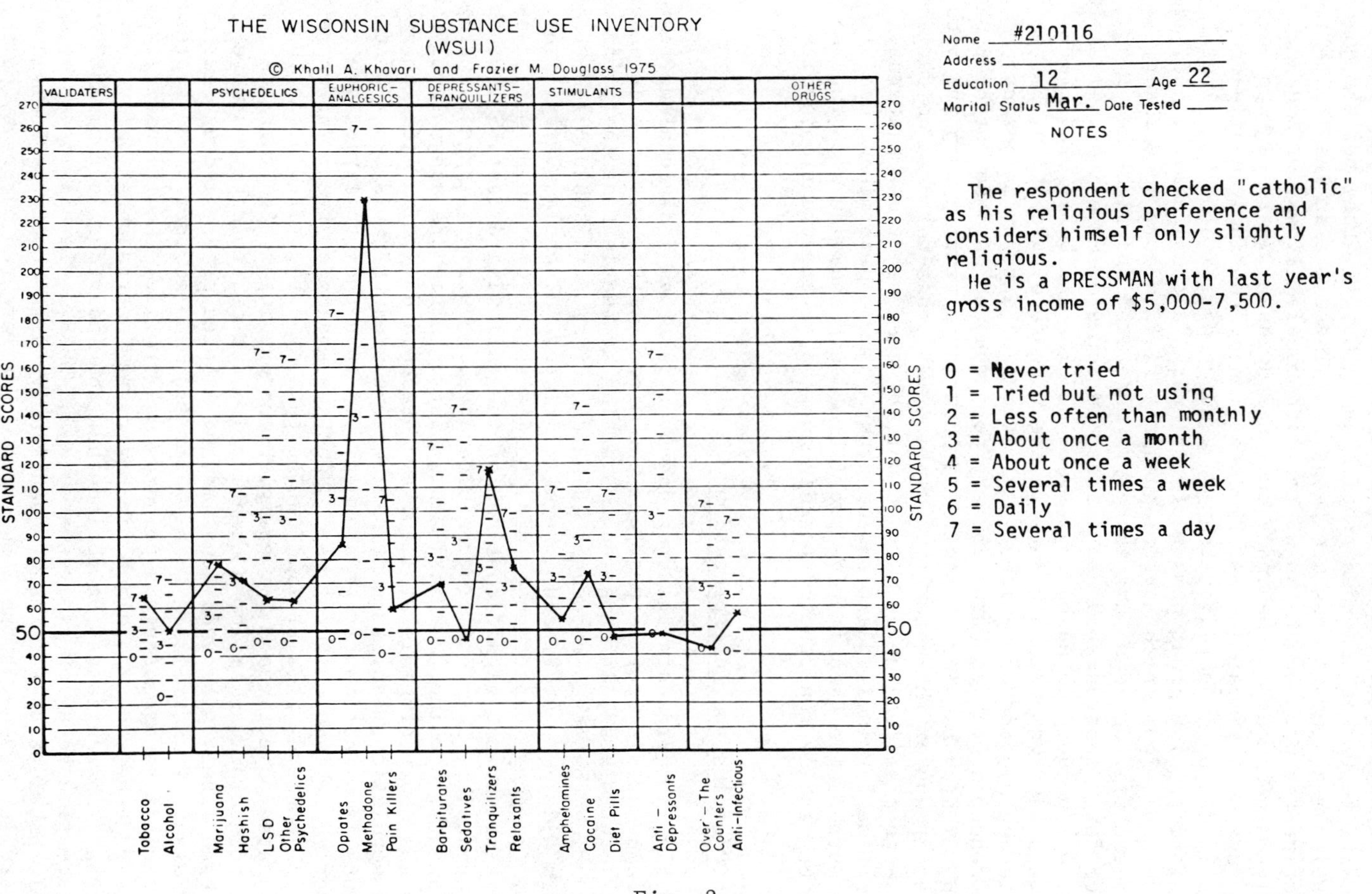
THE WISCONSIN SUBSTANCE USE INVENTORY
(WSUI)
© Khalil A. Khavari and Frazier M. Douglass 1975
VALIDATERS
PSYCHEDELICS
EUPHORIC-ANALGESICS
DEPRESSANTS-TRANQUILIZERS
STIMULANTS
OTHER DRUGS
STANDARD SCORES
Tobacco
Alcohol
Marijuana
Hashish
LSD
Other Psychedelics
Opiates
Methadone
Pain Killers
Barbiturates
Sedatives
Tranquilizers
Relaxants
Amphetamines
Cocaine
Diet Pills
Anti - Depressants
Over - The Counters
Anti-Infectious
Name #210116
Address
Education 12 Age 22
Marital Status Mar. Date Tested
NOTES
The respondent checked "catholic" as his religious preference and considers himself only slightly religious.
He is a PRESSMAN with last year's gross income of $5,000-7,500.
0 = Never tried
1 = Tried but not using
2 = Less often than monthly
3 = About once a month
4 = About once a week
5 = Several times a week
6 = Daily
7 = Several times a day

Fig. 2.

inclusion-exclusion boundaries. Let us suppose that a particular investigation is concerned with examination of individuals who use excessive amounts of sedatives (for example, at least seven SD units away from the population mean) but use no other drug to this extent. Then, in this case we have a highly restrictive inclusion but very lax exclusion criteria and out subject of Fig. 1 is included in the study. Clearly, all kinds of well-defined and objective inclusion-exclusion criteria can be employed to suit the goals of the particular research.

Fig. 2 represents the drug use profile of an individual on methadone maintenance. It is instructive to note that this person, in addition to his daily intake of methadone, uses significant amounts ($p < .01$) of seven other drugs (marijuana, hashish, opiates, barbiturates, tranquilizers, relaxants, and cocaine). Although we do not have the facts, it would seem plausible that this individual's intake of tranquilizers (7=several times a day) is part of his prescribed medication while the remaining six drugs he uses reflect his continued heavy involvement with illicit drugs.

Comment

Currently, we are conducting further studies to increase the data base for calculation of norms and determination of the scale's validity and reliability characteristics. In the meantime, those interested in blank copies of the WSUI profile and the drug use questionnaire can contact one of us (K.A.K.). The WSUI blanks and copies of the questionnaire will be furnished at cost.

REFERENCES

Barter, J.T., Mizner, G.L., and Werme, P.H. 1971. Patterns of drug use among college students in the Denver-Boulder metropolitan area. *Final Report*, BNDD Contract No. J-68-51, Bureau of Narcotics and Dangerous Drugs.

Berg, D. 1970. The non-medical use of dangerous drugs in the United States. A comprehensive view. *Int. J. Addict.* 5: 777-834.

Bucky, S.F., Edwards, D., and Thomas, E.D. 1974. Intensity: The description of a realistic measure of drug use. *J. Consul. Clini. Psychol.* 30: 161-163.

Eells, K. 1968. Marijuana and LSD: A survey of one college campus. *J. Coun. Psy.* 15: 459-467.

Elinson, J. and Nurco, D. (eds.) 1975. *Operational definitions in socio-behavioral drug use research.* DHEW Publication No. (ADM) 76-292.

Glen, W.A. and Richards, L.D. 1974. *Recent survey of non-medical drug use: A compendium of abstracts.* DHEW Publication No. (ADM) 75-139.

Josephson, U. 1974. Studying drug use. In E. Josephson and E.E. Carroll (eds.). *Drug Use: Epidemiological and Sociological approaches.* Washington, D.C.: Hemisphere.

Nail, R.L. and Gunderson, E.K.E. 1973. Physician's Disagreement in Assessing Drug Abuse Risk. Navy Medical Neuropsychiatric Research Unit, San Diego, California: *Report No. 73-7.*

Runyon, R.P. and Haber, A. 1971. *Fundamentals of behavioral statistics.* Reading, Mass.: Addison-Wesley.

Smart, R.G. 1975. Recent studies of the validity and reliability of self-reported drug use, 1970-1974. *Cana. J. Crimin. Correc.* 17: 326-333.

Walizer, D.G. 1975. The need for standardized, scientific criteria for describing drug-using behavior. *Int. J. Addict.* 10: 927-936.

Welsh, G.S. and Dahlstrom, W.G. 1956. *Basic readings on the MMPI in psychology and medicine.* Minneapolis: University of Minnesota Press.

A BEHAVIORAL MODIFICATION TECHNIQUE FOR AN OUTPATIENT METHADONE MAINTENANCE PROGRAM

Richard J. Kestler, Ph.D.
Hope Center (Tucson Southern Counties Mental Health Services, Inc.)

Terrence Hickey, B.A.
Southern Arizona Mental Health Center

INTRODUCTION

A behavioral modification program has been developed to determine eligibility for take-out methadone. Called the "Point System" it assigns point totals to desired behaviors: clean urines, employment, or the equivalent, and counseling. Points are awarded weekly, assuring rapid reinforcement. The system has been operational for five years. Advantages include simple, clear guidelines for clients and staff and reduced conflict level between both staff and clients and among clients and staff.

A questionnaire developed by the authors to assess attitudes toward the system elicited a 56% client return rate. Responses indicated generally favorable attitudes toward the "Point System." Demographic analysis suggests the highest support level is from younger, better educated Anglo population groups.

DISCUSSION

Several studies suggest a direct correlation between the success rate of drug abusers in treatment and the modification of behavior patterns developed before or during their years of drug use. Cheek et al. (1976) found the successful graduates of a methadone maintenance program were those who tired of the drug life style and who, while in treatment, developed " . . . a new and more conventional set of peer groups, familial and occupational roles." Those considered not successful at follow-up were, for the most part, persons who had experienced positive associations with the drug subculture. "The period of treatment did not rescue them from a totally negative situation, but rather threatened a still quite positive one." Success, the authors conclude, probably requires " . . . the pattern of alternation of positive reinforcement and punishing situations." Support for this view comes from Williams and Lee (1975) who describe a positive correlation between a client's length of retention in treatment and his meaningful work history.

Yet an analysis of the literature yields surprisingly few reports of behavior modification techniques being applied as part of the rehabilitation process in out-patient drug treatment programs. Götestam, Melin and Öst (1976) give a detailed analysis of all such programs described in the major journals. Nearly all are either individual and small group case studies or inpatient ward programs. Boudin et al. (1977) describe a drug program using contingency contracting with an outpatient population. The only description we found of a methadone maintenance treatment program utilizing operant conditioning (Melin, Anderson and Götestam, 1976) was small (25 clients) and in an inpatient hospital setting.

The Hope Center has developed a contingency management system for an outpatient program comprising two clinics with a daily methadone maintenance population averaging 275-300 clients. The program began in 1970. In 1973 it experienced rapid growth, ending staff ability to have considerable knowledge of the needs of each client in treatment. The need to develop objective procedures was one of the reasons for developing the "Point System." The other was to help alleviate serious staff conflicts over which clients should be administratively detoxified and which should be awarded privileged medication. Staff tended to polarize medication into "help" vs. "shape up or be expelled" groups, setting administrators against line staff, and ex-addicts against university trained counselors (Heiman, 1977).

Clients were able to exploit these divisions. Schwartzman and Kroll (1977) suggest many addicts are expert manipulators, having learned to pit "authoritarian" against "permissive" family members. Their ability carried over into the clinical setting. Objectifying criteria into the "Point System" reduced the scope of this manipulative activity.

The system specified desired program behaviors (employment or other productive activity, clean urines, counseling, etc.). Performance leads to a reward wanted by nearly every client; the right to take one's methadone dose home. Our expectation was that new habit patterns would gradually supercede those aversive ones established before or during treatment.

The system is designed so a client earning maximum points can gain increasing take-out privileges at a rate consistent with federal guidelines. Points are awarded at a weekly "team meeting" of all clinical staff. Each client's point accumulation total is recorded. These behaviors earn points:

A) Productivity. 5 points for full time work (work status must be verified). 12 or more units of college count as full time work, 5 points may also be earned by homemakers with at least one juvenile or unemployable dependent, by retired persons or by those on SSI or other certified disability. 3 points may be earned for partial work or school.

B) Non-usage of illegal drugs. 5 points are earned for clean urines taken weekly on a randomized schedule and monitored by staff. Testing is for opiates on a weekly basis, for other drugs monthly.

C) 5 points for weekly counseling sessions.

D) 5 points for additional counseling outside the clinic, i.e. marriage counseling groups. One point can be subtracted from the total for failure to pay fees (set on an adjustable scale based on ability to pay, with a maximum of $5.00 per client weekly).

Privileged medication is awarded after attaining 200 points, which earns a one day per week take-out privilege. To earn the remaining privileges a client must acquire: 260 points for weekend privileges, 320 points for tri-weekly privileges and 1000 points for bi-weekly privileges. Up to the 200 point level, a positive urine results in loss of 25% of points earned. Once the client obtains the minimum number of points for privileges (200) the team may choose to suspend points rather than remove them; for example, a client who has earned 500 points and has tri-weekly privileges would have points and privileges suspended for either one positive urine or for failure to pay fees beyond the two week grace period.

Point suspension is for a set period of time. If it is for fees, points and privileges are restored upon payment. Suspension for drug abuse usually is for one month. During that time no additional points are earned, and no privileged medication schedule is permitted. At the end of the suspension period the client's record is reviewed and if the problem is resolved points and privileges are restored. This may be done gradually over a few weeks, or immediately at the option of the clincial team. This permits the staff to exercise flexibility while retaining the point structure. The system provides rapid reinforcement of a return to desired behavior patterns while it penalizes a return to illicit use of narcotics. Should a pattern of illicit drug use become clear during the suspension period, the point total is reduced to 150. The greater a client's point total, the greater the loss.

THE SURVEY: RATIONALE AND RESULTS

Between May and July, 1977, all clients on methadone maintenance were asked to fill out a questionnaire designed by the authors. The "Point System," operational for four years, had become routine. We wished to assess client attitudes and discover why some groups were more comfortable than others with the system. Our null hypothesis, developed subjectively through conversations with clients and staff, was that the highest support level for the system came from the younger, better educated and more middle class oriented clients. Similar to the "Success" category in Cheek et al. (1976), these clients gravitated towards heroin usage as a reaction to problems, but had come to express regret and/or guilt feelings associated with their life "on the streets." They tend to share with the staff the work ethic and have more employability than do the less educated, older addicts who have long considered themselves members of an addict life style. Young Anglo clients not only express these values more frequently than do minority clients, they are also more easily able to change residence and merge into the non-addict community. This group appeared to be the most concerned with earning points and coming to the clinic as little as possible.

At the time the survey was presented in the methadone maintenance component of our program, the clients had the following demographic characteristics (client data is from June and July, 1977): 63% were male, 37% female; 50% Anglo, 35% Mexican American, 11% Black, 2-1/2% Indian and 1% others. Approximately 40% were age 18-24, 30% age 25-30, 20% age 31-40, and 10% over 40. A quarter of the population had an education level of grade 9 or less, half had attended or graduated from high school, and a quarter had some college experience. 40% of the active caseload were employed full time, 3% part time, 7% were students or in training, 6% were retired or disabled, and 19% were homemakers. Only 25% were not receiving points for productivity. 67% of clients tested that month had opiate free urines (on an average of 4 monitored tests per client). 24% of the population had tri-weekly privileges while 13% had bi-weekly privileges.

The survey was distributed by leaving it with the counselor on duty at the dispensary. Counselors asked clients to fill out the survey but no threats or rewards were suggested. Clients were assured their replies would be seen only by Dr. Kestler, who was known to be not directly involved in treatment decisions. A locked box was provided for replies. Clients were asked to include their names or Center ID numbers on the questionnaire.

The survey contains 21 questions each with a five choice scale ranging from A) "I strongly disagree" to E) "I strongly agree." Clients were told to feel free to add comments if they chose (19 did so), and there was a higher response than anticipated (56 percent). Demo-

Table I. Ethnicity, sex and education levels of respondents (R) and non-respondents (N).

	ANGLO n=138		MEXICAN AMERICAN n=97		BLACK n=31		INDIAN n=7		TOTAL	
MALE Grade Level	R	N	R	N	R	N	R	N	R	N
1-8	0	0	6(21%)	11(27%)	0	2(25%)	0	1(50%)	6(7%)	14(18%)
9-11	4(9%)	5(17%)	14(41%)	10(24%)	2(17%)	2(25%)	0	0	20(22%)	17(21%)
H S Grad.	24(56%)	13(45%)	10(29%)	14(34%)	6(50%)	2(25%)	1(100%)	1(50%)	41(46%)	30(38%)
13-15	13(30%)	11(38%)	3(9%)	5(12%)	3(25%)	2(25%)	0	0	19(21%)	18(23%)
Col. Grad.	2(4%)	0	0	1(2%)	1(8%)	0	0	0	3(3%)	1(1%)
TOTAL	43(99%)	29(100%)	33(100%)	41(99%)	12(100%)	8(100%)	1(100%)	2(100%)	** 89(101%)	80(101%)
FEMALE Grade Level										
1-8	1(27%)	0	4(40%)	3(25%)	0	1(17%)	1(100%)	0	6(10%)	4(9%)
9-11	6(14%)	8(33%)	4(40%)	6(50%)	4(80%)	3(50%)	0	2(67%)	14(24%)	19(42%)
H S Grad.	19(45%)	7(29%)	2(20%)	3(25%)	1(20%)	0	0	1(33%)	22(37%)	11(24%)
13-15	16(38%)	8(33%)	1(10%)	0 0	0	2(33%)	0	0	17(29%)	10(22%)
Col. Grad.	0 (99%)	1(4%)	0	0 0	0	0	0	0	0(0)	1(2%)
*TOTAL	42	24(99%)	11(100%)	12(100%)	5(100%)	6(100%)	1(100%)	3(100%)	59(100%)	45(99%)

* Totals do not all = 100% due to rounding.

** Not included in the sample are 1 male and 2 female clients whose ethnicity is other than those above.

graphic characteristics of respondents (hereafter r) were compared to those of non-respondents (hereafter n). 60% of the r's were male, compared with 63% of the n's. Anglo's comprised 57% of the r's but only 42% of the n's. The reverse bias existed for Mexican Americans (30% r, 42% n) and Indians 2% r, 4% n). The greatest difference was for Anglo females who comprised 71% of the r's vs. 51% of the n's. Blacks comprised approximately the same percentage of the r and n populations. Although none of the differences for Anglos and Mexicans from the mean were significant to the 5% level, they all fell between 30% and 8%, indicating a pattern of response to support our null hypothesis (see Table I).

We expected a higher percentage of the r's to be on privileged medication but this was not the case. 37% of both the r and n populations were on either a tri- or bi-weekly schedule. No significant differences within ethnic groups occurred. It is worth noting that 71% of all clients with bi-weekly privileges were Anglos, vs. 50% of the total clinic population ($p<.05$).

Analysis of the questionnaire itself was done in the following manner. Nine questions were pre-coded as positive, i.e., a client highly supportive of the "Point System" would circle E (I strongly agree). Nine questions were pre-coded negative, while three were neutral. Coding for a positive question gave an E response a value of +4; D, +3; C, +2, B, +1, A, 0. The reverse formula was used for negative questions. Neutral questions were not coded. A maximum support score would be 72, while one with all neutral (C) responses would receive a score of 36, and one maximally negative would get a 0.

All clients who responded were separated by ethnicity, sex, and education level. Of 156 responses, 7 were invalid because there was no name or distinguishing ID number, and 5 were invalid because they contained 5 or more blank answers. For the remaining 144 surveys, the high score was 64, the low score 16. The $\bar{x}$ was 42.7, the median 43 and the mode 40, indicating an overall favorable response level.

If we group responses by sex, ethnicity and education level, we get the results found in Table II. Again, while no figure is significant to the 5% level, there is a tendency for a somewhat higher favorable response ratio from the better educated Anglo population.

SUMMARY

For most program staff the "Point System" has proven to be a useful tool. Apparently it is also seen to be one by the majority of clients. Greatest impact is, as predicted, with certain client groups, although the bias is not great.

Table II. Mean levels of response to the survey as a function of sex, ethnicity and education level.

Male Grade Level	Anglo	Mexican American	Other	Total
1-11	40.0 n=4	40.3 n=20	44.0 n=2	40.5 n=26
H.S. Grad.	43.4 n=24	41.6 n=10	40.8 n=6	42.6 n=40
13 or above	46.4 n=13	43.7 n=3	39.4 n=5	44.3 n=21
Total	44.0 n=41	41.0 n=33	41.9 n=13	42.4 n=87

Female Grade Level	Anglo	Mexican American	Other	Total
1-11	41.3 n=7	47.3 n=7	36.9 n=3	42.7 n=17
H.S. Grad.	46.3 n=19	43.0 n=3	50.0 n=1	46.3 n=23
13 or above	43.6 n=15	40.0 n=1	36.9 n=1	43.3 n=17
Total	44.4 n=41	46.0 n=11	38.0 n=5	44.1 n=57

If different contingencies could be introduced it is probably that other behaviors incompatible with using opiates could be reinforced. Götestam, Melin and Öst (1976) suggest that " . . . careful studies must be conducted on the effects of different contingencies in different social contexts that an ex-addict will meet." They further recommend developing individual programs for each client since reinforcing stimuli are highly idiosyncratic. Through this study and additional refinement of our data we are attempting to provide greater individualization of treatment while retaining the underlying behavior modification structure.

REFERENCES

Boudin, H.M., Valentine, V.E., Inghram, R.D., Brantley, J.M., Ruiz, M.R., Smith, G.G., Catlin, R.P., and Regan, E.J. 1977. Contingency Contracting with Drug Abuse in the Natural Environment. *Int. J. Addict.* 12(1): 1-16.

Chambers, K.D., Inciardi, J.A., and Siegal, H.A. 1974. An assessment of the Incidence and Prevalence of Drug Use and Alcohol Beverage Consumption in the State of Arizona. *Resource Planning Corporation*, Arizona Department of Alcoholism and Drug Abuse.

Cheek, F.E., Holstein, C.M., Fullam, F.A., Arana, G., Tomarchio, T.P., and Mandell, S. 1976. From Heroin to Methadone--Social Role Changes and Reinforcement Differentials in Relation to Outcome on Methadone. Part II. The study of Social Role Changes. *Int. J. Addict.* 11(4): 659-679.

Götestam, K.G., Melin, L. and Öst, L. 1976. Behavioral Techniques in the Treatment of Drug Abuse: An Evaluative Review. *Addictive Behaviors.* 1: 205-225.

Heiman, E.M. Submitted 1977. Attitudinal Issues in Methadone Maintenance Programs.

Melin, L., Anderson, B.E., and Götestam, K.G. 1976. Contingency Management in a Methadone Maintenance Treatment Program. *Addictive Behaviors.* 1: 151-158.

Morrison, J. 1972. Final Report: Inquiry into Heroin Use, LEAA Project Grant #72-136-2.

Schwartzman, J. and Kroll, L. 1977. Methadone Maintenance and Addict Abstinence. *Int. J. Addict.* 12(4): 497-507.

Siegel, S. 1956. *Nonparametric Statistics.* New York: McGraw-Hill Book Company.

Williams, W.V. and Lee, J. 1975. Methadone Maintenance: A Comparison of Methadone Treatment Subjects and Methadone Treatment Dropouts. *Int. J. Addict.* 10(4): 599-608.

MULTIPLE FAMILY THERAPY WITH DRUG ABUSERS

Edward Kaufman, M.D.
California College of Medicine

Pauline Kaufmann, M.S.W.
Phoenix House

Multiple Family Therapy (MFT) is a technique which is particularly useful and applicable with drug abusers and their families. This type of therapy can be used in any treatment setting for drug abusers, but is most successful in residential settings where the family is more available and accessible. MFT was initiated as a modality by Lacqueur (1971) in an inpatient unit of a state hospital, but, as used by the authors, has many other roots. These include: Social Network Intervention (Speck, 1971), multiple impact therapy (Ritchie, 1971), the ward or town meeting concept of both the psychiatric and Synanon mode therapeutic community and a host of group and family therapy techniques. Group techniques run the gamut from psychoanalytic through psychodrama, existential, gestalt and encounter. Family techniques include sculpting or choreography (Papp, 1971), structural communications (Minuchin, 1974), and systems approaches (Fogarty, 1974).

This presentation will consist of:

A. Group Composition and Environment
B. The Therapeutic Team
C. Family Dynamics and Techniques
D. Discussion (including literature review)

GROUP COMPOSITION AND ENVIRONMENT

The greater the motivation and involvement of the patient, the easier it is to initiate the family into therapy, particularly in MFT. Thus, it is extremely difficult to generate an MFT in low intervention programs such as most methadone maintenance programs or outpatient psychotherapy clinics. It beomes progressively easier to do so in day programs and inpatient settings. However, the setting which most readily lends itself to the establishment and continuance of a successful MFT group is a residential therapeutic community (TC). One reason that families come to groups in this setting is that they may be required to do so if they are to visit the resident. Also, since the identified patient (IP) is now drug-free for the first time in years, he or she can be related to without chaos and with trust and warmth. It is best to have the IP invite the family to come in, but the therapist may send a form letter or call. A call from the therapist may be necessary in order to involve a reluctant family or family member(s).

One author (Edward Kaufman) recently had the experience of working unsuccessfully for four years with a patient on a methadone maintenance treatment program (MMTP) despite the use of an ancillary mental health clinic and several group psychotherapy experiences. During this time, the patient consistently refused to have his mother or any family member involved in treatment although he lived with his mother throughout the treatment. Federal guidelines about confidentiality, particularly as they are applied to MMTP, and a non-coercive attitude on the part of the N.Y.C. agency sponsoring treatment, prohibited insisting that the family be involved. Finally, after several overdoses and detoxifications from drugs of abuse, the patient agreed to enter a therapeutic community. After two weeks in residence, the patient and his family appeared at the first MFT to which they were invited. The mother's crucial role in treatment was immediately apparent. However, two involved aunts and an older brother, all of whom had never been mentioned previously, attended regularly. This extended family was found to be necessary for restructuring the family system and changing this difficult patient.

In a residence, the group may be composed of all of the families in the TC or separated into several groups of three or four closely matched families. Our experience is with the former type of group, because of lack of primary therapists as well as our view of the group of families as one community.

There are generally as many as forty to fifty individuals in the weekly multiple family group, including ten to fifteen families. The group includes identified patients and their immediate families, as well as any relatives who have an impact on the family. Friends and lovers are included if they are an important part of the addict's network and are drug free. If they are abusing drugs or alcohol, they are excluded from the group until they can control this symptom from being disruptive and destructive to the group. When there are no rigid guidelines about excluding family members and friends, a good deal of meaningful material is produced. One client found that her present drug abusing boyfriend arrived with his "only drug-free friend" who happened to be her ex-husband. Although she was able to ask them to leave before the group began, the feelings which were stirred up between her and her father recreated old unresolved problems which were dealt with in MFT. Families should be oriented and interviewed prior to entry in the group. A genogram can be diagrammed and a family map begun during the initial evaluation. Once it becomes apparent in the group that the troubled family requires individual family sessions, these are provided. Some family secrets may have to stay within the family, but generally, they are encouraged to bring the content of these individual sessions into the multiple family group as this remains the primary modality of family therapy. An experienced family therapist works with counselors in the program as co-therapists. The total group frequently functions as adjunctive family therapists. Usually, family members take their cues from primary therapists and will be appropriately confronting, reassuring and supportive. At times, the family's own needs prevent this and their anger at their child or their possessiveness will spill over to all the former addicts in the group. Families share experiences and offer help by acting as extended families to each other and the residents outside the actual therapy hours. Residents who accompany each other on visits serve as supports in the home and behavioral reporters in MFT groups. Supporting residents may help the IP with their family "homework" on visits, as well as in the MFT.

The group is seated in a large circle with co-therapists distributed at equal distances to provide observation of the total group. We do not use a "fish bowl" with the primary family in the center of the group because this discourages the participation and identification of the other families. Families sit together and their seating arrangements are carefully observed as they usually follow structural patterns. They may be asked to separate if there is a great deal of whispering or disruption. The group begins with everyone introducing themselves by name and role. A group member will describe the purpose of the group, generally stressing the need for families to communicate honestly and to express their feelings openly. At times, the description of purpose emphasizes the importance of understanding and changing the familial forces which have led to drug abuse. The first family frequently is worked with for about an hour.

The conflicts they focus on set the emotional tone and influence the topics discussed in the entire group. Many other families will identify with these conflicts, express feelings, offer support and work on similar conflicts. Generally, three or four families are worked with intensively in a night, almost all families participate verbally, and usually all families are emotionally involved.

The informal contacts which take place before and after a group are crucial. Therapists should mingle and interact during these pre- and post-group sessions. Many pre-session contacts are excellent grist for the therapeutic mill. Post-group interaction may confirm insights and validate feelings or undermine therapeutic work if not monitored. Families of "splitees" are encouraged to continue to attend to maintain their structural shifts and encourage the IP to return to treatment.

THE THERAPEUTIC TEAM

A prerequisite for the therapeutic team is a primary therapist who is experienced in group and family therapy and comfortable in large groups. There must also be several co-therapists who are an integral part of the treatment program and can provide feedback from the group to the program and input from the program to the group. The group should be used to train counselors in the dynamics of families and the techniques of family therapy. All too frequently, younger counselors in TC's tend to over-identify with their client's hostility to parents because of their own unworked-through conflicts. Counselors who are themselves parents may over-identify with the parental system. Thus, counselors should have supplementary training experiences which focus on their own family of origin. Counselors should also have supplementary didactic courses and assigned readings, particularly if they are to become primary therapists. Co-therapists should work together as a team which agrees to disagree so that there is a unity which allows room for individual differences. This provides a role model for a parental system which is unified but not rigid.

Speck (1971) has cited qualities of a good network leader which are also important in an MFT primary therapist. "A sense of timing, empathy with emotional high-points, a sense of group moods and undercurrents and some charismatic presence . . . along with the ability to dominate, the leader must have the confidence that comes with considerable experience . . . the ability to efface himself, to delegate and diffuse responsibility . . . rather than collect it for himself". The role of the MFT therapist, like the network leader, is similar to that of the good theatrical director, but with a greater concern for the therapeutic than the dramatic. Interestingly, Papp (1974) also emphasizes the therapeutic aspects of theatricality in MFT particularly as seen in sculpting. True, theatre strips away superficiality to the bare bones of meaning, reveals a hidden truth, shares a universal experience. Minuchin (1974) also notes that the therapist functions "like the director of a play, setting the stage, creating a scenario, assigning a task and requiring the family members to function within the new sets that he has imposed".

MFT is a stimulating and rejuvenating experience for the therapist and treatment program as well as the family. The therapist becomes the paternal and/or maternal figure for a host of families who become a single family network and, in many ways, a single family. The therapist assumes temporary parental control of all of these families at the same time as he/she is the child of all of them. Thus, the therapist gives and takes in a multitude of parental and childlike roles. At the same time, the therapist must step away from this emotional entanglement and be objective. A therapist may even say to his or her co-therapists and the group, "I am going to join this family system to experience it. Pull me out if I'm getting too involved." The Primary Therapist must always keep in mind that one of his/her primary functions is to be experienced as a supportive ally by every member of the group. He or she must also feel the capability and right to interrupt any communication which is destructive or disruptive.

FAMILY DYNAMICS AND TECHNIQUES

Those dynamics and techniques which are most applicable to MFT will be discussed here. Generally, those techniques which are most familiar to the treatment system are most readily used in MFT. Thus, in MFT's in TC's, confrontation and encounter are used in the service of structural changes. The therapist must join the treatment system as well as each family. New techniques are gradually introduced to MFT and thence to the overall program or vice versa.

In the early phases of treatment, the families support each other by expressing the pain they have experienced through having a drug abuser in the family. The family's sense of loneliness and isolation in dealing with this major crisis is greatly attenuated by sharing the burden with other families. The means by which they have been manipulated are quite similar and are the beginning of a common bond. The addict son or daughter has lied to and stolen from the family. Many families have given the addict money for his/her habit to keep him/her from stealing and resulting in incarceration. Group members express commiseration for the family's suffering. The family is strongly encouraged not to repeat this pattern. The family learns to see hostile aspects of this rather than its benevolence. Many addicts who have difficulty with the demands of a TC will try to convince their families to take them back once again and protect them from the "evils" of a TC just as they had protected them from jail. Intervening in this system helps prevent many early "splitees". We call this "closing the back door". Many families are able to do this merely through group support. Others must learn to recognize and reduce patterns of guilt, provocation and enmeshment before they can close the door to the cycle of symbiotic re-involvement. An initial period of ventilation of anger and resentment may be necessary before strategies for change can be introduced. An atmosphere is created in which all families are encouraged to be open and express everything about everyone. This does not give the family permission to hurt sadistically under the guise of honesty.

The giving of food may be utilized as an important family transaction. Food may also be limited to gifts to the entire group. This helps create a sense of the group as one family. It may also alleviate the families' guilt and permit them to gratify their need to give without infantilizing the IP.

As the therapy progresses, the role of the family in producing and perpetuating the abuse of drugs is identified. Patterns of mutual manipulation, extraction and coercion are identified and negated. The family's need to perpetuate the addict's dependent behavior through scapegoating, distancing, protection or infantilization is discouraged and new methods of relating are tried and encouraged. Families tend to feel guilty when the addict confronts them with their role in the addiction cycle. This has occurred in the home and reoccurs in the early phases of multiple family therapy. If the therapist does not intervene, the family will retaliate by inducing guilt or undermining growth and may ultimately pull the addict out of treatment. Drug dependence is viewed as a family problem; there must be no scapegoats. Parents must be given a great deal of support in family sessions because of their own guilt and the tendency of patients and even counselors to attack them. For some parents, even the admission in public that there is a family problem leads to shame and reactive hostility. Such parents require individual family therapy sessions where they can receive individual support. When they can overcome their embarrassment about expressing feelings in "the public" of the group, they have made a valuable step towards the overt expression of feelings in general. Counselors in the program and families frequently lead the group away from working with the family system to the IP and his/her problems in the TC. When this occurs, the group must be refocussed on the family program. Counselors will often similarly focus on other individual drug abusing family members and should be helped in re-directing the focus to the family system. Frequently, material is revealed in MFT which could lead to severe set-backs and punishments for residents. It is important that the family therapist have input into the program's use of such disciplines lest they undermine the family therapy or recreate destructive family patterns within the program.

Multiple family therapy groups help residents actualize insights about their family which they have achieved in their own therapy. Many families learn to express love and anger directly for the first time in these groups. Deep emotional pain is expressed when appropriate and other family members are encouraged to give support to such expressions rather than nullify them or deny them. Frequently, this is done by the tears of the entire group or by appreciative applause. Kissing, hugging, and rocking are ways that families tend to obliterate pain under the guise of giving comfort. In situations where families are emotionally isolated, encouraging the mutual exchange of physical affection is helpful.

The identified patient acts as the barometer of the total family functioning. When his/her behavior becomes maladaptive or disruptive, it is assumed that the family is under increased stress and may be reverting to former destructive patterns. This is particularly true in adolescent day programs when there is frequent parental contact. The identified patient's behavior is viewed as part of the family stress at the same time as it is handled as his/her individual responsibility by his/her peers in encounter groups within the TC. As an extension of the TC, peers can be quite supportive of each other yet hold each other responsible and challenge each other more effectively than can adults. Therapeutic homework is frequently assigned to reinforce the family's structural changes. Not only are family tasks assigned, but different family roles are also assigned in restructuring the family. Weak ties between family members are strengthened by suggesting joint activities which build closeness and/or identifying the fears and patterns which have led to weak ties. Strengthening weaker ties will diminish other enmeshed ties. Overwhelming family members may be asked to absent themselves for several weeks to strengthen other ties. The family pain at having an addicted member can be used to motivate over-involved family members to begin to separate.

Any number of other therapeutic strategies may be useful in restructuring the family. Much will depend on the individual's style. We consistently focus on dysfunctional communications. We delineate individual boundaries by not permitting family members to speak or feel for each other. We point out nonverbal coercive communications which tend to overwhelm family members, inhibit expressiveness or produce double binds. We ask that messages be stated clearly with underlying meanings made explicit. We also assign tasks to family members to promote individuation. It is most important for mothers who are single parents to find pleasure in their lives and for couples to learn or relearn to enjoy each other. Frequently, grandparents must be brought in and intergenerational patterns demonstrated before parents can change. Interpretations which simultaneously focus on the responsibility of both parties are effective and diminish guilt provocation and scapegoating. They also help maintain the therapist's position as an ally of every family member.

We have found that psychodramas dealing with negotiation and resolution of disagreements, formation of positive subgroups, changing communication style, teaching verbalization of anger, affection, or friendship, are all helpful in changing the dysfunctional system. The "empty chair" technique can be used to tap deep feelings about family members who are not present--generally anger at the withdrawal of the member, but also anguish at loss. Family Sculpture is a technique which is a variant of psychodrama and is very valuable in MFT with drug abusers.

In the latter phases of MFT, families express intense repressed mourning responses which are essential to a healthy family adaptation. This is easiest with feelings about a sibling who has died from an overdose. In still later phases a lost parent can be mourned. Family secrets and myths are also revealed in the later phases of MFT. When the anxiety stirred up by early shifts has been resolved, more advanced tasks can be assigned. In the final phases, the family and IP are separated from the group.

A knowledge of the specific dynamics of a family permits further "family" therapy with the individual alone as well as provides crucial material for individual and group therapy. Frequently, we learn of the resident's overwhelming anger at a parent. At times, it is obvious that this anger is too destructive to be expressed in the presence of the parent, even in a moderate way. The resident is then encouraged to express the anger in his/her own therapy before expression with his/her family is possible.

Videotape can also be used to confront family members with emotions which are denied. By repeated replays it is possible to have family members recognize such patterns as guilt induction and infantilization through enmeshing affection and denigration. Videotapes of MFT are excellent teaching devices and are available from the author (Edward Kaufman).

Families in the MFT act as supports for each other outside the session. They may continue to do so after the therapy has been terminated. These families frequently replace the entire network of the addict with a new, therapeutic network.

DISCUSSION

We would like to summarize the literature on MFT which has influenced our work and discuss it in the context of our experience. Lacqueur's (1971) objectives with schizophrenics are identical to our own, i.e., "improvement of communication between all members of the family and achievement of better understanding of the reasons for their disturbing behavior towards each other." Lacqueur (1971) reminds us that MFT "affords the families an opportunity to learn from each other indirectly, through analogy, indirect interpretation, mimicking and identification" thus "the resources of all family members tend to be exploited more successfully when several families are treated together in one group than when each family is treated as a separate entity." He (Lacqueur, 1971) points out that schizophrenics (like drug addicts) are helpful to transfer the energies of early symbiotic relationships to external objects in such groups. Lacqueur (1971) emphasizes the importance of the therapist not being caught up in the multiple double binds set up by patients and parents in order to demonstrate that emotional distance from these patterns can be developed; thus, the therapist is a role model for individuation. Likewise, the IP may identify with another individuating family member and separate with much less anxiety than is usually associated with such learning (Lacqueur, 1971).

Leichter and Schulman (1974) conduct MFT in an outpatient setting and like Lacqueur (1971) choose three to four families in a manner that creates a homogeneous, yet balanced group. They insist that each family include the entire nuclear family and have at times expanded the group to include the extended kinship where relevant, i.e., grandparents, or divorced spouses. They cite types of families for whom MFT is preferable to work with one family. Such families include those who are isolated or whose system is circulatory and rigidified, especially symbiotic. MFT is also helpful when there is a missing parent (usually the father), since the group provides parental substitutes. Leichter and Schulman (1974) also have found particular dynamics which lend themselves to resolution in MFT. Reality testing is strengthened as distortions within a family are readily apparent to other group members who point them out in a manner which is readily accepted and can then, in turn, correct the reality of their own family. Likewise, transferences to non-family members in the group can be traced to the person's own family. The degree of distortion can be readily pointed out as the person who has been overreacted to is present in the group in his/her own reality. Another goal in these groups is to bridge intergenerational alienation and isolation through experiencing the universality of human needs and emotions. Adults can provide conflict-free parenting to the children of others in MFT and parents and children alike, can gratify their own need to be parented. The unrealness of the "good" child in the family can be pointed out as can the price they pay for their position in the family. At times, even the addict can be the "good child". In general, it is easy for one family to perceive another family's malfunctioning, learn to think in such terms and then apply their new thinking to their own family. Leichter and Schulman (1974) noted that MFT "provides a particularly fertile ground for the emergence of spontaneous and unexpected attitudes and insights" which occur almost as a "byproduct" of ongoing processes.

Papp (1974) has learned that "hopeless" families do better in groups. After six months of trying to match families, she assembled a group composed of the last three families in a waiting list. Thus, she "learned by default that there are no barriers of race, religion, culture, education, psychiatric diagnosis, politics or prejudice which cannot be transcended in a group and used to therapeutic advantage." Papp's (1974) pioneering of sculpting has been very helpful to these authors and is quite useful in MFT. Family sculpting or choreography as she (Papp, 1974) later termed it, can transform a group from an inhibited, intellectualized or anxious rambling body into an "active, alive, highly involved, purposely focussed entity."

The literature on the specific use of MFT with drug abusers is quite sparse. A very early paper was written by Hirsch (1964) which discussed a therapy group composed of parents of adolescent drug abusers at the Riverside Hospital in 1958. This group did not join total families, but did encourage these parents to share their mutual difficulties. Since 1958, parents' groups which excluded the addicted child have not been uncommon, particularly in therapeutic communities. We have not found that including the total family in any way reinforces unwholesome defenses or leads to a deterioration of communication. However, these parents' groups have been more educational than therapeutic and have not dealt with resolutions of underlying conflicts. Berger (1973) and Bartlett (1975) have led MFT's with addict families. Berger's (1973) groups met only monthly in association with the therapeutic community of the Quaker Committee on Social Rehabilitation in New York City. His groups had a code in which they sought to elicit truth in order to examine the past and present without blaming or guilt provoking. There is a focus on patterns which contribute to self-hate and hurting oneself to hurt one's parents. Another emphasis is on nonverbal behaviors which members are not initially aware of through the observations, interpretations and reactions of others. Berger (1973) also uniquely focuses on "crisis creators", "help rejecting complainers", preachers and placaters. A small family group within the larger group is used with two residents assigned as advocates of the "truth", one for the family and one for the resident.

Bartlett (1975) writes of an MFT on a detoxification ward, which is of necessity short term. In the first group or two, the family uses denial about underlying problems. By the third group the therapist confronts the family with their behavior and the underlying dynamics and structure, particularly the family member as "pusher" and the addict roles as scapegoat, interpreter, go-between and emotion supplier. In the fourth through sixth session, a decision is made to implement a treatment plan. This is obviously not a simple decision and underlying issues such as assertions of parental authority and the addict's power to maintain anxiety in others must be dealt with. The therapist supports the parents' capacity to exert change in their own home and "that change is possible."

Bartlett (1975) reports a follow-up study of a group of seven families treated by Kaufmann in which five patients remained drug free a year after termination and no siblings became addicted. Of a group of 45 adolescent substance abusers at Phoenix House, the rate of recidivism after twelve to eighteen months of treatment was over 50%. When the entire family had been part of treatment in another group of 45, the recidivism rate dropped to 20%. Recidivism was evaluated by an hour in-depth interview with the family in addition to a cross check with school and/or employment. Hendricks (1971) compared a group of male narcotic addicts who received multifamily counseling with a control group who did not. He (Hendricks, 1971) found that one year after release, 41% of the treatment group remained in outpatient status as compared to 21% of all male outpatients.

Our approach is that drug addiction is a symptom of family stress that is exacerbated by societal stress. We are involved primarily with the forces within the family that produce and maintain the symptom. We are also involved with those forces within the therapeutic community that serve to maintain these symptoms. The staff and residents constitute their own family system. These also can be dealt with in ways that are similar to those used to deal with family dysfunction. The MFT group frequently acts as a barometer which reveals the overall functioning of the TC and underlines problems of cohesion in the TC family.

Middle class families, particularly Italian, Greek, Irish and Jewish tend to be quite enmeshed. Puerto Rican and Black mothers tend to be over-involved with their sons who are drug addicts. Enmeshed families tend to be the ones that come regularly to therapy and distanced families come rarely, if ever; thus, in our MFT it appears that most, if not all, families of addicts are enmeshed. The multitude of cultures and languages in our families is frequently bridged by the universal aspects of the problems, but also presents many difficulties despite the use of therapists from several ethnic groups.

In some cases, distance between family members is a necessary goal. In many families, the goal is a restoration of the family homeostasis. Certainly, an intact family of origin with appropriate mutuality is curative and preventative of drug abuse. Similarly, so is a healthy nuclear family composed of the addict, a non-drug abusing spouse and their own children.

Multiple family therapy is unique in contemporary society in that families expose themselves to one another and try to have a significant effect on each other's way of life. MFT enriches and stimulates the totality of any therapeutic program which utilizes this technique.

In our experience, MFT reduces the incidence of premature dropouts, acts as preventative measure for other family members, builds a subculture that acts as an extended "good family" and creates and supports structural family changes which interdict the return of drug abuse.

REFERENCES

Bartlett, D. 1975. *The Adolescent in Group and Family Therapy.* New York: Brunner/Mazel, Inc.

Berger, M.M. 1973. Multifamily psychosocial group treatment with addicts and their families. *Group Process.* 5: 31-45.

Fogarty, T. 1974. Evolution of a systems thinker. *The Family.* 1: 26-43.

Hendricks, W.J. 1971. Use of multifamily counseling groups in treatment of male narcotic addicts. *Int. J. Group Psychotherapy.* 21: 34-90.

Hirsch, R. 1961. Group therapy with parents of adolescent drug addicts. *Psychiatric Quarterly*. 35: 702-710.

Lacqueur, H.P., La Burt, H.A., and Morong, E. 1971. *Changing Families*. New York: Grune & Stratton.

Leichter, E., Schulman, G.L. 1974. Multifamily group therapy: A multi-dimensional approach. *Family Process*. 95-110.

Minuchin, S. 1974. *Families and Family Therapy*. Cambridge, Massachusetts: Harvard University Press.

Minuchin, S. 1974. *Structural Family Therapy* in Arrietis, ed. Basic Books.

Papp, P. 1974. Multiple ways of multiple family therapists. *The Family*. 1: 25.

Papp, P. 1973. Sculpting the family. *The Family*. 1: 44-48.

Ritchie, A. 1971. *Changing Families*. New York: Grune & Stratton.

Speck, R.V. 1971. *Changing Families*. New York: Grune & Stratton.

Speck, R.V. 1973. Multifamily psychosocial group treatment with addicts and their families. *Group Process*. 5: 313.

COUNSELOR BURN-OUT: THE REVOLVING DOOR SYNDROME

Catherine E. Jordan, B.A.

Santa Monica Bay Area Drug Abuse Council, Inc.

Counselor Burn-out and low success rates are two inter-related serious problems confronting workers in the drug abuse field. Counselor burn-out is often attributed to the low success rate, but the reverse holds true as well since burned-out counselors lose a lot of their motivation and energy. Counselors become especially discouraged and burned-out when they continually see what is often called the "revolving door" syndrome. This involves addicts coming in repeatedly for detox but not for counseling services. This often creates a vicious cycle as burned-out staff are less apt to reach out and expend the effort necessary to get clients into treatment which only perpetuates the syndrome. This paper focuses on the dual problems of the "revolving-door" syndrome and counselor burn-out.

Unfortunately, drug abuse workers tend to view the revolving-door syndrome as a problem inherent in the addict personality and as something over which they have no control. The research, however, does not support this pessimistic viewpoint. Several studies concerned with the tremenously low rate of addict return to and retention in outpatient treatment programs have reported that program improvements have resulted in higher return rates (Freedman et al., 1963; Brown and Brewster, 1973; Renner and Rubin, 1973). The significant program changes which resulted in improved return rates basically involved increasing the quantity, the quality, and the continuity of staff-client interactions during the detoxification phase. Evidence that an impact can be made on the addicts coming in for detox is exemplified by the experiences of the Santa Monica Bay Area Drug Abuse Council, Inc., an outpatient drug-free clinic. Approximately two-thirds of the population are heroin addicts or abusers. Therefore the most desirable treatment modalities are detoxification facilities. Heroin addicts coming to the agency requesting detox comprise not only the largest subpopulation of clients requesting services from the program, but they are from experience the least likely to return for counseling or aftercare services post-detoxification. Follow-up through site visits to the detox wards has always been conducted in order to provide clients with supportive services during their detoxification. The goal of outreach is to create a bridge between the initial step of physical detoxification to participation in aftercare for psychological, social, and vocational rehabilitation services.

During the first two years of the program's operation, follow-up visits were conducted on a weekly basis. It was believed that by providing clients with this support and by encouraging them to return to the agency for counseling after detox, that they would be more apt to return to the agency. When a small study (Jordan, 1975) was conducted to assess the effect of follow-up on the return rate the results were quite clear: No one was returning for counseling. The clients who did return at some later time inevitably did so to request another detox. It was concluded that only one follow-up visit to clients in detox was not sufficient to affect return rate. Since then several program changes have been made. The detox component was standardized and centralized, follow-up was conducted more frequently, and clients were provided transportation to and from the detox hospital utilized as the primary referral source. The advantage of providing clients transportation was that after they were discharged from the hospital they could immediately be transported to the agency. At that time, counselors encouraged clients to participate in a post-detox counseling session which focuses on their immediate plans, offers them support, and emphasizes the importance of counseling as a means of helping them remain drug-free. It finally appeared that as a result of some of these changes more clients were returning to the agency for counseling after completing detox. The most significant factor was believed to be the increased frequency of follow-up sessions to clients in the detox wards. Jordan (1977) then conducted another study to evaluate the return rate and to assess whether it was related to the number of follow-up sessions clients received while they were in detox.

Data was collected and analyzed on 82 heroin addicts requesting and admitted to inpatient detox at the time of their initial contact with the agency. The most significant finding of this study was that clients receiving three or more follow-up sessions were significantly more likely to return to the agency for counseling sessions whereas clients who did not receive any follow-up or who received only one or two follow-up visits were significantly less likely to return to the agency for counseling. The overall return rate for at least one counseling session other than the post-detox session was 30%. Specifically: 12% of the clients who did not receive any follow-up returned to the agency; 29% returned after receiving only one follow-up session; 23% returned after having two follow-up sessions; and 80% of those clients receiving three or more follow-up sessions returned for counseling. Of those clients who returned to the agency: 28% returned for only one session; 24% returned for two sessions; 8% returned for three sessions; and suprisingly, 40% for four or more counseling sessions.

The study also revealed that those clients who received a post-detox counseling session immediately after discharge from the hospital were significantly more likely to return for counseling than those clients who did not receive a post-detox counseling session. The two groups were compared on a number of demographic variables but no significant differences were found between the two groups.

The results of this study confirm the contention that return for counseling post-detoxification is not related to the addict character, personality pattern, or background variables, but that it is the treatment process itself which is responsible for the low success rate, at least in terms of addict return to aftercare post-detoxification. The area in which outpatient facilities can improve and function to engage the addict into aftercare appears to be during the initial sessions prior to referral, through outreach visits to the detox facilities, and through follow-up after referral. Treatment programs must intensify their efforts to reach the addict during this detox period if we ever hope to make an impact on the heroin addiction problem.

Increasing the quantity, the quality, and the continuity of staff-client interactions decreases the revolving-door syndrome which subsequently should also curb the counselor burn-out, but what other means can be employed to prevent and deal with counselor burn-out? Education, training, and support are additional means to dealing effectively with counselor burn-out. It is crucial that counselors understand the dynamics involved in the transition from addict to "non-addict" and how they can function to facilitate this change for which the addict is typically unprepared. The process of an addict becoming a "non-addict" requires adopting an entirely new role as the addict is entrenched in the addict subculture, its values, its norms, and its way of life. Most of the addict's associations are with other addicts and the addict's identity and way of relating is that of a heroin addict.

Readdiction most frequently occurs during the period immediately following the physical withdrawal from heroin. The addict is most vulnerable to relapse at this point as the old values and meanings experienced as an addict are still immediate and the addict has not yet learned to live and cope without narcotics (Ray, 1961). Wolkon (1974) noted that entering and leaving treatment has the potential for becoming a crisis as a result of the role-change experienced by the individual. Variables which increase the probability of a role change becoming a crisis include: "discontinuity between the roles unexpectedness of the change, lack of preparation for the change. . . the importance or salience of the change to the 'core-self' and heightened feelings of ambivalence and inadeqacy towards the new role". It is all too apparent that addicts are not prepared for this role change and do experience a state of crisis marked by anxiety, ambivalence, and self-doubt. The addict typically responds to anxiety by reverting back into drugs. Therefore, intervention directed to the addict during this crisis period of high anxiety appears to be the most optimal time to motivate and to successfully engage the addict in his/her own rehabilitation. During this period of crisis, the support obtained from counselors is an essential aspect for drawing the addict into the treatment system. As Brill and Lieberman (1969) report, "the addict's pull is generally toward further involvement in the addiction system" and treatment intervention must first focus "on the need to hold the addict in treatment during his periods of crisis". The failure of clients to return for their counseling appointments must be seen for what it is--a manifestation of the very problem in need of treatment.

As counselors often do develop negative feelings and attitudes towards addicts, it is important to have a support system for dealing with these feelings. Clinical supervision groups which focus not only on helping counselors deal more effectively with their clients, but which also deal with the feelings of the counselors, appear to be the best method of support. Counselors also need to understand that abstinence as the exclusive goal of treatment is frustrating to both staff and clients and contributes to the development of negative feelings and discouragement. The emerging community mental health approach views drug abuse as residing in the individual and in the complex relationship patterns of people and their social context (Lennard and Allen, 1973). This approach does not focus on one particular facet of the drug abuse problem as it recognizes a variety of styles in drug-taking and treats each client as an individual. The community mental health approach thus has important implications for the goals of treatment for heroin addicts. Improved functioning in society and habit reduction can then become meaningful treatment goals (Brotman and Freedman, 1968). If we are to deal with the addict's actual life problems, the primary task becomes that of engaging the addict in his/her own treatment and rehabilitation.

There are then a number of things an agency can do to improve services to addicts and reduce both the revolving door syndrome and counselor burn-out. The agency must first establish a liaison with the detoxification facilities treating their clients. This liaison should allow for a free exchange of information regarding the clients placed in detox and should also allow the outpatient treatment staff free acess to the detox wards to visit their clients. Pre-detoxification sessions should be geared towards helping the client prepare for the detox experience and to making realistic plans and preparations for lifestyle changes which are essential if the client is to remain drug-free. Clients should be followed up in detox daily for optimal results and again focus should be on post-detox plans. Clients need a great deal of support and encouragement at this time. Relations between clients and the hospital personnel are often strained as the addict is constantly complaining and seeking relief from tension and anxiety by demanding more medication. Clients need someone to help teach them how to cope with tension and anxiety in ways other than by using drugs as drugs have typically become the addict's means of coping with the world. Clients of the Santa Monica Bay Area Drug Council, Inc. have been overwhelmingly grateful for this contact and support during detox and indicate to the counselors that they feel that someone really does care about them.

Providing transportation to and from the detox facilities reduces the pre-admission drop-outs that result from lack of transportation and also provides a continuity between detox and the aftercare program. Post-detox sessions increase the return rate and are also a means of drawing the client into treatment. A primary counselor should be assigned to each client at time of admission and an appointment scheduled immediately after detox as most addicts relapse shortly thereafter. Continual follow-up contacts are necessary for those clients who do not show for appointments to encourage their coming back for counseling and to let them know that support and help is available to them. Detox follow-up is necessary to get the addict into treatment but continual follow-up efforts are necessary to keep him/her there. Education and training for treatment staff should be geared towards helping the counselors understand the dynamics involved in the role change from addict to "non-addict" and that intervention is most effective at this time. Regardless of the addict's motivation for seeking detox, it is at this time that the addict enters and is visible in the treatment system. Support groups such as small clinical supervision groups provide a means for dealing with negative feelings as well as case management.

The key to improving the success rate lies with the success of our efforts in improving treatment. It is time for research to focus more on studying and devising ways of improving treatment rather than explaining and justifying treatment failures. Too much research has been devoted to studying addict charcteristics and other factors which produce poor results. It is already well-known that addicts are a difficult population to treat. The responsibility for failure does not rest solely within the addict. Studies have demonstrated that the treatment process is also responsible and that we must take greater responsibility for implementing changes which produce better outcomes.

REFERENCES

Brill, L. and Lieberman, L. 1969. *Authority and Addiction*. Boston: Little, Brown and Co.

Brotman, R. and Freedman, A. 1968. *A Community Mental Health Approach to Drug Addiction*. Washington, D.C.: U.S. Government Printing Office.

Brown, B.S. and Brewster, G.W. 1973. A comparison of addict clients retained and lost to treatment. *Int. J. of the Addict*. 8:421-426.

Freedman, A.M., Sager, C.J., Rabinger, E.L. and Brotman, R.E. 1963. Response of adult heroin addicts to a total therapeutic program. *Am. J. of Orthopsychiatry*. 33: 890-899.

Jordan, C.E. 1975. Outreach follow-up on heroin addicts in detox: return to outpatient program after referral. Unpublished manuscipt, California State University, Northridge.

Jordan, C.E. 1977. Motivating the heroin addict: outreach follow-up in detox. Paper presented at the Proceedings of the Annual California State Psychological Association.

Lennard, H.L. and Allen, S.D. 1973. The treatment of drug addiction: toward new models. *Int. J. of the Addict*. 8: 521-535.

Ray, M.B. 1961. The cycle of abstinence and relapse among heroin addicts. *Social Problems*. 9: 132-140.

Raskin, H.A. 1964. Rehabilitation of the narcotic addict. *JAMA*. 189: 956-958.

Renner, J.A. and Rubin, M.L. 1973. Engaging heroin addicts in treatment. *Am. J. of Psychiatry*. 130: 976-980.

Wolkon, G.H. 1974. Changing roles: crises in the continuum of care in the community. *Psychotherapy: Theory, Research, and Practice*. 11: 367-370.

CAN SUCCESSFUL CLIENTS BECOME SUCCESSFUL STAFF IN ADDICTION TREATMENT PROGRAMS?

Jay M. Jackman, M.D.

Salvation Army Addiction Treatment Facility

Honolulu, Hawaii

It has long been recognized that medical and mental health professionals have been limited in what they have had to offer in the field of substance abuse. The treatment of medical complications and of gross psychiatric disorders are those areas they most frequently addressed. Until the late 1950's, little else was offered or available to addicts seeking or forced into treatment for non-federal drug related offenses.

The creation of Synanon (Yablonsky, 1965) in 1959, revolutionized the therapeutic approach to heroin addiction and was the prototype of the therapeutic community which has since evolved in combinations and permutations so numerous as to almost defy description and classification. These residential programs continue to be the heart of many effective treatment programs (Casriel and Amen, 1971; Collier, 1971).

As recreational drug use and then abuse spread, focusing on non-opiate drugs in the early 1960's, a new dimension emerged. For individuals in need of or seeking treatment, an entirely different modality was needed, and thus arose the Haight-Ashbury Free Clinic. This quickly became a model to be emulated of out-patient counselling and treatment services which was radically different from existing out-patient mental health clinics or social service agencies (Smith, 1976).

Like the therapeutic community, the free clinic heavily utilized rehabilitated drug users and heroin addicts to staff and direct the programs. Medical services were provided and may even have been a central service, but the programs did not operate on anything resembling a medical model. Clearly that suited most of the staff and the overwhelming majority of the clientele who had experienced physicians as being notoriously insensitive, hostile and impatient with their problems and concerns.

In the later 1960's methadone maintenance programs emerged as the third of the principal treatment modalities. This element focused exclusively on heroin addicts. While the methadone itself required either pharmacists, nursing or medical involvement, that was clearly distinguished from the rehabilitation elements of the program which once again heavily utilized rehabilitated addicts.

The rationales for the use of rehabilitated drug abusers and addicts were numerous and frequently valid. The most common concern was that staff had to be "relevant" to the community being served or else the potential clients would not appear. Historically, one could say that addicts did not come to traditional institutions because either: 1) Clients did not really exist, or 2) Clients did not feel they could obtain what they needed, or wanted, or else they felt alien, or any of a hundred other reasons why people do not use existing services. Informed people in the field knew the addicts and abusers were there in masses but that existing services were not "relevant." On closer examination, the problems with services really had to do with traditional professional staff.

The difference in attitudes, language, culture, customs, mores, values, judgments, race and socio-economic status added up to insurmountable barriers and the easiest solution appeared to be the recruitment of more "relevant" staff and these in fact were most often rehabilitated addicts (Collier, 1973). By and large, they came initially from Synanon, Daytop Village, Odyssey House, Phoenix House and the Illinois Drug Abuse Program. Later on, each program became the source of its own staffing. There were multiple positive determinants here. Although frequently denied, a major theme running through the more successful therapeutic community programs has been that the therapeutic community is really an alternative community. Synanon has been most candid amongst programs in its claim that Synanon is a life time alternative to the straight world and currently seeks non-drug users to enter and become part of the community.

These therapeutic communities which are successful with their clients teach them how to successfully live in the therapeutic community, not in the outside world. That is why the re-entry phase is so crucial and why the alleged failure rate in re-entry is so high. Being a success in a therapeutic community is not preparation for success in the outside world. Many clients recognize this and are thankful that the therapeutic community has literally saved their life, although it has not prepared them for life "outside" (Freudenberg, 1974; Rachman and Heller, 1974).

At the same time, ideology, better known as "concept" is crucial to the belief system of most therapeutic communities. The powerful belief in ideology probably has a major role in the success of the program. Loyalty to the ideology is probably the central quality of most successful staff. It's often modeled on the ideology of the director, be he recovered addict, physician, minister, administrator, etc. (Goldstein and Dean, 1966).

When looking for staff, the examination of the applicant's ideology is a straight forward process. Predictably, applicants for jobs in treatment programs, who themselves have been successful clients in these programs, meet these requirements most easily. In addition, they are more intimately known to the program seeking additional staff and their loyalty is virtually assured.

For the senior resident of the therapeutic community who is seeking work, a staff position in his own program meets many of his needs. There is an air of security working in "his" program which does not exist in the outside world. He often does not have job skills that are valued in the general market place. And, he is often recruited by his program for the position. The combined power of these forces is very potent in his making such a decision.

While this is readily visible for therapeutic communities, it is just as true for methadone maintenance programs. The reality appears to be that methadone maintenance is a much longer term program than the therapeutic community despite the early wishful thinking that masses of clients could be withdrawn from maintenance fairly easily. That just has not happened. The same issues of ideology emerge in those methadone maintenance programs that sincerely seek rehabilitation. The key element in those programs is the counselor who works with a case load of clients in vocational rehabilitation, negotiating with the criminal justice system and personal counselling. The same issues arise here in selecting staff as occur in the therapeutic communities, and that does not need to be repeated here. Methadone maintenance programs run on a more heavily medical model, do not have this phenomenon as they rely more heavily on medical personnel: doctors, nurses, pharmacists, psychiatric technicians, para-medical assistants, etc. But even these programs use some rehabilitated addicts and often they come from within their own ranks.

The free clinics pose a different problem and do not so heavily rely on former patients, although to some extent, they do. The clientele of such clinics are often transient so there are enormous rates of turnover which is expectable. The demand and expectations of treatment and outcome are different and also less rigorous than in therapeutic communities and methadone maintenance programs and that also is appropriate. However, "street people" who themselves are often heavy drug users, do make up substantial proportions of both volunteer and paid staff. The free clinics are often not nearly as well funded as the other programs, which is truly unfortunate, given how much larger is the pool of clients they serve.

This detailed background was quite lengthy but it was necessary if an observer is to comprehend the principal thesis of this paper which is this: There appears to be very little concern for the risks which the rehabilitated substance abuser faces when he becomes an employee in the same treatment modality from which he was graduated. The risk is a very high likelihood of return to drug abuse, job failure or mental crisis for someone who has successfully completed the treatment aspects of a rehabilitation program. The author in his role as consultant, program director and program administrator, has experienced an alarming rate of recidivism or mental crisis among such staff.

The perceived benefit to the program of having former clients as staff has weighed far more heavily in the minds of administrators and program directors than whatever danger such a position may pose for the former client.

The Salvation Army Addiction Treatment Facility in Honolulu, Hawaii, serves a mixed population of alcoholics and drug abusers and addicts. The program director of the past five years was formerly a heroin addict and a graduate of one of the more respectable therapeutic communities in the East. The treatment philosophy is a modification of a traditional therapeutic approach because a large proportion of treatment staff are candidates for M.S.W. or Clinical Ph.D. degrees and this results in a "dynamic" staff tension in treatment.

In the past six years, approximately 605 clients have passed through the extended care element and some 20 successful clients have been hired on as staff of the agency, 8 in the extended care element, 4 in support services and 5 in an outreach program. Of the 8 staff of extended care who had been clients only one has been successful. 3 returned to drugs or alcohol, two were deemed non-competent and two were unable to adapt to being staff. Virtually, all of the ex-client staff employed in other elements of the program have done well, and when they have left, it has been for better jobs or over conflict with administration.

The author's prior experiences with a series of programs in San Francisco, impressionistically are identical, but the statistics are not currently available. Informal contacts with program directors throughout California suggest similar impressions.

These data suggest that former clients who accept staff positions in their former treatment programs are high risks for failure through return to drug use or inability to adapt to their new positions. This raises some profound ethical problems for program administrators. There is a genuine reluctance on the part of administration or of the individuals involved to recognize that the problem is more a structural, organizational one, rather than one of individual pathology. The process that enables a client to succeed in a treatment program does not prepare him for any of the bureaucratic tasks which a staff member must master in today's atmosphere of cost-accounting evaluation criteria.

Success as a client in a program involves a great deal of role modeling and of acquisition of new behaviors vis à vis other people. This is in effect, an alternative life style to their drug oriented life style. The stresses that successful clients learn to cope with are very much peer related and also authority related. What is healthy behavior is defined by each program and often is related to the program ideology. There is enormous personal reinforcement and it all comes from within the system of the program.

Success as a staff member on the other hand, requires more and more an entire range of skills that are unrelated to being a successful client. There are all kinds of paper work skills and the need to relate to a host of public and private scrutinizing agencies which require an entirely different interactive style than that which they learned in treatment. It is a wholly new "game" from the therapeutic community "game" and the new staff member is rarely informed or instructed in the new rules of the game. Many new staff however do manage to learn these skills.

Significantly, a different organizational dynamic comes into play for these clients who seek to become staff in their own programs as compared with seeking work in some other program. This dynamic tilts the balance more toward failure than toward success (Rachman, 1974).

After the many months that someone has been a client on his way to a successful outcome, he is seen by the agency in the role of client. There is a great institutional resistance to accept a change in his role to staff.

This problem is manifested in several ways:

1) Often many clients in a program who were the peers of the new staff member have ambivalent feelings about the new staff member's role and status change. There is a genuine confusion about how to change one's attitude from that of client-client to the new client-staff attitude. Often there is envy, jealousy, anger and the whole range of feelings which are elicited in any competitive situation where only one or at most, a few people get the available "reward", in this case a staff position. One consequence of this is an excessive hostile distancing of the new staff member from his former client peers. There may even be attempts to sabotage the success of the new staff member by his former peers.

2) The new staff member is beset by his own ambivalent feelings. While he is evidently pleased with this tangible reward of success, he remains for a long period extremely unsure of his role and behavior

expectations. Yesterday, he was a client. Today, he is staff. As far as he can see, he has not changed at all and the real changes he has to make come very slowly and very painfully.

He too is inclined to relate to his former client peers as current peers which undermines his attempt to develop his staff role and the authority that goes with it. He is often unsure of his authority and as yet, it feels unnatural and so he may use it in a heavy handed way which only further distances him from the clients as he appears arrogant, powerful and "higher" than they are. This is a very precarious state that puts into jeopardy all of those qualities which had made the former client a desirable staff member to begin with. The new staff member must rapidly learn that the principal circle of sharing of all sorts has switched from his former client peers to his current staff peers. While this may not pose a great problem for the new staff member, his old peers see this behavior clearly as a distancing mechanism from them and often feel a sense of profound betrayal manifested by angry withdrawal on their own part. This set of bad feelings is probably the most difficult for the new staff member to contend with.

3) Staff members respond with their own ambivalence toward the former client now peer. Many staff have established a relationship with the new staff member when he was a client and that role was heavily determined by authority and power relationships. That imbalance in power must be replaced by a much more power and authority balanced relationship if the new staff member is to be truly integrated as staff. Some staff are unwilling to surrender their prior relationship as it was one sided with their having all of the power. The desire to continue and control and/or manipulate the new staff member is real and should not be underestimated. Privileged positions are often not voluntarily shared even when the well-being of the agency is at stake.

Although these issues are identified as occurring between staff members and clients, they are merely the actors on the larger stage of organizational dynamics. That is why these problems are so common even amongst the best run programs. These dynamics are of critical importance for the success or failure of programs and they are central to this thesis. The operations of these dynamics must be at least recognized and acknowledged prior to any successful intervention. The time is appropriate to acknowledge this reality and either to provide sufficient support to facilitate the transition from client to staff or, not to subject these clients to such risks. At this point, we can only speculate about what such support would involve.

Clearly, however, if the mental health of a rehabilitated client requires separation from the drug scene, we must act on that principle rather than what might be best for staffing the program. This will require a rethinking of staffing needs and patterns.

REFERENCES

Casriel, D. and G. Amen. 1971. *Daytop*. New York: Hill & Wang.

Collier, W. 1971. *An Evaluation Study On The Therapeutic Program of Daytop Village*. New York: Daytop Press.

Collier, W. 1973. A profile study on the residents of Daytop Village. *Journal of Drug Issues*. Vol. 3, No. 1: 10-21.

Freudenberg, H. 1974. We can right what's wrong with our therapeutic communities. *Journal of Drug Issues*. Vol. 4, No. 4: 381-392.

Goldstein, A. and S. Dean. 1966. *The Investigation of Psychotherapy*. New York: John Wiley & Sons. Chaps. 9 and 10.

Rachman, A. and M. Heller. 1974. Anti-therapeutic factors in therapeutic communities for drug rehabilitation. *Journal of Drug Issues*. Vol. 4, No. 4: 392-403.

Smith, D. 1975. The free clinic movement in the United States. A ten year perspective (1966-1976). *Journal of Drug Issues*. Vol. 6, No. 4: 343-355.

Yablonsky, L. 1965. *Synanon, The Tunnel Back*. Pelican Books.

PORTRAIT OF AN ADDICTION TREATMENT AGENT

Lorraine M. Hinkle, A.C.S.W.

CODAAP for the City of Philadelphia

Abstract: Portrait of an Addiction Treatment Agent begins by describing the *"frame"*, or *"work place"* of the Treatment Agent; an optimal "frame" is an agency wherein everyone--from top administration to newest security guard--is convinced that the fight against addiction is necessary, exciting--and that victory is possible. From the "frame", this workshop will move to the *"background"* of the Addiction Treatment Agent: what is his/her chemical substance experience? --what attitudes have developed from this experience? --what kind of addiction education/knowledge does the Treatment Agent have? --Is the Treatment Agent knowledgeable and skillful in the use of treatment techniques? --Is he/she aware of which techniques are most effective when working with addicted people? These questions will be reviewed and answered from the author's extensive experience in the field of Addiction Treatment.

From the "background", the workshop discussion will focus on the *"foreground"* or the Addiction Treatment Agent himself/herself, i.e., what are the characteristics of an effective Addiction Treatment Agent?; what is he/she attempting to do in the role of Change Agent? Emphasis will be placed on both worker and client motivation, as it impacts on the treatment process.

IS GOODWILL THE ONLY QUALIFICATION

Several years ago, in North Jersey I was cooking dinner when the Parish Priest (Pastor pro temp--while the Pastor was in an Alcoholism facility in the West) called to ask if I would help him take care of a woman alcoholic. He had found her in the parish school's boiler room at 10 a.m. The priest with the help of the school nurse had assisted the woman to a cot in the First Aid Room. At 1 p.m. the priest had enlisted the aid of 2 A.A. women, both the mothers of large families. By 6 p.m., when I arrived in the basement room, the 33 year old woman, in an alcoholic stupor, was smoking a cigarette and sipping some coffee. I took a short alcohol-intake history and discovered that the woman had been on a lengthy binge--drinking heavily, with no other nourishment, for several weeks. Realizing that she needed detoxification, I suggested that her parents (who would not admit her to their home) be contacted, and asked if their address could be used to secure admission to our County (and only) detoxification facility.

There was no dearth of goodwill present in that basement room--yet many hours were wasted in offering goodwilled support to a person who had a serious need for detoxification. I believe that Addiction Treatment Agents do need goodwill--most certainly! However, I also think that they need a great many other characteristics if they would be effective in helping people to move away from addiction and toward a happy, healthy life style. The following is my idea of the portrait of an effective addiction Treatment Agent.

The Frame

The Frame, i.e., the setting (agency/institution) wherein the Treatment Agent functions. New, or retreaded, workers in Addiction Programs are not always assured of full administrative encouragement, support, and financial underwriting. The attitude of every staff member, regardless of their job description, is of vital importance to the success of the overall program and the effective functioning of each staff member. None of us ever influences the system we operate in, as we are influenced *by* them. Therefore, it is important that we feel proud of what we are about--if the receptionist calls

Fig. 1. Portrait of an Addiction Treatment Agent.

us in a frightened or disgusted stage whisper to come "get this patient out of the waiting room"--it is not easy to maintain an affect of professional acceptance as one assures the patient that he wants to work with him but cannot in his present high condition. A card bearing the same assurance and requesting a sober call for a new appointment, also given to the person, can be accompanied by disgust and rejection or calm awareness that recovery from addiction is marked by frustration and relapse.

An easily explodable myth is this--that specialized addictive treatment facilities are the optimal frame for Addictive Treatment Agents--that, surely in these settings "everyone from the highest level, on down, sees the fight against addiction as prestigious, remunerative and exciting" (Blum and Blum, 1967). Specialized addiction treatment programs, unfortunately are not always "optimal frames" for the Treatment Agents. Often in these programs we find the same patronizing attitudes, thinly veiled disgust, and quick assumptions and labelling of addicted persons as manipulators/game players that can be seen in General Social Welfare programs. Let us face the fact that--a job is a job is a job! People who need Bread--for bread, are glad to secure employment anywhere! Federal monies given to Addiction Treatment programs--assure only the existence of the program, they, so far, are no surity that such programs will provide the optimal frame to which I have been referring.

I agree with Blum and Blum (1967), who say that the Addiction Treatment Agent is, himself or herself, the most important treatment component. Therefore, it is important that the Addiction Treatment Agent's morale be considered by Agency Administration in the areas of financial/job security and self-esteem, as well as in the other important areas of experimentation and innovation. When Agency Administration label the addicted patient as "uncooperative", i.e., tardy, dirty, dropping out of treatment, not paying bills, etc., and when this attitude is reflected by the spectrum of Agency/Institution Staff, from receptionist to Director, it is not easy to be proud of one's job as an Addiction Treatment Agent. Only when all staff are optimistically oriented--are realistically hopeful--that the exciting, thrilling fight against addiction can be won over and over again, despite discouraging relapses and occasional failures--do we see the optimal frame which becomes a support and assist to effective functioning for the Addiction Treatment Agent.

The Background

When I refer to the "background" of the Addiction Treatment Agent, I mean his/her life experiences, which have developed a set of attitudes; also, his/her education in both addiction and treatment techniques and ability to adapt these techniques to the needs of the addicted person.

It is practically impossible to live in our chemically oriented society and not have some experience, as a child or adult, with alcohol or other drug compounds, and their consequences, as they relate to the human user. Even if no one in our immediate family is an alcohol or other drug, addict, we may have had a relative or neighbor who was hooked on a chemical substance that effected them intra and inter-personally. From actual observations of these individuals, or from hearing our mother or father prohibiting an addicted relative from sharing our home, our attitudes are formed. It is vitally important that we are consciously aware of exactly what attitudes we have developed from our prior life experiences. It is important, because these attitudes, positive or negative, will be transmitted to the people with whom we work. A lack of awareness of negative attitudes toward addicts can be responsible for preventing the formation of a therapeutic relationship. When I lecture, I like to ask treatment agents if they believe that recovered addicts (of whatever persuasion), because they are recovered, make the best Addiction Treatment Agents. I agree with the response which states that only when an addict's recovery is well integrated into his personality, and when he/she has developed personal insight and self-awareness can he/she be an effective Addiction Treatment Agent. This kind of a recovered addict does not see people only in the context of his/her own life experience, but rather is able to exhibit genuine empathy and function as a role-model. Whether recovered or non-addicted, Addiction Treatment Agents must be people who are constantly striving for greater personal insight and self-awareness. They also have to be capable of forming healthy relationships, in a short period of time.

When I supervise Treatment Agents who are involved in psycho-social therapy with epileptics or diabetics, I am appalled when they make no effort to learn about these chronic illnesses. They do not have to become medical experts, but they do have to have sufficient knowledge to enable them to treat the *whole* person with whom they are working. It is essential that the Addiction Treatment Agent be knowledgeable regarding all chemical substances abused by the people he/she is serving. This knowledge must encompass awareness of these chemicals as they individually or in combinations affect the bio-psycho-social facets of the addicted human being. The Addiction Treatment Agent's knowledge must be contemporary and comprehensive. Current research results from Drug and Alcohol Studies should be reviewed and used to supplement or correct his/her theoretical and conceptual fund of addiction information.

Along with this constantly revised addiction-fact-knowledge base, the Addiction Treatment Agent should be aware of a variety of psychological theories and the skills and techniques that flow from each. He/She should understand that the addict is not made of different clay than the rest of humanity; but that he does have extremely high anxiety. This reality demands that the Treatment Agent carefully select techniques for treating the anxiety ridden addict. Treatment techniques that are designed to raise anxiety--such as: silence, confrontation or interpretation, should be used purposefully and carefully. To haphazardly select techniques that will raise addicts' anxiety, is, in my opinion, sadistic!

I am not saying that these three techniques should never be used; they are, in themselves, excellent tools. What I am saying is that in initial treatment sessions (regardless of the treatment modality), *silence* should not be used; and *confrontation* and *interpretation* should be used skillfully and cautiously.

It is easy to remember how early addiction treatment centers, (drug, particularly) leaned havily on encounter confrontations, which were loud, aggressively direct--and often, quite accurate! In the early and middle sixties professionals were loathe to work with addicts (the influx of Federal funds in the late 60's and early 70's changed this attitude) and they were treated, in most centers, by recovered addicts. These paraprofessionals had gallons of goodwill and very little training in treatment techniques. No wonder they used confrontation so extensively; it was a dramatic technique that, not infrequently, resulted in an immediate, dramatic, response. Even to this day, recovered staff really come to life during drug-crisis incidents when their--"Don't B-S me!--I was there!--I know what it's all about!"--rides again!

Confrontations and interpretations, I believe, are most effective when they are used in the context of a healthy, therapeutic relationship, and are appropriately supported. If you are going to hold up a mirror so that someone can see themselves as they truly are (confrontation); you need to help them look at and accept a reality that might be quite difficult to accept. If you are going to take facts that someone has given you and weave them into a picture which they have been hiding from (interpretation); you need to have already helped them to develop a certain amount of trust. I personally believe that the hard, encounter confronations of the early 60's were successful in affecting deep therapeutic change, *only* when they were adequately supported and given by a person whom the addict trusted. I think that these techniques, today and tomorrow, will assist treatment agents to help addicts recover themselves and their lives--only when used, with support, and in the context of an established relationship.

Pseudo-intimacy is not a viable technique in any facet of life. The addict, who so intensely yearns for intimacy is hyper-sensitive to any phony intimation of it. The Treatment Agent who is so quick to put his/her arm around the shoulder of the addicted person; who sets the stage for a paternalistic or pseudo-social relationship, might begin to examine their professional conscience. These signs might well mean that the Treatment Agent sees the addict as a repulsive, or at best, a helpless child--or second class citizen. Pseudo-intimacy blocks the formation of a therapeutic relationship and prevents the addict from receiving the help he/she needs--and deserves!

The Treatment Agent should remember the pain of his/her own anxieties--in order to approach the addicted person, empathetically. Anxiety hurts--excessive anxiety hurts a lot! Therefore, treatment techniques which are aimed at reducing anxiety should be utilized. One of these is: *ventilation*. This is a technique which is often misunderstood and undervalued. Of course, one reason why it is under-used, is the fact that no one who believes they know everything/ have all the answers, is interested in using it. A fair percentage of Treatment Agents (of various professional and paraprofessional orientations) become Treatment Agents so that they can look down from Mount Olympia and give sage advice, recommendations or commands. These Treatment Agents cannot use phrases, such as: "No, I don't know what you mean" (When a person with whom one is working says--"you *know*". It is not easy to admit to one's human limiations); "tell me more about it". "I don't see how that can be"; "What does that mean?" To be able to use ventilation well, one has to be focused off oneself and onto the person one is working with and has to never, never assume that he/she *knows* what the other person's feeling or thinking or why they are acting or speaking in a certain way. An effective Treatment Agent never assumes anything--but is constantly exploring any and all facts offered by the addicted person. Good ventilation opens up the addicted person and causes him to begin to look in corners of his personality and life-style which he has avoided, maybe for several decades. This ventilation may lead to *clarification*--as the addicted person verbalizes he may begin to see certain patterns (with the assistance of the Treatment Agent) of living that add to his addiction problem; strengths as well as weaknesses (underlined by the Treatment Agent) may become clearer to the addicted person; and, problem areas may become more clearly defined.

It is my belief that addicts are in need of a great deal of nurturing; I consider it sadistic to withhold this nurture. Treatment Agents who work with addicted people have to be able to walk the tightrope: between warm, genuine acceptance and sloppy counter-transference; between encouraging autonomous coping and reinforcing a neurotic need to remain infantilized; between feeding enough and over-feeding; and, between giving enough to support growth and giving too much and stunting growth. In order to be successful in choosing warm, nurturing treatment techniques without becoming sucked into an unhealthy, anti-therapeutic situation, an Addiction Treatment Agent should provide himself/herself with ample nourishment--professionally and personally. Lest he/she become--a dry tit--or a burned out resource for the addicted person (Triana and Hinkle, 1973).

The Foreground

The Foreground, i.e., the Addiction Treatment Agent himself/herself. Dr. Joseph Adelstein, several years ago, at Rutgers University Summer School of Alcohol Studies, outlined characteristics of an effective Alcoholism Counselor. I would like to take the four characteristics Dr. Adelstein mentioned and broaden them to include Addiction Treatment Agents in general, as well as adding to/changing them somewhat.

QUALITIES WHICH SHOULD CONCERN
THE ADDICTION TREATMENT AGENT:

1. One's own need for success; this quality is important because recovery from addiction is marked by relapses and failures and a great deal of frustration. If the Treatment Agent is inordinately concerned with being successful, he/she might take each relapse as a personal affront; and take out their anger on the addicted person. They may handle both sobriety and relapses in a subjective, anti-therapeutic manner. If the addicted person achieved a small amount of sobriety, the Treatment Agent may:

a. Act as though this is a personal, delightful gift to himself/herself (then, the first time the addict gets angry--he/she may see getting "high" as a sure way of "getting" their Treatment Agent).

b. Act as though this is the most marvelous accomplishment of the addict (inferring that this is the *only* thing the addict can accomplish).

c. Act as though this is his/her reward for working as a Treatment Agent (a ploy immediately--though usually covertly--suspect by the addict who knows the Treatment Agent gets a salary!)

d. Act as though this is a concrete, irreversible achievement (making it difficult, if not impossible for the addict to return to treatment following a relapse).

e. Act as though the relapse is unpardonable, untenable, unexpected and unaccepted--by him/her (thereby increasing the addict's feelings of guilt, poor self image and often causing rejection of any other treatment approaches).

2. One's ability to meet the demands of extremely dependent persons without becoming either overwhelmed or rejecting, i.e., without presenting the addict with the reflection of society, which either placates or rejects him/her. In our County, dependence has taken on

a negative connotation. God forbid that any of us become dependent! It is as though we make our own shoes and weave the material from which our clothes are made. I think that people should study closely the nature of whatever they work with--automobiles, sheep, numbers--whatever. When one looks at the nature of a human being--one sees an inherently dependent creature. What on earth is more dependent than the human fetus? And, when the baby leaves the womb, he/she does not run down to the A & P for a quart of milk! Along with other professionals (Triana, Busby, and Palmer) in the field of addiction, I think that the addict has never experienced the necessary pre-requisite for forming healthy independent relationships--a healthy, dependent realtionship (Mahler, 1975). It is our job, then, as Treatment Agents to provide this experience--a healthy dependent relationship--to the addict.

This is no little task; the simplicity of its phrasing does not reveal the complexity of the reality. I have a friend, Joe Procaccini, who often says that despite Merton's "No Man is an Island" (1955), we are all "islands"; and we only grow and develop by reaching our hands out and clasping those of other islands. He warns, however, that this is a tricky exercise. Each island has to be very secure in itself, in order to maintain the stability and health of this union and not capsize themselves and others through the unhealthy fusion that either pulls the island on top of theirs, or jumps onto the other, causing mutual capsizing or destruction.

The Treatment Agent who is convinced that the addicted person needs to experience a healthy dependent relationship, is a person who is secure in his/her own "island". This kind of Treatment Agent does not run scared of a transference, but is able to use it to therapeutic advantage. Before anyone can permit the formation of a healthy dependent relationship, that person must love and appreciate himself/herself and be reasonable and consistent in fulfilling his/her own human needs. If so--there will be no situation where a Treatment Agent is available 24 hours a day, 7 days a week. They will be available to, and *for*, the addicted person with rational limits. It is all right if the addict thinks you are his/her mother or father--when you think you are, you have lost your profit as a Treatment Agent.

In this way, they can be a role-model for the addict who is often slow to see his/her needs as *bonafide*. All throughout this talk you can note the focus on the resocialization aspect of treatment. While I work for a salary (which is never enough!) I consider the joy of watching a human adult grow through dependency to independence an unbeatable bonus!

3. One's ability to accept hostility and rejection and still maintain a supportive relationship; it is important to remember that addicted persons, during and for sometime after their substance ab-

use, have extremely low frustration tolerance and poor impulse control. This means that almost everything makes them react in an impatient, acting-out manner. They cannot--wait for anything, tolerate anything they do not like or want; and, they let you know how they feel, by acting it out--screaming, running, fighting, cursing. This is just the reality of the situation. Sure, this kind of reaction resembles an irate infant or todler's behavior--which is not easy to accept even coming from a tiny human. Treatment Agents do not have an easy time working through the din (or storm and stress!) of these reactions. Often, they take the cursing and yelling as a personal affront.

Before my own training at Downstate Alcoholism Clinic (Brooklyn, New York) I had a typical middle-class reaction to anyone who cursed me out, in my office. If they were smaller than me, I threw them out of the agency. If they were larger--I calmed them down, so that the rest of the workers would not realize that I was not in control of the situation! I was indignant that anyone should talk like this to me--a college graduate! Now, I listen to what they are saying, and I work with verbal and non-verbal content as I would if they were not seasoning it with screamed obscenities. I remember that this scream is usually an indication that they are scared and anxious--and their fear and anxiety is up to high G. While not discounting that they might hate me--I remember that it might be other realities that they rant against--such as the system I represent (or other systems of society that have "done them in" as they see it).

It is only when I focus off myself and my own importance that I am able to listen to what the addict has to say--through the orchestration of his/her hurt (Triana and Hinkle, 1973).

4. One's own ability to set realistic goals--I say, *with* the addicted person; I have changed 'for' to 'with', because it is my conviction that no one--unless they are severely psychotic or mentally limited--ever achieves, realistically, the goals that someone else sets for him/her. The only basic differences between the Treatment Agent and the addicted person is that, at this time, in this place, the Treatment Agent is trained, objective and plugged into reality. This is why it is important that the Treatment Agent refers to the differentials of the treatment situation--whether racial, sexual, cultural, or any other variable. This realistic reference can keep the addicted person from using any one of these factors to sabotage the therapy process.

In America we have a multitude of myths--one is that being *open* is a common characteristic of interpersonal interactions. Wouldn't it be great! *Openness*, (as my friend Joe Procaccini is fond of reminding me) demands *trust*; and *trust* is a quality, far from common, in our society. I have worked with couples married 30 years--to each other!--not yet really trusting their mate. *Honesty* demands *integ-*

rity, and people will be as honest with us, as they are with themselves. (Which, may not be all that honest!) In the Bronx, while still a nun, I would begin interviews with clients assigned to me for marital therapy, by asking them how they felt having a Sister as a Marraige Counselor. They would smile and say, "Oh, that's fine, Sister". --When I persisted and asked them if they didn't think it strange that a celibate was going to try to tell them how to do it? --they smiled again and said "Oh no". I never fooled myself that they were open with me; I did let them realize that I was aware of the reality of the treatment situation. When they begin to feel uncomfortable in therapy they had a harder time using that differential, to cop out, because it was an out-in-the-open reality. Months after the initial interview (after a certain amount of trust was established) I have had people say they wonder "Why the h--- the agency gave them a NUN, as a Treatment Agent?"

It is important to note the differences between Treatment Agent and addicted person (or among therapy group members); not because they can prove anti-therapeutic, if ignored. When worked with they (race, sex, culture, etc.) can become a viable, porfitable, part of the whole treatment process. Someone who can work through problems with a therapist of either sex, may concommitantly resolve conflicts, long held, regarding that particular sex.

There is a great deal of controversy, in the field of addiction, over the concept of sobriety. The question reverberating throughout America--is--complete?/or, partial? Then, there are the subdivisions of--can a hard drug user drink alcohol? Can an alcoholic expose himself to powerful medication (even for health reasons?) Personally, I believe that our definitions of addiction and abuse should be more clearly and accurately delineated. Then, surer direction might evolve to solve this dilemma. For example: while the diagnosis of addiction might necessitate eventual total sobriety, that of abuse might indicate some eventual, reasonable, use of a chemical substance.

Regardless of the overall direction of the field--I return to my former conviction. Goals in both sobriety and other treatment issues, are most truly achieved, when the achiever has an active part in designing them. When Treatment Agents deny the addicted person a share in defining his/her areas of concern, setting up goals to resolve this concern and assessing the strengths he/she has to reach the goal--I think they prostitute the very process they claim to utilize.

As I said in the beginning of this session, goodwill is needed to work with, and for, addicted persons, but goodwill alone is practically worthless. If goodwill is to be the wick of the candle that illuminates the terrifying darkness and pain of addiction, it must be supported by talent, training, ongoing supervision and continued education in all aspects of human nature and addiction. I sincerely believe that no one can ever help anyone else to increase his/her insight or self-awareness, unless they (the Treatment Agent) are daily working to increase their own.

Also, to be a genuine Treatment Agent, nothing human should be alien. There should be no act, no phrase that causes a shudder of shock or revulsion to shake us. If we are not intensely aware of the fact that we are *capable* of doing anything the person we are working with is doing (and maybe even with more pizazz!)--we should stop working with people and try machines or plants!

The Portrait of an effective Addiction Treatment Agent is one which shows an *optimal frame*--an agency wherein everyone believes the fight against addiction is thrilling, exciting and hopeful; a *full background*--awareness of attitudes, in-depth education in addiction facts and treatment techniques, and ability and knowledge in adapting techniques to the addiction situation; and a *clear foreground*--a Treatment Agent whose need for success is not excessive, who is able to accept dependent people without becoming overwhelmed, or rejecting them, who is able to accept hostility and rejection and maintain a supportive relationship, and, who is able to set realistic goals *with* the addicted person.

REFERENCES

Blum, E.M. and Blum, R.H. 1967. *Alcoholism.* San Francisco: Jossey-Bass, Inc. (Chapter 17).

Mahler, M.S., Pine. F., and Bergman, A. 1975. *The Psychological Birth of the Human Infant: Symbiosis and Individuation.* New York: Basic Books, Inc.

Merton, T. 1955. *No Man Is An Island.* New York: Harcourt, Brace and Co.

Rev. Joseph Procaccini, Priest Personnel Director, Diocese of Trenton, New Jersey.

Triana, R.R. and Hinkle, L.M. 1973. "Psychoanalytically Oriented Therapy with the Alcoholic Patient". *Social Casework.* 55(5): 285-291.

Robert R. Triana, Founder and Director of Psychoanalytic Training Institute for Social Workers, New York City; Professor Everett Busby, on Faculty of Fordham University School of Social Work, New York City; Juanita Palmer, Founder of Nurses Involved in Alcoholism Treatment, N.C.A. New York City (now Nurse Educator in Albany, Georgia).

IMPLOSIVE THERAPY AS A TECHNIQUE TO HELP DETOXIFY DRUG ABUSE CLIENTS

John A. Heuvelman*, A.C.S.W.

Hope Center, Tucson

La Frontera Center, Inc.

Abstract: This paper was written with the goal of introducing the theory and technique of Implosive Therapy (IT), a relatively new form of behavior therapy, as a technique to help detoxify drug abuse clients. By providing a thorough discussion of the theory, practice, and related issues, it is hoped that more members of the drug treatment profession will implement the implosive procedure in their research and practice. A brief clinical example is also included, and a listing of the experimental evidence relating to the subject of IT and to the subject of detoxification is available for the interested reader.

IMPLOSIVE THERAPY

The implosive technique is based upon the psychological principle of extinction; a decline in responding as a result of non-rewarded repetitions of a response. During the therapy, conditioned stimuli (scenes which elicit anxiety responses in a patient) are presented in imagery without the accompanying unconditioned stimuli (primary aversive scenes which involve physical pain or punishment) really occurring. This procedure is designed to promote in the patient intense emotional reactions, and as such the goal of therapy is to repeatedly present anxiety-provoking scenes until the patient's emotional responses, especially anxiety, are considerably dissipated. Therefore, an emphasis is put on maximizing feelings of discomfort and anxiety within the safety of the therapist's office; so consequently the patient learns that these feelings can occur without physically harmful effects.

Although IT was initially established in 1957, the first formal theoretical discussion was not presented until 1961 when its founder, Thomas G. Stampfl, read a paper at the University of Illinois (Stampfl, 1961). His general intent was to introduce a psychotherapeutic technique based not only on psychological principles of learning, but also on psychoanalytic theories of personality development and treatment. Basically IT was viewed as a synthesis between the Freudian and behavioral approaches to psychotherapy, but its final tenets were recast strictly in terms of learning and extinction principles (Gumina, 1976).

THE NATURE OF EXTINCTION

In response to this theoretical learning framework involving symptom-formation, it is postulated that a sufficient condition for the extinction of a learned fear to occur is that the patient experience anxiety responses to the stimulus compound (i.e., cues associated with trauma) producing the fear; while in reality (i.e., therapist's office), the authentic primary trauma (e.g., pain, punishment, organic deprivation) is absent. Extinction, therefore, involves the reproduction of those stimulus elements (cues) in imagery; particularly those which have been avoided or repressed by the patient in the past. A greater fear (anxiety) reaction, and hence a greater extinction effect, is also hypothesized when the reproduction of cues more closely approximates the original stimulus compound associated with trauma. Then, as a result of multiple and varied repetitions of this procedure, the symptoms and motivating forces (i.e., anxiety reactions) behind the symptoms should be extinguished. The basic premise within this model is that the patient must experience intensely high levels of anxiety as a prerequisite to losing both his fear reactions and symptoms (Gumina, 1976).

RATIONALE FOR IMPLOSIVE THERAPY

Rationales for extinction in general have to do with the Pavlovian or classical model of learning. In the well-known example of Pavlov's dogs, a tone, the conditioned stimulus (CS) shortly precedes the presentation of food powder, the unconditioned stimulus (UCS). After a number of pairings of the tone and the food powder the dog's salivation response, which is the natural or unconditioned response (UCR) to food, occurs to the tone as well as to the presentation of food powder.

Another example, more analagous to the development of clinical anxiety or fear, finds the pairing of a buzzer (CS) with shock (UCS) given to animals, generating the withdrawal and arousal associated with shock (UCR) to the buzzer as well (CR).

Now, two factor theories of how neuroses develop include another step. First, there is the Pavlovian or classical conditioning of fear by repeated pairing of some cue with a painful stimulus. Then defensive behaviors develop on an instrumental or operant conditioning basis; that is, successful avoidance of the fear-producing situation is rewarded through the reduction of fear.

This theoretical explanation of the development of fear may be clearer in a clinical example. Many phobic individuals are afraid of specific items or situations. An explanation of a phobic's development of a fear of elevators might go something like this. Some painful, unpleasant, or traumatic event (UCS) occurs in the presence of an elevator (CS), and the phobic associates the anxiety of the unpleasant situation (UCR) with the elevator and begins to avoid elevators (CR). Each time the phobic encounters an elevator, he avoids it and thus reduces his fear. The association of the elevator with something painful is classical conditioning. The fear reduction resulting from avoidance of the elevator is operant conditioning. (The avoidance may continue for such a long time that the phobic no longer remembers why he fears elevators.)

Stampfl's belief is that the fear must be extinguished by repeated presentations of the conditioned stimulus. Though this initially generates considerable anxiety and discomfort, it is in this way that the client can learn that the conditioned stimulus is no longer followed by or paired with the unconditioned stimulus. The elevator phobic described above, if forced to ride the elevator rather than avoid it, would discover that there was no longer any connection between elevators (CS) and unpleasant experiences (UCS). Stampfl points out that in desensitization approaches, one is often dealing with stimuli other than the original conditioned stimuli. For example, if one had become anxious due to a critical rejecting mother, other critical women or other authority figues may serve as approximations to the original stimulus (stimulus generalization). Thus, a story will be visualized with women criticizing and humiliating the individual.

Stampfl's explanation of what occurs in Implosive Therapy has not gone unchallenged. Other theoretical rationales have been offered. It has been suggested that IT may function as negative practice. That is, the client is satiated with the repeated presentations of the noxious stimulus and abandons the fear through exhaustion. Other theorists suggest that Implosive Therapy works because clients are reassured or given courage when they discover that they don't fall apart when they beome fearful (Storms, 1976).

EXPERIMENTAL EVIDENCE

Since the inception of IT, there have been several experimental and case studies which have attempted to assess the effectiveness of the therapeutic procedures involved in clinical treatment. The evidence, in general, seems to suggest that IT has been relatively effective in dealing with a wide variety of human psychopathological disorders: phobias and phobic-like behaviors (Stampfl, 1967; Hogan, 1967, 1968; Barrett, 1969; Boulougouris, 1971); compulsions (Hersen, 1968); schizophrenia (Hogan, 1966, 1968); test anxiety (Prochaska, 1971); and general inpatients (Boudewyns, 1970) and outpatients (Lewis, 1967). (Gumina, 1976).

DETOXIFICATION

My clinical experience in applying IT as a technique to help detoxify drug abuse clients took place with methadone maintenance clients desiring detoxification. Detoxifying from methadone maintenance will be used as an example throughout the rest of this paper, although the process is applicable to other drug detoxifications.

In Dole's original model and in several other currently operating programs which follow his approach closely, there is no specific provision for detoxification from methadone maintenance. It is assumed that a patient will be on methadone indefinitely and that only if he/she drops out of treatment will it be discontinued. As a result of political pressure, anti-methadone sentiment, and some questioning of "the metabolic disease" model of addiction, the interest and emphasis on eventually removing clients from methadone have increased substantially. In an amendment to the regulations relating to methadone maintenance published by the Food and Drug Administration in December 1972, there is now a requirement that at the end of two years each client's case be reviewed and a specific decision, justified in his/her record, be made to continue him/her on the drug.

Although it may be highly desirable to remove people from methadone or to keep them on the drug for the shortest possible period of time, it appears that a very high percentage of those who were doing well while on methadone fail and return to heroin once they are detoxified, even after a year or more in a program. A minority do, however, seem to be able to achieve complete abstinence and normal functioning (Chushman, 1973) and some would argue that this minority alone would justify a continuing emphasis on detoxification.

CLINICAL PROBLEMS OF DETOXIFICATION

The difficulties of involving methadone maintenance clients in treatment are well documented in the literature. It has been my clinical experience that because of the tranquilizing effect of methadone there are four areas that are more conducive to involving a client in treatment. They are entry to the program, occurrence of crisis, a conscious decision on client's part, and detoxification. The following are common problem areas encountered in detoxification:

1. The ineffectiveness of rational medical information regarding the physical effects of detoxification. This usually creates a counselor credibility gap--"you shouldn't feel like that."

2. Attributing positive life changes to the methadone and not to the person.

3. Hypersensitivity to pain. Possibly related to the active addict not having felt pain because of the tranquilizing effect of most drugs.

4. Underlying problems possibly preconditioning a person to addiction. The most commonly cited are: Inadequacy, Depression, Dependency, Compulsive Personality, A Need to alter Consciousness, Low Frustration Tolerance, Availability, Peer Pressure, and Rebellion.

5. Psychological addiction.

6. A high degree of anxiety and fear related to the detoxification.

GENERAL FORMAT

IT is a technique for overcoming fears and anxieties which involve an apparent paradox. It reduces fear and anxiety by arousing as much fear and anxiety as possible.

The therapist usually conducts two or three diagnostic interviews which focus on specific contexts and feelings from the client's past and present experiences with detoxification. Most psychodynamically oriented therapists are capable of quickly speculating about the significant environmental, interpersonal, or developmental interactions which may have precipitated a particular client's drug addiction. Added support for this contention comes from the clinical observation that therapists' speculations are often substantiated by the client's verbal report following a therapy session. Wolpe and Lang's Fear Survey Schedule may be administered to gather additional information. Next, the client is trained in the technique of visual imagery. The client is asked to play act through imagery an anxiety-provoking scene and then respond emotionally to the scene as if it were real. Implosive treatment usually lasts 4-8 sessions on a crisis intervention frequency of from two to three days between sessions. Clients may also be trained to identify and create their own scenes in an effort to cope with fears and anxieties which may appear sometime in the future, long after IT has been terminated. IT sessions are divided into three sessions: 1) Review (warm-up) evaluate past session and discuss time period between sessions; 2) Implosive session (usually lasting 20-40 minutes); 3) Stabilization-review Implosive Session and gather information for the next session. Usually a client on a lowering dose of methadone has a high anxiety level which is conducive to IT.

CLINICAL CASE EXAMPLE

Mr. J is a 30-year-old Mexican-American single male who has been on methadone maintenance for two years. He has been detoxifying down from a 60 mg dose of methadone and is currently on 10 mg and fearful of completing his detoxification. Mr. J has a strong motivation to detox and has met criteria for a successful detoxification: negative urinalysis for all drugs except methadone, ability to think in long-term goals, stable home life, stability in vocational status.

After explaining the rationale behind the treatment, IT was begun. Following is an excerpt from one of his IT sessions.

> I want you to close your eyes and be aware of the anxiety and fear within your body. You're not ready to detox and you know it. You have always had a crutch all your life and can't make it on your own. The only time you have ever been drug free was when you were in the joint. I want you to be in your cell now and see the bars and feel the sick in the pit of your stomach. You're helpless, dependent, and weak . . . so weak that you've failed at everything in your life. Now you're failing at detox. Feel you're scared, you don't know who or what you are, you'll never make it. Once an addict, always an addict. I want you to visualize taking your last dose of methadone. Feel the goose pimples, the hot and cold spells, the sweating, and now the deep bone aches as you withdraw. This is what you fear the most and without the methadone you'll lose control and go crazy.

A total of eight IT sessions was presented during a two-week period. Mr. J finished his detoxification and follow-up evaluations have shown Mr. J to be a non-user of heroin over the past nine months.

CONCLUSION

IT should be conducted by an experienced therapist. It is a difficult treatment because of the creation of anxiety in both the client and therapist. Weren't we trained to make people more comfortable (thereby ourselves). The therapy is confronting in dealing with both the fear and anxiety of detoxification and the underlying problems of a drug-using person. Change usually involves anxiety. We are faced with the growing concern of drug treatment professionals to use quick and more effective methods of breaking the vicious cycle of drug abuse.

REFERENCES

Barrett, C.L. 1969. Systematic Desensitization versus Implosive Therapy. *J. Abnorm. Psychol.* 74: 587-592.

Boulougouris, J.C. and I.M. Marks et al. 1971. Superiority of Flooding (Implosion) to Desensitization for Reducing Pathological Fears. *Behav. Res. Ther.* 9: 7-16.

Chushman, P. and V.P. Dole. 1973. Detoxification of Well Rehabilitated Methadone Maintained Patients. *Proceedings of the Fifth National Conference on Methadone Treatment.*

Gumina, J. 1976. Unpublished manuscript, Evergreen, Colorado.

Hersen, M. 1968. Treatment of a Compulsive Phobic Disorder Through a Total Behavior Therapy Program: A Case Study. *Psychotherapy: Theory, Research and Practice.* 5: 220-225.

Hogan, R.A. 1966. Imposive Therapy in the Short Term Treatment of Psychotics. *Psychotherapy: Theory, Research and Practice.* 3: 25-32.

Hogan, R.A. 1968. The Implosive Technique. *Behav. Res. Ther.* 6: 423-432.

Hogan, R.A. and J.H. Kirchner. 1967. Preliminary Report of the Extinction of Learned Fears via Short Term Implosive Therapy. *J. Abnorm. Psychol.* 72: 106-109.

Hogan, R.A. and J.H. Kirchner. 1968. Implosive, Eclectic Verbal, and Bibliotherpay in the Treatment of Fears of Snakes. *Behav. Res. Ther.* 6: 167-171.

Johnson, W. 1971. Behavior Therapy: What Place in the Psychiatric Residency? *Proceedings of the 25th Anniversary of the Department of Psychiatry.* Rochester: University of Rochester Medical Center.

Levis, D.J. and R. Carrera. 1967. Effects of Ten Hours of Implosive Therapy in the Treatment of Outpatients. *J. Abnorm. Psychol.* 72: 504-508.

Prochaska, J.O. 1971. Symptom and Dynamic Cues in the Implosive Treatment of Test Anxiety. *J. Abnorm. Psychol.* 77: 133-142.

Stampfl, T.G. In press. Implosive Therapy. In R.M. Jurjevich (ed.)

Handbook of Direct and Behavior Psychotherapies. Chapel Hill: North Carolina Press.

Stampfl, T.G. 1961. Implosive Therapy: A Learning Theory Derived Psychodynamic Therapeutic Technique. Paper read at the University of Illinois, Urbana.

Stampfl, T.G. 1967. Implosive Therapy: The Theory, the Sub-human Analogue, the Strategy, and the Technique. In S.G. Armitage (ed.) *Behavior Modification Techniques in the Treatment of Emotional Disorders*. Battle Creek, Michigan: VA Publication. 12-21.

Stampfl, T.G. 1970. Implosive Therapy: Theory and Technique. In D. Jacobs (ed.) *Behavior Therapy at the Crossroads*. Cleveland, Ohio: Case Western Reserve Press.

Stampfl, T.G. and D.J. Levis. 1967. Essentials of Implosive Therapy: A Learning Theory Based Psychodynamic Behavioral Therapy. *J. Abnorm. Psychol*. 72: 496-503.

Stampfl, T.G. and D.J. Levis. 1967. Phobic Patients: Treatment with the Teaching Approach of Implosive Therapy. *Voices: The Art and Science of Psychotherapy*. 3: 23-27.

Stampfl, T.G. and D.J. Levis. 1968. Implosive Therapy: A Behavior Therapy? *Behav. Res. Ther*. 6: 31-36.

Storms, L.H. 1976. Implosive Therapy: An Alternative to Systematic Desensitization. *Modern Therapies*. 132-149.

*Currently employed by Southern Arizona Mental Health Center.

RURAL/URBAN PROGRAM DEVELOPMENT IN CENTRAL ALBERTA, CANADA

Robert L. Graham, RPN and Val J. Boehme, RPN

The Alberta Alcoholism and Drug Abuse Commission

The need for an effective delivery service of Alberta Alcoholism and Drug Abuse programs to the citizens and communities prompted the development five years ago of a rural/urban philosophy in the Central Alberta Region of A.A.D.A.C.

In looking at the program development and philosophy for the region one must look at several aspects of the geographical area and its population. Some of these factors are the geographic size, the existing population spread and projected growth--the diversity of occupations which includes the full spectrum of industry, farming, ranching, construction and oil.

The region geographically covers 36,000 square miles (one-half the size of Washington State). Population is approximately 200,000; Red Deer contains the largest population of 40,000, the remainder reside in smaller urban and rural communities. One must also consider the broad differences in life styles within the region which necessitate programming to reach the greatest number of the population. The program philosophy that we are using is very basic in that it emphasizes the use of resources that exist in every rural and urban community in the region, thus permitting the full use of the expertise and innovativeness of the Central Alberta Regional AADAC staff. This approach facilitates development of programs that meet the demonstrated community needs as they are seen by the community rather than an overlay of pre-planned programming. This is done by involving our staff throughout the region on a regular basis as consultants, stimulators, trainers, and supporters of existing programs in the communities. The overall guidelines that we are utilizing are as follows:

1. Projects or programs initiated and co-ordinated by AADAC staff with the responsibility for project or program operation solely the responsibility of AADAC (short or longer term programs);

2. Projects or programs initiated by AADAC staff in co-operation with other community resources and with a shared responsibility over the extent of the program (short or long term programs);

3. Projects or programs where AADAC staff may or may not have been involved in the planning and initiation but would act on short term consultant basis either to the program, the individuals or the agencies involved.

The programming approach entails the need for staff mobility and versatility and a high degree of innovativeness and ambition. The

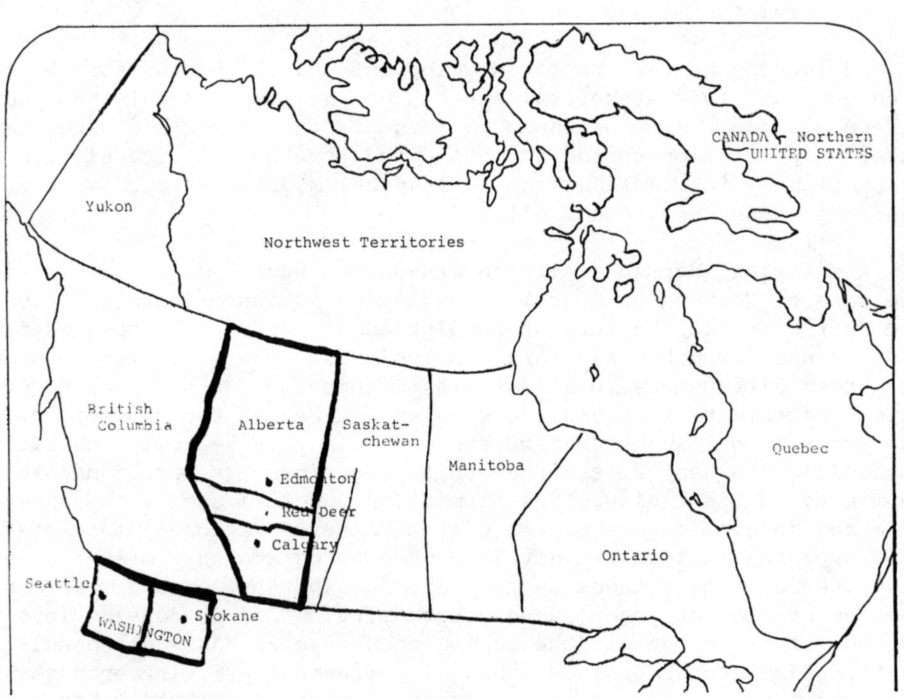

Fig. 1.

staff members must be willing and able to become involved in treatment, education and evaluation of all projects throughout the region. This contributes positively to a high degree of sharing and discussion with regional staff, and willingness of community agencies to co-operate with AADAC staff.

The program thrusts of AADAC provincially are Community Treatment and Community Extension Services. Regionally we use a diversified approach, each field worker functioning in both program areas.

COMMUNITY TREATMENT SERVICES

The regional office in Red Deer houses the non-medical Out-Patient Clinic. Treatment services are available five days per week. Counselors travel from here to area offices in rural Central Alberta, where the same services are made available to the respective communities. Our rural/urban treatment programs contain the following components:

Direct Treatment

Individual, family, and group counselling on an out-patient basis, also with patients in general hospitals at the request of nursing staff and physicians. Weekly Information Series were recently established in Red Deer and Rocky Mountain House. The series in Rocky Mountain House was co-sponsored by AADAC and the Native Friendship Centre. The series of lectures, films, and discussions provides basic information about alcoholism and other addictions. Patients, their families, professionals, students, and the general public participate. Referrals are also made to this program from the Alberta Impaired Drivers' Program.

Referral

To intoxification recovery centres, in-patient and extended care treatment facilities, provincially.

Intoxification Recovery Centres: We have no centres in Red Deer. When detoxification is indicated we are able to utilize the AADAC Intoxification Recovery Centres in Edmonton and Calgary, some 100 miles away. Some 40 miles away is an AADAC funded native detoxification centre which we are able to refer to. Intoxification recovery is available through some general hospitals throughout the region.

In-patient: We are able to refer individuals to AADAC's provincial in-patient programs: Henwood Rehabilitaion Centre is a 28 day, 64 bed, in-patient program, 15 miles north of Edmonton. The David

Lander Centre is a 16 bed in-patient program 65 miles south of Calgary. It is a three phase program: Phase I (12 successive days of residential treatment); Phase II (4-6 weeks return to the community with follow-up by referral source); Phase III (5 successive days of treatment accompanied by spouse when possible). We are also able to refer patients to the 12 bed, 4-6 week alcoholism program at the Alberta Hospital, Ponoka (provincial mental hospital) 30 miles from Red Deer. In referring individuals to each of these in-patient facilities we generally expect a 3-6 week waiting period prior to admission. With regard to long term in-patient facilities we have access to the AADAC Rehabilitation Centre which provides a long term rural residential addiction treatment program to the alcoholic otherwise known as the chronic drunkenness offender. The Centre is located 70 miles from Red Deer, again out of our region.

Consultation

Close working relationship with other private and government agencies and persons. This would include team treatment approaches, individual consultation and referral with professionals and para-professionals in all health related fields. Included in this are probation, courts, Salvation Army, A.A., hospitals, school counselors, occupational health personnel and social service agencies. In addition, we do follow-up consultation and referral.

Follow-up

Is conducted on patients from in-patient treatment programs, jail systems, intoxicant recovery centres and out-patient facilities. Again the utilization of community resources is a very important aspect; it is much more functional and is also more effective in that the client is relating to his own community and is receiving support from his own community.

COMMUNITY EXTENSION SERVICES

We have separated our community services into six (6) categories in an effort to fully outline our programs. They are available throughout the region.

1. Schools
2. Occupational
3. Medical/Professional
4. Special Projects
5. Exhibits
6. Media Presentation

Schools

Seminars/workshops with school administration boards, superintendents and principals are conducted to ascertain and assist in fulfilling their needs as follows:

1. define their needs in regard to alcohol and other related drug problems;

2. clarify the role of AADAC and determine alternatives to meet their needs.

We are also involved in their teacher conferences and teacher training programs (Institutes) as noted below:

1. a general awareness program of AADAC resources;

2. alcohol and other drug use/misuse--identification and associated problems (family, individual and community).

Classroom presentation and discussions are conducted by AADAC personnel to assist the teachers as resource persons in their programming:

1. awareness of drugs and alcohol use/abuse;

2. alternatives to the use of alcohol and drugs;

3. peer group pressure, decision making, etc.;

4. individual responsibilities towards the choice of alternative use;

5. evaluation of programming done by students via questionnaires, etc. for research and evaluation of programs.

In addition, a student-initiated Impaired Drivers' Program has been established in a number of senior high schools throughout the region. This is a preventative program with non-offenders, and is being evaluated in an attempt to establish the effect of preventative education programs.

Occupational

AADAC personnel have been requested to assist in establishing policy and procedures in dealing with employees whose drug or alcohol use is affecting his/her work performance. The occupational consultant in the Central Alberta Region is currently assisting hospitals in setting up policy and procedure. When these policies and procedures are established, AADAC will conduct a workshop for their supervisors and management personnel to assist them in carrying out their policy and procedure.

We are developing with the Behavioral Problems Counseling Unit for the Alberta Government employees, clarification of policy and procedure in order that treatment and referral may be effected. Workshops for supervisory personnel with the Alberta Government Services, Royal Canadian Mounted Police, private business and industry are being conducted.

Medical/Professional

1. Liaison between medical profession and our treatment facilities;

2. From hospitalized treatment to an AADAC out-patient basis;

3. Assist medical profession in furthering in-patient treatment throughout AADAC facilities;

4. Conduct workshops/seminars for the medical, para-medical and supervisory staff of hospitals on supportive elements available to them;

5. Consultations with the medical and other professionals dealing with mutual clients and alcohol or drug related problems;

6. Special Projects: Through these projects we have increased our efforts towards mobilizing community resources in dealing with the alcohol and drug concerns throughout the Region. In some instances AADAC provides short term funding for communities to develop their own projects.

Special Projects

AADAC personnel in the Central Alberta Region have organized or contributed in some manner to the following projects held in this region:

Colleges: We conducted a seminar/workshop for students at the Old's College who were randomly placed into three groups for research and evaluation purposes (requested). This approach to research and knowledge and information on all aspects of drugs and AADAC facilities was also being carried annually in the Camrose College. We provide field experience to the Red Deer College Nursing students and regular workshops for the College's Rehabilitation Program.

Alberta Hospital, Ponoka: Annually, a seminar/workshop on effective means of dealing with alcohol and other drug related problems in the community is presented to the training personnel and students of the psychiatric nursing program (see Appendix "B").

Penhold Base: A cadet program is held over the two month summer period at the base by exposing 2,500 cadets to knowledge and information on alcohol and other drugs. A close working relationship has been established with the Base personnel on this project. Initially AADAC conducted the sessions; however, since 1974 they have been self-sufficient in the program, drawing only on our audio-visual aids and consultant skills.

Student-Initiated Impaired Drivers' Program: Initially five (5) rural high schools took part in this program with 206 students participating. An equal number of students who did not participate were established as a control group. The 412 driving abstracts will be examined in a comparative study in 1980.

In addition, the first Native-Initiated Impaired Drivers' Program was carried on the Sunchild/O'Chiese Reserve west of Rocky Mountain House where 68 persons were in attendance. This was a community initiated program co-sponsored by the Kochitawin Native Society (Native Alcoholism Program--federal and provincial shared funding) and AADAC. There was no deletion on program material and all resource people were used as they would be on any Impaired Drivers' Program for those directed to attend by the Department of Transport or through the courts.

Hole in the Fence: This project can be seen as our being a facilitator to N.M.U.D. (Non-Medical Use of Drugs Directorate--a federal program to coordinate and stimulate a Canada-wide approach to alcohol and drug dependency problems) in putting this program into the hands of the teachers (grades 1 to 3) in the elementary schools throughout our region.

Community awareness: In this region we act as resource personnel, speakers or consultants to interested groups. This would include community groups, A.A. groups, service clubs, planning commissions and committees, especially in areas of industrial and population growth and development; as well we provide general information to these groups re alcohol and drug use and abuse.

Law Enforcement Workshop: Was requested by the local detachment of the Royal Canadian Mounted Police and relates to the defining of roles of the force and our field workers.

Alberta Impaired Drivers' Program: This program is presently carried in nine (9) centres throughout our region. We find this program very useful in detecting early symptoms of alcoholism; it is an excellent referral source for those seeking treatment for alcohol/drug related problems; and a good working relationship has been established with the court system as well as other resource people used on this program.

Courts and corrections: Our involvement with the Nordegg Provincial Minimum Security Institution 120 miles west of Red Deer is one of program development and consultation on existing programs. The Bowden Federal Medium Security Institution has asked for our continued support in developing and delivering programs for training staff and inmates.

Smoke No More: This program was designed to help people to stop smoking--using lectures, discussion and relaxation techniques. The program was designed following a request from the staff nurse at the Red Deer General Hospital for a "Stop Smoking Program" (see Appendix "B").

Pine Lake Elementary Students Living Skills: A program for the Elnora Elementary School students and teachers which was focused on living skills and understanding themselves. It was a three day live-in camp program.

Exhibits

We respond to community requests for public information in setting up display booths at resource fairs with other community agencies and work closely with those agencies in dealing with social problems that may exist in their respective areas.

Media Presentations

Appearance on the broadcasting media or publishing of material in the daily, weekly local press is conducted by our Central Alberta personnel with consultation and co-operation from the provincial AADAC, Department of Public Affairs.

FUTURE PROGRAMMING

The following programs are being considered for development in 1977-1978.

1. Programs in junior high with students and parents to promote a greater awareness in relation to alcohol/other drugs and to achieve positive communication attitudes towards use and non-use of alcohol/other drugs;

2. An industrial program for small businesses and industry in

co-operation with the Continuing Education Department of Red Deer College to assist them in developing policy and procedures to deal with troubled employees;

3. A program for youth group leaders (Girl Guides, Scouts, 4-H Clubs, recreation) to increase their skills in developing programs in their own areas of concern;

4. A treatment program for the new out-patient facilities in Red Deer;

5. Emphasize the consultative, referral, follow-up role in treatment in the rural areas of our region; and,

6. An in-service educational program for senior citizens' facilities.

SUMMATION

In summary, the major points I would wish to leave with you are:

A. The need to clearly define your program themes, i.e.,

(1) programs where you take full responsibility

(2) where the responsibility is shared

(3) programs where you are a consultant, stimulator or trainer.

B. Be aware of and supportive of "locally demonstrated" needs, and be sure the communities are aware of what you are able to provide with "your" resources.

C. Staffing is also of vital importance in that they must be versatile, mobile and innovative.

This program philosophy allows and encourages the full utilization of community resources in its fullest meaning and prevents the kinds of duplication of services that creates confusion to clients and communities, thus increased costs.

The outcome of the past five (5) years has been very encouraging in terms of community response as well as acceptance in AADAC: our field workers are finding that communities are becoming more and more willing to deal with their own problems, develop their own programs and utilize their own facilities efficiently, but also those of specialized agencies such as A.A.D.A.C.

APPENDIX "A"

Red Deer General Hospital Program Presented to Hospital Employees on "Smoke No More"

Co-ordinated by: Bill Ouellet, Director of Personnel, Red Deer General Hospital

Joyce Hagg (Mrs.), Director of In-Service Education, R.D.G.H.

Julie Syrnyk (Mrs.), Staff Nurse, R.D.G.H.

Alex Thomas, Addictions Counselor, The Alberta Alcoholism and Drug Abuse Commission

Manual Prepared by: Alex Thomas, AADAC

Location: Red Deer General Hospital

Dates:
Session 1A - Oct. 20, 1976, Wednesday
Session 1 - Oct. 27, 1976, Wednesday
Session 2 - Oct. 28, 1976, Thursday
Session 3 - Oct. 29, 1976, Friday
Session 4 - Nov. 1, 1976, Monday
Session 5 - Nov. 2, 1976, Tuesday
Follow-up I - Nov. 10, 1976, Wednesday
Follow-up II - Dec. 8, 1976, Wednesday

Table of Contents

APPENDIX "B"

Alberta Hospital, Ponoka Program Presented to School of Nursing

SCHOOL OF NURSING - ALBERTA HOSPITAL, PONOKA--FIRST YEAR STUDENTS

AADAC Co-ordinators: Val Boehme
Don Ward
Val Danielson

February 13, 1978
Monday

GOALS: TO INCREASE THE STUDENTS' UNDERSTANDING OF ALCOHOLISM, AND THE EFFECTS OF ALCOHOLISM ON THE FAMILY

9:00 a.m. : Introduction to AADAC
9:10 a.m. : Distribution and review of reading materials
9:20 a.m. : Students' expectations for the day, consider individually and in groups of 4
9:35 a.m. : Listing expectations and dealing with those which are not already in the plan for the day
10:30 a.m. : Break
10:50 a.m. : Role Play: "I might have a drinking problem" - Discussion
11:15 a.m. : Film: *Conspiracy of Silence*
12:00 noon : Lunch
1:00 p.m. : Role Play: "In the Family" - Discussion; Review of learning so far
1:45 p.m. : Buzz groups of 4 to discuss and list the purposes of the family - Discussion
2:30 p.m. : Break
3:00 p.m. : Film: *The Summer We Moved to Elm Street*; Discussion: (groups of 8) What's going on with each character? Why is the family dysfunctional? Compare this family with the purposes listed--3 groups to observe characters: mother, father and Doreen. General discussion on closed and open family systems. Questions and comments

SCHOOL OF NURSING - ALBERTA HOSPITAL, PONOKA--SECOND YEAR STUDENTS

AADAC Co-ordinators: Val Boehme
Dave McConachie
Guests: Dr. S. Smith and Associates, AHP

February 14, 1978
Tuesday

GOALS: TO LIST STUDENTS' EXPECTATIONS FOR THE WEEK AND BEGIN DEALING WITH THEM--TO INCREASE STUDENTS' UNDERSTANDING OF ALCOHOLISM

9:00 a.m. :	Introduction to AADAC--using slide presentation
9:30 a.m. :	Distribution and review of reading material
9:40 a.m. :	Consider expectations for the week in diads
10:00 a.m. :	Listing expectations
10:15 a.m. :	Break
10:30 a.m. :	Handling questions and expectations not scheduled to be dealt with in the workshop
11:30 a.m. :	Lunch
1:00 p.m. :	Role Play: "Handling Drinking Problems on the Job" - Discussion; Review phases of alcoholism
1:45 p.m. :	Film: *The Secret Love of Sandra Blaine*; Discussion and completing list according to the film
2:30 p.m. :	Break
3:00 p.m. :	Guests: Alcoholism Unit - Dr. S. Smith and Associates; Focus: Effects on the Family; Discussion: Significance of this learning (groups of 8)

SCHOOL OF NURSING - ALBERTA HOSPITAL, PONOKA--SECOND YEAR STUDENTS

AADAC Co-ordinators: Val Boehme
Val Danielson
Ed Boisvert

February 15, 1978
Wednesday

Development

9:00 a.m. :	Introduction to *I'll Quit Tomorrow* film; Mini lecture
9:15 a.m. :	Feeling chart
9:45 a.m. :	Film: *I'll Quit Tomorrow* Reel #1
10:20 a.m. :	Discussion
10:30 a.m. :	Break
11:00 a.m. :	Introspection: John Westlow/General Manager diad; Attitudes and Rehabilitation
12:00 noon :	Lunch

Confrontation

1:00 p.m. :	Introduction to Reel #2 - groups of three (structured questions)
1:20 p.m. :	Element of Intervention (overhead)
1:30 p.m. :	Film - Reel #2
2:00 p.m. :	Discussion group with questions
2:20 p.m. :	Break

Treatment

2:45 p.m. : Introduction to Reel #3
3:00 p.m. : Film - Reel #3
3:30 p.m. : Treatment facilities

SCHOOL OF NURSING - ALBERTA HOSPITAL, PONOKA--SECOND YEAR STUDENTS

AADAC Co-ordinators: Val Boehme
Val Danielson
Ed Boisvert

February 16, 1978
Thursday

GOAL: THROUGH LECTURE AND EXERCISE LEARN A THEORY OF COMMUNITY ACTION PROFILE

9:00 a.m. : Outline program for the morning; Introduce materials and process; Continue as per course outline in packet

12:00 noon : Lunch

GOAL: TO HELP STUDENTS UNDERSTAND THE PROBLEMS SURROUNDING DRUGS OTHER THAN ALCOHOL, SUCH AS: A. PRESCRIBED DRUGS
B. LIFE STYLE ATTITUDES SURROUNDING PILL TAKERS

1:00 p.m. : Presentation on prescription drugs that can cause problems of abuse
2:00 p.m. : Discussion in groups of 5 "When we're dealing with drug problems we're dealing with living problems"
2:30 p.m. : Break
3:00 p.m. : Film: *Up Pill, Down Pill* - (3 groups to observe 1 character each); Discussion

SCHOOL OF NURSING - ALBERTA HOSPITAL, PONOKA--SECOND YEAR STUDENTS

AADAC Co-ordinators: Val Boehme
Barney Winczura
Guests: Norm Lowe, RCMP
Wilson Okeymaw, Director Hobbema Centre

February 17, 1978
Friday

GOALS: TO FAMILIARIZE STUDENTS WITH OTHER PROGRAMS ATTEMPTING TO DEAL WITH ALCOHOL/DRUG ABUSE IN SOCIETY, SUCH AS AN APPROACH TO ALCOHOL/DRUG PROBLEMS AMONG NATIVES

9:00 a.m. : Film: *Highways of Agony*
9:30 a.m. : Discussion: Outlining the sessions as held in the Central Alberta Region
9:45 a.m. : Break
10:15 a.m. : Presentation by RCMP Officer Norm Lowe
--statistics
--procedures of arrest
--use of the breathalyzer
--general questions and answers
11:00 a.m. : Film: *So Long Pal*
11:30 a.m. : Lunch
1:00 p.m. : An overview of Hobbema Centre and Native Alcohol Problems and Programs - Mr. Wilson Okeymaw; General Question Period
2:30 p.m. : Break
3:00 p.m. : Expectations revisited and course evaluation

RITALIN: CONSEQUENCES OF ABUSE IN A CLINICAL POPULATION

Robert M.J. Haglund, M.S.W. and Lynn L. Howerton, R.N., B.S.

Center for Addiction Services

Abstract: A review of the literature suggests that serious complications are to be expected from frequent abuse of Methylphenidate (Ritalin) particularly if administered I.V. Treatment personnel have also suggested complicating factors surrounding the abuse of Ritalin during outpatient treatment for opiate dependency.

A consecutive series of admissions (n=111) to a comprehensive Drug Services program is examined for their use of Ritalin. Special attention is given to history of use, means of supply, and adverse consequences associated with the abuse of this substance. Adverse reactions (client's definition) occurred in nearly 30 percent of the user group. Recommendations for treatment and prevention will be summarized based on the project findings.

BACKGROUND

Seattle once held the dubious distinction of being identified as the Ritalin abuse capital of the United States. That reputation emerged not so much from the prevalence of Ritalin abuse in the Northwest but that it was seldom abused in other parts of the U.S. Today, the abuse of this substance is more widespread, particularly among opiate dependent persons, thus likely to come to the attention of drug treatment personnel.

In 1971, hearings[1] devoted to evaluating federal controls on stimulant drugs were held. Testimony concerning Ritalin focused on the problem in Seattle. Testimony revealed that I.V. administration could result in death to the user, and several such examples were given. At that time, the talc content of the tablets also posed a hazard to the I.V. user. Lung damage resulting from this could be particularly severe.

Ritalin seems to have a variety of actions on the user depending on the perspective of the researcher. Pharmacologists have indicated that Ritalin may inhibit the metabolism of heroin and/or methadone thus extending or "stretching out" the high associated with the use of these narcotics. Inhibition of withdrawal would also occur if the opiate effect was extended. Other investigators have shown that the drug is injected mainly for its "upper" effects typically associated with a stimulant drug. User reports tend to support the latter perspective.

Our interest in examining the abuse of Ritalin was to determine how the user becomes involved with its use, what are the sought after effects, what complications are to be expected, and how these factors relate to the management of the user in a treatment setting. The literature, sparce as it is, seems to concentrate on the acute reactions to I.V. use of Ritalin and other systemic effects. A wider perspective is necessary to effectively deal with this problem in the typical out-patient drug abuse treatment center.

METHOD

Our sample consists of 111 consecutive admissions to the Center for Addiction Services[2] examined six months after completing their initial treatment phase. Structured questionnaires were utilized to gather the self report data. Questions concerning reason for use and adverse reactions were collected in an open-ended format and later coded to preserve the closest approximation to naturally occuring categories of response. The Ritalin data will be combined with that of a larger study to explore possible relationships with background

---

1 Hearings before the Subcommittee to Investigate Juvenile Delinquency of the Committee on the Judiciary United States Senate, ninety-second Congress, first session; July 15 and 16, 1971, Birch Bayh Chairman.

2 The Center for Addiction Services is a private non-profit agency serving as an umbrella to 26 clinics providing drug abuse treatment services to King County, Washington.

Table I. Sample characteristics (N=11).

	Users	Non-users	Total (%)
Male	43	32	75 (68)
Female	20	16	36 (32)
Caucasian	33	40	63 (57)
Black	26	6	32 (29)
Other	4	2	6 (14)
Mean Age	27.4	28.3	--
TREATMENT MODALITY:			
Residential	25	22	47 (42)
Outpatient Drug Free	6	15	21 (19)
Methadone	32	11	43 (39)
TOTAL (%)	63 (57)	48 (43)	

and sociological variables.[3]

RESULTS

Sample characteristics for user and non-user groups are given in Table I. The most notable finding is the high proportion of the total non-caucasian sample who report using Ritalin. Age of first use is considerably older than that typically reported for other drugs, e.g., opiates, amphetamines. Possibly this reflects the popularity of the drug among our sample of addicted narcotic abusers. The average length of time between first use and first continuing use was 2 years. It was interesting to note the respondents tended to either favor or reject the continuing use of Ritalin after the initial trial. This could be due to the initial exposures via I.V. methods with the potential for marked adverse reactions.

---

3 Treatment Porgram Evaluation Project, NIDA Grant 3H81DA01781. Further information available on request from Robert M.J. Haglund, Principal Investigator.

Table II. Characteristics of users.

Mean age First use	22.3	
Mean age First regular use	24	
Frequency of current use (%)		
None	53	(84)
Daily	2	(3)
Weekly	3	(5)
Monthly	2	(3)
Infrequent	3	(5)
Drugs used with*		
Heroin/opiates	46	(63)
Barbiturates	3	(4)
Alcohol	1	(1)
Other	2	(3)
None	21	(29)
Source of Supply (%)		
Dealer	35	(56)
Friend	15	(24)
Prescription	8	(13)
Other	5	(7)
Availability (%)		
Difficult to obtain	5	(8)
Easy	14	(22)
Very easy	38	(60)
Other	6	(10)
Cost per dose (tablet)		
None (free)	13	(21)
<$1	6	(10)
$2-$4	4	(6)
$5	38	(60)
>$5	2	(3)

* More than 1 combination permitted so total >N.

Patterns of use can be evaluated by examining the combinations of drugs used as well as the frequency of Ritalin use. Table II indicates that use with opiates is the most common pattern of abuse in our sample.

Table III. Reasons for use.

REASONS*	N	(%)
Wanted to "get high"; feel good	11	(16)
Curiosity	9	(13)
Availability; peer pressure	7	(11)
Substitute for other drugs	6	(9)
Escape; need "lift"	3	(5)
Remain awake; "speed" effect	7	(10)
Heighten effect of other drugs	17	(25)
Other	7	(11)

* More than one reason permitted so responses > N.

Reasons for use have been summarized in Table III. Getting "high" and other answers typical of amphetamine abuse predominate the responses. Each subject could select up to three primary reasons which accounts for the total number of responses exceeding the number of user respondents. It should be noted that no subject indicated avoidance of withdrawal symptoms or extending the narcotic effects of methadone or heroin as a major reason for use. Ritalin appears to be used for its stimulant effects, particularly in combination with other drugs.

Common adverse reactions and their resolution are listed in Table IV. While the abuse of Ritalin during outpatient treatment seems to present problems for the clinician, it is apparent subjects do not often seek traditional medical services in responding to adverse reactions. Many reactions seem to be of the panic/paranoia type associated with the abuse of other types of amphetamines. Reactions from the injection process itself are infrequent. In reviewing the data, caution must be exercised in interpreting the significance of self-reported physical/psychological symptoms. The interviewer made no medical related evaluation of the respondent.

DISCUSSION

The I.V. abuse of any psychoactive drug can complicate the outpatient therapy of addicts, especially when employing chemo-therapy. The user is seen as desiring a high, speed-ball, or "getting off" in

Table IV. Adverse reactions and treatment.

Frequency of reaction (%)		
None	44	(70)
1	14	(22)
2	2	(3)
3	3	(5)
Major symptoms (n=19)*		
Paranoid, other disturbance	8	(23)
Sweating, chills	4	(11)
Elevated heartbeat, breathing	6	(17)
Stomach irritation	3	(9)
Weak/run down	3	(9)
Abcess/rash	5	(14)
Sleep disturbance	2	(6)
Other	4	(11)
Treatment for reaction (n=19)*		
Self care	12	(41)
Emergency room	2	(7)
Help from friends	1	(4)
Other	14	(48)

* More than one answer permitted so responses >N.

the abuse of Ritalin. It does not seem to be used as a self-medication to deal with withdrawal or other opiate dependence symptoms. At one time physical symptoms concerning the injection of Ritalin accounted for significant accounts of morbidity/mortality. Our findings indicate that reactions similar to those expected from other stimulants are most common among Ritalin abusers.

The late onset of first use implicates the role of opiate dependence in the abuse of Ritalin in our sample. Because this also occurs at a time when opiate dependency therapy is sought, the treatment picture is complicated. Novice users are often unaware of the possible adverse reactions to I.V. Ritalin, possibly accounting for the high proportion of one time-only or infrequent users. The author's own experience in discussing the initial use by the subject confirms the relative uncertainty as to effect that was often apparent during first use. Some education/preventative measures seem appropriate.

While the price of Ritalin has increased rapidly, its availability seems of little concern to the user and non-user alike; it is consistently and readily available. The typical sources of supply (see Table II) given indicate a need to carefully evaluate a request for Ritalin in the medical setting as well as a need to control street dealing.

SELF-SUFFICIENCY: ITS TIME HAS COME

William F. Griglak

Indiana Division of Addiction Services

Whether agreeable or not, substance abuse programming is a business. A service-delivery business, yes; a behavioral service business, yes; *but* a business. As such, programs must have, in the future, less reliance on federal, and/or local government funding bases and more reliance on their own ability to "hustle."

Indiana's Self-Sufficiency Conference last fall provided such a base of consideration for the programs operating in that state. The *sole* emphasis, on a practical level, was to bring programs to the point of seriously viewing their future in terms of a business operation and one which must ensure its own monetary future. As such, profit and loss, funding schema, tricks of the fund-raising trade, and other income-producing methods were important considerations of the conference. It was the contention that, in order to survive, programs had to learn how to "walk through" a series of practical, how-to steps on raising funds adequate to meet their needs.

It was and still is the opinion of the architects of this conference that unless self-sustaining entrepreneurial efforts are engaged in, the substance abuse field will not survive with any vigor (except in a few isolated cases) or will continue to be co-opted by the governmental agencies. Neither prospect holds much promise. Creative, self-sufficient activity on the part of the programs themselves is the only key to service delivery.

The chicken-and-egg dilemma of which priority comes first--service first or fund-raising first--is no dilemma at all. Put rather bluntly, without adequate funding all the well-meaning service intentions will not come to pass. Programs--and their well-meaning staffs--will survive only if there is enough money to pay for both staff and program activities. Without the dollars, the field will pant its way into extinction. *Deus providet* may be valid--but bill collectors and staff demand payment in legal tender, pious protestations notwithstanding.

Self-sufficiency is the only viable alternative for the substance abuse field in the "grey twilight of the seventies" and into the eighties. Self-sufficiency becomes, in truth, the only third party payment plan which will work. We are already seeing massive fall-out from an over-concentration of reliance on federal, state and local governmental units. Energy, ecology, jobs, sewers, roads, and a host of other legitimate social needs are all of much higher priority to government agencies. State of the union messages, congressional utterances--all will not produce adequate funding in the face of massive social unrest in areas which are not considered as "deviant" as alcohol or other drug misuse.

Substance abuse programming--if it is to survive as a delivery system--must "go it" on its own or fall apart. Self-sustaining and energetic attention to fund-raising is the glue that can hold it together.

THE HELPING HANDS PROGRAM: A MODEL FOR THE REHABILITATION OF OLDER ALCOHOLICS

Janet Ashton Glassock, M.A.

Memorial Hospital Medical Center - Alcoholism Treatment and Education Center of Long Beach, California

Alcoholism among the elderly is increasing. As the prohibition youth of the 1930's turn 65 at the rate of over 1000 a day, a much higher number of people are entering old age with well established drinking habits. According to the National Council on Aging, there are 31 million men and women over 60 in the United States. Based on the information that one out of 10 persons is an alcoholic, there are probably well over three million alcoholics over the age of 60.

Current research and observations in regard to substance abuse among the elderly has been thoroughly reviewed in the published Joint Hearings before the Sub-Committee on Aging and Sub-Committee on Labor and Public Welfare, United States Senate Ninety-Fourth Congress Second Session on Examination of the Problems of Alcohol and Drug Abuse Among the Elderly, June 7, 1976.

Examination of the problem reveals that despite the increase in numbers, older problem drinkers are virtually ignored because many are retired and have passed out of the mainstream of society. Their drinking goes largely unnoticed unless they are public inebriates, and few of them are. Elderly alcoholics are frequently uncommon in alcohol treatment centers because severe mental or physical impairments take them to inpatient Psychiatric, Neurologic, or Medical-Surgical centers when problem drinking reaches a point of crisis.

Older problem drinkers have less occasion to be brought face to face with their alcoholism than younger persons. Early detection and crisis prevention can be more difficult. They are more able to avoid identification through family problems, work impairment and driving arrests because they often live in a self-imposed exile from the world, feeling ashamed of themselves and alienated from their families and other potential support.

While signs of alcoholism are more obvious in a crisis situation subtle signs of alcoholism may include legal difficulties, chronic health problems, falls, depression, and housing problems. Questions from health professionals about eating habits, sleep, daily routine, and lack of or kinds of friends may reveal problem drinking patterns. Some older persons genuinely do not know that alcohol is their problem and are relieved to find they can be helped.

There are two types of alcoholism in old age: lifelong and late-life. The former used to result in death long before old age, but because of antibiotics, better nutritional care and hospital management, many alcoholics now live into the later years despite their drinking. Late-life alcoholics turn to alcohol in old age because of life changes, grief, depression, loneliness, boredom, or pain. They may have a life-long habit of drinking which becomes alcohol abuse as physiologic tolerance to alcohol decreases with age.

Older alcoholics seem to drink for reasons different from those of other age groups, and have been found to respond readily to treatment when their aging related needs are met.

The Second Special Report to the U.S. Congress on Alcohol and Health from the Secretary of Health, Education and Welfare in 1974, concluded that excessive drinking in the old is less commonly associated with deep-seated psychological problems. It appears to be more directly related to external factors concomitant with increasing age, such as the loss of a spouse and loneliness, retirement and boredom, loss of status and lower income. Among the causes that turn a middle-aged moderate drinker into an older problem drinker are role changes and crises.

Through therapeutic manipulation of external factors, the geriatric alcoholic can be successfully rehabilitated. According to Zimberg (1974), the alcoholics who develop their drinking problems later in life respond readily when they find a sympathetic ear for their problems and when they feel someone is concerned about them. Once their depression lifts, they discover that they do not need alcohol to adapt effectively to the stresses of aging.

HELPING HANDS

The need for identification, treatment and innovative programs for the physical and social rehabilitation of the older alcoholic stimulated the creation of the Helping Hands Program, in June 1976, as a supplement to the inpatient and outpatient units of the Alcoholism Treatment and Education Center at Memorial Hospital Medical Center of Long Beach, California, where the inpatient population over 60 averages about 15 percent, often runs 25 percent, and is clearly on the increase.

The purpose of the Helping Hands Program is to prevent the cyclical return of older alcoholics to the hospital by the development of new tools for finding new patterns of living such as new roles. Specifically, the goals are: 1) to help the older, recovering alcoholic develop a positive, creative sobriety; 2) to establish and maintain a continuing lifetime support system; 3) to develop and expand the channels between the Helping Hands Program and the Community Senior Services Agencies, Alcohol Treatment and Rehabilitation Facilities and other community resources; 4) to weave the Helping Hands outflow population back into the mainstream of society; and to create a flow of education, prevention, and outreach to the community.

Supporting the need for a community network of inter-agency cooperation between senior services and alcoholism services, the Director of Senior Citizen Affairs in Long Beach, has expressed that the growing problems of alcohol abuse by older people has been neglected. In Long Beach, which has a population of over 72,000 older people, this neglect is resulting in growing harm to large numbers of older people, their families, and the community. It is hoped that this additional community resource will increase the treatability and prognosis of elderly alcohol abuses and reduce the need for hospitalization.

The Helping Hands group is an outpatient group open to any older alcoholic without charge. It meets three times a week--twice a week at Memorial and once a week at the Memorial West Alumni and Friends Clubhouse for recovered alcoholics and their families and friends. Most of the group members are referred from the inpatient Alcoholism Treatment Program at Memorial where there is a special group and counselor for older patients who are retired and have problems associated with aging as well as their alcoholism.

Within this inpatient population treatment is focused not only on the alcoholism and awareness of it, but emphasis in made on environmental and family problems, living conditions, and assessment of whether the patients are able to return home and are able to care for themselves, or if they need the aid of community resources such as Homemaker Services or nutrition programs. Discharge planning includes referral to AA, Women for Sobriety, Senior or other community activities, and Helping Hands. Referrals to the Helping Hands Program come from other alcohol rehabilitation and Senior facilities also.

On any given morning, the Helping Hands population consists of 4 to 12 persons who are outpatients, inpatients who are able to attend, and older persons of long or short-term sobriety from the community. The people come because they want to come: they like the small discussion group where they can talk about retirement difficulties, economic and spousal problems, loneliness, new ideas for using time, bus schedules, social security, taxes, and community resources. Discussion centers about living in the here and now, purposes for living, and making the most of each day. The group is primarily oriented toward reprogramming and replanning their lives, setting new goals, learning new tools, and ways of achieving the goals.

One 70 year old male member, a retired journalist who has always wanted to write feature stories instead of editing news reports found a way to channel his old skills into a new pleasurable pursuit. He is now writing the Newsletter for the Alumni and Friends of Memorial Hospital Alcohol Treatment and Education Center. He also is the facilitator of the Wednesday morning Helping Hands group at the MWA Clubhouse.

The discussion modality varies according to the make-up of the group, but basically it is a mixture of an AA Step Study approach, Women for Sobriety philosophy, and a cognitive approach in a peer group setting.

The most important function of the Helping Hands group is the sharing of common aging and alcoholism-related problems, ways to stay sober, and the rewards of sobriety. During the early period of sobriety, it is sometimes difficult for participants to communicate, even with each other, but once they begin to feel comfortable they open up, gain confidence, and then eventually reach out in to the community and help others. In addition to Helping Hands participation they find AA or Women for Sobriety groups where they are comfortable. They become more at ease with people other than alcoholics and learn to improve their relations at home. The group has been enhanced by the volunteer work of original members. They make telephone calls to the newly sober, do peer counseling, and follow-up. One, as mentioned, facilitates a group.

Throughout the school year Gerontology students from California State University Dominguez Hills assist with the Helping Hands group and the follow-up as a part of their field placement. They also act as group facilitators, help organize a monthly Pot-Luck lunch in conjunction with a group meeting, and arrange for and provide transportation. Very often the participants are not disabled enough to be eligible for funded transportation or escort services, yet they are not quite strong enough to take a public bus. Transportation development and maintenance is one of the major challenges of the Helping Hands program.

RESULTS

We have found that discharge planning, reentry support and continuing lifetime follow-up is an essential part of the recovery process of the older alcoholic. In a review of about 200 patients who have been referred to Helping Hands, we have found that among those who responded to follow-up and participated in the group, the quality of sobriety has been enhanced. Many have long-standing recoveries and have been able to avoid hospitalization.

Although some have had to be rehospitalized, while they are sober they are able to live a much fuller life than before. When they have returned to drinking they usually drink only for a short time ranging from a couple of drinks to a couple of weeks and then get *themselves* back into the hospital. This is attributed to the building block of sobriety that they have developed during their time with Helping Hands. When a relapse occurs, they receive positive, honest, non-judgemental support from the group, focusing discussion on how to avoid sliding into another relapse. The remorse and guilt common at this time is lessened because of the support of the group, and the physical and emotional recovery have often been accelerated.

DISCUSSION

Helping Hands seems to work because it is a small, intimate discussion group. Everyone has a chance to talk. It is a place to become accustomed to sobriety, to interact with others, to develop self-esteem, and to experiment with the realities of life and growing older. It is a place to *grow* among peers who care and understand. Because participants may range from 50 to 95, we focus on adjusting to life changes, not old age itself.

A positive factor in the recovery prognosis of older alcoholics is that they often have the solution to their problem built within. If there is something from a person's past to build upon and redirect, half the battle is won. As with the former journalist, the years of productivity and peer group identification which are the losses being experienced, are the very factors that can be restored and redirected to contribute to the recovery process.

We have found that the families of the older patients have responded from one extreme to the other: from being anxious to work with counselors in discharge re-entry planning, to those who don't care or have given up on the older alcoholic. Some live too far away to be of on-going assistance. Often the family is relieved to find there is help--family counseling and Alanon--and that they have someone with whom to share "their problem".

At the Special Senate Sub-Committee Hearings concerning both Alcoholism and the Aging, Senator Eagleton (D-Mo.) asked, "Do elderly abusers require specialized programs, and if so, should these programs be provided at Alcoholism and Drug Abuse Centers, or should they be a part of Senior Citizen Centers?" Alcoholism rehabilitation facilities afford the anonymity that is often desired--the security and protectiveness of being with other alcoholics during early sobriety. Yet, with increasing education and acceptance of alcoholism as a treatable disease and as America's number one drug abuse problem, perhaps group meetings and counseling will become an accepted part of gerontology service facilities. Both alcoholism and senior facilities must be aware of the problem and work together in the rehabilitation of the older alcoholic.

Referring the older alcoholic to appropriate help--inpatient, outpatient, or a self help group--could make a difference in the treatability prognosis. Likewise, appropriate discharge referral back to community senior services and follow up are important for the long term recovery process.

Experience with the Helping Hands Program has shown that the development of new relationships, repairing of family relationships, new life goals, and peer group identification with other older recovered alcoholics in a small discussion group setting increases the potential for and enhances the quality of, sobriety in the late life years.

REFERENCES

Alcohol and Drug Abuse Among the Elderly. 1976. Joint Hearings before the Sub-Committee on Aging and Sub-Committee on Alcoholism and Narcotics of the Committee on Labor and Public Welfare, United States Senate, Ninety-fourth Congress, second session on examination of the problems of alcohol and drug abuse among the elderly. U.S. Government Printing Office, Washington, D.C., June, 1976.

Second Special Report to the U.S. Congress on Alcohol and Health from the Secretary of Health, Education, and Welfare. June, 1974.

Zimberg, S., M.D. 1974. *The Gerontologist*. 14: 221-224.

A STRESS MODEL FOR DEALING WITH THE ELDERLY SUBSTANCE ABUSER

Earl R. Gardner, Ph.D., Richard C.W. Hall, M.D., and
Sondra K. Stickney, R.N.C.

The University of Texas Medical School at Houston
The Medical College of Wisconsin, Milwaukee
Hermann Hospital, Houston

INTRODUCTION

A review of mental health prevention programs across the country revealed that less than 10% were directed to the elderly and that only 5% dealt with substance abuse in geriatrics. The majority of existing programs rely on one of three primary approaches: 1) self awareness; 2) consumer education; or 3) community referral following recognition of a defined problem.

This prospective study was designed to survey a large geriatric population (i.e., N>600) for the occurrence of: significant psychiatric illness necessitating referral; self reported drug abuse at follow-up and client reported acceptance and improvement following completion of an outreach program which emphasized client responsibility for change and avoided the stigmatization of psychiatric labeling.

RATIONALE FOR THE LIFE STRESS MODEL

Any outreach model which is to be effective must be predicated on a theorem having applicability to the general population which encourages introspection and the analysis of problems in a non-stigmatizing, non-threatening manner. It should provide a method for change, emphasize the affected individual's strengths and facilitate change in an egosyntonic fashion. The failure of the programs mentioned above may relate to their deficiencies in meeting these criteria.

The life stress model provides an outreach system which defines problems as normal, minimizes individual psychodynamics and consequently individual vulnerability, suggests that stress, rather than the reaction to it, is the target of intervention and emphasizes that change is the lynch pin of normal adaptation and is therefore desirable.

This model emphasizes that stress is a dynamic state within each person which continually demands adaptation. (Wolfe, 1968). It assumes that each person has a unique capacity for dealing with stress and that a certain amount of stress (eustress) must be present for the person to function optimally. Too little stress produces inactivity while too much may incapacitate both physically and mentally (Cherry, 1978).

Humans react to both current actual stressors as well as those which are symbolic or anticipated. This is particularly true of the geriatric client who must consider such things as loss of social prestige and role function, diminished future income, loss of spouse, loss of friends, loss of physical health, loss of sexual function, loss of mental functioning, increased dependancy resulting in diminished self concept, possible institutionalization, becoming a victim of crime and their own death (Richitelli, 1972; Clark, 1967; Nordlicht, 1975; Brand, 1974).

In a study of critical life events for individuals over 60 years of age: 22% had lost a close friend; 28% lost a family member or spouse; 13% were forced to go on public assistance; 26% had retired; and 53% had begun to live alone during the preceding year (Smith, 1971).

The stress model is based on the assumption that when stress has exceeded optimal levels and disrupted psychic functioning, the reduction of only one or two stressors can restore homeostasis and psychic equilibrium. If on the other hand stress persists, it is likely to be followed by episodes of clinically significant depression, (Paykel, 1969; Hudgens, 1967) anxiety or the development of physical symptoms. The presence of such anxiety and depression increases the probability of substance abuse following attempts to self medicate.

METHOD

Six hundred and twenty nine individuals of 60 plus years of age were evaluated using an abbreviated format of the Life Stress Schedule developed by Holmes and Rahe and a life history questionnaire which defined demographic, medical, economic, social and drug use data. The subjects were members of senior citizens groups who voluntarily agreed to participate in a data gathering and followup workshop entitled "Stress in the Older American." Each group was seen twice for a period of two hours. A followup questionnaire was administered 6 months after the workshop.

At the initial meeting scales were distributed and each participant scored their Life Stress Schedule and retained a copy. Assistance was available to anyone having difficulty reading or writing. A twenty minute lecture defining the concept of stress and the model for analyzing stressors (Table I) was provided. This was followed by small group discussion (one facilitator with six subjects).

In the small groups, participants were asked to identify their life stressors and define whether each stress was externally, interpersonally or self imposed and whether each could be reduced, modified or eliminated. They were asked to prioritize stressors for elimination and develop long and short term goals for stress reduction. A stress reduction program was written for each member and became part of the data base.

One week after the initial session the second meeting was held. It began with a half hour didactic lecture on specific coping models for stress which emphasized techniques for stress avoidance, stress management, stress resistance and stress reaction management. This was again followed by small group work sessions.

DATA

This report presents only that portion of the data obtained which is directly pertinent to the area of substance abuse. Six hundred and twenty-nine participants (32% male--68% female) with a mean age of 67.3 years completed both parts of the program. Eighty-three percent (83%) had at least one chronic illness about which they were concerned and saw as potentially debilitating. Thirty percent (30%) had lost a close friend within the past year, 32% a family member or spouse, 19% were forced to take some form of public assistance, 35% were fully retired and 60% had begun living alone during the preceding year. Regular use of alcohol was reported by 42% of the subjects. Fifty-eight percent (58%) reported regular use of over-the-counter mood or sleep medication and 39% were using prescribed mood or sleeping medication.

Of the 629 original participants, 42% completed the six month followup questionnaire. Eighty percent (80%) of the respondents reported the stress workshop had significantly effected, for the better, their life style. Five percent (5%) of the total population were referred to community, social or medical agencies following initial contact. Fifteen individuals (2%) were specifically referred for psychiatric evaluation and gladly accepted such referral. Thirty-two additional individuals (5%) indicated at followup that they had self referred to a psychiatrist or mental health agency as a

Table I.

STRESS ANALYSIS STRESSORS	CHANGEABLE	NOT CHANGEABLE	EXTERNAL SOCIAL	INTERPERSONAL	SELF IMPOSED	DECREASED	MODIFIED	ELIMINATED	SHORT TERM	LONG TERM	PRIORITY
1.											
2.											
3.											
4.											
5.											
6.											

direct result of the workshop. Major psychiatric problems defined by these groups were: depression (60%); alcoholism (41%); drug abuse (38%) and organic brain syndrome (12%).

At the first workshop 8% of the subjects defined the use of over-the-counter or prescription medications as a major problem, while on followup 38% of the respondents reported such abuse as significant.

DISCUSSION

The cornerstone of any prevention program is participant acceptance and accessiblity. The geriatric life stress seminar seems to be particularly useful in facilitating both. It has achieved a high degree of accessibility by using well established and accepted senior citizen groups as the vehicle through which the elderly are contacted. A high success rate has been achieved, in that 80% of participants at followup indicated a change in life style in a positive direction. This may be accounted for by the avoidance of the stigmatization of psychiatric labels as evidenced by 90% of all referrals from this program following through with initial contact.

The program is cost efficient in terms of training and manpower and is capable of reaching a large percentage of the needy population. It established a contact point for entry into the health care system should future difficulties arise and provides existing agencies with valuable and immediate feedback.

Depression, drug and alcohol abuse were the primary problems seen in this elderly population. The incidence of such problems and their severity was much higher than generally reported in the literature.

Although one must always be aware of the Hawthorn Effect, the program was thought to have a significant impact upon the participants, as illustrated by the following representative case history.

Bill, a 61 year old white, married male had recently retired from the Air Force after 35 years of service. His wife had glaucoma and was crippled with arthritis. They had moved south becuase of her condition and her need for a warm climate. Bill spent all of his time with his wife, and as a couple since moving, they had developed no social contacts. He had discontinued his hobbies of fishing and bowling which had been life long interests. While in the service he had been a "moderate" drinker but since retirement his drinking had "gotten out of hand."

During the workshop, Bill indicated his greatest stressors were: physical illness of spouse, lack of social relationships, impaired interests and excesssive alochol consumption. He saw these stressors as changeable and self-imposed except for his wife's condition which was something he must learn to live with and that quite possibly, it would get worse. The stressors he chose to work on short-term, were lack of interests and social isolation. A member of his work group, who also felt isolated and lonely, offered to spend time with his wife so that Bill could have some time of his own. He began developing his old hobbies of fishing and bowling. Bill was able to work on outboard engines which led to many contacts for fishing companions as well as supplementing his income. After 3 months, he decided his drinking was still out of hand and began attending AA meetings. His wife joined Ala-non and they began developing social relationships.

At followup Bill had been dry for 3 months, was involved in many activities, had developed a few close friends and reported that he and his wife had much improved social relationships and were considering entering marital counseling. He stated that he could not "find enough time to do all the things I want to do."

Life stress analysis is a valuable adjunct to other methods of dealing with substance abuse in the elderly since it allows them to use their own strengths to systematically evaluate and to change their abuse pattern while serving to improve the overall quality of their lives. To paraphrase Teilhard de Chardin, "The future is in the hands of those who can give . . . valid reason's to live and hope."

REFERENCES

Brand, F.N. and Smith, R.T. Life adjustment and relocation of the elderly. *J. of Gerontology*. 29(3): 336-340.

Cherry, L. 1978. On the real benefits of Eustress. *Psychology Today day*. 11(30): 60-70.

Clark, M. and Anderson, B.G. 1967. Phenomenological description of aging problems. *Culture and Aging*. Illinois: Charles Thomas, Inc. 60-67.

Hudgens, R. et al. 1967. Life events and onset of primary affective disorders. *Arch. Gen. Psych*. 16: 134-145.

Nordlicht, S. 1975. Stress, aging and mental health. *N.Y. State J. of Med*. 75: 2135-2137.

Paykel, E. et al. 1969. Life events and depression. *Arch.·Gen. Psych*. 21: 753-760.

Riccitelli, M.L. 1972. Vitamin C therapy in geriatric practice. *J. of the Amer. Geriatrics. Soc.* 21(1): 34-42.

Smith, W.G. 1971. Critical life-events and prevention strategies in mental health. *Arch. Gen. Psych.* 25: 103-109.

Wolff, S. and Goodwill, H. 1968. The nature of stress for man. *Stress and Disease.* Illinois: Charles C. Thomas, Inc. 3-12.

THE CONFRONTATION-SENSITIVITY (C-S) GROUP

Robert V. Frye, M.S., Mark Friend Hammer, and George Burke III

University of Colorado Therapeutic Community, PEER I

The confrontation group or encounter is basic to most treatment in therapeutic communites for drug abusers. Other types of groups and encounters have also been used with some success in treating people with drug problems. At the University of Colorado Therapuetic Community (PEER I), the staff and senior residents have attempted to bring together a variety of techniques into one comprehensive group, called The Confrontation-Sensitivity (C-S) Group. It may appear evident to the observer that the C-S Group is an eclectic system drawing from such diverse sources as The Synanon and Mendocino Game (Yablonsky, 1965; Brewster and Garrigues, 1974), Daytop Village Encounters, Marathon Group Encounters, Gestalt "Primitive Explosion", Sensitivity Training (Shaffer and Galinsky, 1974), and Reality Therapy (Glasser, 1965).

Confrontation-Sensitivity operates with a social scheme and under a set of social conditions different from those of standard, professional group therapy. The group is flexible and versatile and often functions to make the participant aware of the demands of the environment and the necessity of conforming to those demands. The C-S Group may also assist in setting up altruistic partnerships such as friendship and bonding because it can assess trust worthiness as well as trust unworthiness (Frye, 1978).

REALITY THERAPY INFLUENCE

Reality Therapy is based on the premise that people have problems because they are unable to fulfill their needs in a realistic way and have taken some less realistic way in their unsuccessful attempts to do so. People are encouraged to take responsibility for their own mental health and to take the responsibility to become well (Glasser, 1965). The C-S Group uses the reality approach because it gets totally involved with the individual group member in an attempt to get him to face reality. When confronted with reality by the group, the group member is forced again and again to decide whether or not he wishes to take the responsible path. Reality is often painful and harsh, and it may be dangerous, but it changes slowly. As Glasser (1965) stated, "All any man can hope to do is to struggle with it (reality) in a responsible way by doing right and enjoying the pleasure or suffering the pain that may follow."

THE GAME INFLUENCE

The confrontation part of the C-S Group may be compared to the Synanon and Mendicino Game. Yablonsky (1965) stated concerning the Synanon Game, "No holds or statements are barred from the group effort at truth-seeking about problem situations, feelings, and emotions of each and all members of the group." At times the confrontation process in the C-S Group may sound like a verbal street fight. The "game" part of the C-S Group may take a variety of forms, depending on the purpose and the level of experience of the persons in the group. Some types of confrontation may have a serious, very heavy tone; and some may be noisy and angry, especially if less experienced group members are confronting. Many times the confrontation may seem absurd and hilarious (Brewster and Garrigues, 1974). The Game, according to Enright (1970) is not particularly designed to solve problems but rather to demolish them.

GESTALT THERAPY INFLUENCE

Gestalt Therapy is used within the Confrontation-Sensitivity model and brings with it a "here and now" focus. Gestalt Therapy is used in the "pure way", applying Fritz Perls' style, in which the Group Master or his designate works on a one-to-one basis with individual clients (Perls, Hefferline and Goodman, 1951). The task of the Group Master in this instance is to direct the participant to focus on his own internal self, and the use of special dramatic interventions helps bring out avoided issues. The rest of the group participates, though not directly active, by the process of identification or the group may be called in to assist in the intervention in a structured way.

Gestalt Therapy is used sparingly in C-S to protect the client against becoming too dependent on a therapist-client relationship and the objective of Gestalt Therapy is to spark the individual into future participation in the C-S Group. Gestalt interventions are carefully planned prior to the group. Consideration of the client's drug abuse history (hard or soft core) is basic to the determination of what type of intervention might be appropriate and the type of intervention strongly depends on how high or low his level of trust would be with the Group Master (Burke, Hammer and Milkman, 1978).

The purpose of employing Gestalt Therapy in C-S is to change the pace in the group. It adds intrigue and excitement to the group and is needed because of the past life style of the drug abuser (Frosh and Milkman, 1977). It offers a momentary excitement substitute and provides a variety of potent happenings. Gestalt Therapy offers an experiential model as opposed to the more cerebral, historically-focused dialogue between analyst and patient. The role of the Group Master is to assist the participant to open up and to trust that he can learn from the present.

SENSITIVITY TRAINING INFLUENCE

Another important factor of the Confrontation Sensitivity (C-S) Group is that it draws from the Sensitivity Movement of the early 1960's. T. Groups or Sensitivity Training may be briefly explained as an intensive effort at interpersonal self study, and an attempt to learn from the raw experience (of member participation in a group) how to improve interpersonal skills (Shaffer and Balinsky, 1974). The T. Group was designed to help counteract heirarchically structured and extremely authoritarian systems and the therapeutic community is just that. People learn genuine concern for others from this type of group, and therefore they must incorporate the authoritarian structure of the therapeutic program with sensitivity for themselves and others.

Participants of the group learn not to look for and depend on the so-called "experts" for their sole input of information. People realize, as the length of time in group extends, they have *valid* ideas and observations of themselves and others. The Group Master of the C-S Group assumes the role of the trainer in a T. Group. He doesn't give in to the demands from the group members to assume authoritative control but he will, at times, intervene when a participant's feedback becomes hostile, ill timed or inaccurate. The Group Master is also a participant of the group and therefore needs openness, concern for other's feelings and attention to the behavior to the here and now.

STRUCTURE

The new group member is taught the techniques and rules that govern the C-S Group. The basic rules are taken from the Mendocino Game with some basic modification by PEER I in order to make the technique more eclectic and flexible. The basic set-up for the C-S Group is a room with chairs for between seven to fifteen members who sit in a circle. The chairs should not be too close to each other as it is usually forbidden to touch another person in a C-S Group. Consideration should be given to the location as it is usually quite noisy and might be disruptive to adjacent activities (Brewster and Garrigues, 1974). The C-S Group usually last about two to three hours, but marathons of up to forty-eight hours are not uncommon.

The Group Master's primary function is to start and stop the group and to make sure that the rules are obeyed. He may, at times, assume the role of a T. Group trainer (model role-model behavior), and he may, if skilled enough, direct special interventions. He may function at times as a "Game Master." The Group Master's control may be questioned by the group members and group pressure may sometimes be allowed to take control away from him. The group is highly structured with specific instructions for its use. The following concept is the basic document concerning the Confrontation-Sensitivity (C-S) Group.

CONFRONTATION-SENSITIVITY (C-S) GROUP

I. Rules

A. No violence or threats of physical violence.
B. No chemicals.

II. What is Confrontation-Sensitivity

A. Confrontation is the tool for breaking the criminal mask.
B. It is the safest place in the world.
C. Confrontation is a verbal street fight.
D. The communications are more direct.
E. You can have fun.
F. You have an opportunity to work on your fears and emotional hang-ups through sensitivity encounters.

III. How to Use C-S

A. You work from the outside in.
B. You may request a sensitivity encounter to work on a special problem.
C. By telling the Group Master, "I want to work", you request a special encounter and the regular policies may be suspended for some members. (Decision is up to the Group Master).

IV. Flexible Policies

A. Talk to one person at a time.
B. Support the probe.
C. Spend at least 20 minutes on each person.
D. Break all contracts, verbal and non-verbal.
E. Only one person may leave the room at any given time.

V. Unflexible Policies

A. Don't get into anyone's face.
B. Don't touch anyone.
C. Don't bait anyone.
D. Don't ask to get out of your seat during an encounter or an intensive probe.
E. Don't ask to get out of your seat if you're in the hot-seat.
F. When getting up, first ask the staff; then tell the person on your right, and then the person on your left, loud enough for the whole group to hear you.
G. When in the hot-seat, you may only have a cigarette in your hand.
H. The Group Master starts and stops the group and is generally "in charge."

VI. Tools and Techniques of C-S

A. Catharsis
B. Projection
C. Data running
D. Role-Playing
E. Righteous indignation
F. Engrossment.
G. Belittlement
H. Humor
I. Carom shot, direct and indirect, shotgun
J. Intervention

VII. What You Can Do in the Hot-Seat

A. Remain Silent
B. Give it the barest
C. Listen
D. Defend
E. Dump
F. Go crazy
G. Cop out
H. Tell the truth

The C-S Group may at times be expressed as a Group Marathon of up to forty-eight hours in length. As Stroller (1968) has pointed out, the Marathon Group heightens the experiential nature of what occurs, and may produce singular effects. He points out that intense levels of emotion and involvement are attained which permit much more rapid learning about one's self than generally occurs. The University of Colorado Therapeutic Community, PEER I, has developed a video-tape showing parts of a forty-eight hour C-S Marathon and has developed the following marathon concept to teach the participants the purpose of the marathon:

CONFRONTATION-SENSITIVITY (C-S) GROUP MARATHON

I. What is a Marathon?

A marathon is an endurance contest structured to help you

strip away your criminal-mask or defensive armor.

II. What are the Rules for a Marathon?

Basically the rules of a marathon are the same as a C-S Group with a few necessary changes:

A. There is no sleeping or napping.
B. There is no smoking except at the discretion of the PEER I Staff.
C. A variety of groups may be incorporated in the marathon.
D. The staff, and only the staff, will decide when it shall be terminated. *No time pieces, watches or clocks are allowed in the G.M.*
E. Breaks will be taken when the staff deems it necessary.

III. What Should a Marathon Accomplish?

A. The most obvious accomplishment is that it should aid every *participating* member to grow.
B. It should encourage a bond of brotherhood and sisterhood by constant and prolonged exposure to each other.
C. It aids members to release feelings of anxiety and frustration.
D. It is a real test of a member's stamina and strength.
E. It should encourage real, not faked, feeling and emotions.
F. If a marathon is properly supported, a person should leave at the termination of that marathon with the following:

    1. A sense of feeling "cleansed."
    2. A feeling of being pleasantly exhausted.
    3. A sense of real accomplishment.

IV. The C-S Group Marathon begins with several hours of abstract music to encourage the participants to start feeling emotion.

SUMMARY

The Confrontation-Sensitivity (C-S) Group appears to be especially effective with substance abuse clients in a therapeutic community because it provides a balance between confrontation and sensitivity and a flexibility to meet the individual needs of the group members. The process may involve confrontation to break through the group member's criminal mask or defensive armor, followed by group sensitivity to reinforce the positive changes that occurred and to provide closure. Confrontation-Sensitivity has a definite structure and draws from Gestalt Therapy, Reality Therapy, Sensitivity Training and The Game. It may be expressed as a marathon and provides opportunities for intervention and encounters.

REFERENCES

Brewster, J.T. and Garrigues, C. 1974. The Mendocino Game: Rules, Policies, Modes and Techniques. *DF*. 4(1): 15-29.

Burke, G.H., Hammer, M.F. and Milkman, H.B. 1978. Co-Treatment of Groups of Differentiated Substance Abusers. NDAC proceedings. (conference paper).

Enright, J. 1970. Synanon: A Challenge to Middle Class Views of Mental Health. In Adelson and Kalis (eds.), *Community Psychology and Mental Health*. Chandler Publishing.

Frosh, W.A. and Milkman, H. 1977. Ego Functions in Drug Users. *NIDA Research Monograph*. 12: 142-156.

Frye, R.V. 1978. Sociobiology and The Therapeutic Community. NDAC proceedings. (conference paper).

Glasser, W. 1965. *Reality Therapy*. New York: Harper & Row.

Perls, F., Hefferline, R.F. and Goodman, P. 1951. *Gestalt Therapy*. New York: Dell.

Schaffer, J.B.P. and Galinsky, M.D. 1974. *Models of Group Therapy and Sensitivity Training*. Englewood Cliffs, New Jersey: Prentice-Hall.

Stroller, F.H. 1968. Marathon Group Therapy. In Gazda, G.M. (ed.), *Innovations to Group Psychotherapy*. Springfield: Thomas.

Yablonsky, L. 1965. *Synanon: The Tunnel Back*. Baltimore: Penguin.

HOW DID WE GET WHERE WE AREN'T? A PERSPECTIVE ON NURSING IN OUT-PATIENT DRUG TREATMENT

Marilyn Fisher, R.N. and Kathleen Egan, R.N.

City of Boston Drug Treatment Program

The City of Boston Drug Treatment Program began providing services to drug dependent individuals in 1970, the heyday of federally sponsored methadone programs. The eight-year history of the program reads like a chronicle of change. Satellite clinics were opened. A no take-home policy was instituted. Community pressure resulted in a clinic closing. Funding was lost for the inpatient unit. A new Program Administraiton was created. A vocational rehabilitation program was developed. A recreational therapy program was added. A work experience program was made available for clients. The entries continue on ad infinitum, each change having obvious impact on treatment available to the addict population of Boston. Although nursing has been a consistent presence, its concomitant history is a documentation of passivity; i.e., how these changes affected nursing, rather than how changes within nursing affected the program. It is this phenomenon that we set out to investigate, to determine "How we got where we aren't." and ultimately to propose a means of attaining the role we think nursing should play in substance abuse treatment.

Sharing our understanding of the current status of nursing in the City of Boston Drug Treatment Program (CBDTP) necessitates historically tracing the various influences.

The program began under the auspices of the psychiatry service of Boston City Hospital. The original clinic was located within the outpatient department, with a satellite facility operating in East Boston, a neighborhood five miles away from the main hospital. Nurses working in these units were under the direction of the Psychiatric Nursing Supervisor, who was primarily located on the inpatient psychiatric unit. This unit had been moved to a facility five miles from the main complex in the opposite direction of East Boston. A third methadone clinic was opened in the Brighton neighborhood, again five miles away in a distinctly different direction. The inpatient drug treatment unit and a fourth clinic were later opened on the same grounds as the psychiatric unit. Geography thus influenced the initial phase of nursing's involvement in the program. In essence, the nurses in each unit functioned in isolation of their counterparts in the other clinics. The supervisor made rounds once a week which involved, at most, one hour in each clinic. An hour spent primarily in administrative issues. Most crucial in this design was that so little direction came from the nursing leadership to assist the nurses in defining their role and developing skills in this new speciality area of practice. What resulted was a loss of a strong nursing identity. The phenomenon emerged of a striving toward a collective identity with the various disciplines that comprised the staff. Thereby, the idea of an expanded role was interpreted to mean assuming a more similar role. This meant being a part-time nurse when dispensing medication, and a part-time "counselor" when seeing clients. Functionally, this enabled the nurse to move beyond dispensing methadone, but did not in any way address the issue of how nursing practice was being compromised and clients not receiving the fullest range of nursing interventions. An example of this is evidenced by a discussion with a nurse about a particular client who was pregnant. This woman had multiple medical and psychiatric problems, and was not consistent in receiving prenatal care. After identifying these needs and appropriate nursing interventions, the nurse indicated she would check with the client's counselor to see if it was okay with him that she implement this plan. It appeared to be a critical situation when nurses had to ask permisison to give good nursing care! Their loss of identity resulted in a loss of awareness of their professional responsibility for which they would be held accountable, both by clients and the nursing department.

Although some nurses did see individuals for "counseling", it was contingent on their own initiative and the discretion of the Clinic Director. Their role had not been clearly defined beyond the responsibility for dispensing medication. The nurse in this setting was in an ambiguous situation. She was accountable to a nursing department vis-a-vis a supervisor with whom she had minimal contact.[1] Maximal contact was with a Clinic Director, who provided the bulk of direction to her work, but to whom she had no clear line of accountability. In response to this organizational structure, nurses continued to be perceived as separate or at least unique to the rest

---

1 For purely stylistic purposes single gender references will be used throughout the paper. The feminine gender was chosen as the nurses in the experience being described have always been women.

of the staff. The various degrees of this perception and its consequences depended on the clinic, its staff, and most importantly, the attitude of any particular Clinic Director.

The two major influences then on the early development of nursing were 1) lack of consistent, committed nursing leadership which led to 2) definitions of role and ongoing direction being determined primarily by medicine and administration.

A change in nursing administration's involvement five and one half years ago began a second phase in the development of nursing. A supervisory position was created for the Drug Program. However, the earlier influences had created a model not easily modified. A nursing identity has gradually been restored, aided significantly by an increased number of nurses due to new clinics. This facilitated a sense of a peer group amongst nurses always in a minority within each clinic. Nurses spending time in each other's clinics and meeting as a group weekly supported this process of change.

As the nurses continued to function in their fragmented role, new frustrations emerged from their restored identity. Their preparation for their role was significantly different from that of other members of the staff. Nursing eductionn had provided them with a disciplined, comprehensive approach to meeting client's needs. They were able to view client care in broad perspective, understanding the need for collaborative efforts from the entire staff and recognizing the essential aspects of treatment that went on outside the weekly sessions between client and therapist. The clinic setting could be viewed analogously to the traditional ward setting where most had received their major preparation. In such a setting the nursing staff provided individual care while coordinating the services of various disciplines. However, in the clinic setting they were receiving minimal cooperation in their attempts to replicate this collaborative comprehensive model. Their input did not yield the desired outcome. The root source of this new frustration was that historically nursing had given up (or never exercised) the control of its professional practice and, in general, had assumed a powerless position within the clinic hierarchy.

Having acknowledged this evolution of nursing's dilemma, and being confronted with the consistent frustration and dissatisfaction it yielded, we began to identify ways in which we believed nursing could modify its situation and assume a self-directed role in drug treatment.

Although the exact historical development may differ in parallel nursing situations, we thought it could be useful to share the strategies of intervention utilized to facilitate this movement; our aim being to articulate a role for nursing that would be not only consistent from nurse to nurse, but would be in accordance with national guidelines for nursing practice.

While the role of the nurse is frequently discussed amongst peers and deemed problematic, there is a paucity of available literature. It was our decision, therefore, to utilize standards for Mental Health Nursing Practice of the American Nurses Association as guidelines for the development and implementation of the nursing role within City of Boston Drug Treatment Program.

By educational preparation, the professional nurse addresses his/herself to the bio-psychosocial needs of individuals, families, groups and communities, facilitating movement toward the attainment of health vis-a-vis assessment, intervention, and evaluation of outcome. Given this base of knowledge, it is then both cost-effective and professionally sound to emphasize and develop a role which maximally utilizes these psychotherapeutic skills. Accordingly, our initial strategy of intervention was to reorganize the nursing component in the following three ways: 1) revision of current job descriptions for professional nursing, 2) development of a job description for technical level nursing and 3) implementation of these two levels in the existing nursing positions.

Both these job descriptions evolved out of discussion with the existing nursing staff about their perception of their role and the modification they saw as appropriate and desirable.

Faced with significant staff turnover during the summer months (1977), we designed our proposal before hiring new personnel. This enabled us to move more rapidly in first stage implementation, as we had existing vacancies into which we hired the Licensed Practical Nurses. Our intention was never to cause anyone to lose their job; rather to implement the two level system by attrition.

As the LPN entered the system, further interventions were made to facilitate the shift in the role of the professional nurse. A series of staff development programs were introduced and supervision provided by a nurse consultant. The aims of the consultant were four-fold: 1) to assess the psychotherapeutic skills of the professional nurse group, 2) to identify learning needs, 3) to provide ongoing clinical supervision of client caseloads, and 4) to provide additional nursing leadership and role modeling to augment individual role development. This last aim was seen as pivotal in clarifying the definition and scope of nursing practice versus the "part-time nurse, part-time counselor," identity previously adopted by the nursing staff.

Thus far, we have focused on the specific influences within our program to discuss what we believe to be a prevalent phenomenon experienced by nurses in substance abuse programs. There are additional variables which are perhaps more universal in their influence. These include 1) the lack of involvement by nurse educators, and 2) nursing's identity as an historically female-oriented profession, carrying with it the dilemmas of women in recent society.

This latter variable has emerged as a major impediment to realizing the fullest impact of the strategies already implemented. Having created a viable structure for role change, we recognized that we had superimposed a belief about nursing which, although acknowledged as valid by the staff nurses, was not the popularly accepted view. The women's issue surfaces as it is inextricably tied to the status of the nursing profession overall. As such, consciousness raising as utilized by feminist organizations is a highly applicable and useful vehicle for change. The need exists to foster positive association with the nursing identity, which in current society requires a positive identity as women. We have seen within our program a microcosm of the resistance and negative responses women have experienced when they began to strengthen their collective voice.

A strong negative connotation often accompanies the reference to "the nurses", in lieu of referring to the individuals by name. This is even used in direct address, i.e., "Nurse: vs. Janet, Norma, Sheila, or whoever. How reminiscent of the depersonalizing tactics some men have employed to thwart the individualism of women within our society. The ostracism this reflects can be a discomforting experience and, when effective, leads to the abandonment of the struggle to achieve rightful recognition. It is the very dynamic which contributed to the dilemma we identified in the beginning.

Assertiveness Training is another intervention of which women have found it useful to avail themselves. This is most pertinent to the nursing situation, in which, as we mentioned earlier, nurses generally experienced a sense of powerlessness to assume the role they have defined for themselves, or effect changes in the systems of which they are a part. All too often this sense is born out of a self-fulfilling prophecy accomplished by their passivity.

A demonstration of this was observed in the early phase of our project. In a clinic with a temporary shortage of office space, an office had been secured for the nurses separate from the dispensing area. The new clinical supervisor did not have an office, nor would have one for a month a more. Several weeks following his arrival it was noted that the nurses' office, by appearance, and utilization, would easily be identified as his. The nurses concurred that it was in fact now his office. Exploration of the facts revealed that the Clinic Director had not reassigned the space. The reality the group had to acknowledge was that their office had gradually and easily been usurped without any questioning or challenging on their part. They had quite passively accepted it. By this example it seems essential that each nurse become aware of her participation in maintaining the status quo; being controlled rather than being in control.

Nursing educators, as another group, have influenced the situation for nurses within substance abuse programs by similar omission and inaction.

Recruitment of qualified personnel prepared to deal with drug abuse has been problematic. Traditionally, nursing education departments have spent little time in the preparation of professional nurses for roles within substance abuse. Pamela Burkhalter (1976) states that:

> "Nursing students of all levels should receive instruction on drug abuse and appropriate nursing care interventions. In order for nursing students to gain experience in handling drug abuses, clinical components should be planned as intergral parts of drug abuse education."

Believing that the concept of nursing education and service need to be integrally related, we approached Boston State College School of Nursing to negotiate a clinical placement for students. Beginning in March, 1978, we have two senior nursing students doing a seven-week elective in drug abuse under the preceptorship of two of our staff nurses. Although this phase of intervention is in its infancy, we are hoping that this collaborative venture between education and service will prove useful to both components.

Ultimately, we would like to see an increase in both quality and quantity of substance abuse content within the curriculum and fuller utilization of our program as a clinical placement for students interested in pursuing careers in substance abuse. Such an arrangement has the potential for augmenting the preparedness of nurses in several ways. There are certainly recruitment possibilities inherent in the student placement. Beyond that is the benefit to existing nursing staff to have ongoing exposure to current concepts of nursing education vis-a-vis contact with the students.

In summary, it is apparent to us that nurses have the potential, by definition of their professional practice, to have a major impact on substance abuse treatment and effect changes within the current service delivery systems. For the many reasons discussed here, we have not yet consistently realized that potential. The process of change requires a multi-faceted approach including; communication and support systems for peers on the local, state, and national level; utilization of consciousness raising and assertiveness training to prepare nurses to negotiate existing systems; nursing educators viewing substance abuse programs as valid student placements; nursing administration providing committed leadership and supervision. In addition, job descriptions for administrative and clinical leadership positions must be written to allow equal opportunity for nurses to fill them.

REFERENCES

Burhalter, Pamela 1976. *Nursing Care of the Alcoholism and Drug Abuse Patient*. Philadelphia: W.B. Saunders and Co.

ECONOMICAL EVALUATION: A TRAINING WORKSHOP FOR PRACTICAL PSYCHOLOGICAL TESTING OF DRUG ABUSE CLIENTS

Victor I. Friesen, M.A.

Philadelphia Psychiatric Center Drug Treatment Program

Wurzel Clinic

This paper will concentrate on the practical aspects of evaluation in drug/alcohol abuse programs. It will outline the major areas of evaluation which, although not mutually exclusive, pragmatically have little overlap in terms of testing. Specific test recommendations will be made for each area. These tests have been found to tap into the four areas with maximum efficiency, according to my experience and readings in this area. A discussion of overall use of tests follows, which includes some allusion to a somewhat different perspective on the field of testing when results are to be used for placement and not just diagnostic purposes. Finally, an open-ended report form is presented which can be adapted to any test utilized, in evaluation, either singly or as part of a battery.

Generally, psychological evaluation may be divided into four different areas: (1) Intellectual functioning, (2) Academic functioning, (3) Personality functioning, and (4) Vocational Interest/Aptitude functioning.

INTELLECTUAL FUNCTIONING

The gray area of intellectual functioning may be best approached through attempts to tap into an overall, general intelligence factor; what Spearman had termed the "g" factor (Matarazzo, 1972). Suspending for the moment the controversy between the factor analytic and wholistic approaches to intelligence, breadth and depth of vocabulary and word knowledge has been demonstrated to load high in the "g" factor. The Quick Test (QT) is an individual intelligence test based on perceptual-verbal performance which takes 3-7 minutes to administer. It gives an MA, percentile rank, and tentative IQ. Although sketchy and lacking the wide sampling range of individual intelligence tests such as the W.A.I.S., the QT can give a fairly valid indication of general intelligence functioning. If further exploration in this area is needed, the QT can help indicate this.

ACADEMIC FUNCTIONING

Academic functioning can be measured directly by tapping into three areas: arithmetic, reading level and verbal comprehension. For arithmetical ability, the Employee Aptitude Survey (EAS) Test #2, (Numerical Ability), and the Wide Range Achievement Test (WRAT), Arithmetic Level I, both give valuable information on the basic abilities of addition, subtraction, multiplication and division. Both have a rising complexity format, ranging from problems using whole numbers to those using decimals and fractions. The E.A.S. is somewhat more structured than the W.R.A.T., although both are speeded tests with time limits of 10 minutes each. Both tests have the disadvantage of giving a total score (the W.R.A.T. gives a grade-level equivalent), although the E.A.S. can be broken down into performance indices for the whole number, decimal, and fraction areas. This can be done informally with the W.R.A.T. by scanning results. All tests in this area really require that scanning be done anyway, as it is important to determine areas of weakness for placement purposes. My own preference is for the W.R.A.T., as the administration is somewhat simpler than for the E.A.S.

Reading and pronunciation level can be measured quite efficiently by the W.R.A.T., Word Recognition, Level I. This is a very simple power test which gives a grade-level equivalent indicating the level of difficulty at which words can be pronounced and read. This test generally takes 2-3 minutes to administer.

Word comprehension can be tapped into most effectively using a simple test of vocabulary. The Jastak revision of the Wechsler Adult Intelligence Scale Vocabulary sub-test (Jastak & Jastak, 1964) is a brief, verbally administered test which takes only 3-4 minutes to obtain an accurate word knowledge level. The test is scored using Wechsler's criteria (Wechsler, 1955) and interpreted by doubling the score and utilizing the age-scaled score tables for norm comparisons. Also of course, the results of the Q.T., if administered, can give a critical indication of word knowledge and vocabulary level.

PERSONALITY FUNCTIONING

This is an area of constant controversy, and one which will not be resolved here. There are many tests for personality; almost as many as there are definitions of personality itself. This part of the test battery was constructed with several points in mind, and these points must be understood clearly if the reader is not to accuse the battery of making grossly exaggerated claims. First, the majority of tests in this area, whether projective or non-projective, are oriented towards explicating maladaptive personality features and their dysfunctional behavior manifestations. This is due to the traditional approach to the polemic of defining "normalcy." The Freudian tradition and the medical model (really a reflection of our own language inadequacies) has led to the definition of "normal" as the absence of pathology. The application of this definition leaves much adaptive behavior and its underlying personality aspects undelineated. The investigator in this area can only be referred to the positive attribute sections of the test manuals he or she uses, and comment on those derived from making a conscious effort to note lack of pathology in the results. The second point is that almost all non-projective tests of personality are formulated as extended inventory questionnaires, and since time is at a premium, it was felt best to utilize simple projectives. The third point, which derives from the second, is that results of such tests must be interpreted within the limitations of the tests themselves.

The tests I have chosen for the battery are three graphic paper tests, the Bender-Gestalt Visual Motor Test, the Draw-a-Person Test, and the Kinetic Family Drawing Test. Tests such as the Wagner Hand Test are easily administered non-graphic projectives, but like other similar in format were excluded because of the complexity and open-endedness of the scoring and interpretation.

The Bender-Gestalt Test is a simple drawing reproduction test consisting of nine cards, each with an abstract figure which the client is asked to draw on paper. As reviews in Buros (1972) have suggested, the original intent of the test as one which measures distributions in gestalt-perception has long been relegated to the background. Many writers, chief among them Hutt (1977), have advocated use of the test as an in-depth personality instrument. Evaluative reviews have suggested the test is too limited to warrant such claims, and that interpretation should be limited to organicity/neurosis/psychosis differentiations and gross personality patterns dealing with self-image, needs, conflicts, defenses, and coping methods. For a basic grounding in inferential analysis in the Bender, the work of Hutt (1977), with a strong infusion of common sense, can be quite valuable. The test can be administered in approximately five minutes utilizing the copy-recall method.

The Draw-A-Person Test (D.A.P.) is a graphic task involving the drawing of a "whole person". Instructions are that the person doing the test can draw the figure any way he or she prefers. When a figure has been drawn, the investigator asks that a figure of the opposite sex to the first drawing be done. This test takes about 3-4 minutes to administer. Overall anaysis and interpretation can be based on the works of Machover (1974) and Ogdon (1975), two excellent sources in this area. Machover (1974) has suggested a brief questioning of the person regarding the drawing's sex (if in doubt), age, happiness or unhappiness, current behavior (i.a. "What is this person doing in the picture?"), and other briefly-answered features. The responses can be pencilled in beside the drawing itself, and incorporated into overall interpretation.

The Kinetic Family Drawing Test (K-F-D) is an extension of the D.A.P. which can provide insight into gross family dynamics. The task involves the drawing of the person and his or her family "doing something together." This can be family of origin or current family. Important features here are figure placement, activity, barriers, age of the drawer in the drawing, groupings and separations, etc. This test takes only 2-3 minutes to complete, and there should be a brief question period of identifying who is who in the picture and what they are doing, which can be used as an adjunct to the D.A.P. as comparison and corroboration of salient personality features. There are, as far as is known, no manuals for the Kinetic Family Drawing Test which extend into the adult age group. The work of Burns and Kaufman (1977), taken with a large grain of salt, can be useful. However, most of the dynamics which can be inferred from the K-F-D are based on common-sense intuitions, and the experienced evaluator will gain little from texts in this area.

It is suggested that the Bender, together with the D.A.P. and K-F-D be used as part of a global, overall approach to psychodiagnostic evaluation. A feature-by-feature sign approach is not recommended as these methods simply do not have the validity or reliability required. As Kitay (1972) has pointed out, the incorporation of a global assessment of projective materials with clinical intuition, test behavior and case history is probably the best approach to delineating gross personality functioning when these instruments are used.

A basic understanding of psychobehavioral features and their diagnostic classifications, as an aid to personality evaluation and description, can be obtained from any good text in psychopathology. Also, the American Psychological Association and the American Psychiatric Association both publish glossaries of pathologic nomenclature accompanied by brief descriptions.

VOCATIONAL INTEREST FUNCTIONING

Tests for vocational interests range from the highly structured and high-level (e.g. Strong-Campbell Interest Inventory) to the fully projective and open-ended (e.g. Vocational Apperception Test by Ammons et al), with everything between also represented. Interest testing appears to be a favorite of vocational counselors as it provides access to structured exploration of job possibilities and, parenthetically, provides a delaying mechanism so the counselor can avoid direct confrontation of employment realities and client work readiness issues. Few interest inventories can provide any more information than a verbal interview, and of those tests which do (such as the Strong-Campbell) they are at too high a reading and comprehension level for most drug and alcohol treatment clients. When the general employment picture, client skill levels, general client inability to delay gratification in formal training, and inappropriate job readiness postures are considered, vocational interest testing becomes irrelevant. What is relevant is what jobs, O-J-T slots, etc. are available, what resources the client has, and what makes sense. Given time, investment and the introjection of valued secondary reward systems on a job where ambivalence or conflict of interests exists, this ambivalence will resolve itself through dissonance reduction. This concept is derived from social psychology and is not new to vocational rehabilitation (cf. Neff, 1968), but seemingly its application is not widely practiced. In this area I feel it is important to keep reality in perspective, and vocational rehabilitation based on a static need-reduction model is neither viable nor necessary. It denies the fact of change occurring through dynamic work adjustment, and creates far too much anxiety and frustration for the counselor attempting to supply the "right" job. This allows the counselor to work from a much wider range of options than he or she might utilize.

For sampling the vocational interest area, a brief questionnaire listing major skilled and semi-skilled occupations drawn from the Occupational Outlook Handbook (1978) is recommended. This questionnaire can be verbally administered in 2-3 minutes using a simple like-dislike format. An example appears at the end of this paper. The questionnaire can also be adapted to specific economic and demographic areas, where certain types of jobs may be more predominant than others. The questionnaire can also include space for open-ended questions about occupation most preferred, type of work done most, and most saleable skill. Results can be grouped and used later in follow-up sessions where interview and verbal exchange can establish viable occupational choices.

NOTES ON "CONCLUSIONS AND RECOMMENDATIONS"

In this section of the report it is necessary to coalesce the test results obtained and present a brief write-up of the four areas of functioning which are being evaluated. Generally, the most difficult area to integrate is that of personality functioning. Since this is the primary area on which recommendations are based, it is important that the evaluator strive for simplicity and conciseness of description, outlining areas of conflict or difficulty, and in what manner they might best be approached in treatment (e.g. reality confrontation, insight, concrete suggestion, etc.). This latter is dependent on the client's perceived level of functioning, mode of defense, reality contact, and other salient features. Here it is necessary to rely much on breadth and depth of experience in psychodiagnosis, and these can only be gained over time, thought, and good feedback in supervision.

TEST REPORT

The test report form (see below) has been designed as a brief, two-page fill-in-the-blanks outline which gives brief descriptions of the objective tests, what they measure, how the scores are interpreted, and spaces for the scores. It also provides spaces for a listing of major areas chosen in the Vocational Interest Questionnaire and a note on the responses to the open questions. There is further space provided for brief write-ups of the projective test results and the conclusions and recommendations. This format is flexible in that the report can be easily written to delete tests or add new ones, and provides a model of how the salient features and score-meanings of added tests can be presented. The final saving is in time spent on writing the report, and a stricture to conciseness by its brevity.

CONCLUDING REMARKS

A concise, easily-administered, and brief test battery has been presented which taps a wide range of highly meaningful and relevant material. In some instances, compromises in battery construction were made because of low return which longer and more exhaustive tests might give, when evaluated against the time and effort involved in administering and scoring them. Major sources for interpretation of the tests recommended have been given, and a listing of where these may be ordered is presented below. In addition, other test sources have been listed. A brief, concise and efficient report form has been presented which can provide a flexible framework for individual test preferences. The tests chosen reflect my own experience in the field and my own thinking about what I perceive as useful information when balanced against basic treatment needs. The field of evaluation is flooded with tests which do not measure what they say they measure, and do not matter if they do. I can only urge caution in test selection (a reading of Buros on tests considered can be invaluable), with an eye to the base line treatment needs of the client who is, after all, what it is all about.

REFERENCES

Burns, R.D. and Kaufman, S.H. 1972. *Actions, styles and symbols in kinetic family drawings: an interpretative manual.* New York, New York: Brunner/Mazel.

Buros, O.K. (ed.). 1972. *The seventh mental measurements yearbook.* Highland Park, New Jersey: The Gryphon Press.

Hutt, M.L. 1977. *The Hutt adaptation of the Bender-Gestalt test.* New York: Grune and Stratton.

Jastak, J.F. and Jastak, S.R. 1964. Short forms of the WAIS and WISC vocabulary subtests. *Journal of Clinical Psychology*, Vol. 20: 180-199.

Kitay, P.M. 1972. In Buros, O.K. (ed) *The seventh mental measurements yearbook.* Highland Park, New Jersey: The Gryphon Press, Vol. I: 394-395.

Machover, K. 1974. *Personality projection in the drawing of the human figure: a method of personality investigation.* Springfield, Illinois: C.C. Thomas.

Matarazzo, J.D. 1972. *Wechsler's measurement and appraisal of adult intelligence.* Baltimore: Williams and Wilkins.

Neff, W.S. 1968. *Work and human behavior.* New York, New York: Atherton Press.

Occupational outlook handbook, 1978-1979 edition. Washington, D.C.: U.S. Government Printing Office.

Ogdon, D.P. 1975. *Psychodiagnostics and personality assessment: a handbook.* Los Angeles, California: Western Psychological Services.

Wechsler, D. 1955. *Wechsler adult intelligence scale: Manual.* New York, Psychological Corporation.

APPENDIX A

VOCATIONAL INTEREST QUESTIONNAIRE

Name: ______________________ Date: __________

For the following occupations indicate whether you would like or dislike doing these for a living.

Industrial	Like	Dislike
Machinist	___	___
Sheet-metal worker	___	___
Welder	___	___
Driver (Bus, Truck, Taxi)	___	___
Shipping/Receiving Clerk	___	___
Printer	___	___
Sales		
Cashier	___	___
Salesperson	___	___
Protective		
Policeman/Policewoman	___	___
Guard	___	___
Office		
File/Records clerk	___	___
Receptionist	___	___
Typist	___	___
Social Service		
Recreation worker	___	___
Child-care worker	___	___
Counselor	___	___
Construction		
Bricklayer	___	___
Carpenter	___	___

Service	Like	Dislike
Cook	___	___
Meatcutter	___	___
Waiter/Waitress	___	___
Barber/Hairdresser	___	___
Mail carrier	___	___
Telephone operator	___	___
Gas station attendant	___	___
Building maintenance worker	___	___
Mechanical		
Auto mechanic	___	___
Air conditioning/ refrigeration mechanic	___	___
Appliance repairer	___	___
Radio/T.V. repairer	___	___
Health		
Dental Assistant	___	___
Medical technician	___	___
Practical nurse	___	___
Medical records worker	___	___
Hospital orderly	___	___
Electrical installer	___	___
Carpet layer	___	___
Painter	___	___
Roofer	___	___
Plumber	___	___
Construction Equipment operator	___	___

Type of job most preferred: ______________________________

Type of work best skilled in: ______________________________

Type of work done most in past: ______________________________

TEST BATTERY REPORT

NAME: ______________________ EXAMINER: ______________________

Date of Birth: ______________ Date Tested: ______________

Quick Test

This test provides an overall indication of general intelligence based on verbal ability. It yields a tentative Intelligence Quotent (I.Q. Average of 100, Standard deviation of ± 10), Maturational Age (M.A.), and Percentile Rank (P.R.).

I.Q. ______________
M.A. ______________
P.R. ______________

W.R.A.T. Arithmetic Test

This is a test of arithmetic skills. It assesses the subject's ability to perform computational work. Inspection of the test indicates specific arithmetic strengths and weaknesses as well as the subject's ability to work quickly under the pressure of time. It yields a grade equivalent score expressed in tenths of a grade.

Score: ______________

W.R.A.T. Word Recognition Test

This test gives an indication of the subject's reading ability. It yields a grade equivalent score expressed in tenths of a grade.

Score: ______________

JASTAK Vocabulary Test

This is a test of word knowledge. Vocabulary tests have been found to be positively related to academic success. The score is reported as a scaled score with an average of 10, and a Standard Deviation of ± 2.

Score: ______________

continued...

Test Battery Report
Page two

Vocational Questionnaire

This inventory indicates the extent of an individual's interest in different occupational areas. A preponderance of choices in any one area indicates possible exploration of this area, and the appropriateness of the choices when compared with past work history, current skill levels, etc.

Major areas checked: ________________________________

____

____

Impressions: ____________________________________

____

____

Projective Personality Test Results ______________

____

____

____

____

____

____

CONCLUSIONS, RECOMMENDATIONS

____

____

____

____

____

____

____

MAJOR TEST SOURCES

Western Psychological Services (WPS)
12031 Wilshire Boulevard
Los Angeles, California 90025

The Psychological Corporation (CP)
757 Third Avenue
New York, New York 10017

Psychological Test Specialists (PTS)
Box 1441
Missoula, Montana 59801

Consulting Psychologists Press (CPP)
577 College Avenue
Palo Alto, California 94306

Sheridan Psychological Services Inc. (SPS)
P.O. Box 6101
Orange, California 92667

Quick Test - PTS
W.R.A.T. - WPS, CP
Bender-Gestalt - WPS, CP
Jastak Vocabulary - (From 1964 article)
D.A.P., K-F-D - (Unnecessary to order)

A PRELIMINARY STUDY ON THE USE OF BUDDHIST MEDITATION IN THE REHABILITATION OF DRUG ADDICTS

Saovanee Chakpitak and Bhumirat Chakpitak, with the assistance of Chya Patanacharoen, Sompoon Kritlak, Suthirapan Koralak, and Kovit Prawalpryk

A pilot project conducted jointly by Foundation for the Promotion of Buddhist Meditation in Thailand and researchers from various institutions was undertaken in April, 1974 at one of the Government Rehabilitation Centres for Narcotics Offenders. The purpose was to make a preliminary investigation on the use of Buddhist meditation as one of the effective means in giving mental strength to the addicts in order to prevent or completely stop drug use after they have been cured by medical doctors. In this one-year program of study, 93 volunteer addicts were divided into two groups, the experimental group (E) and the control group (C). Fifty-five addicts in E were instructed in the practice and benefits of meditation. Psychological tests on attitudes and behavior together with detection of drugs in urine specimens by gas chromatography were performed. Though the difference on changes in attitudes and behavior from psychological tests was not statistically significant between E and C, improvement of behavior in E was obvious. Urine tests also revealed striking differences in drug use between the two groups. Follow-up studies to determine the long-term effect of meditation was conducted in April-June, 1977. It was discovered that about 40% of E and 5% of C could remain off drugs completely. It was concluded that the daily practice of meditation was beneficial to the addicts and should be included in the prevention and rehabilitation programs of drug dependence.

Drug addiction is at present a very serious problem in Thailand. It has been estimated that about 300,000-500,000 individuals, mostly teenagers, are being involved (Lanlue, 1973; Shaowanasai, 1972). Numerous different approaches have been attempted to solve narcotics problems but not a single possible method really works so far and the number of drug addicts is alarmingly increasing. Wallace, Benson, and Wilson (1971) found that 36 subjects who practiced transcendental meditation (TM) exhibited physiological changes which were different from those observed during sleep, hypnosis and autosuggestion. This "wakeful hypometabolic physiologic state" involved a decrease in oxygen consumption, carbon dioxide elimination, respiratory rate and minute ventilation without changes in respiratory quotient. The fall in blood lactate concentration which suggested physical and mental relaxation was also observed. On the other hand, there was an obvious increase in skin resistance and the EEG exhibited the greater intensity of slow alpha waves with occasional theta-wave activity. In a study of 1,862 subjects by questionaire method, Benson and Wallace (1972) also reported that "following the start of the practice of TM there was a marked decrease in the number of drug abusers in all categories". A considerable drop in the use of marijuana, hallucinogens, barbiturates, amphetamines and other narcotics were found in those who practiced TM and the longer the practice, the more significant the decrease was. A result of a questionaire survey by Shaft, Lavely, and Jaffe (1972) revealed that "the longer a person had practiced meditation, the more likely it was that he had decreased or stopped his use of marijuana".

As an integral part of Buddhism, the practice of Buddhist meditation in Thailand has existed since nearly 800 years ago. The usual course of mental development in Buddhism is composed of three-step training, namely--Morality, Concentration, and True Wisdom. Morality is the foundation of Concentration or Meditation and meditation is believed to be the only way to free the mind of defilements which would give rise to True Wisdom and finally lead to Supreme Purification and complete freedom from the Wheel of Death and Rebirth. For the development of concentration, the widespread practice is the verbal or mental repetition of Buddho or the observation of breathing. The practice of meditation is considered to be the highest form of merit-making and has many advantages for physical and mental well-being. In the old days and even at the present time, meditation has been used by the monks to give mental comfort and peace to the sick and the dying. Therefore, the present study was designed to make a preliminary investigation on the use of Buddhist meditation as one of the effective means in giving mental strength to the drug addicts in order to prevent or completely stop drug use after they have been cured by medical doctors.

BACKGROUND

A pilot project conducted jointly by Foundation for the Promotion of Buddhist Meditation and researchers from various institutions was undertaken in April, 1974 at one of the Government Rehabilitation Centres for Narcotics Offenders in Thailand. This Centre was established in 1963 for the confinement and treatment of narcotics-addicted prisoners from all parts of the country and has been in operation since February, 1965. At the time of the study, there were approximately 1500 inmates. Those inmates were persons who had violated the Narcotics Acts and were also addicted to narcotics drugs and required treatment. The length of sentence of the inmates confined to this Centre varied from three months to three years. About half of the prisoners were committed for first admissions and a large part of the readmissions were the second or the third ones. Before arriving at the Centre, most prisoners had usually withdrawn from drugs. For those who were heavily addicted and suffering from severe withdrawal symptoms, tranquilizers and sedatives were given. Methadone was used only as the last resort and there were no other alternatives available to save the man during his acute withdrawal phase. A basic academic education program was provided and an on-the-job vocational training program designed to equip the inmates with job skills was offered in the area of carpentry, wood-working, rattan works and farming. The religious program and recreational programs such as music, TV viewing and some outdoor sports were also included in the Centre's total treatment program.

METHOD

In this one-year program of study, 93 volunteer inmates, 18-63 years old, were selected and were divided into two groups, 55 in the experimental group and 38 in the control group. Instruction regarding the practice and benefits of meditation were given to those in the experimental group by Buddhist monks. The training program consisted of three parts. In the first part, instructions and the practice of meditation were made daily for a month. Similar procedure was repeated twice a week for three months in the second part and once a week for eight months in the third part respectively.* The priests visited the Centre for an hour each time. The subjects would have a prayer for 10 minutes, receive instructions for 20 minutes, meditate for 20 minutes and spend the rest of the hour for duscussions with the priests. The subjects were permitted to practice meditation whenever they were free from routine works. In this experiment, the usual practice began with a chanting of the three Refuges and the Sublime Qualities of the Triple Gem. Following the Dhama teachings given by the priests, the subjects sat cross-legged, with eyes closed and developed concentration by means of repeating Buddho, i.e. mentally repeating "Bud" when breathing in and "Dho" when breathing out until the mind was relaxed, concentrated and peaceful.

---

* The training program which was formerly scheduled to close at the end of March, 1975 was changed to terminate about 6 weeks earlier because some prisoners in the project were released in February owing to the Amnesty Act.

Table I. Mean scores and standard deviations of the experimental group (E) and control group (C) in attitude tests.

Attitudes toward		1	2	3	4
1. Self	E	22.23 ± 2.55	21.85 ± 2.11	21.08 ± 3.73	22.77 ± 2.61
	C	22.11 ± 2.81	21.44 ± 2.36	22.56 ± 3.06	20.89 ± 2.73
2. Narcotics suppression	E	17.69 ± 2.49	17.00 ± 3.01	16.69 ± 2.33	17.38 ± 2.17
	C	17.67 ± 2.54	15.67 ± 3.43	18.11 ± 1.85	16.33 ± 2.83
3. Law abiding	E	20.62 ± 2.34	19.08 ± 3.32	19.92 ± 2.76	19.31 ± 3.34
	C	18.22 ± 3.08	19.33 ± 3.59	19.89 ± 3.60	19.11 ± 2.85
4. Families and society	E	21.92 ± 2.37	23.00 ± 2.42	21.62 ± 3.54	21.92 ± 2.62
	C	23.11 ± 2.51	22.56 ± 2.56	24.21 ± 1.55	20.67 ± 4.32

The meditators and the control subjects were interviewed and informations were collected on their personal backgrounds and characteristics such as age, marital status, educational level, previous occupation, types of narcotics addiction, number of admissions and previous criminal records. Psychological tests on attitudes and behavior by questionaire methods together with detection of drugs in urine specimens by solvent extraction and gas chromatography were performed before and after the end of each training period. Follow-up studies to determine the long-term effect of meditation were conducted in April-June, 1977. Each addict was visited and interviewed. In order to locate readmissions and missing cases whose names and addresses had been changed, all fingerprints were checked and identified at the Criminal Records Office, Police Department.

RESULTS

From an interview with the meditators and control subjects, it was obvious that heroin addiction was more prevalent than any other drugs. Most addicts were homeless and poorly educated. The inmates were drawn into drug use owing to their own curiosity or self-initiation or by persuasion of their addicted friends. Most subjects were hired laborers and expressed unwillingness to communicate with the researchers after their release from the Centre. When the experimental group was asked about the meditation program more than half of the meditators reported that they could see the beneficial effects of meditation and many of them tried to practice meditation by themselves on a regular basis. Thirty-four meditators wished to be ordained after they were discharged and twenty-two requested support from the Foundation for their ordination ceremony.

Results of psychological tests on attitudes and behavior are presented in Tables I and II. The questionaires on attitude tests which were answered by the subjects were grouped into the 4 following categories: self-attitudes, attitudes toward narcotics suppression, attitudes toward legal responsibilities which involved law abiding and support, and finally, attitudes toward family and social responsibil-

Table II. Mean scores and standard deviations of the experimental group (E) and control group (C) in behavior tests.

	1	2	3*
E	39.24 ± 5.25	31.13 ± 3.87	31.72 ± 4.39
C	38.13 ± 5.17	30.79 ± 3.81	33.55 ± 3.26

* Change of officials in charge of the units was reported.

Table III. Results of urine tests.

	1	2	3	4
Group E - total	55	55	55	55
Positive findings	54	11	17	3
Heroin and methaqualone	4	1	1	-
Heroin	49	8	7	2
Methaqualone	1	2	9	1
Group C - total	38	38	38	38
Positive findings	38	26	7	10
Heroin and methaqualone	8	2	1	3
Heroin	28	8	4	4
Methaqualone	2	16	2	3

ities. For tests of behavior regarding their work abilities, relation with friends and guardians and their interest in life and surroundings, the scores were judged by the correctional officials in charge of each unit at the Centre. The results on attitude and behavior tests indicated no statistically significant difference between the two groups.

Results of urine examination for heroin and methaqualone are given in Table III. Prior to the meditation program, urine tests of drugs were positive in both groups. After one month of meditation, the results were negative in most subjects of the experimental group while the picture remained nearly the same in the control. In the third examination, a considerable drop of the number of positive cases in the control was noted. Therefore, in the fourth examination, the examiners began to look for other narcotics besides heroin and methaqualone in both groups. The tests revealed the appearance of other tranquilizers in urine specimens of many nonmeditators.

Results of follow-up studies are presented in Table IV. It was found that at the end of two years following their release from the Centre, approximately 40.0% of the addicts in the experimental group remained off drugs completely while only about 5.2% of the control could stop drug use. Seventeen meditators were ordained in 1975 and 9 of them are still in the priesthood and have stopped taking drugs up to the present time. Three control subjects were ordained and 2 of them were again addicted and were forced to leave the priesthood by the police. Many nonmeditators reported false names and addresses and were unwilling to cooperate in follow-up studies.

Table IV. Results of follow-up studies.

	E	C
Total number	55	38
--Death	3	1
--Stop drugs completely	22	2
--Back to drugs but not arrested	4	10
--Readmissions or arrested for taking drugs	17	12
--Commit other crimes	3	3
--Report false names and addresses	5	8
--Unable to follow owing to change of residence	1	2

DISCUSSION

The results of the above study indicated that the daily practice of meditation was essential for the withdrawal of drug dependence. Though the difference on changes in attitudes and behavior from psychological tests was not statistically significant between the two groups, improvement of behavior in the experimental group were also observed and reported by the correctional officials in charge at the Centre. It also appeared that the meditators were more self-conscious and self-confident than the control subjects. These findings conform with Buddhist principles that the practice of meditation helps the meditators to look inward and become more self-oriented. The change in attitudes was also obvious. Following the meditation program, the experimental group was more cooperative and many of them wished to be ordained. These changes agreed with the results of urine examination which indicated much more drug use among the control subjects. Follow-up studies also revealed a striking difference in drug dependence between the meditators and nonmeditators.

The results of this preliminary study described here appear encouraging. The practice of meditation seemed favorable to the addicts and it was recommended that a larger study of the use of meditation in the rehabilitation of drug abusers should be undertaken in other groups of populations. It is interesting to note that in Buddhist culture, ordination and their existence in priesthood serves as one of the best means to provide new environments for the ex-addicts, similar to "Therapeutic Community" in the treatment program of other countries. Besides, their status as priests is more advantageous to them because they are easily accepted by the family and society. Being confined to Buddhist moral precepts for the priests also helps them to behave properly and withdraw from drugs. Finally it should be kept in mind that to achieve "self-control" or "mind-control" in total withdrawal of drugs, the ex-addicts should practice meditation regularly and permanently.

ACKNOWLEDGEMENTS

This work was supported by FPBM. The authors wish to express their deepest gratitude to the Most Venerable Somdet Phra Nyanasamvara, the headpriest of Wat Bovornivesvihara and Chairman of FPBM for his benevolence and encouragement since the beginning of the study. The authors are greatly indebted to Laung Thavil, Managing-Director of FPBM for his valuable suggestions and help which made this work possible. The authors are grateful to Dr. Sudchai Laosunthorn and Police Major-General Pao Sarasin for their many kindnesses and support. The authors wish to thank the priests at Wat Bovornivesvihara, Department of Corrections, Siriraj Hospital, Dhamasart University and Ministry of Education for their participation and cooperation. Appreciation is also extended to the Bangkok Bank Ltd. for the provision of transportation throughout the study.

REFERENCES

Benson, H. and Wallace, R.K., Decreased Drug Abuse with Transcendental Meditation: A Study of 1862 Subjects in Drug Abuse. In Zarofonetis, C. (ed.) 1972. *Proceedings of the International Conference*. Philadelphia: Lea and Febiger.

Lanlue, A. 1973. 300,000 Student Drug Addicts. *Bangkok Post*. (Bangkok, Thailand).

Shaft, M., Lavely, R., and Jaffe, R. 1972. Meditation and Marijuana. *Am. J. Psychiatry*. 131(1): 60-63.

Shaowanasai, A. 1972. Drug Crisis in Thailand. *Bangkok Post*. (Bangkok, Thailand).

Wallace, R.K., Benson, H., and Wilson, A.F. 1971. A Wakeful Hypometabolic Physiologic State. *Am. J. Physiology*. 221(3): 795-799.

REFRAMING: A FAMILY COUNSELING SKILL*

Bruce Fischer, M.A.**

University of Minnesota

Recently the literature on the families of substance abusers has indicated that drug abusers are perceived as powerless, passive, and dependent by their family members (Alexander and Dibb, 1977). Frequently this perspective within a family system is a significant block to positive change. Thus, one strategy to facilitate change within the family of the drug abuser is to reframe or change the family members' perceptions of the drug abuser.

Reframing is an extremely subtle, yet amazingly powerful technique. It may be viewed as both a problem-solving technique and/or a helpful step preceding problem solving. Reframing can be defined as providing a different perspective from which to view the same "problem," thereby eliminating the client's perception of the situation as a problem. Essentially, the problem is eliminated without affecting "reality." What is changed is the client's inner reality. Watzlawick, Weskland, and Fisch (1974) offer a more precise definition:

> To reframe, then, means to change the conceptual and/or emotional setting or viewpoint into which a situation is ex-

---

*This paper is an excerpt from a manuscript entitled "Basic Family Therapy Skills for Drug Abuse Counselors" which was produced while the author was a resident scholar at the National Drug Abuse Center in Washington D.C. Assistance was provided by National Institute on Drug Abuse. Contract # 271-75-4018.

**Bruce Fischer, Instructor, Chemical Dependency Counseling Program, University of Minnesota, 2829 University Avenue, Southeast, Suite 226, Minneapolis, Minnesota 55414; (612) 373-8175.

> perienced, and to place it in another frame which fits the "facts" of the same concrete situation equally well or even better; and thereby changes its entire meaning. The mechanism involved here is not immediately obvious, especially if we bear in mind that there is change while the situation itself may remain quite unchanged and, indeed, unchangeable. What turns out to be changed as a result of reframing is the meaning attributed to the situation, and therefore its consequences, but not its concrete facts--or as the philosopher Epictetus expressed it as early as the first century A.D., "It is not the things themselves that trouble us, but the opinions that we have about these things." (p.95)

The classic joke about the difference between an optimist and a pessimist offers an excellent example of the principle involved in reframing.

QUESTION: What is the difference between an optimist and a pessimist?

ANSWER: The optimist says of a glass of water that it is half full; the pessimist says of the same glass that it is half empty.

Clearly, in this situation, the inner realities of the optimist and the pessimist are different, although, the "reality" of the situation is the same.

For purposes of illustration, let us suppose that the pessimist's awareness that the glass is half empty causes him anxiety. The therapist could attempt to alleviate anxiety in many ways. For example, the therapist could help him realize that his anxiety response is related to his fear of failure or perhaps his long-standing conflict with his father. In short, he would attempt to help the client gain an insight into the problem in hopes of changing the client's anxiety. Another therapist, however, realizing that this man's distress is related to his view of a particular situation--the glass--may simply choose to help the client reframe the situation. In other words, he would help the client view the glass as half full as opposed to half empty. Thus, the "problem" is alleviated without any change in the "real world" whatsoever and without the benefit of insight.

One of the paradoxical characteristics of reframing is that once an object has been reframed, the new "reality" probably will become just as "real" as the old (or problematic) "reality." Simply, the new reality quickly serves a functional role; that is, it works more effectively than did the previous reality. Therefore, the awareness that a new perspective was created is often forgotten. The practical effects of this process are such that after an object has been reframed the old view often seems "silly" or "stupid" to the person who has reframed it.

An example from my experience may further illustrate how reframing may be applied pragmatically in therapy. I was working with a family in which the father was an alcoholic. The mother had entered into a strong coalition with the children against the father. By and large, the father had a peripheral role in the family. He was, however, a strict disciplinarian who continued in this role despite his withdrawal from other roles within the family. His parenting was limited to a strict disciplinarian role. The children saw their father as threatening and imposing. In short, they thought he was "bad" because of his strictness. The children also thought that the father did not care for them because he was so harsh and strict. The wife saw her husband as uninvolved with the children. She thought he did not care about them and was angry because she felt like she was raising the kids alone. In the course of therapy, I found it helpful to reframe the father's disciplinarian role. This was accomplished by discussing how critical discipline is to the parenting process, thereby illustrating that the father was, indeed, involved in the parenting of his children, and that the mother was not raising the children alone. Furthermore, discipline was discussed, and I pointed out to the children that one of the most difficult tasks for a parent is to deny his child something that the child would like, but that parents do this because they know it's best for their children. In other words, the father's discipline was simply an extension of his love and concern for his children, and was not motivated by evil intentions. Thus, the father's evil, distancing behavior was reframed to indicate care and involvement. This allowed the family to begin working together for constructive change, rather than blaming each other for things that had happened in the past.

Lest the reader become too enthusiastic over this seemingly miraculous method of making problems disappear, I would like to add several warnings. To begin, reframing must be based on an accurate understanding of the family's reality and the realities of the individual family members. This includes their views, expectations, motives, and premises. In short, the therapist needs to understand each family member's approach to the world. Should a therapist not understand his client's reality, he could simply reframe the situation from one problematic view to another. In the previous illustration, for example, had I reframed the father's strict behavior as due to his drinking--that is, from bad to sick--the family members probably would have been forced into a position of impotent rage. He was, after all, sick and therefore couldn't help himself. Furthermore, it's not permissible to become angry with someone who is sick. In other words, the family would still see father's behavior as negative, only they would no longer have a legitimate reason to be angry about it. Reframing the family's reality in such a fashion would have only served to perpetuate the problem. Unfortunately, well-meaning therapists frequently use this strategy of reframing from bad to sick in an attempt to reduce blaming and minimize feelings of guilt in families of substance abusers.

To effectively reframe a situation, it is necessary for the therapist to view the problem both from the inside (from the point of view of the family members) and from the outside (from his own "objective" position). While reframing, it is imperative that the therapist "speak the family's language;" if he tries to reframe the situation using a different meaning system, probably he will fail. Using the same language implies more than simply using the same words as the family members use. although using the same vocabulary may be important. Grinder and Bandler (1976) propose that therapists may "use the client's language" by identifying the client's most highly valued representational system. The common representational systems are kinesthetic, visual, and auditory. Once the therapist has identified the client's representational system, he speaks to the client in terms of that specific representational system. That is, the therapist uses the same predicates as the client. For example, if a client's representational system is visual, then the therapist may ask the client how he "sees" a particular situation. The reader is encouraged to consult Grinder and Bandler (1976) for a thorough discussion of this process, as a detailed explanation is beyond the scope of this paper.

In review, reframing involves understanding accurately the client's meaning system associated with a particular problem and then assisting the client, either an individual or a family, to view this situation so that it has a new and different meaning. The therapist's ability to accurately understand the family from both the "inside" and from the "outside" is critical to effective reframing.

REFERENCES

Alexander, Bruce K. and Dibb, Gary S. 1977. Interpersonal perception in addict families. *Family Process*. 16: 1 17-18.

Grinder, J. and Bandler, R. 1976. *The Structure of Magic: Volume II*. Palo Alto: Science and Behavior Books, Inc.

Watzlawick, P. 1976. *How Real is Real?* New York: Random House, Inc.

Watzlawick, P., Weakland, Ch. E. and Fisch, R. 1974. *Change: Principles of Problem Formation and Problem Resolution*. New York: W.W. Norton & Company, Inc.

SEXUAL COMMUNICATION SKILLS TRAINING: A NECESSARY COMPONENT OF AFTERCARE FOR SUBSTANCE ABUSERS

Bruce E. Fischer, M.A. and Marilyn Mason, B.A.

University of Minnesota

INTRODUCTION

While there has been considerable literature devoted to identifying relationship and sexual problems among chemically dependent persons and their partners (De Leon and Wexler, 1963; Gay and Shepherd, 1973; Le Mere and Smith, 1973; Schoener, 1973; Briddell and Wilson, 1973; Simmons, 1977), there has been an obvious lack of treatment programming to meet this need. The purpose of this paper is to briefly describe the goals, philosophy, program format and content of an aftercare program, Sexual Relationship Enhancement Program (SREP) designed to enhance sexual relationships of couples in which one or both partners are chemically dependent.

The impetus for the development of this program has grown from an awareness of the sexual problems frequently encountered by individuals and couples following treatment for chemical dependency. As both authors are practitioners in marriage and family counseling, largely with caseloads of chemically dependent people, they were encountering this problem almost daily. This program was developed to offer a less expensive group format solution for this important client need.

GOALS AND PHILOSOPHY

The goal of the SREP is to improve a couple's sexual relationship. This is accomplished most effectively by an educational format which provides permission to discuss sexuality, information about sexuality, and interpersonal communication. Furthermore, this method emphasizes training in communication skills. This is based on the assumption that many of these couples are not in need of marital or sexual therapy; rather, they possess the ability to resolve their problems with limited information and improved communication skills. Programs of similar design, although not focused on sexual relationships, have been shown to be effective in improving marital communication (Miller, Nunnally and Wackman, 1975; Rappaport, 1976; Guerney, 1976).

Based on prior experience with clients from this population, the authors have found it important to increase clients' awareness in the following ares: sexuality and sensuality, sexual language, gender roles, chemicals and sexuality, sexual dysfunctions, and intimacy, love and sex. Listed below are eleven major assumptions made by the SREP:

1. Sex and sexuality are natural functions.
2. Sexuality is an integral part of everyone's personality and is expressed in all that we do.
3. Human sexual communication is primarily learned behavior.
4. We all have a right to know the range of human sexual behavior and to be acquainted with accurate information about sexuality.
5. We all have a right to our own beliefs and convictions.
6. What is acceptable sexually will vary from person to person.
7. Human sexual behavior *is* a communication process; conversely, all communication has a sexual component.
8. One cannot *not* communicate.
9. One's sexuality is influenced primarily through family rule systems, attitudes and values and behaviors within the family.
10. The ability to discuss sexuality is necessary for a satisfactory sexual relationship.
11. Sexual behavior can be a barrier to intimacy.

PROGRAM FORMAT

The SREP is a short term highly structured training package. It is designed to be delivered in six (6) three hour training modules. Each module attempts to use a variety of learning experiences such as audio-visual presentation, lecture, large group discussion, role play situations, guided fantasy and dyadic exercises. The package places a heavy emphasis on participants' reading assigned materials provided by instructors and completing homework (relabeled "homeplay") assignments (such as self-pleasuring exercise).

The SREP is intended for groups composed of four to seven couples (8-14 people). The focus of this package is strictly on couples in on-going significant relationships. In the authors' experiences, it has not been beneficial to mix couples with singles, because of the differences in needs of these separate populations. Currently, a separate package is being prepared for single persons.

Optimal delivery of this package calls for two co-trainers, preferably one female and one male. Failure to represent both of the sexes may result in a biased training experience, and alienation of one of the sexes in this training group. Furthermore, having a co-trainer of each sex allows for the possibility of trainers to behave outside of traditional proscribed gender roles, thereby forcing participants to examine their values and attitudes towards traditional gender roles. For example, when a male co-trainer is sensitive to others' feelings, shares his own feelings, and asks for help and support from his female co-trainer, group members are forced to examine and deal with the discrepancies between his behavior and the traditional male role.

PROGRAM CONTENT

The program is divided into six modules. The following section will briefly describe each module. The program content is drawn from a variety of sources; some of the major references are McCary, 1973; Annon, 1975; Fischer, 1977; Peele, 1975; McCarthy, Ryan and Johnson, 1975; Dushkin Publishing, 1975; Skolnick, 1973; Dushkin Publishing, 1977; Egan, 1976; Kaplan, 1975; and Williams, 1977.

Module I is designed to introduce the participants and the instructors to each other and begin to build trust within the group. The content presented deals with the similarities and differences between sexuality and sensuality. This is accomplished via lecture, discussion and a sensate focus exercise. The role of non-verbal communication in human sexual communication is also discussed and illustrated in this module.

The second module begins by presenting five types of sexual language. It is emphasized that each type of language represents a distinct set of values and attitudes towards human sexuality. Participants are asked to practice communicating in each of the five styles to illustrate the power of sexual vocabularies, and to help them be aware of which type of language they are most comfortable with. This module ends with a presentation on the skill of being sexually assertive, i.e., asking for what one wants in a sexual relationship. Passive, aggressive and assertive communication styles are discussed and illustrated in the large group. This is followed by participants practicing the skill of sexual assertiveness with their partners.

Module III presents the role of traditional gender roles and their influence on sexual relationships. The advantages and disadvantages of traditional gender roles are illustrated and discussed via a role reversal guided fantasy. Alternatives to traditional gender roles are presented and participants are asked to examine with their partner the roles they play in their personal relationship. The focus is then directed towards sexual myths and misinformation. Common sexual myths are discussed and information is presented to help clarify myths. Some of the myths covered deal with masturbation, menstruation, homosexuality, simultaneous orgasm, and the relative importance of orgasm in sexual relationships.

Module IV outlines the concept of togetherness and apartness in intimate relationships. This concept is then applied to sexual relationships and the role of masturbation is introduced. Emphasis is placed upon the fact that masturbation is an appropriate and normal sexual outlet for singles and people in intimate relationships. A self pleasuring exercise is then used to highlight the importance of individuals allowing themselves this permission. The effects of chemicals on sexual functioning are then presented via lecture. The major points of this lecture are as follows:

1. Few if any chemicals enhance sexual behavior.
2. Many persons have come to believe chemicals can aid their sexual performance.
3. In minor doses, CNS depressants may reduce inhibitions, while larger doses may yield a reduction in sexual performance (this is especially true for males).
4. A state of intoxification can be achieved through sexual relationships.

Couples are then asked to discuss the effects of chemical use on their sexual relationship and to identify if a pattern existed where chemicals served either as an excuse to have sex or a reason to avoid sexual behavior.

The fifth module begins with a description of the types of sexual dysfunctions found in both females and males. Treatments for each of the dysfunctions are then discussed. Couples are asked to consider their sexual relationship and to identify and discuss any existing dysfunctions. Participants are informed that any identified dysfunctions should be considered in the contract setting stage of the next module. The last half of Module V is devoted to the relationships between intimacy, love and sex. This section begins with a lecture and is followed by dyadic discussion. It is requested that the couples consider the level of intimacy in their relationship and how this affects their sexual behavior.

Module VI allows for discussion, clarification and integration of the material previously presented. This is followed by a lecture and role play on negotiating in relationships. Couples are then asked to negotiate a contract for change within their sexual relationship (guidelines for effective contracting are provided). At this point, couples who have identified sexual dysfunction are asked to contact group leaders for referrals. Likewise, couples identifying relationship or intimacy problems are appropriately referred. This module concludes with a verbal and written evaluation of the SREP, and a final wrap-up.

SUMMARY

This paper has briefly presented an after-care program designed to enhance the sexual relationships of chemically dependent persons and their partners. The Sexual Relationship Enhancement Program (SREP) is a highly structured, educationally oriented group training program intended to provide information on sexuality, interpersonal relationships, clarify sexual values and attitudes, and teach communication skills. Some of the content areas covered are: sexuality and sensuality, sexual language, gender roles, chemicals and sexuality, sexuality dysfunctions, intimacy, love and sex. For further information, please contact the senior author.

REFERENCES

Annon, J.S. 1975. *The Behavioral Treatment of Sexual Problems, Vol. I.* Honolulu: Enabling Systems.

Briddell, D.S. and Wilson, G.T. 1976. Effects of alcohol and expectancy set on male sexual arousal. *Journal of Abnormal Psychology.* 85: 2 225-234.

DeLeon, G. and Wexler, H. 1963. Heroin addiction: its relation to sexual behavior. *Journal of Abnormal Psychology.* 81: 1 36-38.

Egan, G. 1976. *Interpersonal Living: A Skills/Contract Approach to Human-Relations Training in Groups.* Monterey: Brooks/Cole Publishers.

Fischer, B.E. 1977. Sexual communication skills for drug abuse counselors. Paper presented at National Drug Abuse Conference, San Francisco, May 1977.

Focus: Human Sexuality. An Annual Editions Reader. 1976. Guilford: Dushkin Publishing Group.

Gay, G.R. and Sheppard, C.W. 1973. Sex-crazed dope fiends! Myth or reality? *Drugs and Youth: The Challenge of Today.* New York: Pergamon Press. 149-163.

Guerney, L.F. 1976. Filial therapy program. *Treating Relationships.* Lake Mills, Iowa: Graphic Publishing Company.

Kaplan, H.S. 1975. *The Illustrated Manual of Sex Therapy.* New York: Quadrangle/The New York Times Book Co.

McCarthy, B.W., Ryan, M. and Johnson, F.A. 1975. *Sexual Awareness: A Practical Approach.* San Francisco: Boyd and Fraser Publishing Co.

McCary, J.L. 1973. *Human Sexuality*. New York: Van Nostrand.

Miller, S., Nunnally, E.W. and Wackman, D.B. 1975. *Alive and Aware: Improving Communication in Relationships*. Minneapolis: Interpersonal Communication Programs.

Peele, S. 1975. *Love and Addiction*. New York: Taplinger Publishing Co.

Rappaport, A.F. 1976. Conjugal relationship enhancement programs. *Treating Relationships*. Lake Mills, Iowa: Graphic Publishing Company.

Readings in Marriage and Family 1977/78. Guilford: Dushkin Publishing Group.

Schoener, G. 1976. The heterosexual norm in chemical dependency treatment programs: some personal observations. *Stash Capsules*. 8:1.

Simmons, R. 1977. Drugs and sexuality. Paper presented at National Drug Abuse Conference, San Francisco, May 1977.

Skolnick, A. 1973. *The Intimate Environment: Exploring Marriage and the Family*. Boston: Little, Brown and Company.

Williams, J.H. 1977. *Psychology of Women: Behavior in a Biosocial Context*. New York: W.W. Norton.

A MODIFIED COURSE IN CRISIS INTERVENTION FOR POLICE OFFICERS

Elena J. Eisman, Ed.D.

Private Practice

Beverly, Massachusetts

This paper describes a modification of the Bard (1972) model for training police in crisis intervention. It will discuss the course in terms of the rationale for the training design, the demographics of the officers involved, the content of the course, and the issues addressed through the training. The paper will also include statements written by the officers during the course of the training.

The major reason the modification of the training was done was to address the unique needs of non-urban police officers. Due to the smaller sizes of these departments, there is less job specialization among the officers. They are expected to perform duties ranging from traffic control to patrol duty; crime investigations to ambulance runs to switchboards. One officer was even called to remove a pet boa constrictor from a toilet. Since the range of training needed is so overwhelming, often the officer is sent with none at all. As one officer wrote, "I can remember my first assignment. I was called to report to work at the evening watch, given a revolver, no holster, a billy club, and a box key. I was told to patrol a certain area, and not to embarrass the department. No other instructions."

In addition, the lower reported incidence of some of the more difficult crisis situations (rape, child abuse, suicide) make on-the-job training through experience a long-term venture.

Finally, the politics of living in a small town where everyone knows everyone else, affect the nature of police work both on a procedural and a personal level. In terms of procedure, in some small towns the officer needing back-up on a domestic call is not allowed to use the radio for his call. He or she must instead gain permission from the disputants (not an easy task) or a neighbor to use the telephone to call the station. The reason for this is that many residents of the town listen to police band radio, and the department is concerned that the disputants might suffer social embarrassment by having their problems broadcast to the whole town. On a personal level, the police officer and their family is identified as such by the community. Their lives are constantly public ones. As one officer wrote, "a cop living in the neighborhood is also looked upon as a resident authority on all neighborhood crises as well as the person to call on to keep the kids in line." An even more poignant statement of the problem of high community visibility for the police family in a small town was written by this ten year veteran:

> "Why did I become a police officer? . . . Sometimes I think I was crazy . . . I never expected to have my house stoned at three in the morning. I never expected to have my girls who are 6 and 7 years old threatened because I was a cop. I never expected to have my wife come home crying because she heard some stupid rumor about me. I never expected my every word and action to be taken apart under a microscope. I never thought I'd be hated by the same people that I took an oath to protect."

The other major reason for modifying the training was a desire on the part of community based, drug treatment agencies to improve relations with their local police departments and hopefully to increase interagency referral between the police and human service agencies. To this end, attention was directed toward building, through the course, a special knowledge of referral sources in the community. Assignments were given to the officers which required them to visit a local human service agency and to interview its staff. They were then required to report on and evaluate this agency in terms of their feelings about the agency (staff, hours, service, and availability) and their likelihood of using it as a resource. These reports were shared in class, and a referral list was compiled for the use of all officers involved.

TRAINEE DEMOGRAPHICS

Most crisis intervention training programs for police are run through local police departments. This program was run as a 15 week, 3 credit elective course through a local community college. The benefit in this approach is that there is a great potential for sharing information and techniques among the officers since the students represent different police departments and different branches of law enforcement such as State police, City police, Town Police, and Prison Guards. The course itself was developed and modified over five semesters. Police participating ranged from inexperienced officers to Captains in the department. One student even went on to become a

Chief of Police. The students also included both men and women officers.

The course was designed to include didactic input, group discussion, and experiential learning through role play. A great emphasis was placed on sharing professional, organizational, and personal issues encountered on the job. This was facilitated by the fact that the class was mixed in terms of department. The officers could feel free to talk without risking departmental sanctions. Newer officers gained a better understanding of the difficulty of policy decisions through listening to the "brass" in the group, and the higher ranking officers learned what their younger colleagues on the force had to deal with as a result of their decisions. Perhaps the most interesting sharing was done between the policemen and policewomen since some of the departments represented in the class had no women on the force. The difficulty of women breaking into police work can be shown by this officer's statement:

> "Being the first woman police officer in our town, I realized that in some functions, I would be subdued, such as in one man cars, barroom brawls, while at the same time, I had an idea that I might be utilized as a Public Relations agent . . . I wasn't too wrong. Gradually as the community becomes more used to my being here, and the more I prove myself, the more my "leash" is let out, in this aspect I feel it is a much harder road for a Rookie woman than for a man."

The fact that the author was a woman and a psychologist (double whammy) created some fascinating and highly predictable dynamics over the course of the semester. Each semester revealed the same pattern:

Week 1 - "Apple polishing" incredulity, summarized with statements like "how could a bright, young, educated professional like yourself know anything about police work?"

Week 2 - Revolved around the cry "we're not ____ social workers."

Weeks 3-6 - "This technique is nothing new, we do it all the time, you just don't know the extenuating circumstances we face."

Week 7 - "I tried it your way, it doesn't work."

Week 8 - "I tried it your way, it does work. Was your father a policeman?"

Weeks 9-13 - "We're really learning a lot about how civilians act in crisis."

Week 14 - "I experience crises too, you know!"

Week 15 - "You didn't *really* teach us anything new. This course should be a required course for every police officer."

The group and educational process is facilitated if the instructor is familiar with police procedure and jargon, and utilizes a great deal of case study material drawn from actual police experiences. Most important however, for facilitating this process is an openness, understanding and acceptance of the pressures encountered in police work. Illustrative of these stresses is the following statement:

"A policeman's role is constantly changing from one minute to the next. In one instance I made (a thirty minute run in eight minutes by ambulance) with a critically ill patient. A year later I arrested the 'patient' for a law violation. Another time I pulled a four year old child from the path of an oncoming train, six years later I took this same child home to his parents a second time for throwing stones at some younger children."

As the course developed, each class became a support system as well as an educational unit. The officers found that debriefing and discussing difficult calls helped them improve their techniques as well as enhancing their emotional well-being. One officer writes:

"I always felt better about my job when I left the class on Friday morning. We talked to you about the things that Police Officers usually keep to themselves."

COURSE DESCRIPTION

The course began with a general introduction which outlined a statement of purpose for the course. This included reduced injury rate among police, decreased repeat calls, and increased effectiveness in handling crisis situations.

The next phase introduced the concepts: what is a crisis, what are the symptoms of a person in crisis, and how does one intervene in a crisis? The manual and paradigm used for this selection was developed by Boxley, Freundl, Rosenthal, and Bradley (1974). The model is a modification of the Bard (1972) training design, focusing on crisis intervention in domestic disputes. The training paradigm covers external and internal assessment skills, defusing techniques, communication skills, data collection, problem solving, referral, and follow-up.

Role plays are drawn from experiences to the officers in the class to keep them relevant and realistic. Roles of both the disputants and the interveners are played by people in the class. Values clarification exercises are built into the curriculum to elicit sexual, racial, and cultural biases and differences. This segment lasts seven to eight weeks.

Following the above segment, individual classes are devoted to focusing on some of the more common crises likely to be encountered on the job. This is done to sensitize the officers to the unique constellation of issues likely to be encountered in victims of these crisis situations. The categories covered are drug and alcohol emergencies, suicide, rape, child abuse, battered women, and adolescent crisis reactions (runaway, truancy, and gangs). Whenever possible, resource people are brought into these sessions, either past victims or human service personnel working with clients facing these crises. This segment lasts for six weeks.

During all phases of the training, officers are encouraged to discuss their personal reactions when dealing with these crisis situations. The next to last class is devoted exclusively to this issue. Often this session revolves around the rewards and frustrations of police work. Much discussion is devoted to the strains faced by the families of the officers. The high divorce and separation rates of police families are discussed, as well as ways that some of the officers have found for maintaining the communication necessary to avert marital discord. Frustrations with departmental decision making are also discussed, especially in light of the difficulty in implementing new crisis intervention procedures and techniques. There is much discussion about the Court system and their feeling that it undoes much of what they work so hard to achieve (one officer called the Court system "Let's Make A Deal."). Other issues raised are the unresponsiveness and unavailability at night of human service resources, and the tendency of legal and medical people to demean their perceptions. By far the greatest amount of time spent in this session is spent reviewing crisis calls they have handled, and the feelings they have engendered in the officer; the family that drinks then fights every Saturday, the experience notifying parents that their small child was killed, the suicide they tried to avert but couldn't, the 14 year old who was gang-raped, and the abused child. The toll that these experiences take on the individual officer is tremendous, especially if there is no person in their life with whom to discuss their feelings. There is the further strain that in dealing with any of these situations, they themselves could be injured or killed. Officers talk, often for the first time about losing control, losing their temper, crying, grinding their teeth in frustration, or drinking too much. The author finds it very helpful at these times to remind the officers that even if their jobs seem to demand super-human qualities, they are still people and still subject to the full range of human reactions. Surprisingly, for many of them, it is the first time that they have been given "permission" to be human. One policewoman writes:

> "I felt a little nervous about situations that might come up, and would I be able to handle them. I've learned that without a small bit of fear and respect of what may be out there, you can become too confident, lose your perspective, and possibly get hurt. Part of this I have learned from this class, and it has made me feel better about those butterflies that sometimes won't go away."

This brings us to the last issue stressed throughout the training, and it is what the author calls a state of "creative alertness." The officer is reminded to stay alert in order to assess the environment for cues of potential physical harm, and also for creative ways to deal with the crisis on hand. In one class an officer reported going on a domestic call with a "chronic" family. This time the wife was threatening the husband with a kitchen knife. Instead of his usual approach, the officer took the knife and brought both parties to the living room piano. On the piano were a doily and two family portraits. He gave each disputant a portrait to hold, sat one on each side of him on the piano bench, lifted the doily, and began playing "Love Makes The World Go 'Round." After they all sang two choruses, things calmed down, and the officer left. The officer in this situation was creatively alert which helped him to assess the situation, protect himself and the civilians from harm, and to find a solution out of this crisis situation with a much better outcome than his usual pattern of arresting one of the parties.

COURSE REQUIREMENTS

There were three written assignments for this course. The officers were informed however, that this was a course that couldn't be learned from a book, so that attendance and participation were also included in their final grade.

The first paper asked the officers to "write a description of a disturbance crisis they had handled." They were asked to describe what they did then, and what if anything, they would have done differently. This paper comes in the middle of the training after the basic crisis intervention technique has been presented. The student is required to answer the following questions:

* Why is it a crisis?
* Who is involved?
* Who is peripheral?
* What are the disputants feeling?
* What are the police feeling? What was the prior call?
* What was the substance of the dispute? Short or long term?
* Describe the initial reponse, preplanning if any.
* External and internal assessment?
* Defusing?

* Resolution?

The second paper was a report on the officer's visit to the human service agency. This was mentioned earlier in this paper.

The third paper asked the officers to focus on a general discussion of the "changing roles of the police officer in contemporary society." This paper was to be largely autobiographical, and to answer the following questions:

What originally attracted you to the job of police officer? What were your expectations concerning your function in the community? How did you think you would be received by the public? How has the role of police officer changed during your tenure as an officer? Did the job conform to your expectations? What do you see as the future of law enforcement? Will it be different from the traditional role of the police?

The papers were useful in giving the officers a structured way to think about the material presented in class and its relationship to their everyday professional lives.

COURSE EVALUATION

Police assessment of the course was uniformly positive. Almost all officers reported that it was useful to them in their work, and that they had entered the course with high initial skepticism. One writes:

> "When I first decided to take this course, I thought it would be interesting, but I never knew it would reach into the real meat of the problems a police officer has to face everyday. At first look I felt that the approach would be too liberal but now I know that this course is a must for all officers."

They reported new learnings:

> "I always thought runaways were fresh little punks. I never considered that they may have been subject to child abuse."

They also reported gaining new tools with which to work:

> "Crisis Intervention has been a rewarding and informative experience for me. This course has shown me that there is always an other (sic) alternative and how to use it effectively. It has opened my eyes as why people act the way they do and how to deal with it effectively. I have already used some of the techniques and to a positive result.

"I enjoyed this experience very much. Thank you."

Most importantly, the officers gained a new level of self-reliance in dealing with all types of crisis situations.

I will close with a rather sad anecdote.

Officer A told the class about the first time he had really lost control with a civilian. It was 14 years ago, and he answered a domestic call where a husband and wife were fighting and drunk. During the fight, the husband had thrown an ashtray and broken a window above where their baby daughter lay in her crib. The officer saw the child laying there covered with snow and broken glass. He saw red, hit the husband, and took the baby back to the station.

Fourteen years later, at the time both officers were in the crisis intervention class, Officer B was manning the phone in the station when a young adolescent called sobbing. He responded to her crisis and convinced her to come down to the station. She came in and threw her mother's works on the table. She said that her mother was a junkie and that she couldn't take it any more. The family had been burnt out of their apartment, and she, her mother and her younger brothers were living in an empty apartment with mattresses on the floor. Her mother was pushing, and there were so many people coming in and out of the apartment, shooting up in the bathroom, that she and her brothers had to wait in line when they had to use it. She had been beaten with a belt and there were angry welts on her back. Officer B took her to the emergency room, and called the mother's probation officer. No one could help. He finally took her to a temporary shelter for adolescents with which the author was affiliated. The staff there finally involved the Welfare Department, but they said that they couldn't remove her from the home because she was 14 and the welts weren't "big enough" to constitute a life threatening situation.

The mother said that she wanted the reluctant girl to come back home. Officer B was then assigned to escort the mother to the temporary shelter to bring the frightened, hysterical girl back home. Officer B writes:

> "As you can see, I'm not satisfied with the way things are going. And, Dear Elena, as I am done with college I doubt if our paths will cross again. I want you to know . . . I enjoyed your classes very much. I'm just sorry I couldn't teach you anything."

You did.

REFERENCES

Bard, M. and Shellow, R. 1976. *Issues in Law Enforcement*. Reston, Virginia: Reston Publishing Co.

Bard, M., Zacker, J., and Rutter, E. 1972. *Police Family Crisis Intervention and Conflict Management*. LEAA.

Bittner, E. 1970. *The Functions of Police in Modern Society*. Rockville, Md.: NIMH Center for Studies of Crime and Delinquency.

Boxley, R., Freundl, P., Rosenthal, K., and Bradley, R. 1974. *Family Crisis Intervention*. Boston: Police Training Reference Series, Boston Police Department.

Siegel, M. 1978. Workaday world of the cop: stress, unending stress. *Parade*. March 12.

URINE ANALYSIS FOR METHADONE PROGRAMS: A CLIENT-CENTERED COMPUTER-BASED AUTOMATED TECHNIQUE

Donald J. Egan, M.D., David Owen Robinson, Ph.D., J. Richard Pearson, Ph.D., and Jann Wagner, R.N.

University of Colorado Medical Center

Urine surveillance of opiate addicts in methadone treatment programs is an important feature of contemporary treatment techniques. The purposes of urinalysis were defined by Nightingale, Michaux and Platt (1972):

* To yield clinical information to the counselor.
* To encourage the client's abstinence from illicit use.
* To symbolize the counselor's interest in the client's progress.

To these we may add that urinalysis results are used:

* As a measure of treatment program efficacy.
* To comply with Federal regulations.

This last factor--compliance with Federal regulations--is possibly the most important to many treatment programs since it delineates the minimum urine testing which must be done. Presently, (under 21 CFR 130.44), programs are required to test a client's urine samples at least once a week for opiates and at least once a month for methadone, barbiturates and amphetamines and for other drugs of abuse as indicated.

GOALS OF URINE TESTING

Once the Federal regulations are satisfied, the primary goal of urine testing should be *to provide clinically useful information*. Urine testing is useful when it provides:

* Information which is not available from other sources.
* Information which is believable to counselor and client.

Catlin (1973) outlined the necessary attributes of urinalysis for these goals to be achieved:

Sensitivity

Urine tests must be able to detect drug use when it occurs. If a client gives a urine specimen when he knows that he has recently used and the urine testing fails to detect this use, then the client's confidence in the urinalysis (and hence the treamtent program) is diminished.

Specificity

Urinalysis must not suggest that drug abuse has occurred when it has not. For example, the appropriate use of over-the-counter medications should not result in a "false-positive". A confrontive discussion between counselor and client which is based upon erroneous information is destructive to the therapeutic process.

Turn-around-time

Urinalysis has little clinical significance if a large delay exists between the time a specimen was collected and the time when the results of the program are known to the counselor. When a delay of several days exists before results are known, all the information becomes moot. A client sees little connection between his behavior (drug use) and the consequences (his counselor becomes aware of the drug use).

Costs

Even if urinalysis is an extremely useful adjunct to methadone maintenance treatment (and this point is debatable, see Goldstein, 1974), its value is diminished if its actual cost is so high that it interferes with other functions of a treatment program. When testing costs are high, clinical staff will fail to make maximum use of urinalysis.

A CLIENT-CENTERED APPROACH

Even when the attributes outlined above are satisfactorily achieved, urinalysis can be made more valuable by adopting what we call a "client-centered approach". In the system presented in this paper, both the frequency of testing and the type of tests which are ordered are determined for each client by his counselor to maximize the therapeutic potential of testing. A number of examples will clarify this point:

1. A stable methadone maintenance client, who is employed and has been "clean" for a year, would have only the minimum number of tests ordered to comply with regulations.

In this first example, additional negative urine results (i.e., no detection of illicit use) would give no useful information to either counselor or client. The client knows that either a relapse into heroin addiction or occasional "chipping" will be detected by random analyses so the goal of encouraging the client's abstinence is achieved at the smallest cost to the program.

2. A maintenance client comes to see his counselor in a crisis state; he says he has left his wife and has used heroin once, two days ago, and he feels sure that he will be able to go to work and stay clean this week. After discussion between counselor and client it is agreed that the client will give urines three times a week and that the urines will be screened for opiates.

In the second example, the client expresses the hope that he can stay clean even though he is under emotional pressure. The counselor can be supportive and express confidence in the client's ability. The increased costs are worth it if they help the client to avoid a relapse.

3. A 21-day detoxification client says on interview that she took Darvon for a number of years before recently switching to heroin. The counselor makes a note to screen all urines for propoxyphene (Darvon).

In the third example, the additional expense of the extra test is justifiable because information about Darvon use would be clinically valuable. The success of the methadone detoxification would be limited if heroin use were merely translated into re-addiction to Darvon. For this reason it is important for the counselor to know whether that particular drug is being used; however, its use in the clinic as a whole is rare and routine screening for Darvon of all clients would be an unwarranted expense.

4. A methadone maintenance client has had several positive urinalyses for cocaine. He makes it clear to his counselor that he has no intention of discontinuing his cocaine use which he sees as recreational. The counselor no longer orders cocaine tests on this client.

In this example, the client's continual cocaine use is a matter of common knowledge to client and counselor. Additional testing adds no new information to the treatment situation. The counselor decides to eliminate the special test for cocaine because the costs are not justified by clinically relevant information. If at some later time the client and counselor agree to work on eliminating cocaine use, testing for this drug could be reinstated.

To summarize, the client-centered approach to urinalysis aims at maximizing the clinical utiliy of the results while minimizing the costs of testing.

DISSATISFACTIONS WITH THIN LAYER CHROMATOGRAPHY

Thin layer chromatography is widely used as a method or urinalysis for drug treatment programs. Its advantages are that it is a broad screening technique, testing materials are relatively cheap, and the demands for specialized equipment are small; however, in our experience it has a number of disadvantages:

* Sensitivity is low, especially for synthetic opiates such as Dilaudid.
* Specificity is poor; cocaine may be hidden by methadone and nicotine and even Mexican food may lead to spurious unidentified positives.
* Turn-around-time may be high because each test must be interpreted by hand.
* Costs depend largely on salary costs since the method is labor-intensive. Personnel costs may make this an expensive technique.
* The same test is run on each client regardless of the clinical utility of the information gained.

A COMPUTER-BASED TECHNIQUE

We have developed a system for urine drug screening which overcomes the disadvantages of thin layer chromatography. The system uses the Enzyme Multiplied Immunoassay Technique (EMIT) (Syva, 1976) and an automated laboratory analyzer (American Monitor KDA) in a computer-controlled (IBM 1800) laboratory. The details of the equipment and laboratory techniques are described in a companion paper (Pearson, 1978).

Urine specimens are collected at the treatment clinic in accordance with usual practice. The counselor who collects the specimen consults the client's sheet in the urine record book which indicates the tests to be ordered for the individual client and which shows the date and specimen number of all samples taken. The counselor makes out a lab order slip (shown in Figure 1) and attaches the same sequence number to all three copies of the slip, the specimen test-tube, and the client's sheet in the record book. One copy of the order slip is kept at the program and the other two copies accompany the specimen to the laboratory. Specimens are taken to the laboratory by the program on a regular schedule agreed upon by the laboratory and program staff.

When specimens arrive at the laboratory, they are assigned a laboratory accession number. That number, the client number, and the test instructions are then entered by keypunch into the laboratory's master computer. The computer then prepares a work list which instructs a laboratory technician how to load the KDA analyzer. The location of calibrating specimens is specified and specimens from applicants are automatically placed at the head of the queue. The KDA then performs the analyses ordered for each specimen by consulting the information in the laboratory master computer. Results are displayed on the KDA so that the technician can observe the progress of the run and they are simultaneously automatically entered into the master computer. The technician then telephones the program with results of analyses on applicants' specimens. The computer collates the results for each client and sorts them into client number order. Two copies of results are then printed: a daily list for the entire clinic in which positive results are flagged and an individual result slip for each client (shown in Figure 2). These are then delivered to the treatment program where clerical staff interpret client numbers to client names and file the individual result slips. A final stage in the implementation of this system will be to install a computer terminal at the treatment program which will print results on site.

EXPERIENCE

After some minor adjustments (to printed forms and to schedules) the system has been developed to the satisfaction of both treatment and laboratory staff. The system has apparently higher sensitivity than the thin layer chromatography technique and the rate of "positives" with EMIT has been approximately 20% higher than with thin layer chromatography, a difference which is highly statistically significant. Our clinical experience suggests that this represents a genuine increase in sensitivity (rather than an increase in the number of false positives) since there has not been a concomitant increase in disputes between clients and counselors over the accuracy of urine results.

With some exceptions (which will be discussed below) the specificity of the EMIT technique has been good and both clients and staff have developed confidence in the results which the system gives. This confidence is aided by rapid turn-around-time. Results are obtained

	U. C. M. C.
01	FULL SCREEN INCLUDES OPIATES, AMPHETAMINE, METHADONE AND BARBITURATE
02	ROUTINE OPIATE
03	ALCOHOL [ETHANOL, ETOH]
04	DIAZEPINES [VALIUM]
05	COCAINE
06	AMPHETAMINE
07	METHADONE
08	BARBITURATE
09	PROPOXYPHENE [DARVON]
	ADDICTION RESEARCH AND TREATMENT SERVICES

SEQUENCE #

CLIENT #

SITE #

DATE COLLECTED — MONTH DAY YEAR

LAB ACCESS #

ART
15276
ART
15276
ART
15276
ART
15276
ART
15276

Fig. 1. Urinalysis order slip and sequence numbers (self-adhesive strip of five numbers). One sequence number is placed on each of the three copies of the slip, one on the urine specimen test-tube and the fifth is used to record the collection of the sample.

```
THIS REPORT PRINTED ON   3/28/78 AT   8.09 HRS                WARD     1
          CLIENT #     0725           ART  AGE      HOSP NO.  15123.
          DR.     0    0
SAMPLE   UR     RECVD ON   3/27/78 AT   0.00 HRS ACCESS NO  2381  03/26/78
C                  OPIATE              POSITIVE
E                  AMPHETAMINE         NEGATIVE
N                  METHADONE           WILL RPT
T                  BARBITURATES        NEGATIVE
R
A
L

L
A
B
S
```

CHART COPY - ORIGINAL   PHYSICIANS COPY - DO NOT CHART - PINK STRIPE

Fig. 2. Urinalysis result slip. This becomes part of the client's permanent medical record after a clerk has interpreted the client number to a client name.

in under 24 hours for applicant urines (for which the results are telephoned to the clinic) and the results of routine analyses are typically available within 48 hours. The staff time-load on the laboratory is small: 40 specimens may be tested for each of four drugs of abuse by one technician in a total time of $2\frac{1}{2}$ hours, including the calibration and loading of the machine.

The costs of operating this system are quite acceptable to the program. The services provided to the treatment program by the laboratory on an inclusive annual contract (rather than a cost-per-test basis). Present estimates are that urinalyses for 298 clients (262 in outpatient treatment and 36 in residential treatment) can be met at an average cost of $1.68 per client per treatment week. It is estimated that the on-site computer printing of results would increase this figure to approximately $1.75. Two points about this cost estimate should be emphasized. First, this figure could not be achieved without continual re-evaluation for the need for special tests. If staff are encouraged to order extravagant numbers of tests on each client, the materials' cost alone for one specimen could exceed $5. Second, this system depends on a pre-existing computer-controlled laboratory with expensive automated analyzing equipment whose capital costs are not amortized into this cost estimate.

The attitude of clinical staff toward urinalyses has been favorably altered by the introduction of this system. Urinalysis is no longer viewed as an unfortunate bureaucratic intrusion into the treatment process, but is viewed as a valuable therapeutic tool.

RESERVATIONS

The only problem area which we have to report concerns the specificity of certain Syva reagents used in the EMIT process. We did not at first realize that the Syva "opiate" reagent effectively picks up the use of synthetics such as Dilaudid and Percodan. This particular incidence of lack of specificity is convenient from the point of view of a methadone treatment program; it does not really matter whether illicit use is of heroin itself or some synthetic opiate. A more perplexing problem was that a number of clients who are maintained on LAAM (and not methadone) had positive methadone results which we believe are spurious. We understand from Syva that the specificity of the methadone reagent has been improved and that this problem will not occur with future shipments of reagent.

REFERENCES

Catlin, D. 1973. *A Guide to Urine Testing for Drugs of Abuse.* Special Action Office Monograph, Series B, Number 2.

Goldstein, A. and Judson, B.A. 1974. Three critical issues in the management of Methadone programs. In P.G. Bourne (ed.) *Addiction*. New York: Academic Press.

Nightingale, S.L., Michaux, W.W., and Platt, P.C. 1972. Clinical implications of urine surveillance in a Methadone maintenance program. *Int. J. Addict.* 7: 403-414.

Pearson, J.R. 1978. Drug screening by enzyme immunoassay with the American Monitor KDA. In preparation.

Syva, 1976. *Package insert: EMIT drug abuse urine assays*. Palo Alto: Syva.

PREDICTING METHADONE TREATMENT RETENTION AND SUCCESS WITH THE MINI-MULT

Bruce Duthie, Ph.D. and J. Ray Hays, Ph.D.

Texas Research Institute of Mental Sciences

Treatment success for methadone patients can be approached in various ways but might best be defined as the complete rehabilitation of the addict. The four general areas which should be assessed by any evaluation of the rehabilitation of the opiate addict are reduced drug use, reduced legal involvement, improved psychological functioning, and improved vocational circumstance. There are many approaches to this rehabilitation such as vocational adjustment (Meyer, 1972; Schoolar, Winburn and Hays, 1973), legal involvement, and illicit drug use (Winburn and Hays, 1974). Before an addict may reach the point of successful treatment the addict/prospective patient must undergo a rather severe transition in lifestyle and severe internal changes, psychological and physiological. The changes are best illustrated in a pioneering study which examined a cohort of eight individuals in their transitions from opiate addiction to methadone patient (Pablant and Hays, 1976). The drop-out rate for programs remains quite high with the general rate of only 25% pursuing treatment after the initial contact with the clinic. This statistic is based upon informal analysis of data on contacts, intakes, and those admitted to treatment. The number of drop-outs creates difficulties for the administration of programs; it may not represent treatment failures but is at least a failure on the program's part to deal effectively with that erstwhile patient.

In order for the program to be successful with a particular individual that patient must remain in contact with the clinic long enough to evaluate each of the four areas of functioning, listed above, whether or not that individual is receiving direct services from that program.

The most illusive of the four areas cited is the change in psychological functioning. Such change can be measured on various instruments which relate to the underlying psychology theory of human function and can range from dynamic or analytic theories measured by projective instruments such as the Rorschach to the behavioral or interpersonal theories measured by behavioral contingency analysis or adjective checklists. The plethora of instruments available, the abundance of theories, and the unique orientation of the individual care provider combine to make comparison among studies on this dimension of improved psychological functioning difficult if not impossible.

One of the most commonly used psychological assessment instruments which was empirically developed upon psychiatric populations and is useful for description of the intrapsychic processes of individuals is the Minnesota Multiphasic Personality Inventory. It has also been used extensively in the drug abuse population. A sub-scale of the MMPI has been developed which identifies the unique responses of the heroin addict: the heroin addiction scale (He) (Cavior, Kurtzberg, and Lipton, 1967).

The heroin addiction scale has been used successfully to differentiate an addict group from a physical rehabilitation group (Kwant, Rice, and Hays, 1976). A number of other studies using the MMPI clinical scales to examine the personality of the addict have found evaluations on psychopathic deviate (Pd) (Hill, Haertzen, and Glaser, 1960) and hysteria, depression, psychopathic deviate, masculinity-femininity, and psychasthenia (Henriques et al., 1972).

Several studies have examined demographic variables associated with success in the methadone treatment of heroin addiction (Levy and Tortelli, 1974; Rosenberg, Davidson, and Patch, 1972; Babst, Chambers, and Warner, 1971). Several factors were found to be associated with successful retention of clients on a methadone program. These factors were single drug abuse, a short history of addiction, employment, and living with a family.

Zuckerman et al. (1975) utilizing the MMPI found that clients staying in treatment had elevations on the psychopathic deviate and hypomania scales. Clients who terminated treatment early had high scores on psychasthenia, schizophrenia, and hypomania. This was interpreted to mean that personality disordered and neurotic profile types had better success in therapeutic communities than did clients with psychotic profile types. Bess, Janus, and Rikkin (1977), also using the MMPI, found that recidivist addicts when compared to ex-addicts scored higher on the F, psychasthenia, and schizophrenia scales.

This evaluation was conducted to determine whether or not differences exist between successful and unsuccessful methadone outpatients on the Mini-Mult (Kincannon, 1967) which is a short version of the

MMPI. Retention on the program for 6 months, and urine records are the criteria of success.

METHOD

Subjects

All clients initiating treatment from May 1977 and January 1978 at the TRIMS Substance Abuse Research Clinic were included (n=50) in this study.

Measures of Success

Two criteria of therapeutic success were examined: retention on the program for six months versus termination prior to six months, and urine records for January and February 1978 for clients who had been on the program at least 3 months. Those with opiate positive urines were compared to those with opiate negative urines. Each comparison was made using multiple discriminant analysis.

Instruments

The Mini-Mult was administered as part of the routine admission procedure and scored using the standard techniques. The standard clinical MMPI scales were used to examine the psychological differences between the groups.

RESULTS

Visual profile differences from each set of outcome criteria were obtained. The mean MMPI T-scores for the groups are presented in Table I. Multiple discriminant analysis of the client profiles on the program six months versus those terminating before six months was not significant ($p > .05$). Multiple discriminant analysis of the profiles of clients with opiate positive and opiate negative urines also was not significant ($p > .05$). However, when the sign test is applied to the eight clinical scales of the MMPI profiles of those dirty versus those clean there is a significant difference between the profiles ($p < .05$) with the profile of those who are clean being more elevated than those who have dirty urines. This is an extremely surprising result which has far reaching implications for the therapeutic work conducted by treatment programs.

Although not significant, the main difference between the clients who stayed on the program when compared to the clients who quit the program before six months was obtained on the Depression scale of the Mini-Mult. The Hypochondriasis and Psychasthenia scales were also slightly higher for the clients who quit the program before six

Table I. MMPI T-scores of patients with "clean" versus "dirty" urines and those who terminate versus those who stay in treatment.

MMPI Scale	Clean n=11	Dirty n=18	Stayers n=30	Quitters n=21
L	50	56	53	53
F	66	55	60	60
K	51	57	53	49
Hysteria	72	72	70	72
Depression	77	75	75	82
Hypochondriasis	75	73	73	71
Psychopathic deviate	76	74	74	71
Paranoia	70	62	66	68
Psychasthenia	73	69	69	73
Schizophrenia	88	71	73	73
Hypomania	60	55	59	60

months. These results suggest that differences may be related to a generally more neurotic trend in the clients who left the program prior to six months. Clients with opiate positive urine test results seem to show less overall psychopathology than clients with opiate negative urine test results.

DISCUSSION

Further research is needed to clarify findings of this evaluation. Two hypotheses have emerged clearly from this study: (1) Methadone clients who terminate methadone treatment before six months have generally more neurotic psychopathology than those remaining in treatment longer than six months, and (2) clients on a methadone program with opiate positive urines have less general psychopathology than those with opiate negative urines.

Those who terminate treatment prior to six months tend to have less ego strength and tend to be more depressed than those individuals who remain in treatment. The pattern of an individual who has less than the usual amount of self esteem and who shows clinical symptoms of depression is less likely to remain in treatment than the average prospective patient. Such individuals could be approached with this fact at the outset of treatment as a therapeutic strategy in order to reduce the loss to the program of this type of individual. A paradoxical strategy such as that used by Watlawick could be employed. In any event the foreknowledge on the part of the intake and treatment staff of the tenuous nature of the commitment of such individuals is important.

Replication with a larger sample is definitely in order. A design examining urine records over a year's period of time employing a regression analysis might be more productive in examining "clean" versus "dirty" urines.

REFERENCES

Babst, D.V., Chambers, C.D., and Warner, A. 1971. Patient characteristics associated with retention in a methadone maintenance program. *British Journal of Addiction*. 66: 195-204.

Bess, B., Janus, S., Rikkin, A. 1972. Factors in successful narcotics renunciation. *American Journal of Psychiatry*. 128: 861-865.

Cavior, N, Kurtzberg, R.L., and Lipton, D.S. 1967. The development and validation of a heroin addiction scale with the MMPI. *International Journal of the Addictions*. 2: 129-137.

Henriques, E., Arsenian, J., Cutter, H., and Samaraweera, A.B. 1972. Personality characteristics and drug of choice. *International Journal of the Addictions*. 7: 73-76.

Hill, H.E., Haertzen, C.A., and Glaser, R. 1960. Personality characteristics of narcotic addicts and criminals. 62: 127-139.

Kincannon, J.C. 1968. Prediction of the standard MMPI scale scores from 71 items: the mini-mult. *Journal of Consulting and Clinical Psychology*. 32: 319-325.

Kwant, F., Rice, J.A., and Hays, J.R. 1976. Use of heroin addiction scale to differentiate addicts from rehabilitation clients. *Psychological Reports*. 38: 547-553.

Levy, N.M. and Tortelli, J.A. 1974. Methadone dropouts in a drug free therapeutic community. *Drug Forum*. 3: 225-231.

Meyer, R.E. 1972. *Guide to Drug Rehabilitation--A Public Health Approach.* Boston: Beacon Press.

Pablant, P. and Hays, J.R. 1976. Descriptive study of behavior during transition from heroin addict to methadone patient. *Psychological Reports*. 39: 607-620.

Rosenberg, C.M., Davidson, G.E., Patch, V.D. 1972. Patterns of dropouts from a methadone program for narcotic addicts. *International Journal of the Addictions*. 7: 415-425.

Schoolar, J.C., Winburn, G.M., and Hays, J.R. 1973. Rehabilitation of drug abusers--A continuing enigma. *Rehabilitation Literature*. 34: 327-330.

Winburn, G.M. and Hays, J.R. 1974. Methadone: The carrot at the end of the stick. In J.M. Singh and H. Lal (eds.), *Drug Addiction Volume 4*. New York: Stratton Intercontinental Medical Book Corporation.

Zuckerman, M., Sola, S., Masterson, J., and Angelone, J. 1975. MMPI patterns in drug abusers before and after treatment in a therapeutic community. *Journal of Consulting and Clinical Psychology*. 43: 286-296.

SMALLTOWN, NEW MEXICO: AUTOPSY OF A FAILURE

Peter V. DiVasto, Ph.D.
University of New Mexico

David L. Ryther, B.S.
New Mexico Health Education Coalition

Abstract--The Focus Program: "I should have been able to take a course on raising children for 20 years before I even had them!"

A father in family counseling was almost desperate when he made that statement. Oddly enough, it is true that we receive little if any training for what is probably the most important job of our lives--raising children.

Project Focus is a training program which provides parents and teachers with important skills in communication, child management, and values clarification in order to help them help their children to grow. The underlying assumption of Project Focus is that children become involved in deviant, self-destructive, and destructive behaviors (such as truancy, vandalism and drug abuse) because they have poor self-concepts, undeveloped decision-making skills and poor communication abilities. By helping children develop stronger self-images and accept responsibility for their behavior, we can help prevent these destructive behaviors.

Local Project Focus training groups are conducted by trained local counselors and teachers and meet once a week for eight, ten or fifteen weeks at a time and place convenient to the participants.

TRAINING PROGRAM CONTENT AREAS

Communication

Non-verbal: Much of what we say to children and what they say to us we say without words. How can we become more aware of our non-verbal behavior so we don't send mixed messages and how can we become more sensitive to children's non-verbal messages?

Verbal: How to talk to children so that they'll listen, how to listen to children so that they'll talk and how to solve problems so that no one has to lose.

Values Clarification

How do we help children discover and act on their own system of values, guiding beliefs for their lives based on their own needs and experiences?

Child Management

Rule setting: Some rules are necessary and important for children. How can we set rules that are reasonable to children and possible for us to enforce?

Democratic families: How to solve family conflicts constructively, understanding the goals of children's misbehavior and helping children develop responsibility.

Rural communities in New Mexico wishing to conduct Project Focus Substance Abuse Prevention efforts may receive assistance through the State Project Focus Coordinator in the following areas:

1. Community development. Identifying key individuals and institutional personnel in the community and soliciting their support and cooperation in developing a local community prevention effort.

2. Program planning. Assisting the local group in planning a prevention effort based on the Focus Training model and other prevention approaches.

3. Training. Assisting local program in identifying potential local program facilitators. Providing initial training in the skills area of the Project Focus model (30-32 hours) to approximately 20 potential facilitators and school personnel. Providing 16 hours of follow-up training designed to assist the local facilitators in the implementation of the local

parent and/or teacher study groups. Providing local program monitoring.

PROJECT IMPLEMENTATION

In September 1976, the Drug Abuse Education Center in Albuquerque received a telephone call from Ed Bailey, the community drug program coordinator from Smalltown, New Mexico. Smalltown is a ranching and farming community of 12,000, located some 200 miles west of Albuquerque. The gist of Ed's call was as follows: "We're still havin' some drug problems here in Smalltown, even though I've been showin' films in the high school and two of our deputy sheriffs have been to the F.B.I. narcotics school. The sheriff and the high school principal are really gettin' uptight and want me to do something about it. I heard that you folks up in Albuquerque have a bang-up program that you've been doin' that might help us. Something called Project Prevention."

The call was taken by the coordinator for Project Prevention, Ann King. Ann, after hearing Ed out, told him that she would be glad to come right down to Smalltown by the end of the week and tell him about the project. Ann drove down and spent an afternoon describing the project and how it had originated. She described the way the project has been embraced by the public schools as an excellent way for teachers to earn recertification credit. She was careful to point out that, since the model was aimed at primary prevention of drug abuse, it was applicable to all of the potential abusers in Smalltwon's schools. Ann left Ed wildly enthusiastic and anxious to get started as soon as possible. Her parting instructions to Ed were: "Get some folks together and have a meeting. I'd be glad to come down and talk to them."

In a few weeks, Ed called back, saying: "I'm having some trouble getting a group together. I'm kind of new in town and haven't gotten to know some people real well." He finally managed to round up two teachers from the high school, a sheriff's deputy, a nurse from the local hospital and three concerned parents.

When Ann came to speak to the group, they were somewhat skeptical. However, her enthusiasm for the project, coupled with the fact that it would initially cost Smalltown nothing, soon won them over. They agreed to begin Project Prevention training as soon as possible. Ann had just met a psychologist and his wife, newly relocated from the East, who impressed her as being witty, bright and attractive, so she assigned them as trainers for Smalltown.

In November, the trainers, Dr. Mize and his wife, Minnie, were sent by Ann to Smalltown to talk with Ed about beginning the training. Both Dr. Mize and his wife thought Smalltown to be utterly horrid, but carefully concealed that from Ed. Dr. Mize was careful to point out to Ed that it was crucial to have very receptive people at the initial workshops. He vetoed some of Ed's choices and inserted some of his own, based on his knowledge of psychological types.

After the training had begun, it was discovered that the second Saturday workshop had been scheduled on the same day as the Smalltown High School's Homecoming game. Ed informed Dr. Mize that the attendance would be affected. The consultant replied: "It can't be rescheduled, people will just have to make choices." Four people, including the consultants, attended the workshop. Dr. Mize commented that this seemed indicative of "passive-aggressive resistance" on the part of Ed and the others.

After the initial training had been completed, the question of recertification credit for teachers had to be addressed. The superintendent said he would look into it, but as the deadline for the workshops drew near, no policy statement was forthcoming. At the same time, the school system advised Ed that they could not assume any responsibility for funding the program in the future, since they had not been appraised of the need to contain the program within the school system. "We thought it was a drug program and therefore should be housed in Mr. Bailey's area," was the superintendent's statement.

At this point, Ed expressed great reservations about continuing Project Prevention. His hesitancy was based on his feeling that Smalltown folk would find it impossible to open up in a group setting. "This isn't Albuquerque; we have to see our neighbors every day," Ed said. Dr. Mize was not at all surprised by Ed's reluctance, as he had grave misgivings about presenting the program in Smalltown at all. In January 1977, the Drug Abuse Education Center received a polite letter from Ed Bailey, thanking them for the consultation. He was not heard from again.

APPENDIX

Flowchart

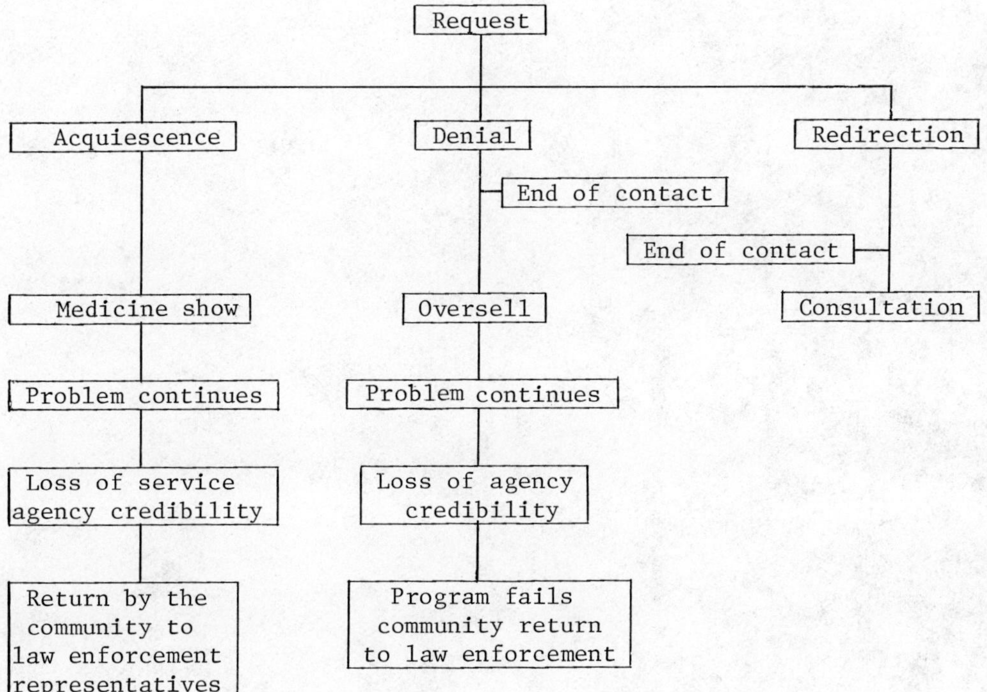

CLINICAL DECISION MAKING IN A CENTRAL INTAKE UNIT FOR DRUG ABUSE TREATMENT

Don C. Des Jarlais and Alan Kott

New York State Office of Alcoholism and Substance Abuse

Central intake units for drug abuse treatment offer the possibility of greatly improving treatment outcomes. In theory, the central intake unit matches the client with the most appropriate treatment modality within a specific area, leading to improved treatment success rates and more efficient use of the scarce resources available for drug abuse treatment. The effectiveness of any central intake unit, however, is limited by the quality of its clinical decision-making. If the clients cannot be appropriately matched with available treatment programs, there is little reason for having a central intake unit. The following evaluation study was undertaken to examine in detail the clinical decision-making process of one particular central intake unit.

The unit was considered to be "functioning well". It had established good relationships with the local court system; drug involved defendants are sent to the central intake unit for determination of their specific treatment needs. The unit had also established good relationships with a variety of treatment agencies throughout the country to which it referred its clients.

There were three major questions that served to focus this study of the clinical decision-making process of the central intake unit: 1) Determining the comprehensiveness and quality of the information used in making the treatment referral decisions; 2) Determining how that information was processed to decide upon specific referrals; 3) Determining the existence of any major limitations on the effectiveness of the treatment referral process.

METHODS

Interview

In order to determine how data on clients is obtained by the unit staff, an hour interview was conducted with one of the counselors. Data from this interview was supplemented by data obtained from numerous short interviews with other staff conducted during our frequent visits to the unit.

Observation

The case conferences at which the referral decisions are made were observed and recorded. The unit staff was very conscious of the observer and the tape recorder and occasionally asked questions about the ongoing research. These questions were answered to the best of the researcher's ability without divulging the content of the research. When asked, the unit staff stated that the presence of the researcher did not arouse anxiety. Staff were quite friendly and cooperative at all times. It appeared that they rapidly became acclimated to the presence of the observer and the tape recorder and that the content of their decisions was not affected.

Coding the Data

Each case conference was divided into two distinct phases. In the first phase the counselor in charge of the case read the "Case Summary and Evaluation" report on the subject.

The second phase consisted of a group discussion to evaluate the information presented and decide upon an appropriate treatment referral. These decisions were made when a consensus of opinion was reached by the group. In the Bales and Strodtbeck (1951) phases of group decision-making, reading the Case Summary and Evaluation report corresponds to the orientation phase, while the group discussion corresponds to the evaluation and control phases.

In order to better understand the decision making process, a standard form was constructed to record references to different aspects of the case made during this discussion. The different aspects of the case represent different types of information to be considered in making the decisions. Each reference to a particular aspect of the case was recorded on the form, though the sequencing of the references was not recorded.

The form consists of ten specific variables and an "other" category to capture references made by the unit staff to factors not included among the first ten. Several of the variables were subdivided in order to make fine distinctions, though these were later collapsed in the tabulations. The variables were selected from interviews with the intake unit staff regarding their case conference decision-making and from observation of several case conferences prior to developing the standard form.

The definitions of each variable is as follows:

1. Legal considerations--includes all references made to any crime charged to this client, any references made to a court ordered determination of whether client is addicted, any general references made to pending court cases, lawyers, etc.

2. Client's drug and/or alcohol use history--includes any references made to the client's past or present use of alcohol and/or drugs.

3. Client characteristics--includes any mention of a client's personality traits or his psychological functioning, and any references made to the degree of his motivation for treatment.

4. Client's family relationships--includes any mention of the makeup or status of the client's family situation and any reference to a particular family member or family problem.

5. Client's socio-economic characteristics--includes any reference to the client's financial position, his educational attainment, goals or plans or his vocational situation.

6. Characteristics of judge--includes any references to a judge's personality, preferences, sentencing patterns, requests or name.

7. References to referral source--includes any mention of any referral source (other than the court) including probation, aftercare, The Department of Social Services, etc.

8. References to treatment facilities--includes any references to or characteristics of a treatment facility.

9. Needs of the intake unit--includes any mention of actions or procedures which would best suit the needs or requirements of the unit.

10. Client's medical characteristics--includes any references to the client's past or present medical conditions.

11. Other--this category included those variables which did not fit into the first ten. Four variables were recorded. They were: a) Residence--includes any reference to a client's

place of residence and/or transportation problems related to commuting to a treatment facility; b) Social relationships--includes any references to the client's friends or acquaintances; c) Age--includes any references to the client's age; d) Race--includes any references to the client's race.

For twenty-six case conferences, an investigator equipped with a tape recorder and an ample supply of rating inventory forms, observed and coded each case. Both the tape recorder and the investigator were in full view of the group.

Subsequently, the second author listened to and coded five randomly selected cases. A Pearson product moment correlation of .85 was obtained for the two coders on the scores for 11 variables x 5 cases. Disagreements between the raters appeared related to the occasionally poor quality of the tape recording due to telephones ringing and other interrupting noises.

RESULTS

Information About Clients

Once a client is referred to the unit, the staff takes several courses of action simultaneously. Any information regarding the client's legal history, other agency contact and treatment history is obtained from the referral source and any other agencies involved. This information is usually obtained by telephone.

The client is asked to come to the unit office for an intake interview. He is asked to bring any close family or friends if, in the judgment of the counselor, important data can be elicited from them. For those clients being referred while still in jail, arrangements are made for the counselor to interview them in situ. Their families can be seen at another time at the unit office. In the cases where the client is able to travel to the unit, the interviews are held in a specific manner. First the client is interviewed individually. This is followed by the counselor interviewing the client's family or friends without him present. Finally, a joint interveiw is held. In most instances, only one set of interview sessions is needed. However, in approximately 10% of the cases additional interviews may be required as a result of the acquisition of new important information received subsequent to the original interviews.

In most cases the interview usually lasts from one to three hours. The variations are due to differing degrees of cooperation on the part of the client and his family, the amount of information needed and the particular counselor who is conducting the interview. When all the necessary information is obtained from various sources, the counselor must evaluate it and put it into a concise, intelligible form. A report is written or dictated and submitted to the secretarial staff for typing. The counselor then notifies the supervisor that he or she has a case for conferencing. In some instances if the case is not an urgent one and/or if the counselor has little time available to write the report, it will not be written up until after the case conference.

Even before the case conference is held, the case has usually been discussed with other staff members. Opinions or possible recommendations have been solicited, particularly if the counselor is faced with problems in an area of another staff member's expertise. These interactions occur throughout the day but intensify around four o'clock when most of the staff return to the unit office, having completed their daily field work. At this point, the counselors exchange information on their daily experiences in a manner which helps familiarize each counselor with certain aspects of the cases which are scheduled for conferencing.

The report is typed on a standard "Case Summary and Evaluation" form. This form is divided into the following sections: Basic Identifying Data, Legal Status, Drug Abuse History, Deviant or Criminal Behavior, Interfamilial and Interpersonal Relationships, Client's Perception of Problem, Urinalysis Results and/or Spectroscopic Analysis, and Recommendations for Treatment.

As mentioned previously, the speed with which information must be gathered and the report written and delivered varies with the client and his circumstances. When the court notifies the unit that it has a client to be evaluated, it is not unusual for the report to be needed within a day or two. In contrast a self-referral or a referral from a social service agency may allow the counselor up to several weeks for preparation of the evaluation report.

The average caseload per counselor is eight per week and may run as high as twelve. Approximately one to three hours are spent in the interview, one to one and a half hours for writing and/or dictating the report, with telephone contacts and staff consultations extra.

A typical case conference begins with the counselor's reading of the "Case Summary and Evaluation" report. Once this has been completed, the floor is opened to any comments, questions and/or suggestions from other staff members. During this process, possible treatment alternatives are discussed and appraised on their relative merits. The conference is completed when a consensus is reached as to the most appropriate treatment plan and the means to implement it.

The case conferences usually begin between 2:30 p.m. and 3:30 p.m. All clinical staff members are required to attend if they are not in the field or involved with other business of a special nature. The variation in starting time is due to the desire to have as many of the ten staff members present as is possible. On the average, there were six staff members present at the conferences; however, the number fluctuated between four and nine. The length of each case conference varied by counselor and by client. Some lasted only ten to fifteen minutes, while others required nearly an hour to complete. The average length appears to be from twenty-five to thirty minutes.

There were obvious differences in preparedness among the counselors. Some regularly had their reports typed or had them written and ready for typing. Others simply had pages of notes which often resulted in long pauses as they searched for necessary information. The latter was observed relatively infrequently, and when it occurred it was usually a consequence of the short period of time which had passed since the client's referral.

In addition to the differences in the preparedness among the counselors, there was considerable variation in the thoroughness of their reports. While some counselors routinely included psychological and sociological data as well as extensive historical information, others included little information other than the minimum required by the report format.

Generally the reports were presented in a smooth, precise manner. The degree of individual participation in the discussion following case presentations also varied from counselor to counselor. Some rarely contributed while others were involved in every discussion.

Coding of the Decision-Making

Results of the tabulation of the coding forms are presented in Table I. A variable was considered "primary" with respect to an individual case if it was the most or second most frequently mentioned in the discussion of the case (ties were included). It appears that thirteen of the fourteen variables can be divided into the following three categories:

1. Those variables that are frequently mentioned and are often "primary" in the individual case. This category includes client's drug and/or alcohol use, client's family relationships, and references to treatment facilities.

2. Those variables which are frequently mentioned but are rarely "primary" for an individual case. This category includes legal considerations, client (personality) characteristics, client socio-economic characteristics, and references to referral source.

3. Those variables which are infrequently mentioned and are rarely "primary" for individual cases. This category in-

Table I. Factors mentioned in twenty-six client case conferences.

Category	Total # of times each variable is mentioned in all 26 cases	# of times each variable is mentioned at least once in a case	# of times out of 26 cases where a variable was mentioned most or second most frequently
Legal considerations	48	18	5
Drug and/or alcohol use	62	24	15
Client characteristics	54	20	7
Family relationships	53	17	10
Socio-economic relationships	24	11	4
Characteristics of Judge	4	3	1
Reference to referral source	27	18	3
Reference to treatment facility	78	22	16
Needs of intake unit	1	1	0
Medical characteristics	15	7	3
Other:			
Residence	18	11	3
Social Relationship	5	5	0
Age	1	1	0
Race	1	1	0

cludes characteristics of the judge, needs of the unit client's residence, client's social relationships, client's age and client's race.

The one variable that does not fit into this classification is the client's medical characteristics. A possible explanation for this is presented in the discussion section.

DISCUSSION

The methods used by the unit staff to obtain information about the clients appear close to optimal given the time constraints on the process. Gathering data from the client's family members and friends, and from other agencies is certainly to be commended. The use of separate interviews followed by a joint interview in which interaction between the client and his family may be observed can provide much valuable information regarding such issues as the need for family involvement in treatment.

Preparing evaluations within a single day would appear to be a severe constraint on their quality. It is not certain, however, whether the information missed in such rapid evaluation outweighs the need to present an evaluation when it will be used.

The variability in the amount and quality of the information obtained in the psycho-social interviews by different counselors suggests that some form of standardized testing might improve the quality of the client evaluation process.

It is important to note that case conference decision-making is based on neither explicit theories of drug dependence nor on systematic empiricism. There is no formal theory of drug abuse that is invoked during the discussions. There is also no systematic data base used in making the referral decisions. Several hundred referral decisions have been made by the unit, but there has been no study of the outcomes of those decisions that could provide an empirical base for future decision-making.

The data from the coding forms gives an outline of how the various types of information available are used in making referral decisions. It should be emphasized that this data collection procedure measures only how frequently references were made to the different types of information. This data collection did not include any of the "implicit" understandings regarding referral decisions, nor the weight of these factors in case decision-making, but focused on what needed to be discussed. Within these methodological limitations, there are some clear indications of how information is explicitly evaluated in referral decision-making.

The types of information were divided into three categories. The first included those variables that were mentioned at least once per case. These were often the most frequently mentioned on an individual case. Client's drug and alcohol use, family relationships and treatment program characteristics were included in this category, with client's personality characteristics on the border between this category and the second category. These variables seem to define the nature of the referral decision task: to select a treatment program for a person with a substance abuse problem that is linked to his or her family relationships and psychological functioning.

In relating the client's substance use, family relationships and pscyhological functioning to the different treatment programs available, the unit staff did not utilize any formal theory of "differential diagnosis." The staff had no formula that weighted the importance of the different aspects of the particular case, or even an exhaustive list of the treatment implications of the different types of substance abuse. Rather decisions were made by what can be characterized as using "ideal types" (Weber, 1964), or "stereotypes" (Brown, 1966). Several examples of the types used by the staff in their decision-making will help to clarify this type of decision-making.

1. The "young kid whose problem is not really drug abuse, but how he or she gets along with his or her family." This type of client was typically a late adolescent, who used alcohol, marijuana and other "soft" drugs. Evidence of conflict with the parents was usually found, and the treatment program recommendation was to ambulatory where family counseling would be provided.

2. The "system manipulator." This client typically had a long history of either heroin or alcohol abuse, with many previous treatment episodes. The motivation for treatment was considered questionable, with either staying out of prison or obtaining eligibility for welfare benefits seen as the "real" reason for entering treatment. Such clients would be referred to "tough" treatment programs, particularly those that involved encounter groups that would challenge the client's motives and any lack of progress in treatment.

3. The "immediate danger to society." This type of client had an extensive criminal history, and was believed to be likely to commit additional crimes even when he did not "need" money for drugs in the sense of a physiological addiction--avoiding withdrawal symptoms. This criminal involvement was seen as an additional factor to the client's drug abuse, but one that required treatment to occur within a secure (locked) treatment program. (Such programs involved a civil commitment process.)

The essence of the clinical decision-making process, then, was to compare the given client to the variety of "types" of clients, and to follow the recommended course of action for the type that the given client most closely approximated. There was no exhaustive comprehensive listing of the different types of clients. Rather the types appeared to have evolved from the work experiences of the staff and be subject to continued modification and addictions.

The second category of variables discussed in case conferences were usually mentioned at least once per case, but not frequently mentioned within a given case conference. These variables were legal consideration, (client) socio-economic characteristics, and references to the referral source. The client's residence was on the border between this second category and the third. This second category seems to include the restrictions on the referral process. It represents pragmatic, "non-therapeutic" considerations, such as constraints imposed by the law or the agency referring the case to the unit.

The third category included those variables that were mentioned relatively infrequently both across individual cases and also within individual cases. These appear to be "special considerations" that are occasionally used in the decision-making, and include the characteristics of the judge, the needs of the unit, and the age, race and social (non-family) relationships of the client.

As noted earlier, medical characteristics did not fit into any of the three categories. It was mentioned in only seven of the twenty-six cases. Severe medical complications appear to re-define the nature of the referral decision process, so that it becomes a problem of finding appropriate medical treatment, with the drug dependence problem as a secondary consideration.

Three questions were noted in the introduction as forming the core of this descriptive evaluation of a central intake unit for drug abuse treatment. With respect to the quality and comprehensiveness of the information used in the clinical decision making, the unit appears to be gathering as much good information as it is capable of processing. The variability among the staff in the quality of their "Case Evaluation and Summaries" is to be expected when standardized instruments are not utilized.

With respect to the information processing, the staff make referral decisions by comparing the individual client with a variety of "ideal types" of drug abusers. This type of decision making is the norm for diagnosis in the mental health field and its limitations have been well assessed (Meehl, 1955). The chief limitation is that humans are relatively poor at aggregating information across the many dimensions involved in ideal type comparisons (Edwards, Lindman, and Phillips, 1965).

With respect to the effectiveness of the central intake unit, first we must state that it appears much more effective than the alternative of having the many treatment agencies with the county doing recruitment within the court system. The greatest single limitation upon the effectiveness of this central intake unit was the lack of a feedback loop that would have informed the staff how the various clients had done in the treatment programs to which they had been referred. In the absence of a formal theory to use in making the referral decisions, feedback to shape future decisions would offer the greatest chance of improving the effectiveness of the central intake process.

REFERENCES

Bales, R.F. and Strodtbeck, F.L. 1951. Phases in group problem solving. *Journal of Abnornal and Social Psychology*. 46: 485-495.

Brown, R. 1966. *Social Psychology*. New York: The Free Press.

Edwards, W., Lindman, H., and Phillips, L.D. 1965. Emerging technologies for making decisions. In *New Directions* in *Psychology II*. New York: Holt, Rinehart and Winston.

Meehl, P.E. 1954. *Clinical versus Statistical Decision Making*. Minneapolis: University of Minnesota Press.

Weber, M. 1964. *The Theory of Social and Economic Organization*. New York: The Free Press.

ASSERTIVE TRAINING WITH PRESCRIPTION DEPENDENT DRUG USERS*

Kenneth L. DeSeve, Ph.D.

Spokane Community Mental Health Center

The inability to engage in a straightforward expression of feelings is a behavioral deficit characteristic of many clients who use prescription medications to cope with anxiety. Assertive Training is considered the treatment of choice when such a deficit occurs in the context of interpersonal relationships (Wolpe and Lazarus, 1966; Wolpe, 1969).

Consistent with Wolpe's reciprocal-inhibition framework (Wolpe, 1958), assertiveness and anxiety are viewed as incompatible responses that do not usually occur concurrently. Recent research has indicated that many patients have never learned appropriate assertive responses, and are greatly in need of acquiring verbal coping skills (Eisler and Hersen, 1973; Laws and Serber, 1971; Lazarus, 1971). The acquisition of such skills presumably results in a reduction of interpersonal anxiety, and facilitates self-directed and self-initiated responding.

One of the consistent findings regarding the effectiveness of assertive training is the high degree of success and, more importantly, specificity in dealing with target maladaptive behaviors. In the perjorative sense, there is little generalization or transfer of learning from one assertiveness demanding situation to another (Lawrence, 1970; McFall and Marston, 1970; McFall and Lillesand, 1971). Rimm and Masters (1974) state that "Assertiveness does not manifest itself as a broad 'trait,' and elevating assertiveness in one class of situation should not be expected to increase assertive-

---

* This study was supported by Contract Number 271-76-2220 from the Department of Health, Education and Welfare.

ness in situations markedly different (p. 117)." Indeed this would be the case if the content of the training was restricted to role playing specific, narrowly defined situations that required assertiveness.

In this vein, Baer (1968) argues that generalization should be programmed into the training rather than simply hoped for. In their concluding remarks on transfer of training of assertive responding, Hersen et al. (1973) suggest that attention to the common elements of assertiveness (verbal and non-verbal) would facilitate generalization of learning.

It would appear that there are common denominators of assertive behavior that could be identified and programmed into a training program. Evaluation of such a program would require an instrument that was sensitive to change in self-directed responding, as well as general in scope (as opposed to situation specific).

In a present study, outpatient clients in a Community Mental Health Center Drug Treatment Program served as subjects in an assertive training group. The focus of the sessions was to develop general assertive coping skills to deal with interpersonal anxiety (as opposed to learning to deal with specific situations requiring assertiveness). Smith (1975) has delineated several verbal skills that mitigate against feelings of anxiety, ignorance or guilt. These techniques are general in nature and are purported to be applicable to any assertiveness demanding situation. The present study attempts to evaluate the effectiveness of such skill acquisition in altering the degree of self-directedness. In addition, by selecting an evaluation instrument that is broad in scope, this study attempts to shed some light on the degree of generalization offered by such an approach.

METHOD

Subjects

Fourteen clients of the Spokane Community Mental Health Center Drug Treatment Services served as subjects. All subjects were direct referrals from primary therapists working in the agency, or from private psychiatrists in the community. With respect to diagnostic classification, the group included nine personality disorders (passive dependent, passive aggressive and inadequate), four neurotics and one marital maladjustment. All but two of the clients were using some form of psychotropic medication, predominately antidepressants or antianxiety agents. The average age of these eleven females and three males was 33.30 years, with a range from 21 to 53 years of age. Nine of the clients were married, five were single, and the mean number of years of education was 12.21 with a range from nine to 16 years. The average length of involvement with the Mental Health Center from the date of intake until termination of the group was 13.71 months, with a range from four to sixty months.

Measurement

Rotter's Internal-External Locus of Control Scale (1966) was used to assess the degree of self-directedness, both pre- and post-treatment, for each subject. This instrument was selected primarily because of the degree to which its internal variance is accounted for by the one factor deemed important in general assertiveness training --personal control. Franklin (1963) reports 53% of the variance attributable to this variable in his factor analysis of the Rotter IE Scale.

In addition, convergent validity is substantiated by the fact that over 50% of the IE locus of control studies have employed the Rotter Scale (Joe, 1971). One month test-retest reliability coefficients have been reported as r=.83 for females and r=.60 for males (Rotter, 1966).

Rotter's Test consists of 29 forced choice items, including six filler questions. The range of scores is from 0 (most internal control) to 23 (most external control). Normative data reported by Rotter (1966) suggests a population mean score of 8.3 (SD=3.9).

Procedure

The Rotter IE Scale was administered to all *Ss* prior to the first session of a ten week training group. *Ss* met one day per week at the Spokane Community Mental Health Center for a two hour session with two therapists. The content of the sessions was as follows:

Session 1. Focus on non-assertive assumptions that restrict freedom of expression and self-directedness. *Ss* were also given an Assertive Bill of Rights (Smith, 1975) and a mini-lecture on how individuals come to give up these rights.

Sessions 2 and 3. During these sessions, *Ss* were engaged in learning the basic communication skills of making self-disclosing statements and responding to the free information others offered. This was done in the context of a "getting acquainted" exercise. Also during these sessions *Ss* began defining the areas in which they most often experienced assertive response deficits.

Session 4. The major theme of this session was the "process" (as opposed to content) of assertiveness. Nonverbal exercises were employed to emphasize the importance of such factors as volumes, posture, eye contact, etc.

Sessions 5 and 6. These sessions focused on increasing *Ss* tenacity by having the *Ss* role play various commercial transactions with the therapist. *Ss* task was to persevere in their request for something from the role-playing therapist, who tried to wear them down or make them feel guilty, ignorant or anxious for making the request.

Sessions 7, 8 and 9. Several techniques for dealing with criticism were covered in these sessions. *Ss* spent much of the time practicing these general coping skills with one another. The therapists began each session by role-playing with each other (modeling) and demonstrating the general context in which each technique was most effective.

Session 10. This session focused on any assertiveness difficulties the *Ss* were experiencing or feared they might experience in the future. In addition, the post therapy assessment test was administered and individuals were referred back to their private therapists for follow-up or termination.

At the outset of each session the *Ss* were given handouts that covered the essentials of the exercise. In addition, at the end of sessions 5, 6, 7 and 8, the *Ss* were given a written dialogue that demonstrated the general nature of the upcoming exercise.

RESULTS

Eight of the initial 22 *Ss* terminated involvement prior to the last session (36%). The pretreatment IE mean score for these *Ss* was 19.37 (SD=5.06). The pretreatment mean for *Ss* who completed all ten sessions was 11.35 (SD=4.53). These means were not found to be significantly different from each other. They are, however, in a direction that suggests that individuals with less internal control are less likely to complete treatment in an assertive training group.

A major finding was that the combined *Ss* pretreatment, mean scores (12.59) on Potters test, were significantly higher than that of the general population ($P < .001$).

DISCUSSION

Major findings of this study involve the development of perceived control over one's life, and a concurrent reduction in feelings of external control. The gain in internal control was the main dependent measure, and was reflected by scores on the Rotter IE scale. These test taking changes corroborated the therapist's subjective impressions of reduced anxiety on the part of internal control gaining clients. Descriptively the data supports that such an acquisition did not occur for *Ss* completing a ten week "course" in assertive training. However, due to limitations of the study (lack of control groups), it is difficult to make inferential statements of causality.

It would appear that the Rotter IE Scale was sensitive to a characteristic many therapists viewed as nonassertiveness. This is substantiated by the significantly higher than normal ($p<.001$) pretreatment scores for clients referred by experienced therapists for assertive training. It would be of interest to assess the overall IE elevation for clients not referred to this type of treatment modality as another reference point.

The loss of *Ss* clearly limits the generalizations that can be made regarding the effectiveness of the assertive training as conducted in the described procedures. Based on a small sample, then, we cautiously interpret the significant increase in internal control for *Ss* completing treatment. The author believes that this concept (locus of control) is general enough to satisfy the requirements for assessing generalization of learning. Given this assumption, it would appear that assertive training of general coping skills enhanced the *Ss* perception of control they felt they had over their lives. This was not a situation specific feeling of control, but rather a general pervasive one.

Obviously, more extensive research is necessary before conclusive evidence will specify the nature of the relationship between locus of control and assertiveness. The present study suggests that such a relationship does exist.

REFERENCES

Baer, D.M., Wolf, M.M., and Risley, T.R. 1968. Some current dimensions of applied behavior analysis. *J. of App. Behav. Analysis.* 1: 91-97.

Eisler, R.M., and Hersen, M. 1973. Behavioral techniques in family oriented crisis intervention. *Archives of Gen. Psychiatry.* 28: 111-116.

Franklin, R.D. 1963. Youth's expectancies about internal versus external control of reinforcement related to N. variables. *Dissertation Abstracts*. 24: 1684.

Hersen, M., Eisler, R.M. and Miller, P.M. 1973. Development of assertive responses: clinical, meaurement and research considerations. *Behavior Research and Therapy*. 11: 505-521.

Joe, V.C. 1971. Review of the internal-external control construct as a personality variable. *Psychol. Reports*. Monograph Supplement 3-V28: 619-640.

Lawrence, P.S. 1970. The Assessment and modification of assertive behavior. Unpublished doctoral dissertation, Arizona State Univ.

Laws, D.R., and Serber, M. Measurement and evaluation of assertive training with sexual offenders. Paper presented at Association for Advancement of Behavior Therapy. Washington, D.C.

Lazarus, A.A. 1971. *Behavior Therapy and Beyond*. McGraw-Hill, N.Y.

McFall, R.M. and Ullesand, D.B. 1971. Behavior rehearsal with modeling and coaching on assertion training. *J. Abnor. Psychol.* 77: 313-323.

McFall, R.M. and Marston, A.R. 1970. An experimental investigation of behavior rehearsal in assertive training. *J. Abnor. Psychol.* 76: 295-303.

Rimm, D.C. and Masters, J.C. 1974. *Behavior Therapy*. Academic Press, N.Y. and London.

Rotter, J.B. 1966. Generalized expectancies for internal versus external control of reinforcement. *Psychol. Mono.* 80: (1, whole No. 609).

Smith, M.J. 1975. *When I say no, I feel guilty*. The Dial Press, New York.

Wolpe, J. 1958. *Psychotherapy by Reciprocal Inhibition*. Stanford University Press. Stanford, Ca.

Wolpe, J. 1969. *The Practice of Behavior Therapy*. Pergamon Press, New York.

Wolpe, J. and Lazarus, A.A. 1966. *Behavior Therapy Techniques*. Pergamon Press, New York.

LENGTH OF TREATMENT AS A PREDICTOR OF OUTCOME

Kenneth DeSeve, James Loudermilk, and Frank Cannata

Spokane Community Mental Health Center

Abstract: This study investigated the point of diminishing returns for one specific program (drug treatment) along one specific objective (reduction of drug usage). The methodology involved having each client's drug type and extent of usage rated at the time of intake and termination by the primary counselor. The rating scale ranged from one (no present use) to six (daily consumption). The results indicate that most clients receiving treatment in an outpatient setting make their greatest drug consumption during 9-18 weeks of therapy. Excluded from this group are the young, non-prescribed depressant abusers who make their drug pattern changes later in treatment.

In custodial environments such as 1950 era state hospitals, lengthy episodes of care are usually associated with poor prognosis. On the other hand, Luborsky et al. (1971) have reviewed 22 outcome studies conducted in community mental health centers and found length of treatment positively related to outcome in 20 of them. Although length of treatment is most often related to number of sessions, Lorr (1962) found that the duration of treatment was more predictive of a positive outcome than was the sheer number of sessions.

There are several possible explanations for the above findings, besides the obvious one that if therapy is beneficial, the more the better. Indeed, as Luborsky (1971) points out, it may be that patients who are getting what they want simply stay in treatment longer, or therapists may be inclined to overestimate positive changes for those clients in whom they have invested much over a long period of time. Regardless of the explanation, it appears well documented that lenth of time in treatment is a good predictor of a positive outcome.

Given that extended lengths of treatment cost more, and heavy investments of resources into one client necessarily draws away from the pool of resources available to others, clinicians and program directors are forced to make some very difficult decisions about how much time they can afford to invest in a given client. At the heart of the decision is a conflict between effectiveness and efficiency, the former being an assessment of how much was accomplished, the latter a measurement of benefit per unit of investment.

In the business world it is a well known fact that the most effective solution to a problem may not be the most efficient one. Business managers often must operate on the principal of diminishing returns, whereby they desist from engaging in a profit making activity at some point where benefits no longer keep pace with costs. It would seem imperative in this time of scarcity, that mental health centers examine the efficiency of their operations. Looking for a point of diminishing returns is one approach to the assessment of efficiency. Potter, Halpern, and Binner (unpublished) contend that while costs for therapy rise at a constant rate, the benefits from such increase at a decreasing rate. Ultimately there is a point at which the additional investment of resources exceeds the benefits derived from continued treatment. The difficulty lies in the definition of benefit. Although an assessment of outcome measures is not within the scope of this paper it seems reasonable that the specialized treatment programs of a comprehensive mental health center would operate on more-or-less specific objectives. In many cases these objectives define the intended benefit or value the program offers to its clients.

The purpose of this study is to ascertain the point of diminishing returns for one specific program (drug treatment) along one specific objective (reduction of drug usage). It is recognized at the onset that there are many other benefits resulting from treatment in a drug program, e.g., occupational, educational, personal adjustment, etc. For the purposes of this study, however, we will define benefit in a drug treatment program as decreased usage of drugs.

Thus the major dependent variable, reduction in drug consumption, will be related to the length of time in treatment. In addition two other factors, drug type and age of the client will be assessed. The rationale for looking at these two factors and relating them to drug usage is that the type of drug used and the age of the user may relate directly to how long it takes an individual to profit maximally from treatment.

METHOD

Subjects

The sample included all clients admitted into the drug treatment program of the Spokane Community Mental Health Center during calendar year 1976. Of the total 272 clients, complete assessment information was available on 221 of them, resulting in a 16% subject loss. The loss was randomly distributed across age, sex, drug type and length of treatment.

The subject pool included 55% female and 45% male clients with an average age of 28.2 years and 12.2 years education. Of these clients 77% were voluntarily involved in treatment, with the remaining 23% having been referred by the criminal justice system.

Procedure

The individuals involved in this study received treatment under a "psychological model". Primary emphasis was on individual and group psychotherapy. Additional treatment methods available included assertiveness training, biofeedback, autogenics and activity therapy. Focus of treatment was on problem solving and alternative coping skills rather than symptom relief.

The major dependent variable was the "change" in the extent of drug usage from intake to discharge. At the time of intake and termination from the program each client's drug type and usage was assessed by a counselor using rating a rating scale ranging from one (no present use) to six (daily consumption). For purposes of this study, a change in extent of drug usage constituted an index of benefit derived from treatment.

The length of treatment, the major predictor variable, was broken down into the following groupings: (A1) 0-8 weeks; (A2) 9-18 weeks; and (A3) 19 weeks or longer. This breakdown was intended to separate those individuals involved in early terminations (A1), brief treatment (A2), or long term therapy (A3).

For purposes of this study drug types were grouped into three categories: (B1) nonprescribed depressants including heroin and other opiates, methadone, alcohol, barbiturates and over-the-counter medications; (B2) stimulants/hallucinogens including amphetamines, cocaine, marijuana/hashish, hallucinogens and inhalants; (B3) prescription tranquilizers/sedatives/hypnotics.

Clients were grouped according to age, into two specified ranges: (C1) 16-25 and (C2) 26-99 years of age.

Table I. Changes in drug usage.

Source	SS	df	MS	F
A (Weeks in therapy)	10.66	2	5.33	8.62*
B (Drug type)	.13	2	.06	.11
C (Age)	.06	1	.06	.10
A x B	12.32	4	3.08	4.98*
A x C	12.39	2	6.19	10.02*
B x C	22.92	2	11.46	18.54*
A x B x C	12.26	4	6.13	9.92*
Error		203	.62	

* $p < .01$.

The experimental design allowed for an assessment of main effects and interactions among the three factors under study as they related to changes in drug usage.

As can be seen from Table I, the only significant main effect related to changes in drug usage was the length of time an individual was in treatment. Neither the type of drug used nor the age of the user were found individually to have a significant bearing upon changes in drug consumption from pre- to post-treatment.

In addition to the main effect findings, the factorial analysis found all interactions of variables reaching statistical significance. Figure 1 shows the interaction between weeks in therapy and drug type as they relate to changes in drug consumption.

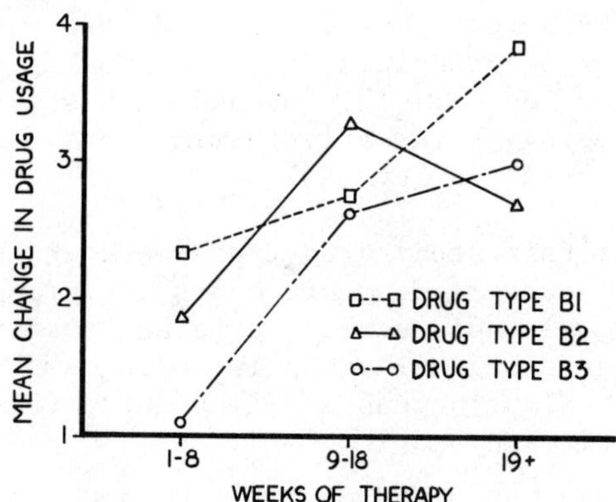

Fig. 1. Mean changes in drug usage as a function of weeks in therapy and type of drug used.

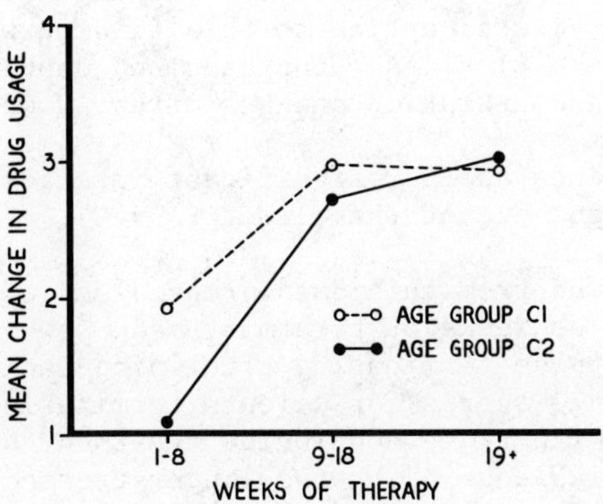

Fig. 2. Mean changes in drug usage as a function of weeks in therapy and age of client.

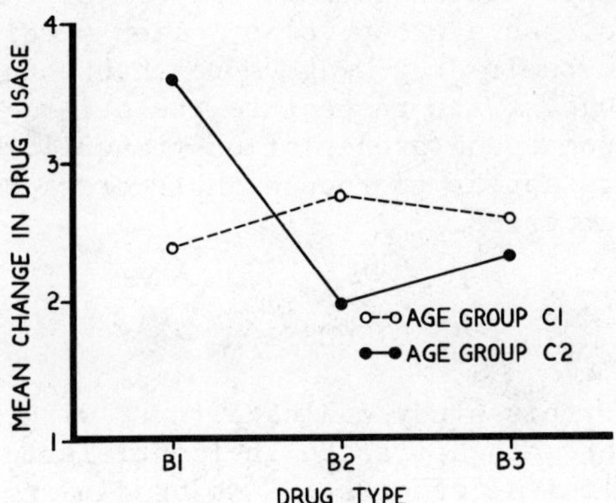

Fig. 3. Mean change in drug usage as a function of drug type and age of user.

As can be seen from this data, clients terminating treatment during 1-8 weeks of treatment (A1) made significantly less changes in their drug consumption than did those who terminated during 9-18 weeks (A2). Clients continuing treatment beyond 19 weeks (A3), however, did not make significantly greater changes than those reported for the A2 group. The exception to this finding was for clients using drugs from the B1 class (nonprescribed depressants) who made most of their drug consumption changes after 19 weeks in therapy.

Figure 2 presents another significant interaction: that between weeks in therapy and age of the client (A x C).

As can be viewed from this data, regardless of age, clients terminating before nine weeks of treatment made fewer changes in drug consumption than those terminating after nine weeks of treatment. In comparing the change scores for clients terminating during the 9-18 week period (A2) with those continuing treatment beyond 19 weeks (A3), we find no appreciable gain for the longer treated clients of either age grouping.

Another significant interaction is presented in Figure 3 and shows the relationship between drug type and age of user (B x C).

This data indicates that the changes in drug consumption, regardless of how long the client remains in treatment, are significantly influenced by an interaction between the age of the user and the type of drug being used. As a group, older clients using nonprescribed depressants reduce their drug usage more than younger clients (C2) using the same drugs. With respect to the other classes of drugs (stimulants, hallucinogens/prescription tranquilizers and sedatives), the younger clients appear to reduce their drug consumption somewhat more than the older group.

DISCUSSION

The results of this study indicate that, with respect to changes in drug usage, there appears to be an identifiable point of diminishing returns for clinical efforts. A major finding of this study consistent with previous research was a positive outcome associated with length of treatment.

The interaction between weeks in therapy and drug type, as shown in Figure 1, was found to be significant. When contrasting group A1 with A2 we find that regardless of drug type, clients terminated during 9-18 weeks of treatment reduced their drug consumption more than those terminated prior to that period.

In comparing groups A2 and A3 the data indicates that the longer term clients (A3) who were terminated after 18 weeks of treatment, did not reduce their drug consumption more than clients receiving 9-18 weeks of treatment (A2). This is not to say that other benefits do not continue to accrue over time; however, when looking solely at changes in drug usage this data leads one to question the investment of resources beyond 18 weeks. The one exception to this finding is for those individuals using nonprescribed depressant drugs. Even then, Figure 3 suggests it is primarily the younger clients in this category who make the greatest reduction in drug consumption.

One can only conjecture about the reasons clients using non-prescribed depressants continue to make gains beyond 19 weeks in treatment. Indeed, many of the more recalcitrant character disorders are found among this group of drug users. It would be reasonable to assume that in dealing with this population a therapist would meet with an initially higher level of resistance that may take longer to work through.

This in part may be expected in light of the fact that many were court ordered and were previous treatment failures. Higher therapeutic resistance may account for the later changes for this group; however, further research in this area is indicated.

The interaction between client age and weeks in therapy (Figure 2) further supports the 9-18 week interval as being the optimal length of treatment for bringing about changes in drug usage. Regardless of age, clients appear to profit maximally during nine to eighteen weeks of treatment. Beyond that interval continued effort brings diminishing returns in terms of changing drug usage patterns.

In summary, it would appear that most drug using clients receiving treatment in an outpatient setting make their greatest drug consumption changes during 9-18 weeks of therapy. Excluded from this group are the young, nonprescribed depressant abusers who make their drug pattern changes later in treatment.

ACKNOWLEDGEMENTS

This study was supported by Grant 10 H 000060 02 and Contract Number 271-76-2220 from the Department of Health, Education and Welfare.

REFERENCES

Lorr, M. 1962. Relation of treatment frequency and duration to psychotherapeutic outcome. In H. Strupp and L. Luborsky (eds.), *Research in Psychotherapy*. Vol. 2. Washington, D.C.: American Psychological Association.

Luborsky, L., Chandler, M., Auerbach, A.H., Cohen, J., and Bachrach, H. 1971. Factors influencing the outcome of psychotherapy: a review of quantitative research. *Psychological Bulletin*. 75: 145-185.

Potter, A., Halpern, J., Binner, P. 1971. Predicting resource utilization in a comprehensive center: an evaluation of three alternative methods. Unpublished manuscript. Fort Logan Mental Health Center.

SOME ADVANTAGES OF INTEGRATING JOB TRAINING AND TREATMENT

Marc Carpenter

Project DARE, Foundation for Research and Community Development, Inc.

OVERVIEW

This paper proposes to outline the job of training and treatment experience of Project DARE and point out some programmatic advantages of treatment philosophies embracing and delivering job training and placement services.

It has been my view that the drug treatment community has perceived job training, placement, re-entry, CETA, economic development, the business community, outcome evaluation, cost accountability, etc. with apprehension and distrust. This is not inconsistent with NIDA's feeble efforts to deal with the subject nor with the resulting lack of emphasis by local, state and national planning and priority setting bodies.

It seems the point has been made many times that vocational services are necessary and can greatly enhance treatment outcome; however, progress has been slow and serious national attention and leadership continues to be absent. It is my hope that this paper will stimulate ideas, discussion and an awareness of potential strategies and nerve centers that will activate treatment providers to more vigorously pursue vocational resources on behalf of their clients.

PROJECT DARE - BACKGROUND AND EXPERIENCE

Project DARE is sponsored by the Foundation for Research and Community Development, Inc., a non-profit private corporation in the state of California. Project DARE is located in San Jose, Santa Clara County in the Bay area of northern California. Santa Clara

County has a population of approximately 1,200,000 and has major industries in electronic manufacturing services and real estate within its boundaries; as well as a complex political make-up consisting of fifteen (15) cities, seventy-five (75) special districts (fire, water, school, etc.).

Since its inception in 1974, DARE has served a population that is approximately 60% Chicano, 30% Caucasian and 10% Black. 95% were unemployed or underemployed and 70% had criminal records and all had some dehabilitating drug or alcohol problem.

DARE was originally funded by the Office of Economic Opportunity in 1974 on a one-time only research and demonstration basis. Spawned from the idea that drug treatment efforts were largely concerned with immediate treatment and had little capability to attend to vocational needs, and that traditional manpower training programs avoided anyone with known drug problems, DARE was designed to integrate job training and outpatient counseling while clients received money for participation. Shortly after DARE's implementation, NIDA assumed contract monitoring responsibilities and wasted no time in telling us that we would not be considered for any future NIDA funding, no matter what our results were.

DARE struggled through its first year of operation, providing job training (and counseling) in four occupational subjects and paying clients $100 per week for active participation. As there existed little reference material for operating this kind of model, we viewed ourselves as pioneers and invented new approaches and adapted traditional concepts to our needs. As a result of the experience, DARE began to learn that client motivation and interest were visibly stimulated and the rewards of "cleaning up" became real and tangible via monetary incentive.

Our initial experience was productive in many ways; however, our acceptance as a legitimate provider of an essential service was still shrouded by suspicion, primarily because our structure and philosophy were different from traditional approaches. Even the treatment community withheld their cooperation because we didn't embrace their rigid attitude toward total abstinence.

Our initial grant was coming to a close and even though we had been productive, NIDA's position had not changed. We were told by local and state drug abuse funding sources that since we were a job training program we should pursue job training funding sources. The job training funding sources told us that since we were a drug program we should pursue drug program funding sources!

We eventually derived partial funding from both the drug treatment and the job training sources after months of hassles and enlisting the support of every known elected official on behalf of our program. The political environment proved to be more responsive than the rigid "cover your ass" bureaucrats or the "we've got all the answers" treatment community. One suspects this was a result of our presentations which, out of absolute necessity, stressed every imaginable angle of our value. The concept that has always been central to DARE and apparently rang a bell with the "politicos" is economics; an addict learning a job skill and going to work at a cost of $1400 (for training, counseling and placement) is a wise investment of the taxpayers' money.

Today, Project DARE enjoys the prosperity of operating a multiple-service, multiple-funded program. In a comprehensive environment DARE provides outpatient heroin detoxification with emphasis on vocational planning; PCP treatment, primarily for adolescents; CETA Title I job training, counseling and placement; and CETA Title III youth services providing basic education and job training (See Figure 1--Client Flow Chart).

An Advisory Board representing business, education, labor, criminal justice agencies and community programs aggressively supports the efforts of Project DARE. This group has been of invaluable inspiration, guidance and support over the last two years. Problems and progress are discussed honestly and openly, and their participation is genuinely integrated into our entire program.

DARE maintains a philosophy that promotes independence from drugs, welfare, jails, parole officers, etc., and offers the client the opportunity to become self-reliant. The client gains confidence from learning and achievement and having options as to where they market their labor bolsters their self-esteem. In this way the client's transition to the straight world is transmitted through the social and economic catalyst of work. Even as clients begin detoxification in the treatment component, vocational skills are inventoried and discussed. This emphasis is reinforced throughout all dimensions of the program.

To focus our discussion, the remainder of the paper will deal with the CETA Title I job training and employment experience of Project DARE since 1974.

A BRIEF DESCRIPTION OF DARE'S TITLE I PROGRAM

The Department of Labor from which CETA is administered is concerned with quantifiable numbers. Most Title I job training programs share this common concern since their perpetuation is based on numerical success. The following describes how DARE as one Title I program is structured and has achieved success on Department of Labor (hereinafter D.O.L) terms.

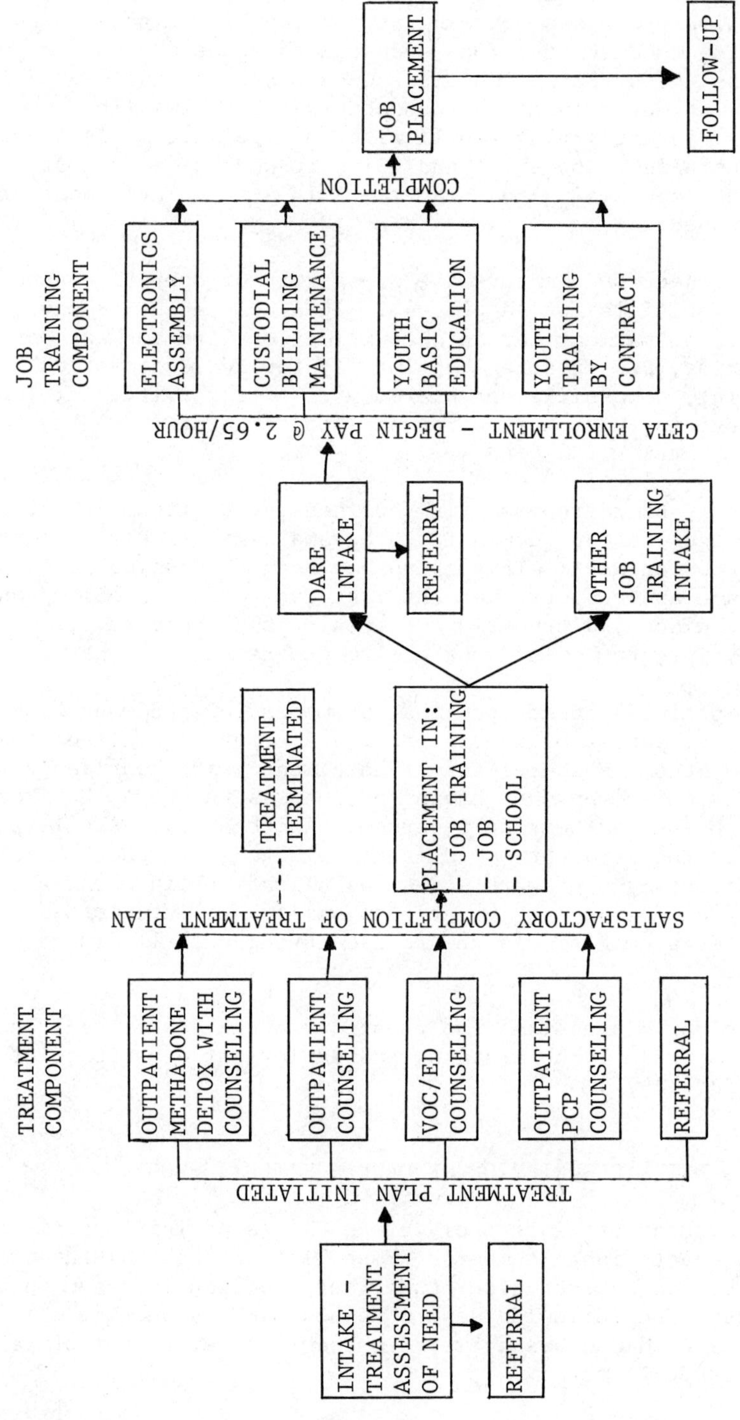

Fig. 1. Project DARE client flow chart.

Project DARE's Title I program basically consists of:

Intake and assessment of client needs, skills and motivation. This function is essential to the ultimate success of our effort. We are required by CETA to report outcome data monthly. A selection process increases the potential for success. No client earnestly wanting our services is denied. Negative terminations are greatest in the first four weeks of the client's participation.

Counseling is based upon an Employment Development Plan and behavior contract, that serve as client policies and procedures and are mutually agreed upon by the client and staff. Counseling works toward the client developing self-reliance and the problem solving skills needed upon placement into a job. Acceptance in the program is based upon a desire to go to work, hence the focus of behavior change is toward what needs to be done to succeed in securing and maintaining employment.

Job training takes place in a simulated work setting. We have 25 training slots in Electronics Assembly and 15 in Custodial Building Maintenance. The training is both classroom lecture and practical experience in a 12 week open entry/open exit model. Each client is paid $2.65/hour while participating in training.

Job placement - Preparation for this essential phase begins at intake. The job developer initiates pre-employment seminars throughout the client's participation. He is also responsible for establishing working relationships with employers and the business community.

Follow-up - With the limitations of funding, follow-up is done by all staff, no one having full-time responsibility for this function. Clients are encouraged to return to the program for whatever services are necessary.

This is one use of Title I funds that has been relatively effective in providing the numbers necessary for continued D.O.L support. There are many approaches to utilizing CETA that will be discussed later. For the results of our last two years of operation, see Table I.

SOME ADVANTAGES DERIVED FROM PROVIDING TITLE I SERVICES

In providing job training service for drug abusers for the past 3½ years, several benefits for both clients and staff have been derived.

* Upon entering treatment, the client is aware of the realistic opportunities for jobs, resulting in greater motivation to complete treatment.

* The credibility of participation in school or job training can relieve the continuous pressure placed on the client by parole officers, social workers, family, etc.

* The client earns while he learns, reducing the need to hustle and place him/herself in high risk situations for money.

* With job training as an option, the staff has more tools to work with.

* Seeing clients going to work and "making it" has been inspirational for the staff, who has come to accept the futility of the revolving treatment door.

* Job placements are quantifiable in tax dollar terms and provide a cost effective information base for enlisting public support of your effort.

* Demonstrated credibility in provided job training and placement services opens the door to service expansion through CETA. This results in more diverse services for your clients and increased job opportunities for your staff.

Table I. Project DARE two year performance CETA Title I.

	1975-1976	1976-1977	Total
Total enrollments	132	156	288
Total terminations	97	119	216
Completed job training	74	95	169
Placed in jobs	65	74	139
Placed in school	3	-0-	3
Drop-outs	20	24	44
Carried out	35	29	64
CETA Placement Rate*	69%	62%	
Average starting salary (yearly)	$6,136	$6,739	$6,437
Approximate taxes paid per person	$1,227	$1,347	$1,287
Potential return by client working	$79,755	$99,740	$179,495

* The placement rate is the number of terminations divided by the number of placements. However, if this were computed by dividing the number of training completions (those actually having marketable skills) by the number of placements, the rate would be much greater.

CETA

The Comprehensive Employment and Training Act (CETA) was established to serve unemployed, underemployed and economically disadvantaged persons. The enacting legislation has seven (7) titles, each having corresponding amounts of annual funding. Any governmental unit with a population of 100,000 or more qualifies as a Prime Sponsor and would be eligible to contract with the Department of Labor to administer CETA funds. The decision-making structure of the Prime Sponsor consists of a governing board made up of local elected officials and an advisory council open to business, labor, consumers and the general public.

Spending approximately 3.2 billion dollars annually, CETA offers a variety of potential resources. In Santa Clara County, drug treatment programs operate on an annual budget of about 1.5 million dollars; while the local CETA Prime Sponsor distributes over 50 million dollars annually.

WHAT CETA MONEY CAN FUND OR SUPPORT

* A job placement component in your treatment program.
* A basic education class in your T.C. and pay clients to learn.
* A job training class in your facility.
* On-the-job training for your clients.
* A job referral and placement center for your clients.
* Vocational counseling for your clients.
* Occupational testing and skill assessment.
* Some residential services for clients in job training.
* Additional staff and special projects through Title VI.
* Overhead and supervision costs for these services.

In most CETA activities clients are paid to participate, adding an additional resource to your program.

PROBLEMS TO ANTICIPATE WITH CETA

As mentioned previously, the D.O.L and CETA are greatly concerned with accountabilities; hence there is a good deal of *paperwork* involved and regulations to observe.

Being concerned with self-perpetuation, there has been *limited interest* in CETA *serving drug or alcohol abusers.* They will not come looking for you, so you must advocate persistently on behalf of your clients having equal access to these vast resources. Demand accountability. CETA knows who they are and aren't serving.

Existing providers of manpower services have generally been successful by *avoiding high risk populations.* You may enlist their cooperation or pursue services administered through your own organization. Any community based organization is eligible to contract with a Prime Sponsor. Starting your own training and placement program requires multiple levels of expertise.

SUGGESTIONS FOR APPROACHING CETA

Know the players - Who are the governing board members of the Sponsor and who are their constituents. Who does the advisory council to the Prime Sponsor represent and are the needs of the drug abusers being articulated.

Know the process - Each year the Prime Sponsor must submit to D.O.L. an annual plan of their priorities.

Get involved - Drug abusers are legitimately eligible for CETA services. Request membership on the advisory council. Request drug abusers be considered a priority population which would require each provider of CETA services to designate a set proportion of their services for drug abusers. Handicapped drug abusers fulfill a national priority.

Be persistent - This is a giant bureaucratic maze and can be extremely frustrating. It took Project DARE three years to establish credibility with our local CETA Prime Sponsor. Our services have tripled in the last two years as a result of our work.

RECOMMENDATIONS FOR AN ACTION PLAN

* Aggressive grassroots advocacy of client vocational needs to NIDA and CETA.

* A national policy statement should be developed for a clear commitment of NIDA and CETA's responsibilities to the vocational needs of drug abusers.

* NIDA advocate to CETA to include drug abusers as a national target population and mandate all Prime Sponsors to serve drug abusers in Title I programs.

* NIDA initiate a national training and technical assistance effort to equip drug treatment providers with the capability to pursue CETA funding.

* NIDA fund semi-annual meetings of vocational service providers to develop a long-term strategy and monitor CETA and NIDA efforts to improve services to drug abusers.

* More research and development money be available to measure the impact of existing efforts.

* The next National Drug Abuse Conference provide a greater forum for participation of vocational rehabilitation issues.

LIFE CHANGES AND HEALTH IN TWO DRUG TREATMENT MODALITIES*

Marcia Andersen, R.N., Ph.D.

National Women's Drug Research Coordinating Project

INTRODUCTION

Change is an integral part of life; researchers in the health sciences have found relationships between the occurrence of life changes (measured in different ways) and:

onset of physical illness (Rahe and Arthur, 1968)
onset of mental illness (Paykel et al., 1969)
perceived severity of illness (Andersen and Pleticha, 1974)
occurrence of traffic accidents (Seltzer, 1975)

One group of people with multiple health problems, who are undergoing a great deal of life change, are drug dependent clients in treatment (Brill, 1972; Byrd, 1970; Reed and Herr, unpublished; Rowland, 1975; Smith, 1972).

BACKGROUND

The literature has shown that illness is associated with a high number of recent changes in one's life, such as moving or changing jobs, divorce, or death of a spouse (Andersen and Pleticha, 1974; Bell, 1977; Dohrenwend and Dohrenwend, 1974; Holmes, 1970; Lundenberg,

---

* Supported in part by National Institute on Drug Abuse Grant # 5H81-DA-01496. Grant awarded to the Wayne County Department of Substance Abuse Services With sub-contract to the University of Michigan School of Social Work.

Theorell, and Lind, 1975; Marx, Garrity, and Bowers, 1975; Rahe, 1968, 1974a, 1975; Seltzer and Vinokur, 1974). Recent changes are defined as those occurring within the past six months.

LIFE CHANGES AND HEALTH IN DRUG DEPENDENT CLIENTS IN TREATMENT

Since poor health accompanies a high number of changes in one's life, drug addicts can be expected to have a high incidence of illness because of their life changes. Numerous investigators have documented the multiple health problems of drug-dependent individuals (Atlee, 1972; Bloom and Butcher, 1970; Challenor and Garner, 1975; Cherubin, 1975; Christakis et al., 1973; Felton, 1975; Harris, Yeoh, and Brown, 1975; Kilcoyne, 1975). Heretofore, however, it has been assumed that the high numbers of illnesses in this population were due primarily to withdrawal from drugs (Baden, 1975; Louira, Hensle, and Rose, 1967; Kreek, in press; Stimmel and Kreek, 1975).

Rowland (1975), in discussing health problems of drug-dependent individuals, states: " . . . much more than physical illness is involved. With drug abuse, the interface between medicine and social factors involved in the health of people . . . is blurred (p.xv)." He points out that as long as the relationship between health and social factors remains blurred, medical problems will persist for this group of individuals.

Reed and Herr (unpulbished) note the large number of life changes associated with drug treatment admission for an individual. They point out that entire careers, life patterns and associations often change. Entering drug treatment is also associated with different amounts of life change, depending on the drug treatment modality (Langrod et al., 1972; Glasscote et al., 1972).

Changes are associated with outpatient methadone programs admission, but only a few are required. They are: frequent attendance, forming a relationship with a counselor, and becoming and remaining drug free. While clients are often counseled to seek and obtain employment or education, and to form new relationships apart from their drug-related acquaintances, these are not requirements of treatment in an outpatient modality.

Residential therapeutic communities on the other hand, have more required changes at admission. Some of these changes are: change of residence; isolation from family and friends; change in eating habits; change in sleeping habits; changes related to sexual activity; required participation in program work assignments; required meeting attendance; coping with confrontations; forming new relationships; becoming financially dependent upon the program; remaining drug free; leaving drug-related careers.

The purpose of this investigation was to examine:

1. The amount of change reported by drug-dependent clients in two drug treatment modalities.

2. The relationship between the clients' life changes and their health.

3. The relationship between the clients' life change scores and their adherence to the treatment goal of remaining drug free.

MEASURING INSTRUMENTS

Adjustment to Change

Holmes and Rahe (1967) theorized that recent life changes were sources of stress. They felt that a life event, regardless of desirability, evokes an adaptation response. It was the stress embodied in this adaptation response that they sought to measure. They developed a list of life events empirically observed to cluster at the time of disease onset (p. 213). This tool was called the Social Readjustment Rating Scale (SRRS) and can be seen in Table I. Not all the events were negative or stressful in the conventional sense (i.e., socially undesirable (p. 217). Rather, there was identified "one theme common to all these life events. The occurrence of each usually evoked or was associated with some adaptive or coping behavior on the part of the involved individual" (p. 217).

The mean number of life change units (LCU's) assigned to the event by a group of subjects was transformed into the LCU value for the event. A measure of adjustment to life changes was constructed from these readjustment ratings by summing the readjustment ratings for all events reported in a given period of time Dohrenwend, 1973).

The classic set of norms for the SRRS is the result of all data collected from several thousand Navy and Marine Corps subjects in early studies by Rahe (1972, 1974b). They are:

85 LCU's - Healthy baseline for past six months
150 LCU's - Healthy baseline for past one year
150-300 LCU's - Half of these reported illness in the next year
300 LCU's - 70% reported illness in next year

Rahe (1975) revised the SRRS and called the new version The Recent Life Change Questionnaire (RLCQ):

In addition to adding events he has designed this version of the SRRS to collect subjective data, i.e., he wanted the respondents to be given a chance to score their own readjustment to the life change

Table I. The Social Readjustment Rating Scale (Holmes and Rahe, 1967), p. 216

Rank	Life Event	LCU Value
1	Death of spouse	100
2	Divorce	73
3	Marital separation	65
4	Jail term	63
5	Death of close family member	63
6	Personal injury or illness	53
7	Marriage	50
8	Fired at work	74
9	Marital reconciliation	45
10	Retirement	45
11	Change in ehalth of family member	44
12	Pregnancy	40
13	Sex difficulties	39
14	Gain of new family member	39
15	Business readjustment	39
16	Change in financial state	38
17	Death of a close friend	37
18	Change to different line of work	36
19	Change in number of arguments with spouse	35
20	Mortgage over $10,000	31
21	Foreclosure of mortgage or loan	30
22	Change in responsibilities at work	29
23	Son or daughter leaving home	29
24	Trouble with in-laws	29
25	Outstanding personal achievement	28
26	Wife begin or stop work	26
27	Begin or end school	26
28	Change in living conditions	25
29	Revision of personal habits	24
30	Trouble with boss	23
31	Change in work hours or conditions	20
32	Change in residence	20
33	Change in schools	20
34	Change in recreation	19
35	Change in church activities	19
36	Change in social activities	18
37	Mortgage or loan less than $10,000	17
38	Change in sleeping habits	16
39	Change in number of family get-togethers	15
40	Change in eating habits	15
41	Vacation	13
42	Christmas	12
43	Minor violations of the law	11

on a scale of 1-100. No real norms have yet been developed using this new scale. One study of arthritic patients found those subjects' mean score to be 438 SLCU's (McEwan and Stevens, 1977).

Health Status Indices

This investigator has chosen three dimensions of health for study as dependent variables. They are:

1. numbers of illnesses
2. subject's perception of the severity of the illnesses
3. types of illnesses

Numbers of illnesses has been used as a dependent variable in numerous studies (Marx, Garrity, and Bowers, 1975; Rahe, McKean, and Arthur, 1967; Rahe et al., 1971). In addition to numbers of illnesses, Andersen and Pleticha (1974) found that life changes were related to emergency room patients' perception of severity of their illnesses (t=2.3, p<.02).

Heroin Use

Heroin use, while differing somewhat from region to region, proved to fit with remarkable similarity onto a chart devised to compare drug usage across programs and cities. Though heroin, for example, was sold in packs, bags, balloons, or chunks, usage could be converted into teaspoons or McDonald's® spoons (from McDonald's® Restaurants). These were remarkably universal measurements of heroin across the country. An estimate of heroin quality was obtained by asking the client to give an estimate of the heroin purity they used. Data on cost per day was also collected.

Adherence to Treatment Goals

While adherence to treatment goals is a multi-variate situation, one easily measured variable that suggests such adherence, is the lack of illicit drugs in clients' urine samples. Urines free of illicit drugs are said to be "clean," and those with evidence of illicit drugs are said to be "dirty." Clean urines then, are a sign that clients are remaining drug free and adhering to that one treatment goal.

METHOD

Program Selection

Ten heroin treatment programs in four urban areas in four regions of the United States were selected for participation in this study. They were chosen because of prior connections to other phases of The National Women's Drug Research Coordinating Project, and because procedures for confidentiality and entry had already been established. While this limits the generalizability of the findings to these ten programs, significant trends for future study can be identified.

Subject Selection

Absolute randomness was not possible, but the sample closely represents the clients in the sample programs. An attempt was made to approach the first client to arrive or arise, and every third client thereafter. This procedure did not always work as planned.

While the sample can only be considered a convenience sample, trends for future life change research can emerge from this study.

Data Collection Procedure

Each respondent was given a stack of cards. An event was written on each card. Respondents were asked to place them in two piles:

"Yes, it has happened in the past two years" or

"No, it has not."

After the respondent had separated the cards into two piles, this interviewer went through the "yes" pile of cards asking for a score representing the amount of adjustment required by each event on a scale of 1-100; and, in what time period the event had occurred.

Data was thus collected and coded in two ways. Utilizing the subjective scores assigned by the subject, and by using the preassigned scores of the SRRS.

RESULTS

Data were collected from 47 drug-dependent subjects (19 males, 28 females). Therapeutic communities contributed 24 subjects and outpatient methadone clinics, 23. The sample was 55% Black, 41% White, 2% Hispanic, and 2% Asian.

The sample represents clients in ten sample programs located in the Midwest, Northeast, West, and Southeast. In each city there was at least one therapeutic community and one outpatient methadone clinic. In addition, in the Midwest, there was an all-womens' therapeutic community. This accounts for the fact there were more women than men in this sample.

The education distribution shows over 50% of the subjects graduated from high school and 25% completed some college.

In this sample, only ten subjects, or 21%, supported themselves by working at legal occupations.

Study Question One

How do the readjustment scores of these drug-dependent clients compare to other groups noted in the literature?

The range of the life change scores was from 52 - 1,420 LCU's. The mean was 432 SLCU's (sd=236). This mean was statistically significantly higher than the "healthy baseline" quoted in the literature, of 85 LCU's (one sample t=10.10, $p<.001$). It was even statistically significantly higher than 300 LCU's, the high reported in the literature. (One sample t=3.80, $p<.001$).

Using the subjective life change units, the range was 160 - 2,200. The mean was 955 SLCU's (sd=500). This mean differed significantly from the only mean SLCU available at the time of analysis, the mean of 438 SLCU's obtained from arthritic patients, by McEwan and Stevens (1977). (One sampe t=7.09, $p<.001$). Thus it appears that the readjustment scores of drug-dependent clients in treatment programs are very high, higher than those reported in the literature to date.

There were no differences among races, cities, or gender on adjustment scores.

Study Question Two

What is the association between life change scores and numbers of illnesses?

Figure 1 shows the types of illnesses reported by this sample during the past six months. Respiratory conditions accounted for the largest number of illnesses reported. Diseases of the digestive system include dental disorders.

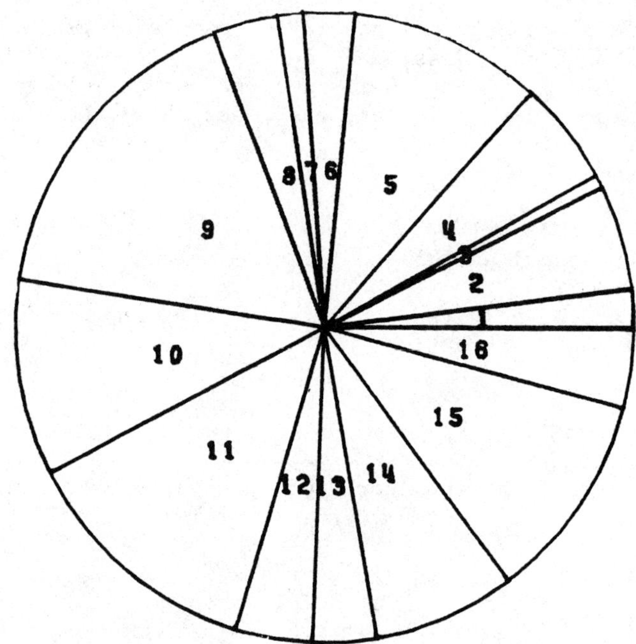

1 INFECTION
2 METABOLIC
3 BLOOD
4 MENTAL
5 NERVOUS SX
6 EYE
7 EAR
8 CIRCULATORY
9 RESPIRATORY
10 DIGESTIVE
11 G.U.
12 PREG. COMP.
13 SKIN
14 MUSC-ORTHO
15 INJURY
16 MISC.

Fig. 1. Types of illnesses reported in the past six months.

Types of changes reported by subjects in the two modalities are shown in Figure 2. It can be seen that outpatient subjects have considerably more financial changes than do residents of a therapeutic community.

Subjects with high adjustment scores, measured using either LCU's or SLCU's, have statistically significantly more illnesses than those with low scores. Table II shows these correlations. It is interesting to note that while there were no statistically significant differences among race or cities or health variables, there were differences between sexes. Women in this sample had statistically significantly more illnesses than males in the past six months (t=2.3, df=45, p<.02).

Illness severity. It is interesting to note in Table II that the correlations between life change scores (LCU's) and numbers of serious illnesses, in a six month period of time, are higher than those between LCU's and total numbers of illnesses, in the same time period.

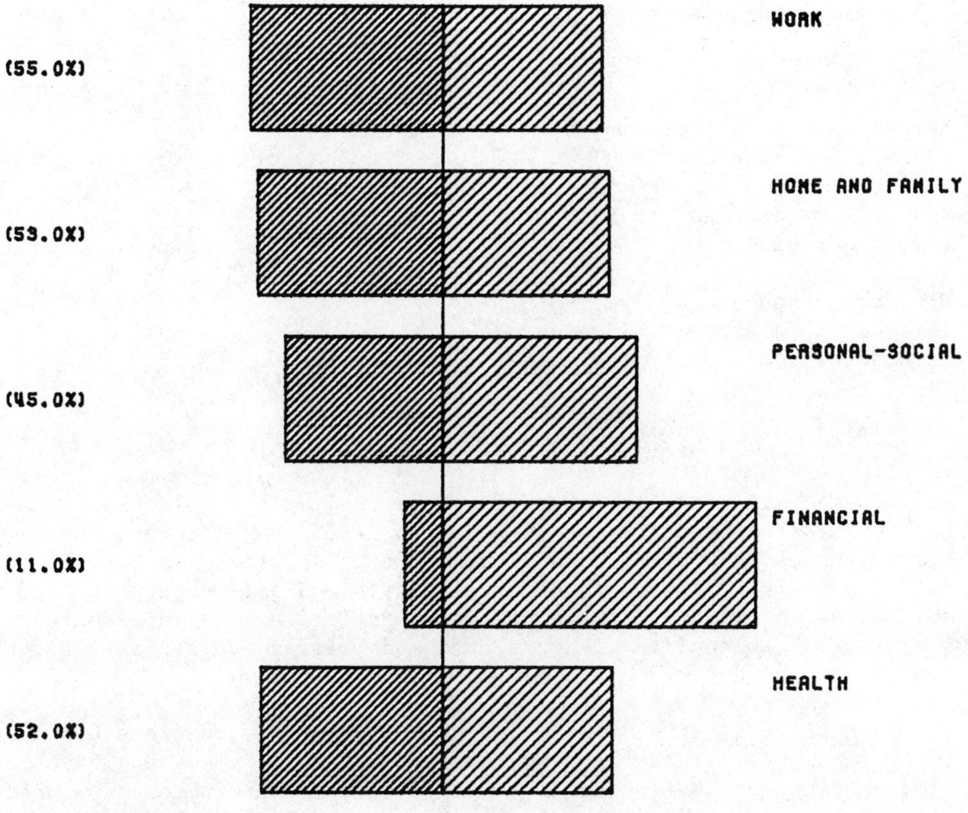

Fig. 2. Changes reported by subjects in two modalities.

Table II. Correlation between adjustment scores and illnesses in past six months.

	r	p<
SLCU and numbers of illnesses	.36	.01
SLCU and numbers of serious illnesses	.38	.01
LCU and numbers of illnesses	.46	.001
LCU and numbers of serious illnesses	.56	.001

Table III. Association of illnesses and life change scores (LCU's).

	p	r	r^2
Mental disorders (depression, anxiety, suicide attempts)	.01	.38	.14
Respiratory diseases (colds, flu, pneumonia)	.02	.33	.11
Diseases of the digestive system (dental disorders, liver disorders, vomiting, diarrhea)	.03	.31	.10

Types of illnesses. Table III shows the types of illnesses statistically significantly associated with life change scores. While illness onset is clearly a multivariate concept, there is a respectable association between types of illnesses and life change scores in this group of clients. There are more statistically significant associations between life change scores and illness types than would be expected to be due to chance.

Illness prediction. Table IV shows regression equations based on these findings. It can be seen that the original Holmes and Rahe Social Readjustment Rating Scale is still the best predictor of numbers of illnesses for this group of people.

Drugs and illnesses. Heroin quantity, quality or cost was not statistically significantly associated with either numbers of illnesses or numbers of illnesses perceived as serious. There were some few statistically significant associations between heroin use and types of illnesses, but no more than would be expected to be due to chance alone.

Table IV. Illness prediction.

Number of illnesses = .004 (LCU score for last six months) + 1.34

($p<.001$, $r^2 = .21$, se = 2.01)

Number of illnesses = .002 (SLCU score for past six months) + 1.7

($p<.01$, $r^2 = 13$, se = 2.1)

Table V. Differences between two treatment modalities.

	Outpatient Means	Residential Means
Numbers of changes in last six months	13.6 changes	12.5 changes
Changes within six months of admission	24 changes	17.0 changes
LCU scores in last six months	459 LCU's	406 LCU's
SLCU scores in last six months	1016 SLCU's	898 SLCU's
Number of illnesses in last six months	3.3 illnesses	3.1 illnesses

Study Question Three

What is the association between life changes and treatment modality?

Table V shows the differences between therapeutic communities and outpatient methadone clinics. Although outpatient subjects had more changes, more changes within six months of program admission, higher adjustment scores and more illnesses, the differences between modalities was not statistically significant.

Study Question Four

What is the relationship between life changes and clean urines?

Clean urines were not correlated significantly with life change scores. Dirty urines, however, were correlated with life change ($\underline{r}$=.50, $p<.001$). These findings support the similar findings of Chorens and Dempsey (1977). Dirty urines represent non-adherence to the treatment goal of remaining drug free.

DISCUSSION

Adjustment Scores

The findings of this study support the notion that drug-dependent clients are adjusting to a great deal of life change. Their life changes are statistically significantly higher than the norms reported in the literature. Caplan (1975) suggests that norms need to be established for each study population that differs from the population upon which the currently accepted norms were derived. Data on females, drug-dependent clients and low socio-economic groups need to be gathered and compared to the findings of this study, for a more accurate comparison of drug-dependent clients versus non drug-dependent clients.

The scores in the subjects in this sample are so high, however, that in spite of the lack of generalizability of the findings based on data collection procedures, and the lack of adequate norms for this population, it can be safely said that these individuals are adjusting to a great deal of life change.

Illness Prediction

For this group of subjects, the Holmes and Rahe Social Readjustment Rating Scale is still the measurement of adjustment that is the most predictive of illness. This is contrary to the opinion that the subjective version would be a better measuring instrument.

It may be that as an organism, we have a certain amount to adjustment to change allotted to use before illness may occur. When that amount of adjustment has been reached, regardless of our perception, physiological manifestations occur. This is in line with Holmes' and Rahe's original theory of the cumulative effect of change, and Hans Selye's (1956) General Adaptation Response to Stress Theory.

Treatment Modalities

The results of this study also show that while all subjects have very high scores, subjects entering residential programs have fewer changes and lower scores than do outpatients living in the community.

Two factors, structure and social support, may account for these results. Living in a low socio-economic community may be highly associated with large numbers of changes. Entering the structured environment of a therapeutic community actually protects clients from those changes. Secondly, social support and structure are guaranteed with entrance to a therapeutic community.

Outpatient subjects often only stop in daily for medication and see a counselor once a week. For most of their day they have no structure. The past structure of obtaining money and drugs is gone. Social support may also be absent as evidenced by the fact that 70% of outpatient subjects, on whom marital status data are available, live with no mate. These variables need to be measured in any future studies attempting to study change and illness in this population.

Adherence to Treatment Goals

The strong correlations between life change scores and dirty urines lead to the conclusion that adjustment to change must be taken into account by program staff, if they are interested in increasing adherence to treatment goals. Clients who are undergoing stress and reverting to old coping mechanisms, that of drug dependence.

RECOMMENDATIONS

Based on these findings, that life changes are associated with illnesses and particularly illnesses perceived as serious, clinical interventions such as life change assessment need to be implemented to evaluate the impact of life changes on drug-dependent clients in treatment. These interventions could be instrumental in preventing some health problems.

One way to evaluate the impact of life changes on clients entering treatment would be to employ a tool such as the one used by health care providers at the Kaiser Permanente Medical Facility in San Jose, California. While it is not a drug treatment program, their intervention would work well in all types of drug treatment modalities. They have provided a "Golden Rod" assessment tool for clients, consisting of several questionnaires among which is a *Schedule of Recent and Anticipated Events*. Each client is given a short booklet on life change philosophy and a self-administered life change questionnaire. A concise detailed model of how each client can monitor and affect his/her own life changes over the next year is included. Also included in the booklet are norms such as those provided in your handout, and a method for provision of assistance in coping with changes, if desired. Clients are encouraged to keep an eye on their risk levels and if they feel they are having trouble coping, they are encouraged to use the method for provision of assistance. This self-management intervention allows clients to have some control over their own future health.

CONCLUSION

In conclusion, I would like to draw your attention to a phrase coined by Alvin Toffler (1970) in his book *Future Shock*. He described future shock as the disease of change. He says it is no longer a potentially distant danger, but a real sickness from which large numbers of people already suffer. We as drug-treatment providers must recognize the presence of future shock in our clients and begin to take it into account as we plan their treatment.

REFERENCES

Andersen, M.D. and Pleticha, J.M. 1974. Emergency unit patients' perceptions of stressful life events. *Nurs. Res.* 23: 378-383.

Atlee, W.E. 1972. Talc and cornstarch emboli in eyes of drug abusers. *JAMA*. 19: 49-51.

Baden, M.M. 1975. Pathology of the addictive states. In R.W. Richter (ed.), *Medical Aspects of Drug Abuse*. Hagerstown, Maryland: Harper & Row.

Bell, J. 1977. Stressful life events and coping methods in mental-illness and wellness behaviors. *Nurs. Res.* 26(2): 136-141.

Bloom, W.A. and Butcher, T. 1970. Methadone side effects and related symptoms in 200 methadone maintained patients. *3rd National Conference on Methadone Treatment Proceedings*, 44-47.

Brill, L. 1972. *The De-Addiction Process*. Springfield, Illinois: Thomas Publisher.

Byrd, O.E. 1970. *Medical Readings on Drug Abuse*. Reading, Mass.: Addison-Wesley Publishing Co.

Caplan, R.D. 1975. A less heretical view of life change and hospitalization. *J. Psychosom. Res.* 19: 247-250.

Challenor, Y.B. and Garner, A.L. 1975. Hand problems in the addict. In R.W. Richter (ed.), *Medical Aspects of Drug Abuse*. Hagerstown, Maryland: Harper & Row.

Cherubin, C.E. 1975. Liver disease in addicts. In R.W. Richter (ed.), *Medical Aspects of Drug Abuse*. Hagerstown, Maryland: Harper & Row.

Chorens, J.A. and Dempsey, G.L. 1977. Life stress events, a dimension of recidivism. Paper presented at the *National Drug Abuse Conference*, San Francisco.

Christakis, G., Stimmel, B., Rabin, J., Fintanaes, R., and Archer, M. 1973. Nutritional status of heroin users enrolled in methadone maintenance. *5th National Conference on Methadone Treatment Proceedings*.

Dohrenwend, B.S. 1973. Events as stressors: A methodological inquiry. *J. Health Soc. Beh.* 14: 167-175.

Dohrenwend, B.S. and Dohrenwend, B.P. 1974. A brief historical introduction to research on stressful life events. In B.S. Dohrenwend and B.P. Dohrenwend (eds.), *Stressful Life Events: Their Nature and Effects.* New York: John Wiley & Sons.

Felton, D.P. 1975. Pulmonary infections in the addict. In R.W. Richter (ed)., *Medical Aspects of Drug Abuse.* Hagerstown, Maryland: Harper & Row.

Glasscote, R.M., Sussex, J.N., Jaffe, J.H., Ball, J., and Brill, L. 1972. *The Treatment of Drug Abuse.* Washington, D.C.: American Psychiatric Association.

Harris, P.D., Yeoh, C.B., and Brown, J. 1975. Endocarditis in addicts. In R.W. Richter (ed.), *Medical Aspects of Drug Abuse.* Hagerstown, Maryland: Harper & Row.

Holmes, T.H. 1970. Psychological screening. In *Football Injuries.* Paper presented at a workshop, sponsored by the Subcommittee on Athletic Injuries, Committee on the Skeletal System, Division of Medical Sciences, National Research Council, February, 1969. Washington, D.C.: National Academy of Sciences.

Holmes, T.H. and Rahe, R.H. 1967. The Socail Readjustment Rating Scale. *J. Psychosom. Res.* 11: 213-218.

Kilocyne, M.M. 1975. Heroin-related nephrotic syndrome. In R.W. Richter (ed.), *Medical Aspects of Drug Abuse.* Hagerstown, Maryland: Harper & Row.

Kreek, M.J. Medical complications in methadone patients. *Annls. of N.Y. Acad. Sci.* (in press).

Langrod, J., Brill, L., Lowinson, J., and Joseph, H. 1972. Methadone maintenance from research to treatment. In L. Brill and L. Lieberman (eds.), *Major Modalities in the Treatment of Drug Abuse.* New York: Behavioral Publications.

Louira, D.B., Hensle, T., and Rose, J. 1967. The major medical complications of heroin addiction. *Annls. Int. Med.* 67: 1-22.

Lundberg, U., Theorell, T., and Lind, E. 1975. Life changes and myocardial infarction: Individual differences in life change scaling. *J. Psychosom. Res.* 19: 27-32.

Marx, M.B., Garrity, T.F., and Bowers, F.R. 1975. The influence of recent life experience on the health of college freshmen. *J. Psychosom. Res.* 19: 87-98.

McEwan, S.E. and Stevens, R.F. 1977. An investigation of life change events in a sample with arthritis using the Social Readjustment Rating Questionnaire. Master of Science in Nursing Thesis. University of Michigan.

Paykel, E.S., Meyers, J.K., Dienelt, M.N., Klerman, G.N., Lindenthal, J.J., and Pepper, M.P. 1969. Life events and depression. *Arch. Gen. Psych.* 21: 753-760.

Rahe, R.H. 1968. Life-change measurement as a predictor of illness. *Proc. Roy. Soc. Med.* 61: 1124-1126.

Rahe, R.H. 1972. Subjects' recent life changes and their near future illness reports. *Annls. Clin. Res.* 4: 250-265.

Rahe, R.H. 1974. The pathway between subjects' recent life changes and their near-future illness reports: Representative results and methodological issues. In B.S. Dohrenwend and B.P. Dohrenwend (eds.), *Stressful Life Events: Their Nature and Effects.* New York: John Wiley & Sons.

Rahe, R.H. 1974. Subjects' recent life changes and their near-future illness reports: Previous work and new directions of study. In J.P. Riehl and C. Roy (eds.), *Conceptual Models for Nursing Practice.* New York: Appleton-Century-Crofts.

Rahe, R.H. 1975. Epidemiological studies of life change and illness. *Internat. J. Psych. Med.* 6(1): 1-14 (a).

Rahe, R.H. 1975. Life change and near-future illness reports. In L. Levi (ed.) *Emotions - Their Parameters and Measurement.* New York: Raven Press, 511-529 (b).

Rahe, R.H. and Arthur, R.J. 1968. Life change patterns surrounding illness experience. *J. Psychosom. Res.* 11: 341-345.

Rahe, R.H., McKean, J.D., and Arthur, R.J. 1967. A longitudinal study of life change and illness patterns. *J. Psychosom. Res.* 10: 355-366.

Rahe, R.H., Bennett, L., Romo, M., Siltanen, P., and Arthur, R.J. 1973. Subjects' recent life changes and coronary heart disease in Finland. *Amer. J. Psych.* 130: 1222-1226.

Rahe, R.H., Pugh, W.M., Erickson, J., Gunderson, E.K.E., and Rubin, R.T. 1971. Cluster analyses of life changes. *Arch. Gen. Psych.* 25: 330-339.

Reed, B.G. and Herr, J.N. 1977. Deviant social careers and heroin addicted women: Issues in treatment. Unpublished paper. Women's Drug Research Coordinating Project. School of Social Work, University of Michigan.

Rowland, L.P. 1975. Foreword. In Richter, R.W. (ed.), *Medical Aspects of Drug Abuse*. Hagerstown, Maryland: Harper & Row.

Seltzer, M.L. 1975. Role of life events in accident causation. *Ment. Health Soc.* 2: 36-54.

Seltzer, M.L. and Vinokur, A. 1974. Life events, subjective stress, and traffic accidents. *Amer. J. Psych.* 131: 903-906.

Selye, H. 1956. *The Stress of Life*. New York: McGraw-Hill Book Co.

Smith, R.C. 1972. Community approaches to drug treatment. In C.J.D. Zarafonetis (ed.), *Drug Abuse Proceedings of the International Conference*. Philadelphia: Lea & Febiger.

Stimmel, B. and Kreek, M.J. 1975. Pharmacologic actions of heroin. In B. Stimmel (ed.), *Herion Dependency: Medical, Economic, and Social Aspects*. New York: Intercontinental Medical Book Corporation.

Toffler, A. 1970. *Future Shock*. New York: Random House.

CLIENT EVALUATION OF COUNSELING SERVICES IN A METHADONE MAINTENANCE CLINIC

Lewis P. Bennett, M.A.

Drug Abuse Services

Wilmington Medical Center

OBJECTIVES

Although regulatory agencies set standards for counseling and counselors themselves have various conceptions of their job function, there is less information on how methadone clients view counseling. Therefore, this survey was conducted to provide counselors with feedback on how their efforts are received by clients, to provide clients with a way to express their views and feelings through a measurement device, and to give program directors a chance to examine their expectations of how counseling services are perceived by clients.

A secondary objective of this survey was to provide a clear demographic picture of the clients who answered the questionnaire. The survey included questions on demographic information as well as questions on counseling evaluations.

METHOD

The survey was conducted at the Wilmington Medical Center Methadone Clinic. The clinic is the only methadone clinic in a city of approximately 90,000 people on the Eastern seaboard. The subjects are ethnically mixed and include both inner city dwellers and farm workers from nearby Maryland and Pennsylvania. Formal education among clinic members includes those with a third grade education and those in graduate school.

The survey was administered by placing the forms at the dispensing station. When each client came in to receive medication, the nurses were instructed to tell the clients that the survey would be used to help evaluate the counseling they received and that they did not have to put their names on it or worry that information could be used against them. The nurses were instructed to tell clients to answer as honestly as they could and to feel free to make additional comments in the space provided.

Out of a total clinic population of 140, 131 forms were turned in. If more than two questions in a row were unanswered, the form was discarded. Seven forms were discarded for this reason leaving a total of 124 completed survey forms. In part, the questionnaire on evaluation of counseling services is based on a similar but more extensive device by Orlinsky and Howard, reported by Lazarus (1971). The survey was administered in the spring of 1976.

RESULTS AND DISCUSSION

As can be seen on the sample questionnaire (see Appendix), the survey included two main areas: Demographic and Evaluation of Counseling. Demographic self-report was compared to demographic information collected during intake procedures.

Age

On the question of age, 56 clients reported they were between 18 and 25 years of age; 62 clients were between 26 and 39 years of age; and 6 stated they were over 40. This information is accurate when compared to clients' ages reported on intake forms.

Sex

Eighty-seven males and 37 females responded to the survey. While there are almost twice as many males as females, the females tended to be younger than the males. There were 36 males between 18 and 25, 45 males between 26 and 39, and 6 males over 40. Twenty females were between 18 and 25, and 17 females were between 26 and 39. No females in the clinic were over 40.

Race/Ethnic Group

The question on Race/Ethnic group elicited more resistance than the two previous categories of age and sex. Seventeen persons refused to answer this item. Several people refused to be classified by answering "human," "person," or some other general category. Others gave responses that could not be scored such as "male" or "Baptist". Of those who answered, 54 stated they were Black (16 between 18 and 25, 33 between 26 and 39, and 5 over 40). Forty-seven people reported they were White. This figure includes five individuals who stated they were Italian. With this exception, no other clients listed themselves as members of specific ethnic groups. Of the 47 Whites, 24 were between 18 and 25, 22 were between 26 and 39, and one was over 40. For both Blacks and Whites, self-report correlated with clinic intake records.

Four people reported they were Puerto Rican (three between 18 and 25 and one between 26 and 39). According to intake records, there are nine Puerto Ricans (Hispanics) on the clinic. When questioned later several of the Puerto Ricans stated they listed themselves as White. Two people reported they were American Indians. Clinic intake records show one Indian. Later on a client said he and his cousin, also a client, answered "Indian" since they are part Cherokee and identify with that category rather than White. While not statistically significant, it is interesting to see how people view themselves in this area.

Time on Clinic

The final demographic question asked about number of months the individuals were on the clinic. Thirty-nine individuals reported they were on the clinic six months or less (16 between 18 and 25, 21 between 26 and 39, and two over 40). On the clinic between six months and one year were 31 individuals (16 between 18 and 25, 15 between 26 and 39, and none over 40). Forty-six individuals stated they were on the clinic between one and five years (17 between 18 and 25, 27 between 26 and 39, and two over 40). Of those clients on the clinic more than five years, none were between the ages of 18 and 25. On the clinic over five years were seven clients. Five of those clients were between 16 and 39; the two remaining were over 40. This question yielded a total of 123 responses; on one form this question was not answered. Again this information is in agreement with clinic records.

Except for the questions on Race/Ethnic group, the clients' responses were in agreement with clinic records. On the question of Race/Ethnic group, the answers seemed emotionally honest which is perhaps even more important. There seemed to be no noteworthy trends associated with age except that no one under 26 years old was on the clinic for five years or more. The accuracy of the clinic information given by clients is a good indication that most clients responded in an honest way.

Counseling Evaluation

The first item related to counseling asked how often the client saw a counselor. Out of 111 responses to this item, 27 clients saw a counselor once a week or more; 22 saw a counselor several times a month; 57, the largest group, saw a counselor only when specifically requested to contact a counselor; five clients reported they were not interested in counseling. Slightly less than half the respondents were involved in on-going counseling.

Under the item on types of counseling services requested, eight services were listed, and clients could mark as many as they felt applied. Forty-seven respondents marked all eight which probably reflects a tendency to simply mark all categories. The remaining 64 clients of a total of 111 who answered this item marked the services selectively. The three greatest areas of concern were drug problems (including alcohol) with 36, problems with personal feelings 31, and vocational counseling 31. Of least concern were problems related to housing; seven clients checked this item.

Categories in between were legal problems, 22 responses; interpersonal problems, 19; family problems, 18; and problems with other agencies (Welfare, Probation and Parole, etc.), 18. Since clients could mark more than one item, there are more responses than the total number of persons answering this item.

Besides ascertaining what services clients request and how often they see counselors, several items were asked on counselor effectiveness.

Out of 106 responses, 84 stated that their counselors were available when needed. Nineteen clients reported that sometimes their counselors were available, and three responded that counselors were not available when needed.

On the question "do you trust your counselor," there were 107 responses. Eighty-two said they trusted their counselor; 18 said sometimes they trusted their counselors; and seven said they did not trust their counselors.

"Do you feel your counselor cares about your problems" elicited 103 responses (64 felt counselors cared; 36 thought counselors sometimes cared; and three did not think their counselors cared).

The final question in this area only received 82 responses which could be because the question appeared unclear to the clients answering the survey. The question was, "Does your counselor understand your thoughts and feelings when you talk with him?" In the margin one person wrote, "Ask the counselor." Someone else wrote, "How should I know what the counselor thinks?" These comments probably show the confusion caused by the wording of this item. Of the 82 responses, 62 stated "yes;" 14 said "sometimes;" and six said "no, the counselor does not understand."

This survey successfully demonstrated several things about this particular group of methadone clients:

1) Clients will give accurate information on surveys of interest to them. This statement is demonstrated by comparing their answers on the demographic section to that gathered on intake forms.

2) Counseling for drug problems, personal problems (those dealing with how the client feels), and vocational problems were the most frequently requested services, although other services were in demand.

3) The vast majority of clients believed their counselors cared about them. Also counselors were available when needed and understood the information clients expressed to them.

4) While it is impossible to generalize on particular responses given in one clinic, this study demonstrates the feasibility of using measurement devices of this type. It is the belief of the author that feedback from clients is an important source of information for those making policy decisions.

Hopefully this survey will provide positive feedback to counselors and program planners. Also, for the clients who shared their feelings with us by answering the questionnaire, it gave them a chance to look at how they felt about counseling and consider what services they consider important.

REFERENCES

Lazarus, A.A. 1971. *Behavior Therapy & Beyond*. New York: McGraw Hill Co. 253-271.

APPENDIX

This questionnaire is designed to evaluate or "check out" what happens in counseling sessions. Hopefully, your answers will give the counselors a better idea of what they are doing right as well as what they could do better. If you want to make any additional remarks, please feel free to put them at the bottom of the page in the placed marked "Comments".

It is not necessary to put your name on this sheet.

Age:

Sex:

Race or Ethnic Group:

How Long have you been on the clinic?

a. six months or less

b. six months to one year

c. one year to five years

d. over five years

How often do you see a counselor:

a. once a week or more

b. several times a month

c. only if the counselor asks to see me

d. I am not interested in counseling

Check the types of problems that you feel you can talk to your counselor about. (Check all that apply to you.)

1. legal
2. family
3. jobs
4. housing
5. problems with Welfare, Probation and Parole, and other agencies

6. drug problems (including drinking)

7. problems with other people

8. personal problems and feelings

Is your counselor available to see you when you need him/her?

Yes Sometimes No

Do you trust your counselor?

Yes Sometimes No

Do you feel your counselor cares about your problems?

Yes Sometimes No

Does your counselor understand what you are thinking and feeling when you talk to him?

Yes Sometimes No

Comments (particularly anything your counselor could do to help you that he/she is not doing now):

CO-TREATMENT OF GROUPS OF DIFFERENTIATED SUBSTANCE ABUSERS*

George H. Burke III, Harvey B. Milkman, Ph.D., and Mark F. Hammer

University of Colorado
Therapeutic Community, PEER I

There is a continuing controversy among treatment personnel over how to treat whom among drug users. Understanding differences in personality structure and function may provide clues which will permit careful delineation of the variety of treatment programs designed to meet the needs of particular groups of drug users. Traditional approaches to differentiating among users rely on behavior definitions of the type of drug, frequency, and the intensity of use. We propose a psychodynamic definition of hard and soft core drug use based on the careful assessment of levels of trust and denial. Treatment strategies are geared to meet the user at his or her level of personality organization with outcome measures providing an assessment of the relative effectiveness of differentiated treatment. The therapeutic community provides a vehicle through which investigation of this type will be conducted.

At the present time, there is little rationale for assigning addicts to treatment. With the exception of patients who are directed to specific programs as an alternative to incarceration, most drug addicts are not assigned to treatment (Pittel and Hofer, 1974). "Both traditional and ad hoc agencies tend to group together patients with varying patterns of drug involvement. They often abnegate their responsibilities in determining how each patient might receive the most appropriate and effective treatment." According to Aron and Daily

---

* The opinions expressed in this paper are those of the authors and are not necessarily those of the University of Colorado, Addiction Research and Treatment Services.

(1976), "A greater number of drug abusers are seeking treatment and as drug rehabilitation program resources become more and more scarce, selection processes should become based on scientific research rather than on the intuition of a few influential personnel."

Increasingly, therapeutic communities (T.C.'s) will be called upon to work more intensely and accomplish what we must in a shorter period of time (Freudenberger, 1976). Many T.C.'s remain of the old wave of drug abuse treatment. Problems exist that have to be looked at, and dealt with, for effective treatment. Looking back from today's research, the early days of T.C.'s had a primitive approach to treatment. Rigid rules were established within the treatment modality. Stress-inducing situations, such as enforced haircuts, were utilized, ego shaming devices were employed; and close attention to the minutia of daily behavior was considered essential to ultimate personality change (Freudenberger, 1976).

Existing problems have detrimental effect on treatment approaches. "If it worked for me, it will work for you," is the standard shot-gun approach used by many addict clinicians. Freudenberger (1976) observes that, "Many staff of therapeutic communities are made up of graduates from other programs. Each staff person will use the concepts which worked for him and her and these previous program concepts must be the only ones that make sense to him or her. The result is confusion of treatment terms and treatment philosophers." Many addicts appear to resist change. Rigidity, it seems, is the solution to this problem. This mode of operating is still incorporated in today's T.C. This presents a problem because effective treatment no longer dictates the hard attack encounter. Residents have become sophisticated in defenses, conning, manipulative and evasive techniques so they can survive encounters without being changed by it.

The population of today's T.C.'s are made up of the younger, more sophisticated, not hard core, poly drug abusers (Freudenberger, 1976). Addicts seem to have the mix and match method of getting high. They might be using heroin and alcohol or speed and alcohol or many other formulas to help them cope with the pressure of being who they are and they not liking it. The personality of the poly drug abuser seems to be multi-faceted. Increasing numbers of researchers have developed theoretical positions about the psychodynamics of drug dependence. Blaine and Julius (1977) edited a collection of analytically oriented articles, each of which addresses the problem of what constitutes realizable and adequate treatment. There was consensus that analytic insight, rigorously applied to the problems of addiction can make a major contribution to the treatment of individuals in trouble with drugs. The specific elaborations of treatment issues were summarized by Khantzian and Treece (1977) in terms of four basic constituents: 1) Multiple modality approaches to do justice to the complexity and multiplicity of determinants in pathological drug use; 2) bi-phasic therapeutic strategy, consisting first of interim measures for keeping patients available and intact until the second phase of longer range work in psychotherapy can be firmly established; 3) the use of treatment personnel who are fully trained and who hold the particular qualities needed for this work; and 4) specificity of diagnosis and treatment planning.

Our method of developing a more effective treatment strategy, in the therapeutic community, incorporates the above constituents and focuses on the increasingly reported differences in trust and denial among users. Wurmser (1977) observes that there may be two basic types of compulsive drug users; one far more afraid of engulfment (lack of trust) and the other fearing more isolation and loneliness. "In the first type, the distance needs to be preserved far more and far longer for trust to arise. In the second type, far more direct support is needed." Milkman and Frosh (1973, 1977) report differences in the personalities of long-term preferential users of heroin and amphetamines. "The heroin user's mood of depression and despair is contrasted with the amphetamine user's denial of depression and compensatory optimism." More severe psychopathology was observed in heroin users as indicated by significant differences in ego functions.

CLINICAL OBSERVATIONS TOWARD A DEFINITION OF HARD AND SOFT CORE USERS

In the past, practitioners have had to rely on their intuition in determining treatment. Clinical observation and interpretation of the client's behavior, past and present, has been the method for writing and implementing treatment plans. The problem with this approach is that there is a lack of systematized knowledge regarding differential treatment strategies.

Hard core individuals, from our observations, have a "so-what syndrome". A person is told that he or she will eventually die or end up in jail for a good part of life. A typical response is, "yes, I know that, but so what?" We interpret that this person has limited denial of a difficult life situation but does little, if anything, to change. As the client progresses in treatment they continue with behaviors that are detrimental to themselves. Residents continue to "game" at living life. Self-esteem is contingent upon how they have "run a number", (munipulate), and who they have "run a number" on. Discipline is given to them; the clients understand what they did, but continue to demonstrate these behaviors. We interpret this as lack of trust in themselves and even less trust in the information and support of the community. The hard core individual has learned not to trust himself or others over the years of abuse. His job was to make sure that he had enough money, drugs, or both, to help make it through that day. He had to employ many techniques and rationalizations to reach this state. While the resident was living in the community he became "partners" with people of the same type. This person knew, after a while, that these people were after the same state as he, being high. These residents were involved in extensive criminal activities but could not tell anyone for fear of being arrested. These clients learned not to trust anyone or anything for their own survival. When this individual enters treatment, this primary way of thinking and reacting is still employed. Time and consistency of information is a factor for developing their trust.

Conversely, there are individuals in treatment who have a high rate of observable denial. Clients enter treatment due to court or parental pressure. When confronted about their life's situations, i.e., being in treatment or getting arrested on drug related charges, they start denying why they are in treatment. The most common response is, "I wouldn't be here if it wasn't for court or my parents because I really don't have a drug problem." Clients have a tendency to blame externals for their present situation. Another form of denial is comparing themselves to individuals who might have done more drugs or have been in more trouble. A common response is, "Well, I'm not as bad as so and so." We have observed these clients depending on others for their sense of self-esteem. These clients appear to seek and need extensive approval from their externals. Becuase of this they seem to have more trust. The soft core person, behaviorally, has not been in as much trouble with their family or community support systems (police, schools, etc). These residents have not used drugs as long; their street survival skills are not as refined, as the harder core. With these observations, and theoretical positions, we have operationally defined soft-core users on two levels: Level 1, high denial, high trust; Level 2, low denial, high trust. Hard core users are in two levels also: Level 3 is low denial, low trust; Level 4 is high denial, low trust.

HYPOTHESIS

This study will test the hypothesis that within a therapeutic community, more effective treatment can be achieved by differentially treating hard and soft core users, independent of drug preference, frequency or intensity of use.

METHOD

Treatment-site

The University of Colorado Therapeutic Community, PEER I, is an adult program aimed at the single and multiple substance abuser. Residents who have an alcohol problem are accepted provided that this problem is secondary to drug abuse. Specifically, it is an intermediate-term program (five to nine months), flexible in its philosophy and emphasizes individualized treatment plans. The present capacity of the program is 36. Treatment is intensive, voluntary and drug-free and both men and women are treated interactionally. A subjective staff appraisal is made of the applicant to determine suitability for admission to the Program, is based on expressed motivation, history of drug abuse, observed character personality traits, milieu dynamics, and census.

Subjects

The average age of the client-population is 27 years. Women comprise 19% of the total population. Sixty percent (60%) of the women are white, 20% are black, and 20% are Spanish American. Ninety-five percent (95%) of the men are white and 5% are Spanish American. PEER I total client population is roughly 8% Spanish American, 4% Black and 88% Anglo.

Existing Treatment

PEER I offers an eclectic approach to treatment; both traditional and non-traditional tools are offered. The Program incorporates discipline, verbal haircuts, signs, work monads, essays and loss of privileges or job status. A unique group, the Confrontation-Sensitivity Group (C.S.) was started in order to meet some of the needs of the multi and single substance abuser. The group draws from such diverse sources as Attack Therapy, The Synanon Game, Daytop Village Encounters, Gestalt "Primitive Explosion" as well as Gestalt exercises and Reality Therapy (Frye, Burke, and Hammer, 1978).

PROCEDURE

Instrument

Leary (1956) devised a theory of personality whereby interpersonal behavior is classified into sixteen mechanisms or reflexes. These mechanisms operate to different degrees in 5 levels of personality ranging from the level of public communication to the level of deeply internalized values of which the individual has no conscious awareness. He developed the Interpersonal Checklist as a means of quantifying behavior at two of these levels: 1) Level II--the individual's conscious descriptions of self and others; 2) Level V--the individual's conceived ideal self.

The checklist comprises 128 items--eight for each of the sixteen interpersonal variables. The sixteen variables can be combined to form a group of eight octant variables, which are coded by the use of letters: 1) A P--Managerial--Autocratic; 2) No-Responsible--Hypernormal; 3) C M--Cooperative--Over-Conventional; 4) J K--Docile--De-

Table I. I P C categorization.

Adaptive Interpersonal Diagnostic Categories	Letter Code	Maladaptive Interpersonal Diagnostic Categories
Managerial	A P	Autocratic
Competitive	B C	Narcissistic
Critical	D E	Sadistic
Skeptical	F G	Distrustful
Self-effacing	H I	Masochistic
Docile	J K	Dependent
Conventional	L M	Over-conventional
Responsible	N O	Hypernormal

pendent; 5) H I--Self-effacing--Masochistic; 6) F G--Rebellious--Distrustful; 7) D E--Agressive--Sadistic; 8) B C--Competitive--Narcissistic. An intensity dimension is built into the checklist. The items in the left column are the most moderate aspects of the interpersonal trait, and the items in each of the rows increase in intensity as they move towards the right. Example: "Well thought of" is a moderate item for the variable "P" (Prestige), and "Tries to be too successful" is an intense or maladaptive "P" item. Table I indicates the categories employed for interpersonal diagnosis of adaptive and maladaptive behavior at all levels.

There are a number of ways in which the data from the Interpersonal Checklist can be treated statistically. A simple procedure is to count and graph the number of raw score items falling in each octant. The octant we are concerned with is "Skeptical-Distrustful."

In this study the Interpersonal Checklist will be given for the subject's description of his perceived "self". Test-retest reliability of this use of the Interpersonal Checklist correlates on an average of .78. These correlations suggest that Interpersonal Checklist scores have sufficient stability to be useful in personality research of this type.

Implementation

The Interpersonal Checklist (I P C) will be given to all new clients after a 10-day period. Their scores will be graphed accordingly with attention being given to the Distrustful scale. Simultaneously, two (2) staff and one (1) senior client, who have become close in contact with the new resident, will independently complete the I P C inventory, for each new resident. A staff caucus will be held to reach a consensus about the I P C ratings. Staff assignments are randomly determined. The combined caucus description reflects reality. The difference between caucus and client I P C rating is defined as the degree of denial, i.e., the difference between self-description and objective reality. Degree of trust is determined by combined client and caucus ratings. A Control Group, comprised of an equal number, will be matched for age, race, sex, I.Q., social class, psychopathology, and I P C ratings. Staff will devote an equal amount of time and effort in writing and implementing treatment plans for experimental and control groups.

Treatment

Computation of Trust and Denial I P C ratings permits subdivision of the experimental group into four levels. The level categories are presented in Table II. Depending on where the individual scores on the I P C rating, differential treatment strategies are employed.

Level 1 (soft core): High Denial, High Trust; The client is saying, "I don't think I have a problem, but I trust you." The primary clinician may write a more direct treatment for this resident. Group Confrontation is recommended from the start to help break through the denial system. The initial treatment factor for this group is that neither therapist nor residents need to consume time in building trust. Direct interpersonal encounters are indicated with rapid implementation. Gestalt and Primal techniques are appropriate because trust is already established and deep self-disclosure may be relatively rapid. Peer pressure and weekly discipline tasks continue to be part of this client's treatment. Individual counseling can be direct and confrontive.

Level 2 (soft core): Low Denial, High Trust; The resident who scores in this level says, "I know I have a problem and I trust you to help me solve it." A treatment plan will be written to keep confrontation at a minimum. Trusting Interventions, relying on high levels of trust, e.g., Gestalt, Primal, Sensitivity, are used from the beginning. Weekly discipline for negative behavior is not as severe. Instead, it is task oriented to facilitate continued high levels of trust. Distance need not be preserved by the staff. Peer pressure remains an effective tool.

Table II. Level categories for hard and soft core.

Soft Core	Level 1 High Denial High Trust	Level 2 Low Denial High Trust
Hard Core	Level 3 Low Denial Low Trust	Level 4 High Denial Low Trust

Level 3 (hard core): Low Denial, Low Trust; This resident is saying, "I know I have a problem but I don't trust myself or anyone else to help me solve it." In this case, treatment is of longer duration. A supportive environment needs to be maintained until trust begins. Residents begin with small, success-oriented tasks designed to initiate self and milieu trust. Very little confrontation is recommended at the beginning of treatment. The clinician needs to maintain a greater distance for a longer period of time. Pressure needs to be applied but not to the point of forcing them out of treatment. Discipline, when needed, should not lend itself to humiliation but towards the goal of self-trust.

Level 4 (hard core): High Denial, Low Trust; This resident is saying, "I don't have problems and I don't trust you." Treatment consists of supportive confrontation to begin breaking through defensive armor. Simultaneously, discipline is designed to help develop internal and external trust. Staff is careful to offer greater support in developing rapport. While the trust level is low, these residents do not participate in any Gestalt work or other interventions requiring high levels of trust. This group constitutes the most difficult treatment category in the therapeutic community.

Clients in the experimental group are retested at 3-month intervals to indicate change in denial and trust levels. Treatment plans will be reviewed and up-dated accordingly. The data will indicate the effect of individualized treatment vs. regular treatment in a therapeutic community. Standard evaluation procedures will be employed, i.e., degree of drug use, criminality, employment, recidivism.

Discussion

There are many advantages to this differentiated and systematic method of treatment. The Caucus forces the staff to recognize and make concrete differences among individuals in the community. This calls for staff to upgrade their skills in order to become more flexible in their treatment approach. In this design, research and treatment become "married" in a manner that is acceptable to staff and clients. The interpersonal nature of our approach is consistent with the overall thrust of the therapeutic community. Clients may find themselves in a group atmosphere that promotes and preserves individuality.

ACKNOWLEDGEMENTS

The authors gratefully acknowledge the clerical assistance of Helen Rash and Bonnie Boex in the preparation of this paper. In addition, the research assistance of Kirstin Strand was invaluable and imperative for the successful completion of this paper.

REFERENCES

Aron, W.S. and Daily, D.W. 1976. Graduates and splitees from therapeutic community drug treatment programs. *IJA*. 11(1): 1-18.

Blaine, J.D. and Julius, D.A. 1977. Psychodynamics of Drug Dependence. *NIDA Research Monograph*.

Freudenberger, J.J. 1976. The Therapeutic Community Revisited. *AJDA*. 3(1): 33-43.

Frye, R.V., Hammer, M.F., and Burke, G.H. 1978. The Confrontation-Sensitivity Group. NDAC Proceeding (Conference paper).

Khantzian, E.G. and Treece, C.J. 1977. Psychodynamics of Drug Dependence: An Overview. *NIDA Research Monograph*. 12: 11-25.

Leary, T. 1957. *Interpersonal Diagnosis of Personality*. New York: Ronald Press.

Frosh, W.A. and Milkman, H. 1977. Ego Functions in Drug Users. *NIDA Research Monograph*. 12: 142-156.

Pittel, S.M. and Hofer, R. 1974. A Systematic Approach to Drug Abuse Treatment Referral. *JPD*. 6(2): April-June.

Wurmser, L. 1977. Mr. Pecksniff's Horse? *NIDA Research Monograph*. 12: 36-71.

RELAXATION TRAINING AS AN ADJUNCT TO DRUG ABUSE TREATMENT

Floyd A. Aprill, MSW, ACSW

Milwaukee County Mental Health Complex
Drug Abuse Program

Anxiety is a major psychological symptom associated with narcotic dependency (Cheek, 1976; Lesser, 1967; O'Brien et al., 1972; Cheek et al., 1973); Alcoholism (Kraft, 1971; Eno, 1975; Cheek, 1976); and other Drug Abuse (Kraft, 1968; Kraft, 1970, Bergland and Chal, 1972). There is strong evidence that chronic and/or severe anxiety are major causative factors in the onset of certain patterns of drug taking behavior. Certainly anxiety plays a major role in sustaining and reinforcing patterns of compulsive drug use. Development of effective non-chemical mechanisms for coping with stress and anxiety in daily living is an important part of drug abuse treatment.

Relaxation training is a learned motor skill (like driving a car) that can be taught to clients to provide them with an active non-chemical coping skill for alleviating or managing tension and anxiety. Clients are taught to identify feelings of muscular tension and how to consciously relax away tension and anxiety. The process involves systematically tensing and releasing various muscle groups and learning to perceive and differentiate the resulting sensation of tension (muscle sense training) and relaxation. By progressively and systematically training all muscles in this procedure, deep relaxation can be obtained.

Progressive Relaxation is similar to meditation, yoga and biofeedback in that each provides a viable non-chemical method of handling stress and each is capable of producing states of deep relaxation with accompanying physiological changes that can be objectively verified. Progressive relaxation offers the advantage of providing a non-mystical, non-religious method of obtaining these results that can be utilized without expensive instruments or sophisticated equipment.

Relaxation training as opposed to chemotherapy with anti-anxiety medication maximizes client control and responsibility for anxiety management and facilitates the growth of client mastery skills. Positive success experiences in the self-management of anxiety frequently lead to other therapeutic gains such as a more positive self-concept and improved ego strength.

Little research has been done in the area of applying relaxation training in substance abuse treatment. Most studies combine relaxation training with systematic desensitization aversion training, assertive training or other behavior modification techniques. The effect of relaxation training by itself is rarely reported except in anecdotal studies involving limited numbers of patients. Although solid research in this area is sorely needed, the available evidence suggests that relaxation therapy (either by itself or in combination with other behavior modification techniques) is a useful adjunct to traditional drug abuse treatment.

Given the sparsity of research on relaxation training in the drug abuse literature, one might logically conclude that this technique is largely unknown to drug treatment personnel. Contrary to this expectation, many drug and alcohol programs are utilizing progressive relaxation as an adjunct to traditional drug abuse treatment. The verbal reports from these practitioners closely conform to the author's own experience. The general concensus is that progressive relaxation is no panacea, but that it can be a useful adjunct to a traditional problem focused treatment approach. Progressive relaxation can be used either by itself or more frequently in combination with other behavioral modification treatment techniques in an overall educational approach, stressing more effective ways of dealing with stress. Progressive relaxation can be especially useful in methadone maintenance and other medical model programs by providing a non-medical alternative to prescribing minor tranquilizers and sleeping pills.

Clinical staff can readily learn how to instruct clients in relaxation training and relaxation techniques are easily learned by drug abuse clients.* Although it is desirable for treatment

---

*Deep states of relaxation are often accompanied by physical sensations of warmth, heaviness, tingling, numbness, or a floating sensation. It has been this author's experience that drug treatment clients instructed in deep muscle relaxation frequently report these symptoms and liken it to the feeling of a drug high. This same experience has been reported by Lesser (1967), O'Brien et al. (1972), Bergland and Chal (1972). Cordeiro (1972) reports that, "Most patients spontaneously and rapidly discover the similarity between the sensual impression of deep relaxation and the phenomena previously experienced only by drug taking." Similar observations have

staff to learn relaxation techniques from an experienced instructor, it is possible to learn relaxation procedures on your own by utilizing a variety of training manuals and audio tapes which are available at a very modest cost.

This article contains a comprehensive working bibliography which the reader can utilize for self study. The substance abuse bibliography refers the reader to specific applications of relaxation training to alcohol and drug abuse clients. The clinical bibliography includes those journal articles, books, manuals, and audio tapes which focus upon the clinical skills required to instruct clients in progressive relaxation. These materials include step-by-step instructions, written relaxation scripts, and/or prerecorded audio tapes. The most useful material is marked with an asterisk.

It is hoped that this article will stimulate drug abuse researchers and clinicians to learn more about progressive relaxation and its potential applications in drug abuse treatment. This author encourages individuals with expertise in this area to contribute the results of their experience to the professional literature.

---

also been noted with alcoholics learning deep muscle relaxation (Burtle et al., 1974). This phenomenon may explain the generally positive response of most substance abusers to progressive relaxation training. Clients continue relaxation training because they enjoy the physical sensation of deep relaxation.

BIBLIOGRAPHY

Bernstein, D.A. and Borkovec, T.D. 1973. *Progressive Relaxation Training: A Manual for the Helping Professions.* Champaign: Research Press.

Ferguson, J.M., Marquis, J.N. and Taylor, C.B. 1977. A Script for deep muscle relaxation. *Disease of the Nervous System.* 38(9): 703-709.

Gershman, L. and Clouser, R.A. 1974. Treating insomnia with relaxation and desensitization in a group setting by an automated approach. *J. Behav. Ther. and Exp. Psychiat.* 50: 31-35.

Goldfried, M.R. 1976. *Behavioral Management of Anxiety: Self-Modification of Anxiety: Client Instructions* (audio tapes). New York: Biomonitoring Applications, Inc.

Goldfried, M.R. and Davidson, G.C. 1976. *Clinical Behavior Therapy.* New York: Holt, Reinhart and Winston.

Goldfried, M.R. and Trier, C.S. 1974. Effectiveness of relaxation as an active coping skill. *Journal of Abnormal Psychology.* 83(4): 348-355.

Goldfried, M.R. 1973. Reduction of generalized anxiety through a variant of systematic desensitization in M.R. Goldfried and M. Merbaum (eds.). *Behavior Change Through Self Control.* New York: Holt, Reinhart, and Winston.

Goldfreid, M.R. 1971. Systematic desensitization as training in self-control. *Journal of Consulting and Clinical Psychology.* 37(2): 228-234.

Hartman, C.H. 1976. *Mixed Scanning Relaxation Training.* (audio tape) New York: Biomonitoring Applications, Inc.

Haugen, G.B., Dixon, H.H., and Dickel, H.A. 1958. *A Therapy for Anxiety Tension Reactions.* New York: Macmillan.

Israel, E. and Beiman, I. 1977. Live versus recorded relaxation training: a controlled investigation. *Behavior Therapy.* 8: 251-254.

Jacobson, E. 1964. *Anxiety and Tension Control.* Philadelphia: Lippencott.

Jacobson, E. 1938. *Progressive Relaxation.* Chicago: University of Chicago Press.

Lazarus, A.A. 1970. *Daily Living: Coping with Tensions and Anxieties.* Relaxation exercises (audio tapes). Chicago: Instructional Dynamics, Inc.

Lomont, J.F. and Edwards, J.E. 1966. The role of relaxation in systematic desensitization. *Behav. Res. and Therapy.* 5: 11-25.

Paul, G.L. 1969. Inhibition of physiological response to stress-full imagery by relaxation training and hypnotically suggested relaxation. *Behav. Res. and Therapy.* 7: 249-256.

Paul, G.L. 1966. *Insight vs. Desensitization in Psychotherapy.* Stanford: Stanford University Press.

Paul, G.L. 1969. Physiological effects of relaxation training and hypnotic suggestion. *Journal of Abnormal Psychology.* 74(4): 425-437.

Paul, G.L., and Trimble, R.W. 1970. Recorded vs. "live" relaxation training and hypnotic suggestion: Comparative effectiveness for reducing physiological arousal and inhibited stress response. *Behavioral Therapy.* 1: 285-302.

Rachman, S. 1965. Studies in desensitization-1: The separate effects of relaxation and desensitization. *Behav. Res. and Therapy.* 3: 245-251.

Wolpe, J. and Lazarus, A.A. 1966. *Behavior Therapy Techniques: A Guide to the Treatment of Neurosis.* New York: Pergamon Press.

Wolpe, J. 1970. *The Practice of Behavior Therapy.* New York: Pergamon Press.

Wolpe, J. 1958. *Psychotherapy by Reciprocal Inhibition.* Stanford: Stanford University Press.

Yates, D.H. 1946. Relaxation in psychotherapy. *Journal of General Psychology.* 34: 213-238.

SUBSTANCE ABUSE BIBLIOGRAPHY

Anderson, L., Lubetkin, B., and Alpert, M. 1973. Comparison of relaxation methods for alcoholics: differential relaxation vs. sensory awareness. *Proceeding 81st Annual Convention, APA.* 8: 393-394.

Bergland, B.W. and Chal, A.H. 1972. Relaxation training and a junior high behavior problem. *School Counselor.* 19(4): 288-293.

Blake, B.G. 1965. The application of behavioral therapy to the treatment of alcoholism. *Behav. Res. and Therapy.* 3: 75-85.

Blake, B.G. 1967. A follow-up of alcoholics treated by behavior therapy. *Behav. Res. and Therapy.* 5: 89-94.

Burtle, V., Whitlock, D. and Franks, V. 1974. Modification of low self-esteem in woman alcoholics: a behavior treatment approach. *Psychotherapy: Theory, Research, and Practice.* 11(1): 36-40.

Cheek, F.E. 1976. Behavior modification for addicts on methadone maintenance. *Current Psychiatric Therapies.* 16:223-232.

Cheek, F.E., Tomarchio, T., Standen, J., and Albabary, R.S. 1973. Methadone plus-a behavior modification training program in self-control for addicts on methadone maintenance. *Intern, J. Addiction.* 8(6): 969-996.

Cordeino, J.C. 1972. A new perspective in the treatment of drug addicts: relaxation. *Annales Medico--Psychologigues.* 1(1): 11-17.

Eno, E.N. 1975. A comparison study of the level of state-trait anxiety and muscle tension of alcoholics when treated by electromyograph biofeedback relaxation training and other clinical techniques. *Dissertation Abstracts International.* 36(4-B): 1914.

Kraft, T. 1971. Social Anxiety model of alcoholism. Perceptial and motor skills. 33: 797-798.

Kraft, T. 1968. Successful treatment of a case of drinamyl addiction. *Brit. J. Psychiat.* 114: 1363-1364.

Kraft, T. 1970. Successful treatment of drinamyl addicts and assoiated personality changes. *Canad. Psychiat. Ass. J.* 15: 223-227.

Lesser, E. 1967. Behavior therapy with a narcotics user: a case report. *Behav. Res. and Therapy.* 5: 251-252.

Lesser, E. 1976. Behavior therapy with a narcotics user: a case report. Ten year follow-up. *Behav. Res. and Therapy.* 14(5): 38.

Miller, M.M. and Barlow, D.H. 1973. Behavioral approaches to the treatment of alcoholism. *J. Nervous and Mental Disease.* 157: 10-20.

Miller, P.M. 1973. Behavioral treatment of drug addiction: a review. *Inter. J. Addictions.* 8(3): 511-519.

Mc Farlain, R.A., Mielke, D.H., and Gallant, D.M. 1976. Comparison of muscle relaxation with placebo medication for anxiety reduction in alcoholic inpatients. *Current Therapeutic Research.* 20(2): 173-176.

O'Brien, J.S., Raynes, A.E., and Patch, V.D. 1972. Treatment of heroin addiction with aversion therapy, relaxation training and systematic desensitization. *Behav. Res. and Therapy.* 10: 77-80.

Paynyard, C. and Wolf, K. 1974. The use of systematic desensitization in an outpatient drug treatment center. *Psychotherapy: Theory, Research and Practice.* 11(4): 329-330.

Steffen, J.J. 1973. Tension reduction in chronic alcoholics during prolonged experimental drinking: effects of electromyographic feedback training. *Dissertation Abstracts International.* 34 (10-B): 3212.

Steffen, J.J. 1975. Electromyographically induced relaxation in the treatment of chronic alcohol abuse. *J. of Consulting and Clinical Psychology.* 43(2): 275.

CLINICAL BIBLIOGRAPHY*

Bernstein, D.A. and Borkovec, T.D. 1973. *Progressive Relaxation Training: A Manual for the Helping Professions.* Champaign: Research Press.*

Ferguson, J.M., Marquis, J.N., and Taylor, C.B. 1977. A script for deep muscle relaxation. *Disease of the Nervous System.* 38(9): 703-709.*

Goldfried, M.R. 1976. *Behavioral Management of Anxiety: A Clinician's Guide and Self-Modification of Anxiety: Client Instructions* (audio-tapes). New York: Biomonitoring Applications, Inc.

Goldfried, M.R. and Davidson, G.C. 1976. *Clinical Behavior Therapy.* New York: Holt, Reinhart, and Winston.*

Hartman, C.H. 1976. *Mixed Scanning Relaxation Training.* (Audio-tape) New York: Biomonitoring Applications, Inc.*

---

* The clinical bibliography includes those journal articles, books, manuals, and audio tapes which focuses upon the clinical skills required to instruct clients in progressive relaxation. These materials include step by step instructions, written relaxation scripts, and/or prerecorded audio tapes. The most useful material is marked with an asterisk.

Jacobson, E. 1964. *Anxiety and Tension Control*. Philadelphia: Lippencott.

Jacobson, E. 1938. *Progressive Relaxation*. Chicago: University of Chicago Press.

Lazarus, A.A. 1970. *Daily Living: Coping with Tensions and Anxieties*. Relaxation exercises (audiotapes). Chicago: Instructional Dynamics, Inc.*

Paul, G.L. 1966. *Insight vs. Desensitization in Psychotherapy*. Stanford: Stanford University Press.

Wolpe, J. and Lazarus, A.A. 1966. *Behavior Therapy Techniques: A Guide to the Treatment of Neurosis*. New York: Pergamon Press.*

RELAXATION TRAINING IN AN OUTPATIENT DRUG DEPENDENCE PROGRAM

Howard F. Blumenfeld, M.A. and Craig V. Showalter, M.D.

Substance Abuse Services, Inc.

Chicago, Illinois

Relaxation training sessions were offered to predominantly methadone maintained patients in a large urban outpatient setting who presented symptoms such as anxiety and insomnia. We investigated (1) the development of a non-pharmacological approach to treatment of stress, (2) methods of handling patient and staff resistance to the introduction of an unfamiliar approach to therapy, and (3) patient responses and treatment progress after relaxation training. While many articles on relaxation responses have appeared in recent years there is a sparsity of literatue related to the use of relaxation techniques with outpatient methadone maintained populations. A review of the literature revealed that the majority of relaxation programs have involved research studies with college students not abusing drugs, relaxation techniques in private practice therapy situations, or relaxation programs for hospital inpatients. The patients in this study were predominantly urban, black welfare recipients in an outpatient treatment facility. They were on low dosages of methadone with treatment oriented towards detoxification.

Since no format was available for outpatient relaxation treatment with methadone maintained populations we selected three relaxation exercises which we believed would be effective. One was the use of a calm scene, a visualization technique (Samuels and Samuels, 1975), the second was an autogenic type of exercise (Schultz and Luthe, 1959), and the third was a progressive muscle relaxation exercise (Jacobson, 1929; Bernstein and Borkovec, 1973). Staff and patient resistance to this unfamiliar approach to treatment was anticipated. A staff relaxation training program was designed to develop a functional skill that could be utilized in the midst of daily routines as well as to introduce the staff to our approach. The patient training program offered a modified relaxation approach that would hopefully attract our patients and keep them interested in attending relaxation sessions.

Of the 35 participants in this study (20% of our total clinic population), 58% returned for more than one relaxation session. Of these returning patients 65% reported on the effects of relaxation training on insomnia. Twenty-five percent reported no change in their sleep pattern while 35% stated that they were able to relax somewhat better than before relaxation training. Five percent reported that they were able to relax very well before going to sleep since becoming involved in relaxation training. Anxiety was also a frequent presenting symptom with such manifestations as feeling out of control, rushed or upset in general. Fifty-nine percent of those attending more than one relaxation session reported the effects of the training on their anxiety. Fourteen percent of this group said that there were no changes in their control of anxiety, 28% reported some control of anxiety and 17% stated that they were now able to get significant control over anxiety with relaxation exercises.

Our findings in this initial program indicate that for outpatients in this socioeconomic group, delivery of a modified relaxation training is feasible and thereby can provide a non-pharmacological alternative to the reduction of stress. Continued evaluation and refinement of the most effective relaxation training approach in outpatient methadone treatment programs is necessary.

REFERENCES

Bernstein, D. and Borkovec, T. 1973. *Progressive Relaxation Training: A Manual for the Helping Profession*. Champaign, Ill: Research Press.

Jacobson, E. 1929. *Progressive Relaxation*. Chicago: Univ. of Chicago Press.

Samuels, M. and Samuels, N. 1975. *Seeing With the Minds Eye*. New York: Random House, Inc.

Schultz, J.H. and Luthe, W. 1959. *Autogenic Training: A Psychophysiologic Approach to Psychotherapy*. New York: Grune and Stratton.

TEAM APPROACH TO TREATMENT OF ADDICTION

Emizie Abbott, B.S.

Cleveland Treatment Center, Inc.

The Cleveland Treatment Center, Inc. is a multi-modality center using methadone as a therapeutic tool in the treatment of addiction. There has been much controversy about methadone since it was first used in treatment. Some of the critics say it is detrimental to one's health because it is an addicting drug. Methadone isn't the ultimate answer, nor is it a cure-all for heroin addiction, but it is the only drug we have at the present time which helps combat drug addiction. Methadone, however, plays a small part in the overall rehabilitation of an addict. Methadone enables an individual to improve his physical health because it is no longer necessary to shoot adulterated street heroin and thus become subject to an enormous range of health destroying and life threatening hazards. Further, methadone improves one's legal status in that the addict on the street is subject to arrest or prolonged imprisonment and this constant threat of imprisonment for possession of drugs or drug-taking equipment has a negative effect on his entire life pattern. When one is in treatment and on methadone, he has very little reason to be involved in criminal activity. Methadone also improves his economic status in that an addict in treatment can secure a job and thus satisfy his basic human needs. If he does have a tendency to use "black market" heroin, as some methadone clients do, he will not buy as much as he would if he were solely on street heroin. Methadone also improves employability. An addict is able to obtain and maintain a job, which is one of the main goals of the program's efforts.

Methadone also restores an individual's freedom of choice. On heroin, he is the slave of the supplier and must devote his whole life to raising money for drugs and maintaining his supply line. After enrollment in a methadone program, he can spend his time and money as he pleases.

What is success in a methadone program? Success is defined by the Cleveland Treatment Center, Inc. as a positive intervention into an addict's illegal use of heroin, thereby enabling him to become a healthy and productive individual in society and giving him an alternative life style; free from criminal activity, and free from health destroying and life threatening hazards.

Addiction is the physical and mental dependence on an opiate drug. During a person's addiction, all energy and resources are spent attempting to find a means to obtain money to continue using drugs. This is a daily situation. There is no real medical cure for addiction because it is a recurring disease. After a person has become drug-free, he may have tendencies to relapse and may maintain a craving to get high. During the relapse, an addict needs a great deal of support in order to keep his mind occupied and keep him away from drugs or the thought of using drugs. While there is no cure for addiction, it can nevertheless be treated. Methadone is a major therapeutic tool which we have at present. A more ideal drug would be a non-addicting drug that would (1) stop withdrawal and relieve whatever the mechanism is in one's body that trigger's withdrawal, (2) eliminate relapse and the craving for drugs, and (3) eliminate all desire to use drugs. It would preferably be long-acting and non-addictive.

The Cleveland Treatment Center, Inc. is a flexible program in which patients can move at their own optimal rate from methadone to total abstinence and freely back to methadone if relapse occurs. We feel that if you can gain the confidence of the addict, he will feel free to turn to the program when in need of help. The addict is a person who has made a mistake. He is an individual who needs guidance.

TEAM APPROACH

The Cleveland Treatment Center, Inc. adheres to one basic operational philosophy which it utilizes in the treatment of its clientele. This philosophy, simply stated, is that we believe that the most positive results are obtained when we, as a Clinic, attempt to rehabilitate clients using the team approach. The team approach involves all personnel and begins with the Board of Trustees and the Director.

Where there is major concern which affects the total program, the Director, his administrative staff, and the Board negotiate in order to define general policy. The policy is interpreted and implemented as a team to the entire staff. If necessary, the Client Committee is involved.

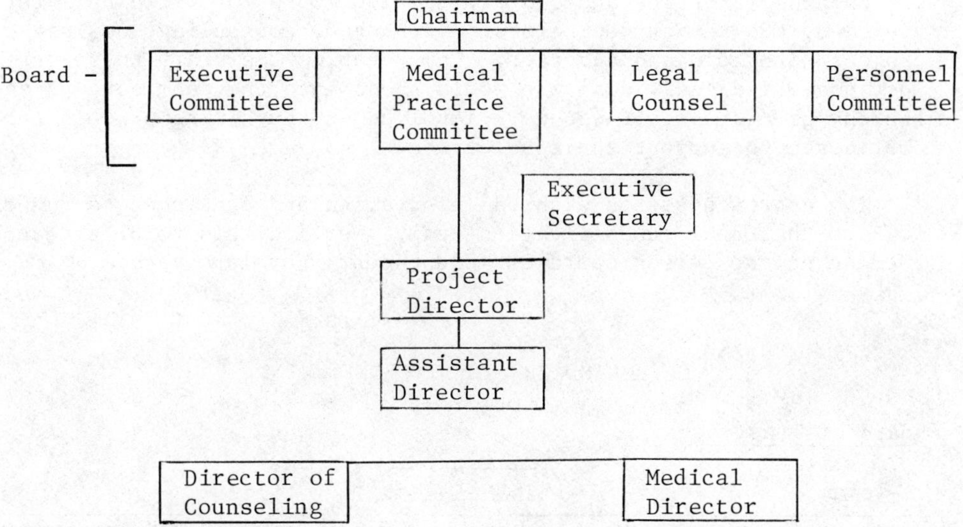

Figure 1. Administrative segments of the Cleveland Treatment Center, Inc.

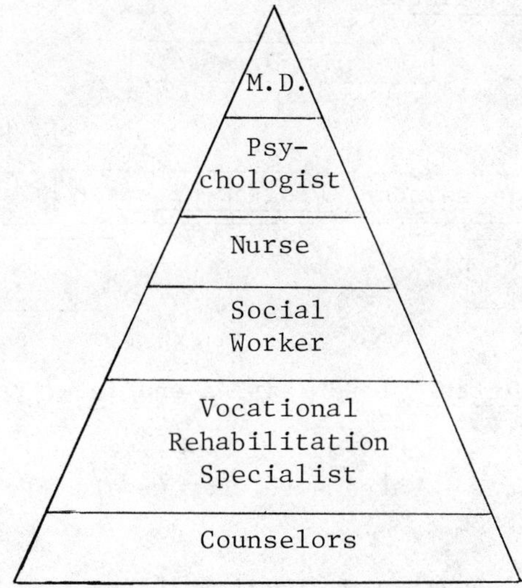

Figure 2. Traditional staff structure (example).

Staff hiring is also done as a team. An individual is interviewed by the senior staff of the particular department to which he is applying. For instance, a prospective counselor would be interviewed by the senior counselors, director of counseling, assistant project director, and director. This enables the staff to select the most qualified individual who would blend into the team concept and thereby give the staff who were involved in the hiring process the confidence to support their selection.

The approach starts with administration and continues to funnel down to the total rehabilitation team. We find this to be effective because no one person operates independently or in a vacuum.

CLEVELAND TREATMENT CENTER, INC.

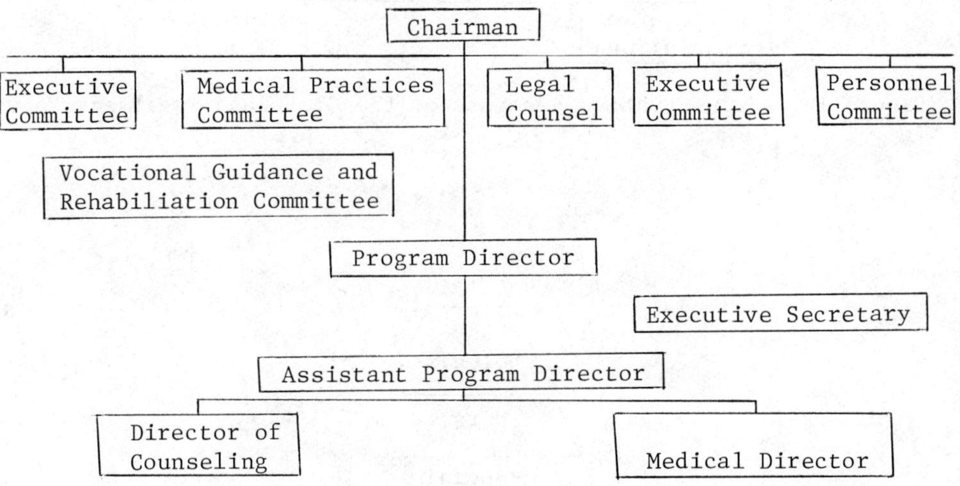

Executive Committee. Fund raising and interpretation of Board policy and procedures.

Medical Practices. Makes sure that we are adhering to FDA, DEA, and State regulations.

Legal Counsel. Advises on legal matters.

Personnel Practices. Develops personnel policies and handles grievances by staff.

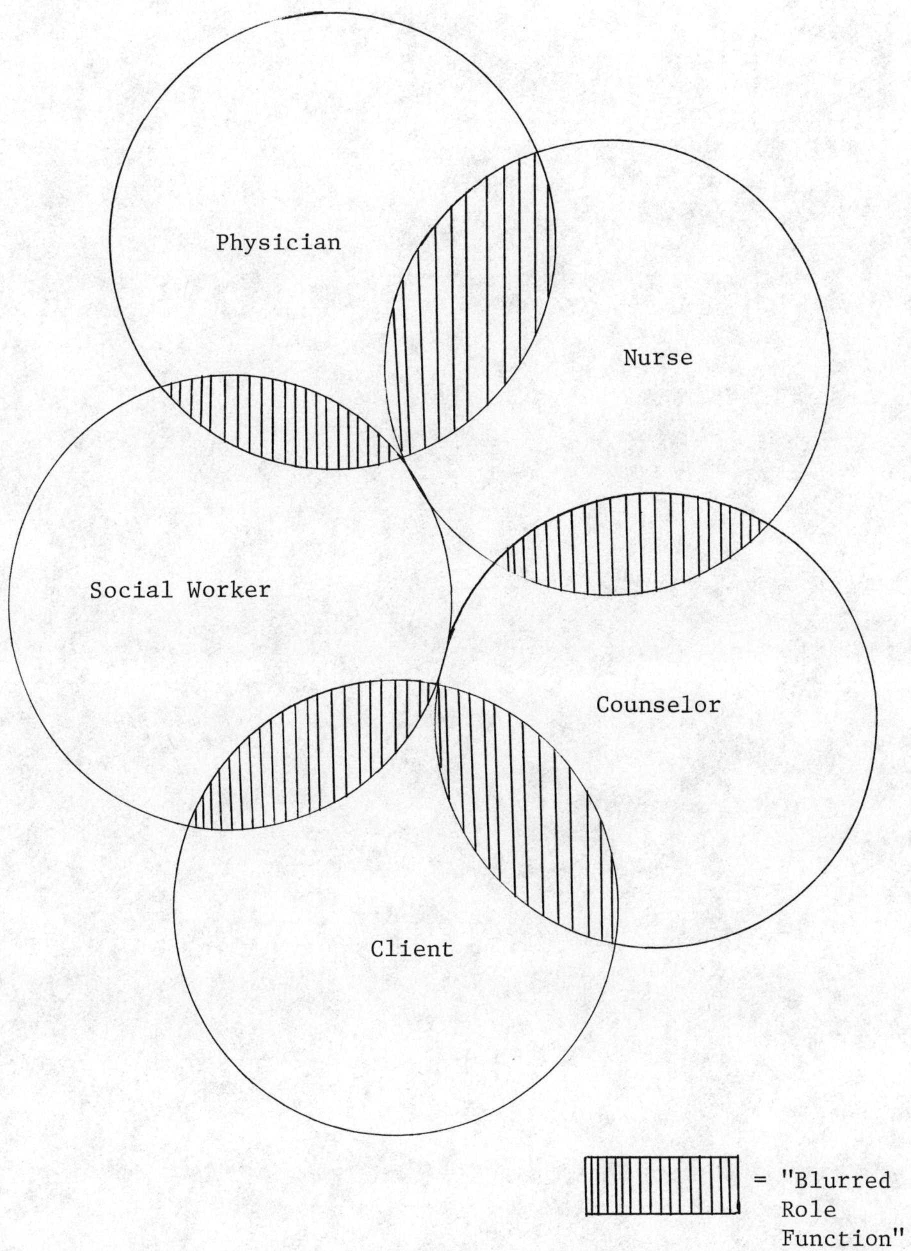

Fig. 3. Typical team model illustrating "Blurred" Role Functions.

SUPERVISION AND THE PSYCHOLOGY OF SUPERVISION AND THE MULTI-MODALITY DRUG ABUSE TREATMENT PROGRAM USING THE TEAM APPROACH

Emizie Abbott, B.S.

Cleveland Treatment Center, Inc.

CONDITIONS AFFECTING LEADERSHIP

All leaders in an organization, from chief executive officer to the first line supervisor, perform such functions as planning and organizing, communicating orders and instructions, motivating and controlling the performance of subordinates, resolving conflicts between individuals and groups, settling grievances, and taking disciplinary action. These functions, however, may be performed in a variety of ways, depending upon the nature of the organization, the level of the leader's position and his specific job requirement. For example, a supervisor in a treatment program will have different conditions under which to perform than one in a hospital. Similarly, the supervisor of a group of college graduates will be required to use an approach that is different from what would be used in supervising a group with minimal formal education.

The first line supervisor may supervise a few or a large number of subordinates, depending largely upon the type of work for which his unit is responsible. The nature of his duties may also vary considerably according to objectives, policies, functions, and other characteristics of the organization. The supervisor may be thought of as a linking pin who belongs to two groups within the organization. He is the superior in one group and a subordinate in the other. The supervisor is the linkage between these two groups. In order to function effectively, a supervisor must have sufficient influence with his superior. The amount of influence he exerts upward, as well as laterally, determines the amount of influence he exerts on his subordinates. When his subordinates perceive him as having great influence with his superior, they, in turn, give him more authority. Unless higher management and staff departments, such as the Personnel Department, help to strengthen the influence that supervisors have, the organization will suffer. This simply means that supervisors have to get the support of higher management.

Probably the most important functions performed by the supervisor are those of motivating and controlling employees' performance. It is the supervisor's responsibility to create conditions and incentives that will motivate employees to achieve the objectives established for their jobs and for the Department. He must then determine the extent to which the goals are being achieved and whether or not prescribed quality standards are being maintained. He must also determine that employees are conforming to the policies, procedures, and regulations prescribed by management, by law, (FDA, DEA, State Authority), or by any professional or technical societies that may have a legitimate voice in the establishment of standards of determining of procedures.

The supervisor must be able to promote good human relations with individual employees, as well as with his work group and, at the same time, he must insure that they do what is expected according to job functions.

The promoting of effective human relations is frequently one of the more difficult aspects of supervision since effective human relations depend upon attitudes as well as skills.

Employees are quick to sense whether the supervisor's skills in human relations are based on sincerity or whether they are management devices. Management tools for promoting good human relations on an insincere basis are likely to be recognized by subordinates. This suspiciousness concerns two major issues. The first is that the supervisor is out for himself and will do anything to advance his own interest. Therefore, the employee learns to become a "yes" man or politician so as not to do anything that would be perceived by the boss as threatening to his own personal goal-seeking. (Counselor having more influence on client's supervisor being threatened.) The second basis of suspiciousness is a fear of manipulation. Managers who try to force good human relations may be thought of as attempting to mislead and deceive by pretending interest that they do not feel. So, in order to promote good human relations among staff, supervisors need to be on guard not to increase the amount of suspiciousness that already exists. (Lay everything out on the table.) It is particularly important that supervisor's good human relations practice not be one of attempting to win friends and use people.

Effective human relations depend on sincerity and a concern for ethical values. While it is not always easy to obtain agreement as to what is right and what is wrong, supervisors should be encouraged through examples by top management to give careful attention to ethical values in the performance of their duties. Particularly in motivating and controlling the performance of subordinates, supervisors should make a special effort to consider the human values that are involved. One way to emphasize its importance is to have a formal code of ethics:

1. Set an example of what you expect from others.
2. Emphasize the future rather than the past or present.
3. Look for, and deal with causes rather than symptoms.
4. Admit, and learn from, making mistakes.
5. Don't pass the buck.
6. Consider both the long and short term results.
7. Everyone involved should benefit.
8. Legal and ethical means should be used to achieve legal and ethical ends.
9. The dignity of every individual should be respected.
10. Try to understand others and make yourself understood by them.

Also, a part of effective personnel planning is the establishment of sound work rules and other controls to govern the behavior of all employees in a reasonably uniform manner. (Everyone must play by the same rules.)

Another important role in the Team Concept of a supervisor is to provide employees with the experience of participating in the making of decisions. As a supervisor learns to trust his group and to recognize that it, too, can make good decisions, he is usually willing to let the group participate in decisions about important matters.

The supervisor of a group of employees has two roles when the group is involved in the decision-making process. In one role, he is the discussion leader and has the job of conducting a good discussion about the problem under consideration. In the other, he is an expert who has certain information about the problem that should be made available to the group. The skills needed for the democratic type of discussion must be consistent with a permissive approach in which each member of the group is encouraged to present his opinions and feelings. No one should be permitted to dominate or to utilize the discussion for his own selfish purposes. Before attempting to use the group-decision approach, the supervisor should develop some competency in leading discussion, including the ability to:

1. State what the problem is in such a way that the group does not become defensive, but instead approaches the issue in a constructive way.
2. Supply essential facts and clarify the areas of freedeom without suggesting a solution.
3. Draw persons out so that all members will participate.
4. Wait out pauses.
5. Restate accurately the ideas and feelings expressed, and in a more abbreviated, more pointed, and more clear form than when initially expressed by a team member.

6. Ask good questions so that problem solving behavior is stimulated.
7. Summarize as the need arises.

FACILITATING EMPLOYEE ADJUSTMENT

From the time that an employee is recruited until he is separated from the job, a large part of his life is influenced by the personnel policies and procedures of management, the supervisor for whom he works, and the relationship he has with his employees. All of these influences, together with his past and present life experience, determine his adjustment to his job as well as to other areas of his daily life. Because emotional adjustment is very important to his employees, their families, the organization, and to society at large, it should be one of the areas of primary concern to the supervisor. The supervisor should be able to recognize changes in behavior, such as excessive absenteeism, tardiness, hostility, moodiness, withdrawal, and decline in job performance as indicators that the individual requires understanding and assistance.

Since the supervisor is usually in the best position to observe changes in a subordinate's behavior and interacts with him frequently, it is only logical and efficient that he assumes the initiative in helping the employee to identify the cause of his change in behavior and to assist him in resolving his problem. If the relationship between supervisor and subordinate are harmonious, the supervisor's ability to be of assistance to the employee will be enhanced. The supervisor, however, must have some knowledge of skill in counseling and above all, know when and how to make referrals to professional persons if he is to perform his counseling role satisfactorily. This depends on active listening and letting the subordinate determine the course of the interview. Free expression reduces some of the obstruction in getting to the core of the problem, such as tension and frustration to enable him to look at the problem more objectively and with a problem-solving attitude.

The morale of workers, their interest in and enthusiasm for doing their work, is vital to the success of a Center and how it progresses. The supervisor must be observant of staff morale.

The supervisor must be able to motivate and get people to do their jobs because they want to.

AFTERCARE: POST-TREATMENT SUPPORT FOR SUBSTANCE ABUSERS

Richard M. Scott, M.S.
Veterans Administration Medical Center
Salt Lake City, Utah

Larry J. Lantinga, Ph.D.
Veterans Administration Medical Center
Salt Lake City, Utah
and
University of Utah

INTRODUCTION

The substance abuse treatment process may be thought of as a long continuum emcompassing a number of necessary activities. The list of activities usually includes: 1) intake and assessment; 2) treatment plans; 3) intervention and progress notes; 4) discharge; 5) aftercare and 6) followup. Of these, we single out aftercare as being relatively new in concept and design. This novelty often leads to misunderstanding. Yet aftercare is receiving increasingly more attention by important groups such as the Joint Commission on Accredition of Hospitals (JCAH).

At the Salt Lake City Veterans Administration Medical Center's Outpatient Substance Abuse Clinic we have long been aware of the need to perform follow-through activities with our clients. In short, we have viewed such activities as a necessary ingredient of program success. However, in spite of our awareness we did not fully comprehend the value of follow-through activities until we brought our program into compliance with the JCAH standards for Drug Abuse Treatment and Rehabilitation Programs (1975). Specifically, the JCAH Aftercare Standards have caused us to provide more viable, effective post treatment supportive care to each of our clients. We wish to share our experiences with you by discussing aftercare--what it is, its goals and objectives, and how it can be done.

OBJECTIVES AND METHODS

It is the intent of aftercare services to provide post treatment care to clients who have progressed sufficiently through primary care to a point in their rehabilitation where they no longer require the intensive level of treatment offered in the priamry treatment component. Aftercare then is post treatment care specifically designed to support and increase the gains made to date by the client in the treatment process. Aftercare assists the client in maintaining lasting behavioral, emotional and intellectual changes in his life. In so doing, it also insures that continuity of care through primary treatment and into community readjustment is achieved for each client.

It is the primary objective of aftercare services to insure that continuing supportive and referral services are available to each client to maintain the gains made in primary treatment. To accomplish this we suggest treatment programs include a number of specific aftercare objectives as follows:

1. Aftercare services are provided to *each* client being discharged from direct service care. The services are either provided directly by primary treatment staff or indirectly by referral to other community agencies for services unavailable within the primary treatment components. Whether aftercare services are provided by primary treatment staff or community agency staff depends on the individual needs of the client.

2. Each client participates with the treatment staff and significant others in the development of a written individualized Aftercare Plan completed upon discharge from primary care.

3. Each client benefits from regularly scheduled aftercare follow-through contacts; the frequency of aftercare contacts is determined on an individual basis and specified in the written Aftercare Plan.

4. Each individualized Aftercare Plan is reviewed as scheduled, but at least once quarterly in contacts initiated by the client or aftercare treatment providers. Each reviewed Aftercare Plan is updated when and where needed; documentation of the review and update of the Aftercare Plan is included in the client's clinical record.

5. The duration of aftercare services varies for each client, but is provided for as long as deemed necessary by the client, staff and significant others. As a general guide, aftercare services are provided to each client for at least ninety (90) days, with one year of aftercare considered optimal for most clients. However, for those clients whose problems are particularly severe and debilitating, special efforts will be made to keep the client involved much longer in aftercare.

Progress toward achievement of aftercare service objectives is assessed on a regular basis by aftercare staff. This is accomplished through the development and periodic evaluation of performance indicators directly related to the above objectives. However, the most important indicator of attainment of aftercare objectives is the client's continued success in the community as evidenced by his stable, self-sufficient, and productive life-style. Thus, the ultimate indicator of aftercare service delivery success, is a reduction in the number of clients needing or seeking readmission to primary treatment components. Therefore, recidivism rates are also utilized as an indicator of aftercare service success.

Aftercare is provided by that primary treatment component with which the client has most recently completed treatment. It is here that the client has established a therapeutic relationship with his primary therapist and with other staff members. Therefore, the primary treatment staff are in the best position to reach out to the client and deliver the supportive and referral services which provide the best continuity of care for that client.

Aftercare planning begins with initial assessment following admission. Aftercare begins with the development of treatment plans prepared by the client and the primary therapist. These plans are goal centered and identify the objectives and methods by which the client can achieve resolution of his problems. These plans are revised as the client proceeds through treatment, culminating in a thorough and complete aftercare plan developed just prior to discharge. Thus, aftercare plans are derived from those treatment plans begun and revised during primary treatment. The written Aftercare Plan includes the following elements:

1. Goals and Objectives, i.e., specific goals and objectives are identified as related to the individual needs of the client and, as appropriate, may include the following:

    a) continued abstinence from alcohol and illicit drugs.
    b) continued avoidance of illegal activities.
    c) continued vocational-educational stability.
    d) continued improvement in family relationships.
    e) continued improvement in interpersonal relationships.
    f) continued improvement in physical health.
    g) continued improvement in the client's affect, cognition, and self-esteem.
    h) continued financial stability.
    i) continued stable, independent living arrangements.
    j) continued pursuit and participation in recreational and avocational activities.

2. Methods and Procedures, i.e., therapeutic approaches to be utilized to meet the needs of the individual client either

through direct treatment services or supportive referral services. Examples here include individual counseling, group counseling, vocational rehabilitation, education groups, interpersonal skills training, legal assistance, medical treatment, recreational therapy, crisis intervention, urinanalysis, blood alcohol content analysis, Anatbuse, etc.

3. Scheduled Aftercare Contacts with Staff, Family or Significant Others, i.e., the scheduled frequency of contacts, identification of who (client, staff, family or significant others) will initiate the contacts, identification of who will participate in the contacts, how the contacts will be accomplished (personal visits, groups, telephone, collateral sources, etc.), and specifically when the contacts will be made.

4. Criteria for Re-entry into Primary Treatment, i.e., urges to use, actual use, medical emergencies, personal or social circumstances which require return to treatment, etc.

5. Specific Point of Contact to Facilitate Obtaining Needed Treatment Services, i.e., where and with whom aftercare contacts are made and who is contacted in a crisis situation.

6. Criteria for Termination of Aftercare Services, i.e., demonstrated drug and/or alcohol abuse free functioning for an individually specified period of time, demonstrated vocational stability, demonstrated avoidance of illegal activities, and demonstrated success in dealing with the individual's past incapacitating problems.

7. Referrals, i.e., referral resources individualized to the client's need, and designed to provide either primary support or additional support to the client.

8. Programmed Review and Update of Aftercare Plan, i.e., the individualized aftercare plan is reviewed and updated at periodic, scheduled intervals. The dates for review are scheduled at the same time as when the plan is formulated. During review, both the client and the advocate initial the plan attesting to its validity.

9. Signatures. To gain maximum participation in the development of and compliance with the plan, the plan will be formulated and signed by the client, the primary staff member, and significant others (prior treatment providers, family).

The individualized written aftercare plan is an important part of aftercare service delivery. It is the vehicle which identifies specific treatment objectives for each client and provides for timely aftercare contacts with the client to review and, if necessary, revise the plan. The aftercare contacts enable progress toward attainment of objectives to be measured. This provides an opportunity for the plan to be refined consistent with the client's continuing treatment needs.

Once the Aftercare Plan is completed, aftercare service delivery commences. The methods, procedures and activities specified in the plan are now implemented to bring about the successful achievement of individualized goals and objectives. The aftercare staff coordinate the delivery of aftercare services with the client and significant others consistent with the parameters established in the Aftercare Plan. Justification for termination of aftercare services is identified in the Aftercare Plan and is dependent upon the successful attainment of the client's individualized goals and objectives.

Occasionally individuals in aftercare may require or desire services unavailable through the program. Since no one program can meet all the needs of its clients, aftercare services are closely coordinated with other community service providers. This is done primarily through supportive referral activities specified in the individualized Aftercare Plan. To insure client confidentiality, referrals are only initiated after obtaining the client's permission as demonstrated by his signature on appropriate release of information forms.

DISCUSSION

Certainly not all primary treatment programs are in a position to provide aftercare services directly to their clients. Limitations on staffing and funding may cause primary treatment providers to consider other approaches to aftercare service delivery. For instance, they may want to consider having one staff member oversee an aftercare program staffed principally with properly trained volunteers. Or another alternative would be the use of alumni groups in aftercare service delivery to reinforce the gains the client has made in primary treatment. And still yet another alternative is contracting with another health care provider for the provision of aftercare services.

Regardless of the method followed, however, the primary treatment program retains the responsibility for development of the original Aftercare Plan, which must be recorded in the client's clinical record. Additionally, the primary treatment program has responsibility for insuring that necessary and appropriate care is being provided to the client and that such care is documented in the clinical record. Updates and revisions to the Aftercare Plan are also documented in the clinical record and are attested to by client and staff signatures.

Treatment programs, to be most effective, should provide aftercare services. Aftercare services are a natural continuum and an integral part of the treatment process. Treatment programs may call aftercare what they please; they may even wish to incorporate it into one of their existing program components; or they may wish to provide aftercare through referral or by contracting with other health providers. But, however they choose to provide this service, the standard has been clearly set that: 1) post treatment supportive service be provided and 2) documentation exist in the client record of the provision of this service.

Aftercare, as discussed in this paper, is a relatively new concept. It is assumed that clients will respond better if they are given the benefit of post-treatment support. But, it is important to realize that this is merely an assumption. Little information is available concerning the true effectiveness of aftercare. We have implemented aftercare because JCAH says we must. We believe it is an effective procedure. However, methodologically sound research is necessary to truly demonstrate the efficacy of aftercare.

REFERENCES

Joint Commission on Accreditation of Hospitals 1975. *Standards for Drug Abuse Treatment and Rehabilitation Programs.*

AN EXAMINATION OF THE REDUCTION OF CLIENT POPULATION IN TREATMENT PROGRAMS

Gail Shortell, B.A.

State of Alaska, Office of Alcoholism and Drug Abuse

About two years ago, Mr. Leon Hunt, Statistician, published through the Drug Abuse Council a lag formula for estimating admissions into heroin programs. Admissions into Anchorage programs though not numerous had been healthy for some time. We were surprised when the formula told us that we might anticipate a tapering off of admissions. Our feeling at the time was that the formula was not appropriate for our needs, or that we had too little data to run an accurate computation. Some time later, the Drug Abuse Council did some consulting in Anchorage; they also ran the formula and came up with substantially the same estimation. And, in fact, experience has borne out the predictions.

Admissions to methadone maintenance in Anchorage have dropped in the last year. Admissions to the therapuetic community have risen but we have seen an increase in the poly-drug use versus straight heroin use. Beyond our local experience, NIDA has spoken recently to a grave concern of theirs regarding national decreasing admissions and decreasing levels of utilization. At a training conference in Portland last January, Mr. Roberton spoke with conferees about this subject asking the group for insights into this growing issue. NIDA's Executive Report of April, 1977--their printout of the NDATUS data--also reveals some possible trends that perhaps bear out Mr. Roberton's concerns: rates of utilization of budgeted capacity were down in 13 of 32 indicator SMSA's cities from 1975 to 1976 and down in 15 of the same 32 cities from 1976 to 1977. The overall utilization was slightly higher in 1977 over 1976 but in the same report we see a drop in 28 states in number of clients in treatment per 100,000 of population from 1976 to 1977. Additionally, the Central and South regions showed a drop in utilization of two out of three major modalities from 1975 to 1977. The Western region reflected a utilization drop in all three modalities.

Undeniably, these trends have serious consequences from a national to local level. If we as planners or as service providers are to be able to continue to meet the population in need, we must attempt to analyze the reasons for this trend and develop appropriate strategies to meet the current needs.

Our major belief is that fewer folks have become addicted to heroin in the 70's than did in the 60's. This may be explained by the overwhelming increase in the use of cocaine and the movement away from a primary drug of abuse and towards poly-drug abuse. This has resulted in fewer new heroin addicts and possibly a weaning away from heroin as an addiction for existing addicts.

Additionally, a case may be developed that in the 60's, a large population of folks from 18 to 25 years old became heroin users. They were defined as "addicted" and entered treatment programs. It seems possible that many of these clients were only marginally addicted, experiencing identity crises and, subject to the vogue of the day, used heroin to help them through this difficult time. Subsequently, they "matured-out"; rejoined, if only the fringes, society; and or adopted new drug use patterns that no longer qualified them for heroin treatment.

Finally, there is a good deal of evidence to suggest that much less heroin is available nationally, and what does exist is of much lessor quality. As a result, it would be logical that fewer people are sufficiently strung out to require treatment.

Overall, heroin programs seem to be experiencing a downward trend in admissions and utilization. Simplistically speaking, this may be accounted for by the fact that there are fewer heroin addicts in the U.S. at this time. It is not axiomatic, however, that there are no longer folks in need of treatment. Several years ago Dr. Joel Forte of San Francisco was fond of reiterating that there were not drug problems, only people problems. Certainly, this continues to be the case.

Social problems like other issues in the U.S. seem to be the subject of vogues or trends. Twenty years ago programs' dollars went to juvenile delinquency, 20 years ago welfare fraud occupied lots of public concern. Since the late sixties drugs have provided one major focus.

Today there is an increasing amount of concern over alcohol, women, and native American problems. The foci shift frequently. Programs as specialized as ours may find themselves under-utilized as the trends shift.

Possibly, it is time that we approached the people problem from precisely that perspective. It appears that a broader view and a more holistic approach is called for. Now that substantial progress has been made at solving so many diseases perhaps we can begin to look towards promoting health with an approach which breaks down barriers that have separated alcohol, mental health and drug abuse service providers in the past.

PROPOSAL FOR A STRUCTURAL REFINEMENT IN OUR TREATMENT METHODOLOGY: USING THE SHORT-TERM TREATMENT CONTRACT

Joel L. Smith, M. Ed.

The Hartford Dispensary

GOAL

This proposal is predicated on the belief that methadone should be a tool for the client to use to make meaningful adaptations in his/her life system. Methadone should not be an end in itself. (Functionally it is not, because of the incidence of polydrug abuse). This proposal makes the assumption that the client must assume a responsible approach toward re-entering the social superstructure. If we are going to take the responsibility for addicting/maintaining the client on methadone, then we also share the responsibility for helping the client to make reasonable efforts at developing a more successful and socially acceptable life style. In short, we have the mandate, and the responsibility, to expect that the client will make these efforts.

The proposed adaptation is predicated on resources that already exist, in some form, within the program. The short-term treatment contract represents an additional refinement.

THE CONTRACT

The case contract has generally been an implied or assumed tool within the counseling relationship. The client-worker contract is potentially a tool that can contribute substantially to the effective outcome of the counseling relationship. When formally introduced, it can be used to clarify objectives and encourage clients to participate in the entire process of intervention. An aspect of the changing theory and practice of the use of contracts is the conviction that the client has an important role in formulating policy in planning his/her own therapeutic model. One model of service delivery proposes that the client has a choice in what services are provided, and in how far in the therapeutic relationship he/she wants to go. What this exacts is a real participation on the part of the client. It is logical then to assume that the responsibility for therapy and for change lies with the client. The case worker then also becomes a tool. The ultimate goal of therapy is one wherein the client assumes a contractual relationship with his/her life situations, so that he/she deals with the experiences encountered in a realistic and responsible manner, and looks for healthful ways of coping with those situations. Essentially the contract represents a conscious agreement between the participants in the therapeutic process to work towards certain goals. These goals are agreed upon. What is achieved then is a method for adaptation and measurement. What we are looking for is a means by which the client sets him/herself a pattern, looks at that pattern, determines that the pattern is appropriate, and identifies a step-off point for his/her next pattern. The contract itself provides the therapist, or the person who is responsible for objectively looking at the contract, a means of measuring the effort that is made by the client. It also provides the therapist with a tool for measuring the steps that he/she has taken in helping the client achieve those goals. The contract itself becomes very limiting in one sense, in that in order for a client to use it to measure his/her achievement, it must be time-limited. On the other hand, the contract series can be almost cyclical, in that each contract builds upon those preceding it, to develop a therapeutic program, ultimately leading to the achievment of those goals. Goals appear to be, generally speaking, those things that we can (1) achieve, or (2) aspire to. What we are trying to do by using the contract is to analyze the broader goals, and set up smaller packets of realistic goals that are more immediately achievable.

Wolberg (1965) identifies a series of briefer goals, including: (1) relief of symptoms, (2) restoration to the level of functioning that existed prior to the problem, (3) some understanding of the forces that precipitated the current upset, (4) recognition of some of the personality problems that interfere with a better life adjustment, (5) some partial understanding of the origin of these circumstances in past experiences, (6) some degree of awareness of the relationship between prevailing personality problems and the current problem, and (7) comprehension of those measures that can remedy current environmental difficulties. These goal packets would developmentally lead to larger goal satisfaction.

The contract relieves the client of his passive orientation and his passive involvement in the therapeutic process. It almost demands that the client assumes the responsibility for the goal and direction and time-frame of the therapy. The counselor then becomes essentially a navigator, helping the client to achieve those goals within the set time-frame. Traditionally, the responsibility for change has fallen on the therapist, in that he/she is the one who determines when a change is made, how much is made, and so forth. What the contract model does, then, by putting the responsibility back in its rightful place, onto the client, is assume that the client will accurately, or acceptably, develop a series of goals for him/herself. The counselor's function is to moderate the achievement of those goals, and to test for the accuracy of those goals.

To briefly define contract, we can say that a contract is a promise or set of promises that must be achieved in order to satisfy the demands placed by both partners. The contract suggests mutuality of participation, and it suggests a certain amount of action. Essentially what is being said here, then, is that a contract be defined as the explicit agreement between the counselor and the client concerning those problems they want to target on, the goals, strategies, and time-frame involved, and the roles and tasks of those people involved (i.e. the counselor and the client). The primary demand of developing the contract is mutuality. Without the mutuality the contract, by definition and in actual practice, cannot succeed. Therefore, the client has to buy into the plan of developing the contract. The contract has to be realistic for the client. If the client is to believe and accede to achievement of those goals, they must be realistic, and although they may provide frustration, they must not be so frustrating that they turn the client off completely. It is entirely possible for the client and the counselor to have almost mutually exclusive contracts going, or even concomitant contracts going. The counselor may perceive the situation as needing additional specificity or measures, and thereby set up a contract almost with him/herself in working with the client (often called a "hidden agenda"). What we are looking for here is an overt contract between the program (i.e. the counselor) and the client to help the client better develop and assess his/her own growth.

The contract itself must be rooted in reality. If the client provides several goals that are false, or unobtainable, the contract then becomes useless as a tool for achievement of those goals (although the tool still exists for dealing with the client in terms of analyzing why those goals were chosen). The therapist must help root the client in reality as goals are identified. They must be the client's goals and not the goals of the therapist.

Oftentimes in therapy, the client tries to provide the counselor with the information that he/she perceives the counselor as needing (i.e. suggesting goals that the therapist might feel are appropriate). As counselors we must be very careful not to fall into that trap, because ultimately what then happens is that the contract is our contract. It becomes very difficult for us to help the client work toward unrealistic goals. It might be helpful to point out that the worker and the client have essentially "equal but different" roles. This is not simply a theoretical concept. It is almost a prerequisite to the process, because what it says is that although I work with my clients, they also work with themselves, and so we are equals, and are working to achieve a common goal. We are also different, however, in that we are approaching the goals and the process in different ways.

The therapist has a major responsibility to delineate with the clients the unique aspects of the participation. It is my responsibility to help the client to see how he/she is working toward those goals. The contract provides me with that tool. The contract provides both the client and the counselor with a measure of accountability, and that accountability is reciprocal. The contract can help make both parties as aware as possible of those reciprocal obligations. The client's responsibility must be emphasized. Insufficient attention to it may partially account for the limited involvement of some clients in the helping process, or their withdrawal from it, and therefore the failure of the process. The contractual alliance brings into sharp focus the role of the client. At the same point, considering the reciprocal agreement, the client knows exactly what to expect from the counselor, and gets a measure of consistency from the counselor, over and above that which he/she provides as a person, because the client has in the contract an understanding of what he/she is going to be held to, what the counselor will expect, and how the counselor will respond to the achievement or the attainment of those goals, or the attempt to achieve or attain those goals. It is my belief that the contract is dramatically important to the positive outcome of the therapeutic process, in particular to bringing focus and meaning to the inherent values and principles implicit in the therapeutic process, and in making them work. If the counselor has conviction about the contract and implements it, he/she can help the client participate more actively in dealing with his/her own situation. In doing so he/she can affirm the client's preeminent role in the therapeutic process. The contract has the potential to serve as an active instrument in the process. The contract is derived from their shared experience in exploring the situation, and in reaching an agreement on goals and tasks. It gives both the client and the counselor a sense of immediate involvement in meaningful agreement and participation. It signifies a mutual commitment to the process of achieving those goals. It provides a baseline for periodically reviewing those accomplishments, and for assessing whether or not the goals are appropriate or adequate.

BACKGROUND

The development of this proposal began simply as an analysis of those tools that already exist within the program. The contract itself exists in the form of the treatment plan. What I am suggesting by using this contract, however, is a slightly modified approach with several controls added. What I tried to do is look at what happens in methadone maintenance, within the Hartford Dispensary, and determine those situations that need to be refined a little bit. There were several different characteristics of the program that I considered.

First and foremost, perhaps, is the question of whether or not the true opiate addict exists any longer. It has come to our attention that most of the clients that we have tend to be polydrug abusers; polydrug also including alcohol. The life style, the quality of drugs on the street, the existence of social pressures, all seem to minimize those users who are only opiate addicts. Those who do tend to be strictly opiate addicts are likely to be older people, at least in our experience. The question then is raised: What does methadone maintenance treat in terms of the polydrug abuser? The clear answer is that methadone itself deals only with the addiction to drugs in the opiate family. We then need to look at how in methadone maintenance can we treat those people who are using drugs along with or other than opiates.

The second characteristic, then, is what is the philosophy of the Dispensary in particular, and methadone maintenance in general at this point. The indicators seem to show that what we are is a delivery system of methadone to those people who are interested in changing their life situations in regard to their addictive pattern. This does not necessarily mean that they are interested in changing their life style. In fact, oftentimes quite the opposite exists, whereby they are interested in and satisfied with their life style, but merely need to have changes in their addictive pattern. Some of the reasons they may need to change that pattern could be drug cases, other legal cases, the degree to which they are strung out, and their ability to provide themselves with enough drugs. If in fact all we are asking is that a client picks up his/her methadone, and lives up to the minimal behavioral rules of the clinic, then there is very little need for any program refinement. If, on the other hand, we are interested in helping the clients make adaptations in their life styles, and in fact if we are making demands on the clients that they make adaptations in their life, we need to help them adapt and develop behaviors that are more appropriate and acceptable to society (society being represented as a generalized set of norms).

That brings us to the third point, which is how methadone itself relates to the set of social norms. As a broad overview, methadone would seem to fit in, in terms of the social superstructure, with other almost-renegade institutions such as prisons, reform schools, and mental hospitals, in that those people who have problems tend to be an eyesore for the rest of society. Society in fact does not deal with them very well. So it would seem that methadone maintenance is an idea that is still very germinal, in that it provides methadone to people who are drug-addicted, but it oftentimes is not equipped in a programmatic way to help them with those problems that their initial drug addiction may be the symptom of.

Another point that needs to be considered is the counseling process within the methadone maintenance treatment program. What is the role of the counselor? Is it in fact the role of the counselor to help the client become able to deal with those problems encountered within the program; or, on a larger scale, is it the role of the counselor to help the client make those adaptations that he/she needs to make in his/her life style? It seems to me that the counselor has a larger role than just helping facilitate ease within the program. The counselor, must, if he/she is truly involved with his/her clients, help the clients make those changes that will make the social superstructure less threatening, and help the client fit into that superstructure more easily. What is that nebulous superstructure? Well, oftentimes it is an ideal, an artificial ideal; but we all do need to develop coping mechanisms, and those coping mechanisms are measured in terms of what society expects of us. Is the client going to be on welfare for his/her whole life? Will he/she need a Medicaid card? Will he/she be considered a junkie for his/her whole life? Is the prisoner always a prisoner whether he is released from prison or not? If someone has spent time in a mental hospital, will he/she have to deal with a related stigma as well as those situations that may have precipitated the hospitalization? These are questions that society does not deal with particularly, and I think they hold true for methadone also, so what we are looking at is how a counselor can help a client cope with well before, so that he/she can function more successfully in the larger society outside the drug culture.

If that is to be the role of the counselor, we have to look at the controls that are available to the counselor. Many of the behaviors that our clients exhibit lend themselves to treatment by some of the behavior modification techniques, and in fact we employ those techniques with in-house behavior. There are few realistic controls in our program; those that do exist really have to be manipulative. One of them is the take-home medication. Take-home essentially is to be provided those clients whom we feel would not divert the medication, and who will live up to their program expectations. But, in order to get the client to look at his or her life, we oftentimes need to use that as a reward. We need to provide the take-home, or withhold the take home, on the basis of how they are functioning. The only other real tool that we have here is the relationship that develops with the client. As a subculture, those people who are addicted, tend to be non-trusting people. They feel they have been put-upon, and wronged, and the rest of us who represent society at large owe them something. Also they feel that we are not part of their culture, so that the trust that is the foundation of a counseling relationship is something that happens over a long period of time with the clientele that we serve.

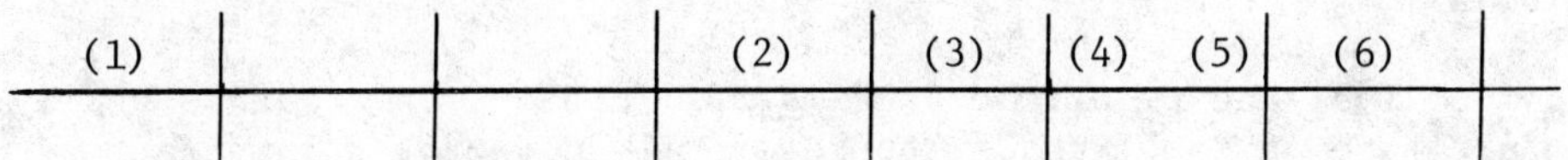

(1) Entrance into treatment; development of a treatment plan agreeable to the client and to the program.

(2) Five-month period; to satisfy the parameters of the treatment plan.

(3) Beginning of the sixth month; analysis of the satisfaction of the treatment plan - options:
   a. request additional six months in treatment;
   b. client requests detox;
   c. program requests detox.

(4) Option b and c: Medically supervised detox.

(5) Sixth month - used to complete treatment goals and to develop a treatment plan for the next treatment sequence.

(6) Process begins again.

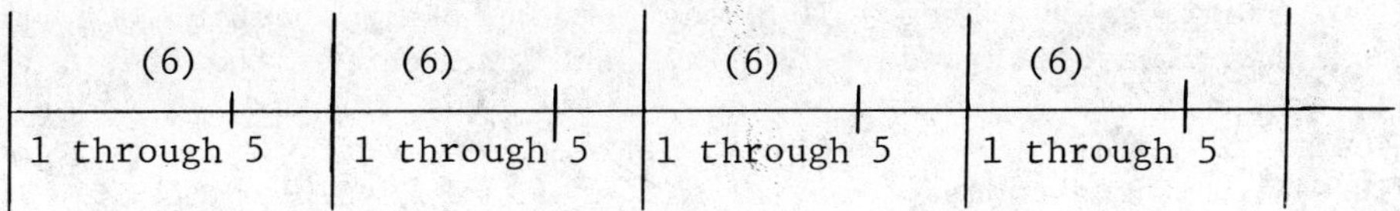

Fig. 1. Short-term treatment contract: A six-month renewable program adapted for the Hartford Dispensary.

We have identified too that there seems to be a treatment dichotomy, between those clients who work and those clients who do not work; often we say that those clients who work are doing well. This may, in fact, not be the case at all. They just are out of our sight and out of our control. What we need to do then is look at what they are doing with their lives, in terms of what they want to do, what they expect to do, what would help them enhance their coping mechanisms. The added controls that this proposal suggests would seem to do that.

The final question is that of staff burnout, and that is really the ultimate of the first six points. If in fact the frustration factors built on the lack of programmatic controls are so great, then the staff tends to become frustrated and finds it very difficult to implement any kind of change. Additional control would be beneficial for the client and the staff, in that there would be a reciprocity of agreement as to what expectations are.

METHODOLOGY

The methodology is designed to offer the client an explicit six-month period of methadone therapy, rather than having the therapy be open-ended. The six-month period of treatment would be consistent with our goal of having the client take additional responsibility for his/her treatment. The methadone itself would then be part of an all-over treatment program, rather than be the entire treatment. The process would be renewable, so that essentially the client would be in treatment six-month period after six-month period, without any break or interruption in his/her treatment. The process would expect, upon entrance into the program, development of a serious treatment plan, that is acceptable to both the client and the program. For example, client X comes into the program, and identifies his personal goal of becoming drug-free. That may be completely acceptable to both the program counselor, and the client and that can be the treatment plan for the first six-month period. That gives all parties an opportunity to look at how the client progresses toward the avowed goal. Obviously, the treatment plan will differ for different clients. The initial treatments may have some basic similarities, in that all people generally are coming on the program to clean up their street drug habit. For some clients, however, that may not be enough even in the beginning. It may well be that a client who has certain responsibilities, such as a business, or a case pending, would have to do certain other things, such as get to work, deal with the lawyer, and so forth. For certain people, the tasks may be simply improving their appearance, or changing their vocabulary.

Again, there are three agreements necessary to the treatment plan: that which would have to be approved by the client, that which would have to be approved by the clinic staff, and that which would have to be acceptable to the counselor in terms of his/her ability in working with the client. Ultimate acceptance of the treatment plan has to be with the client, in that it has to be real for the client, so that it has meaning, and the client takes responsibility in working towards it. The treatment plan itself represents a negotiable process, and that in and of itself is part of the process of having the client assume responsibility, as the client has to make realistic decisions. It is the counselor's job to help guide the client; if the client indicates from day one, the first treatment plan, that he/she would like to be President of the United States, the counselor knows that to all intents and purposes the goal is an unrealistic one, and must help the client develop a series of more realistic goals, perhaps ultimately leading to becoming President. The process is negotiable too, in that at some point within the six-month period of time the client may feel a need to renegotiate, saying he/she can achieve more, or is having trouble achieving the original goal. The main determinant for the program and for the counselor should be the client's progress. We all have expectations for ourselves, and make demands on ourselves, sometimes they are not very realistic, regardless of the process we go through in setting them up. So it is the process--if the client wants to become the president, how does he or she go about it--is he or she being realistic, making appropriate steps, talking with people in the political party. Those are the kinds of things that must be considered when analyzing the process.

Our intake procedure would remain virtually the same as it is now. The difference perhaps would be the heightened role of the staff in the procedure, and the utilization of the mental health specialist. By this I mean that the staff would have greater input in determining how therapy is started, who the most appropriate counselor might be, whether or not the goals are determined realistically, etc. The mental health specialist would be a key person, responsible for helping staff develop an understanding of where the client is coming from, in order to help the client get to where he/she wants to go.

The process itself has been limited to six months in a very arbitrary fashion. The thinking behind this is that at the very least a client would have six months of freedom from the street-addict cycle even if he/she did not qualify for continued treatment. The process divides the six months into two periods, essentially a five-month treatment period, and a one-month period for analysis of progress, and potential detoxification from treatment. The first five-month period is designed to help the client satisfy the demands of the treatment plan that he/she helped to devise. On the first day of the sixth month, the client would analyze his/her treatment plan with the counselor, to determine whether or not there has been goal satisfaction. Naturally, this process will be undertaken throughout the five month period, but at the beginning of the sixth month it would be a formal procedure. There would be three options available at that point: (1) if there was acceptable progress in terms of the initial treatment plan, the client could request an additional six months in treatment, and if he/she was accepted for an additional six months, would design a new treatment plan; (2) at that point the client might determine he/she is not interested in any additional treatment, and might request a detox; or, (3) based on the client's performance in the initial five-month period, the program would request a detox. Essentially these parameters exist within the program, but in a less formal way. At the point where the client requests an extension of treatment, the counselor would present a progress report at staff meeting, and the process would again be repeated: has the client done what he/she expected of him/herself, or what the program expected; how in that process has he/she fared; have there been any success experiences; how has he/she dealt with the failures experienced; what are the counselor's perceptions; what are the rest of the staff's perceptions. If in fact all of these things indicate an additional period of treatment, then with the updated therapeutic treatment plan, an additional treatment segment is instituted. If a detox is ordered either by the client or by the medical staff, then, as now, this would be a medically-supervised detox. A one-month period of time is specified for this, but this is certainly up to the medical staff of the clinic, rather than in terms of a formalized treatment plan. If the program orders the detox, the client cannot be reconsidered for methadone maintenance for a minimum of six months.

The therapeutic process then could become repetitive over a long period of time, perhaps years, allowing the client to be in treatment for as long as is necessary. What happens, though, is that this gives the client and the clinic more formal options as to how to deal with one another. It takes a serious look at expectations, but more than that it places more of the responsibility on the client for the maintenance of his/her treatment. It is also somewhat more clear-cut when a client violates the agreement, in terms of the strict code of clinic rules and regulations, or in terms of the more negotiable set of goals that the client has set up.

POTENTIAL RESULTS OF IMPLEMENTATION OF THE METHODOLOGY

In analyzing how this methodology would benefit the program, I can see five potential areas for improved services. The first would be putting increased responsibility on the client for his/her own therapy. It would seem that in any therapeutic process this is very important, and in terms of the drug-addicted clientele, if there are any generalities to be drawn, it would be that they have been irresponsible toward themselves, and in their interaction with society.

The second gain would be a programmatic one, and that would be a greater sense of control over the direction of therapy and those involved in therapy. As evasive as we often want to be about this, dealing with the clientele that we have, and the medium that we have to offer, we need certain kinds of controls, in terms of counseling relationships, in terms of providing a therapeutic environment, and in terms of being able to have certain expectations for the clients.

The third benefit would be greater use of staff resources, as the staff would have a greater opportunity to help direct clients towards realistic and achievable goals, and interact with the clients in terms of the achievement of those goals, or the acceptable dealing with the resultant frustrations. A clear-cut programmatic benefit (four) would be in terms of a measurement tool for client development. This is something that programs always seem to be looking for, and what we would have here is a truly understandable measure of whether or not the client lives up to the expectations that he/she sets, and that the program sets. At the same time, the fifth benefit would be a better measurement tool for the clinic's ability to deliver services. For example, if many clients are asking for job-related services, it would be measurable whether or not the counselors had been able to provide those services. This would then help us to determine the directions we as a clinic need to be going in, that as professionals we need to be working towards. The sum total of the implementation of this methodology is a heightened sense of responsibility of the counselors, and what this does is ask that the counselors become intermediaries between a program methodology, and the human beings that methodology is designed to serve. I think this is a little more specific than what now exists; moreover I think it demands certain things from the counselors. Because it is clear-cut, it provides the counselors with the tools to satisfy those expectations.

POTENTIAL PROBLEMS

I can anticipate some difficulties implementing this methodology, the first of which would be how to deal with the existing clientele, in terms of changing over to this system. I think part of the problem would be for those clients who are doing well. But a rejoinder would be that if, in fact, they are doing well, this program would essentially have no impact on them whatsoever. It would be entirely possible to reach a point where the treatment program would be "maintain progress as per last treatment plan". For someone who was functioning at a relatively optimum level, he/she may not have any additional expectations, as such that would not be a problem. Those clients on the program now who are maintaining a non-progressive status quo, or are doing very little, might have some reaction to this kind of a program, but this is the kind of client that we are trying to reach anyway, in terms of developing a more acceptable treatment tool.

Initially an additional problem may be a census fluctuation caused by something new, something that can be essentially stressful, or demanding. I think traditionally new treatments that are implemented in any kind of environment cause some fluctuation, but ultimately the kind of service that we provide serves an essentially captive audience, and I think history has shown that fluctuation within a norm is entirely acceptable.

REFERENCES

Maluccio, A.N. and Marlow, W.D. 1974. The case for the contract. *Social Work: Journal of the National Association of Social Workers,* 19 (1).

Mann, J., M.D. 1973. *Time Limited Psychotherapy*, Harvard University Press, Cambridge, Mass.

Small, L., Ph.D. 1971. *The Briefer Psychotherapies*, Brunner/Mazel Publishers, New York.

Wilson, R.S. 1937. *The Short Contract in Social Casework*, Volume 1- Theory. National Association for Travelers Aid and Transient Service, New York.

Wilson, R.S. 1937. *The Short Contract in Social Casework*, Volume 2- Cases. National Association for Travelers Aid and Transient Service, New York.

Wolberg, L.R. .965. The Technique of Short-term Psychotherapy, in Wolberg, L.R. (ed.): *Short-Term Psychotherapy*, Grune and Stratton, New York.

ART AS AN ADJUNCT THERAPY IN DRUG TREATMENT: SELF-ACTUALIZATION THROUGH NON-VERBAL COMMUNICATION

Joan E. Standora, M.A.

Addiction Research and Treatment Corporation

Brooklyn, New York

To begin, I would like to describe a scene which stands as an extended metaphor for the substance of this paper: art therapy, at its best, is, for the client, a non-verbal work in progress, not a completed work or a product.

Ms. J. stalked into the art studio with anger on her face. She was extremely overweight, her hair was unkempt, her clothing haphazard. Not only was she unhappy about being there, she complained that she never could "do" art and would not put her hands in "that nasty Mud". After watching other members of the group work and, being cajoled by them to try, Ms. J. accepted a small lump of clay and began to knead it with her fingers. In a few minutes she rose and helped herself to a bit more of the "nasty stuff". She rolled it on the table, poking it with her fingers, pulling it apart, and mashing it back together. Again, she arose, returning to the table with two fistfuls of clay. This time she remained standing, forming the clay into a large ball which she then beat down with her fists, threw on the floor, and reformed into another ball. She began to pour water over the clay, squeezing the huge lump so that it oozed between her fingers. She carried it to the pottery wheel head. Her hands caressed the smoothness of the clay. As she pushed the wheel faster, she added more water to the clay until it became almost liquid, dripping down the sides of the wheel. It spattered, covering Ms. J.'s arms and clothing, the floor and the wall behind her. Since her opening complaint, Ms. J. had not uttered a word. She appeared unaware of the activity around her, immersed as she apparently was in sensation and fantasy. It was only when the spattering clay fell on several other group members and their protests became audible that Ms. J. ceased her frenetic activity. She good-naturedly cleaned her, by now, clay-soaked work space, and she laughed with other clients who teased her about her play with the "nasty mud". As the session ended, she begrudgingly admitted it had been fun. In subsequent sessions, her childlike manipulation of the clay gradually developed into an interest in creating objects. Although she remained essentially non-verbal throughout her art therapy sessions, her approach became less erratic and her work showed increasing integration and organization of form. Likewise, her interaction with her peers became more cooperative and relaxed. Ultimately, she was able to express satisfaction and pride in the work she was doing and, most importantly, in herself.

Amongst addicted individuals like Ms. J., personal growth and self-awareness are often blocked by feelings of worthlessness, unacknowledged self-hatred, and a fragile ego-structure. Finding meaningful, positive alternatives to the use of drugs and its attendant lifestyle, and the development of healthy coping mechanisms, in general, are two very major and difficult tasks for clients seeking rehabilitation. Growth toward self-actualization, i.e., a basic acceptance and expression of the "inner core" or self, is thwarted by the unfulfillment of basic needs: love, respect, safety, belongingness, and self-esteem (Maslow, 1972). The stages of personality development in which these needs are met are mirrored within the stages of artistic development. From initial manipulation of tactile materials to the production of well-organized and integrated products, an individual can experience himself, his environment and his interactions with that environment. On a non-verbal level, therapeutic art treatment can provide a means toward satisfying and transcending the basic needs, thereby fostering growth toward self-actualization through innovation and creativity. This is of particular importance in working with clients for whom verbalization has been a means to maintain a pathological lifestyle.

Verbal manipulation and "game playing" enable mechanisms of rationalization, denial, displacement, and projection to serve the addicted individual in his evasion of self (Foulke and Keller, 1976). Non-verbal art therapy; however, does not reinforce these patterns of behavior but provides, instead, a form of spontaneous expression which circumvents the usual defensive coping mechanisms to put the addicted client in touch with his own feelings and needs. For those non-verbal clients who have difficulty in communicating generally, an outlet is offered through which expression is always acceptable.

Important, too, is the sense of mastery which develops through the learning of basic artistic skills, although primary emphasis in our program was on the expressive process. For many clients, the sense of accomplishment in an area of constructive and independent pursuit is a novel experience, particularly when supportive feedback from peers is forthcoming. Working in the group art therapy setting enables clients to feel a sense of cohesiveness through common endeavor, through learning about oneself via the interaction of the group process.

In December, 1973, art therapy was introduced as an adjunctive treatment modality in ARTC's therapeutic community. It became an important part of the treatment structure formulated for in-house residents, a structure which included methadone maintenance, individual counseling, education classes, and other, more traditional, group counseling modalities. For two years, residents were involved in a thrice-weekly group art therapy program which stressed self-awareness and positive self-expression through non-verbal communication. Ms. J. was a resident of this program.

Our therapeutic community, Quincy Village, accommodated a maximum of twenty-eight residents, male and female, who were in various stages of treatment. The census fluctuated as clients graduated from the program or terminated treatment for various other reasons. The ages of participants ranged from 21 to 53. The majority of these clients were maintaining on methadone. Each client progressed through four treatment phases from entry to termination.

Participation in art therapy paralleled the treatment phase structure of the program. Each client in the initial phase of treatment, for example, attended the group for a minimum of six hours per week. Those in the last phase attended voluntarily.

The studio in which sessions were conducted was equipped with facilities for clay work: pottery, ceramics, and sculpture; as well as for drawing, painting, hand-printing and other graphic activities. When the studio was not being utilized for scheduled sessions, it was avalilable for self-initiated individual work as the clients wished.

Reaction to the inception of this program was somewhat hostile and fearful. Clients for whom past school experiences had been negative perceived the group initially as a type of "class" or school-related activity and were not disposed to place themselves in a similar anxiety-provoking situation. For this reason, emphasis was placed on supportive but essentially non-directive techniques. In areas of practical information (e.g. methods of joining clay pieces, missing of paints, etc.) the clients were given general instruction and supervision; however, at no time were judgmental and controlling directives imposed on a client's expression. As a result, personal experience was enhanced and validated. Artistic expression was viewed as uniquely reflective of an individual's affective state and, as such, contained a high potential for self-discovery (Betensky, 1973).

In addition to working in groups, several clients found single, individual art sessions meaningful. Such sessions were initiated by the individual to discuss feelings about his/her work which served often as a catalyst for self-revelation and indicated a budding sense of trust in the client's acceptance of personal emotions.

For some clients, art activity was particularly helpful in dealing with certain critical periods in treatment. For example, clients whose detoxification from methadone was nearing the final stages could be found in the art studio in the late hours of night, working through the anguish of withdrawal discomfort and insomnia by relieving and translating their anxiety into visual representations. Most frequently, clients' work reflected stages of treatment response as disorganized and eratic approaches to expression, illustrated by Ms. J., became integrated and functional.

Gradually, group members began to move away from a narcissistic preoccupation with self. Group effort surfaced as clients began to work in concert with one another, helping in the resolution of practical problems, sharing project ideas, and, significantly, maintaining an atmosphere of non-conflict through peer pressure on occasionally disruptive members.

To further the growing sense of self-esteem experienced by most participants in art therapy, community exhibits and sales were implemented twice a year. The therapeutic community was located in a residential neighborhood with which program interaction had been positive. In addition, exhibits were presented on an in-house basis to the corporation's other staff and clients in ARTC's out-patient clinics. Exhibits were also hung in a city college and a part of a major New York City museum's community art program. The response to these exhibits was extremely positive. Clients, who were given their respective monies through sales, often expressed extreme surprise and pride in such acceptance of their work. The difficulty of assimilating into society seemed, perhaps, less insurmountable as a result of such interaction.

Various studies have been done on the effectiveness of artistic expression as a tool of the psychotherapist. More recently, however, art therapy as a viable approach to personal problem-solving has become a branch of psychotherapy in its own right (Naumberg, 1966). The primary focus has been on patients in psychiatric institutions or with "problem" children and adolescents; nevertheless, the utilization of art therapy in drug treatment has been increasing. The art therapy approach to the client population in our therapeutic community can be differentiated from the traditional psychoanalytic approach of symbol interpretation of the product; "process" was deemed of primary importance in the development of realistic self-perception.

A questionnaire was formulated when this treatment modality had been in effect for 18 months. The information was collected to elicit clients response to the program. Additionally, it represented a beginning line of investigation into the relationship between the art experience, itself, and its meaning to each individual in terms of personal growth and self-awareness. At the time the questionnaire was administered, fifteen clients were in residence. Five of these were working or attending school and were, therefore, rarely involved in the therapy sessions. Eight men and two women were avaliable to answer the questionnaire, ranging in age from 21 to 53. Four had been in residence for one year or more, four from six to nine months, and two from one week to one month. All shared past experiences of heavy drug abuse as well as alcohol dependencies in varying degrees. With three exceptions, none of the clients had finished high school. The exceptions were two with GED's and one with two years of college. There was some hesitation in answering the questionnaire because a lack of writing skills inhibited the respondents. When assured that this was not a "test", however, most were able to proceed without concern.

For the purposes of presentation, certain results have been selected as pertinent to the essential concept of this paper. The relationship between attendance in group and attitude toward artistic activity indicated that the generally positive response to the group's work was a factor in consistent attendance. The primary motivating factor in positive response to the art experience, itself, was self-satisfaction. The five respondents who ranked this as foremost explained their opinions on the basis of self-fulfillment, self-awareness, increasing artistic ability, and the pleasure of producing something by oneself. A majority of the clients (nine) answered affirmatively to questions involving the development of self-confidence and personal growth through self-expression. While the motives for chronic drug abuse are complex and involve physical as well as psychological and social needs, one characteristic seems to be the need to ease "certain anxieties and tensions" and "to feel more confident as a person" (Waldorf, 1973). The positive reactions to this set of questions which were answered affirmatively by nine of the residents may be interpreted as a sign of the therapeutic value of artistic activity in assisting the growth of positive self-image through an alternative means of behavior.

Administration of a follow-up questionnaire, as planned, was not possible because of the phasing out of the therapeutic community from the overall treatment structure of the agency. However, the fact that eight of the ten respondents had no experience with art prior to the art therapy sessions and found it a valuable mode of activity, through which positive feelings of self emerged, suggests the potential for success which art therapy offers.

In conclusion, Ms. J. and her fellow clients in ARTC's therapeutic community were able to experience themselves in a highly specialized and supportive environment. Through the bonding of the group in its common concern for expression, the individual was safe: he was respected for his individual creativity; he was able to feel genuine self-esteem, and he "belonged". The conditions for growth toward self-actualizating behavior had been established. The process of "becoming" was facilitated by the non-verbal process of artistic self-expression.

REFERENCES

Betensky, Mala, 1973. *Self-Discovery Through Self-Expression.* Illinois: Chas. C. Thomas.

Maslow, Abraham, 1972. Some basic propositions of a growth and self-actualizing psychology. *Perspective on Personality.* (Salvatore Maddi, (ed.)). Boston: Little, Brown & Co. 216-229.

Naumberg, Margaret, 1966. *Dynamically Oriented Art Therapy: Its Principles and Practice.* New York: Grune & Stration.

Waldorf, Dan, 1973. *Careers in Dope.* New Jersey: Prentice-Hall, Inc.

Foulke, E. and Keller, T.W., MD, 1976. The art experience in addict rehabilitation. *American Journal of Art Therapy.* 15: 75-80.

APPENDIX A

ART QUESTIONNAIRE

Male ______
Female ______

Age: 20-30 ______
30-40 ______
40+ ______

1) How long have you been working in the art studio?

____

2) How would you describe your attendance?

Very often ______ Enthusiastic
Average ______ Not Interested
Rare ______ Depends on mood

3) Did you have any chance before entering this T. C. to work at art activities?

Yes ________ No ________

4) If yes,

Where? ______________________
For how long? ______________________
What kind of art work did you do? ______________________
____

5) Do you feel comfortable working in the art studio?

____

6) Are you more comfortable working in the art studio by yourself?

Yes ________ No ________ Doesn't matter ________

7) If you are more comfortable working alone, why?

____
____

8) Have you completed any projects of which you are proud?

Yes ________ No ________

9) Do you feel relaxed when working with art materials?

Always ________ Never ________ Sometimes ________

10) When you are working in the studio, do you find yourself thinking about problems or situations not related to your art?

____

____

____

11) If so, is it helpful to work on these projects when thinking about such things?

____

____

____

12) Have you ever felt excited by something you were working on in the studio?

____

13) Do you feel free to come down to the art studio and work whenever you like?

____

14) What kind of art activity do you like best in art and/or what would you like to see added to the available materials?

____

____

15) Have you ever felt that you can think more clearly after working in the studio?

____

____

16) Are you sometimes surprised at what you produce artistically?

____

17) Do you relate better to your peers when you are working in the art shop with them?

____

____

18) Do you feel now that you have some artistic ability which you weren't aware of before?

____

____

19) What is more important to you in relation to your art work? Check items below from 1 (most important) to 6 (least important).

Self-satisfaction____
Peer-satisfaction____
Money possibilities____
Learning new skills____
Staff Approval____
Fulfillment of T. C. Phases ____

20) Explain your most important item above:____________

21) Have you ever entered the studio in a particular mood or frame of mind and found that mood changed when your work in the studio was done?

Yes__________ No __________ Explain ________________

22) Do you feel free to express your own ideas in relation to your art work?

23) Would you like to see more involvement between staff and residents in the art program? In what way?______

24) Do you feel there is any change in your <u>own</u> attitude toward yourself from working in the studio?

Yes __________ No __________ Explain________________

25) Would you like to discuss your work with your counselor?

26) Do you feel able to ask for help in your art work? Do you get enough attention from the group leader when you are faced with a problem in your work?

27) Is it possible to work out certain problems you may have about yourself through artistic activity?

Yes __________ No __________ Maybe __________

Explain your answer______________________________________
____
____

Thank you for answering these questions-----------------
J. S.

CRISIS OUTREACH WITH DRUG ABUSERS: UTILIZING THE EMERGENCY MEDICAL SYSTEM

Steven B. Robbins, B.A. and Michael D. Decaria, Ph.D.

Drug Referral Center

Abstract: The Emergency Medical System (EMS) comes into contact with a multitude of substance abusers. Emergency rooms frequently serve as "family doctors" for drug abusers who seek health care when in crisis. Paramedic and ambulance personnel make initial contact with drug abusers experiencing both medical and psychological crisis. Because substance abusers in crisis are prone to considering alternative behaviors, the EMS, if appropriately utilized, can become a fertile referral source for drug treatment programs. Emergency medical system personnel have been found to be resistant to facilitating a linkage between drug treatment staff and substance abusers unless given adequate training with regard to the importance of psychotherapeutic crisis intervention and referral counseling.

INTRODUCTION

The treatment of drug abuse has been defined in many different ways, but, in its broadest sense, drug abuse treatment consists of those activities which will change the abuser's drug use pattern in the direction desired by the treatment personnel (National Polydrug Collaborative Project, 1976). Drug treatment programs have seen the enormous difficulty in impacting the behavioral patterns of drug abusers. Heroin users, who have a well entrenched lifestyle (Gay, 1973), are not the only population who abuse drugs. Polydrug abusers, diverse in nature, constitute a high percentage of the drug abusers being treated in the United States today (National Institute on Drug Abuse, 1976). It would seem apparent that drug treatment programs would be interested in identifying and treating a wide variety of drug abusers who are prone to be influenced by drug treatment personnel.

The EMS, if appropriately utilized, is a system which can provide a potential source of motivated drug abusers.

THE APPEAL OF CRISIS INTERVENTION

Bloom (1968) has stated that the appeal of crisis intervention is due to its primary theoretical postulate, i.e., a person in a crisis is unusually receptive to help and can be influenced in an unusually brief period of time. Crisis intervention is an important aspect of mental health care (Darbonne, 1967). Community mental health centers have made crisis intervention an integral part of community mental health (Federal Legislation, 1976). Finally, Wesson et al. (1974) pointed out that many drug treatment agencies list crisis intervention as one of their services.

DRUG CRISIS INTERVENTION

Traditionally, drug crisis intervention has referred to the medical and nonmedical management of adverse drug reactions (Wesson et al., 1974). Much has been written on the medical side of crisis intervention (Foreman and Zerwekl, 1971; Chapel, 1973; and Bourne, 1975) which has resulted in a high level of sophistication in the medical management of drug abusers. Unfortunately, the psychological side of crisis intervention with drug abusers has not kept pace with the medical side. Yet, it is the authors' opinion that drug abusers frequently experience crisis states without acute medical difficulties and that these abusers often seek out emergency medical care as another means of coping.

DRUG ABUSERS IN THE EMS

McCarrial and Skudder (1960) found that emergency room visits have increased as much as 600 percent since the 1940's. They also found that about half of these cases were not true medical emergencies. Bartolucci and Drayer (1973) speculated that the average drug abuser in the emergency room is seeking an immediately available and secure environment, i.e., the hospital, for the treatment of his ills and problems. Those two researchers further stated that the emergency room staff have become prime deliverers of social services as well as medical services.

The emergency room is frequently a place where drug abusers encounter organized health care (National Polydrug Collaborative Project, 1976). The authors' own experience is that paramedic and ambulance services also come into contact with drug abusers who are never transported to a hospital becuase they do not need further medical care. In summary, then, many or most drug abusers who enter the

EMS have nonemergent problems and are ranked low in social worth (Jacobs, 1976); and, although they are inappropriate for emergency medical care, they are prime candidates for drug abuse treatment programs.

THE IMPORTANCE OF TRAINING EMS PERSONNEL IN PSYCHOLOGICALLY-ORIENTED HEALTH CARE

The authors have found that, in agreement with Roth (1972), the EMS usually does not consider the psychological care of a patient as its role. Consequently, EMS personnel are often resistant to referring patients from the medical field to the mental health field. In all fairness, it must be admitted that when drug abusers are referred they often do not keep their appointments; but, while EMS personnel cannot be held responsible for the characteristics or lifestyle of their patients, the EMS personnel can be instrumental in getting the drug abusing patients to accept referral into treatment. Experience has shown that once EMS personnel realized the importance of psychological treatment of drug abusers, they are more likely to refer drug abusers into appropriate treatment programs. Caring and nonjudgmental EMS personnel may actually enhance a drug abuser's chances of seeking and completing treatment.

TEACHING PSYCHOLOGICALLY-ORIENTED DRUG CRISIS INTERVENTION TO EMS PERSONNEL

Following is an outline of the training model the authors have developed in response to training EMS personnel to better establish relations between the "drug treatment system" and the "emergency health care system."

I. Differentiating Functional Disorders from Organic Disorders

   A. Psychopharmacology: An overview of commonly abused substances (Brecher, 1972)
   B. Nonemergent versus emergent medical problems; the medical management of drug abuse (Thorn et al., 1972)
   C. Prolonged versus acute abstinence syndrome (National Polydrug Collaborative Project, 1976)
   D. The mental status examination: Identifying underlying psychological problems (Solomon and Patch, 1974)

II. Importance of Crisis Intervention

   A. Potential impact of a crisis state on a person: Crisis theory (Parad, 1965)
   B. Phases of drug treatment; crisis intervention (medical/psychological) (National Polydrug Collaborative Project, 1976)

C. Psychic trauma and stress
D. The role of EMS personnel in crisis situations (Committee on Allied Health, 1977)

III. Attitudes Towards Drug Abusers

A. Legitimate versus illegitimate clients in the EMS (Roth, 1972)
B. Social worth of the client
C. Identifying personal feelings and attitudes toward the client

IV. Gathering a Baseline of Data for the Inital Assessment

A. Identifying the precipitants of the crisis
B. Common precipitants; drugs, law enforcement, self or others
C. Establishing the extent of resources; financial, psychological, interpersonal, etc. (National Drug Abuse Center, 1977)

V. The Referral Process (National Polydrug Collaborative Project, 1977)

A. Establishing rapport and clarifying roles (e.g., health care staff are *not* law enforcement)
B. Intervention with family and friends
C. Appropriate utilization of drug treatment programs

VI. Followup and Feedback

A. Responsibility of drug treatment personnel to followup
B. Establishing communications channels for problems that may arise
C. Developing goals and objectives common to both the EMS and drug treatment system

REFERENCES

Bartolucci, G. and Calvin, C. 1973. An Overview of Crisis Intervention in the Emergency Rooms of General Hospitals. *American Journal of Psychiatry*. 130:9.

Bloom, B. 1968. Perspective on Crisis Intervention. *International Journal of Psychiatry*. Vol. 6: 380-382.

Bourne, P. (editor). 1975. A Treatment Manual for Acute Drug Abuse Emergencies. *U.S. Department of Health, Education, and Welfare Publication* (ADM). 76-230.

Brecher, M. (editor). 1972. *Licit and Illicit Drugs*. Boston: Little, Brown and Company.

Chapel, J. 1973. Emergency Room Treatment of the Drug Abusing Patient. *American Journal of Psychiatry*. 130: 3.

Committee on Allied Health. 1977. *Emergency Care and Transportation of the Sick and Injured*. Menasha, Wisconsin: Banta Co.

Darbonne, A. 1967. Crisis: A Review of Theory, Practice, and Research. *Psychotherapy: Theory, Research, Practice*. Vol. 4, No.2.

Federal Legislation 1975. Community Mental Health Centers Amendments Public Law 94-63.

Foreman, N.J. and Zerwekl, J.V. 1976. Drug Crisis Intervention. *American Journal of Nursing*.

Gay, R. George, et al. 1973. The Pseudojunkie; Evolution of the Heroin Lifestyle in the Nonaddicted Individual. *Drug Forum*. Vol. 2 (3).

Jacobs, P.E. 1976. Emergency Room Drug Abuse Treatment. *Journal of Drug Education*. Vol. 6 (1).

McCarrol, J.R. and Skudder, P.A. 1960. Conflicting Concepts of Function Shown on a Rated Survey. *Hospital*. 24: 35-38.

National Drug Abuse Center. 1977. *Assessment, Interviewing and Treatment Planning*. Washington, D.C.

National Institute on Drug Abuse. 1976. Statistical Series Quarterly Report.

National Polydrug Collaborative Project. 1976. Medical Treatment for Complications of Polydrug Abuse. HEW Publication No. (ADM) 76-336.

National Polydrug Collaborative Project. 1977. Referral Strategies for Polydrug Abusers. Health, Education and Welfare Publication No. (ADM) 77-15.

Parad, Howard J. 1965. *Crisis Intervention*. New York: Grune & Stratton.

Roth, J.A. 1972. Some Contingencies of the Moral Evaluation of Control of Clientele: The Case of the Hospital Emergency Service. *American Journal of Sociology*. Vol. 77, No. 5.

Solomon and Patch. 1974. *Handbook of Psychiatry*. Los Altos, California: Lange Medical Publications.

Thorn, G.W. et al. 1977. *Harrison's Principles on Internal Medicine*. New York: McGraw-Hill.

Wesson, D. et al. 1974. Drug Crisis Intervention: Conceptual and Pragmatic Considerations. *Journal of Psychedelic Drugs*. Vol. 6, No. 2.

A FUNCTIONAL MODEL FOR INCREASING THERAPEUTIC SUCCESS AND COST EFFECTIVENESS IN SUBSTANCE ABUSE TREATMENT PROGRAMS

A. Thomas McLellan, Ph.D. and Michael D. Thumme, B.S.

Philadelphia Veterans Administration Hospital
and
University of Pennsylvania

Abstract: In order to increase the therapeutic success and cost effectiveness of a substance abuse treatment network involving five different inpatient and two outpatient programs, a computer assisted evaluation model has been developed.

The evaluation model offers a quasi-experimental analysis of the matching of patients with treatment programs. The specific goals of the model are:

1. The development of a generalizable evaluation process to efficiently match patients with appropriate programs.

2. An overall increase in treatment success through specific improvement of treatment assignment procedures.

3. A general reduction in cost per unit of treatment effectiveness, patient drop-out, and treatment overlap.

INTRODUCTION

It seems clear that the addictive process is a complex syndrome of psychological, medical, legal, social, and family problems. Our experience shows that these problems combine in various ways to support substance abuse and create, particular treatment needs (e.g., psychotherapy, marriage counseling, job counseling, etc.). Given a large population of substance abuse patients, with a wide range of treatment needs, it seems likely that any single therapeutic program will be successful with only the sample of patients whose particular treatment needs correspond with the specific therapeutic approach of the program. Thus, while a program may be only 35 to 40% effective with the total patient population, each individual program may be much more effective in treating that particular sample of patients for which it is best suited. By determining the specific treatment needs most successfully addressed by each therapeutic program, and then assigning patients with those needs to the appropriate program, it seems possible to improve the effectiveness and efficiency of the entire treatment network.

At the Coatesville Veterans Administration Hospital, a Substance Abuse Treatment Unit was organized in July of 1975 to combine all the individual alcohol and drug treatment programs into one coordinated network. At present this substance abuse treatment network incorporates five distinctly different inpatient programs based at CVAH and two large outpatient programs based at the Philadelphia Veterans Administration Hospital. This treatment network treats approximately 1,800 patients each year, at an annual cost of over three and one-half million dollars (1975 figures). The wide range of therapeutic programs, and the large and varied patient population offers a promising opportunity for development of a specific model designed to improve and evaluate the most beneficial matching of patient to treatment program without disrupting normal hospital procedures.

The need for such an analysis has been widely documented (Pattison, 1969; Brill and Lieberman, 1969; Gottheil et al., 1971; and Roman, 1976). However, the experimental designs and methodology typically required for these studies, have been prohibitive in terms of time, cost, and disruption of delivery of health services. Furthermore, the meager results from these studies have not been robust or generalizable for the most part (Luborsky and McLellan, 1978).

The present paper presents a quasi-experimental model developed to be both effective in matching patients with treatment programs, and particularly applicable to an ongoing treatment setting. In the remaining sections of this paper the model and methodology will be presented.

METHOD

Our evaluation model is designed in three different stages. Stage I requires 12 months and is designed to collect patient data at the time of the patients' admission, during treatment and after discharge. Stage II requires 15 months and is concerned with analyzing Stage I data, and determining program assignment changes to best suit patient needs. Stage III requires 15 months and evaluates the changes that were made during Stage II by comparing the Stage III performance against the Stage I baseline.

The model suggested is a quasi-experimental design (Campbell and

Stanley, 1963). This design was chosen because it permits evaluation of ongoing treatment, does not interfere with staff or patients, does not require random assignment of patients to programs, or untreated control groups, and still permits sufficient power to interpret results.

Stage I

In Stage I three types of data will be collected from each patient treated in the existing system: Pre-treatment (admission), within-treatment (treatment progress), and post-treatment (follow-up). The cost and effectiveness data collected during Stage I, serves as a baseline from which we can assess the results of subsequent changes.

An Addiction Severity Index (McLellan et al., 1978) was developed to aid in collection of both pre- and post-treatment data. This Index focusses on the patient's need for treatment in six areas potentially important within the addiction syndrome. These areas include: medical, means of support, substance abuse history, legal problems, family problems, and psychological problems. Within this index, objective and subjective data are collected on each of the six problem areas and serve as an index of the duration and severity of each problem.

In addition to objective data on the patient's life pattern with regard to each problem area, several items are included to determine the patients' recent status 30 days prior to admission. This will permit a time-based comparison with the 30-day period prior to the post-discharge follow-up interview. Finally, the subjective items are included in each problem area to assess the patient's perception of his problems and his need for specific treatment.

Within-treatment (treatment progress) data is collected monthly and at the time of transfer or discharge from any of the treatment programs. Staff of the assigned program are responsible for recording objective and subjective measures of within-treatment effectiveness. Objective measures include number of group sessions, individual counseling sessions, disciplinary actions, etc. Subjective measures include staff ratings of the patient's mood, motivation, and benefit from program. This data will serve as a within-treatment measure of patient-program compatibility and treatment effectiveness.

Post-treatment data will consist of a repeat administration of the Addiction Severity Index six months following admission. This is designed to permit exact comparisons of the patient's status in each of the six problem areas before and after treatment. All pre-treatment and post-treatment data will be collected by technicians trained in psychological interviewing and follow-up techniques. The technician has primary responsibility for all admission, treatment progress, and follow-up data collected on his assigned patients.

Technicians will channel their data to a central collecting point where scanning for errors and keypunching takes place. After the data is reduced to IBM cards, it is introduced to computer storage. Thus by the end of Stage I a data file will be established for each patient and will consist of his admission, within-treatment and post-treatment data.

Data manipulation and filing requirements for an information system comparable to this demand both computer hardware and software. Basic hardware needs consist of access to an IBM 360-370 or similar system. This size unit provides tape storage capability, disk storage capability, on-line storage large enough for data manipulation, and most commonly used periperhal equipment (e.g., card punch, card reader, high speed printer, card sorter, etc.). Human considerations should be accessibility, convenience, quick turn around time, minimum downtime, established back-up system, and any needed special equipment. Basic software will need to be generated either by a programmer on your staff from a canned data management program. The canned programs (e.g., Mark IV, IMS, or CICS) will provide for all data manipulation and filing.

Stage II

The second stage of the overall design is accomplished in two parts. During the first part, data collected under Stage I is analyzed. It is expected that analysis of the Stage I admission data will provide factor combinations which differentially predict treatment outcome in each of the treatment programs. The within-treatment data will indicate how a patient performs during treatment and may predict post-treatment outcome. Post-treatment data will show if any benefit was achieved and maintained from the program. The actual analyses will be performed in two phases:

1. Data reduction

2. Discrimination and prediction

In order to reduce the data to a more relevant and manageable form, and to facilitate the accomplishment of subsequent predictive analyses, multiple factor analysis will be employed. The general purpose of factor analytic techniques is to find a way of condensing (summarizing) the information contained in a number of original variables into a smaller set of new composite dimensions (factors) with a minimum loss of information--that is, to search for and define the fundamental constructs or dimensions assumed to underlie the original variables. The use of factor analysis will therefore enable us to identify appropriate variable clusters for subsequent regression, correlation or discriminant analysis.

The initial factor analysis may produce a series of "factors" describing the different aspects of the admission data. The next step will be to subject these "factors" to a discriminant analysis. Discriminant analysis is useful when the analyst is interested either in understanding group differences or in correctly classifying individuals into groups or classes. This analysis can be considered either as as a type of profile analysis or as an analytical predictive technique.

The discriminant analyses will determine which of these factors differentiate between patient successes and failure in each of the programs. For example, the analyzed data for a program may suggest that factors such as history of previous psychiatric problems and history of previous employment problems, may effectively differentiate between those patients who demonstrated good outcome and those that did not. Similarly, other factors may prove to be useful in differentiating success from failure in other programs. These factors will also be subjected to a multiple regression analysis; a procedure which can be used to determine what combination of factors predict treatment success with each program.

In the second part of Stage II, the results of the analyses will be presented to the program coordinators, unit chief, and hospital administration for judging modifications to existing treatment assignment procedures. Finally, the recommended modifications in treatment assignments will be instituted.

Stage III

During Stage III the results of the treatment assignment modifications (Stage II) will be directly compared for cost and treatment effectiveness wtih the Stage I baseline. This will be accomplished by collecting admission, within-treatment, and post-treatment data using the same instruments and procedures as in Stage I. Thus, Stage I and Stage III are identical in design and procedure.

DISCUSSION

The evaluation model is expected to provide information on the factors within the addictive syndrome of each patient which relate to treatment outcome and which differentiate the population into subgroups of patients with common addiction related problems. In addition, the process described will determine the specific therapeutic strengths of each treatment program and will permit assignment of patients to programs having proven success in dealing with their particular addiction related problems. Similarly, it should be possible to determine the types of patients not currently receiving help, and to develop new programs to treat these patients. Finally, by developing specialized treatment approaches and eliminating or modifying ineffective ones, it should be possible to reduce waste and cost overlap.

This model has other benefits than treatment matching. Admission data can indicate the duration and severity of patient needs in medical, psychological, legal, social, and family areas. These indications can demonstrate needs for additional staff time and involvement, special procedures, and/or specific therapies. For example, with regard to medical problems, patients that demonstrate a significant number, and long duration of problem symptoms may require additional medical coverage. Similarly, a patient demonstrating severe psychological problems may require extended psychological evaluation and special treatment. Further, program analyses of the patient population, in terms of their specific treatment needs, may permit more effective utilization of staff and facilities.

The within-treatment reports may be used for quick and efficient evaluation of patients' therapeutic progress. This allows the program to determine if a patient is receiving sufficient benefit from the treatment offered, and will permit rapid transfer of patients who are not receiving adequate benefit, to a more appropriate program. The collected data base is also an excellent start of a management information system. In our model we have the basic ingredients of a management information system (e.g., monthly the unit chief and program directors could obtain information describing patients in treatment, types of treatment offered to patients, treatment outcome, etc.).

Finally, total benefits are not limited to direct model applications. The model suggested may serve to effectively generate additional evaluation studies, experimental projects, and important management information in a variety of treatment settings. Thus, the model has obvious benefits for the health care delivery system and a wide range of other potential usages.

REFERENCES

Brill, L. and Lieberman, L. 1969. *Authority and Addiction*. Boston: Little, Brown and Co.

Campbell, D.T. and Stanley, J.C. 1963. *Experimental and Quasi-Experimental Designs for Research*. Chicago: Rand McNally.

Gottheil, E., Corbett, L.O. and Grasberger, J.C. 1971. Treating the alcoholic in the presence of alcohol. *Amer. J. Psychiat.* 128: 475-480.

Luborsky, L.B. and McLellan, A.T. 1978. Our surprising inability to predict the outcome of psychological treatment: with special reference to treatments for drug abuse. *Am. J. Drug and Alcohol Abuse.* 5: 387-398.

McLellan, A.T., Erdlen, F.R., LaPorte, D.J., Crits, K., Intintolo, V., and Thumme, M.D. 1978. *Instruction Manual for the Administration of the Addiction Severity Index*. Philadelphia, Pa.: Veterans Adminstration Press.

Pattison, E.M. 1969. Evaluation of alcoholism treatment: A comparison of three facilities. *Arch. Gen. Psychiat*. 20: 478-488.

Roman, P.M. 1976. Prospects for behavioral science research on alcohol abuse and alcoholism. *Annals of New York Academy of Science*. 273: 555-557.

MEGASCORBATE: A DETOXIFICATION ALTERNATIVE

The San Francisco Drug Treatment Program, Inc.

Since its inception in 1968, the San Francisco Drug Treatment Program, Inc. has responded to the mental and physical health needs of narcotic addicts. The search for safer, cost-effective detoxification alternatives led to an exploration of non-traditional methods of detoxification which would address the poor nutritional habits of narcotic addicts. The present pilot study investigated the use of megadoses of ascorbic acid as sodium ascorbate, multi-vitamins and mineral supplements in the treatment of drug withdrawal symptoms. Fifty-eight clients were compared; seventeen clients using the sodium ascorbate detoxification procedure; thirty clients using symptomatic relief medications and eleven clients using symptomatic relief medications for three days and the mega-scorbate procedure for the remainder of the detoxification period. The preliminary data presented indicates that the sodium ascorbate or megascorbate detoxification procedure is at least as effective as symptomatic relief medications in alleviating narcotic withdrawal symptoms, and that the procedure can be adapted successfully to an outpatient setting. Results of the present pilot study clearly indicate a need for further testing of the megascorbate procedure by pointing to its acceptance and success rate with clients, its cost effectiveness, and its holistic approach to the process of drug rehabilitation.

INTRODUCTION

The abuse of drugs and alcohol has become recognized as a major health problem to millions of Americans. The appearance of drug abusers on all socioeconomic and racial levels has made this problem evident to clinicians, physicians, judges and employers alike. Thus, many efforts have been made in recent years to treat the drug abuser more effectively and holistically.

One area of concern of the addict or drug abuser is the process of drug withdrawal. Withdrawal symptoms differ according to the individual and the drug(s) of abuse, but are typically painful and anxiety producing. Methods commonly used in the past by clinicians and physicians to help alleviate drug withdrawal symptoms include a combination of symptomatic medications such as Darvon, Librium, Belephen and Chloral Hydrate, Methadone, Codeine and Valium, and intensive psychotherapy. While these and other chemical combinations have long been employed as drug detoxification alternatives, many of them present harmful or uncomfortable side effects of their own. (Methadone is indeed a powerful, addictive substance.) Thus, individuals working with drug addicts continue to search for more effective, safe methods to help make the withdrawal process an easier one, and to encourage a healthier lifestyle.

San Francisco Drug Treatment Program, Inc.'s current study of the use of megavitamin therapy for the treatment of narcotic withdrawal began in December, 1977, in response to our clients' need to find a more healthful alternative to standard chemical detoxification procedures. Megavitamin therapy using sodium ascorbate, ascorbic acid, calcium and other mineral supplements was seen as a cost-effective, convenient, *safe* way to detoxify narcotic addicts, and is also a way to address the poor nutritional habits of our client population.

BACKGROUND AND RATIONALE FOR USE OF MEGA DOSES OF ASCORBIC ACID

Since the discovery and synthesis of ascorbic acid in the early 1930's, a vast amount of medical research has been carried out on the physiologic effects of this substance and its possible uses as a therapeutic agent in various disease states. In addition to being identified as Vitamin C, the substance is known as ascorbic acid (and its salt sodium ascorbate). Ascorbic acid in small amounts (10-60 mg.) is required in the human diet to prevent the endstage deficiency state known as scurvy, but both ascorbic acid and sodium ascorbate have been shown to be of benefit in large doses (megadoses) in a great variety of pathologic conditions.

Medical research and clinical experience have yielded results which tend to support the following possible therapeutic effects of ascorbic acid: preventing heart disease by lowering blood cholesterol and preventing deterioration of arterial walls; counteracting the side effects of many drugs; alleviating the effects of several kinds of toxic substances; decreasing disease of the circulatory, digestive and nervous system in the elderly; combating the effects of hay fever and various allergic conditions; helping relieve pain and decrease healing time in burn patients; helping control leprosy; decrease fractures and increase physical activity in patients with Osteogenesis imperfecta; promoting spontaneous resolution of rectal polyps;

combating urinary tract infections and inflammations; preventing and/or ameliorating symptoms of the common cold; promoting increased survival in terminal cancer patients with improvement in symptoms; treating many viral diseases, (especially hepatitis, mononeucleosis and viral pneumonia); enhancing the effectiveness of antibiotics in bacterial diseases.

These therapeutic effects have been more effectively demonstrated using doses of ascorbic acid many times in excess of the daily doses required in human nutrition to prevent scurvy. For example, in the treatment of viral hepatitis and pneumonia as much as 150 gram doses have been used orally and intravenously without evidence of adverse effects.

THE APPLICATION OF ASCORBIC ACID TO THE DETOXIFICATION OF NARCOTIC ADDICTS

More pertinent to the San Francisco Drug Treatment Program, Inc. Pilot Study is the recent work by Dr. Alfred Libby and Irwin Stone (published in the December 1977 Journal of Orthomolecular Psychiatry), on the use of mega-doses of ascorbic acid to detoxify heroin addicts. They compiled 100 case reports of heroin addicts whom they detoxified using ascorbic acid and/or sodium ascorbate in doses of 25-85 grams per day for the first few days, gradually tapering to a holding dose of about 10 grams per day. In addition, on the theory that addicts are malnourished in general and protein deficient in particular, most of these patients were given high levels of multivitamins and minerals and a pre-digested protein preparation.

The patients in this study almost uniformly reported a loss of craving for drugs while taking megascorbate. These patients were treated for 1-2 weeks in a residential setting and then patients were discharged on holding doses of ascorbic acid for out-patient follow-up.

POSSIBLE SIDE EFFECTS

The possible side effects of mega doses of ascorbic acid are remarkably few, and no serious side effects have been documented even in patients receiving as much as 150 grams per day for treatment of viral diseases. They can be divided into short term and long term complications. Of the short term problems, the most common and well-documented is gastrointestinal distress with symptoms of heartburn, stomach-ache, diarrhea and excess gas. These are generally dose dependent and are influenced by the concept of "bowel tolerance", meaning that the dose that will cause gastrointestinal symptoms increases in proportion to how sick the patient is. (A person who might normally get diarrhea on 2 grams per day will be able to tolerate 50 grams per day if he has the flu). In fact one way to tolerate the dose of ascorbic acid for an individual is to stop at just below that dose that will cause diarrhea.

Other short term side effects that have been reported include: a rash in the rare patient who proves allergic to ascorbate; worsening of active peptic ulcer disease (may be avoided by using only sodium ascorbate, which has a high pH); overloading a patient having hypertension or congestive heart failure with too much sodium (avoided by using only ascorbic acid); bitter taste of ascorbic acid (ascorbate is tasteless).

Possible complications of long term use of mega-doses of ascorbic acid that have been suggested but not proven include: formation of oxalate or uric acid stones in the urinary tract; destruction of vitamin B-12; pentosuria and possible increased chances of infertility or abortion, and dependency (people taking 10-15 grams per day for years might get sick if they stop). None of these suggested complications have been confirmed by researchers and clinicians who have been using mega-doses of ascorbic acid in humans for up to 30 years. However, these researchers do suggest that due to the chelating effects of ascorbic acid supplemental multivitamins and minerals be taken along with it.

GOALS OF SFDTP, INC. PILOT STUDY

With such an absence of serious complications, and the encouraging results of Dr. Alfred Libby, San Francisco Drug Treatment Program, Inc. initiated (Dec. 1977) a pilot study of the use of ascorbic acid/sodium ascorbate in the treatment of narcotic withdrawal symptoms.

Goals of the SFDTP, Inc. Pilot Study were identified as follows:

1. to compare the effectiveness of the ascorbic acid/sodium ascorbate detoxification procedure with symptomatic medications in alleviating withdrawal symptoms;

2. to determine whether maintenance on ascorbic acid/sodium ascorbate relieves the craving for drugs after detoxification;

3. to determine the appropriateness of this detoxification approach to an outpatient setting.

METHODOLOGY

In order to more easily compare the effectiveness of the two detoxification procedures (ascorbic acid and symptomatic medications) three detoxification groups were established: 1) clients using only the ascorbic acid procedure, 2) clients using symptomatic relief medications, and 3) clients using symptomatic relief medications for three days followed by the ascorbic acid procedure for the remainder of the detoxification period.

At the point of intake, clients were given the option of participating in one of the three detoxification groups and were assigned a counselor who dealt with clinical issues of concern during the client treatment period.

All clients received a routine physical examination from the Medical Director upon admission to the program. Following the examination a symptom checklist was completed for each client (see Appendix). Symptom checklists were then completed daily on all clients. Each client met regularly with an assigned counselor who dealt with clinical issues. Clients in all groups followed a 21-day detoxification protocol.

Group #1 - Ascorbic Acid Procedure

The ascorbic acid program consisted of the following: sodium ascorbate or ascorbic acid (in crystalline form) dispensed in packets containing 24-48 grams per 24 hours for 5-7 days, tapering to 8-12 grams per day for 14 days; multivitamins and multimineral tabs (one or two) for 21 days; calcium complex and magnesium tabs three times per day; liquid protein (20 oz.) three times per day for 3-5 days. Individual client dosages were dependent upon client reports of active withdrawal symptoms, observable symptoms and the symptom checklist report. Dosage sheets were maintained on each client depicting daily dosage levels and supplements given (see Appendix).

Group #2 - Symptomatic Medication Procedure

Symptomatic relief medications administered to clients in this group consisted of the following: Propoxyphene hydrochloride 65 mg. per day for 21 days; Belephen #3 three times per day for 14 days; Libritabs 10 mg. three times per day for 21 days; Chloral Hydrate 1,000 mg. per day for 7 days. Symptomatic medications were routinely administered in individual medication packets and dosage levels were lowered gradually as withdrawal symptoms decreased during the 21 day detoxification period. Medication sheets were maintained on each client depicting daily dosage levels (see Appendix).

Group #3 - Combination Procedure

Clients participating in this group were administered routine doses of symptomatic medications for three days. Beginning with the fourth day clients were given ascorbic acid/sodium ascorbate in doses determined by subjective and objective withdrawal symptoms, multivitamins as needed and minerals as needed. Dosage levels were ascertained by the Medical Director and/or Program Nurse and were tapered during the remaining 18 days of the 21 day detoxification period according to client need.

RESULTS

To date, 58 clients have participated in the SFDTP, Inc. Pilot Study. Table I illustrates that the majority of clients chose to detoxify with symptomatic medications, and that these clients were younger, on the average, than clients in the other groups.

The highest dropout rate is among clients in Group #1 utilizing the ascorbic acid procedure. It is also apparent that clients in Group #1 reported having smaller drug habits than clients in the other groups, as evidenced by the amount of money spent on drugs per day.

Table I. Client background information.

	GROUP 1	GROUP 2	GROUP 3
	Ascorbic Acid	Symptomatic Medications	Combination
# of Clients in Group	17	30	11
Average Age of Clients	32 yrs.	28 yrs.	31 yrs.
Ave. # yrs. used heroin	9 yrs.	6.5 yrs.	9 yrs.
Ave. Amount Used ($)	$55.00 day	$70.00 day	$80.00 day
Ave. Length of Run	2 yrs.	1 yr.	2 yrs.
Ave. # Days in Program	6 days	11 days	9 days
Sec of Clients	5 Females	7 Females	5 Females
	12 Males	23 Males	6 Males

It can also be noted that the majority of male clients opted for the Group #2 detoxification procedure using symptomatic relief medications, as was true with female clients, but to a lesser degree.

Table II illustrates the average number of symptoms reported by all clients in each group for the second and fourth days of the detoxification period.

The data from Table II indicates that clients in Group #1 utilizing the ascorbic acid only experienced the most marked decrease in symptoms over this two day period. Clients in Group #2 utilizing the symptomatic medications only reported no decrease in symptoms over the two day period.

It is interesting to note that although clients in Group #2 reported no decrease in symptoms due to the symptomatic relief medications utilized during the detoxification period, the average length of stay in the program was longer than that for any other group.

SUMMARY AND CONCLUSIONS

It must be understood that the data presented in this paper is incomplete. In order to obtain an adequate sample size the Pilot Study should be continued for several months. Conclusions drawn from the data presented must therefore be made cautiously, and are necessarily based upon both raw data and the experiences of the San Francisco Drug Treatment Program, Inc. staff. With these thoughts in mind, a discussion of this study is in order.

The preliminary data of the ascorbic acid pilot study is encouraging. The results presented indicate that the ascorbic acid detoxification procedure is at least as effective as symptomatic medications in alleviating narcotic withdrawal symptoms. In some cases,

Table II. Client symptoms.

	GROUP 1	GROUP 2	GROUP 3
AVERAGE # OF SYMPTOMS	Ascorbic Acid	Symptomatic Medications	Combination
DAY #2	6	8	7
DAY #4	4	8	5.5

according to client reports, relief was found to be more complete with ascorbic acid and supplemental vitamins and minerals, and also provided a new sense of well being to the client. Several clients are currently being maintained on doses of 8-10 grams of sodium ascorbate per day and have remained drug free. The data also shows that clients utilizing the ascorbic acid procedure, on the average, dropped out of treatment sooner than did clients on symptomatic medications. This could point to the effectiveness of the ascorbic acid procedure in alleviating withdrawal symptoms (thus eliminating a client's need to continue the detoxification phase), to a client's frustration with the more complicated procedures of this form of detoxification, or to a client's satisfaction with the results. It is the author's opinion that the second alternative listed above was the primary reason for the earlier dropout rate in this group. The lack of standardization in administering the ascorbic acid procedure to clients and the subsequent lengthy time involved during each client visit drew many complaints from clients in this group. It is hoped that with more experience, increased standardization and continued staff support, the ascorbic acid procedure will be utilized by more clients who will remain in treatment longer. However, even the preliminary data presented points to support for the hypothesis that the ascorbic acid procedure is at least as effective in alleviating withdrawal symptoms as symptomatic medications and that this alternative was an acceptable one to the other form of detoxification. Considering the cost effectiveness of ascorbic acid and supplements and the health benefits to clients, this is indeed an encouraging find.

Due to its short duration thus far, the study presented here does not include data concerning clients who are being maintained on sodium ascorbate and supplements. Thus, no conclusions may be drawn at this early date regarding this procedure's ability to relieve the craving for drugs after detoxification. Initial client reports, however, suggest that at least two individuals find this to be the case and have remained drug free.

The data presented suggests that the ascorbic acid detoxification procedure could, indeed, be included in outpatient drug treatment. The fact that more than half of the clients presented in this pilot study opted for one of the two ascorbic acid detoxification procedures indicates its acceptance potential. As was mentioned earlier, however, unless the procedures involved are standardized to minimize the length of each client visit, the dropout rate will continue to be high. A high degree of staff support for this treatment alternative is extremely important in an outpatient setting. Much of the data collection and monitoring of client progress in an outpatient setting depends upon continued staff interest and support. The primary problem in presenting this detoxification alternative in an outpatient setting is the need for close monitoring of clients during the first week of treatment. In the present study, clients were asked to come to the program two or three times daily so that doses could be altered as needed. Initially, doses were only given for a four or five hour period, necessitating a client's return. It was evident, however, that this was not acceptable to clients and interfered with counseling practices within the program. With more experience, the Medical Director and Program Nurse were able to more easily and accurately match dose levels to demonstrated withdrawal symptoms, and clients using ascorbic acid and supplements received a 24-hour dose once per day. This streamlining of the ascorbic acid procedure was seen as a necessity for an outpatient setting, which tends to attract clients who can not (or do not wish to) enter a more demanding treatment environment. It was also found that written directions explaining doses and the dosage schedule greatly aided clients who could not otherwise have such questions answered immediately.

It can be speculated that the ascorbic acid detoxification procedure is a potentially successful one. The data presented here indicate a need for further research, including a research design which allows for a greater control of the many variables listed. Data from this Pilot Study suggests that the older drug addict, who has used drugs for a longer period of time and who may have tried other forms of detoxification without success, would be more likely to opt for this type of treatment. A streamlining of the administration procedure for ascorbic acid would be a necessity, as would a closer monitoring of client symptoms and for a longer period of time. A follow-up period of at least two months but not more than three months should be included in the research design as well.

It is the opinion of the San Francisco Drug Treatment Program, Inc. staff that the field of drug abuse is largely unexplored. The search of drug treatment programs for innovative ways to treat the problems of heroin addiction must continue and be encouraged if the problems of drug abuse are to be eradicated. The presented treatment alternative utilizing ascorbic acid/sodium ascorbate should be seen as a pioneering effort which has a high potential in treating narcotic withdrawal symptoms. The fact that it views treatment as a holistic approach to drug abuse is also one of its positive contributions.

REFERENCES

Gerson, Charles P. Ascorbic Acid Deficiency in Clinical Disease Including Regional Enteritis. *Ann. N.Y. Acad. Sci.* 483, 490.

Libby, Alfred F. and Stone, Irwin. 1977. The Hypoascorbemia Kwashiorkor Approach to Drug Addiction Therapy: A Pilot Study. *T. Orthomolecular Psychiatry.*

Klenner, Frederick R. 1971. Observations on the Dose and Administration of Ascorbic Acid when Employed Beyond the Range of a Vitamin in Human Pathology. *J. of Applied Nutrition.* 23(3 & 4): 61-87.

Stacpoole, Peter W. 1975. Role of Vitamin C in Infectious Disease and Allergic Reaction. *Med. Hypothesis.* 1(2): 43-45.

Stone, Irwin. 1972. *The Healing Factor: Vitamin C Against Disease.* New York: Grosset & Dunlap.

Stone, Irwin. 1966. The CSS Syndrome: A Medical Paradox. *J. of the Northwest Academy of Preventative Medicine.*

APPENDIX

Table I. Vitamin "C" Project Medication Sheet.

CLIENT # ____________

D.O.B. ____________

DATE																							
MEDICATION	1	2	3	4	5	6	7	8	9	10	11	12	13	14	15	16	17	18	19	20	21		
SODIUM ASCORBATE																							
ASCORBIC ACID																							
50-50: Sodium ASCORBATE/ASCORBATE ACID																							
MULTI VITAMIN																							
DISPENSED BY:																							
URINANALYSIS RESULTS																							

KEY: A-amp. B-barb. C-codiene N-negative D-darvon K-cocaine PR-procaine UFDS- Urine for drug screen M-Morphine Q-Quinine METH=methadone I= Insufficient

Tine Test:
Date Tested ____________
Checked: Results:

Blood Drawn ____________

Notes:

Date: ____________

I REQUEST THAT THIS PRESCRIPTION NOT BE DISPENSED IN A CHILD RESISTANT CONTAINER.

____
Signature

Table II. Nurses' Medication-Urinalysis Record.

____

D.O.B. ____________________

DATE ____________________

MEDICATION	1	2	3	4	5	6	7	8	9	10	11	12	13	14	15	16	17	18	19	20	21
DISPENSED BY																					
URINALYSIS																					

TINE TEST:
DATE TESTED ____________________
DATE CHECKED ____________ RESULTS ____________

Urinalysis Key: A - Amphetamines, B - Barbituates, C - Codiene, D - Darvon, K - Cocaine, Pr - Procaine, M - Morphine, Q - Quinine, Meth - Methadone

NOTES:

Table III. Symptom Checklist/Vitamin C Project.

CLIENT # __________

Please check if symptom is present, 2 checks for severe symptoms.

During the last 24 hours did you experience:

DATE:																								
Subjective/Objective	S	O	S	O	S	O	S	O	S	O	S	O	S	O	S	O	S	O	S	O	S	O	S	O
1. Running eyes																								
2. Running Nose																								
3. Sweating																								
4. Chills																								
5. Muscle Aches & Pain																								
6. Abdominal Cramps																								
7. Nausea																								
8. Feeling sick																								
9. Diarrhea																								
10. Craving for Drugs																								
11. Anorexia																								
12. Insomnia																								
13. Daily Dose																								
ANSWER YES OR NO:																								
14. Did you use heroin																								
15. If so, usual effect?																								

Interviewer:

S=Subjective, ie client report O=Objective, observed by interviewer

BROAD SPECTRUM BEHAVIORAL THERAPY IN THE TREATMENT OF SUBSTANCE ABUSE

Harvey Weiner, D.S.W.
Addictive Behavior Programs, Hahnemann Medical College

Sonja Fox
Walk-In Clinic of the Addictive Programs

INTRODUCTION

The primary objective of this paper is to expose clinicians to the use of broad spectrum behavioral therapy in the treatment of substance abuse. Those interested in a fuller description of the theory and strategies which underlie this approach are encouraged to review the items listed in the bibliography.

Evidence suggests that traditional psychotherapeutic modalities are of limited usefulness in the treatment of substance abuse. Behavioral therapy has proven useful as an alternative and/or adjunct to traditional approaches, as will be discussed below. The central assumptions which underlie the cognitive-behavioral approach include:

(1) Evaluating the client's perceptions of the effects of abused substances as a critical part of the treatment planning;

(2) Eliciting the client's reasons for abuse, to explore alternatives and challenge misconceptions;

(3) Mutually establishing specific "targets" of treatment, so that client and therapist can objectively measure progress. A "target" is defined as a problem area which is related to the substance abuse behavior.

Therapeutic methods include cognitive appraisal and restructuring, relaxation and assertiveness training, and activity scheduling. Heavy emphasis is placed on the daily performance of appropriate task assignments. The cognitive approach is based on the fact that a person's behavior is determined in part, by the way in which he structures his experience. If he believes that he is an incurable addict with no way to control discomfort other than through the use of drugs, he will probably continue such use as the "only solution." Or, if he believes that all of his actions will result in failure, it is unlikely that he will initiate any positive action. The therapist must work with these faulty cognitions, because treatment efforts will prove futile if the client's underlying assumptions remain unchanged. This may help to explain the extraordinarily high relapse rates associated with inpatient detoxification programs, for example. While physiological detoxification can occur in a relatively brief period of time, little has been accomplished if the client retains the faulty cognitions which originally led to his addiction.

In the field of substance abuse there are some common cognitive distortions which consistently occur in both staff and clients. Corrections of such distortions might significantly affect treatment outcome. These common cognitive distortions are as follows:

(1) Distortion--Addicts are manipulative and manipulative behavior is totally "bad" and must be extirpated. Correction--Maniputive behavior can be viewed as shrewd, innovative, problem-solving activity when channeled in a socially adaptive way, and in fact, is practiced intensively and ubiquitously by "successful" people in our society.

(2) Distortion--Addicts are not goal oriented. Correction--The lifestyle of the addict demonstrates goal-directed activity rarely achieved by the average individual. Approximately 80% of the activities of the addict are directed towards the goals of obtaining and using drugs. The task of therapy is to change goals--not to convince the individual that he has no skill or experience in this area.

(3) Distortion--Addicts are weak, dependent individuals. Correction--The activities which an addict must engage in daily to satisfy his habit, and the sacrifices and hardships he must endure to maintain his habit, demonstrate amazing strength, initiative and perseverance.

(4) Distortion--One must always confront an addict and preferably in the group situation. Correction--Most of the addict's interpersonal interactions with peers and authorities (police, probation officers, treatment personnel, etc.) are confrontive. He has become an expert at this because his deviant life style has demanded defensiveness. The average therapist is not even in the same league with him in terms of this skill. One to one relationships which are supportive, unprejudiced and rational are rarely within the experience of the addict and thus represent an area of deficient skills, as well as a novel approach against which he has prepared few defenses except avoidance. To summarize, many thera-

pists see the addict as totally lacking in resources. In fact, their resources are already well developed, but misdirected. The consequences of the therapist's distortion is to convince the addict that he has nothing going for him, that all of his coping behaviors are without value, that he has learned nothing which can be useful to him in "straight life", and, consequently, that his rehabilitation must start from zero. No wonder he gets depressed.

Clearance of these distortions enables the therapist to approach the addict with a positive feeling, and a more accurate assessment of his strength and potential. This in turn, leads to more hopeful feelings in the client, thus working to lift depression early in the treatment process, rather than adding to it.

The following case study illustrates the use of broad spectrum behavior therapy in the treatment of substance abuse.

CASE HISTORY - MR. D.

Background

Mr. D. is a 27 year old black male, married, with two children (ages 5 and 8), referred by the therapist of his 8 year old daughter. The therapist felt that Mr. D.'s behavior was largely responsible for his daughter's psychotic symptoms. Mr. D. has been addicted to alcohol since age 14, and he also abuses Valium on occasion. His father and stepfather were both alcoholics. When Mr. D. was 17, he stabbed his mother's lover to death during a quarrel.

Mr. D. dropped out of school in the 10th grade but he has some training as a machinist. He is an intelligent, articulate individual with well developed social skills who could best be described as an "operator" or "con artist". Mr. D. is a binge drinker; he drinks steadily for 5 or 6 weeks until he is too sick to continue, then he stops until the next binge. He had developed pancreatitis and a fatty liver by the age of 24. He always felt that he was powerless to control his irresistible urge to drink. The drinking binges always began at his mother's home, and usually involved his mother and/or sister. He would return to his own home only after becoming too ill to continue.

Mr. D.'s wife is a stable, responsible woman who works hard to maintain the family and pay the bills. The marriage was almost destroyed by Mr. D.'s drinking behavior.

Treatment Strategies

(1) Marital Counseling. Contracts were made which clearly delineated household responsibilities for Mr. and Mrs. D. Rules for argu-

ments were developed, and each kept records of the other's deviations from these rules. These deviations were then discussed in the presence of the therapist. Mr. and Mrs. D. also agreed that she would contact the therapist immediately if Mr. D. showed any signs of drinking and/or drug use, rather than waiting until a binge was underway.

Mr. D. agreed that he would deposit all of his money in a joint bank account which would require Mrs. D.'s signature for withdrawal. He also agreed to undergo hospital detoxification (which he particularly detested) if he went on a binge. In addition, a schedule was prepared for the ongoing involvement of Mr. D. in recreational, educational, and disciplinary activities with the children.

(2) Meeting with Mr. D. and His Mother. During this session, the therapist clearly identified mother's role in maintaining Mr. D.'s drinking. Mother has also had problems with alcohol and Mr. D. discovered that drinking functioned as a ritual between himself, mother, and sister. This ritual included events such as celebrations, as well as mutual commiseration.

(3) "Controlled" Drinking. Mr. D. initially chose the "controlled" drinking treatment option, since he recognized that drinking stimuli were ubiquitous in his environment. Self-monitoring records revealed the cyclical nature of the drinking, and the fact that money in the bank was always accompanied by sexual contact with women. After learning so much about his drinking behavior, he eventually opted for abstinence.

(4) Physical Symptoms. Drinking records and physical symptoms suggested the presence of hypoglycemia. Examination and laboratory tests revealed that Mr. D. was borderline hypoglycemic, and a proper diet was instituted. New cognitions replaced old ones, i.e., "This is not just anxiety or the need for a drink--this is a physical symptom and I can deal with it by eating a handful of peanuts and relaxing for a few minutes. The feeling will pass."

Relaxation training was introduced to control massive anxiety. Mr. D.'s cognition that he was feeling anxious led to immediate exacerbation of the anxiety. The relaxation training proved successful, and imagery was later used along with the relaxation for deconditioning purposes.

(5) Examining Distorted Cognitions About Women. Mr. D. at first thought that he "loved" women, and that this accounted for his promiscuity. Therapy led to the awareness that after casual sex the real message to himself was, "Well, I got over on that bitch." He was able to associate this with the fact that his mother has always manipulated him for her own purposes, and tried to infantilize him. Mr. D. was

projecting the hatred for his mother onto all women. He was also able to acknowledge other reasons for extramarital sex, which included: a) to spite his wife; b) to distract himself from anxiety; c) to prove that he could do something well when he was feeling low self esteem; d) a way of acting out his contempt for women; and e) to compensate for homosexual fears. Mr. D. recalled that he developed the "Mr. Hyde" part of his personality to overcome his fear of fighting back when somebody called him a "faggot" during his early teens.

(6) Other Distorted Cognitions. Homework assignments and tests were used as part of the educational process involved in helping Mr. D. look at his distorted cognitions. This enabled him to recognize that the technology for behavioral change was under *his* control, and not some magical knowledge or quality possessed by the therapist. For example, vocational/educational counseling enabled Mr. D. to deal with his cognition that his wife was superior to him because she had completed high school and was engaged in further training. Similarly, written assignments helped to provide delays in regard to impulsive behavior, and to gain "distancing" from his own behavior so that he could objectively understand and evaluate it.

RESULTS

(1) The marriage is stronger, and Mr. and Mrs. D. feel more secure about their future together.

(2) Mr. D.'s daughter is doing very well, and he is proud of the fact that he contributed to her recovery.

(3) Mr. D. has had only two binges in two years, and each provided him with a learning experience.

(4) Mr. D. has a much healthier relationship with his mother and he is able to confront her directly with negative feelings.

(5) Mr. D. is unashamedly able to ask for help when he needs it, instead of trying to pretend that everything is all right and ending up on an extended binge.

(6) Mr. D. has made significant gains in controlling his impulsive behavior.

(7) Anxiety is under complete control and no longer poses a problem.

(8) Mr. D. has been working steadily, and he has been able to pay off many of his back debts.

SUMMARY

This paper has presented a discussion of the theory and practice of broad spectrum behavioral therapy in the treatment of substance abuse. An attempt has been made to define the ways in which cognitive distortions can lead to addiction, and prolong the addictive cycle.

Common cognitive distortions which are often shared by both clients and therapists have been noted, and emphasis has been placed on the importance of correcting these faulty cognitions. A case study was then presented to illustrate the use of these techniques.

Addiction is a complex disorder characterized by frequent relapse. The frustrations inherent in working with the addicted often stem from feelings of futility due to limited and ineffective treatment options. It has been the experience of the authors that the techniques discussed in this paper help provide viable and effective treatment alternatives to the more traditional psychotherapeutic approaches. It is hoped that other clinicians will report similar results.

REFERENCES

Beck, A.T. 1976. *Cognitive Therapy and the Emotional Disorders*. New York: International Universities Press.

Beck, A.T. 1970. Cognitive Therapy: Nature and relation to behavior therapy. *Behavior Therapy*. 1: 184-200.

Beck, A.T. Measuring depression: The depression inventory. In T.A. Williams, M.M. Katz and J.A. Shield (eds.), *Recent Advances in the Psychology of the Depressive Illnesses*.

Beck, A.T., Weissman, A., Lester, D. and Trexler, L. 1974. The Measurement of Pessimism: The Hopelessness Scale. *Journal of Consulting and Clinical Psychology*. 42: 861-865.

Callner, D.A. 1975. Behavioral treatment approaches to drug abuse: A critical review of the research. *Psychological Bulletin*. 82: 143-164.

Droppa, D.C. 1973. Behavioral treatment of drug addiction: A review and analysis. *International Journal of the Addictions*. 8: 143-161.

Jacobson, E. 1938. *Progressive Relaxation*. Chicago: University of Chicago Press.

Khantzian, E.J., Mack, J.E. and Schatzberg, A.F. 1974. Heroin use as an attempt to cope: clinical observations. *American Journal of Psychiatry*. 131: 160-164.

Kovacs, M. Beck, A.T. and Weissman, A. 1975. Hopelessness: An Indicator of suicidal risk. *Suicide*. 5: 98-103.

Miller, P.M. 1973. Behavioral treatment of drug addiction: a review. *International Journal of the Addictions*. 8: 511-519.

Shaw, B.F. and Beck, A.T. The treatment of depression with cognitive therapy. In A. Ellis and R. Grieger (eds.), *Handbook of Rational Emotive Theory and Practice*. New York: Springer Publishing Co. In press.

Taintor, Z. and D'Amanda, C. 1973. Multiple drug abuse in heroin addicts receiving outpatient detoxification. *Proceedings of the Fifth National Conference on Methadone Treatment*. New York. 1003-1009.

Wolpe, J. 1965. Conditional inhibition of craving in drug addiction: a pilot experiment. *Behavior Research and Therapy*. 2: 285-288.

Wurmser, L. 1972. Methadone and the craving for narcotics: observations of patients on methadone maintenance in psychotherapy. *Proceedings of the Fourth National Conference on Methadone Treatment*. San Francisco. 525-528.

BIOFEEDBACK AS AN ADJUNCT TO TREATMENT WITH HEROIN ADDICTS

Royal H. Tyler, M.A. and Mark Disorbio, M.A.

Project Reality

Salt Lake City, Utah

Drug addiction continues to be very difficult to treat effectively. Therefore, a continuous flow of new ideas is useful for generating and exploring alternative approaches. One treatment method which may have interest arose from a combination of factors.

Initially, interest was generated from the response of clients in a private clinical setting who repeatedly responded to their increased ability to control some physiolgical functions they had previously believed to be autonomous. This process was the focal point for the initiation of a relaxation therapy group. Volunteer drug addicted clients were trained in relaxation techniques, basically autogenic training. Although no systematic physiological monitoring was conducted on these clients, there was a consistent positive subjective response, which centered around the clients' increased ability to put their body at rest. Clients also reported a remission in many sleep disturbances.

The combination of the positive acceptance of the relaxation training and response of clients in the private clinical setting generated the idea of using biofeedback equipment with the addicted population. Another factor suggesting this approach was the intrinsic interest shown in the visual display of the biofeedback equipment itself.

The use of biofeedback has been shown to be a useful treatment approach for a variety of psycho-physiological disorders, such as migraine headaches, cluster headaches, and Raynaud's disease (Green and Green, 1977; McQuade and Aikman 1974; Alder, 1976; Stroebel, 1976). This then raises the issue of the rationale for the use of biofeedback with the drug addicted population. There are several reasons why biofeedback training may be an effective treatment adjunct for drug addicts. While addicted to heroin, abusers have responsibility for medicating themselves, which gives them at least some measure of self control over their lives. At the beginning, and during treatment, clients express fear of losing this control because of the program's need to regulate medication as well as other regulatory functions. Biofeedback training gives an individual a degree of control over some physiologic states, which is reinforcing in terms of regaining a sense of control. Second, addicts focus heavily on somatic complaints and respond to these complaints with medication. This response mode contributes to the complexity of the addiction problem. Biofeedback training provides the drug abuser an alternative response. The successful use of the biofeedback training reduces the general stress level which may help to limit any magnification of physiologic responses. Third, one common aspect reported by addicts during treatment is a general increase in nervousness and sleeplessness. This appears to be especially true during periods of detoxification. Physically this may be due to the removal of the constant state of central nervous depression which occurs during heroin or methadone addiction. Biofeedback training provides a systematic way to deal with these arousal problems. The activation of parasympathetic activity, characteristic of successful biofeedback training, should act to improve sleep problems as well as general nervousness. Fourth, many clients report that their drug use is in response to anxiety and nervousness which they experience in social interpersonal situations. Teaching addicts to use their biofeedback training to reduce at least the concimmittant physiological response levels in these situations may decrease their desire to respond by medicating. The specificity of this situational anxiety is perhaps less pertinent to addicts alone than the other reasons. It does, however, represent a repeated theme in clinical interventions. With this philosophical rationale and the experience mentioned earlier, the trial use of biofeedback with an addicted population appears to represent a viable approach.

Biofeedback as a treatment modality might be employed in several manners. In either a methadone or drug free setting, biofeedback training can be used as the main form of therapy. Clients involved in such a regimen would participate in one or two treatment sessions per week. The biofeedback sessions would last a maximum of two months with periodic refresher sessions available.

More effective might be the employment of biofeedback as an adjunct to other treatment modalities. Once again this may be employed in either methadone or drug free situations. The regimen would include biofeedback training as part of a larger therapy package. Clients may be involved in individual group therapies in addition to the biofeedback. One advantage such a multi-modality approach may have, is to engage clients more quickly by responding to their individual needs. The biofeedback training itself also has inherent advantages over some more traditional approaches.

Biofeedback training produces some changes which the client can experience immediately. Most persons begin to obtain at least some degree of control in the first session. This, coupled with the intrinsic interest in the biofeedback equipment itself, tends to engage the client very quickly. This is a critical aspect of any treatment program due to the difficulty that is often experienced in attempting to engage clients at the onset of their treatment.

The use of biofeedback is certainly not viewed as a panacea for drug treatment. It does appear to have some advantages, as previously mentioned, over more traditional modalities. It should be emphasized that it is most useful as part of a more global treatment system.

The actual biofeedback training that is envisioned would consist of a variety of systems. Clients involved in such a system would receive training on two biofeedback instruments. The first training would take place on a biofeedback temperature monitor and the second on a biofeedback EMG (electromyograph) monitor. Temperature monitoring appears to be a reflector of autogenic nervous system activity and EMG of skeletal muscle tension. Together they give some idea of how the body is responding and the manipulation of these functions by the client gives a sense of self-control.

These two response modes are selected for several reasons. The instruments used for these modes tend to be very sensitive and therefore reflect small changes. This is very important for the client to be able to see, as control may come in small increments initially. EMG and temperature are selected as they produce the most accurate and documented feedback in both the autonomic and muscle feedback systems.

The biofeedback training or therapy sessions herein described are not limited to only the attempts to gain physiologic control. Clinical experience indicates that much fruitful information surfaces during these sessions. Therefore the biofeedback technician, who also should be a trained counselor, will watch for and encourage the client to elaborate on these areas of concern. This focus, or allowing of the free flow of these cognitive concerns, becomes very important for the successfulness of the biofeedback training. As the client begins to have awareness of the areas that seem to be related to physiological responses, especially those that are more extreme, he can examine the cognitive processes involved. The therapist must help the client look at these concommitant internal dialogues if the training is to have impact on the client's everyday behavior. To the extent the client is able to improve his day to day functioning in interpersonal situations the process is a success. This improved functioning may also be important for reducing the dependence on drugs, which these clients use as a means of coping. The aim is for the client not only to be able to control his physiological responses without external aid but to recognize these cognitive ruminations which appear to be part of the impaired functioning process.

There have been very few attempts to employ biofeedback training in the treatment regimen of drug addicted clients and even fewer controlled studies involving its use. Some investigators have attempted to use biofeedback with a wider abuse population including alcoholics. Dr. Robert Schmitz, for example, believes that biofeedback training may be a valuable tool for the treatment of alcoholics (Schmitz, 1966). He cites two effects of biofeedback which may be helpful: 1) the use of biofeedback training to reduce the high tension levels experienced by alcoholics during detoxification, and 2) the re-establishment of the patient's sense of identity and control over his environment as well as himself.

Dr. Paul Kurtz, director of the Alcoholism and Chemical Addiction Unit of St. Cloud Hospital, Minnesota, has expressed support for the use of biofeedback techniques with drug addicts (Kurtz, 1973). Dr. Kurtz believes that the use of biofeedback allows the addict to gain a sense of independence and self control necessary for rehabilitation.

Barbara Schneider of the Chemical Dependence Unit of Gladman Hospital, Berkeley, California, supports the use of biofeedback with heroin addicts. Ms. Schneider, (by phone on 10-27-77) who has treated addicts with biofeedback training for several years, believes that this approach offers the client a sense of power and self-assurance which becomes a positive part of long term rehabilitation.

A less promising research study was conducted by Cohen, et al. (1977) at the Mid-West Research Institute of Kansas City, Missouri. Although Cohen's research was more specifically designed to investigate double-blind methodology, he did use biofeedback with addicts and found no difference in outcome under such conditions.

It is somewhat unclear as to why there is a lack of research data in this area. It may be it is a rather novel approach, or that drug treatment centers have limited budgets and cannot afford the necessary equipment, or even a general lack of enchantment with the addicted population itself. However, this still is a viable research approach with heroin addicts.

REFERENCES

Alder, C. 1976. Biofeedback and psychotherapy in the treatment of combined tension, migraine headache syndrome. *Handbook of Physiological Feedback*. Berkeley: Autogenic Systems, Inc. Vol. II.

Cohen, H., Graham, C., Fotopoulos, S., Cook, M. A double blind methodology for biofeedback research. *Psychophysiology*. Vol. 14, No. 6.

Green, E. and Green, A. 1977. *Beyond Biofeedback*. San Francisco: Delacorte Press.

Kurtz, P. 1973. Turning on without chemicals. *Journal of Biofeedback*. II: 88-103.

McQuade, W. and Aikman, A. 1974. *Stress*. New York: Bantam Books.

Schmitz, R. 1976. "Relaxation Training." A paper presented at the Sixth Annual Kentucky School of Alcohol Studies.

Stroebel, C. 1976. EMG and thermal feedback in the treatment of a complicated case of chronic recurrent migraines. *Handbook of Physiological Feedback. II*. Berkeley: Autogenic Systems, Inc.

Stroebel, C. Biofeedback in the treatment of primary idiopathic Raynaud's disease. IBID.

PATIENT SELF-REGULATION OF METHADONE DOSAGE: A TWO-YEAR PRELIMINARY REPORT

Carl I. Thistel, A.C.S.W., William E. Abramson, M.D., F.A.P.A., and Mary Louise Miller, R.N.

Comprehensive Drug Abuse Program (COMDAP)

The Sheppard and Enoch Pratt Hospital

The Sheppard and Enoch Pratt Hospital is a 288-bed private, not-for-profit, psychiatric hospital located in Towson, Maryland. The Comprehensive Drug Abuse Program (COMDAP) is one outcome of the hospital's expansion from the traditional psychiatric inpatient service, into the broader approach of provision of a range of community-oriented mental health services, and as such is an outpatient program with a capacity of 145 patients.

Our interest in a patient regulation of dose policy dates back to the National Drug Abuse Conference in 1974, when Dr. Avram Goldstein and co-authors presented the results of the first comprehensive study in this area. Early in 1975, we designed a six-month pilot study aimed at partial replication of the Goldstein model. Our own study was for a six-month period from June through December 1975. The departure of the principal investigator before the study was completed did not permit a full presentation of findings. However, we felt that the data we had collected was of sufficient validity to warrant a basic program change, implementing the policy of patient regulation of dosage for all methadone maintenance patients in our program effective June 1976 to the present. What we are presenting here is a two-year preliminary report based on our experience to date, with reference to four basic assumptions underpinning both our original study in 1975 and the patient regulation of dosage policy now in effect: (1) that given the option to determine their own methadone dose, maintenance patients will stabilize themselves along a wide spectrum of dose levels and not cluster around the maximum dose permitted; (2) that a self-determination of dose policy would liberate the counseling relationship from preoccupation with methadone, and would permit greater focus on specific treatment objectives; (3) that the rate of patient-initiated long-term detoxification under a self-determination of dose policy would at least equal the rate of patient-initiated detoxification under a physician-determined dose policy; and (4) that the rate of retention in treatment of new admissions would significantly increase under a patient self-determination of dose policy.

LITERATURE REVIEW

The clinical value of dosage self-regulation in drug abuse treatment is now amply documented in the literature. Three pioneer studies (Wikler, 1952; Aldrich, 1969; Schuster, Smith, and Jaffe, 1971) provided the first evidence that narcotic patients could be permitted to regulate their own dosages without dire consequences, so long as adequate medical safeguards were in place. Renault (1973), reporting on a study to evaluate the effects of patients knowing their dosage, concluded that, contrary to staff expectations, dosage knowledge did not promote competition for higher dosages. Other unpredicted findings in this study were that knowledge of dosage enhanced rather than impaired the therapeutic relationship, and did not result in an escalation of somatic complaints.

Of special relevance to the widely-expressed concern within the field regarding the high dosage selections presumed to be the inevitable outcome of any dosage self-regulation policy, were the dosage comparison studies of Goldstein et. al (1971, 1972, and 1973). These studies established no significant correlations between dosage level and treatment performance; highly important findings with respect to a self-regulating protocol. One year later, Goldstein et al. (1974) presented their landmark paper at the NDAC entitled "Control of Methadone Dosage by Patients". This study, which incorporated much of the valuable groundwork which predated it, has functioned as a basic frame of reference for many patient regulation of dosage studies since presented at the NDAC. In 1977 alone, four papers were presented. Burglass, Renner, and Greenfield reported the preliminary results of a study comparing the single blind and the patient-controlled dosing models used in the Boston Drug Treatment Program. The patient-controlled model yielded favorable results with regard to patient and staff acceptance, patient retention, frequency and outcome of detox attempts, and overall treatment performance. Scher, O'Connor, and Childers gave a favorable report on an ongoing policy in effect since 1972 at the Methadone Maintenance Institute of Chicago, where patients are routinely informed of their dosage and participate in any decisions regarding increase or decrease of medications. A study by Havassy (1977) in San Francisco comparing two self-regulation of dosage groups with a standard control group, showed that the self-regulation groups did opt for higher dosage levels but showed less evidence of supplementing their medication with other drugs.

In both the Havassy and Goldstein models the element of contingent take-homes was used as a counterincentive to high dosage selection, although neither study showed any significant correlation between dosage level and treatment performance.

The Fulwiler and Bortman study, also presented at the 1977 NDAC, found that self-regulation of dosage for patients in detoxification effectively involves the patient in his own treatment and inculcates feelings of responsibility for discontinuance of herion use.

DESCRIPTION OF POLICY

Consistent with the Goldstein model, the COMDAP patient regulation of dose policy (hereinafter referred to as PRD) has permitted COMDAP methadone maintenance patients, after the first 21 days in treatment, to increase or decrease their dose by 5 mg. per week up to a maximum of 80 mg. A departure from the Goldstein protocol is that no restriction with regard to take-home privilege has been placed on patients who have selected a dose above 50 mg. On a specified day each week, patients wishing to modify their dose submit a dosage change slip to the nurse. These changes are then recorded in the medical chart, reviewed and signed by the program physician. The patients are given to understand that self-regulation of dosage is a privilege which can be removed at medical discretion, when urine results or clinical observation suggest a dangerous level of illicit drug use. Patients with two or more consecutive clinic no-shows are temporarily removed from dosage self-regulation privilege as a safeguard against tolerance loss.

FINDINGS TO DATE

Assumption #1: Given the option to determine their own methadone dose, maintenance patients will stabilize themselves along a wide spectrum of dose levels and not cluster at the maximum dose permitted.

The data source examined with reference to this assumption was the dosage order sheets on all active methadone maintenance patients for the period covered, and eight quarterly reports submitted by the chief nurse over the two-year period. The comparison or representative months during the staff regulation of dose period (May 1975) and the patient regulation of dose period (March 1977) shows that in March 1977 45% of our patients stabilized themselves at a high-dose range (61-80 mg.); 38% at a medium-dose range (31-60 mg.); and 17% at a low-dose range (5-30 mg.). This compares with 26% high dose, 59% medium dose, and 15% low dose, during the staff regulation of dose month of May 1975 (see Table I).

Table I. Dose range comparison under Staff Regulated Dose policy (SRD) and Patient Regulated Dose policy (PRD) for 2 sample months (May 1975 and March 1977).

	SRD Policy May 1975		PRD Policy March 1977	
	# Pts.	%	# Pts.	%
Low Dose Range 5-30 mg.	12	15	19	17
Medium Dose Range 31-60 mg.	46	59	43	38
High Dose Range 61-80 mg.	20	26	50	45
TOTAL	78[1]	100	112[2]	100

[1] Census of MM Pts. May 1975.
[2] Census of MM Pts. March 1977.

The average dose under the PRD policy after 2 years was 62.5 mg. compared with 54.3 mg. average under the SRD policy (see Table II).

Although these data depict a measurable preference for high dose by 45% of the population, this hardly bears out the prevailing concern of several years ago, to the effect that methadone maintenance patients granted a self-regulation of dose privilege would promptly escalate to the maximum dose allowable, without built-in incentives to keep the dose below 50 mg. To the contrary, our experience to date shows that 55% of the methadone maintenance population have opted for medium or low dose, despite the absence of the incentive built into the original Goldstein protocol restricting take-home privilege to the medium- and low-dose patients at 50 mg. or less.

From July through December 1977, under the PRD policy, 12.9% of the patients increased their dosage, 5.4% decreased their dosage, 61.3% did not change and 20.4% varied their dosage in both directions (see Table III). Those who selected a higher dosage were slightly older, had a longer history of drug abuse, and were more stabilized in treatment as evidenced by a higher number of earned take-home days (see Table IV).

In any one week over the period, an average of less than 10% of the population submitted change requests. We had anticipated a far greater volume of such changes, on the basis of the high degree of manifest enthusiasm with which the patients greeted the announcement of a self-regulation policy.

Table II. Methadone dosage averages 1975-1977.

	SRD period[1]	SRD period based on Quarterly reports							
	5/75	6/1/76[2]	9/1/76	12/1/76	3/1/77	6/1/77	9/1/77	12/1/77	3/1/78
Average dose all MM pts.	54.3	53.3	56.3	56.1	58.4	56.8	59.9	60.7	62.5
Average dose of MM Pts. with TH med.	60.1	52.4	69.3	58.2	60.8	61.4	61.3	61.2	63.2
MM census	78	96	97	101	103	105	104	116	113

1 Baseline month prior to 6 month study period of 6/1/75 through 12/1/75.

2 Beginning of PRD policy.

Table III. Patterns of dosage changes for 93 methadone maintenance patients under a self-regulation of dosage policy from July 1977 through December 1977.

Patient Group[1]	# Patients in group	% of total	# Meds Changes	% of all Meds Changes
Increasers	12	12.9	60	28.4
Decreasers	5	5.4	21	10.0
Constants	57	61.3	7	3.3
Variables	19	20.4	123	58.3
TOTAL	93	100.0	211	100.0

[1] Groupings derived from the Goldstein study presented in 1974.
Increasers = those whose dosage was 5 mg. higher than starting dosage.
Decreasers = those whose dosage was 5 mg. lower than starting dosage.
Constants = those who did not make more than 1-5 mg. change during period.
Variables = those who made multiple change in both directions.

Assumption #2: A self-determination of dose policy would liberate the counseling relationship from patient preoccupation with methadone, and a corollary assumption that a counseling relationship unencumbered by dose level concerns would permit greater focus on specific and measurable treatment and rehabilitation objectives.

That the PRD policy has, in fact, "liberated" the counseling relationship and permitted massive redirection of clinical time, is an assumption about which virtually no COMDAP staff member now has the slightest doubt. Staff attitudes which were somewhat negative about the policy in the beginning, are now uniformly positive. It has, indeed, removed a major battleground in treatment, and freed up significant amounts of time to address issues more pertinent to basic treatment objectives such as employment, stability of primary relationships, reduction of illicit drug use, etc. Supporting these staff impressions with data, however, has presented some formidable problems. In hard fact, we know that staff-processed medication change requests numbered 73 in the calendar year 1974, in contrast to 11 such changes in the year 1977, reflecting a sizeable 564% reduction. Quantifying these changes with reference to staff time utilization, however, will need to be addressed in a more detailed report. Assuming, however, that the PRD policy has indeed liberated much staff time from preoccupation with the chemical aspects of treatment, to what extent has this time been diverted to more constructive and measurable objectives related to basic lifestyle changes? The data

Table IV. Comparative Data: Age, length of drug history, days of take-home, and dosage level of methadone maintenance patients under the COMDAP self-regulation of dosage policy.

Variable		High dosage[1]	Low dosage[1]
Age on admission (in years)			
Total N		40.00	34.00
Mean		31.00	28.29
Std. Deviation		5.09	4.18
Std. Error		.80	.72
	T=2.52	P<.01	
Length of drug history (in years)			
Total N		40.00	34.00
Mean		11.15	9.97
Std. Deviation		5.16	4.01
Std. Error		.82	.69
Days of take-home			
Total N		40.00	34.00
Mean		2.50	1.53
Std. Deviation		1.50	1.65
Std. Error		.24	.28
	T=2.63	P < .01	

1 High dosage=61-80 mg.
Low dosage = 0-60 mg.

presented in Table V suggest a positive correlation between the PRD policy and 5 of the 9 parameters by which treatment is evaluated in COMDAP (see Treatment Evaluation Parameters appended), although the N we used was much too small to establish significance levels (see Table V). Probable gains in treatment performance parameters were suggested in scores for frequency of illicit drug use, compliance with program rules, regularity in clinic appointments, as well as regularity in fee payments.

Assumption #3: The rate of patient-initiated long-term detoxification under the patient regulation of dosage policy would at least equal the rate of patient-initiated detoxification under the staff regulation of dosage policy.

The same sample groups were used for evaluating this assumption as with Assumption #2, namely, 25 admissions randomly selected for a

Table V. Comparative treatment performance scores of first admission methadone maintenance patients under staff regulation and patient regulation of dosage policies.

COMDAP treatment evaluation parameters	Scores during 7th, 8th, & 9th month following admission to Meth. Maint. SRD N=9[1]	PRD N=18[2]
1. Drug use	2.15	1.56
2. Compliance with program rules	1.41	1.11
3. Regularity and promptness in clinic appointments	1.85	1.67
4. Motivation for therapy (in context of need)	1.85	1.85
5. Quality of participation in therapy	1.63	1.61
6. Regularity in fee payments	1.26	1.15
7. Stability of primary relationships	1.38	1.46
8. Arrest/conviction rate	1.00	1.02
9a. Self support or 9b. School and/or vocational training	1.37	1.37
Means for all 9 parameters	1.54	1.42

[1] Randomized sample of admissions for period 1/74 through 3/75 under staff regulation of dosage policy.
[2] Randomized sample of admissions for period 6/76 through 8/77 under patient regulation of dosage policy

15-month period during the SRD period, and 25 new admissions selected from a comparable 15-month period under the PRD period. Among the 50 patients comprising the two samples, 12% initiated long-term detoxification requests under the SRD policy, and 28% initiated such requests under the PRD policy, reflecting an increase of 133%. It should be noted, however, that COMDAP stresses long-term treatment for methadone maintenance patients, and uses 18-24 months as a bench mark for considering detoxification. The 15-month period from which these samples were drawn is not an adequate time frame for evaluating the effectiveness of the PRD policy, with respect to the long-range goal of discharge in a drug-free detoxified state. An additional year of program experience under the PRD policy will be necessary to evaluate patient-initiated and successfully completed detoxification. On the basis of the data we have thus far, however, there is certainly some indication that the frequency of long-term detoxification initiated by patients under the PRD policy is at least as high if not significantly higher than the frequency of detoxification under the SRD policy.

Assumption #4: The rate of retention in treatment of new admissions (i.e., those active in treatment 6 months after admission) would be significantly higher under the PRD policy.

Retention data were derived from the two sample groups of 25 patients from which evaluative data on Assumptions #2 and #3 were derived. At the end of 6 months following admission, 15 patients from the SRD group were still in treatment compared with 20 patients still active after 6 months among the PRD group. Significance levels here will need to be determined in a subsequent evaluative report.

SUMMARY OF FINDINGS

Given the option to self-regulate their dose, COMDAP methadone maintenance patients did not promptly escalate to the maximum allowable dose. After nearly a two-year period, 55% of the methadone maintenance population had opted for dosages of 50 mg. or below, in the absence of any counterincentives to high dose such as take-home medication restriction. Staff time formerly invested in dose-related interactions with patients has, we believe, been redirected to address more fundamental and long-range treatment objectives. Patients under the dose self-regulation policy show measurable gains in treatment performance, in comparison to patients whose performance was assessed under the staff regulation of dosage policy.

Although we have been unable to establish a statistical comparison of successfully completed detoxifications between the two groups, it is noteworthy that the number of patient-initiated long-term detoxifications has been higher under the patient regulation of dosage policy. Perhaps our most significant finding in comparing the two groups, is that those under the PRD policy are more likely to remain in treatment during that critical period following the first six months after admission.

SOME PHILOSOPHICAL OBSERVATIONS

Having now observed a variety of patients during the two-year period of our policy of patient regulation of methadone dose, we have made a number of observations related to patient change and patient growth under this policy. These are observations and impressions, rather than quantifiable data, but they lend a great deal of support and evidence for the usefulness of this policy in the treatment and rehabilitation of narcotic addicts. Many of our patients are coerced into treatment, come to us reluctantly, or enter into the treatment situation under covert protest. As a treatment program, we represent both the "straight" society and establishment, and figures of authority, that most of our patients have come to distrust as part of their lifestyle as narcotic addicts. The policy of patient regulation of dose serves very well to overcome some of this patient resistance, by setting up a therapeutic alliance between the patient and the program. The patient becomes an active participant in his own treatment. This has a significant effect in reducing both patient resistance to therapy, and the various forms of passive-aggressive behavior often seen with the reluctant or coerced patient. It is our feeling that with this policy, we are also teaching and training our patients in developing responsibility for themselves, and adding the factor of self-determinism to their treatment. We feel that this also leads to the growth of autonomy and independence, which goes a long way toward helping our patients in the development of emotional maturity. The resistance of some patients to acceptance of autonomy and responsibility is demonstrated by anecdotes of patients resisting the actual writing out of the dosage change slip, presented to them by the nurse. Some patients have indicated a wish to reduce their dosage, but at the final moment have asked the nurse to do the actual writing out of the dosage change slip. To us, this typifies how foreign it is to some of our patients to take responsibility for themselves, and function as autonomous and mature individuals. The patient regulation of dose policy confronts this issue with them, and moves them into more mature patterns of behavior.

Under this policy, the patient becomes an involved active participant in his treatment, as contrasted with his being a passive recipient who can act out with passive-aggressive behavior. After years of hearing the typical statement of patients, "My methadone is not holding me," we have often felt that our patients are actually expressing their unconscious passive-dependent need to be "held" and made to feel secure, somewhat like the need of the child for being held by the mother. With this policy, the patient is now responsible for "holding" himself, in terms of his dosage, and maintaining his feeling of security is now in his own hands.

Another "street" legacy which our patients bring to us when they come into treatment is their established pattern of being mistrustful of others who have any influence or control over their lives (parents, police officers, parole and probation officers, courts and judges, etc.). With self-regulation of dose, the patient now has to learn to trust himself and his own judgment. He is less able to use this built-in mistrust, as a passive-aggressive mechanism for resisting or sabotaging or attacking the authority represented by the program and his counselor.

The final advantage of this policy is that it fits in quite well with the current emphasis in delivery of all health care on "consumerism." By being responsible for their methadone dose, our patients have to be more knowledgeable and educated and aware in their treatment, since they are now part of their own treatment. This helps to make them a more respectable patient population, since they are now more like all other patients, who are also being encouraged to be aware and knowledgeable about the treatment they receive, as consumers of the services of health care systems.

In summary, our overall impression is that the policy of patient regulation of methadone dose adds significantly to the difficult process of psycho-social rehabilitation of narcotic-addicted patients, in moving them toward a role as autonomous and self-responsible mature individuals.

REFERENCES

Aldrich, C.K. 1969. Experimental self-regulated readdiction to morphine. *Int. J. Addict.* 4: 461-470.

Angle, H.V., Knowles, R.R., Marrazzi, A.S., and Sletten, I.W. 1973. Method for self-administering methadone by heroin addicts. *Int. J. Addict.* 8: 435-441.

Angle, H.V. and Parwatikar, S. 1973. Methadone self-prescription by heroin addicts in an inpatient detoxification program. *Psychological Record.* 23: 309-314.

Burglass, M.E., Renner, J.A., Jr., and Greenfield, D.P. 1977. Comparison of dose regulation protocols in an outpatient methadone treatment program. *Proc-NDAC* (in preparation).

Fulwiler, R.I. and Bortman, R.A. 1977. Self-regulation of methadone dose in outpatient heroin detoxification. *Proc-NDAC* (in preparation).

Garbutt, G.D. and Goldstein, A. 1972. Blind comparison of three methadone maintenance dosages in 180 patients. Proc. Fourth Nat. Conf. on Methadone Treatment. *Nat. Assoc. Prevention of Addiction to Narcotics*, New York, pp. 411-414.

Goldstein, A. 1971. Blind controlled dosage comparisons in two hundred patients. Proc. Third Nat. Conf. on Methadone Treatment. *U.S. Public Health Service Pub. No. 2172,* U.S. Govt. Printing Office, Washington, D.C., p. 31.

Goldstein, A. and Brown, B.W., Jr. 1970. Urine testing schedules in methadone maintenance treatment of heroin addiction. *JAMA.* 214: 311-315.

Goldstein, A., Hanstren, R.W., Horns, W.M., and Rado, M. 1974. Control of methadone dosage by patients. *Proc. First NDAC*, Chicago.

Goldstein, A. and Judson, B.A. 1973. Efficacy and side effects of three widely different methadone doses. Proc. Fifth Nat. Conf.

on Methadone Treatment. *Nat. Assoc. Prevention of Addiction to Narcotics*. New York, pp. 21-40.

Havassy, B. 1977. Increased freedom and control for methadone maintained heroin addicts: self-regulation of dose and contingent take-homes. *Proc-NDAC* (in preparation).

Renault, P.F. 1973. Methadone maintenance--effect of knowledge of dosage. *Int. J. Addict.* 8: 41-47.

Scher, J., O'Connor, H., and Childers, B. 1977. Methadone maintenance: patient participant known dosage setting, dosage changing, and self-determined detoxification scheduling, a four year study. *Proc-NDAC* (in preparation).

Scher, J., Rice, H.M., and Baig, K. 1977. Methadone maintenace: a national comparative study of dosage, dirty urines and dropouts. *Proc-NDAC* (in preparation).

Schuster, C.R., Smith, B.B., and Jaffe, J.H. 1971. Drug abuse in heroin users. An experimental study of self-administration of methadone, codeine, and pentazocine. *Arch. Gen. Psychiat.* 24: 359-362.

Wikler, A. 1952. A psychodynamic study of patients during experimental self-regulated re-addiction to morphine. *Psychiat. Quart.* 26: 270-293.

APPENDIX

THE SHEPPARD AND ENOCH PRATT HOSPITAL

TREATMENT EVALUATION PARAMETERS – COMPREHENSIVE DRUG ABUSE PROGRAM (COMDAP)

White - COMDAP
Canary - Research

Name: ______ COMDAP No.: ______ Period of ______ to ______

Worker: ______ Circle: Admission - Termination - Other

Circle *one number*: At the present time, how motivated is the client to involve himself/herself in the treatment program? (On termination: How involved was the client in the treatment program?)

Not at all 1 - 2 - 3 - 4 - 5 - 6 - 7 Very

On admission, complete only items 1, 7, 8, and 9 below. At other times, complete all items:

PARAMETERS	RATING (check ✓)			
	1	2	3	4
1. Drug Use	____ No apparent use of non-prescription drugs	____ 1 or 2 self-reported or lab-reported +'s	____ 3 or 4 self-reported or lab-reported +'s	____ Apparent habitual use of non-prescription drugs.
2. Compliance with Program Rules	____ No recent infractions	____ Occasional minor infractions	____ One or more major infractions	____ Multiple infractions
3. Regularity and Promptness in Clinic appointments	____ Highly regular and prompt	____ Periodic no-shows with calls	____ Frequent no-shows with calls	____ 2 or more no-shows without calls
4. Motivation for Therapy (in context of need)	____ Optimal or sufficient	____ Good	____ Fair	____ Poor
5. Quality of Participation in Therapy	____ Appropriate Involvement	____ Mostly goal-directed with minor conscious resistance	____ Periodic manipulation and/or conscious resistance	____ Frequent manipulation and/or conscious resistance
6. Regularity in Fee Payments	____ Consistent	____ 1-2 no-pays with notification	____ 1-2 no-pays without notification	____ Chronically in arrears
7. Stability of Primary Relationships	____ Relatively Stable	____ Occasional upheavals	____ Cyclical upheavals	____ Tumultuous in constant flux
8. Arrest/Conviction Rate	____ No arrests, no convictions	____ Arrest (s) No conviction (s)	____ One or more arrests with conviction (s)	____ Multiple arrests and/or convictions
9a. Self Support *OR* 9b. School and/or Vocational Training	____ Gainfully and/or constructively employed *OR* ____ Enrolled and performing well	____ Periodically employed *OR* ____ Enrolled, marginal performance	____ Seldom employed; motivation questionable *OR* ____ Enrolled, poor performance	____ Not employed; clearly unmotivated *OR* ____ Not enrolled

881-796-174

AN ANALYSIS OF METHADONE PATIENTS INCARCERATED WHILE IN ACTIVE TREATMENT*

Peter L. Vogelson, M.A. and Lee Koenigsberg, M.B.A.

Community Treatment Foundation

New York, N.Y.

INTRODUCTION

The Community Treatment Foundation (CTF) is a not-for-profit Corporation established in 1973. Prior to the establishment of CTF, the organization operated under the aegis of Rockefeller University where it was referred to as the Methadone Information Center at Rockefeller University. The continuity of the Information Center was maintained as one division of the Foundation. Its central task is the registry and clearance of all methadone patients in New York State. The collection of certain identifying data for the purpose of admission screening and retention of this data for both active and inactive individuals has placed the organization in a unique position. We presently have data elements on approximately 85,000 distinct individuals of whom some 33,000 are presently in active treatment.

In 1973, the New York City Department of Health asked Dr. Vincent Dole of The Rockefeller University to establish a system for detoxification of all opiate addicts incarcerated in the city's prison system. The detoxification schedules for those individuals who were currently in methadone maintenance treatment had to be different from those people addicted to heroin or other opiates on the street. The Methadone Information Center instituted a verification procedure to determine if an incarcerated person was active on a program. In doing this, we

---

* This publication is based upon research activities supported through a grant from the New York State Office of Drug Abuse Services. The conclusions stated herein are not necessarily those of that office.

also obtained certain medical information for use by the prison health personnel in treating these people. Such information as present methadone dose, last date of pick-up, and ancillary medical problems are obtained and communicated to the prisons in order to facilitate good health care.

Many articles in the popular press and professional journals have discussed criminal behavior among methadone maintenance patients. The data presented has often been fragmentary, based on either anecdotal material or on a very small sample within one or two small programs. The data presented here encompasses all individuals arrested and incarcerated in New York City while awaiting arraignment or trial during 1977, who were verified as being actively in methadone maintenance treatment at the time of arrest. We are confident that our information encompasses virtually all such individuals. This viewpoint is based on the realization that upon verification of treatment the individual will be medicated differently than the street addict. In addition, patients generally want to notify their clinics of their incarceration in order to be kept on their active rolls and to avail themselves of legal services provided by clinics.

DATA PRESENTATION

The following tables and discussion present a composite picture of the New York City methadone patient who was incarcerated during 1977. Where appropriate, an attempt has been made to compare this data to that of the total New York treatment population.

Table I shows the number of arrests per individual. The 2,971 arrests involved 2,496 specific individuals with 16% of them having been arrested more than once. Many articles in the popular press attribute a great deal of criminal behavior to methadone patients but this arrest data does not seem to confirm this concept. Approximately 7.7% of the total methadone maintenance and methadone-to-abstinence patients in New York City engaged in criminal activity leading to incarceration for at least one night.

Table I. Distribution of arrests by frequency.

Arrest episodes	Number of patients	%	Number of arrests	%
1	2,108	(84.5%)	2,108	(70.9%)
2	320	(12.8%)	640	(21.5%)
3	52	(2.1%)	156	(5.3%)
4	13	(.5%)	52	(1.8%)
5	3	(.1%)	15	(.5%)
TOTAL	2,496	(100.0%)	2,971	(100.0%)

Table II. Demographics--Age distribution.

Age	Number Arrested	%	Total methadone* population N.Y.C.	%
Under 21	59	(2.4%)	489	(1.5%)
21-25	750	(30.0%)	8,099	(25.1%)
26-35	1,267	(50.8%)	16,475	(51.0%)
36-45	353	(14.1%)	5,039	(15.6%)
over 45	67	(2.7%)	2,205	(6.8%)
TOTAL	2,496	(100.0%)	32,307	(100.0%)

* as of 6/30/77

As shown in Table II, 32.4% of those arrested were under the age of 25 at the time of arrest. It is significant to note that this is disproportionate to this age group in the total MMTP population, where it is 26.6%.

A disproportionate share of the incarcerated patients are male as compared to the total population, as shown in Table III. This finding is not particularly surprising in light of the types of crimes typically perpetrated by males.

The ethnic distribution shows whites to be the smallest group and significantly lower than their representation in the total patient body. This is most probably a reflection of the sociological parameters of our society. Whites generally have an easier time obtaining employment and tend not to be arrested as readily as other groups.

Ethnicity data was not available for all patients included in the study. Table IV contains what we believe to be a representative sample of both the incarcerated and total population.

Table V provides a breakdown of the arrests by type of crime. Crime against property accounts for 45% of the total charges. Of particular interest is the fact that only 1% of the arrests pertain

Table III. Demographics--Distribution by sex.

Sex	Number arrested (%)	Total methadone population N.Y.C.	%	Arrested patients as a percent of total population
Male	2,116 (85%)	23,780	(74%)	9%
Female	380 (15%)	8,527	(26%)	4%
TOTAL	2,496 (100%)	32,307	(100%)	

Table IV. Demographics--Ethnicity distribution by sample data.

Ethnicity		Sample of arrested patients	(%)	Sample total methadone population N.Y.C.	(%)
White		45	(15%)	5,723	(29%)
Black		131	(44%)	7,703	(40%)
Hispanic		108	(37%)	5,770	(30%)
Other/Unknown		12	(4%)	171	(1%)
	TOTAL	296	(100%)	19,498	(100%)

to sale of methadone. Given the large anecdotal attention to methadone diversion, this low incidence rate is illuminating.

Table V. Analysis of charges.

	Number of arrests		(%)
A. Drug crimes			
--possession of a controlled substance	205		
--sale (other than methadone)	201		
--possession of paraphernalia	21		
--sale of methadone	30	457	(15%)
B. Crimes of violence		488	(16%)
C. Crimes against property		1,333	(45%)
D. Minor violations		266	(8%)
E. Other			
--warrants	226		
--prostitution	71		
--violation of probation or parole	58		
--gambling	18		
--child abuse	1		
--harboring a fugitive	1	375	(13%)
F. Charges unknown		92	(3%)
TOTAL ARRESTS		2,971	(100%)

Table VI. Treatment-related data: Distribution by time in treatment.

Length of treatment	Incidence of arrests	(%)	Total population*	(%)
Less than 6 months	807	(27.2%)	3,088	(9.2%)
6-12 months	510	(17.2%)	2,630	(7.8%)
1-2 years	624	(21.0%)	5,056	(15.0%)
2 years or more	1,030	(34.6%)	22,961	(68.0%)
TOTAL	2,971	(100.0%)	33,735	(100.0%)

* as of 12/30/77

Table VI reveals that 44.4% of the incarcerated group were in treatment less than one year, whereas only 17% of the total methadone population falls into this category. The disparity is even more striking in the comparison of those in treatment less than six months. This data provides clear evidence that the likelihood of arrest decreased the longer a patient remains in treatment.

The dosage analysis presented in Table VII suggests that those on high, or blockade, doses, are less likely to engage in criminal activity. Only 8.3% of the arrested population were receiving 80 or more milligrams per day, as compared to 29.4% of the larger sample population. A further analysis showed no relationships between dosage level and types of crimes.

SUMMARY AND CONCLUSIONS

While this study does not purport to be an exhaustive survey of methadone patient crime, it represents a comprehensive analysis of those crimes in which arrests resulting in incarceration occurred in

Table VII. Treatment-related data: Distribution by dosage level.

Dose level	Arrested N=2496	Sample of total population* N=10,004
Less than 40 mg.	32.0%	24.8%
40-79 mg.	58.0%	45.8%
80 mg. or more	8.3%	29.4%
Unknown	1.7%	
	100.0%	100.0%

* as of 12/30/77

New York City during 1977. The most significant facts revealed in the analysis are:

--7.7% of the patient population are involved in this incarcerated group.

--Males and individuals under the age of 25 are overrepresented in the incarcerated group, while whites are underrepresented.

--Methadone sales accounted for only 1% of the arrests, while 45% related to crimes against property.

--People in treatment less than one year accounted for 44% of the arrests and those on high dosages were very underrepresented in the arrested group.

--There were no significant relationships between the types of crimes committed and the dosage level of the individual.

We recognize that not all criminal activity results in arrests and further, that all arrests do not lead to incarceration. However, the data upon which this study is based is of enough significance to support some overall conclusions.

The data clearly suggests that methadone patients exhibit far less criminal behavior than is generally attributed to this group, particularly by the popular media. The evidence relating to the low incidence of arrests for the sale of methadone should give rise to major questions regarding the promulgation of treatment policies which place too much emphasis on the fear of diversion.

We believe that the phenomenon of patient criminality requires more in-depth study, particularly because so much of governmental policy is framed in this context. It is hoped that future policy planning will be influenced by the findings of this and other studies.

SEX ROLE IDENTIFICATION OF DRUG ADDICTS

Robert W. Weiss, Ed.D. and Don Russakoff, M.B.A.

Samaritan Halfway Society

Societal norms regarding the differential appropriateness of various behavior patterns and psychological characteristics for each sex are learned at an early age, so that by age six the interests, activities, and attitudes of boys and girls diverge substantially (Guttentag and Bray, 1976). Generally, masculinity is associated with a task orientation - a focus on getting things done; while femininity is associated with an orientation toward emotional expressiveness and concern for others (Bem, 1974). In American society, it is more desirable and appropriate for a man to act as a leader, be assertive and analytical; it is more desirable for women to be affectionate, understanding and compassionate.

The concept 'sex-role identity' relates to whether a person has internalized the socially desirable sex roles for one or both sexes. Thus, a male describing himself in masculine terms and avoiding characteristics appropriate for the opposite sex would have a 'sex-typed' role identity. If his description of himself is predominantly in terms of feminine rather than masculine characteristics, he would have a 'sex-reversed' role identity. Similarly with women: those ascribing mostly to the appropriate characteristics of females are 'sex-typed,' and those adhering primarily to masculine characteristics are 'sex-reversed.' Within both sexes, there are people who are neither in the sex-typed or sex-reversed categories. They do not adhere to the standard sex role for either sex, but have flexible self-concepts containing neither a predominance of masculine nor feminine traits. They describe themselves using the preferred characteristics for both sexes equally, and are, therefore, considered to have an androgynous sex-role identity.

In recent years, there has been increased interest in sex-role identity research. This is probably due, in part, to the growth of the women's rights movement. There has been a concern with the extent to which the stereotyped female sex-role leads to discrimination against women, with a concomitant effort to reduce the divergence between approved male and female societal roles. There is some evidence that people who have either a masculine or feminine sex-role identity restrict their range of behaviors, thereby reducing their capacity to effectively adapt to varying environmental demands. Those who are androgynous are more likely to display a wide range of adaptable behaviors across situations (Bem, 1975).

Sex differences in behavior are determined by a combination of biological predispositions and social learning. Since parents are an important source for learning sex roles (Maccoby and Jacklin, 1974), there are reasons to believe that sex-role identity distributions for drug-abusers may differ from the general population, and that this variable may also relate to the motivation for drug-taking behavior. In a summary of numerous studies examining the family of the addict (Salmon and Salmon, 1977), a general finding was that there is a lack of strong father relationships, with the father either a detached figure or absent from the household. Addicts tend to be strongly attached to the mother, who is likely to be the dominant figure in the family. Because the mother is more available as well as more powerful, her behavior rather than the father's behavior, would more likely be modelled by the child. Therefore, drug-abusers of both sexes may have sex-role identities that are in a more feminine direction in comparison with groups characterized by a cohesive family and dominant father. In addition, sex-role identity may be a determinant of motivation for drug use among males, with the more feminine-typed males more likely to use drugs to overcome passivity.

An exploratory study of sex-role identity among drug-abusers was conducted by Samaritan Halfway Society; a multi-modality agency serving the New York City area. Samaritan has both drug-free and methadone-to-abstinence programs available in residential and ambulatory environments. The current clients as well as the new admissions over a 5-month period were administered the Bem Sex-Role Inventory (Bem, 1974). This questionnaire contains twenty personality characteristics that have been judged to be more desirable for a man than a woman, and twenty characteristics judged more desirable for a woman than a man. (The twenty items measuring social desirability response set were omitted, and there were some synonym additions to the presented items to make them more comprehensible for clients with limited vocabulary.) Clients indicated the extent to which each of the forty characteristics described themselves. Both a Masculinity and a Femininity score were obtained. The sex-role identity score used in the data analysis is the *difference* between the average Masculinity and Femininity scores, divided by a term reflecting the amount of response variation for the individual items (Student's t).

In addition to the Bem Inventory, clients answered questions relating to: 1) the extent to which they use drugs for various reasons, such as relief from anxiety, to get high, etc.; 2) the composition of their families; 3) attitudes toward each parent and each sex; and 4) basic demographic factors such as age, ethnic group, religion. For most current clients, MMPI results were available from a prior administration and selected scales were included in the analysis.

A total of 246 clients are in the sample: 203 males and 46 females. This ratio reflects the preponderance of males entering Samaritan programs. The data analysis consisted of: 1) A categorization of sex-role identity scores into five categories - Masculine, Near Masculine, Androgynous, Near Feminine, Feminine - based upon the criteria suggested by Bem (1974). This distribution for clients was compared to a normative sample of undergraduate students (Bem, 1976). 2) Correlations between the sex-role identity scores for males and the other data. The raw continuous scores for sex-identity were correlated with each of the other variables using Pearson correlation, point biserial correlation, or F - scores as appropriate.

A summary of the results follows:

1. Most female drug-abusers are sex-typed; a minority are androgynous, and very few are sex-reversed. As predicted, they are more often sex-typed than the normative sample. (See Table I.)

2. About one-half of male drug-abusers are sex-typed; a smaller number are androgynous and relatively few are sex-reversed. Contrary to predictions, they are not less often sex-typed or more often sex-reversed than the normative sample. In fact, differences in sex-role identity distributions for the groups are minor. (See Table I.)

3. Male drug-abusers report they use drugs most often for hedonistic reasons - to get high and feel good. The second most frequent reason for drug use is relief from aversive stimuli - to get away from feelings of anxiety, stop worrying. Drug use for social reasons or to create an image is not as frequent as the other reasons. Sex-role identity scores for male drug-abusers are related to the extent of reported drug use for social reasons and for relief from aversive stimuli. Clients with sex-role identities in a more feminine direction are more likely to use drugs to be sociable and to relieve anxiety and worrying. (See Table II.)

4. Male drug-abusers prefer and feel closest to their mothers rather than fathers by a 4:1 ratio, although they rate each

Table I. Comparison of Sex Identity Scores of Drug Abusers With Bem's Normative Sample

	Drug Abusers (N=203)	Normative Sample (N=375)*	Drug Abusers (N=23)	Normative Sample (N=290)*
Feminine	2%	8%	37%	20%
Near Feminine	10%	11%	28%	18%
Androgynous	39%	30%	30%	41%
Near Masculine	24%	21%	5%	13%
Masculine	25%	30%	0%	9%

* Stanford University undergraduates.

Table II. Percentage Distributions of Response Relating to Reasons for Drug Use (N=203 Males)

	Never or Almost Never True	Usually Not True	Sometimes But Infrequently True	Occasionally True	Often True	Usually True	Always or Almost Always True
Go Along With My Friends, Be Sociable: (1)	20.1%	13.8	10.6	15.9	15.9	11.1	12.2
Get Away From Feelings Of Anxiety, Nervousness: (2)	13.2%	6.3	6.8	14.7	16.3	20.5	22.1
Impress People With My Ability To Use Drugs: (5)	42.3%	21.2	8.5	9.5	7.9	5.8	4.8
Get High: (5)	11.1%	2.1	3.2	4.7	10.0	18.9	50.0
Be Like My Friends: (3)	30.9%	23.4	14.4	10.1	8.0	5.3	8.0
Stop Worrying About Myself: (4)	20.8%	15.3	6.0	12.0	14.8	14.8	16.4
Create An Image Of Myself That I Like: (5)	23.1%	16.5	9.3	11.0	12.6	11.5	15.9
Feel Very Good: (5)	6.5%	5.9	7.0	7.5	14.0	17.2	42.0
Stop Aches And Pains: (5)	25.3%	11.5	9.9	12.6	9.9	12.6	18.1

(1) Clients with sex-role identity score in more feminine direction would more often use drugs to be sociable: Pearson $r=.20, p<.01$

(2) " " " " " " " " " " " " " " " " relieve anxiety: Pearson $r=.19, p<.01$

(3) " " " " " " " " " " " " " " " " be like friends: Pearson $r=.22, p<.01$

(4) " " " " " " " " " " " " " " " " stop worrying: Pearson $r=.21, p<.01$

(5) No significant relationship between sex-role identity score and this item.

Table III. Percentage Distributions for Personal Background and Family History Variables (N=203 Males)

Variable		%
PEOPLE LIVED WITH MOST OF TIME UP TO AGE 12:		
Both parents	-	67.5%
Mother alone	-	21.0
Father alone	-	3.0
Other relatives	-	4.5
Foster parents	-	2.5
Others	-	1.5
NUMBER BROTHERS:		
None	-	29.4%
'1'	-	24.9
'2'	-	18.3
'3-4'	-	15.7
5 or more	-	11.7
NUMBER SISTERS:		
None	-	24.9%
'1'	-	29.4
'2'	-	18.3
'3-4'	-	18.8
5 or more	-	8.6
POSITION IN FAMILY:		
Oldest child	-	24.8%
Middle	-	38.4
Youngest	-	29.8
Only child	-	7.1
HAVE BEEN IN SERIOUS RELATIONSHIP WITH OPPOSITE SEX:		
Yes	-	78.3%
No	-	21.7
WHICH SEX SPENDS MORE TIME WITH:		
Same sex	-	21.7%
Opposite sex	-	25.3
Both equally	-	53.0
SCHOOL GRADE COMPLETED:		
5th Grade or less	-	1.0%
6th - 8th	-	14.2
9th - 11th	-	48.7
High School Graduate or GED	-	19.8
More than High School		16.2
PARENT CLOSEST TO:		
Mother	-	57.0%
Father	-	13.0
Both same	-	18.0
Cannot tell	-	12.0

Table III. Percentage Distributions for Personal Background and Family History Variables (Continued) (N=203 Males)

Variable		%	Variable		%
PARENT LIKED MORE:			SAMARITAN PROGRAM:		
Mother	-	49.0%	Methadone-to-Abstinence	-	36.0%
Father	-	12.5	Drug-Free	-	64.0
Both same	-	26.0			
Cannot tell	-	12.5	RELIGION:		
PARENT CONSIDERED MORE SUCCESSFUL:			Protestant	-	26.3%
			Catholic	-	58.1
Mother	-	36.7%	Jewish	-	8.6
Father	-	35.2	Other	-	7.1
Both same	-	16.8			
Cannot tell	-	11.2			
AGE:					
Less than 17	-	11.3%			
17 - 19	-	18.7			
20 - 22	-	12.8			
23 - 25	-	18.2			
26 - 28	-	17.7			
29+	-	21.2			
ETHNIC GROUP:					
White	-	53.7%			
Black	-	33.5			
Hispanic	-	12.8			

parent as more successful with equal frequency. There is not a significant relationship between sex-role identity scores and responses to these questions, nor to other questions relating to family composition and behavior toward the opposite sex. (See Table III.)

5. There is not a significant relationship between sex-role identity scores and demographic variables such as ethnic group, age. (See Table III.)

6. On the MMPI scales, clients with sex-role identity scores in a more feminine direction score higher on the Mf scale, indicating some preference for traditionally feminine interests and occupations. They also have higher scores on the Social Introversion scale (Si); indicative of a tendency to withdraw in social situations. There is no relationship between sex-role identity scores and the MMPI pathology scales - Hypochondriasis, Depression, Hysteria, Psychasthenia. (See Table IV.)

The results of this exploratory study provide some evidence that there may be a sub-group of male clients typified by a less masculine sex-role identity than male clients as a whole, who tend to withdraw in social situations, and use drugs more often than other males for the purposes of being sociable and reducing anxiety. For these clients, assertiveness training may be an effective auxiliary treatment method for increasing social competence and reducing anxiety. Male role models may be preferred for these clients to effect behavior change. A cautionary note in interpreting these results: The relationship between sex-role identity scores and drug use for anxiety relief could merely reflect less honesty in male sex-typed clients, since it is not considered masculine to admit to experiencing anxiety.

A substantial number of male clients were found to be sex-typed, contrary to expectations based upon research about drug-abusers' families. This could indicate the pervasive influence of other factors, such as peer group, in determining or changing the sex-role identity for this population. The lack of relationship between sex-role identity and family characteristics and attitudes towards parents would seem to provide some support for this view. However, it seems unlikely that family characteristics are totally unrelated to the sex-role identity of male drug-abusers. Perhaps the development of sex-role identity is not related to the simple and objective family characteristics, such as number of brothers and sisters, used in this study. If the focus were on more complex psychological characteristics of the family and the family members, such as which parent made important family decisions and the responsibilities of older siblings, then the results may have been positive.

Table IV. MMPI Results: Mean, Standard Deviation and Correlation With Sex-Role Identity Score (N=83 Males)

MMPI Scale*	Mean	Standard Deviation	Correlation With Sex-Role Identity Score
Lie (L)	3.48	2.04	-.18 NS
Hypochondriasis (Hs)	60.20	13.70	.11 NS
Depression (D)	73.47	13.62	.16 NS
Hysteria (Hy)	61.72	11.02	.01 NS
Interests (Mf)	64.16	10.18	.22, $p < .05$
Psychasthenia (Pt)	69.47	12.67	.19 NS
Social (Si)	57.40	9.04	.39, $p < .001$

* T-Scores (with K correction where applicable) were used in calculations, except for Lie scale where raw score was used. Only selected scales were analyzed.

REFERENCES

Bem, S. 1974. The Measurement of Psychological Androgyny. *J. Consult & Clin. Psych. 42:* 155-162.

Bem, S. 1975. Sex-role Adaptability: One Consequence of Psychological Androgyny. *J. Pers. & Social Psych. 31:* 634-643.

Bem, S. 1976. *Scoring Packet: Bem Sex-Role Inventory.* Stanford University Psychology Department Mimeograph.

Guttentag, M. and Bray, H. 1976. *Undoing Sex Stereotypes.* New York: McGraw Hill.

Maccoby, E. and Jacklin, C. 1974. *The Psychology of Sex Differences.* California: Stanford University Press.

Salmon, R. and Salmon, S. 1977. The Causes of Heroin Addiction - A Review of the Literature. Part 1. *Int. J. Addict. 12:* 679-696.

THE CULTURAL FIT FACTOR IN DEVELOPING DRUG-FREE THERAPEUTIC COMMUNITIES FOR PUERTO RICAN ADDICTS IN THE UNITED STATES

Efren Ramirez, M.D.

National Coordinator, Hogar CREA, Inc., 14 East 60th Street, New York, New York 10022

The purpose of my paper is to share with you my views on the particular cultural aspects that have to be taken into consideration when promoting rehabilitation services for Puerto Rican drug addicts and prevention services for Puerto Rican communities where addiction is endemic. My views are based on 18 years of experience in the design and implementation of the psychiatrically supervised drug-free therapeutic community system in Puerto Rico, it's implementation in the United States in the last 12 years and in the Dominican Republic in the last 3 years.

The Puerto Rican Therapeutic Community Model was developed by myself and by a mixed staff of professionals and former addicts between 1961 and 1965. It was first described at a conference on New Developments in the Rehabilitation of the Narcotics Addict, held in Fort Worth, Texas, on February 1966. In the same year it was published by the Vocational Rehabilitation Administration of the Department of Health, Education and Welfare under the title of "The Mental Health Program of the Commonwealth of Puerto Rico" (Ramirez, 1966).

The Puerto Rican Therapeutic Community Model is a professionally supervised prevention, treatment and rehabilitation unit, a *Dome* (Ramirez, 1973a), that includes outreach Community Orientation Centers in endemic areas to engage the street addict in long term treatment, Day-Night Care Centers for additional motivation and partial detoxification, a residential therapeutic community for personality reconstruction, and a reentry program for resocialization and vocational training. The reentry program has 5 stages or levels from which the ex-addict staff is selected. In addition to the previously described services, the ex-addict staff runs the prevention and orientation services rendered to the community, schools and prisons, and the follow-up services of discharged residents. A dome was designed to engage about 700 addicts in all the phases of the rehabilitation process at a total cost of under $1,000 a year per treated addict.

Between 1966 and 1968 I participated in an attempt to develop the Dome model in a large scale in New York City and in Puerto Rico.

In New York City, I initiated a long range plan for the management of endemic urban drug addiction which provided the design and staff training of Odyssey House, the creation of the Addiction Services Agency of the City of New York, the establishment of the Phoenix House Program, the development of penal therapeutic community programs in Riker Island and Hart Island, the designing and the securing of five years' funding for the Horizon Project of the Lower East Side, and the development of a prevention program based on an original method of Attitudinal Skills Training implemented by a mixed professional and para-professional training staff in schools and community groups such as RARE and AWARE (Ramirez, 1973b). I also created the Addiction Specialist Job Line series in the city civil service system which opened job opportunities for qualified former addicts. The development of the therapeutic community movement in New York since that time is well known to all of you.

In Puerto Rico, I encouraged and supervised a group of my former patients headed by Juan Jose Garcia, in developing the Hogar CREA program. Today, Hogar CREA has 19 domes, 18 in Puerto Rico and one in the Dominican Republic, with a total of 65 residential therapeutic communities and a population of about 3,200.

During 17 years as a participant observer of the Therapeutic Community movement in Puerto Rico, I have been able to identify a cluster of culter-related characteristics which I believe have to be carefully considered by programs in the United States that are providing or planning to provide drug-free rehabilitation services to addicts of Puerto Rican descent. I call this cluster of culture-related program characteristics the cultural-fit factor.

1. COMMUNITY ACCEPTANCE OF THE SUBSTANCE ABUSER

Although resented because it leads to parasitism, unproductivity and even dangerous social behavior, substance abuse is by and large accepted by Puerto Ricans living under conditions of poverty as a natural outcome of their social and economic deprivation. This acceptance, which in untreated neighborhoods has led to a rapid escalation of substance abuse levels, has been utilized in Dome catchment areas as a source of neighborhood support for the program.

The therapeutic community program is understood not as a mechanism to remove addicts from the neighborhood, or exclusively as a treatment for addicts, but as a system to treat the roots of the social and moral illness of the local community of which addiction is a prime symptom. Local community people are able to see that they too have the same moral, social and psychological problems of the addict and therefore participate in the local program support system in exchange for social and psychological services for themselves.

2. THE THERAPEUTIC ALLIANCE

Most of the professionals who join the former addicts to form the therapeutic alliance in the Puerto Rican therapeutic community movement come from the lower socio-economic classes themselves. We all have known poverty, we speak and think the same language, we share equal racial mix, the same ethnic background and the same history. In this alliance, economic and educational differences are quickly forgotten and people relate in a more genuine parity. The addicts are more likely to see the professional as a role model to admire and emulate rather than as a superior, condescending being to resent.

3. ABSENCE OF GRANTSMANSHIP

The Puerto Rican therapeutic community pilot project as CISLA (Centro de Investigaciones Sobre la Adiccion), provided services at less than $1,000 per addict per year. Its off-shoot, Hogar CREA, operated during its first four years without government grants, entirely supported by the local community. The first 10 therapeutic communities were developed this way. Since 1972, at the recommendation of a study by the University of Puerto Rico, the government has been helping CREA with grants for clinical and administrative professional services. With a total yearly budget of $3.0 million and a resident population of about 3,200, the program is still providing services at less than $1,000 a year per addict. The government funding is only 25% of the total budget.

The spartan quality of the Hogar CREA budget has had a profound impact on the morale and collective motivation of the movement. Top salary is $12,000 a year. My own is $1.00 a year. They owe me $10.00. There are no frills in this program, no fat, no waste. We recycle everything, produce as much as we can, and impart to our residents and staff the sense of dedication and sacrifice that makes our program a social revolutionary movement striving for individual autonomy and collective self-sufficiency (Ramirez, 1977).

4. PROGRAM EXPANSION RATE

With 60 therapeutic communities, we have achieved 30% of our projected goal of 200 therapeutic communities by the year 2000. Opening 6 Hogares CREA every year, one every two months, maintains a momentum of growth and the creation of new job opportunities that is essential for program development. The multiplicity of identical programs provides a healthy opportunity for rotation and a variety of work and rehabilitation settings unmatched anywhere.

5. COMMUNITY EMPLOYMENT FACILITIES

Our graduates have a higher rate of employment than comparable groups in the open community in spite of their lower average educational level. The most likely explanation, other than the higher degree of self-assurance and competitiveness acquired in the therapeutic experience, is that employers make up a substantial percentage of the community volunteers who serve in the local therapeutic community steering communities, friends of CREA, all of whom are members of our human resources bank. The CREA Human Resources Bank is a register of local community members who have pledged their cooperation to the program in donations, time, technical assistance or job opportunities.

6. COOPERATIVISM

The cooperative movement is strong in Puerto Rico - 30% of the population belongs to a cooperative association of one kind or another. We have organized a CREA Savings and Loan Cooperative which is open to all CREA residents, staff, relatives and friends. Besides serving as an additional control in the process of socialization, the cooperative binds the CREA movement in a huge extended family structure.

7. RESIDENTS' GOALS AND EXPECTATIONS

By and large, the residents and graduates of CREA do not use their program experience as a springboard to joining the culture of ostensible consumption characteristic of upwardly mobile groups in a free enterprise society. Their goals are more directed a living within their means, developing a sense of social responsibility and reserving some of their time and energy for helping others. Individually and collectively, they are becoming a role model for socially and economically deprived people in Puerto Rico, a living proof of an alternative way of coping with the conditions prevalent in the countries of the third world.

8. SEPARATION BY AGE AND SEX

CREA has separate residential facilities for women, adolescent males and adult males. This segregation at the personality reconstruction stage of the process has been found to be essential to the moral and social rehabilitation of the residents. Social contact between the different groups is resumed at the reentry stage, when the residents are capable of mutual respect and consideration and are least likely to fall into acting out behavior.

9. RELIGIOSITY

Ecumenism has been a consistent feature of the Puerto Rican Therapeutic Community movement since its inception. Residents, staff and volunteers participate in religious services of their choice, invocations mark the beginning and end of important meetings and every day behavior guidelines are patterned along the lines of universal religious and moral principles. Religious faith, including faith in Espiritismo, is encouraged and expected in everyone.

THE CULTURAL FIT FACTOR HYPOTHESIS

The nine items I have mentioned collectively constitute what I call the Cultural Fit Factor of the Puerto Rican Therapeutic Community movement. Traditionally, culture is defined as the attitudes and patterns of behavior that constitute the distinctive achievement of a human group (Kroeber and Kluckhohn, 1952). In the development of the Puerto Rican Therapeutic Community experience, we see an autochthonous attempt of our people to cope with some of the consequences of colonialism, rapid industrialization, runaway population explosion and the head-on collision with the Anglo-American culture. This combination of factors has produced high levels of poverty, 28% among Puerto Ricans in the United States (U.S. Labor Department, 1978), and over 60% in the island (Commonwealth of Puerto Rico, 1977). A culture of poverty (Lewis, 1966) has emerged, a collective attitude characterized by carelessness toward the environment, apathy toward civic life, hostility toward fellow human beings, cruelty towards children and the elderly, lack of concern for health and hygiene, lack of moderation in pleasure seeking behavior, disdain for property, tardiness, limited capacity for commitment, chauvinistic machismo in males, self-debasing submission in females, dishonesty, banality and anti-intellectualism.

In the Puerto Rican Therapeutic Community movement we see the implementation of the existential principles of encounter with reality as it is and personal responsibility applied as an antidotal frame of reference to deal with our bleak social situation (Ramirez, 1968; May, 1969). In the Puerto Rican Therapeutic Community we see the consistent application of the Alderian tools of uncovering of fictitious goals and life style, imparting of courage and retraining toward community orientation (Ellenberger, 1970).

It is my considered opinion that to cope with the problem of substance abuse among Puerto Ricans in the United States, a therapeutic community movement similar to the one being developed in the island is urgently needed. The drug-free therapeutic community programs that are providing services to Puerto Ricans at present must consider adding the cultural fit factor to their present approach. More importantly, local, state and federal authorities must be educated on the nature and the application of these cultural considerations so that local community efforts to develop Puerto Rican rehabilitation and prevention programs are not strangled by alien clinical and administrative guidelines.

A national strategy is required to achieve this goal, yet not a single high ranking Puerto Rican represents our point of view in the relevant federal organizations. This omission must be corrected soon. In the meantime I propose that we organize our own Substance Abuse Policy Council that would recruit all the available Puerto Rican expertise to undertake the following tasks: 1) Develop a research branch to document the incidence and distribution of substance abuse among Puerto Rican communities; 2) Utilize this information to request federal seed funding allocated in proportion to need through the single state agencies and earmarked for programs specifically designed to meet the needs of our people; 3) Develop a nationally available technical assistance staff to design and aid in the development of these programs; 4) Coordinate at local, state and national levels with bilingual education, urban policy, youth employment, community mental health, social services and related programs to aid these agencies to provide relevant support to our program development efforts; 5) Develop a Puerto Rican position paper on certification and licensing; 6) Evaluate existing program development and training resources and make this information available to interested local groups; 7) Launch a national Spanish mass media educational campaign for the Puerto Rican communities to mobilize their volunteer services, their financial support and their political influence in behalf of our national strategy; and, 8) Create our own foundation to finance our activities.

REFERENCES

Commonwealth of Puerto Rico. 1977. *Department of Social Services Annual Report*.

Ellenberger, H. 1970. *The Discovery of the Unconscious*, Chapter 8. New York: Basic Books.

Kroeber and Kluckhohn. 1952. *Culture, 47 (1)*. The Papers of the Peabody Museum.

Lewis, O. 1966. *La Vida*. New York: Random House.

May, R. 1969. *Existential Psychology*. New York: Random House.

Ramirez, E. 1966. *The Mental Health Program of the Commonwealth of Puerto Rico*. Rehabilitating the Narcotics Addict. Vocational Rehabilitation Administration. U.S. Department of Health.

Ramirez, E. 1968. The Existential Approach to the Management of Character Disorders. *Rev. Exist. Psych. and Psychiatry, 8:* 43-53.

Ramirez, E. 1973a. The Dome Model in the Management of Addiction. *Major Modalities in the Treatment of Drug Abuse*. New York: Behavioral Publications.

Ramirez, E. 1973b. The Addiction Services Agency of the City of New York. *Major Modalities in the Treatment of Drug Abuse*. New York: Behavioral Publications.

Ramirez, E. 1977. Self-Sufficiency: A Survival Approach from a Puerto Rican Perspective. *Proceedings, NDAC,* San Francisco.

U.S. Labor Department. 1978. *Study by Bureau of Labor Statistics*.

THE ROLE OF SOCIAL WORKER SERVICES WITHIN A THERAPEUTIC COMMUNITY

Richard Pruss, Rabbi

Samaritan Halfway Society

The original concept of a therapeutic community for drug addicts, as exemplified by Synanon, did not use professionals as providers of clinical treatment. Among the probable reasons for this exclusion are the following:

1. Modeling was an important therapeutic tool, and ex-addicts were considered more effective models for clients than people without drug abuse histories. The high-level position of the ex-addict within the organization provided evidence of the rewards for conforming to therapeutic community regulations.

2. The therapeutic community was a supportive and cohesive "family" environment, as opposed to the institutional environment with rigid staff - patient separation that typified the relationship of professionals to their clients.

3. The therapeutic community required adherence to community norms with peer pressure used to control deviance. Clients not conforming were subject to threats of banishment, and strongly worded accusations to induce guilt and shame. This was radically different from the approach of most professionals who would be expected to be critical of these methods.

4. The clinical positions provided a career opportunity in a protected environment for many clients who had poor job prospects partially due to discrimination against addicts, or who could not succeed away from the therapeutic community. In addition, promotion of addicts into clinical positions may have had therapeutic value in keeping them drug-

free. In order to fulfill their responsibility to rehabilitate clients, they had to remain abstinent.

5. Finally, professional treatment methods, whether in an institution or in private therapy, had failed with the heroin addict.

In recent years, some therapeutic communities have been incorporating professional treatment providers into the clinical program, as part of a general tendency towards diversifying services. At Samaritan, a multi-modality agency serving the New York City vicinity, a Casework Department was established in 1977, which is staffed by three social workers. This unit is independent of the Clinical Department, which is responsible for most clinical treatment and remains staffed primarily by ex-addicts successfully completing a therapeutic community program. Social workers have become an integral part of the treatment process, although they function primarily in a consulting and supportive role. The addition of social workers did not reflect a policy change denying the validity of therapeutic community methods. Indeed, the components of the therapeutic community have remained mostly intact since the program began. Social work services were instituted primarily due to new treatment needs resulting from changes in characteristics of clients entering in recent years. There are increasing numbers of polydrug users, more adolescents, and apparently more clients with psychiatric disorders. Many of them can benefit from a therapeutic community, but they often require professional auxiliary services to succeed in the program.

The integration of social workers within a therapeutic community must be conducted in a manner that does not detract from the therapeutic value of the concept. If social workers were used to provide comprehensive counseling to all clients as per the psychotherapy model, the effects of the group-oriented treatment activities could be substantially diluted. Clients might become dependent upon their counselors rather than clinical staff and their fellow clients. The result could be reduced adherence to community norms, and a lessening of the effects of peer pressure. However, social workers can perform at least four functions within a therapeutic community setting, that strengthen rather than detract from the program. The four major services they provide at Samaritan are: 1) Intake diagnosis and referral, 2) Problem-oriented brief counseling, 3) Coleadership in groups, 4) Family therapy and liaison services. A brief description of these services follows:

1. INTAKE DIAGNOSTIC AND REFERRAL SERVICES

Clients entering any Samaritan modality go through an evaluation process during their initial 2 - 3 weeks. Data on the new client is obtained from various sources: a physical examination, laboratory tests, interviews by a clinical psychologist and a social worker, and psychological tests. In addition to these professional evaluations, the ex-addict staff conduct a biographical interview as well as the traditional clinical interview to assess motivation for treatment and to obtain a commitment from the client. All this client data is presented at a multidisciplinary case conference coordinated by the Casework Department. The goals of the conference are to determine whether the client is appropriate for a Samaritan program, and to construct a treatment plan. Decisions are made on a cooperative basis. The treatment plan is implemented primarily by clinical staff at the treatment facilities, with social workers available to provide prescribed services for certain clients and consult with clinical staff when needed. If the case conference decision is that the client is inappropriate for Samaritan, social work staff are responsible for appropriate placement. No client is discharged onto the street.

2. PROBLEM-ORIENTED INDIVIDUAL COUNSELING

Clients may have difficulties functioning within a therapeutic community which could be remedied by individual counseling. The goal of this time-limited counseling is to enable the client to obtain maximum benefits from the primary program. If the client requires extensive counseling rather than a therapeutic community environment, the client would be referred to a counseling-oriented agency. Examples of the interfering problems that are amenable to time-limited counseling are: moderate depression, social anxieties which hamper interactions with clients and staff. In addition, certain crisis situations may be handled by social work staff, particularly those that require assistance with a situation outside the program. An example would be obtaining care for a client's child while the client is in treatment.

The counseling is conducted according to rules that are different than those of a counseling center. The interaction between counselor and client cannot be kept entirely confidential. Clinical staff must be informed of relevant discussions which impinge upon treatment, and clients enter counseling with this understanding. The counselor does not allow the client to continue sessions indefinitely, since this would produce dependence on the counselor which should be avoided. The counselor does not obtain special privileges for his clients, and all decisions about changes in treatment procedures for a client are made in conjunction with clinical staff. The clinical staff must be perceived by clients as maintaining power, otherwise their effectiveness is diminished.

3. GROUP LEADER ASSISTANCE

On occasion, a social worker will co-lead a group therapy session with a member of the clinical staff. There are a variety of new group procedures which have been developed since the encounter group methods used by the clinical staff. The social worker has the opportunity to demonstrate other procedures in the group, providing both therapy and a training opportunity for staff and clients. Social workers can use whatever group methods they can handle adequately, providing the procedures do not conflict with the therapeutic community's philosophical orientation. Methods that extensively explore clients' past history or that could be used by the client to avoid responsibility for his behavior and blame others, would generally be avoided. Another benefit derived from the occasional presence of professional staff in a group session is that it provides an opportunity for clients to explore their attitudes towards "squares." Since many of them experience anxiety with people outside of their peer group, this can be a valuable experience and excellent preparation for their eventual return to the outside community when they will be interacting with a new peer group of non-drug abusers.

4. FAMILY THERAPY AND LIAISON SERVICES

About one-third of Samaritan clients are adolescents, and most of them will be returning to their families after graduation. In many cases, the family was an important determinant of the child's drug-abusing behavior. Although it may not be desirable for the child to return home, it often is the only practical alternative. Therefore, in order to reduce their detrimental influence on the returning client, a social worker may provide services to the family while the client is in the program. It is sometimes adequate for the social worker to provide liaison services for the family with public agencies: assisting them in obtaining adequate housing, psychiatric care, legal services, employment. It may also be necessary for the social worker to provide family therapy. To some extent, this can be done at family association meetings which occur periodically at a program facility. In addition, individual families are provided services separately. The therapeutic community concept assumes that the current social environment can be a crucial determiner of client behavior; without this belief a therapeutic community would be irrelevant. Therefore, it is reasonable for the treatment program to include services that will positively affect the environment that the client enters after graduation. This is what social work staff attempts to do. They are uniquely qualified for this assignment, due to their extensive knowledge of community resources and their familiarity with family therapy procedures.

The four primary job functions for social workers describe the situation at Samaritan during the past 15 months. Although the implementation of an independent Casework Department has been a bumpy road with conflicts between casework and clinical staff, the two departments now function reasonably well together. The social workers have enabled the program to have greater impact on many clients, particularly those who, without social worker intervention, would have split early in treatment. In addition, the agency has been able to more quickly identify admissions not appropriate for a therapeutic community, so that they could be referred before harming themselves or the program. Although successful, incorporating social worker staff and giving them the discussed responsibilities is not necessarily a desirable direction for all therapeutic communities. Among the other possibilities for dealing with the problem clients who need additional services are: to provide clinical staff with the expertise for handling both the problems now referred to a social worker as well as the other social worker services; to restrict admissions to clients who can succeed in a traditional therapeutic community. Regarding the former, it would seem an ideal solution to equip clinical workers with additional expertise, but perhaps not a realistic answer. Clinical staff already have substantial skills acquired from their experiences at various levels of a therapeutic community, usually starting at the client level. Many of them are not amenable to academic learning, and tend not to return to school for extensive periods. There are few staff people around with a combination of ex-addict background, therapeutic community experience, and graduate level mental health training, who could perform both traditional therapeutic community functions as well as social worker tasks. Therefore, the division of labor seems necessary considering current manpower availability. However, social work services do not eliminate the need for training clinical staff to deal with some problems they now avoid. An example would be homosexual behavior. Regarding the restriction of admissions to clients suitable for a traditional therapeutic community, this approach seems detrimental to the therapeutic community movement. The type of client that seemed most suitable for a traditional program was the adult heroin-only addict. However, they are now only a small percentage of the New York City area drug-abusing population. If admissions were restricted to only these clients, therapeutic communities would not make a substantial contribution towards controlling the drug-abuse problem. There are reasons to justify the expansion, rather than contraction of programs. A therapeutic community is probably the current treatment of choice for most sociopathic individuals. While the degree of success is not remarkable, who has done better? It would seem that the therapeutic community movement should expand client populations to eventually include criminals without drug histories, and youthful delinquents. This expansion might be more effective if professional treatment services as described are incorporated into the concept.

A STRUCTURED INCENTIVE SYSTEM FOR RE-ENTRY CLIENTS

Don Russakoff, M.B.A. and Robert W. Weiss, Ed.D.

Samaritan Halfway Society

Within those therapeutic communities that are divided into the traditional three phases - Orientation, Main treatment phase, Re-Entry; the last phase concentrates on assisting clients to establish themselves in the community as productive citizens. Vocational rehabilitation is an important part of the re-entry process, since most clients do not have the prerequisites for success in the labor market. At Samaritan, which serves the New York City vicinity, entrants generally are high-school drop-outs without job skills who have either a sporadic work history in unskilled jobs or no work history at all. Educational remediation and high school equivalency preparation begin early in treatment, while vocational evaluation, counseling, and placement occur primarily during the 3 - 12 month period in ReEntry.

Placement is the culmination of the educational/vocational rehabilitation process, and is considered among the most important aspects of treatment. The Samaritan treatment philosophy recognizes that clients who can remain vocationally productive are more likely to remain drug-free than those who have frequent periods of unemployment. Therefore, placement is oriented toward establishing clients in a long-term career. In the past, two problems frequently occurred during placement: 1) Clients accepted dead-end positions rather than enter training or education programs leading to more substantive careers. The salary for these jobs would be higher than stipends received in training, which provided a financial disincentive to enter training. 2) Clients entering training/education programs dropped out prior to completion. They would not adhere to the demands of these programs in order to receive the long-term benefits. As a result of these problems, graduates often did not have sufficient skills to maintain a career, although they usually did have a job at the time of graduation.

In November, 1976, Samaritan established an experimental financial incentive system to provide incentives to clients for entering and remaining in vocational training. The system is called VEFAP: Vocational, Educational, Financial Assistance Program. Clients in the residential programs, both the drug-free and methadone-to-abstinence modalities, who are at least 18 years old become eligible for VEFAP after completing a one-month probationary period in ReEntry phase. If they enter an approved training/education program, a weekly payment is accrued for them during their attendance. The payment is given to the client after training has been successfully completed (or a specified phase of training completed for lengthy programs). Clients not entering approved programs do not receive accrued payments, while those who drop out of approved training have their entire accruals cancelled. This system reduces the incentive for accepting an unskilled job, since clients receive about the same total payment by pursuing training. Also, an incentive is provided for remaining in training, because termination leads to loss of accrued money.

Clients in VEFAP receive $40 - 45 per week on accrual. Total costs to the agency are substantial: about $35,000 for 1977; and an estimate of $50,000 for 1978. In order to determine whether these costs are justified, a preliminary evaluation was conducted of the initial VEFAP clients. The basic design of this evaluation was a comparison of training program attendance and ReEntry termination rates for two time periods. One time period was the first 14 months of VEFAP operation (November, 1976 - December, 1977); the other time period was the 10 months prior to the start of VEFAP (January, 1976 - October, 1976). A summary of results follows (also see Appendix):

During the 10-month period without VEFAP, 13 clients entered training programs who did not have the opportunity to participate in VEFAP. By the end of the period (October, 1976), ten clients had dropped from training - seven lasting not more than two months. Two trainees had completed programs, and one was deceased. During this same period, 23 of the ReEntry clients had terminated, an average of more than two per month. Overall, the record of the ReEntry phase for this period was unsatisfactory.

During the succeeding 14 months when VEFAP was in effect, 38 clients entered approved programs and received accruals. By the end of this period (December, 1977), seven had graduated, and only nine had dropped out. Of the 22 continuing in training, most had completed at least two months. During this period, the termination rate from ReEntry was exceedingly low: only three terminated; an average of about one per five months. A comparison of this period with the preceding non-VEFAP period indicates an increase in the average amount of time clients remain in training; a reduction in the percentage dropping from training; and a substantial decline in the termination rate for all ReEntry clients.

This preliminary evaluation does not conclusively demonstrate the favorable aspects of VEFAP. Only two time periods were compared, and factors other than the introduction of the VEFAP program could have accounted for the changes in client performance. However, we are not aware of other factors that could have resulted in the improvement; indeed there were non-controllable factors that could have created a decline in client performance during the VEFAP period. For example, the local unemployment rate declined which could have led to an increase in terminations from clients able to obtain jobs. We are inclined to attribute the favorable effects primarily to VEFAP.

The financial incentive system is continuing in 1978, although it presents a strain on the current financial resources of the agency. So far, VEFAP has been sufficiently beneficial to clients to justify the expense. We are pursuing private sources of financial support to meet the projected budget increases for the program as greater numbers of clients enter.

Table I. Comparison of Training Program Progress for Clients Entering Training Programs During a VEFAP Period (11/76-12/77) Versus the Prior Non-VEFAP Period (1/76-10/76)*

VEFAP (11/76-12/77) 14 month period		Non-VEFAP (1/76-10/76) 10 month period	
Number entering VEFAP (includes one client entering twice)	38	Number placed in training program	15
		Number transferred into VEFAP (these are not counted in the data below, but are in the VEFAP column)	2
Number completing training program (as of 12/77)	7[1]	Number completing training program (as of 10/76)	2
Number above who were Samaritan trainees	4[2]	Number above who were Samaritan trainees	0
Number of drop-outs	9	Number of drop-outs	10
Number continuing	22	Number continuing	0
		Deceased	1

Average time spent in training program (VEFAP):

months	<1	1-2	2-4	4+
Drop-outs (9)	2	1	2	4
Continuing (22)	4	0	16	2

Average time spent in training program (Non-VEFAP):

months	<1	1-2	2-4	4+
Drop-outs (10)	1	6	3	0
Continuing (None)				

*Only residential clients age 18 and above are included for both periods.
[1]One client lost VEFAP account, although completed scool.
[2]In-house trainees prepared for a therapeutic community position.

Table II. Types of Training Program Entered

	VEFAP (11/76-12/77) 14 month period	Non-VEFAP (1/76-10/76) 10 month period
Samaritan trainee	12	None
Technical trades	9	7
Clerical	7	2
Community college	5	1
Dietetics/cooking	3	1
Beautician	None	1
Four-year college	1	None
Non-specific	1	1
	38	13

Table III. Clients Leaving ReEntry During Period*

	VEFAP	Non-VEFAP
TERMINATED (Split or Discharged)	3	23
RE-CYCLED TO ANOTHER STAGE	10	6
AVERAGE POPULATION DURING PERIOD	26.5	16.7

* All clients in this stage are included.

THE EFFECTIVENESS OF TWO THERAPEUTIC TECHNIQUES IN THE TREATMENT OF DRUG ABUSE PATIENTS

Bernard G, Rosenthal, Ph.D.

Evanston Comprehensive Drug Treatment Program, Inc.

Previous studies (Gorsuch, Butler and Sells, 1975) of drug abuse treatment programs have shown that methadone maintenance programs are the most effective of all outpatient programs, and, of the former, change-oriented approaches in contrast to adaptive approaches are better, if only slightly so.

However, within the methadone maintenance change-oriented framework, no studies have yet been carried out directed to which specific types of counseling would produce the greatest clinical improvement or rehabilitation.

The present study, now planned for undertaking at the Evanston Comprehensive Drug Treatment Program, Inc., is directed to determine which of two counseling (therapeutic) modes dealing with two specific psychological types of clients will result in the most effective outcomes.

Of the two types of counseling involved, the first is nurturant counseling which is or approximates supportive, affectionate or TLC counseling. Though this would not preclude probing, insight-oriented penetrating approaches, it would be done in a context of kindliness, acceptance, and concern for the client. Firmness would be in order but, again, in a context of cordiality or warmth and with evident concern for the client's feelings and state of mind. If specific ultimatums, severe admonitions or substantial toughness is in order, this again would be done with evident concern for the client's feelings and in a context of cordial sensibility.

The second type of counseling to be employed is confrontational counseling which is more directed to reality, candor and intrusion in regard to the dynamic issues or etiology of the client's drug condition rather than mollification, "sweet talk," and exclusive support. This does not mean that this should be done callously, though it connotes a certain dispassionate and "hard" orientation in contrast to the "softer" nurturant approach. When necessary, however, it means coming down hard on the client, persisting obdurately without consideration of the niceties so that the client can achieve insight and awareness of his motives as well as a recognition of reality which he heretofore denies, dissimulates about or otherwise circumvents. It means, then, a pushing or forcing of reality into the client's consciousness. If this can be done without harshness or cruelty, fine. But, if not, it may require severe and penetrating measures, tumult, surely unequivocal persistence in the combating of evasion and, if necessary, attack and abuse to make the client aware of denial facts and his own motivations.

The two psychological types of clients who the study will investigate are: 1) dependent, passive, weak types, and 2) dominant, strong, tough types. We have developed a number of methods to differentiate and identify these client types.

The basic nature of the design is of a 3 x 2 factorial type in which equal numbers of both types of clients will be administered either nurturant or confrontational therapy during the year's treatment process. Thus a rigorously controlled design will be employed with clients being matched between each of the four client-and-therapy-type groups for a number of background and demographic variables. In addition, a matched control/contrasting group would be employed which would involve casual, informal and unsystematized treatment. Counselors who have had intensive training in each of these therapeutic modes will participate in the study.

The instruments used to assess the nature of the results and of the processes that occur during the treatment year are the following: 1) conventional performance criteria, including drug use, criminal activity, occupational status, etc.; 2) a newly devised psychological-behavioral scale to be used bi-weekly to assess the client's behavioral and psychological condition through the treatment year. This scale has been used extensively and its inter-rater reliabilities approach 95% agreement for judgment of identical scale points; 3) a newly developed environmental scale to be used bi-weekly which assesses the client's environmental situation, his friendships, family support, group associations, diversions, economic support, and the like; 4) finally, both a therapist and client scale to be used after each weekly counseling session which gauges the therapist and client's experiences, feelings, perceptions and other reactions to these sessions. Among the modalities assessed by these scales are: therapeutic changes, client progress and improvement, estimate of processes and events that transpired during the session, the perception by both client and counselor of what each conveyed to the other and how each acted toward the other, each participant's estimate of how nurturant, confronting, warm, indifferent and hostile the therapist was to the client, what psychological needs the client had to fulfill during the session and whether the therapist helped him to do so, and numerous other dimensions of the counseling experience and relationship. One critical purpose of these scales is to ensure that the designated nurturant and confrontational sessions which the experimental design calls for will be just that. Thus, if discrepancies occur from the prescribed mode of treatment as uncovered by use of these scales, such deviations may be corrected in subsequent sessions. Further, these scales, especially the clients', will afford evidence that what the therapist thinks is happening to the client--and his therapy--in these sessions is, in fact, according to the client, actually happening.

Analysis will cover the following issues: 1) changes in the performance criteria, psychological-behavioral variables and therapy-scale variables over the course of treatment so that the course of change in the two counseling and the two client-type groups can be followed and the sources of change discovered; 2) change in the environmental scales over the course of study will be followed to determine if any environmental change is related to performance, psychological-behavioral and/or therapy-session change. The foregoing changes will be studied both independently for each separate scale and also for prospective relationships between them, e.g., relations between therapy-session changes and performance criteria changes; relations between environmental scale changes and psychological-behavioral scale changes; relations between therapy-session changes and performance criteria changes and psychological-behavioral scale changes; relations between environmental scale changes and performance criteria changes; etc. Thus, a number of potentially significant relationships will be fully explored weekly or bi-weekly from the study's inception to its end.

Thus, this will be the first, rigorously monitored and controlled study to determine whether a specific mode of change-oriented counseling in a methadone maintenance program will be more effective than another mode when two types (psychological) of clients, equally matched except for a crucial independent psychological variable, are investigated. The results of the study will thus be important for the following reasons: 1) it will demonstrate whether either or both of the experimentally-designated therapies are more effective than a control group with only limited, casual or unsystematized amounts of therapy; 2) it will demonstrate whether nurturant therapy is more effective than confrontational therapy--or vice versa-- and, if so, with one or both types of clients. If the study demonstrates that neither of the experimentally-designated types of therapy is more effective than the other, it will be clear that there is no advantage in using either confrontational or nurturant therapy to the exclusion or minimization of the other in a methadone maintenance program.

Similarly, if the study demonstrates that both these types of therapy are no more effective than the therapy employed in the control group, then there is no gain in using these more complex and demanding therapies over the simpler, less taxing and less sophisticated ones.

Thus, whatever its results, the study will have a significant practical and effective value for it will demonstrate that either nurturant and/or confrontational therapy--in terms of the study's context--should be used to obtain more effective treatment outcomes in methadone maintenance programs or, on the other hand, are no more effective than relatively unsophisticated, casual therapies which can be used with much less difficulty and cost in time, training and money to obtain comparable results.

REFERENCE

Gorsuch, R.L., Butler, M.D. and Sells, S.B. 1975. Evaluation of treatment in DARP Cohort 3 based on behavioral outcomes during treatment. *IBR Report No. 75-17*. Institute of Behavoiral Research, Texas Christian University.

COMMUNITY RELATIONS FOR THERAPEUTIC COMMUNITIES

Gary Perkins

The New Arizona Family, Inc.

Phoenix, Arizona 85003

All of us involved with T.C.'s and other drug programs have one common goal and one common problem aside from our treatment process. Nearly all T.C.'s have a goal, or at least a fantasy, of becoming self-sufficient and independent of bureaucratic government funding sources. A problem we all share is that in order to complete the mountains of paper work, proposals and other red tape to comply with government funding criteria we don't have half the time we need to spend with our clients--let alone time to pursue business and community resource possibilities.

From time to time I'm sure we feel that counseling drug abusers is merely secondary and our primary job function is pushing pencils over tons of paperwork that no one but the auditor ever reads.

If your program is to grow and progress in the areas of treatment, quality and education of staff, matrix, as well as the physical aspects such as your facility and transportation, it is essential to develop and maintain credibility within the criminal justice system and the political arena as well as the various publics in your community.

A well structured P.R. program can be an invaluable tool in convincing your community that your program is a much needed good neighbor.

This presentation will attempt to pass on some simple procedures in setting up a low-budget public relations program. Emphasis will be placed on planning, research, setting goals, defining target publics, basic P.R. tools, and of course how to use the media. For the purpose of brevity this presentation can only point out the basics of public relations. Anyone interested in incorporating an effective P.R. program is urged to take advantage of the numerous P.R. and communications courses available in many colleges and universities.

PLANNING A P.R. PROGRAM

Defining Goals

The first step in setting up a P.R. department is to evaluate the needs of your T.C. and define them on paper in order of priority. If the Executive Director is the initiator of the P.R. department then the needs, problem and goals are close at hand.

However, if a subordinate staff member designs the P.R. department he/she must study the T.C. program's mechanics, philosophies history, problems and goals in order to know the program inside and out.

The most efficient P.R. departments operate with a policy of being directly responsible to the top management level, or the executive director in the case of therapeutic communities. Get to know the director. Strive to gain a thorough understanding of his beliefs, goals and plans for the future.

Keep in mind that the community relations or P.R. department is a representative of both your program and your director. It is vital that he is aware of all the plans and moves that the department makes.

Choosing a Staff

After defining the needs, problems and goals of your program it is time to choose a staff suitable for addressing these. Few therapeutic communities can afford to hire a community relations staff. There is a vast resource to draw from in the residents of your T.C.

Utilizing residents serves a dual purpose in that aside from saving monies for salaries it gives residents an opportunity to learn in areas of publications, media, speaking engagements and to gain experience in dealing with a more positive and successful level of society than they have in the past.

Most residents in the upper levels of program have developed verbal skills and are responsible, honest individuals. Those qualities, along with any printing, writing, layout and editing skills will be helpful.

DEFINING GOALS AND PROBLEMS

After the planning, staffing and defining of goals/problems have been completed it is time to start addressing the most urgent problem. Since community support is vital to all T.C.'s, that might serve as your first problem/goal.

Before a therapeutic community can gain the support from the public, the public must be convinced that the T.C. is needed and credible.

This means the public must be educated concerning the detrimental impact of drug abuse.

DEFINING PUBLICS

Whatever your goal, keep in mind that you have many different kinds of publics to reach and it is important to communicate information that affects the individual publics the most.

Some major target publics for T.C.'s are:

1. The judicial system
2. Parole boards
3. Parole and Probation officers
4. Local and state politicians
5. Corporations
6. Referral agencies
7. Businessmen
8. Community clubs and organizations (i.e., Optimists Club, Rotary Club, Kiwanas, etc.)
9. The media.

Various residential areas should be divided according to the media vehicle used. For example, if you use the newspaper the public would be divided in relation to the circulation of the paper, if you use radio the public would be in relation to the listening audience of the particular station.

Choosing your target public is simple. Just put yourself in the position of each individual public. Then ask yourself if you would be interested or affected by the message. For example, if your goal is obtaining more referrals from the courts then the average housewife would not be interested--your communication would zero in on judges, P.O.'s and attorneys.

RESEARCH

In order to most effectively reach your defined publics, research in certain areas if necessary.

A) How, where and why did the problem/goal originate?

B) Has anyone else dealt with the same or similar problem/goal effectively? If so, how did they deal with it?

C) What facts and information will interest your target publics the most?

D) What media and P.R. tools are most likely to reach the target publics?

E) Gather as much information concerning your subject as needed, from several credible resources, to insure that your message is factual.

COMMUNICATION TOOLS

It is important that you use the media and P.R. tools within your budget that will most effectively reach your target public. If you are inexperienced in the field of public relations and communication skills, there is advice available. Many professsional public relations practitioners are willing to assist non-profit organizations in setting up P.R. programs. Teachers in those fields are another resource for advice.

Here are some communications "tools"--vehicles to get your message across:

- News releases (TV, radio and newspaper)
- publications (newspapers, brochures, magazines, posters, flyers)
- articles in newspapers and magazines
- photographs
- speaking engagements
- press conferences
- Public Service Announcements
- talk shows
- documentaries
- gimmicks (benefit ball games, contests, blimps, balloons, parties)

UTILIZING MASS MEDIA

Television

Without question, television is one of the most powerful and effective media today. You don't have to spend $2,000 per minute to take advantage of this powerhouse.

A good rapport with major local TV stations is invaluable. If you can convince a few key figures of a station that your organization provides a valuable service to the community, TV will help you convince the community.

There is no law forcing TV stations to air public service announcements. However, all stations must renew their license regularly with the F.C.C. in order to broadcast, and F.C.C. requires a certain amount of broadcast time provided for P.S.A.'s. This is 20, 30, and 60 seconds air time for worthy organizations. And it is free.

News broadcasts are another source of free air time. Any happenings within your organization that are newsworthy are candidates for news stories. Annual reunions, benefit ball games, large donations from corporations, litter clean-up campaigns, and Christmas caroling for hospitals are all examples of newsworthy possibilities.

TV documentaries are another source of publicity.

Many local talk shows are interested in what treatment programs have to say.

Key figures of your local stations to approach are the News Assignment Editor, Public Relations Director, news broadcasters and reporters.

Radio

The procedures and concepts for receiving radio air time are basically the same as television. Radio stations are similarly licensed and regulated as far as P.S.A. air time is concerned.

Key figures in radio are all the same as television. Many radio stations have talk shows about community problems, and drug abuse is a major problem of every community--even though some do not realize just how big a problem.

Newspapers

The opportunities for newspaper coverage vary somewhat from radio and TV. Larger newspapers are less likely to print articles about your program as such, but there are ways to work coverage on your program in with more newsworthy articles.

Through research you can come up with some startling statistics on the social, economical and public safety impact that drug abuse has on your community. The more recent and startling the statistics, the more newsworthy the articles will be. You can work followups concerning what is being done to curb the problem and what your program is doing into these stories.

Social events and Women's Sections are other possibilities, i.e., women in treatment, nutritional experts advising your program about meals and menus, employers supporting graduates of your program, etc.

Smaller publications. If your program is located in a large community, don't discount the smaller newspaper, journals and college newspapers. These smaller publications focus more on community concerns than on national news. For this reason, it is usually easier to obtain coverage from them than large circulation newspapers.

OBTAINING CONTACTS IS VITAL

You will find that developing a rapport and even friendships with a wide variety of community figures will be one of the most valuable tools you can obtain. It does require a great deal of legwork, playing politics and constant communications, but then that is what P.R. is all about, and the benefits are more than worth the effort.

Developing contacts in the following areas will be very rewarding to your community relations campaign:

- Superior Court as well as City Court Judges
- State Board of Pardons and Parole
- City, County and Federal Probation and Parole Officers
- Public Defender's Office
- Private attorneys
- City, County, State and Federal Law Office
- All non-profit organizations and referral agencies
- Television Stations
- Radio Stations
- Newspapers, large and small
- Public Relations practitioners and teachers
- Printing and publishing companies
- City Councilmen
- State Senate and House

- Businessmen, corporations, car dealers
- Banks
- Hospitals
- Community organizations and clubs
- Other T.C.'s and drug and alcohol programs
- Churches

There is an old saying, "what goes around comes around." Be consistent, be honest, and be dependable. If you expect help and support from the areas mentioned above, you must first get across your good will. This entails a lot of personal contact and participating and helping people in their areas of concern. Participating and helping out on other organizations' boards as well as their campaigns will open the door to a vast resource.

If you have been involved with the intake work of your T.C., you know that a great deal of communications and personal contact is necessary with judges, P.O.'s and referral agencies. The same applies for any other areas of the community that you intend to solicit support from.

EDUCATION IS IMPORTANT

Again, anyone interested in developing an effective community relations program is urged to participate in P.R. seminars and college courses in this field. The valuable information that can be obtained will save you time, money and headaches. It will also teach you techniques and shortcuts to more successfully use the media.

Suggested Courses

- Public Relations
- Mass Media and Communications
- Newswriting
- Journalism
- Public Relations writing

Many TV stations and Media organizations sponsor public relations seminars that explain how to most efficiently utilize the services of the media. You may also want to consider joining the Public Relations Society of America.

Finally, it is important to remember that you are a representative of your therapeutic community. As far as the public is concerned, what you do and say is what your program does and says.

ASSERTIVENESS, FEARS, AND PREFERRED REINFORCEMENTS OF PATIENTS AND STAFF IN A THERAPEUTIC COMMUNITY FOR DRUG ABUSERS

Louis A. Moffett, Ph.D. and Richard N. Bale, Ph.D.

Veterans Administration Hospital, Palo Alto, Ca.
Stanford University School of Medicine

Clinical descriptions (e.g., MacKinnon and Michels, 1971) of drug addicts portray them as impulsive, passive, dependent, having low tolerance for frustration and discomfort including anxiety, and reduced pleasure in living, with especially impoverished interpersonal relationships. Many drug addicts are often indentified as sociopathic, borderline, or narcissistic characters (e.g., Kernberg, 1975; Vaillant, 1975). These clinical descriptions are generally more consistent than the results of standardized measuring instruments. Beyond finding that addicts are clearly more maladjusted than normals, there has been little evidence for a single addictive personality or even for specific personality traits that are characteristic across all populations of drug abusers (Platt and Labate, 1976). Furthermore, the manifest personality of drug addicts may change dramatically when seen in an inpatient rather than an outpatient setting (Vaillant, 1975). In any case, the common stereotype of the drug addict depicts him as a socially undesirable character.

How do the patients and staff in our program compare in assertiveness, fears, and sources of pleasure? Satori (Zarcone, 1975) is a residential therapeutic community for drug abusers, mostly young heroin addicts.

METHOD

Subjects

In the first stage, all members of the Satori community participated. The clinical staff (N=17) consisted of a psychiatrist, psychologists, nurses, nursing assistants, an ex-addict drug counselor, psychiatric residents, and a vocational counselor. The 15 residents (patients) were all men, aged 21 to 48 years old (Mean=30.8; Median=29); most (12) of them were Caucasian, two were Black, and one was Mexican-American. Their primary drug of abuse had been mainly heroin (12); a few (3) were polydrug abusers. They had been in the program from 9 to 46 weeks (Mean=20.2; Median=20). Thus, these two groups represented the community at a given point in time.

A second group of residents, 25 consecutive admissions, completed the questionnaires within their first two days. These men ranged in age from 22 to 41 years (Mean=27.8; Median=27); most (17) were Caucasian, some (5) were Black, and a few (3) were Mexican-American. Again, most (19) had abused heroin primarily, and the rest (6) were polydrug abusers. This group represented the entering resident.

Measures

Assertiveness was measured by the Rathus Assertiveness Schedule (1973); fears were measured by the Fear Survey Schedule (Wolpe and Lang, 1964), and preferred reinforcements were measured by the Reinforcement Survey Schedule (Cautela and Kastenbaum, 1967). Reviews of the reliability and validity of these instruments are available (e.g., Kanfer, 1976 ; Krasner, 1975).

Procedure

After completing the questionnaires in a staff meeting, each of the 12 day staff members predicted whether the current group of 15 residents would score higher, the same, or lower than the staff on each of the questionnaires. Evening staff filled them out privately. Current residents completed the questionnaires at one sitting in small group sessions. Entering residents completed the forms privately.

Statistical Analyses

For every item on each of the schedules two-tailed *t* tests for significance ($p \leq .05$) between means were conducted between the entering residents and the staff, between the entering residents and the current residents, and between the current residents and the staff.

RESULTS

Ten of twelve staff predicted that the current residents would be less assertive than staff and two predicted that residents would be more assertive.

For fears, seven staff predicted no differences between the groups; three predicted that residents would be more anxious, and two expected residents to be less anxious.

For reinforcements, seven staff predicted that the residents would be lower; three predicted no differences, and two expected the residents to be higher.

Entering Residents vs. Staff

Generally, the entering residents were less assertive, more afraid, and reported less pleasure than staff.

Out of 30 RAS items the residents were less assertive on 11 items (1,7,8,9,15,20,21,23,24,29, and 30). These residents reported that they were less likely than staff to express their feelings in a variety of situations. For example, entering residents highly endorsed "There are times when I just can't say anything," "I tend to bottle up my emotions rather than make a scene," and "I often have a hard time saying "No"."

Out of 76 FSS items, the entering residents, compared to staff, were more afraid on 16 items (4,5,9,10,12,15,16,17,23,27,30,31,32, 52,57, and 64); staff were more afraid on 3: "Receiving injections" (22), "Weapons" (41), and "Sight of fighting" (44). Both entering residents and staff rated similar interpersonal fears as their ten highest fears:

Entering Residents	Staff
Failure[1]	Feeling rejected by others
Looking foolish	Prospect of a surgical operation
Feeling rejected by others	Failure
Feeling disapproved of	Parting from friends
One person bullying another	Feeling disapproved of
Making mistakes	Looking foolish
Enclosed places[1]	Making mistakes
Prospect of a surgical operation	One person bullying another
Being ignored	Sight of fighting[2]
Parting from friends	Being criticized

---

1 Residents more afraid than staff ($p \leqslant .05$).
2 Staff more afraid than residents ($p \leqslant .05$).

Out of 140 RSS items the residents reported less pleasure than staff on 29 items (1d,1e,2e,5,7a,8,13,14,16a,16b,17a,17b,17c,17d,17e, 19h,21,23,30f,30g,30i,31,32,33,34,41,42,45, and 48); staff reported less pleasure on 13 items (2b,11b,11i,12n,19a,19j,19h,191,19n,19o, 20f,36, and 53). Staff indicated a preference for active interpersonal events, e.g., being praised by others, talking, flirting, singing, and dancing. Residents preferred more aggressive sports, e.g., football, boxing, judo, and hunting, "Babies," and "Having somebody pray for you." The ten highest sources of pleasure were:

Entering Residents		Staff	
1.	Making love	1.	Being praised about your understanding of others[3]
1.	Being close to an attractive woman	2.	Being praised about your mind[3]
3.	Beautiful women	2.	Making love
4.	Nude women	4.	Looking at beautiful scenery[3]
5.	Making somebody happy	5.	Talking with people who like you[3]
6.	Being right about your work	6.	Being praised about your work
7.	Listening to music, Rock and Roll	6.	Being praised about personality[3]
7.	Talking with people who like you[3]	8.	Making somebody happy
7.	Happy people	9.	Having people seek you out for company[3]
10.	Being praised about your work	9.	Being close to an attractive woman
		9.	Talking to friends[3]

<u>Entering Residents vs. Current Residents</u>

Compared to the current residents, entering residents were slightly less assertive, slightly more afraid, and nearly identical in reported sources of pleasure.

Out of 30 RAS items the new admissions, compared to the current residents, were less assertive on 5 items (6,8,10,25, and 30).

Out of 76 FSS items, the new admissions were more afraid than the current residents on 7 items (17,27,54,62,63,64, and 68) including "Entering a room where other people are already seated," "People in authority," and "Feeling disapproved of."

Out of 140 RSS items, the new admissions reported more pleasure than the current residents on 8 items (1b,7f,11b,11c,12n,15,20f, and

[3] Staff reported more pleasure than residents ($p \leq .05$).

29c); new admissions preferred candy and more passive recreational interests, e.g., watching T.V., movies, and sports, and listening to music. Current residents indicated more pleasure on 2 items (2e, 8): coffee and nude men.

Current Residents vs. Staff

There were few differences between the current residents and staff except in reported sources of pleasure.

Out of 30 RAS items the residents were less assertive than staff on 2 items (9,23): "To be honest, people often take advantage of me," and "I often have a hard time saying "No"." Residents were more assertive on one item (6): "When I am asked to do something, I insist upon knowing why."

Out of 76 FSS items the residents were more afraid than staff on 5 items (5,9,10,26, and 57) including "People who seem insane," "Feeling angry," and "Enclosed places." Staff were more afraid on 4 items (22,44,54, and 62) including "Receiving injections," "Sight of fighting," and "Feeling disapproved of."

Out of 140 RSS items the residents reported less pleasure than staff on 26 items (1d,5,8,16a,16b,17b,17c,19h,29b,29c,30a-g, 30i,31, 32,33,34,41,42,45, and 48); residents reported more pleasure on 10 items (7c,11e,11i,11j,19k,19l,19m,19n,36, and 53). Again, staff preferred active interpersonal events, e.g., being praised, talking, flirting, singing, and dancing. Residents preferred more aggressive sports, "Babies," and "Having somebody pray for you."

DISCUSSION

Assuming that this self-report data accurately measures the actual assertiveness, fears, and preferred reinforcements of patients and staff, the common stereotype of the drug addict is both confirmed and qualified. Drug abusers were more passive, more anxious, and received less satisfaction from interpersonal transactions than the staff. On the other hand, staff and residents shared many very similar social anxieties and a few similar interpersonal values, e.g., "Talking with people who like you."

This study has important implications for research, program design, therapeutic processes, and staff development and training.

Validity

Measures validated with other populations in other contexts are not necessarily valid with drug abusers in a treatment setting. There are clear indications of validity in our findings. For example, heroin addicts would presumably be less afraid than a psychiatric staff of receiving injections, weapons, and the sight of fighting while being more afraid of people who seem insane. Similarly, patients entering a therapeutic community are presumably more afraid than current residents of entering a room where other people are already seated and people in authority. Still, residents' responses such as "When I am asked to do something, I insist upon knowing why" and their fear of "Enclosed places" may also be messages from patients to therapists regarding their relationship and conditions of treatment rather than a simple or accurate self-description.

Program Design

These instruments can identify patients' assertion deficits, anxieties, and values, thus constituting the first step in the development of programs in assertion training (Bower and Bower, 1976; Moffett and Stoklosa, 1976), systematic desensitization (Wenrich et al., 1976), and values clarification (Simon et al., 1972). For example, in Satori one goal is for social approval to become a more salient reinforcer for residents.

Staff's assertiveness, fears, and values can also be identified. In Satori, the staff's need for praise about their understanding of others, their work, their minds, and their personalities along with their fears of failure, making mistakes, and criticism require that we build in opportunities for staff support and positive feedback, especially when such gratification is not forthcoming from patients. These same needs and fears may account for staff's reluctance to adopt innovative treatment techniques. In addition, these questionnaires identify staff's vulnerabilities, personal sensitivities that may be used as levers for manipulation by certain patients.

Finally, specific fears regarding anger and angry transactions (e.g., bullying, fighting) must be considered in designing anger management programs (Pursell and Moffett, 1978).

Research as Therapy

The collection and feedback of group research data can be used therapeutically. Feedback of this study's results to Satori's residents demonstrated the communality of many hopes and fears, thus increasing self-acceptance and facilitating further self-disclosure. Feedback of outcome data helps patients re-evaluate their own unrealistically high or low change goals.

For individual patients, these questionnaires offer skills training in self-assessment of problems, with repeated measures providing feedback on their progress.

Research in Staff Development

Staff are often anxious about research because of its evaluative implications. Yet, open discussion and analysis of staff behavior is necessary to provide effective treatment, especially in a therapeutic community for drug abusers. In such settings, staff are prone to extreme misperceptions of themselves, other staff, and patients (Adler, 1973). It is essential that staff differentiate their accurate perceptions of others, which establishes therapeutic empathy, from their inaccurate projections, which often reflects an antitherapeutic externalization process (Reed et al., 1974). Active participation in relevant research can facilitate this corrective self-examination process. More importantly, a given staff can develop norms of questionning their assumptions, exchanging compliments and constructive criticisms, and experimenting with different treatment modalities.

Staff who are inexperienced with drug abusers can benefit substantially from exposing their stereotypes of drug addicts. At our hospital, clinical psychology interns prematurely dismiss drug patients as desirable clients because of erroneous stereotypes. It is important to present an accurate image of drug abusers in order to recruit competent staff for drug abuse treatment programs.

REFERENCES

Adler, G. 1973. Hospital treatment of borderline patients. *Am. J. Psychit.* 130: 32-36.

Bower, S.A. and Bower, G.H. 1976. *Asserting Yourself.* Menlo Park, CA: Addison-Wesley.

Cautela, J.R. and Kastenbaum, R.A. 1967. A reinforcement survey schedule for use in therapy, training, and research. *Psych. Rep.* 20: 1115-1130.

Kanfer, F.H. 1975. Report on outcome measures in behavior therapy. In I.E. Waskow and M.B. Parloff (eds.): *Psychotherapy Change Measures.* Rockville, MD: NIMH.

Kernberg, O.F. 1975. *Borderline Conditions and Pathological Narcissism.* New York: Aronson.

Krasner, L. 1975. Techniques of assessment in behavior therapy. In I.E. Waskow and M.B. Parloff's (eds): *Psychotherapy Change Measures.* Rockville, MD: NIMH.

MacKinnon, R.A. and Michels, R. 1971. *The Psychiatric Interview in Clinical Practice.* Philadelphia: Saunders.

Moffett, L.A. and Stoklosa, J.M. 1976. Group therapy for socially anxious and unassertive young veterans. *Int. J. Grp. Psychother.* 26: 421-430.

Platt, J.J. and Labate, C. 1976. *Heroin Addiction.* New York: Wiley.

Pursell, S. and Moffett, L.A. 1978. Management of anger in a therapeutic community for drug addicts. NDAC paper. Seattle, WA.

Rathus, S.A. 1973. A thirty-item schedule for assessing assertive behavior. *Beh. Ther.* 4: 398-406.

Reed, J.P., Bale, R.N. and Zarcone, V.P. 1974. Clinical processes in a residential drug dependence treatment program. *N. Am. Con. Al. and Drug Prob.* Paper. San Francisco.

Simon, S.B., Howe, L.W. and Kirschenbaum, H. 1972. *Values Clarification.* New York: Hart.

Vaillant, G.E. 1975. Sociopathy as a human process. *Arch. Gen. Psych.* 32: 178-183.

Wenrich, W.W., Dawley, H.H. and General, D.A. 1976. *Self-Directed Systematic Desensitization.* Kalamazoo, MI: Behaviordelia.

Wolpe, J. and Lang, P.J. 1964. A fear survey schedule for use in behavior therapy. *Beh. Res. and Ther.* 2: 27-30.

Zarcone, V.P. 1975. *Drug Addicts in a Therapeutic Community: The Satori Approach.* Baltimore: York.

GETTING CLOSER TO THE GROUND--DRUG ABUSE AND GROUNDING

Christine E. Miller, M. ed.

Tucson Awareness House, Inc.

The language of the substance abuser tells the story of their body.* This paper will discuss the relationships between street talk and grounding concepts.

Getting grounded is the state of having two feet rooted on the ground where people can feel support from the earth. Getting grounded is part of the bodywork process used in Bioenergetic therapy. This therapeutic process, originating with Wilhelm Reich and Alexander Lowen, engages the client in deep breathing, vocal expression, and physical stress exercises to facilitate energy flow and thus emotional understanding. The body speaks for itself about its life experience.

While the client is under stress in a basic grounding exercise, the therapist notices the variation in age of different parts of the body. The hands may be the hands of a forty-year old and the face that of a young man of twenty. The therapist also notices body armor on the client, such as a huge middle area, large burly legs, rounded shoulders and back. The lack of body support is evident in spindly legs, flat feet, and a caved in chest. All people have various walls or defense mechanisms manifested physically in the body. Defense mechanisms or body blocks shield people from that which they fear.

Drug abusers are ungrounded individuals. Their reality exists off the ground in space--"spaced out." "Getting high" is a means

---

* Pronouns "they" and "their" are used throughout the paper to include both male and female persons.

for a substance abuser to avoid coping with everyday reality. "Getting high" implies leaving the ground, the place where they must deal with the stress of family, peers, job, and school life.

This reality is visible in the addictive lifestyle which denies natural bodily functions, hence, deadening bodily feeling. Improper diet, lack of adequate rest, little physical exercise, decreased sexual desire--being "strung out"--are all signs of a body lacking vital energy. Living for a better and better "head trip" (drug experience) eventually leads to the end of a body trip (health failure or death). The abuser no longer experiences bodily feeling and acts without feeling in daily life, no longer caring for self, family, or friends.

Substance abuse disables a person from feeling grounded by creating illusory bodily and mental sensations. Grounding enables a person to recognize the physical/emotional strengths and weaknesses--a recognition that is necessary to respond appropriately (with feeling) in stressful situations.

"Falling" in drug treatment terms means failure. When a former heroin user shoots heroin after a period of abstinence or "staying clean," news spreads that John Doe "fell."

Ungrounded	"High"	"Shooting Up"
Grounded	"Clean"	"Straight"
Ungrounded		
Falling	"Fell"	Dirty Urine

As diagrammed above, both "shooting up" and "falling" are states of ungroundedness. The former user therefore "shoots up" becoming ungrounded, as a means to cope with stress. Why does someone "fall" in street terms? Bioenergetically, falling is an anxiety producing action. When a client is asked to fall into a pile of pillows on the floor, various fears may emerge, such as: falling into hell (punishment), falling into a pit of nothingness (self), falling into a pain or anger that lies within. Some attributes of falling in the literal sense are vulnerability, humiliation, and physical hurt. These are qualities that a drug abuser does not want to feel. Rather than finding out what feelings lie beneath the fear by falling to the ground (symbolically), rather than expressing the feelings in time of stress (literally), the addict copes by falling into dope. So when John Doe "fell," his base of support or ground was not stable enough to rely on. He became ungrounded by shooting dope. Grounding exercises and Bioenergetic therapy provide a framework for building a base of support within the body.

A RECOMMENDATION

Traditional drug treatment modes see drug addiction as *the problem*. After the client is drug free, the treatment focus is environmental problems: family, poverty, racism, education, sexism, etc. Bioenergetic therapy provides a missing link to this treatment model. The focus of therapy is on the client's ability to cope with the stress that they experience. Once a client becomes grounded, their coping ability and tolerance for stress increase as their body energy is tapped and feelings are expressed. This process establishes the client on firm ground, so that working on environmental issues is viable as the client becomes self-supporting.

REFERENCES

Lowen, A. 1970. *Pleasure: A creative approach to life.* New York: Penguin.

Lowen, A. 1958. *The language of the body.* New York: Collier MacMillan Publishers.

Lowen, A. 1975. *Bioenergetics.* New York: Penguin.

Lowen, A. and Lowen, L. 1977. *The way to vibrant health.* New York: Harper and Row.

LONG-TERM FOLLOW-UP OF THERAPEUTIC COMMUNITY CLIENTS*

Sherry Holland

Director of Research
Gateway Houses Foundation, Inc.

This is the first in a series of reports on a large-scale, long-term outcome evaluation of Gateway Houses, a residential drug-free treatment program (therapeutic community) for drug abusers in Illinois. Follow-up subjects (N=648) are a sample of all first admissions to a Gateway treatment facility between July, 1968 and June, 1974 (N=1418). Time between discharge from treatment and date of follow-up ranges from one to nine and one-half years.

Two earlier Gateway follow-up studies (Slotkin, 1972; Slotkin and Senay, 1973) presented post-treatment interview data concerning drug abstinence and employment for admissions during the program's first two years (July 1968-June 1970). The average follow-up period was one and one-half years. A third Gateway study (Holland, 1978) examined official arrest records one year prior to admission and one year following discharge for a sample of admissions between 1968 and 1974. Results indicated that pre-post decreases in arrests occurred as a function of time in treatment.

The primary objectives of this most recent follow-up are 1) to evaluate the independent contribution of the Gateway treatment ex-

* Funding for this project was provided in part by the Joyce Foundation, Chicago, Illinois. The author gratefully acknowledges the perseverance of the Policy Research Corporation in conducting the field work, the enthusiastic support of the Gateway Houses staff, and the special assistance of Michael Darcy, Sandie Swieca, and Amy Barron.

perience in facilitating both short- and long-term behavioral change; 2) to examine the relationships among pre-treatment (client) variables, treatment variables, and post-treatment outcomes and to identify those types of clients who appear to benefit from treatment in Gateway; and 3) to identify those features of the treatment experience which are associated with positive behavioral change.

The current report presents initial tabulations of the follow-up data and focuses on the issue of treatment effectiveness. Because the randomized experiment is not feasible at Gateway, a quasi-experimental design utilizing a contrast group of short-term residents and three treatment groups defined by longer stays in treatment is employed. Hypothetically, the comparison subjects provide change norms for the untreated population. Some data concerning the initial equivalence of the groups are presented here. Major multivariate analyses of client characteristics and of the relationships among client, treatment, and outcome variables, however, will not be presented in this report, but in subsequent reports in this series.

METHOD

Subjects

The follow-up sample consists of 648 former Gateway Houses residents. A sampling strategy was employed in order to obtain approximately equal numbers of follow-up subjects (Ss) in each of four time in treatment groups within three bi-yearly admission cohorts. Each first admission to a Gateway residential treatment facility between July 1, 1968 and June 30, 1974 (N=1418) was identified as to admission cohort (1968-1970; 1970-1972; 1972-1974); and time in treatment (one day to less than three months [Group I]; three months to less than nine months [Group II]; nine months and over but did not complete treatment [Group III]; and completed treatment [Group IV]). Varying proportions of subjects were randomly selected from each of the four time in treatment populations, i.e., approximately one out of four Group I Ss, one out of two Group II Ss, two out of three Ss, and all Ss in Group IV.

A Gateway "admission" is defined as a person who has been individually interviewed by a counselor concerning treatment and the individual's goals, has completed intake forms and received a unique identification number, and has entered a residential facility. Elapsed time between first contact with a program counselor and admission to treatment varies. Some persons are admitted to a facility immediately following the intake interview. Others may be requested to attend outpatient groups at a Gateway storefront facility for three to five days either because of lack of bed space in a residential facility or in order to test the prospective resident's motivation for treatment. A smaller number make contact with Gateway staff while in jail, hospital, or other treatment programs and may enter the program after a considerable period of time. It should be noted that time in treatment includes only time spent in a residential facility and does not include pre-admission contacts. By the same token, pre-admission losses (drop-outs during the intake process) were not eligible for inclusion in the follow-up sample.

Approximately 18% of admissions who leave treatment prematurely subsequently return to the program after absences ranging from 24 hours to several years. For persons who split and return, time in treatment is defined as their single longest treatment stay. This definition of treatment time is in conformity with clinical practice. Clinical staff members generally assume that regression occurs during the split and treat the readmission as a new resident. Irrespective of number of splits and readmissions, the individual is counted as a first admission only once and retains his/her unique identifier.

Cases in the original follow-up sample were substituted when the clinical record contained no last known address (N=3); and when the time between date of discharge and date of follow-up was less than one year (N=8).

Table I presents selected pre-admission characteristics of 395[1] interviewed follow-up Ss by time in treatment group. The table also shows chi-square and F values and the corresponding level of significance of difference among the four groups with respect to each characteristic. The follow-up Ss are predominantly male and white. Their mean age was 23 years when they entered Gateway. Two-thirds had never been married. They had completed between 10 and 11 years of school. Only 11% had a full-time job at admission, although another 62% had had some full-time employment prior to their admission. Mean number of months employed at the longest job was 13. Three-quarters were living in an apartment, house, or hotel prior to admission, while the remaining 25% came to Gateway from jail, hospital, other residential treatment programs, or had no stable living arrangements.

Average age at first use of any illicit drug (including marijuana) was 15 years. The primary drug for half the sample was opiates. For one-third, barbiturates, amphetamines, hallucinogens, inhalants, or cocaine was the primary drug. Fifteen percent reported that marijuana was their primary drug.

Over 90% admitted to having participated in illegal activities, other than illicit drug use, at some point before entering Gateway.

---

1 Five additional completed interviews were not available at the time of data compilation for this report.

Table I. Selected characteristics of follow-up sample at admission by time in treatment group.

Characteristic	Time in Treatment Group				Statistic	Significance Level
	I	II	III	IV		
Sex						
Male	74%	72%	77%	67%		
Female	26	28	23	33	X^2= 3.20	n.s.
Total %	100%	100%	100%	100%		
Total N	77	92	109	117		
Race						
White	64%	54%	61%	67%		
Black	29	40	37	23		
Spanish	6	4	2	9	X^2=12.97	n.s.
Other	1	1	0	1		
Total %	100%	99%	100%	100%		
Total N	77	92	109	117		
Age						
Mean (in years)	21.5	23.0	23.7	23.1		
S.D.	6.0	7.5	7.4	8.4	F = 1.49	n.s
N	77	92	109	117		
Education						
8 years or less	16%	18%	14%	8%		
9-11	45	56	48	47		
12	24	13	24	29	X^2=13.05	n.s.
13 or more	16	12	15	16		
Total %	101%	99%	101%	100%		
Total N	76	92	109	117		
Mean (in years)	10.7	10.3	10.9	11.0		
S.D.	2.2	1.9	1.9	2.1	F = 2.13	n.s.
N	76	92	109	117		
Marital Status						
Never Married	75%	65%	65%	67%		
Married	9	15	9	8	X^2= 6.38	n.s.
Separated, Divorced, Widowed	16	20	26	25		
Total %	100%	100%	100%	100%		
Total N	77	92	109	117		

Table I (cont'd.).

Characteristic	Time in Treatment Group				Statistic	Significance Level
	I	II	III	IV		
Living Arrangement at Admission						
Apartment, House	75%	61%	63%	68%		
Hotel, Rooming or Boarding House	13	8	5	9		
Jail	7	23	23	15	X^2=21.28	n.s.
Hospital	4	2	5	5		
Treatment Program	1	0	1	0		
Other	0	6	3	3		
Total %	100%	100%	100%	100%		
Total N	77	92	109	117		
Full-Time Job at Admission						
Yes	9%	5%	14%	15%		
Previous Only	57	66	65	60	X^2= 9.21	n.s.
Never	34	28	21	25		
Total %	100%	99%	100%	100%		
Total N	77	92	109	117		
Months Employed Longest Job						
Mean (in months)	13.9	10.6	14.0	14.1		
S.D.	21.4	17.3	18.9	20.6	F = 0.72	n.s.
N	77	90	108	116		
Age First Used Illicit Drug						
Mean (in years)	14.8	14.8	14.9	15.4		
S.D.	3.7	3.1	3.7	4.7	F = 0.42	n.s.
N	76	90	102	115		
Primary Drug						
Opiates	50%	55%	56%	44%		
Other Drugs, excluding marijuana	36	33	31	33		
Marijuana	13	11	13	21	X^2= 7.30	n.s.
None	1	1	0	2		
Total %	100%	100%	100%	100%		
Total N	75	87	109	115		
Ever Engaged in Illegal Activities (other than drug use)						
Never	9%	7%	3%	12%		
Yes	91	93	97	88	X^2= 7.20	n.s.
Total %	100%	100%	100%	100%		
Total N	76	92	106	115		

Table I (cont'd.).

Characteristic	Time in Treatment Group				Statistic	Significance Level
	I	II	III	IV		
Mean age first engaged in illegal activities	15.1	15.6	14.9	15.6		
S.D.	4.7	4.5	3.7	4.3	F = 0.56	n.s.
N	68	84	102	100		
First Engaged in Illegal Activities						
Prior to onset of drug use	33%	34%	33%	26%		
Same time as onset of drug use	38	19	29	31	X^2=10.08	n.s.
After onset of drug use	29	47	37	43		
Total %	100%	100%	99%	100%		
Total N	69	85	102	101		
Arrests						
None	29%	20%	9%	18%	X^2=11.53	.01
1+	71	80	91	82		
Total %	100%	100%	100%	100%		
Total N	73	89	108	115		
Mean age at first arrest	16.4	17.0	16.9	17.2		
S.D.	4.1	4.8	3.7	4.5	F = 0.48	n.s.
N	51	71	95	94		
Times in Jail						
None	46%	29%	31%	30%		
1	16	22	17	22	X^2= 8.14	n.s.
2+	38	49	52	48		
Total %	100%	100%	100%	100%		
Total N	76	90	108	116		
Mean times in jail	2.8	3.2	5.7	5.2		
S.D.	5.9	4.7	9.5	10.3	F = 2.86	.05
N	76	90	108	117		
Treatment Experiences Prior to Gateway						
None	57%	49%	50%	44%		
1	27	20	21	20	X^2=10.50	n.s.
2+	16	31	29	37		
Total %	100%	100%	100%	101%		
Total N	77	91	109	117		
Mean	0.9	1.1	1.2	1.7		
S.D.	1.8	1.4	2.1	2.6	F = 2.57	n.s.
N	77	91	109	117		

Table I (cont'd.).

Characteristic	Time in Treatment Group				Statistic	Significance Level
	I	II	III	IV		
Ever Treated for Emotional Problem						
Yes	32%	38%	33%	47%		
No	67	62	67	53	$X^2 = 5.74$	n.s.
Total %	99%	100%	100%	100%		
Total N	77	92	109	116		
Legal Status at Admission						
None	66%	42%	43%	47%		
Some	34	58	57	53	$X^2 = 12.73$	.01
Total %	100%	100%	100%	100%		
Total N	77	91	109	117		
Type of Admission						
Voluntary	77%	69%	71%	74%		
Legal commitment to treatment, not necessarily Gateway	1	1	0	1		
Legal commitment to Gateway	21	30	29	25	$X^2 = 2.52$	n.s.
Total %	99%	100%	100%	100%		
Total N	75	91	109	117		

Of those, thirty-one percent became involved in illegal activities prior to the onset of drug use, while 29% started using drugs and engaging in other illegal activities at about the same time. Forty percent did not become involved in illegal activities until after they began using drugs. Ss were an average of 15 years old when they began illegal activities.

Eighty-two percent of the sample had been arrested at least once prior to admission. The time in treatment groups are significantly different on this measure. Percent never arrested is highest for Group I (29%) and is lowest for Group III (9%). Average age at first arrest is 17 years.

Two-thirds had been jailed at least once. Mean times in jail significantly differentiates the time in treatment groups. (It will be noted that arrests, times in jail, and the third characteristic on which the groups differ, legal status at admission, are not independent but correlated measures.) Ss who remained longer in treatment had been in jail more frequently on the average than Ss who left the treatment program early. The function relating times in jail to time in treatment has a significant linear component ($F=18.26$, $p<.001$).

Half of the follow-up Ss had had at least one treatment experience for drug or alcohol abuse prior to their admission to Gateway. Although mean number of prior treatment experiences increases across the four groups, from 0.9 for Group I to 1.7 for Group IV, the differences are not significant. Thirty-eight percent had also sought assistance with emotional problems before entering Gateway.

As mentioned above, the time in treatment groups are significantly different with reference to legal status at admission. "Legal status" includes warrants pending, being on probation or parole, under supervision, or under a deferred sentence. Whereas 53-58% of Groups II-IV entered treatment with some form of legal status, only 34% of Group I did so. Legal status, however, does not necessarily entail involuntary committment to treatment. Almost three-quarters of the follow-up Ss were voluntary admissions.

In summary, then, the four time in treatment groups show significant pre-treatment differences only in the area of criminality. Increased time in treatment is associated with higher pre-treatment numbers of arrests, times in jail, and percent who entered treatment with legal status.

Instrument and Other Data Sources

The major source of client and outcome data is the follow-up questionnaire. Self-reported drug use (current), arrests and convictions, and employment are validated by means of urinalysis, official arrest records, and Social Security Administration records. Death certificates are obtained for deceased Ss. Treatment variables, including admission and split dates, type(s) of discharge(s), and assigned facilities are retrieved from Gateway clinical records. Only interview data are presented in this report.

The criterion outcome measures include type and frequency of drug use; type, quantity, and frequency of alcohol use; employment; education; sources of support; treatment for drug and alcohol abuse; treatment for emotional problems; and criminality. Other social, demographic, and attitudinal data are obtained, such as marital status, living arrangements, family background, attitudes toward the Gateway Houses treatment program, self-rating of behavioral improvement over time and its relationship to treatment, and the like. Questions are asked with reference to five distinct time periods: 1) pre-treatment ever: the subject's life up to one month prior to entering Gateway; 2) pre-treatment current: the one month prior to date of admission; 3) during treatment in Gateway; 4) post-treatment ever: from date of discharge to one month prior to the follow-up interview; and 5) post-treatment current: one month prior to the follow-up interview. For the outcome criteria, lifetime measures pre- and post-treatment as well as year-by-year data for the five years preceding admission and for each year following discharge up to five years are obtained. Thus, pre-post comparisons of criterion behaviors can be made over equal time periods for each S up to five years after discharge from the treatment program.

In an effort to identify whether the order of the questions interacts with reported behavioral improvement over time, two versions of the questionnaire are randomly administered to Ss. In Version I, the interview begins with questions concerning post-treatment activities. Version II of the questionnaire begins with pre-treatment questions.

Procedure

The field work was carried out by the Policy Research Corporation (PRC) under contract to Gateway Houses Foundation, Inc. Identifying and locating data for the follow-up sample were obtained from Gateway clinical files. PRC recruited and trained a professional field staff, and assumed full responsibility for contacting, locating, and interviewing former residents. While Gateway and PRC staffs worked closely together during all phases of the field work, PRC maintained a policy of strict confidentiality regarding the whereabouts of follow-up Ss and of the content of the follow-up interviews.

As of March 1978, 497 of the 648 follow-up Ss (77%) had been interviewed or otherwise accounted for. Table II shows the current interview status of the sample by time in treatment group. Percent interviewed ranges from a low of 42% for Group I (short-term) Ss to a high of 83% for Group IV Ss (graduates). Seventy-eight percent of the interviews were face-to-face. The remainder were obtained by telephone, usually from out-of-state Ss.

For this preliminary assessment of the effectiveness of treatment, four outcome measures are examined: employment, drug use, criminality, and treatment for drug and alcohol abuse. The employment, criminality, and treatment measures refer to the two-year period prior to admission to Gateway and the two-year period following discharge. For drug use, the measure selected is drug used most frequently during the month before admission and during the month two years following discharge.[2] In the event S was incarcerated, in

---

[2] Eleven Group III Ss and 18 Group IV Ss who had been out of treatment less than two years as of follow-up were omitted from the analysis.

Table II. Interview status of follow-up sample by time in treatment group.

Interview Status	Time in Treatment Group				Total
	I	II	III	IV	
	%	%	%	%	%
Interviewed	42	58	69	83	62
Personal Interview	(73)	(74)	(75)	(87)	(88)
Telephone	(27)	(26)	(25)	(13)	(22)
Deceased	9	12	6	4	8
Out of Country	2	3	1	*	2
In Jail, Not Interviewable	*	0	*	0	*
Refused	7	4	4	6	5
Not Located	39	23	20	7	23
Total %	99	100	100	100	100
N =	182	163	162	141	648

* Less than one percent.

residential treatment, or otherwise institutionalized during the index months, frequency of drug use during the last full month S lived "in the street" was obtained. All measures are presented for time in treatment groups.

RESULTS

Figure 1 shows mean number of months employed full-time during the two years preceding admission to and the two years following discharge from the Gateway Houses program by treatment group. All groups show increases in post-treatment employment over pre-treatment, although the increase for Group I is very small.

Table I lists the pre- and post-employment means for each group, as well as F statistics, which test the extent to which there are differences among the groups before and after treatment; and t statistics, which test the extent to which pre- and post-treatment scores differ for the same group.

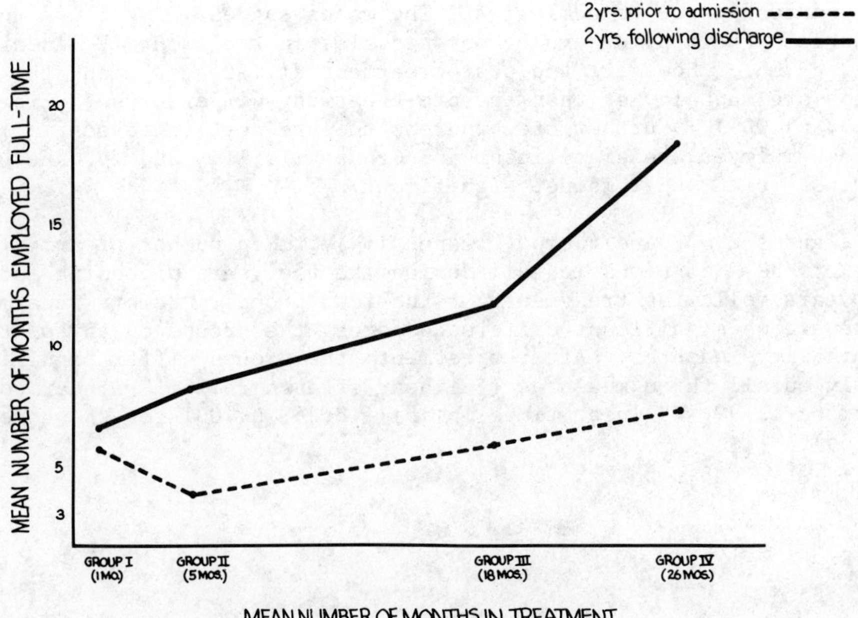

Fig. 1. Mean number of months employed full-time two years prior to admission and two years following discharge by time in treatment.

Table III. Mean number of months employed full-time two years prior to admission and two years after discharge by time in treatment group.

Group	N	Mean Number Months Employed: Two Years Prior to Admission	Mean Number Months Employed: Two Years After Discharge	Value of t	Significance Level
I	77	5.64	6.52	0.89	n.s.
II	91	3.75	8.14	4.48	.001
III	98	5.61	11.54	5.91	.001
IV	98	7.03	18.25	10.98	.001
Value of F		2.65	33.49		
Significance Level		.05	.001		

There is significant variability among the groups on the employment measure prior to admission. The means suggest a positive relationship between prior employment and time in treatment (F linear= 3.91, p<.05). However, the post-treatment differences among the groups are much larger than the pre-treatment differences (F linear= 93.39, p<.001). Furthermore, whereas the pre-post increases in mean months employed are significant for Groups II, III, and IV, the small increase for Group I is not significant.

Figures 2, 3, and 4 show, respectively, mean number of arrests, convictions, and times in jail during the two years preceding and the two years following treatment for the four groups. Before treatment, there are no significant differences among the groups on any of the criminality measures. After treatment, the groups differ significantly on all three measures (F linear arrests=25.15; F linear convictions-12.99; F linear times in jail=18.26; p<.001 for the three F ratios).

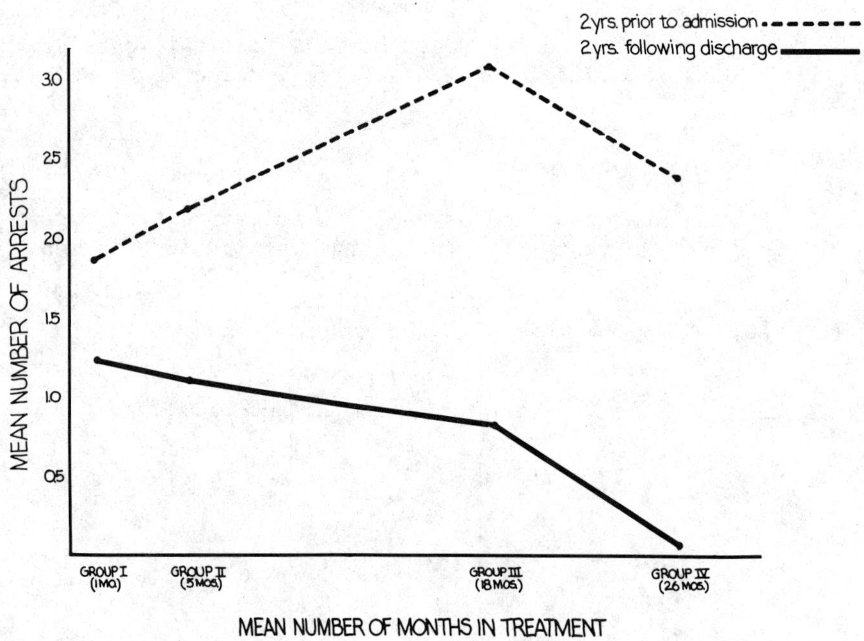

Fig. 2. Mean number of arrests two years prior to admission and two years following discharge by time in treatment.

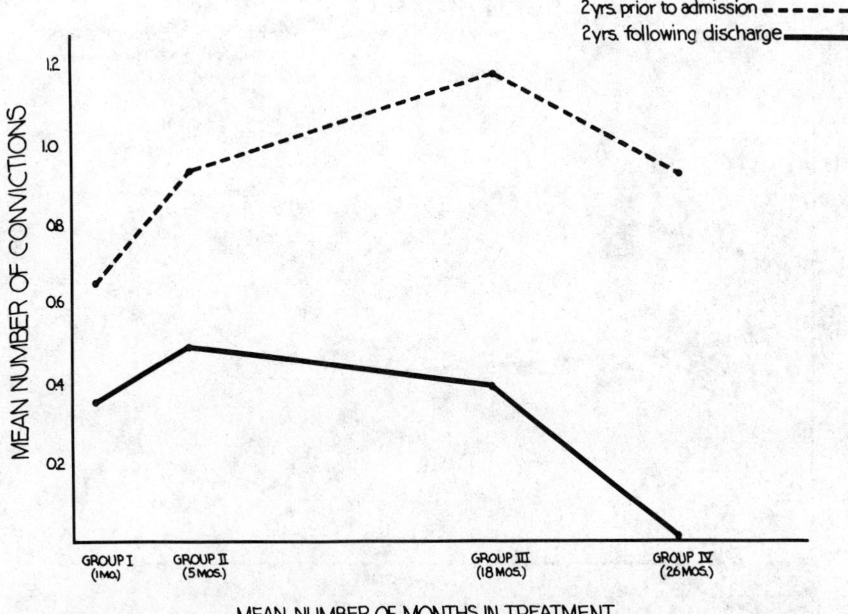

Fig. 3. Mean number of convictions two years prior to admission and two years following discharge by time in treatment.

The t statistics presented in Tables IV, V, and VI reveal no significant pre-post differences in arrests and times in jail for Group I, although Group I does show a significant decrease in mean number of convictions. Groups II-IV show significant pre-post decreases on all three criminality measures.

Figure 5 shows percent in each treatment group using opiates, other drugs (barbiturates, amphetamines, hallucinogens, inhalants, or cocaine), marijuana, or no drugs during the month prior to admission and the month two years after discharge. In general, each S's single most frequently used drug at each of the two points in time is shown. The exception is marijuana. Marijuana use is shown only when there is no other drug use. If marijuana is the most frequently used drug but S also is using opiates and/or other drugs, the second most frequently used drug is shown. Thus, Ss in the opiate and other drug classes may be using more than one drug. Ss in the marijuana class are using only marijuana.

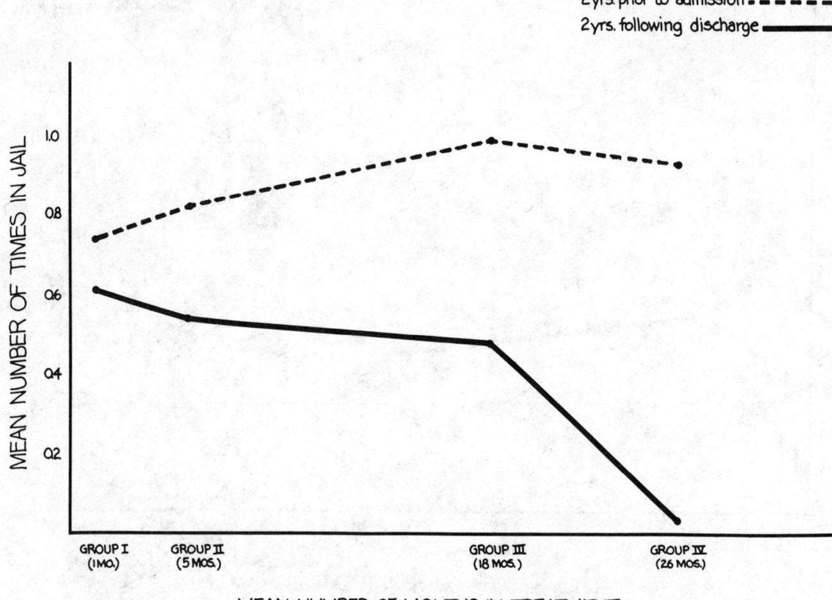

Fig. 4. Mean number of times in jail two years prior to admission and two years following discharge by time in treatment.

Table IV. Mean number of arrests two years prior to admission and two years after discharge by time in treatment group.

Group	N	Mean Number of Arrests: Two Years Prior to Admission	Mean Number of Arrests: Two Years After Discharge	Value of t	Significance Level
I	71	1.86	1.23	1.40	n.s.
II	87	2.18	1.10	3.22	.01
III	94	3.06	0.81	5.31	.001
IV	95	2.37	0.05	6.73	.001
Value of F		1.82	9.40		
Significance Level		n.s.	.001		

Table V. Mean number of convictions two years prior to admission and two years after discharge by time in treatment group.

Group	N	Mean Number of Convictions		Value of t	Significance Level
		Two Years Prior to Admission	Two Years After Discharge		
I	71	0.65	0.35	2.16	.05
II	87	0.93	0.49	2.65	.02
III	94	1.17	0.39	3.64	.001
IV	95	0.92	0.01	5.26	.001
Value of F		1.42	7.74		
Significance Level		n.s.	.001		

Table VI. Mean number of times in jail two years prior to admission and two years after discharge by time in treatment group.

Group	N	Mean Number of Times in Jail		Value of t	Significance Level
		Two Years Prior to Admission	Two Years After Discharge		
I	77	0.74	0.61	0.89	n.s.
II	92	0.82	0.54	2.29	.05
III	97	0.99	0.48	3.38	.01
IV	98	0.93	0.04	5.73	.001
Value of F		0.71	7.84		
Level of Significance		n.s.	.001		

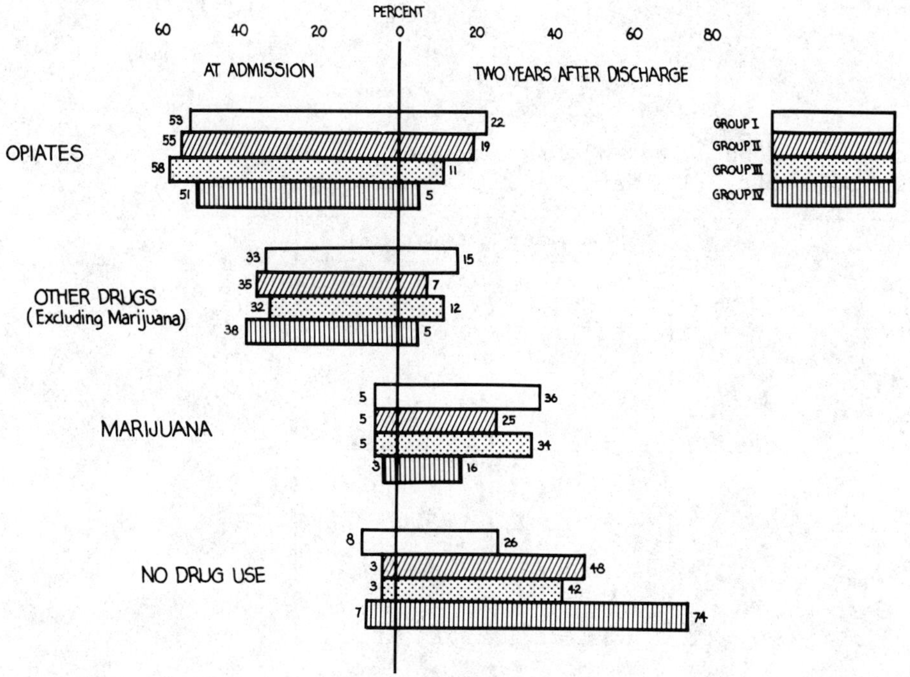

Fig. 5. Percent using opiates, other drugs (excluding marijuana), marijuana, or no drugs prior to admission and two years after discharge.

The distributions of Ss using opiates, other drugs, marijuana, or no drugs at admission are approximately equal for the four groups (X^2=5.10, n.s.). Two years after discharge, the drug use distributions are significantly different (X^2=47.31, $p<.001$). While all groups report reduced drug use after treatment, reductions are greatest for Group IV, somewhat less for Groups II and III, and least for Group I.

For clarity of presentation, frequency of use is disregarded in Figure 5. Table VII shows the percentages in each group using daily, weekly, or monthly within each drug class. The order of the groups remains relatively unchanged with the introduction of frequency of use.

Figure 6 shows mean number of times treated for drug or alcohol abuse two years before and after treatment. The groups are not significantly different with respect to number of prior treatment experiences, although the mean is highest for Group IV. There is significant variability among the groups after treatment. However, the

Table VII. Most frequently used drug type and frequency of use at admission and two years after discharge by time in treatment group.

Drug Type and Frequency	At Admission Group				Two Years After Discharge Group			
	I	II	III	IV	I	II	III	IV
	%	%	%	%	%	%	%	%
Opiates								
Daily	46	50	49	45	13	13	5	2
1-6 days/week	7	4	8	5	9	6	3	3
1-3 days/month	0	1	1	1	0	0	3	0
(Subtotal)	(53)	(55)	(58)	(51)	(22)	(19)	(11)	(5)
Other Drugs (Excluding Marijuana)								
Daily	16	12	15	18	7	1	4	0
1-6 days/week	16	19	15	13	7	2	4	3
1-3 days/month	1	4	2	7	1	4	4	2
(Subtotal)	(33)	(35)	(32)	(38)	(15)	(7)	(12)	(5)
Marijuana								
Daily	0	1	3	0	9	9	7	0
1-6 days/week	4	4	2	2	15	13	22	10
1-3 days/month	1	0	0	1	12	3	5	6
(Subtotal)	(5)	(5)	(5)	(3)	(36)	(25)	(34)	(16)
No Drug Use	8	3	3	7	26	48	42	74
Total %	99	98	98	99	99	99	99	100
Total N	74	90	97	99	74	90	97	99

pre-post decreases for Groups II and III are not significant. Only the decrease in mean treatments for Group IV is significant (see Table VIII).

DISCUSSION

The object of this preliminary analysis of selected outcome measures for a sample of former therapeutic community clients is to try to determine whether observed changes in behavior can be attributed to treatment. In the absence of a randomized design, it is difficult to estimate the extent to which the observed changes may be associated with systematic differences among clients. Nevertheless, we are willing to conclude that observed pre-post differences in behavior

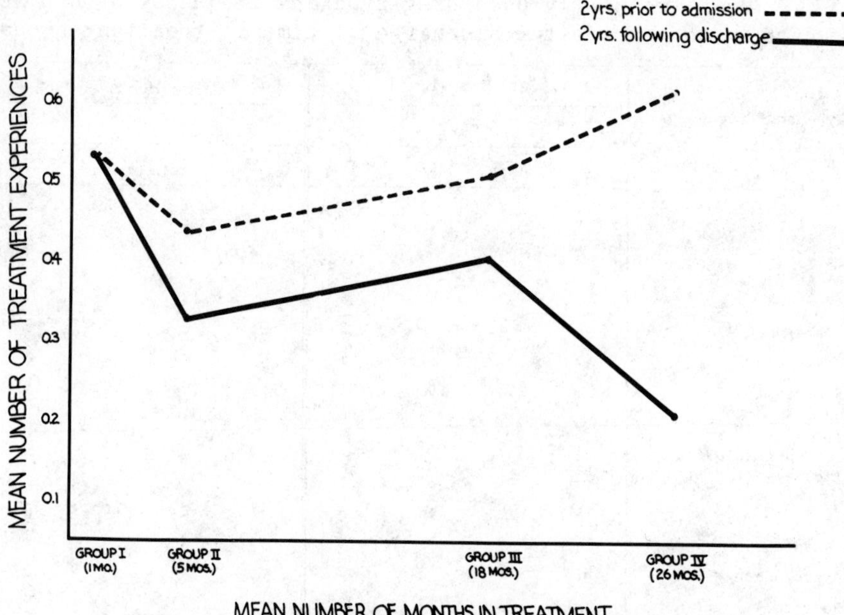

Fig. 6. Mean number of treatment experiences two years prior to admission and two years following discharge by time in treatment.

Table VIII. Mean number of treatment experiences for drug and alcohol use two years prior to admission and two years after discharge by time in treatment group.

Group	N	Mean Number of Treatments: Two Years Prior to Admission	Mean Number of Treatments: Two Years After Discharge	Value of t	Significance Level
I	77	0.57	0.56	0.07	n.s.
II	91	0.47	0.35	1.16	n.s.
III	98	0.52	0.41	0.83	n.s.
IV	98	0.63	0.22	2.84	.01
Value of F		0.29	3.48		
Significance Level		n.s.	.025		

are compatible with a treatment effect hypothesis when 1) Ss who receive varying amounts (i.e., duration) of treatment do not appear to be significantly different before treatment; 2) these Ss do show significant differences after treatment; and 3) there is a positive relationship between amount and direction of change and amount (duration) of treatment.

The data presented in this report provide powerful, if not unequivocal, evidence of a treatment effect. With respect to the outcome criteria, treatment groups showed significant pre-treatment variability only on the employment measure, whereas there were significant post-treatment differences among the groups on employment, arrests, convictions, times in jail, drug use, and treatment experiences. The contrast group (Group I) showed significant pre-post change only on the convictions measure. Groups II, III, and IV showed significant changes with regard to employment, arrests, convictions, times in jail, and drug use. Group IV also had significantly fewer treatment experiences post Gateway. Furthermore, degree of change increased as a function of time in treatment.

With respect to pre-treatment client variables, the group differences in criminality and legal status suggest differential motivation for treatment. A number of interview questions deal with the issue of motivation, and these will be explored in subsequent reports.

Despite the proportionately few significant client differences evident in this analysis, it is possible that clustering of more reliable composite measures will differentiate client types who evidence initial differential probability of successful outcome. It should also be possible to identify short-term clients who benefit from treatment, as well as long-term clients with negatives outcomes. These and other more detailed analyses of client-treatment-outcome interactions will be undertaken and should serve to broaden our understanding of treatment dynamics.

REFERENCES

Holland, S. 1978. Gateway Houses: Effectiveness of treatment on criminal behavior. *Int. J. of the Addictions*. 13(3), 369-381.

Slotkin, E.J. 1972. *Gateway, the First Three Years*. Unpublished manuscript.

Slotkin, E.J. and E.C. Senay. 1973. *Gateway's Success in the Rehabilitation of Drug Abusers*. Unpublished manuscript.

NEW DIRECTIONS HOUSE: DAY TREATMENT PLUS ACTIVITIES

Richard G. Greer, M.Ed. and Samuel R. Patterson, M.Ed.

New Directions House - PPC Drug Treatment and Rehabilitation Program
Philadelphia, Pa.

INTRODUCTION

The rehabilitation of drug abusers involves many diverse treatment approaches. It is essential that we recognize, identify and gain knowledge of other less known treatment modalities. This is particularly true in the case of the New Directions House program, which combines activity therapy in a day treatment environment. The three year experience in an urban setting is objectively described, pointing attention to the developmental process, philosophy of service, techniques, and our team approach.

Our focus on Day Treatment vs. Therapeutic Community as a treatment alternative to incarceration, is designed to provide realism and new insights for criminal justice system personnel.

The imaginative steps to find alternatives to frustration, hopelessness, and loneliness is documented to serve program administrators as a model for future planning and implementation in the continued struggle against drug abuse.

DEVELOPMENT

New Directions House was officially opened March 4, 1974. It was conceptually established in September of 1972 as part of a comprehensive Drug Abuse Treatment and Rehabilitation Program. The project funded by NIDA H-80 grant was designed by the Project Director, Dr. Alfred Friedman, Ph.D. The project provides drug abuse treatment to catchment area #4 located in West Philadelphia (see Figure 1). Modalities included are methadone maintenance, out-patient drug-free,

COMMUNITY MENTAL HEALTH AND MENTAL RETARDATION CENTER OF THE PHILADELPHIA PSYCHIATRIC CENTER

CATCHMENT AREA #4

POPULATION: 169,000

Note: Code numbers on map designate service locations. Match with code numbers in the following service description. Code letters refer to Specialized Agency locations.

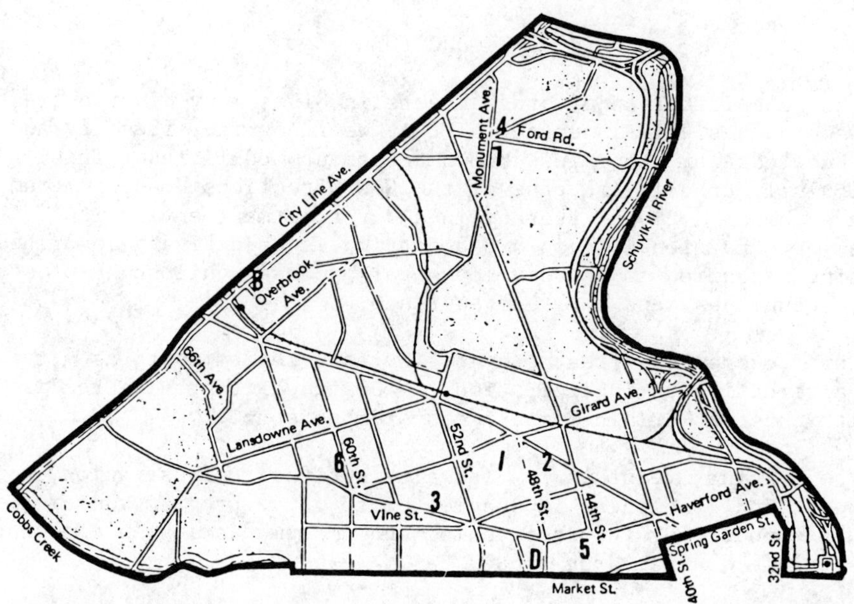

Fig. 1.

in-patient detoxification and drug-free day care. The latter being the subject of this presentation.

New Directions House became a reality after a series of delays based on locating a suitable site and difficulty in gaining the acceptance of the immediate community.

The location ideally should be as close to the clients as possible. The clinic designed as a Social Rehabilitation Center, that would provide aftercare services, defined its target population based on arrest records, by precinct; emergency room records, existing drug treatment programs and records from other social institutions.

The resistance of the immediate community to the proposed site resulted from several factors:

1. Hostile parents,
2. Close proximity of churches and schools,
3. Zoning board bureaucracy,
4. Negative opinions and publicity about addicts and treatment programs,
5. Political and personal goals of influential community members.

The process used to extinguish or resolve these factors involved a series of community meetings and a Zoning Board hearing. This is more easily stated than done. The process is one that allows for the discharge of a great deal of hostility. This cathartic effect came more as a phenomena than the result of careful design. However, subsequent interactions with the community were less hostile and more concilatory.

The Zoning approval was granted, but with provisos regarding staffing and the hours of operation. These provisos did not require any major alterations to the program's design, therefore, the criteria for a day care facility was fulfilled.

The approval of the community was not immediately measurable. Community members were invited to join the existing Community Advisory Board of the Philadelphia Psychiatric Center Drug Treatment Project, few did accept and are active participants. Further indicators of increasing community support are shown in various ways:

1. Total lack of vandalism, i.e., graffiti, broken windows, etc.,
2. Cooperative agreements and interagency affiliations,
3 The use of community recreational facilities by New Directions House groups,
4. The participation in community cultural and aesthetic events,
5. Police support.

The facility originated from two consecutive row houses that first appeared as prime candidates for condemnation. Through extensive renovations, the houses were developed into a most attractive facility. The fact that it has remained so reflects the respect the clients and the community hold for the clinic.

ORGANIZATION

New Directions is one of five (5) drug treatment facilities of the Philadelphia Psychiatric Center (PPC). The Philadelphia Psychiatric Center is a private non-profit multi-services organization. Among its components are psychiatric hospitals, community-based mental health programs, a research unit and vocational skills development unit. Aside from the NIDA funding, additional revenues are generated from medical assistance payments on a fee for service basis and from client fees.

New Directions uses the Day Care modality. The flexibility of this modality enables the clinic to use activities as a therapeutic approach within a milieu therapy environment. The activities used are those which were determined as being significant in the majority of our clients' lives at one time or another. The activities are music, art, recreation, Domestic Arts and Horticultural therapy.

The static capactiy of New Directions House is ninety-two (92) clients. The staff consists of:

	% Treatment	% Administration
1 - Clinic Director	50	50
1 - Vocational Rehabilitation Specialist	75	25
1 - Secretary		100
1 - Billing Clerk		100
2 - Security/Maintenance	10	90
1 - Art Therapist	100	
1 - Domestic/Horticultural Therapist	100	
1 - Music Therapist	100	
1 - Recreation Therapist	100	

New Directions House operates in a manner that allows for a creative utilization of its personnel. Without specifically speaking, this is necessary in order to meet the myriad of needs and concerns presented on a daily basis within a day care modality. All of the staff are involved with the clients around a variety of experiences. This is consistent with the reality of seeing individuals in multidimensional roles. It also provides for variety of interaction, resulting in growth and alleviation of the Burnout syndrome.

The decision making within the clinic is by consensus. We may not all agree at any one time but all must be willing to state their opinion and work them through. Divisiveness is counter productive and cannot be tolerated. The entire clinic operates a management by objectives method whereby all staff participates in identifying major areas of responsibility in their work, sets standards for performance and measures results against those standards (Drucker, 1959).

In taking into account all of the problems affecting the addict treatment is designed to include:

Vocational rehabilitation
Interface with the criminal justice system
Research, Demonstration and Evaluation
Education

PHILOSOPHY

The basic New Directions House philosophy is that individuals that abuse drugs have problems in life which they try to solve with the use of drugs. These problems can be generally categorized as feelings of inadequateness or avoidance conflicts. New Directions House is committed to sharing its concern, skills, knowledge and experience with each individual.

Another aspect of the New Directions House philosophy is that the client should not be seen as completely responsible for his difficulties. The problems of the poor, the Black and the needy are the products of the systematic conditioning of one segment of society by another. With an economic motive clearly in view, minority groups have been systematically conditioned to internalize inferior attitudes and behavior.

New Directions House recognizes the incredible diversity of its clients. However, there is one theme that persists constantly in our striving to get each client to confront his problems. That is the doctrine of free will. Branden (1969) states, "Free Will"--in the widest meaning of the term--is the doctrine that each client is viewed as capable of performing actions which are not determined by forces outside his/her control. Free will consists of a single action, a single basic choice. To think or not to think. One is not allowed to escape this choice. It is not made by the gods, the stars, the chemistry of the body, the structure of the family constellation or the economic organization of society. To recognize that each client is free to think or not to think is to recognize that in a given situation he or she has the ability to focus on what is really concerning him in a real sense, and has the freedom to adopt any attitude regarding this concern. This freedom of choice is part of the story and half the truth. Being free is but the negative aspect of the whole phenomena whose positive aspect is being responsible (Frankl, 1970). In addition (Glasser, 1976) the typical . . . addict is a person who is severely frustrated in his own particular search for love and worth. He made the first choice to give up usually when he was very young. If little or no praise or reinforcement of positive actions from the family is received, then the individual develops a poor self image with no clear idea of self worth or potential for achievement and success (Satir, 1972). A person's view of himself plays a crucial role in value choices (Branden, 1969). Each individual can only absorb just so much anxiety, pain and frustration before reaching out for release and relief if only momentarily. Fulfillment, pleasure, recognition, a sense of personal value, a sense of worth, the enjoyment of loving and being loved are not optional, they are the facts of life. Each individual finds this in his or her own way, but in general everyone finds them through (1) love that is through loving and being loved and (2) by doing something one believes is worthwhile (Glasser, 1976).

We refuse to define the problem as "drug abuse". When drug abuse is considered as the problem, one tends to think people are problems and that drugs are major problems and that we are compelled to provide an alternative, which is only a relief of the symptom rather than leave the client free to define what the real problem is.

New Directions House holds a non-pathological view of its clients. Because someone is experiencing difficulties in life does not mean he is sick or crazy. However, it must be stated that we will not do a disservice to anyone who after extensive diagnosis it becomes obvious that he is severely incapacitated and cannot be helped by New Directions House, or might possibly regress--a referral is made in the appropriate manner. In the world of individuals there is no "best".

DESCRIPTION OF CLIENTS

Our participants ages range from 18 to 55, are urban, varied educational backgrounds, poor, with histories of: broken homes; unemployment; illegal activity; anti-social behavior patterns; conflicts with their family, peers and community; frequent contacts with the criminal justice system; prior treatment experiences; and multi-drug absue. The typical New Directions House client is a Black male, 29 years of age, living within four (4) city blocks of the clinic and is currently on probation or parole. His drug abuse history involving heroin use at age 17 and his entrance into treatment occurred at age 23.

A review of the literature with respect to demographic information from treatment services in other sections of the United States, I found a striking similarity (Ball, 1965; Chein, 1968; Gerard, 1955; Newmeyer, 1972; Saloway, 1973; Waldorf, 1973).

We recognized early on the possible advantages of treating clients who receive methadone treatment elsewhere, with "drug free" clients. Certain members of this latter grouping have served as a positive influence on others who are reportedly involved in multi-drug aubse. It is of equal importance to note however, that case management must be coordinated between the respective services to avoid counter treatment approaches.

We have had greater success in attracting Black narcotic abusers than whites. The wide variance might be attributed to our environmental setting, or as indicated by O'Neal (1975) that over one half of Black male drug abusers are unemployed, compared to less than one third of white males. The social/economic implications are pronounced, that is, the unsolved problems of prejudice, discrimination and segregation remain for the masses.

TREATMENT

New Directions employs an eclectic treatment approach that emphasizes each individual's own reality. The levels of empathy, positive regard and congruence provided by the therapists in their actual activity procedure relates positively to the cognitive growth of their clients. The development of the concept of Milieu therapy using activities suggests that client movement is, to a large extent, a function of the therapist. Given particular interaction patterns of relevant variables, a variety of counseling approaches may be employed to produce favorable results.

ART THERAPY

Much interest has been maintained in drawing analysis as a projective technique. Anna Freud was one of the first child analysts to use drawing in her psychoanalytic work. Beginning in the 1930's, Paul Scheider and Lauretta Bender wrote extensively about art work in relation to diagnosis and therapy. Bender (1968) stresses the meaning of art productions as creative activity and as projective phenomena and symbolism. Many diagnostic drawing techniques have been developed. The House-Tree-Person test by Buck (1950) as a means of exploring areas of intellect and personality. Karen Machover's innovative work, Personality-Projection-in-the-Drawing-of-the-Human-Figure have been two (2) of the forerunners in the use of art productions for diagnostic assessments.

The use of spontaneous art as an adjunct to other forms of therapy in institutions has been an established form of psychotherapy for over twenty-six (26) years.

Art therapy, both on an individual and group level, by means of pictorial projection encourages a method of symbolic communication between client and therapist. In a therapy setting the personal relationship between the client and the therapist (or the client and the group members in group therapy) is a key factor in the treatment program. The art activity may be primarily for the purpose of promoting such a relationship or it may be viewed as the main feature of the therapy process itself. As a method of treatment, art therapy is a here-and-now action oriented process which involves learning through doing.

The art activity may be seen as a means to encourage social interaction in order to develop the individual's sense of trust or acceptance and his communication skills. Group members are encouraged to share materials and to discuss each other's work, but not critically in terms of esthetic quality. The way that the art work is used by the client is an important method of assimilating and mastering trauma as well as facilitating the expression of pent up feelings, and speeds up the process of gaining insight.

In a number of instances, a sense of mastery gained in experience with art seemed to turn the tide of hopelessness and self-destructive behavior. Through the art projections and discussion the client can discover where some of their strengths lie, giving the client a sense of confidence and achievement.

Art therapy also contributes to rehabilitation by helping people express inner conflicts and examine them realistically. Through their spontaneous art work clients objectify their subjective experiences, project their problems outside of themselves, and thus can begin to perceive and define their concerns. When the process is successful, a person has more of himself available for purposeful thought and action. Also symbolic solutions to problems can be tried as a way of accustoming oneself to them and perhaps trying them in reality.

The art therapist functions as a catalyst and sometimes as an expert. The therapist should help and encourage clients to complete works they could be proud of, and gain information about each client's perceptual awareness, self-concept, ego-strength, motor control, manipulation of tools and material, developmental level, anxiety level, and attitudes towards peers and authority figures.

Art therapy can be conducted on a reality oriented supportive basis to actively work upon problems in reduction of overwhelming anxiety, to give clients another way of expressing feelings not yet expressed verbally, to help nonverbal or intellectualizing clients to communicate their feelings symbolically, to help clients not in touch with their feelings to gain insight into their problems, to help clients whose low self-esteem hindered their progress in achieving goals and to provide opportunities for social interaction in a nonthreatening environment. Individual art therapy can be recommended for those who could not tolerate being in a group and for those who needed individual attention. Art therapy emphasizes the value of the art experience in fostering self-awareness.

Art therapy can be conducted on a reality oriented supportive basis, analytical and diagnostic, making recommendations and referrals to appropriate agencies to supplement or continue treatment as well as creating motivational techniques which encourage the client's participation and attendance.

DOMESTIC ARTS/HORTICULTURE THERAPY

Domestic Arts consists of sewing, cooking, handicraft (crochetting, knitting, and quilting), as well as Horticulture therapy. In addition, there are several survival skills that are related to domestic activities such as money management, meal planning, consumer protection, and weight-watching.

While cooking, sewing, and housekeeping are perceived as a female responsibility, there is a definite need for males to learn such skills as well. Many male participants at New Directions House do live alone, are unattached, and may need to fend for themselves. Also, fathers who are supportive to their spouse in times of illness, childbirth, or family emergency can contribute concretely to the household. In a therapeutic sense, an individual can enhance his self-esteem through the personal involvement of self-care. The Domestic Arts group's therapeutic focus has been cohesion and cooperation through interaction. Participants have been primarily female, single heads of families who encounter difficulties because of their stressful lifestyle peculiar to the poor and minorities in this society. The frustration engendered by these daily struggles of living results in low self esteem, feelings of inadequacy, and the inability to establish and sustain interpersonal relationships. As a result, participants are often isolated, lonely, and depressed. While homemaker services have been a vital part of the helping professionals and have provided insight for the need of continuity within the home when a significant adult is unable to function adequately, there is little in the literature about homemaking or domestic arts as a therapeutic activity.

Cooking as a therapeutic focus can be stated as a vehicle for individuality, self-expression, and creativity. Cooking provides an opportunity for a participant to explore personal skills, increase self-confidence, and receive positive strokes within a comfortable supportive setting. Personal interaction takes place during the activity of giving, and receiving instructions, debating about the importance of certain ingredients and finally consuming the meal in a family milieu. In addition, the less liked task of cleaning up the kitchen, is shared because the activity itself was initially rewarding. A significant part of the domestic arts activities is sewing. Several group members possess sewing skills, and are willing to share ideas with those who are less knowledgeable. As it often develops within the group, an individual who for example, sews sleeves well, does not have the same success with buttonholes or darts. Consequently, the activities bring about a natural learning process and provides a great opportunity to realize therapeutic goals such as sharing of experience, cooperation and establishing interpersonal relationships.

Horticulture therapy has a long history of success. This activity can motivate participants who previously displayed no interest in plants by actively providing plant materials and demonstrating simple procedures. Usually Horticulture therapy is conducted in two groups, a primary group of approximately eight (8) to twelve (12) members and an advanced group of the same number. Clients usually suited for plant therapy are those with a short attention span, who have difficulty in delaying gratification, and a history of under achievement.

The activities of the primary group may consist of identifying plants, starting cuttings, and growing plants in small containers such as milk boxes. In this first group there is structure and control. Rewards are readily available, for example, plant growth that can be observed based on the amount of proper care provided by the client. Treatment goals can be realized and the participant is evaluated for progress before advancing to the advanced group. Within the group setting the focus is on interaction, clarifying established techniques, and reality testing. In addition, the safe environment allows the acceptance of individual differences and values. Frustration tolerance levels are raised and the feeling of significant achievement is gained by observing the natural guidelines formulated by the plants themselves.

MUSIC THERAPY

At New Directions House music therapy is part of the therapeutic milieu and total team approach (Tyson, 1973; Cox, Vahil, Mayer, and Steele, 1973).

The task of the music therapist is to design a music activity to aid the client in achieving goals which have been set in the treatment plan. This treatment may take place in an individual setting (ex. learning to play an instrument) or a group setting (ex. rhythm band). The real value of music therapy lies in its flexibility. It provides a vehicle for social integration and a stage on which an infinite number of relationships can be explored (Alvin, 1974). More importantly this modality provides the client with a relaxed and non-threatening atmosphere (Hoffer and Ulrich, 1974) and an opportunity for self-expression.

Besides its use as a treatment tool, music can be used as an evaluative tool. Through group and individual sessions it is possible to observe the client's:

1. ability to remain "on task" for specific times
2. ability to work towards long and short term goals
3. gross and fine motor-skills
4. interpersonal skills
5. leadership qualities
6. self concept and/or ego strength
7. drug abuse

These observations can then be used to redefine and develop the treatment plan.

As a purely treatment modality music can be used effectively to: improve self-esteem and interpersonal relationships, and to provide a positive means for utilization of leisure time (Margolis, 1972; Alvin, 1975; Keller, 1971).

Objectives in group and individual psychotherapies are related whenever possible to the music activity. In turn, any behaviors or feelings generated during a music group can be dealt with immediately and/or later, in the regular therapy group or individual sessions. In this way therapy becomes an ongoing, daily process.

A client with an interest in music might be given lessons on a particular instrument. Through this individual session we can work on: self-esteem, relationship with an authority figure and negotiating skills, and self-expression. If the therapist has reason to believe that the client's self-concept is particularly low, he can select an instrument with the potential for immediate success (ex. a percussion instrument). This success will raise self-esteem and motivate the client to futher participation and possible involvment in a group.

An example of a possible group music session would be a rhythm band. Each participant is given a different percussion instrument and a different part to play. The players must concentrate on individual parts while trying to blend with the group. This exercise is obviously analogous to our everyday existence, i.e., we try to maintain our individuality while still a member of society.

Since a pleasing sound can be produced with very little effort in this activity, interpersonal relationships, leadership skills, self-expression and group cohesion can be developed. Again, these skills can be integrated into the regular group therapy sessions and hopefully, into the client's life. Finally, it is important to note that the goal of the music therapist is behavioral change, not musical achievement (Cox, Vahil, Mayer, and Steele, 1973).

RECREATION THERAPY

Recreational activities are generally looked upon as providing participants with the opportunity for socialization, feelings of personal accomplishment, physical conditioning and varying degrees of skill development. However, from a drug rehabilitation point of view, client-participants are provided with the opportunities to develop "alternative" ways in which to meet social needs, feelings of accomplishments, etc.

When the addicted individual fulfilled the same needs within the drug culture, he or she generally did so under negative self-destructive circumstances. For example, social needs were met while congregating on a street corner where drugs are sold, which increases the probability for police involvement. Further, personal accomplishment needs might be fulfilled by an individual who develops sound burglary or shoplifting skills, which naturally are used to support a habit and so on. Countless other examples can be pinpointed.

At the very foundation of recreational activities therapy lies the idea of image-building which enables individuals to feel better about themselves through developing recreational skills, the participation and the accompanying feeling of self accomplishments. In my opinion, self-esteem gradually improves the individual'a ability to be introspective while examining personal problems. As individuals feel better about themselves, they have less fear of the unknown or hidden personal concerns. Therefore, one of the first steps in the recreation therapy process is to address the individual's skilled areas and areas of interest. Individuals with limited skills are introduced to new, non-threatening activities. Non-competitive activities, such as individualized physical conditioning programs, can measure an individual's progress against his or herself.

Within a group setting there is also an opportunity for peer learning as individual group members share recreational skills and interests with other members of the group. This sharing of information and skills, in turn, encourages trust and the development of a support system. Both are necessary initiatives within an activities and group psycho-therapy day treatment program such as New Directions House.

In short, recreational activities therapy is no guaranteed cure-all for drug rehabilitation, but is a treatment tool which can be effectively used in conjunction with other supportive services, i.e., psychiatric consultation-evaluation, vocational counseling, etc. At

New Directions House a great deal is also dependent upon the therapist's skills as well as the client's commitment to making a change in his or her own lifestyle and life's direction. Recreation therapy is a *medium* through which individuals can be given the opportunity to learn alternative ways of fulfilling personal needs; increasing self-esteem; developing self-discipline and control; and a more optimistic outlook on life through healthier participation.

DAY CARE VS. THERAPEUTIC COMMUNITY

The topic seems to suggest a possible confrontation between an established major treatment modality, against a lesser known approach. This is not our intent, we choose rather to simply acknowledge the differences and in so doing generate wider prospectives on Day Care treatment in the community as an alternative to incarceration.

In the Pennsylvania State Plan for the prevention, treatment and control of drug and alcohol abuse, under the provisions of Pennsylvania Act 63, guidelines are set forth for treatment and rehabilitation services for male and female juveniles and adults who are charged with, convicted of, or serving a criminal sentence. The provisos have a tremendous impact on the system with reference to the drug abuser. It identifies drug dependence as a health problem, provides for the confidentiality of patient records and offers the criminal justice system means to promote rehabilitation et al.

The courts, parole/probation officers have historically shown a high degree of preference in the placement of drug abusers in therapeutic community programs. The rationale is based primarily on the assumption that this highly structured modality will provide:

1. Close supervision in a controlled environment;
2. Means to prevent recidivism through therapy, getting in touch with feelings, dealing with anxiety and depression, etc.;
3. Former addicts are given the opportunity to serve as change agents;
4. A financial investment/saving to the taxpayer/community.

However, an examination of the literature (DeLeon, 1973), points out that the therapeutic communities have become costly ventures to federal, state and local funding sources. It was estimated that it cost $5,000 to maintain a patient slot for one year in a therapeutic community, while it cost $2,500 for a drug free day care slot. On a national level (DeLeon, 1973) therapeutic communities have acknowledged their special problems, i.e., credentialing, accountability, ethnic issues, and high recidivism rates. Further documentation locally, by the City of Philadelphia Coordinating Office for Drug and

Alcohol Abuse Programs (CODAAP) indicated, discharges since November, 1975 showed a 14.63:1 ratio of drops to graduates with Residential Therapeutic Community to a 3.76:1 ratio for our day care program. The vast amount of social pressure brought to bear on therapeutic community participants is frequently too much for the client to tolerate, resulting in the take flight syndrome. Day Care provides a greater variety of treatment approaches, enabling the client to establish realistic short/long term goals and self (growth/actualization) through supportive counseling in a real environment as opposed to the artificial therapeutic community climate. The reason for greater flexibility and availability of treatment approaches is because the client is not removed from the community but rather remains in it. Then all the community resources are brought to bear to assist him/her in the recovery.

Based on our progressive (supported by Helms, 1975) work experience with: probation/paroles; incarcerated adult males and females; work release clients; and criminal justice personnel at all levels, it is our professional view that day care treatment is a more viable means of serving the offender population.

SUMMARY

New Directions helps those who want to help themselves. We offer no quick and easy method to banish drug abuse. However, those who have practiced the self-help coping methods and have participated regularly in the activities have proved that our treatment approach works! New Directions House is also very aware of the fact that for every successful graduate of the program there are four (4) failures. This fact is not overlooked, but in order for one to maintain a facilitative position, success is measured in the smallest of increments.

We have made a statement concerning day treatment, milieu therapy, the use of activity or the alternative process. It is because the phenomenon of narcotic addiction treatment is a conviction that no single explanation will suffice. Narcotic addiction, like happiness, is different things to different people and hence will resist any unitary hypothesis. Many promising models remain to be more fully explored. There should not be the eventual triumph of one approach but the weaving together of whatever values that may be within each. The test of clinical skill is the assemblage of an appropriate mix for a particular client (Lazare, 1973).

By and large every program in order to function as it should, it must be understood it is wholly a function of the people involved: it's no better than the combination of the clinic staff and the client. Without the dedication of the first and motivating of the second, it's no more than a junkie's hype.

APPENDIX

Table I. Demographic (Intake census data as of March 30, 1978).

Sex			
Male	53		
Female	28		
Race			
Black	80		
White	1		
Other	0		
Age	Male	Female	Total
18-19	0	0	0
20-21	3	0	3
22-23	10	1	11
24-25	9	4	13
26-27	9	1	9
28-29	11	8	19
30-31	8	3	11
32-33	2	3	5
34-35	1	3	4
36-37	1	0	1
38-39	0	0	0
>40	3	1	4
Highest Grade Completed	Male	Female	Total
<9	7	2	9
10	18	4	22
11	5	7	12
12	24	11	35
13	1	0	1
14	1	0	1
15	0	0	0
16	1	0	1
>17	0	0	0
Current Marital Status			
Never married	40		
Married	20		
Living together	6		
Separated	14		
Divorced	1		
Widowed	0		

Table I (cont'd.). Demographic (Intake census data as of March 30, 1978).

Living Arrangement	
Alone	9
With parents	18
With spouse and/or children	26
With others	11
Unknown	17

Current Employment Status	
Full time (35 or more hours per week)	5
Part time (less than 35 hours per week)	4
Retired	0
Unemployed--has sought employment in last 30 days	70
Unemployed--has not sought employment in last 30 days	2

Number of Prior Treatment Experiences

	Total
0	5
1	10
2	14
3	12
4	5
5>	2

REFERENCES

Alvin, Juliette. 1975. The Identity of a Music Therapy Group: A Developmental Process. *Brit. J. of Music Therapy*. 6(3): 9-17.

Alvin, J. Music. 1974. Music Therapy in Psychiatric Social Clubs. *Anali Klinicke Bolnice Dr. M. Sotjanovic*. 13(3): 239-243.

Ball, John C. 1965. Two patterns of narcotic drug addiction in the United States. *J. of Criminal Law, Criminology, and Police Science*. June.

Bender, L. 1938. *A Visual Motor Gestalt Test and Its Clinical Use*. American Orthopedic Assn.

Branden, Nathaniel. 1969. *Psychology of Self-Esteem*. Los Angeles: Nash.

Buck, J.N. 1950. *Administration and Interpretation of the H-T-P*. Calif.: Western Psychological Services.

Chien, Isidor. 1968. *Psychological Functions of Drug Abuse*. Paper read at Symposium on the Scientific Basis of Drug Dependence, London, April 8-9.

Cox, Patricia, Vakil, Rama, Mayer, Morris F., Steele, Louise. 1973. Music Therapy as an Integral Part of Total Treatment in a Residential Center. *Am. J. of Orthopsychiatry*. 43(2): 257.

DeLeon, G. and Beschner, G. 1976. The Therapeutic Community. Proceedings of Therapeutic Communities of American Planning Conference. DHEW Publication No. (ADM) 77-464.

Ellis, Albert. 1973. *Humanistic Psychotherapy - The Rational Emotive Approach*. New York: Julian.

Fink, P.J., Goldman, M.J., Levick, M.F. Art Therapy - A Diagnostic and Therapeutic Tool. *J. of the Hahnemann Medical College*.

Gerard, Donald and Kornetsky, C. 1955. Adolescent Opiate Addiction. *Psychiatric Quarterly*.

Glasser, William, M.D. 1976. New York: Harper and Row.

Helms, D.J., Scurra, W.C., and Fisher, C.C. 1975. *Treatment of the Addict in Correctional Institutions*. New York: Harper and Row.

Hoffer-Ulrich, E. 1974. Musiktherapie Im Konzept Der Psychotherapie. *Analie Klinicke Bolnie "Dr. M. Stojavnovic*. 13(3): 276-279.

Keller, U. Techniques with Active Individual Music Therapy in Schizophrenics, Depressives and Psychosomatic Patients in the Psychiatric University in Basel. *Proc. of the 3rd Int. Congress of Social Psychiatry*. V4.

Lazare, Aaron. 1973. Hidden Conceptual Models in Clinical Psychiatry. *Eng. J. Med.* 288: 345-51.

Machover, K. 1949. *Personality Projection in the Drawing of the Human Figure.* Springfield, Ill.: Thomas.

Margolis, Phillip M. 1972. Community Mental Health: Harmony or Cacophony? *J. of Music Therapy.* 9(3): 123-129.

Naumburg, M. 1966. *Dynamically Oriented Art Therapy: Its Principles and Practice.* New York: Grune and Stratton.

Newmeyer, J. and Gay, G. 1972. The Traditional Junkie, the Acquarian Age Junkie and the Nixon Era Junkie. *Drug Forum.*

O'Neal, Thomas R. Patterns of Drug Abuse Among Blacks: A Preliminary Analysis. National Urban League Research Department, Washington, D.C. (Paper presented at the National Drug Abuse Conference, New Orleans, La. April 5-7, 1975).

Saloway, I. 1973. Methadone and the Culture of Addiction. *J. of Psychedelic Drugs.*

Saodap: *Outpatient Drug-Free Treatment Manual.*

Taylor, Richard. 1963. *Metaphysics.* Englewood Cliffs, N.J.: Prentice-Hall.

Tyson, Florence. 1973. Guidelines Toward the Organization of Clinical Music Therapy Programs in the Community. *J. of Music Therapy.* 10(3): 113-124.

Waldorf, D. 1973. *Careers in Dope.* Englewood Cliffs, N.J.: Prentice Hall.

SOCIOBIOLOGY AND THE THERAPEUTIC COMMUNITY*

Robert V. Frye, M.S.

University of Colorado Therapeutic Community, PEER I

In a 1977 meeting of the American Association for the Advancement of Science, the controversy surrounding sociobiology was the culmination of arguments raging since publication two years ago of *Sociobiology: The New Synthesis* by Harvard Zoologist, Edward O. Wilson, a work regarded as definitive in the field (Gustaitis, 1977). This paper will attempt to define the sociobiological system and apply some of the basic principles inherent in that system to the therapeutic community, which is to some extent a microcosm of society, combining the elements of "family" with cultural attributes and survival skills. It will also attempt to suggest some possible areas where modification of the community to correspond with sociobiological theory might lead to a more effective utilization of the community as a therapeutic tool.

EVOLUTION AND SOCIOBIOLOGY

Charles Darwin's (1871) theory of evolution holds that the diversity of living forms has evolved in response to natural selection or the principle that the best adapted variants have the best chance of survival. It implies that orderly changes in species have arisen by the operation of genetic mutations with the survival of the best-adapted mutants, and he believed that the *individual* organism was the

---

* The opinions expressed in this paper are those of the author and are not necessarily those of the University of Colorado Addiction Research and Treatment Services.

primary unit of selection. Darwin could not fully explain why some organisms help other members of their species. Some birds, for example, risk their lives for the flock by crying out to warn others of the presence of a predator; thus increasing their chance of attracting the enemy's attention and being singled out for attack. Sociobiology attempts to explain this apparent contradiction by speculating that behavior which promotes survival of the reproductive winners is passed on by their genes.

Wilson (1975) stated, concerning sociobiology, "Not only has new light been shed on the mechanisms of social behavior, but the structure of societies has been laid open to biological analysis for the first time . . . " He speculates that such human emotions as love, hate and fear, as well as acts of heroism and brotherhood may well be rooted in man's genes and that human strengths and weaknesses are conditioned genetically rather than shaped entirely by environment. He points out that human beings are social in a way that is basically mammalian yet distinctive among the mammalian species. The human constraints, or unique differences in our social behavior which make us distinctive, according to Wilson, appear to have evolved from a mammalian plan over the past several millions of years during which our ancestors lived in a hunter-gatherer economy (Barash, 1977).

Sociobiology has already provided great insights into the behavior of animal sub-species and there appears to be optimism in regarding its applications to human animals. Sociobiology searches for the biological foundations of social behavior. Certain traits have developed (evolved) that are individually disadvantageous, but beneficial to a larger social unit. Such a trait is called altruism and the presence of a genetic altruistic tendency would have to contribute greatly to the reproductive success of the group. Human altruism may have been necessitated by the existence of biological tendencies for personal selfishness; tendencies that must somehow be overridden if complex human societies are to function smoothly in the modern world. All animal social systems may well represent neither more nor less than the sum of behaviors of individuals, each of whom is acting to maximize *its* inclusive fitness (genetic contributions to future generations). Even though an individual member of a society does not reproduce, if its behavior assists the group's gene pool to continue, then the unreproductive member's gene traits will be passed on in the gene pool. Sociobiology relies heavily upon the biology of male-female difference and upon the adaptive behavioral differences that have evolved accordingly; however, sociobiology does not imply that either sex is better (Barash, 1977).

Barash (1977) asks, "What behaviors do we find satisfying and why? Sex, good food, rest, the respect of others, physical comfort, personal power and autonomy, coordinated and successful movements (athletics, dancing), the accomplishments of ourselves and our offspring, all these pleasures contribute eventually to our own fitness, and therefore, we have been selected to engage in them. We find them sweet."

At the heart of sociobiological theory is the concept of *inclusive fitness*. Inclusive fitness implies that the group carries the totality of the genes of its members. Helping one's kin survive and reproduce maximizes one's inclusive fitness. One member's behavior may not necessarily cause him to reproduce, but it will help the group to reproduce successfully. Thus the tendency for that behavior will be genetically continued by the group's gene-pool and altruism (self-sacrifice) will continue to occur throughout succeeding generations through kin reproducing more successfully (Barash, 1977).

THE THERAPEUTIC COMMUNITY

The therapeutic community, as initiated by Maxwell Jones and the drug-treatment model developed by Synanon et al., gives maximum responsibility for rehabilitation to members of the group. Members' "disorders" are treated with a variety of tools; Behavior Theory, Gestalt techniques and Existentialist philosophy often pervade the therapy. The group tends to become a "family" with overtones of kinship and is a miniature model of the real world. It is often authoritarian in structure with a definitive hierarchy for both staff and residents. Its stated objectives, at least in the community for drug abusers, is to help its members obtain a drug-free and more "meaningful" life. It is called a "self-help" operation within the group matrix. This philosophy is reinforced with the motto, "You alone can do it, but you can not do it alone." The group confronts "bad behavior" in its members, provides discipline for "errors" and promotes change in individuals by the use of powerful peer pressure. The community teaches its members to act by permission and not by impulse (Yablonsky, 1965).

People who come into a therapeutic community are people with *problems, problems caused by laboring under stresses that are novel to our biology*. Individuals unsuccessfully attempted to cope with these stresses by using and abusing drugs. Most residents are products of a nuclear family, or less; when a more primitive family with aunts, uncles, grandparents and cousins may be more in line with their evolved biological propensities. The concept of family in the therapeutic community is more related to kin: The structure and hierarchy of the therapeutic community provide many brothers and sisters, parents and grandparents (senior peers and staff), uncles and aunts (graduates), and cousins (supporters and persons in other T.C.'s). There is an opportunity to both expend and receive kin investment and the family member has an opportunity to fulfill a major biological satisfaction, that is the feeling of being part of a family. This satisfaction is the major factor in reinforcing peer (kin) support and control and modifying the behavior of individual family members.

SOCIOBIOLOGICAL IMPLICATIONS IN THE THERAPEUTIC COMMUNITY

Senior family members (those in authority positions) and staff are parents of the family. These "parents" have an investment in the "fitness" of younger family members. In return, younger family members (brothers and sisters) must compete for the resources of the community in an attempt to marshal those resources to increase their own fitness (to lead a drug-free and more "meaningful" life, i.e., to later genetically reproduce). Expulsion from the family means death (often literal death from drug over-dose) or at least a loss of meaningful life. Aggression (selfishness) in obtaining these resources pays off only on a cost-benefit ration and too much aggression leads to expulsion and death, and it then becomes too costly to the individual. Aggression pays off only when the cost does not exceed the benefit and therefore aggression may often be very subtle. Members would lose fitness if they are too selfish and aggressive in their demands for resources. Aggression is expressed in the first part of the motto, *"You alone can do it*, but you can not do it alone."

Since family members are presumed to share in the same genetic pool, it is in the interests of the group that resources also be shared, thus increasing the group's reproductive chances (kin selection). This biological tendency retards extreme selfishness on the part of the individual and promotes altruism in the group. Altruism increases the chance that members will survive and propagate. We then see altruism expressed in the therapeutic community family as loyalty to the group, helping each other and sharing resources, and altruism is expressed in the last part of the classic motto, "You alone can do it, *but you can not do it alone."*

Senior family members and staff show evidence of parental investment and "prospects" for the family members are less valued than older family members. If a new member leaves (splits), there is little grief over his "death", but if an older member leaves, there is a great sense of loss and grief. Sociobiology would speculate that this may be an evolved tendency to maximize genetic investment in those most likely to reproduce (Trivers, 1972). Older program members are given more resources (better living conditions, more privileges, extra attention from staff) in an attempt to maximize their fitness and protect the parental investment.

"To love and be loved" is one of the objectives of a therapeutic community, and it is part of developing a "meaningful life." Brotherly and sisterly love is encouraged in the therapeutic family. Love is a "feeling" that encourages selflessness and altruism, and thus helps protect the family and maximizes its inclusive fitness. Love may have been selected by evolution to do this and it is part of *human behavior.* Love brings a "sweetness" to life and makes life "meaningful" for many.

Altruism is subject to reciprocity and may be given only when the *benefits* (in terms of inclusive fitness) are greater than the cost. Thus the altruistic act should carry a low risk for the altruist and a high benefit for the receiver. Altruism would not have evolved if the altruist did not get enough back for the cost (survival and reproduction) (Wilson, 1975). This makes the act subject to cheating, i.e., those who accept the altruism of others but fail to reciprocate. Evolutionary selection may have encouraged procedures for identifying the cheaters and ostracizing them or withholding altruism from them, and these procedures may be interpreted as human feelings such as indignation, guilt, gratitude, sympathy and moralism (Trivers, 1971). These feelings are powerful factors in the therapeutic community, where one is encouraged to be "open and honest" and where deceit is considered a "felony offense." If a member does not reciprocate the support of the group, he is ostracized and subject to considerable peer pressue, and then either begins to reciprocate or is forced to leave (death). Some unconsciously turn to "subtle cheating" to help them survive and display involuntary self-deception such as self-pity and martyrdom which gives individuals a more convincing display of honesty (Barash, 1977). The therapeutic community may counter the subtle cheater by saying, "Pity kills and self-pity kills most of all."

SOCIOBIOLOGICAL GROUP PROCESS

The confrontation group or encounter is basic to most treatment in therapeutic communities for drug abusers. In the Confrontation-Sensitivity (C-S) Group, it is evident that this type of group is an eclectic system drawing from such diverse sources as "attack therapy," the Synanon and Mendocino Game (Brewster and Garrigues, 1974). Daytop Village encounters, Gestalt "primitive explosion," sensitivity movements and Reality Therapy. Confrontation-Sensitivity operates with a social scheme and under a set of social conditions different from those of standard, professional group therapy. The group is flexible and versatile and often functions to make the participant aware of the demands of the environment and the necessity of conforming to those demands.

A possible sociobiological interpretation of the interpersonal psychodynamics of this type of group (C-S) may be that it provides a test of the degree of group members' individual tendencies toward altruism or cheating (cheating is used to denote failure to reciprocate altruism: no conscious intent or moral connotation is implied. According to Trivers (1971), human reciprocal altruism includes helping in time of danger, sharing food, helping the sick, sharing implements and shring knowledge. Altruism, however, is subject to both gross and subtle cheating. In gross cheating, the cheater fails to reciprocate at all, and the altruist suffers the cost of whatever he has given without getting anything back. Subtle cheating, however, involves reciprocating, but giving back *less* than one received. This still benefits the giver (altruist), but not as much as if the situation were reversed. The C-S Group gives an opportunity for the members to assess the relative altruism vs. cheating potential for individual group members. Since it often pays to cheat, especially when the partner will not find out, or when he will not discontinue his altruism even if he does find out, it is imperative for group members to be able to assess their potential for getting away with cheating as well as assessing the potential for reciprocal altruism in individual members. Thus the C-S Group spends much time testing and discussing ways to detect the subtle cheater: talking and confronting each other regarding trust, trust worthiness, and suspicion. Guilt is often an important factor in individual group response to confrontation because guilt motivates the cheater to compensate for his misdeed and to behave reciprocally in the future, and helps prevent the rupture of reciprocal relationships (Trivers, 1971). The C-S Group often shows moralists aggression and indignation in the group as a whole or in individual group members because altruists are in a vulnerable position as cheaters have been selected to take advantage of the altruist's positive behavior. Indignation and moralistic aggression is a protective mechanism used by the group altruists to withhold support from the individual if he will not reciprocate. This process educates and frightens the cheater and in extreme cases will isolate or exile the cheater (cause him to leave the community). The C-S Group deals with injustice, fairness, and lack of reciprocity (cheating) in this manner and individual group members may show strong verbal aggressiveness when the cheating tendency is uncovered (physical aggression is taboo). The C-S Group also assists in setting up altruistic partnerships such as friendship and bonding because it can assess trust worthiness as well as trust unworthiness.

SOCIOBIOLOGY AND SEX ROLES

Women have the primary child-care roles in all human societies and men are universally less concerned with infants and children. This implies certain biologically generated (selected) tendencies. Women may be more interested in *families* and men more interested in *vocations*. This sociobiological speculation reinforces the "biology of the double standard" and may be the root of the current controversy over "male chauvinism." Women may tend to be selected to look for resources (support) for care of her young and men may be selected to provide these resources through male-to-male competition. Vertebrate aggressiveness is largely the concern of male hormones while prolactin and progesterone are concerned with child care. Sociobiology predicts that human males may be more *intolerant* of infidelity by their wives than wives are of their husband's infidelity. If a man philanders, but still provides well for his family, then his fitness (for evolutionary success) is probably high; but a cuckolded husband's fitness is probably low and he then becomes the subject of ridicule (Barash, 1977). Responses to this selected biology seem to appear in the therapeutic community, where women are often confronted more than men concerning their previous sexual experiences in an apparent attempt to assess their fidelity potential.

Women generally appear to try to relate closely with only one male at a time in the community, while a subjective observation of the males leads one to speculate that (in general) the men are not as specific in their brotherly (male) attention to individual women. Women also seem very preoccupied with concerns centering around their menstrual cycle and are not beyond using biologically selected menstrual stresses to manipulate more support and attention (subtle cheating?). Interpersonal sexual activity within the therapeutic community itself is taboo. Men usually far outnumber the women in a therapeutic community and a male-bond usually develops (brotherhood) which frees the men from the contamination of sexual jealousy, and they can then get on with "male business" (Tiger and Fox, 1971). Women in the therapeutic community do not seem to bond together in this way and seem more competitive with each other for the attention of the males (resources to support potential offspring). Sisterhood may seem more competitive in the therapeutic community than brotherhood. Men have historically been called upon to act together in an all-male activity, particularly a dangerous one and have at times been biologically selected to affirm male solidarity at the expense of male-female bonding. As Tiger and Fox (1971) pointed out, "In some circumstances, no woman is as important to any man as men are to one another." This selected behavior appears to influence the therapeutic community where women often appear to be more individually isolated than the men. Male-female bonding is discouraged in the T.C. and women appear to have difficulty forming woman-to-woman bonds. Women therefore seem to be less secure in this treatment modality. Aron and Daily (1976) state that success in treatment for women in therapeutic communities is correlated strongly with the variable measuring strength of cohesion of self-image. Assuming that strength of cohesion of self-image and completion of treatment may somehow be equated with chances of reproductive success, then the therapeutic community, using sociobiological concepts, should attempt to improve the fitness of its female members by developing techniques to strengthen the self-image of women and to prevent its further fragmentation.

The sociobiological view of homosexuality is that it may be genetically selected. Homosexual relatives (kin) did not reproduce, but may have helped support the raising of the children in the more primitive extended families and tribes. Homosexuals, therefore, may have had inclusive fitness and through kin selection, the genetic tendency may have continued through succeeding generations. Bisexuality and homosexuality (as a sexual orientation) is accepted in most therapeutic communities and a variety of sexual orientations do not seem to impede bonding (here again, all interpersonal sexual *activity* is taboo). A family member may be encouraged to accept his sexual orientation rather than reject it. Freudenberger (1976) states that in the therapeutic community the homosexual needs to be given the opportunity to come out and that gays are still too much closeted in programs and are often the victims of jokes, derision, and confusion. Aron and Daily (1976) state that gay or bisexual males tended not to finish the T.C. program because they tend to feel less comfortable, being minority members (12.5% of their sample), and therefore tended to leave the program early. Trivers (1974) speculates that individuals who choose not to reproduce are not necessarily acting counter to their genetic self-interest and a non-reproducer may increase his *parents'* inclusive fitness while lowering his own, because his parents would be expected to value more the increased reproductive success of kin (relatives). Such a non-reproducer (homosexual) would, according to Trivers, be expected to show conflict over his adult role and to express ambivalence over the past, particularly over the behavior and influence of his parents (parents unconsciously encouraging or molding offspring toward non-reproduction in order to increase the parent's own inclusive fitness). Homosexuals and bisexuals may, therefore, have certain special problems and may require more individualized treatment in the therapeutic community: He or she must come to terms with his or her adult role and resolve the conflicts concerning such a role. Evidenced ambivalence concerning the past and the behavior of parents should be examined and resolved. The homosexual should be accepted as part of the group and his minority status should be minimized.

SOCIOBIOLOGICAL APPLICATIONS

The expectation of the amount of individual change possible in therapeutic communities may be limited by the knowledge that people are not infinitely maleable and are only able to exercise free will within the limits set by their genes. Human social behavior may be largely controlled by what sociobiologists call "facultative genes", those that can be influenced by environment to change their effect (Galvin, 1977).

The facultative genes may have a major responsibility in shaping human *culture*. Culture and environment may always be ahead of evolved biology and may produce stress and health problems. Humans may employ a strategy that their genes tell them is right (adaptive), but modern culture may have rendered that behavior obsolete if not suicidal (one needs but mention the genetic tendency to reproduce and over-populate). The major conflict in the human experience today appears to be a conflict between genotype (genetic composition of the individual); phenotype (any physical characteristic - behavior, physiology, anatomy, etc.); and environment. The phenotype is produced by the interaction of genotypes and the environment (Barash, 1977). Often the phenotype and genotype have not caught up with changes in environment. Biological evolution takes much time and environment can change much more quickly than evolutionary genetic change.

Diagram #1

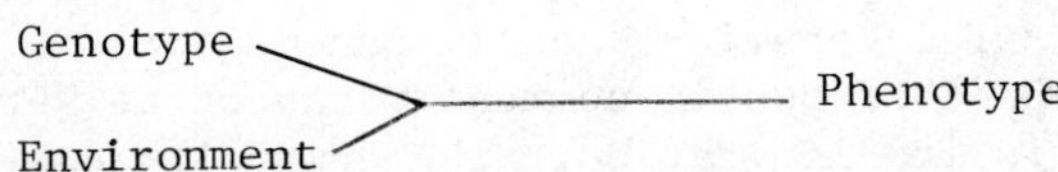

Heroin addicts make up a large proportion of applicants for treatment in a therapeutic community for drug abusers. Using the flow specified above, let us speculate as to a possible interpretation of heroin abuse behavior. Heroin use and abuse behavior would be the *phenotype* (including possible criminal, i.e., lawbreaking behavior), and this behavior may possibly be the result of environmental factors plus the genetic predisposition or the affinity of brain cells for opiates. The environmental factor might be (among many variables) societal stress and drug accessibility and the genotype might possibly be the gentic predisposition to produce human brains with an affinity for opiates (or a similar chemical substance naturally produced in humans). Thus the model for this interaction might look like this:

Diagram #2

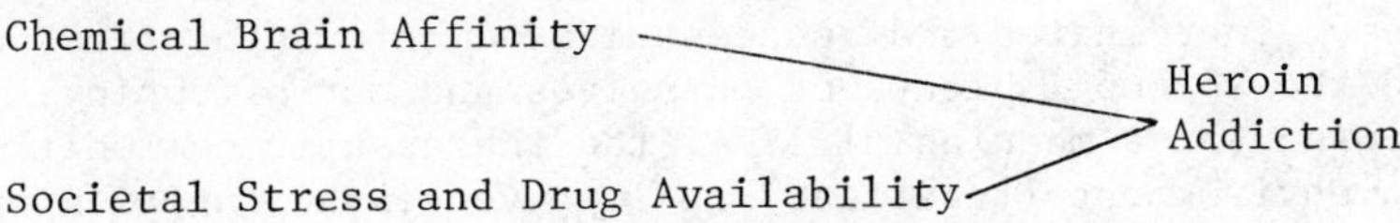

Heroin addiction or phenotype then, may possibly be successfully treated (removed) by changing the genotype or the environment, or both. A chemical antagonist might be administered to block the brain affinity for heroin and heroin might also be taken out of the environment so that its availability is curtailed. Societal stresses might be reduced. The reader will probably recognize that all of these approaches have been used, either individually or in combination with some successes. In addition, society might want to advocate the sterilization of addicts to prevent this characteristic being selected for future generations. The therapeutic community takes away the availability of heroin and attempts to reduce societal stresses. The concept of a drug-free environment is constantly reinforced in the therapeutic community and new community members are usually not allowed to communicate with the "outside world" for an extended period of time. Techniques are used to reduce outside societal stress and the community provides support to equip the member to deal more effectively with society. This support may take the form of vocational training, relocation upon graduation, meaningful employment or direct intervention with family and friends.

The therapeutic community attempts to *change* its members. It attempts to remove unproductive (bad) cultural behavior and replace it with productive (good) cultural behavior, and it can manipulate the environment to do this. The therapeutic community cannot as yet change genotypes and this is its limitation. The community can only work with phenotype and environment. It cannot, at least as yet, manipulate evolved genetic predispositions for certain behaviors. Wilson (1975) speculates that the genes need not always be obeyed and notes that man has "a genetically inherited array of possibilities. Some of these possibilities set limits on man's aspirations, others do not, and the search should be for where biology pushes mankind and where man can resist the push."

The therapeutic community must find those areas where its residents have an "array of possibilities" and concentrate on choosing the most appropriate possibilities that might lend themselves to what the community calls, "a meaningful life". A meaningful life should, according to sociobiological theory, be a life that would maximize one's inclusive fitness, i.e., how fit one is as measured by personal reproductive success and *that of relatives* (kin, tribe, community). So a *meaningful life may be one that contributes to the community's success*. The things that contribute to our own fitness are those things that we find satisfying. Barash (1977) specifies sex, good food, rest, the respect of others, physical comfort, personal power and autonomy, coordinated and successful movements (athletics, dancing), and the accomplishments of ourselves and our offspring. Using this criterion for a meaningful life, the therapeutic community could provide an environment (treatment) to improve sexual functioning through education and psychotherapy. It should teach nutrition and provide nourishing and flavorsome food. It should provide an opportunity for rest and recuperation, and a chance for physical comfort (warmth, good furniture, good beds, etc.). It should give the resident a chance to obtain the respect of others through opportunities to accept responsibility and fulfill tasks successfully. The therapeutic community should provide an opprotunity for the member to achieve personal power and status within the group and a chance to make autonomous decisions and accept responsibility for those decisions. The community should provide an opportunity for some freedom of movement and coordinated physical activity and it should give the member an opportunity to experience substantial accomplishment. Since the time a resident is a member of the community is limited in most instances, he should be *taught* how to be able to obtain (manipulate) and have an opportunity to experience these satisfying behaviors (phenotypes) later in an independent living condition. In this case the formula might be interpreted:

Diagram #3

Genetic Composition of the Individual → Behavior to promote a meaningful life

Environment (manipulated) → Behavior to promote a meaningful life

Environment must be manipulated by the individual in order to satisfy biological (genetic) needs if one is to be successful in obtaining and maintaining a meaningful life and the therapeutic community resident must be taught this "positive manipulation" if he is to be successful in independent living. Positive manipulation is altruistic because even though it is assertive, it is so at a minimum of expense to others and promotes the inclusive fitness of individuals in the community. The therapeutic community must provide an environment as compatible as possible with the genetic needs of its residents and must teach the resident to manipulate his environment in a way to make it more compatible with his genetic needs.

Diagram #4

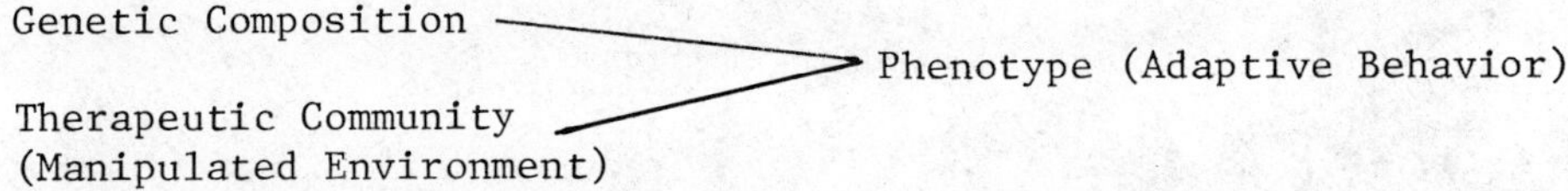

Genetic Composition → Phenotype (Adaptive Behavior)

Therapeutic Community (Manipulated Environment) → Phenotype (Adaptive Behavior)

The therapeutic community must therefore respond to the positive manipulation of its residents and support genetically satisfying (adaptive) behaviors (within the cultural environment) if it is to be successful in helping people. Adaptive behavior is behavior which helps the individual interact more effectively with his environment. Behavior today, which appears to be non-adaptive, seems so because it is assessed in an environment far different from the one in which it evolved. As Barash (1977) has stated, "There should be a sweetness to life when it accords with the adaptive wisdom of evolution."

The therapeutic community should help its residents make choices regarding appropriate behavior and help them select those behaviors that appear to be most adaptive to their environment. These choices, if carefully selected, should provide residents with the foundations for a meaningful life.

SUMMARY

Sociobiological theory can be helpful in interpreting the therapeutic community and can suggest possible options for change in order to increase the community's effectiveness. A meaningful life for residents should be a life that improves the resident's inclusive fitness, one that often helps others to be fit, and behaviors that encourage this concept should be supported by the community. Techniques for assessing trust and trust worthiness (such as the confrontation group) are essential to promote individual growth and fitness for the community members. Sex roles may be biologically determined and individual sex role orientations should be accommodated in the community and used as a basis for possible individualized treatment plans. Therapeutic communities should provide its residents with an opportunity to achieve personal power and status within the group, a chance to make autonomous decisions and to accept responsibility for those decisions. The member should be taught how to obtain satisfying behaviors.

ACKNOWLEDGEMENTS

The author gratefully acknowledges the clerical assistance of Helen Rash and Bonnie Boex in the preparation of this paper. In addition, the research assistance of Kirstin Strand was invaluable and imperative for the successful completion of this paper.

REFERENCES

Aron, W.S. and Daily, D.W. 1976. Graduates and splitees from therapeutic community drug treatment programs: a comparison. *IJA*. 11(1): 1-18.

Barash, D.P. 1977. *Sociobiology and Behavior*. New York: Elsevier.

Brewster, J.T. and Garrigues, C. 1974. The Mendocino Game rules, policies, modes and techniques. *DF*. 4(1): 15-29.

Darwin, C. 1871. *The Descent of Man, and Selection in Relation to Sex*. New York: Appleton.

Freudenberger, J.J. 1976. The Therapeutic Community Revisited. *AJDA*. 3(1): 33-43.

Galvin, R.M. 1977. Sociobiology: A new theory of behavior. *Time*. 110(5): 54-63.

Gustaitis, R. 1977. Watch Out! Here come the sociobiologists. *Straight Creek Journal*. 6(30): 1.

Tiger, L. and Fox, R. 1971. *The Imperial Animal*. New York: Holt.

Trivers, R.L. 1971. The evolution of reciprocal altruism. *QRB*. 46: 35-57.

Trivers, R.L. 1972. Parental investment and sexual selection. In *Sexual Selection and the Descent of Man* (B. Campbell, ed.) Chicago: Aldine.

Trivers, R.L. 1974. Parent-offspring conflict. *AMZO*. 14: 249-264.

Yablonsky, L. 1965. *Synanon: The Tunnel Back*. Baltimore: Penguin.

Wilson, E.O. 1975. *Sociobiology: The New Synthesis*. Cambridge: Harvard.

THE BOARD OF DIRECTORS: A MICROCOSM OF THE THERAPEUTIC COMMUNITY

Ching-piao Chien, M.D., William P. Rockwood, Ph.D. and Lewis F. Krupka, M.A.

Hope House, Inc.
Albany, New York

INTRODUCTION

Traditionally the role of the Board of Directors in agencies providing human services is that of "rubber stamping" the policies and action set forth by the directors of agencies. The prime concern is often in the fiscal stability and is usually not directed in depth to the day to day operation of the agencies. If the drug problem is a common concern of our society, then the Board of Directors involved in the drug program should be the microcosm of this concerned society. The objectives and goals for the drug addicted that the conventional society desires can be theoretically implemented by the "therapeutic" Board of Directors through the action of the program staff. The current status of the drug treatment program is still in a nebulous state searching for effective modalities in terms of who, what and how to achieve the goals. The literature is mostly oriented in psychology, sociology, criminology, medicine and training programs, etc. There is hardly, if any, reporting regarding the effective usage of the Board of Directors as a necessary component of a total therapeutic operation. Since the Board of Directors closely represents the concerned community, the acceptance or rejection of a drug program in a community depends significantly on the degree of participation of the Board of Directors in the program. The present paper presents a unique attempt in this direction that has taken place in Albany, New York.

HISTORY

In 1967 Father Howard Hubbard, a street priest and now the youngest Bishop in the nation, together with a small group of concerned citizens of Albany, New York became concerned about the increasing use of drugs within their community. A drug program called Hope House was established initially as a storefront in the inner core of the city where drug problems prevailed.

As with most private, community-based programs, Hope House did not originate from a governmental source. The initial beginnings of the program were from members of the Albany community who, themselves, actively participated to bring the problem to the forefront. This led to the opportunity to receive small financial assistance from the then Narcotic Addiction Control Commission of New York State. Along with the birth of this drug program, a Board of Directors was established.

EVOLUTION OF THE ROLE OF THE BOARD OF DIRECTORS

From 1967 through 1968, the Hope House Board of Directors consisted of fifteen individuals. These people represented different aspects of society such as banking, education, medicine, clergy, business, engineering, etc. The role of the Board during this period of time was that of a "rubber stamp" primarily because the Board did not truly understand their role in the overall operation of the program. Brief monthly meetings were held wherein the Board usually accepted the reports from the staff without any intensive interaction. During these initial stages, the program appeared to be very successful perhaps because of the fact that it was the only drug program at the time in upstate New York. From these original program developments similar type programs developed in surrounding counties which were modeled after the Hope House program. In 1971, the New York State Narcotic Addiction Control Commission provided funds so that a residential program could be established in Glenmont, New York, a rural suburb of Albany. Therefore, in 1971, we had the original storefront and the residential "therapeutic community" for eleven individuals. At this time, it would appear that all was running well, the program now had diversified to residential care as well as day treatment and outreach. Problems developed as implied earlier, primarily because of lack of communication between the Program Director and the Board. In many cases the Board did not want to look at things that were occurring with the program for fear that if the Board exerted any type of active involvement at this time, the effects would be detrimental to the program. This concern is commonly shared by most Boards involved with community-based programs, that is, a feeling that there is a certain mystique about treatment. As a result, the program started a steady decline through 1972, and the early part of 1973. This decline had its effect on the number of clients that were being served, the quality of services provided, and the amount of involvement of Board members. Drastic action had to be taken if the program was to survive and continue with original goals and objectives formulated by the original Board. After a careful review, the Board felt that two elements were essential in improving the situation: a)the Board must adequately represent the community at large; b) the Board had the ability to coordinate and communicate its objectives and goals to the staff of agencies. Without these two elements, the program would likely not achieve its maximum efficiency even in the presence of competent staff. In 1973 a new slate of officers were elected and the committee structure which previously existed on paper but was never activated was initiated. These committees were Personnel, Finance, Housing, Program, and Public Relations. Each one was chaired by an individual having the expertise within this particular area. After this reactivation took place, the Board, as a total functioning unit, felt that the program also had to revamp itself in terms of its existing staffing pattern. In 1973 the Board voted to hire two consultants in drug abuse treatment from the state of Michigan to make recommendations to the present Board as to what route should be undertaken. From this evaluation, a significant change occurred in terms of the role the Board would play. The Board saw itself as a policy making group who needed someone to carry out their particular goals and objectives. The Board had decided that they were a functioning unit who would determine what kinds of services would be provided by Hope House rather than leaving it to the discretion of a Program Director alone. The result was the hiring of a person for a newly created position called Executive Director to which the Board formed a partnership so that future activities were done in joint concert rather than in isolation. With this new direction, the Executive Director was given the freedom to develop or alter the program with close consultation with the Board of Directors. This clarification of role diminished the amount of friction which had existed between the Board and the staff. The communication improved because there was openness on the part of both parties and a willingness to share and learn from one another.

As stated earlier, the original composition of the Board was made up of individuals who had a desire to help the persons having drug problems in the community. During the last five years with the maturing of the program there has been a definite selection process developed for replacement of outgoing Board members. This selection process is based on the particular needs of the agency rather than on the status of an individual within a community. Hope House has divided the Board into specific areas directly related to the committee structure presented earlier. Therefore, the Chairman of the House Committee is a contractor, the Chairman of the Personnel Committee is a director of a social service program, the Chairman of the Finance Committee is a financial expert, etc. In addition, the Board has maintained key representatives from various needed professions. For example, from the Criminal Justice System, the District Attorney of Albany County has become a member of the Board. From the Federal System, the District Director of the Internal Revenue Service participates as a Board member. From the medical profession, we have a psychiatrist, from the public relations area we have the president of a local public relations firm as well as representatives from two T.V. stations. When one of these positions is vacated, a person having the same general occupational background is appointed to the Board. In this way, the Board becomes functional rather than a Board of persons who are well known in the community but are unable to satisfy the needs of the agency. This selection process, with its emphasis on utilization of each Board member by giving each specific tasks and responsibilities and making them accountable for these tasks, has created a spirit of service.

PROLIFERATIVE BOARD ACTIVITIES

One of the unique aspects of the relationship between the program and the Board is the operation of a "Board of Review" where members of the Board of Directors meet with individuals receiving treatment within the program to review their progress. The psychiatrist who is active as a Board member assisted in the proper interviewing and evaluation of the patients by the lay Board members. This psychiatrist also introduced the drug program to the training of psychiatric residents in the local medical college to highlight their consciousness of the drug problem.

Another Board member who was a former president of the Board, through his experience and empathy of the program felt the need to develop special courses in substance abuse counselling at the graduate level in the local college where he is a faculty member. This has led to the creation of a first of its kind counselling program in the northeast part of the country.

Another Board member, who was a consumer of the program many years ago and then became a staff member, is currently the Program Director of a local T.V. station. With her own sympathy and empathy to the drug problem, she was able to utilize the news media to help the public to become more aware of drug problems in general and our program in particular. As an example, our second annual radiothon resulted in donations five times the amount received in the first radiothon. As a consequence of these concerted efforts together with many others, Hope House has become a prestigious organization where Board membership is held in high esteem by the community. It is not a surprise to find many state and local dignitaries and community leaders actively participating in many of our events. This is gratifying to us in view of the national stigma still attached to many drug programs. During the last ten years Hope House has directly helped more than 800 men, women, and children and indirectly affected the lives of more than 50,000 persons through family conferences, school presentations, and individual consultation. Cumulative statistics from July, 1975 through January, 1977 show marijuana use down from

100% to 70%, other drug use down from 100% to under 10%, legal involvement down from over 80% to 20%, and employment and educational involvement up from 40% to 45%.

Our program has clearly demonstrated that Board of Directors in a drug program can be a very active element in a total therapeutic approach, particularly in a non-metropolitan area where the Board members have more visible representation in the community.

MANAGEMENT OF ANGER IN A THERAPEUTIC COMMUNITY FOR DRUG ADDICTS

Sandra A. Pursell, M.A. and Louis A. Moffett, Ph.D.

Veterans Administration Hospital

The major focus of this paper is on the problems, solutions and benefits we found in implementing Novaco's (1977) systematic program for anger management in an inpatient therapeutic community for drug addicts. We became interested in implementing such a program in our community because anger and angry transactions are emotional and behavioral patterns frequently encountered among our patient population. In general, it seems that among this population conflict and frustration are handled predominately in one of two ways: either the individual is passive, experiences little or no anger and totally withdraws from conflict, or becomes provoked very easily, experiences intense anger and behaves aggressively. Novaco's program is designed particularly for the latter response pattern while assertion training and fair fight techniques (Moffett, 1976) are interventions designed for individuals on the passive end of the anger continuum.

TWO CASE STUDIES--INTERNAL AND EXTERNAL ANGER PROBLEMS

Two residents, X and Y, illustrate the types of anger problems with which Novaco's program seemed especially suitable to deal.

X was a young man who was easily provoked, especially by criticism or losing competitive games. He usually responded with insults, name-calling, aggressive gestures, throwing objects, beating walls, etc. He escalated confrontations almost to fist fights. His behavior drew much criticism from fellow residents and jeopardized his membership in the community.

Y's problem with anger was more internal. He retained good control except under intense provocation to which he responded with deeply personal insults, but his behavior troubled him less than his intense and painful feelings of anger. He reported extreme physical discomfort accompanying intense anger including a burning sensation in his legs. The only outlet he found satisfying was to break things. Since he could not destroy the ward, he suffered great discomfort.

Their anger caused both men pain: one felt the pain chiefly as a result of responses from his external environment and the other experienced pain due to internal pressures. Each one, however, also experienced rewards from his anger, and this is a very crucial point. X reported that he felt satisfaction in seeing people become intimidated and back down. On a more unconscious level the anger served as a defense against less acceptable feelings of hurt, rejection and inadequacy. For Y anger also served a defensive function against feelings of inadequacy and self-rejection.

It can be obvious to the client that his anger is causing him pain and is dysfunctional, and yet he will resist changing this emotional and behavioral pattern because it does fulfill some positive functions from his perspective. Implementation of the anger management program, therefore, can be difficult for the therapist.

NOVACO'S ANGER MANAGEMENT PROGRAM

The actual program we used will be described only briefly here, since extensive written descriptions, including manuals for both the therapist and client, can be obtained by writing to Dr. Raymond Novaco, Social Ecology Program, University of California, Irvine.

The first step in Navaco's program is to assess the extent and conditions under which anger occurs. This is accomplished through the use of a daily journal in which the client records the frequency, intensity and duration of his anger, the situations in which it occurs, the persons with whom it occurs and the mode in which the anger is expressed. External factors such as abuse and frustration are noted. Internal factors such as expectations and interpretive self-statements are included in the assessment.

Through discussion of his journal entries, the client begins to understand how anger functions for him. The next step, and a very crucial one, is for the client to decide which anger responses are adaptive and therefore should be retained, and which responses are maladaptive (e.g., too frequent, too intense, too lasting, inappropriately expressed, negatively affecting performance or relationships).

To deal with inappropriate responses, the treatment interventions include use of a hierarchy of anger situations paired with relaxation as a desensitization procedure, and cognitive rehearsal of anger situations in which the client employs coping self-statements which decrease his anger and restructure the anger-inducing event as a task which needs a solution. The strongest emphasis of the program is on this cognitive component of anger management. These are some examples of the types of coping self-statements clients learn to make:

"Remember, stick to issues and don't take it personally."
"Try not to take this seriously."
"I'll let myself relax. Easy does it."
"What he says doesn't matter. I'm on top of this situation and it's under control."
"Time to take a deep breath."
"Try to reason it out. Treat each other with respect."
"I'll get better at this as I get more practice."
"It could have been a lot worse."
"I actually got through that without getting angry."

As treatment progresses, role-playing is used to enhance control and to help the client see events through his antagonist's eyes, thus increasing his empathic ability. Assertion training is also emphasized so that the angry client does not fear that he will be turned into an ineffectual Caspar Milktoast. We have found that it is particularly important to stress that managing anger includes strong assertive behavior that will bring the positive results that anger did *without* its negative effects.

IMPLEMENTATION OF PROGRAM - PROBLEMS AND SOLUTIONS

The first major difficulty we encountered in using the anger management program occurred in the assessment stage when the client was deciding which anger responses were inappropriate and hence needed replacement. It was at this point that ambivalence about change emerged most strongly.

As mentioned previously, anger has hitherto served some positive functions from the clients' perspective and he is loath to replace this manner of coping with stress even though he has become aware of the drawbacks to his responses. In addition, many of our residents resent authority and react very oppositionally to suggestions for behavior change. Therefore, we have found it helpful in the initial sessions to use a paradoxical approach if there is any sign of hesitancy or resistance. For example, in the course of discussing X's anger journal, he realized that when he was criticized by anyone he generally reacted by countering with personal insults that frequently resulted in a major angry confrontation. While evaluating the consequences of this response, it became apparent to the therapist that handling criticism in any other way was not appealing to him. In this instance, rather than attempting to demonstrate the maladaptive effects of his anger, the therapist took a paradoxical position and argued that actually this was a response he probably should *not* change. After all, it probably would discourage people from criticizing him in the future. He might succeed in intimidating onlookers as well: they would learn to keep a healthy distance from him and he would not have to interact with them either. As the therapist aligned herself with the part of him that was afraid to change, he was then able to oppose the therapist and argue that actually his response was less than ideal since he disliked the subsequent angry confrontation, did not enjoy fighting and even felt lonely as people kept their distance from him.

Taking the paradoxical stance and arguing *for* resistance rather than *against* it mobilizes the client to move into the program and take it on his own terms. In the case of Y, who disliked his status as a patient and deeply resented authority, resistance appeared even earlier. He expressed dissatisfaction with his anger, but was highly skeptical that any treatment could relieve him of it. At this juncture, rather than attempting to persuade Y of the efficacy of Novaco's program, the therapist mentioned its existence in an offhand way but hastened to add the she was sure he would not be interested in the program. It was added that his anger served a useful function for him and that he probably needed to retain it. After several therapy sessions in which the anger program and its efficacy were mentioned casually and in a paradoxical context, Y expressed a desire to try it. The therapist was especially careful to maintain a paradoxical stance throughout the use of the program.

After the therapist and client agreed on which of the clients' transactions could be handled more effectively without anger, the next problem we encountered in implementing the anger management program was in helping the client to do the homework required to effect a behavior change. The program requires that the anger journal be maintained daily and that the client set aside fifteen minutes each day to practice relaxation and cognitive rehearsal of coping with anger situations. Despite good intentions, it was very difficult for our clients to do the daily homework. Our solution to the homework problem was to suggest that the client write the anger management homework into the behavioral contracts which determine his progress through various levels of privilege and responsibility in the Sartori community (Zarcone, 1975). This step immediately increased the regularity with which homework was done. In the absence of a similar structure in an inpatient community, a system of reinforcements should be arranged which are given by another resident or staff who checks to see that the homework is being done each day. Novaco does not stress the use of material reinforcements in his program, but for our client population it appears to be an essential factor.

ADVANTAGES OF ANGER MANAGEMENT PROGRAM

One of the advantages that the program provided was that in addition to helping the client manage his own anger, he in turn became a model for his fellow residents. One of our clients, for instance, was observed giving instructions to another resident on how to handle his anger. Another benefit of the program for the clients was that they became aware of their internal processes, and came to the realization that they had a choice whether or not to respond with anger. Previously they had viewed their anger as an inevitable response to stress.

The program also has benefits for staff as well as for the patients. Various assessment measures (Moos, 1974; Moffett and Bale, 1978) indicate that staffs of therapeutic communities consistently prefer less angry behavior than do patients. Given this disparity, an anger management program can give the staff a sense of potential control over anger transactions and thereby allay their anxiety in dealing with chronically angry patients.

STAFF TRAINING

As experienced administrators are well aware, the introduction to staff of a new treatment program is not to be undertaken without a good measure of tact and forethought. Resistance to change is a phenomenon found in clinical staffs as well as in patients.

The approach we used was successful and might provide ideas for others. Initially, one staff member expressed an interest in working with the anger management program. The staff was presented with a very brief description of the program and their consent was requested in allowing the staff member to proceed on an experimental basis. When the patient's individual therapist was another staff member, the anger specialist was careful to keep the therapist apprised of his client's progress and to discuss any issues which might arise.

As staff and patients saw the results of the anger program, their interest in the program was stimulated. Only after six months of work with patients and after several requests were made by staff, was a formal training presentation given.

In summary, we think the major lesson we learned in implementing an anger management program is that even maladaptive anger serves useful functions for clients and they will find it difficult to replace this emotional response. They need reassurance, therefore, that the new skills will make them even more effective. It is also important to communicate to the client that the behaviors he alters are his choice and not yours and, when appropriate, side with his resistance rather than against it.

REFERENCES

Novaco, R.W. 1977. Stress inoculation: A cognitive therapy for anger and its application to a case of depression. *J. Consulting and Clin. Psych.* (in press).

Novaco, R.W. 1976. The functions and regulation of the arousal of anger. *Am. J. Psychiatry.* 133: 1124-1128.

Novaco, R.W. Anger management procedures: Therapist manual. Unpublished materials.

Novaco, R.W. Anger and coping with provocation: An instruction manual. Unpublished materials.

Moffett, L.A. 1975. Fair fight training in a therapeutic community for drug abusers. NDAC paper. New Orleans, LA.

Moffett, L.A. and Bale, R.N. 1978. Assertiveness, fears and preferred reinforcements of patients and staff in a therapeutic community for drug abusers. NDAC paper. Seattle, WA.

Moos, R.H. 1974. *Manual for Word Atmosphere Scale.* Palo Alto: Consulting Psychologist Press.

Zarcone, V.P. 1975. *Drug Addicts in a Therapeutic Community: A Sartori Approach.* Baltimore: York.

ARE DIFFERENCES IN THE DISULFIRAM-ALCOHOL REACTION THE BASIS OF RACIAL DIFFERENCES IN BIOLOGICAL SENSITIVITY TO ETHANOL?

Arthur R. Zeiner, Ph.D.

The University of Oklahoma Health Sciences Center

Ingestion of alcohol is a prerequisite for becoming an alcoholic. However, neither do all people drink alcohol equally nor are they all equally at risk for becoming alcoholics. There are large individual differences in both alcohol use and abuse. Such data leave the researcher with perplexing questions. Why is it that some people choose not to drink (for other than financial or religious reasons), why are some others indifferent to alcohol and why do some people prefer to drink alcohol? What factors govern the self-selection of alcoholic beverages? In partial answer to some of these questions, it has been proposed that there exist differences in biological sensitivity to alcohol which may modulate the self-selection process (Kalow, 1962; Wolff, 1972). Implicit in the biological sensitivity hypothesis is the notion that some people experience predominantly pleasant effects from alcohol ingestion whereas others experience mainly unpleasant effects. Further, it is assumed that the experience of pleasant-unpleasant effects has a biological basis.

The purposes of this presentation are (1) to delve further into the biological sensitivity hypothesis and (2) to propose that the basis for the biological sensitivity hypothesis relates to the ability of the body to handle acetaldehyde.

Pharmacologically, alcohol is classified as a depressant (Ritchie, 1975). Yet contradictory results have been found in studies where physiological responses have been monitored after an acute dose of alcohol. Thus, in Caucasian subjects, Zeiner (1974) found heart rate increases to ethanol ingestion; Ewing, Rouse and Pellizzari (1974) did not find heart rate increases and Hanna (1977) found intermediate heart rate increases. Peripheral vasodilation to alcohol ingestion has been found in Japanese (Ijiri, 1974), as well as various

Oriental groups (Wolff, 1972, 1973; Ewing, Rouse and Pellizzari, 1974) and Tarahumara Indians (Zeiner, Paredes and Cowden, 1976) but not in a matched group of Caucasians (Zeiner, et al. 1977) Decreases in blood pressure among both Orientals and Caucasians have been demonstrated to alcohol ingestion (Zeiner, Paredes and Christensen, 1977; Hanna, 1977).

In the past such physiological changes observed after ingestion of an acute dose of ethanol have been ascribed to ethanol (Gillespie, 1976; Naitoh, 1972; Ritchie, 1975). We are proposing here that these effects attributed to ethanol (increase in heart rate, peripheral vasodilation, decrease in blood pressure, increased cardiac output, increased rate and depth of respiration, and increase in skin conductance) may actually be due, not to ethanol, but to its metabolite, acetaldehyde or due to an interaction between alcohol and acetaldehyde on central nervous system (CNS) structures. Thus, the basis for individual and/or racial and ethnic differences in the ability to handle acetaldehyde.

EARLY ANIMAL ACETALDEHYDE DATA

Eade (1959) studied the sympathomimetic actions of the acetaldehydes in cats. He compared control cats and cats pretreated with resperine, a catecholamine depletor on nictitating membrane and blood pressure pressor responses to a variety of substances. In the control cats cocaine potentiated the responses to aldehydes and ganglionic blockade with hexamethonium did not prevent the sympathomimetic actions of the aldehydes. In cats pretreated with resperine, neither nictitating membrane nor blood pressure responses could be elicited by the aldehydes. Eade's data suggest that the catecholamine releasing activity of the sympathomimetic aldehydes appears to be a direct effect on the structures retaining the catecholamines. (See extensive review by Akabane, 1970.)

EARLY HUMAN DISULFIRAM-ALCOHOL DATA

In 1948 a group of Scandinavian experimenters embarked on a series of studies to characterize the alcohol disulfiram reaction and to describe the physiological concomitants. In the first study, Hald, Jacobsen and Larsen (1948) demonstrated that pretreatment of human subjects with antabuse resulted in facial flushing and an increase in facial temperature after ingestion of alcohol and that the duration of the alcohol sensitivity was maintained for at least three days. On the average flushing became evident 7-8 minutes after alcohol ingestion. The maximum effect was seen after 30 minutes. The face became a scarlet red with a tinge of blue. The flushing could extend to the neck, chest, arms and in a few cases to the abdomen. Simultaneously with the flushing, the heart rate would accelerate as high as 120-140 bpm and the blood pressure would show a small drop. With higher doses of alcohol nausea might set in and the blood pressure might drop to 65 mm Hg systolic and zero diastolic. About 5 grams of alcohol would produce some mild symptoms in individuals. After about 10 grams symptoms were always seen. In the second experiment, Asmussen, Hald, Jacobsen and Jorgenson (1948) studied pulse rate, cardiac output and depth and rate of respiration before alcohol intake, after alcohol intake and after alcohol intake pretreated with antabuse. They showed reliable increases in pulse rate, cardiac output and both depth and rate of respiration to alcohol after antabuse pretreatment but not after alcohol without antabuse pretreatment. In the third experiment, Hald and Jacobsen (1948) showed that the concentration of acetaldehyde was increased about seven-fold in breath to ethanol intake after pretreatment with antabuse. They proposed that at least some of the symptoms observed after antabuse and alcohol ingestion could be explained as a result of the increased formation of acetaldehyde. In the fourth experiment, Asmussen, Hald and Larsen (1948) infused acetaldehyde as a 5% solution into the venous circulation of human subjects. They noted a marked increase on pulse rate and ventilation. They also noted a marked vasodilation in the face similar to that seen after antabuse and alcohol in previous experiments. The marked flushing was evident even at low concentrations of acetaldehyde having only a small effect on pulse rate and ventilation. Some of their subjects who had been in some of the preceding experiments, spontaneously reported that the effects of acetaldehyde were very similar to the effects experienced after antabuse plus alcohol.

In the early 1950's Raby carried out an extensive series of investigations aimed at: (1) further characterizing the clinical course of the disulfiram-alcohol reaction, (2) documenting the associated EKG changes (size of T wave and S-T segment depression), and (3) relating the clinical symptoms of disulfiram-alcohol reactions to blood acetaldehyde concentrations. In the first study, Raby (1953a) tested thirty-nine subjects with varying doses of alcohol (≃40 gms absolute alcohol in water) after varying doses of disulfiram pretreatment (.5-7.5 gms). All thirty-nine subjects showed a facial flush response which appeared on the average 7.6 minutes after alcohol intake. Thirty-eight out of thirty-nine subjects showed an average heart rate increase occurred 25 minutes after alcohol ingestion. Blood pressure effects were biphasic with a small initial increase followed by a larger decrease. The average systolic decrease (maximum change of each subject) was 45 mm Hg. The group average diastolic maximum decrease was more pronounced at 61 mm Hg. Besides these changes the author observed a dry cough in 20% of the cases. Over half of the subjects reported headaches. Nausea and vomiting were present in about one half of the cases.

In his second experiment, Raby (1953b) proceeded to record EKG, heart rate and blood pressure both before and during the disulfiram-alcohol reaction. Four control groups were run to further delineate the cardiovascular effects. Some disulfiram-alcohol subjects received oxygen inhalation in the course of the reaction. Some subjects received alcohol only. Some subjects were exposed to hyperventilation only. Neither oxygen inhalation nor amyl nitrate inhalation altered the disulfiram-alcohol reaction. The group which received only alcohol showed less of an effect in terms of flattening of the T waves than did the disulfiram-alcohol group. Hyperventilation had a very slight effect on T wave flattening in a few cases. Raby also noted that there was a greater systolic blood pressure drop (53 mm Hg) in subjects showing both flattening of T waves and depression of the S-T segment than in subjects showing only flattening of T waves (31 mm Hg.)

In the third experiment, Raby (1954) explored the relationship between acetaldehyde level and physiological symptoms. Blood was sampled before ethanol ingestion and at 10-15 minute intervals after ethanol ingestion (varying doses of absolute alcohol in water).

One group received only alcohol. A second group received varying doses of disulfiram and was then given alcohol. Group I, the alcohol intake group, reached substantial blood acetaldehyde concentrations. Group II, the disulfiram pretreated group, had higher acetaldehyde concentrations. In Group II, the physiological signs were related to acetaldehyde concentration. Larger heart rate increases and blood pressure decreases were related to higher acetaldehyde concentrations even though the blood alcohol concentrations did not differ between sub-groups. The higher acetaldehyde values, in turn, were related to pretreatment with larger doses of disulfiram.

More recently Sauter, Boss and von Wartburg (1977) reevaluated the disulfiram alcohol reaction in man. Sixteen alcoholics were tested after various doses of disulfiram pretreatment or no pretreatment with an acute dose of alcohol (13 ml pure ethanol in 350 ml of beer). Ethanol and acetaldehyde concentrations were determined from venous blood by gas chromatograph analysis. Disulfiram did not interact with either peak BAC attained nor with ethanol elimination rate. There was no difference in acetaldehyde concentration between ethanol only and low dose disulfiram pretreatment groups. However, the high dose disulfiram group attained reliably higher acetaldehyde concentrations. Both systolic and diastolic blood pressures decreased in the high dose disulfiram pretreated group (down to 97/71 mm Hg) and heart rate increased about 35-40 bpm to 110 bpm. The magnitude of diastolic blood pressure decrease was highly correlated with blood acetaldehyde concentration (r=.95) in the high dose disulfiram pretreated group. Significant correlations were also demonstrated between acetaldehyde and systolic blood pressure, pH, pCO_2 and plasma potassium level. Large individual differences were found on the disulfiram alcohol reaction.

The most replicated finding of the disulfiram-alcohol studies is the demonstration of facial flushing. Reliable increases in heart rate have also been demonstrated as well as sometimes precipitous drops in blood pressure. A few studies have also shown increases in contractibility of the heart and in cardiac output. Increased depth and rate of respiration has also been demonstrated. Increased acetaldehyde concentration in either breath or blood correlates with the physiological changes described above. However, there are large individual differences in the size of the observed reactions to the same dose of disulfiram and alcohol or to acetaldehyde by itself.

ETHANOL SENSITIVITY STUDIES

It has been suspected for some time that there are individual ethnic and racial differences in biological sensitivity to ethanol (see review by Kalow, 1962). Investigation of the phenomenon is a relatively recent development. Morikawa, et al. (1968) performed one of the early studies on the problem. They tested about 160 Japanese college students with a dose of alcohol. About 85% of their subjects were poor drinkers who showed reddening of the face, large increases in heart rate and changes in the dichrotic notch index.

Wolff (1972) tested various groups of Orientals and Caucasians with an acute dose of alcohol. Eighty-three percent of the Orientals showed vasomotor flush responses whereas only six percent of the Caucasians showed such responses. The Orientals also demonstrated more or greater dysphoric symptoms than did the Caucasians.

Wolff (1973) followed up with a second study which extended the results to American Indians. The Indians showed similar responses to the Orientals.

Ewing, Rouse and Pellizzari (1974) replicated and extended on the Wolff findings by also recording blood pressure and assaying for acetaldehyde from blood samples. They found more flushing in Chinese than in Caucasians. The Chinese showed a 10 beat per minute heart rate increase. The Caucasians did not show any heart rate change. Further, more Chinese subjects showed a blood pressure drop than did Caucasians. There was a trend for higher acetaldehyde levels among Chinese than among Caucasians although the effect was not statistically reliable.

Ijiri (1974) performed a very interesting experiment with Japanese subjects. On the basis of the facial flush response after ethanol ingestion, a flushing and a nonflushing group were formed. The groups did not differ on baseline measures or on peak alcohol concentration attained nor in the clearance rate of ethanol. Acetaldehyde concentration after alcohol ingestion was double that of nonflushers in the group which flushed. Similarly there was a large increase in catecholamine excretion in flushers whereas nonflushers did not change their excretion rate after ethanol ingestion. Flushers also demonstrated large increases in heart rate, skin temperature and digital pulse wave amplitude whereas such changes were minimal or nonexistent in the nonflushing group.

Seto, et al. (1978) replicated the flush response findings of Wolff and Ewing. They did not find differences in acetaldehyde concentrations between Oriental and Caucasian groups.

Zeiner, Paredes and Cowden (1976) demonstrated digital pulse wave amplitude changes, heart rate increases and increases in frequency of nonspecific skin potential responses between groups of Tarahumara Indians given alcohol and placebo.

Zeiner, et al. (1977) demonstrated that Tarahumara Indians were less reactive physiologically to an acute dose of ethanol than were matched groups of Caucasians in all measures except digital pulse wave amplitude increase.

Zeiner, Paredes and Christensen (1977) demonstrated breath acetaldehyde concentrations after alcohol ingestion in Chinese which were three times as large as among Caucasians. The degree of heart rate increases was correlated with acetaldehyde concentrations. Both groups showed decreases in systolic and diastolic blood pressures after ethanol ingestion. Not all Chinese flushed. Those who flushed the most showed the greatest drops in blood pressure and the largest heart rate increases. Those Chinese who did not flush, showed the minimal or nonexistent Caucasian type physiological reactions to ethanol.

Hanna (1977) compared Chinese, Japanese and Caucasian subjects on cardiovascular reactions to an acute dose of ethanol. A greater percentage of Orientals showed facial flush responses than did Caucasians. Likewise, the Orientals demonstrated greater heart rate increases to ethanol ingestion than did Caucasians. The Oriental subjects also showed larger blood pressure decreases than did Caucasians after ethanol ingestion.

The biological sensitivity to ethanol data indicate that the physiological measures which delineate the sensitivity are facial flushing, an increase in heart rate, a decrease in blood pressure and an increase in acetaldehyde concentration. At present it is not clear exactly what factors are causal in the sensitivity. However, it is clear that it is not a racial difference. That is, not all members of a race demonstrate the phenomenon. It is more likely an individual differences effect. What makes it seem racial is that a greater percentage of, for example, Orientals possess the sensitivity than do Caucasians. However, there are Caucasians who are as sensitive to the effects of ethanol as are Orientals.

Several factors seem to be involved. First, there seem to be racial differences in absorption of ethanol. Both Chinese (Zeiner, Paredes and Christensen, 1977) and American Indians (Farris and Jones, 1978) absorb ethanol more quickly than do Caucasians. Second, there seem to be metabolic differences which may have to do with atypical liver enzymes (von Wartburg and Schurch, 1968; Stamatoyannopoulos, Chen and Fukui, 1975). If a group metabolizes more quickly because of these enzymes, then more acetaldehyde will be found which can lead to larger dysphoric physiological changes. Third, there may be differences in organ sensitivity such that the same amount of acetaldehyde might release a larger physiological reaction. Studies are under way to further elucidate the basis for the reaction.

SUMMARY

The literatures relating to the disulfiram-alcohol reaction and the biological sensitivity to alcohol reaction were reviewed. The data indicated that under both conditions the physiological reactions observed can be characterized by: (1) fcaial flushing, (2) increases in heart rate, (3) decreases in blood pressure, (4) increases in cardiac output, (5) increases in rate and depth of respiration, (6) increases in digital pulse wave amplitude changes, and (7) increases in both skin conductance and in frequency of nonspecific responses. It has also been demonstrated that the size of the obtained physiological effects correlates strongly with either blood or breath acetaldehyde concentrations. The results suggest that the biological sensitivity to ethanol observed in some individuals and in some racial groups may be related to acetaldehyde concentrations in those individuals. The exact mechanism for the sensitivity has not been worked out yet. However, studies are under way to further elucidate the mechanisms involved.

REFERENCES

Akabane, J. 1970. Aldehydes and related compounds. In: J. Tremolieres (Ed.), *International Encyclopedia of Pharmacology and Therapeutics*. Oxford: Permagon Press, Volume II, Section 20, pages 523-560.

Asmussen, E., Hald, J., Jacobsen, E. and Jorgensen, G. 1948. Studies on the effect of tetraethylthiuram disulphide (Antabuse) and alcohol on respiration and circulation in normal human subjects. *Acta Pharmacologica Toxicol.* *4*, 297-304.

Asmussen, E., Hald, J., and Larsen, V. 1948. The pharmacological action of acetaldehyde on the human organism. *Acta Pharmacologica.* *4*, 311-320.

Eade, N.R. 1959. Mechanism of sympathomimetic action of aldehydes. *J. Pharmacol. and Exp. Ther. 127*, 29-34.

Ewing, J.A., Rouse, B.A. and Pellizzari, E.D. 1974. Alcohol sensitivity and ethnic background. *American Journal of Psychiatry. 131,* 206-210.

Farris, J.J. and Jones, B.M. 1978. Ethanol metabolism in male American Indians and Whites. *Alcoholism: Clinical and Experimental Research. 2*, 77-81.

Gillespie, J.A. 1976. Vasodilator properties of alcohol. *British Medical Journal. 2*, 274-277.

Hald, J., and Jacobsen, E. 1948. The formation of acetaldehyde on the organism after ingestion of Antabuse (Tetraethylthiuram disulphide) and alcohol. *Acta. Pharmacol. 4*, 305-310.

Hald, J., Jacobsen, E. and Larsen, V. 1948. The sensitizing effect of tetraethylthiuram disulphide (Antabuse) to ethyl alcohol. *Acta Pharmacologica Toxicol. 4*, 285-296.

Hanna, J. 1977. Cardiovascular response of Chinese, Japanese and Caucasians to alcohol. Paper presented at the NATO International Conference on Behavioral Approaches to Alcoholism, Bergen, Norway, August, 1977.

Ijiri, J. 1974. Studies on the relationship between concentrations of blood acetaldehyde and urinary catecholamine and the symptoms after drinking alcohol. *Japanese Journal of Studies on Alcohol. 3*, 35-39.

Kalow, W. 1962. Pharmacogenetics, Heredity and the Response to Drugs. Philadelphia: B.W. Saunders.

Morikawa, Y., Matsuzaka, J., Kuratsune, M., Tsukamoto, S. and Mikisumi, S. 1968. Plethysmographic study of effects of alcohol. *Nature. 220*, 186-187.

Naitoh, P. 1972. The effect of alcohol on the autonomic nervous system of humans: psychophysiological approach. In: B. Kissin and H. Begleiter (Eds.), *The Biology of Alcoholism.* New York: Plenum Press.

Raby, K. 1953a. Investigations of the Disulfiram-Alcohol reaction. *Quarterly Journal of Studies on Alcohol. 14*, 545-556.

Raby, K. 1953b. Electrocardiographic changes following the ingestion of disulfiram and alcohol. *Quarterly Journal of Studies on Alcohol. 14*, 557-567.

Zeiner, A.R., Paredes, A. and Cowden, L. 1976. Physiological responses to ethanol among Tarahumara Indians. *Annals of the New York Academy of Sciences. 273*, 151-158.

Zeiner, A.R., Paredes, A., Musicant, R. and Cowden, L. 1977. Racial differences in psychophysiological responses to ethanol and placebo. In: F.A. Seixas (Ed.), *Currents in Alcoholism*, Vol. I, New York: Grune and Stratton, 271-286.

Zeiner, A.R., Paredes, A. and Christensen, H.D. 1977. Relationships between acetaldehyde blood alcohol concentrations and physiological reactivity to an acute dose of ethanol among different racial groups. Paper presented at the NATO International Conference on Behavioral Approaches to Alcoholism, Bergen, Norway, August.

ORGANIZATIONAL ISSUES AND FACTORS IN COMBINED DRUG AND ALCOHOL PROGRAMMING

Maureen A. Young, M.S.W., Executive Director

Lighthouse/McLean County Alcohol & Drug Assistance Unit, Inc., Bloomington, Illinois

Lighthouse/McLean County Alcohol and Drug Assistance Unit, Inc. is a private nonprofit agency which provides services to people with alcoholism and drug abuse problems. Lighthouse is a community oriented program which develops services for the target population primarily within McLean County. McLean County is located in central Illinois and is the site of the Bloomington-Normal SMSA. Total county population is approximately 110,000.

From its inception in July of 1970, the program has dealt with alcohol and drug problems on a combined basis. The original impetus for this decision was that the combination would be more economical and more feasible in terms of community support than separate systems. In order to communicate a definition of the problems and the goals and objectives of the combined services, an economic model is used. The health and social problems resulting from alcoholism and drug abuse are described as following a progressively disabling course during which the abuse of alcohol and/or other mood-altering drugs leads to some or all of the following conditions:

--acute and chronic health problems which necessitate repeated medical/hospital care for the same condition unless the individual ceases the abuse of alcohol or drugs

--decreased earnings and productivity during work years

--earlier dropout or retirement from the labor force

--extended period of dependence through welfare or disability benefits-average 10 to 15 years in advanced cases

--earlier than average death due to disabling conditions, acute intake of alcohol or drugs or accidents while intoxicated

--extended dependence, personal distress and family breakup due to alcoholism or drug abuse of one or both parents, including placement of children, neglect and abuse

--increased community costs due to efforts to control negative social behavior such as public intoxication, disorderly conduct, petty theft, vandalism, driving while intoxicated, and family disturbances

This approach to the definition of the problem has resulted in a high level of community support as indicated by increasing amounts of local tax dollars contributed to the program. Approximately one-third of the current agency budget is generated through local tax sources and provides the match for state and federal dollars.

Our starting point was the development of a comprehensive plan for services for both alcohol and drug abuse. We chose to organize and plan for both areas because we were acutely aware of the need for services and believed that in a county of our size it did not make sense to develop separate organizations. In addition, we felt that similarities between the types of services needed made it feasible to combine developmental efforts. Finally, we believed that we could address the differences between the various alcohol and drug dependent populations by programming service elements for specific target populations.

Since January of 1973, the McLEAN COUNTY ALCOHOL AND DRUG ASSISTANCE UNIT has provided both alcohol and drug-related services through LIGHTHOUSE under a combination of local and state tax revenues. The full range of LIGHTHOUSE services includes: information and referral, crisis intervention, non-medical detoxification, residential treatment, individual and group counseling, community education, in-service training to cooperating and referral agencies, consultation and coordination of treatment resources. In-patient hospital services are provided through affiliation with local hospitals.

Our service programs are operated in accordance with a basic policy on treatment goals and methods which provides orientation and direction for all of our services. Progress in treatment, regardless of service or specific substance involved, is identified by achievement in five specific steps:

--Admission that self-destructive behavior exists.

--Realization that the destructive behavior is related to the abuse of alcohol or drugs and willingness to explore the effects of this abuse on personal and family functioning.

--Acceptance that living without these chemicals is necessary for survival.

--Realization that continued personal growth socially, emotionally, and spiritually, is desirable and often necessary for sustained success.

--Realization that productive behavior and positive participation within the community is desirable and often necessary for sustained success.

We have obtained support for our comprehensive plan from a wide range of community groups, organizations, and individuals. Formal community sanction for our plan comes from being designated by the county as the sole agency to plan and deliver alcohol and drug related services.

Our relationship to the county law enforcement and judicial system is a contractual arrangement which provides alternatives for both the legal agencies and the individuals involved. At the same time, our status as a private, not-for-profit, corporation provides us with the necessary autonomy to develop and plan a wide variety of services for populations not coming into contact with the law enforcement system. Participation in all programs and services is on a voluntary basis.

The following sections will describe the essential features of the organizational model currently supporting the combined drug and alcohol treatment provided by Lighthouse and go on to identify the critical factors and issues confronting programs that want to assess their own capacity for implementing combined programming.

I. ORGANIZATION OF THE SERVICE DELIVERY SYSTEM:

A. All clients enter and receive services according to the same process:

1. Intake
2. Evaluation
3. Intervention
4. Follow-up

B. Intervention or treatment is on an individualized basis which makes any needed adjustments necessary due to specific drugs and unique characteristics of the client. This is documented in the case record.

C. Information is collected on drug using history of all clients including alcohol usage.

D. All staff work with clients regardless of presenting drug problem.

E. An administrative review system tracks all clients by drug using history and monitors services provided in order to assure that quota by drug and funding source is met. Adjustments are made in terms of new referrals and waiting lists for specific services.

F. Staff organization and individual job descriptions are according to service modality and functional level of competency.

II. ACCREDITATION AND LICENSURE:

A. Separate standards and accreditation require separate documentation being submitted according to the required format. Applicable standards for our program are:

1. Joint Commission on Accreditation of Hospitals
   a. Alcoholism Program Standards
   b. Drug Program Standards

2. State of Illinois Licensure Standards
   a. Department of Public Health Licensure for Alcoholism Facilities in conjunction with Illinois Department of Mental Health and Developmental Disabilities.

   b. Illinois Dangerous Drugs Commission Licensure for Drug Abuse Programs.

   c. Illinois Department of Mental Health Guidelines and Regulations for Grant-In-Aid Agencies.

B. Administration needs to translate all requirements in a coherent manner down to the operating level in such a way that services can function effectively in terms of client flow and the treatment process. In our agency, one policy handbook describes the total treatment process. Administrative and supervisory staff have responsiblity for assuring that services operate in accordance with the applicable standards.

C. Multiple standards and regulations within the same program increase the need for adequate in-service training in order to interpret and explain the rationale for various procedures.

D. Problems creating additional financial resources arise when different standards and regulations conflict in terms of requirements. In general, this usually means that the more "stringent" standard will have to be followed or that required review mechanisms will have to apply to all cases. Examples from our program are as follows:

1. Non-medical or social setting detoxification physical facility standards are more stringent in terms of physical facility fire protection than alcohol or drug residential program standards. The state fire marshall determined that our entire physical facility would have to meet the detox requirements even though only three of seventeen beds were used for that purpose.

2. CODAP admission and discharge reports are required for all drug clients. However, in order to accomplish this on a reliable basis, all clients seen by the agency are screened. Prior attempts at having all staff attempt to screen out drug clients for reporting purposes resulted in a very low reporting rate.

3. Utilization review and patient care audit studies are now required in all alcoholism treatment programs, but not in drug programs. Utilization review and audit are technically defined terms for prescribed ways of evaluating program or treatment effectiveness. They are a more complex type of evaluation than what is usually required for drug programs. In our agency, we are currently using utilization review and audit for all cases because it would be even more complicated to set up separate evaluation systems for alcohol and drug clients.

III. CLIENT RECORD REQUIREMENTS

A. There is a need to track all clients according to multiple drug usage, including alcohol usage as a drug.

B. Tracking has to be done on a continuous basis because drug usage patterns change with clients over time and with acute episodes.

C. The tracking system has to be able to identify changes in client drug using patterns and take necessary action when required in order to maintain "balance" or "quota" set by funding sources.

D. Internal system monitors client drug usage whenever any significant change occurs and whenever client makes a change in terms of treatment modality, i.e. going from detox to residential, from emergency care to outpatient, from outpatient to detox, from outpatient to followup status.

E. Classification of clients into categories necessary for funding sources, i.e. primary drug, primary alcohol needs to be done by administrative team on basis of reported drug usage in order to assure consistency of classification. Experience with having different counselors make classifications resulted in inconsistent classifications of clients with similar drug usage patterns.

F. There is a major problem with differentiating usage of alcohol as a drug and usage of alcohol which is symptomatic of alcoholism. No clear consensus on this issue exists in either field. Consequently, internal classification criteria need to be specific and defensible if challenged. Our agency solution is as follows:

1. Specify quota of clients according to modality and primary drug or primary alcohol.

2. Classify clients with most extensive drug history first as primary drug and most extensive alcohol history as primary alcohol. Classify remaining clients according to successively broader criteria into either primary drug or primary alcohol.

IV. INTERFACE OF THE SYSTEM WITH GRANT AND THIRD PARTY PAYOR FUNDING MECHANISMS:

A. Interface with grant sources requires ability to track clients by type as described above, or ability to keep service modalities completely separate in terms of staff and clients and program costs.

B. Interface with grant sources requires that costs be identifiable for program costs and that specific client activities be identifiable for each program.

C. There are many similarities between alcohol and drug programs relative to potential third party payments. Very few are available at present except for direct medical services or for employed individuals with private insurance. The future advantage is probably with alcoholism programs in terms of potential for collecting health insurance payments for nonhospital and nonmedical services, especially residential based programs because there are more precedents in a number of states and because alcoholism treatment is easier to measure in terms of cost benefit to the employer and thus easier to sell. One advantage in combined programming is that more employers seem to be willing to talk about including other drug abuse in a program along with alcoholism than are willing to consider a solely drug abuse program.

V. ISSUES RISING FROM COMBINED PROGRAMMING WHICH AFFECT TREATMENT PROCESS:

A. If combined programming is initiated with inadequate client record keeping, funding sources may force the program into assignment of clients to staff by primary drug problem rather than need for the skills of that particular staff member, or into restriction of certain funded modalities to clients with particular substance problems, i.e. outpatient services only for drug clients, nonmedical detox only for alcoholism clients, residential services only for alcoholism clients. This results in poor staff morale, vast confusion among community referral sources, and loss of potential clients. All of these situations were experienced at one time or another in our efforts to organize and run combined treatment.

B. Funding source regulation for accountability and monitoring purposes always conflict to some degree with the optimum treatment process as viewed by staff and client. Combined programs place the treatment process and staff in "double jeopardy" because now two different sets of values, regulations, and accountability standards are imposed upon the treatment process with the added problems created by the synergistic interactions between the multiple parties.

C. Combined programs also face difficulties in finding and training staff to deal effectively with all clients regardless of presenting alcohol or drug usage problem. However, our experience indicates that staff who have difficulties in working with particular types of substance abusers may also be limited overall in working with other problems of the client.

This situation represents an especially difficult problem with existing staff if a "transition" or changeover is contemplated from single substance to combined programming. Our recommendation is that if an effective changeover is to be made, that all staff who stay have to agree to positively support change. The changeover will probably not be effective if demands of individual staff members are allowed to take precedence over a functional system.

VI. KEY DECISIONS INVOLVED IN GOING TO COMBINED TREATMENT:

The following questions have been formulated based upon our experience over the past 8 years with the development and implementation of combined programming. These questions are designed to assist programs assess their own capacity for implementing combined programming.

A. Will additional income from combined programming be sufficient to absorb increased and possibly "hidden" costs associated with the changeover?

B. Is the existing information and cost accounting system sufficient to support information and monitoring requirements of all funding sources if combined programming is initiated?

C. Will existing requirements be "escalated" if combined programming is initiated?

In other words, what is thought to be adequate by a funding source if the program is a single substance agency may not be acceptable if the funding source subsequently feels that their money may now be used to treat clients for whom they are not responsible.

D. If the existing information and cost accounting system is not adequate, does the added income for the combined program include sufficient resources to develop and operate an adequate system?

E. Will initiation of a combined program result in a net loss or gain in terms of the number of clients served by the program?

Loss or gain of clients after going to a combined program format can occur as follows:

1. Gains due to the fact that an increased number of clients can now be served.

2. Gains due to the fact that a wider variety of clients, particularly those with multiple alcohol and drug problems can now be served.

3. Losses due to the fact that the existing client group is not compatible with the new clients being served in a combined program.

4. Losses due to the increased administrative and record keeping required in a combined program which cuts down on time available for direct service.

F. Is there sufficient demand and acceptance from referral sources to use the combined program?

The program may experience either a net gain or loss in terms of the number of referrals received before and after combined programming. Our experience indicates that some referral sources simply will not accept combined programming, particularly as it relates to the usual position of such programs on alcohol usage. On the other hand, some referral sources will be more likely to use the combined program because alcohol problems are less threatening than drug problems in some sectors of the community. Each program's referral network will present a different set of circumstances.

G. Are existing staff able to accommodate additional demands of combined programming?

A program which has outstanding success with a single substance or alcohol approach may not experience the same level of success with a combined program, particularly if existing staff are opposed.

H. Do additional funds include necessary training and consultation resources?

Even if the addition of combined programming includes new staff, these new staff members will have to interact with existing staff and potential or actual value conflicts must be resolved.

J. Does the program have adequate internal leadership to resolve conflicts on value issues?

Value problems reduced to the simplest terms revolve around the following: For alcohol professionals, attitudes of drug professionals toward use of alcohol are often thought to be too "liberal." For drug professionals, attitudes of alcohol professionals toward the use of any drugs is often thought to be too "conservative."

VII: EVALUATION OF COMBINED PROGRAM EXPERIENCE AT LIGHTHOUSE:

Our experience with combined treatment has been positive in that both alcohol and drug clients have been successfully treated within the same programs by the same staff. The primary problems encountered have come at the administrative level in terms of coordinating multiple funding sources and tracking clients by primary drug of abuse. These problems have been resolved through the introduction of machine based record keeping and a more sophisticated data retrieval system.

Operation of combined programming has probably resulted in more and better services being available to drug populations than would have been otherwise possible in this community.

Costs are thought to be higher than for straight drug programs at present in terms of stricter physical facility requirements and higher demands for evaluation. However, expenses for drug programs may soon be "inflated" by the same factors at work in the alcohol field - i.e., demand for greater protection of the client and greater accountability to the funding source.

ROLE OF DIET IN PEOPLE-WORK: USES OF NUTRITION IN THERAPY WITH SUBSTANCE ABUSERS

Mark Worden and Gayle Rosellini

Douglas County Council on Alcoholism

Box 1121, Roseburg, Oregon 97470

> *For as the distraction of the mind, amongst other outward causes and perturbations, alters the temperature of the body, so the distraction and distemper of the body will cause a distemperature of the soul; and 'tis hard to decide which of these two do more harm to the other.*
>
> -Robert Burton, 1628

INTRODUCTION

From surveying the treatment literature, it is clear that nutritional factors and other related physiological considerations are generally overlooked or neglected in treatment planning and implementation. Instead, psychological treatment dominates alcohol and drug rehabilitation. There is a substantial bias in individual and group counseling that emphasizes intrapsychic conflict, problems of socialization and coping, relationships, authenticity, and other psycho-social variables. This practice has been aptly described by Wallace (1972):

> ...The symptomatology of the illness under scrutiny is assumed to be motivated behavior expressive of psychological conflicts and to some degree effective in reducing tension and anxiety; the symptoms are "interpreted" in terms of some deductive schema intended to lay bare the (usually assumed to be unconscious) conflicts...This procedure almost completely neglects the victim's body; or,

> rather, it attributes to the victim's psyche a virtually magical ability to control the state of its body, by uncritically assuming that almost any somatic expression can be satisfactorily explained merely by asserting a plausible concomitant intrapsychic conflict...Thus, even with regard to syndromes familiar to Western clinicians and conventionally (if not invariably) conceived as functional in etiology, the assumption that biological determinants are negligible is becoming an increasingly hazardous one to make. (p. 364)

In this paper we intend to review biological, diet-related factors frequently neglected in treatment, and we will present examples of how nutrition is integrated into the counseling process in a small, semi-rural alcohol and drug treatment program.

BACKGROUND

Inborn Errors of Metabolism

Nutrition and other diet-related metabolic processes have long been known to affect physical, cognitive and emotional functioning. In 1908 A.E. Garrod introduced the concept of "inborn errors of metabolism", which referred to those congenital disorders characterized by an enzyme defect generally resulting in physical and intellectual impairment (Stanbury, Wyngaarden and Fredrickson, 1966). Phenylketonuria (PKU) is a fairly common example of an inborn error of metabolism, a "missing enzyme disease" (Jackson,1973). In normal metabolism the enzyme phenylalanine hydroxylase converts the amino acid phenylalanine into tyrosine. In PKU this conversion is incomplete because the enzyme is missing. The result is a form of autointoxication in which the body produces various chemicals having a deleterious effect on the brain. If detected early enough, very often this condition can be controlled by scrupulous attention to diet (Knox, 1966). PKU is an instructive model of the potent effects of metabolic "errors". Many other enzymatic defects have been discovered, serving as reminders that microscopic physiological processes often have profound behavioral consequences.

The early work of Garrod and subsequent work on biochemical individuality (Williams, 1965) have set the stage for Pauling's (1968) concept of "optimum molecular environment of the mind". Without too much distortion, one may summarize Pauling's view by noting that where there is a disorder in thought, behavior, or emotion, there will be a concomitant disorder in the molecular environment of the central nervous system. While the implications of this line of thought are far-reaching, the importance of the internal molecular environment and kindred metabolic processes in the etiology and management of behavioral dysfunction has generally been neglected by most counselors and therapists. The primary exceptions are those therapists who have become identified with the controversial orthomolecular movement (Williams and Kalita, 1977; Hawkins and Pauling, 1973). As will be seen, there is convincing evidence that certain kinds of emotional and behavioral problems are related to various kinds of nutrient deficiency or food intolerance, and these latter conditions may result from inborn errors of metabolism.

Malnutrition

With few exceptions, starvation and frank vitamin deficiency diseases are now relatively uncommon. Yet up until World War II it was not at all uncommon to find large segments of the population suffering from the debilitating physical and psychological effects of vitamin deficiency. Pellagra is a classic example of a deficiency disease that affects mental and emotional behavior, even in the early stages of niacin deficiency prior to the manifestation of gross symptoms (Spies, 1947). Although pellagra was epidemic in the southern states and in parts of the north after 1908, the widespread fortification of flour and other changes in dietary patterns made pellagra a rarity in the U.S. following World War II (Etheridge, 1972; Roe, 1973). When it was discovered that niacin could alleviate the suicidal depression and confusion which accompanied the bizarre behavior in pellagrins, many physicians began to reexamine their psychogenic frameworks: biological factors were important in many kinds of problems formerly regarded as strictly "mental problems."

In a lengthy review article, Bell (1958) wrote: "that the nutritional status of an individual may affect his psychological well-being appears to be demonstrated by experimental evidence which has accumulated during the past two decades" (p.47). Others were not so reserved and circumspect in their observations. Watson (1956) in a paper rhetorically titled "Is Mental Illness Mental?" draws two conclusions from a summary of experimental evidence on the relation between nutritional factors and personality disturbances:

> (a) Some states which are psychologically diagnosed as functional mental illness may originate from nutritional deficiences, and (b) Some states which are psychologically diagnosed as functional mental illness may be relieved by appropriate nutritional therapy (p. 326).

Yet only in the treatment of the more severe stages of disability, during detoxification, does one typically find nutritional problems routinely being addressed (Leevy, Valdellon, and Smith, 1971; Vitale and Coffey, 1971; American Medical Association, 1977; Seixas, 1971).

It would appear as if treatment and rehabilitation staff assume (a) that once detoxification takes place the client's ordinary patterns of eating will automatically fill all nutritional needs; and (b) that one's customary dietary practices are irrelevant to one's mental and emotional behavior in the post-detox treatment phase. Both of these assumptions are erroneous and represent the conventional view that special attention to nutrition is important only in catastrophic cicumstances (Kalita, 1977).

Quite possibly the controversy surrounding the role of nutrition in the etiology of alcoholism has generalized and cast doubt on the role of nutrition in treatment (Jellinek, 1960; Popham, 1954). Yet there can be no doubt that it is well substantiated that many conditions reflecting complex biochemical individuality contribute to the development of vitamin and mineral deficiencies which affect behavioral functioning.

Similarly there can be little doubt that the casual and chronic use of alcohol and other drugs leads to widespread vitamin and mineral deficiencies. Problems of nutrient metabolism, absorption, and utilization occur as frequent side effects of many of the commonly prescribed pharmaceuticals, OTC preparations, and recreational drugs. Roe (1976) has compiled an extensive summary of these nutrient/drug interactions and concludes that drug-induced nutritional deficiencies may be the single most frequent cause of malnutrition in America today.

Hypoglycemia

The hypoglycemic response was first identified by Seale Harris (1924). He called it "hyperinsulinism" and noted that the various symptoms accompanying this blood-sugar anomaly could be alleviated by dietary treatment.

During the period subsequent to Harris' report, hypoglycemic symptoms have been clinically and experimentally linked to several diseases, including pancreatic cancer (Freinkel, 1975; Meiers, 1973). However, the condition known as relative (or functional) hypoglycemia, low blood-sugar occuring in the absence of organic disease, is subject to much controversy. In the opinion of many, if not most, physicians, functional hypoglycemia is a popular "non-disease", a condition resulting from a rather neurotic "misattribution". According to this interpretation the patient accounts for the symptoms by means of an incorrect system of explanation, a misattribution usually derived from friends, relatives, or magazine articles which feature the disease of the month. "Rather than endure a 'psychologic' or otherwise stigmatizing condition, the patient may suffer a respectable metabolic illness and enjoy the corresponding status and privileges" (Yager and Young, 1974, p. 907). Cahill and Soeldner (1974) agree with this view and offer an alternative diagnosis, "clinical pseudo-hypoglycemia," but they offer no clear criteria to differentiate the two phenomena.

While there is controversy surrounding the prevalence of "true" hypoglycemia in the general population, several clinicians and investigators have reported unusually high rates of hypoglycemic conditions in alcohol-dependent populations. Meiers (1973) states: "There are some conditions that are not purely psychiatric in which there is an extremely high incidence of hypoglycemia. A few of these are alcoholism, peptic ulcer, and asthma. In my own experience, it occurs in 95% if alcoholics" (p.454). Milam (1974) claims to find a high percentage of alcoholics with hypoglycemia in his in-patient treatment program, and such claims are widely reported in the popular literature (Fredricks & Goodman, 1969).

Unfortunately, these clinical observations are not supported by an appreciable body of research on hypoglycemia in alcoholics or other drug abusing populations. Indeed, research on hypoglycemia in these populations is virtually non-existent, with the single exception of ethanol-induced hypoglycemia, a well-documented phenomenon (Arky, 1971; Freinkel, 1975). In a recent report Poulos, Stafford, and Carron (1976) compared 50 out-patient alcoholics and 50 halfway house patients with a control group of nurses and teenagers. Of the 100 alcoholics, 96 were diagnosed as hypoglycemic. Despite these suggestive findings, the study has serious flaws. The authors omitted details of experimental procedure, quantitative comparisons of blood sugar determinations, or other experimental data, making interpretation quite limited and inconclusive.

One leading proponent of the orthomolecular movement considers hypoglycemia to have been misclassified: "Relative hypoglycemia has been considered a disease, but I believe it is more appropriate to consider it an abnormal laboratory test indicative of disturbance of carbohydrate metabolism. I consider it a symptom of the saccharine disease". (Hoffer, 1977, p.15) The saccharine disease stems from the consumption of excessive amounts of sugar and has been linked to obesity, diverticulitis, cancer of the colon, peridontal disorders, cardiovascular disease, in addition to disturbances of carbohydrate metabolism (Cleave, 1975).

Randolph (1976) adds a precautionary note to the interpretation of functional hypoglycemia. He finds "more consistently satisfying results from treating the manifestations of hypoglycemia as an allergic response to multiple specific foods" (p.114). Glucose tolerance tests and food allergy ingestion challenges often employ sugar glucose. Both procedures may provoke hypoglycemic reactions, but the reaction may be due to the individual's special susceptibility to corn. "In other words, symptoms occuring in the course of performing glucose tolerance tests may not be specific for hypoglycemia" (Randolph, 1976, p. 115).

Food Allergy and Intolerance

According to Corwin (1976), "There is no such thing as a universally safe food" (p. 122). If one is sensitive to the varied manifestations of allergic reactions to foods, it becomes both simplistic and misleading to imply that all "health foods" are good and curative, while nonorganic foods are uniformly noxious. On the contrary, so-called natural, unrefined foods are not necessarily salubrious or curative; nor are refined foods totally debilitating. "To a given individual, any food in any of these categories may be severely damaging" (Corwin, 1976, p. 122). To contradict a current advertisement, every body does not need milk, particularly since milk and milk products are high on the list of foods provoking intolerances and allergic reactions (Rowe, 1972). Other widely-used allergenic foods include wheat flour and other gluten-containing grain, corn, eggs, and the chocolate-cola-caffeine triad (Speer, 1970).

Some food allergies and intolerances appear to be inherited and have much in common with inborn errors or metabolism. They may contribute to a constellation of chronic symptoms once thought to be "psychosomatic". As a recent textbook on nutritional therapy points out, many people formerly diagnosed as hypochondriacs or neurotics are in fact allergic to common foodstuffs (Robinson and Lawler, 1977).

Caffeinism

There is growing consensus that daily overdoses of caffeine by law-abiding adults constitute one of the most widespread and least-acknowledged forms of drug abuse. Long term heavy use creates tolerance, and coffee addicts regularly report withdrawal effects when deprived of the drug (Brecher, 1972). Many people are allergic to caffeine (Dickey, 1976), and it is well documented that even in small doses, caffeine stimulates the adrenals, increases hormonal activity, raises the blood sugar level and impairs psychomotor performance (Robertson et al., 1978; Nash, 1966).

Greden (1974) identified caffeinism as an ubiquitous clinical syndrome characterized by intensified feelings of anxiety, apprehension and irritability and by physical symptoms of tachycardia and tremor. Greden suggests that reducing or eliminating coffee intake in such cases might minimize the tendency to prescribe another drug to offset the effects of caffeine. In a more recent study, Greden et al., (1978) conclude:

> The message of this study for clinicians is that caffeinism probably can be found among a fairly large percentage of patients with psychiatric symptoms, especially those presenting mixed anxiety/depression profiles. Such subjects will only be identified, however, by history taking.

> Without inquiry there will be no diagnosis; without diagnosis there will be no relief (Greden et al., 1978).

But this notion is evidently not an idea whose time has come. Considering the fact that caffeine is a physiological stressor which causes or augments physical and psychological symptoms, it seems odd that one seldom finds discussion of this diet-related factor in reports on counseling processes. Perhaps this omission should be expected in view of the fact that the coffee intake of staff often equals or surpasses that of the patients (Winstead, 1976). As frequently noted, we tend to protect our own favorite addictions.

Ruminations

It is curious that none of the foregoing diet-related factors is discussed in any detail in textbooks on alcoholism, abnormal psychology, adjustment, diagnosis, techniques of counseling and psychotherapy. With few exceptions (e.g., Robinson & Lawler, 1977), the role of diet in mental and emotional problems has been equally ignored in nutrition textbooks. At the same time, as we have seen, there has been a plethora of reports scattered throughout the literature for the past 70 years, clearly describing the diagnosis and treatment of nutrition-related intellectual, emotional and behavioral dysfunction. We maintain not only that attention to diet-related factors might enhance the therapeutic process, but that in some cases "psychological" therapy alone is futile when behavioral problems stem from biochemical processes.

THE COUNSELOR'S ROLE

In our treatment program, diet is an indispensable part of a biosocial framework for treatment. It is frequently difficult for the client to believe that his drinking and emotional discomfort may be related to his eating patterns. Before most clients are willing to alter strong entrenched dietary habits, they must be convinced that the change will be worth the effort. During the initial sessions, it is the counselor's role to explore the possibility that there may be a connection between the client's eating habits and his life problems. This can be done by reviewing the client's hereditary background, present diet, and presenting symptoms.

Heredity

During the initial evaluation, it is important to ask if, to his knowledge, he has a close relative (mother, father, sister, brother, aunt, uncle, grandparent), who suffers from alcoholism, diabetes, mental illness, or allergies. There is growing evidence that all of these disorders may occur in individuals who have genetic susceptibility to metabolic dysfunction. The presence of such family history is not conclusive proof that the client's current problems are related to errors of metabolism or other genetic factors. But such a family history is an indicator of high risk. In our clinical practice, it is unusual to find an alcohol or drug abusing client who does not report a family history of at least one of the hereditary conditions.

It is obvious that nothing can be done to change a person's heredity, but it is possible to manipulate the client's diet and other contributory environmental stresses to compensate for possible metabolic errors.

Diet

Alcohol and drug abusing clients usually report many of the symptoms listed in Table I (a) even when they are alcohol and drug free. These symptoms may become overwhelming. We have found that clients suffering from these symptoms usually have a diet similar to that described in Table I (b).

When evaluating a client's eating patterns, it is important to ask specific questions such as:

-What did you have for breakfast today?
-How many cups of coffee have you had today?
-How many times did you eat yesterday?

General questions such as "how is your diet" usually elicit a one word response: "fine".

TABLE I

Symptoms and Typical Diet

I(a) Symptoms		I(b) Typical Diet	
Depression	Weight problems	No breakfast or high sugar breakfast	Heavy consumption of:
Nervousness	Tiredness-weakness	Skipped meals	Sugar
Anxiety	Dizziness-faintness	Light eating during day	White flour
Craving for sweets	Morning nausea	Heavy eating at night	Caffeine
Craving for alcohol	Blurred vision	Refined carbohydrate snacks	Salt
Irritability	Transient muscle aches		Alcohol
Rages	Transient joint pain		Tobacco
Feeling of doom			Junk food
Insomnia-nightmares			Packaged food
Headaches			

Clients who report symptoms similar to those in Table I (a) and who also eat irregularly and consume sugar, coffee, alcohol, and junk food, are asked to follow a diet as outlined in Table II.

The recommended diet is not a high protein, low carbohydrate diet, or a natural, organic diet. It does not specify portion size, which fruits and vegetables to eat, or fat intake. However, clients report it is easy to understand and follow and it is usually a vast improvement over previous eating habits. Food preferences are learned and difficult to alter. Consequently it is not usually a good idea to suggest to a client that he begin eating foods that are totally different from what he is used to, because he probably will not do so for any length of time. Changes can be suggested within the framework of the client's present eating pattern. Generally a new diet must be implemented gradually. Clients who are overwhelmed by the idea of totally altering their eating habits may be persuaded to experiment with eliminating coffee for two weeks, or having three balanced meals a day for a month.

Over a period of several weeks, a client will be given nutritional information and will be encouraged to follow the complete dietary program. The counselor helps to devise a diet program tailored to each individual's specific needs. It may be important to help the client to plan menus and to also provide pertinent reading material. The counselor also asks the client to keep journals of everything she eats for a period of one to two weeks and to also make short notes concerning how she feels during this time. Together, they review the food intake journals at the beginning of each counseling session, and the client is encouraged to stick with the diet.

Dietary counseling must take into account not only the beneficial nutrient aspects of food, but also the possibility of food intolerances or allergies. Food allergies can produce not only physical symptoms, but confusion, irritability, neurosis, behavioral dysfunction, even psychosis.

As noted in the previous section, it is possible to develop a sensitivity to any food, but usually allergies develop to foods that are consumed frequently. Clients who fail to respond to the recommended diet may be suffering symptoms that are produced by allergies. Eliminating sugar will not alleviate distress caused by sensitivity to wheat or milk.

Several methods for determining food allergies are, (1) referral to a physician for skin testing; (2) a four day fast followed by challenge with the suspected food; (3) elimination diets (Rowe, 1974). Foods that are most likely to cause an allergic response are sometimes those consumed in large quantities (Randolph, 1956). The client's food journals are invaluable at this time.

TABLE II
Recommended Diet

BASIC RULES:

1. EAT AT LEAST THREE EVENLY SPACED, WELL BALANCED MEALS PER DAY.
2. CONSUME ADEQUATE PROTEIN DAILY (RULE OF THUMB TO DETERMINE PROTEIN NEEDS: DESIRED BODY WEIGHT ÷ 2 = GRAMS OF PROTEIN DAILY). PROTEIN MAY BE OF ANIMAL OR VEGETABLE ORIGIN.
3. CONSUME FRESH FRUITS AND VEGETABLES DAILY.
4. USE ONLY WHOLE GRAINS. INCLUDE LEGUMES AND NUTS.
5. TOTALLY ELIMINATE: SUGAR (WHITE, BROWN, RAW TURBINADO, SYRUP, HONEY, MOLASSES, ETC.) WHITE FLOUR, WHITE RICE, ALCOHOL.
6. USE SPARING: SALT, DRIED FRUIT, COFFEE, TEA, TOBACCO.
7. SUGGESTED: FRUIT, VEGETABLE, OR PROTEIN SNACK BETWEEN MEALS AND BEFORE BED.

FOODS TO FAVOR

LEAN MEAT
FISH
POULTRY
EGGS
MILK
CHEESE
PLAIN YOGURT
WHOLE GRAINS
FRESH FRUITS
FRESH VEGETABLES
LEGUMES
NUTS, SEEDS
HERBAL TEAS

FAVORITE FOODS TO USE MODERATELY

SWEET FRUIT AND VEGETABLE JUICES, I.E. CARROT, GRAPE, APPLE, OR ORANGE JUICE
DRIED FRUIT
SALT

FOODS TO AVOID

SUGAR-WHITE, BROWN TURBINADO, RAW
HONEY. MOLASSES
CORN SYRUP
WHITE FLOUR
CAKES. COOKIES. PIES. PASTRIES. DONUTS, CANDY
WHITE BREAD
BREAKFAST CEREAL, COMMERCIALLY MADE GRANOLA
ALL SOFT DRINKS
FRUIT FLAVORED DRINKS
ICE-CREAM
FLAVORED YOGURT
CANNED FRUIT
CANNED VEGETABLES
PROCESSED OR PREPACKED FOOD
COFFEE, TEA
ALCOHOL

COMMON SENSE SUGGESTIONS

1. OVERWEIGHT? FOLLOW BASIC RULES, BUT LIMIT FAT INTAKE AND PORTION SIZE.
2. BALANCE MEALS WITH PROTEIN FOODS, FRUITS, VEGETABLES AND UNREFINED STARCHES.
3. OBSERVE HOW YOU FEEL. DON'T EAT ANYTHING THAT LATER MAKES YOU FEEL BAD.

WHAT IS A GRAM OF PROTEIN?

1 EGG = 6 GRAMS PROTEIN
1 OZ. MILK = 1 GRAM PROTEIN
1 OZ. MEAT = 6 GRAMS PROTEIN (APPROX.)
1" CUBE CHEESE = 4 GRAMS PROTEIN (APPROX.)

Because the biological substrate has been neglected in treatment, we have purposely avoided discussion of other therapy techniques. Obviously dietary factors are not the only significant elements in the therapeutic smorgasbord, where one man's meat may be his neighbor's poison. One may drink quarts of coffee, smoke cartons of cigarettes and eat pounds of sugar and, despite these insults to the body, still function in society. Nutritional counseling is not the sole salvation for the substance abuser. There are many ways to stay sober, many paths to a drug free existence.

If one uses nutrition as an adjunct to counseling, one should keep in mind that there is no simple, quick, magic nutritional cure for alcoholism, drug abuse, and emotional problems. However, there is much evidence to suggest that attention to dietary factors may help the client more adequately deal with problems and render the counseling process more efficient, productive and rewarding. As one client said, "I used to feel crazy all the time. Now I only feel that way when I cheat on my diet. I know how to control that. It sure makes it a whole lot easier to stay sober."

REFERENCES

American Medical Association. 1977. *Manual on Alcoholism.*

Arky, R.A. 1971. The effect of alcohol on carbohydrate metabolism: carbohydrate metabolism in alcoholics. In Kissin and Begleiter (Eds.), *The Biology of Alcoholism*, Volume 1. New York: Plenum.

Bell, E.C. 1958. Nutrional deficiencies and emotional disturbances. *J. of Psychology.* 45: 47-74.

Brecher, E.M. 1972. *Licit and Illicit Drugs.* Boston: Little, Brown.

Burton, R. 1628. *The Anatomy of Melancholy.* F. Dell and P. Jordan-Smith (Eds.), New York: Farrar and Rhinehart, (1927 edition).

Cahill, G.F. and Soeldner, J.S. 1974. A non-editorial on non-hypoglycemia. *The New England J. Med.* 291: 905-906.

Cleave, T.L. 1975. *The Saccharine Disease.* New Canaan: Keat's Publishing.

Corwin, A.H. 1976. The rotating diet and taxonomy. In Dickey, L.D. (Ed.), *Clinical Ecology.* Springfield: C.C. Thomas.

Dickey, L.D. (Ed.) 1976. *Clinical Ecology.* Springfield: C.C. Thomas.

Etheridge, E.W. 1972. *The Butterfly Caste*. West Port: Greenwood Publishing Company.

Fredericks, C. and Goodman, H. 1969. *Low Blood Sugar and You*. New York: Grosset & Dunlap.

Freinkel, N. 1975. Hypoglycemic disorders. In Beeson and McDermott (Eds.), *Textbook of Medicine*.

Greden, J.F. 1974. Anxiety or caffeinism: a diagnostic dilemma. *Am. J. Psychiatry*. 131: 1089-92.

Greden, Fontaine, P., Lubetsky, M., and Chamberlin, K. 1978. Anxiety, depression and caffeinism among psychiatric inpatients. In press, *Am. J. Psychiatry*.

Harris, S. 1924. Hyperinsulinism and dysinsulinism. *J. Am. Med. Assoc.* 83: 729-734.

Hawkins, D. and Pauling, L. (Eds.) 1973. *Orthomolecular Psychiatry*. San Francisco: Freeman and Company.

Hoffer, A. 1977. Supernutrition. In Williams and Kalita (Eds.), *A Physician's Handbook on Orthomolecular Medicine*. New York: Pergamon.

Jackson, J.F. 1973. Genetic factors in disease. In Brunson and Gall (Eds.), *Concepts of Disease*. New York: Macmillan.

Jellinek, E.M. 1960. *The Disease Concept of Alcoholism*. New Haven: Hillhouse.

Kalita, D.K. 1977. Orthomolecular medicine. In Williams and Kalita (Eds.), *A Physician's Handbook on Orthomolecular Medicine*. New York: Pergamon.

Knox, W.E. 1966. Phenylketonuria. In Stanbury, Wnygaarden and Fredrickson (Eds.), *The Metabolic Basics of Inherited Diseases*. New York: McGraw-Hill.

Leevy, C.M., Valdellon, E. and Smith, E. 1971. Nutritional factors in alcoholism and its complications. In Israel and Mardones (Eds.), *Biological Basis of Alcoholism*. New York: John Wiley and Sons.

Meiers, R.L. 1973. Relative hypoglycemia in schizophrenia. In Hawkins and Pauling (Eds.), *Orthomolecular Psychiatry*. San Francisco: W.H. Freeman and Company

Milam, J.R. 1974. *The Emergent Comprehensive Concept of Alcoholism*. Kirkland, Washington: ACA Press.

Pauling, L. 1968. Orthomolecular psychiatry. *Science*. 160: 265-271.

Popham, R.E. 1953. A critique of the genetotrophic theory of the etiology of alcoholism. *Quart. J. Stud. Alc.* 14: 228-237.

Poulos, J., Stafford, D., and Carron, K. 1976. *Alcoholism, Stress and Hypoglycemia*. New York: Sterling Publications.

Randolph, T.G. 1956. The descriptive features of food addiction: addictive eating and drinking. *Quart. J. Stud. Alc.* 17: 198-224.

Randolph, T.G. 1975. Biologic dieletics. In Dickey (Ed.), *Clinical Ecology*. Springfield: C.C. Thomas

Robinson, C.H. and Lawler, M.R. 1977. *Normal and Therapeutic Nutrition*. New York: Macmillan.

Roe, D.A. 1973. *A Plague of Corn: The Social History of Pellagra.* Ithaca: Cornell University Press.

Roe, D.A. 1976. *Drug-Induced Nutritional Deficiencies.* West Port, Connecticut: AVI Publishing.

Rowe, A.H. 1972. *Food Allergy*. Springfield: C.C. Thomas.

Seixas, F.A., (Ed.) 1971. *Treatment of the Alcohol Withdrawal Syndrome*. New York: National Council on Alcoholism.

Speer, F. 1970. *Allergy of the Nervous System.* Springfield: C.C. Thomas.

Spies, T.D. 1947. *Rehabilitation Through Better Nutrition.* Philadelphia: W.B. Saunders Company.

Stanbury, J.B., Wyngaarden, J.B. and Fredrickson, D.S. 1966. *The Metabolic Basis of Inherited Disease*. New York: McGraw-Hill.

Vitale, J.J. and Coffee, J. 1971. Alcohol and Vitamin Metabolism. In Kissin and Begleiter (Eds.), *The Biology of Alcoholism,* Volume I. New York: Plenum.

Wallace, A.F.C. 1972. Mental illness biology and culture. In Hsu (Ed.), *Psychological Anthropology*. Cambridge: Schenkman Publishing Company.

Watson, G. 1956. Is mental illness mental? *J. of Psychology*. 41: 323-334.

Williams, R.J. 1956. *Biochemical Individuality*. Austin: University of Texas Press.

Williams, R.J. and Kalita, D.K. 1977. *A Physician's Handbook on Orthomolecular Medicine*. New York: Pergamon.

Winstead, D.K. 1976. Coffee consumption among psychiatric in-patients. *Am. J. Psychiatry*. 133:12.

Yager, J. and Young, R.T. 1974. Non-hypoglycemia is an epidemic condition. *New Eng. J. Med*. 291: 907-908.

ALCOHOLIC SUPPORT SYSTEMS IN AN ALASKAN COMMUNITY

Barbara Stuckel

Salvation Army Comprehensive Alcoholism Services

It is generally accepted that alcohol has a major impact on not only the drinker, but on the total family. The abuse and misuse of this drug affects interactions with all significant others--with all social relationships. However, despite this knowledge of the impact of alcohol misuse on all who come in contact with the alcoholic and/or problem drinker, there is not much information available to the public concerning the social support systems that tend to perpetuate this addiction.

Support systems can simply be defined as anything that reinforces the ongoing behaviors. Caplan (1976) refers to these supports as "continuing social aggregates that provide individuals with opportunities for feedback about themselves and for validation of their expectations about others, which may offset deficiencies in these communications within the larger community context." Support system implies a pattern of continuous or intermittent acts that perform a significant part in maintaining the psychological and physical well-being of the individual over a time (Caplan, 1974). These supports usually include at least one significant other who helps the individual to use his psychological resources and carry his emotional load, supplementing money, goods, skills and cognitive guidance to aid in the handling of his circumstances, and in the sharing of any labors. These systems can be both spontaneous and/or organized. They are usually a combination of professional and formal community agencies coupled with natural systems such as families and person to person care-giving efforts. Most communities develop integral social systems to support the alcoholic. These contribute to the alcoholic's instability as well as stability. These systems are negative as well as positive.

One of the basic reasons for small success rates with alcoholics is that the majority of the existing supports that have been built around them are negative, immediate and on-going. To combat this, the action-reaction effect of these negative systems needs to be broken; and to fill the void, immediate, long-term, positive behavior modification must be the replacement. With this focus on supports, committees of appropriate individuals should be formed to view their own community and the alcoholism problem. A state that is no stranger to the growing alcoholism problem and that is trying innovative measures is Alaska.

Alcoholism is Alaska's number one health, economic and social problem. This problem of alcohol abuse is distributed throughout the state, in rural and urban areas alike. It is known that whole villages in the "bush" areas decide when there is to be a drinking binge. When these "social" gatherings first began, the older relatives would take the responsibility for the children. (This aspect of extended family care is an accepted cultural practice with Alaskan natives.) As these drinking binges continued to occur, more members of the older age group became active drinkers resulting in less care given to the children. Numerous cases of child neglect occur. Due to role modeling, as well as alcohol becoming more plentiful in the home, children begin to drink at younger ages.

Some Alaskan villages consider themselves "dry". In the villages where prohibition of alcohol is the law, bootleggers become prosperous. Alcohol continues to come in, by carrier, and the people do not abstain. A reason why drinking continues may be that the accepted urban definition of having a problem with alcohol is not the same in rural native villages. In these "bush areas" alcoholic behaviors have become accepted and are condoned. In most rural areas if one does not drink with someone, it is thought that he considers himself "too good" to drink with the regular crowd and may risk being ostracized. A lack of agreement on what constitutes alcoholic behavior as well as lack of apporpriate alcohol legislation and enforcement of this legislation would seem to both reflect and contribute to the absence of strong social controls over drinking in the settlements.

The Alaskan natives' rural way of life and daily routines that have been passed down through generations are in a state of crisis. The natives find their roles changing as subsistence rural lifestyles evolve into more hectic urban routines governed by Western regulations. The rural native is accustomed to seasonal work (i.e., a fishing season, a hunting season, etc.). Their concept of time is different from the city individual. They make the move to an urban setting or their rural system begins to modernize, and there is little need to take the time to learn or to practice skills of subsistence living. They replace these practices with new and convenient ones. Cultural and traditional behaviors decrease, alcohol consumption increases. An article written in 1955, "Drinking Patterns of the

Aleuts" (Berremen), emphasizes the relationship between drinking and anxiety within a community. It states that when a village disintegrates as a community the drinking will increase causing greater adverse effects upon the community and its individuals. Berreman focuses on the statement that if economic security and independence could be achieved, it would lower the level of anxiety and the social situation and drinking patterns would become less of a problem.

Many bush areas do not offer proper academic curricula. This creates a double edged sword. Native parents find integral family members gone when older children are sent to boarding schools hundreds of miles away and are rarely seen by the family. The parent may begin to drink more.

In a recent study (Kleinfeld & Bloom 1977) of the social and emotional problems of Alaskan Eskimo adolescents during their freshman and sophomore years in various boarding school environments, it is noted that drinking problems involved only 8% of the freshman class of a public boarding school, but increased to 40% during the following year. It is suggested that heavy drinking and violent drunken behavior were related to social stress, largely caused by racial tension between Eskimos and whites. Wives and mothers are prone to "cabin fever" during winter when daylight hours are gone as soon as they come. Some villages have approximately two hours of daylight in midwinter months. "Cabin fever" is a state of depression, similar to the effects one can suffer when living in a climate that offers continuous rain, long droughts or constant sandstorms. Alaskan women are more susceptible to the loneliness and isolation housewives frequently feel due to the remoteness of rural communities as well as out of the way "subdivision" layouts of urban centers.

When rural Alaskan natives make trips to the Anchorage area, there are two generally acceptable areas where natives can congregate, feel comfortable and meet other natives, the Alaska Native Medical Center and Fourth Avenue. If they are found drinking or drunk while on the premises of the medical center, they are terminated from the hospital. This increases their chances of drinking. One of Anchorage's biggest tourist attractions is Fourth Avenue. It is a mixture of smart shops, attractive flower beds and skid row alcohol/drug abusers. They can be found crawling out of dumpsters, sleeping in the downtown library or under bushes, interrupting weekly church services as well as asking for handouts. They are commonly referred to as the Fourth Avenue Native Population.

There are members of this subculture that are non-natives. The "native" population includes Eskimo, Aleut and Indian. They represent approximately four to five percent of Alaska's total alcoholic populace. The rest have the support systems that include homes, jobs, and families. However, the condition of the Fourth Avenue public inebriate is one aspect of alcoholism peculiar to Alaska. The Fourth Avenue public inebriates of Anchorage form their own subculture, the style of living forms a recognizable pattern in which a high degree of inter-connecting and mutual dependency is the rule. Shapiro (1971) found that, despite the painful basis of these rejects of the larger culture, a core with their own health standards emerges. They have built a private culture that affords them a network of social supports. In this world, some are maintaining "quasi-families", having learned to safeguard each other from the excesses of their own pathology. There is a kindred spirit between these street people. When someone new wishes to join the group, there is a certain period of initiation, to see if they make the grade. As they pass muster, their individual feelings and attitudes are assimilated into the group. They are also taught "the community" ropes, made aware of the support systems the district has to offer, both positive and negative, and avail themselves of the ones they desire depending on whether they want sobriety or inebriety.

The Anchorage area has in the past provided a walk-in center for the derelict alcoholic. Services offered included a place to gather, sit and socialize, to keep warm and dry, coffee, bouillon, sandwiches, and counselors who in reality were monitors. An average number of clients seen in this center were approximately 750 different individuals each month, some of them visiting the center up to 40-50 times each month: a shining example of a negative support system.

Granted, public inebriates and other social service problem cases had a place to find aid in maintaining necessary minimal health needs. If necessary, transportation was given to a hospital for emergency medical attention or to a detox unit. When necessary, a shower and change of clothing was provided, giving the alcoholic just enough support to maintain comfort, reinforcing existing drinking patterns. This was help that was available on a 24 hour, seven day a week basis, help given by counselors acting as significant others, helping the alcoholic to utilize available resources, helping to alleviate emotional loads, handling their problems by supplementing goods and guidance.

This walk-in center closed its doors in November of 1977. In its place a central screening unit was developed by January, 1978. Services included a community service patrol for pick-up and delivery of intoxicants to hospitals or appropriate community agencies. Medical triage service is offered. There is no sleep off room, no place to sit with a buddy and chat. Clients are also requested to actively do something for themselves by participating in an intake interview, a medical exam and an appointment or referral to an appropriate agency.

It is important for community members, an agency and its counselors to recognize the potency of generating enough force during a time of crisis in a client and utilizing these forces in the interest of client change.

The average number of clients seen in this screening unit since January 1, 1978 has been approximately 250 transports on the patrol van and approximately 150 walk-ins (in three months). Certainly this is a marked difference from the 750 different individuals previously seen each month in the Walk-In Center. Where have all the clients gone? The derelict alcoholic seems to always find a way to maintain an existence on the street. It clearly points out that when positive support is offered, not every alcoholic really wants sobriety.

In the past, the Alaskan Federation of Natives (AFN) and the Bureau of Indian Affairs (BIA) were two of the main operations offering supports to natives. Due in part to the number of native clients and due to bureaucratic blanket policy statements, natives came face to face with a system that fostered frustration and anxiety. The present plan for native corporations in Alaska is to form smaller agencies that are ethnically defined. Past cultural practices of the natives point out that this differentiation is an important aspect. Alaska is divided into 12 native regions. When someone identifies themself to another native as an Eskimo, the first question asked is: "Where is your village?" "Are you a Northern or Southern Eskimo?" Smaller agencies can offer a support system conducive to ethnic and regional problems and hire natives from similar backgrounds, setting up trust and rapport in aiding the alcoholic. The relation of drinking to acculturation has been frequently noted. This disruption of the native culture and the marginality of the individuals existing between two cultures, often leads to individualistic drinking patterns and problem drinking.

Another problem area that needs to be cited as offering negative support is the one of legislation. The state excise tax on liquor has remained unchanged for fifteen years while consumer price indexes for many other items continue to rise significantly. It is cheaper to purchase wine than to purchase milk or soft drinks. Present legislation subsidizes liquor by making it a relatively cheap commodity, resulting in increased consumption; drinking as an activity is promoted providing social reinforcement for alcoholics.

The Alaska State Statutes provide for arrest of operating a motor vehicle while intoxicated. It does not state that it is illegal to drive with a BAC of .10%. Since the State Statute does not allow conviction on BAC level alone, a great amount of resources are spent on trials to determine conviction. The drunk in public law was erased from the statutes. The remaining law as written is ambiguous and ineffective. There are no specific enforcement procedures for chronic repeaters.

State statutes create barriers for effective legislative action to be implemented on a local level. Common carriers are not prohibited from transporting alcohol into "dry" communities. Special local liquor taxes are prohibited, preventing local government from obtaining a source of revenue for local treatment and prevention.

Licensed liquor outlets are numerous in Anchorage and have little difficulty with license renewals. Local legislation controls the hours of sale. In Anchorage, bars are closed only from 5:00 a.m. to 8 a.m.--barely enough time to be swept out and to replenish the stock.

A range of treatment programs for the alcohol abuser are not available in the rural areas. These communities have no persons trained to identify and refer behavioral health problems nor to offer basic counseling techniques which can provide a model for development throughout the bush areas.

On the positive side, the National Health Planning and Resources Development Act established a nationwide system of regional health planning organizations. The primary responsibility of these agencies is to develop and implement plans for health improvement. South Central Health Planning and Development, Inc., a health system for South Central and Western Alaska, is aware of some of these problem areas and has set up goals and objectives for the next year to aid in the desired State/community changes necessary in alleviating the problems created by alcohol abuse throughout Alaska.

Other major problems of alcoholism service programs, especially native programs, that inhibit the best possible positive supports, is the high turnover rate of counselors due to low pay, political pressures, lack of program directors with organizational experience and fragmentation of services. Psychometric tests offered in treatment often have cultural biases, giving erroneous and/or insufficient information resulting in the inability to offer the best possible treatment to natives. These are only some of the areas that contribute as negative supports to Alaskan alcoholics, many of these just as relevant to other communities throughout the United States.

The alcoholic (public and private) receives supports from health services. Public assistance programs provide enough funds to sustain on-going drinking behaviors. Frequently, a patient with a history of alcoholism is a target for multi-drug misuse, as physicians may prescribe amphetamines for weight loss and sedatives to calm nerves.

Third party payors do not provide incentives for abusers to seek treatment services. Alcoholism is not allowed as a diagnosis for reimbursement. Health insurance programs as well as the Veteran's Administration do not pay for the coverage of out-patient treatment services without prior inpatient care.

Rehabilitation facilities that have no or very little interagency contact provide convenient havens when the alcoholic chooses to use them. This creates a "revolving door" syndrome, offering temporary supports and care between drunks. Too often social service agencies employ counselors who suffer from the "bleeding heart" syndrome, initiating too much assistance, and over-nurturing, giving the alcoholic just one more person to become dependent upon. This is mainly due to the counter-dependent need of the counselor. Treatment is geared for the counselor's personal needs, and not for what the client can obtain from it. To transform this negative support into a positve awareness, the concept of "tough love" or confronting with caring needs to be taught.

As this paper addresses community support systems conducive to reinforcing drinking habits, I have not taken the time to discuss supports given by immediate as well as extended family systems. I have mentioned only a few of the positive alternatives that can and should be offered by a community. The two major problem areas tend to be a lack of knowledge and experience on the part of the service providers with respect to the kinds of support systems they create, and a lack of knowledge and available information on alcoholic support systems given to the general public. This information should include the subjective beliefs, attitudes, and feelings of society concerning "helping" the alcoholic.

It is important to note that the divorce rate is 5 times greater for recovering alcoholics than for the national average. The suicide rate is twelve times greater for recovering alcoholics than for the national average. This should tell us without question that the total social environment needs to be treated. I have picked out and picked upon specific systems in Alaska. These are the ones I am most familiar with. I am sure that these systems are not alien to other communities throughout this country and can be related to both alcohol and other substance abuse.

It is the hope of this writer that this paper will act as a catalyst for service providers to develop and to distribute materials identifying negative supports to the community focusing more attention on the impact these supports have on the substance abuser, and outlining specific steps to take corrective measures.

REFERENCES

Berreman, G., 1956. Drinking Patterns of the Aleuts: *The Quarterly Journal of Studies on Alcohol*. 17: 503-514.

Caplan, G. 1974. *Support Systems and Community Mental Health*: Lectures on Concept Development. New York: Behavioral Publishers.

Caplan, G. & Killilea, M. 1976. *Support Systems and Mutual Help Multidisciplinary Extrapolations.* New York: Graine and Stratton.

Kleinfeld, J., & Bloom, J. 1977. Boarding Schools: Effects on the Mental Health of Eskimo Adolescents, *American Journal of Psychiatry.* 134: 411-417.

Lookout, F.M. 1977. Alcohol Drinking and the Indian Predicament, *Alcohol Technical Report.* Oklahoma City. 6: 13-16.

Shapiro, J.H., 1977. *Communities of the Alone*, New York: Association Press.

Somers, M.L., 1963. The Small Group in Learning and Teaching: *Learning and Teaching in Public Welfare*, Report of the Cooperative Project on Public Welfare Staff Training I. Bureau of Family Services, Welfare Administration, U.S. Department of Health, Education and Welfare.

A "SHARP" APPROACH TO ALCOHOLIC TREATMENT

Peter Stead, M.S.W., Director, SHARP Program
Veterans Administration Hospital, Palo Alto, California

Judith Viders, M.S., Assistant Program Director
Veterans Administration Hospital, San Francisco, California

Abstract: This paper describes a hospital based, total abstinence, self-help alcohol recovery program (SHARP) which has an active social support system in the community called CORK. This self-help program engages chronic alcoholics in assuming major roles in governing the program and in helping one another. In SHARP's approach, new and meaningful social roles are created for patients so they can actively participate in their own recovery. Patients are helped to develop skills in assertiveness, organization and negotiation. SHARP also includes a work-for-pay program (Payday) which re-establishes dignity, promotes camaraderie, and provides a financial grubstake. SHARP provides a setting for mutual planning to re-enter the community with an improved self-image and health. Because we believe follow-up support is essential to maintain sobriety, SHARP created a community program, CORK. CORK sustains the values and skills of SHARP which are a common bond, sobriety, and mutual assistance. Since its inception in October 1975, CORK has grown from 8 to 162 members discharged 6 months or more, plus their families and friends. Of these members, 104 are totally abstinent and 38 are drinking, but still occupied and in charge of their lives. The evidence of SHARP's success is shown by results of the 1976-1977 follow-up study which surveyed former patients' levels of drinking activity and life functioning. Results show 37% totally abstinent, 20% drinking controlled and infrequently, and 14% drinking but still in charge of their lives.

An alcoholic treatment for long term chronic alcoholics with impoverished social relationships has created a subculture which starts in a hospital and extends into the community. By providing a social support system, this subculture aids the recovering alcoholic in maintaining sobriety and remaining productively occupied.

INTRODUCTION

The Share Help Alcohol Recovery Program (SHARP) based in the Palo Alto Veterans Administration Hospital, Menlo Park, California approaches alcoholic recovery by involving the patient and his family in an ongoing social support system offering fellowship, a sense of belonging, life-related problem solving, and recreation. The main goals of SHARP are (a) to support the alcoholic in his goal of attaining and maintaining total abstinence and (b) to provide a setting for mutual planning to re-enter the community with an improved self-image, health, and a constructive discharge plan for maintaining sobriety.

The SHARP program is a 28-bed, self-help, total abstinence program for veterans. The patient spends 45 days in the hospital and afterwards is involved in a community program component for Counseling, Organization, Recreation, and Knowledge (CORK). This program component functions to support recovering alcoholics in staying sober through social gatherings, recreation, problem-solving meetings, and frequent informal interactions among its members. This community effort covers three San Francisco Bay area counties in northern California and includes SHARP graduates, their families and friends.

CHARACTERISTICS OF SHARP PATIENTS

A sample of 90 patients admitted over a 3-month period was taken in 1976. The average age of the SHARP population was 47 and 84% were Caucasian. Ninety-seven percent had a primary diagnosis of alcoholism. According to the patients, they had not been able to control their drinking for an average of 13 years. The staff believes this period is much longer. Sixty-three percent had been in previous inpatient alcoholic treatment programs and 67% had gone to Alcoholics Anonymous. Fifty-four percent lived alone, while 46% lived with someone (wife, other family, girl friend, or significant other). Eighty percent received income from some source, such as employment, disability pension, retirement pension, or welfare, whereas 20% had no income from any source.

Although the majority of SHARP residents are middle-aged veterans with at least a 13-year history of excessive drinking and marginal functioning, there are also two distinctively different sub-groups within the patient population, (a) the 50-plus retired group, and

(b) the recently emerging under-35 (Vietnam War) group. Many of these people are seeking help for the first time.

The Older Alcoholic

Retirement is a life-situation that takes a heavy toll on the once-employed, work-oriented persons who come into SHARP. Prior to retirement, the degree of their social functioning varied from marginal to highly functional. Independent of their place on this continuum, they either were--or appeared to be--in control of their lives. Following retirement, their social role functioning deteriorated. Without a clear role to fill and with time on their hands, there were significant changes in their daily lives. They found their lives now characterized by a pervasive sense of purposelessness and lack of direction. Their drinking accelerated while their sense of worth decreased. With this increase in drinking, they became embroiled in legal, medical, and family problems which finally brought them to our attention.

A good example of the older alcoholic is Mr. R. J., a 64-year-old, retired, married, father of two grown children. When Mr. J was admitted to SHARP, he was so debilitated that he required a wheel chair. His family rejected him because of his drinking, and he had nothing to do except clean the house--because other family members were employed. Through the support of SHARP and CORK, Mr. J. has been totally abstinent over 2 years. He is happy and is actively involved in community activities, including CORK.

Vietnam Veterans

The under-35 patients in this program are products of the Vietnam War drug-culture period. They have a history of experimenting with--or even being addicted to--such hard drugs as heroin, amphetamines, barbituates, and psychedelics. More recently, their use of these drugs decreased and their alcohol intake increased. Often they have merely given up a drug habit for an alcohol habit. The authors note that these men's lives are characterized by transience, a lack of direction, and difficulty formulating and following through on life goals.

A good example of a SHARP Vietnam War patient is Mr. D. L., a 26-year-old, married, unemployed veteran with a 4-year history of heroin addiction and multiple drug use before becoming an alcohol abuser. Mr. L. was a patient in the Menlo Park VA Hospital's drug detoxification unit seven times without any change in his behavior. When he was admitted to SHARP, alcohol abuse had become his main problem. After going through a stage of hostility, suspiciousness, and indifference to SHARP, Mr. L. came to like and be liked by the other patients. He became friendly, took an interest in the program, and made constructive discharge plans which included abstinence from all drugs. He has not been rehospitalized since his discharge on January 10, 1977, and has been staying sober.

THE PROGRAM CALLED SHARP

In response to 7 years of experience with 1500 habitual excessive drinkers, the staff has evolved the current model. This model has four main components:

1. Share help (self-help) concept
2. Social systems model
3. Discharge planning, including the CORK family and community program

Share Help (Self-Help)[1]

The major purpose of this system is to provide a social climate contrary to the destructive one in which the patient has been mired, a social climate wherein the patient discovers he can function as an effective participant and decision-maker in a positive social role. Patients assume major roles in governing the program and in helping one another to change. Leaders, chosen by the patients conduct all meetings, supervise the patient-run committees, and make major decisions. Staff act as consultants to the leaders.

The share-help model creates new and meaningful social roles, enabling patients to actively participate in their own recovery. In the course of dealing with day-to-day problems that arise, they develop skills in assertiveness, negotiation, and collaboration. In this setting patients experience a sense of responsibility, renewed self-worth, and an esprit de corps.

Social Systems Model

Staff in many alcoholic-treatment programs tend to view excessive drinking as a problem characteristic of a particular individual that can be alleviated by a time-limited exposure to a medical or psychological treatment setting. SHARP regards alcoholism as part of a pathological social system, rather than simply a disease of the indi-

---

[1] Recently the name was changed from "self-help" to "share help" because the patients continued to see self-help as meaning helping oneself only, whereas, to them, share help meant to help others.

vidual. Therefore, the group process in the patient-run, share-help program is regarded as an important part of the treatment. Furthermore, SHARP puts much emphasis on constructive discharge planning, including a supportive community environment to maintain sobriety. CORK, the postdischarge phase of the program, is SHARP's way of establishing this helpful community environment.

Involvement of family members in the program is mandatory for those who live in the Bay area, in that the drinking is seen as a symptom of a pathological family system.

An example is Mr. L. S., a 43-year-old father of six children who had a long history of alcoholism and an unsatisfactory marriage. Prior to his admission to SHARP, Mr. and Mrs. S. had received marital counseling, and Mrs. S. had received psychiatric treatment, without any changes in their behavior. SHARP gave Mr. S. support after the separation from his wife and helped him to make a new life for himself. His new life includes total abstinence from alcohol since July 28, 1976 and full-time employment.

Role of Staff in a Self-Help Program

The Sharp staff is composed of only 6 full-time clinical employees working with 28 inpatients and over 200 outpatients. The Staff maintains a specialized setting in which a self-help program can occur. This setting entails being responsible to the hospital administration for running the program and abiding by hospital policies. The staff also deals with the outside world, including referaal agencies, other hospitals, detox units, community agencies, etc. Furthermore, the staff maintains continuity of beliefs in a program in which individual patients remain for a maximum of 45 days. New patient leaders are being continually trained to take over.

Because of staff's long experience, professional training, and psychiatric and historical information on new applicants, staff has considerable influence in the decision making of whether or not to accept an applicant to the program, even though patients could outvote staff. Staff provides medical care, nursing care, and psychiatric evaluations when needed.

The patients, especially the patient leaders, receive consultation from the staff. This consultation usually covers policy, scheduling, and how to deal with program and patient problems. The patients have the choice of whether or not to take the staff advice and have a good deal of freedom in handling program and patient matters in their own way. Staff monitors the progress of each patient through the program by observation and consultation with program leaders. If a patient has a personal problem or a feeling to share and he comes to a staff member, he is usually encouraged to seek help from his peers.

However, patients most often seek help from each other. The day a patient graduates we ask him what feature in the program helped him the most. Almost all the patients respond that their evening rap sessions, when no staff is present, helped them most.

Other Program Features

The SHARP program includes daily problem solving, relaxation therapy, vocational counseling, recreation, and a weekly AA meeting. An important feature of SHARP is assertiveness training and the encouragement of all residents to be more open, direct, and outgoing. Specific program assignments are designed to help patients become more assertive. For example, the most passive patient is given the cleanup chairman assignment. In this capacity, he is not allowed to do any menial work himself, but he supervises the daily cleanup. In short, he acts as a top sergeant. This particular assignment has succeeded in giving several cleanup chairmen self-confidence and a higher self-esteem.

One example is Mr. C. K., a 49-year-old, single, unemployed veteran receiving a disability pension for depression. Prior to his admission to SHARP, Mr. K. had been admitted to VA Hospitals 21 times for depression. Between hospitalizations, he lived a marginal, isolated life style and drank excessively. When he was screened for SHARP, he admitted to only one psychiatric admission for fear he would be rejected. Since his admission to SHARP, Mr. K. came to realize his main problem was drinking, not depression. With the help of his new responsibility as cleanup chairman and the group support he received, he changed from a passive, uninvolved person to an assertive, self-confident leader.

Payday

Although many alcoholics tend to be impoverished in their interpersonal relationships, they do form close relationships when they perform tasks together. Useful work in groups seems to reestablish a feeling of dignity, self-confidence, and camaraderie with these men. It also provides the grubstake they need to move out into the community. Therefore, our work-for-pay program called "Payday" was initiated and continues to be an important feature of the SHARP program.

Payday is a voluntary program for SHARP residents and graduates who want to earn money. It generates a mutual support system which starts in the program and, for some residents, continues in the community. Payday has included a wide variety of work, such as house painting, carpentry, cleaning, construction work, gardening, laying tile, cutting trees, selling firewood, and the making and selling of such items as rugs and planters, cat trees, etc. The patients solicit the business, work when the program is not in session, and handle the money. Handling the money includes negotiating the distribution of earnings and dealing effectively with those who do not carry their weight in work projects.

Payday brings out the creativity, talents, and skills of the patients which tend to be overlooked because of the more immediate problem of alcohol abuse. When these talents surface, they are found to be varied and impressive. For example, in 1974 a man whose skill was vinyl repair trained three other residents and secured contracts with a restaurant and other establishments for vinyl restoration. While in the program, this man earned $1,200, and each of the three men who worked with him also earned several hundred dollars. These four men continued to work together for one year after they completed the program. They obtained a contract with the San Francisco Municipal Railway to repair bus seats. This job and others maintained them for a whole year. Having formed close interpersonal relationships, they were part of the nucleus of CORK.

CORK

One of the primary principles of SHARP is having a brief (45-day) inpatient stay, followed by a long-term, supportive outpatient program to maintain total abstinence. SHARP was originally a much longer program--90 days or more, but in the longer program, the men tended to become institutionalized in the VA Hospital system and neglected to attend to their real problems in the community. Therefore, the program time was cut to 45 days with a minimum of a 6-month outpatient follow-up. In the daily SHARP problem-solving meetings, the focus is on discharge planning, including moving to a supportive living environment and staying active. While they are in the program, the patients are referred to alcohol recovery homes, AA, and SHARP's own follow-up program called CORK (Counseling, Organization, Recreation, and Knowledge)--the extension of SHARP.

CORK was founded October 10, 1975 by Joe Archambault, a successful SHARP graduate, who became an effective community organizer in the field of alcoholic treatment. In 1974, prior to the official starting of CORK, he developed a nucleus by organizing an informal sober group of eight SHARP graduates. He continued to develop and expand CORK into the much larger, well-organized program it has now become.

Because SHARP believes that to stop or control one's drinking is not a single event but, instead, a process, CORK was devised to provide an ongoing support system to reinforce this process. CORK sustains the skills and values learned in SHARP. These skills include shared help, organization, negotiation, and assertiveness. CORK also sustains the same value system, including a belief in sobriety, taking responsibility for one's actions, helping others, fellowship, and maintaining a common bond. CORK's main functions are socialization and solving real life problems such as job finding, housing needs, and resource assistance. Members of CORK, their families, and friends are given an opportunity to socialize in a nondrinking environment. This type of socialization is very important because most alcoholics have difficulty socializing without alcohol.

Mr. S. J. is a 43-year-old carpenter and bartender, married, the father of three children, whose drinking had been out of control at least 2 years before he was admitted to SHARP. Due to drinking he was separated from his wife, and she had contacted a lawyer to file for divorce. He had not worked for months and was physically debilitated. Mr. J. did well in SHARP, took Antabuse faithfully, and indicated an intention to remain totally abstinent. After completing SHARP, Mr. J. joined CORK and established a close relationship with the other members. In that special setting, he confessed that he had never really intended to give up drinking but planned to resume it on a controlled basis. His CORK friends strongly advised him against this plan and also involved Mr. J.'s wife in CORK functions. As a result of the support from CORK, the establishment of close personal friends there, and the involvement of Mrs. J. in CORK, Mr. J. changed his mind and has been totally abstinent since December 1, 1975. His marriage is unusually close, he is employed full time as a bartender, and has become secretary-treasurer of one of the CORK chapters.

CORK and the Co-alcoholic

The co-alcholic includes the spouse, girl friend, or anyone who is significantly involved with the problem drinker. SHARP has found that helping co-alcoholics with their problems facilitates reaching the alcoholic. Consequently, the co-alcoholic group and the couples group, which are led by staff, are a very important part of the CORK program. Since the inception of these groups in October, 1976, 78 co-alcoholics, 62 couples, and 12 teenage family members have been seen. The couples group has averaged eight couples per week. Some of the results of the CORK co-alcoholic and couples' group are as follows: The reconciliation of four couples, the working through of a reasonable separation of two couples, and the referral to the community for continued counseling of five couples.

Mr. E. B.'s 27-year-old son drove 300 miles round trip every Thursday to attend CORK meetings while his father was in the inpatient SHARP program. The son, who had a chronic asthmatic condition, worked with his father on a long standing unresolved conflict in their relationship. After six weeks, the son and father had greatly improved their interaction and the son's asthma had improved enough for him to discontinue routine, daily medication.

EVALUATION AND RESULTS

The evaluation consisted of two parts, both lasting from July 1976 through June 1977. The first part was keeping an up-to-date count of the level of drinking of CORK members by staff report from observation and interviewing them at intervals.

In the second part, 50% (108) of our SHARP ex-patients were contacted on a random sample basis by one of four trained staff persons. All patients who entered the program, regardless of whether or not they completed it, were told they would be included. The random sample patients were contacted 6 months or later after completing the program. Of this group 44 (40.8%) were seen in face-to-face interviews by staff. These ex-patients returned to the hospital for the follow-up interview or they were seen at CORK functions, although the majority of those randomly sampled were not CORK members. An additional 24 (22.2%) were interviewed by telephone or information was obtained by correspondence with the ex-patient. Information on 31 (28.7%) was received from sources such as probation officers, VA Hospital medical charts, or staff in other programs. Information on 5 (4.6%) was obtained through information received from friends or family of the subject (the "grapevine"). Three (2.8%) ex-patients could not be located and one died (.9%).

The following information on each ex-patient was obtained by staff using an informal interview format:

1. Extent of drinking
2. Level of activity
3. Level of interpersonal relationships

The most important information sought was whether or not the subject had been drinking since he left the program, and if so, to what degree. Any substantial deviation from the SHARP goal ot total abstinence would indicate that other goals, such as employment, high level of community involvement, and close friend or family relationships would be in jeopardy.

No systematic reliability study was done; however, the 44 (40.8%) seen in person were seen by more than one staff person, who corroborated the information. We believe the information obtained is valid for the follwing reasons: (a) the staff has extensive training and long experience in assessing the real situation of our patients; (b) information received from probation officers, medical charts, and staff from other programs is accurate; (c) in our experience we have found that the grapevine system is quite accurate. In this system, if someone falls off, the word spreads quickly.

This evaluation was done without sponsorship from any outside staff after initial consultation assistance from two VA Hospital staff employees who helped to create the design.

The first part of the evaluation, the up-to-date count of the extent of drinking of 162 CORK members 6 months or more after discharge, showed 64.2% totally abstinent--including 16% abstinent 2 years or more, 30.9% up to 2 years, and 17.3% abstinent from 6 months up to 1 year. This figure clearly demonstrates that an organized social support system, such as CORK, helps maintain a very high rate of total abstinence.

The second part, the random sample study, shows 37% totally abstinent and 20.4% drinking moderately and very infrequently,[1] or 57.4% in complete control of their drinking. Another 13.9% were marginal drinkers.[2]

The 57.4% of SHARP's ex-patients either totally abstinent or controlled drinkers reveals a considerably higher success rate than most alcoholic treatment programs and by far greater than any other reported figures with samples over 100. Emrick's (1974) review of 271 alcoholic treatment programs showed one-third of all ex-patients in the programs reviewed remained abstinent during follow-up and only one-twentieth or fewer were controlled drinkers. Our results also compare favorably with the Lemere and Voegtlin (1950) study which showed an overall abstinence rate of 51% with patients from more favorable financial circumstances who paid a substantial fee. Indigent, inadequate, psychopathic, or extremely neurotic patients only occasionally responded to their conditioned-reflex method. SHARP accepts these types of patients.

The next question to which the staff addressed itself was the relation between the extent of drinking and extent of occupying one's time. Of the 40 subjects totally abstinent, 85% were occupied[3] full time and of the 22 controlled, infrequent drinkers 72.7% were occupied full time. This outcome confirms SHARP's belief in focusing on the self-help concept, sharing, and the patients' learning to plan, organize, and be productive and assertive. Maintaining employment or satisfactory involvement in some personal effort upon graduation from SHARP is a program goal constantly stressed through many activities in

---

1 Controlled, infrequent means the subject drinks one drink, one night per week before dinner, or has an occasional glass of wine with dinner, etc.

2 Marginal means excessive drinking but the subject functions in some areas of his life, such as working, paying rent, keeping up his personal appearance.

3 Work, school, volunteer work, or structured time.

our program, such as the productive and self-image enhancing work done in the share-help involvement and the Payday effort. To see these goals being carried over into our patients' life style after graduation indicates a significant sign of program success.

The final question to which the staff addressed itself was how the extent of drinking was related to the level of interpersonal involvement. Ninety percent of the totally abstinent group actively and frequently interacted with others[1] and 91% of the controlled, infrequent drinkers did likewise.

This outcome shows that there is a statistically significant relationship between the extent of drinking and the extent of interpersonal relationships of our ex-patients. SHARP empahsizes social interaction and it is evident that many ex-patients do socialize, whether they abstain, control their drinking, or become marginal drinkers. The quality of ther interpersonal relationship must be considered. It appears that abstainers and controlled drinkers have more lasting and intimate relationships. Many marginal and out of control drinkers go to bars and interact with others where relationships tend to be transient.

REFERENCES

Emrick, C.D. June 1974. A review of psychologically oriented treatment of alcoholism. *Q. J. Stud. Alcohol* 35:523-549.

Lemere, F., & Voegtlin, W.L. 1950. An evaluation of the aversion treatment of alcoholism. *Q. J. Stud. Alcohol* 11:199-204.

---

1 Actively and frequently indicates involvement and contact with others daily.

COERCION OR CAREGIVING? CLIENT AND CAREGIVERS' PERSPECTIVES ON COURT MANDATED DISULFIRAM USE

Walter P. Shepherd, M.A. and P. Clayton Rivers, Ph.D.

University of Nebraska - Lincoln

The past decade of legal action in this country has reflected an increased concern by members of both the legal and mental health professions for those individuals caught in the interface between the legal and mental health systems (Wyatt v. Stickney, 1972; Wexler, 1975). As might be expected, the concern for the civil rights of these individuals has been unevenly distributed with some areas, e.g., the rights of the mentally ill in the commitment process, receiving considerable attention and other areas being almost entirely neglected. The focus of the present paper is on one of these neglected areas, the practice of including disulfiram (Antabuse) as part of a client's probation for a legal system offense that is related to alcohol use and abuse.

Disulfiram is a drug which chemically alters the metabolism of ethyl alcohol and causes the accumulation of toxic acetaldehyde in the body. Much of the clinical literature has described disulfiram as being relatively non-toxic when taken by itself. The side effects included skin rash, slight drowsiness, lethargy, and headaches (Hald and Jacobsen, 1948; Fox, 1967). More recent research, however, tends to question the extent of these side effects when disulfiram is administered over a prolonged period of time. Several investigations (Kane, 1970; Rainey and Neal, 1975) have described cases of carbon disulfide poisoning after prolonged use of disulfiram. Fried (1977) notes that carbon disulfide, a metabolite of disulfiram, can lead to a deficiency of pyridoxine (vitamin B_6) and that it is probable that disulfiram can either cause or aggravate a pre-existing pyridoxine deficiency. Since vitamin deficiencies are often found in chronic alcoholics this observation becomes all the more crucial.

If alcohol is ingested with the concurrent presence of disulfiram, an adverse physical reaction occurs. The general symptoms comprising the disulfiram-alcohol reaction begin within seven to twenty minutes after the person has taken a drink. The symptoms are generally related to peripheral vasodilation and hypotension and include the following: a feeling of warmth in the face, intense flushing, especially in the face, but also in the neck, chest, and upper arms, dilation of the scleral vessels resulting in a reddening of the eyeballs, tachycardia, palpitations, headaches, slight rise in blood pressure, rapid breathing, precordial pain, sweating, thirst, and blurred vision. After ten to fifteen minutes, the blood pressure decreases markedly, and the following symptoms may be noted: pallor, nausea, vomiting, and feelings of dizziness and weakness. The entire reaction usually lasts between two to six hours and is often followed by sleep and a return to normal (Hald and Jacobsen, 1948; Fox, 1967).

The intensity of the reaction varies from person to person and depends on the amounts of alcohol and disulfiram consumed. Disulfiram is excreted slowly from the body and can produce a strong reaction with alcohol several days after ingestion of the drug. Some symptoms can occur for as long as two weeks after taking disulfiram. Tolerance to these effects does not take place even after prolonged use.

While drugs designed to treat various disorders having psychological ramifications are routinely dispensed by physicians, it is unusual for the court system to be involved in the indirect "prescription" of such drugs as a part of an alternative to loss of driving privileges and/or imprisonment. The failure to focus on the civil rights aspect of the courts using disulfiram treatment has corresponded to the dearth of research on the relationship of disulfiram to patient behavior in general. While a considerable body of literature has been developed in the four decades during which disulfiram has been used in alcohol treatment, the bulk of this research has been concerned with either the drug's pharmacological aspects or with contraindications and complications involved in its use (e.g.: Cahol, 1972; Hotson and Langston, 1976; Gardner-Thorpe and Benjamin, 1971).

When research has focused on the therapeutic aspects of the drug in treating alcoholism, the emphasis has been primarily on outcome studies where the criteria for success has been either narrowly or vaguely defined (Hoff, 1967; Hoff and McKeown, 1953; Billett, 1968). The evidence to show the efficacy of disulfiram over other modes of treatment is not convincing. Clinical studies which conclude that disulfiram is useful for clients suffer from lack of adequate control groups, failure to explore the interaction between disulfiram and other treatment variables, and inadequate followup procedures to assess effectiveness over an extended period of time. This latter point is especially relevant to programs involved with the court system where information on clients is often not available after their probation periods have concluded.

Little has been mentioned in the literature about how the clients themselves perceive disulfiram and whether or not they see it as being useful for them (Lubetkin, Rivers and Rosenberg, 1971; Victor, 1973). In particular, little has been said concerning the perception of the clients in the legal system when it includes disulfiram as a part of the client's probation for an alcohol-related offense. If the assumption can be made that a person's perceptions are related in some way to his or her eventual behavior, then it would be useful to have a better understanding of how these clients perceive taking disulfiram. In particular it would be helpful to know if these perceptions are related to the treatment effectiveness of disulfiram.

The purpose of the present research is to gain some initial impression of the attitudes and opinions of clients who are using disulfiram. A second, but not secondary, purpose is to establish what, if any, differential perceptions exist between the clients who receive disulfiram and the alcohol treatment workers who are connected with the program that dispenses the drug. Third, the study is an attempt to better understand some of the possible variables involved in establishing disulfiram as an effective treatment modality for alcoholism. Finally, we would like to explore some of the ethical and legal issues that may be associated with the courts' involvement with disulfiram.

Fifty clients volunteered to be interviewed on their views concerning disulfiram. The clients were not selected on any specific set of criteria except they came to the clinic to take their disulfiram during one of several times the interviewer was at the clinic. Forty-three of these clients had been placed on disulfiram as part of their probation for alcohol related legal offenses and the present paper will focus on these clients.

The age range for the clients was 20-56, with the mean age being 31. Forty-one of the clients were males and two were females. Thirty-five were working full-time, with only five neither working nor going to school. Seventeen were single, fifteen were married for the first time and nine were divorced. Almost half (20; 46%) were earning in excess of $800.00 a month. The educational range was 8-17 years with the mean being 12.4 years. The average time on disulfiram was 5.76 months with the range being one week to two years. For thirty-nine of the clients it was their first experience in taking disulfiram. Twenty-eight of the clients reported being involved in some form of concurrent treatment (e.g., group therapy, family therapy, A.A.) while taking disulfiram.

A modified form of the interview was also administered to ten staff members (alcohol service workers and registered nurses) at the short term detoxification center where the disulfiram was dispensed. These individuals were all full-time employees at the center who had worked there for at least three months prior to being interviewed. Since the outpatient disulfiram clinic was housed within the center, all the staff members had the opportunity to dispense disulfiram to the clients as part of their normal workday. There were seven males and three females in this group and their ages ranged from 21 to 35 years. Educational level ranged from the 12th grade to the completion of college.

Interview Format

The interview format employed was a structured one which focused on the clients' perceptions of disulfiram as well as their thoughts and feelings about taking the drug. The court-referred clients were also asked their opinions about the courts' involvement in using disulfiram as a treatment agent. Other questions were designed to elicit the clients' plans for continuing the use of disulfiram after the end of the probationary period. Attempts were also made to establish any resistance to taking the drug (e.g., drinking while taking disulfiram or pretending to take the drug while surreptitiously disposing of it).

Procedure

The interviewer (the first author) was stationed near the room where the disulfiram was dispensed. After the client had taken his disulfiram he was approached by the interviewer who identified himself as a graduate student in psychology at UN-L who was doing research on "Antabuse" (the trade name for the disulfiram most frequently used by clients). The interviewer then asked the client if he could obtain his or her views on Antabuse. Any client who refused to cooperate was not approached again by the interviewer (only one of 50 clients initially refused, and he later changed his mind without further intervention by the interviewer and agreed to participate in the study).

Once the clients agreed to be interviewed, they were told the interviewer's status and informed that he was not associated with the alcohol center in any way. It was stressed that they were under no obligation to participate in the study and could refuse to be involved without repercussions. They were told that the purpose of the study was to find out how people taking Antabuse feel about taking the drug.

Clients were informed that the interview would take about 30 minutes and that they had not been singled out in any way to participate. They were also assured that no information given to the interviewer would be passed on to the center's staff or any other people not directly related to the research. Strict anonymity was maintained. Clients were also instructed that they could refuse to answer any questions they were asked and could withdraw at any time without penalty. If the clients had questions at this point, they were answered and the clients were then asked to sign a research consent form. No one withdrew from the study or refused to answer any of the questions asked.

Results

The results of the present study are presented in a descriptive rather than a statistical format. The results are organized into (a) perceptions of the probationary clients; (b) the perceptions of the alcohol treatment workers and a comparison of the client and alcohol workers' perceptions.

a) Perceptions of the Probationary Clients. Twenty-one of the clients (49%) saw disulfiram as being helpful to them when asked to give a yes or no response. Two indicated they were unsure (5%). When asked to rate the helpfulness or harmfulness of the drug on a 7 point scale, 30 found it to be helpful to some degree (70%), 6 found it to be harmful to some degree (14%) and 7 found it to be neither (16%). In response to a subsequent open-ended question, those who found it helpful stated that it assisted them in getting through difficult periods when they might have taken a drink had they not been on disulfiram. Those individuals who found it not helpful generally commented that in the long run it was up to the individual to remain off alcohol, and that the drug was just forestalling the time when they could establish whether or not they could maintain sobriety on their own.

A majority of the clients (35; 81%) said they would not have taken disulfiram if the court had not made it part of their probation. Only nine of the clients (21%) would have continued its use at the time of the interview had they been given the opportunity to stop. Because it was felt that a positive or negative decision to continue taking disulfiram might be related to length of time in the program, clients in the first 2 months (N = 16) were compared with clients in the program from 5-12 months (N = 16). There was no difference in opinion as a function of length of time in the program (3 in the 0-2 months group and 4 in the 5-12 months group would continue). These results suggest that the majority of clients would not continue taking disulfiram even after they had been exposed to it for an extended period of time. It would appear, then, that time in the program is an irrelevant issue when deciding whether or not to continue taking the drug. Perception of helpfulness may be relevant, however, since all of the 9 clients who wanted to continue taking disulfiram saw it as being helpful (yes-no response). Perception of helpfulness may be a necessary but not sufficient condition, however, since 12 other clients who saw disulfiram as being useful stated they would discontinue taking the drug if given the opportunity to do so at the time of the interview. Since we wondered if time in the program was related to acceptance by the client of an alcohol problem, we compared the 16 clients in the first two months of treatment with the 16 clients who had been in the program for 5-12 months on this question. We found that 7 (44%) clients in the early group indicated that they had an alcohol problem while 11 (69%) of the clients in the 5-12 month group gave a similar response.

Returning to the 43 clients, only 6 (14%) reported drinking while on disulfiram, while 7 (16%) stated that they had pretended to take their disulfiram without actually doing so. The most frequently cited reason for this latter behavior was a need by the clients to do things on their own, without help of the drug. Another reason frequently cited was a general rebelliousness against being told to do things and seemed less specifically related to disulfiram than a possible characterological response on the part of the clients involved.

When asked to respond in open-ended fashion to questions involving pressure on them to take disulfiram and their opinions about the courts directing them to take the drug there were several strong opinions. Many probationary clients (19; 44%) discussed what they perceived to be an "unfairness" on the part of the legal system in not offering them alternatives to treatment which did not include disulfiram and some believed that the court was stepping out of its boundaries to make disulfiram part of their probation. Other clients noted that the court was well within its guidelines to do this and that although they resented it at first, they gradually recognized the need for their being placed on disulfiram. Twenty-five (58%) of the clients felt that they had nothing to say in the decision that they be placed on disulfiram. Of those who did feel that they had a choice, some (8; 19%) saw their only other alternative as accepting a straight sentence if they didn't agree to the disulfiram as part of their probation. Some saw this as being a legitimate choice, while others commented on the possibility that such a choice amounted to nothing more than concealed coercion. Thus, 33 clients (77%) perceived that they had either no choice or a potentially negative choice when they accepted disulfiram as a part of their probation.

b) Perceptions of the Alcohol Treatment Workers. The alcohol treatment center personnel were asked to make two ratings: 1) They were to give their own opinions as to the utility, etc. of disulfiram and 2) they were to attempt to match the ratings the clients gave to the same questions. For this second rating, the staff was asked to base their responses only on the disulfiram clients who came to the center as part of their probation.

When asked what percentage of the clients were helped by the disulfiram, the workers' average response was 42.5%. When asked how many of the clients believed disulfiram was helpful to them, the workers' average response was 37.5%. If you'll recall the clients' actual responses, 49% saw it as helpful on the yes-no question. Using the 7 point scale of "helpfulness-harmfulness," the alcohol workers gave disulfiram an average rating of 5.30 for the clients (slightly helpful). (The clients gave it an average rating of 5.12.) However, when asked to respond as they thought the clients had on this scale, the workers' average rating was 3.6 (between slightly harmful and neutral). These findings suggest that the clients' and workers' perceptions of how helpful disulfiram is are quite close, but the workers misperceive the opinions of the clients by a rather wide margin.

This tendency to misperceive how clients respond to critical questions is also found on the question relating to alcohol abuse. The workers were asked what percentage of the clients had alcohol problems. Their average response was 75%. However, on the average, they saw only 41% of their clients as actually saying that they had an alcohol problem. The clients themselves responded affirmatively 55% of the time, again indicating that the workers' perception of how the clients see a relevant issue may be substantially different from the actual perceptions of clients.

Finally, workers estimated that 14.4% of the clients would have taken disulfiram if they had not been placed on probation, while 19% of the clients indicated that they would have done so. The workers believed that 18.6% of the clients would continue to take disulfiram if they were taken off probation at the time of the interview, while in actuality 30% of the clients indicated a desire to do so.

DISCUSSION

When we began this study, we had few research studies on which to base predictions about how the clients would perceive taking disulfiram. What we discovered was that the majority would not have taken disulfiram on their own, many indicating they clearly saw the court as playing a coercive role in their taking of disulfiram. Despite the perceived coercion, many clients did see the disulfiram as being helpful, and while this perception tended to be related to whether clients would continue taking the drug if released from probation, the majority of clients would clearly not keep taking the drug if given the opportunity to discontinue it. Also, the majority would not have initiated taking the drug. Their strong feelings about taking disulfiram decrease over time, even though they seemingly do become more accepting of the fact that they have a drinking problem.

These findings raise a serious question about what effect being placed on disulfiram under conditions perceived as coercive may have on the continuum of care for the alcohol abuser in general and in particular for the client who is court-referred. Despite the fact that almost half (49%) of the clients saw disulfiram as helping them, 79% of them would have stopped taking it immediately had they been given the chance. While this appears to be contradictory, it is possible that the issue of how the clients came to be placed on disulfiram has something to do with the discrepancy. The clients' perception of being coerced (77% felt that they had no choice or a limited choice in taking disulfiram) may have remained with them and influenced their decision to discontinue the drug.

The comparison of results between the perceptions of the alcohol workers and the perceptions of the clients also raises a question as to what effect the alcohol workers' perceptions have on the perceptions of the clients. In general it seemed that the alcohol workers saw the clients as perceiving disulfiram in a more disfavorable light than the clients actually did. It is possible that the alcohol workers might have changed their style of interacting with the clients based upon such perceptions, the results of which could have been a more pessimistic or cautious response to the client concerning the outcome of the client's treatment.

One of the most critical issues raised by various researchers regarding the treatment of the alcohol abuser is the relationship between the alcohol client and the caregiver. Chafetz, Blane and Hill (1970) have noted the need for a positive relationship between the caregiver and client and indicate that a positive relationship is necessary before more involved treatment can be undertaken. They also note that certain attitudes, such as self-righteousness or punitiveness, may serve to reinforce characteristics commonly seen in the alcoholic, i.e., a need to be rejected and a fear of becoming involved with others. Similar comments are made by Gerrein, Rosenberg and Manohan (1973) who believe that alcholics are especially responsive to any increase in care that is provided by the personnel who dispense disulfiram.

These conceptions regarding treatment of the alcoholic are particularly germane to the discrepancies between the perceptions of the two groups. What seems to be happening is that the alcohol workers are being too negativistic in their estimation of how the clients are perceiving both disulfiram and their alcohol problem. Such incorrect perceptions are likely to affect the development of a positive relationship between client and caretaker. A more serious possibility is that the alcohol workers may adopt a more confrontive "policeman's" role in their relationship with the clients. That is, many of their actions toward court-referred clients may focus on being sure they take the drug rather than their being open and supportive toward the client. One might predict that when caregivers have inaccurate and less positive notions about a client's perception of his or her own problem, the caretaker is less likely to be helpful to the client. The questions are then, why do the caregivers have these misconceptions, what can be done to change them, and what effects do they have on the client's behavior.

We would like to turn at this point to what we believe is a civil rights-ethical issue involved in this type of treatment. As mentioned earlier, the issue of the court system requiring some clients to take disulfiram for alcohol related offenses has received little attention. The courts operate on the basis of a cost-benefit system which tries to take into account both the individual and society, ensuring that neither has to bear the burden of a particular action to an unreasonable degree. The courts continually use coercion in their system, and almost all alcoholics who enter the treatment system are there as a result of some family, job-related, or external agent pressure. The issue of coercion alone is not what is being objected to; however, it is the type of coercion. The courts are presented with legal offenses that involve alcohol in some way but instead of dealing with the offense in particular, they are dealing with the entire concept of alcoholism. Since the courts have adopted the general medical concept of alcoholism as a disease, and the treatment of choice as abstinence, they allow themselves to become involved in dispensing disulfiram. Disulfiram deals with alcoholism, not legal offenses. It is here that the courts may have overstepped their boundaries, making a decision which does a great deal of good for society (e.g.: it keeps intoxicated drivers off the road) but which diminishes the rights of the individual. Taking disulfiram affects other aspects of the individual's life besides the legal system offense which he or she has committed.

Two suggestions for change in the present system evolve out of the issues of perception of choice by the clients and coercion to take disulfiram. The first suggestion is based on the assumption that disulfiram will continue to be used by the court system, while the second does not. First, if procedures could be changed so that clients perceived more freedom of choice in taking disulfiram, it might increase the number of people who choose to stay on it. A possible strategy would be that disulfiram be offered to the client on a three-month trial basis. During the three months the usefulness of the drug would be evaluated by the client and his or her probation officer. The client would also be asked to do some reading on disulfiram and to attend several classes where its action was fully explained. At the end of the three months, the client's progress would be reassessed and he or she would have the option of continuing on disulfiram for nine months. Another option would be to discontinue the drug and to have a suspension of his or her driver's license for four months instead of the six months that would have happened at the time the initial decision about disulfiram was made. It is our contention that this method of placing clients on disulfiram involves more choice points for the client and yet rewards him or her for at least trying the drug. Such a contention is supported by the clients in our study who were not court-referred. Though four of the seven were awaiting formal court action on an alcohol related legal offense and were taking disulfiram to increase their chances of receiving probation, six of them felt disulfiram was helpful to them (yes-no alternative). This reinforces the position that perception of choice may have been an important variable in attitudes towards the helpfulness of disulfiram since these six felt they had something to say in the decision that they take disulfiram.

The second suggestion is that the courts decrease their concern with alcoholism per se and focus on the behavior which brought the individual before the court. In the case of a charge of driving while intoxicated, it is the behavior of driving that has brought the individual before the court. If that individual had the same blood-alcohol content but had remained in his or her own home and had not ventured out onto the streets in an automobile, chances are that he or she would not have gotten involved in the legal system.

What conclusions can be drawn from what has been said to this point? First, from the viewpoint of the client, entering the caregiving system with a perception of being coerced into taking disulfiram may mitigate against the continued use of the drug. Second, caregivers see the clients as being more against taking disulfiram than they actually are. The implications of this include a possible decrease in the efficacy of disulfiram treatment. Finally, the courts may be overstepping their boundaries by dealing with the clients' alcoholism rather than the specific behaviors which have brought the client before the court. While concrete solutions to this problem are difficult to achieve, future investigations into the general area of court-mandated disulfiram should make some attempt to develop new strategies.

REFERENCES

Billet, S.L. Antabuse Therapy. In R.J. Cantanzaro (ed.) *Alcoholism: The total treatment approach*. Springfield, Illinois: Charles C. Thomas, Publisher, 1968.

Cahol, C.A. Safety of disulfiram. *New England Journal of Medicine*, 1972, 287, 935-936.

Chafetz, M.E.; Blane, H.T. and Hill, M.J. (eds.) *Frontiers of Alcoholism*. New York: Science House, 1970.

Fox, R. Disulfiram (Antabuse) as an adjunct in the treatment of alcoholism. In R. Fox (ed.) *Alcoholism: Behavioral, Research, Therapeutic Approaches*. New York: Springer Publishing Company, 1967, 252-255.

Fried, R. Comments on Antabuse therapy of alcoholism. *Clinical Experimental Research*, 1972, 1, 275-276.

Gardner-Thorpe, C. and Benjamin, S. Peripheral neuropathy after disulfiram administration. *Journal of Neurology and Neurosurgery*, 1971, 34, 253-259.

Gerrein, M.S., Rosenberg, C.M. and Manohan, V. Disulfiram maintenance in outpatient treatment of alcoholism. *Archives of General Psychiatry,* 1973, 28, 798-802.

Hald, J. and Jacobsen, E. A drug sensitizing the organism to ethyl alcohol. *Lancet,* 1948, 255, 1001-1004.

Hoff, E.C. The use of disulfiram (Antabuse) in the comprehensive therapy of a group of 1020 alcoholics. In R.J. Cantanzaro (ed.) *Alcoholism: The total treatment approach.* Springfield, Illinois: Charles C. Thomas, Publisher, 1967.

Hoff, E.C. and McKeown, C.E. An evaluation of the use of TETD in the treatment of 560 cases of alcohol addiction. *American Journal of Psychiatry,* 1953, 109, 670-673.

Hotson, J.R. and Langston, J.W. Disulfiram-induced encephalopathy. *Archives of Neurology,* 1976, 33, 141-142.

Kane, F.J. Carbon disulfide intoxication from overdosage of disulfiram. *American Journal of Pyschiatry,* 1970, 127. 690-694.

Lubetkin, B.S., Rivers, P.C. and Rosenberg, C.M. Difficulties of disulfiram therapy with alcoholics. *Quarterly Journal of Studies on Alcohol,* 1971, 32, 168-171.

Rainey, J.M., Jr. and Neal, R.A. Disulfiram, carbon disulfide and atherosclerosis. *Lancet,* 1975, 284-285.

Victor, C.A. Psychological factors that influence attitudes and feelings of alcohol clients to Antabuse. *Smith College School for Social Work,* 1973, 49, 69-70.

Wexler, D.B. Token and taboo: Behavior modification, token economies and the law. In D. Rosenhan and P. London (eds.) *Theory and Research in Abnormal Psychology*, New York: Holt, Rinehart and Winston, Inc., 1975.

Wyatt v. Stickney, 344 *F. Supp.* 373 (M.D. Ala. 1972) (Bryce and Searcy Hospitals).

A COMPARISON OF TWO QUICKLY ADMINISTERED EVALUATIVE INSTRUMENTS USED TO PREDICT A PREDISPOSITION TO ANXIETY IN ALCOHOLIC MALES

Stephen G. Shafer, M.A.

Community Mental Health Center

Spokane, Washington

BACKGROUND

The problem of anxiety as a human emotion can be traced back into history to at least the time of ancient Egypt where the attribute of fear is recorded (Cohen 1969). As early as the eleventh century, a major work entitled, *A Philosophy of Character and Conduct* by the Arab philosopher Ala ibn Hazm, purported that anxiety was basic to the human condition. (" . . . no one is moved to act or moved to speak a single word who does not hope by means of this action or word to release anxiety from his spirit" (Kirtzeck 1955).

In terms of philosophers who have influenced the direction of psychological inquiry into the nature of anxiety, Spinoza has often been quoted. For Spinoza, "fear was essentially a state of mind or attribute, a subjective condition of uncertainty in which there was the expectation that something painful or unpleasant might happen" (Ed. 1963). The thought of Pascal, particularly his belief that the "perpetual restlessness" (anxiety) he observed in himself and his fellow man could not be overcome completely by the powers of reason, also had its effect on modern psychology (Ed. 1963). Pascal obviously recognized the power of emotions to direct or overwhelm human behavior.

In the nineteenth century the observation of fear and anxiety was noted by biologists such as Darwin from whom we have a rather classic description of the physical manifestations of fear. This description included the typical manifestations as, "rapid palpitations of the heart, trembling, increased perspiration, erection of the hair, dryness of the mouth, change in voice, dilation of the pupils and the like" (Ed. 1963).

The work of Sigmund Freud led to an investigation of anxiety which included a distinction between a neurotic anxiety and neurasthenia as early as 1894 (Ed. 1963). Freud later defined anxiety as "something felt" (Ed. 1963). He submitted that this "something felt" this "unpleasant emotional state" (affective) was, "a nodal point, linking all kinds of most important questions: a riddle of which the solution must cast a flood of light upon our whole mental life" (Freud 1969). When one compares this statement with that of the philosopher Ala ibn Hazm it is easily recognizable that two of the great minds of recorded history are almost in total agreement on the importance of anxiety to the development of a comprehensive theory of human behavior.

The nature of anxiety and its corresponding measurement covers a wide range of theorists. Likewise the definition of anxiety ranges from a distinction between fear and anxiety to a seeming equation of the two concepts. For example, Izard defines anxiety as "a 'pattern of emotions,' a complex emotional reaction that includes fear as well as other fundamental emotions" (Spielberger 1972). Likewise Lazarus and Averill define anxiety as "emotion based on the appraisal of threat, an appraisal that entails symbolic, anticipatory and other uncertain elements" (Spielberger 1972). It is the cognitive appraisal of the stressful stimuli which separates fear from anxiety for these two researchers. In contrast, anxiety is pictured as the result of chemical reactions such as hypothalamic--hypophyseal in other words endocrine controlled reaction (Spielberger 1972), or a high concentration of lactate ions in the patient's blood (Spielberger 1972).

In a time of history which has been labeled the age of anxiety it is not surprising to find that anxiety is suspect as a cause of many evils. Tension and worry, two possible results, are by-products of anxiety, and are thought to contribute to heart diseases. The problem of alcohol abuse is also suspected as having anxiety at its base (Swerling, Rosenbaum 1966).

Swerling and Rosenbaum have characterized alcohol addiction in behavioral terms as follows: "the pattern of anxiety, alcohol, relief from anxiety, is practiced repetitively and becomes highly over learned. The reinforcement, which is the relief of anxiety, is immediate in contrast with the delayed reinforcement of other alternatively adaptive maneuvers for handling the anxiety" (Swerling, Rosenbaum 1959).

Alcoholism has been seen in another light as follows: "Their premise seems to be that to face real fears such as arrest, lack of 'necessary' food, shelter, running out of something to drink, the delirium tremens, et cetera is preferable to daily facing the free-floating fear and existential anxiety inherent to so-called 'normal living' in American society" (Rose 1970). Considering the evidence above, it would seem that anxiety plays at least some role in alcohol addiction.

In Spokane as in most large cities there is a section of town which has become known as Skid Row. In spite of recent attempts at eliminating the environment of cheap housing, small bars and the pawn shops that are indigenous to any skid row, Spokane skid row stretches from Division Street to Washington Street, East and West, and from Trent Avenue to Sprague Avenue, running North and South. It is estimated that 90% of all Spokane's Skid Row population lives in this twelve square block area. The Spokane Alcoholic Rehabilitation center (SPARC) is currently one of the major organizations attempting to deal with the homeless, rootless type of person found in this section of town. The SPARC program includes three half-way houses and a work ranch. The clients admitted to the work ranch made up the population for this study. This study was done following a pilot study which produced statistics concerning the length of sobriety after discharge from the ninety day program. Following the pilot study it was apparent that some descriptive analysis of the population other than biographical data only was important.

In this study fear was contrasted with anxiety, with anxiety being seen as "unspecific, vague, and useless" (May 1950). As such, anxiety is pictured as attacking "the foundation (core, essence) of the personality" (May 1950). This centrality and at the same time vagueness are the qualities which distinguish anxiety from fear. "Fear, as opposed to anxiety, has a definite object (as most authors agree), which can be faced, analyzed, attacked and endured" (Tillich 1952). In not having a specific object, anxiety leaves the person in a position where one "cannot 'stand outside' the threat, cannot identify it and thereby is powerless to take steps to meet it" (May 1950).

> In fine, the objectless nature of anxiety arises from the fact that the security base itself of an individual is threatened, and since it is in terms of this security base that the individual has been able to experience himself as a self in relation to objects, the distinction between subject and object also breaks down (May 1950).

This statement of anxiety formation and operation is in keeping with the author's thoughts in as far as it goes. It does not seem, though, to suggest the possible mechanism by which "the security base" (May 1950) of a person is threatened. In other words, what possible object, person or thoughts can attack a person's "core or essence"? Being of a somewhat nebulous nature, one's essence or core are probably not capable of being attacked by physical objects or persons. Thoughts or ideas, however, are constructs of the same or similar nature as one's core or essence and therefore may be possible sources of influence for producing anxiety.

This day and age of what has been labeled the age of anxiety can be contrasted with the time in history known as the Middle Ages. It is suggested that during the Middle Ages, community served as a stronger rallying point than during the Renaissance and preceding ages. "Each person theoretically knew his place in economic structure of the guilds, in the psychological structure of the family, and the hierarchy of feudal loyalties, and in the moral and spiritual structure of the church" (May 1950). During this time, when culture was in such a unified state, the manifestation of anxiety was much less than it is today.

The origins of the increase in the expression of anxiety can be seen toward the end of the Middle Ages, the fourteenth and fifteenth centuries. As Huizinga pointed out, "the hierarchical forms of church and society, previously serving as ways of channeling emotions and experiences, had become methods of supressing individual vitality" (May 1950). This beginning of disunity in the socio-psychological makeup of the culture gave rise to "feelings of depression, melancholy, skepticism and much anxiety" (May 1950). The anxiety took the form of excessive dread of death and pervasive fears of devils and sorcerers (May 1950).

The following period in history known as the Renaissance became symbolized by the power of the individual and the waning of the power of the socio-psychological unity of the culture. One of the results of this shifting emphasis is that social mobility increased. The medieval family cast was broken down and, "by courageous action" one "could now achieve eminence regardless of the level of his birth" (May 1950). The growth of this concept of the powerful individual is basic to the understanding of the anxiety creating patterns of modern culture.

This increased emphasis of the individual in Renaissance life was not a valuation of the person as such, but rather the valuation of the strong individual (May 1950). "Success was the standard by which acts were judged; and the man who could help his friends, intimidate his enemies, and carve a way to fortune for himself by any means he chose, was regarded as a hero" (May 1950). "With this emphasis, man's freedom was increased and with it came an increased feeling of strength and at the same time an increased isolation, doubt, skepticism and resulting from all of these, anxiety" (May 1950).

It is at this point that the author would like to take issue with the ideas of those who profess that anxiety is the result of early childhood experiences beginning as early as the separation of the mother and child. The capacity for anxiety as well as all other emotions is present but what are manifested at first are not emotions but reflex (innate) reactions to separation from the only other environment that the child has known, mother. It is not for months that good recognizable emotions are displayed and the first of these is fear, not anxiety.

Anxiety has its basis in *doubt*. The lack of anxiety expressed during the Middle Ages is the result of the carefully constructed culture. The author does not wish to say that anxiety did not exist, simply that the situations that could cause doubt and thereby anxiety were limited. There was little reason to be doubtful about one's place in the social world when one's birth determined that, when psychological concerns were met by a close community structure and family ties. The greatest potential for doubt was that for one's spiritual future. It was precisely at this point that the greatest anxiety as a result of doubt was to be seen, in the excessive doubt as to the presence of devils, sorcerers and the relationship to what was to become of one in death.

For the developing person it is the same. One's first experiences within the family are well defined. One's needs, physiologically, sociologically, and psychologically are met in an atmosphere of unity. The structure within the family is relatively standard and one's place in the structure is assured, at first. Later, though, as the person's awareness of the disunity of the culture in the society in which the family is located and even the disunity of the family unit grows, doubt arises. This doubt, a mental process, gives rise to anxiety, and emotion. This is in agreement to the author's early thinking that persons or objects, even situations cannot threaten a person's "core or essence". One's essence can only be attacked and threatened through thought, namely the person's own thoughts and in the case of anxiety, thoughts of doubt. Certainty does not give rise to anxiety.

Rollo May (1950) suggested that the Western world's competitive structure could be developed as follows: "competitive individual striving (CIS)→intrasocial hostility (IH)→isolation (I)→anxiety (A) →increased competitive striving (ICS)". This author suggests that the addition of doubt is a necessary inclusion for a complete equation. In other words CIS→IH→I→*doubt*→A→ICS. This inclusion of doubt rounds out the equation to reflect more accurately the basis of anxiety in our modern society.

INSTRUMENTATION

There are several instruments used today to measure anxiety. The two instruments considered in this study are the *Luscher Color Test (LCT)* and the *State-Trait Anxiety Inventory (STAI)*. The *LCT* is a test based on people's choice of colors. In his theory, Dr. Max Luscher states that "the structure" of a color remains the same for everyone (Luscher 1971). This is so, he suggests, because the colors have an evolutionary basis which is the same for all people (Luscher 1971). There are two versions of the test: the complete test which employs seventy-three color patches and the short version which uses eight color patches (French, Alexander 1972). It is the shortened version of this test, utilizing eight color patches, which was used in this study. The background material and administration are described in a book translated and edited by Ian Scott (1971). The *LCT* itself represents more of the ideal test as for being quick, easily administered, as it requires no reading ability, and is easily scored. Responses are recorded as eight pairs of numbers. Even color blindness is not supposed to effect the results (Pickford 1971). The other measure of anxiety used in this study is the *STAI* developed by Charles D. Spielberger and his associates. The *STAI* requires the handling of two pages of results for each client. It takes between fifteen and thirty minutes to administer, and one is expected to have at least a sixth grade reading ability. This instrument is claimed to measure anxiety somewhat differently than other instruments in that it divides the concepts of anxiety into two constructs, state anxiety, or A-state, and trait anxiety, or A-trait. A-State is described as a non-pleasant emotional state or condition which is characterized by "subjective, consciously relieved feelings of apprehension and tension which are accompanied by or associated with activation of the autonomic nervous system" (Ed. 1973). A-Trait is defined as, "an acquired behavioral position based primarily on residues of past experiences and reflected in behavior by relatively stable individual differences in an anxiety proneness in response to stress" (Ed. 1973).

The population on which this study was conducted consisted of only male clients admitted to the SPARC program. The average age of the clients was 45.04 years with a range from 22 to 64 years of age. The age groups which appeared most frequently were forty to forty-four fifty to fifty-four and fifty-five to fifty-nine years. The racial characteristics included 94 percent Caucasian, 7 percent blacks, 4.7 percent Indians. In terms of economic levels, better than half, 58.1 percent were unskilled while 41.9 percent had a marketable skill. Nearly 65 percent (64.9%) were veterans of the armed services with 27.8 percent of all clients having served prison time for major crimes other than jail time for drunkeness. 44.9% of the clients had been in contact with alcoholism programs such as half-way houses, half-way alcoholism wards and in-resident treatment programs. 60% had attended Alcoholics Anonymous one or more times. The average length of the stay at SPARC between the testing dates was one month, fifteen days for all clients and three months, thirteen days for all clients shown to still be sober. This sample consisted of 44 clients which represents those who stayed two weeks or more from May through the following March.

During each bi-weekly period of the 40 weeks involved in this study the researcher was informed of new entries into the SPARC program. Testing at two week intervals was used as it allowed the subjects time to sober up, and adapt to their new surroundings. On Friday evenings at the end of the two week period the new members of SPARC were tested. The new members were tested in groups of four or less.

At each session, form X-1 of the *STAI* was administered first. The directions at the top of the form were called to the attention of the group. The sentence and the directions which makes the X-1 form different from the X-2 form was read aloud. Following completion of the form X-1, the *Luscher Color Test* was administered to each subject. Directions concerning the *LCT* were given to the group as a whole and repeated to each member as they began the procedure of turning over the color cards. After all subjects had completed form X-1 and the first selection of the color cards, they were given form X-2 of the *STAI*. Again the directions at the top of the page were called to the attention of the subjects. When this form was completed the subjects were asked to respond to the color cards of the *Luscher Color Test* as they had before. At this time the observation was made that this *LCT* was not a test of memory, therefore it is not an attempt to see how many of the colors can be turned over in the order that they had been before.

RESULTS

The results of the study were grouped into two major categories: the correlational relationship between the anxiety scale of the *LCT* and *STAI* either as a whole or in its two parts, the possibility of using either as a whole or in its two parts, the possibility of using either of the three measurements in the prediction of the aspects of length of stay or continued sobriety. The methods of examination of these categories consist of application of Pearson's Product Moment Correlation Test of statistical significance and chi square test of independence of categorical variables.

Only one result was found to be statistically significant by either method. The -.32 correlation coefficient between the *TRAIT* anxiety scale of the *STAI* and the length of stay at SPARC demonstrated a statistically significant negative correlation at the .05 level. Some slight indications of predictive ability were suggested in the chi square results for the *TRAIT* anxiety scale and continued sobriety 3.67. It was also interesting to note the extremely small Pearson r results for the *LCT* anxiety scale and the complete *STAI* (-.063) and the *LCT* anxiety scale as compared with the individual parts of the *STAI* (state anxiety -.1032 and the TRAIT anxiety .0286).

The highest correlation between two aspects of this study was found to be the negative .32 correlation between the *TRAIT* anxiety scale and the *STAI* and the length of stay in the program. This may be a direct indication of one aspect of one of the conditions necessary in staying with the program such as SPARC. The negative correlation obtained in this study between the *TRAIT* anxiety scale and the length of stay suggest that the lower the *TRAIT* anxiety is scored, the longer the involvement in the program. According to the test theory, the lower the score the less they manifest anxiety. The 3.67 chi square score obtained for the *TRAIT* anxiety scale and the length of stay relationship was not significant at the .05 level. It was significant at the .20 level. Adjustment for the limitations for the non-randomness and small N may increase its reliability as a predictor of continued sobriety.

DISCUSSION

The results showed that state anxiety seemed to play a lesser role in length of stay and continued sobriety. One possible reason for the depressed scores was that most testing was done immediately following the evening meal at the program's work ranch where the clients had been located long enough to feel at home, i.e., the surroundings were not foreign to them. The researcher feels that these facts would tend to lessen any state anxiety present. Also, the testing was structured to be as non-threatening as possible. Again, this would reduce state anxiety as theorized by creators of the test.

It is interesting that the two instruments, both claiming to measure a phenomenon called anxiety, could score such very low correlation scores. It would seem that the "anxiety" being measured is very dissimilar indeed. Granted the size of the sample was not large. Even so, the mean value of 1.977 closely resembled the reported averages of the *LCT* of 2, and the mean score and standard deviation for this study on the *STAI* closely resembled those obtained with college freshmen and even more closely resembled neuro-psychiatric patients.

The researcher believes the current confusion concerning just what anxiety is, the extremely low correlations between the *LCT* and the *STAI* only being examples of this confusion, is the result of seeing many different manifestations of a central problem. This confusion is the result of viewing anxiety in terms of Freud's thinking that anxiety is the nodal point of personality problems. In effect such thinking suggests that anxiety is the very basis of the problem. The reason this thinking has led to many theories and few positive solutions is that there is still one step further down. The basis of anxiety is what should be dealt with. The anxiety, in this researcher's thoughts, is the manifestation of doubt. They are not synonymous. Anxiety is an outgrowth of doubt. It is doubt, not anxiety, which is the very basis of the problem.

The effects of treatments of anxiety are stop-gap at best. It is the underlying doubt that must be dealt with. Some research has suggested this already. In some research electrode-shock failed to elicit increased anxiety in human subjects. One of the suggested reasons is that the students did not think the researchers would do anything to actually harm the subjects. They did *not* become *anxious* because they *did not doubt* the ethics of the researchers.

Doubt is ubiquitous and very basic to our lives. It is necessary and impossible to avoid. It is harmless. But what one does with it is what creates the problems. If one dwells upon it one becomes anxious. If one acknowledges it, and continues to be hopeful, growth does not stop.

After reviewing the findings of this study, the researcher concluded the following: 1. The use of the *LCT* anxiety scale in place of the *STAI* in part or in whole is highly questionable; 2. Prediction of continued sobriety or length of stay in the SPARC program with any of these instruments is questionable; 3. Further research with the *LCT* anxiety scale is necessary to establish its possible usefulness. Further research using the *STAI* with the alcoholic population in alcoholic abuse programs should be pursued.

REFERENCES

Cohen, J., 1969, *Personality Dynamics*. Chicago: Rand McNally.

French, C., and A. Alexander, 1977, The Luscher Color Test: An Investigation of validity and underlying assumptions. *JPA* 8:361-365.

Freud, S., 1969, *A General Introduction to Psychoanalysis*. New York: Simon and Schuster.

Luscher, M. (I. Scott, translator), 1971, *The Luscher Color Test*. New York: Pocket Books.

May, R., 1950, *The Meaning of Anxiety*. New York: Ronald Press.

Pickford, R., 1971, Review Article: The Luscher Color Test. *OP* 45:15-7.

Rose, T., 1970, A Descriptive Study of the Skid Row Alcoholic in Houston, Texas. *ERIC* No. Ed. 072-193 (Criminal justice monograph).

Spielberger, C. (Ed.), 1963, *Therapy and Research on Anxiety*. New York: Academic Press.

Spielberger, C. (Ed.), 1972, *Anxiety and Behavior*. New York: Ronald Press.

Swerling, I. and M. Rosenbaum, 1959, *American Handbook of Psychiatry*. Ariesti, S. (Ed.). New York: Basic Books.

Tillich, P., 1952, *The Courage To Be*. New Haven, Connecticut: Yale University Press.

A STUDY OF RETENTION OF COMMUNICATION SKILLS AFTER COMPLETION OF A COMMUNICATION SKILLS TRAINING COURSE FOR ALCOHOLISM COUNSELORS

D. Rowden, Ph.D., D. Sansbury, Ph.D., K. Roberts, M.A., B. Hettinger, Ed.D.

National Center for Alcohol Education, Arlington, Virginia

INTRODUCTION

During 1976, the National Institute of Alcohol Abuse and Alcoholism directed the National Center for Alcohol Education (NCAE) to develop a training program in communication skills for alcoholism counselors. This training program entitled *Counseling Alcoholic Clients* (CAC) was completed in late 1976 and pilot tested in early 1977.

The CAC course is a 30-hour training program designed to assist practicing alcoholism counselors in improving their application of basic communication skills in one-to-one interactions with clients. The program was developed in collaboration with a team of practicing alcoholism counselors from the State of West Virginia's Division of Alcoholism and Drug Abuse.

The training program is designed for those whose major job responsibility is counseling clients with alcohol-related problems, and who have had less than three years' experience and little or no previous training in counseling.

Eight basic communication skills are addressed in the training: attending, paraphrasing, reflection of feeling, summarizing, probing, counselor self-disclosure, interpreting, and confrontation. The course is designed in twelve sessions which range in length from one to two-and-a-half hours and can be delivered either in consecutive sessions or by modules--whichever is required to meet local training constraints. Alternatives for course delivery are described in the materials.

The program can be conducted for up to 21 participants; the

total number of the group should facilitate working in triads, i.e., be divisible by three.

Methods employed in training include structured exercises, observation of skill demonstrations on videotape, skill practice, reading, written quizzes, and group discussion.

The program requires a trainer with experience in counseling clients with alcohol-related problems and with training delivery skills in short lecture, role-play, structured exercises, and group discussion techniques. It is recommended that a training team of two or more conduct the training and that trainers have the ability to co-train.

To evaluate this training program three states were selected for pilot testing--West Virginia, Wisconsin, and Maryland. The training programs in the first two states were residential; the third state conducted the program on a commuting non-residential basis for the trainees. In addition, a comparison group of 23 alcoholism counselors was randomly selected from a list of 180 certified alcoholism counselors in Texas.

DATA COLLECTION

The instruments employed in generating evaluation data for the pilot tests and follow-up study were: 1) a standard trainee profile; 2) an alcohol crisis hot line audio tape which was used to generate the data for the communication skills analysis; and 3) during the follow-up, a semi-structured guide was used by the interviewers in the telephone discussion with the participants. The evaluation design for this phase of the project addressed two major questions.

* To what degree have the trainees been able to maintain the communication skills they learned in the program?

* What impact (if any) has the training program had on counseling activities?

In order to answer these questions, a total sample of 30 counselors were to be contacted. Twenty of this number were drawn from the pilot test participants and ten from the comparison group of counselors in Texas who had not participated in the training. The ten comparison group members were to have completed the pretest but not the posttest. Administration of the follow-up test to this group would identify any long term learning of skills which was neither a function of the training program nor a function of multiple testings of comparison group participants. All of the appropriate Texas counselors were contacted. However, when the cut off point for the collection of data was reached, complete data on only 14 trainees had been obtained.

Thus a total of 24 participants were available for the follow-up study.

The procedure used to gather the necessary data was as follows: First, participants were mailed their standard profile, which they had completed six months prior to the current contact. The participants were asked to examine their profile and note any changes in the information which had occurred and return it to NCAE.

A time was arranged when the participants could be contacted by telephone to complete the rest of the data gathering procedures. The first part of the telephone interview was the administration of audio cassette test procedure for recording data on the basic communication skills covered in training. The interviewer then focused on any changes in the professional life or agency situation of the participants which had occurred since the training program. The participants discussed such things as change in position, major policy changes at their agency, counseling style, how much time they spend in a counseling session. When significant changes were noted, an attempt was made to determine if such changes were a function of the training.

Data from all these sources were then analyzed to provide the necessary information to answer the questions posed for this stage of the evaluation. Because of the relatively small number of cases involved in the follow-up, information from the profiles and telephone interviews were content analyzed by two raters with previously established reliability in the rating procedure developed for the audio tape. Data from pre, post, and follow-up testing were used to assess stability of behavioral gains over the last six months.

FINDINGS

Analyses of pilot test data indicated the program successfully taught basic communication skills to Alcoholism Counselors. Specifically, the following findings were noted:[1]

---

1 For a complete and detailed discussion of the evaluation design for the pilot tests and the results of those tests see: *Counselor Skills Project: Proposed Evaluation Design*, National Center for Alcohol Education, December 30, 1976; *Evaluation Report on Pilot Tests of Listening, Processing, Feedback: A Training Program in Basic Communication Skills*, National Center for Alcohol Education, July 12, 1977; and *Addendum Report on Comparison Group Findings for Analysis of Pilot Test Results of Counseling Alcoholic Clients Training Package*, National Center for Alcohol Education, September, 1977.

* At all three pilot test sites (West Virginia, Wisconsin, and Maryland) the CAC program was successful in reducing the participants' felt need for training in counseling and increasing the participants' perceived skill in counseling clients.

  - Participants at all three sites exhibited significant gains in their ability to employ basic communication skills in a simulated counseling situation.

* Analysis of selected characteristics among the trainees indicated these behavioral gains were not a function of age, previous training, experience in counseling or in the alcohol field or education.

* Analysis of comparison group data indicated the behavioral gains exhibited were not a function of the testing procedures employed in the evaluation.

This report focuses on the follow-up of the trainees which took place approximately six months after completion of the training program. In this phase of the evaluation effort the purpose was to measure the impact and/or influences of the training program on the trainees after they had returned to their home work settings.

First, the follow-up analysis of the comparison group data will be presented. The focus of this analysis was demonstration of communication skills. Using a dependent t-test, no significant differences between pretest scores and follow-up scores were found. ($t=.38$, $df=9$, $p>.05$). Thus, for the comparison group, no significant skill learning had occurred over the last six months. Next, the question of skill retention for the trainees will be addressed.

Skill Retention

The first question is "to what degree have the trainees been able to maintain the behavioral gains observed at the conclusion of the training programs?" The answer to this question, while somewhat complicated, in general indicates skill retention did occur. First, if only statistically significant gains and losses of the 14 people contacted are considered only one had regressed in demonstrated communication skills to original pretest levels. An additional six persons had regressed on the follow-up test to levels below their posttest performance but still superior to their pretest performance. Thus, a total of seven persons had follow-up scores lower than posttest scores. Statistical analysis of this group indicated that the differences between pretest scores and follow-up scores was marginally significant ($p=.06$). This is interpreted as possible through week evidence of a maintenance of behavioral gains in spite of the observed posttest to follow-up test declines in scores. Another five respondents exhibited gains from the posttest ot the follow-up test scores. This suggests some continued learning since conclusion of the program. Finally, the remaining two respondents exhibited no change between pretesting, posttesting and follow-up. These findings are displayed in Table I.

Behavioral gains as measured during the pilot test phase were not based upon statistical significance alone. While that analysis was encouraging there was another assessment procedure employed which was more rigorous and perhaps more useful in judging the true effects of the training program. Specifically, an *a priori* criterion for success was established as a gain of at least two skill demonstrations between pretesting and posttesting. Using this criterion as a basis for analysis of follow-up scores, the following operational definitions were developed for outcomes in this follow-up study:

* No Change - an individual exhibits no gain in skill demonstration at posttesting and at follow-up.

* Skill Maintenance - an individual gained at least a two unit skill differential from pretesting to posttesting and maintained this same differential at follow-up.

* Skill Gain - an individual demonstrates a gain of at least two skill demonstrations pretest to follow-up test and a gain of at least one unit posttest to follow-up.

* Skill Loss - an individual demonstrates at least a two unit gain pretest to posttest, but less than a two skill unit differential pretest to follow-up test.

Using these definitions five participants experienced meaningful loss of ability to demonstrate communication skills during the six month period between posttesting and follow-up testing. Four participants who exhibited gains pretesting to posttesting maintained these gains at follow-up testing even though three of the four had follow-up scores that were lower than their posttest socfes. Three participants exhibited gains at the conclusion of pretest-posttest and follow-up administrations. Two of these three participants did not register a meaningful gain between pretesting and posttesting, but did gain at least two units over their pretest scores at follow-up testing. The remaining two respondents did not demonstrate any gains at posttesting or at follow-up testing. To summarize, of the 14 participants' scores analyzed, seven either maintained their original pretest-posttest differential at follow-up or exhibited a meaningful increase in skill demonstration between pretesting and follow-up testing. The remaining seven participants either exhibited no gain at any stage of testing or exhibited a gain pretest to posttest but subsequently regressed to pretest performance levels at follow-up testing. In short, 50 percent of the sample maintained the skill gains that resulted from the training program.

TABLE 1

COMPARISON OF TEST SCORES ON PRETEST, POSTTEST AND FOLLOW UP

ADMINISTRATIONS OF THE COMMUNICATION SKILLS TEST INSTRUMENT

Group Analyzed	Mean Scores*						Mean Scores*					
	Pre Test	Follow Up	$\overline{D}$	t	df	p	Post Test	Follow Up	$\overline{D}$	t	df	p
All Participants (N=14)	1.85	3.80	-2.2	-2.03	13	.005	5.20	3.80	+1.1	2.03	13	.03
Participants Posttest > Follow-Up (N=7)	2.20	3.40	-1.2	-1.78	6	.06**	6.20	3.40	+4	6.38	6	.0003
Participants Posttest < Follow-up (N=5)	1.5	5.80	-4.3	-2.9	4	.02	4.70	5.80	-1.1	-3.8	4	.009

*Scores based on average for all respondents within the category on the number of responses to 10 client statements classified as demonstrated on the eight skill areas.

**p .05 not significant using traditional rejection level, however close enough to be given cautious consideration.

The observation that approximately one-half rather than a substantial majority of the participants were able to maintain their skill levels after six months raises the question of which differences among the participants might predict maintenance or loss of skill level over time. Examination of participant characteristics such as age, education, previous experience, and length of time in the field were not related to posttest performance. However, some of these factors may be related to skill retention over time. Examination of profiles of those who lost skill capability in comparison to other trainees indicates that some difference in background did exist.

First, five of seven participants exhibiting loss of skills were from the Maryland pilot test. The Maryland pilot test differed from the others in that this was a non-residential program. Further, the Maryland participants as a group had somewhat lower educational levels and less counseling experience than the other participants. At the time of the pilots, analysis of these factors indicated they were not associated with posttest outcomes. Results of a Fisher's exact probability test comparing those participants in Maryland who lost skill ability and those who maintained these factors was based upon a comparison of those participants who demonstrated a loss of skill at follow-up to those who maintained or increased their original training program performance. Statistical significance was determined by application of the Fisher's exact probability test. A median was determined for each independent variable (education, months in counseling, percent of time spent in counseling, etc.). The number of participants in the groups that either lost skills or maintained/gained skills was tallied for each independent variable. The participants were divided on the basis of whether they fell above or below the computed median.

In addition, a more sensitive independent t-test was applied to the data for each of the variables of interest. In this case the groups for the t-test were defined as 1) those who maintained or gained skills at follow-up and 2) those who demonstrated loss of skills. The scores for each of these groups were defined as the values for each of the variables of interest.

To summarize, the maintained/gained and lost skill groups were compared on the following variables:

* Education,
* Months of Counseling Experience,
* Months in Alcohol Field,
* Number of Hours Training in Counseling, and
* Percent of Time Spent in Counseling.

Table 2 presents a summary of the analyses.

None of the above analyses attained statistical significance using the traditional rejection criterion of p=.05. However, both counseling experience and percentage of time spent in counseling were close enough to significance to warrant some discussion. First, both of these variables relate to the practice of counseling rather than education and training in counseling. In the case of previous experience in counseling, the group that maintained or gained in demonstration of skill had slightly more experience than the group which lost on skill demonstration at follow-up. However, all the participants in the pilot tests had less than three years previous experience. This may suggest that while the program was successful with trainees who have less than three years experience, it is more successful with those trainees who are closer to the 36 month maximum of previous experience. The reason for this is not as obvious as it might appear. The first inclination would be to attribute the above trend to skill learning on the job. That is, more experienced counselors have already learned some of the necessary communication skills by virtue of their contacts with clients in counseling situations. If this were true, this more experienced group should have higher pretest scores than the less experienced group. This was not the case. Another explanation is that those counselors with some experience are better able to use the communication skills they learn. That is, with experience in counseling situations, they are better able to apply the skills, thus reinforcing the use of those skills and subsequently leading to a trend of increased skill retention in relation to less experienced counselors. If future information should support this hypothetical explanation, it may indicate that the package is best suited for counselors with at least one and one-half years experience but less than three years total experience. At this point such a presumption would definitely be premature.

In the case of the percentage of time spent in counseling, the finding suggests that practice and/or usage helps in skill retention. This is not surprising. If, in fact, it is valid, does increased communication skills lead to increased percentage of time spent in counseling? Or, perhaps, spending a significant portion of time in "one-to-one" counseling allows for the reinforcement and development of use of communication skills. As was the case with previous experience in counseling, at this juncture it is not possible to establish which of these alternatives is the best explanation. Further, the reader is cautioned that the observed relationships between performance and counseling experience and performance and percentage of time spent in counseling are marginal at best. In short, these findings are intriguing and warrant further investigation. However, at this point, on the basis of the five variables examined, none appear to significantly influence the outcomes of the follow-up study.

To summarize the findings of this six month skill retention study:

TABLE 2

RESULTS OF ANALYSIS OF SELECTED VARIABLES HYPOTHESIZED AS POTENTIALLY INFLUENTIAL IN FOLLOW-UP OUTCOMES

Variable	t value	P	Fisher P	Significant? (p=.05)
Education	-1.14	.14	.14	No
Counseling Experience	-1.50	.08	.14	No
Experience in Alcohol Field	.13	.45	.50	No
Training in Counseling	- .06	.52	.50	No
Percent of Time Counseling	-1.39	.09	.29	No

* Fifty percent of the trainees contacted were able to either maintain or increase the use of skills learned in the training program.

* The other half of the 14 trainees contacted had skill usages which were no better than their original pretest assessments.

  - Two of this number had also not registered meaningful gains in the original pretest-posttest assessment.

* Two of the seven trainees who gained in ability to use the communication skills at follow-up did not register a meaningful increase on their pretest-posttest assessments.

* Thus, of the 14 trainees evaluated, five who had originally gained from the pretest to posttest, regressed to pretest performance levels. Four trainees maintained their original gains at follow-up. Three trainees increased their skill capability between posttest and follow-up. Moreover, two of these three trainees registered a meaningful gain at follow-up which was not present after pretest-posttest evaluation. Finally, two trainees showed no improvement at any stage of testing.

The literature on the subject of skill retention in microcounseling programs is sparse. For those studies examined by the authors the findings are somewhat mixed. Haase, DiMattia and Guttman (1971) reported a systematic one year follow-up study of training of support personnel in human relations skills. Specific skills which are of interest to this report were: 1) attending, 2) reflection of feeling and 3) expression of feeling. A total of 13 trainees were contacted one year after completion of their training. Attending skills, as measured by eye contact, remained constant at follow-up. Ability to demonstrate reflection of feeling had decreased and ability to express one's feelings had increased. An overall rating of counselor effectiveness was also employed. On this measure, the trainees as a group had regressed to base line levels of competence.

Another study by Guttman and Haase (1972) investigated the use of microcounseling skills in counseling situations after completion of training in those skills. Two groups of ten each were employed in this study. The first group consisted of counseling students specifically trained in attending, reflection of feeling, and summarization skills. The remaining ten students served as the control group and received no training in these skills. Skill measurements were obtained prior to training, immediately after training, one week after training and two weeks after training. The students who received the training exhibited higher skill levels on reflection of feeling and summarization than did the control group. There was not a statistically significant difference between the groups on the skill of attending. At follow-up there was some regression in ability to demonstrate reflection of feeling and summarization. However the trainees as a group did not regress to their pretraining skill levels in these areas.

Further work by Parker (1972) focused on retention of microcounseling skills over time and a comparison of this trainee group to another group which went through a standard one-semester practicum in counseling. This training program was similar to CAC in that it consisted of 1) receiving a didactic presentation 2) viewing video counseling models 3) counseling pseudo clients 4) critiquing the sessions with the pseudo clients. There were ten students assigned to this experimental condition. Another group of ten students were exposed to a placebo activity during this segment of the practicum. A series of 30 isolated videotaped client statements were administered to students in both groups immediately after training and again at the end of the semester. Training effectiveness was measured by the affective and understanding dimensions of the Counselor Verbal Response Scale (CVRS) and the Truax Accurate Empathy Scale (TAES). Comparison of the two groups immediately after training indicated the experimental group scored significantly higher than the control group on the measures of affect employed for this study. By the end of the practicum semester, the experimental group maintained the same level on the affective dimension noted after training. However, by this time the control group had also achieved this same level. This indicates that the non-specific learnings of the practicum were also able to transmit counseling skills to the students outside of a skill specific microcounseling model.

The current study seems to mirror to some extent the above findings. For example, in this follow-up, gains after training and skill retention at follow-up were observed. The measurement approach and training program in this follow-up which generated it are somewhat similar to those described in Parker's work. The outcome measures and trainees seem fairly close to those employed by Guttman and Haase. Both the Parker and Haase studies employed long term follow-ups, which indicated skill retention. Thus the findings and approach of this follow-up activity are generally consistent with the research reported on similar programs.

Impact on Job Performances

To carry out this analysis, four areas of work activity were identified.

* Change in jobs,

* Change in case loads,

* Change in amount of time spent with clients, and

* Change in counseling style.

In addition, a fifth area which focused on changes in agency organization and functioning was examined to determine when an observed job related change was due to the training program and when such changes were agency generated. To facilitate the analysis of each of these factors that might be associated with loss of skills or maintenance of skills, each of the job performance variables were analyzed with reference to two groups: Skill Loss Group and Skill Maintenance Group.

With reference to job changes, only two of the 14 participants reported any change in jobs since completion of the training program. One person in the follow-up sample, had advanced to the position of Alcoholism Counselor from a lower support staff position. He stated this advancement was a result of attendance at the training program. Another trainee had also changed jobs, however, this was a lateral move which was not attributed to the training program. Both of these changes were in the Skill Loss Group.

Seven of the 14 participants reported increased case loads since the training program. One person reported a decrease and two others reported a shift from previous clients to a new type of client. Specifically, one trainee reported a greater emphasis on women clients and the other reported a shift toward court referred cases. Two of the seven who reported increases were in the Skill Loss Group and the remaining five were in the Skill Maintenance Group, as was the trainee who reported a decreased case load. None of the changes in either composition or size of case load was attributed to the training program. This is not surprising since the impact of the training program would more likely be apparent in the amount of time spent with clients rather than the number of clients seen.

Six of the participants indicated that they now spend more time with each client. All of these participants indicated that this increased time was a function of their CAC experience. The average amount of time spent with a client for this group was 57 minutes, while the group that stated no change had occurred in time spent with clients averaged 58 minutes. Four of six persons who reported an increase on this variable were in the Skill Maintenance Group. Thus, it appears that the training program was moderately successful in increasing contact time between counselors and clients. Given that the group which reported no change in contact time reported approximately the same amount of contact as the group that increased the amount of time spent with clients (about one hour for each group), a tentative explanation of program impact in their area would be that given one hour as the usual maximum amount spent with a client, the training program assisted trainees in reaching that one hour upper limit.

Eight of the participants reported changes in their counseling style since the training program. All eight attributed these changes to the training program. Five of the eight participants were in the Skill Maintenance Group. The types of changes reported were as follows:

* More Attending (3 participants)
* Increased Reflection of Feeling (1 participant)
* More Confrontation (1 participant)
* More Feedback and Empathy (1 participant)
* More Interpretation (1 participant)

The last factor examined was the change in the agency. One agency switched from public sponsorship to private sponsorship. Two others instituted policy changes dealing with such things as third party payment, NIDA policies, and Federal and State regulations. None of these changes were the result of the training program or had an influence on the counselor's activities with regard to the application of skills learned in the training program.

This concludes the major factors examined to assess the impact of the training program on participant job performance. Two other analyses were carried out and even though they provided no new information they are worth some discussion.

Other Considerations

One measure of success in counseling, which is confounded by other factors, is the return rate of their clients after the initial intake session. Therefore, prior to training and at follow-up, an estimate of the percentage of their clients who returned for second, third and fourth counseling sessions was requested. Updated information on this question was not received from four participants. Of the ten who did respond, only three indicated any change in the return rate and one of those indicated a slight decrease. In short, this turned out not to be a particularly useful measure.

Additionally, in the original design for the follow-up, the possibility was raised as to whether or not the training program had any impact on the participants' levels of confidence in their counseling ability. Two questions were asked of each participant to generate information for this inquiry. The questions were:

* How confident are you of your ability to effect change in an alcholic client using only individual counseling?

* How confident are you of your ability to establish and maintain a climate for counseling that includes getting clients to talk freely, to discuss problems and their feelings about those problems, and to feel willing to participate actively in the treatment process?

These two questions were asked of each participant at pretesting, posttesting and follow-up testing. Answers were recorded on a Likert type five point scale in which a value of "one" indicated the respondent was very unsure of his or her ability and a score of "five" indicated the respondent was very confident of his or her ability. Analysis for change in pre to post scores and post to follow-up scores indicated no increases in confidence for either of the two questions examined. Separating the total group into the Loss and Gain in Skills Groups also yielded no differences. Examination of pretest scores provided an explanation for this lack of growth in confidence in the participants' ability. The modal response for both questions of the pretest was a "four" (somewhat confident). There were also several respondents who rated both questions a "five", the top category. Thus, there was a topping out effect. The participants were so confident in their skills at the beginning of training, that there was little, if any, room for improvement in individual confidence assessments. In short, this inquiry did not yield any information of importance, other than that the trainees were quite confident in their abilities both prior to and following training.

To summarize, the data on impact on job performance indicated the following:

* The CAC program produced only one job change among the 14 participants queried.

* While there was a large number of increases and change in case load among the participants, none of these changes were attributable to the training.

* Six of the participants (43%) reported they had increased the amount of time they spend in a counseling session with a client. These increases were attributed to the training program.

* Eight participants (57%) reported changes in counseling styles which were attributable to the training program.

SUMMARY AND CONCLUSIONS

A total of ten comparison group members and 14 trainees were contacted and interviewed during this follow-up activity. The general findings are as follows:

First, no significant skill learning took place among the comparison group members during the period between pretesting and follow-up testing.

Fifty percent of the trainees contacted were able to maintain their posttest skill levels or to increase those levels.

There is some indication that this group had more experience in counseling and report a higher percentage of their time in one-to-one counseling than the other trainees who did not maintain their skill levels. Because of the marginal relationship of these factors, their influence is considered weak at best.

Two of the seven trainees who exhibited gains in skill from pretest to follow-up testing had not demonstrated a meaningful gain pretest to posttest. This is evidence of continued significant learning during post training for these trainees.

The CAC training program produced only one job change among the 14 participants contacted.

Six of the 14 trainees reported they now spend more time with their clients and that this increase was due to the CAC training they received.

Eight of the participants (57%) reported changes in counseling styles which were attributable to the training program.

Thus, as was indicated in the pilot test findings, this appears to be a successful program. The follow-up findings are positive though not conclusive support of the validity of the program. There is slight evidence that counselors with two to three years experience may benefit more from this type of program. Perhaps, as more data accumulate on this subject, a firm conclusion might be reached which will help focus the program on the best target audience. In any case, for the pilot test version of this package, a 50 percent success rate for skill retention coupled with the other positive changes noted is acceptable evidence of the validity and utility of this training program.

REFERENCES

Guttman, M.A. and Haase, R.F. 1972. Generalization of microcounseling skills from training period to actual counseling setting. *Coun. and Sup.* 12: 98-108.

Haase, R.F., DiMattia, D., and Guttman, M.A. 1971. Training of support personnel in human relations skills: A systematic one year follow-up. *Coun. and Sup.* 11:194-199.

Parker, R.D. 1972. *The effects of microcounseling with counseling practicum students*. University of South Dakota: unpublished doctoral dissertation.

Rowden, D., Maloney, H., Roberts, K., Hettinger, B., Stepney, T., and Freeman, K. 1977. *Evaluation report on pilot tests of listening, processing, feedback: A training program in basic communication for alcoholism counselors*. Arlington, Virginia: National Center on Alcohol Education.

Rowden, D., Sansbury, D., English, C., Maloney, H., Roberts, K., and Freeman, K. 1976. *Counselor skills project proposed evaluation design*. Arlington, Virginia: National Center for Alcohol Education.

Rowden, D., Roberts, K., Maloney, H., Stepney, T., and Freeman, K., 1977. *Addendum report on comparision group findings for analysis of pilot test results of the Counseling Alcoholic Clients training package*. Arlington, Virginia: National Center for Alcohol Education.

CONDITIONED ETHANOL TOLERANCE

Lorne F. Parker, Ph.D. and Joseph D. Skorupski, M.S.

University of Washington

While it is evident that conditioning is involved in the development of drug addictions, the specific etiological role (or roles) of conditioning has not yet been clearly delineated. Some believe that addiction is maintained via addicts learning to avoid withdrawal stress through self-administration (Beach, 1957), while others have suggested that the reinforcing effects of drugs are sufficient to maintain operantly conditioned self-administration (Schuster & Villarreal, 1968). Wikler (1971) has further pointed out that environmental situations that regularly surround drug administration or withdrawal stress may also gain conditioned positive or negative reinforcement qualities, respectively.

Siegel (1975;1976;1977) has recently expanded this view of conditioning even further by demonstrating that tolerance to morphine analgesia is acquired through conditioning in rats. In a large series of experiments Siegel found that rats showed tolerance to morphine's analgesic effects only when it was administered in conjunction with environmental stimuli that previously had been paired with morphine. If the drug was administered without these conditioned stimuli (CSs), the animals showed no evidence of tolerance. Siegel also showed that morphine tolerance is mediated by a compensatory, conditioned hyperalgesic response, that like other conditioned responses, is susceptible to the decremental effects of partial reinforcement, latent inhibition and extinction contingencies.

Since it has not been possible to account for ethanol tolerance entirely in terms of physiological adjustments (Kalant, Le Blanc & Gibbins, 1971), the following experiment was conducted to ascertain whether classically conditioned compensatory responses might be involved. We decided to examine the development of tolerance to ethanol's depressant effects because (1) it is a major effect of the drug, and (2) we felt that the development of tolerance to drug-induced narcosis would eliminate the possibility of operantly conditioned tolerance (Le Blanc, Gibbins & Kalant, 1973), since the narcotized animal presumably cannot perform overt behaviors.

The development of tolerance to ethanol's depressant effects on the central nervous system (CNS) was clearly demonstrated by Newman and his colleagues (Newman & Card, 1937; Newman & Lehman, 1938), but little substantial work has been done since then. Newman et al.(1937; 1938) found that the depressant effects of ethanol on reflexes, respiration, coordination, etc. became less severe after repeated exposures to the drug, and they suggested that a development of cellular tolerance within the CNS was responsible. It is possible, however, that a compensatory "arousal" response may have been conditioned to the stimuli of the experimental setting (i.e., the administration ritual, etc.) and served to combat the depressant effects of ethanol during their tests.

We tested this hypothesis by employing a variation of the "sleeping-time" test to monitor the narcotic efficacy of ethanol in both the presence and absence of stimuli that had previously been paired with ethanol's depressant effects. The "sleeping-time" test was used because it provides a continuous measurement scale of depressant potency, and it is sensitive over the effective dose ranges of many depressant drugs (Aston, 1965; Ebert, Yim & Miya, 1964; Milner, 1968).

GENERAL METHOD

If tolerance to ethanol's depressant effects is mediated by a conditioned compensatory response (i.e., a conditioned "arousal" response), then it follows that if ethanol is administered in the absence of the CSs that elicited the "arousal" response, the organism should show a decrement in tolerance. This line of reasoning underlies the design of the present experiment. The conditioned response, which is referred to as an "arousal" response for lack of a better descriptive term, was not observed directly. Rather, its existence was verified by noting the decrement in ethanol tolerance that occurred when ethanol was administered in the absence of CSs.

In general, the experiment consisted of pairing the effects of ethanol with a unique set of environmental stimuli on repeated trials, and subsequently measuring their "sleeping-times" after a large dose of ethanol was administered either in the presence of these stimuli (CSs) or in an unconditioned, neutral situation. Thus, the effects of ethanol were always paired with a separate experimental room and clear, plastic chambers (CS Chambers) that were saturated with odor of "Mentholatum." After many such pairings the animals were given large, anesthetizing doses of ethanol in either the CS Chambers or in their home cages, and the durations of their "sleeping-times" were recorded. On the following day the test was repeated in a counterbalanced manner using the opposite test situation. Comparisons between the "sleeping-times" obtained in the CS Chambers and in the neutral, home cages revealed the extent to which a conditioned compensatory response mediated tolerance.

Subjects

The subjects were 36 male, albino rats derived from the Wistar strain. The animals were obtained from Simonsen Laboratories, Inc., Gilroy, California, and were between 90 and 110 days old when they arrived at our laboratory. All animals were allowed one week to habituate to their individual, stainless steel, home cages prior to initiating the experiment.

Conditioning Procedure

Beginning on Day 1 and on every other day thorugh Day 19, all animals were removed from their home cages during the night portion of their 12 hr/12 hr day/night cycle, and were individually transported to a separate room in a carrying cage. The animals were then given either intraperitoneal (ip) injections of isotonic saline (n=18) or 4.0 gm./kg. (15% v/v) ethanol (n=18) and immedately placed into individual CS Chambers. While the dose of 4.0 gm./kg. ethanol is severely debilitating when administered to rats parenterally, it was employed because it is possible to produce maximal behavioral tolerance to ethanol by administering such dosages every other day (Wallgren & Lindbohm, 1961). The CS Chambers were constructed of clear plastic and measured 30.0 x 8.5 x 16.0 cm. A gauze pad soaked in "Mentholatum" was attached to the inside of each chamber. All animals remained in the Chambers for six hours, after which they were returned to their home cages.

On the days that intervened between conditioning trials all animals were left in their home cages. At the time of their usual injections in the CS Chambers, however, they were given injections of isotonic saline and immediately returned to their cages. This sham injection procedure was employed to diminish the likelihood that the rats would use the administration ritual as a predictive cue for the onset of ethanol's effects. Since the injection procedure was used in the tolerance tests in both the CS Chambers and the neutral home cages, it was felt that the sham injections were important to reduce the probability of introducing part of a compound CS into the supposedly neutral home cage test situation.

Tolerance Tests

After ten conditioning trials (20 days) it was evident that the ethanol treated animals had developed tolerance and were recovering earlier than they were initially. Thus, two consecutive sleeping-time tests were administered to each animal on Day 21 and Day 22. The tests consisted of administering a 4.0 gm./kg. (ip) dose of ethanol and, beginning 30 minutes after the injection, checking every 10 minutes to see if the "righting reflex" was present. The criterion for "righting" was that the rat, after being placed on its back, would roll back to an upright position, with all four feet touching the floor, within 15 seconds. If "righting" was not obtained, the rat was returned to an upright position and checked again 10 minutes later. For half of the animals in each group the first test took place in the CS Chambers and the second test took place in their home cages. For the remaining animals the tests took place in the opposite order.

Results

The 4.0 gm./kg. dose of ethanol was very debilitating, and eight animals died during the conditioning and testing portions of the experiment, and an additional four animals were dismissed from the experiment because of poor health. Thus, the data from 10 experimental animals and 14 control animals was used to determine whether ethanol tolerance had been elicited by the CS Chambers.

Figure 1 shows a summary of the "sleeping-times" of both groups in the CS Chamber and home cage tests. As depicted, the experimental group, which had received 10 large doses of ethanol, showed no evidence of tolerance in the home cage situation. That is, their mean "sleeping time" was identical to that of the saline treated control group. In the presence of the CS Chambers, however, the experimental group showed a greatly reduced "sleeping time" duration as compared to the control group ($U=40.5$, $p<.05$).

From Figure 1 it is evident that, while the experimental group showed a decrease in "sleeping-time" when tested in the CS Chambers, the control group showed an increase in "sleeping-time". The reasons for the increase in ethanol's baseline depressant effects in the CS Chambers are not clear, but one reason might be that poor ventilation in the CS Chambers produced prolonged unconditioned ethanol narcosis. Many other differences between the two situations may have also been involved.

The raw "sleeping-time" data were fairly variable between individuals within a group and between the two testing situations, so the data were normalized with respect to these factors by computing the percentage of each animal's home cage "sleeping-time" that was spent in the CS Chamber. Figure 2 shows these percentage scores and, as evi-

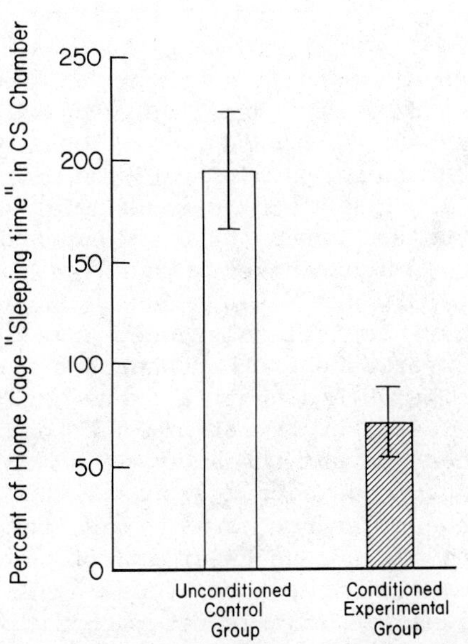

Fig. 1. Percentage of Home Cage "Sleeping-time" in CS Chamber

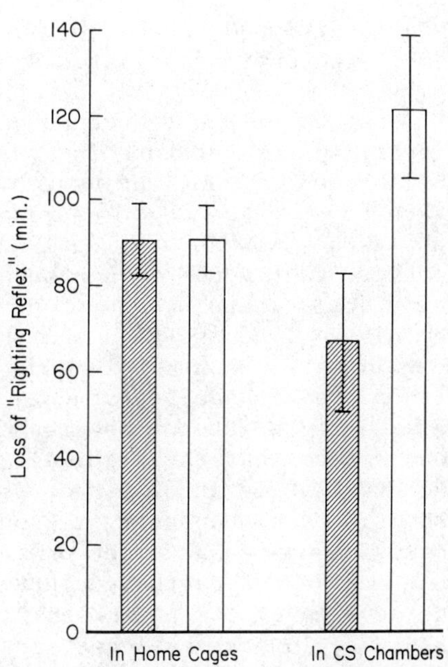

Fig. 2. Loss of Righting Reflex" (min.)

dent, the effects of conditioning are much more obvious in the transformed data. The unconditioned control group showed a 95% increase in the CS Chambers, whereas the conditioned experimental group showed a 20% decrease in the CS Chambers. A Mann-Whitney U test revealed that this difference was highly significant ($U=17.0$, $p<.01$).

DISCUSSION

The finding that the conditioned experimental group and the unconditioned control group did not differ in their sensitivities to the narcotic effects of ethanol in the home cage situation suggests that the ethanol treated animals did not develop any dispositional or physiological tolerance during the course of the repeated ethanol injections. Rather, tolerance to ethanol was only evident in the presence of stimuli (i.e., the CS Chambers) that had been previously paired with ethanol's effect, and thus it appears that all of the ethanol tolerance was mediated by conditioning. The nature of the conditioned response that mediated ethanol tolerance is obscure, since it was not noticed in the present study by its absence, but it might be described as a conditioned "arousal" response since it functionally served to combat the depressant effects of ethanol.

That tolerance to ethanol can be conditioned is an important consideration for interpreting previous studies of ethanol tolerance. Until recently, an improvement of behavioral performance was usually taken to be evidence of neuronal adaptation or physiological tolerance (e.g., Newman et al., 1937; 1938). Chen (1968) demonstrated, however, that much of such behavioral tolerance can be attributed to individuals learning how to perform while intoxicated. Thus, Le Blanc and his colleagues (Le Blanc et al., 1973; 1975; 1976), in an attempt to delineate learned tolerance from physiological tolerance, have dissociated ethanol treatments from performance of the tolerance index task during tolerance development. While such a control procedure does ensure that their animals cannot learn how to perform the tolerance task while intoxicated, it does not prevent the acquisition of conditioned compensatory responses (e.g., conditioned arousal) that could in turn enhance performance on the tolerance task. Thus, tolerance manifested by their control groups did not necessarily reflect the development of physiological tolerance, but rather it may have been due, in part at least, to conditioning. In this view, conditioning mediates ethanol tolerance by (1) allowing the individual to acquire relatively complex external responses while intoxicated, and (2) by providing compensatory internal responses that serve to diminish the drug's disruptive effects on homeostatic balances.

It is clear that much more work must be done to delineate the importance of conditioning in the overall picture of drug tolerance, but the initial findings with morphine (Siegel, 1975) and ethanol suggest that such an understanding might provide a crucial piece to the larger puzzle of drug addictions and their etiologies.

REFERENCES

Aston, R. 1965. Quantitative aspects of tolerance and post-tolerance hypersensitivity to phenobarbital in the rat. *J. Pharmac. Exp. Therap.* 150: 253-258.

Beach, H.D. 1957. Morphine addiction in rats. *Canad. J. Psychol.* 11: 104-112.

Chen, C.S. 1968. A study of the alcohol-tolerance effect and an introduction of a new behavioral technique. *Psychopharmacologia* 12: 443-450.

Ebert, A.G., Yim, G.K.W. and Miya, T.S. 1964. Distribution and metabolism of barbital-^{14}C in tolerant and nontolerant rats. *Biochem. Pharmacol.* 12: 1267-1274.

Kalant, H., Le Blanc, A.E. and Gibbins, R.J. 1971. Tolerance to and dependence on, some non-opiate psychotropic drugs. *Pharmacol. Rev.* 23: 135-190.

Le Blanc, A.E., Gibbins, R.J. and Kalant, H. 1973. Behavioral augmentation of tolerance to ethanol in the rat. *Psychopharmacologia* 30: 117-122.

Le Blanc, A.E., Gibbins, R.J. and Kalant, H. 1975. Generalization of behaviorally augmented tolerance to ethanol, and its relation to physical dependence. *Psychopharmacologia* 44: 241-296.

Le Blanc, A.E., Gibbins, R.J. and Kalant, H. 1976. Acquisition and loss of behaviorally augmented tolerance to ethanol in the rat. *Psychopharmacologia* 48: 153-158.

Milner, G. 1968. Modified confinement motor activity test for use with mice. *J. Pharmaceut. Sci.* 57: 1900-1902.

Newman, H. and Card, J. 1937. Duration of acquired tolerance to ethyl alcohol. *J. Pharmacol. Exp. Therap.* 59: 249-252.

Newman, H.W. and Lehman, A.J. 1938. Nature of acquired tolerance to alcohol. *J. Pharmacol. Exp. Therap.* 62: 301-306.

Schuster, C.R. and Villarreal, J.E. 1968. Psychopharmacology: A Review of Progress. Washington, D.C., PHS Publ. 1836, U.S. Gov't. Printing Office.

Siegel, S. 1975. Evidence from rats that morphine tolerance is a learned response. *J. Comp. Physiol. Psych.* 89: 498-506.

Siegel, S. 1976. Morphine analgesic tolerance: Its situation specifically supports a Pavlovian conditioning model. *Science* 193: 323-325.

Siegel, S. 1977. Morphine tolerance acquisition as an associative process. *J. Exp. Psych: Animal Behav. Processes* 3: 1-13.

Wallgren, H. and Lindbohm, R. 1961. Adaptation to ethanol in rats with special reference to brain tissue respiration. *Biochem. Pharmacol.* 8: 423-424.

Wikler, A. 1971. Some implications of conditioning theory for problems of drug abuse. *Behav. Sci.* 16: 92-97.

A GRADUATE TRAINING PROGRAM FOR ALCOHOLIC REHABILITATION SPECIALISTS

May Palacios, Ph.D.

Ball State University
131 N. Washington
Marion, Indiana 46952

The purpose of this program is to provide intensive specialized graduate training in the area of alcoholic rehabilitation for selected graduate students who are interested in a career in the drug abuse field. This training was a joint effort of the departments of General and Experimental Psychology and Physiology and Health Science at Ball State University and the Alcoholic Treatment Units at the Veterans Administration Hospital in Indianapolis, Indiana. A balanced program of scientific, theoretical and applied knowledge in this specialized area will prepare the selected graduate students at the Master's level for a career in this field.

THE TRAINING PROGRAM

The graduate training of the alcoholic rehabilitation specialist consists of two parts: the academic and the clinical. The academic aspect of the training is done at Ball State University in the departments of General and Experimental Psychology and Physiology and Health Science. The Department of General and Experimental Psychology presently offers an MA degree in pre-clinical psychology. The selection criterion are Graduate Record Examination - Verbal Score 400, Quantitative Score 400, a grade point average in undergraduate work of 2.5, and a grade point average of 3.0 in psychology. The above criteria will generally be applied to the selection of students who are primarily interested in the alcoholic rehabilitation specialty. Special consideration will also be given to minority groups and graduate students who have had a drug problem in the past.

The academic training will consist of the successful completion of 40 credit hours selected in a predetermined manner from the following courses:

Motivation and Emotion
Sensation and Perception
Psychology of Intelligence
Psychology of Learning
Theories of Personality
Behavior Disorders of Children
Abnormal Psychology
Introduction to Clinical Psychology
Introduction to Psychodiagnosis
Psychodiagnostic Aspects of Intellectual Dysfunction
Psychological Bases of Behavior Modification
Introduction to Psychotherapy
Experimental Psychology
Experimental Clinical-Personality
Physiological Psychology
Seminar in Personality
Seminar in Clinical Psychology
Psychological Investigations

The minor will consist of 12 hours credit to be taken in the Department of Physiology and Health Science, namely alcohol problems and drug dependence and abuse, in addition to an elective from the following courses:

Health in the Family
Health Quackery
Public Health Practice
Environmental Health

GOALS AND TRAINING OBJECTIVES

The primary objectives of this training program are as follows:

1. To provide basic knowledge and information about alcohol and substance abuse, and to assist the student in his ability to distinguish between facts and myths.
2. To provide factual information on the etiology and dynamics of addiction.
3. To familiarize the trainees with effective use of resources in the prevention, treatment and rehabilitation of the alcoholic.
4. To help increase students' clinical skills in the following areas:
   a. More effective clinical communication with clients.
   b. Skills in problem assessment.
   c. Skills in the therapeutic or helping relationship.
   d. Skills in problem management.

e. Skills in interviewing and record keeping.
f. Skills in effective team work.
g. Skills in case management.
h. Skills in the dynamics of current group procedures such as encounter groups, trans-actional therapy, gestalt therapy, synanon and other confrontation group procedures.

5. The ethics of therapy, importance of confidentiality and the basic respect for the integrity of the client will be stressed.

The student completing this program will earn a master's degree, M.A., in pre-clinical psychology, with special emphasis on substance abuse.

The clinical part of the training consists of 400 hours to be spent in the Alcoholic Treatment Unit in the V.A. Hospital in Indianapolis, Indiana under the direct supervision of a Ph.D. in clinical psychology.

The V.A. Hospital in actuality has one building that specializes in the treatment of alcoholism and substance abuse. Following is a brief description of the alcohol treatment program.

Since October, 1970, the Alcoholism Treatment Unit (ATU) of the Veterans Administration Hospital, Indianapolis, Indiana, has offered a comprehensive, combined inpatient-outpatient rehabilitation program to eligible veterans suffering from alcoholism. The ATU has a multidisciplinary team approach to treatment with a braod range of treatment techniques in use. The treatment philosophy is essentially a belief that no one has all of the answers to effective treatment of the alcoholic, and therefore, the staff will use as many different treatment techniques as possible, providing that each technique has some evidence available as to its effectiveness.

ADMISSION PROCEDURES

Individuals accepted for the ATU program must be veterans eligible for VA hospitalization. An eligible veteran comes to the Unit in one of two ways. (1) He may be seeking help on his own and applies to the hospital Admissions Office. The admitting Physician for Psychiatry Service will evaluate the veteran to determine if his problem is alcoholism and if he might benefit from the ATU program. If the veteran is accepted, he is placed on Pre-Bed Care (PBC) status and begins the ATU program as an outpatient. (2) A veteran already hospitalized for medical, surgical or psychiatric treatment may be found to have a drinking problem by his ward physician. If the ward physician (or other staff) judges the veteran to have a significant alcohol problem, he sends a Consultant Request to ATU. An ATU staff member (usually a

Social Worker) will see the veteran while he is still an inpatient for the purposes of evaluating the patient's motivation for and ability to make use of the ATU program. If the veteran is accepted for the ATU program, he normally completes his hospital stay and is released to out-patient status with ATU. Occasionally the veteran is transferred directly from his ward to the ATU in-patient program (depending on bed availability).

CRITERIA FOR ACCEPTANCE

In general, eligible veterans will be accepted in the ATU program if they meet two main criteria.

1. The veteran should have some awareness that alcohol has created some significant life problems and should show some willingness to accept treatment for his alcohol problem. ATU staff who evaluate the veteran attempt to motivate him for treatment, but the veteran must voluntarily agree to the program before he is accepted.

2. The veteran must be physically and mentally able to participate in the active and verbally-oriented ATU program. Thus, the individual with a seriously disabling physical handicap or medical problem would not be accepted until his physical condition improves. The ATU does not have sufficient medical and nursing coverage to care for the bedridden patient or the individual with acute medical problems beyond simple alcohol withdrawal. Likewise, the veteran who is overtly psychotic or shows a serious organic brain syndrome could not benefit from the program and would not be accepted until such time as he has been treated elsewhere and has improved to the point that he could participate and gain from the program.

PRE-BED CARE PROGRAM

Both walk-in applicant and the inpatient seen on consultation generally follow the same treatment program when accepted for ATU. Both types of applicant spend from two to six weeks on the PBC program. During PBC, the veteran has regular counseling appointments, attends two group meetings (with his family) involving an orientation to alcoholism and to ATU, is introduced to, and encouraged to attend both hospital and community AA meetings, and has a contact made with his family to encourage them to begin regular attendance in the ATU Family Program.

The PBC period is seen as the beginning of the whole treatment process which involves both inpatient and outpatient care of at least 14 months duration. Two of the primary goals during PBC are (1) to give the veteran a gradual introduction to the ATU program so he is familiar with the staff and program to the point he has no fears about being admitted to the inpatient unit, and (2) to attempt a "social detox" approach to the problem drinker so that he either has little to no withdrawal, or is sober when he is admitted to inpatient treatment. The attrition rate from the PBC program is fairly high, at about 35 percent. Of the 65 percent who do attend PBC and are admitted to the inpatient program, more than one-half are sober or show little or no physical withdrawal on admission. With rare exception, all full-term inpatient admissions are ambulatory and begin attending treatment activities the day after admission. These data tend to support the effectiveness of the "social detox" approach used during the PBC period.

INPATIENT PROGRAM

The ATU inpatient unit has 24 beds. Generally, 21 beds are used for new admissions; 2 beds are reserved for ATU outpatients who experience a "slip" and require a short-term inpatient detoxification; and 1 bed is used as a "floater" for former patients in crisis but not drinking, for short-term detox, or for a full-term patient depending on current needs. The inpatient and aftercare components of the program are described in the green folder entitled "Information About the Alcoholism Treatment Unit." The "Daily Schedule" lists the variety of treatment activities which the inpatients attend. The lecture schedule lists the topics covered in the daily lectures. Some lectures are given by ATU staff, some by staff from other parts of the hospital, and some by other professionals participating under an informal staff-sharing relationship between ATU, a state hospital program, and a private hospital for alcoholics. Daily therapy groups with 8 patients and 2 co-therapists each are oriented toward working on "here and now" problems, exploration of feelings together with learning more effective ways of expressing emotions and becoming more effective in developing and maintaining significant interpersonal relationships. An understanding of the basic structure and purpose of Alcoholics Anonymous is developed through three weekly discussion sessions on the 12 Steps and Traditions of AA. Attendance at an open speaker meeting and a closed men's group of AA also helps prepare the individual for making a knowledgeable choice about whether or not AA will become a useful part of his continued treatment after release from the inpatient program. Inpatients receive regular individual counseling and meet in small groups for weekly counseling with nurse-counselor teams.

AFTERCARE (OUTPATIENT) PROGRAM

During the final week of the inpatient program, each patient spends some time with his counselor developing an "Outpatient Contract." From the attached sample Contract, it is apparent that there are a number of options available in the aftercare program. Most patients receive the "standard" agreement which involves continuing use of Antabuse, twice weekly AA attendance, and regular outpatient individual counseling. Other activities are added depending on the individual's needs and problems. Counseling is usually on a weekly basis at first with time between appointments gradually lengthened as the patient does well. By the end of one year on aftercare, appointments may be down to once per month.

FAMILY PROGRAM

For all ATU patients, there is a strong attempt made to involve their significant others in the ATU family program. The family program is open to anyone with whom the patient has a close relationship whether it be a spouse, parents, siblings, girlfriends, friends, etc. If a family member accompanies the patient when he applies for the program, the family is seen by an ATU staff member and has the program explained by them. If the family does not accompany the patient on his first visit, the ATU staff attempts to make an appointment for an interview. At the very least, there is phone contact, and an explanatory letter and family program schedule is sent to the family (or others).

The main elements of the family program are the two PBC orientation meetings, the twelve week ATU family night program (topic schedule attached), and a weekly Alanon meeting. In addition, selected couples may receive con-joint counseling or couples group therapy. On occasion, wives are counseled individually or are placed in a wives' group.

EFFECTIVENESS OF THE PROGRAM

The ATU maintains an ongoing evaluation as a part of the total program. At three months after release from inpatient, approximately 65 percent of all outpatients are improved, 8 percent are unimproved, and 27 percent have lost contact with ATU. By 12 months on aftercare, those patients out of contact have risen to 47 percent of all patients (including 27 percent who have been totally abstinent for 1 year). The main effort recently has been to improve the ATU outreach and follow-up techniques in order to reduce the number of veterans who drop out of the aftercare program.

STAFF

Full-time ATU staff includes a clinical psychologist as Program Director, a staff Psychologist, two Social Workers (one is half-time), five Alcoholism Counseling Technicians, an Evaluation Coordinator (counselor) and eight Nursing personnel. Psychiatry Service staff physicians provide medical coverage on a rotating basis. They also provide psychiatric consultation as needed.

To provide adequate coverage for both inpatient and outpatient program activities, the ATU staff all work some irregular hours of duty. In addition to 24-hour nursing coverage, other ATU staff are on duty and offering treatment activities from 8:00 a.m. to 9:00 p.m. Monday through Thursday and 8:00 a.m. to 4:30 p.m. on Friday and Saturday. Most Family and outpatient activities are scheduled in the evenings and on Saturday.

The students participating in this program will take an active part in *all* phases of the Alcoholic Treatment Programs under the direct supervision of a clinical psychologist. They will also be required to study the literature on alcoholism, and will participate in the ongoing training program at the V.A.

Supervision and evaluation of the students will be a joint effort of the Department of General and Experimental Psychology at Ball State University and the V.A. staff who are directly involved in training the students.

ADMINISTRATION OF THE TRAINING PROGRAM

The coordinator of this training program will keep a file on each student. In the file will be information on his academic performance and progress, clinical experiences in the V.A. Hospital and pertinent evaluations from his supervisors. After graduation, contact will be maintained with the student regarding placement and follow-up. Detailed questionnaires regarding the adequacy of training and clinical experience will be systematically collected from each graduate who has been in the field working. This feedback will serve as a guide to improve the program and to provide the types of training the graduates feel will better prepare the student to perform at a higher level on the job. The following is a brief description of the procedures and processes of administering and evaluating the training program.

1. Training Manual. A detailed manual describing the aims, purposes and procedures of this training project was prepared. This manual is given to every trainee and to interested graduate students who are not as yet ready for their internship. The manual also includes a detailed description of all the courses offered in psychology. It specifies required and elective courses, both in the major and minor areas of specialization.

The goals of this training program are outlined. The skills required for an effective internship are detailed. This is specifically designed to guide the potential trainee in planning her/his curriculum and course of study.

The Alcohol Rehabilitation Treatment program at the Veterans Administration Hospital in Indianapolis is completely described. The names of the clinical supervisors are included. The contractual agreement between the V.A. Hospital and Ball State University is part of this Student Manual. This document is designed to give the trainee and potential trainee a working knowledge of and the guidelines for the whole training project.

2. Application Form. A detailed application form was developed for use in this project. This form seeks demographic, educational, experiential, and motivational information from the applicants. The applicant is also required to take the MMPI. The data from this form will be used to select trainees and it will also be used to generate a personality profile of the Alcohol Rehabilitation Specialist.

3. Bibliography. A comprehensive bibliography "Addiction References 1971-1976" was compiled. This was given to every interested student and all trainees.

4. Trainee Supervision and Evaluation Form. Every trainee is evaluated by her/his clinical supervisors at the end of his/her internship period. The trainee was evaluated in the following areas: professional conduct, work habits, personal adjustment, work performance and clinical skills.

   Supervision of the trainee is a joint effort of the clinical staff at the V.A. Hospital and the academic staff at the University. Weekly evaluation sessions are held with the trainee to assess progress and training.

5. Procedures for Trainee Placement and Supervision. Once the trainee has been accepted to this program, a form letter is sent to the Director of the Alcohol Treatment Unit at the V.A. Hospital in Indianapolis, informing him of the trainee. Basic data as to name, age, and dates of internship are included in the letter. The trainee is then instructed to make an appointment with the Alcohol Treatment Unit for the initial interview and orientation.

6. Alcohol Rehabilitation Internship Evaluation Form. An evaluation form was developed and a pilot study instituted to determine its effectiveness in evaluating the program from the trainees' point of view. This evaluation procedure is designed to gain systematic, orderly, and meaningful information from all trainees who have completed this training program. The general areas of academic training, clinical

skills acquired, basic information about alcohol and substance abuse and the clinical internship are assessed. An effort is also made to evaluate the working skills of the graduate and the impact this training grant has in the area of alcohol rehabilitation. Special effort is made to determine the type of jobs, responsibilities and duties the graduate students are involved in.

7. Personality Profile of the Alcohol Rehabilitation Specialist. Systematic information was gathered from every trainee regarding their motivation, education, clinical training and personality dynamics. It is hoped that all of this information will be used to generate a profile describing the alcohol rehabilitation specialist.

8. Impact. In determining the impact of this training grant the following variables will be examined:

   a. The number of graduate students who have completed this internship and are now actively working in the area of alcohol rehabilitation and are delivering services.

   b. The populations served.

   c. The number and types of programs developed by these graduates in terms of education, prevention and/or remediation.

   d. The types of jobs held by the graduates will also be examined.

THE PREVALENCE OF ALCOHOLISM AMONG GENERAL MEDICAL PATIENTS IN LARGE *MUNICIPAL HOSPITALS*

Roger Mazze, Ph.D., Theodore Feldman, B.A., and Edward Julie, B.S.

Albert Einstein College of Medicine

The diagnosis and treatment of alcoholism plays a major role in the dynamics of health care delivery in every general hospital in the United States. Yet, there exists a significant disparity between the number of alcoholics treated in a general hospital and the actual number of patients who are alcoholic. Estimates of the actual number of alcholics range from 10 to 45% of the adult medical population based on various sources of information and different criteria of what constitutes alcoholism.

This study, applying three different techniques, determined the prevalence of alcoholism among general medical patients at four municipal hospital centers which were selected from the eighteen facilities that comprise the New York City public hospital centers. Using physician discharge notes, a point prevalence study and an in-depth interview protocol, we were able to provide an accurate count of the number of alcoholic patients seen in these facilities on general medical services.

The first element of the study consisted of a review of all general medical patient's records (approximately 15,000) at the four hospitals in the year 1976 to identify those patient's charts coded 303 (alcoholism according to the International Classification of Diseases). This coding is based on the physician's discharge notes constituting a discharge diagnosis of episodic excessive, habitual excessive or alcohol addiction. The second element of the study involved a one day point prevalence procedure in which all of the general medical patients on randomly selected hospital wards (male and female) charts were reviewed by a team of investigators and matched against the following criteria: (a) evidence of alcohol withdrawal syndrome, (b) drinking patterns, (c) social dysfunctional behavior in family and employment, (d) psychological factors of dependence, craving and disorientation and (e) presence of alcohol associated disorders (liver disease, pancreatitis, gastritis, etc.). The presence of any criteria in categories a, b and e, or the presence of a medical problem with evidence of alcohol ingestion provided basis for the identification of the alcoholic patient. In total, 125 patient charts were reviewed in this manner.

An in-depth interview with the chief of the medical service investigated, constituted the third element of the study. Although subjective in nature, this data provided information about the perceptions of the primary medical house officer and may effect both diagnoses and treatment of alcoholism. Each chief was asked to estimate the percentage of alcoholic patients on his service for the previous year and to characterize in general terms the profile of the modal alcoholic patient in terms of age, sex, employment, life style and prognosis for successful treatment.

FINDINGS

The results of the three part study show, overall, that the discharge diagnoses identification was significantly lower than the point prevalence study which in turn was lower than the chief's estimate. At hospital I, the ICD code indicated .05% in contrast to 47% and 50% from the point prevalence and chief's study respectively. At hospital II, the ICD produced 14%, the point prevalence revealed 16% and the chief estimated 50%. This same pattern existed at hospital III, where 14% were identified alcoholic by the ICD method, 50% by the point prevalence procedure and 60% by chief's estimate. Finally, at hospital IV both ICD and point prevalence identified approximately 34% of the adult medical patients as alcoholic, while the chief estimated 75% were alcoholic.

If the ICD 303 figures are pooled and averaged, a mean score of 13% is reached. The mean for the point prevalence studies is 37% and the mean for chiefs is 59%. It is therefore clear that both within individual hospitals and among hospitals, a serious discrepancy exists between the hospital-based ICD method, the point prevalence study and the chiefs estimate technique.

The uniformly low number of alcoholics reported by discharge diagnosis data represents an inability or unwillingness to identify both the alcoholic and the problem drinker. In two of the hospitals, the ICD percentage was found to be significantly lower than the percent of alcoholics (10-12%) found in a "healthy population." Subsequent review of these charts revealed that, in general, unless a patient presented with overt manifestations of alcoholism or alcohol abuse (intoxification or withdrawal) or voluntarily admitted to being alcoholic, physicians were reluctant to screen for and eventually identify the alcoholic patient. This reluctance may be attributable to such factors as a bias toward the alcoholic patients, poor skills and knowledge about alcoholism and frustration over poor prognosis.

The point prevalence study, which, through the application of the one criteria-set, by one research team, at four different hospitals, enabled us to compare on the same dimensions the extent of alcoholism among general medical patients. While the range of prevalence suggests some difference between facilities, it should be noted that at hospital II with 16% alcoholism, an active in-patient detoxification unit existed which was not the situation at the other facilities that were examined. More important, however, was that the records review procedure which enabled us to identify the number of alcoholic patients above provided us with insight into the diagnostic and treatment procedures at these hospitals. As the data indicated, the majority of these patients were not formally identified as alcoholic by the house staff. Fewer than 20% had received any type of psychosocial consultation. The majority had incomplete social and personal histories which generally omitted questions concerning alcohol use. When such questions were asked they were usually limited to the amount of alcohol consumption.

Finally, the perceptions of the chiefs of service provide insight into the enormity of the problem. In-depth interviews revealed that generally no formal education about alcoholism occurred on their service, and that house staff were expected to have, as medical students, already learned about alcoholism. Additionally, it was found that while the chiefs wanted to do something about the problem a general ambivalence existed. This was expressed in terms of poor prognosis, limited resources and "more deserving patients need our attention."

This study, although limited in scope, indicates that the identification and consequently the treatment of the alcoholic has been both inadequate and ineffective. Suggested, is the need for a consistent set of criteria to identify the alcoholic and a consistent standard for his treatment to assure proper management. Only in this way can this leading cause of morbidity and mortality be properly assessed and provided with sufficient resources for its prevention, detection and treatment.

STATISTICAL DATA RELEVANT TO ALCOHOLISM INCIDENCE IN ALASKA

Cheryl Mann

Salvation Army, Comprehensive Alcoholism Services

In this paper, I intend primarily to discuss some of the factors that cause unique problems in working with the alcoholic population in Alaska. While accepting that there are many areas of commonality we share with most service providers in alcoholism treatment programs, I intend to point out some dissimilarities encountered in Alaska with specific emphasis placed on the problems faced by the Alaskan native population.

Alcoholism has been identified by both the non-native and Alaskan native populations as the number one public health problem in the state. In the Rowan Group Report Study conducted in 1972, 148 people were interviewed in nine villages and the data showed overwhelmingly that rural Alaskans consider alcohol to be not only the major social problem in their communities, but also in the entire state. Also, studies have indicated (Kelso, 1975) that the conservative estimate of the economic cost of alcohol abuse and alcoholism in Alaska was $131.2 million in 1975 and the cost rises steadily each year.

While the mere fact of consumption of alcohol, even in large quantities is not necessarily presumptive evidence of alcoholism, there is a definite correlation between increasing consumption or al-related problems. There is no question that Alaska can be considered a high consuming population. The average per capita consumption of alcohol exceeds the national average by 57%. What is more alarming is that in the eighteen years following statehood there has been an 80% increase in the average annual per capita consumption of alcohol. The corresponding increase for the rest of the country during the same period was 46%, just about half as much.

One interesting fact has emerged regarding the per capita consumption of absolute alcohol. While the general trend in the rest of our country, and most other countries has been a gradual but marked shift away from distilled spirits toward consumption of the so called lighter alcoholic beverages, i.e., beer and wine, the continued and definite preference in Alaska is towards consumption of distilled liquor rather than wine or beer. This also contributes to the image of the hard drinking Alaskan.

One of the major factors involved in this increasing consumption is, of course, availability. Compared to the rest of the country, Alaska has nearly twice as many retail liquor outlets per 1,000 population. Added to this is the fact that we have lengthy hours of sale. Many bars do not close until 5:00 a.m. to re-open again at 8:00 a.m. Furthermore, the tax structure is very favorable for alcohol purchase. In Alaska, there has not been an increase of the excise tax on beverage alcohol since 1961. Thus, while the average per capita income has increased steadily and the cost of other consumer goods has kept pace with the rising income, beverage alcohol prices have remained fairly stable and in relation to other rising prices the cost of alcohol has decreased, therefore making it all the more accessible and resulting in increased excessive consumption.

Currently, it is estimated (NCA, 1976) that there are more than 30,000 alcoholics in Alaska. This is about 15% of the drinking age population. While only about 17% of the states population are Alaskan Native, this group represented nearly 2/3 of the admissions to Alcohol Treatment programs in 1975. Thus, Alaskan Natives are highly impacted by problems of alcohol and alcoholism.

According to data gathered in 1975, (Kelso) the alcoholism mortality rate in Alaska had jumped to 418% greater than the national average. These are deaths due directly to alcohol. However, the rate of violent deaths, homicides and suicides, related to alcoholism is also extremely high. In this category Alaskan Natives, who constitute only 17% of the total state population, were involved in nearly half of both the homicides and suicides in the state. This group also represented more than 50% of all deaths directly related to alcoholism. Thus, the alcoholism mortality rate is much higher for natives than non-natives.

One of the major legal problems related to alcohol prior to the passage of the Uniform Alcoholism and Intoxication Treatment Act in 1972 which decriminalized public drunkenness and set up mechanisms for developing treatment, education and rehabilitation programs was the public inebriate.

In 1968, while Alaskan Natives constituted only 4% of the total population in Anchorage, they represented 66% of the drunk in public arrests. This equals 372 arrests per 1,000 Alaskan Natives compared to 8.4% per 1,000 in the non-native population. Thus, the Alaskan Native population had an arrest rate 49 times greater than the non-native population. This resulted in a very high recidivism rate with some individuals being arrested numerous times. There seem to be many cultural factors at work here, some of which include:

1. Inadequate education and vocational preparation by Alaskan Natives to enable them to compete with non-natives for good jobs in the city.

2. There are few opportunities in Anchorage for native people of the same ethnic background to meet socially outside the bars.

3. Many individuals coming from small, closely knit village communities upon arriving in the metropolitan areas suffer from feelings of loneliness, homesickness, isolation and a great deal of frustration.

4. Due to having unskilled low-paying jobs, or no jobs at all, many people in this group suffer economic insecurity, resulting in feelings of severe anxiety.

5. Therefore, with any or all of these or other factors involved, the Alaskan Natives will often go to bars seeking companionship or release. Then, they would often get drunk, be arrested and be caught up again in the whole self-perpetuating cycle.

Although a large percentage of violent crimes in Alaska are considered to be alcohol related, as high as 64% in the incidence of homicides, the majority of the alcohol related crimes are of the misdemeanor type that constitute about 40% of all the arrests in the State. In a sentencing data study conducted in 1974 by the Division of Corrections, it was reported that alcohol offenses accounted for almost half of the sentences received by Natives. However, in the case of drinking driving arrests, it is indicated that these are more predominant in the non-native population.

In Alaska, alcohol was the major factor involved in traffic fatalities for 1976. The figure for the entire state indicates (NCA, 1975) that 75% of all traffic fatalities were alcohol related, the rate climbing to 81% in Anchorage and an astronomical figure of nearly 100% in Fairbanks.

I think it is very important to mention when discussing Alaskan Natives that this is not a culturally homogeneous group, but a very rich mixture of many different cultural and ethnic backgrounds. In fact, there are more than ten different cultural groups with over sixteen different language groups to accommodate them. However, the major groupings are Eskimo, Aleuts, and American Indians. It is also important to mention that geographically, the state is divided into twelve Native Regional Corporations that deal with many social, health, education, and vocational problems of Alaskan Natives.

Many of the problems facing the Alaskan Native population are cultural in nature. The last 30 to 40 years have seen the change in lifestyle of thrusting people who have lived in small communities virtually unchanged for centuries into the jet age. This abrupt movement has caused many changes in their traditional cultures. Recently, during the past 30 years, there has been a high and steadily increasing amount of family disruption and emotional upset. Families which have been the foundation for their lifestyles have been split up, often in the following ways:

1. Increased incidence of illness and diseases, especially tuberculosis, causing long hospitalizations.

2. Enforced separations when one or more of the family's children had to leave home to be sent to a boarding school or to the city because the village lacked educational facilities. This also resulted in a very high incidence of attempted suicides among the youth. Many of these young people are sent to high schools in Anchorage where their placement is often in non-native foster homes and they frequently develop a high percentage of emotional problems. This in turn can lead to the use of alcohol and drugs and in 1970 resulted in a *reported* suicide attempt rate of 9% (Aleut League Report, 1975) of the Alaskan Native boarding students in Anchorage.

3. Increasing and alarming incidence of alcoholism and all the social, psychological and health related problems that accompany it.

4. Increasing numbers of individuals being incarcerated and spending time in jail especially for offenses related to alcohol.

In conclusion, I want to reemphasize the major factors that make the alcoholism problem unique in Alaska:

1. Extremely high consumption of beverage alcohol per capita and the steadily increasing consumption over the last 20 years.

2. Ready availability and relative low cost of beverage alcohol.

3. High incidence of alcoholism in state compared to the rest of the nation.

4. Alcohol mortality rate of 418% greater than the national average.

5. The percentage of traffic fatalities is much greater than the national average.

6. The very special cultural problems faced by a heterogeneous Alaskan Native population who while constituting only a small portion of the state's population bear the brunt of alcohol related problems especially in alcohol related arrests and mortality.

REFERENCES

Geffin, B. 1975. Dimensions of Alcoholism and Personality Within a Population of Alaskan Inpatient Alcoholics. Anchorage, Alaska.

Kelso, D. 1975. Working Papers: Descriptive Analysis of the Impact of Alcoholism and Alcohol Abuse in Alaska. Volume IV. Social Systems Indicators of Alcoholism and Alcohol Abuse in Alaska, 1975. Volume V. Executive Summary: Descriptive Analysis of the Impact of Alcoholism and Alcohol Abuse in Alaska, 1975.

N.I.A.A.A., 1975. Alaska Fact Finder on Alcohol Abuse and Alcoholism.

Peterson, W.J. 1975. A Social Systems Analysis of Alaskan Alcoholism.

State of Alaska, 1974. Division of Corrections, Sentencing Data Report.

State of Alaska, Department of Health and Social Services, Division of Family and Children's Services, 1973. Synopsis: Alaska State Plan for the Reduction of Alcoholism and Alcohol Abuse. Juneau 1973.

HEALTH EDUCATION NEEDS OF ALCOHOLISM SERVICES

Marian V. Hamburg, Ed.D., Sanford A. Weinstein, Ed.D., and Harold N. Weiner, A.B.

New York University

When New York University's Department of Health Education was funded in 1975 by the National Institute on Alcohol Abuse and Alcoholism to develop an Alcohol Studies specialization as part of our Community Health Education masters program, we had no certain knowledge about the job market for the people we would be preparing. We believed that there was an immense prevention job to be done in the alcoholism field, and we had a pretty good idea that trained health educators were a relatively scarce resource in that field. Beyond that, we felt that our Project had a sound theoretical base, namely that people prepared at the graduate level in community health education and with knowledge of alcohol problems would have a good deal to offer the alcoholism field.

Our first inkling of the realities of the job market came when we graduated the first six persons in June 1977 from our Alcohol Studies program. All but one found a spot in the alcoholism field, although the majority assumed responsibilities that were more administrative than educational and more concerned with treatment than prevention. This experience with our first crop of graduates, as well as our curiosity about the usefulness of our evolving curriculum to the real needs of alcoholism agencies, moved us to conduct a survey in the New York City area to get a better sense of that market for persons with graduate preparation combining health education and alcohol studies.

We developed and sent a 55-item questionnaire to 72 administrators and program directors in 66 organizations concerned with treating and/or preventing alcoholism. They were asked to rate the importance of a variety of health education tasks and areas of knowledge about alcohol problems to the purposes of their programs. In addition, they were asked to indicate whether a salaried position responsible for health education tasks existed in their organizations and the degree to which they supported the existence or creation of such a position. They were also asked how they felt about hiring a person trained in community health education but not in alcohol studies, since we wanted to find out how important the knowledge of alcohol problems is to the employability of community health educators in the organizations polled.

Responses were solicited from representatives of organizations listed in the "Directory of Alcoholism Resources and Services for New York City" and the "Directory of Social and Health Agencies of New York City." Additional respondents were selected by the faculty and staff of our Department of Health Education. While the vast majority (85%) represent New York City organizations, the remainder represent organizations from surrounding communities, notably Long Island.

The initial mailing of the questionnaire and cover letter brought in a 50 percent response. A second mailing of the questionnaire and cover letter was made to those who did not respond to the first mailing. This was then followed by a postcard reminder and phone calls until an 80 percent (N=55) response was achieved.

The questionnaire contained 28 items representing community health education tasks and 20 items representing knowledge about alcohol problems. Respondents were asked to rate the importance of these items to the purposes of their programs on a five-point scale ranging from "totally irrelevant" to "indispensable." The questions about the degree of support for retaining or acquiring health education positions, and for hiring persons trained in health education but not in alcohol problems, were also put on a five-point scale, this one ranging from "no support" at the bottom to "very strong support" at the top of the scale.

The health education tasks and alcohol knowledge items were examined as summated scales and were found to have a high reliability. This indicates that respondents tended to rate the items within each scale in a consistent fashion. The two scales were then analyzed on an item by item basis. In order to establish relative importance of the items, they were rank-ordered on the basis of the average ratings for each.

Upon inspection of the responses, it appeared they had clustered in their rankings around a number of themes and could be categorized by those themes. The most highly rated health education tasks, according to the respondents, concern the conceptual aspects of program design and development, such as planning and evaluation. We categorized the next most highly rated group of tasks as mobilizing resources and executing educational programs. This group included such tasks as developing educational policies, staff training, and public speaking. On the five-point scale, the average ratings for these two top-ranked categories of health education tasks approach the overall rating of "important" to the purposes of the responding alcoholism agencies.

The third-ranked category of health education tasks combines activities that play a supportive role in the delivery of educational services. This category includes such tasks as writing reports, organizing committees, and conducting research. Finally, those tasks considered least important by the responding alcoholism agencies fall into a category we have called fiscal and program accountability. Tasks in this category include budgeting, grantsmanship, and writing professional articles. These third and fourth-ranked categories earned ratings on the five-point scale predominantly in the range of "moderately important."

It should be noted that as one moves downward through the rankings, there is a trend toward increasing variance in the ratings as evidenced by standard deviations for the items. This suggests that the uppermost ranks are not only ascribed greater importance but also have greater levels of agreement about their importance.

One perspective from which this analysis of the data may be viewed is that of defining the health educator's role in the alcoholism agency in terms of the reported educational needs of the agencies. Based on these findings, the greatest educational need seems to be for health education personnel who will assume an authoritative and directive role in program design, development, execution, and evaluation. Of lesser importance, apparently, is a range of activities which we have characterized as "support" and "accountability" tasks.

The rankings of the alcohol knowledge items were found to cluster around four themes. The first and most highly rated category is composed of concerns about alcoholic persons and their treatment. The average rating on the five-point scale for the items in this category indicates that respondents consider knowledge about them almost "indispensable" to persons being considered for community health education positions. The second-ranked category of alcohol knowledge items embraces concerns about alcoholism as an illness and about methods of intervention. Items in this group were rated above "important" on the five-point scale. The third-ranked category of items has to do with prevention in special populations, notably the young, and was rated just below "important" but high above "moderately important." The category of items with the lowest rank, the themes of which concern such things as the social, economic, and political aspects of alcohol problems, received a rating just above "moderately important" but considerably below "important."

It should be noted here, too, that the increase in standard deviations in the lower ranking areas of knowledge suggests less agreement about the importance of the items contained in them and is similar to the trend among the health education tasks mentioned earlier.

In answer to the question about hiring a community health educator who has no specialized knowledge about alcohol problems, most respondents were less than enthusiastic. The mean rating of the response to this item fell between "little" and "moderate" support. It is also worth noting that the majority of the respondents indicated having a position in their organizations responsible for community health education tasks, although we do not know whether the persons in those positions were trained in health education. The respondents very strongly support continuation of that position and strongly support creation of one or more additional positions. Those indicating the absence of such a position in their organizations also expressed strong support for the creation of a position with community health education responsibilities.

Based on the data gathered, in summary, the following implications seem reasonable:

1. Many of the community health education tasks that define the role of the health educator in health or education agencies generally are also important to the needs of alcoholism agencies. Those tasks considered of greatest importance by alcoholism agencies, namely program design, development, execution, and evaluation, are also highly valued by the overall health field.

2. However, generalized community health education training by itself appears currently to be insufficient background for employment in the alcoholism field. This training must also be supplemented by a number of areas of special knowledge about alcohol problems, especially knowledge about alcoholic persons, about the nature of their illness and its treatment.

3. Finally, the expressed health education needs of the alcoholism agencies in the New York City area and the experience to date of our Alcohol Studies graduates are evidence that there is a job market for people who combine health education skills with knowledge about alcohol problems. And although jobs with education and prevention titles and responsibilities continue to be a relatively scarce commodity in the alcoholism job market, I am optimistic that the forces of education and prevention will gain a greater foothold there, as they must inevitably expand in the mainstream of health services.

ALCOHOL ABUSE IN A METHADONE MAINTENANCE TREATMENT PROGRAM

Joyce B. Goodale, M.S.W., Jayne Burton, B.A., Gregory Carr, B.A., Patricia LePage, and Orlando P. Orfitelli, M.D.

INTRODUCTION

This paper is the completion of a study undertaken at the Hartford Dispensary Methadone Maintenance Treatment Program in 1976-1977. The original purpose of the study was to determine whether the actual extent of alcohol abuse among our client population was as high or higher than subjectively estimated by clinical staff. In addition, we also hoped to identify patterns or trends in order to alter treatment approaches, referrals and techniques with an eye toward prevention. The study was conducted in five phases, three of which were completed, written and presented at the 1977 National Drug Abuse Conference. The five phases are:

Phase One

The development and completion of a questionnaire to determine clinical staffs' (supervisors, counselors, nurses and control clerks) impressions and attitudes toward alcohol abuse.

Phase Two

A survey of the eight counselor/nurse teams to identify clients on their caseload considered to be alcohol abusers or problem drinkers. The definition of alcohol abusers/ problem drinkers included: 1) those having sought or received treatment for alcoholism and/or related problems, 2) admission of alcohol abuse, 3) obvious frequent intoxication, 4) those having documented problems due to alcohol involving their social, economic or physical well-being.

Phase Three

The research included unannounced random testing of clients with a breathalyzer and back-up full drug screen urine testing on the same day. This phase was conducted within an eleven week period.

Results of the first three phases were reported in a paper entitled "Alcohol Abuse in a Methadone Maintenance Treatment Program."

THIS PAPER WILL PRESENT PHASE FOUR AND FIVE, THE FINAL TWO PHASES OF THE STUDY

Phase Four

We conducted an anonymous survey of client-stated alcohol use and attitudes. The questions asked actually constituted an abbreviated version of the Michigan Alcoholism Screening Test (Fig. 1).

Although 192 clients participated in the study, a small percentage did not answer each question. Therefore, the numbers do not add to 192.

Of a total of 338 clients in treatment at the time of the survey, approximately 57% (N=192) answered the questionnaire. The results are as follows:

Question #1 -- 106 (55%) stated they believed themselves to be normal drinkers; 79 (41%) answered no.

Question #2 -- 98 (51%) stated friends or relatives believed them to be normal drinkers; 88 (46%) felt they were not considered normal drinkers by friends or relatives.

Question #3 -- 59 (31%) have attended one or more meetings of Alcoholics Anonymous; 130 (68%) said no.

Question #4 -- 33 (17%) admitted to having lost friends due to drinking; 153 (80%) said no.

Question #5 -- Alcohol-related problems at work were stated by 23 (12%). 159 (83%) had no difficulties.

Question #6 -- 18 (9%) admitted to neglecting obligations for two or more days due to drinking; 169 (88%) answered no.

Question # 7 -- 19 (10%) stated yes to experiencing DT's, shaking or hallucinations as a result of drinking; 167 (87%) answered no.

TO: All clients
SUBJECT: Alcohol research

In completing our research study, we are asking for your help with this questionnaire. It is for statistical purposes ONLY in order for us to recognize the extent of the problem of alcohol abuse in our program. The results will NOT be seen or used by counselors or nurses.

Please take one minute to answer the questions.

Thank you.

		YES	NO
1.	Do you feel you are a normal drinker?	( )	( )
2.	Do friends or relatives think you are a normal drinker?	( )	( )
3.	Have you ever attended a meeting of Alcoholics Anonymous (AA)?	( )	( )
4.	Have you ever lost friends or girlfriends/ boyfriends because of drinking?	( )	( )
5.	Have you ever gotten into trouble at work because of drinking?	( )	( )
6.	Have you ever neglected your obligations, your family, or your work for two or more days in a row because you were drinking?	( )	( )
7.	Have you ever had delirium tremens (DT's), severe shaking, heard voices or seen things that weren't there after heavy drinking?	( )	( )
8.	Have you ever gone to anyone for help about your drinking?	( )	( )
9.	Have you ever been in a hospital because of drinking?	( )	( )
10.	Have you ever been arrested for drunk driving or driving after drinking?	( )	( )

Fig. 1.

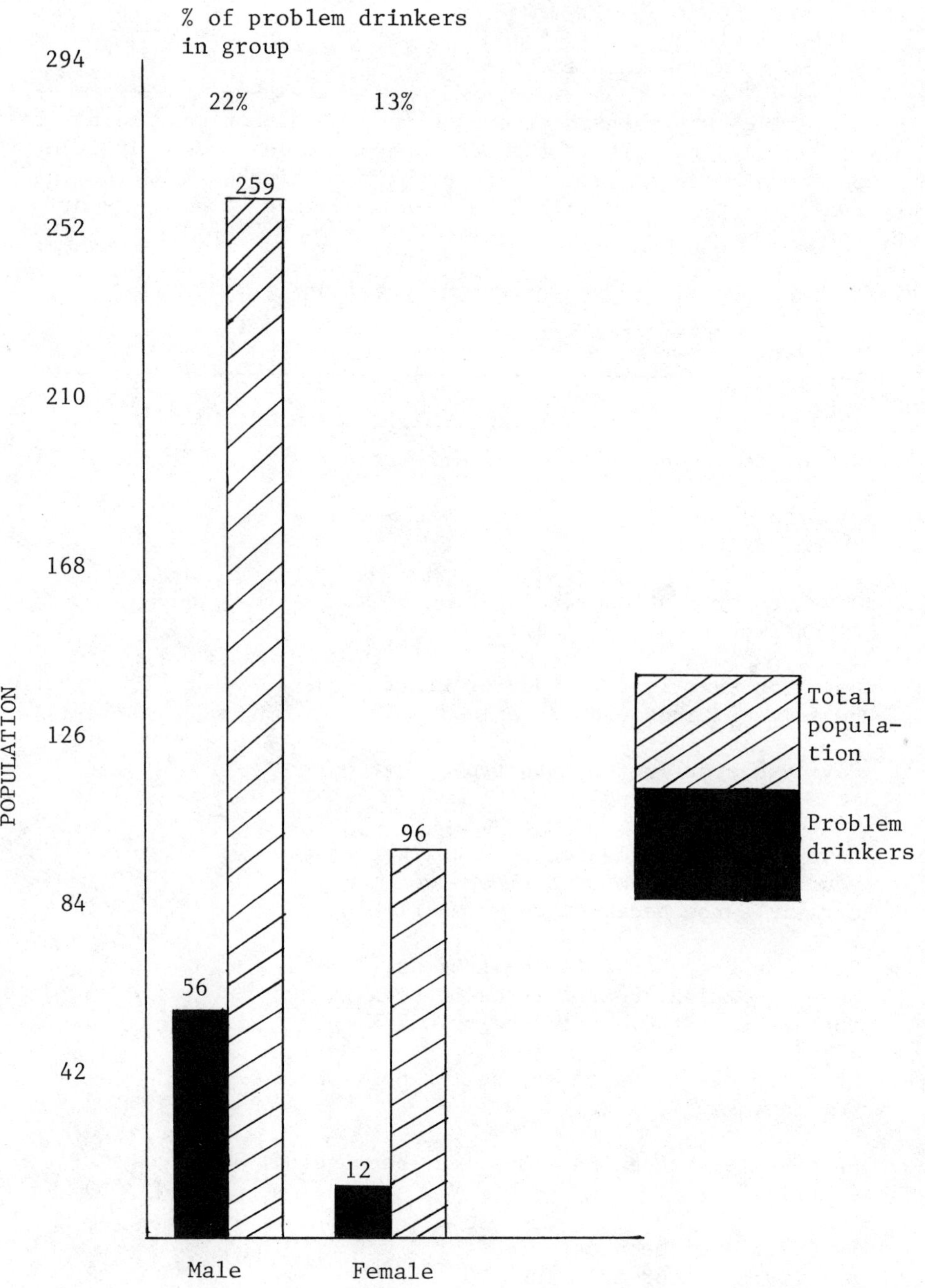
% of problem drinkers
in group
22%
13%
294
252
210
168
126
84
42
POPULATION
259
96
56
12
Male
Female
Total
popula-
tion
Problem
drinkers

Fig. 2. Sex.

Question #8 -- 21 (11%) have sought help for drinking; 166 (87%) have not.

Question #9 -- 18 (10%) have been hospitalized; 168 (88%) have not.

Question #10 -- 19 (10%) stated yes to having been arrested for drinking and driving; 168 (88%) have not.

Phase Five

This was the correlation of various demographic and clinical data of problem drinkers as compared to non-problem drinkers. The analysis of variance was tested using X^2 with $p<.05$. Listed below are the results:

1. As shown in Figure 2, male problem drinkers comprised 22% of the total male clients, while 13% of the total number of female clients were alcohol abusers. The differences were not statistically significant ($X^2=3.75$, d.f.=1).

2. Program status--Figure 3 shows of all clients in the build-up phase (first ninety days), 11% were seen as problem drinkers, compared to 21% in the maintenance phase and 12% in the detoxification state. The differences were not significant ($X^2=3.74$, d.f.=2).

3. Dosage--Dosage levels were separated into six categories. As dosages are adjusted in increments of 5 mg., each category is mutually exclusive. The percentages of problem drinkers in each category (see Figure 4) is:

0-25 mg.	30-40 mg.	45-55 mg.	60-70 mg.	75-80 mg.	85+ mg.
20%	19%	29%	17%	12%	0%

   Differences were not statistically significant ($X^2=$ 7.88, d.f.=5).

4. Ethnicity--Figure 5 indicates that of 109 black clients, 28% were considered problem drinkers, compared to 18% of white clients and 7% of Hispanics. The differences were statistically significant at the .01 level ($X=12.68$, d.f.=2, $p<.01$).

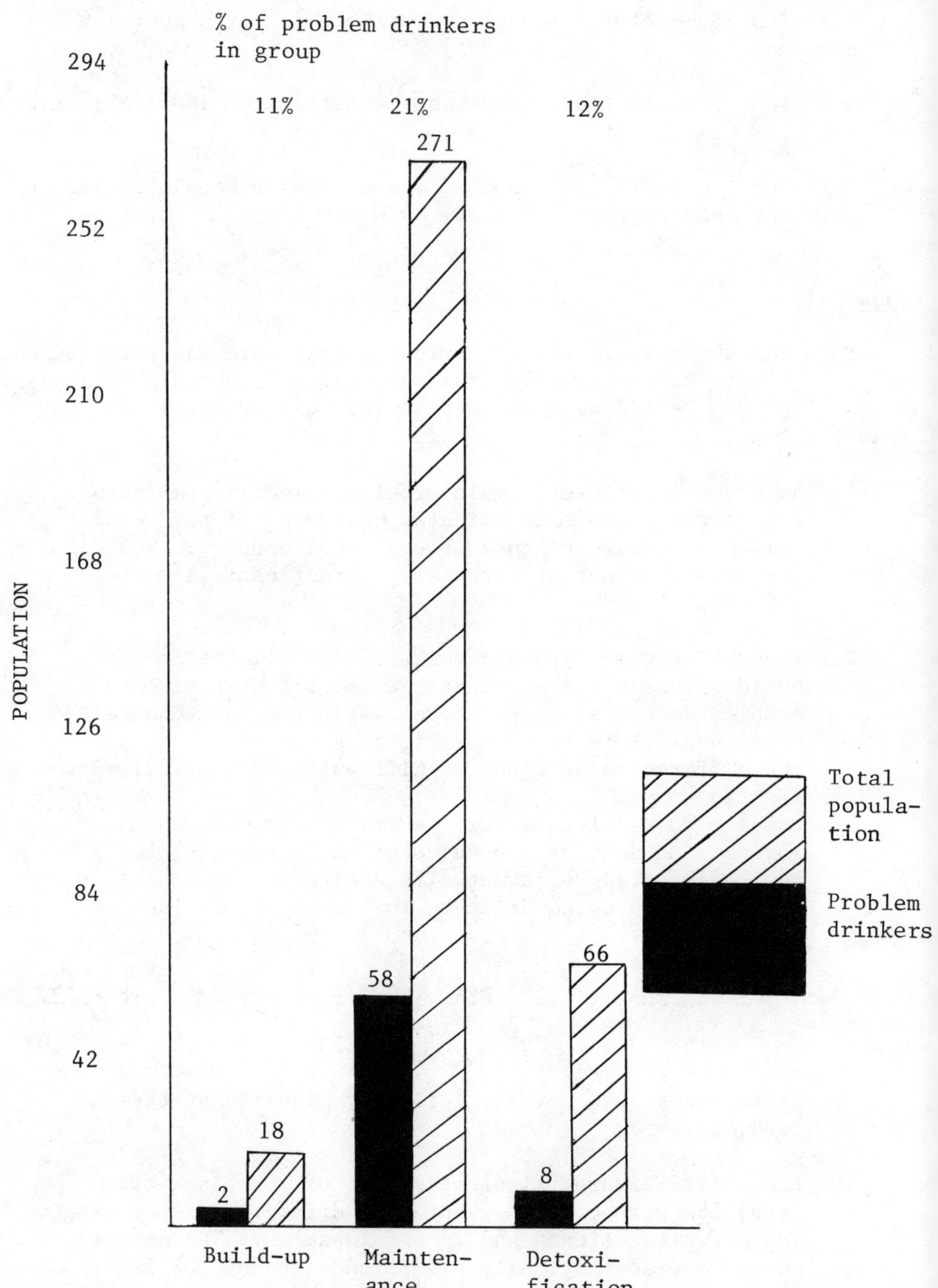

Fig. 3. Program status.

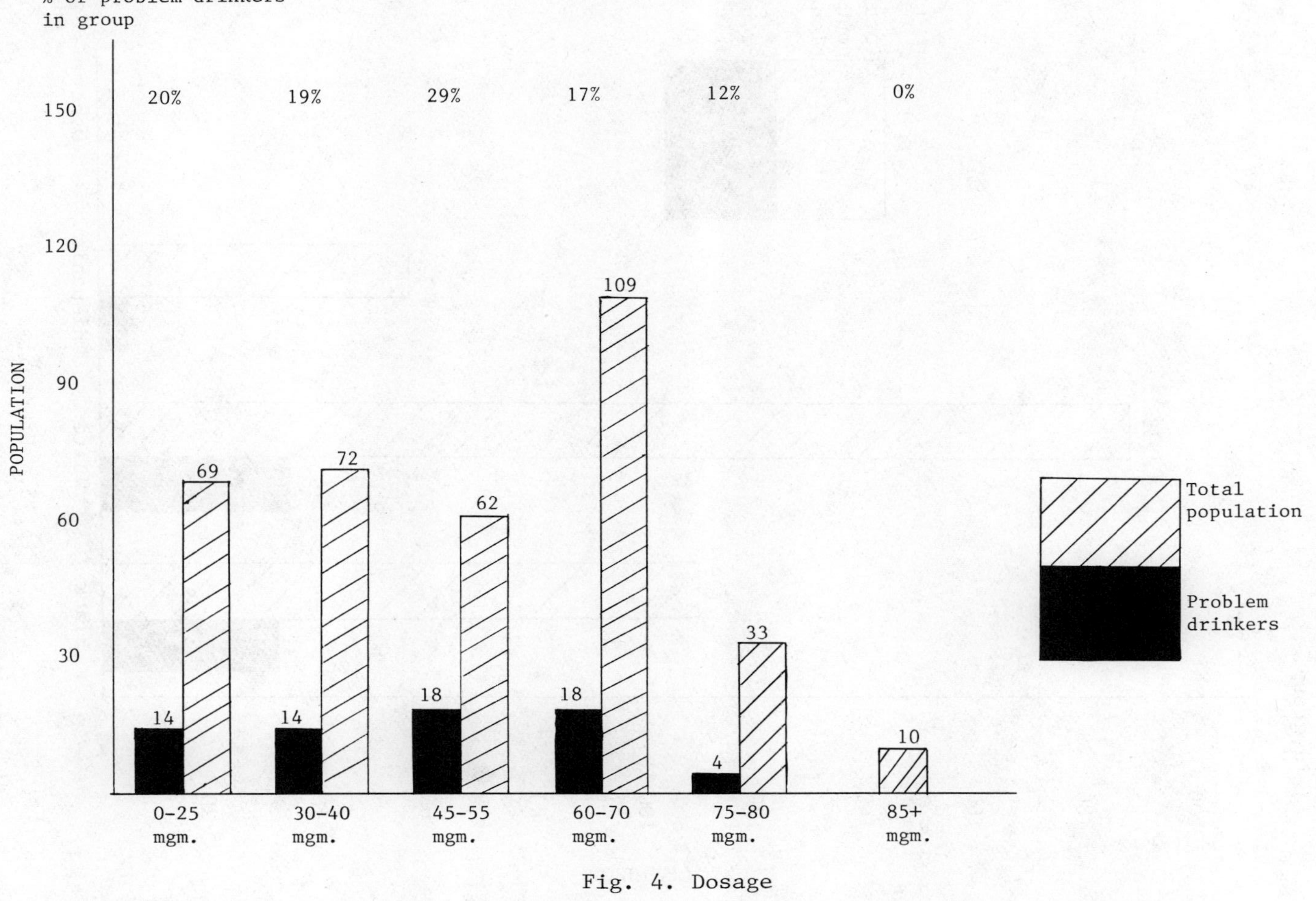
% of problem drinkers
in group
20%
19%
29%
17%
12%
0%
POPULATION
150
120
90
60
30
14
69
14
72
18
62
18
109
4
33
10
0–25 mgm.
30–40 mgm.
45–55 mgm.
60–70 mgm.
75–80 mgm.
85+ mgm.
Total population
Problem drinkers

Fig. 4. Dosage

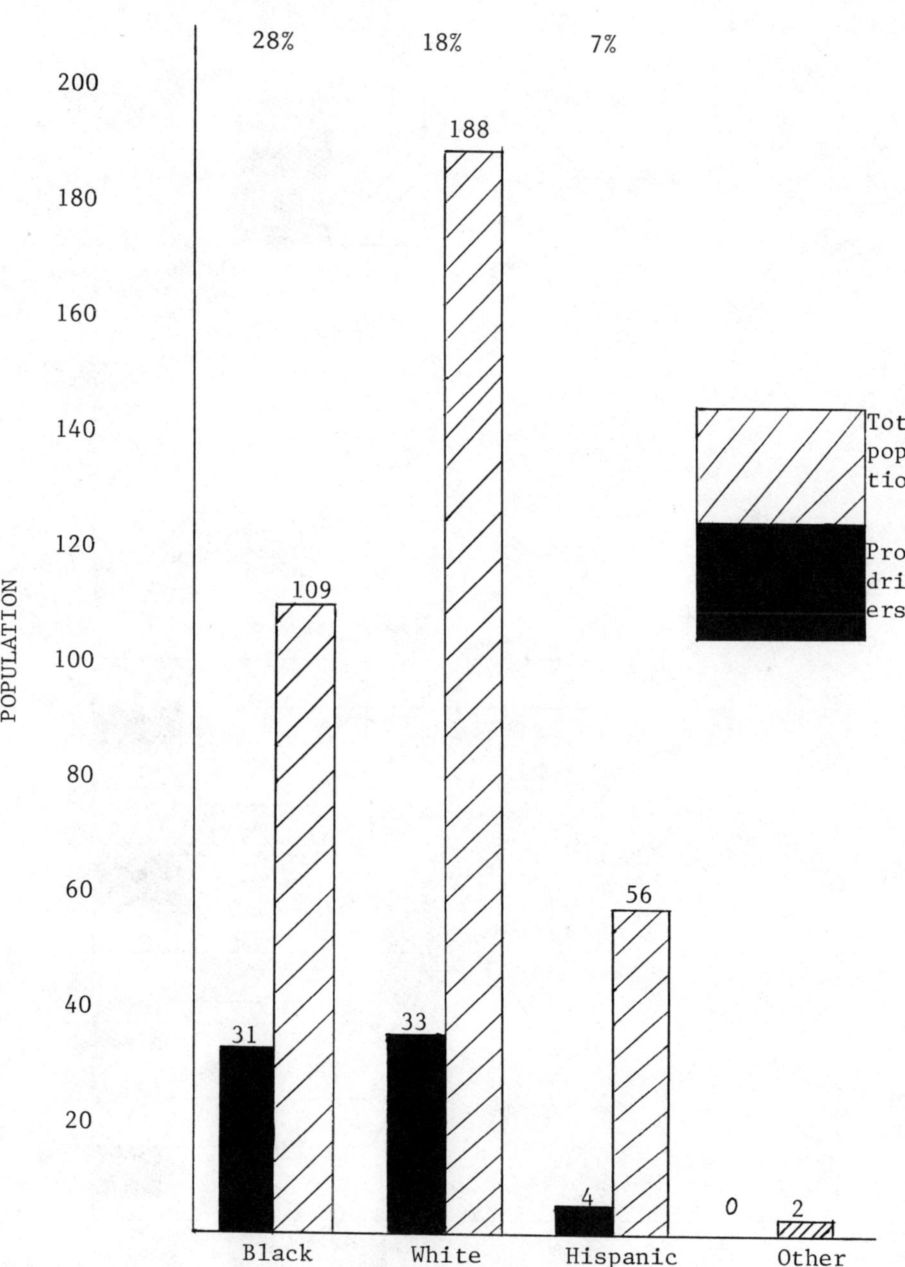

Fig. 5. Ethnicity.

5. Age--Seven age groups were delineated; age 18 (minimum age of eligibility) to 54+.

Age Group

18-23	24-29	30-35	36-41	42-47	48-53	54+
10%	13%	28%	41%	13%	43%	50%

The differences were statistically significant (X^2= 23.34, d.f.=6, p<.01). (See Fig. 6).

6. Months in treatment were separated into six categories. The percentages of problem drinkers in each are as follows:

0-6 mos.	7-13 mos.	14-20 mos.	21-27 mos.	28-34 mos.	35+ mos.
14%	19%	13%	18%	29%	9%

The differences were not statistically significant (X^2= 4.40, d.f.=5). (See Fig. 7).

DISCUSSION

Phase Four

Because the survey was anonymously answered, it is impossible to know how many of the clients who answered the questionnaire were suspected or known problem drinkers. In general, however, the results indicated a higher incidence of alcohol-related problems than we expected. Although the sample constituted 57% of the total population, it was not a random sample. Therefore, we are not able to generalize the results to the total population.

Referring to Phase Three of our study, *objective data vs. subjective data in determining alcohol abuse*: essentially, our objective data does not support greater incidence of alcohol abusers as determined by our subjective observation. However, interestingly enough if we look at the Phase Four questionnaire answered by 55% of our clients (N=192), we see that some (46%) of those clients do *not* consider themselves normal drinkers. Additionally, 41% of this population feel that friends and relatives do not consider them normal drinkers.

Regarding these two questions, however, it should be noted that some clients felt they were not normal drinkers in that they drink less than they considered normal, or not at all. However, we also find that some (31%) of these people have attended AA meetings which may connote some concern they are, in fact, problem drinkers. What

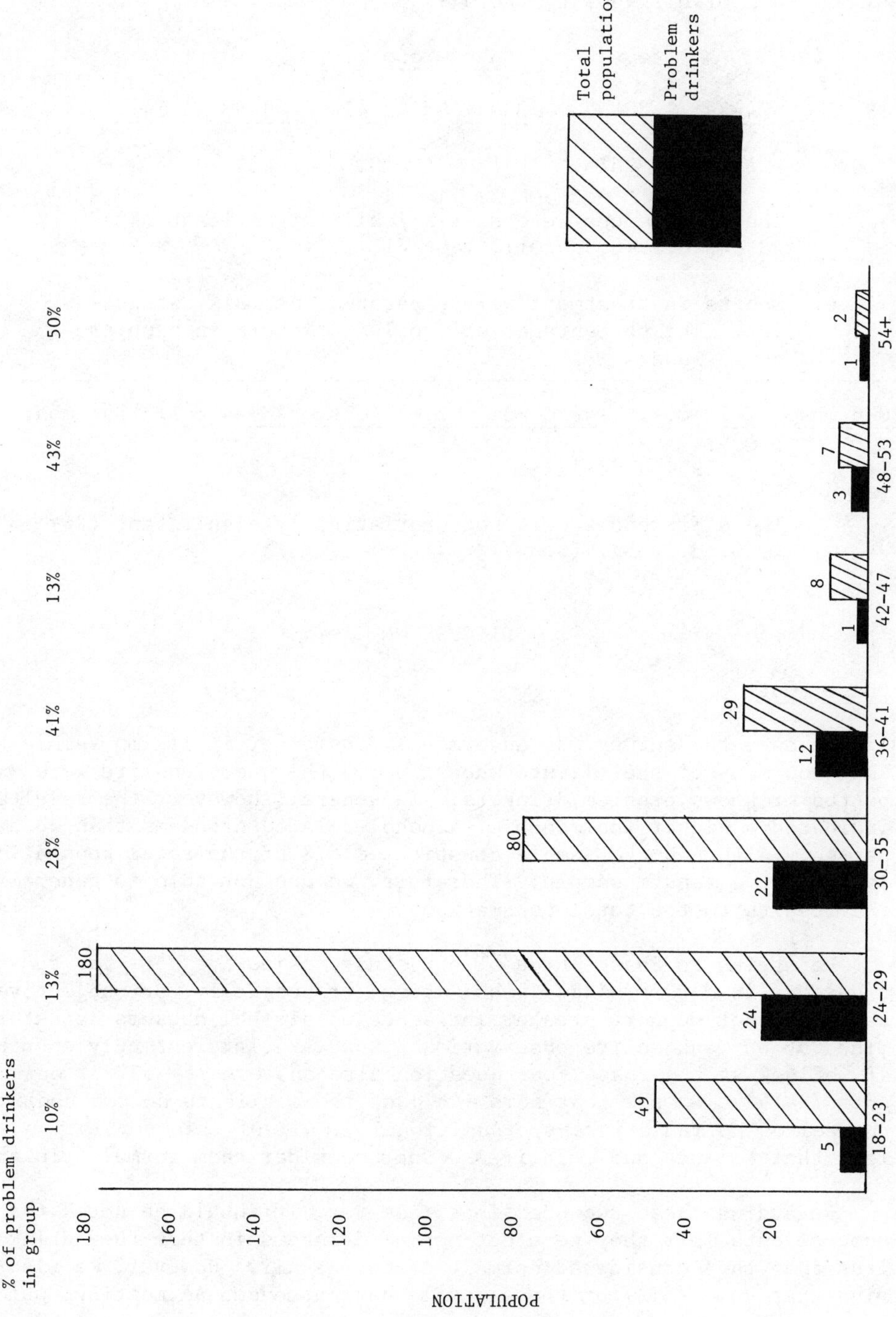

% of problem drinkers in group
10%
13%
28%
41%
13%
43%
50%
POPULATION
180
160
140
120
100
80
60
40
20
5
49
24
180
22
80
12
29
1
8
3
7
1
2
18-23
24-29
30-35
36-41
42-47
48-53
54+
Total population
Problem drinkers

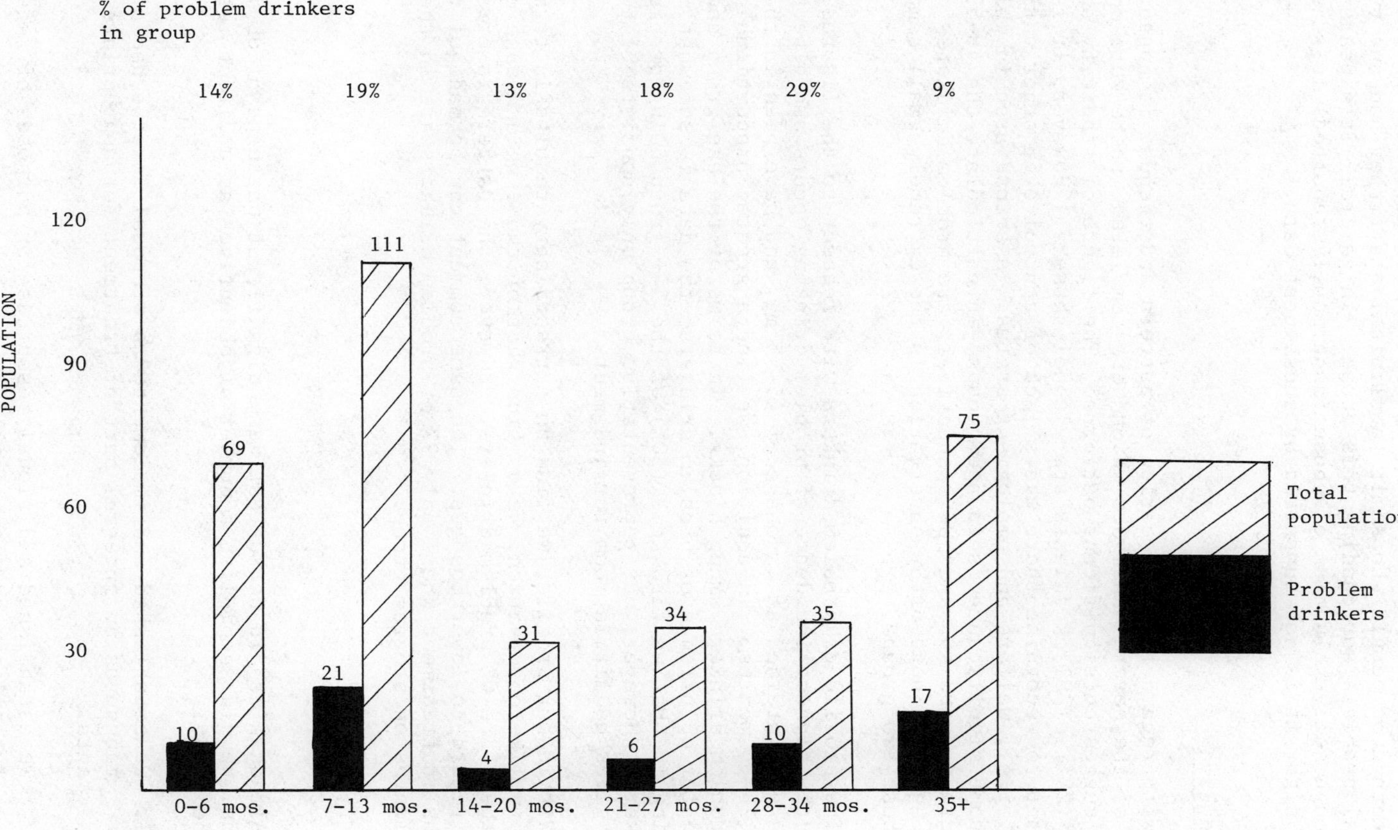

Fig. 7. Age (differences not statistically significant)

we do not find supportive of this sense of alcohol abuse is the low percentage of people who admit to clinical staff they had any type of social problems surrounding this abuse. In general, it appears that although self-evaluation of abuse (anonymously reported) is reasonably high, it is not supported by subsequent data.

Phase Five

With regard to our question of pattern and trends, it appears that of the several analyses which were completed, there were only two that suggested significant patterns. The figure of ethnicity shows some (28%) of black clients are alcohol abusers. However, it is important to recognize there was a lack of control over other variables, i.e., welfare and employment status. A second area in which we see some significance is in age groups, in particular, the 48-53 and 54+ group where 43% and 50% respectively are problem drinkers. Once again, a noteworthy observation is the extremely small sampling of these two groups.

The profile of problem drinkers with respect to the figures shows that the highest incidence of problem drinking occurs among people who are maintained at a dosage of 45-55 mg. and have been in the program 28-34 months. In addition, we find a greater proportional group of problem drinkers among blacks. In terms of the age group showing a greater occurrence of problem drinkers--the data is somewhat misleading in that we find some (50%) of the 54+ group being problem drinkers. However, our representative group only consists of two and this gives us little room for judgment.

Within our profile, we find no extraordinary results in terms of pattern-like observation. We do find that of those on a dosage of 85+ mg. there are 0% problem drinkers. However, the pattern is erratic with regard to lower dosages. The same comment can be made of the months in treatment where only 9% of problem drinkers are in the 35+ months on program group.

SUMMARY

Although we did not detect many significant patterns of obvious trends, we did accomplish a number of objectives as a direct result of the study.

1. We established and implemented a proposal for the use of Antabuse as an integral part of treatment for appropriate clients.

2. We established additional resources for purposes of referral for clients abusing alcohol.

3. In-service training sessions were held regarding the subject with emphasis on the seriousness of the issue of alcohol and methadone maintenance.

In summation, our study indicated a significant problem of alcohol abuse among our clients. The study proved valuable to clinic staff in heightening awareness of the problem in general as well as indicating its seriousness. As staff members become more aware of the clients' propensity for alcohol abuse, we expect that in individual cases the problem might be identified earlier and successful treatment approaches introduced.

REFERENCES

Pokorny, A., Miller, B., and Kaplan, H. 1972. The Brief Mast: A Shortened Version of the Michigan Alcoholism Screening Test. *Am. J. of Psychiatry*. 129:3.

THE NEED TO FREE CLIENTS OF LABELS OR BEYOND THE NEED TO LABEL BEHAVIOR ALCOHOLIC

Thomas Goldston

Salvation Army Comprehensive Alcoholism Services

Two words stand out predominantly in the vocabulary of both the therapist and the client in treatment: these two words are alcoholic and alcoholism. The commonly accepted definition of these denotes reference to sets of behaviors that are associated with a diseased state of an individual's total being, both physiological and psychological (Block 1956). Validation of this disease process has not been substantiated through clinical research. (Pomerleau, Pertschuk & Stinnett 1976) It is accepted that certainly what is called alcoholism does have similarities to other disease processes.

In most literature used in the programs that I have been employed with, emphasis is placed on getting the client to accept these labels (Tiebout 1963). I have observed that even when a client did actually show positive behavior changes relative to family relationships, improved employment records, or improved community social functioning and the individual did not accept the accompanying label of alcoholic, the discharging diagnosis was that the client was not improved. As a counselor and an observer of other counselors and therapists, I have seen the emphasis that is given to getting the client to accept that he is an alcoholic with the disease of alcoholism: a disease that has not been proven to exist and consequently there is a forced acceptance of a label alcoholic which is derived of the nonexistent disease alcoholism.

The labels were first employed to gain recognition and acceptance of a problem (Jellinek 1960), yet long after this objective had been accomplished the labels seem to serve another purpose, that being to serve as a convenient classification for all behavior that does not fit into other behavioral description systems. As professionals in treatment services, we all, I'm sure, recognize a need to have commonly understood terms. Unfortunately, I discovered that all too often as a counselor if I did not know the basis for a particular set of observed behavior in a client, I could classify it as alcoholic behavior and quickly dig out another handout showing the client how this particular behavior was descriptive of alcoholism.

In listening to clients speak of themselves I keep hearing a constant reference to their being alcoholics with the disease of alcoholism. This proclamation is used to explain any behavior that is undesirable including poor employment, marital discord, anxiety reactions, etc. I've noted that even during periods of non-drinking for up to two years, still the reason most often shared by the client for what he viewed as personal problems was his being an alcoholic. What has occurred to me is that if alcoholic and alcoholism as descriptive words are imprinted on the minds of the client in treatment then would it not follow that he will also consciously or unconsciously accept the feelings of guilt associated with the terms? Both terms, alcoholic and alcoholism, were born out of a need to remove the phenomenon of abusive consumption from the moral stigma and condemnation of the early temperance movement in this country. I see there was a need to use the labels then, and now I see a need to go beyond these labels.

The formation of AA lay groups of people, experienced in both drinking and non-drinking, allowed for individuals to share their personal experiences with drinking and recovery with other individuals still seeking a means of stopping their own destructive use of alcohol. The individualized mutual help and assistance that allowed the originators of AA to learn to deal more effectively with their life problems and to cease all drinking practices have been shown to be an effective means of dealing with the problem of alcohol, with the ultimate objective being total abstinence. For purposes of clarity and as a means of identification the terms alcoholic and alcoholism were used by Alcoholics Anonymous as reference points to ensure that public and personal reactions could be filtered through and accepted as medical terms of classification (Jellinek 1960). It's interesting that the terms alcoholic and alcoholism take on a very different meaning to both client and counselor in treatment. For the client, the acceptance of the label alcoholic seems to have a predetermined effect on his subsequent behavior and attitude towards self and others. This predetermined attitude I see as stemming largely from the emphasis placed on the labels of alcoholic and alcoholism by the treatment staff or the program philosophy, whichever happens to be the case. The problem as I see it, is that the therapist and client start confusing the classification labels alcoholic and alcoholism with the true self or real self of the client in treatment, rather than working with the real person who demonstrates definite problems with managing his own life: we are working with the unreal phenomenon of the alcoholic, with the unreal or unproven disease of alcoholism, rather than saying we're working to help a person come to grips with his pattern of over-consumption of alcohol first and a feeling, experiencing, loving person second.

One of the first things that is to be considered here is where the terms came from. How long have we used alcoholic and alcoholism? How did we refer to these same alcoholics prior to the investigation of these labels? What advantages existed then in creating new terms to describe old behaviors? Often when discussing these terms, one is referred to the disease concept of alcoholism hypothesized by Dr. Jellinek. It has often been said that the most important thing occurring as a consequence of the Jellinek classification system has been the removal of the moral stigma and condemnation of drinking practices from the influence of the early temperance movement (Pomerlau, Pertschuk & Stinnett 1976). This statement does merit attention, for it is truly one positive result of the early efforts of some professionals to work with a common occurrence, that being alcohol abuse. Alcohol abuse is not alcoholism in a true sense, given that alcoholism as a true disease state has not been proven to exist. (Verden & Shatterly 1971) Most of you, I hope, have become acquainted with these various concepts and theories about causes of alcoholism and what it is. Consequently, it is now taken for granted that it will just be a matter of time before we do have conclusive proof of the disease state (Jellinek 1960). In our assumption or belief in the disease concept we begin the treatment process for the client, despite clinical research and other evidence that says conclusively that no such disease process has been shown to exist. Most programs in Alaska still use the disease concept of alcoholism as the only basis of program emphasis.

As you can see, we are treating an unestablished disease and we are calling the clients that we work with "alcoholics". So as counselors and therapists we are classifying clients into an uncertain category of behavior called alcoholism, and alcoholics are persons afflicted with this disease. The contradiction is that we are treating real people for an unreal disease: rather than working with observable behaviors, we are working with non-observable behaviors. Rather than working with a client that has been shirking his personal responsibilities for self respect, self care, self love, we are working with alcoholics, sick people, diseased organisms. We say on the one hand that using the terms alcoholic and alcoholism has removed the moral stigma attached to alcohol abuse, yet in the very necessity for using the terms there is an implied moral stigma. It can be argued that the creation of the terms alcoholic and alcoholism have no relationship to that early temperance movement. Yet, again, looking into the literature, there is no information to substantiate the relations between the early temperance movement and the latter formation of the terms alcoholic and alcoholism (Jellinek 1960).

Recently, in my role as a counselor, I asked the client group to define themselves without using terms associated with alcoholism, i.e., length of time sober, how long they have been struggling with alcohol, etc. Unremarkedly enough the group was stymied--this is something that no one had ever presented to them before. They really did not know how to speak about themselves without reference to their drinking practices. Even where the client was able to report extended periods of sobriety, still they wanted to refer to some symptom of alcoholism for the cause of their behavior. The point I'd like to make here is that the alcoholic is not the whole person, and it seems that in the treatment process that's the only way we want to look at the individual. It has been supported in various clinical studies that the philosophy of the program tends to influence the attitude of the client in therapy (Verden & Shatterly 1971). As a consequence, if a treatment center places emphasis on the client's acceptance of the label alcoholic, and labels whatever behavior happens to be going on with the individual as alcoholism then the clients will consciously or unconsciously follow along with such labeling practices.

As a beginning counselor, my emphasis was on getting the person to accept that he was an alcoholic, yet I was not also creating a space in which the client could see himself as not an alcoholic (Moore & Murphy 1961). This ability to see himself as not an alcoholic is essential if the client is ever going to be a fully functioning whole person again. Given this reality, then it becomes clear that my role as counselor, to allow the client to see himself as not an alcoholic in addition to being an alcoholic, stems from the fact that relative to continuation of the treatment process he will encounter other individuals who work under the assumption that alcoholism does exist. In that the terms are used to describe behaviors associated with types of alcohol abusers it becomes appropriate for me as a counselor to do various educational presentations on terminology and the history of those concepts used. Given that the individual in treatment also exhibits behaviors not consistent with what is called alcoholic, I have a responsibility to provide educational information and the history of other classification terms, i.e. self-esteem, depression, self-actualization, etc. I see a need for we as treatment personnel to clarify that the labels are for increased understanding and communication and that they are not to be confused with the person's real self. The whole self is really what we are attempting to get the client to acknowledge and do something about. The alcoholic picture itself that we try to communicate to the client was, or is, that behavior which we classify as their abuse of alcohol. I've seen in practice that the client often sees only the alcoholic self partly due to the influencing factor or our emphasis on it. I do not know if it has been done and I would be interested in knowing if any program uses the labels alcoholic and alcoholism. My question is, have the labels come to stand for justification of the program's existence rather than being valid descriptions of prior behavior on the part of the client in treatment?

In reality, the environment that we grow up with in this country is that alcoholics are not O.K. people. Even in employment practices, the trend is to stay away from recovered alcoholics due to an assumed unreliability to perform the job. We work in treatment to get clients to gain a higher measure of self-esteem, self-assurance, and appropriate use of their emotions: to see things as they really are and not as they have assumed, thought, or believed them to be. Yet we as professionals are not willing to accept the reality that there may be no such thing as the disease called alcoholism. In presenting education presentations I do cover the disease concept of alcoholism, not as a factual existent disease, but rather as a hypothesis about the causes of alcoholism. I also present other information on personal responsibility, self expression, nutritional deficiencies, etc. I've placed my primary emphasis on the clients discovering their own true self without the need to rely on the words alcoholic and alcoholism. Being in Alaska has allowed me this unique opportunity to start moving away from my need to use the alcoholic label. The treatment of alcohol abuse is just coming of age in Alaska; we have a unique opportunity to put forth a new attitude, a new philosophy of recovery and cure for the clients in treatment.

Ideally, we would all like to know that there is no stigma attached to being an alcoholic or a recovered alcoholic, yet we know that there is. The clients that seem to do the best in the recovery process are those who have sufficient supports to see that alcoholism is only one aspect of their behavior and that they really are beyond the need to be alcoholics. I extend this idea one step further by postulating that the individual can be taught to go beyond the need to classify himself as a recovered alcoholic, and reach a point of recognizing the desirability of simply seeing himself as a person assuming a more responsible role in the management of his life. It is my observation that the alcohol abuser has always taken care of his needs by virtue of being an alcoholic. We tell the client that alcoholism is a progressive disease and once an alcoholic always an alcoholic, consequently we provide the client with a perpetual vehicle by which he defines his needs. I see a need to communicate to the client that I will do whatever I can to assist him whether he accepts the label or not. I am essentially concerned with teaching clients better methods of managing their life situations. I no longer have a need to fit them into a classification of alcoholism in order to provide this service.

Alcoholic and alcoholism, then, can be seen as intellectual constructs used to explain a specific set of behaviors that varies with the individual. As a counselor, I'm dealing with the client's psychological need to be an alcoholic and this need has extended so as to encourage this person to learn exactly what to do to exhibit this behavior. Even when the manifest drinking behavior has stopped we see a continuation of the need for the client to be an alcoholic. Intellectualizing is a defense mechanism that most of us are familiar with. It is used when the individual is unable to deal with the realities around him. Are we, as alcoholism counselors and therapists, intellectualizing about alcoholism because we are not able to deal effectively with the reality that some people make life choices that are not consistent with what we feel are "right" choices? Father Martin has said that we know nothing about alcoholism. We do know a lot about what people do when they over-imbibe. With this simple truth can we get past our need to work with alcoholics and begin work with those people who come to us for assistance in making their lives more manageable. No one is powerless over alcohol. Some people do choose to give up their power to alcohol. No one is born an alcoholic and many do choose to be alcoholics. We, as the workers in the field, can take the lead in freeing clients of their need to be alcoholics. If we set out in Alaska to cure alcoholism then we will, if we continue to only work for arresting the problem and not for curing the problem, keep ourselves in business forever.

Not until alcohol abuse treatment facilities expand services to effectively meet other problem needs of the clients will there be a significant reduction in alcohol abuse. Treating alcoholism involves getting the client to stop drinking, treating the whole person means providing total support of an immediate and long range nature. The needs manifested by a majority of alcohol abusers extend beyond just treating drinking behavior; many are not prepared to meet their individual needs for love, respect, physical safety, and effective interpersonal relationships. Identifying themselves as alcoholics has met these needs in the past and will continue to meet their needs until we professionals free the client of the need to label themselves alcoholics.

REFERENCES

Block, M.A. 1956. Alcoholism is a disease. *Today's Health*. 34: 36-39.

Block, M.A. 1963. Approaching the problem of alcoholism. *MT*.

Giffen, B.M. 1977. Personality Assessment and Classification. *USIU*. Human Behavior, 704.

Jellinek, E.M. 1960. *The Disease Concept of Alcoholism.* New Jersey: Hillhouse Publishers.

Miller, W.R. & Caddy, G.R. 1977. *JSA*. Vol. 38, No. 5.

Moore, R.A. & Murphy, T.C. 1961. Denial of Alcoholism as an Obstacle to Recovery. *QJSA*. 22: 597-609.

Pattison, E.M. 1972. Proceedings of the second annual alcoholism conference. *NIAAA*. 211-221.

Pomerlau, O., Pertschuk, M., Stinnett. 1976. A critical examination of some current assumptions in the treatment of alcoholism. *JSA*. Vol. 37, No. 7: 850.

Tiebout, H.M. 1971. The act of surrender in the therapeutic process. *NCAI*.

Tiebout, H.M. 1963. Normal drinking in recovered alcohol addicts. *QJSA*. 24: 109-111. Comment by D.L. Davies.

Verden, Paul & Shatterly, D. 1971. Alcoholism research and resistance to understanding the compulsive drinker. *QJAS*. Vol. 55, No. 3: 331-336.

TYPES OF INTERACTION IN ALCOHOLISM COMPLICATED MARRIAGES AND THEIR RELATIONSHIP WITH EXISTING PATTERNS OF SEXUAL DIFFICULTIES.

I. PRELIMINARY DATA AND METHODOLOGICAL CONSIDERATIONS

Irene Gad-Luther, M.D.
Family Research Institute

Herbert Laube, M. Div.
University of Minnesota

Many hypotheses have attempted to explain the clinical behavior of the alcoholic and his interpersonal (Hore 1971) and marital difficulties (Orford et al. 1976), but they usually do not build one upon the other and are not integrated in a unifying model or theoretical framework (Steinglass 1971).

Although sexual dysfunctions and alcoholism have been associated in countless anecdotal observations and clinical reports (Paredes 1973; van Thiel 1974; Viamontes 1974, 1975; Renshaw 1975; Lemere 1976) there are no hypotheses which attempt to explain their interrelation. This is unfortunate considering that sound theoretical guidelines, based on identified typological interactions have been found important indicators on which to choose the elements most likely to be effective in a particular therapeutic approach (Olson 1970).

The tragic reality of repeated relapses in alcoholism is obvious in statements about treatment such as: ". . . Entering at the superficial level and attempting to suppress behavior with legal measures has obviously failed. Entering on the level of life-style through counseling and vocational rehabilitation *may be at times successful* in calming and quieting *temporarily* the feelings of anger, boredom and anxiety . . . " (Wurmser 1972). " . . . However complex the intra-individual determinants, behavior treatment programs limited only to this dimension are fragmentary, inadequate and non-comprehensive ..." (Glasscote et al. 1967). Furthermore, Rae and Forbes (1966) have pointed out that the personality of the spouse is at least as important as that of the patient in maintaining sobriety. This was also confirmed by the finding that very "hostile" spouses are an omen for poor prognosis for their alcoholic partner (Ritson 1970) and that marital therapy not only helps marital interaction but also improves the probability that there will be less chances of relapse into drinking (Burton and Kaplan 1968; Rae and Drewery 1972).

Data on the effect of sexual therapy are so far missing although scattered information emerges here and there, connecting the spouse's sexual withdrawal with relapse into drinking (Orford et al. 1975) which would indicate that sexual therapy may have a beneficial effect.

We have come to believe that both individual traits and interpersonal dynamics play a major role in the sexual dysfunction of the alcoholic. At the intra-personal level the parallelism between some of the characteristic personality traits of the alcoholic and of the sexually dysfunctional person is striking.

We have only to look first at some of the trait clusters of the alcoholic as described by Blane (1970): low frustration tolerance, low self-esteem, fearfulness, dependency conflicts, rigid cognitive styles and add to this the often mentioned lack of assertiveness and social skills (Johnson 1974), anhedonia (Greaves 1974) or self-imposed sensory deprivation (Jones 1974; Zuckerman 1972) to see the following picture emerging: an individual with low frustration tolerance and instability, who expects to have his needs met but lacks assertiveness to ask for what he wants. All this is allied with a rigid avoidance of interdependence, rigid need to control others while denying, blaming and lacking responsibility for self. Such an individual will be incapable of an in-depth commitment and therefore will not achieve any satisfactory interpersonal relationship.

We can now address ourselves to what Kaplan (1977) has described as the trait cluster of the "failures" of short-term sexual therapy--the patients with low libido. "As a group these patients tend to be more emotionally injured, more vulnerable to psychic pain, so they tend to set up more rigid psychological defenses. The greater the vulnerability, the greater the defensive rigidity, the more deeply rooted the problem, the more tenacious the sexual anxiety and conflict . . . Sexually disruptive anxiety may be produced by fear of rejection, of intimacy, of commitment to another person . . . "

It was therefore with no surprise that we noted in the data reported by Jensen (1977) that the most frequent form of sexual dysfunction observed in post-treatment during sobriety, among other indications of increased severity of sexual malfunction, was the decrease in sexual appetite. Although seemingly paradoxical, the predominance of a more severe symptom of sexual dysfunction during periods of sobriety makes sense if we look upon it as a result of a psychodynamic imbalance: as alcoholism is alleviated, the sexual malfunction is aggravated. This would confirm our speculation that the basic personality configuration that led to alcoholism, also leads to sexual dysfunction and to other interpersonal difficulties. It seems reasonable to speculate that, possibly, both the chemical dependency and the sexual dysfunction are related in some fashion to interpersonal competence and may be directly or indirectly caused, precipitated or facilitated by lack of interpersonal effectiveness. Given that one's attitudes, cognitive capacities and behavioral skills combine to determine how effective we will be in our interpersonal interactions, those with low interpersonal skills will be more likely to engage in drug abuse. The lack of interpersonal skills will further result in isolation, loneliness and weak and transient relationships (Johnson 1974).

Among the interpersonal determinants, one of the most obvious patterns contributing to the alcoholic's difficulties is represented by his relationships to the opposite sex (Ward and Faillace 1970). Moreover a circular quality in the "complementary relationship" established by the alcoholic and his spouse, in which one partner's behavior provokes and at the same time presupposes the other partner's response, is maintained by an essential incapacity to break the chain of interaction presumably in order to maintain homeostasis (Giovacchini 1967).

Furthermore, as the functions carried out at different stages of the marriage differ, the state of the marriage has to be judged in relation to it. An interaction between the stages of alcoholism and the developmental stages of the marital relationship has been described by Jackson (1954); we hypothesize that the sexual maladjustments will vary in relation to these developmental stages. It is for this reason that we have chosen to investigate the stage of alcoholism, the type of marital interaction and the range of sexual interactions.

The Subjects

Our subjects were couples married ten years or less, in which at least the male, if not both are alcoholics in treatment for at least 10-14 days. We arbitrarily decided on a ten year time span because on one hand the Inventory of Marital Conflicts we used, had been validated on such couples, and on the other hand Jackson (1954) seemed to suggest that after a longer period the marital relationship may be irretrievably broken and we were interested to examine relationships with intact areas to build upon. The 10-14 days time span after onset of treatment for alcoholism was based upon empirical data suggesting that the hostility of the spouses is so great at the very beginning that they would be unwilling or even resistant to cooperate. After two weeks of treatment this was no longer the case.

METHODOLOGY AND PRELIMINARY RESULTS

Alcoholism

An alcohol questionnaire was devised in order to identify the stages of alcoholism. It focused on the three stages proposed by Jellinek, but reorganized by Park (1973): 1) prodromal, 2) crucial, 3) chronic and the questions refer to specific symptoms within each stage. The questionnaire also asks several questions addressing themselves to symptom clusters appearing before the prodromal stage is established. Another questionnaire for the alcoholic's spouse duplicated some of the questions asked from the identified alcoholic but included also some areas in which the spouse's perception of the alcoholic's behavior was focused upon. This provided a check on the veracity of the answers of each of the individuals and some insight into the couple's dynamics.

The Marital Relationship

a. The Locke-Wallace Marital Adjustment Test. Significant differences found between non-alcoholics, cured alcoholics and active alcoholics on this test, showed that while social functioning improved in cured alcoholics, sexual adjustment and ways of handling anxiety remained impaired (Negrete and MacPherson 1966). We, therefore, decided to use it also as an overall estimate of the relationship. Our results show so far a modest deterioration, the scores ranging from 50-100.

b. The Pair Attraction Inventory (PAI). Shostrom and Kavanaugh (1971) used a battery of self-report items referred to as the "Pair Attraction Inventory" to identify major types of marriage relationships. They differentiated the couples along two basic psychological dimensions based on each spouse's capacity to express anger-love and strength-weakness. The seven types of couple relationships are:

(1) The nurturing relationship: mothers and sons
(2) The supporting relationship: daddies and dolls
(3) The challenging relationship: bitches and nice guys
(4) The educating relationship: masters and servants
(5) The confronting relationship: hawks
(6) The accommodating relationship: doves
(7) The rhythmic relationship

The first four types of relationships are called complementary because the individuals involved are trying to find in the other what is missing within themselves. The authors believe that these relationships can be taken as support for the notion that "opposites attract"--the idea of complementary needs. The other two types of relationship--hawks and doves--are referred to as symmetrical because both spouses struggle for control. The hawks exaggerate their strength or anger and deny their weakness or love, the doves exaggerate their love or weakness and deny their strength or anger. These couples appear to avoid each other by mutual manipulation and support the idea that "like attracts like". The seventh type of relationship is called the rhythmic relationship and it describes how actualizing a couple is in their capacity to express all four aspects of the test--anger, love, strength and weakness. We used the PAI because the categorization system based on the two polarities does identify couple relationships; it is a convenient self-report instrument and the impressionistic categories are potentially useful for a clinical setting.

We expected to find most of the couples in the challenging relationship category--bitches and nice guys--and this was confirmed by our results so far. The percentage of couples falling into this category, as compared to the matched couples without alcoholism complicating their marriage is significant.

c. The Inventory of Marital Conflicts (IMC). Based on three empirical dimensions: task leadership, opinionated struggle and affective coping, the IMC developed by Olson and Ryder (1970) was able to identify nine types of relationship. These are:

(1) wife-led disengaged
(2) wife-led congenial
(3) wife-led confrontive
(4) cooperative
(5) husband-led disengaged
(6) husband-led expressive
(7) husband-led confrontive
(8) husband-led conflicted
(9) husband-led conventional

This classification is empirically strong in that the criteria are specified, the procedure for assigning couples to types is objective and the reliability is known.

The spouses are given nine short vignettes of which six present conflict items; for these the husband and wife received slanted information to create potential conflict. This constitutes a relatively standardized task by which the couple is given the opportunity to resolve the artificially created discrepancies. A code system is utilized to identify the behaviors exhibited during conflict and non-conflict situations. The couples seemed to be less flexible and less capable of solving the artificially created conflict. We expect to find most couples in the wife-led disengaged or wife-led confrontive categories but the computer data were not yet available.

The Sexual Interaction

a. Sexual attitudes. We investigated the sexual attitudes of the couple by using a two page paper and pencil test, the Minnesota

Sexual Attitudes Survey (MSAS). As compared to matched couples we found more rigid sexual attitudes in our sample.

b. Sexual behavior. We utilized the Sexual Interaction Inventory (SII), formerly called the Oregon Sex Inventory, devised by LoPiccolo and Steger (1974). The instrument is a self-report behavioral inventory of 17 sexual activities focusing on the couple's sexual interaction. For each of these 17 activities the same six questions are asked.

(1) how regularly the activity forms a part of the sexual repertoire
(2) how often he/she would like it to
(3) how pleasurable he/she finds the activity
(4) how pleasurable he/she thinks his/her mate finds the activity
(5) how pleasurable he/she would like the activity to be
(6) how pleasurable he/she would like it to be for his/her mate

The scores on these six questions are combined in various ways in orde to generate eleven scales:

1. Frequency dissatisfaction, male
2. Frequency dissatisfaction, female
3. Self acceptance, male
4. Self acceptance, female
5. Pleasure mean, male
6. Pleasure mean, female
7. Perceptual accuracy, female of male
8. Perceptual accuracy, male of female
9. Mate acceptance, male of female
10. Mate acceptance, female of male
11. Total disagreement

Scales 1 and 2 assess the satisfaction with the range of activity and its frequency; scales 3 and 4 compare actual self with ideal sexual self, i.e., how the patient would like to be; scales 5 and 6 depict how much they are enjoying what they are doing; scales 7 and 8 measure the difference between the sexual message sent and the one received; scales 9 and 10 assess the degree to which the perceived sexual self of the mate is accepted by each spouse. Scale 11 is simply the total disagreement between husband and wife regarding sex. According to LoPiccolo and Steger (1974) it is the best overall measure of the sexual adjustment of the couple. We were expecting to find a great dissatisfaction with the alcoholic mate, provided the non-alcoholic partner is sexually functional, not inhibited and without a pathology of her own (Parigi 1970). We found less frequent sexual behavior, less flexible sexual behavior and less satisfaction with the sexual behavior as compared to the matched sample.

c. Sexual pathology. We used the Sexual Difficulties Behavior Checklist, modified to fit our sample. The initial self-report checklist was devised by Bruce Fisher (1976). There is a form for each spouse in which the male and the female list the number of times that a difficulty has occurred for self and partner in the last four weeks. We had to modify it including difficulties having occurred since drinking began and those having occurred in the last year. Usually there was no sexual interaction of any kind in the last four weeks, so that category of questions was left unanswered in our sample. All in all there was more male pathology reported by our subjects as compared to the matched couples.

Matching

We administered a check list which included such general information as highest grade completed in school, present occupation, income bracket, religious affiliation, degree of involvement with religious activity, etc. This enabled us to compare our couples with couples used as subjects for the Doctoral Dissertation of one of us (H.L.) and who presented sexual dysfunctions without having alcoholism complicating their marriage.

Final considerations

Largely not talked about during treatment for alcoholism, or minimized by the assumption that sobriety will cure any sexual problem occurring during drinking (Lemere 1976) it is now becoming increasingly clear, as more data are gathered, that two separate pathological entities may exist in the complex field of sexual/alcoholism interaction. Indeed the episodic impotency leading to situational anxiety and to the self-perpetuating dysfunction of the social drinker (Masters and Johnson 1970; Sackett 1977) is essentially different from the complex psycho-sexual malfunction of the chemically dependent.

The increased severity of sexual dysfunction (Jensen 1977), the persistence of impotency after years of sobriety (Mills 1975) as well as the data presented herein, seem to indicate a deep seated personality conflict and a complex, mutually reinforcing neurotic system.

REFERENCES

Blane, H.T. 1970. The Personality of the Alcoholic. In Chafetz, M.E., Blane, H.T. & Hill, M.J. *Frontiers of Alcoholism.* New York: Science House. 16-29.

Burton, G. & Kaplan, H.M. 1968. II. The Correlation of Excessive Drinking Behavior with Family Pathology and Social Deterioration. *Brit. J. Addict.* 63: 161-170.

Fisher, B. 1976. Sexual Difficulties Behavior Checklist. Department of Family Social Sciences, University of Minnesota.

Glasscote, R.M., Plaut, T.F.A., Hammersley, D.W., O'Neil, F.J., Chafetz, M.E. & Cumming, E. 1967. *The Treatment of Alcoholism: A Study of Programs and Problems.* Washington, D.C.; Amer. Psychiat. Assn. & Natl. Assn. for Mental Health, Joint Information Service.

Giovacchini, P.L. 1967. Characterological Aspects of Marital Interaction. *Psychoanal. Forum.* 2:10-14.

Greaves, G. 1974. Toward an Existential Theory of Drug Dependence. *J. Nerv. and Mental Dis.* 159: 263-274.

Hore, B.D. 1971. Factors in Alcoholic Relapse. *Brit. J. Addict.* 66: 89.

Jackson, J.K. 1954. The Adjustment of the Family to the Crisis of Alcoholism. *Quart. J. Stud. Alc.* 15: 562-586.

Jensen, S.B. 1977. Sexual Behavior and Sexual Problems Among Alcoholics. *Ugeskrift for Laeger.* 139: 35-40.

Johnson, D.W. 1974. A Theory of Social Effectiveness. In *Why Drugs. The Psychology of Drug Abuse*, University of Minnesota.

Jones, H.B. 1974. The Effects of Sensual Drugs on Behavior: Cues to the Function of the Brain. *Adv. Psychobiol.* 2: 297-312.

Kaplan, H. 1977. Hypoactive Sexual Desire. *J. of Sex & Marital Therap.* 3: 3-9.

Lemere, F. 1976. Sexual Impairment in Recovered Alcoholics. *Med. Asp. of Human Sex.* 10: 69-70.

Locke, H.J. & Wallace, K.M. 1959. Short Marriage - Adjustment and Prediction Tests: Their Reliability and Validity. *Marriage and Family Living.* 21: 251-5.

LoPiccolo, J. & Steger, J.C. 1974. The Sexual Integration Inventory A New Instrument of Sexual Dysfunction. *Archives of Sexual Behavior.* 3: 585-95.

Masters, W.H. & Johnson, V.E. 1970. *Human Sexual Inadequacy.* Boston: Little & Brown.

Mills, L.C. 1975. Drug Induced Impotence. *Amer. Family Physician.* 12: 104-106.

Negrete, J.C. & MacPherson, A. 1966. Comparative Study of Matrimonial Adaptation of "Active" and "Cured" Alcoholics. *ACTA Psiquiatrica y Psicologica de America Latina.* 12: 251-256.

Olson, D. 1970. Marital and Family Therapy: Integrative Review and Critique. *J. of Marriage and the Family.* 32: 502-38.

Olson, D.H. & Ryder, R.G. 1970. Inventory of Marital Conflicts: An Experimental Interaction Procedure. *J. of Marriage and the Family.* 32: 443-448.

Orford, J., Guthrie, S., Nicholls, P., Oppenheimer, E., Egert, S. & Hensman, C. 1975. Self-reporting Coping Behaviour of Wives of Alcoholics and Its Association with Drinking Outcome. *Journal of Studies on Alcohol.* 36: 1254-67.

Orford, J., Oppenheimer, E., Egert, S., Hensman, C. & Guthrie, S. 1976. The Cohesiveness of Alcoholism-Complicated Marriages and Its Influence on Treatment Outcome. *Brit. J. Psychiat.* 128: 318-339.

Paredes, A. 1973. Marital-Sexual Factors in Alcoholism. *Medical Aspects of Human Sexuality.* 7: 98-115.

Parigi, S. 1970. Considerazioni sulle alterazioni ipotalamiche e talamiche nella insorgenza del comportament di gelosia degli alcoolisti. *Rivista di Neurobiologia.* 16: 255-66.

Park, P. 1973. Developmental Ordering of Experiences in Alcoholism. *Quarterly Journal Stud. Alcoholism.* 34: 473-488.

Rae, J.B. & Drewery, J. 1972. Interpersonal Patterns in Alcoholic Marriages. *Brit. J. Psychiat.* 120: 615-621.

Rae, J.B. & Forbes, A.R. 1966. Clinical & Psychometric Characteristics of the Wives of Alcoholics. *Brit. J. Psychiat.* 112: 197-200.

Renshaw, D.C. 1975. Sexual Problems of Alcholics. *Chicago Med.* 78: 433-36.

Ritson, B. 1971. Personality and Prognosis in Alcoholism. *Brit. J. Psychiat.* 118: 79-82.

Sackett, V.A. 1977. Sexual Dysfunctions in Alcoholics. Paper presented at the NADPA Conference, Detroit, September.

Shostrom, E. & Kavanaugh, J. 1971. Pair Attraction Inventory. San Diego: Educational & Industrial Testing Services.

Steinglass, P., Weiner, S., Mendelson, J.H. & Chase, C. 1971. A Systems Approach to Alcoholism. *Arch. Gen. Psychiat.* 21: 401-8.

van Thiel, D.H. & Lester, R. 1974. Sex and Alcohol. *New England Journal of Medicine*. 291: 251-3.

Viamontese, J.A. 1974. Alcohol Abuse & Sexual Dysfunction. *Med. Aspects of Human Sex*. 8: 185-6.

Viamontes, J.A. 1975. Sexual Depressant Effects of Alcohol. *Med. Aspects of Human Sex*. 9: 31.

Ward, R.F. & Faillace, L.A. 1970. The Alcoholic & His Helpers: A Systems View. *Quart. J. Stud. Alc*. 31: 684-691.

Wurmser, L. 1972. Drug Abuse: Nemesis of Psychiatry. *Amer. Scholar*. 41: 393-407.

Zuckerman, M. 1971. Dimensions of Sensation Seeking. *J. Consult. & Clin. Psychol*. 36: 45-52.

AN EDUCATIONAL PROGRAM TO INCREASE ELECTIVE USE OF DISULFIRAM TREATMENT

Anne-Marie Deutsch, M.A. and Vitali Rozynko, Ph.D.

V.A. Hospital

Palo Alto, California

As a widely used pharmacological treatment for alcohol abuse, disulfiram (commonly called by its trade name, Antabuse) has been demonstrated to be effective in temporarily helping alcoholics refrain from drinking (Lundwall and Baekeland, 1971). The value of disulfiram is due to the adverse physical effects which occur when the drug is combined with alcohol; a patient who takes disulfiram regularly knows that drinking any form of alcoholic beverage will cause feelings of nausea, lowered blood pressure, palpitations, headache, and intense anxiety. When disulfiram is administered on a voluntary basis, it has the potential of being a powerful self-control device because it gives the problem drinker the freedom to make only one decision a day; that is, to take his or her disulfiram tablet. Unfortunately, the effectiveness of disulfiram is limited by the fact that patients choose not to use it or neglect to take it after it has been prescribed. It has been suggested that patient education be the preferred method of increasing patient use of disulfiram (Lubetkin, Rivers and Rosenberg, 1971). The present paper is a description of one such educational approach being developed at a Veterans Administration Hospital alcohol treatment program.

DISULFIRAM USE AT *OUR* PROGRAM

Our Program at the Menlo Park Division of the V.A. Hospital in Palo Alto is, in many ways, typical of V.A. alcohol treatment facilities. It is a 30 bed inpatient unit for males, following a therapeutic community model. Being a self-help program, the unit has no nurses or physicians on full-time staff. One physician is available on the unit for one hour a day; medications, however, are kept to a minimum, with disulfiram being one of the few drugs which are typically prescribed. One of the contingencies for a patient's remaining in the 90-day treatment program is that he must remain totally abstinent from alcohol during his length of stay. Until this past year, this condition was insured by requiring that all patients receive a daily dosage of disulfiram. As program goals were re-evaluated, however, it was decided that increased control would be given to patients, that disulfiram would only be prescribed at a patient's request, and that the drug would be taken voluntarily (i.e., a nurse would not be present to watch that the pill was actually swallowed).

Although this decision resulted in fewer patients being in disulfiram treatment, there were several potential advantages to the voluntary method: first of all, it decreased the view that disulfiram is a drug which is literally being pushed down patients' throats. Second, it encouraged patients to make their own decisions (in part) about their treatment plans, and finally it increased the possibility that patients would choose to continue the use of disulfiram when they were discharged from the hopsital.

It became apparent after several months that patients were not, on the whole, choosing to use disulfiram as part of their treatment, and this reluctance seemed to be based on numerous rumors, myths and fears regarding the effects of disulfiram as well as the implications of its use. Some of these misconceptions are as follows:

1. An alcohol-disulfiram reaction will result in death.

2. Some people can drink alcohol after taking disulfiram, with no effect.

3. If one uses disulfiram as treatment, it will be a lifetime commitment to the drug.

4. Using disulfiram means that the user is weak and has no self-control.

5. Disulfiram has dangerous side effects, such as long-term impotence.

6. Disulfiram is a "crutch."

It was decided that because of the potential usefulness of disulfiram in the control of alcohol abuse, an educational program was needed to train patients in the use of the drug--not only in the specific pharmacological aspects of disulfiram, but also uses of disulfiram as a way of exerting self-control and "buying time" to increase the therapeutic value of other alcohol programs. Thus, the purpose of the class was *not* to convince patients that disulfiram was good for them, rather that their decision to use or not use disulfiram treatment could best be made on the basis of accurate information and careful consideration. Obviously, it was hoped that this approach would increase patients' electing to use disulfiram treatment.

DISULFIRAM AS A FORM OF SELF-CONTROL

The notion of self control is defined, for the purposes of this program, as the manipulation of environmental variables in such a way that one can increase the likelihood of one's own desirable behaviors occurring, or decrease the likelihood of undesirable behaviors. This is essentially a behavioral definition of self-control and successfully sidesteps the concept of "willpower", an amorphous term which often is used to explain situations in which the source of control is unclear.

In conducting a behavior analysis of the drinking problem, it becomes clear that one of the most immediate and powerful discriminative stimuli for drinking is the presence and appearance of an alcoholic beverage. The time when an alcoholic is most likely to say "yes" to a drink is when he or she has a bottle present or when the situation is one in which other people are drinking. Self-control then can be aimed at *reducing* the probability that a "yes" response will be made.

The decision to use disulfiram is usually made at a time when alcohol appears to have less of a reinforcing value, e.g., while the patient is actually involved in an alcohol treatment program, when he or she is away from bars, liquor stores or social situations containing alcohol. At such times the probability of saying "no" to alcohol is high. Once the decision is made, it does not need to be made again, particularly in high-risk situations with alcohol present. This aspect of disulfiram, as well as the fact that it is ineffective unless the patient agrees to use it faithfully, creates a chain of responses which can clearly be labelled as behavioral self-control.

THE CLASS PRESENTATION

Information about disulfiram treatment is presented to patients in a 1½ hour class discussion format. Because the class is still in a process of development, the results of each class discussion have been used to shape future classes. Classes have essentially consisted of four sections, discussed below:

1. *Free discussion of disulfiram: experiences, myths, feelings.* Patients are encouraged to offer all they know about disulfiram; those who have used it discuss how it has or has not been helpful, and those who have not used it may discuss their reasons for not doing so (e.g., fears and myths).

2. *Specific information about disulfiram.* Patients are offered information about how disulfiram creates a reaction with alcohol, dosage information, side effects, disulfiram's effects with other medications, etc.

3. *Discussion of disulfiram as self-control tool.* The rationale for use of disulfiram as a self-control device is presented to patients. Emphasis is placed on re-arrangement of environmental stimuli, thus debunking the traditional notion of "willpower."

SUMMARY

To increase the voluntary use of disulfiram, a class in the uses and values of disulfiram treatment has been developed for patients on a residential self-help alcohol treatment unit. The important role which disulfiram holds in the treatment of alcoholism makes it critical that patients be given a detailed introduction to its use. Finally, alcoholic patients are frequently making attempts to gain back their self-control, to get away from a drug (alcohol) which they perceive as a crutch; it is therefore necessary that the self-control aspects of disulfiram be emphasized, the hope being that patients will see disulfiram as a powerful tool in their endeavor to regain control of their own behavior.

REFERENCES

Lubetkin, B.S., Rivers, P.C. and Rosenberg, C.M. 1971. Difficulties in disulfiram therapy with alcoholics. *Quarterly Journal of Studies in Alcoholism.* 32: 168-171.

Lundwall, L and Baekeland, F. 1971. Disulfiram treatment of alcoholism: a review. *Journal of Nervous and Mental Disease.* 153: 381-393.

TO ABSTAIN OR NOT ABSTAIN--AN EAGLEVILLE SAGA

Jerome F.X. Carroll, Ph.D.

The events leading up to an institution's decision to examine a very controversial issue are sometimes as intriguing as the process of inquiry and its outcome. This was especially true of Eagleville Hospital and Rehabilitation Center's (EHRC) decision to compare and contrast the assets and liabilities of the traditional total and permanent abstinence treatment objective with those of the controlled substance use treatment objective.

The roots of this decision can be traced to an earlier EHRC decision to initiate a combined treatment program for *both* alcoholics and drug addicts in 1968 (Ottenberg and Rosen, 1971). Prior to this, EHRC had treated only alcoholics. Like many other substance abuse programs, EHRC hired former drug addicted and alcoholic patients as staff, some of whom had successfully completed the EHRC program.

Over the intervening years between 1968 and the present, EHRC developed a special recognition ceremony for those recovered staff who had once been patients at EHRC and had achieved one year of *total* sobriety (meaning *no* use of any psychoactive substances, including both alcohol and other drugs). At this annual ceremony, recovered alcoholic staff members and recovered drug addicted staff members received a Post Hospitalization Discharge (PHD) degree. Relatives, friends, staff, and even the hospital's Board of Directors usually attended this ceremony.

In 1976, a minicrisis arose around the awarding of the PHD degree. Several recovered drug addicted staff members scheduled to receive their PHD degree were known to be drinking (but not abusing) alcohol. This was a "violation" of the *total* abstinence ethic which was the principal criterion for awarding the PHD degree. Although these staff members were working satisfactorily on their jobs, seemed generally adjusted in terms of their social relationships, and were abstaining from any drug use/abuse, nevertheless, they were not awarded a PHD. Some staff members believed this decision was wrong and demanded that the decision be reevaluated in terms of future PHD degree ceremonies.

I don't think this demand would ever have been made in 1970 when I began working at EHRC, as then staff members were not expressing any doubts about the wisdom of the traditional total and permanent abstinence model of rehabilitation. Over the last decade, however, a growing number of research and clinical findings have been published which have challenged the *absolute* validity/merit of the toal and permanent abstinence model of substance rehabilitation (Davies, 1962; Kendell, 1965; Gerard, 1966; Merry, 1966; Pattison, 1966; Bailey and Stewart, 1967; Pattison et al., 1968; Paredes et al., 1969, Cutter et al., 1970, Lovibond and Caddy, 1970; Mello and Mendelson, 1970; Nathan et al., 1970; Cohen et al., 1971; Engel and Williams, 1972; Gottheil et al., 1972; Gottheil et al., 1973; Marlatt et al., 1973; Paredes et al., 1973; Sobell and Sobell, 1973; Drewery, 1974; Sobell and Sobell, 1974; Sobell and Sobell, 1975; Armor et al., 1976; Gottheil, 1976; Miller and Munoz, 1976; Popham and Schmidt, 1976; Miller, 1978). These findings and opinions have been challenged by other authors who still insist that total abstinence is the proper path to successful rehabilitation (e.g., Fox, 1967; Davis, 1976; Ewing and Rouse, 1976). Thus it is not so surprising that the total abstinence ethic was officially challenged at EHRC in 1976.

The demand to reevaluate eligibility for the PHD degree was referred to EHRC's Therapy Review Committee[1] for further deliberation in June 1976. The Therapy Review Committee is responsible for overseeing the quality of therapy at EHRC. This committee made six recommendations:

1) all staff were advised by memo to come prepared to discuss the "abstinence issue" at the next general staff meeting;

2) an all day conference be planned to discuss the issue;

3) that EHRC's Learning Community (a staff development program) develop a special seminar series on the subject;

4) that each department develop a special educational program for its staff;

---

[1] Now the Treatment Review Committee.

5) a formal debate be planned on this topic to include outside experts; and

6) that the Policy and Planning Committee[2] (EHRC's principle decision-making body) appoint a staff member to plan and implement these recommendations.

Unfortunately, after this initial burst of concentrated interest in the abstinence-controlled use/PHD degree issues, no further effective action was taken for a period of nine months. Then in February 1977, a special committee consisting of both recovered and non-recovered staff began meeting over a period of several months in order to examine the issue in depth and to make recommendations to the community. At its initial meeting, a number of concerns and grievances were aired:

1) the lack of adequate information about the implications and effects of non-abstinence (controlled substance use) after treatment, e.g.,
   a) who can use what substances?
   b) under what conditions?
   c) how?
   d) when?
   e) and how is the success of such use to be determined?

2) a wish to see the community discuss and explore the issue of a non-abstinent (controlled substance use) approach, i.e., the possibilities of moderate, controlled use over time; a need to seek greater acceptance for a non-abstinence controlled use philosophy that would allow people to make their own decisions; a need for adequate information to help people make such decisions;

4) the question of guilt feelings generated among recovered persons who engage in controlled use in a community that officially advocates abstinence;

5) the discrepancy between Eagleville's official policy of abstinence and the fact that some recovered staff do use (as opposed to abuse) substances in a controlled fashion;

6) the problems and anxieties raised by the wide variation in treatment philosophies extant among Eagleville staff members;

---

2 Now the Policy and Review Committee.

7) the variations in approach taken by different program elements within the EHRC complex;

8) a need to arrive at mutually acceptable definitions and understandings of terms; and

9) a need to explore the political and legal implications for both Inpatient and Outpatient Programs of a controlled substance abuse approach to substance abuse rehabilitation.

The committee eventually evolved a plan for a series of seminars which was to feature both outside experts and selected EHRC staff who would both provide information and debate the issue. It was decided that these seminars would be open only to staff and not residents and candidates.

One decision reached by this committee was to have an "Abstinence/ Controlled Use Day" for EHRC staff in order that the staff could *experience* this issue more fully and personally. On that day, those staff that wished to participate could totally abstain from or use in a controlled manner some substance or activity that they typically "abused" (e.g., coffee, cigarettes, TV watching, etc.). April 28, 1977 was designated as Abstinence/Controlled Use Day at EHRC.

Those staff who volunteered to participate in the Abstinence/Controlled Use Day program were asked to report their commitment to total abstinence or controlled use at the morning community meeting prior to Abstinence/Controlled Use Day (in the same manner that our residents report their commitments at the mike). They were also asked to report to the community on the following day regarding their experiences. In addition, several staff agreed to wear small signs with their abstinence/controlled use commitment depicted in order to solicit community support in adhering to their commitment.[3]

I might add parenthetically that the "controlled use" aspect of this exercise was added, because several staff didn't wish to *totally* abstain from the abused substance (e.g., cigarettes) or activity. They insisted that if they could reduce the level of their use/activity of whatever they were abusing, this would represent "success."

---

[3] Since then, the use of signs as a therapeutic intervention has been discontinued at EHRC.

THE SEMINAR SERIES

First Seminar

Following Abstinence/Controlled Use Day, a series of four seminars was initiated (May 11, 18, 25, and June 1, 1977). The first of these seminars entitled *"Total Abstinence--Is It the Only Way?"* featured three presenters: Dr. Donald J. Ottenberg, Executive Director, Eagleville Hospital and Rehabilitation Center; Dr. Edward Gottheil, Professor of Psychiatry of Human Behavior, Jefferson Medical College Associate Director, Division of Drug & Alcohol Services, Thomas Jefferson Community Mental Health/Mental Retardation Center; and Mr. Thomas Zampanis, Director, Gaudenzia House, Palmyra, Pa. Dr. Jean Feinberg, Chairperson of EHRC's ad hoc Abstinence Committee, served as moderator for this session.

Dr. Donald J. Ottenberg: Dr. Ottenberg opened this session by reviewing EHRC's history concerning its present commitment to a total and permanent abstinence model of substance rehabilitation. He also noted that EHRC's decision to examine this commitment more closely was due to several factors:

1) a number of narcotic addicts who had graduated from the EHRC program and continued to do well in their recovery have resumed "social" drinking without any apparent adverse effects; similarly we have heard of recovered alcoholics who have maintained sobriety and stable living while using marijuana "socially" or "recreationally."

2) a number of recent research and clinical reports (e.g., the Rand Report) have suggested that some alcoholics have been able to resume drinking without drinking alcoholically; and

3) various substances continue to be used in the treatment of addicts (e.g., methadone and naltrexone) which even includes a growing interest in the prescriptive use of heroin as a treatment method.

Dr. Edward Gottheil: Dr. Gottheil stated that in order to control excessive drinking, it is first essential to understand the factors which enhance or decrease the need to drink. For this reason, he maintained that it is necessary to study alcoholics drinking in a controlled, experimental environment.

To accomplish this objective, Dr. Gottheil and his associates developed an *experimental* treatment approach which they called the Fixed Interval Drinking Decision (FIDD) program. This program was begun in 1970 at the Coatesville VA hospital. In the FIDD, alcoholics who volunteered to participate could decide to drink or not to drink one or two ounces of 40% ethanol each hour on the hour from

9 a.m. to 9 p.m., Monday through Friday. No drinking was allowed on weekends. This schedule provided 13 decision points daily and a maximum intake of 26 ounces. The FIDD's eclectic and multidisciplinary treatment program focused on trying to help patients understand why they drank at one time and not another.

Without being rewarded or punished for drinking or not drinking, patients exhibited a broad range of drinking behavior: 45% never took a single drink; 20% began drinking, then stopped; and 35% drank in some degree throughout the four drinking decisions weeks of the six week program.

Those that drank, drank in every conceivable fashion. A few drank almost every drink that was available, some drank heavily on some days and not on others, some drank in moderate amounts, and some took an occasional drink with meals.

Gottheil reported that those patients who drank and then stopped, in the presence of freely available alcohol and other drinking patients, did not suffer as a result of stopping but felt better--even if they had been drinking as much as a fifth a day for two or three weeks. He also noted that his patients had been able to withstand the dry weekends and cessation of drinking at 9 p.m. without exhibiting much in the way of craving or withdrawal symptoms. Gottheil also noted that patients who did not drink in the program did better on the followup than those who did.

While Gottheil was careful to point out that his findings were based on a restricted sample of volunteer patients under controlled conditions, he did, nevertheless, observe alcoholics abstaining, drinking then stopping, and drinking moderately in presence of available alcohol and other drinking patterns.

In a six months followup study of 23 to 25 patients, Gottheil reported findings that only two patients had been completely abstinent since discharge (total abstinence had not been presented as an objective of the program). On the other hand, 12 persons (52%) reported drinking less than twice a week; 14 people (61%) reported that they had never been intoxicated during the six months post discharge period, while 16 persons (70%) reported they had not been intoxicated more than once during the last 30 days. Those who were drinking, moreover, reported drinking much less than prior to entering the program.

Gottheil stated that if we set our treatment goals at total and permanent abstinence, we are doomed to failure. Such a goal is unrealistic and creates feelings of hopelessness among both patients and staff.

While he concluded that the most practical treatment goal at present is *abstinence*, he emphasized that not everyone can achieve that goal. He argued, therefore, that it is crucial not to condemn or view as a failure the patient who achieves something less than total and permanent abstinence, that is, who learns to successfully control his drinking.

Mr. Thomas Zampanis: Mr. Zampanis began his presentation by providing a brief overview of Gaudenzia's drug abuse treatment program. Gaudenzia's therapeutic community serves approximately 130 addicted men and women. The program includes a short-term (90 day) cycle and a long-term (18 months) cycle. The short-term cycle is directed to abstinence from all chemicals, including alcohol.

The long-term cycle, however, did permit drug addicts to attempt to learn to drink "socially." This cycle had three phases. The first, an exposure phase, consisted of having drug addicts go out to places where they would be exposed to people drinking.

In phase two, introduction to alcohol, the drug addicts could request permission to drink, and if their request was granted, they were permitted to drink--as long as they were accompanied by a staff member or an older, responsible patient who had already demonstrated the ability to drink responsibly. In phase three, the drug addicts were permitted to drink alone.

Since its inception, only three out of 130 graduates have become alcoholics, although a considerable number of other graduates have gotten into difficulty because of alcohol abuse. As a result of this experience, the program has taken a more critical view of drinking. Newly admitted patients are now, for example, routinely administered an alcohol abuse test whose results are later used to evaluate a patient's request to drink.

The notion of *complete* abstinence is obsolete, stated Zampanis. He maintained that it is more productive to focus on the patterns of abuse and to teach abusers how to gain control over the chemicals they have abused. In some cases, this would necessitate abstinence, while in other cases, controlled use was a possibility.

He argued that a drug abuser (e.g., a heroin addict) who learned to smoke pot, yet did not abuse drugs or engage in criminal activities, has gained a large measure of successful control over his life. Abstinence oriented therapy, which views the *drug* as the problem, is a perspective which he believed would not help many substance abusers. The best way to help substance abusers, according to Zampanis, is to assist them in making constructive changes in their lives.

The Second Seminar

The second seminar, entitled "*Alternative Approaches to Abstinence*," featured four presenters, including Dr. Christopher D'Amanda, Chief Medical Officer, Coordinating Office for Drug and Alcohol Abuse Program, City of Philadelphia; Dr. Genevra Driscoll, Director of Psychiatry, EHRC; Mr. Grant Freeman, Director of the EHRC Candidate Program; and Dr. Jerome F.X. Carroll, Director of Psychological Services, EHRC. Mr. Avrom Soifer, Utilization and Review Coordinator, EHRC, served as the moderator for this panel.

Dr. Christopher D'Amanda: Dr. D'Amanda presented an overview and critique of the Rand Report (Armor, Stambul, and Polich, 1976). He noted that the Rand Corporation had been awarded a grant from NIAAA to review the status of the Alcoholism Treatment Centers (ATC) reporting to NIAAA's national data collection system (ACTMS). The report was based on data obtained from 14,000 admissions to 44 ATC's, plus data from four successive Harris Surveys.

The basic purposes of the study were to ascertain what had worked and is working in alcohol abuse rehabilitation; to test the validity of various theories of why people start and/or continue to drink excessively; and to determine whether or not various types of treatment result in differential outcomes of success or failure.

The Rand Report created quite a shock wave, since it produced evidence which cast doubt on the validity of the disease model of alcoholism and seemed to indicate that some alcoholics could resume normal drinking.

Some of the specific findings of the Rand Report which bear on the abstinence vs. controlled use issue (just one of 12 basic research questions studied) were:

1) the presence of long-term abstinence (not having consumed alcohol for a period of six months or longer) was rare; only 17% of the respondents at six months post treatment and 24% of the respondents at 18 months post treatment were totally abstinent;

2) relative improvement, however, was reported by 70% of the sample; for example, the percentage of persons drinking in excess of five ounces of alcohol per day dropped from 51% to 13% at the six months followup and from 53% to 16% at 18 months followup;

3) close to 50% of the total population reported consumption of *less* than one ounce per day at the time of followup;

4) totally abstaining patients appeared to be just as likely to

relapse as those who pursued "normal drinking" (less than three ounces consumed a day; a typical quantity of consumption at any one time less than five ounces; absence of any tremors; and absence of any serious symptomatology; and

5) some alcoholics were able to return to some pattern of drinking without necessarily exhibiting alcoholic symptoms.

The authors of the Rand Report did *not* advocate a normal drinking policy in the clinical treatment of alcoholism. In fact, where physical complications had occurred (e.g., liver disease), they indicated that abstinence was imperative. They did, however, question the wisdom of teaching alcoholics that one drink was as bad as ten, since this might cause the recovering alcoholic not to try to stop drinking after one or two drinks.

Mr. Grant Freeman: Grant Freeman began by noting that he too was not advocating doing away with abstinence as a treatment goal. He did state, however, that he believed it was healthy to question basic assumptions.

He noted that the literature he had reviewed seemed to stress that abstinence continues to be the most practical treatment goal, especially since the greatest number of successes have been achieved through total abstinence.

At the same time, he expressed concern that the total and permanent abstinence model did seem to set up a self-fulfilling prophesy. He suggested that substance abuse is a complex problem requiring complex solutions.

He noted that more alcoholics, according to the literature, than previously suspected were able to resume drinking in a normal manner. He expressed the hope that rather than allow the issue of abstinence vs. controlled use to polarize us, the discussion of the issue would result in advancing knowledge in the field which ultimately would help addicted men and women everywhere.

Dr. Genevra Driscoll: Dr. Driscoll reviewed the results of a number of studies that indicated that some alcoholics were able to resume drinking in a non-alcoholic manner. Some of the factors that seemed to distinguish those former alcoholics who were able to resume drinking normally were: post treatment improvement in occupational and/or social/family conditions and pre-treatment drinking levels (those who drank less were more likely than those who drank more to drink normally).

She noted that those who had challenged these results had questioned whether or not those who had resumed normal drinking were "true" alcoholics. She indicated that some critics had referred to such patients as "pseudoalcoholics." She also noted finding that what constituted "normal" drinking varied from one study to another and that some "normal" drinking levels seemed rather high.

Dr. Jerome F.X. Carroll: Dr. Carroll indicated that the literature he had reviewed seemed to lend support to his position that the medically oriented disease model of addiction is invalid and obsolete, as are the labels "alcoholic" and "addict." He cited the following liabilities associated with the disease model:

a) overemphasizes intrapsychic (within the person) causes while underestimating/ignoring social, political, and economic causes;

b) overemphasizes medical causes and role of medicine in administering and providing treatment;

c) makes it more difficult for the addicted person to accept his role/responsibility in choosing to abuse substances;

d) the underlying assumptions of the disease model have *not* been supported by recently published facts:

   1) uncontrollable urge/impulse to continue use has not been confirmed--some alcoholics didn't continually drink at every opportunity to do so

   2) some addicted persons were able to resume substance use without continually and totally abusing the substance.

He proposed a "dis-ease" model as an alternative to the traditional disease model (Carroll, 1976). According to this model, substance abuse represents a maladaptive habitual mode of coping with life and its problems to the extent that the person's life style becomes centered around substance abuse. Medicine in this regard plays an ancillary role in rehabilitation, while social, political, and economic factors receive major emphasis as interacting, causative agents in addiction. In addition, this model encourages a wider and more equitable distribution of decision-making power among the many and diverse disciplines which presently work in the addictions fields.

Carroll also noted that since most substance abusers are multiple substance abusers (especially those below the age of 40)--abusing both alcohol and other drugs--the issue of abstinence and controlled use is much more complex than many have indicated. To demonstrate his contention, he cited findings from the National Drug/Alcohol Collaborative Project (Carroll, 1977) which indicated that many substance abuse patterns have a developmental history involving several substances and/or reflected interactive or concurrent substance abuse of several substances (e.g., where one substance of abuse is substituted for another or used with another substance to get a better high or mitigate the discomfort of withdrawal symptoms).

He also agreed with those critics who believed that the total and absolute abstinence criterion was too rigid and unrealistic a criterion of "successful" rehabilitation for all substance abusers and proposed that new and additional criteria be developed to supplement the total abstinence criterion.

Third Seminar

The third seminar, *"Advantages and Disadvantages of a Total Abstinence Model for EHRC's Inpatient and Outpatient Programs,"* featured six EHRC staff who prepared formal statements on this topic. This session was chaired by Dr. Jerome F.X. Carroll, Director of Psychological Services, EHRC.

The participants were Dennis Deal, Counselor-Therapist, Candidate Program; Jeanne Pleis, Team A Administrator; Sidney H. Schnoll, M.D., Ph.D., Medical Director and Clinical Director; Avrom Soifer, Utilization and Review Coordinator; Murray Synigal, Assistant Director of Inpatient Division and Director of Therapy; and Joseph Watson, M.S.W., Director Mid-County Clinic. Following is a brief synopsis of each participant's formal statement.

S. Schnoll: stressed the limitations of the disease model of addiction; strongly objected to the present process of teaching recovering persons that one use will absolutely lead to a total relapse--his own experience as a rehabilitative specialist did not support this contention, since he personally knew recovered people who had been able to use one substance without becoming addicted again; he also stressed the need for various support systems for those treated substance abusers who wished to attempt controlled substance use, especially in view of the many relapses and remissions observed among addicted people; he estimated that 10% of addicted men and women were able to use substances formerly abused without becoming addicted again; finally he stressed need for a spectrum of treatment modalities, with better diagnoses--such a spectrum would allow for controlled substance use as well as abstinence; until such treatments and quality diagnoses were available, then substance abuse rehabilitation and treatment would continue to be a hit or miss type of situation.

M. Synigal: argued strongly that absolute and permanent abstinence were essential for substance abuse rehabilitation--that the greatest gains had been realized by programs (e.g., AA) that followed this model; he cited various authorities in the addictions problems field who had expressed such views (e.g., Davis and Fox); he cited his own personal experience as a recovered alcoholic who had attempted "controlled" drinking and failed; he stated categorically that a "chronic alcoholic" can not drink in a controlled fashion--that the first drink that enables the former alcoholic to "relax" will assuredly lead to the resumption of compulsive drinking.

D. Deal: argued that treatment facilities need to help addicted people to consider that some patients may be able to learn to use without abusing substances; he cited his personal experience with 10 recovering addicts who had drank and/or smoked pot after leaving the EHRC abstinence program--two had failed totally; one was borderline; but seven were coping successfully despite their drinking and/or use of pot; he cautioned against introducing a less than total abstinence model into EHRC's Inpatient and Candidate Programs for fear the patients would "play games" with such a model; he also shared with us how he had felt as a "social outcast" when he had decided to attempt controlled drinking, even though his addiction was heroin.

J. Pleis: spoke of the value of the disease model in that it had afforded her a significant degree of relief when she was attempting to recover from alcoholism; she also praised AA's emphasis on sobriety, since it had proven very helpful to her; she saw some merits in the "dis-ease" model too; she noted that until researchers could devise the means of identifying which addicted persons could and couldn't successfully manage controlled substance use, it would seem wiser to stay with the total abstinence approach to rehabilitation; she did not question the abstinence "forever" phase, in that her sobriety was built around the AA one-day-at-a-time approach; she also stated that her "self research" with controlled drinking had led to her resumption of alcoholic drinking.

J. Watson: stressed the ultimate goal of self sufficiency, behaving and choosing responsibly--this emphasis allows for both abstinence and controlled substance use; he did advocate a period of total abstinence following the completion of treatment (from one to two years) before attempting controlled substance use; he declined attempting to structure any guidelines for staff (recovered and nonrecovered) re their use of substances--he asserted that what staff do on their own time was their responsibility.

A. Soifer: argued that an abstinence approach for the present was the most practical path to pursue, especially since no known characteristics distinguished those who could and those who could not handle controlled substance use; from his review of the literature, he estimated that from 15-25% of all recovered addicts could successfully handle controlled substance use; he urged the substance abuse field to seriously explore new alternative treatment modalities and goals--including controlled substance use, although *not* as a substitute for abstinence; he acknowledged that mistakes would be made during this period of exploration, but that the end result justified the risk; he supported a one year post-treatment period of abstinence before any efforts to experiment with controlled substance use; he stressed the need for developing sanctioned support systems for controlled substance users who formerly had abused substances; he declared that it was *unethical* to deny help and support to those recovering addicts who had decided to attempt controlled substance use.

Fourth Seminar

The last seminar, entitled "*A Community Day Discussion of Recommendations and Implementations*," was chaired by Dr. Jean Feinberg, chair of the ad hoc Abstinence Committee. This program began with a brief review of the preceding seminars by Dr. Feinberg.

Following this review, the audience was directed to read each of four issue statements concerning abstinence and a brief description of a situation relating to each of the issues. The audience then was allowed to form discussion groups around whichever of the four issues that they were most interested in discussing.

The discussion groups were allowed approximately ½ hour to discuss the issue and accompanying situation they had chosen and to develop three recommendations based on their discussion.

Below are the four issue statements, with their accompanying situations, and the recommendations made by the discussants.

#1 Issue Statement: What do we do about residents who say they want to use again, when they leave Eagleville?

Situation: Jack Jones, a 20 year old heroin addict, who is finishing the final week of in-patient treatment, tells his therapist that he intends to drink socially and smoke a little pot after leaving the program (he never abused either of these two substances). He states that he doubts he would ever again get involved with heroin and seems very sincere about this. Jack has shown excellent progress and growth in his therapy group. Jack's therapist, Marty, is a recovered heroin addict who just recently began drinking socially without any detrimental effects following one year of total abstinence.

Recommendations: 1) *Open communication* so that the resident will be open to explore *why* s/he made the decision.

2) Introducing various alternatives by utilizing *several* staff models (recovered drug addicts) who have begun to drink following different periods (lengths of) abstinence.

3) Provide supports for decision-making around using and for alleviation of guilt and "failure."

#2 Issue Statement: What about recovered staff members who seek support for their controlled use of a substance?

Situation: Martha Howard, a recovered alcoholic social worker at EHRC, has been smoking pot for nearly six months. She is presently smoking one joint a day, usually in the evenings. She is very confused and conflicted because she gets good feelings from smoking, and she is afraid that her enjoying smoking may mean she'll "slip" back into alcoholism again. She would like to discuss these concerns with another recovered staff person who is attempting to use some substance without abusing that substance, but she doesn't know where to go.

Recommendations: 1) Use informal social network to gather information to make her own decision (include non-recovered staff).

2) An institutional statement/policy put forth that recovery is a process, and that in advanced stages, alternatives to total abstinence are possible.

3) Formalized support group.

#3 Issue Statement: What about the policy of the institution regarding recovered staff members who begin to use--and control their use?

Situation: At the conclusion of a meeting of department heads, the director of department X informally tells the director of department Y that one of his recovered staff members (a former drug addict) got drunk at a party over the weekend and had to be driven home in someone else's car. The director of department Y knew that this staff member had been drinking socially for at least six months after two years of total abstinence. The staff member's work has been very satisfactory to date.

Recommendations:

1) Approach problem in two distinct ways--one for the welfare of the individual and one for the institution. Look at the performance, attendance, etc., and the objective evidence over time. Offer support, help to the individual.

2) Minimize inherent conflict between policy that allows recovered staff to use in a controlled manner and abstinence in therapy by having a specified period of time during which complete abstinence is required in the treatment, following which each individual makes his/her own decision. Be out in the open re this policy with staff and residents.

3) Re institutional policy at EHRC. We have no detailed written policy--there are some precedents.

#4 Issue Statement: How can/should we present information on controlled use to residents and candidates?

Situation: Howard Thomas, a staff member at EHRC, has been approached by a number of candidates and residents seeking information about the Rand Report which indicated that some alcoholics who had undergone treatment were able later to resume drinking in a non-alcoholic fashion. Howard is familiar with the Rand Report and other research data, but he is not sure whether or not he should share this information with residents and candidates. A group of residents and candidates have asked if this information could be dealt with in a patient education program.

Recommendations:

1) Information should be shared due to the following: the information has already been shared and published; residents and candidates have a legitimate right to the information; and there is no reason to hide the information.

2) The information should be shared in a most digestible fashion--constructive manner

(choices/changes/guidelines/responsibilities.

3) To carefully explain the "why" of Eagleville's commitment to abstinence, i.e., to gain some initial control over "TOTAL LIFE."

1977 EHRC CONFERENCE

The staff development seminar proved to be such a positive experience for our staff that we decided to share our experience with other workers in the drug and alcohol field by devoting the 1977 EHRC Conference to the abstinence-controlled use issue. For this purpose, a select panel of experts, representing all points of view on this topic, were invited to address the conference. The panel included:

Jerome F.X. Carrol, Ph.D., Director of Psychological Services, Eagleville Hospital and Rehabilitation Center
Edward Gottheil, M.D., Ph.D., Professor of Psychiatry and Human Behavior, Thomas Jefferson University Hospital
Frank Seixas, M.D., Medical Director and Director of Research and Evaluation, National Council on Alcoholism
Edward C. Senay, M.D., Professor, Department of Psychiatry, University of Chicago; Executive Director, Substance Abuse Services, Inc., Chicago, Ill.
Mark B. Sobell, Ph.D., Associate Professor, Director of Graduate Studies on Alcohol Dependence, Vanderbilt University, Nashville, Tenn.
Ms. Ariel Winters, Founder/Director, Association of Drinkwatchers International, Haverstraw, N.Y.

Dr. Joel Fort, Executive Director of the National Center for Solving Special Social and Health Problems was the conference's featured Donovan Day Memorial Lecturer.

CONCLUSION

Subsequent to the above discussion/presentations, EHRC's two most influential committees, Policy and Review and Treatment Review, met in February 1978 to finally resolve the issue of abstinence vs. controlled use as a condition for awarding the PHD degree. The results of their discussions have been summarized as follows:

Arguments for Abstinence as a Criterion for Award of the PHD:

1. As illustrated by the work of AA, there is considerable evidence that the total and permanent abstinence model works.
   a) countless numbers of addicted people have attempted to "experiment" and/or use socially and have failed--including Eagleville recovered staff.
2. Within any complex program, such as Eagleville's, it would be very difficult to promote philosophies or values which seem to be at variance; for example, total abstinence and controlled use:
   a) the patients would seize on these apparent contradictions in a manner that would impede their rehabilitation.
3. There has to be a minimum period of abstinence from all chemicals (at least one year) in order to enable the patients to rid himself/herself of chemical dependency and lifestyle.
4. The experimental and clinical evidence in support of controlled use is both scant and of questionable quality/relevance.
5. Presently, it is impossible to distinguish which substance abusers can resume substance use without abusing substances; therefore, it is too risky to support a less than total criteria.
6. Total abstinence is essential in order to generate evidence of the patient's having established internal controls.

For Less than Total Abstinence:

1. Clinical and experimental evidence does exist to support a less than total abstinence model of rehabilitation.
2. It is a well-known fact that there are many failures in total and permanent abstinence programs.
3. There are living, honest-to-God Eagleville recovered staff who use but don't abuse substances.
4. There is a significant portion of the drug and alcohol field which sanctions a less than total abstinence approach to rehabilitation; this includes methadone maintenance programs which permit patients long term use of psychotropic medications, and programs like Drinkwatchers which permit people to choose their self-defined goal of rehabilitation.
5. Since a significant portion of patients will elect to attempt social use, it is better that they be permitted to discuss this decision and to obtain support for it while in treatment; this support would take the form of assisting the patient to think through the decision and to monitor carefully its consequences.
6. A controlled use approach to rehabilitation will generate useful evidence of internal controls.

Arguments for Abstinence as a Criterion for Award of the PHD: (cont'd.)

7. Eagleville would be blackballed by other drug and alcohol programs and referral sources if it acquires the "image" of an institution that favors controlled use, that is, which teaches people to drink or use drugs socially. In a related vein, other agencies might not understnad and/or accept the controlled use notion of successful rehabilitation. For example, doubt was expressed whether controlled substance users could qualify for receipt of the Bureau of Vocational Rehabilitation's "certificate of rehabilitation."
8. Staff objected to a "watered down PHD," meaning that allowing controlled users to receive the award would diminish the significance of the award for those who abstained. It was also argued that allowing both abstinence and controlled users to receive the PHD would change the "meaning" of the award for those who had already received the award.
9. While total abstinence as a criterion of successful rehabilitation has its limitations, it is a neat and clean (no nonsense) criterion which everyone can understand/appreciate.
10. Total abstinence will benefit those substance abusers who are cross-addicted and/or who have abused different categories of substances, e.g., alcohol and other drugs.

For Less than Total Abstinence: (cont'd.)

7. Allowing controlled users as well as abstainers to qualify for the PHD does not mean Eagleville "favors" controlled use, rather it would simply represent an acceptance of another variety of successful rehabilitation. Other agencies *can* understand and appreciate this duality of successful rehabilitation. Besides, Eagleville has committed itself to a pursuit of truth and cannot be intimidated by the negative judgments of others regarding this pursuit.
8. Awarding the PHD to both abstainers and controlled users will in no way diminish the significance of the abstainers' achievement of one year of sobriety nor would it in any way affect the significance of the achievements of past recipients of the award.
9. Total abstinence is not equivalent to personal growth and self-fulfillment--in fact total abstinence may actually mislead the former substance abuser into thinking he/she has made more progress than he/she actually has achieved, thus increasing the probability of that person relapsing.
10. Total abstinence fails to distinguish between substances of abuse and substances of use. It contributes to a life-long stigmatization of the person as a diseased/crippled/different (inferior) person.

Arguments for Abstinence as a Criterion for Award of the PHD: (cont'd.)

11. Those controlled users who really "have it together" will not lie to get a PHD. If they have made "real" progress, they'll know it, and they won't need a PHD to prove it to anyone.

For Less than Total Abstinence: (cont'd.)

11. The total abstinence criterion may encourage dishonesty among recovering staff who have achieved significant personal growth and have not abused any substance(s) who believe they are as deserving of formal recognition for their efforts as their abstaining, recovered co-workers. At present, there is no alternative to the PHD degree.

On February 16, 1978 the issue was resolved, first by Treatment Review, then by Policy and Review of Feb. 23, 1978. Both committees agreed to sustain the total abstinence rule for the award of the PHD. In recognition of the points raised by those who favored allowing controlled users to qualify for the PHD, a proposal was made to create a parallel award. This award would be given to those recovered staff who have demonstrated personal growth while continuing in treatment on an outpatient basis during the 12 months following their discharge from our inpatient program, but who elected to attempt controlled use. This proposal is under active consideration, although no final decision has been taken to date on its implementation.

Although this paper dealt with the granting of a particular award at a particular institution, I think all drug and alcohol specialists can appreciate the generality and significance of the issue we have addressed at EHRC. My own personal prediction is that more will be heard on the issue of abstinence-controlled use in the years ahead--both at EHRC and elsewhere.

REFERENCES

Armor, D.J., Stambul, H.B., and Polich, J.M. 1976. *Alcoholism and Treatment*. Santa Monica: Rand Corporation.

Bailey, M.B. and Stewart, J. 1967. Normal drinking by persons reporting previous problem drinking. *Quarterly Journal of Studies on Alcohol*. 28: 305-315.

Carroll, J.F.X. 1975. "Mental illness" and "disease": Outmoded concepts in alcohol and drug rehabilitation. *Community Mental Health*. 11(4): 418-429.

Carroll, J.F.X. (ed.) 1977. *National Drug/Alcohol Collaborative Project (NDACP) Final Report* (NIDA Grant H81 DA 01113). Rockville, Md.: National Institute on Drug Abuse.

Cohen, J., Liebson, I.A., Faillace, L.A., and Speers, W. 1971. Alcoholism: Controlled drinking and incentives for abstinence. *Psychology Reports*. 28: 575-580.

Cutter, H.S.G., Schwab Jr., E.L., and Nathan, P.E. 1970. Effects of alcohol on its utility for alcoholics and nonalcoholics. *Quarterly Journal of Studies on Alcohol*. 31: 369-378.

Davies, D.L. 1962. Normal drinking in recovered alcohol addicts. *Quarterly Journal of Studies on Alcohol*. 31: 369-378.

Davis, F.T. 1976. Abstinence: Goal for Rehabilitation. *American Journal on Drug and Alcohol Abuse*. 3(1): 7-12.

Drewery, J. 1974. Social drinking as a therapeutic goal in the treatment of alcoholism. *Journal of Alcoholism*. 9: 43-47.

Engle, K.B. and Williams, T.K. 1972. Effects of an ounce of vodka on alcoholics' desire for alcohol. *Quarterly Journal of Studies on Alcohol*. 33: 1099-1105.

Ewing, J.A. and Rouse, B.A. 1976. Failure of an experimental treatment program to inculcate controlled drinking in alcoholics. *British Journal of Addiction*. 71: 123-134.

Fox, R. 1967. A multidisciplinary approach to the treatment of alcoholism. *American Journal of Psychiatry*. 123: 769-778.

Gerard, D. and Saenger, G. 1966. *Outpatient Treatment of Alcoholism: A Study of Outcome and Its Determinants*. Toronto: University of Toronto Press.

Gottheil, E. 1976. Advantages and disadvantages of the abstinence goal in alcoholism. *American Journal on Drug and Alcohol Abuse*. 3(1): 13-23.

Gottheil, E., Corbett, L.O., Grasberger, J.C., and Cornelison Jr., F.S. 1972. Fixed interval drinking decisions I. A research and treatment model. *Quarterly Journal of Studies on Alcohol*. 33: 311-324.

Gottheil, D., Alterman, A.I., and Skoloda, T.E. 1973. Alcoholics' patterns of controlled drinking. *American Journal of Psychiatry*. 130(4): 418-422.

Kendell, R.E. 1965. Normal drinking by formal alcohol addicts. *Quarterly Journal of Studies on Alcohol*. 26(2): 247-257.

Lovibond, S.H. and Caddy, G. 1970. Discriminated aversive control in the moderation of alcoholics' drinking behavior. *Behavior Therapy*. 1: 437-444.

Marlatt, G.A., Demming, B., and Reid, J.B. 1973. Loss of control in alcoholics: An experimental analogue. *Journal of Abnormal Psychology*. 81: 233-241.

Mello, N.K. and Mendelson, J.H. 1970. Experimentally induced intoxication in alcoholics: A comparison between programmed spontaneous drinking. *The Journal of Pharmacology and Experimental Therapeutics*. 173: 101-116.

Merry, J. 1966. The "loss of control" myth. *Lancet*. 1257-1258.

Miller, W.R. 1978. Behavioral treatment of problem drinkers: A comparative outcome study of three controlled drinking therapies. *Journal of Consulting and Clinical Psychology*. 46(1): 74-86.

Miller, W.R. and Munoz, R.F. 1976. *How to Control Your Drinking*. Englewood Cliffs, N.J.: Prentice-Hall.

Nathan, P.E., Titler, N.A., Lowenstein, L.M., Solomon, P., and Rossi, A.M. 1970. Behavioral analysis of chronic alcoholism. *Archives of General Psychiatry*. 22: 419-430.

Ottenberg, D.J. and Rosen, A. 1971. Merging the treatment of drug addicts into an existing program for alcoholics. *Quarterly Journal of Studies on Alcohol*. 32(1): 94-103.

Paredes, A., Hood, W.R., Seymour, H., and Gollob, M. 1973. Loss of control in alcoholism: An investigation of the hypothesis with experimental findings. *Quarterly Journal of Studies on Alcohol*. 34: 1146-1161.

Paredes, A., Ludwig, K.D., Hassenfeld, I.N., and Cornelison Jr., F.S. 1969. A clinical study of alcoholics using audiovisual self-image feedback. *Journal of Nervous and Mental Disorders*. 148: 449-456.

Pattison, E.M. 1966. A critique of alcoholism treatment concepts with reference to abstinence. *Quarterly Journal of Studies on Alcohol*. 27(1): 49-71.

Pattison, E.M., Headley, E.B., Gleser, G.C., Gottschalk, L.A. 1968. Abstinence and normal drinking: An assessment of changes in drinking patterns after treatment. *Quarterly Journal of Studies on Alcohol*. 29: 610-633.

Popham, R.E. and Schmidt, W. 1976. Some factors affecting the likelihood of moderate drinking by treated alcoholics. *Journal of Studies on Alcohol*. 37(7): 868-882.

Sobell, M.B. and Sobell, L.C. 1973. Individualized behavioral therapy for alcoholics. *Behavior Therapy*. 4: 49-72.

Sobell, L.C. and Sobell, M.B. 1974. Time to acknowledge reality. *Addictions*. 21(4): 2-29.

Sobell, L.C. and Sobell, M.B. 1975. Legitimizing alternatives to abstinence--implications now and for the future. *Journal of Alcoholism*. 10: 5-16.

ADAPTATION OF A TIME TESTED TREATMENT MODALITY FOR ALCOHOLISM TO A HEARING IMPAIRED POPULATION

Hank Berman

Catholic Social Service/Hearing Impaired Program

The last few years have seen an upsurge in attempts to provide various services to handicapped individuals. In the area of mental health, professionals have been taking a closer look at the special needs of the disabled. In examining the mental health needs of the hearing impaired, many unique and difficult issues arise, especially with regard to the crucial factors related to communication. Therapy, for example, will have a greater chance of success when there is full communication and a common understanding of the nature of the "enterprise". This is especially true in the case of alcoholism, in which an appreciation of the existence of the problem is considered by most a prerequisite for effective therapeutic intervention. This paper will examine problems faced in the establishment of an Alcoholics Anonymous group for the hearing impaired as well as efforts made to overcome these problems.

Before proceeding, some definitions are in order. The alcoholic is considered "a person whose excessive use of alcohol results in serious medical, social, domestic, vocational, or legal problems" (Woodruff, Goodwin & Guze 1974). For use in this paper, a hearing impaired person (used interchangeably with deaf) is one who is unable to hear and understand speech (Schein & Delk 1974) and for whom sign language can be helpful.

Estimates of the extent of alcoholism range from five to nine million Americans (S. Cain 1970). Assuming this percentage holds true for the deaf, since deafness cuts across cultural, racial, age and sexual groupings, we would expect to find between 45,000 and 81,000 hearing impaired alcoholics in the United States. Given these numbers the question arises as to where these individuals have received help in the past and what is the extent of our knowledge of this population with regard to the problem of alcoholism. A review of the literature is startling in its paucity and can be explained best by looking at some of the cultural components of the deaf community (to be undertaken in a subsequent part of the paper). It should be noted that while "denial" is an integral part of the alcoholic's repertoire, it functions for the professional and many significant others in the alcoholic's life. Charles Hill, in his attempt to seek out the deaf alcoholic, found that counselors with the State Department of Rehabilitation tend to "shy away from a person with an alcohol problem, especially if he were deaf, because they didn't want their recovery caseloads to be way down. They accepted those easy cases. They didn't want to deal with something they didn't know about. It was easier to say he has an employment problem, he can't get up in the morning" (Boros & Sanders, eds. 1977).

The issue of denial of the existence of the problem of alcohol abuse among the deaf by hearing professionals working with the deaf was the first encountered by Jack Gorey of the Alcohol Project for the Deaf. For those wanting to serve this population, denial of the problem can be quite disheartening, especially when it comes from those one would expect to have the information and perhaps be in a position to refer those in need of help. Boyce R. Williams, himself hearing impaired, refers to paternalism as "that albatross of deaf people . . . that degrades, imprisons, stifles and defeats at every turn" (Rainer & Altschuler 1970). Paternalism often manifests itself as a desire to see the deaf population as untainted by behavior considered aberrant by the hearing majority. There seems to be a need, especially among those traditionally involved in advocating for the deaf, to "picture the deaf in a good light". Unfortunately this sort of paternalism draws attention and services to those who need it least and away from those who need it most.

In the case of hearing professionals, the message was mixed: the explicit message was "we will help all we can", while the implicit message was "there is no problem". There were also those unfamiliar with deafness who felt that deafness must be so isolating and frustrating an experience that periodic alcohol abuse could be seen as somehow more acceptable among the deaf. After many frustrating encounters with agencies already serving the deaf, efforts to assess the nature and depth of the problem focussed on the deaf community itself.

Leaders of the deaf community play an important part in disseminating information to the community and from the community to hearing agencies. Unfortunately not enough deaf leaders themselves work for these agencies. Many of those who were approached seemed to have introjected elements of paternalism toward other members of their community. Again, denial of the problem was pervasive; there seemed a desire to present only the healthy attributes of the deaf to those concerned with helping. Here the denial appeared to have almost an "ego splitting" function (i.e., those who do drink to excess are not for us; since they look bad to the outside world we cannot consider them part of the deaf world). There was also an element of protectiveness toward the deaf which implied that since the hearing world has been the source of much frustration and false hopes for the deaf, they needed to be protected from further misguided efforts at helping. Other efforts at assessing the problem have met similar forms of resistance (Boros & Sanders, eds. 1977).

It was felt at this point that if the problem of alcoholism did indeed exist in the deaf community, the best approach was to undertake a discussion of what might be the best treatment modality and begin its implementation with the expectation that word of mouth and the deaf "grapevine" would eventually bring those in need of treatment. Such a discussion took place between the author (currently a marriage and family counselor with Catholic Social Service/Hearing Impaired Program) and Jack Gorey (Director of the Bay Area Hearing Society's Alcohol Project for the Deaf), the former having at that time only a cursory knowledge of the parameters of deafness. The author undertook a review of the literature in determining the value of Alcholics Anonymous (hereafter referred to as A.A.) as a therapeutic tool in the treatment of alcoholics. What follows is a brief critique of the A.A. modality.

Raymond Corsini (1957) posits that "In terms of numbers A.A. is the most important group engaged in psychotherapy". By way of classification, A.A. is seen by most to be an "inspirational supportive" group in which one of the "prime therapeutic agents is the idea that the individual has potential goodness in him and that others just like him have achieved resolution of problems similar to his" (Kaplan & Sadock, eds. 1971). Thimann (1966) sees the therapeutic approach as "religious conversion and an individual 'older brother-kid brother' relationship between an older member and one who has just joined A.A."

Concerning the efficacy of the group approach versus individual treatment, Kaplan and Sadock (1971) feel that "with the essential or primary type of alcoholic, the usual type of individual psychotherapy is almost impossible because of the patient's marked ambivalence and rage if any frustration is encountered." They see the phenomena of transference as being diminished and better handled in a group situation. "The patient does not change within himself but feels changed because the supposedly powerful therapist acts benevolently" (Kaplan & Sadock 1971). For Bales (144) the "whole process of relating past experiences and comparing notes" that occurs in A.A. is a means of laying bare the ruses, rationalizations and techniques which enable the alcoholic to continue to use drinking as a means of adjustment. He feels the "moral principles advocated by a closely knit solidarity group, can only be internalized and made effective against self-centered, satisfaction directed impulses by an involuntary feeling of belongingness and allegiance to such a group". Hence, the A.A. group fosters identification with other members of the group in their common struggle and as such offers ego support to its members. Suggestions are shared as to ways of avoidance as well as methods of coping with the frustrations of everyday life without the use of alcohol. The message which gets transmitted over and over again is that it can be done! The sponsor (one by whom the alcoholic is introduced to A.A.) acts as a positive role model and guide. His/her example stands in direct opposition to the sense of hopelessness and aloneness which characterizes the alcoholic before treatment. Entry into the group is seen as a positive step and subsequent efforts at remaining sober are reinforced. Analysis of the etiology of alcoholism is deemphasized and instead the group acknowledges that causation may forever remain a mystery; understanding is not a prerequisite for control. There seems to be little coercion to participate in discussion. The decision of when to share thoughts and feelings is left to the individual. This allows as much time as is needed by the incoming individual to begin to identify with other members of the group by listening to their stories. The new member does however immediately feel the concern of the group.

Cain (1970) criticizes the dogmatism of A.A. He takes umbrage with the "anti-analytical dogmatism" of A.A. which he calls the "utilize don't analyze" message. Cain feels it is a mistake for A.A. to see sobriety as an ultimate goal in life and for that "glorious end to justify any means". In spite of this criticism, Cain goes on to say that he still believes that "A.A. provides the best possible way, at present, for most alcoholics to get sober and start a new life without alcohol". Fox (1957) goes as far as to prescribe that "No matter what other form of treatment is used, each patient should be urged to take part in the group life of A.A. as well."

Since the emphasis placed on anonymity by A.A. precludes any statistical material being available in terms of success rate, the author had to content himself with observation and reading of the judgments of others in the field of alcoholism. The author responded enthusiastically when asked to act as interpreter/facilitator for an A.A. group for the deaf. The importance of securing cooperation between someone familiar with the A.A. process and someone having an understanding of the dynamics of the deaf community cannot be understood when considering outreach into this community. Ideally the person having familiarity with the deaf community should him/herself be deaf. If such a person is unavailable, a hearing person with a fluency in sign language is preferable. It is the author's experience that a fluency in ASL (American Sign Language) is the key necessary to gain entry into the deaf community. It is this fluidity in sign language that indicates to the prelingually deaf person that his/her handicap will be understood and communication insured. Indeed it would be difficult to imagine a sponsor-neophyte relationship existing (given the prelingual deafness of the neophyte) without the sponsor having some knowledge of sign language. The distrust between "cultures" is a reality best understood if one compares the feelings of other minority groups as an analogy. A history of being misconstrued and misunderstood sows seeds of distrust. It is interesting that the author noticed an increase in the effectiveness of the group leader as role model within the group as his sign language improved. A complication arose as resistance was found to the use of sign language in the group by members who had become deaf after many years of functioning in the hearing world. This resistance was in many instances part of the process of denying their deafness. The message given these members was consistent with that given to all A.A. participants: "you are here to deal with your alcohol problem. If sign language is not helpful in communication you have a right to ignore it but it will continue to be used for the benefit of those who use it as their only means of communication." Both members who complained about the use of ASL now use it as an adjunct to lip reading in the group. One member alternates between using sign language and denouncing those who teach it.

Once a time and place are chosen for A.A. meetings, the information should be given to any organization providing services to the deaf. It should also be given to local police stations, jails and the courts. Unfortunately most referrals come from the latter three sources. However, some do come from within the community itself and it is here that the "deaf grapevine" is most important. The deaf grapevine is the means by which news of the community and its members is shared among the deaf. If information is made available to several members of the community, one can expect this information to be disseminated throughout the community. More formal avenues such as newsletters and news programs aimed at the deaf community should also be utilized.

The group should secure the services of an interpreter skilled in sign language for every meeting, ideally the same person at every meeting. The interpreter should be made aware of the nature of confidentiality as it is observed in the group as well as its specific applications to the deaf community. A professinal or certified interpreter should already be aware of the importance of confidentiality in the deaf community. It should be noted that while A.A. itself offers its members the cloak of anonymity, that cloak does not work as well for members of the deaf community since they may realistically assume that their participation in A.A. will be known throughout the community if another member shares that information with any deaf person outside the group. It is wise therefore to explain the concept of anonymity at the beginning of each meeting, especially if a new member is present.

Previously mentioned were the issues of denial and protectiveness as they related to hearing professionals working with the deaf. These same issues are operational for the deaf individual as well. There is some evidence to show that prelingually deaf individuals are more prone to the use of the more infantile defense mechanisms (Rainer and Altschuler 1969). The mechanism of denial becomes a bit clearer when viewed in relation to the prevailing attitude of the deaf to alcoholism. The deaf community views alcoholism as a moral problem. Those unable to control their drinking are seen as lacking the will power to stop (Corsini 1957). Hence if one knows that community awareness of a drinking problem will only earn moral castigation among his/her peers, there is a greater tendency to deny the problem. The denial operative in all alcoholics is then potentiated by membership in a valued social group which has a negative image of alcoholism. Indeed, potential members of the deaf A.A. group have expressed a fear of ostracism or ridicule by the deaf community if their alcoholism became known.

Another problem affecting the delivery of services to the deaf alcoholic is the cultural differences perceived by the deaf within their own subculture. Many times those deaf individuals with more education and verbal skills seek to separate themselves from those who are "low-verbal". Since many of the "high-verbal" deaf hold professional positions in which they deal with other members of that community, they are often more reluctant to enter a group treatment situation. The terms "high" and "low" verbal are used here to refer only to English skills and have no correlation with intelligence (Jacobs 1974). A cultural consideration unique to the deaf population is that many cities have "clubs for the deaf" which usually function as social gathering places for members of the deaf community. There clubs are traditionally run by the membership. In some of these clubs financial hardships force the members to augment dues with profits from the bar, if one exists on the premises. This set of circumstances makes for a built-in resistance to cooperation with a program leading to a reduction in drinking.

Our experience with deaf A.A. groups in the San Francisco Bay Area has shown the importance of establishing a cooperative relationship between one familiar with the A.A. process and one knowledgeable about the deaf community. Familiarity with the range of social services in the community is essential. One hundred percent of the hearing impaired clients who have entered the deaf A.A. group in San Francisco have presented ancillary problems congruent to their entry into the group. In many cases addressing these problems was the only way to insure the clients' gaining a foothold in the process. For example, two deafened adults ["individuals who lost their hearing after they were ill through their educational programs" (Jacobs 1974)] presented the dual problems of excessive drinking and frustrations connected with their inability to receive communication clearly. Both reported frustrating attempts to benefit from "hearing" A.A. groups due to their lack of lipreading skills and the inefficacy of well-intentioned notetakers. It was evident that these men would need instruction in communication skills before they could benefit from the use of an interpreter. Their sponsor was able to help them see the need for this and was able to secure them this service. Other kinds of ancillary problems include financial or vocational difficulties relating to alcoholism, deafness or both. There is a therapeutic advantage in being able to offer help directly or to refer appropriately: this helps to build a supportive relationship, overcoming distrust and beginning a process that stands in opposition to longstanding isolation.

A significant criticism of A.A. is that while it deals with the dysfunctinal behavior, it neglects the underlying psychodynamics which may have led to the behavior. It has been helpful in our experience to look at A.A. as the central component in a multidimensional approach, including individual counseling, vocational training, etc. While this could be seen as contravening the orthodoxy of A.A., it is the author's feeling that many alcoholic clients can benefit from individual or family counseling as an adjunct to the group. Such counseling can provide an opportunity for the client to deal with issues which s/he may feel uncomfortable about raising with the group, or which the group is ill equipped to address.

Historically A.A. has answered the criticism of those who see the family of the alcoholic as needful of help by establishing Al-Anon. Al-Anon has been extremely useful in the family to understand and cope with the behavior of the alcoholic. In many instances the alcoholic has been led to treatment as a result of what family members have learned at Al-Anon meetings. This also applies to wives and husbands of alcoholics who happen to be deaf. Since the numbers have not yet warranted separate meetings for deaf spouses, integration into a hearing Al-Anon group has been successfully tried with the use of an interpreter. Until this integration was feasible, one spouse attended the deaf A.A. group to get an idea of the nature of the problem.

Lacking a clear set of criteria for success in the field of alcohol abuse, it is difficult to make specific statements in this regard. It should be noted that during the two years of the group's existence, three participants have died. Others, who might have died if not for their participation in the group, are continuing in their attempts to achieve sobriety. Two members referred by the prison system have begun to lead productive lives, unencumbered by a compulsion to drink. One deaf participant in an Al-Anon group reports a decrease in her guilt feelings surrounding her husband's alcoholism. Two deafened members have shown incredible progress in learning sign language and are making tentative steps toward utilizing their newly acquired skills to find a place in the deaf community. Most important perhaps is the sense one gets from the deaf community that the deaf A.A. group has achieved a level of acceptance as a valid helping agent. American Sign Language has no single word for "alcoholic", but in the Bay Area the word that one sees repeatedly for this affliction is "A.A." This is but one indication that a chink in the attitudinal armor has been made.

This leads to thoughts of the future of the Alcohol Project for the Deaf. A continuing struggle to change the attitudes of both professionals working with the deaf and the deaf community itself is necessary. The fight will be carried on most successfully by recovered deaf alcoholics themselves. Links to the community are continuously being strengthened. Prelingually deaf adults will be reached more effectively if the A.A. literature (which plays an important part in the therapy) is made accessible to the conceptual level of most deaf adults. As more groups arise, more judges will be willing to divert individuals in the criminal justice system to programs such as the Alcohol Project for the Deaf. The author has already seen the need for a Narcanon project for the deaf as there are many young deaf people involved in drug abuse who have nowhere to turn for help. In general we expect to see a rise in the number of deaf people participating in such programs as awareness catches up with the problem. We are hopeful that as schools for the deaf realize the extent of the problem, they will welcome preventative programs which can provide the information so vital in overcoming prejudices and fears which hold those troubled in check.

REFERENCES

Altschuler, D.Z. & Rainer, J.D. 1969. *Mental Health and the Deaf: Approaches and Prospects.* U.S. Dept. of Health, Education and Welfare.

Anonymous. 1957. *Alcoholics Anonymous Comes of Age - A Brief History of A.A.* Alcoholics Anonymous World Services, Inc. New York: American Book.

Bales, R.F. 1944. "The Therapeutic Role of Alcoholics Anonymous As Seen By A Sociologist", *Quarterly Journal of Studies on Alcohol.* 2: 267-278.

Boros, A. & Sanders, E., eds. 1977. *Dimensions in the Treatment of the Deaf Alcholic.* Cleveland: Northeast Ohio Deaf Development Organization, Inc.

Cain, S. 1970. *The Treatment of Alcoholics - An Evaluative Study.* New York: Oxford University Press.

Corsini, R.J. 1957. *Methods of Group Psychotherapy.* New York: McGraw Hill Book Co., Inc.

Di Carlo, L.M. 1964. *The Deaf.* Englewood Cliffs, New Jersey: Prentice-Hall, Inc.

Griffith, J., ed. 1969. *Persons with Hearing Loss.* Springfield, Illinois: Charles C. Thomas, Publisher.

Hardy, R.E. & Cull, J.G. 1974. *Educational and Psychosocial Aspects of Deafness*. Springfield, Illinois: Charles C. Thomas, Publisher.

Himwich, A.E., ed. 1957. *Alcoholism - Basic Aspects and Treatment*. Washington, D.C.: American Assn. fo the Advancement of Science.

Jacobs, L.M. 1974. *A Deaf Adult Speaks Out*. Washington, D.C.: Gallaudet College Press.

Kaplan, H. & Sadock, B.J., eds. 1971. *Comprehensive Group Psychotherapy*. Williams and Wilkins, Publishers.

Rainer, J.D. & Altschuler, K.Z. 1970. *Expanded Mental Health Care for the Deaf: Rehabiliation and Prevention*. Washington, D.C.: U.S. Dept. of Health, Education and Welfare, Social Rehabilitation Service.

Rainer, J.D., Altschuler, K.Z. & Kallmann, F.J., eds. 1969. *Family and Mental Health Problems in a Deaf Population*. Springfield, Illinois: Charles C. Thomas, Publisher.

Schein, J.D. & Delk, M.T. Jr. 1974. *The Deaf Population of the United States*. Silver Spring, Maryland: National Assn. of the Deaf.

Schlesinger, H.S. & Meadow, K.P. 1972. *Sound and Sign - Childhood Deafness and Mental Health*. Berkeley: University of California Press.

Thimann, J. 1966. *The Addictive Drinker - A Manual for Rehabilitation*. New York: Philosophical Library, Inc.

Trice, H.M. 1956. *A Study of the Process of Affiliation with Alcoholics Anonymous*. Ph.D. Dissertation, University of Wisconsin.

Wallerstein, R.S., ed. 1957. *Hospital Treatment of Alcoholism - A Comparative, Experimental Study*. New York: Basic Books, Inc.

Woodruff, R.A. Jr., Goodwin, D.W., & Guze, S.B. 1974. *Psychiatric Diagnosis*. London: Oxford University Press.

ESTIMATING THE PROPORTION OF PROBLEM DRINKERS IN AN EX-ADDICT POPULATION: USE OF CAGE AND DRINKING FREQUENCY MEASURES[1]

Steven Belenko, Ph.D.
Mathematica Policy Research, Princeton, New Jersey

Sandra Kellum, M.A.
San Francisco, California

A number of studies have presented estimates of the extent of alcohol abuse among heroin addicts. Although most researchers and treatment personnel have agreed that it is a major problem or potential problem in this population, the extent of the problem has been somewhat difficult to determine, since there has been a wide variation in the reported proportion of alcohol abusers. In addition, the findings are difficult to interpret due to differences in methodology, in the populations studied, and in the definition of alcohol abuse. It has become increasingly clear, however, that problem drinking presents a major barrier to the successful treatment of heroin addiction.

There remain important barriers to dealing with alcohol-abusing addicts in treatment: it has been difficult to get a precise estimate of the prevalence of problem drinking in this population (Belenko 1979), and accurate, manageable means of screening active and potential problem drinkers at the time of admission to treatment need to be developed.

The quantity of alcohol consumed and the frequency of use are often used as a measure in identifying the alcoholic or problem drinker. However, reporting how much and how often one drinks is influ-

---

1 The research for this paper was conducted at the Vera Institute of Justice in New York City, where the authors were Senior Research Associate and Research Analyst respectively. Supported by grants from the National Institute on Drug Abuse and New York City Department of Employment.

enced by a number of factors: one of these is the drinker's perception of his/her drinking. A recent study on the reliability of the quantity/frequency measurement showed a tendency for men to underestimate the amount and frequency they drink, while women were more likely to overestimate their drinking (Garrett and Bahr 1974). Anxiety about one's drinking may play a role in the perceived need to defend or camouflage excessive drinking. In addition, quantity and frequency of use may be a poor identifier of the "binge" drinker who frequently has long periods of abstinence between binges.

In recent years a number of questionnaires have been designed specifically to identify the problem drinker without asking quantity/frequency questions. For example, the Michigan Alcoholism Screening Test (MAST) has been shown to adequately discriminate alcoholics from nonalcoholics in both its standard form (Zung and Charalampous 1975) and in a shortened, self-administered version (Selzer, Vinokur, and Van Rooijen 1975). Another self-administered alcohol screening test developed by Swenson and Morse (1975) was recently tested for validity in a methadone maintenance population (Cohen et al. 1977).

The CAGE questionnaire (Ewing and Rouse 1970) is a brief four-question scale which has been shown to be a viable, rapid screening instrument for identifying alcoholics among hospital patients (Ewing and Rouse 1970; Mayfield, McLeod, and Hall 1974). The items in the CAGE questionnaire are as follows:

CAGE QUESTIONS

C - cut down	Have you ever felt you should cut down on your drinking?
A - annoyed	Have people annoyed you by criticizing your drinking?
G - guilty	Have you ever felt bad or guilty about your drinking?
E - "eye opener"	Have you ever had a drink first thing in the morning to steady your nerves or get rid of a hangover?

Because of its brevity and ease of administration and understanding, the CAGE questionnaire is a potentially useful tool for identifying problem drinkers among persons entering treatment for heroin addiction.

The present study describes a test of the ability of CAGE to identify problem drinkers in an ex-addict population, in comparison to questions on the frequency of drinking.

METHOD

A group of 227 ex-addicts from New York City participated in this study. These individuals represented a randomly selected subsample of ex-addicts participating in an experimental demonstration program of transitional employment of ex-addicts and ex-offenders, called "supported work". Participants in the present study were predominantly male (95%), minority (65% Black and 30% Puerto Rican) and referred from methadone maintenance programs (85%). Their average age was 30 years. In comparison to similar ex-addicts in treatment in New York in 1974, they probably represent a "middle" group of treatment program clients: judged work-ready by their counselors but not yet prepared to hold jobs in the regular marketplace.

The CAGE, drinking frequency, and drinking problem questions were administered as part of the first annual follow-up interview given to participants in the three-year longitudinal supported work experiment. The entire interview lasted about one hour, and was given in person by a member of a trained interviewing staff comprised primarily of minority ex-addicts or ex-offenders.

CAGE Criterion

Prior work with the CAGE questions has shown that positive responses to at least 2 of the 4 items are indicative of alcoholism. In an early study with the instrument, Ewing and Rouse (1970) found that 95% of an alcoholic population answered yes to 3 or 4 of the CAGE items, while 82% of a similar but nonalcoholic group answered no to all 4 items. Mayfield, McLeod and Hall (1974) found that a 2 or 3 positive response criterion was a sensitive screener for alcoholism. In the present study, therefore, a criterion of two or more yes responses to the CAGE questions was utilized.

TABLE 1

RESPONSES TO CAGE QUESTIONS

Number of positive responses	N	%
0-1	176	78%
2	14	6%
3	28	12%
4	9	4%
(2 or more)	(51)	(22%)
Total Sample	227	

TABLE 2

PERCENTAGE OF CAGE-SCREENED RESPONDENTS REPORTING CURRENT OR PAST DRINKING PROBLEMS

Number of Positive CAGE Responses	Reported Drinking Problems N	%
2	11	79
3	23	82
4	8	89
2 or more	42	82%

RESULTS

Of the 227 participants in the study, 51 (22%) had positive responses to at least 2 of the CAGE questions (Table 1); most of these persons (42) answered 2 or 3 of the items positively.

An analysis of the responses of these 51 individuals to survey questions related to the effects of their drinking on employment, health and family indicated that 82% had a current or past alcohol problem. The probability of having a problem increased with the number of positive CAGE responses (Table 2).

The drinking patterns of the CAGE subgroup during the year covered by the interviewer indicated that nearly all of the 51 individuals were drinking daily (61%) or 2-3 times a week (29%). This high rate of drinking suggests that the drinking problems discussed in the interview represented current problems.

In addition to the CAGE questions, all respondents were asked questions on the frequency, quantity, and type of alcoholic beverage consumption during the preceding year. Examination of the data on drinking frequency indicated that the CAGE questions did not select out all the problem drinkers in the sample; a substantial portion of the respondents who drank daily during the year, but did not answer "yes" to two or more CAGE questions, had current or previous drinking problems.

---

1 Minimum of three consecutive months of daily drinking during the year.

TABLE 3

PROBLEM DRINKERS NOT IDENTIFIED BY CAGE OR DAILY DRINKING MEASURES

	N	% of Non-Screened Subgroup
Total not screened	149	--
No drinking problems reported	135	91
Reported drinking problems	14	9
a. Drank 2 to 3 times a week	8	5
b. Drank less than 2 to 3 times a week	6	4

Twelve percent of the sample (N=27) reported daily drinking,[1] but fewer than 2 "yes" responses to the CAGE questions. Of these 27 daily drinkers, 21 (78%) indicated current or past drinking problems. This rate is similar to that of CAGE-screened respondents who indicated a drinking problem (82%), and represents 9 percent of the sample of 227 respondents. Thus, daily drinking, without positive CAGE responses, is also highly suggestive of the existence of a drinking problem.

Combining both the CAGE and daily drinking measures, it is apparent that a large proportion of this ex-addict sample have a current or past drinking problem. A total of 78 respondents (34%) were screened by these two measures, of whom 81% also specifically indicated that a drinking problem existed; this represents 28% of the full sample of 227 respondents.

In order to determine how comprehensively all problem drinkers are screened by the CAGE and daily drinking measures, drinking data from the remaining 149 respondents were also examined. Table 3 summarizes these data. Of the non-screened sample, only 14 (9%) indicated a drinking problem on the interview; the majority of these persons (8 out of 14) reported drinking 2-3 times a week during the year.

In summary, out of the full sample of 227 ex-addicts, there were 77 persons with current or past drinking problems (34%) of whom 63 (82% of the problem drinkers) were identified by the CAGE (55%) and daily drinking (27%) questions.

DISCUSSION

The data from this sample of ex-addicts suggest that a substantial percentage (34%) have current or past drinking problems and that the CAGE questionnaire is a necessary but not sufficient instrument for identifying these problem drinkers. Nearly half of the self-identified problem drinkers in the sample were not screened by the standard CAGE criterion score; most of the others were identified by the drinking frequency measures. By itself, the daily drinking criterion was also not sufficient to identify a reasonable proportion of the problem drinkers.

It appears that the combination of CAGE and daily drinking measures screen most of the problem drinkers (82%). However, there are some possible problems in these data. Given that the existence of drinking problems was self-reported, it is possible that those who under-report their drinking frequency also under-report the existence of drinking problems; this would result in an exaggeration of the ability of the frequency measure (and therefore the combined CAGE and frequency measures) to discriminate problem drinkers. On the other hand, it is possible that some individuals who were screened by the CAGE and drinking frequency criteria but did not report drinking problems actually had past or current problems; this would tend to understate the sensitivity of the measures.

Given the exploratory nature of this study, and the potential attraction of a simple, valid screening device such as CAGE to identify problem drinkers entering treatment, further validation studies should be carried out on other ex-addict samples. At the present time, however, it seems that multiple complementary screening devices are needed to ensure adequate identification of problem drinkers.

REFERENCES

Belenko, S. 1979. Alcohol abuse by heroin addicts: Review of research findings and issues. *Int. J. Addict.* 14(6).

Cohen, A., W. McKeever, M. Cohen, and B. Stimmel. 1977. The use of an alcoholism screening test to identify the potential for alcoholism in persons on methadone maintenance. *Amer. J. Drug Alc. Abuse.* 4(2): 257-266.

Ewing, J.A. and B.A. Rouse. 1970. Identifying the "hidden alcoholic". Paper presented at 29th International Congress on Alcohol and Drug Dependence, Sydney, Australia.

Garrett, G. and H. Bahr. 1974. Comparison of self-rating and quantity frequency measures of drinking. *Q. J. Stud. Alc.* 34: 1294-1306.

Mayfield, D., G. McLeod, and P. Hall. 1974. The CAGE questionnaire: Validation of a new alcoholism screening instrument. *Amer. J. Psychiatry*. 131: 1120-1123.

Selzer, M., A. Vinokur, and L. Van Rooijen. 1975. A self-administered short Michigan Alcoholism Screening Test (SMAST). *J. Stud. Alc.* 36: 117-126.

Zung, B. and K. Charalampous. 1975. Item analysis of the Michigan Alcoholism Screening Test. *J. Stud. Alc.* 36: 127-132.

PREVALENCE OF ALCOHOLISM AMONG METHADONE PATIENTS

Roger Mazze, Ph.D. and Alice Kornblith, Ph.D.

Albert Einstein College of Medicine

A recent prevalence-of-alcoholism study carried out at a New York City hospital-based methadone maintenance program provided data concerning the number of patients addicted to ethanol and an opportunity to test different intervention strategies for polydrug users. There has been, over the past decade, increasing interest in the phenomenon of polydrug use/abuse among methadone maintained patients. Alcoholism has become especially prevalent and given rise to a number of studies attempting to determine both the number with and etiology of this disorder. Current theories range from the use of alcohol to supplement methadone (suggesting that this practice begins after entry in a methadone program) to alcoholism preceding methadone treatment (suggesting that the patient was alcoholic prior to program participation.) The estimated prevalence of alcoholism has ranged from less than 1% to greater than 70% among this patient population.

To assess the prevalence of alcoholism among methadone maintained patients, a study was designed using a single facility whose patient population, in general, reflected the sexual, racial and ethnic characteristics of most urban centered programs. The counseling staff, comprised of both paraprofessional and professional personnel, were asked to complete an open-ended form for each patient in their case load for three points in time: September 1977, October 1977, and March 1978. The form required the counselors to provide reasons for patients they identified as either alcoholic or suspected alcoholic. Throughout the prevalence study period, a special educational program on alcoholism was undertaken by faculty from the affiliated medical school.

FINDINGS

Eleven counselors participated in the first prevalence study. Of the 252 patients reviewed in September, 23.8% (N=60) were identified as alcoholic, 13.1% (N=33) were suspected of being alcoholic. A rank ordering of the reasons for characterization of "alcoholic" revealed that medical problems were most often cited (34 cases). Within this category, convulsions, liver disease, kidney failure, ulcer, diabetes, pancreatitis, and gastritis were predominant. The second reason most given by counselors was voluntary admission by patients (found in 23 cases). History of hospitalization and history of drinking ranked as the third most mentioned reason. Among other listed causes for identification were attendance on the service under the influence of alcohol, general depression and reports from patient's spouse or friends. In contrast, the "suspected alcoholic" category relied almost entirely upon both drinking habits and social dysfunctional behavior. Cited twelve times was excessive alcohol use, followed by patient seen under the influence of alcohol, episodic drinking binges, accidents, physical appearance and violent behavior.

In the October prevalence study, ten counselors identified 27% (N=60) as alcoholic and 12% (N=29) as suspected alcoholic for the 236 patients reviewed. The alcoholic category consisted of medical problems identified 29 times, history of alcohol abuse 28 times, and admission of alcoholism 22 times. The medical problem dimension was noted to be more specific than the September study. Edema, seizures, heart murmur and enlarged liver were cited in addition to the disorders noted in the earlier study. Overall, this category included many more specific conditions which were mentioned by the counselors. Among these were physical indicators of weight loss, swollen abdomen, tremulousness and skin pallor. Additionally, blackouts and patient denial were more often cited. The suspected category showed a marked change in comparison with the September study with the inclusion of specific medical problems (6 times) and patient self-admission to daily drinking (8 times). Again, the pattern of more reasons and greater specificity as shown in the alcoholic category persisted.

The March prevalence study showed no significant difference in the number of patients identified as alcoholic or suspected alcoholic. Cited as alcholic were 24.2% (N=62) and suspected were 12.5% (N=32) of the 256 patients evaluated. What was significant, however, was the continued pattern of citing more specific evidence for identification as alcoholic. Medical problems were noted in 36 cases with hypertension and cardiomyopathy added to the list and a greater frequency of blackouts, and liver disease mentioned. One-third of those identified as alcoholic had liver disease as a reason for this characterization. Self report of excessive alcohol use was noted in 25 cases with history of alcoholism (15), hospitalization for detoxification (15) and overall physical appearance (11) mentioned in descending order. The suspect category generally followed the latter category with identification based on alcohol abuse, patient self-admission, medical problems and dysfunctional behavior (family and friends) mentioned most often.

The data suggest that alcoholism among the methadone patient population ranges from 25% to 40% (if the suspected category is included) and presents medical and social problems requiring immediate as well as long term remediation.

The prevalence study marked the beginning of an educational program for the methadone maintenance counselors. This was the strategy selected by the center's administration and staff to deal with alcoholism among the patient population.

Over the course of eighteen hours, counselors were taught about the causes of alcoholism (genetic, social and psychological), the natural history of the disease (liver disorders, cardiomyopathy, gastrointestinal and neurological problems), management of the alcoholic from both medical and psychosocial aspects and various treatment modalities. Mini-simulations designed to reflect the types of encounters counselors would find themselves in were included in the course. To insure feedback following the course, semi-monthly case conferences were held. Each conference included a case presentation by a counselor plus discussion of possible intervention strategies. It was through this format that realistic and effective treatment modalities could be evaluated.

A number of conclusions may be drawn from the educational course and the case conferences. The data presented earlier indicates that, while the number of identified alcoholics did not increase, knowledge about alcoholism as measured in open-ended lists of "reasons" showed a marked improvement. What is perhaps more revealing, however, is that there exists a fundamental difficulty in treatment of the alcoholic which was demonstrated both in the sustained number of alcoholics and the case conferences. During seven case conferences, the patients presented revealed numerous and complex social and medical problems. Each patient had significant physiological involvement due to alcoholism ranging from early liver disease (fatty liver) to severe cirrhosis and cardiomyopathy. Social problems ranged from marital and employment dysfunction to violent behavior. Many of the patients were candidates for detoxification from alcoholism. The majority were between 20 and 30 years old with almost equal representation of males and females. Intervention strategies for these patients were severely limited due to their own intransiency to treatment, few polydrug in hospital beds in the New York metropolitan area and a lack of an organized program within the clinic for alcoholic patients. None of these problems seemed easily remedial.

Patient intransiency to treatment for alcoholism presents a more substantial problem among polydrug users than among those drug abusers limited to ethanol. The denial process among this patient population seems exacerbated by methadone use. The strategy of withholding methadone until treated for alcoholism has not yet been successful. There also appears to be a reluctance of alcohol detoxification units to accept methadone patients. Since this requires attention to both drugs, the reluctance seems understandable. Finally, the absence of a formal alcoholism program at methadone maintenance units is most probably due to a lack of knowledge about this disease and a serious question as to the efficacy of alcoholism treatment programs. Clearly, new specific programs will be required if this problem is to be managed.

TREATMENT EVALUATION OF ALCOHOLIC PATIENTS IN AN ALCOHOLISM PROGRAM VS. THOSE ON GENERAL MEDICAL SERVICES

Roger Mazze, Ph.D., Edward Julie, B.S., and
Theodore Feldman, B.A.

Albert Einstein College of Medicine

More than 20% of the adult medical patients treated in New York City municipal hospitals were discharged with either a primary or secondary diagnosis of alcoholism during the one year period December, 1976 - December, 1977. These patients were treated either on a general medical service or a special alcohol detoxification unit. In general, patients with complicated medical sequelae of alcoholism were admitted to general medical services (GMS), while patients who voluntarily sought treatment for alcoholism were admitted to detoxification units (DU). These individuals included primarily lower socioeconomic groups, racial and ethnic minorities, predominantly males, and ranged in age from 17 to 67 years (with a mean age of 48).

A treatment outcome study was designed at one of the large New York City municipal hospital centers to determine whether management for alcoholism and its medical complications on a general medical service or a detoxification unit lead to a significant difference in prognosis for the alcoholic patient. Two patient samples were randomly selected from 730 patients discharged with the diagnosis of alcoholism, to represent the population of patients treated on general medicine and on detoxification units. Each subject who participated in the study was administered a one hour interview protocol designed to measure, six months following discharge, abstinence from alcoholic beverages, social stability, and medical status. These three dimensions together formed a continuum of diagnostic criteria (CDC). Abstinence dimension gauged the patient's drinking habits prior to hospitalization, during the six months post discharge period and at the time of interview. Social stability was measured in accordance with pre-tested and weighted questions from the Minnesota Multiphasic Personality Inventory which were designed to gauge the social status of the alcoholic in terms of family, employment, criminal and general social behavior. The medical status dimension evaluated the subject's current medical problems, recent hospitalization(s) and physician visits. Additionally, it monitored thirteen specific medical complications/symptoms of alcoholism and whether medical or psychosocial intervention had ameliorated the complications/symptoms.

A base-line profile of each study participant was established from discharge data included in the hospital chart. This data consisted of an evaluation of the medical status and social condition of the patient at discharge. Although the two patient populations differed in terms of medical diagnosis and alcohol related psychosocial problems at entry, at discharge these two dimensions had been significantly stabilized making it possible to compare the two populations.

The two samples initially consisted of 66 patients of whom 35 were drawn from the alcohol detoxification unit and 31 were drawn from the medical unit at the same hospital center. Patients were contacted by letter with an enclosed return postcard, to which six subjects responded favorably and 51 did not reply. The latter group was followed-up with a telephone call which eventually reached 30 patients of whom 7 agreed to participate in the study. The remainder of the samples were accounted for by false names and addresses, deaths, and people moving. A replacement sample of 16 was randomly selected of whom 9 participated in the study.

FINDINGS

For the purpose of this study, the abstinence dimension was graded on a continuum from no alcohol ingestion (1) to chronic alcohol intake (7). A combined score derived from this continuum for the period prior to and proceeding hospitalization was developed for each subject. A social problem inventory consisting of a list of discrete but related problems (family, friends, job, criminal) formed one element of the social stability dimension. If a patient exhibited none of the problems he/she received (0), if all of the problems the score was (5). A second variable concerning specific dysfunctional social behavior prior to and after hospitalization was combined with the social problems inventory to form the social stability dimension. Medical status was evaluated using two variables: (1) an open-ended question concerning current health status and (2) a check off list of complaints, symptoms and disorders. Scores ranged from 1 to 14.

The two samples scored significantly different on the pre and post hospitalization parts of the abstinence dimension. General medical unit patients had a mean score of 5.3 pre hospitalization and 3.4 post hospitalization. Alcohol unit patients scored 6.7 pre hospitalization and 1.9 post hospitalization. This data indicates that, while both samples drank less after hospitalization, the patients treated in the DU substantially improved in terms of the abstinence dimension.

The social stability dimension comprised three separate scores. On the social problem inventory the GMS scored 2.1 and the DU scored 2.1. For the social dysfunction variables, the scores were: pre hospitalization GMS 2.3 and DU 3.6; post hospitalization GMS 1.7 and DU 2.0. It was noted that post hopitalization DU patients improved to a more significant degree than did their GMS counterparts.

For the current health status portion of the medical status dimension, the GMS mean was 3.0 and the DU mean was 2.0. For the complaints' list the scores were: GMS 8.4 and DU 7.6. It should be noted that the groups did not significantly differ in terms of medical status indicating that no major change in medical status had occured for either sample since discharge six months earlier.

The data suggest, overall, that no significant difference exists six months following treatment in the patient population managed for alcoholism in a general medical service compared with those on an alcohol detoxification unit. Important differences do, however, exist on certain dimensions of the CDC which we believe should be noted.

On the abstinence dimension, the DU patients appeared to score significantly better following discharge from the unit. This may be explained by the large numbers of patients from the DU sample continuing rehabilitation in the OPD of the alcohol unit. However, this contrasts sharply with the data concerning social stability and medical status which indicate less of a difference between groups.

While the data show no significant difference between the two modalities, hospitalization on either service for the alcoholic appears palliative. On all dimensions, the patients showed improvement following hospitalization which lasted at least six months. The lack of impact by the particular modality may be explained by the chronicity of the disease, the nature of the population at risk, the short hospitalization and the state of art of treatment.

We by no means have drawn the conclusion that treatment for alcoholism by a specialized unit is ineffective. This study was designed to conceptualize and pretest a method for evaluating alternate treatment modalities which encompass the multidisciplinary aspect of the symptoms of alcoholism. The three dimensions of social stability, abstinence and medical status still require refinement and synthesis. We believe that we have accomplished this to some extent by creating out of these dimensions the concept of a continuum of diagnostic criteria (CDC). Multidisciplinary in its construction we hope to be able to eventually apply it to alcoholic patients at entry, during treatment and following discharge, as a gauge of the progress of the patient and the impact of medical and psychosocial intervention.

ASSESSING THE TREATMENT OF THE ALCOHOLIC PATIENT IN AN URBAN HOSPITAL

Roger S. Mazze, Ph.D., Samuel M. Rosen, M.D., Edward Julie, B.S., and Theodore Feldman, B.A.

Albert Einstein College of Medicine

During the six month period June - December, 1976, 365 of 4,500 adult medical patients were discharged with either a primary or secondary diagnosis of alcoholism at a large New York City medical center. These patients were treated on either a general medical service or a special alcoholism unit within the center complex. Generally, those patients with alcoholism secondary to another problem (e.g., pneumonia) were admitted to general medicine, while those who voluntarily sought treatment for alcoholism or whose medical problems were minor or easily controllable were admitted to the alcoholism unit.

A "quality of care" protocol was developed to encompass the medical and psychosocial aspects of diagnosis, treatment, and discharge planning for alcoholic patients on either a general medical service or special alcoholism unit. Although general medical patients were considered to be suffering from more severe medical problems than alcoholism unit patients, nevertheless, both groups were expected to meet similar standards in terms of the management of alcoholism. Since both services existed within the same medical complex and both had access to the same medical, psychosocial and nursing personnel, it was hypothesized that treatment for alcoholism should be similar on both services, as should treatment for medical problems related to alcoholism.

Whether admitted to a general medical unit or an alcoholism unit, diagnosis of alcoholism was to encompass: (1) complete and comprehensive personal and family social history, gauging the extent and nature of drinking of the patients, and members of the patient's family; (2) specific dysfunctional behavior (marital, social, financial and criminal) related to drinking; (3) summary of past treatment modalities; (4) history of blackouts, minor withdrawal and delirium tremens; (5) complete general physical examination; (6) neurological evaluation; (7) follow-up at entry with EKG, liver function tests (SGOT, SGPT, Phosphotase), hematocrit and, if unconscious, lumbar puncture and skull films.

During hospitalization patients were expected to be managed for detoxification through the use of minor tranquilizers (librium or valium), provided with multivitamin therapy for accompanying malnutrition and evaluated for medical complications--specifically liver disease, gastrointestinal disorders, neurological problems and cardiomyopathy. Each patient was also expected to receive a social service consultation.

Discharge criteria included (1) control of medical complications; (2) off tranquilizers for 24 hours preceding discharge; (3) management of malnutrition; and (4) referral to an in-patient or ambulatory alcoholism program for rehabilitation services.

The study group against which the quality of care criteria was applied consisted of two samples randomly drawn from 365 patient charts with primary or secondary diagnosis of alcoholism. These charts had been coded 303 (alcoholism according to International Classification of Diseases) by records' clerks based on the discharge notes of the patients' physicians. Twenty general medical ward (GM) patients were selected from 160 charts and twenty-five alcoholism unit (AU) patients were selected from 205 charts (the respective portions of the 365 patients).

Each subject's chart was initially reviewed by a records evaluator. The purpose of this review was to identify the criteria *not* met by the physician at entry, during treatment and at discharge. Each criteria not met was then examined by an internist and social scientist. This second step was designed to determine whether there existed any justification for not meeting the criteria. To insure complete accuracy, all evaluative data was then turned over to the chief-of-service for comment. This three step mechanism insured that the evaluation was fair and comprehensive.

FINDINGS

At entry, it was found that 95% of the general medical unit sample had undergone a comprehensive physical examination which included a complete system review (head, ear, nose, throat, chest, heart, gastrointestinal and urinary/reproductive). In contrast, 40% of the patients admitted to the alcoholism unit (AU) had incomplete and inadequate physical examinations. Six AU cases (24%) had no funduscopic examination. Incomplete documentation for neurological examination was discovered in one-third of the AU cases. Both samples scored 100% on required tests: EKG, liver function, hematocrit, SMA-6, lumbar puncture and skull films (when patient presented unconscious or with head trauma).

Meeting quality of care standards on the psychosocial dimension presented a completely different picture. The general medical patient sample scored very low on completing psychosocial admission criteria. Only 20% of these patients were given an adequate personal and family history. None of these charts contained information about previous blackouts, minor withdrawals or delirium tremens. In contrast, the alcoholism unit patients met or surpassed every category of the psychosocial dimension. A complete drug history for the patient, his/her parents, spouse, siblings and children was found in 100% of the cases. Past treatment experience and past dysfunctional social behavior was elicited for each AU patient.

Therapy for alcoholism, the second category of the study, was found to be absent or inadequate in 55% of the general medical unit patients. No social service consultation had occurred for any patient in the GM sample. Thirty percent of the GM cases had not been given any minor tranquilizers to manage withdrawal. Although it was deemed appropriate for all patients in the GM sample to be given multivitamin therapy, only 30% met this criteria. Management of the medical complications of alcoholism (pneumonia, gastritis, hepatitis, cirrhosis and cardiovascular disease) was found satisfactory in each case. This contrasted significantly with the alcoholism unit patients. This sample scored 100% on therapy for alcoholism and less than 60% on treatment of medical complications. The latter may have been due to extremely poor documentation regarding treatment for medical complications for patients on the alcoholism unit.

For the final category, discharge status and planning, the general medical patient sample was rated low on three study dimensions. Only three patients had been referred for rehabilitation; 20% had been discharged on tranquilizers; no patient received social service counselling. In comparison, all alcoholism unit patients were referred to an ambulatory rehabilitation program; 100% had been discharged off tranquilizers for 24 hours; all patients had received a a social service consultation. A sharp contrast appeared in terms of discharge planning in terms of medical complications. General medical service discharge notes were found to be comprehensive, detailing follow-up procedures with referral to the outpatient clinics. Alcoholism unit patients charts left it unclear as to the plan for follow-up management of medical complications.

Two specific problems were found in the alcoholism unit patient charts. The comprehensive psychosocial notes were maintained physically separate from the general medical progress notes. In-hospital daily records did not include nor reflect the extensive notes made by the triage counsellors, social workers and psychiatrists. Only after discharge were the medical and psychosocial notes united. The second problem noted was that 50% of the AU sample exceeded the normal 7 to 9 days length of stay. These patients remained in the hospital for 14 days. While there may have been a specific reason for the extended length of stay, no definitive explanation could be found (none of the general medical patients exceeded the length of stay criteria without adequate documentation).

There appears to be a marked contrast between admission, treatment and discharge management of the alcoholic patient on the general medical service as compared with the special alcoholism unit. In neither case was the management of the alcoholic considered totally adequate. Complete psychosocial evaluation and care found on the alcoholism unit was consistently lacking at entry, during treatment and at discharge for the majority of the alcoholic patients admitted to the medical service unit. Evaluation and care of the medical problems associated with alcoholism was found inadequate in a significant number of those cases reviewed from the alcoholism unit, while these elements were found acceptable for the general medical service patients sample. The lack of complete neurological and funduscopic examination among the alcoholism unit sample and similar insufficiencies in psychosocial management of the general medical unit patients, both of which groups were treated in the same medical complex, points to a problem which we believe can be generalized to other medical centers. It is apparent from the data collected that poor communication exists between these types of services and, equally apparent, that with adequate communication many of the insufficiencies could be resolved.

The quality of care medical audit which focuses on a particular disease, irrespective of where treatment takes place permits us to carefully evaluate the management of the patient in various settings. Under such circumstances, we are presented with both the opportunity for discovery of insufficiencies and methods of remediation. In general, special alcoholism detoxification units are better equipped to manage the psychosocial elements of alcoholism than the general medical service. This factor is sustained in our data. Equally predictable is that the medical complications due to alcoholism are better managed on a general medical service which is more familiar with the disorders. The uniting of the two services in the case of the management of alcoholism (and other diseases with major psychosocial components) would, of course, be the best approach. At the urban hospital center in which this study took place the merging of the two services was not considered a viable solution for the uneven and poor treatment of the alcoholic. What did prove possible, however, was the implementation of a plan by which adequate provision for psychosocial evaluation and consultation by alcoholism unit staff for general medical service patients would be provided. In turn, a closer liaison with the general medical staff for consultation on the alcoholism unit was also planned. Both moves are directly attributable to the results of the medical audit, which enabled us to bring the inadequacies of management to the attention of the respective chiefs of service.

HOSPITAL-BASED EMPLOYEES ALCOHOLISM PROGRAM

Roger S. Mazze, Ph.D., William Schneider, M.D. and Ernest Drucker, Ph.D.

Albert Einstein College of Medicine and Montefiore Hospital and Medical Center

Contemporary industrial alcoholism programs have moved away from the "punitive" model of identifying and subsequently terminating the alcoholic employee to a "caring" model in which an offer of treatment is made following discovery of the alcoholic worker. This approach is often largely dependent upon the discovery of alcoholism through work-related dysfunctional behavior, e.g., accident, mistake, absenteeism and irresponsibility. It relies heavily on the *threat* of treatment as its rationale for prevention; it provides little help for the potential alcoholic, since treatment could only start after discovery. But perhaps the most striking criticism, is the basic and fundamental contradiction of the entire approach--*in an effort to maintain efficiency and productivity and to avoid accidents, the companies adopt a plan that would force the individual to be inefficient, unproductive and accident prone before he/she could be treated for alcohlism.* By making it necessary for a crisis to be precipitated before an individual could be "forced into treatment" (though threatening him with dismissal) the strategy fosters chronic alcoholism for it is only at this stage in the alcoholic's history that dysfunctional behavior of the type discovered in an industrial/institutional setting would manifest itself. In general, this approach is used by most corporations and institutions termed to have modern industrial alcoholism programs.

Recent experience at Montefiore Hospital and Medical Center in an employee alcoholism program suggests that identification need not depend upon supervisory intervention nor need it await dysfunctional work performance. A program based in the hospital's Employee's Health Service (EHS), utilizing third person intervention strategy and an on site counselor trained in the management of alcoholism was instituted in 1975. Vigorous educational efforts were employed to train selected personnel representing all employment levels (physicians, administrators, social workers, engineers, kitchen workers, etc.) to identify the early manifestations of alcoholism and to assume the role of "referral counselors." This educational program was also designed to promote a hospital-wide change in attitude, recognizing that alocholism is a disease and that appropriate treatment can be of benefit.

The program was jointly sponsored by the hospital administration and by Local 1199 of the National Union of Hospital and Health Care Employees. During the first year of the program three teaching cycles were completed. Each cycle lasted one month and consisted of four two-hour sessions. Recruitment of participants was left to labor and management each selecting an equal number (18) for each cycle.

Members of the Department of Social Medicine at the Montefiore Hospital Center, outside consultants and the Employees Health Service counselor comprised the teaching staff.

The schedule for the first session was based on the need to simultaneously introduce the staff, promote a group concept and detail the program. Each participant was greeted by a staff member and introduced to other participants. At the same time participants were registered and told something about the general program. Next, the program, its goals and specifically the role to be assumed by the group in which separate smaller groups were then asked to develop a "model" of the typical alcoholic. To guide their deliberations, special forms listing physical, social, economic and psychological characteristics were prepared. This exercise encompassed the representative of each group of the "typical" alcoholic. This exercise also served a second purpose pretesting the attitudes of the participants towards alcoholics.

Using these typical characterizations as a starting point, the second session focused on what we know in general about alcoholism. Presented by the staff, discussion included: who are alcoholics, the definition of the disease, stages of the disease, causes and prognosis. The second half of the afternoon was a series of mini simulations between staff and staff, staff and participants and participants and participants. These simulations reenacted the contact between referral counselor (the role participants would eventually assume) and the co-worker with an alcohol related problem. Each encounter was followed by discussion centering on the communicator's reaction and outcome of the contact. Suggestions about the most effective method, when to stop and how to proceed were encouraged from the audience.

The third session focused on forms of treatment and further counselling practice. Various treatment modalities were presented by the staff. Representatives of A.A. were engaged to simulate an "open meeting". The employee alcoholism counselor explained his approach (ambulatory detoxification and group counselling). This was followed by two exercises designed to help the referral counselors feel more comfortable in that role: triad simulations and counselling demonstrations. The triad simulation divided the group into three person units (an alcoholic, a referral counselor, an observer). Situations were designed by the participants and roles were exchanged to facilitate self and peer evaluations.

The last session focused entirely on the role of referral counselor repeating earlier exercises. Questions and concerns were addressed by staff. A system of continued reeduction and communication was established to insure feedback after the "referral counselors" were back at work.

As conceptualized, the program in the work setting begins with initial contact between the referral counselor and co-workers. This contact may occur in one of several forms: a worker may seek out the referral counselor, or a counselor may initiate the first contact, or alternatively, a third party may contact the counselor who, in turn, may initiate a relationship with the individual. Once communication is opened, the purpose is to ascertain the extent and nature of the problem and to bring the individual to the attention of either the professional counselor or the medical staff at the Employees Health Service clinic. At this point, professional staff would take responsibility for treatment and the referral counselor would remain the co-workers "buddy". These steps avoid the necessity of waiting for a precipitating crisis to force a confrontation and insist on mandatory treatment. The likelihood of crisis occurring and awaiting detection until the chronic alcoholism is manifested in dysfunctional behavior is reduced in favor of early identification, quick treatment and security for one's position. To assure client cooperation, it is important that this procedure remain confidential between the employee and the counselor/physician.

There exist multiple consequences from such a program. Employees trained as voluntary worker advocates, themselves, become the target of primary prevention. Having been familiarlized with the problems associated with alcoholism, they are less likely to abuse alcohol. Second, such a program is likely to attract a number of alcoholics as participants, who through this form of intervention, may voluntarily seek treatment. Third, a cooperative effort to alter attitudes concerning alcoholism begins in these sessions and is later promoted by participants among co-workers. Fourth, reaching all levels of employment, the program makes clear the universality of alcohol abuse.

The medical center employs approximately 7,000 individuals. During the first four months of the program, 52 persons exhibiting various manifestations of alcoholism were seen by the counselor. Twenty-four were referred by EHS physicians, 20 by supervisors, 5 were self-referred and 5 by the "referral counselors". Various approaches to treatment have been utilized; these have included in-patient detoxification, personal and group sessions with the counselor, and referrals to Alcoholics Anonymous. Of the 52 employees seen in 1975, 31 remain actively employed in the hospital at present.

This data contrasts sharply with the one year period prior to the program. During that time fewer than 15 employees were considered alcoholic. Additionally, no hospital-wide prevention program had existed; consequently, prevention and early identification of alcoholism among this worker population was absent. During the six months of the program, 118 employees were trained as referral counselors. Until the program, standard practice at the hospital center was for a superior to confront a worker (at a lower level) to force the individual into treatment. The superior may have no knowledge, or worse, misinformation about alcoholism and thereby may cause irreparable damage to the potential patient. If, instead, a trained co-worker (of equal status, and thereby, non-threatening) were to approach the suspected alcoholic and convince him to seek help with his assistance, the result may be both avoidance of crisis and voluntary acceptance of treatment. Because the pressure to seek treatment initiates with a peer and does not include any overt or covert threat of dismissal, the necessity of precipitating a crisis is removed. Additionally, since the rationale behind this approach is a human concern for the well-being of a fellow employee, there is no need to wait until serious dysfunctional behavior to precipitate contact. Further, because the individual is reached earlier in his/her alcoholic career, referral to treatment by a co-worker advocate on a voluntary basis is more likely to be successful.

An education and treatment program can serve to increase general awareness of alcoholism as an illness and to direct significant numbers of individuals with this problem to effective therapy. The overall impact of this program is presently being evaluated. An important question remains as to the consequences of an educational program in preventive and early identification. Measurement of the phenomenon has yet to be devised for a hospital setting in which productivity among the professional staff has never been quantifiable.

TREATMENT PROGRESS OF SUBSTANCE DEPENDENT MOTHERS AND THEIR CHILDREN AT FAMILY HOUSE

Malcolm West, M.A., Barbara Frankel, Ph.D., Jeanne Dalton

Eagleville Hospital and Rehabilitation Canter

INTRODUCTION

Women and their children live together at Family House, Eagleville Hospital and Rehabilitation Center's residential treatment program founded in 1975. Family House was created to provide a necessary treatment alternative for the female addict with children.

Although women account for up to 40% of the addicted and 50% of the alcoholic population, female oriented halfway houses in the United States are limited (Kaubin, 1974). Treatment centers which additionally address the need of the recovering parent are even scarcer. In 1975 only one other residential parenting program, Odyssey House, N.Y., was in existence. Women who would not or could not abandon their family responsibilities were denied the advantage of an intensive, inpatient treatment experience.

The recovering mother brings her own special needs into treatment. The additional responsibilities and added stress of parenting, and the identity conflicts accompanied by resentment and subsequent guilt feelings toward her children are unique areas which must be recognized and addressed if treatment is to be successful. In the therapeutic milieu of a parenting program a woman must confront those problems which have supported her addiction.

The addicted mother frequently lacks the parenting skills and ability to nurture that is essential for healthy development of the child. Outpatient seminars designed to teach mothering skills have been unsuccessful because they do not interact directly in the mother-child relationship (Faires, 1977; Haley, 1976). If a mother seeks residential treatment her children are placed with relatives or in foster care, further establishing the pattern of upheaval and instability in their lives. In a parenting program the child's addiction related problems are also treated. Physical, emotional and educational needs are given priority.

Poor parenting and subsequent developmental problems render a child more vulnerable to substance abuse (Copolillo, 1975). Through a "grass roots" treatment modality we hope to minimize the risk of second generation addiction.

Family House admitted its first residents in October, 1975. At that time the program was housed in a single-family home that could accommodate only three women and their children. Efforts to locate a larger facility had been plagued by difficulties with zoning ordinances, and the reluctance of landlords to offer properties on a three year lease.

The decision to begin on a small scale was fortunate. The initial staff was comprised of the Director, a research evaluator, three full-time mother/child counselors (two female and one male) and evening and weekend coverage workers. The small caseload (one family per counselor) allowed us the opportunity to acquire additional necessary skills, such as diapering, home maintenance, lawn care, and kickball.

During the shake down period we re-examined many of the previously untested policies and procedures. Often staff disagreed on basic treatment issues. A fundamental dispute was the degree of control that staff should exercise over residents. In general, recovered staff favored more sanctions and the concept of "earned privileges". Others felt that we should offer support and counsel and limit our rules to those which protect the sobriety, health and safety of the mother and child. The philosophy that has emerged incorporates both positions to some extent, and the once salient distinction between "recovered" and "non-recovered" has blurred.

Family House originally relied heavily on the use of group and individual therapy services of a local community clinic. It became evident that the residential support unit and the primary therapy provider must not be separate programs. Residents were occasionally torn between conflicting messages, and often the two programs priorities were at odds. Family House was able to negotiate an agreement which allowed our residents to receive therapy in separate groups at the clinic, under the supervision of our own staff.

Six months after its inception Family House moved to a larger facility in Norristown that accommodates nine women and their children. The luxury of being able to indulge in long and frequent discussions all but disappeared as the census tripled. Policies began to emerge based on practical as well as philosophic perspectives.

The issue of staff qualifications and training has been crucial. The aforementioned mix of "recovered" and formally educated individuals lent a combination of experiential and theoretical grounding. Staff brought skills in early education, addiction therapy, child care and research and administration. Consultants on child development, psychology, problems of women and program development are utilized to upgrade staff skills through regular training programs.

APPROACH TO TREATMENT

Our major treatment goal is to return the recovering woman and her children to the community as integrated family units requiring minimal supports. To meet the varying needs of our families we have developed a holistic approach to treatment, making major use of community resources in addition to the talents of our own staff.

The most salient aspect of "treatment" at Family House is the atmosphere of concern and support. The knowledge that one has friends who will help and understand is an important part of recovery. Most counseling is spotaneous and informal, although structured sessions are regularly scheduled.

Women learn housekeeping skills by doing; chores are rotated so that each has the opportunity to practice cooking, decorating, meal planning, cleaning and child care. Practical experience is supplemented by regular classes on nutrition, parenting, budgeting and child development. Classes on self-assertiveness and sexuality help women to define themselves as responsible, valuable people.

Recreational activities ranging from ski outings to arts and crafts for mothers and children, separately and jointly, are designed to help our residents experience joy and learn to make good use of their leisure time.

Parenting techniques are taught by demonstration. Positive rewards are emphasized and mothers are forced to find alternatives to physical punishment.

Therapy, psychiatric evaluation and counseling, vocational/educational services and group parenting seminars are provided through the Eagleville Hospital Community Clinic or Community Services Unit. To break down the pattern of isolation all women are urged to utilize local recreational and support services and to become actively involved in community social, educational or political organizations in the latter phases of treatment.

For children, the Family House staff provides initial diagnostic testing, child care, counseling and treatment planning services. In addition, all children participate in arts and crafts, music and physical activities at the house or local recreational centers. Older children have the opportunity to participate in scouting programs and to develop special physical or creative skills at the community YMCA or YWCA. Regular visits to the local library, zoo and parks, special field trips and daily opportunities to explore the community are considered essential experiences for the child. Health care and special development needs are provided by the local Visiting Nurse's Association. Emotional problems and learning disabilities beyond staff treatment scope are referred to the County Mental Health/Mental Retardation Center for further treatment.

All services provided to mothers and children are coordinated through a formal, individualized treatment plan. Counselors meet with each resident to discuss the residents treatment needs and special problem areas. This counseling involves a frank review of test data, referal information, psychological and social histories and the clients own insights. The Treatment Plan Worksheet is then used to document these needs and to establish objectives. Program and community resources are identified which will help the resident achieve the mutually endorsed goals.

Residents proceed through four phases of treatment, beginning with orientation and culminating with re-entry. Bi-monthly progress notes indicate the degree to which a client is able to attain goals. The treatment plan is evaluated and updated at the end of each phase. The final treatment plan recommends a pattern of aftercare for mother and child.

EVALUATION OF TREATMENT AND PROGRESS

Two types of evaluation are conducted at Family House. The first is based on an anthropological model and involves the content analysis of daily logs, minutes and progress reports by a cultural anthropology consultant. The second type of research involves a more traditional evaluation design. Mothers are rated according to parenting skills, self image, motivation and other aspects which bear on their continued sobriety. Ratings, based on standardized test and measures developed here, are established at entry, discharge, and on a 6-12 month follow-up basis. The children are similarly rated on social, emotional and physical growth and development.

We hope the most salient benefits of treatment at Family House will be long-term. Since these effects can only be accurately measured over the life-time of our residents (both mothers and children) and our resources are limited in this matter, we have concentrated our evaluation on more immediate treatment effects.

We have attempted to make our study sample as representative and inclusive as possible. However, we are able to include only those women who were in treatment after the program had been in operation for six months. For our follow-up study we can necessarily include only those women who we able to locate and test. Following is a summary of our preliminary findings:

Sobriety: Approximately 30% of our residents remain in our program the entire six to twelve months necessary to complete treatment. Failure to complete does not necessarily indicate failure to maintain sobriety or to make progress in other areas, as our findings demonstrate. Of all former residents surveyed 84% were judged to be alcohol and drug free. Of these, 25% had experienced temporary relapses, but had regained sobriety through the support and assistance of individuals associated with Family House and Eagleville Hospital.

Living Situation: Of all former residents, 86% are judged to have adequate, fairly stable housing at discharge, compared with 54% at entry. An additional focus has been the living situation of the children. Follow-up interviews showed that 77% of the children who had gone through Family House were living with their mothers. Only 35% had lived with their mothers immediately prior to entry. This suggests that the goal of returning mother and children to the community as an integrated family unit is being met.

Legal Status: The legal status of the mother is an important progress indicator. Family House staff works to resolve pending legal problems, and often negotiates with justices and parole and probation officers on behalf of clients. Six women who were awaiting trial were released with either suspended sentences or parole only. Four successfully completed parole, probation, or work release terms. No women received jail terms subsequent to admission to Family House, and none were arrested for crimes committed subsequent to admission.

Education: Two-thirds of the women entering Family House participate in one or more educational programs during their stay here, and the average grade level for clients at discharge is one full year above that of clients at admission. Four women have acquired their General Equivalency Diploma and six of our former residents have gone to college for the first time. At follow-up, approximately 70% of our clients were either working or participating in an educational program.

The Wide Range Achievement Test is administered to see whether the educational emphasis of the program impacts on basic skills such as math, reading and spelling. An average increase of 12% was noted in math scores, with slight increases in reading and spelling. It is also interesting to note that the average scores of women who complete treatment is 13% higher at entry and the discrepancy is greater at discharge. The greater achievement of those who complete treatment may be a function of higher motivation, more time in program, or a combination of both.

Parenting Skills: The Assessment of Mother/Child Relationships is used to determine the impact of our program on the parenting techniques of the mother. Ratings are made by staff members in four general areas: Attention to Physical Needs, Attention to Emotional Needs, Play Techniques, Discipline and Education. Women who are going through treatment at Family House show a noticeable increase in their ability to be effective and affective parents. Those who complete treatment show a much higher increase than those who drop out early. A possible explanation is that mothers who come to adopt the parenting values of Family House are more highly motivated to complete treatment.

Self-concept: One of the most fruitful areas of research has been that of self-concept. Our staff members early noted the overall low sense of self-worth which our clients felt. A review of the literature (Lindbeck, 1972) supported our observations that women entering treatment often have experienced a negative sense of self originally conveyed by parents and later reinforced by husbands, family and society. Kaubin (1974) explains that substance abusing women carry a heavy load of guilt "about their addiction itself and, even more, about all the things they have done or haven't done because of it...".

It follows then that a significant part of treatment must be aimed at increasing self-esteem if rehabilitation is to be successful (Vasanti, Whitlocks and Franks 1974). We decided to test whether our clients' self-concepts could be enhanced, and used the Tennessee Self Concept Test as a measure. (See Fitts and Richards 1971 for further information on this topic.) The profile obtained for our clients, who are primarily alcoholics with cross-addiction patterns, was strikingly similar to that obtained by Robinson (1973) for drug abusing women. In all liklihood, drug-abusing women and alcoholic women have similar patterns of self-concept (Carroll, Santo, Klein 1976). Our research indicates that the average T.S.C.S. scores of our clients increases markedly (5%) between entry and follow-up. A comparison of scores between women who completed treatment and those who did not demonstrates that those completing treatment scored higher both at entry and follow-up.

Child Development: We have not experienced the degree of emotional instability, developmental retardation, or social ineptitude on the part of children which we had anticipated (Frankel et al. 1978). We have been impressed by the independence, adaptability and intelligence of our children. There are occasional incidents of squabbling, sibling rivalry and jealousy. In short, our children encounter the same problems living at Family House that they would in a large family. But they also experience a closeness and a sense of mutual support that lightens the child care burden. The Slosson I.Q. test has been administered to children. This instrument was chosen because it is applicable to persons of all ages. Children entering our program do slightly higher on the average than the general population. Their scores tend to increase about 3 points during their stay here.

Younger children (6 and under) enter our program with Slosson I.Q. scores considerably higher (15 points) than older children (7 and up). While this may be a quirk in our sample it certainly merits further investigation. The fact that children's scores increase during their stay at Family House is an encouraging indication that a high degree of attention and stimulation can contribute to a child's growth. Measuring the emotional stability of a child is difficult but fundamental to Family House. The Behavior Rating Scale for children is administered several weeks after discharge, when the child has readjusted to her new environment. The scale includes scores for Activity Level, Relationship with Adults, Relationships with Peers, Emotional Well-Being and Special Symptoms. The results show modest improvement in all areas with the exception of Physical Functioning. The greatest improvement is in Emotional Stability.

SOME PROBLEMS IDENTIFIED BY OUR RESEARCH

Our most conspicuous problem has been that about two-thirds of the residents leave before completing treatment. To learn why, we analyzed Discharge Summaries and have accumulated some preliminary data, which, if confirmed, has implications for treatment.

The living situation at Family House may create interpersonal problems for some residents. Eight to ten women and their children live together in a single dwelling and share the responsibility of cleaning, cooking and caring for the children. It is not unusual for arguments to occur and to be aggravated by these circumstance. Women who are lax in their responsibilities or have a relapse of substance abuse face harsh confrontation by other residents. Staff must intervene frequently to prevent scape-goating and the development of factions. The solution to this problem lies in employing approaches to conflicts which relieve tension and reduce hostilities.

Unlike many therapeutic communities, we do not always advocate open expression of anger, because it may become uncontrolled and thus counter therapeutic, especially when antagonists are living together. Role playing out difficult situations has assumed increasing importance in our approach. We also continue to teach methods of responsible assertiveness.

Considerable evidence points to the fact that many of the resident's difficulties are situational. Women who were sloppy or careless about their rooms here maintain well kept apartments after discharge. It is also apparent that parenting techniques, educational tools and social skills acquired at Family House are often practiced most consistently after discharge. We hope to alleviate the situational difficulties by finding a facility which will provide more personal space for our residents.

In searching for factors which are predictive of successful outcome we can apparently rule out some of the obvious possibilities on the basis of research to date. Race accounts for no variance at all. It is also intriguing to note that privious treatment experience is not a determinant. We cannot test for the effect of pre-admission detoxifcation, since it is required of all residents.

We have noted that most of our residents are cross dependent on alcohol and other drugs. Our observations are reinforced by similar findings, nationwide (Carr, 1975; Driscoll and Barr, 1972). Women who are purportedly alcoholic with no secondary addiction are less likely to complete treatment and have an overall shorter length of stay. A tentative conclusion is that the alcoholic does not identify as closely with other residents who shared common experienced procuring illicit drugs and participating in other "street activities". We are also considering the possibility that women who use illicit drugs as well as alcohol are, in a sense, adaptable to non-traditional life styles. This may prepare them for the communal and somewhat hectic ambience of Family House. It seems unlikely that the substance chosen for abuse is of much significance. Even the "pure alcoholics" in our program displayed a general dependency syndrome, relying heavily on over-the-counter and legally prescribed drugs while in treatment. In any event, we consider it important to further study this phenomenon.

Our data indicates that unhealthy intimate personal relationships pose one of the greatest obstacles to recovery (and to well being subsequent to discharge). Several authors (Wilsnack, 1973; Gamberg, 1975, to name two) have found that women entering treatment have seldom experienced a beneficial sustained personal relationship. Our residents' relationships have been primarily sexual in nature and have fostered extreme dependency. Crises in the relationship have led to bouts of drinking or drug use. It is difficult to interrupt this pattern, but necessary for continued recovery (Driscoll and Barr, 1977; Sandmeir, 1977). Our approach includes sexuality courses, personal counseling and life skills courses which help increase independence. While it may be possible to eliminate damaging relationships in favor of healthy ones, it does appear possible to reduce the ensuing emotional and psychological trauma.

Mothers are often concerned that the staff's ultimate authority undermines the respect they receive from their children. Family counselors also stress the danger of subjecting a child to several different authority figures (Haley, 1976). Since direct intervention in discipline (or lack of it) is sometimes necessary we do, in fact, disrupt the mother-child relationship from time-to-time. In order to minimize this problem we take extra care to make it clear that mothers have primary responsibility for the care of their children.

When a residential treatment program prepares to accommodate both children and parents it must be ready to face series of complications, distractions and general confusion.

Maintenance of the health and welfare of the residents is a crucial challenge. Contagious diseases such as colds, flu, rashes and bacterial infections can easily reach epidemic proportions. It is difficult, at best, to maintain a hygienic atmosphere and isolation of afflicted individuals is all but impossible. We have the services of a pediatrician who consults with us on steps toward disease prevention and sanitation. Safety is a similar concern. Some residents tend to be lax in precautions that are necessary around infants and small children. Matches, razor blades, tacks, knives and other hazardous objects are confiscated when left within reach of small hands. Cleaning fluids, medications and similarly dangerous ingestibles must always be secured. Any door left ajar or window left unlatched suggests an adventure to some precocious youngster.

Another inherent complication is that kids are everywhere. While a delight to have around, they present distractions at the most inconvenient times.It is hard to convey the atmosphere which pervades. The confusion reaches a tumultuous crescendo at about 3:30, as youngsters awaken from naps, older kids arrive from school and mothers return from the afternoon activity. It then gradually subsides sometime after dinner.

Rivalry among residents exists in any number of combinations. Sibling rivalry often develops among children and women rival for the attention of staff and other residents. Women accuse staff of caring more for the children than for them, and the children envy the "adults only" treatment sessions. When fights break out among children, it is difficult to predict how the opposing teams will be formed. Often the incident will provoke ill-feelings among the adults after it has been long forgotten by the children.

Staff must face agonizing decisions when there is a conflict between the needs of a mother and the best interests of the children. Often, children need the attention and care of their mother at a time when the mothers would benefit from some time to themselves.

Many of the problems which have precipitated a need for treatment among women substance abusers surround the family situation in general, and mothering in particular (Dolgren, 1975; Lindbeck, 1972). Family House provides the opportunity to observe parenting behavior first hand and to help mothers cope with difficult situations as they arise, with direct intervention when necessary. Our staff emphasizes that while motherhood is a responsibility and a difficult job, it can also be a source of satisfaction and joy.

REFERENCES

Carroll, J.F.Y., Santo, Y., Klein, I.M., 1976. A Comparison of the Similarities and Differences in the Self-Concept of Male Alcoholics and Addicts. Report to N.I.D.A. Grant H 81 DA 01113

Copolillo, H.P., 1975. Drug Impediments to Mothering Behavior. *Addictive Diseases: An International Journal.* 2(1): 201-208.

Dalgren, L., 1975. Special Problems In Female Alcoholism. *British Journal of Addiction.* Vol. 70, #1: 18-24

Driscoll,G.Z., Barr, H.L., 1972. Comparative Study of Drug Dependent and Alcoholic Women. Proceedings of the 23rd Annual Meeting of Alcohol and Drug Problems Association of North America.

Faires, T.M., 1976. A Group Experience To Foster Mothering Skills In Drug-Using Mothers. *Drug Forum.* Vol. 5(3): 229-235.

Fitts, W.H., Richards, W.C., 1971. Self-Concept, Self Actualization and Rehabilitation: An Overview. *Self Concept and Self Actualization.* Nashville: Dede Wallace Center.

Frankel, B., West, M., Dalton, J., Lapp, L., 1978. Children of Addicted Women: Some Suprises and Hypotheses. Paper Presented at the Meetings of the Society for Applied Anthropology.

Fontaine, D., Undated. Halfway House Programs for Women: A Chance and a Choice. Publication of the Association of Halfway House Alcoholism Programs of North America.

Gomberg, E.S., 1975. Alcoholism and Women: State of the Knowledge Today. Special Report of the National Council on Alcoholism.

Haley, J. 1976. *Problem Solving Therapy: New Stategies For Effective Family Therapy.* San Francisco: Jossey-Bass Publishers.

Kaubin, B., 1974. Sexism Shades the Lives and Treatment of Female Addicts. *Contemporary Drug Problems.* Vol. 3: 471-484.

Lindbeck, V., 1972. The Woman Alcoholic, A Review of the Literature. *The International Journal of the Addictions.* 7(3): 567-580.

Robinson, J.S., 1973. *The Self-Concept of Drug Abusers.* Nashville: Dede Wallace Center.

Vasanti, B., Whitlock, D., Franks, V., 1974. Modification of Low Self Esteem In Women Alcoholics: A Behavior Treatment Approach. *Psychotherapy: Theory, Research and Practice*. Vol. 11, #1: 36-40.

Wilsnack, S., 1973. Sex Role Identity in Female Alcoholism. *Journal of Abnormal Psychology*. Vol. 82, No.2: 253-261.

A DESCRIPTIVE AND COMPARATIVE ANALYSIS OF THE SOCIAL SUPPORT STRUCTURE OF HEROIN ADDICTED WOMEN

M. Belinda Tucker, Ph.D.

University of Michigan

The aim of the present study was to explore various aspects of the social support structure for addicted women as compared to addicted men and a group of socioeconomically similar women (presumed to be unaddicted) in an attempt to identify those critical elements that might be usefully attended to in treatment.

This study is part of a much larger effort, funded by NIDA, which examined a number of other psycho-social and demographic aspects of the lives of drug dependent women. Our sample consists of women entering heroin addiction treatment programs in Detroit, Los Angeles, and Miami; the men entering those programs at the same time; and a group of socioeconomically similar women obtained from a state employment office that served the neighborhood in which several of our treatment centers in Detroit were located.

The analyses first compare addicted women in Detroit with the non-addicted comparison women in Detroit, and secondly compare all addicted women with all addicted men. (The limitations of the research did not permit us to obtain such comparison samples at all three sites.)

While there is relatively little empirical work done on addicted women generally, virtually no examination of their support structures is available. One notable exception has been the network analysis undertaken in 1976 by Neva Wallace, an anthropologist, in Detroit. Interviewing male and female addicts, she observed that the women were *not* loners (although they tended to have smaller networks than the men), that most of their relationships were with non-drug addicted persons, and that most maintained relationships with their families of origin. Furthermore, the women's own mothers and husbands and boyfriends figured prominently in their support systems. While a landmark study in the field, it unfortunately suffered from an extremely small sample, and the lack of a nonaddicted comparison group.

Other drug research related to the issue of social support is for the most part either demographic (with an emphasis on the marital status of drug abusing women) or fixated on pathology (often describing partners and parents of drug abusers as "neurotic" or "insecure" and largely responsible for the addicted state of the spouse or child).

The social support literature in general has stemmed from two rather distinct theoretical orientations--the sociological/anthropological tradition which tends to be more descriptive and comparative in nature (cf. Barnes, 1972; Bell, 1957) and the social psychological and psychiatric fields which were more focused on the impact of social support on various psychological and physiological states (cf. Cobb, 1976; Dean and Lin, 1977; Pinneau, 1975).

Borrowing from both directions, this study attempted to address six issues:

1) the perceived adequacy of the respondent's friendships in general,
2) the nature and amount of support received from relatives,
3) the degree of closeness to, nature and amount of support received from partners,
4) the degree of closeness to, nature and amount of support received from best same-sexed friend,
5) the respondent's coping styles (that is, given specific problems, who respondents tend to go to for support and the extent to which non-social outlets are used, and
6) the extent to which the respondent has to contend with adversive relationships. (We felt that such negative influences, when perceived as such, could be just as important as supportive relationships.)

The results presented today are the first descriptive and comparative analyses of the data. More intensive examinations are presently underway.

You have before you a packet of tables. I'll summarize the findings and you can read for yourself the exact statistical results.

GENERAL FRIENDSHIP PATTERNS

Importantly, over 90% of the respondents in every group declared that they had at least a few good friends. However, as shown in Tables I and II, addicted women in Detroit were found to be signifi-

Table I. Friends in neighborhood--response distributions.

Groups	n	Responses: None	Few	Many	Nearly everyone
Detroit addicted women	73	34.2%	50.7	9.6	5.5
Detroit comparison women[a]	175	15.4	49.7	19.4	15.4
All addicted women	146	32.9	52.1	9.6	5.5
All addicted men[b]	202	14.9	67.3	9.6	7.9

a X^2 (3) = 15.81, $p<.01$.
b X^2 (3) = 16.27, $p<.01$.

cantly more likely than comparison women to report having "no friends" in their neighborhoods and overwhelmingly complained of being lonely some or most of the time. The same pattern held in the comparison between all female addicts and male addicts with the women evidencing greater isolation.

PARTNER-RESPONDENT RELATIONSHIPS

We felt that the critical variable here was the presence of a *relationship* rather than the legal status of that relationship. Therefore, we asked questions about a given respondent's present romantic involvement, and, if they had none, the respondent's husband/wife or boyfriend/girlfriend. We learned that in many cases, having a spouse did not mean that one had a romantic involvement!

Table II. Feelings of loneliness--response distributions.

Groups	n	Responses: Most of time	Some of time	Hardly ever	Never
Detroit addicted women	73	24.7%	61.6	9.6	4.1
Detroit comparison women[a]	175	12.6	42.3	30.9	14.3
All addicted women	146	28.8	52.1	13.7	5.5
All addicted men[b]	202	16.8	52.0	26.2	5.0

a X^2 (3) = 22.89, $p<.01$.
b X^2 (3) = 11.93, $p<.01$.

The basic difference between addicted and comparison women was in the existence of a relationship--only 56.2% of addicts compared to 78.9% of non-addicted women reported being involved with someone, X^2 (1) = 13.21, $p<.001$. Beyond this, however, the most striking feature of the data is the similarity between the groups. The women did not differ in length of relationship, degree of togetherness, how well the couple got along, or satisfaction with the relationship.

Similar patterns were observed in the comparison between all female addicts and all male addicts. Significantly more men than women reported involvements--74.3% versus 62.1%, X^2 (1) = 5.88, $p<.05$. Furthermore, as shown in Figure 1, women more often than men were associated with individuals who had used drugs.

The predominant aspect of these data is the absence of "disorder" and "abnormality" as forwarded in the literature. Relationships are fairly long lasting and stable. Problems encountered are similar to those reported by non-drug users in similar environments. Most im-

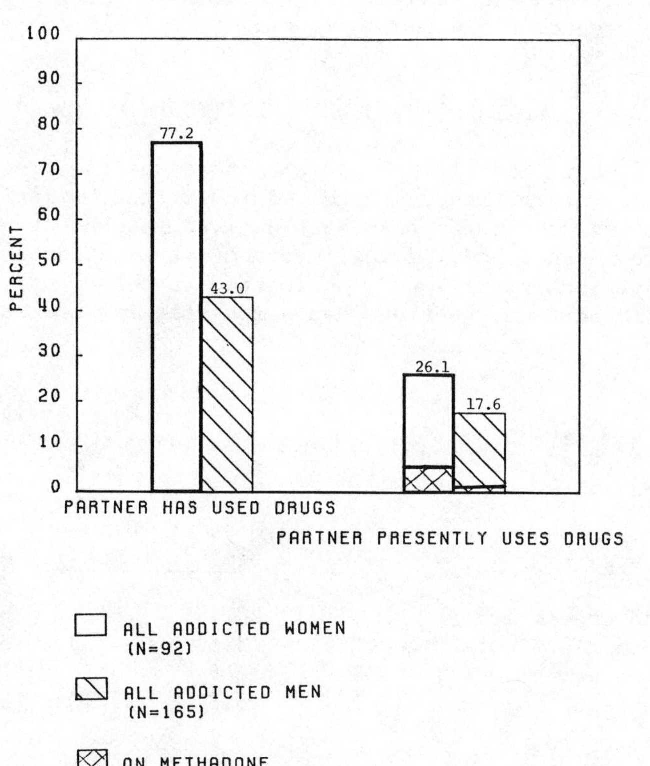

Fig. 1. Proportions of addicted respondents with drug using partners.

portantly, women as well as men were overwhelmingly supported in treatment attempts by their partners--73.2% and 78.0% respectively said that partners were "very happy" (for them)--glad that (they were) trying to get off drugs."

The data indicate then, that relationships for addicted women are not unusual, given our measures. Nevertheless, addicted women are less likely than either the addicted men or the comparison women to be involved in relationships.

BEST FRIEND-RESPONDENT RELATIONSHIPS

We attempted to examine the relationships between the respondent and her or his best friend of the same sex. At first glance, the most striking aspect was the apparent similarity between the groups. However, the two differences that did exist were fundamental and potentially more important than all other elements. Detroit and all addicted women were less likely than comparison women and addicted men respectively to have best friends--49.3% vs. 86.2%, 48.6% vs. 62.9%, X^2 (1) = 37.6, 7.01, $p < .01$. Furthermore, as you can see in Table III, all addicted respondents tend to give and receive practical help (e.g., financial, housework, childcare) from best friends while comparison women predominantly exchange emotional help (e.g., love, respect, confidence, sharing activities) with best friends.

This finding could be artifactual. That is, the practical needs of addicts may be so great that they overshadow emotional needs and fulfillments. In terms of immediate needs only, practical help is possibly viewed with greater salience and thus more likely to be elicited by direct questioning. If the responses do reflect a real deprivation in terms of emotional support, addicted women are truly in an unenviable state since they are *more likely* than the other samples to have *no friends* in the neighborhood, to have *no romantic involvement*, to have *no best same-sex friends*, and to receive *less emotional support* from friends.

DYSFUNCTIONAL RELATIONSHIPS

As I mentioned earlier, the existence of aversive or dysfunctional relationships can potentially be as disturbing as a lack of supportive relationships. The drug lifestyle may indirectly contribute to the development of such linkages (through unkept commitments, criminal activity, etc.). While all groups were equally likely to know people who disliked them or caused them trouble, 43.5% of addicted women, as opposed to only 20.6% of comparison women in Detroit, knew more than one such person, X^2 (2) = 6.11, $p < .05$. Since no such sex differences emerged in the gender comparison, the difference is probably a function of the drug abuse lifestyle.

Table III. Types of help exchanged by respondents and best same-sex friend.

Groups	n	General	Emotional: Love, respect, etc.	Emotional: Other emot.	Emotional: Sharing of activities	Practical: Financial help	Practical: Housewk/ children	Practical: Oth. prac.
Detroit Addicted Women	36	33.3%	36.1	25.0	19.4	58.3	41.6	19.4
Detroit Comparison Women	150	22.0	53.3	10.0	52.6	44.6	34.0	23.3
All Addicted Women	69	43.4	31.8	31.8	24.6	42.0	27.5	23.1
All Addicted Men	127	33.0	27.5	14.9	34.6	61.4	2.3	33.8

Note: Percentages do not sum to 100 because more than one response was allowed.

COPING MECHANISMS

Through analysis of the respondents' actual responses to the stress invoked by specific life difficulties we can better understand the nature and depth of their support structure. Our exploration took several forms:

1. Given the specific aversive but common emotional conditions of depression and anger, how does the respondent cope?
2. To what extent is the respondent confronted by specific practical problems?
3. Given actual practical difficulties, who, if anyone, does the respondent go to for help?

In the interests of time, I will briefly summarize our results on coping mechanisms. As shown in Table IV, when coping with emotional stress, there were a few clear differences between the groups, with addicted women overall more likely to choose internally oriented, non-social strategies for stress alleviation than the comparison women (i.e., they were more likely to "get away", "just stick things

Table IV. Reported behavior when upset or angry.

Responses	Groups			
	All Addicted Women[a]	Detroit Comparison Women[b]	All Addicted Women[c]	All Addicted Men[d]
lose temper and yell	60.3%	50.9	61.6	49.0
talk over things . . . (female)	58.9	64.0	54.1	53.5
talk over things . . . (male)	42.5	33.3	43.8	35.6
get away . . .	87.5	69.9	81.4	76.1
talk over things . . . (partner)	67.1	65.7	67.8	76.2
talk over things . . . (friend)	60.3	65.9	56.2	56.2
just stick it out	80.8	54.1	71.9	68.5
take out on children . . .	18.1	6.9	16.6	2.5
drink alcohol	17.8	16.6	26.0	34.7
take drugs	67.1	9.1	76.7	79.6
other	19.4	38.9	26.9	32.2
Reported behavior when upset or angry				
go to bed	39.7%	36.0	45.2	27.7
talk over things . . . (female)	54.8	56.9	47.6	56.4
talk over things . . . (male)	38.4	32.9	38.4	35.8
get away . . .	89.0	73.6	85.6	78.7
talk over things . . . (partner)	58.9	62.3	61.6	74.8
talk over things . . . (friend)	58.9	62.6	52.1	53.7
just stick it out	79.5	62.1	73.8	72.1
take out on children . . .	16.7	5.7	15.2	3.5
drink alcohol	20.5	14.9	28.8	32.2
take drugs	75.3	8.0	82.2	83.7
other	16.7	33.7	26.2	19.8

[a] n=72-73.
[b] n=172-175.
[c] n=145-146.
[d] n=200-202.

out", or "take drugs"). Interestingly, the men tended to have a more balanced set of coping strategies. At the risk of prematurely suggesting causation, it seems clear that one result of an insufficient support structure is the development of non-social coping styles (although the alternative is minimally plausible). For example, does the lack of primary social support contribute to the high incidence of or continuation of drug use for the alleviation of common emotional crises among addicted women? The fact that men and women did not differ on reports of drug use as a coping mechanism, does not necessarily negate this argument. That is, given apparently equal types of use, causal factors may be quite diverse. Since, as Chodorow (1974) has asserted, women seem to be more dependent on external supports than men, the lack of such structures may be potentially more devastating.

INCIDENCE OF PROBLEMS

Figure 2 shows the percentages of each group reporting the occurrence of the listed problems during the month preceding the interview. The results are fairly astonishing. While addiction clearly relates to an increase in financial, health, and interpersonal problems (as indicated by the higher percentages for all addicts), female addicts are particularly and disturbingly prone to health problems. While this has been generally asserted by those particularly concerned with addiction among women and while serving as a fundamental issue among those involved in the establishment of specialized treatment centers for women, empirical support has only recently been presented (Andersen, 1977).

When problem occurrences were summed (dropping drug and child related items to form a comparable index) addicted women in Detroit and as a whole had significantly more problems than comparison women and addicted men, $\underline{t}$ $(246)=3.13$, $\underline{p}<.01$; $\underline{t}$ $(346)=2.46$, $\underline{p}<.05$. Though the addict sex difference is due primarily to the overwhelming health problems of the women, the reality is that this group is being subjected to significantly more potentially stressful situations than either of the other groups.

RESPONSES TO SPECIFIC PRACTICAL DIFFICULTIES

We also asked respondents who they want to help solve their problems. Several clearly defined patterns were evident in these results:

1. Sources tend to differ radically by problem type. That is, certain people are perceived among our samples to be more able to effectively deal with certain crises than others.

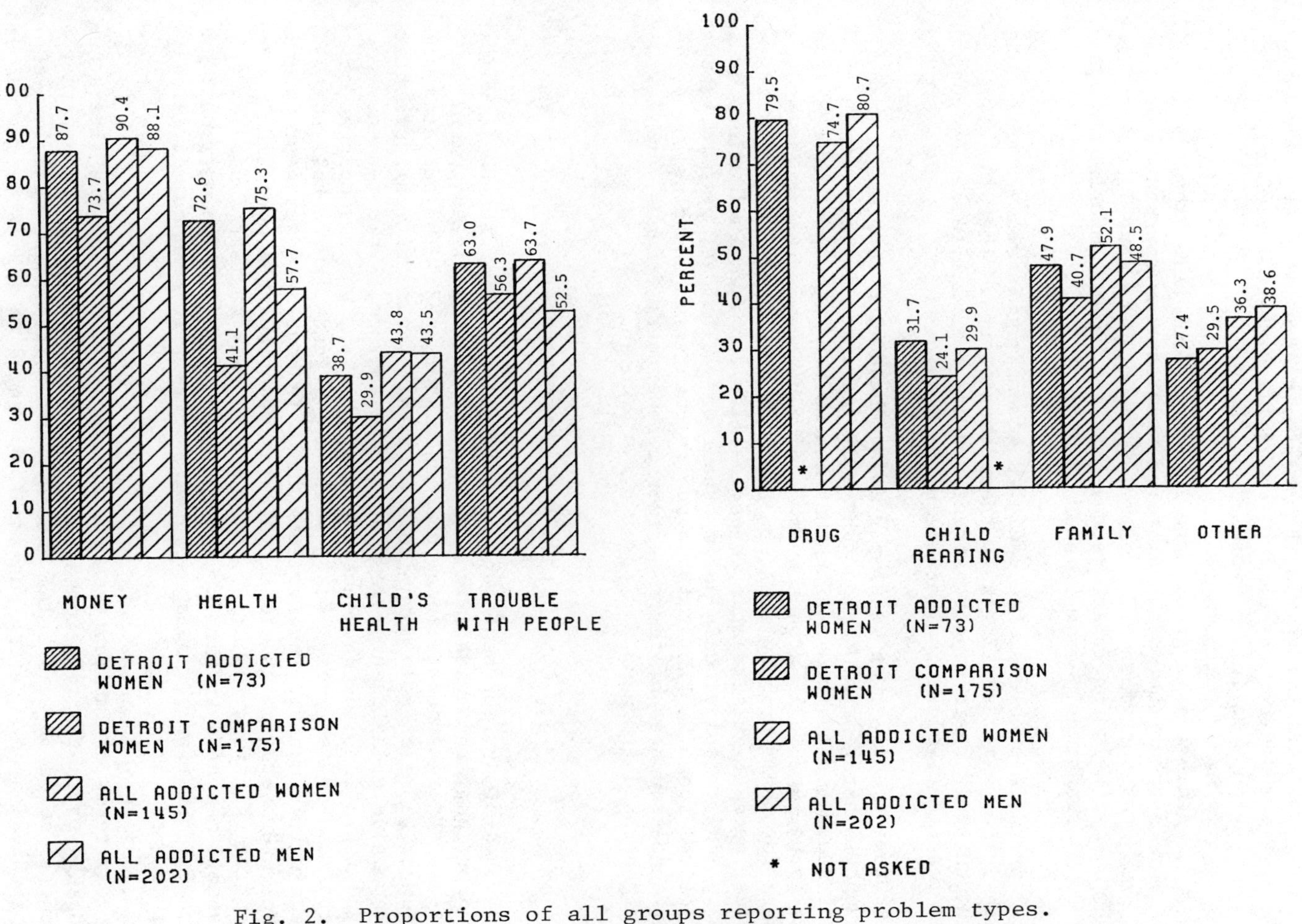

Fig. 2. Proportions of all groups reporting problem types.

2. Certain problems elicit common solutions. This, thankfully, was most evident in the health related problems. When faced with either personal or child illness, most members of all groups solicited professional help.

3. Addicts as a group were likely to seek help from similar sources that often differed from sources approached by comparison women. For example, addicts are more likely to seek *no* help for interpersonal problems while comparison women sought help from friends and family.

Over all problems, the addicted women in this study are no more likely to resist seeking help than other groups. That is, even though their basic support structures are more limited, they continue to reach out, seeking alternate resources.

Other results that we simply do not have time to present today, support that general notion that addicted women in this sample are not the outcasts from family that some of the literature would lead us to believe. All groups overwhelmingly depended on family and partners for their primary socio-emotional needs as well as their practical needs. This relates to our earlier reported indications of female addict isolation in an important way. That is, primary support received from the family may serve to counteract the possibly negative consequences of fewer partners, fewer best friends, and fewer neighborhood friends.

CONCLUSIONS AND IMPLICATIONS

This study represents an initial exploration of the basic social support patterns evident in a select sample of addicted women. A rather straightforward modal pattern has emerged. The addicted women seem to be *relatively* more isolated with certain critical potentially supportive relationships less available to them. This tendency toward isolation, though, does not mean that the women should be *characterized* as isolates. Nearly all have friends, about half have meaningful romantic involvements and half have same-sex best friends. The interpretation is clearly a relative one. When potentially supportive relationships exist, they are largely similar (based on the aspects measured in this study) to those of the two comparison groups. Similarly, coping strategies, overall, were not strikingly different.

In terms of implications for treatment, there is some literature to suggest that a relationship between social support and a rather general sense of well-being exists. Anthropological, social-psychological, and psychiatric research has tended to indicate that people, in general, are better off when they maintain critical supportive ties with others (e.g., Caplan and Killilea, 1976; Cobb, 1976; Farkas, 1976; Schneider and Smith, 1973; Stack, 1974). In the real world, however, few phenomena are so straightforward and easily understood. One problem, particularly relevant to drug abuse studies, is that of element definition. Specifically, is addiction a *consequence* of perceived inadequate support (either by virtue of observable reality or a tendency among addicted women to be more sensitive to such needs) or a *cause* of insufficient support by driving existing and potential significant others away? In all probability, both occur in an interesting, but disturbing interplay.

This distinction matters less if we focus on the state of addiction as a crisis or stress situation. The literature as a whole, according to Cobb (1976) seems to indicate that social support *protects* people in crisis from a wide variety of negative states including arthritis, alcoholism, depression, social breakdown, death and, as Lowenthal and Haven (1968) note, widowhood. Gerald Caplan (1974) has for some time emphasized that social and mutual supports, particularly from family, are indispensible in community health treatment and the handling of individual crises, in general.

The issue of support then becomes relevant from two aspects for women in treatment: first, how supportive is the treatment environment itself? Second, is there a support structure outside of the treatment center for women? As to the former, this research has shown that, when available, individuals will approach the treatment center for specific needs. While health and drug problems were, understandably, most amenable to aid by centers, it could conceivably extend to a whole host of areas--e.g., child care, legal advocacy, emotional stress. The facilitation of the development of adequate and meaningful social support structures for women without them could become a valuable treatment service.

The present research attended more specifically to the second issue of outside support. Clearly, there are broad implications for our findings. When women had best friends or romantic involvements, these significant others were overwhelmingly supportive of the women's treatment attempts. What of those lacking this basic support? Moreover, mothers often had the child care supports critical to their recovery. This could mean that mothers lacking these supports are less likely to enter treatment--an assertion made by advocates of child care components for treatment centers some time ago.

While more complex analyses of relationships between the study variables are planned for the future, the trends in the data suggest that treatment personnel should become cognizant of the client's support structure as soon as possible, ascertaining the degree of support for her treatment attempt, providing potential support in case of competing crises (e.g., child, financial, or family emergencies) and general emotional outlets. We have seen here that addicted women are less likely than even addicted men to have such supports which may partially explain reported lower "success" rates for women. It remains that, for women in particular, addiction is just one problem out of many and that the whole woman (mother, partner, sister, worker, daughter, addict, etc.) presents herself in treatment.

RERERENCES

Andersen, M. 1977. Medical needs of addicted women and men and the implications for treatment. *Women's Drug Research Project Report* to the National Institute on Drug Abuse.

Barnes, J.A. 1972. *Social Networks*. Reading, Mass.: Addison-Wesley.

Bell, W. and Boat, M.D. 1957. Urban neighborhoods and informal social relations. *AJS*. 62: 391-398.

Caplan, G. and Killilea, M. (eds.) 1976. *Support systems and mutual help*. New York: Grune and Stratton.

Caplan, G. 1974. *Support systems and community mental health*. New York: Behavioral Publications.

Chodorow, N. 1974. Family structure and feminine personality. In M. Rosaldo and L. Lamphere (eds.), *Women, Culture, and Society*. Stanford: Stanford University Press.

Cobb, S.C. 1976. Social support as a moderator of life stress. *PM*. 38: 300-314.

Dean, A. and Lin, N. 1977. The stress buffering role of social support. *JNMD*. 165: 403-417.

Farkas, M.I. 1976. The addicted couple. *Drug Forum*. 5: 81-87.

Lowenthal, M.F. and Haven, C. 1968. Interaction and adaptation: intimacy as a critical variable. *ASR*. 33: 20-30.

Schneider, D.M. and Smith, R.T. 1973. *Class differences and sex roles in American kinship and family structure*. Englewood Cliffs, N.J.: Prentice-Hall.

Stack, C.B. 1974. *All our kin*. New York: Harper & Row.

Wallace, N. 1976. Support networks among drug addicted men and women. *W.O.M.A.N. Evaluation Project Report* to National Institute on Drug Abuse.

AN INVESTIGATION OF WAYS IN WHICH THE DRUG TREATMENT PROCESS AFFECTS WOMEN AND MEN DRUG ABUSERS DIFFERENTIALLY

Barbara Sowder, Ph.D.

Burt Associates, Incorporated

During 1977, we at Burt Associates, Incorporated surveyed clients and staff of six geographically dispersed programs to investigate:

- Staff views of female and male clients, including staff attitudes and beliefs regarding posttreatment functioning and expectations.

- Staff behaviors toward female and male clients regarding their expectations, attitudes and behaviors regarding posttreatment functioning and the treatment process.

- Female and male client's views of staff and treatment and client attitudes and beliefs regading posttreatment functioning and expectations.

The questionnaires for clients and staff were similar in wording and content. Questions centered on the client's reasons for using drugs and enrolling in the program, their goals for "getting themselves together," services needed to reach goals and ratings of counselors and program services. Clients were asked to answer each item from their own experience or perspective; staff were asked to answer each question as they believed it pertained to certain percentages of or to three fourths or more of their female as well as their male clients.

Random sampling with replacement was used to select the 226 female and 260 male clients. The 39 female and 31 male staff represented persons most involved in the client respondents' treatment. Clients completed their questionnaires in group sessions where questions were read aloud by a female and male interviewer. Staff questionnaires were mailed for self administration.

CHARACTERISTICS OF RESPONDENTS AND PROGRAMS

The majority of client respondents were black or white, non-Hispanic, were between the ages of 21 and 30, were parents and had at least a high school education. Between 18 to 47 percent of the 12 client groups had some college education. Most clients were heroin abusers and had been enrolled previously in drug treatment. There were no statistically significant differences between female and male clients on these characteristics. Use of barbiturates and minor tranquilizers was more prevalent among females while marihuana and alcohol use were more common among males; however, each of these drug usage sex differences was statistically significant in no more than two sites (p range $= < .05 - < .01$). In three programs males were significantly more likely than females to have had drug problems for eight or more years (p range $= < .05 - < .01$) and in four programs males were significantly more likely to be employed than females (p range $= < .05 - < .001$). Finally, there were significant differences ($p < .001$) across all programs in sexual preference: 21 percent of the female clients as compared to 7 percent of the male clients reportedly preferred a homosexual or bisexual relationship.

Among staff participants, female staff were more likely than males to be white, under age 30 and to have completed college.

The programs constituted four methadone clinics, one therapeutic community (T.C.) and one multimodality program. All offered a range of counseling and other services and two provided special services for women (e.g., a pregnant mothers program and self awareness group). The programs had been in operation from four to nine years and served average monthly caseloads ranging from 185 to 320 clients. The representation of females in the average caseloads was 15 percent at the therapeutic community and from 30-39 percent in the five other programs.

The programs employed 150 full time and 19 part time staff; 56 percent were female. Twenty six percent of both the male and female staff were counselors. In all programs the ratio of female staff to female clients was higher than the male staff/client ratio; this pattern also held for counseling staff in five programs. The total staff/client ratios ranged from 1:5 to 1:13.

FINDINGS

Reasons for Using Drugs and Entering the Program

According to clients, many factors contributed to their drug use, including intrapsychic problems, interpersonal relationship problems and employment problems. Females were significantly ($p < .01$) more likely than males to report using drugs because they felt suicidal, hated themselves and/or were pressured by a spouse, lover or friend to use drugs. Males were significantly ($p < .001$) more likely than females to say they used drugs to "improve" their "sex life".

These statistically significant differences in reasons for using drugs did not statistically differentiate females and males when clients checked these same items as reasons for entering the program. For example, even though females were significantly more likely to say they were pressured into using drugs by a spouse, lover or friend, there was no indication that these "significant others" were more likely to pressure women than they were to pressure men into entering treatment. There were, in fact, no statistically significant differences between females and males on reasons for enrolling in the programs. The majority of all clients reportedly entered treatment because of drug related reasons, e.g., their drug habit was too costly, was "messing up" their minds and emotions or was hurting their family too much. Thus, the majority reported being "tired of drugs".

Staff responses to the questions of why clients used drugs and entered treatment were essentially the same as the clients' responses. Staff rather accurately perceived the presence as well as the lack of male and female differences on the various "reasons" for using drugs and entering treatment.

TREATMENT GOALS

Female and male clients were in general agreement as to the goals they had for "getting themselves together" and there were no statistically significant differences between the groups on these goals. Major goals were to never touch drugs again, to feel healthy, to think clearly, to have self respect, to set realistic goals for themselves, to get along well with and have the respect of significant others (e.g., spouse, parents, children), to do well on the job or in school and to stay out of trouble with the law.

Staff rather accurately perceived the clients' stated goals. However, male and female staff differed with regard to the importance of two goals for *all* clients. Male staff gave more importance to the client goals of "never touching drugs again" and "helping others with their drug problems" than did female staff and they were closer to both the male and female clients' ratings of these goals than were female staff.

Services Needed to Reach Goals

The therapeutic community clients perceived methadone-related services as "harmful" to attainment of their goals. With this exception, clients of both sexes rated all major modalities and all forms of counseling as services that were highly important in helping them achieve their goals. High ratings also were given to child care services, transportation to the program, help in obtaining housing and public assistance, vocational training and assistance in getting a job. There were no statistically significant differences between females and males on this set of questions and client opinions were rather accurately perceived by staff respondents. However, male staff gave significantly ($p<.01$) higher importance ratings than female staff to the provision of treatment in a therapeutic community while female staff gave significantly ($p<.05$) higher ratings to providing clients transportation to services. These differences in staff ratings pertained to the importance of these services to the goals of *both* female and male clients.

Ratings of the Quality of Services and Programs

In rating the quality of their present services on a 4-point scale ranging from "very bad" to "very good", clients varied more within programs than across programs. In three methadone clinics, males rated 10-12 of the 14 services listed higher than females rated them; this pattern of ratings by sex was reversed among T.C. clients. There was no consistent pattern of ratings by sex in the remaining programs. The mean ratings given by all client groups to most services ranged from "fair" to "very good". Especially high ratings were given to individual counseling by all groups. However, mean ratings given social services ranged between "very bad" to "poor" at all programs and "poor" ratings were given to methadone and detoxification services by T.C. clients. Very few statistically significant differences were found between male and female ratings and the few that appeared did not represent any consistent pattern of ratings by sex group.

Staff tended to minimize the few significant male-female client differences found within programs. Staff and especially male staff, rated the 14 services listed more highly than the clients rated them.

In addition to asking for ratings of the typical services provided in drug treatment programs, respondents were asked to rate how well the program was serving working clients and persons of their own sex, age group, ethnic group, educational level and type of drug problem.

Neither male nor female clients gave high ratings to the programs' services for working clients. Clients did give rather high ratings to services for persons of their own age, ethnic group, educational level and drug problem.

There was no indication in the clients' responses of any substantial feelings of discrimination on the basis of sex in response to a 4-point scale ranging from "a lot" to "none at all". However, females in four programs were more likely than males to indicate a "little" discrimination toward persons of their own sex but these differences were statistically significant ($p < .01$) in only one site. Female staff were significantly ($p < .001$) more likely than male staff to indicate "a little" bias in services for persons of their own sex. Mean ratings given by male staff were close to zero (.2) indicating a perceived lack of discrimination toward persons of their own sex; the mean female staff rating was close to one (.9) which signified "a little" bias. This mean rating by male staff was very close to the mean rating of male clients at five programs. The mean rating of the female staff was close to the mean rating (.8) of female clients at only one program; at all other programs the mean ratings of female clients were lower than those of female staff (ranging from .08 to .5).

Overall, staff--like clients--rated services for different groups highly. Male staff consistently rated services to different groups higher than did female staff but the male-female staff differences were statistically significant only for the variables of educational level ($p < .05$), users of opiates other than heroin ($p < .01$) and, as noted, persons of their same sex.

Counselors and Counseling Services

Although individual programs varied in counseling emphases, there was no evidence of any significant relationships between sex of clients and the provision of different counseling services. There was considerable variation among the 12 client groups in their desire for more hours of different forms of counseling. Between 20-60 percent of all six female groups and four of the six male groups wanted more family counseling. Across all programs, between 20-60 percent of the 12 different groups wanted more individual counseling but more group counseling was desired by 20-60 percent of the client groups on only four programs. There were no statistically significant differences between female and male groups in their desire for more hours of different types of counseling.

When asked about the issues they would like to have "more", "less" or the "same" emphasis upon in their counseling, some significant differences were found between males and females. In four programs, females were more likely than males to cite the need for more emphasis on "help with mental or emotional problems" and these differences were statistically significant ($p < .05$) in three programs. In two programs males were significantly ($p < .05$) more likely than females to cite a need for more emphasis on employment-related assistance. The highest ratings given by both males and females for additional counseling emphasis centered on help with job plans or employment, vocational training and help in getting an education. High ratings were given also to more counseling emphasis on "help with drug problems". Other concerns with lesser rankings centered around personal and interpersonal relationship problems, help in being more assertive, help in building a strong male or female identification, etc.

In rating their current counselors on a 4-point scale ranging from "poor" to "excellent," clients' responses yielded mean ratings that ranged between "fair" to "good". In the one program where females gave significantly lower ratings than males to how well the program was serving persons of their own sex, females rated their counselors significantly ($p < .05$) lower than male clients on availability, fairness, sensitivity to their feelings and being respectful of them. At one other program women rated their counselors significantly ($p < .05$) lower than did males on being understanding of their problems and on focusing on their past in counseling sessions.

This list of counselor characteristics was repeated in a later section of the questionnaire and clients were asked to rate these characteristics in terms of their perceptions of the "ideal" counselor. In general, the ratings of the characteristics of the "ideal" counselor were higher than the ratings given current counselors on these same characteristics. A few statistically significant sex differences were found within programs but there was no consistent trend by sex on these differences. The "ideal" counselor characteristics given the highest ratings by all client groups were being understanding, fair, respectful of the client, trustful and believing in the client and focusing on the future and on present problems during the counseling sessions.

When asked their preference regarding the sex, age, race-ethnicity, education and former drug use of counselors, the majority of both female and male clients indicated they had no particular preference on any of these characteristics except the counselor's drug history. The majority of clients preferred a counselor who had had problems with drugs in the past. No statistically significant differences were found between females and males on these particular counselor preferences.

Staff appeared to perceive the clients' preferences and opinions accurately on a number of the questions pertaining to counselors and counseling services. A few staff-client discrepancies were noted, however. Staff inaccurately felt that female clients would be less likely than male clients to prefer a counselor with a drug use history; however, they were correct in assuming that most clients would prefer a counselor with a drug history.

Also, staff were more likely than clients to give a high rating of importance to an "ideal" counselor being firm and a good role model. They were less likely than clients to rate availability of the counselor and focusing on the client's past as important "ideal" characteristics. However, they rather correctly perceived the ratings clients gave of their current counselors but in some cases predicted slightly lower ratings than the clients gave on specific counselor characteristics.

Staff did underestimate both the male and female clients' wish for more counseling emphasis on vocational training and on assistance in getting a job, an education and improving their personal appearance.

DISCUSSION

Many results of this study were unexpected. For example, we expected women respondents would report substantial feelings of discrimination by sex since the drug abuse literature contains several accounts of actual sexual exploitation as well as subtle forms of discrimination rooted in societal sex role ideologies (e.g., Edwards & Jackson, 1975; Levy & Doyle, 1974,1976; Peak & Glankoff, 1975; Schultz, 1975; Soler, et al., 1975). Yet, women clients and staff in this survey reported only "a little" discrimination toward women in all six programs. Too, we expected considerable discrepancies between staff and client attitudes and perceptions, particularly in relation to women, since some such discrepancies were reported by Levy and Doyle (1974; 1976) and Levy and Broudy (1975). Instead, staff rather accurately perceived the clients' opinions and stated needs--including the few female-male client differences and the many similarities in the two groups' responses. Perhaps the most important client-staff discrepancies were the greater values placed on services related to employment, vocational training and education by clients. The staffs' somewhat lesser importance ratings of these services were not related to the clients' sex, however, but pertained to both females and males.

The question arises as to why our findings differed from others in regard to sex discrimination and client-staff discrepancies in opinions about treatment.

Our data cannot answer this question. However, several factors possibly influenced our results. One is the fact that programs were not randomly selected. To perform statistical testing within programs, we sought programs serving at least 40 women.[1] Few programs contacted met this criterion; most that did participated with great interest. This interest and caseload size may indicate that these

---

[1] The universe of T.C. female clients was only 22; efforts to locate a larger program were unsuccessful. Sample sizes for all other groups ranged from 34-61.

programs are attuned to women's needs, even though they were not perceived as "exemplary models" by the staff. In other words, selection "biases" may have been operating.

It is possible also that the results reflect changes in the field brought about by the relatively recent increased interest in and knowledge about treating drug abusing women. Studies, such as those of Levy and Doyle, may have raised the consciousness of staff and resulted in positive efforts to better understand and meet the needs of women.

It is possible also that some women surveyed simply do not recognize subtle forms of discrimination that seem apparent to women attuned to the messages of the feminist movement. Although we can only speculate as to why our findings differ from other accounts on the literature, we have provided a medium through which women in six programs could express their treatment needs. Whether these findings are generalizable to women in other programs cannot be determined. However women in this study reported the following treatment needs as being especially important to them: support and understanding from staff and family; help in getting better housing; help in getting an education, a job or job training; medical services; transportation to the program; and individual counseling. The need for counseling services is further supported by the fact that women clients often reported intrapsychic problems as reasons for using drugs (e.g., "I hated myself", "I felt suicidal").

The data suggest also that client-staff meetings would help staff gain more insight into clients' needs. Further, male and female staff should exchange views about the importance of different services so that the strengths of their different opinions can be incorporated into treatment.

In terms of research, several types of projects are recommended: 1) a survey of client-staff opinions and expectations regarding treatment *after* both groups have left a program; 2) a survey of dropouts to determine reasons for leaving treatment; 3) the development of new techniques that will gather indepth data from women clients and decrease the possibility of obtaining "socially desirable" answers; and 4) pre-post studies of women clients' attitudes and behaviors before and after the introduction of special services for women (e.g., consciousness raising groups, assertiveness training, pregnant addicts' groups).

ACKNOWLEDGEMENT

This survey was part of a larger study on drug abusing women conducted for the National Institute on Drug Abuse, Contract No. 271-76-4401. The paper was adapted from a report by B. Sowder, T. Glynn, and N. Yedlin, *An Investigation of Ways in Which the Treatment Process Affects Women and Men Drug Abusers Differentially*. Bethesda, MD.: Burt Associates, Inc., 1977.

REFERENCES

Edwards, E.D. and Jackson, J. 1975. Rehabilitation services provided women versus men in a substance abuse treatment program. In E. Senay, V. Shorty, and H. Alksne (eds.) 1974. *Developments in the Field of Drug Abuse. Proceedings 1974 of the National Association for the Prevention of Addiction to Narcotics*. Cambridge, Massachusetts: Shenkman Publishing Company.

Levy, S.J. and Broudy, M. 1975. Sex role differences in the therapeutic community: Moving from sexism to androgyny. *Journal of Psychedelic Drugs*. 3: 291-296.

Levy, S.J. and Doyle, K.M. 1974. Attitudes toward women in a drug abuse treatment program. *Journal of Drug Issues*. 428-434.

Levy, S.J. and Doyle, K.M. 1976. Attitudes towards women in methadone maintenance program. In A. Bauman et al. (eds.) 1976. *Women in Treatment: Issues and Approaches. Resource Manual*. McLean, Virginia: National Drug Abuse Center, Materials Distribution Center.

Peak, J.L. and Glankoff, P. 1975. The female patient as booty. In E. Senay, V. Shorty, and H. Alksne (eds.) 1975. *Developments in the Field of Drug Abuse. Proceedings 1974 of the National Association for the Prevention of Addiction to Narcotics*. Cambridge, Mass.: Shenkman Publishing Company.

Schultz, A.M. 1975. Radical Feminism: A treatment modality for addicted women. In E. Senay, V. Shorty, and H. Alksne (eds.) 1975. *Developments in the Field of Drug Abuse. Proceedings 1974 of the National Association for the Prevention of Addiction to Narcotics*. Cambridge, Mass.: Shenkman Publishing Co.

Soler, E., et al. 1975. *Trick or Treatment: Women in Drug Programs*. This report was supported by a grant from the Drug Abuse Council, Incorporated.

CHILDREN OF ADDICTS: A FORGOTTEN ASPECT IN DRUG ABUSE TREATMENT?*

Barbara Sowder, Ph.D. and Marvin Burt, D.P.A.

Burt Associates, Incorporated

INTRODUCTION

The basic research question addressed in this study is:

> Are the children of heroin addicts at greater risk than other children for health, learning, behavioral, socioemotional and/or adjustment problems?

The first phase of this study was a literature review focused on children of addicts aged 3 through 18. This review indicated that very few studies existed that addressed children beyond the neonatal stage. Those which have been conducted have mainly been followup studies of children born to women addicted to heroin or maintained on methadone during pregnancy. They have been focused on preschool aged children and provide little information about the child rearing practices and home environments of these children. The measures used to assess the developmental progress of these children have varied, the sample sizes were extremely small and some were conducted without the use of control groups.

The findings, however, are intriguing. Followup results of children born to mothers addicted to heroin during pregnancy generally point to growth, perceptual, motor, socioemotional and/or learning or attentional deficits among these children. The extent to which drug use during pregnancy accounts for these developmental

---

* This study was conducted for the National Institute on Drug Abuse under Contract No. 271-76-4416.

problems remains unclear, especially since the provision of better prenatal and health care to women maintained on methadone during prenancy seems to be associated with a better developmental prognosis. (For a review of these studies see Sowder, 1977.)

These results of previous investigations, while intriguing, prove to be inconclusive and limited largely to a small segment of the population of children of addicts. The literature review clearly supported the need for the second step taken in this study, an actual survey of children who are being reared by parents with heroin abuse problems.

The second study phase, which is the subject of this report, was designed to accomplish two objectives:

1. To examine and attempt to determine whether or not a significantly larger proportion of the 3 through 18 year old offspring of heroin addicts exhibit behavioral and other difficulties when compared to their same aged peers from similar families with no evidence of heroin abuse.

2. To examine the above question with a view toward understanding the potential usefulness of family intervention techniques for the heroin abuser and his or her children to contain and/or treat the latter's behavioral or other problems.

The survey was conducted during 1977; the major results are summarized briefly in this paper. More complete reports of the findings are currently in press for publication by the National Institute on Drug Abuse (Sowder & Burt, 1978a; Sowder & Burt, 1978b).

METHODOLOGY

Sample Selection

The investigation focused on five geographically dispersed urban areas selected on the basis of the following factors:

1. Geographic dispersion in five different areas of the country

2. Cooperation of drug treatment programs in each site that served a large heroin addicted clientele from which to sample 40 addict parents per site who maintain responsibility for a 3-7 and/or 8-18 year old child.

Random sampling with replacement was used to select Index parents from the universe of eligible addict parents at each site. The Index parents were clients of eight drug treatment programs in these five sites. The Index children were randomly chosen (by age group) from each Index family.

Comparison parents were chosen from the neighborhoods in which the Index families resided by WATS line interviewers using the Hanes Criss Cross Directory and, where necessary, by field interviewers using door-to-door sampling procedures. Random sampling with replacement was used. Comparison parents were selected to be as equivalent as possible to the participating Index respondents on three variables: age group of the child (e.g., 8-10, 11-13), family structure (one vs. two parent households) and sex of parent respondents.

The final sample totaled 734 parents and children. With the consent of the parents, information was obtained from the children's teachers (although few questionnaires were received because the survey coincided closely with summer vacation). Other information was collected from social service agencies, mental health agencies, alcohol and drug treatment programs, courts and police for the purpose of validating parents' responses to certain questions.

The Index and Comparison samples at Site 1 consisted of parents maintaining responsibility for at least one 3-7 year old child (n=34 in each parent-child group); the samples at Sites 2-5 consisted of parents maintaining responsibility for at least one 8-18 year old child (n=126 in each parent-child group).

Instrumentation and Data Collection

Three basic methods were used in this survey to gather information on children of heroin addicts and their parents (the Index Group) and nonaddict parents and their children (the Comparison Group): (1) developmental testing (Site 1 only); (2) personal interviews and (3) a mail survey of schools, other agencies that serve children and youth and the drug treatment programs serving the Index parents. The developmental testing was conducted by trained clinicians; the interviews with parents and children were conducted by professional interviewers.

Data Analysis

A number of different mathematical and statistical methods were used to analyze the data collected. Appropriate statistical tests were used to determine whether differences between Index and Comparison groups were statistically significant. The data from standardized tests were scored using standardized clinical procedures and appropriate age norms for each test. Since it is well known that performance on one test is not a valid measure of the developmental status of very young children (Bender-Gestalt, Draw-A-Person IQ,

Peabody Vocabulary and Stanford-Binet Vocabulary). Children were classified into risk categories based upon their performance on the combined tests. Figure drawings of children were rated independently by three clinical psychologists with similar training and experience; reliability coefficients were obtained to determine inter-rater reliability (which was determined to be quite high).

Discriminant analysis, factor analysis and covariance analysis were used for the purpose of determining what variables distinguish the Index and Comparison children, what variables are related and the extent to which the results change when certain variables are covaried out.

FINDINGS

Table 1 summarizes the demographic characteristics of the parent respondents. The results of special statistical analyses show that the differences between groups on income, employment and other demographic characteristics of parents and/or their spouses accounted for little or none of the differences found between Index and Comparison children.[1]

Arrests and Drug Use Histories

Since becoming a parent, 65 percent of Site 1 parent respondents and 30 percent of their spouses had been arrested; for Sites 2-5 the comparable figures were 56 percent and 32 percent for respondents and spouses respectively. Few respondents in the Comparison groups indicated arrests for themselves or their spouses.

There is no evidence of nonmedical drug use among Comparison parent respondents or their spouses since assuming the parent role. Nor were there any reports on use of drugs during pregnancy with the target children by the Comparison group. Use of drugs by Index mothers during pregnancy with the target child was particularly high in Site 1 --29 percent used heroin, 26 percent used other drugs and 44 percent of the target children were exposed to drugs in utero. For respondents in Sites 2-5, 5 percent of the mothers used heroin during pregnancy with the target child and 13 percent other drugs. However, this drug use during pregnancy was not statistically related to child outcomes.

Family Structures

The average number of children was about three in both Index and Comparison families. Family stability was generally less for the Index families as measured by the percentage of legally married

Table I. Selected Characteristics of Parent Respondents (Percent)[1]

CHARACTERISTIC	SITE 1		SITES 2-5	
	I	C	I	C
Male	50	29	38*	24*
Age < 31	64	52	53***	12***
Race/Ethnicity				
Black	15	23	70	64
White	82	71	24	29
Other	3	6	6	6
Years of School				
< 9	5	3	19	19
9-12	74	71	62	55
> 12	21	27	18	26
Respondent Employed	24***	53***	30	41
Family Income 1976				
< $5,000	47	15	51***	25***
n =	34	34	126	126

1. Figures rounded
* Significant @ .05
*** Significant @ .001

respondents and by the spouse/partners' average length of residence in the household. Index families were considerably more likely to have the child's grandparents or other relatives help raise the child and Index parents were more likely to spend less than 12 hours per week with the child.

Two thirds of the Index parents indicated that their addiction has prevented them from being the kind of parent they would like to be. Two thirds of the Index parent respondents also indicated that it would help their parenting if drug treatment programs provided additional services. The types of services they felt would be helpful included family counseling, drug education for their children, recreational services for the family, medical counseling and/or psychiatric services for the parents and/or children, vocational or employment services for the parent and financial support services for the family.

The Children

There were no statistically significant differences between Index and Comparison groups in terms of sex, age or school status of the children. There were no statistically significant differences in parental attitudes toward school or parent's perception of school performance. However, teachers reported significantly more behavioral problems for a subsample of Index children and more academic problems (e.g., 24 percent of 46 Index children repeated one or more grades vs. 7 percent of the 49 Comparison children).

Test Performance, Health Status and Utilization of Mental Health Services

The scores obtained from the testing of the 3-7 year old children at Site 1 strongly indicate that a substantial proportion of these Index children are at risk for learning and school adjustment problems. Index children performed significantly lower than Comparison children on three out of four tests. When assigned a "risk" score based on their performance across four tests (Draw-A-Person, Peabody, and Stanford-Binet Vocabulary, Bender-Gestalt), 56 percent of Index children vs. 29 percent of Comparison children were judged to be "at risk." Forty two percent of Index children vs. 20 percent of Comparison children were judged to be at "high risk" (i.e., fell below the tenth percentile on three or four of the four tests). Based on clinical assessments of figure drawings, 35 percent of Index children and 12 percent of Comparison children were judged to have a "severe socioemotional problem"; 39 percent of Index and 18 percent of Comparison children were judged to have a "moderate socioemotional problem."

There was little difference between Index and Comparison children in terms of their physical health status and their utilization of medical and dental services. However, it should be noted that Index children at Site 1 and Sites 2-5 experienced high rates of low (< 6 pounds) birthweight (21 percent and 19 percent respectively) while comparable rates for Comparison children were 9 percent and 18 percent. Both the Index and Comparison groups experienced high rates of current receipt of mental health services (2 and 5 percent for the Index and Comparison groups respectively) when compared with the 1 percent reported nationwide. Teachers reported that 37 percent of 46 Index children were in need of family counseling and that 46 percent were in need of individual counseling compared to 26 percent of 49 Comparison children (for both types of counseling). Also, Index children were considerably more likely than Comparison children to report "ever" having received counseling in school (28 percent vs. 19 percent).

Abuse and Neglect

Index parents were considerably more likely to report the target children having ever been abused or neglected (about four times as likely as for Comparison parents). Based on official reports to community agencies, the incidence of abuse and neglect during 1976 for both the Index and Comparison children was extremely high. The rates for Index children were 20 percent at Site 1 and 9-14 percent at Sites 2-5; for Comparison children the rates were 12-18 percent at Site 1 and 11-14 percent at Sites 2-5. Rates of abuse and neglect for the Index children during 1976 were <u>at least</u> 10.2 to 15.9 times higher than rates reported for the U.S. (Nagi, 1975). These differences could be partially attributable to reporting laws among states and other factors, but it is unlikely that the combined effects of these factors could account for a very large proportion of the difference. The conclusion would still be unambiguous: <u>these children are at very high risk of abuse and neglect</u>.

Delinquent Behavior Among the 8-18 Year Old Children

More Index than Comparison parents said that their target child had appeared before a Juvenile Court before 1976 (5 vs. 2 percent respectively) and during 1976 (3 vs. 1 percent respectively). The 11-18 year old Index children were also more likely than their Comparison peers to report serious encounters with the police (e.g., robbery and burglary). The community validation data yielded similar results. Although small cell sizes prohibited statistical testing, the data do suggest that the Index group is at greater risk of delinquency than the Comparison group.

With respect to alcohol and drug use, the community validation shows that 37 percent of the 14-18 year old Index Children (n=16) received alcohol treatment in 1976 while none of the Comparison group (n=22) received such treatment. With respect to drug treatment, 31 percent of the 14-18 year old Index children vs. 14 percent of their Comparison peers received drug treatment in 1976. There was little difference in the smoking behavior reported by the two groups; smoking appeared to be less prevalent than among other children of the same age in the U.S. population.

DISCUSSION

Any conclusions drawn from this study of children of addicts must be viewed in light of the limitations of the study (Sowder & Burt, 1978a, 1978b). Because of these various limitations, the conclusions are best viewed as tentative ones that require additional research before they can be generalized to the universe of children of heroin addicts.

Based on the samples included in this survey, it appears that children of addicts are at risk for:

1. Learning disorders
2. School adjustment problems
3. Health disorders
4. Mental health problems
5. Child abuse and neglect
6. Delinquency and/or behavioral problems
7. Alcohol abuse
8. Drug abuse.

In most of the above problem areas, children of addicts appear to be at greater risk than children of nonaddicts from the same neighborhoods.

Implications for Research

The findings of this study clearly require further study. It is recommended that:

1. A similar study be conducted in which developmental testing is done on large samples of addicts' children in different areas of the country. This testing should include older children (8-18) as well as younger children (3-7). Since there are existing norms for these tests, it is not particularly necessary that this testing include a comparison group composed of children of nonaddicts.

2. A large scale study be conducted to obtain additional data on school, health, mental health, abuse/neglect, delinquency, alcohol and drug problems among children of heroin addicts. Such a study should provide for a longer research time frame

than this current study in order that investigators may have time to carefully "validate" parent and child responses via contacts with schools and relevant community agencies.

3. A study be conducted to determine whether the developmental status of children of addicts is better, the same, or worse when these children are placed in surrogate care than when they are maintained by their addict parents, and the conditions under which surrogate care may make some difference (positive or negative).

4. NIDA fund pilot projects with both research and treatment components to determine the impact of providing coordinated, comprehensive services both directly and through referral networks to children of addicts who are maintained in the care of their addict parents (see below).

The above recommended research strategies should be based on sound methodologies including adequate sample sizes and, where appropriate, carefully selected comparison groups. The drawing of equivalent comparison groups, however, may prove difficult. Addicts have unique "lifestyles" (in some respects) which may prohobit true equivalency between addict and nonaddict parent groups. The problem of selecting the most relevant dimensions on which to match addict and nonaddict parent groups should, perhaps, be the first issue addressed in any research.

Implications for Treatment

Under current mandates, Federally funded drug treatment programs can do little to assist addicts' children except to provide them referral services and/or include them in family counseling. These services are not well developed in many programs.

Based on findings of this study, we recommend that NIDA fund pilot projects designed to provide comprehensive and coordinated services to addict parents and their children through (1) referral networks with relevant community agencies, (2) direct service components within drug treatment programs, and (3) combinations of direct and indirect services. The services should include family therapy, parent education and child development components. It is further recommended that some projects be based on the "ideal" model of a comprehensive community mental health center, i.e., a service providing for the "total" needs of the client (including family needs), and that they include research components (using pre/post measures where possible) to assess the effectiveness and efficiency of each "model" used. More specific recommendations related to R & D projects for different developmental problems and stages of childhood are proposed elsewhere (see Sowder, 1977).

Drug treatment programs now provide some services which, according to addict parents in this study, may assist addicts in their parenting role by helping them "provide better financial support" for their children. These include educational and vocational rehabilitation services and help in obtaining public assistance. Enhancing these services for addict parents may well enhance family functioning and child development outcomes as well as clients' treatment outcomes.

NOTES

1. "Spouse" is used in this study to denote unmarried parents as well as legal spouses.

REFERENCES

Nagi, S.Z. 1975. Child Abuse and Neglect Programs: A National Overview. *Children Today*, 4: 13-17.

Sowder, B.J. 1977. *Children of Addicts and Other Problem Families: A Review of the Literature*. Bethesda, Maryland: Burt Associates, Incorporated.

Sowder, B.J. and Burt, M.R. 1978a. *Children of Addicts and Non-addicts: A Comparative Investigation in Five Urban Sites*. Bethesda, Maryland: Burt Associates, Incorporated. (To be published by NIDA.)

Sowder, B.J. and Burt, M.R. 1978b. *Children of Addicts and Non-addicts: A Comparative Investigation in Five Urban Sites. Summary Report*. Bethesda, Maryland: Burt Associates, Incorporated. (To be published by NIDA.)

THE MANY FACES OF ADDICTED WOMEN: IMPLICATIONS FOR TREATMENT AND FUTURE RESEARCH*

Beth G. Reed, M.D., Judith Kovach, Nancy Bellows, Ph.D. and Rebecca Moise, Ph.D.

Women's Drug Research Coordinating Project

Until fairly recently, research investigating the characteristics of addicted women has been quite sparse. Women were often not included in research samples or represented in such a small percentage of the total sample that any description of their distinctive characteristics and related needs was difficult. Data were rarely analyzed and reported by sex. During the 70's, coinciding with a general societal reappraisal of theories and knowledge about women and their service needs, substance abuse researchers began to address these omissions. This new research has identified key differences between addicted men and women (e.g., Aron and Daily, 1976) and some important value similarities between addicted and "straight" women (e.g., Miller et al., 1973).

While such research begins to delineate important gender differences, this paper will argue that research which treats addicted women as a single group obscures key subgroup differences that have important treatment and policy implications. Evidence of such subgroups is beginning to accumulate. Studies that have analyzed their data by both race and sex, for instance, have found different patterns on a number of variables for men and women in different racial groups (Chambers, Hinesley, and Moldestad, 1970; DeLeon, 1974). File (1976); File, McCahill, and Savitz (1974); and Weissman and File (1976) have described different roles and patterns of criminality among addicted women. Other studies have focused on the differences between "rebellious" young middle class drug users and "traditional"

---

* Supported by the National Institute on Drug Abuse Grant number 5H81 DA 01496 awarded to the Wayne County Department of Substance Abuse Services, Detroit, Michigan, with a subcontract to the University of Michigan School of Social Work, Ann Arbor, Michigan.

black lower class addicts (Heller and Mordkoff, 1972; Baldinger et al., 1972; Miller et al., 1973). Berzins et al. (1974) used the MMPI for a typological analysis of personality differences among female and male addicts.

The Women's Drug Research Coordinating Project (WDR) is engaged in an effort to identify important dimensions that can be utilized to characterize key subgroups of women addicts. Although analyses are still underway, key areas have emerged in which subgroup identification seems to have some utility. Data will be presented as examples of:

* pervasive racial and cultural differences among female addicts;

* distinct subgroups within racial groups that do not fit the overall pattern for the racial group; and

* subgroup differences in patterns of drug usage which are related to other key variables.

These data raise many more new questions about the types of information and analyses necessary to describe addicted women adequately. Some of these questions will be identified and implications for treatment and future research discussed.

SAMPLE DESCRIPTION

The data to be reported here were collected by the WDR and trained program staff, from female (and some male) clients, at the time of admission into 26 drug abuse treatment programs of two modalities (methadone dispensing and therapeutic communities). By means of a three page, pre-coded admission form, rather extensive information was collected including demographic data, living arrangements, source of support, number of pregnancies, births and abortions, criminality, family problems, marital and cohabitation status, previous treatment admissions, etc. An extensive section on drug use was adapted from an instrument used by the National Polydrug Collaborative Project. Some of the variables were included specifically because research on women in other areas and experts in the field indicated that they would be especially useful in describing key treatment-related issues and needs of women.

Of a total of 656 women in the total sample, 193 (29%) were white, 373 (57%) were black, and 90 (14%) represented several other racial groups. Only the two larger groups of blacks and whites will be discussed in this paper since the size of each of the other racial groups is too small for meaningful statistical analyses at this time. Some of these women were clients in demonstration programs designed to meet the special needs of women, while others were clients in more traditional male-female programs. Located in five major U.S. cities, the programs represent a variety of urban areas and geographical regions.

In reporting all comparisons, an .05 level of statistical significance was adopted. Some similarities among groups are reported along with key differences when these seemed to have research and/or treatment implications.

RESULTS

Racial Differences Among Addicted Women

Race and sex are two major components of an individual's identity and society differentially reacts to and reinforces individuals according to their racial group membership and sex. Culturally defined sex role behaviors often differ across racial groups as well, so it is not surprising that women (and men) in different racial and cultural groups might differ considerably on a number of dimensions. Only a few of the many differences related to race among this sample of addicted women will be reported here to illustrate the pervasiveness and complexity of the effects of cultural differences and other factors associated with race in developing an understanding of addicted women.

Choice of treatment modality. Although the majority of women of both races entered outpatient (methadone) programs as opposed to inpatient or residential programs, black women (72.7%) enter outpatient programs more frequently than white women (55.7%) (X^2 (3) = 17.32, $p<.001$). Whether this pattern reflects a preference for outpatient or methadone services by black women (or residential by whites), a societal tendency to make drug-free and more expensive residential services more available to whites, or some other factors is not known. There are also regional variations in this pattern.

Utilization of environmental resources. As depicted in Table I, white women are more likely than black women to utilize private professional services, e.g., lawyers, and private physicians, for both physical and mental health needs. Black women are more likely to use outpatient medical facilities. This may reflect general cultural preferences, differences in economic resources, or some other factors not known at this time. Whatever the reasons, understanding these patterns has treatment implications, especially given the level of medical problems found among addicted women who must be referred to medical resources outside the drug treatment program for appropriate care (Andersen, 1977). In addition, 51.7% of blacks compared to 30.6% of whites utilize the Welfare Department.

Table I. Use of professional services broken down by race.

Category of Professional Services	Race	
	White	Black
Lawyer	63** (32.6%)	83 (22.3%)
Physician for physical health needs	95** (49.2%)	115 (30.8%)
Physician for mental health needs	21** (10.9%)	15 (4.0%)
Clergy	11 (4.8%)	16 (4.3%)
In-patient mental health facility	12 (6.2%)	17 (4.6%)
Out-patient mental health facility	26 (13.5%)	38 (10.2%)
Out-patient medical facilty	45** (23.3%)	126 (33.8%)
Welfare	59** (30.6%)	193 (51.7%)
Employment services	15 (7.8%)	49 (13.1%)
Other social servie programs	24 (12.4%)	67 (18.0%)

* Racial differences are significant at the .05 level.
** Racial differences are significant at the .01 level.
*** Racial differences are significant at the .001 level.

Marriage and cohabitation. Although the two racial groups were equally likely to be involved in an informal cohabitant relationship, they differed in the kinds of formal or legal relationships they had with men. As seen in Table II, the percentage of women who have never been married was higher for blacks than for whites. Among those who had been married, but the marriage disrupted, whites were more often divorced and blacks were more often separated. Although the numbers are small, proportionately more blacks were widowed, often at a young

Table II. Marital status broken down by race.

Race	Marital Status: Married	Widowed	Divorced	Separated	Never Married
White women	42 (21.8%)	4 (2.1%)	39 (20.2%)	26 (13.5%)	82 (42.5%)
Black women	52 (14.0%)	12 (3.2%)	24 (6.5%)	80 (21.6%)	203 (54.7%)

age. Whites were more likely to be living with a dealer (20% for whites, 9.7% for blacks; X^2 (2) = 10.525, $p<.01$).

Children. More black women than white women have children (51.8% for whites, 76.4% for blacks; X^2 (1) =35.11, $p<.001$), and among those who have children, the total number of children is higher ($\bar{x}$ = 1.78 for white women, $\bar{x}$ = 2.20 for black women, t (382) = 2.65, $p<.01$). When the total number of children is held constant, black women have proportionately more of their children living with them in their home ($\bar{p}$ = .44 for whites, $\bar{p}$ = .63 for blacks; t (382) = 3.62, $p<.001$). (The welfare contacts noted earlier may well be related to ADC support for these children.) White women report that proportionately more of their children are in the care of their fathers ($\bar{p}$ = .10 for whites, $\bar{p}$ = .04 for blacks; t (375) = 2.67, $p<.01$).

Again, whatever the reasons for this pattern, a substantial number of these women have children, and black women appear to have ongoing responsibility for a larger number of children. These responsibilities can both enhance or be a detriment to treatment depending on the support arrangements the women have and how the program chooses to deal with child-related issues.

Family of origin. White women report that their mothers had psychiatric problems more often than black women (10.8% for whites, 3.4% for blacks; X^2 (1) = 12.05, $p<.001$). More whites also reported alcohol problems among their parents, although the difference between blacks and whites was statistically significant only for alcohol problems among mothers (19.4% for whites, 7.8% for blacks; X^2 (1) = 15.59, $p<.001$). Since mother-daughter relationships are very significant developmentally and since psychiatric and alcohol problems might be expected to interfere with these relationships, this finding may have important treatment implications. Comparisons between the group of white women reporting these parental behaviors with other groups on other key variables would help identify other important related characteristics.

Stress indicators. The measures in this area are somewhat problematic since it is known that black and white women express and cope with stress quite differently. Thus, some measures are likely to be higher for all groups of white women; others will be higher for black women. These relationships may be quite independent from their addiction status. One of the measures in the WDR sample, number of suicide attempts, is an example. Suicide attempt rates for white women are higher than black rates across all societal levels (Hendin, 1969). In this sample of drug addicted women, the percentage of those who had attempted suicide one or more times was substantially higher among whites as well (29.2% for whites, 10.1% for blacks; X^2 (1) = 31.29, $p<.001$). More work must be done to determine if these rates differ significantly from general population rates.

Consistent with this finding, however, are the comparative overdose rates. Whites report more overdoses for the first drug ever used (X^2 (7) = 25.31, $p<.001$), for their most preferred drug (X^2 (7) = 32.38, $p<.001$), and for other drugs used in addition to the most preferred drug (X^2 (7) = 24.15, $p<.01$; X^2 (7) = 22.82, $p<.001$; X^2 (7) = 19.39, $p<.01$).

To better interpret the meaning of this data, one would need data on stress indicators likely to be higher for blacks and would need to understand the relationship of the rates among addicted women and general population rates. Even without this information, however, these findings have implications for treatment programs. Suicide attempts and overdoses make different demands on program resources and require utilization of different emergency medical service agencies than other "coping" strategies.

Summary and implications of racial differences of selected variables. While one should be cautious about drawing too may inferences from this partial report of these data, some speculation may be warranted. At least a substantial subgroup of the white women appear to be reporting a constellation of indicators of family disturbance and personal stress reactions. This group may use drugs in response to familial and personal stress. The greater likelihood that white women's cohabitants are dealing drugs (and that white women themselves report more drug dealing (52.44% for whites, 40.8% for blacks; X^2 (1) = 6.77, $p<.01$), suggest that black and white women relate differently to the various drug subcultures and may need different program interventions as a result. Direct intervention into the relationship with the cohabitant may be more necessary in order to involve both in treatment, or to encourage the women to sever the relationship in order to pursue her own treatment. Black women have more on-going responsibility for children and may be more in need of child-related services if they are to pursue their own treatment.

Subgroups Within Racial Groups

Differences related to education. When comparing overall educational levels, black and white women differed very little in the mean

Table III. Racial differences with respect to education.

Highest education level completed	Race	
	White	Black
Less than 11th grade	70 (36.6%)	101 (27.2%)
11th and 12th grade	80 (41.9%)	228 (61.5%)
Some college	41 (21.5%)	42 (11.3%)

X^2 (2) = 21.28, p .001.

number of years of education. Patterns of educational attainment are quite different for each racial group, however. Table III presents the numbers of black and white women attaining three different levels of education: 1) completing 10th grade or less, 2) completing 11th grade or graduating from high school, and 3) attending some college. Relatively speaking, whites were somewhat evenly distributed across the three levels. Most of the black women, however, fell in the middle category, having completed the upper years of high school but going no further. In other words, there were proportionately more whites than blacks not only in the same-college group but also in the group of early high school drop-outs. Interestingly, as depicted in Table IV, white women with less education are found disproportionately in the women's demonstration programs.

As compared to white women with more education, the less educated white women were more likely to have had a marriage disrupted by separation or divorce (X^2 (2) = 10.465, $p<.01$) (see Table V), more likely to have been involved in prostitution (X^2 (1) = 9.92, $p<.01$), and more likely to have attempted suicide (X^2 (2) = 6.20, $p<.05$) (see Table VI). Although not statistically significant, the incidence of reported problems within their families also tended to be greater. As a group, they have had more children than white women at other levels of education (although these children are not necessarily living with them (t (187) = 3.12, $p<.01$). In contrast, black addicted women who dropped out of high school early do not seem to be appreciably different from other black women addicts. The small group of black college-educated women do seem to be different from other blacks, although the group's size precludes using statistical techniques. These were more likely to use professional services (e.g., private physicians), more likely to be employed, and also more likely to be drug dealers than less educated black women

Table IV. Addicted women: Program differences with respect to education group by race.

Education group	Program: Women's Demonstration Programs	Comparison Programs
a. White		
Less than 11th grade	43 (82.7%)	9 (17.3%)
More than 11th grade	56 (66.7%)	28 (33.3%)

Note: x^2 (1) = 4.17, $p<.05$.

Education group	Program: Women's Demonstration Programs	Comparison Programs
b. Black		
Less than 11th grade	68 (76.4%)	21 (23.6%)
More than 11th grade	176 (75.2%)	58 (24.8%)

Note: N.S.

Table V. Addicted women: Marital Status Differences with respect to education group by race.

Education group	Marital Status: Married	Divorced/ Separated	Never Married
a. White			
Less than 11th grade	15 (21.4%)	34 (48.4%)	21 (30.0%)
More than 11th grade	27 (23.1%)	31 (26.5%)	59 (50.4%)

Note: x^2 (2) = 10.47, $p<.01$.

Education group	Marital Status: Married	Divorced/ Separated	Never Married
b. Black			
Less than 11th grade	11 (11.0%)	28 (28.0%)	61 (61.0%)
More than 11th grade	41 (15.8%)	76 (29.3%)	142 (54.8%)

Note: N.S.

Table VI. Addicted women: Number of suicide attempts per education group per race.

Education group	Suicide attempts in last 2 years		
	None	One	More than one
a. White			
Less than 11th grade	40 (59.7%)	12 (17.9%)	15 (22.4%)
More than 11th grade	85 (77.3%)	11 (10.0%)	14 (12.7%)

Note: x^2 (2) = 6.20, $p<.05$.

Education group	None	One	More than one
b. Black			
Less than 11th grade	84 (90.3%)	6 (6.5%)	3 (3.2%)
More than 11th grade	235 (89.7%)	17 (6.5%)	10 (3.8%)

Note: N.S.

The socioeconomic status of the family-of-origin of the lower education white group is unknown. They may have always been among the urban poor or they may have been middle class runaways from the suburbs. Whatever their origin, they seem to have fewer resources and more disruptions than other white women. This may be a result of a greater degree of alienation from original reference and support groups, or having fewer psychological or practical coping skills. The data are insufficient at this point to speculate about these issues with any confidence, but it is likely that treatment issues would be quite different for this group of women. One might expect particularly difficult self-esteem problems, perhaps.

Differences related to religion. Nearly half (49.4%) of the white women were Catholic, concentrated most heavily in the large urban areas, but distributed through all regions at 32% or higher. Among the black women there was a small but noticeable group of Catholics (9.6%, N=34). White Catholics, in preliminary analyses, do not seem to be distinguished from other white women addicts. The black Catholics, however, did look quite different from other blacks. Although their numbers are small, and the use of statistical tests problematic, they are more likely than other black women to be married or living with a cohabitant, less likely to have children, and more likely to be supported by a cohabitant or through drug dealing.

Although not quite statistically significant, she reports fewer problems among members of her family of origin.

Patterns of Drug Usage--A Typology

Although nearly all in this sample were either addicted to heroin or had been addicted in the past, there were some who were addicted to other kinds of opiates, and women differed generally in the kinds of drugs they used in combination with the opiate. In order to examine any relationships between drug usage and other variables, a drug use typology was developed which allowed women addicts to be distinguished according to drug usage at least on a preliminary basis. The typology outlines the following groups:

1. Heroin only. For this group, the only drugs which had been used in the last two years were heroin and methadone. For this preliminary typology, we did not attempt to separate licit from illicit use of methadone.

2. Heroin plus recreational drugs. In addition to using an opiate, which was nearly always heroin, this group also used cocaine and/or marijuana. These drugs have the property of being "recreational", i.e., associated with a pleasure seeking orientation, (generally) less dangerous, have less abuse potential, and finally are widely available in many addict populations.

3. Other opiates. All drugs used by this group were opiates. In nearly all cases, heroin plus another opiate was being used, but there were a few cases in which the use of other opiates entirely replaced the use of heroin.

4. Heroin plus non-opiate drugs. Women were included in this group if they reported using heroin and some other drug from one or more of the general categories: barbiturates, amphetamines, hallucinogens, pyschotropics, inhalants, and other drugs apart from cocaine, marijuana, and other opiates. Included here, then, is some use of drugs that have abuse potential and/or that may be significantly dangerous. In our population, there was relatively little use of hallucinogens, inhalants, or other of the more destructive "thrill seeking" drugs. There was, however much use of "self-medicating" drugs, especially valium and barbiturates.

5. No opiates. Although this was a very small group, drug usage for this group was limited to drugs which were not opiates.

Examination of the drug typology breakdowns indicates major differences between white and black female addicts in drug use patterns.

Table VII. Race differences with respect to the drug typology.

Race	Heroin only	Heroin and recreational	Other opiates	Heroin and others	Non-opiates
White	48 (26.5%)	35 (19.3%)	19 (10.5%)	69 (38.1%)	10 (5.5%)
Black	132 (37.1%)	154 (43.3%)	7 (2.0%)	62 (17.4%)	1 (.3%)

Note: x^2 (4) = 78.73, $p < .001$.

The drug types themselves, however, also seem to be associated with particular characteristics within the sample, independent of race although the patterning is somewhat different within race.

As seen in Table VII, black women were more likely to report using *heroin only* or *heroin and recreational drugs*, marijuana and especially cocaine. White women were more likely to report *other opiate* use or *heroin plus non-opiate drugs*. The small group of non-opiate users is almost entirely white. Women *heroin only* users of both races are more likely to seek treatment in outpatient methadone programs (X^2 (4) = 35.57, $p < .001$ for whites; X^2 (4) = 14.51, $p < .01$ for blacks). The reasons for this are unclear and in need of further exploration. Perhaps those using heroin only are more severely addicted, or that using *heroin only* represents a later stage in the addiction process.

Some of the users of *heroin only* may be women who for one reason or another prefer not to "mix" their drugs. Table VIII suggests another possible rationale for differences in the variety of drugs consumed. The literature usually assumes that drug dealers have access to more and a wider variety of drugs. In fact, women who were dealers, in both races, were more often characterized by multiple drug use. Apparently, those who have access to a wider variety of drugs use more. For the white women, the dealers tended to fall in the *heroin used with other non-opiate* drugs group, while black women dealers used recreational (primarily cocaine) drugs with heroin. Another explanation would postulate that these women may deal in order to support their preference to use multiple drugs, but whatever the reason, this connection between dealing and use of multiple drugs needs further study.

There are also program differences that seem to be related to race and drug type. White women entering the women's demonstration programs are more likely than white women in other programs to report using *heroin only*. These women are also less likely to be in-

Table VIII. Racial differences in the relationship between dealing and drug use typologies.

Race	Involvement in drug dealing	Drug group				
		Heroin only	Heroin with recreational	Other opiates	Heroin and Other	No Opiates
White						
	Dealers	15 (15.2%)	19 (19.2%)	15 (15.2%)	48 (48.5%)	2 (2.0%)
	Not dealers	32 (40.0%)	16 (20.0%)	3 (4.0%)	21 (26.0%)	8 (10.0%)

Note: x^2 (4) = 26.86, p<.001.

Race	Involvement in drug dealing	Heroin only	Heroin with recreational	Other opiates	Heroin and Other	No Opiates
Black						
	Dealers	46 (30.0%)	76 (50.0%)	3 (2.0%)	25 (17.0%)	1 (1.0%)
	Not dealers	85 (42.0%)	78 (38.0%)	4 (2.0%)	37 (18.0%)	0 (0.0%)

Note: x^2 (4) = 7.35, N.S.

volved in dealing than white women in other programs (X^2 (1) = 10.12, p<.001) and are twice as likely to be involved in prostitution (X^2 (1) = 6.39, p<.02). Black women in demonstration programs are also less likely to be dealing, but there are no differences in prostitution rates or rates of *heroin only*. Future analyses will explore these differences and attempt to describe this group of women which seems to be found primarily in the women's demonstration programs.

SUMMARY AND CONCLUSIONS

The examples just described illustrate, at least in a preliminary way, some of the diverse profiles that seem to exist within the population, "female addict". Until such profiles become more clear, the authors are reluctant to make any generalizations about the causes and correlates of female addiction, and would caution other researchers to be similarly careful. There is no single entity called "female addiction"; causes and correlates are likely to vary considerably from subgroup to subgroup. WDR data analysis is still underway to refine some of the typology classifications, describe more fully the subgroup categories, and then begin to interrelate them more fully. As noted earlier, this report is not intended to present a thorough description of the WDR sample of addicted women, but rather to raise issues and questions about data collection and analysis strategies. From the work completed thus far, a number of such implications have become clear.

1. Different racial groups *must* be treated separately at some point in the analysis. Strong differences in one racial group can obscure differences or even trends in other directions in other groups. Considerable evidence has been presented that such strong racial differences occur, and that they are interactive with sex-related differences. In addition, other key demographic variables must be examined for important subgroupings which are obscured in many typical forms of reporting results. For instance, the education clusters summarized here are related to a number of other client group characteristics that were not obvious in earlier analyses. Preliminary attempts to investigate the impact of age within the WDR sample have also yielded no obvious relationship, although one's age is very likely to be related to numerous other variables.

2. Discrete patterns of drug taking must be developed, including the type(s) and mixtures of drugs taken and also, if possible, historical and situational information about the ways drugs are taken. This may be especially important for developing a better understanding of women's drug use patterns and their relationship to other variables. While such a pattern is not clear in the WDR data thus far, other data systems suggest a wider and more varied pattern of drug use among women. If drug use patterns are at all related to demographic or lifestyle variables or have any implications for treatment planning, documenting the types of patterns that exist among women will be important.

3. Studies concerned with women must routinely collect information about life areas important for understanding women, the responsibilities she carries and the forces operating that can either promote or inhibit progress in treatment. This paper deliberately presented some variables that are key demographic variables and appear in most major data systems, e.g., race, education, religion. Others, e.g., numbers and living arrangements of children, are not routinely collected and are important when discussing treatment needs and variations in subgroups of women.

4. Different types of treatment programs attract clients with different characteristics. This may be a result of many forces: client choice, agency selection criteria, community norms,

etc. Whatever the reasons, careful attention must be paid to identifying key client characteristic differences by program, especially if any attempt is to be made to link client and program characteristics to assess treatment effectiveness. Also, regional differences can be substantial and need to be checked.

All of the above suggest the need for fairly large, diverse samples and/or the regular use of a common core of items so that data from separate studies can be more meaningfully compared. Given that women are still the minority in most treatment programs, special efforts must be expended to acquire adequate samples to be studied. Especially if minority and other subgroups are to be understood better, adequate sample sizes will be important. Perhaps some of the current national data systems can be revised to include more of the women-relevant information. Independent researchers interested in doing research on women could also collaborate informally to develop a set of common items for inclusion into separate studies.

There are many problems with the data just described and much work remaining to be done even to identify all the problems. This approach to data analysis is certainly a less common one and thus far is identifying more questions than it has provided "answers". Hopefully, these questions can help define the directions and more specific hypotheses that can guide future research on the many types of addicted women.

REFERENCES

Andersen, M. 1977. Medical nees of addicted women and men and the implications for treatment: focus on women. *WDR Report #4*, National Institute on Drug Abuse.

Aron, W. and Daily, D.W. 1976. Graduates and splittees from therapeutic community drug treatment programs: a comparison. *Int. J. of the Addictions*. 11: 1-18.

Baldinger, R. et al. 1972. Pot smokers, junkies and squares: a comparative study of female values. *Int. J. of the Addictions*. 7: 153-166.

Berzins, J.I. et al. 1974. Subgroups among opiate addicts: a typological investigation. *J. of Abnormal Psych*. 83: 65-73.

Chambers, C.D., Hinesley, R.K., and Moldestad, M. 1970. Narcotic addiction in females: a race comparison. *Int. J. of the Addictions*. 5: 257-278.

DeLeon, G. 1974. Phoenix House: psychopathological signs among male and female drug-free residents. *Addictive Diseases*. 1: 135-151.

File, K.N. 1976. Sex roles and street roles. *Int. J. of the Addictions*. 11: 263-268.

File, K.N., McCahill, W., and Savitz, D. 1974. Narcotics involvement and female criminality. *Addictive Diseases*. 1: 177-188.

Heller, M.E. and Mordkoff, A.M. 1972. Personality attributes of the young, nonaddictive drug abuser. *Int. J. of the Addictions*. 1: 65-72.

Hendin, H. 1969. *Black Suicide*. New York: Basic Books, Inc.

Hughes, P.H. et al. 1971. The social structure of a heroin copping community. *Am. J. of Psychiatry*. 128: 551-558.

Miller, J. et al. 1973. Value patterns of drug addicts as a function of race and sex. *Int. J. of the Addictions*. 8: 589-598.

Weissman, J.C. and File, K.N. 1976. Criminal behavior patterns of female addicts: a comparison of findings in two cities. *Int. J. of the Addictions*. 11: 1063-1077.

SERVING WOMEN: THE IMPACT OF PROGRAM STRUCTURE AND RESOURCES*

Andrea Savage-Abramovitz, Suzanne Rinaldo and

Beth G. Reed

Women's Drug Research Project, Ann Arbor, Michigan

INTRODUCTION

One of the components of the Women's Drug Research Coordinating Project (WDR) is a study exploring organizational and environmental characteristics of treatment programs which are likely to impact on the quantity and quality of services provided to all clients and to women in particular. The human services organizational literature suggests that a better understanding of program services and activities is likely to help explicate the determinants of proximate and long-term outcomes for drug addicted clients.

This paper will focus on program environment, boundary mediating structures, program autonomy and resources as they related to the level of services and activities provided for women and other clients. These services include those obtained directly from the program and those which are made available to clients through interorganizational linkages. Of particular importance is the postulated relationship of program scope to outcome for women clients.

---

* Supported by the National Institute on Drug Abuse, Grant # 5H81-DA 01496, awarded to the Wayne County Department of Substance Abuse Services with a subcontract to the University of Michigan, School of Social Work, Ann Arbor, Michigan. The work involved ongoing collaboration with Marjorie Chandley of the Dynamics of Therapeutic Communities, Norman Washburne, Principal Investigator (Newark, NJ and Atlanta, GA), and Pam Schwingl and Laura Sperazzi of Women, Inc. in Boston. Theresa Hluchyj of WDR worked closely with the authors on all phases of design and implementation.

Program scope, a concept derived from Etzioni (1961), is a measure of organizational productivity and has two aspects:

1. quantity - the number of activities within the organization in which clients participate, and

2. pervasiveness - the degree to which the organization seeks to intervene in the lives of clients. This component could include the numbers and types of required activities and the range of life areas into which the program seeks to intervene.

The literature on treatment agencies and our field research suggests that programs may optimize their client outcome by increasing the "pervasiveness" component of scope. This involves widening the targets of service interventions beyond the three traditional goals for clients: 1) attaining drug abstinence; 2) decreasing criminal activity; and 3) increasing social productivity (e.g., obtaining employment).

This broadening of program focus should be useful for all clients but may be essential for intervention models likely to be effective for women. The literature suggests that a wider range of services is likely to be necessary to assist women for a number of reasons, e.g., as a result of the greater stigma associated with female drug use (Reed and Herr, 1977) and a greater range of reality problems that women face compared to men (e.g., more medical problems (Andersen, 1977), on-going responsibility for children and a poorer work history (Ryan, 1978)).

Exploring scope and other key organizational variables requires adopting a sociological focus by viewing drug abuse treatment programs as organizations whose task is changing clients' behaviors. The theoretical framework to be utilized is an open system approach which views the treatment program as an energy transforming system. Through interaction with its environment, it garners energy (resources) and transforms it into outputs: 1) services and activities and 2) rehabilitated clients (Katz and Kahn, 1965). Our research model, described more thoroughly elsewhere (Rinaldo and Reed, 1977), attempts to identify those aspects of the programs' critical environment, linkages and resources which have some relationship to several measures of program scope.

This paper will present both interview and observational data obtained from two key drug treatment modalities: 1) residential therapeutic communities (TC) and 2) outpatient methadone programs (MM). While these two program types might be expected to differ in the quantity measure of scope, the pervasiveness measure is less likely to be affected by program type.

The individual program or facility was chosen as the unit of study, much as research on industrial organizations focuses on the manufacturing plant as a unit of investigation. This allows for easier comparability of different programs. In addition, with this definition of program boundaries, one can examine the costs and benefits of intermediary structures such as parent or umbrella programs and coordinating agencies.

DATA COLLECTION, KEY MEASURES AND SAMPLE DESCRIPTIONS

Two phases of standardized interviews with program directors and managers were conducted at each program site. Each interview schedule focused on a broad range of organization research areas, including ideology, internal structure, environment and service delivery. Some questions and subscales were adapted from organizational research in other settings for use in drug treatment programs (Inkson, Pugh and Hickson, 1970) while others were developed from the very exploratory first phase interview through the collaboration of a number of independent researchers. Data were collected from 24 (12 MM and 12 TC) programs in geographically diverse urban areas. Of these programs the bulk are located in urban neighborhoods. Eight (33.3%) are in primarily residential areas, 4 (16.7%) in non-residential areas, and 12 (50%) in settings which are both residential and commercial. The programs were chosen primarily because of accessibility (most were already cooperating with WDR in client information research) and the possibility of additional linking to client data. In order to get a sizeable sample of women clients, programs were selected in each region which attracted larger proportions of female clients. Several additional screening factors were utilized so that the programs represent a range of client and staff sex distribution. The intent of this study is exploratory; since the sample is not random (see Rinaldo and Reed, 1977 for more detail) great care should be taken not to assume universality in the findings.

Several measures of size (a structural feature) are utilized to describe the programs. The static capacity of the programs ranged from 12 to 280 clients. The mean static capacity of the TC's was 57 with only two programs reporting capacities above 75 clients. The mean static capacity for MM programs was 177 clients with 5 (41.7%) reporting capacities of more than 200 clients. The number of clients actually served during the year (dynamic capacity) was another measure of size or throughput (in system terms). The TC's served a mean of 164 clients during the year while MM programs served a mean of 488 clients each during the year.

The programs identified the services and activities available to their clients from a standardized list of service activities. This list included everything mentioned as a program activity during the first interview and some additions from the literature on addicted women. The respondents also indicated the frequency of the activity, to what extent it was required, and the nature of client and staff participation.

To develop our preliminary measure of scope quantity of service we summed the number of activities offered for each program. The results ranged from a low of 15 activities to a high of 51. We also computed a tentative measure of pervasiveness of asking which activities were required for at least some of the clients. The range of required activities was from 2 to 33. Both of these ranges were categorized (cut at the medians) to develop dichotomous scores on activities, high and low. As one would expect, TC's (probably because they are interacting with clients for a longer time frame each day than MM programs) comprise 10 of the 12 programs in the high activities category; only two of the MM programs also fell in this category. This same pattern holds for required activities, high and low.

In connection with our concern for the outcome of women clients, we identified from the master list of program activities those which could be considered important for female clients based on a thorough literature review and the opinion of experts in the field. These included some female-stereotyped activities such as sewing, homemaking training, etc. and a wide range of other activities designed either to address women's issues directly (women's groups, assertiveness training, sex role education), particular areas of need for women (health self-help, family intervention, prenatal care, etc.) or those related particularly to women's key roles (parenting, hygiene and birth control, etc.). Child-oriented services were also included. A possible total score of 26 activities and services was identified. For each program a sum of women-focused activities was calculated. The range was 2 to 19 with the overall mean of 10.54 women-focused activities per program.

RESULTS

The Nature and Range of Women's Services

TC's, in general, provided more women's services (75% of programs higher in women's services were TC's). Only two MM programs provided a substantial range of women's services, and both were women's demonstration programs. A number of the MM programs had essentially no services for women and one-third did not even do a Pap smear as part of the entry physical. Only the two TC's that were women's demonstration programs could be considered high on women's services.

Eighty-three percent (5) of the freestanding TC's fall in the group with a higher range of women's services. The three embedded programs which report higher numbers of women-related services all have some history of struggling to adapt their models and orient their staffs to be more sensitive to the needs of and biases about women. The umbrella organizations, in all three cases, have mandated special attention to programs for women residents. (This was reported to be a result of either an influential woman and receptive men within the upper hierarchy, or of responsiveness to the results of special research studies investigating the status of women in the program.)

The pattern and pervasiveness of women-oriented services varied considerably across programs although WDR has not yet developed a specific pervasiveness index. Family and couple intervention and attention to hygiene education were particularly common and about half the programs had instituted women's groups. The eight programs highest in woman-focused activities all provided some services for children; none of the others did. Services for children ranged from referral and consultation to providing on-site day care services and intensive parent education. About half the TC's provided extensive program resources for the care of these children. The MM programs higher in women's services reported that much staff time was expended in advocacy activities with community agencies on behalf of their clients. They also were very concerned about training their women clients to advocate more effectively on their own behalf.

Environment and Program Boundaries

Several hypotheses concerned aspects of the programs' environment and boundaries as they related to the utilization of resources and their impact on program scope. These aspects include: 1) the numbers and types of demands from regulatory agencies and funding sources; 2) structures and mechanisms developed for coping with these demands from external sources; and 3) procedures for maximizing linkages to the environment and responding to shifting needs of the client population.

> Hypothesis 1: Drug treatment programs operate under multiple demands for interaction and accountability from funding sources and regulatory agencies and program resources must be diverted to meet these demands.

We found that one-half of our sample programs were accountable to no fewer than six regulatory agencies simultaneously. Three programs in our sample were accountable to no fewer than eight regulatory agencies at the federal, state, regional, county and city level. The requirements and expectations of most of these agencies were viewed by the programs as ranging from moderate to very extensive.

We asked our respondents to identify the sources of funding which required program-funding source interaction. Eight (33.3%) reported up to two sources, 9 (39.1%) reported 3 to 4 sources of funding and 6 (25%) identified in excess of five sources of funding. The maximum number of sources reported was 16.

> Hypothesis II: Drug abuse programs have adopted a number of internal strategies to minimize the energy drain of coping with environmental demands.

Program energy and resources are expended to respond to demands from the environment. Organizations seek to minimize the diversion of their energies from task accomplishment (client rehabilitation in this case). Drug abuse programs have a number of strategies available to them to accomplish this. One is to adjust the internal operation of the program to the expectations of the environment and to reduce variability in the program operation. This may include development of a more bureaucratic structure and the minimizing of uncertainty through formalization, e.g., using written procedures. Our data suggests in fact that some programs are quite formalized. Responding to a thirteen item scale tapping a number of areas in which standardized procedures may occur (adapted from Inkson, et al. 1970), 66% of the programs had a score of 10 or more.

Another defensive strategy may be to hire staff whose professional status, legitimacy and personal history may not incur additional scrutiny from the environment. Although our data is not historical, we have found that nearly 30% of the programs report that in excess of 50% of their staff possess professional degrees. An additional 26% of the programs report that at least 25% of their staff possess professional degrees. Also the level of employment of ex-addicts seems lower than one would expect traditionally. Fifty-four percent of the programs report that they have staffs with 0 to 24% ex-addicts.

A third strategy that programs might use would be to organize themselves so as to have specialists with primary responsibility for tending to environmental demands. This was not extensively used by the programs in our sample although some utilized specialists located at the parent program level who served similar functions for all the component programs in the umbrella structure.

> Hypothesis III: Drug abuse programs, like other organizations, use certain structures which lie outside their boundaries, but with which they are formally linked to "buffer" the programs and conserve their internal resources in dealing with the environment (Thompson, 1967). Boards of directors, advisory boards, coordinating agencies and parent organizations may be utilized in this way.

Twenty (83%) of the programs report having a board of directors which have a significant influence on some aspect of the program. Only four have advisory boards that have significant influences on the programs. When asked if their program was part of an umbrella organization or was independent, seventeen of the 24 programs reported being embedded in a parent organization. The parent organizations of the embedded programs included hospitals, mental health services and organizations which specialize in drug abuse treatment programs. All but one of the MM programs were embedded. Seven of the TC's were embedded, with the remaining five, freestanding.* Since MM programs were largely embedded, much of the following analysis will focus on TC's.

After our first wave of data collection we had theorized that embeddedness would have both costs and benefits. We anticipated that the benefits of embeddedness would include: 1) buffering of the program from environmental demands through the provision of supportive program services; 2) increasing activities and services for clients through the utilization of parent program resources; and 3) increasing the linkage to the community's human service network with a central unit or some outside structures facilitating referrals of program clients to needed services unavailable within the program.

Programs indicated whether they obtained certain services and benefits through their parent organizations, coordinating agencies, boards of directors or other sources. Table 1 lists the types of program services available that can facilitate the management of the program and responses to environmental demands for accountability. Nearly all the TC's which are embedded in a parent organization receive assistance from either their parent organization or other sources in all the areas of service. The largely freestanding TC's were less likely to receive such benefits. Receiving services in these areas is likely to increase the program resources which are available for client-centered services. Embedded programs clearly would have more freed resources in this fashion than their freestanding counterparts.

Respondents also identified those services and benefits directly available to clients from the parent organization. Table 2 lists these services and program responses. At least 57% of the embedded programs received such services in all but two areas (lab procedures and employment) while no more than two of the freestanding TC's (40%) received any of the services. The bulk of MM programs received all the services. Eleven (92%) of the MM programs received psychiatric consultation which was the most widely received service.

---

* As our sample was not drawn randomly it is not appropriate to conclude that this is the typical distribution of embeddedness in drug programs.

Table 1. Number of Programs Receiving Services From a Source Outside the Program

	Therapeutic Community Freestanding		Therapeutic Community Embedded		Methadone	
Program Services	Number	%	Number	%	Number	%
Financial Mgmt/Budgeting	1	(20)	7	(100)	10	(83)
Legal	3	(60)	7	(100)	8	(66)
Fundraising	1	(20)	7	(100)	6	(50)
"Political" Help (re: zoning)	2	(40)	7	(100)	9	(75)
Management Consultation	2	(40)	7	(100)	10	(83)
Record Keeping/Statistics	2	(40)	6	(86)	10	(83)
Interaction With Regulatory Agencies	2	(40)	6	(86)	10	(83)
Staff Development/Training	2	(40)	6	(86)	12	(100)
Treatment Consultation	1	(20)	7	(100)	10	(83)
Information re Developments in the Field	3	(60)	7	(100)	9	(75)
Grant Preparation	---	---	7	(100)	9	(75)

Table 2. Number of Programs Receiving Services From a Source Outside the Program

Services Provided Directly To Clients	Therapeutic Community Freestanding		Therapeutic Community Embedded		Methadone	
	Number	%	Number	%	Number	%
Medical (excluding Intake)	1	(20)	6	(86)	8	(66)
Legal Services	2	(40)	6	(86)	9	(75)
Intake	2	(40)	5	(71)	8	(66)
Employment	1	(20)	3	(43)	6	(50)
Vocational Counseling	2	(40)	4	(57)	8	(66)
Psychiatric Consultation/ Psychological Testing	2	(40)	6	(86)	11	(92)
Lab Procedures (Non-Intake)	1	(20)	2	(29)	8	(66)
GED Classes	2	(40)	4	(57)	7	(58)

The pattern emerging suggests direct benefits through embeddedness for program clients.

Besides those services that the parent organization provided, the programs also obtained services for their clients through the use of referrals. We asked the respondents to identify the categories of services referrals which they made for 5% or more of their clients in 1976. Table 3 lists referral areas and program responses. Most programs had made referrals to medical facilities, the social welfare department and to vocational training agencies. From a possible 11 categories, the mean number of categories of referrals made from MM programs was 7.4; for TC's which are embedded the mean was 8.58 categories; for the freestanding TC's the mean was 5.6 categories. Though the data is based on small numbers, it suggests that embeddedness facilitates more effective linkage to the existing human service network.

Relationship of Embeddedness to Scope

Given all these available services, programs embedded in parent organizations ought to provide more services than those programs which are freestanding. The mean activity score of the embedded TC's was 43.7 while the mean activity score of these freestanding was 36.6. In examining the scores for activities required for at least some of the clients, the pattern is repeated. Embedded TC's have a mean score of 24 required activities; freestanding, 16.6 activities, and MM programs a mean score of 8.1 activities.

Women's Services and Embeddedness

We examined the woman-focused activities scores to determine if this same pattern was continued. We found essentially no difference in the mean female-focused activity scores for embedded and freestanding therapeutic communities. Thus, while total services are increased in embedded programs, those for a special population which might be considered to demand more innovative programming, do not seem affected by embeddedness.

Program Autonomy

Though being embedded within a parent organization seems to afford certain service advantages to programs and their clients, there may also be some disadvantages. A further hypothesis was that those programs embedded in parent organizations sacrifice power and autonomy. We suspected, based on findings in other organizational research, that autonomy is a key factor in the programs' adaptability to changes in their operant fields. This might suggest that autonomous drug abuse programs would be more adaptable to identifying

Table 3. Programs Reporting Referrals for 5% or More of Their Clients In 1976

Type of Referral	Program Type		
	Therapeutic Community Freestanding	Therapeutic Community Embedded	Methadone
Social Services	5	6	12
Vocational Training	4	7	12
Non-Vocational Training	4	4	7
Legal Aid	–	4	8
Employment Agencies	3	3	9
Medical Facilities	5	7	11
Drug Treatment Agencies	1	3	8
Criminal Justice	1	2	5
Mental Health	1	4	9
Child Care	3	5	4
Child Health	2	5	4
Mean Number of Referrals Made by Programs for 5% or More of Their Clients	$\bar{x} = 5.6$	$\bar{x} = 8.58$	$\bar{x} = 7.4$

treatment needs of special populations, in this case women, and be able to mobilize in order to innovate and meet those needs.

Using an adaptation of previous work (Inkson, Pugh and Hickson, 1970) on autonomy and locus of decision-making, we asked respondents to identify who had the final decision in each of 21 areas.* On the whole, freestanding programs made their own decisions, with three programs identifying their boards of directors as making some final decisions (in the area of spending unallocated funds, and in moving the facility).

In contrast the embedded programs had complete autonomy in few areas. In 19 of the 21 decision areas, the majority of programs reported that the parent organization either made the decision or the decision was made jointly by the program and its parent organization. We constructed a dichotomous variable based on the responses of the programs to the decision-making schedule. Not surprisingly, all freestanding TC's were high on autonomy; all embedded TC's were low. The relationship in the MM programs is less clear: one-third of the MM programs fall in the high category and two-thirds in the low.

One might speculate that the similarity of quantity of women-focused services for both freestanding and embedded TC's, despite the marked differences overall in services, might be attributable to the greater autonomy of action which freestanding TC's can exercise in order to innovate in response to client population needs. More work must be done to understand the autonomy findings for MM programs. Perhaps the coordinating regulatory agencies have different requirements in different geographical regions.

Financial Resources, Scope and Women's Services

Each program provided WDR staff members with whatever budget information they were able to compile. Total budgets for comparable fiscal years with a rough breakdown of allocations were computed for each program component. Budget amount per client served was computed for each program and correlated with the programs' scores on scope (quantity) and women's services. As expected, financial resources were positively related, both to total activities and to women's activities scores (Total Activities Score: $Tau_B = .5822$, SE = .1806, Sig. = .0009; Women's Activities Score: $Tau_B = .5250$, SE = .1831, Sig. = .0036). These relationships were somewhat less strong when modality is controlled (TC's, in general, were much higher in both scope measures).

---

* Respondents often identified multiple loci of decision-making, suggesting that the authority structure for many programs may not be well articulated.

The programs' ability to increase their scope and, therefore, women's-focused activities seems to depend heavily on the level of available financial resources. Programs providing only minimal services are unlikely to be providing special services to women. Interest in providing such services will, by definition, increase program scope, probably both in quantity and pervasiveness. Such an increase requires more resource availability.

SUMMARY AND CONCLUSIONS

Our findings suggest that drug abuse treatment programs operate in environments with high demands for accountability and interaction. Such environments may encourage the diversion of program energy from provision of services to clients and client rehabilitation to system maintenance - responding to environmental demands. Many mechanisms which programs can utilize to minimize this diversion occur in this sample of programs. Being embedded in a parent organization seems to have some clear advantages for the TC's. Further exploration of the disadvantages is necessary. Indications of decreased autonomy and perhaps limited flexibility might suggest that embedded organizations may be very efficient for administering routine treatment but less in developing and implementing more innovative programming. Unfortunately the state of the treatment art, especially for women, suggests that routinizing treatment may be quite premature. A disturbing trend in certain programs was the adoption of minimum formulas required by regulatory agencies as the standard of treatment for all clients. In essence the required minimum became the maximum treatment, and was applied to all clients regardless of individual needs.

Provision of women-oriented services seems to be related to higher program scope, in general, and to a higher level of program fiscal resources. Women-oriented services are not necessarily higher in embedded programs, although most report that much higher levels of both program and direct client services are available from the parent organization. Whether these additional resources result in more women-oriented programming seems to depend heavily on the degree of sensitivity to women's needs at the umbrella level and the level of fiscal resources available to a given program.

The measures of program scope and program productivity appear to have some utility in understanding characteristics of service delivery systems. Client outcome studies could use these concepts to focus more extensively on the quantity and quality of treatment services and the context in which they are provided. Policy makers must become aware of the energy drain on treatment programs that environmental demands create. Mechanisms must be developed to assist programs to manage these demands while minimizing the development of program structures and buffering mechanisms that inhibit innovative programming. Understanding the special service needs of addicted women is in its early stages and the development of effective programming will continue to require experimentation and program flexibility.

Our data suggest that it will also be more expensive. Policy makers and funding agents might want to consider incentive funding to programs demonstrating both a high quantity and pervasiveness of scope in their treatment services if they wish to stimulate more effective service delivery related to women's needs.

REFERENCES

Andersen, M.D. 1977. Medical Needs of Addicted Women and Men and The Implications for Treatment: Focus On Women. WDR Report to the National Institute on Drug Abuse. Special Treatment Projects Section, Services Research Branch, Div. Resource Development.

Etzioni, A. 1961. *Comparative Analysis of Complex Organizations*. New York: Free Press.

Inkson, J.H.K., Pugh, D.S. and Hickson, D.J. 1970. Organizational Context and Structure: An Abbreviated Replication. *Adm. Sci. Quarterly, 15:* 518-529.

Katz, A. and Kahn, R.L. 1965. *The Social Psychology of Organizations*. New York: John Wiley.

Lefton, M.R. and Rosengren, W. 1966. Organizations and Clients: Lateral and Longitudinal Dimensions. *Amer. Soc. Rev., 12:* 802-810.

Reed, B.G. and Herr, J.N. 1977. Deviant Social Careers and Addicted Women: Issues in Treatment. WDR Report to the National Institute on Drug Abuse. Special Treatment Projects Section, Services Research Branch, Div. Resource Development.

Rinaldo, S. and Reed, B.G. 1977. The Organizational Context of Treatment for Female and Male Heroin Abusers: Preliminary Observations on Service Delivery Determinants From a Systems Perspective. Presentation to the National Drug Abuse Conference, San Francisco, California.

Ryan, V. 1978. Differences Between Males and Females in Drug Treatment Programs. Presentation to the National Drug Abuse Conference, Seattle, Washington.

Thompson, J.D. 1967. *Organizations in Action*. New York: McGraw-Hill.

HOSPITAL TREATMENT OF PREGNANT ADDICTS[1]

Virginia S. Ryan, Ph.D.

INTRODUCTION

The treatment of addiction during pregnancy has become an overwhelming problem in many hospitals; some report drug abuse among as many as 15% of their delivering patients. Recent research suggests that pregnant addicts have many problems. These may include:

1. Physiological consequences of drug use and life style, for example, anemia, malnutrition, venereal disease, hepatitis, and numerous types of infections (Pelosi et al., 1975; Perlmutter, 1974);

2. Obstetrical complications, for example, spontaneous abortion, placenta previa, abruptio placenta, and post-partum hemorrhage (Pillari, 1975);

3. A reticence to request medical treatment until late in labor if at all (Pelosi et al., 1975); and

4. A need for psycho-social support to prepare the addicted woman for the coming child and to help her deal with the possible problems of caring for a baby suffering from subacute withdrawal following in utero associations with drugs (Brazelton, 1970; Wilson, 1977; Wilson, 1975).

---

1. Undertaken as part of the Women's Drug Research Project, funded by the National Institute of Drug Abuse Grant #5H81-DA 01496, awarded to the Wayne County Department of Substance Abuse Services with a subcontract to the University of Michigan School of Social Work.

Failure to address these problems has many serious consequences, not only for the woman herself, but also for her baby. However, many of the problems of pregnant addicts may be dealt with using special treatment protocols (Zarin-Ackerman, 1977; Ostrea, 1975; Lief, 1977). Consequently several hospitals have instituted pregnant addicts programs to deliver specialized and comprehensive treatment to these women. These programs have all reported treatment success which may be attributed to the fact that they offer extensive services by sensitive professionals in the areas of obstetrics, pediatrics, drug treatment and social services (Davis, 1977; Finnegan and Mcnew, 1974). These elements are typically coordinated so that the addicted women are given comprehensive care by personnel with whom long-term relationships of trust are established.

Unfortunately, the majority of cities with large numbers of addicts do not have such special programs. Heretofore, it has not been clear what kinds of treatment are available for pregnant addicts in most cities. As a result, the research that I will describe to you today was designed to review the kinds of treatment that are typically available to addicts, problems inherent in this treatment, an overview of treatment needs, and ways in which current treatment may be improved.

The research was conducted in six major U.S. cities with distinctly different characteristics. They represent different population types, geographical locations and treatment protocols for pregnant addicts. Half of these cities have pregnant addicts programs and half do not. In each city information was gathered from hospitals that have the primary responsibility for caring for pregnant addicts and their babies. Most cities have only one hospital with primary responsibility for this kind of treatment. Only one city had two major hospitals delivering treatment and in this city both hospitals were visited. The chief of the obstetrical department, pediatric department, and the head of social services, or their designated alternates were interviewed at each hospital. I also spoke with the directors of the special programs and other individuals that interviewees felt were pertinent, including administrators, researchers, nurses and other physicians. An average of seven interviews of an hour or more each were completed at each of the hospitals. In addition, at least one drug treatment program was visited in each city, providing the opportunity to get an overview of problems of addiction as a whole in the city, determine the extent to which drug programs liaison with hospitals, and to get some idea of the socio-political milieu in which drug treatment programs operate. I would like to use the limited amount of time available today to review some of the overriding problems in treatment reported by interviewees and to describe some issues which appear to be of highest priority in delivering adequate treatment to pregnant women.

IDENTIFICATION OF ADDICTION

The most outstanding impediment to delivering good hospital treatment to pregnant addicts is the inability of most treatment deliverers to identify women who use drugs. This problem has three distinct aspects. The first of these involves the inability of hospitals to identify abusers of non-opiate substances. This is so problematic that hospital personnel are unable to provide any specifics of treatment of this group. Research suggests that as many as 5% of delivering females abuse non-opiate substances (Hill, 1973). All the physicians that were interviewed in the present research felt that this is a serious problem in their hospitals. However, abusers of non-opiate substances were rarely identified as such, and nothing was known about these women.

The second problem in identification involves the inability of many hospitals to identify methadone-maintained women. I was able to ascertain this by utilizing a very simple measure. Methadone programs were randomly selected in the cities visited and it was determined that an average of 14% of their female clients become pregnant each year. Then the figure that represents 14% of all women in methadone programs in the area served by each hospital was computed. This estimate of local pregnancy of methadone clients was compared to the number of methadone-maintained pregnant addicts identified by each hospital. The results of this comparison suggest that hospitals with pregnant addicts programs tend to identify more than the number estimated, presumably because they treat newly recruited heroin addicts who have not previously been in treatment, as well as transfers from traditional methadone programs. However, hospitals without special programs identify less than half of the number of methadone-maintained women estimated as entering their system during the period of a year.

These, and other data strongly suggest that a third problem in identifying addicted women exists. That is, the failure to identify heroin addicts is even greater than the failure to identify methadone-maintained women. It is most likely that the majority of heroin addicts go undetected in all of the hospitals visited, including those with special programs. It is very difficult to ascertain the size of this "hidden population." However, drug urinalysis screening suggests that 2/3 to 3/4 of the addicted obstetrical patients are heroin users. Hospitals typically report that only half of the addicts they identify use heroin.

It must be noted that although hospitals are generally unable to identify their addicted patients, occasionally a given service or individual makes a concerted effort to do so. However, this information is not generally shared by those who make identifications. For example, in one site the obstetrical department identifies only half as many addicted women as the number of babies in withdrawal identified by the pediatric department. In another situation the social service department identifies 13 times more pregnant addicts (4%) than the obstetrical department (.3%) or pediatric department (.3%).

Clearly, problems in identification of addiction must be resolved if addicted women are to be: 1) encouraged to seek special care; 2) monitored for the typical problems of this group, many of which may affect the fetus (e.g., anemia and venereal disease); 3) given counsel relevant to the impact of drug use on the fetus and 4) given information about hospital services, how to receive help, and what to expect as an in-patient.

A first step in identification requires instituting a procedure which dependably screens all in-coming patients. Two methods were identified in the present research: 1) in depth interviews with all patients and 2) drug urinalysis. The urinalysis is undoubtedly the cheapest and easiest method, and is the approach least likely to be biased by personality factors. However, there are problems in this method relevant to confidentiality issues, and these must be addressed prior to instituting the practice. The one hospital visited which utilizes this approach was allowed to do so only after implementing the test as a standard procedure. It is part of a battery of tests performed on all in-coming patients regardless of sex or other variables, and is a condition of admission to the hospital to which all patients must agree.

ATTITUDES OF TREATMENT DELIVERERS

The attitudes of hospital based professionals is another important area of treatment, because attitudes may be expected to have a significant impact on the ability of the staff to deliver optimal care. While many respondents appear deeply concerned about pregnant addicts, it was clear during my interviews that many others do not have a positive attitude towards addicted women. Some respondents report shaking their patients, screaming at them, feeling they should all be sterilized or killed. It was not possible to quantify these attitudes in a rigorous unbiased manner. Nevertheless, I attempted to compare the attitudes of professionals in a very simple way. Three basic attitudinal categories were constructed including the classification "hostile", "intermediate" and "sympathetic." These may be defined as follows:

1. hostile - actively engaged in attempts to decrease available services or verbalizing punitive attitudes or behavior as previously described;

2. sympathetic - actively engaged in research or in attempts to increase available services; and

3. intermediate - not exhibiting overt behavior which is either hostile or sympathetic by these classifications.

Not surprisingly, all but one of the respondents in pregnant addicts programs were classified as sympathetic, because they reported being involved in attempts to increase the kinds of services they were able to provide, and many were also actively engaged in research.

However, when the attitudes of all treatment deliverers are tabulated by these classifications, the results are not overwhelmingly positive. 44% are classified as intermediate. Although 39% may be classified as positive, 24% of these are from special programs while only 15% are not. 17% of the respondents are classified as hostile. In attempting to determine the impact of that 17% who are overtly hostile towards pregnant addicts, the data were analyzed by sex and profession of treatment deliverers. 86% of those with hostile attitudes are males and 70% are male obstetricians. Although there are not enough subjects to analyze the interaction between profession and sex, it is possible to make individual comparisons. These analyses indicate that women in treatment are significantly more sympathetic towards pregnant addicts than are their male counterparts ($X^2 = 17.11$, df = 2, $p < .01$, Table I), and that individuals delivering social services are the most sympathetic, obstetricians are the least sympathetic, and pediatricians are between these two ($X^2 = 19.87$, df = 4, $p < .01$, Table II).

Table I. Attitudes of Treatment Deliverers By Sex

Sex of Respondents	Attitude			
	Sympathetic	Intermediate	Hostile	Total*
Males				
Physicians	2	5	5	12
Soc. Work.	1	1	0	2
Others	1	2	1	4
Subtotal	4	8	6	18
Females				
Physicians	4	3	1	8
Soc. Work.	9	0	0	9
Others	1	10	0	11
Subtotal	14	13	1	28

* $X^2 = 17.11$, df = 2, $p < .01$.

Table II. Attitudes of Treatment Deliverers By Profession

Profession	Attitude			
	Sympathetic	Intermediate	Hostile	Total*
Obstetrician	2	5	5	12
Pediatricians	4	3	1	8
Social Workers	10	1	0	11

* $X^2 = 19.87$, df = 4, $p < .01$.

One would expect women treatment deliverers to be especially sympathetic towards the problems of other women. One would also expect that social workers, who have been trained to handle social-emotional aspects of patients, to be more comfortable and accepting of addicted women, many of whom do present serious management problems. On the other hand, emotional problems may interfere with delivery of medical care, and consequently physicians may resent the extent to which addicted patients disrupt their schedules and hospital protocol. Furthermore, obstetricians may find treatment of pregnant addicts more of a problem than do pediatricians, because obstetricians have more extensive contact with these patients.

Nevertheless, the punitive attitudes of many physicians are unconscionable and contrary to the ethics of the medical profession. Unfortunately, most hospitals do not have a staff person who may act as an advocate for addicted women or protect them from unfair or biased treatment. However, it is clear from this study that such a person is needed in most hospitals, not only to maintain a high level of professionalism, but also to help addicted women in other ways. For example, such a staff person is usually required if services are to be coordinated and the pediatric department is to be appraised of a mother's drug use in preparation for: 1) potential problems in neonatal withdrawal; and 2) requests for premature discharge of the baby. Because most hospitals have small social service staffs that cannot begin to meet the needs of pregnant addicts, addicted women often spend a great amount of time talking to nurses. An advocate could also help both social workers and nurses by accepting responsibility for the psychosocial needs of pregnant addicts. An advocate may attend to many patient needs including: 1) helping patients negotiate within the hospital system; 2) preparing the woman to deal with the potential withdrawal of her baby from drugs, and for the problems of motherhood; and 3) providing referrals for out-patient social services or drug treatment. Such a staff person would not only relieve hospital staff of time-consuming activities but, in many cases, provide services that are imperative in providing optimal treatment and currently not available. These services would be of particular importance to heroin addicts who do not have the support of drug treatment programs, are often alienated from our cultural mainstream, and may have few options for receiving necessary help.

Many hospitals are aware of the problems some of their staff have in dealing with pregnant addicts. Some hospitals have begun to provide in-service training, teaching staff what to expect of addicted patients, and how to cope with these women. This tact was reported to reduce many of the problems. However, some respondents indicate that not all physicians are able to learn to deal effectively with addicted patients. Consequently, most of the people interviewed feel that treatment of pregnant addicts should be provided by only a limited number of staff, those who feel comfortable in treating these patients. It was also felt that addicted women respond best when they are always treated by the same physician, one whom they have come to know and trust.

AVOIDANCE OF HOSPITAL CONTACT

Another major problem was found in delivering hospital care to pregnant addicts. Many of these women are reticent to approach the hospital, and consequently receive little hospital care. At some sites pregnant addicts are reported to avoid hospital contact altogether, preferring to deliver their babies at home, without medical intervention.

It was impossible to determine how prevalent this phenomenon actually is. One can only assume it is a problem as a result of emergency room activity related to postpartum obstetrical complications following home deliveries. Although most hospitals do not keep records relevant to this type of emergency room activity, one social worker has tabulated this information. This individual reports that she sees six times more addicted women requiring postpartum obstetrical care in the emergency room than she sees delivering within the hospital.

Many addicts will opt for hospital delivery although they will not seek prenatal care. Specifically, although prenatal care was reported to be received by 66% of the methadone maintained women, only 27% of the heroin addicted women were reported as having received prenatal care. Even when a special pregnant addicts program is available within a city, many addicts will not opt for these services. One methadone program visited has established liaisons with a special pregnant addicts program and reports that more than 1/3 of the women that they refer for special treatment never follow-up on these referrals.

Women who opt to deliver within a hospital are described as often entering the hospital when very high on drugs and late in labor. This makes them particularly difficult to manage and often results in narcoticized babies at birth. Unfortunately, many of these women choose to leave the hospital soon after delivery against medical advice and often want their babies to be prematurely discharged.

In speaking with treatment deliverers it became clear that there are several reasons for these problems at delivery: 1) the punitive behaviors and negative attitudes of physicians, which have already been discussed, may make the hospital stay unpleasant; 2) the discomfort of obstetrical complications and labor which precipitates medical contact by most women may be handled by addicted women through self-medication; and 3) drug maintenance and analgesia of these women may be inadequate. Although all of the hospitals visited make every attempt to provide women in methadone programs with appropriate methadone doses, physicians often tend to be conservative in providing additional drugs. The doses of methadone which are delivered daily to patients are geared to the individual's tolerance level. They do not serve to sedate or anesthetize a patient during labor or postpartum distress. Unfortunately, many addicted women require greater than normal doses of analgesia, because of cross tolerance for drugs, greater physical problems, or a decreased pain threshold. However, only half of the hospitals visited were sensitive to this problem and responded with larger than average analgesic doses when treating addicted women.

A most unfortunate problem in medication was found in the treatment of heroin addicts, who make up the majority of the addicted population. These women are generally offered methadone to deal with their opiate dependence. A few hospitals also provide non-opiate substances, and one hospital does not provide any medication. Many heroin addicts refuse methadone when it is offered, presumably because they fear addiction to it and subsequent withdrawal. Only one of the hospitals visited utilizes morphine (the opiate from which heroin is derived) to manage the drug maintenance needs of heroin addicts. However, the hospital which utilizes this treatment is unique in reporting that their patients respond positively, are comfortable, do not demand premature discharge, and there is no drug trafficking in the wards.

ATTRACTING ADDICTS FOR TREATMENT

It is clear that a most important goal in improving the treatment received by pregnant addicts involves developing mechanisms that motivate these women to seek and continue to receive treatment. The methods utilized must be sensitive to the diverse qualities of the addicted population, and the rights of each addict relative to making personal choices and maintaining confidentiality. The limitations of many of these women must also be considered when designing treatment models. Procedures which increase the probability of hospital contact should rely heavily on information sharing, including informing addicts of their medical needs, the consequences if appropriate treatment is not received, and the kinds of treatment that are available. The mechanisms employed in such processes should protect the addict's identity until she is comfortable in revealing her drug use to selected individuals, and should allow the addict free choice in opting for various types of treatment. Treatment models should also be sensitive to the addicts' personal needs. Many addicts do not respond well to tight schedules. These patients must be cared for using an "open door" policy and not be expected to conform to firm appointments. In addition, many addicts lack a sense of personal worth and self-determinism. Consequently, hospital treatment models must enhance these patients' positive self-concept if contact is to be maintained.

It is clear that the greatest challenge involves increasing the involvement of heroin addicted women in health delivery systems. This group represents the majority of addicted women; it is the group which is most likely to have the greatest physical and psychosocial needs, and has the fewest resources. Unfortunately, this group is also the most difficult to reach because of several irregularities which preclude early diagnosis of pregnancy. Second, when these women discover they are pregnant they have few reliable sources of information if they are not receiving social services. And they generally cannot receive treatment in special pregnant addicts programs as there are few of these and most will not deliver treatment to addicted women unless they are willing to accept methadone. Finally, treatment within hospitals may be feared, either because of previous experiences or because of the potential impact of revealing illegal drug use and criminal activities.

Most probably the most effective way of bringing these women into treatment would involve the institution of new treatment models in many hospitals. In addition, attempts to appriase heroin addicts of their needs, alternatives, and legal rights during pregnangy may be effective. General public information campaigns might have some effect. However, it is likely that information would be viewed as more trustworthy if it came from community based social workers and street level workers who have liaisons with drug-dealing networks.

Because as many as 15% of the women who become pregnant each year may be abusing drugs, and because neonatal withdrawal is the first or second most common neonatal complication at many hospitals, it is not unrealistic to suggest more comprehensive measures. For example, drug urinalysis might be performed whenever pregnancy tests are made, or whenever any individual opts for medical attention. However, information gleaned from such tests must be carefully guarded and used only in ways which will not violate the individual's rights. It might be more reasonable to supply all women testing positive for pregnancy with a fact sheet which summarizes information relative to local services available, and basic treatment needs during pregnancy, especially if they are using drugs.

Women in methadone programs are, theoretically, easier to bring into treatment. They already have program support and have established liaisons with traditional systems which may serve to reduce their fears of "the establishment." Nevertheless, many of these women receive little or no medical care and are maintained on high levels of methadone during pregnancy. There are several things that drug treatment personnel may do to help these women, and it is important to underscore some of the responsibilities of drug treatment programs to these clients. Drug treatment programs are in a unique position because they are most usually the only outside agency that is aware of both their clients' drug use and obstetrical status. This information may be used to motivate the female to seek obstetrical care, as well as to determine appropriate methadone goals. However, providing any drug, including methadone, to a pregnant woman who is not under an obstetrician's care is irresponsible in view of the potential danger to the fetus, particularly when high doses are administered. Those involved in the delivery of maintenance drugs should interact with obstetricians as soon as pregnancy is determined to: 1) establish appropriate methadone treatment goals for the pregnant addict, and 2) determine schedules of prenatal clinic visits to be encouraged by the program. Programs may effectively go one step further to help these women. They may refuse to treat any pregnant client who is not under an obstetrician's care. This tact is used successfully in special pregnant addicts programs delivering methadone, and there is no reason why it would not be effective in traditional programs as well.

CONCLUSIONS

It is the overwhelming opinion of people who have extensive contact with addicts that the period of pregnancy is a critical time in the lives of these women. During this period women are open to making positive changes in their lives and, for the first time, may view their lives in long-range terms. Many addicts experience a sense of success and self-worth during pregnancy which may be used as a positive substrate for treatment intervention. They may reassess their patterns of drug use, and consider the impact of addiction on their family and the coming baby.

Unfortunately, health delivery systems have not generally responded to these challenges, and many have not begun to recognize their roles in rehabilitation. It is imperative that efforts be expended at all levels to legitimize the treatment of pregnant addicts and address the fact that early (prenatal) intervention is a critical factor in solving a problem which is currently self perpetuating, and socially debilitating by all standards.

REFERENCES

Brazelton, T. 1970. Effect of Prenatal Drugs on the Behavior of the neonate. *Am. J. Psychiat., 129:* 95.

Davis, M. 1977. The Quality of Prenatal Treatment of Narcotic Addicted Women as a Factor in Neonatal Outcome. *Symposium on Comprehensive Health Care for Addicted Families and Their Children.* Services Research Report, National Clearing House on Drug Abuse.

Desmond, M., et al. 1972. Maternal Barbituate Utilization and Neonatal Withdrawal Symptomatology. *J. Ped., 80:* 190.

Finnegan, L. and Mcnew, B. 1974. Care of the Addicted Infant. *Am. J. Nurs., 74:* 685.

Hill, R. 1973. Drugs Ingested by Pregnant Women. *Clin. Pharmacol. Ther., 14:* 454.

Lief, N. 1977. Some Measures of Parenting Behavior For Addicted and Nonaddicted Mothers. *Symposium on Comprehensive Health Care for Addicted Families and Their Children.* Services Research Report, National Clearing House on Drug Abuse.

Ostrea, E.M., Chavez, C.J. and Strauss, M. 1975. A Study of Factors That Influence the Severity of Neonatal Narcotic Withdrawal. *Add. Dis., 2:* 187.

Pelosi, M., et al. 1975. Pregnancy Complicated by Heroin Addiction. *Ob. Gyn., 45:* 512.

Perlmutter, J. 1974. Heroin Addiction and Pregnancy. *Ob. Gyn., 29:* 439.

Pillari, V. 1975. Special Problems and Management of the Pregnant Drug Addict. *Ped. Ann., 4:* 377.

Wilson, G. 1977. Management of Pediatric Medical Problems in the Addicted Household. *Symposium on Comprehensive Health Care for*

Addicted Families and Their Children. Services Research Report, National Clearing House on Drug Abuse.

Wilson, G.S. 1975. Somatic Growth Effects of Perinatal Addiction. *Add. Dis., 2:* 333.

Zarin-Ackerman, J. 1977. Developmental Assessment of All Infants Born to the Family Care Program, 1975-1976. *Symposium on Comprehensive Health Care for Addicted Families and Their Children*. Services Research Report, National Clearing House on Drug Abuse.

A FEMINIST COUNSELING MODEL BEYOND THE MYTH OF THE PERPETUAL CHILD

Christine E. Miller, M.ed.

The Matrix, Tucson Awareness House, Inc.

A feminist counseling model strives to empower women. Empowerment means a rediscovery of strength from within that has been lost in female socialization. A woman is socialized to be a perpetual child and the struggle for personhood is traumatic. The counselor validates and encourages a woman in her struggle to become herself. This paper presents the perpetual child myth as a major stumbling block to personal self-actualization. The feminist counseling model contains necessary techniques for moving beyond this non-productive myth.

THE MYTH OF THE PERPETUAL CHILD

The female is socialized to be a child throughout life. She depends upon parents throughout adolescence, and then transfers the dependency to an adult man as she comes of age. As a child, she learns a passive orientation to life. Waiting for events to happen, waiting for people, waiting for life to become meaningful, all reinforce the passive role. She needs parents for food, shelter, clothing, and emotional support. Later, her survival skills consist of finding the right man to fulfill these same parental needs for her. Her training lacks both tools and incentives for learning how to survive on her own. Some manifestations of this training are: girls should look pretty, girls should help mother in the house, girls should speak softly, girls should be ornaments of beauty, mommy's helper, passive in social situations and physically underdeveloped.

Adulthood carries the same message for females: beauty is a woman's most marketable trait, a woman's place is in the home, speak softly, the man makes the important decisions. Throughout life, the social messages given females are self-limiting. Living out the expected female role has stifled the growth and creativity of many women. The idea of living with independence, and the opportunity to openly declare one's personhood, is frightening when contrasting the comfort of routine with the exploration of the unknown.

Feminists analyze the socialization of females to redefine the concept of woman. For feminists, a healthy concept of woman means independence, female identity, self-acceptance, self-understanding in the female perspective and, most importantly, growth beyond strict gender role behaviors.

The myth of the perpetual female child, like all stereotypes, is both true and false. In truth, women are not children. However, all women know they are supposed to conform to the stereotype. Pressures to conform cause women to live with fears of sex-atypical behavior. Women regulate their lives based on this stereotype, straying not too far from it and yet never perfecting the stereotype either.

The stereotype as it exists is falsified by acting differently. Even those women who conform to the wife and mother role have strength and inner resources attributed to adult, not childlike, behavior.

The socialized female role is changing and women today carry both the lessons of childhood and the benefits of an expanded role with newly opened possibilities. All women are in transition today. With more opportunity comes more responsibility. The desire to live differently from past generations leads women to experiment with new lifestyle options. However, many do not have the skills to live life in a different way. Making new choices can be fulfilling or it can lead to a crisis. One important determination of the difference is the extent to which a woman is surrounded by people who have a strong interest in her remaining in a tightly proscribed role. Feminist counseling provides an atmosphere of support while a woman experiences her changes.

A FEMINIST COUNSELING MODEL

Feminism is a social/political/cultural movement working toward the elimination of discrimination based on sex. Feminism also strives to equalize relationships between people. It is believed that the origins of social injustice began with the oppression of women.

The traditional counselor/client relationship mimes many characteristics of the male/female roles. Prior to the humanistic psychology movement, most counselors were male and most clients were female. Success in treatment reflected the degree to which a woman adjusted to her role of wife and mother (Chesler, 1972). The counselor labeled the client based upon how far she deviated from the norm. The norm for the health adjusted female once meant nurturing, passive, and feminine.

The feminist counseling model considers both the client and the counselor equal participants in the counseling process. It differs from other psychotherapeutic models in that its focus is on sex roles and female identity. The feminist counselor relates to the client as a peer who has skills to share. One of these skills, which is valued highly, is the ability to articulate personal experience.

Sex roles perpetuate a false assumption that males are strong and females are weak. These roles prohibit the growth and development of full personality traits. People long to be complete and search for that completeness in self and others. The feminist counseling model assumes each person can be a whole person, able to express the yin and yang, positive and negative, the strong and the weak aspects of mental, emotional, and physical endeavor.

THE COUNSELING PROCESS

Validation

Women enter counseling feeling sick, bad, stupid or crazy (Shaeff, 1975). The origins of these labels are in the continual messages denying female daily reality. The media portrays the rosy picture of woman as seductress, playboy bunny, elegant career woman, and earth mother. When her feelings conflict with these images, she may feel crazy, or significant others may suggest she is crazy.

The counselor validates a woman's experience by carefully listening. Many of the experiences told by the female client have also been experienced by the counselor. The counselor identifies with this experience and lets the client know she understands by the recognition. Validation means giving support to another woman. The counselor also encourages a woman to validate her own experience. For example, discussing her body in detail and how she feels about her body is an act of self-validation. A woman is stating "I am". This step initially begins the process of a woman defining herself on her own terms as opposed to society's image of who she should be.

Expressing Rage

A woman's exploration into herself takes her back to all the times she waited for others to do things for her. The exploration takes her back to memories of being told to act like a lady, when she wanted to play vigorously. She retraces various life experiences realizing the inequity in her development compared to male peers. Rage results from recognition of opportunities gone by due to an unjust social order. Before a woman accepts herself as a self-identified woman, this rage, held in for years, must emerge.

The rage is expressed in several ways. In counseling, pent up frustrations come out through pillow pounding exercises, role playing past situations, Bioenergetics and Gestalt work. Expressing rage is very liberating. The physical and emotional release enables a woman to see her own strength.

Feelings of rage can occur in cycles, depending on a woman's daily life experience. Sometimes harassment by construction workers sparks rage feelings in women. Power dynamics of male/female relationship patterns can bring out a sense of rage. Both the initial emerging of rage and its cyclical bouts are viewed as valid in the feminist counseling model. The release of rage constitutes a cleansing rite that heals women from the wounds of internalized hostility from the environment.

Assertiveness

> "Apology and powerlessness have characterized the lives of many women for generations. A woman's traditional social role has been a dependent, submissive one. Women have been expected to react rather than act, to have decisions made for them rather than make decisions for themselves. What women seek now is the power to determine the cause of their own lives without apology, to make their own decisions, and to be free from the absolute authority of others" (Phelps and Austin, 1975).

Women have basic human rights. However, women may come to counseling thinking they have no rights. A productive counseling technique is listing basic rights on butcher paper, which encourages a woman to state what is truly hers. Commonly mentioned rights include: the right to express anger, the right to have opinions, the right to be confused, the right to say "no", the right to get a good job, the right to privacy, the right to be myself, the right to control my body, the right to be free.

Assertiveness teaches a woman that she comes first. Self-esteem increases as a woman realizes her needs and wants and effectively expresses them to others. Techniques introduced are: assertive body language, direct speech, statements including feelings and wants and the use of fair fighting techniques. Rigorous practice of these techniques, weekly homework, assertive readings, and role-playing reinforce assertive behavior through which a woman learns to experience more control over her life.

Empowerment

Empowerment is the process of realizing the strength within (Riddle, 1973). All people have power. Sometimes people give their power away to others. Women give power away to others by being silent, speaking softly, apologizing, giggling in a serious situation, smiling when they feel angry, playing dumb, acting coy, avoiding direct eye contact and choosing subservient tasks. Often, women are in power under/positions in relationships. The counseling setting offers practice sessions of alternate behavior that tap the strength and wisdom from within. There are many ways a woman empowers herself. Being honest about personal strengths and weaknesses, spending time alone, self-nurturing and getting involved in women's centers and self-help clinics are some examples of empowering activities.

An exercise in empowerment is to have the counselor ask the client, "How do you give your power away?" The client responds by saying, "I give my power away by smiling", for example. Repeat this question to elicit answers as many times as possible. Then the client asks the counselor the same question. Following the series, the counselor asks, "How do you empower yourself?" The client responds with, "I empower myself by taking a stand", for example. This continues and then they reverse roles.

Squirreling

This technique is taught to women who want to get out of the situations they are in, but have no resources to call upon. The technique is predicated upon whether the client needs to take what available goods she can and flee, in time of crisis, or whether she can stay and put resources in reserve, until a safe time for leaving arises.

In particular, women living in abusive environments squirrel until they have stored enough rent money, food money, and other necessities for survival on their own. When a woman decides to change her living situation, she discusses this in counseling. The counselor explains the squirreling technique and how essential this forethought can be to the success of her change. Some women return to abusive home environments because they could not survive in any other situation. Squirreling is preparation for a possible cold winter.

Counselor Focus

The counselor gives support and unconditional positive regard to the client. The counselor is a resource person on matters of sexuality, child care, abortion, pregnancy, rape, consciousness raising groups, gay resources, welfare, and divorce. Through feminist literature, support and education are acquired which adds to a positive self image in a sexist society. Feminist literature deals with historical and cultural implications of sexism and offers tools to deal with sexism as a group member and individually. The counselor and other group members share information within the group as equals.

Client Focus

A woman expresses a need for change and engages in a contract to change. Her awareness of social and political forces empower her. She learns to make decisions and to be independent. She engages in a process of growth. During this process, she explores various life-styles and new personal definitions.

CONCLUSION

The feminist counseling model utilizes many techniques. The combination of strong support for the female life experience and humanistic counseling skills together form a counseling modality useful to female clients.

REFERENCES

Chesler, P., 1972. *Women and Madness.* Garden City: Doubleday and Company, Inc.

Chicago, J., 1975. *Through the Flower: My Struggle as a Woman Artist.* Garden City: Doubleday and Company, Inc.

Phelps, S. and Austin, N., 1975. *The Assertive Woman.* Fredericksburg: Impact Publishers.

Riddle, D.,1973. Power, powerlessness and empowerment. (Unpublished article).

Schaeff, A., 1975. The white-male system. (Unpublished lecture).

COUNSELING LESBIAN WOMEN

Christine E. Miller, M.ed.

The Matrix

Tucson Awareness House, Inc.

This paper discusses three topics important to counseling lesbian women: the social context within which the lesbian creates her lifestyle, specific therapeutic issues related to lesbianism, and lesbian survival skills. It is my intention to facilitate questioning the myths and stereotypes of lesbians that therapists have accepted as truth. Therapists will gain an awareness of counseling lesbian women from a viewpoint different from the traditional psychotherapeutic approach.

THE SOCIAL CONTEXT - BACKGROUND

Knowing the background information of the lesbian lifestyle presents a context enabling the therapist greater understanding when problematic issues unfold with the client. Lesbians define themselves by choosing to relate to other females as sexual partners. They consist of a group of women from all classes and strata of society: all religious backgrounds, ages, occupations, ethnic groups, and political persuasions. The lesbian lifestyle is created by a group of women who are socially stigmatized by existing either visibly or invisibly outside the WHASP (white-heterosexual-Anglo-Saxon-Protestant) mainstream. National power is held by the WHASP mainstream. Legal rights, child custody, property laws, income tax, employment rights, and housing codes are all based on an assumption that everyone should be heterosexual. Anyone straying from this assumed norm is labeled mentally sick or criminal.

Often when people hear that Jane Doe is a lesbian, all other information about the woman is somehow forgotten. Her sexual preference becomes paramount. Her career, her hobbies, her personality, and her family seem secondary to the fact that she is "queer". People do not refer to the next door neighbor, Mrs. Alice Goodwin, as "Alice the Hetero", yet it is common to hear such comments as "Jane the Lesbian."

Women loving women fulfills a basic need to love and be loved. Both the laws of God and State qualify who has the right to love whom, and this excludes love of same sex individuals. Current research shows the existence of lesbianism throughout time, along with the consistent oppression of this lifestyle. Lesbian love receives no social validation and must originate and maintain support from a strong sense of self-acceptance. The therapist facilitates the process of self-acceptance (unless she opposes this lifestyle). Homophobia, or the fear of one's own same-sex love feelings, prevents many from affirming the experience of lesbians. Therapists who refuse to accept and support a client's lesbianism need to be honest with themselves and make a referral. Lesbians have the right to a therapeutic atmosphere of validation and understanding to freely explore personal issues in the context of the explained social reality. This background material is not a part of the client "problem" or a counseling issue, but is a part of the counseling technique to be considered during therapy with lesbians.

THERAPEUTIC ISSUES

Clearing the Air

Through personal experience as a therapist and a lesbian, I find common threads in the issues discussed in counseling. Often lesbians initially need to vent remaining feelings from past therapy, which they found to be destructive. Lesbians report statements made subtly or overtly by former therapists to the client's detriment. Common examples are: being gay is the problem, if they stop shooting dope they will get their sexuality straightened around, lesbians hate men, lesbians are afraid of men, lesbians have a hormonal imbalance, lesbians really want to be men, lesbians have authoritarian fathers and weak mothers, lesbians have weak fathers and strong overbearing mothers. This element of the therapy process I call *clearing the air*.

Coming Out

When a woman identifies herself as a lesbian, she is making a statement that she loves women. Before a lesbian "comes out" to a therapist, she experiences a feeling of secrecy which is uncomfortable. Many women need to discuss "coming out" in therapy. Some elements involved in this process are risk-taking, self-acceptance, and practicing. "Coming out" is an ongoing process in which a woman shares her lesbian identity with people. The process is a continual one which may include family, co-workers, friends, and acquaintances.

Acknowledging one's lesbianism is a choice. Some women are not sure if they are lesbians. They may be involved with another woman or possibly desire intimacy with another woman. We are all sexual beings with a variety of exploration and variation of sexual experiences. When a woman chooses to "come out" as a lesbian, she openly declares her love of women.

Women who align themselves with the Feminist Movement are more willing to "come out" publicly and identify themselves as lesbians. Often these women work independently or in privately owned businesses. Non-political lesbians are much more willing to hold 8-5 jobs, where their identity is not known. "Coming out" represents becoming a part of lesbian culture. This creates fear in women who have an investment in others' (parents, children, co-workers, and/or employers) acceptance of them. When a lesbian "comes out", the relationship between her and the other person may feel precarious. A lesbian never knows what the response will be from her parents, children, or peers.

"Coming out" to one's parents presents the highest degree of risk. You can get another job if co-workers or an employer rejects your lesbianism, you can find accepting friends, but you cannot get different parents. "Coming out" to one's parents is somewhat analagous to telling them of your desire to marry someone of a different ethnic group, i.e., inter-racial marriage (although even with inter-racial marriage there is a male/female relationship which is socially sanctioned). The issue of acceptance stands above all. Will parents accept their child following a different path from friends' and neighbors' daughters? Lesbians, as daughters, struggle with the desire to self-disclose to parents and/or close members of the family. They want to share the joy and happiness of knowing themselves in a deeper way. They may want to include their partner in the family. In both cases, the fear of family rejection exists. Homophobia is so deeply ingrained that parents may choose to uphold norms rather than lovingly accept their daughter.

Lesbian mothers may experience similar anxiety when they want to come out to children. Younger children see mother in a day-to-day interaction, and if they have a loving relationship, the children accept the love they see in adults, be it between man and woman or woman and woman. Older children exposed to the homophobia and heterosexism of society may find acceptance of their mother's lifestyle in conflict with their already preconceived notions of love. Talking openly with children is important. Lesbian mothers often face the added pressure of a possible custody suit if there are negative feelings from the other parent. These suits generally fall in favor of the father taking custody of the children. The lesbian mother must decide how important it is for her to have her identity known. For many lesbians, the desire to be themselves publicly is important and they are willing to risk the possible legal consequences.

Career

The more a woman identifies as a lesbian, the less approval she receives from the larger society. The lesbian sub-culture, sometimes referred to as "the community", has unique lifestyle definitions unto itself, including a strong network of support. Lesbians are not waiting for the "white knight" to take care of them and protect them from the world. They face the issue of survival in a direct way. With the need for self-sufficiency, work aspirations are a constant element of life. As the desire for professional or career goals increase, the need for social approval increases in order to maintain one's status in the work system. Conflict arises since the more open a woman is about her lesbianism, the less likely she is to get that needed social approval to maintain her work status. Hence, many women who are professionals are "in the closet" about their lesbianism. Other lesbians, angry at the "system" which opposes their lifestyle, refuse to work so they can be "out of the closet". Both ways are at the expense of personal integration of social and personal goals. Being self-employed provides an answer to this conflict for some.

Based on these considerations, therapists often spend a great deal of time clarifying with the client what she really wants. The combined personal/social/political realities weigh heavily on lesbian decision making.

RELATIONSHIPS

Society at large knows only the heterosexual model for intimate relationships. Lesbian relationships do not fit this model. The Feminist Movement and Gay Liberation discuss at length the oppressiveness of male dominant/female submissive roles. Lesbians are very conscious of the potential of falling into roles and struggle to incorporate role-free patterns in the relationship. For some, monogamous coupling is not desirable. For others, short term or long term relationships are desirable. Lesbians have stepped out of the traditional female role simply by virtue of female identification versus male identification in marriage. This gives lesbians added room to explore many possibilities for intimate affiliation. This opportunity also presents the dilemma of choosing from an extremely wide range of options, which often becomes problematic.

Lesbians make no legal vows for life. There is no dating, engagement, wedding sequence as in heterosexual associations. Because there is no formal label for the status of a lesbian relationship, outsiders tend to view lesbian relationships as fleeting. This is not true. The potential and destiny of the relationship is entirely up to the two women involved. Each couple needs to be taken seriously in the context of their loving affinity for one another. There is little support for the lesbian couple in the outside world, therefore, learning self-validation is essential.

DRUGS AND ALCOHOL

Many therapists see lesbians who are abusing drugs. Given the aforementioned social issues and personal pressures, many women get high as a primary means of coping. Needing to be strong, independent, O.K. (so people won't say, "No wonder . . . she's a lesbian") can be too awesome for a woman trying to be recognized for who she is. Over and over again, I see lesbians who feel fine about their sexuality, but feel totally beaten down by the response of others. Thus, the therapeutic issue here is how to find positive coping mechanisms to be able to survive in an unsupportive environment.

SURVIVAL SKILLS

Lesbian survival skills add further support. There are many ways lesbians survive and have survived over time in a heterosexist society. Brainstorming and sharing these skills with lesbian clients aid in building a needed self-support system. Some lesbian survival skills are:

(1) finding places where lesbians gather, i.e., women's center, women's coffee house, lesbian rap groups, gay metropolitan church, Daughters of Bilitis, women's bookstores, and gay activist groups;

(2) making a list of all service and professional people supportive of the lesbian lifestyle, i.e., church leaders, doctors, lawyers, professors, counselors, politicians;

(3) making social contact with other lesbians, i.e., women's parties, political meetings, gay bars;

(4) establishing a support group of friends who know and understand who you are;

(5) reading feminist literature and gay literature.

CONCLUSION

The unique elements of counseling lesbians mentioned here offer the therapist background information. It is our responsibility as therapists to be honest about our limitations, values, and expectations regarding homosexuality. Only then can lesbians who go into therapy receive the services that will enable them to reach their full potential.

REFERENCES

Abbott, S. and Love, B. 1972. *Sappho was a right-on woman: A liberated view of lesbianism.* New York: Stein and Day.

Chafetx, J.S. et al. 1976. *Who's Queer? A study of homo and heterosexual women.* Sarasota: Omni Press, Inc.

Chesler, P. 1972. *Women and Madness.* New York: Doubleday.

Firestone, S. 1970. *The Dialectic of Sex.* New York: Morrow and Company, Inc.

Klaich, D. 1974. *Woman plus woman: Attitudes toward lesbianism.* New York: Morrow and Company, Inc.

Martin, D. and Lyon, P. 1972. *Lesbian Woman.* California: Glide Publications.

Tobin, K. and Wicker, R. 1972. *The Gay Crusaders.* New York: Coronet Publishers.

Wysor, B. 1974. *The Lesbian Myth.* New York: Random House.

WOMEN WHO RELATE TO DRUG ABUSERS

Christine E. Miller, M.ed.

The Matrix

Tucson Awareness House, Inc.

The Matrix, a part of Tucson Awareness House, Inc., is the drug intervention and education program in the treatment network. The Matrix offers supportive services to programs in the City of Tucson involved in drug treatment.

In February 1975, several referrals were made to The Matrix of parents of addicts. Several months later, a group was formed composed of mothers of addicts and women relating to someone abusing drugs. The group was facilitated by Ellen Litman and the author. During the group, various themes emerged from the experiences and problems the women faced stemming from their relationship to the addict. The group was convened out of an expressed need for mothers of heroin addicts to get support to "cut the cord" between themselves and their children (children in this paper refers to adult addicts who are children of the group members; the word children will be used to clarify the viewpoint of the group members).

Members of the group, all white, middle-class women, supported each other by sharing their experience. An understanding of their behavior patterns was gained through the sharing of common problems.

GROUP DYNAMIC

The group evolved through three phases. During the first phase, the women were child-focused. The second phase consisted of an internal group focus. Women focused on themselves, primarily, and those issues directly affecting their personal life, during the third phase.

Phase One

Each woman told her story, unfolding the social history of the family. The culmination of the story was the child's present abusive behavior. Women reported coming to the group out of desperation. They had remained silent for years, discussing their problems with a spouse or no one at all. Essentially, each woman experienced isolation by this continual silence.

One group member spent the entire session on her difficulty in asking for help. She feared group members blaming her for her unhappiness. She blamed herself for what happened to her daughter and felt uncertain of the support from others. Self blame prevented many of the women from seeking help earlier. They labeled themselves because their children failed.

Mothers in this society receive most of their own validation on the basis of the success or failure of their children. Traditionally, a mother's primary job is to tend to the home and children. The literature pays attention to the problems of addicts. No one pays attention to the mothers of addicts who suffer the same personal feelings of failure as the addicts themselves do. They have spent their entire lives investing in their child's success, which is the mother's promise of success.

Mother-loving, a drug treatment term, refers to over-nurturing and protecting a child from the world. The overprotection disables a child from standing on its own two feet, learning to cope with everyday stress. The term reflects negatively on the job of mothers, which is to love and care for their children. Group members continued to protect their children throughout this phase of the group. Reports from members included bailing a son out of jail for the sixth time, covering up a household theft, ignoring lies and giving money away when it was known the money would be spent on heroin.

Overprotection of the child and taking responsibility for the child's actions keep women in a powerless position. They invest so much in everything the child does, they lose touch with the self. At the point where women did reach out for support and feedback, a commitment to themselves began. By the very act of reaching out for help to another woman experiencing the same struggle, they make a positive statement of "I am" and "I need", something women have denied themselves for centuries.

Therapeutically, Phase One focused on the Victim-Rescuer-Persecutor triangle. Women entered the group as rescuers wanting to find a way to help their son or daughter. Defending the addict's behavior, bailing them out of jail, protecting them from police, were all rescue actions. They experienced victimization when addict sons or daughters stole from them, lied to them, and when they shot dope. They saw themselves victims of another's actions. The persecutor role meant blaming oneself and the child for all wrong doing. The triangle continues and each person plays all three roles at various times: rescuer, victim, and persecutor. The victim triangle originates in a state of powerlessness.

Women experience a state of powerlessness in this society. Men have power over women and women have power over children. The degree to which male society dominates women is the degree to which women dominate children.

Facilitators confronted group members when examples of the victim triangle were expressed by the women. Group members analyzed their behavior in the context of this triangle and, with new self-understanding, began exploring other behaviors.

Phase Two

The second phase was characterized by a strong group identity. Group members telephoned each other for support between group meetings. In group, women were discussing many areas of their lives, particularly interpersonal relationships with spouse and other family members. The group supported and encouraged each other to explore themselves. An atmosphere of friendship and closeness developed.

We discussed group dynamics and how each person fits into a role within the group. A group member requested assertiveness training. They spent a month learning the theory and practice of assertiveness behavior. Members then made contracts to be assertive in daily life. Positive and negative feedback was encouraged. Women began confronting each other on self-defeating behaviors. Members stated what they wanted to work on and took greater responsibility for the direction of the group.

The Feminist Counseling Model techniques were utilized during the group. After the group stabilized, facilitators became participants and also worked on various issues.

A theme throughout the second phase was expressing anger. Some women began to express anger during various assertiveness exercises. This led to discussions about playing the Supermom who is never angry. Women are not encouraged to show anger. Thus, they experience depression, which is anger turned into the self.

In ensuing sessions, women expressed deep anger and were encouraged by other members to do so. Anger was directed to a number of areas: at the addict son or daughter, at the family for taking mom for granted, at the self for always being there to smooth things over.

Once the anger was mobilized outward, the women spoke freely of resentments toward their children. They realized the difference between the way they feel about themselves and the way they feel about their children. During this period, the symbiotic nature of these relationships changed. Actions taken by group members showed signs of a de-investment in the addict's behavior. Toward the end of the second phase, discussion arose about changing the name of the group. Some of the ideas discussed were older women's group, second forty's group, women's group, and second start group. The desire to change the name of the group pointed to the significant change in focus, which became more apparent in the third phase.

Phase Three

Phase Three can be characterized by a strong focus on the self. Group members began talking about topics solely related to themselves, such as body awareness, sexuality, menopause, death, and careers. Women started communicating assertively with children for fear of being rejected. At this point, the desire to be true to one's own feelings outweighed any fear of rejection. The more they succeeded in asserting themselves, the more they explored new definitions of themselves. One woman at the age of fifty-two decided to go to college full-time.

The pain over the child's addiction to heroin never disappeared. Throughout the group there were moments of sadness experienced by someone who finally accepted the fact that she could not change her son or daughter's lifestyle. The women learned to take responsibility for their own feelings and allowed room for their children to now take responsibility for themselves. For some the greatest awareness of the group was knowing that each of the women, too, has her own life to lead.

TWO PERSONAL ACCOUNTS

Ruth

Ruth was one of the original group members. She entered the group during a time in her life when she felt she was "at the end of the rope." A mother of twelve children, two of whom were addicted to heroin, she carried most of the family responsibilities. Her husband, a career man in the Air Force, was a fine husband and father, but was away from home frequently. The addict daughter lost custody of her child so Ruth and her husband raised the granddaughter. While Ruth was in the group, both children were in and out of jail and also in and out of drug treatment programs. Both children have been addicts for nearly ten years.

Ruth was absorbed in guilt and self-pity at the beginning of the group. She judged herself harshly. She received a great deal of encouragement from group members. Her children were continually asking for money. We role-played situations as they arose, and Ruth saw herself through another group member's words. She eventually said "no" to their requests for money. This breakthrough marked the beginning of Ruth standing on her own two feet. She gained a sense of respect for herself that had been lost for many years. She now has firm limitations on what she will do with the two children. She feels they know her limits, and they have no choice but to acknowledge her strength. She reassures the children of her love for them, and yet, she will not be a cover-up for their lifestyle. Six months after the group terminated, one of Ruth's children stole a tool set worth several hundred dollars from her home, and she reported this to the police. Two years ago, she would have made excuses for this kind of action.

She is now active in the community and hopes to start a women's group herself.

Jan

Jan was the pleaser of the group. She would never say anything that might hurt someone. She spoke in a very high pitched girlish voice. Her marriage was an unhappy one. She asked permission for grocery money and other household necessities. Throughout her thirty years of her married life, she never once enjoyed sex. Yet, she had intercourse with her husband upon demand as part of her wifely duties. Jan had a strong belief in God without any belief in herself.

She found retreat in the group telling many tearful experiences of her unhappy life. One of her sons had been in to therapeutic communities and several other treatment centers. He had been using heroin for twelve years. Jan also blamed herself.

Jan was confronted about her passive approach in life. Over the months, she started making demands on her family. She got a job and requested that some household tasks be shared. She saved up enough money to take a trip to Europe with a music group--something she had always dreamed about.

She always gave in to her son's request for money. During the group, she had great difficulty setting any limitations on her son's relationship with the family. Eventually, she told him to stop calling collect, not to call when he was in trouble, and to stop asking for money. This may sound like a simple request for a mother to make of a thirty-five year old son. For parents who have given everything in hopes that this will be the last time, saying "no" is a recognition that their help is no aid.

CONCLUSION

Women Who Relate to Drug Abusers groups had a profound effect on several group members. They came to the group looking for a way to help someone else and learned how to help themselves.

WOMEN HELPING WOMEN: THE EVALUATION OF AN ALL FEMALE METHADONE MAINTENANCE PROGRAM IN DETROIT

Jeanne C. Marsh, Ph.D.

The University of Chicago

In recent years, there has been increased concern about the female addict. This concern has been fed by a limited amount of research about the female addict which overwhelmingly indicates women do not fare as well as men in the treatment process. A number of factors have been proposed to account for the poor prognosis for women in treatment. Among the factors cited most frequently are those which suggest that women are less likely to be employed (Eldred and Washington, 1975; Ryan, 1976), less likely to be living with their families (Eldred and Washington, 1975; Maglin, 1974), and more likely to be discouraged from pursuing treatment as a result of a steady income from prostitution (Maglin, 1974; Suffet and Brotman, 1976), or as a result of child care responsibilities (Eldred and Washington, 1975). Clearly, most of these factors place the major responsibilities for failure in treatment upon the woman client--her particular set of personal characteristics or life events.

A second approach to understanding the inaccessibility of women to rehabilitation seeks explanations with the traditional approaches to drug treatment. These explanations suggest that treatment programs originally designed for a largely male addicted population are not geared to meet the special needs of women (Levy and Doyle, 1974; Schultz, 1974; Chesler, 1972). Further, expectations of treatment staff typically are not as high for women as they are for men in traditional programs (Doyle and Levy, 1975; Ryan, 1976). Proponents of the programmatic explantion for the traditional failure of women in treatment have proposed all female programs as alternatives to traditional programs and as avenues for reaching women who have not been served adequately or *at all* by traditional programs.

In 1974 NIDA funded several all female treatment programs, one of which--the WOMAN Center in Detroit--is the focus of this study (see

Appendix). Today I'm going to present data from the evaluation of the WOMAN Center program designed to investigate whether the frequent failure of women in treatment is related to the characteristics of clients or the characteristics of male-oriented programs. I will present two kinds of evidence to examine whether all female programs can achieve anything beyond what is achieved by mixed sex programs. First, I will compare characteristics of clients who elect service from an all female program with characteristics of clients in mixed sex methadone maintenance programs in Detroit. To the extent that all female programs are serving clients who are different from those served by traditional programs, i.e., to the extent that they reach clients not otherwise reached, then these programs are making a unique contribution. Secondly, I will examine the effectiveness of the WOMAN Center program as compared with mixed sex methadone maintenance programs in Detroit in improving the functioning of addicted women. To the extent that outcome evidence indicates all female programs to be more effective with female clients, there will be some indication these programs have something positive to offer.

The demographic and psychological data I am going to present come from admission forms completed by 67 clients admitted to the WOMAN Center between September 1976 and April 1977 and 50 clients admitted to 5 mixed sex methadone maintenance programs in Detroit during this same period. Data on level of client functioning derive from a behavioral checklist completed by counselors who rated adequacy of client functioning on 14 dimensions (e.g., housing, legal needs, health, education/vocational training, primary care of children and drug use). At the WOMAN Center, adequacy of client functioning was rated by counselors for each client each month. For comparison clients, client functioning was rated at termination only.

While the data I present today will not determine conclusively the source of difficulty in treating heroin addicted women, it provides information on the value of providing the alternative of an all female facility. We seek this information in the answers to two questions:

(1) *Does the WOMAN Center serve a female heroin addicted population that is different from the population served by more traditional programs?*

Age. The mean age of the woman admitted to the WOMAN Center and to comparison methadone maintenance programs in Detroit is approximately the same (28.4 years for the WOMAN Center clients and 27.8 years for the comparison clients). 62.7% of the WOMAN Center sample and 76.2% of the comparison sample are between the ages of 20 and 29. There are slightly more WOMAN Center clients 30 or older.

Race. Well over half of the women admitted to treatment at the WOMAN Center and at the comparison program are black Americans, 64.3%

and 84.4%, respectively. 32.8% of the women admitted to the WOMAN Center and 13.7% of the women admitted to the comparison programs are white. The relatively larger white population in the WOMAN Center accounts for the significantly different racial composition of these two programs ($\chi^2_{(2)}$=6.03, $p<.05$).

Marital Status. A comparison of the marital status of women admitted to the WOMAN Center and comparison programs shows WOMAN Center clients are less likely to have married and more likely to be divorced (although these differences are not significant). Further, when we examine the clients' "cohabitation" i.e., whether they live with a spouse or a non-legally sanctioned partner, we see WOMAN Center clients are more likely than the comparison groups to live *alone*. However, when they do have a cohabitant, it is significantly more likely to be a non-legally sanctioned sexual partner ($\chi^2_{(3)}$=9.07, $p<.05$).

Employment. In the city of Detroit, where the unemployment rate is extremely high, the employment picture for women admitted to the WOMAN Center and to the comparison programs is bleak at best. 83.3% of WOMAN Center clients and 88.0% of comparison program clients are unemployed when admitted to the program. Our findings show that clients derive support primarily through benefits or assistance, i.e., welfare, social security or unemployment insurance. In the WOMAN Center sample, a somewhat larger proportion support themselves through illegal means, primarily prostitution, and a somewhat smaller proportion through public assistance or pensions.

Criminal activity. What are the criminal activities in which these women report they are involved and how do they differ for women from the WOMAN Center and from comparison programs? A significantly larger percentage, 61.5% of WOMAN Center clients report they have engaged in prostitution as opposed to 36.7% of clients from the comparison programs ($\chi^2_{(1)}$=6.21, $p<.01$). For drug dealing, the trend is reversed: more comparison clients (60.4%) than WOMAN Center clients (40.8%) report involvement in drug dealing. These differences between the two groups are significant ($\chi^2_{(1)}$=3.73, $p<.05$). Successively fewer women in both groups are involved in property crimes, non-violent victim crimes (any crime involving a victim where the victim does not sustain physical harm) or violent victim crimes (any crime including physical aggression directed toward a human victim).

Children. The relationship of a woman addict to her children has been a concern of the WOMAN Center program since its inception. And it is interesting to note that clients admitted to the WOMAN Center have had significantly more live births than women from comparison programs ($t_{(116)}$=1.71, $p<.01$). However, WOMAN Center clients are less likely than comparison program clients to live with their children ($t_{(115)}$=2.53, $p<.01$). Comparison program clients, while they have given birth to significantly fewer children, are significantly more likely to live with these children.

Drug Use. Both groups first used a drug around the age of 17 and they first used heroin at the average age of 20. For both groups there was very little difference in the initial source of heroin. Gifts and illegal street purchases were the initial sources of heroin for women from both groups. So, in terms of early patterns of drug use, these women were quite similar. However, they do show some differences in their patterns of drug use at admission to their respective programs. WOMAN Center clients used drugs significantly more frequently at admission ($\chi^2_{(4)}$=10.8, $p<.05$). 46.3% of WOMAN Center clients versus 25.5% of comparison clients used heroin daily; only 16.5% of WOMAN Center clients and 33.4% of comparison clients used heroin three times a week or less. Further, although the differences are not significant, WOMAN Center clients are more likely than comparison program clients to use drugs alone rather than with other people. Finally, for both groups, the source of drugs at the time of treatment was more likely to be illegal street purchases rather than gifts.

Thus, the patterns and frequency of drug use varies to some extent between these groups, with WOMAN Center clients using heroin more frequently and more often using it alone.

Previous Treatment. A significantly larger proportion of WOMAN Center clients were readmissions. Indeed, WOMAN Center clients had a mean of 6.4 previous drug treatment experiences and the comparison programs had a mean of 3.2 previous treatment experiences ($t_{(116)}$= 6.43, $p<.001$). The higher percentage of readmissions at the WOMAN Center is consistent with the finding that more clients are over 30 in this program.

Psychological Characteristics. Among the psychological characteristics measured in this study are some that have been measured before with addicted populations: self-esteem, depression and anxiety. Among these scales not previously used with addicted populations and of particular interest in this study were those measuring the subject's perception of herself on a masculinity-femininity continuum, and her stereotypes about male and female attributes.

Although there was no difference between the WOMAN Center and comparison program clients on the self-esteem measure, the WOMAN Center clients were significantly *more* depressed and *more* anxious than the comparison program clients ($t_{(112)}$=9.21, $p<.001$ and $t_{(91)}$=15.73, $p<.001$, respectively). The WOMAN Center clients also viewed themselves as significantly *more* feminine than the comparison program clients viewed themselves. That is, WOMAN Center clients do not report the kinds of masculine skills needed to effectively manipulate the environment (e.g., independence, decision-making abilities) when describing themselves. Furthermore, WOMAN Center clients viewed other women and men as significantly *more* traditional than comparison clients ($t_{(113)}$=2.32, p<.05). WOMAN Center clients clearly buy into the traditional role for women much more strongly than comparison clients.

To summarize these findings briefly, it appears that the WOMAN Center is serving women who are different from those served by more traditional methadone maintenance programs in Detroit in several ways. There are a larger percentage of white women; they are older and more likely to have been in treatment before. They are more isolated to the extent that they are *not* likely to live with partners or with children. Their incomes are more likely to come from an illegal source--typically prostitution. They use drugs more frequently and more frequently use them alone. Psychologically, they are more depressed and more anxious, they are more likely to buy into traditional sex role stereotypes both for themselves and for others. In other words, most of the characteristics which distinguish these women from those in comparison programs suggest that they have a large number of problems, i.e., they will be more difficult to treat. In the discussion which follows, possible explanations for these findings are explored.

(2) *Does the WOMAN Center have a more positive influence on the functioning of female clients than more traditional programs in Detroit?*

To address this question I will present findings on several measures of program impact. First, program effectiveness is measured using four general outcome measures that are frequently employed in evaluations of drug treatments. The measures examined are (1) reason for closing (whether the program was completed or whether the client or staff initiated early termination); (2) counselor evaluation at termination (whether the client was recovered, improved or not improved); (3) client employment at termination (full or part-time employed or unemployed); and (4) whether or not client was detoxified from heroin. The two client groups are not significantly different on any of these measures. Indeed, the two groups look very similar on all measures with one noteworthy exception. Examining reasons for closing or termination in both groups, we see fully one half of the comparison program clients left the program on their own initiative, while only one third of the WOMAN Center clients chose to do so. At the WOMAN Center, if a client was terminated before completing the program, it was more likely to be a staff than a client decision. In other words, more WOMAN Center clients were engaged by the program, i.e., were satisfied enough with the service they were receiving to decide to remain in the program.

The extent to which participation in the WOMAN Center program resulted in more adequate functioning in specific areas of client's lives is a central concern of this study. Change in client functioning from the beginning to the end of the program was considered to be another general measure of the program impact. As described previously, client functioning was assessed on behavior rating scales completed by counselors for each client each month. Since clients remained in the program for different lengths of time, program exposure could vary for each client from 1 to 6 months. Using data at 6 months or last recorded value as termination data, the mean functionality score for WOMAN Center clients was 6.8 at admission and 8.7 at termination, a difference of 1.8, ($t(50)=5.78$, $p<.001$). That is, as we look across all areas of client functioning, we find WOMAN Center clients are significantly more competent at termination than they are at admission.

While these positive findings suggest drug abuse intervention can have a positive impact on the functioning of heroin addicted women, they provide no information that a special program for women, specifically the WOMAN Center, is any more effective than more traditional co-sex porgrams in treating women. The WOMAN Center has identified several areas of functioning which are particularly problematic to the female addict. Among the most problematic are health, mental health, status with the criminal justice system, housing, education, vocational training, employment, and drug use. These were areas in which WOMAN Center clients were particularly deficient and areas in which the program focused resources through inhouse services and a strong advocacy and referral program. A comparison of the proportion of clients from each group functioning adequately in each of these areas as judged by their counselors at termination shows the two groups look very similar with one exception. WOMAN Center clients were somewhat more likely to leave the program with education and vocational skills judged to be adequate by counselors.

In summary, the data show that WOMAN Center clients show significant improvement in functioning during treatment, and they leave the program functioning as adequately as clients leaving the comparison program. We do not have functionality ratings for the comparison clients at admission. However, given the characteristics which distinguish WOMAN Center clients from comparison clients, it is possible WOMAN Center clients had to change more to reach the same functional level as the comparison clients.

In conclusion, I would like briefly to mention some implications of this study. The clients attracted to the WOMAN Center represent a different population from those attracted to mixed sex programs. A significantly larger proportion of these women had been in treatment before, and in terms of their drug habits, legally, socially and emotionally, they appear to be in "worse shape" than comparison clients. WOMAN Center clients represent the types of female clients who have traditionally "failed" in mixed sex programs. These clients may have chosen the Center for its special attention to the needs of women or, for some, it may have been the only alternative left. While WOMAN Center clients were not significantly different from comparison clients at termination on several measures of effectiveness, they had significantly improved their level of functioning since admission to the program and they appeared more satisfied with the program as they were less likely to drop out. The findings show an all female program provides a treatment option that is of value for seriously addicted women. Indeed, for some women it may be the only viable treatment option. To the extent that WOMAN Center findings are replicated, i.e., that other all female programs are serving a special, perhaps more seriously addicted, population of women, it is imperative this option be provided.

REFERENCES

Chambers, D.C., Hinesely, R.K., and Moldestad, M. 1970. Narcotic addiction in females: A race comparison. *International Journal of the Addictions*. 5: 257-278.

Chesler, P. 1972. *Women and Madness*. New York: Doubleday.

Doyle, K.M. and Levy, S.J. 1975. The female client: How treated in drug abuse programs? Paper presented at the meeting of the American Psychological Association, Chicago, August 1975.

Eldred, C.A. and Washington, M.N. 1975. Female addicts in a city treatment program: The forgotten minority. *Psychiatry*. 38: 75-85.

Ellinwood, E.H., Smith, W., and Vaillant, G.E. 1966. Narcotic addiction in males and females: A comparison. *International Journal of the Addictions*. 1: 33-45.

Levy, S.J. and Doyle, K.M. 1974. Attitudes toward women in a drug treatment program. *Journal of Drug Issues*. 4: 428-434.

Maglin, A. 1974. Sex role differences in heroin addiction. *Social Casework*. 55: 160-167.

Ryan, V.S. 1976. The female client in several drug treatment modalities. Unpublished paper, Women's Drug Research Project, Ann Arbor, Michigan.

Schultz, A. 1975. Radical Feminism: A treatment modality for addicted women. In E. Senay, V.T. Shorty, and H. Alksne (eds.) *Development in the Field of Drug Abuse: Proceedings for the National Association for the Prevention of Addiction to Narcotics*. Cambridge, Mass.: Schenkman.

Suffet, F. and Brotman, R. 1976. Female drug use: Some observations. *International Journal of the Addictions*. 11: 19-33.

Sutker, P.B. and Moan, C.E. 1972. Personality characteristics of social deviant women: Incarcerated heroin addicts, street addicts, and non-addicted prisoners. In J.M. Singh, L.H. Miller, and H. Lai (eds.) *Drug Addiction: Clinical and Sociological Aspects* (Vol. III). New York: Futura Publishers.

APPENDIX

The WOMAN Center

In addition to the all female composition of the WOMAN Center (staff and clients) the design of the program reflects the concern for the special needs of the *female* addict. Specific components of this program which distinguish it as a women's program, different from mixed sex methadone maintenance programs are the following:

1) Individualized program planning. Each client and her counselor work out individualized treatment plans for service and referral to address her unique medical, legal, educational, employment and child care needs.

2) Medical service. A physician is retained by the program to provide clients with physical examinations, diagnosis and emergency medical care at the program facility.

3) Legal service. An attorney is also retained by the program to provide legal advice, to represent clients in court, and to help clients resolve long-standing legal problems resulting primarily from prostitution activity.

4) Child care. A child care center provides structured and free play, arts, crafts, outdoor play, reading hours and trips to places of interest for children as well as consultation with mothers focusing on improving parent-child interactions.

5) Referral network. Services provided by the program are expanded by referral linkages to a range of diverse services including social services, vocational rehabilitation.

6) Client advocacy. Program staff serve as client advocates to help clients become aware of the services available to them and to insure clients receive services. Advocacy takes a variety of forms: (a) identifying appropriate service delivery agencies for clients, (b) providing transportation to services, (c) accompanying clients to the care facilities, (d) confronting agencies about discriminatory practices, (e) lodging complaints in appropriate plans when adequate services were not received. Program staff have identified advocacy as a key element of the program since many services are otherwise unavailable to clients due to overt and covert discrimination precipitated by fear, incorrect information and lack of exposure.

Thus, the program focuses its resources on the often acute medical, legal and child care needs of female addicts using advocacy to help clients gain access to services outside the program.

MYTHS AND STEREOTYPES OF THE SEXUAL MINORITY WOMAN

Rosemary Madl

> "At one moment we are children playing and the next moment powerful women. Our soft sides and strong sides at once."
>
> - from *Women Loving* by Ruth Falle

Women as a group have been allowed to express affection in a fairly open manner. The kissing on the cheek, hugging and walking arm and arm are seen in a "sisterly" way by the American culture. All is acceptable as long as the demonstration does not go beyond the boundaries that have been subtly and at times overtly imposed over the centuries. The writer's goal in speaking about the Lesbian Woman is to allow a discussion of the myths and stereotypes that have been imposed without the need to judge the "rightness" of the process. The writer will state that she has the belief that the lesbian woman has the innate right to choose whom she wishes to establish an intimate relationship with and that this decision rests on her and the person with whom she is sharing a relationship. The writer does not see homosexuality as a mental disorder in need of curing, as a perversion or a "lack of" in relation to the heterosexually oriented persons in our society. Due to the choice made by the woman to be with another woman in an intimate relationship, the lesbian woman has special needs that differ from the heterosexually oriented woman. These needs tend to be both emotional and legal. These will be explored and qualified as the discussion continues.

When women began to sense that being in love with the same sex was a viable and valid choice, they began to look for role models to aid them in establishing a support system and a sense of community. As expected, the only overt role models were that of the heterosexual culture - the man was allowed to express only his male principle, the woman was allowed to express only her female principle. Thus, lesbian women began to intuitively sense a need to establish their own sense of culture. To do this, they first had to examine and play out the already existing role modeled culture. From this exploration, the lesbian woman became aware of her female-male principles in a manner that the heterosexually oriented woman could not. The stereotyping of the lesbians, the tough aggressive butch, and the general stereotyping of "feminine" and "masculine" roles have been prime targets for the lesbian liberationists. If two women cohabitated together, there would definitely be a need to discuss who would provide the financial support and who would aid in the house duties. This would sound legitimate in a heterosexual household. In this household, the man would be the financial provider and the woman would nurture the children, etc. In a household of two women, the discussion centers around the sameness of the women and the "sharing" of the lifestyle. This means that neither woman needs to stay home and neither needs to work. The decision rests on whether or not the women are work oriented or home oriented as far as their energy make-up. Thus, there may be two women working and one of the women cooks and the other cleans. There could also be one woman willing to work and the other wanting to stay at home. The importance is that the women see a choice in their lifestyle rather than an accepted indoctrination of "shoulds". Thus the woman with energies around the working world and who was assertive was called "dyke," "butch" by the culture to define her male principle energy and the woman who stayed in the home or who was less assertive was labeled the "fem." To further elaborate, lesbianism is defined as a homosexual relationship between women. This phenomenon has existed back as far as the written word and prior to this. She is in all myths, cultures and beliefs. Do we negate this truth or do we begin to define her as she sees herself and not through the eyes that wish her to be "cured"? It appears that she has had a sense of herself. It is the role modeling that is limited and ill defined. One reason for this can be attributed to the not-okayness given to her by her heterosexual counterparts.

As the lesbian woman begins owning her own personal power, the heterosexual role play model is no longer as unquestionably valid to her. The lesbian woman is into her own and has begun to address her needs. The lesbian woman, after centuries of living out the female/male energies within her, has begun to combine and integrate the energies to form an androgynous person, totally owning her female/male principles. The female (anima) and male (animus) energies were now seen as a combination to be worked with and not denied and separated. This is to say that the female principle embodies feelings, intuition, moods; and the male energy principle embodies thought, logic, opinions. It is important to note that the male principle in the woman is not the same as the male man but is the energy associated with the masculine. Jung states that the "animus in the woman, is the counterpoint of the anima in man...derived from three roots: 1) the collective image of man which woman inherits; 2) her own experience of masculinity coming through the contacts she makes with the men in her life; 3) the latent masculine principle in herself." This masculine principle in women was allowed or given permission to express itself during the wars when women filled positions that had been previously reserved by men, who were now in combat. So far, the writer is intimating that the lesbian woman happens to be overt in her "living out" of her female/male principles. If this is so, then what is all the fuss about? What is the need to talk about her behind closed doors, ridiculing her through jokes which are degrading and often an expression of rage? What makes people so quick to point a finger at her? Could it be that by living out her life through choice she is telling others that sexuality cannot be so rigidly defined and categorized? That everyone has a choice to make and need to own their own personal power in their life? When the lesbian woman in the United States in the late fifties, sixties, and now the seventies owned her ability and right to choose, she then had no further need to define herself in the categories laid out by society and she began to develop and believe in herself as a woman into loving women. With this came a non-verbal challenge to the rest of the people to reexamine their own beliefs and structures.

Roles and behaviors were established by the lesbian community to create a sense of the following:

Intimacy - rules on relationships, how to deal with the female/male principle energies, expression of affection.

Territory - Since the heterosexual community scorned and mocked the lesbian woman, meeting places were established where the woman could go and socialize with other women. Dress codes helped to identify the lesbian woman from the heterosexual woman and often aided in the meeting. Most meeting places were not in the most desirable part of town - boundaries were seen as less rigid. Thus, the lesbian woman had to learn to express aggressiveness and defend her personhood.

Personal Boundary Setting - The recognition of "roles" allowed the couple to be seen in an overt manner. The roles also aided the woman who was seeking to establish a relationship. Her dress and manner was a non-verbal language to others who would possibly be complementary to her personhood. Otherwise, the woman could possibly experience verbal and/or physical abuse from her "straight" sisters or brothers.

Community - The lesbian woman oftentimes cannot share her lover relationship with her family. Oftentimes the parents

> give non-verbal cues not to tell them. There is also the possibility of disownment, rage and guilt. The other women now become family. Support systems are established and nurturance is given by the auxilliary family system. All understand the price of their choice in establishing lesbian relationships. The pain arises out of the rigid rules set forth by the heterosexual community on the ways one "ought to act and exist." The internal pain stems out of a knowing what is right for her and a sense of helplessness to the dogma which condemns her.

The lesbian of today can choose to look feminine and/or masculine. Most often she is a secretary, high school teacher, nurse, executive, social worker, maid, cook, mother. There are no boundaries on her. No income is too low or too high to exclude her. She is comfortable with her womanhood. She has learned the difference between her assertiveness and her aggressiveness. She does live a schizophrenic existence. In the daytime she must "put up with" the heterosexual pressure of marriage, children and/or the male passes. She is attempting to establish herself in the working world and is labeled as easy prey because she is not married or is the object of sexual passes and come ons. She has learned to say "he" instead of she and has learned to censor her speech prior to talking about what she did the evening before. She has learned to like herself when she laughs at the "gay" jokes or pretends that the male pass was sexy. She has learned to sleep with men to protect her job and/or defend against public ridicule. She has become a master at dodging her parents' inquiries as to what man she is dating. In the evening, she owns her lover is a woman. There is no need to censor or laugh fakedly. She also knows that in the morning she must again go back to the other world. Maybe today she will tell them all...maybe today she will be able to stand her silence. Through her owning of choice, she also owns the rage that knows no limits, the rage of non-acceptance by others. More painfully, the denial of choice by her other sisters.

Most lesbians will state that it takes more than sleeping with a woman to make you gay. There is an emotional bonding that occurs and it is this bonding that makes the difference. Women have the capacity to choose multi-levels of interacting. They can see themselves as emotionally bonded with women, yet sexually fulfilled with men. There can be a sexual bonding with women and an emotional bonding with men. There can be both a sexual and emotional bonding with women and vice versa. In other words, both the sexual and emotional bondings must be explored and given permission to co-exist together. The bondings do not occur out of a lack of but are on a continuum of growth and energy levels. To further elaborate on the bondings (see Table I):

The lesbian woman recognizes her maleness and her femaleness.

Table I. Bonding

	HETEROSEXUAL		PAN SEXUAL		HOMOSEXUAL	
	male	female	male	female	male	female
HETEROEMOTIONAL						
female/male						
PANEMOTIONAL						
female/male						
HOMOEMOTIONAL						
female/male						

These two energy principles are related both sexually and emotionally. This then means that the woman has two energies which can bond with the other woman's two energies in relation to sexuality and also emotionality.

The male principle in the woman may bond with the female principle in the other woman while both may bond emotionally out of their female energy. There could also be two women bonding emotionally, one female, the other male, and sexually both out of female energies. Another alternative would be a sexual bonding female to male and an emotional bonding male to female.

In the series "All in the Family," Archie bonds emotionally male to male with his men friends. Archie is seen as heterosexual and homoemotional. He is heterosexual with Edith, yet his emotional bonding is with men.

Webster's fourth definition of lesbian is "one addicted to lesbian." If the person ponders the female addict and her pain and then the lesbian and her pain, we may begin to see similarities in a phenomenon. The addicted woman knows rage and contempt. She also knows that she denies herself for the sake of her addiction. The woman does not own her personal power and in doing so denies her ability to choose. Instead, she makes love to her addiction and plays helpless in the ever revolving triangle. In her heart, she is aware that there is an alternative, yet she fears the price. Thus, she hides her overt rage and denies her right to feel and cope. She begins to believe her own cons and in doing so, begins to lose sense of herself. Is this knowing so different from the woman who knows that she has a choice in regards to emotional and sexual bonding and that, if she follows her choice, it may mean being seen as a second class person much like the female addict? By denying her choice she also denies herself. One does it through addiction, the other through a denial of feelings and choice.

PEER COUNSELING: TREATMENT MODALITY FOR FEMALE NARCOTIC ADDCITS ISSUES AND OUTCOMES

Linda Gurrister, M.S.W.

Project Reality

Salt Lake City, Utah

The problem of treating the female addict is gradually emerging as an issue of interest in the drug treatment field. Recently, awareness of the lack of literature dealing with this concern is becoming apparent (Cooperstock, 1971; Levy and Doyle, 1975; Thure and Moore, 1977). Of particular interest is the finding that female addicts are not being effectively treated when they do seek help. An ASA follow-up study found that "20% more men than women are retained for 30 days after referral . . . drop-out rates of women from treatment programs is twice that of males" (Schwartz, 1976). In another study geared to look at sex-role differentials in treatment, it was discovered that "over half the men are working while only one-fifth of the women work. Concomitantly more women than men are job hunting" (Levy and Doyle, 1975).

A preliminary review of statistics taken from CODAP and treatment records at Project Reality methadone clinic corresponds with the data cited above. Project Reality is a methadone program which provides methadone maintenance, methadone 21 day detox and drug free counseling services. Approximately 260 patients are treated through this clinic, 37% of whom are female. As of July 1975 it was noted that of female clients who were non-homemakers, 28% were employed, while 54% of the male clients were employed. Females tended to remain in treatment for a shorter period of time than male clients. Females tended to receive more individual counseling and less vocational counseling than did their male counterparts. The implications of this data point to a desire on the part of female clients to engage in treatment; however, lack of needed responses and resources from staff may be the reason for eventual disengagement from the program.

Utilizing this information, a group of female staff members from the Project Reality program began exploring alternatives for treatment of female substance abusers. It was determined that further data about women in treatment would be beneficial in identifying the scope of the problem. It was also suggested that the use of trained female clients as a resource to help other women on the program might be a viable treatment approach.

Support for a peer-counseling approach was generated partly as an outgrowth of the general success of women's consciousness-raising groups during the past few years. Literature on women and drug abuse is beginning to consider the idea of women helping women as a means of enhancing self-esteem, coping skills, identity and other variables crucial to the restoration of the female addict. Some researchers (Thure and Moore, 1977) contend that the alienation considered to be a factor in women's drug dependency can be penetrated by female peer/self help groups that facilitate sharing of experiences and provide opportunities for self understanding and reciprocal nurturing. Preventative, as well as remedial, approaches to treatment are including peer and feminist counseling as alternatives to traditional therapeutic endeavors (Weston and Pancoast, 1974). A third consideration in selecting peer counseling as a treatment approach was the shortage of available staff members and the hope that training clients in counseling skills might prove a viable adjunct to the use of professional counselors. Attention specifically would be focused on the effectiveness of peers as counselors and on the impact of the training on the trainee.

In order to gather data on the special needs of women in the Project Reality clinic, a needs assessment questionnaire was compiled by the female staff of the program. The questionnaire addressed the following areas: self-esteem, support systems, knowledge and utilization of resources, employment opportunities, sexuality and physical appearance. Questions were answered by the use of a rating scale and yes/no responses. Questionnaires were passed out in January 1978 to approximately 80 women and 38 were returned. I will briefly summarize the relevant data in the following paragraphs. In the area of support systems for female clients, the majority of respondents (77%) felt they got emotional support from Project Reality counselors. They next chose spouse/partner (74%) and parents (59%) as sources of support. Less often did the respondents turn to friends (43%) or other clients (12%) for emotional support. The women also ranked the needs of spouse/partner and children as taking priority over attending to their own needs.

In the area of employment, 44% were employed at the time the questionnaire was filled out. Fifty-seven percent of the respondents felt it was important to work at a job they enjoyed, but 39% felt they could not adequately support themselves with their current job skills. Eighty-nine percent said that it was important to secure some type of employment. Interestingly, 43% of the respondents said that they learned this information from Project Reality personnel. If job training were available 94% said they would utilize the opportunity. Fifty-seven percent said they contributed a great deal of income to the family unit.

When responding to questions about physical appearance, 94% said that it was very important, but only 49% were satisfied with their own physical appearance. However, 64% of the women said they felt others saw them as attractive. When asked to rank the source of their models for physical attractiveness the majority of respondents chose men in general first and a certain man second.

In the area of sexuality 90% of the respondents said they did not derive pleasure from sex, but 57% said they went for long periods without sex. Sixty percent of the respondents seemed to feel that women are generally uninformed about their bodies and sexual desires and 51% also felt that males were uninformed in this area. Eighty-one percent said they felt physical affection and touching were important. And 76% said that at times they had sex with a partner simply to touch and be touched.

When asked to choose a label family members might give them, 78% chose "black sheep"; however, 89% of the respondents said they felt "good" about themselves. When asked to choose the roles that currently applied to them, those of homemaker, mother, working and wife were most frequent and in the order listed, with working and mother ranking equally.

Pearson's correlations were run to see if any further significant relationships could be established between the way questions were answered. Women who said they felt good about themselves also saw others' perceptions of them as favorable ($p<.004$) and seemed to seek emotional support from parents ($p<.002$). Also, women who felt that others perceived them favorably were able to seek emotional support from friends ($p<.004$) and felt it easier to find employment ($p<.01$). Women who said they sought support from other clients also seemed to feel able to get support from other clients ($p<.001$). Respondents who felt they got support from Project Reality counselors also said they felt they got support from the staff nurses ($p<.003$), secretaries ($p<.004$) and the doctor ($p<.002$). Those women who seemed to feel they could deal with their environment also ranked high in their perceptions of the amount of support they could seek and get from clients and friends, partners and Project Reality counselors. Clients who placed value on more education also rated positively their feelings about themselves ($p<.001$) and said they did not turn to friends or clients for emotional support ($p<.006$).

Generally, the data from the questionnaire suggests that women place value on employment but do not seem to feel they are able to get the training necessary. The data also suggests that the women tend to be other-directed rather than meeting personal needs. Also, they tended to take models for physical attractiveness from a source outside themselves. Pertaining to sexuality, the respondents felt that women are uninformed about sexual needs and perhaps enjoy sex more for the sake of intimacy and physical affection. Self-esteem seems to play an important part in the ability of the respondents to seek and utilize emotional support and resources in their environment. In particular, clients who did seem to feel good about themselves were generally more able to perceive and seek support from other clients and friends. All in all it was felt that the data indicated some special needs for the women responding, as well as some potential in training other clients or peers to begin meeting these needs.

The remainder of this paper will concentrate on the process of setting up the peer counseling program at Project Reality and my personal observations and experience throughout this time. By way of review, the literature on peer counseling maintains that its major advantage is that it occurs in the "real world environment rather than in the artificial environment of the professional's office" (Wasserman, McCarthy, Ferree, 1975). It is generally regarded as a mutually beneficial process that is not impeded by some of the barriers inherent in the usual patient/therapist relationship (Leibowitz and Rhoads, 1974; Allen, 1973; Edgar and Kotrick, 1972). However, there are numerous considerations in structuring such a program. Crucial to the success of the program is the impact of peer counseling on the peer counselors themselves, particularly with drug dependent women, who may and probably will be wrestling with their own personal issues. The success or failure of such an endeavor rests primarily in the type of supervision and preparation given the peer counselors. According to a recent study (Allen, 1973) "The lack of clear role definition and job performance criteria has nearly wrecked some programs because of the paraprofessionals' uncertainty about what they would be accountable for."

Initial criteria for recruiting potential peer counselors was similar to that used in determining eligibility for receiving methadone take-outs, i.e., illicit drug abstinence, legal status and some involvement in therapy. In addition, the women considered had demonstrated some ability to utilize community resources, were able to engage in non-drug related activities and had expressed interest in working with people in some counseling capacity. It should be noted that at first staff preferred recruiting women who were employed, in a stable personal relationship, skilled, and somewhat more personally attractive. In a nutshell, we chose women who were not personally needy but did appear interested. However, as the time to begin training approached some of these women suddenly began to have unanticipated excuses for not participating. One common resistance appeared to be coming from the partners or spouses who feared what the renewed "association with other addicts" might precipitate. Whether this fear was that of the partner or the woman herself was difficult to ascertain. The result was that the women being trained were generally less self-assured, not involved in a stable intimate relationship, and not as skilled as the first women recruited. However, the group participants, four women, are not presently involved in drug abuse, have been involved in therapy and have some initial skills or educational involvement. They are women who seem to be ready to move to a less introspective level and begin extending themselves to others. All have verbalized the sense of needing a training program such as this one to help them improve skills in dealing with people and their environment, as well as a desire to be helpful to others.

The formal training program consisted of seven three hour sessions and will continue with a weekly one and a half hour supervisorial group. The women were sensitized in the areas of values clarification, problem solving and decision making. They were then taught a transactional analysis model of personality functioning, a communication model, empathic levels of responding, basic interviewing skills and how to use community resources. Role-play was fundamental throughout the training course as a means of practicing and integrating new skills. The women were critiqued and given feedback about their skill development on an on-going basis.

One area that needed constant monitoring and sensitive handling was focusing the interactions of the group on their commonalities as women rather than as drug abusers. Initially the women tended to focus on drug "war stories" but as the training progressed they appeared to begin sharing related to content and experiences in sessions, i.e., personal examples about relationships, utilizing resources, day care concerns, etc. Eventually the peer counselors were able to begin expressing their own reservations about working with new clients who might still be involved in drugs, and asked for methods of intervention they could utilize in contending with this situation.

Another crucial issue was that of clarifying for the peer counselors the limits of their roles. The role-playing exercises generated a great deal of anxiety concerning handling problem areas that could stimulate too much personal unrest on the part of the women counseling. The women were trained in how to make appropriate referrals and encouraged to use the co-facilitator and myself for in-depth supervision. Emphasis was placed on learning to own personal feelings and recognition of individual limits in each peer counselor. The women seem to have a clear perception of themselves as advocate-big sisters rather than junior therapists. The women have been encouraged to ventilate their fears and concerns in the group and to the facilitators as they occur.

The intensive training program has been completed and the second phase of the training is ready to begin. Weekly group sessions will be continuing and a projection for the latter part of April has been established as the time for the peer-counselors to begin working with individual women. Content for the on-going group sessions will revolve around more practice of role-play situations, and information about resources and referrals. The peer counselors are not being rushed into working with clients at this point. They are in the initial phases of group cohesiveness and the development of the group is viewed as essential to the support the counselors will need to do effective work.

In examining this first phase of the peer counseling program salient elements to consider in treatment come to mind. The first ingredient appears to be consistency of facilitators in their approach. Membership for training groups was not stable or punctual. However, the training sessions were regularly held. Often they started late but the entire content was covered. The women were rewarded for the efforts they extended. Feedback to the facilitators was that even if things were unstable in a peer counselor's life, they were still given the message that they were acceptable and valuable to the group.

Flexibility in terms of facilitators' willingness to take the individual needs of the peer counselor into account was another important variable. Home visits and extra sessions were arranged to provide make-up training sessions for absences. Encouraging the women to participate as much as they could, and providing additional time if necessary, seemed valuable in keeping the group going.

Facilitator role was designed to model a women-helping-women format, rather than patient-therapist roles. The facilitators participated in the exercises, role-plays and sharing. When one facilitator taught didactic content the other participated in the experience. Feedback from the peer-counselors indicated that this was significant for them in enhancing openness among the group and trust in the facilitator's understanding and commitment to the concept of the group.

Currently I cannot report on a peer counseling program that is functioning. However, despite trial and error and a few rough periods of discouragment, I feel the initial phase of the process is a success. My expectation is that the women in training will need a great deal of support as well as group and individual supervision once they begin work as peer counselors. They all have numerous fears about performance and idealized notions about "success". Yet they are excited and open to learning about themselves. Perhaps the most valuable aspect of this experience for me has been to see this group of women get enthused about the possibilities and really attempt to learn new skills that will be valuable to them as women whether or not they are peer counselors. At minimum, we hope to have a women's group that functions supportively and enhances self-esteem for the participants. The real test of self-esteem and learning will come in the next few months when the selection of clients for the counselors is complete and the peer counseling has begun.

REFERENCES

Allen, Dean A., Ph.D. 1973. Peer Counseling and Professional Responsibility. *College Health*. Vol. 21: 339-342.

Cooperstock, Ruth. 1971. Sex Differences in the Use of Mood Modifying Drugs: An Exploratory Model. *Journal of Health and Social Behavior*. Vol. 12: 238-244.

Edgar, K.F. and Kotrick, Cathy. 1972. The Development of a Peer Counseling Center. *Psychotherapy: Theory and Practice*. Vol. 9 No. 3: 256-258.

Leibowitz, Zandy and Rhoads, David J. 1974. Adolescent Peer Counseling. *The School Counselor*. Vol. 12: 280-283.

Levy, Stephen J., Ph.D. and Doyle, Kathleen M. 1975. Attitudes Toward Women in a Methadone Maintenance Program. Presented at New Jersey Medical School, January.

Schwartz, Bryna. 1976. The Female Addict. *Women in Treatment: Issues and Approaches*. National Drug Abuse Center for Training and Resource Development.

Thure, Karen Lee and Moore, Bettye Ann. 1977. Women in the Age of Aquarius. NIDA, Division of Resource Development, Prevention Branch. Contract No. 271-75-4014.

Wasserman, Craig W., McCarthy, Barry W. and Ferree, Elizabeth H. 1975. Student Paraprofessionals as Behavior Change Agents. *Professional Psychology*. Vol. 6(2): 217-223.

Weston, Lynda Martin and Pancoast, Ruth. 1975. Feminist Psychotherapy. *Social Policy*. Vol. 9: 54-62.

THE RESULTS OF DATA ANALYSIS ON 50 PREGNANCY AND ADDICTION STUDIES 1965-1977*

Susan Dickey
Burt Associates, Incorporated

Margruetta Hall, M.S.
National Institute on Drug Abuse Services Research Branch
Division of Resource Development

PURPOSE AND METHODOLOGY

During 1977, Burt Associates, Incorporated conducted a study of pregnancy and addiction for the National Institute on Drug Abuse. The major research questions addressed were:

* What are the differential effects of heroin and methadone on the mother and her offspring?

* How does polydrug abuse in conjunction with heroin and/or methadone affect pregnancy outcomes?

* What are the risks to the fetus/infant and mother as a result of drug taking behavior?

The sources of data were published and unpublished reports on pregnancy and addiction written between 1965 and 1977. Two hundred and fifty articles were qualitatively analyzed; 50 which contained

---

* This study was part of a larger project on drug abusing women conducted for the National Institute on Drug Abuse, Contract No. 271-76-4401. The paper was adapted from a report by S. Dickey, *Pregnancy and Addiction*, Rockville, Maryland: National Institute on Drug Abuse, in press. We wish to thank and acknowledge Mr. Charles A. Darby, M.A. for his outstanding contribution on the data analysis for this paper and the larger report.

numerical data were selected for quantitative analysis. Forty independent and dependent variables were selected for analysis (see Table I); examples of these variables are polydrug use, drug abuse treatment, complications in labor and delivery, postpartum maternal mortality, infant abnormalities and complications other than drug withdrawal, and placement of the infant. Data on each variable were extracted across studies to investigate trends and relationships as well as gaps and conflicts in the literature.[1]

The numerical treatment of data originally reported in the literature varied greatly (i.e., ranges, percentages, and averages); therefore, the data could only be summarized in the form of descriptive statistics and statistical testing was precluded. Variability in data reporting also occasionally limited the number of studies that could be used in the data analysis of a specific variable; for example, studies reporting "up to" and "greater than" could not be used. Although summarizations of data are given here in the form of percentages and averages, originally the numerical values in the literature often ranged widely for each variable both within and across studies.[2] Thus the results of the data analysis must be viewed with caution and considered suggestive rather than conclusive.

FINDINGS

Although most findings reported here are taken from the quantitative analysis of 50 studies, results from the qualitative analysis from 250 studies were given when insufficient data on a variable precluded numerical summarization. Findings on the addicted women are presented first followed by data on their infants. In each case, where appropriate, the data are reported for pregnant addicts who were in drug treatment vs. those who were not. Where this occurs, the groups are referred to as treated and untreated.

---

1 Originally statistical tests on at least subsamples of the literature data were planned in order to investigate the relationship between variables in different samples of mothers and infants (e.g., by ethnicity, methadone dosage, and nontreatment) and between different outcomes for both the mother and infant. However the data were not reported in ways amenable to use of statistical tests nor were they, in some cases, readily summarized statistically.

2 Where results of data analysis are derived from a large number of studies, individual study references cannot be given because of space limitations. The reader is referred to the original report.

Table I. Number of studies presenting numerical data on each variable out of 50 studies.

A Characteristics of Pregnant Addicts	B Prenatal Care	C Treatment and Complications in Labor and Delivery	D Outcomes for Mother	E Outcomes for Fetus/Infant	F Neonatal Withdrawal Symptomatology	G Neonatal Withdrawal Symtomatology	H Treatment for Infant
Ethnicity 20	Medical Services 21	Anesthesia 5	Portpartum Medical Complications 8	Apgar Score 23	Symptoms 23	% of Infants Manifesting Withdrawal 28	Medication for Withdrawal 14
Age of Parent 29	Social and Psychological Services 2	Complications in Labor and Delivery 17	Postpartum Psychological Complications 0	Scales Used 12		Time of Onset of Symptoms 15	Dosage 9
Family Status 20	Drug Treatment 25	Treatment in Labor and Delivery 9	Length of Hospitalization 4	Neonatal Mortality 27		Duration of Symptoms 8	Duration of Treatment 11
Health Status in Pregnancy 24			Postpartum Medical Treatment 5	Abortion 10		Relationship of Symptoms with Maternal Drug Intake 11	Special Nursing 2
Primary Drug of Abuse 24			Postpartum Social Services 1	Stillborn 19			Hospitalization 19
Polydrug Use 16			Postpartum Maternal Mortality 4	Abnormalities Complications Other Than Withdrawal 25			Followup Medical 16
Sociological Psychological Condition 2				Birthweights 45			Followup Services 6
				Gestation 32			Placement 8

Addicted Mothers

Data on ethnicity show that women in both the untreated and the methadone maintenance groups are most likely to be black, although there was a higher percentage of whites in methadone treatment groups than in the untreated groups. These data, however, were derived from urban hospital studies where the normal patient population was black.

The average age for the untreated addicted woman is approximately 24 years while the woman in methadone treatment tends to be about one year older. Many of the mothers in both groups, however, are likely to be under 21 years of age.

Data on family status suggest that the woman enrolled in treatment is more likely to be either married or living with a partner than the untreated woman. However, the number of women maintaining these relationships appears to decline as treatment, particularly psychological support, is intensified. Women in both the treated and untreated group are likely to have had an average of three previous pregnancies and two living children.

The ten most frequently mentioned health complications for the pregnant mother in both the treated and untreated groups are, in rank order: 1) Venereal disease; 2) Anemia; 3) Hepatitis; 4) Toxemia; 5) Hypertension; 6) Infection; 7) Malnutrition; 8) Diabetes; 9) Epilepsy; and 10) Preeclampsia. It should also be noted that two studies reported physical abuse to the mother; one case resulted in miscarriage (Claman and Strang, 1962; Mondanaro, 1976).

Both the untreated and treated pregnant woman are likely to have a history of approximately four years of heroin use with heroin as the primary drug of abuse. Illegal methadone also may be frequently used. Dosages of drug intake were reported in so many different ways that meaningful data summarization was not possible.

Barbiturate use in conjunction with heroin and/or methadone maintenance was reported twice as frequently as any of the other drugs. Amphetamines, alcohol, and illegal methadone ranked second, third, and fourth respectively for both the untreated and treated groups. Women in methadone maintenance programs appear likely to use the above drugs sporadically during treatment.

Pregnant women receiving methadone maintenance are not likely to have completed high school. They generally lack specific job training or employable skills. No data on education or employment are available on the untreated group.

The pregnant addict not receiving treatment is likely to register at a prenatal clinic in the last trimester or term and never return for additional treatment until labor. The treated woman, however, will probably make an average of seven visits to a prenatal clinic, beginning treatment fairly early in her pregnancy.

Methadone maintenance was reported as the most viable treatment approach for the heroin addicted population; the average daily methadone dosage for pregnant addicts was approximately 42 mg.

Generally, the pregnant addict is treated as any other patient in labor; however, staff may take special care to locate veins for intravenous medication early in the labor and to avoid hepatotoxic medication. The untreated pregnant heroin addict is likely to deliver early by breech presentation, and to have premature rupture, placental abnormalities and very rapid labor. The average labor time for the untreated pregnant addict is approximately 9½ hours for primigravidas and 5½ hours for multigravidas. However, mothers who have received methadone maintenance and prenatal care approach more normal times for labor and delivery; average time for general population is approximately 13 hours for primigravidas and 8½ hours for multigravidas. The treated woman has decreased amounts of delivery complications such as breech delivery, premature rupture of membranes, abruptio placenta and amnionitis. This appears to be directly related to the amount of prenatal care received (e.g., Finnegan et al., 1972; Connaughton et al., 1975). Both treated and untreated groups are more likely to have low and mid-forceps delivery and less likely to have cesarean section compared to women in the general population.

Postpartum hemorrhage appears to be the single most frequent medical complication occurring in the postpartum period; however, this complication occurred much more frequently in the untreated group than in the treated group.

The average length of maternal hospitalization for the woman receiving drug treatment is approximately three days. The untreated patient however may sign out against medical advice even if she has received "narcotic" maintenance since delivery (Clark et al., 1974). Very little data on maternal mortality are available.

The addicted mother appears to experience depression after delivery and, if she is on methadone maintenance, she may request an increase in her daily methadone dosage.

Postpartum data pertaining to the mothers is extremely sparse and thus represents an area for future research.

The Infant

The infant of the untreated mother is likely to be premature and weigh approximately 5 lbs. The Apgar score for this infant may range from 7 to 8 at one minute and about 9 at five minutes.

The infant of the woman enrolled in methadone maintenance is likely to be delivered closer to term and weigh approximately 6 lbs. The Apgar score for these infants may range from 6 to 10 at one minute and from 7 to 10 at five minutes.

Analysis of the infant health complications, other than drug withdrawal, suggests that approximately 8 percent of infants of drug addicted mothers have some type of additional complication at birth. The possibility of polydrug use in the methadone maintenance group may explain the slightly higher rate of additional infant complications (about 11 percent) in comparison to the infants of the untreated group (about 5 percent). Infection, particularly intestinal infection, was reported more frequently in the untreated group than in the methadone maintenance group. Meconium aspiration, jaundice and asphyxia were reported frequently for the infants of the treated mother group.

Approximately 1.1 percent of the infants born to both groups were reported as having congenital malformations. Malformations of the urogenitals and digestive system were reported more frequently across studies for infants of untreated mothers while infants of women in treatment were reported more frequently across studies as having malformations of the heart, chest, and limbs. Again, the possibility of polydrug use among the methadone group may explain this difference and is an issue that needs further study.

Across studies, mortality rates among newborns of addicted women appear to be substantially higher than among newborns in the general population. Mortality rates for infants of untreated mothers were higher than rates for infants of treated mothers. Whether this higher mortality rate is a result of heroin per se, lack of prenatal care or a combination of both, cannot be determined from the literature. Prematurity ranked as the most frequently mentioned cause of death for the untreated group while Hyaline Membrane Disease ranked first for infants of mothers receiving methadone.

Spontaneous abortion was reported as occurring more frequently among the women who were untreated. The mean percent for this group was approximately 6.6 while the mean percent for women receiving treatment was about 5.5. The mean percent of stillbirths (3.2) was also higher among the untreated group than among the treated group (1.6 percent).

Infant withdrawal symptoms were ranked by order of times reported in the literature. The five most frequently mentioned symptoms for infants of untreated mothers in order of frequency were: 1) Tremors or myoclonic jerks; 2) Irritability; 3) Hypertonicity; 4) Shrill continuous crying; and 5) Rapid breathing. The five most frequently mentioned symptoms for infants of methadone maintained mothers in order of frequency were: 1) Irritability; 2) Tremors or myoclonic jerks; 3) Rapid breathing; 4) Generalized convulsions; and 5) Vomiting. The percentages of infants manifesting withdrawal sympomatology ranged from 39 to 100 percent for infants of untreated mothers and from 58 to 100 percent for infants of mothers who had received methadone maintenance. Because of the wide variability in data reported from these studies, average percentages for the time of onset and duration from symptoms were not computed.

The near absence of studies relating the effects of multiple drug abuse in conjunction with heroin or methadone treatment to infant outcomes was striking. Since barbiturate use in conjunction with methadone maintenance was frequently reported, it is surprising that its effects in relation to infant withdrawal have not been addressed more directly by investigators. Descriptions of infant barbiturate withdrawal are suspiciously like descriptions of the delayed, more severe, longer lasting methadone withdrawal symptoms. In view of this, it is hypothesized that some of the reports on the seriousness of infant methadone withdrawal may, in actuality, be descriptions of symptoms caused by CNS depressants. If this hypothesis is true, it would explain the diverse and often conflicting data surrounding neonatal methadone withdrawal.

Medications used for infant withdrawal were ranked according to most frequent mention. Phenobarbitol, Paregoric, Chlorpromazine and Diazepam were the four most frequently mentioned medications. Dosages were reportedly individualized. The average duration of treatment among the infants of treated and untreated mothers could not be ascertained because of the variability of data reported and because this variable was not delineated between the two groups.

Data analysis showed that the average duration of hospitalization for infants in treated and untreated mother groups did not greatly differ. The average hospital stay for infants of untreated mothers was approximately 13 days compared to 14 days for infants of the treated group. Prenatal care reportedly shortened the duration of hospitalization. The difference between hospitalization of the mother (who averaged 14 days) can be a major factor in disrupting effective maternal infant bonding.

Most medical followup data pertain to infants of mothers who receive drug treatment during pregnancy. Few such data are available for infants of untreated mothers.

One month followup revealed that the babies of treated mothers often continued to manifest subacute withdrawal symptoms at home; they showed unresponsiveness, restlessness, irritability, regurgitation and a need for frequent feeding and constant attention. Their scores on the behavior and mental scores of the Brazelton Neonatal

Behavioral Assessment Scale have also been reported as differing from infants in the general population.

There were several reports of Sudden Infant Death. Seven cases were reported out of 258 infants. This is considerably higher than the normally expected rate of 2.5/1,000 live births in the general population. It is not clear, however, as to whether the mothers of these infants were taking heroin or methadone or both or other combinations of drugs. Further research on this is needed.

One case of child abuse was reported (Ramer et al., 1973). However, in view of the emotionally deprived and battered backgrounds of mothers, the frequency of immediate separation of the mother and infant after birth, and the difficulty in caring for these sick babies, it is surprising that more cases of abuse were not reported. Research on the relationship between birth factors and child abuse and neglect indicates that birth factors such as low birthweight, prematurity and infant illness are associated with child abuse and "failure-to-thrive." Such research suggests that abuse and neglect may be prevalent among the addict population (Fanaroff, Kennel, and Klaus, 1972; Klein and Stern, 1971).

Mothers receiving methadone maintenance reportedly kept their babies twice as often as the untreated mothers. Infants of the untreated group were most likely to be placed in foster care. Apparently neither group of mothers was likely to place their babies for formal adoption.

Parenting classes which teach child rearing knowledge and skills and emphasize parental enjoyment of child care, have been reported as a highly successful approach in the treatment of both the mother and child.

IMPLICATIONS FOR TREATMENT AND FUTURE RESEARCH

Although there is a plenitude of studies on pregnancy and addiction, this area of drug research is clearly in its infancy particularly with regards to polydrug abuse.

The issue of multiple drug use is possibly the most crucial research and treatment issue pertaining to pregnancy and addiction. The lack of documentation of multiple drug use in the literature is striking. Barbiturate use in particular was frequently mentioned as commonly occurring among pregnant clients; yet little consideration was given to the effect of this drug on maternal and infant outcomes. It may be that many of the conflicting findings in this literature may be explained by multiple drug use among the mothers. In other words, the occurrence of multiple drug use may be an invisible independent variable which has confounded the majority of study outcomes.

Greater efforts on achieving excellent experimental and statistical design in pregnancy and addiction research are needed. Investigators must: 1) use comparable controlled study groups which control possible relevant independent variables (i.e., parity, ethnicity, socioeconomic status); 2) use noncontaminated study groups; 3) use adequate sample sizes to allow for statistical testing; 4) report samples sizes clearly and correctly; 5) report on important variables including the value of zero in the absence of occurrence; and 6) delineate outcome data by study groups and independent variables.

Additionally, it appears that the pregnancy and addiction literature has primarily focused on the consequences of heroin abuse or methadone maintenance to the infant rather than to the pregnant client. Research pertaining to the mother, particularly during the postpartum period, is greatly needed. For example, postpartum psychological conditions, retention treatment rates related to particular services and mortality rates are areas of needed research. The effects of dietary habits and nutritional status on the infant *and the mother* also needs further study. Effective maternal-infant bonding and parenting skills are crucial to this population. Methods, including followup care, which enhance these interactions need furdevelopment and longitudinal evaluation.

Documentation on the incidence of Sudden Infant Death and child abuse and neglect among the infants of drug abusing women is urgently needed. This also includes documentation of the *absence* of these factors in a study population.

There is an urgent need for cooperative standardization of reporting among investigators. A minimum set of independent and dependent variables and a consistent way of reporting data need to be established. This is not to suggest that research would be limited to these variables, but rather that a limited number of standard variables would be controlled for and reported in all studies addressing pregnancy and addiction.

REFERENCES

Claman, A. and Strang, R. 1962. Obstetric and gynecologic aspects of heroin addiction. *American Journal of Obstetrics and Gynecology*. 83: 252-257.

Clark, D., Keith, L., Pildes, R., and Vargas, G. 1974. Drug-dependent obstetric patients. *Journal of Obstetric, Gynecologic and Neonatal Nursing*. 3(5): 17-20.

Connaughton, J.F., Finnegan, L.P., Shut, J., and Emich, J.P. 1975. Current concepts in the management of the pregnant opiate addict. In: R. Harbison (ed.), *Perinatal Addiction*. New York: Spectrum Publications, Inc.

Fanaroff, A., Kennel, J., and Klaus, M. 1972. Follow-up of low birth weight infants--the predictive value of maternal visiting patterns. *Pediatrics*. 49: 287-290.

Finnegan, L.P. Connaughton, J.F., Emich, J.P., and Wieland, W.F. 1972. Comprehensive care of the pregnant addict and its effect on maternal and infant outcomes. *Contemporary Drug Problems*. 1(4): 795-809.

Klein, M. and Stern, L. 1971. Low birthweight and the battered child syndrome. *American Journal of Diseases of Children*. 122: 15-18.

Mondanaro, J.E. 1976. Women: Pregnancy, children and addiction. In: A. Baumen (ed.), *Women in treatment: Issues and approaches research manual*. National Drug Abuse Center for Training and Resource Development, 1901 North Moore Street, Arlington, Virginia, 22209.

A DESCRIPTIVE AND COMPARATIVE ANALYSIS OF SELF PERCEPTIONS AND ATTITUDES OF HEROIN ADDICTED WOMEN

Mary Ellen Colten, Ph.D.

Institute for Social Research--University of Michigan

In 1974, NIDA funded several demonstration programs to provide services for heroin addicted women and to gather information on the utility of special approaches to treatment for female substance abusers. A year later, the Women's Drug Research Group was given a mandate to collect data on the psychosocial characteristics of women addicts. We went about this in several ways. The data I'm going to present here today comes from structured intensive interviews with 146 women in mixed sex treatment programs in Los Angeles, Miami, and Detroit, and from interviews with 202 men from similar programs. Additionally, we interviewed 175 non-addicted women from the inner city of Detroit for comparison purposes.

Today I'm going to present some of the more interesting data about the self perceptions and attitudes of our respondents. The two areas I'm going to discuss are sense of well-being and attitudes toward addicted women. While this is certainly not a comprehensive portrait of the female addict, the data speak to many of the old myths about addicted women and give us some very good clues concerning treatment for heroin addicted women.

First, sense of well being. Among the instruments given to our respondents were questions that had to do with self esteem, feelings of depression and feelings of anxiety. All three of the scales used were relatively standard ones, chosen because their validity has been established and because they appear to be quite effective for use with low education populations. For each of them, I will give our results and then our interpretation and the consequences for treatment.

We know that it has been demonstrated repeatedly that heroin addicts have poor self images--that their regard for themselves, their sense of self worth, is considerably lower than that of non-addicts. It is also generally acknowledged that women have lower self esteem than men. Accordingly, we would suspect that substance abusing women would be lower in self esteem than other groups, both because they are addicted and because they are women. Our data, not surprisingly, confirmed this expectation. The addicted women as a group were significantly lower in self esteem than the addicted men ($t(346)=4.12$; $p<.001$) and lower in self esteem than the non-addicted comparison sample of women ($t(159)=6.81$; $p<.001$). Additionally, the addicted men scored lower in self esteem than the non-addicted women ($t(174)=3,86$; $p<.001$).

There has been a tendency to attribute the comparatively low self esteem of addicted women to particular aspects of their situation--to claim that it is a consequence of prostitution, for example. It has also been claimed that a woman must be extremely deviant, much more so than a man, to use heroin. Repeated references are made to the self-hatred of women addicts, to their repressed aggression and to guilt they may have over their children.

Since this is the only study we know of showing lower self esteem in any group of men compared to any group of women, the finding that male addicts are lower in self esteem indicates that the condition of addiction is related to self esteem. It also suggests that the addiction process itself contributes more heavily to low self esteem than prostitution or particular aspects of the situation of the female addict.

Although addicted men have lower self esteem, it still must not be ignored that women addicts display comparatively lower self esteem. Clearly, the self images of women addicts suffer a double burden from the female role and from substance abuse. For example, treatment programs using confrontation techniques may exacerbate rather than alleviate this problem for women. A woman having little sense of her own self worth may also feel that she is not worth the effort of treatment.

The drug culture is at least as, if not considerably more, male oriented as the dominant culture. Women and the roles relegated to them are of secondary status. Lower self esteem of the women addict might best be viewed as an understandable response to the situation in which she finds herself.

Next, the feelings of depression.

The measure of depression in this study has been used in national surveys (e.g., Americans View Their Mental Health, 1957, 1976). It has been shown to be highly reliable and valid, is considered to be good for use in a variety of populations (Radloff, 1977).

It was selected because it is *not* considered to be a diagnostic tool or a measure of pathology. Any of the symptoms in the scale could be experienced and reported by psychologically healthy individuals, with more depressed persons tending to report more of the symptoms at any one testing time. It is also not diagnostic because it is oriented only toward the present; respondents are asked if they have felt this way *in the past week*.

Women, generally, report being more depressed than men. This has been attributed to response biases--the greater willingness of women to admit to depressive symptoms (Cooperstock, 1971; Phillips and Segal, 1969) and to the realities of women's lives. That is, given their lot in life they *ought* to be more depressed.

The addicted women in this study are, in fact, considerably more depressed than both addicted men ($t(345)=4.05$; $p<.001$), and non-addicted women, ($t(159)=4.74$; $p<.001$). Addicted men and non-addicted women also differ significantly in reported mean levels of depression, with the addicted men reporting greated depressions than non-addicted women ($t(173)=2.20$; $p<.05$).

As with the self esteem results, we must conclude that addiction itself results in reports of more or greater depressive symptoms. If this were not the case, we would expect the addicted males to show less, rather than more, depression than the control group women.

The addicted women are overwhelmingly more depressed than the other groups. This finding should not be taken lightly, particularly since the individuals comprising the other two groups have quite serious life problems--the men are substance abusers just entering treatment and the comparison women are all unemployed and actively looking for work. Response biases may be involved in this result but it is far-fetched to accept the notion that women addicts are simply more willing to complain. Again, we must look to their life circumstances. The data on their social relationships and on their life problems indicate that they have more about which to complain. This finding may also be linked to the data on self esteem and the causes of both findings may be somewhat similar, if not identical. The addicted women in this sample have all recently entered mixed sex treatment programs. Critiques of traditional treatment programs indicate that staff are less likely to expect women to successfully complete treatment and may, in fact, implicitly communicate this to women. The women may be at a low point when they enter and confrontation and forced self examination in conjunction with weakened defenses may leave women feeling discouraged and hopeless. They may have been forced to relinquish their children, and have relatively fewer social supports (Tucker, 1977). It also has been suggested that women addicts are treated like "whores" whether or not they ever prostituted to earn money to support their habits (Schultz, 1974). They are also less likely to be employed (or to have hopes of employment) than male addicts. As we shall see later they are well aware of the negative societal response to women substance abusers and over half of the addicted women agree that "women addicts are worse than men addicts." So a good number of these women may have internalized this negative appraisal and their lower self esteem and greater depression may reflect awareness and/or acceptance of these attitudes. In all, their situations are dreary and depressing. If they did not appear more depressed than male addicts we might wonder what was wrong with them. Less depression could constitute an unnatural response to a difficult situation.

In one study (Fisch, et al., 1973) it was found that more highly depressed clients in a methadone program tended to show greater improvement than low depression clients; in fact, the low depression clients tended to get worse in conjunction with therapy. On the other hand, highly depressed clients tended to be more likely to drop out from treatment. This suggests that the greater depression of the female addicts may make them good candidates for successful therapeutic intervention, provided that the program makes intensive efforts to engage them and keep them in the program during the initial phases of treatment. So, their heightened depression may be a hopeful sign if it is dealt with appropriately.

Now, we turn to feelings of anxiety.

Women addicts are more anxious than men addicts ($t(346)=4.45$; $p<.001$) and more anxious than non-addicted women ($t(159)=3.46$; $p<.001$). Addicted men also report greater anxiety than non-addicted women ($t(174)= 1.99$; $p<.05$).

As with the depression results, it appears as though the process of addiction and/or entering treatment affects reports of anxiety, since both male and female addicts report significantly more anxiety than the control group. The addicted women again are reporting more or greater symptoms than the addicted men. This should be taken as a sign that they are experiencing greater discomfort.

The greater anxiety of women addicts may also be due to their fears that they will be unsuccessful in the program. Their lower self esteem and lower sense of control over their lives supports this notion. Additionally, we do not know what treatment programs do (or do not do) to make women feel that they can be successful. It is possible that the initial phases of treatment in most programs have not been designed to alleviate the fears and anxieties of women clients.

Very high levels of anxiety have been shown to cause decrements in performance (Sarason et al., 1960; Sarason, 1961). We suspect that addicted women may get trapped in the proverbial vicious cycle, with heightened anxiety leading to poorer performance and lower self esteem which then again lead to greater anxiety.

Part of the negative feelings women addicts have may be better understood in light of some of our findings on attitudes that male and female addicts have toward addicted women.

We asked our respondents how much they agreed with the following three statements:

1. Women addicts are worse than men addicts.

2. Men look down on women addicts more than they do on men addicts.

3. Women look down on women addicts more than they do on men addicts.

Overall, addicted women, more than addicted men, report that women addicts are worse and looked down on more ($t(345)=2.61$; $p<.01$).

A look at the individual items enhances our understanding of these results. Addicted women are significantly less likely than addicted men to agree that "Women addicts are worse than men" ($X^2(3)=19.7$; $p<.001$). Still, more than half of the women (55.5%) and more than two thirds of the men (67.8%) agree. (This does not mean that those who disagree necessarily think men are worse; they may see them as equally bad.)

Although there are no significant differences between the responses of men and women to the item "Men look down on women addicts more than they do on men addicts," 89.9% of the women and 88.6% of the men agreed with this statement. So, while a slight majority of the addicted men and women actually feel that women addicts are worse, a whopping majority feel that men look down on women addicts more.

Also, the vast majority of addicted women (69.2%) and addicted men (72.8%) agree with the statement: "Women look down on women addicts more than they do on men addicts." Although both figures are high, women are significantly less likely to agree ($X^2(3)=12.23$; $p<.01$) and more likely to strongly disagree (15.1% of women vs. 6.9% of men).

Many writers in the field have referred to the self-hatred of women addicts. These results indicate that what may have been called "self-hatred" is simply a recognition of the fact that others, particularly men, more strongly disapprove of addiction in women.

There is, in reality, no reason why female addicts should be condemned more than male addicts; they share the same problem of addiction. And yet, they are treated as if they are worse. The data show that this is not paranoia on the part of these women--the men also acknowledge the situation.

The extreme isolation of addicted women (Tucker, 1977), their low self esteem, and their high depression, may all be linked to these reports of attitudes toward women addicts. Both men and women say that women addicts are viewed with disfavor--rejects among a community of rejects. It appears that addicted women are censured because they are women.

These data suggest that any treatment program which does not confront this disparity between attitudes toward male and female addicts is shirking its responsibility to its female clients, and will probably not help them to overcome the low self esteem and heightened anxiety and depression which may well stand in the way of treatment goals.

Before concluding, I would like to point out that there is a bright side to the picture: we did find that addicted women have many social skills, that they evidence a willingness to get involved with other people and appear to be no more manipulative or mistrustful than comparison women. In other words, they do enter treatment with some positive and solid characteristics on which to build.

These data remind us that addicted women are not exactly like other women--and, more importantly, they are not exactly like addicted men. Their problems and perceptions of themselves should be considered and treated as unique.

REFERENCES

Cooperstock, R. 1971. Sex differences in the use of mood-modifying drugs: an explanatory model. *J. of Health and Social Behavior.* 12: 234-244.

Fisch, A., Patch, V.D., Greenfield, A., Raynes, A., McKenna, G., and Levine. 1973. *Depression and self concept as variables in the differential response to methadone maintenance combined with therapy.* Paper presented at the Methadone Conference, Washington, D.C.

Phillips, D. and Segal, B. 1969. Sexual status and psychiatric symptoms. *American Sociological Review.* 34" 58-72.

Radloff, L.S. 1977. The CES-D scale: a self report depression scale for research in the general population. *Journal of Applied Psychological Measurement*. 1.

Sarason, I.G. 1961. Test anxiety and the intellectual performance of college students. *Journal of Educational Psychology*. 52: 201-206.

Sarason, S.B., Davidson, K.S., Lightfall, F.F., Waite, R.R., and Ruebush, B.K. 1960. *Anxiety in elementary school children*. New York: Wiley.

Schultz, A. 1975. Radical feminism: a treatment modality for addicted women. In E. Seray, V.T. Shorty, and H. Alksne (eds.): *Developments in the Field of Drug Abuse*. Cambridge, Mass. Schenkman.

Tucker, M.B. 1977. A descriptive and comparative analysis of the social support structure of heroin addicted women. *Women's Drug Research Project Report*. National Institute on Drug Abuse.

ISSUES SURROUNDING THE WOMAN ADMINISTRATOR IN DRUG ABUSE

Rebecca Brownlee, Executive Director

Drug Action of Wake County, Inc.
Raleigh, N.C.

I am interested in both the theoretical issues that prevent or discourage women from moving into administrative positions and the practical problems which seem to surround the woman who tries to be an administrator. I shall deal briefly with facts about women in the field before focusing on the theoretical and practical issues.

In 1973 a small group of women formed the first women's caucus on drug abuse at a methadone conference in Washington, D.C. The primary concern expressed by this group was that the special needs of women were not being met in drug abuse treatment. These women and others reconvened a year later. At that time they asked the National Institute on Drug Abuse to take a stand with regard to the absence of women in treatment, as well as to work in other areas affecting women, such as prevention, research, and education. The response to the second caucus was the implementation of a program for Women's Concern by N.I.D.A.[1] The result of this early groundwork has been a new emphasis on the woman as a client in the drug abuse system. This emphasis has taken place with special concern given to bringing the female client into treatment, to reaching hidden abusers through intervention programs, and to developing prevention models for women in the home. Drug abuse agencies have been encouraged to target women as special populations. Funds are becoming available to deal specifically with the female client and demonstration projects are being set up to develop models. A lot of time and energy has been spent on the wom'n who may or should enter the client system. Despite all of this emphasis on the female client, little attention has been given to the people in these agencies who make decisions concerning the female client. For the most part those who remain in positions of authority are men.

1. *Drugs, Alcohol, and Women: A National Forum Source Book*, p. 1.

I made a national survey to determine how many top administrators are female in drug specific agencies (Table I and II). I was interested in how many women actually head agencies in the field. I did not seek information about women who may be in middle management positions. The data which I cite is my own compilation from a 75 percent response on a letter to all Single State Agency Directors requesting the names of female administrators in their respective states. My request referred specifically to top executives, but I find that some people interpreted that to mean any administrative position. Therefore, the data may understate the actual number of women in top positions. It is interesting to note that 19 percent list no female administrators and only 21 percent report ten or more women administrators. With 37 states responding and a range from no women administrators to 76 in one state, the overall average number per state is 9.5. In the state which reported 76 women administrators, the same state lists 301 men administrators. This comparison shows that even in states with a seemingly large number of women administrators, the women administrators comprise only 20 percent of the total. In the five states which reported total administrative make-up, only one position in five was filled by a woman. It would

Table I. Chart on Women Administrators

Women Administrators Reported	Number of States	Percentage of Total States
0	7	19%
5 or less	18	49%
6 to 10	4	11%
10 or more	8	21%
TOTAL REPORTING	37	100%

Table II. 5 States Reporting Total Administrators

Number of Total Administrators		Number of Women Administrators	Percentage of Women to Total
I	126	25	19%
II	246	39	16%
III	377	76	20%
IV	29	10	34%
V	8	1	13%

seem that, although a great deal of pressure has been put on bringing the female client into the system, the advancement of women into high level positions of authority remains considerably behind.

Why in a field which has been so diverse and innovative have women's positions remained traditional? Drug abuse seems to have followed the pattern of other social service agencies: the women do the counseling and casework while the men move into management positions. "... Women are said to 'have intuition' and a gift for handling inter-personal relations and are encouraged to become social workers. Yet, men who work well with people are defined as 'good diplomats' and are given important political-executive posts which are closed to women and their imputed skills."[2] Women seem to have moved more easily into professions which require nurturing, such as teaching and social work. In these fields women often outnumber men significantly and the professions are referred to as 'female'. The few men who enter these fields, however, move into executive positions, while the women continue in the lower ranks.[3] What happens, even in less traditional fields like drug abuse, that deters women from what might be considered a natural progression?

Theoretical issues are significant factors to consider as we examine the question. In terms of theoretical issues a first step is to look at traditional expectations for women. Woman in our society is most commonly viewed in a maintenance role. She is most easily identified with the mother image. Man on the other hand is seen as the objective, factual being. It is the man from whom we expect firmness and the woman from whom we expect softness. The maintenance role is one of support and giving. The mother of the infant is viewed as an all giving, all sacrificing person. She sublimates her own needs and her own wishes for the purpose of supporting the infant. Whether a woman actually becomes a mother or not, femininity is closely associated with the characteristics of motherhood. Because of the soft nature, women are thought of as being emotional and are thought of as approaching decision making irrationally and impulsively. Interestingly enough, as Elizabeth Janeway points out in her book *Man's World, Woman's Place,* that role of motherhood is also a very powerful position: "The grown child remembers the mother as a slave, as loving nurturer who tends and watches and serves. But the mother is also the master... she now has the power to create [the child] as a social being, a member of the community; and without her this creation will not take place...".[4] Mother who is all giving is also all controlling. She has the power to take away as

---

2. Cynthia F. Epstein, *Woman's Place,* p. 22.
3. Eli Ginzberg, "Challenge and Resolution," in Ginzberg and Yohal Yohalem, eds., *Corporate Lib: Woman's Challenge to Management,* p. 142.
4. Elizabeth Janeway, *Man's World, Woman's Place,* p. 53.

well as to give. Motherhood may evoke fear, as well as respect from the small child, as the little person begins to learn limits from that primary figure in its life.

These stereotypical images, even if subconscious, come into play when a woman is considered for a position of authority. And it must certainly be true that women themselves harbor these images just as much as men. "Moving into management is a much more daring step for women to take than management men understand. The life of a role breaker makes constant psychological demands."[5] The view then of woman as a soft, nurturing, supportive, giving person must conflict with our expectations of administrative positions and fits easily into what we think of as the more traditional job situation for women in our society. In the field of drug abuse, for instance, we have found that the majority of women hold positions such as secretaries, educators, counselors, or crisis intervention specialists. In all of these areas a woman can fit more naturally into the stereotype of being a maintenance-oriented person. Edward Robie points to the fact that "even where women occupy higher level jobs...women tend to have helper oriented titles, such as executive assistant or assistant manager."[6]

A study of staffing patterns in 1975 and 1976 within the Division of Drug Rehabilitation in Massachusetts revealed that, although 60 percent of the employees were female, only 38 percent of the supervisory positions were occupied by women. Of the 10 supervisory positions only 3 (30 percent) were occupied by women. On the other hand, of the non-supervisory professionals, 55 percent were held by women.[7]

When we consider the traditional female role image, we must recognize the hinderances of such an image in terms of an administrator. The traditional woman is not considered a leader. The generalized view of a person with all-giving characteristics is not a person who would be first chosen to direct an agency. When we think of administration, we want someone who can be firm, not necessarily giving. We also tend to think of an administrator as someone who can make clear, unemotional decisions. Traditionally, women are not in situations which demand objective decision-making. "When we are told that women are, by nature, bad at making decisions, we might reflect that they have usually had little practice at choosing

---

5. Janeway, "Family Life in Transition," in Ginzberg and Yohalem, eds., *Corporate Lib,* p. 123.
6. Edward A. Robie, "Challenge to Management," in Ginzberg and Yohalem, eds., *Corporate Lib,* p. 15.
7. Joy Camp and Della Hardy, *Study on the Special Treatment and Prevention Needs of Women Drug Abusers in the Commonwealth of Massachusetts,* p. 36.

consciously to initiate overt action."[8] In essence, in any kind of agency, especially in an agency with the complexities of drug abuse, an administrator must be a person who can provide strong direction in all aspects. This image is not the one which comes to mind when we think of "woman."

The power of the traditional female must also be a consideration. Although this characteristic is more subtle, the type of power which motherhood possesses is one which may be frightening to all of us. It is she who dominates the environment and atmosphere of the child. In the traditional role she may well assign the task of punishment to another source, e.g., the father figure; yet it is she who exercises control. The strong matriarch is usually not one whom we admire. This aspect of motherhood must affect the consideration of women for positions of authority. Both men and women would shy away from selecting a person for a leadership role who might suggest to them such totality of power and control.

These role images evoke psychological factors which we all deal with, consciously or not. The positive and negative aspects of the traditional woman's role become difficult to mesh with what we think of as significant characteristics for a good administrator to possess. But, in fact, many women do not fit the traditional model. Women who succeed in top positions are able to function effectively as decisive, objective managers, even though they may be warm, responsive people. We have a very strong administrator when we find a woman who balances the maintenance role with objective rationality. William Goode even defines the female stereotype in positive terms for managers, stressing the human relations and maintenance skills as strengths which women bring to management positions.[9]

I have contended that the theoretical issues which we have discussed may deter or discourage women from progressing naturally into top management positions. I think practical issues are equally significant.

Let us now look at these practical issues that also make it less likely for a woman to become an administrator. First, consider the application process for an administrative position when the applicant is a woman. Often women do not seriously consider themselves applicants for positions, nor are they considered as serious applicants when they apply. The initial challenge becomes the daring to apply for the top position in an agency. I found myself in such a predicament when I discovered that the incumbent Executive Director

---

8. Janeway, *Man's World, Woman's Place*, p. 86.
9. William Goode, "Family Life of the Successful Woman," in Ginzberg and Yohalem, eds., *Corporate Lib*, p. 99.

was planning to resign. I had been employed in the agency for five years and was serving as Deputy Director at the point of his resignation. I realized early on and he later shared with me that he had much difficulty in even considering recommending a woman to replace him. Perhaps, the idea of a woman being able to do the same job as a man is a threatening one. My employment record was fine in a second place position, as long as a man was around to make final decisions. This particular man did make a decision to recommend me without qualification for the position. The indecision, the self-doubt that women are put through, even when they know they are qualified for such positions, is a strain. After the first hurdle of having been recommended for the job, the next hurdle is competing for the position. In my case a Board of Directors made the hiring decision. Many of these Board members had supported me when I had moved into a position as Administrative Officer some years earlier and had supported my application for Deputy Director. Although I came equipped with more experience than the previous Directors and with a Masters degree, there were still many questions in the minds of these men and women who were to hire a new Executive Director.

I think realistically speaking that a Board of Directors in such a situation must consider how the agency itself relates to the community. The separateness of drug abuse agencies actually becomes another reason for them to use in not installing a woman as the head of such an agency. Some people might think that a woman administrator would intensify the differences and impede the agency's relations with the community at-large. I see this consideration as one which the applicant may have to deal with directly.

These beginning obstacles are some of the ones which discourage women from applying for such positions. Women, as I have noted earlier, are not moving up into top positions of leadership in a natural sequence. For many the difficulties of placing themselves in competition with men for such positions are simply not worth the effort. "There is no real tradition of women in leadership roles... Thus, while there is little doubt that many women qualify for management employment, it is an open question whether large numbers will move toward managerial careers without special encouragement and support."[10] My data indicate that women are certainly not moving toward top managerial careers in the field of drug abuse.

The woman who finds herself as the first female administrator in an agency will be in a pioneer position. Equality of pay will be only one of many issues. If men on the Board of Directors and men in governmental positions are not accustomed to working with a female administrator, there will be many trying situations. Some

10. Ginzberg, "Challenge and Resolution," p. 142.

may welcome a woman during a time when equal opportunity is getting so much attention. One official greeted me warmly saying that I was the token woman in the state.

Still others may not place full confidence in a woman's ability to handle the responsibility and may not feel comfortable unless a man continues to be involved in major decisions. One way that a Board of Directors may insure that a man is close to the top position of authority is to request that the next person in line be a man. My Board of Directors strongly encouraged me to hire a man as Business Manager to assuring having one man in administration. In other words, with a woman in the top position it continues to be important to have a man close at hand. It is interesting to note that the reverse is not true: most male administrators are not encouraged to designate a woman as next in charge.

Since women are expected to be emotional beings, some unexpected difficulties may arise. I have found, for instance, that I must be more diplomatic than my male colleagues. It appears that women who confront directly and try to fight the system boldly are labeled "difficult to deal with." My male counterparts are admired for "blowing off steam". I have not stopped fighting for things that my agency needs, but I have learned to temper my demands so that I do not appear in a traditionally emotional role of a woman who approaches problems irrationally.

A woman may not be able to move initially with the same freedom as a male predecessor, I discovered quickly. Suggestions even to meet for lunch may be misinterpreted, particularly if the woman initiates such a meeting. I was once rebuked (in a facetious way) by a Board officer who told me, "This is not going to be a social relationship." I never suggest twosome luncheons now.

Since drug abuse is considered a female social service occupation, the woman administrator may also have to deal with the old-fashioned perception of the "do-gooder" social worker. I think particularly with criminal justice officials this image must be confronted. Judges, for instance, may view a woman in drug abuse as someone who is trying to save the world and has no clear understanding of how "bad" an addict really is. I won over a notoriously tough superior court judge once by simply saying, "I should have told you when I first came in that I'm not a missionary social worker."

The other extreme is also possible: a criminal justice official may associate a professional woman in drug abuse with the female clients in the field and expect passive, seductive behavior. Credibility is more difficult for a woman to achieve in these situations, but consistent factual directness results in gains. The increasing number of female attorneys will certainly facilitate the woman administrator's ability to relate effectively to those in criminal justice, as professional women become more accepted by the system at-large.

The theoretical and practical issues which have been discussed tend to hinder the woman who seeks a top position in drug program administration. These problems seem to have resulted in fewer female administrators than one could expect in a field which has been innovative in other ways. Despite these problems women can become outstanding administrators dealing with the special demands that face drug abuse agencies.

I think there are unique challenges that set drug program administration apart from other kinds of administration. The process of funding alone demands special flexibility in planning. Personnel matters differ in that few other agencies or businesses have the range of expertise from degrees in street experience to degrees in psychology. Community relations remain complicated, as our society continues to place a stigma on drug use other than alcohol. The field needs strong leadership combined with a concern for the human elements involved. Capable women can provide this particular kind of leadership. I agree with William Goode's conclusion that, "... the skills and behavior they acquire in becoming a woman are exactly those of good managers: they are trained in human relations, not test tubes and machinery; in insight; in the organization and maintenance of a social unit, the family; in command not through arbitrary orders, but through persuasion and participation in taking care of subordinates and serving their needs so that they will produce better."[11]

These qualities are certainly those needed by an administrator in the ever-challenging field of drug abuse. While there are outstanding men in the field, given an opportunity, many women, too, will prove to be exceptionally competent leaders.

REFERENCES

Camp, Joy, and Della Hardy. 1977. *Study on the Special Treatment and Prevention Needs of Women Drug Abusers in the Commonwealth of Massachusetts*. Unpublished manuscript. Boston, Massachusetts: Division of Drug Rehabilitation.

Epstein, Cynthia Fuchs. 1970. *Woman's Place*. Berkeley, California: University of California Press.

---

11. Goode, p. 99.

Ginzberg, Eli and Alice M. Yohalem, eds. 1973. *Corporate Lib: Women's Challenge to Management*. Baltimore, Maryland: John Hopkins University Press.

Janeway, Elizabeth. 1971. *Man's World, Woman's Place: A Study in Social Mythology*. New York: Dell Publishing.

Lynch, Edith M. 1973. *The Executive Suite - Feminine Style*. New York: Amacom.

Nellis, Muriel, ed. 1975. *Drugs, Alcohol and Women: A Source Book*. Washington, D.C.: National Research and Communications Associates.

Pogrebin, Letty Cottin. 1970. *How to Make It In A Man's World*. Garden City, New Jersey: Doubleday and Company.

THE ROLE OF FEELINGS IN MEN FOR THE TREATMENT OF WOMEN

David W. Boots and Barry Nazar

Within our society there exist two gender-based subcultures: a "world" for men, a separate "world" for women. A pervasive set of social norms maintain these subcultures through extensive socialization of our children and through institutionalizing sex-role stereotypes. The two subcultures specify different dress codes, different value systems, different aptitudes, and, most significantly for this discussion, differing ego functions, that is, the concepts and percepts of self.

Some of these differences were brought out in a series of research projects which examine "Sex Role Characteristics" as perceived by both male and female psychotherapists (Broverman et al., 1970, 1972; Fabrikant et al., 1973, 1974). All of the therapists were asked to respond to an adjective check list describing Sex Role Characteristics as applied to either male or female.

There was significant agreement between male and female therapists in describing the male as:

> "aggressive, assertive, bold, breadwinner, chivalrous, crude, independent, virile"

Male therapists added the following in their descriptions of the male:

> "achiever, animalistic, attacker, competent, intellectual, omnipotent, powerful, rational"

Female therapists disagreed with the above but added:

> "exploiter, ruthless, strong, unemotional, victor"

In describing the female, both male and female therapists agreed on the following:

"chatterer, decorative, dependent, dizzy, domestic, fearful, flighty, fragile, generous, irrational, nurturing, overemotional, passive, subordinate, temperamental, virtuous"

Male therapists added to the list:

"manipulative, perplexing"

Female therapists added:

"devoted, empathetic, gentle, kind, sentimental, slave, yielding"

These studies further evaluated the word descriptors. The words were grouped with respect to the positive and negative values placed on them by society. The male therapists rated 70% of the female words as negative, as contrasted to rating 71% of the male words as positive. Surprisingly, or perhaps it is not surprising in the face of the contemporary scene, the female therapists were very close, rating 68% of the females words as negative, and 67% of the male words as positive.

The attributes of the two subcultures, then, take on very unbalanced connoted value by the total society, and more specifically, are held in unequal esteem even by psychotherapists. The import of this inescapably comes to bear on the counselor-client relationship. Fiedler and Senior (1951) stated that the patient or client is very sensitive to the therapists' feelings about themselves as well as about the client. Stevens (1971) wrote:

> "But the patients psych out therapists too and fairly rapidly. How therapists feel about themselves, how they relate to others, what kind of behavior makes them uncomfortable or they approve of, are readily seen by the patients. Moreover, they provide extensive clues for deciphering the therapist's values and beliefs about women."

Obviously, holding that masculine traits are positive and that feminine traits are negative, will certainly pose severe problems when attempting to relate to the female client. If the therapist is male (which most therapists happen to be), the likelihood for difficulties of this nature is greater. The male therapist has a lifetime of reinforcement for masculinity behind him. His successes are typically the result of exercising characteristically male virtues; *viz.*, assertiveness, achievement, competence, independence, intellect, rationality, victory, etc. So, if the male therapist has a positive self-esteem, it is likely that (1) society has rewarded him well for playing out his masculinity adequately, and (2) he has internalized these masculine traits so that societal reinforcements for these traits apply to his own concept of self or ego. Needless to say, the female client will readily perceive this in the counselor, and a constrained dynamic will ensue based upon those perceptions. Clearly, a double bind exists for male therapists dealing with female clients.

At the Women's Training and Support Program in Pennsylvania, we provide training for both male and female counselors regarding the needs of women in treatment. For the male counselors, particular emphasis is placed on dealing with this double bind and on eliminating the unwanted dynamics that can proceed. It is also not accidental that the solution to this problem is also the resolution to a life of greater fulfillment for the participants in the program. The solution comes from understanding the concept of *Androgyny* (Singer, 1976). "Andro" meaning male, "gyny" meaning female, combined as androgyny, meaning the harmonious blend of male and female within one body. Each human being is both "male" and "female" regardless of their physical gender. That is to say, men, beneath the prestigious "macho" exterior, are also fearful, fragile, irrational, dependent, nurturing, empathetic, etc. And females, beneath their virtuous feminine exteriors, are also intellectual, competent, strong, powerful, victorious, etc. We all possess both male attributes and female attributes alike --both positive and negative of each.

The crux of the problem lies with the nature of our society more so than with any given individual. Our societal system is decidedly masculine. Our capitalistic economy spawns competitiveness, aggressiveness, technological intellectualization, achievement, exploitation, and quest for power. The masculine is positive because it turns profits and gains power. Unfortunately, the masculine nature also abhors ambiguity. Consequently, what is masculine is relegated only to males, and what is feminine is relegated only to females. As we "buy in" to success in our society, we obligatorily buy in to a strict dichotomy of maleness and femaleness as well as the "recognized" positive and negative values associated with these characteristics. We deny our truly androgynous nature in order to hold our respective places in society. We debilitate the fullness of our human heritage--and ultimately the survival of our species. For what is humankind without compassion, generosity, and sentiment, but a reptilian viper with a technological venom? In less poignant terms, there exists a need to balance competition with cooperation, to balance the male principle with the female principle.

In our society men are not permitted to express feelings. The preferred male image is still essentially the autonomous male, the independent strong achiever who can be counted on to be always in control. From early childhood the male population learns that masculinity means not depending on anyone. Their needs are bottled up and disowned. The inner pressure to deny fear leads many males to various kinds of self-destructive activities. The ring of "big boys don't cry" checks the flow of tears and the full experience of sadness. The combination of underlying competitiveness, fear of homosexuality, the need to maintain a masculine image, allows very little safe territory for personal involvement among men, other than impersonal discussions about such things as automobiles, sports, politics, and business.

Male counselors come to the Women's Training and Support Program in Pennsylvania to learn about the female addict. Their initial expectations are, that they will acquire a counselor's "bag of tricks". What they soon discover is, the way to understand women is to find harmony with their own anima/animus. The greatest resource for understanding the female client lies within themselves, *if* they are willing to transcend society's strict concepts of male and female. The transcendence is an experiential process. The experience is both joyful and painful. Joyful are the experiences of the richness of their emotionality and the intuitive powers that arise from free emotions. Painful is the experience of oppression. As women are oppressed, so is the anima of female principle in men. As men experience their own female, they also begin to feel the oppression that femininity suffers through devaluation by our culture/society.

A DESCRIPTIVE COMPARISON OF THE FAMILY OF ORIGIN OF WOMEN HEROIN USERS AND NON-USERS

Victoria J. Binion, M.A.

Institute for Social Research
University of Michigan

The family, the basic unit of human organization in most societies, perpetuates human and societal existence by performing the classic functions of socialization and biological reproduction. Because families are part of such a powerful agency in the formation of adult attitudes, familial relations are a subject of great interest. Familial relations are also critical in the development of active, productive individuals. William B. Goode (1959) found "emotional maintenance" to be important in modern family dynamics. Seldin (1972) reviewed the family of the addict and found emotional maintenance to be highlighted in the family interaction of urban populations.

The family of origin (the family one is born into) of drug addicted people is presumed to deviate from normative socialization patterns exhibited by families of non-drug users. There is a substantial body of literature to support the notion that discontinuity, disorganization and pathology exist disproportionately in the families of origin of drug abusing persons. The general conclusion is that the family of origin does not adequately equip some of its members to assume adult responsibilities thereby contributing to drug addiction. Chein, Gerard, Lee & Rosenfeld (1964), found that drug users' families have greater weaknesses than do other families. Studies by McCord (1965), Aron et al. (1976), Ellinwood et al. (1966), Chambers et al. (1968) and Wolk & Diskind (1961) all show links between family disorganization and addiction.

Chein et al. (1964), in their study of young male addicts, wanted to understand why some individuals in marginal communities became addicted and others did not. They found the critical factor to be "the degree of family emotional health, with the mother's relationships especially crucial." Other authors have also found the addicts relationship with their mothers to be of special importance. Mason (1958), in his clinical work with adolescent and young adult drug addicts in New York City, found certain factors to be repetitious, stating that "The father of the addict is usually either physically absent through death, separation, or work away from home; or he represents a shadowy background figure... The mother, on the contrary, is the "boss," and is always present -- if not in person -- exerting her influence upon the patient even when removed from him physically." To the addicts seen as patients by Mason, the mother was always the preferred and the important parent.

The mother is seen as the central character in the family of the addict. Chein et al. (1964) categorized the mothers of female adolescent drug addicts as "insecure...judgemental, rigid, authoritarian...punitive or indifferent in regard to their daughters' sexual functions and development." The mother of the drug addict is described as the domineering parent in Valliant, 1966a & 1966b; Laskowitz (1961), Frazier (1962), Fort (1954), Nyswander (1956), Johnson (1960), and Hirsch (1961). Most of the drug literature refers to mothers of male addicts and a "peculiar" mother-son relationship.

The female figure is frequently seen as dominant because the family of the drug addict is usually a one-parent or father absent household where the mother must take the lead to survive. Johnston (1968), in a study of 100 convicted female narcotic residents, found that 65% of the sample had parents who had separated during their formative years. Aron (1975) found nearly one-half of the sample (44.5%) came from homes where one or both of the biological parents were absent due to death, separation or divorce. Other studies have reported sex and race differences in rates of parental separation for drug addicts. In the Lexington sample, Ellinwood et al. (1966) found that separation of the parents of women occurred at an earlier age than that of males. Chambers, Hinesley, & Moldestad (1970) found the majority, 54.8% of the female addicts in their sample had been reared in a home which had been broken prior to age 16. Collier et al. (1966), Baer & Corrado (1974), Osmos & Laskowitz (1966), Merry (1972) and Willis (1969), also found broken homes to have an influence on the development of drug addiction.

Personal trauma, which includes incest, rape, violence, death by drug overdose and attempted suicides, are also hypothesized in the drug treatment literature to be more prevalent in families of drug addicts. Aron in his 1975 study makes a case for a greater number of occurrences of traumatic events in the families of drug addicts. Benward and Densen-Gerber (1975) suggest that mothers in families of drug addicts will rarely take action against their husbands to protect their children from incestuous advances.

Drug or alcohol use in the family of origin is also seen as a precipitating factor to later drug addiction in the child. Robins' et al. (1969) showed a correspondence between parental drug dependency (alcoholism) and sociopathy and alcoholism in the offspring. Cahalan (1970) found that high risk drug users came from families who use and value alcohol and where there is a permissive and supportive climate for other drug use. Blum & Associates in 1970 and 1972 found parents' drug use to have an effect on their children's later drug use.

Few studies have compared drug users with non-drug users. Craig and Brown (1975) looked at youthful male heroin users and non-users and found users were significantly less likely than non-users to have both parents available to them in formative years. This was the major difference between the two groups in terms of their early years. Glaser, Lander and Abbott's 1971 study of siblings in low income areas of New York City indicated that addicts were not in conflict with their parents more often nor were they more alienated from their parents than their non-addicted siblings. However, they (Glaser et al.) also found that addicts did poorly in school and had more difficulty holding jobs than their non-addicted siblings and were more involved in street life, gang activities, arrests and a hustler life style.

This present research is a more in-depth look into differences in psychosocial functioning of the families of origin of heroin users and nonusers. A better understanding is needed of the differences in family dynamics that contribute to certain people using drugs, while others from similar backgrounds do not. Most of the drug treatment literature, especially in the area of background characteristics, looks only at male addicts. This study is an attempt to bridge a serious gap in the literature by focusing on background characteristics of female heroin users and a similar comparison group. Little is known about the female drug user as compared to her non-drug using peer and until recently many public and professional assumptions have been tempered by myths and assumptions. This research focuses on psycho-social aspects of the family of origin of female heroin users and non-users. Her family structure, her relationship with her parents, her attitudes about her childhood and her attitudes about herself while growing up will be explored.

METHODS

Subjects

The addicted women ($\underline{n}$=73) were enrolled in drug treatment programs in low-income areas in Detroit, Michigan. A comparable socioeconomic sample of comparison women ($\underline{n}$=175) was recruited from a Michigan Employment Security Commission (MESC) branch office located in similar low-income inner-city communities in Detroit. The average age of the women in both the addicted and non-addicted sample was 25.0 years. Approximately 80.8% of the addicted sample was Black or Afro-American and 19.2% White or Caucasian. The non-addicted sample was 70.7% Afro-American, 26.4% White and 2.9% other.

Procedure

Questionnaires containing demographic, situational and psycho-social information were administered in personal interviews to women in drug treatment programs and comparison women in Detroit. The social history section of the questionnaire was designed to cover six general areas; living arrangements, perceptions of significant others, family interaction patterns, child rearing experiences, religious experiences, and self-perceptions as a child. Other areas covered in the questionnaire were parenting, social history, social support, sex-role, and personal and attitudinal states (e.g., self-esteem, depression) (Tucker et al., 1976).

CONCLUSION

The results of this research indicate that the drug treatment literature has grossly overstated the differences in the family dynamics of heroin users and non-users. The differences in the psycho-social milieu of the families as demonstrated in this study are subtle rather than the pathological explanations presented in prior drug literature. This research would suggest that the family structure of addicted women is more heterogeneous than past studies might show. The addicted women in this sample were just as likely as the comparison women to be reared in a two-parent household. The comparison women were more likely to be reared with only a mother, while the addicted women were more likely to have lived with a mother and a stepfather. This research further indicates that a one-parent household is not more likely to influence later drug use. Gorush and Butler (1976) found that family intactness is only an indirect measure of many of the important family variables.

The addicted and comparison women were just as likely to describe their family life as happy while growing up. The families of both groups of women were also able to provide basic economic necessities of life and did not consider themselves poor. The conditions of the childhood of the addicted women were not marked by unusual poverty. The parents of both groups had jobs that would be classified as "blue-collar." The parents of the addicted women were more likely to have only some high school education, but the majority were still able to provide a "steady income" for their families. The parents of the comparison women were more likely to have some college, but were not more likely to provide a better economic level. Somehow the parents of the addicted women have been able to compensate in the market place by building a stable economic environment for their families.

The description of their family life, the closeness of both groups to their parents and siblings, in addition to the amount of time spent in large family groupings would suggest that the women in both groups were reared in extended domestic networks. As children, both groups of women got together frequently with their families, including aunts, uncles, grandparents and cousins. Gans (1962) and Berger (1960) among others, reported that residents of working class communities were especially likely to be involved in kinship networks. Stack (1970), Meadow (1960) and Blumberg and Bell (1959) have also found that most Blacks and other urbanites have relatives living in their vicinity and that a majority interact with their relatives regularly. Feagin (1968) hypothesizes that "informal networks may provide an organized context in which many, if not most, ghetto dwellers are able to cope..." It would seem that family solidarity and the stable neighborhood surroundings played a part in the creation of the relatively stable economic and home situation of these two groups of women.

In describing their parents both the addicted and comparison women slightly favor their mothers. It isn't the neurotic, dependent relationship characterized in the drug literature (Mason, 1958; Chien et al., 1964; Wolk & Diskind, 1961). In fact, both the addicted and comparison women perceive their mothers as being helpful, loving, strict people who were good mothers and were easy to get along with. This research would suggest that the mothers of both groups of women played a central role in their childhoods. The mothers of the addicted women did not seem to reject their daughters after they became addicted. Wallace (1976) found that heroin addicted women most frequently mentioned their own mother as "the person I would miss most if they were no longer around." The mothers of the addicted women in this sample also continue to support their daughters after their addiction. More importantly, the mothers of both groups were seen as warm-acting, relaxed, supportive parents, who provided a happy home situation (see Table I).

Each group of women were just as likely to see their fathers as being helpful, loving people, but were likely to have more positive perceptions of their mothers. The impact of sex-role identification and family ideals may also be operating in these perceptions. "Wives, mothers and sisters are all focal figures in American family life" (Cumming and Schneider, 1961). Girls are also socialized to be like their mothers and consequently spend a great deal of time with them (Chodorow, 1974). The women in both groups were more likely to say that they engaged in a variety of activities with their mothers, than they did with their fathers. The father's role in the family has been underemphasized in American life and it has only been in recent years that social scientists have paid some attention to his presence (Pleck, 1975).

Table I. A Pairwise Comparison of Semantic Differential Descriptions of Mother and Father

Variable	Mean	Std. dev	t-stat
Addicted Women[a]			
Mother description	72.719	29.520	2.6517**
Father description	62.351		
Comparison Women[b]			
Mother description	73.300	23.378	2.4149*
Father description	68.529		

a n=57.
b n=140.
** p < .02.
* p < .01.

The women in both samples describe their fathers a little less positively than their mothers, but other indications of rejection by or rejections of the father are absent from this research. Instead the perceptions of these two groups of women take into account the realities of their different family constellations. The less positive description of the father only serves to highlight the central importance of the mother-daughter relationship for women. It may also suggest that the addicted and the comparison women do not know their natural fathers as well as their mothers.

Socialization techniques and the perceptions of them during childhood differed for the two groups of women. Some of the addicted women felt they were punished "much more than other children," but a similar number of them felt they were punished "not nearly as often as other children." Neither group of women felt they were punished and disciplined differently than the comparison women. The addicted women were more often made to do extra work, given a lecture on what they had done wrong, not allowed to go someplace to do something and they were screamed and yelled at more often. However, the parents of the addicted women did not physically punish them any more than the parents of the comparison women. These differences, considered with the fact that addicted women in this sample were much more likely to run away from home, indicate discord in their family of origin.

The attitudes and perceptions of their parents and family life would suggest that the behavior of the addicted women when they were adolescents might possibly explain some differences in punishment and discipline. Besides running away from home, the addicted women were more likely to "hustle" money from adults and were also more likely to go to a friend's home instead of a relative's when they did leave home. The onset of their drug use began before they left their family of origin and while they were an adolescent. The majority of the addicted women first tried heroin when they were between 16-18 years.

The addicted women were more likely than the comparison women to leave home before or around 18 years old. Drug use seems to be an important variable in the separation of the addicted woman from her family. It may also be related to the increased likelihood of the addicted women leaving school without a high school diploma, since the overwhelming majority of the addicted women also viewed themselves as "good or average" students. The new friends and environment of drug users may have also influenced the addicted women in their decision to leave high school. Although drug use was one of the critical differences in why addicted as opposed to comparison women left school, being pregnant or having children and being bored or tired of school were the most important reasons for both groups.

The results of these data may provide some support for Cloward and Ohlin's (1960) theory that blocked aspirations lead to heroin use. Apparently those who aspired to upward mobility and found their way blocked were more likely to become addicted than those who did not have such aspirations (Luckoff & Brooks, 1974 and Kleinman et al., 1975). Possibly because that extra bit of encouragement from a special teacher was lacking for the addicted women or because they began using drugs in high school, the addicted women lost interest in getting a diploma and instead turned to the adult world. Some may say the lesser educational attainments of their parents was the cause of their dropping out but very few addicted women reported that their family was a reason for their leaving school.

The peers of the addicted women seemed to have greatly influenced them during adolescence. The addicted women in this sample indicate that "being around people who use drugs, being in that environment" was one of the more important reasons why they used heroin. It is difficult to determine if the hassles with their parents preceded their drug use or if the drug usage of the addicted women in adolescence caused their discontent with their family. One of the three most important reasons the addicted women use heroin is "because of hassles with their parents." These addicted women thought highly of their parents, and probably wanted to please them but the lure of the drug life style seemed overpowering.

The differential use of alcohol in the family of the addicted and comparison women may be a factor in the hassles, disciplinary techniques, and the discontent that the addicted women expressed. The family members of the addicted women were more likely to "drink a lot." Interestingly, a large percentage of comparison women also indicated that "there had been some problems in their family because a family member drank alcohol." The atmosphere often created in the home by excessive use of alcohol may have certainly influenced the addicted women's drug use. This research also shows that even though the comparison women's family members also had drinking problems, the addicted women were significantly more likely to perceive heavy drinking as a problem.

Heroin addiction arises from a complex interaction of powerful multiple forces and the family of origin of the addicted person has been identified as a significant factor in the interplay. The role of the family in the later use of drugs cannot be diminished, but this research suggests that the notion of the multigenerational transmission of pathology in the family of heroin users is a myoptic and inaccurate view (Carr, 1975 and Distasio, 1973). Heroin users are a heterogenous group like others in society, and only subtle differences exist between the family life of female heroin users and non-users of the same socio-economic background.

REFERENCES

Aron, W.S. September 1975. Family Background and Personal Trauma Among Drug Addicts in the United States: Implications for Treatment. *The British Journal of Addiction, 70:* 295-305.

Aron, W.S. and Daily, D.W. 1976. Graduates and Splitees From Therapeutic Community Drug Treatment Programs: A Comparison. *The International Journal of the Addictions, 11 (1):* 1-18.

Baer, D.J. and Corrado, J.J. 1974. Heroin Addicts Relationships With Parents During Childhood and Early Adolescent Years. *Journal of Genetic Psychology, 124:* 99-103.

Benward, J. and Densen-Gerber, J. February 19, 1975. Incest as a Causative Factor in Antisocial Behavior: An Exploratory Study. Paper presented at the American Academy of Forensic Sciences, Chicago, Illinois.

Berger, B.M. 1960. *Working-class Suburb.* Berkeley, California: University of California Press.

Blum, R.H. & Associates. 1970. *Students and Drugs.* San Francisco, California: Jossey-Bass.

Blum, R.H. & Associates. 1972. *Horatio Alger's Chindren.* San Francisco, California: Jossey-Bass.

Blumberg, L. and Bell, R.R. 1959. Urban Migration and Kinship Ties. *Social Problems, 6:* 328-333.

Cahalan, D. 1970. *Problem Drinkers: A National Survey.* San Francisco, California: Jossey-Bass.

Carr, J. 1975. Children Addicts and the Multigenerational Transmission of Social Pathology. Family Life Problems of Drug Abusing Women. Presented at a meeting of New York City's Addiction Services Agency, September 17.

Chambers, C.D., Moffett, A. and Jones, J.P. Fall 1968. Demographic Factors Associated With Negro Opiate Addiction. *The International Journal of the Addictions, 3(2):* 329, 343.

Chambers, C.D., Hinesley, R.K. and Moldestad, M. June 1970. Narcotic Addictions in Females: A Race Comparison. *The International Journal of the Addictions, 5(2):* 257-278.

Chein, I., Gerard, D.L., Lee, R.S. and Rosenfeld, E. 1964. *The Road to H.* New York: Basic Books, Inc.

Chodorow, Nancy. Family Structure and Feminine Personality. In M.R. Rosaldo and L. Lamphere (Eds.), *Women, Culture and Society.* Stanford, California: Stanford University Press.

Clinard, M.B. 1968. *Sociology of Deviant Behavior.* New York: Rinehart.

Cloward, R. and Ohlin, L. 1960. *Delinquency and Opportunity: A Theory of Delinquent Gangs.* Glencoe, Illinois: Free Press of Glencoe.

Collier, M.V. 1972. *Daytop Villages Drug Abuse Programs: The 1972 Report.* New York: Daytop Village, Inc. Mimeographed.

Craig, S.R. and Brown, B.S. 1975. Comparison of Youthful Heroin Users and Nonusers From One Urban Community. *The International Journal of the Addictions, 10(1):* 53-64.

Cumming, E. and Schneider, D.M. Sibling Solidarity: A Property of American Kinship. In B. Farber (Ed.), *Kinship and Family Organization.*

Distasio, C.A. 1974. Multigenerational Familial Pathology in Opioid Addiction. *Proceedings of the 5th Annual National Drug Abuse Conference,* p. 515.

Ellinwood, E.H. Jr., Smith, W. and Vaillant, G.E. June 1966. Narcotic Addiction in Males and Females: A Comparison. *The International Journal of the Addictions, 1(2):* 33-45.

Feagin, J.R. 1968. The Kinship Ties of Negro Urbanites. *Social Science Quarterly, 49:* 661-665.

Feldman, H.W. 1968. Ideological Supports to Becoming and Remaining a Heroin Addict. *Journal of Health and Social Behavior, 9:*131-139.

Fort, J.P. 1957. Heroin Addiction Among Young Men. *Psychiatry, 17:* 251-259.

Frazier, T. 1962. Treating Young Drug Users: A Casework Approach. *Social Work, 7:* 94-101.

Gans, H.J. 1962. *The Urban Villages.* New York: The Free Press.

Gerard, D.L. and Kornetsky, C. 1954. Adolescent Opiate Addiction - A Case Study. *Psychiatric Quarterly, 28:* 367-380.

Glasser, D., Lander, B., and Abbott, W. Spring 1971. Opiate Addicted and Nonaddicted Siblings in a Slum Area. *Social Problems, 18(4):* 510-21.

Goode, W.J. 1959. The Sociology of the Family. In R.R. Merton, L. Broom, and L.S. Cottrell (Eds.), *Sociology Today.* New York: Basic Books.

Gorush, R.L. and Butler, M.C. 1976. Initial Drug Abuse: A Review of Predisposing Social Psychological Factors. *Psychological Bulletin, 83(1):* 120-137.

Hirsch, R. 1961. Group Therapy With Parents of Adolescent Drug Addicts. *Psychiatric Quarterly, 35:* 702-710.

Holzner, A.S. and Ding, L.K. 1973. White Dragon Pearls in Hong Kong: A Study of Young Women Drug Addicts. *The International Journal of the Addictions, 8:* 253-263.

Johnson, A., et al. 1960. The Transmission of Superego Defects in the Family. In N.W. Bell and E.F. Vogel (Eds.), *A Modern Introduction to the Family.* Glencoe, Illinois: Free Press.

Johnston, C.W. Summer 1968. A Descriptive Study of 100 Convicted Female Narcotic Residents. *Corrective Psychiatry, 14:* 230-236.

Kleinman, P. and Lukoff, I.F. July 1975. Generational Status, Ethnic Group and Friendship Networks: Antecedents of Drug Use in a

Ghetto Community. Mimeographed. New York: Center for Socio-Cultural Studies on Drug Use, Columbia University School of Social Work.

Laskowitz, D. 1965. Psychological Characteristics of the Adolescent Addict. In E. Harm (Ed.), *Drug Addiction in Youth*. London: Pergamon Press, Inc.

Lukoff, I.F. and Brook, J.S. April 1974. *A Socio-Cultural Exploration of Reported Heroin Use*. Mimeographed. Columbia University.

Mason, P. 1958. The Mother of the Addict. *Psychiatric Quarterly, 32:* 189-198.

McCord, W.M. November 21, 1965. We Ask The Wrong Questions About Crime. *New York Times Magazine*.

Meadow, K.P. 1962. Negro-White Differences Among Newcomers to a Transitional Urban Area. *Journal of Intergroup Relations, 3:* 320-330.

Merry, J. 1972. Social Characteristics of Addiction to Heroin. *British Journal of Addiction, 67(4):* 322-325.

Nyswander, M. 1956. *The Drug Addict as a Patient*. New York: Grune.

Occupational Classification System. 1970. *Classified Index of Industries and Occupations*. Census of Population, U.S. Department of Commerce, pp. x-xiv.

Osnos, R. and Laskowitz, D. 1966. A Counseling Center for Drug Addicts. *Bulletin on Narcotics, 18(4)*.

Pleck, J.H. August 1975. Work and Family Roles: From Sex-patterned Segregation to Integration. Presented at the American Sociological Association, San Francisco.

Robins, L.N. and Murphy, G.E. September 1967. Drug Use in a Normal Population of Negro Men. *American Journal of Public Health, 57 (9):* 1580-96.

Rosenberg, C.M. 1968. Young Drug Addicts: Addiction and Its Consequences. *Medical Journal of Australia, 1:* 1031-1033.

Rosenfeld, E. 1962. Teenage Addiction. In W.C. Bier (Ed.), *Problems in Addiction*. New York.

Seldin, N.E. 1972. The Family of the Addict: A Review of the Literature. *The International Journal of the Addictions, 7(1):* 97-107.

Stack, C.B. 1970. The Kindred of Viola Jackson: Residence and Family Organization of an Urban Black American Family. In N.E. Whitten Jr. and J.F. Szwed (Eds.), *Afro-American Anthropology*. New York: The Free Press.

Staples, R. February 1971. The Myth of the Black Matriarchy. *The Black Scholar*, p. 9-16.

Tucker, M.B., Colten, M.E. and Douvan, E. 1976. Report on the Development of Psycho-Social Measures Appropriate for Use With Low Income/Low Education Women. *Women's Drug Research Project Report* to National Institute of Drug Abuse. Institute for Social Research, University of Michigan.

Valliant, G.E. 1966. Parent-Child Cultural Disparity and Drug Addiction. *Journal of Nervous and Mental Disease, 15:* 599-609 (b).

Wolk, R. and Diskind, M. 1961. Personality Dynamics of Mothers and Wives of Drug Addicts. *Journal of Crime and Delinquency, 7 (2):* 148-152.

Wallace, N. August 1976. Support Networks Among Drug Addicted Women and Men. *Woman Evaluation Project Report* to National Institute of Drug Abuse.

Wiener, J.M. and Egan, J.H. 1973. Heroin Addiction in an Adolescent Population. *Journal of the American Academy of Child Psychiatry, 12 (1):* 48-58.

Willis, J.H. 1969. Drug Dependence: Some Demographic and Psychiatric Aspects in U.K. and U.S. Subjects. *The British Journal of Addiction, 64 (2):* 135-146.

MEDICAL NEEDS OF ADDICTED WOMEN AND MEN AND THE IMPLICATIONS FOR TREATMENT; FOCUS ON WOMEN*

Marcia DeCann Andersen, R.N., B.S.N., M.S.N.

National Women's Drug Research Coordinating Project

INTRODUCTION

This study was begun amid increasing concern about treatment omissions and inadequacies for women, and increasing anecdotal reports from programs that addicted women's medical needs were more extensive than those of addicted men.

One of the main methodological problems in the drug abuse-related literature addressing the issue of medical complications of heroin and methadone use has been the often extremely unequal numbers of males and females. Much of what is know about medical problems of addicts in treatment has been based on largely male samples, the only notable exception being the literature regarding the pregnant female addict. While strategies can be devised to take the occurrence of unequal group size into account, the discrepancy in numbers is usually not even mentioned (Reed and Herr, unpublished).

PURPOSE

The purpose of this study is to document in a preliminary way the nature and scope of medical problems and their treatment in heroin-addicted females.

---

* Supported in part by National Institute on Drug Abuse Grant # 5H81-DA 01496. Grant awarded to the Wayne County Department of Substance Abuse Services with sub-contract to the University of Michigan School of Social Work.

BACKGROUND

Health Problems Associated with Heroin Use

Health problems related to parenteral use of heroin are well documented (Cherubin, 1967; Louria, Hensle, and Rose, 1967; Stimmel and Kreek, 1975; White, 1973). "In New York City heroin is reported to be the single leading cause of death in young men between the ages of 15 and 30" (Stimmel et al., 1973).

Drug-related health problems have been identified in a largely male heroin-addicted population. In many cases this was because the treatment programs from which subjects were selected were composed mostly of males.

In drug-related medical studies that do specify their female sample, results are usually not reported by gender (Christakis et al., 1973; Cushman, 1973; White, 1973; Yaffe, Strelinger, and Parwatiker, 1973). This is an important omission because it does not address directly the issue of any differentiated medical risks for males and females.

However, studies have found differences between males and females on measures of illness (Roskies, Iida-Miranda, and Strobel, 1975; Uhlenhuth and Paykel, 1973; Uhlenhuth et al., 1974). Uhlenhuth and Paykel (1973) and Uhlenhuth et al. (1974) found that females respond in a manner that is higher than males on a symptom intensity scale. Roskies, Iida-Miranda, and Strobel (1975) found a strong relationship (rho=0.42, p=0.00) between sex and illness using Wyler's Seriousness of Illness Scale, with females being more "sickness prone" (p.238). Differences between males and females, if they exist in drug program populations, have implications in planning treatment interventions. These differences cannot be utilized in planning interventions if they are not reported.

Health Problems Associated with Methadone Use

There has been some controversy in the literature regarding the presence of health problems associated with methadone. Some authors report some side effects (Bloom and Butcher, 1970; Cushman, 1973; Espejo, Hogben, and Stimmel, 1973; Goldstein and Judson, 1973; Yaffe, Strelinger, and Parwatiker, 1973). Kreek et al. (1972) suggest on the other hand, that drug dependent individuals enter methadone programs with multiple health problems. They have found that the client's liver function tests were abnormal at admission. While they remained largely abnormal throughout the study, they did not appear to worsen. Kreek (1973) does report some side effects of methadone, such as excessive sweating and constipation. Like the heroin studies, these studies often had a small proportion of female subjects.

METHOD

Sampling Procedures

Ten heroin treatment programs, some therapeutic communities and some out-patient methadone clinics, in four urban areas in four regions of the United States were selected for participation in this study. They were chosen because of prior connections to other phases of the Women's Drug Research Coordinating Project, and the fact that procedures for confidentiality and entry had already been established. It should be noted that this program selection process, though an asset to the data collection phase, does limit the generalizability of the findings of the study to clients in these ten programs. However, significant trends have been identified.

An archival research design was selected utilizing a random sample of twenty client records in each of the ten programs.

Measurement of Drug and Health Variables

Drug use, while differing somewhat from region to region, proved to fit with remarkable similarity onto a chart devised to compare drug usage across programs. Though heroin, for example, was sold in packs, bags, balloons, or chunks, usage could be converted into teaspoons or McDonald's spoons (from McDonald's Restaurants). These were remarkably universal measurements of heroin across the country. While purity varied, even estimates of heroin purity could be identified from many records.

Measures of health abnormalities were measured at two points during a client's participation in a drug treatment program. The problems identified on the initial physical examination constituted the first point. The number of subsequent medical problems developing after the initial physical constituted the second. For clarification, the problems identified on the initial examination will be called initial problems. Those developed later will be called subsequent problems. The combination of the two will be referred to as total abnormalities.

System classifications for initial abnormalities were developed using systems routinely examined in physical examination (e.g., skin, eyes, ears, lymph nodes, abdomen, etc.). Therefore, system classifications for subsequent and total abnormality categories were developed based on systems commonly cited in the literature reviewed. As a classification check, it would be good to recode the illnesses using a universal coding tool such as the H-ICDA Classification Index (Commission on Professional and Hospital Activities, 1973).

RESULTS

Health Differences in Male and Female Addicts

Initial problems. Table I shows the initial medical problems of this sample. Women had more problems in 11 of 16 systems and statistically significantly more in 5 systems. The mental system was the only one in which males had statistically significantly more problems than females.

Table I also shows that 41% of the females had dental problems on admission. 33% had eye problems. Caution is needed when looking at the large number of eye problems as physicians noted miosis and mydriasis (a side effect of drug usage) on many charts. This could account for some of the eye problems and needs to be recoded and separated out on the next analysis. It can be seen that 43% of examined females had genital problems at admission. While males also had many dental and eye problems, they had no genital problems.

Table I. Initial medical problems.

	Total			Female		Male				
System of abnormality	examined N	N	%	Female N	%	Male N	%	M1	df	p
Skin	69	11	16	9	17	2	13	-		
Dental	125	53	42	35	41	18	46	-		
Ear	121	4	3	2	2	2	6	-		
Nose	117	13	11	12	15	1	3	4.0	1	.05
Eye	134	51	38	32	33	19	45	-		
Lymph nodes	76	11	15	11	20	0	0	7.4		
Breast	80	6	8	6	9	0	0	-	1	.01
Chest	126	9	7	8	9	1	3	-		
Heart	125	16	13	15	17	1	3	6.1	1	.01
Abdomen	127	17	13	14	16	3	8	-		
Spine	60	5	9	5	12	0	0	-		
Extremities	119	17	15	10	12	7	19	-		
Genital	68	18	27	18	43	0	0	21.2	1	.00
Rectal	50	3	6	2	7	1	4	-		
Neurologic	102	7	7	7	10	0	0	6.2	1	.01
Mental	61	8	13	2	6	6	23	4.0	1	.05

(Note: Percentage developed from numbers of clients examined in each category (examined N). Clients often had more than one abnormality.)

Total abnormalities. Fifty-six percent of the females had one or more gynecological problems while in drug treatment, and over one-third had dental problems.

Statistically significant differences between males and females were noted in the following systems of total abnormalities developing during treatment (initial problems plus subsequent problems) with females having had more problems in every case.

Endocrine and metabolic ($t=2.61$, $df=193$, $p<.01$; $u=3569.0$, $p<.01$)
Gynecology ($t=5.44$, $df=193$, $p<.001$; $u=2038.0$, $p<.001$)
Respiratory ($t=2.23$, $df=193$, $p<.03$; $u=3405.0$, $p<.02$)
Urinary system ($t=2.23$, $df=193$, $p<.05$; $u=3385.0$, $p<.00$)

Even excluding gynecology, three significant relationships out of sixteen possibilities are more than can be accounted for by chance alone.

Differences in treatment variables. Figure 1 shows the difference on treatment variables between males and females. The only variable initially on which males and females differed significantly was physical examination completeness ($t=2.5$, $df=166$, $p<.01$), with males having more complete examinations. This was probably due to the fact that gynecologic examinations for women were not often done on this initial exam.

During drug treatment, however, females developed statistically significantly more problems, had more treatments, and had more of the treatments performed outside the program than did males.

In a preliminary attempt to look at cost inside and outside medical interventions, the study director enlisted the aid of one of the medical staff in a sample program. Together they developed an approximate and decidedly conservative cost estimate for both inside and outside care. Inside care was estimated at $5.00/visit and outside care at $20.00/visit. Utilizing this rudimentary estimate, females' treatment were found to be significantly more expensive than males'.

Variables Related to Medical Problems in Females

Heroin. Aspects of heroin past use are related to numbers of abnormalities in females, both initially and in those developing during treatment. Table II shows heroin to be associated with urinary and endocrine/metabolic problems. Heroin is also marginally associated with dental problems and nervous system problems in females.

Legal methadone. Legal methadone past use was found to be associated with the following health problems in females: number of initial medical problems; number of subsequent medical problems; liver abnor-

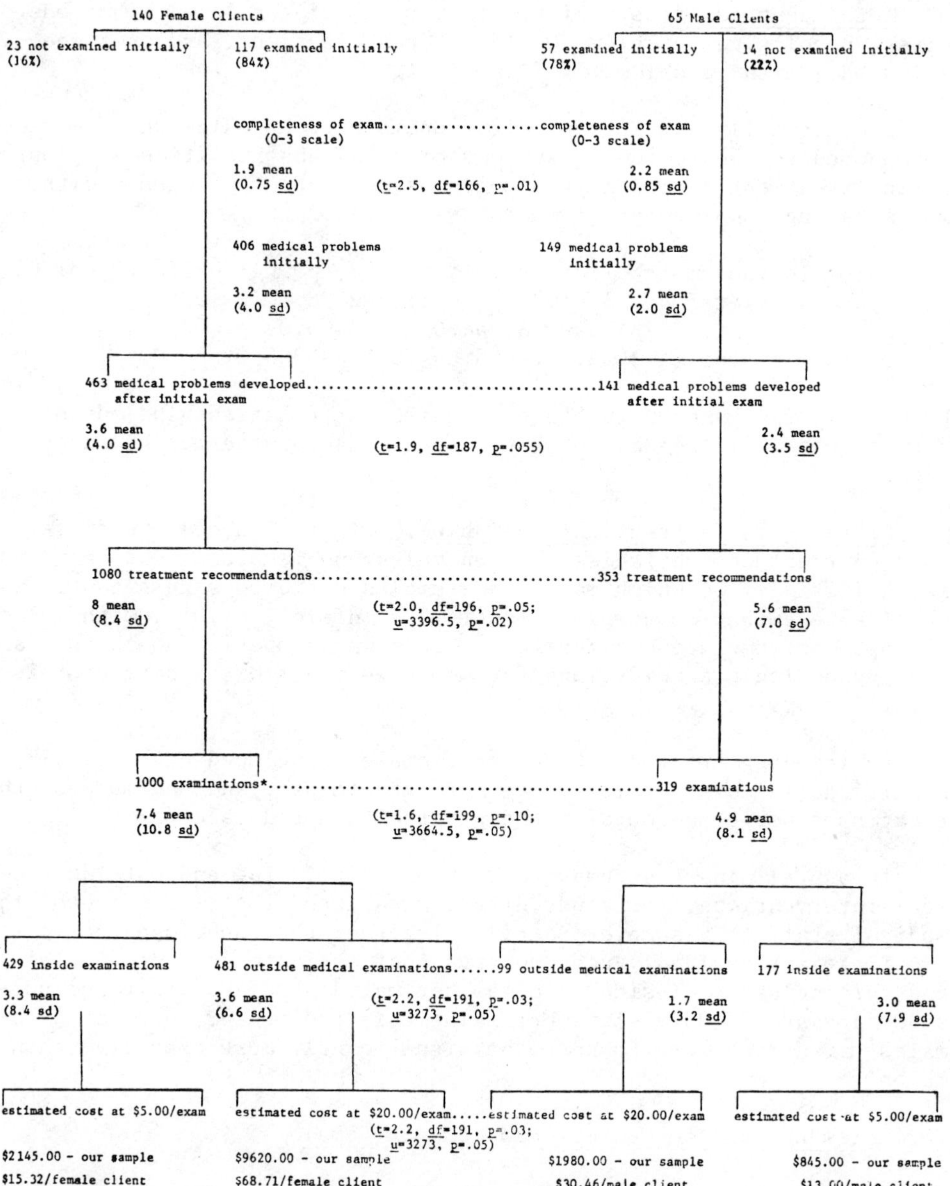

Fig. 1. Overview of medical differences between male and female addicts in treatment. (* There was some missing data with regard to whether treatments were done inside or outside the facility.)

(Note: These same difficulties hold when the clients in two all women's programs are not used in the computation (N=100 females, 65 males).

Table II. Correlations between heroin use and health abnormalities in females.

Health variables	Heroin variables			
	Heroin quality	Heroin dose	Heroin past use	Heroin cost
Number of initial abnormalities	r=.29, p=.00			
Total number of abnormalities	r=.25, p=.01	r=.33, p=.05		
Total urinary system abnormalities		r=.20, p=.05	r=.20, p=.02	
Total dental abnormalities			r=.16, p=.068	
Total nervous system abnormalities				r=.22, p=.06

malities; and dental abnormalities. Table III shows these relationsihps.

Drugs other than heroin and legal methadone. Part of the variance in a variety of health variables in females was accounted for by drug variables in different ways. Opiates other than heroin, sedatives/tranquilizers, and barbiturates are associated with ten or more of twenty possible health variables measured. Table IV shows these relationships.

Prostitution. Finally, it is important to note that 41 (29%) of the females in this sample supported at least part of their drug habit with prostitution. Table V shows that prostitution contributed to the variance in the following measures of health variables for fe-

Table III. Association between time on legal methadone and medical variables for women.

Health variables		
Number of initial medical problems	r=.20	p=.04
Number of subsequent medical problems	u=568.0	p=.01
Liver abnormalities	r=.20	p=.04
Respiratory abnormalities	r=-.19	p=.04
Dental abnormalities	r=.22	p=.02

Health Variables	Alcohol	Amphetamines	Barbiturates	Cocaine	Hallucinogens	Illegal Methadone	Marijuana	Opiates (other than heroin)	Other drugs (i.e., glue, Darvon)	Sedatives/ tranquilizers
Initial medical problems	p=27.6 p=.00	p=.31 p=.060		u=1165.5 p=.00		p=3.6 p=.05	u=1409.5 p=.05	u=1343.5 p=.05	r=.40 r^2=.16 p=.04	u=1277.0 p=.01
Subsequent medical problems								p=7.0 p=.00		u=1193.0 p=.00
Total number of medical problems		p=4.1 p=.03				p=3.4 p=.056	p=4.6 p=.01	p=6.2 p=.00		u=1090.5 p=.00
Number of treatment recommendations		p=5.8 p=.01	u=1555.5 p=.00					u=1119.5 p=.00		u=1036.5 p=.00
Number of treatments	p=3.2 p=.00	p=3.8 p=.03	u=1772.5 p=.03					p=12.7 p=.00	u=1011.0 p=.04	u=1024.5 p=.00
Systems of medical problems										
Circulatory	F=3.7 p=.00									
Dental			F=3.9 p=.00				F=3.0 p=.056	u=1279.0 p=.00		
Dermatology			F=81.9 p=.00	F=5.4 p=.01	F=4.3 p=.03	F=5.5 p=.02	F=34.2 p=.00	F=39.0 p=.00		
Ear			F=82.8 p=.00	F=42.7 p=.00			F=131.2 p=.00			F=243.5 p=.00
Endocrine and metabolic				F=2.9 p=.02						
Eye								u=1431.5 p=.01		
Gynecology	F=2.3 p=.01		F=4.3 p=.02					F=2.5 p=.060		u=1432.0 p=.02
Liver			F=4.0 p=.00	F=3.6 p=.00				F=19.2 p=.00	u=1256.6 p=.05	
Mental			F=4.6 p=.02				u=1763.5 p=.00		u=1087.0 p=.02	u=1441.5 p=.00
Muscle	F=3.4 p=.00					F=3.3 p=.061				u=1743.0 p=.05
Nervous system										u=1505.5 p=.02
Orthopedic			F=3.1 p=.00		F=25.1 p=.00	F=23.4 p=.00				
Respiratory			F=5.0 p=.01				F=2.0 p=.062	F=2.6 p=.057		u=1500.0 p=.03
Urinary system	F=2.1 p=.02								F=3.9 p=.02	

Table IV. Relationship of drugs (other than heroin) to health variables in females. (Note: The Mann Whitney u test results (u) compare drug variables (i.e., whether a client ever (or never) took the drug) with health variables. The Analysis of Variance test results (F) compare past or present frequency of drug use categories with health variables. The Pearson Correlation Coefficient test results (r) compare the association of the numbers of "other" drugs to health variables.)

Table V. Correlations between number of years of prostitution and health variables.

Health variables	Years of prostitution		
	r	r^2	$p<$
Number of initial medical problems	.59	.35	.02
Number of circulatory abnormalities	.83	.69	.001
Number of nervous system abnormalities	.80	.64	.001
Number of urinary system abnormalities	.66	.44	.001

males: number of initial medical problems; number of circulatory abnormalities; number of nervous system abnormalities; and number of urinary system abnormalities. It is interesting to note that nearly 70% of the variance in circulatory abnormalities in this female sample can be accounted for by number of years of prostitution.

Interrelationships and Clinical Implications

It is clear that many of the variables measured in this study are interrelated. These interrelationships can be utilized in clinical settings for prediction purposes. Regression equations were attempted with several variables measured in an analytical fashion. The resulting equations represent a rudimentary attempt to predict medical variables in women from drug and prostitution variables. While these equations have margins of error, they could be useful clinically as prediction tools. They are more accurate, even with the error margins, than the current prediction tool, a best guess.

Number of circulatory abnormalites=2.5 (years of prostitution)-.25
(r^2=.69, se=.44, p<.001)

Liver abnormalities=.39 (months on methadone)-.19
(r^2=.04, se=.26. p<.04)

Nervous system problems=.3 (years of prostitution)-(heroin cost/day) +.29
(r^2=.93, se=.42, p<.02)

Urinary system problems=.24 (years of prostitution)=.1 (heroin quantity)+ .1 (heroin dose)+ .13
(r^2=.59, se=.62, p<.025)

For example, looking at the third equation, a staff person could predict numbers of nervous system problems a client might have during drug treatment by knowing her years of prostitution and the cost of her heroin habit per day. This knowledge could aid in planning staffing patterns and health resource alliances.

Limitations of the Study

While the findings of this study are only generalizable to participants in the ten sample drug treatment programs, the trends uncovered in this study are significant and warrant further study.

The variance in health problems needs further analysis. For example, tests should be run to partial out the amount of variance in health problems that can be accounted for by methadone, when examining the relationship between heroin use and health.

Treatment Recommendations

Female examinations. In lieu of the high percentage of gynecological abnormalities in this population, gynecological examinations need to be made a routine part of all initial physical examinations. Either an internal capability to conduct such exams needs to be developed or close ties with outside facilities providing gynecological treatment need to be established. Mechanisms for insuring completion of "deferred" examinations must be implemented and enforced with appropriate chart notation when completed.

Direct questions about prostitution activities should also be routine on initial physical examinations. If the questioning is done in a sensitive, nonintrusive manner, accurate data may result.

Dental care. Dental abnormalities are a major problem for clients in drug treatment due to lack of available dental facilities. Most clients have very little money and most dentists do not take clients on Public Assistance Medical Insurance.

Since this is such an overwhelming problem in this group and facilities to provide dental care are so few, innovative ways to develop treatment resources need to be discovered. One such thought that occurred to this investigator would be to contact colleges of dentistry and offer clinical placements for dental students, if appropriately supervised by the college. Dental students would get great experience and drug treatment clients would get dental care. This does not exhaust the innovative possibilities for obtaining dental care.

Health liaison positions. Currently, overworked program staff individually arrange health care for clients outside the treatment program. It would prevent duplication of efforts and resource burnout if programs were funded at a level to permit hiring a part-time health liaison person. This individual would become responsible for arranging, planning, transporting and following up all outside medical appointments.

CONCLUSION

This report confirms the prevailing notion that female addicts in treatment have more medical problems than male addicts in treatment. Health needs must be met along with the psychosocial needs not being addressed in drug treatment programs. If medical needs are not met, the resultant stress often leads back to drug use as a means of coping with discomfort and frustration. The heartbreak felt by one female client in our sample who left a therapeutic community before her treatment was complete is described in a note she left behind: "I'm in a lot of pain and will be ill for a long time." This woman had been to see physicians twenty-six times during her four month stay in the drug treatment program. She no longer felt able to cope with this physiological stress without drugs. She also felt uncomfortable creating the obvious financial and energy burden she put on the staff providing her with transportation and medical care. In a closely related study, this investigator found that there are many clients like this one who return to their former drug patterns when the stress of unsolved medical problems becomes too great (Andersen, unpublished 1978).

REFERENCES

Andersen, M.D. 1978. The relationship between adjustment to life change and illness in two drug treatment modalities. Unpublished dissertation, University of Michigan.

Bloom, W.A. and Butcher, T. 1970. Methadone side effects and related symptoms in 200 methadone maintained patients. *3rd National Conference on Methadone Treatment Proceedings*, 44-47.

Cherubin, C.E. 1967. The medical dequalae of narcotic addiction. *Annls. Int. Med.* 67: 23-33.

Christakis, G., Stimmel, B., Rabin, J., Fontanaes, R., and Archer, M. 1973. Nutritional status of heroin users enrolled in methadone maintenance. *5th National Conference on Methadone Treatment Proceedings*, 494-500.

Commission on Professional and Hospital Activities 1973. *H-ICDA*. Vol. I & II. Ann Arbor, Michigan: CPHA.

Cushman, P. 1973. Significance of hypermacroglobulinemia in methadone maintained and other narcotic addicts. *5th National Conference on Methadone Treatment Proceedings*, 515-522.

Espejo, R., Hogben, G., and Stimmel, B. 1973. Sexual performance of men on methadone maintenance. *5th National Conference on Methadone Treatment Proceedings*, 490-493.

Goldstein, A. and Judson, B.A. 1973. Efficacy and side effects of three widely different methadone doses. *5th National Conference on Methadone Treatment Proceedings*, 21-44.

Kreek, M.J. 1973a. Medical safety and side effects of methadone in tolerant individuals. *JAMA*. 223:b: 665-668.

Kreek, M.J., Dodes, L., Kane, S., Knobler, J., and Martin, R. 1972. Long-term methadone maintenance therapy: effects on liver function. *Annls. Int. Med.* 77: 598-602.

Louria, D.B., Hensle, T., and Rose, J. 1967. The major medical complications of heroin addiction. *Annls. Int. Med.* 67: 1-22.

Reed, B.G. and Herr, J.N. Deviant social careers and heroin addicted women: Issues in treatment. Women's Drug Research Coordinating Project, School of Social Work, University of Michigan (unpublished).

Roskies, E., Iida-Miranda, M.L., and Strobel, M.G. 1975. The applicability of the life events approach to the problems of immigration. *J. Psychosom. Res.* 19: 235-240.

Stimmel, B. and Kreek, M.J. 1975. Pharmacologic actions of heroin. In B. Stimmel, *Heroin Dependency, Medical, Economic and Social Aspects*. New York: Stratton Intercontinental Medical Book Corp.

Stimmel, B., Lipski, J., Swartz, M., and Donoso, E. 1973. Electrocardiographic changes in heroin, methadone, and multiple drug abuse: A postulated mechanism of sudden death in narcotic addicts. *5th National Conference on Methadone Treatment Proceedings*, 706-710.

Uhlenhuth, E.H., Lipman, R.S., Balter, M.B., and Stern, M. 1974. Symptom intensity and life stress in the city. *Arch. Gen. Psych.* 31: 759-763.

Uhlenhuth, E.H. and Paykel, E.S. 1973. Symptom intensity and life events. *Arch. Gen. Psych.* 28: 473-477.

White, G. 1973. Medical disorders in drug addicts: 200 consecutive admissions. *JAMA*. 223: 1469-1471.

Yaffe, G.J., Strelinger, R.W., and Parwatiker, S. 1973. Physical symptom compliants of patients on methadone maintenance. *5th National Conference on Methadone Treatment Proceedings*, 507-514.

A NEW SOCIOCULTURAL-CLINICAL APPROACH TO THE USE OF NARCOTIC ANTAGONISTS IN REHABILITATION OF OPIATE DEPENDENT PERSONS

Arnold Schecter, M.D.

New Jersey Medical School

In the 1960's two chemotherapeutic approaches to opiate addict rehabilitation were introduced. Opiate maintenance clinics were once again tried--this time using the orally effective long acting synthetic opiate, methadone (Dole and Nyswander, 1965). At almost the same time Martin and colleagues at the Addiction Research Center, Lexington, Kentucky suggested the use of the synthetic opiate antagonist cyclazocine, an opiate blocking agent, as a useful chemotherapeutic adjunct in rehabilitation. It too was orally effective, potent and long acting. As a slightly agonistic antagonist, it caused dysphoric reactions during induction in some patients (Martin et al., 1965 and 1966; Martin, 1968). Methadone maintenance became enormously popular at this time and the antagonists were relegated to the role of a limited use research instrument.

In the early 70's Martin and colleagues reported a new pure semisynthetic opiate antagonist EN 1639A now known as naltrexone (Martin et al., 1971, 1973 and 1973). By adding the cyclopropylmethyl group of cyclazocine to the naloxone molecule an orally effective long acting almost completely pure opiate antagonist was prepared and found useful in man. It blocked the effects of opiates, had few agonistic or other side effects and offered hope for those seeking a "pure" antagonist. Schecter and also Resnick, working independently, introduced naltrexone into clinical use (Schecter et al., 1974, 1974, 1974 and 1975; Resnick et al., 1974 and 1974). Both found the antagonist to have few side effects, but noted that average patient retention was short, usually under one month. A second series of clinical trials established the same findings in other settings (Lewis, 1975; Curran and Savage, 1976; Hurzeler et al., 1976; Haas et al., 1976; Taintor et al., 1975; Curran et al., 1976).

It is a tribute to the impressiveness of methadone that most workers became discouraged by such short retention periods. Kissin in Brooklyn (personal communication) and Crawford in St. Louis (Parwatikar et al., 1975) suggested a slight agonistic or opiate like antagonist such as cyclazocine or oxilorphan might give an addict slight euphoria--in a similar manner to methadone and should be used instead It was not noted by many clinicians that most programs rejected few if any applicants for antagonist therapy, that therapeutic communities, which are considered by most authors to have at least some clinical efficacy, also lose most patients during the first month of treatment, and that Sells methadone retention figures nationwide indicated a 50% loss of patients during the first year in treatment. It did not occur to some observers that an opiate maintenance program such as methadone maintenance, was for many patients similar to the "street" drug addiction scene and ought, therefore, initially retain clients longer than the antagonist therapy, which was but one step removed from total opiate abstinence, one of the final goals of treatment.

At about this time Goldstein and others suggested a stepwise treatment progression to be repeated as needed (Goldstein, 1975 and 1976). This envisioned heroin to methadone to detoxification followed by narcotic antagonist treatment and finally on to abstinence without the aid of an antagonist. It was thought by many in the field that such a step ladder progression, with entry wherever appropriate and repetition as often as needed made more clinical sense than rigid long term maintenance with either methadone or narcotic antagonists.

Despite such enlightened insights into potentially fruitful approaches to the clinical application of narcotic antagonists with opiate addiction rehabilitation programs an attempt was made to force the use of naltrexone into a prospective double blind placebo-naltrexone random assignment clinical trial. The Committee to Evaluate Narcotic Antagonists of the National Research Council of the National Academy of Sciences was forced to accede to the desires of the Food and Drug Administration and ran one such cooperative study (Hollister et al., 1977). The results were disappointing to some of those concerned with the study. *Safety* for short term use was again confirmed: *efficacy* in blocking the effects of opiates was already known. Clinical efficacy as demonstrated partially by long term retention, like the methadone model was, of course, not seen. Those hoping for such results were again disappointed with the so called limited usefulness of the antagonists, especially as compared with methadone.

At this time, in April of 1978, with the phase III Naltrexone clinical trials hopefully soon to begin, it seems appropriate to review what we know of the natural history of opiate dependence in the United States, the efficacy of methadone in the rehabilitation of opiate addicts and consider reasonable expectations for the antagonists in rehabilitation therapy.

In 1962 Winick first described the "maturing out" of opiate addiction seen in many addicts as they grow older (Winick, 1962). Robbins and Murphy, in a survey of drug use in black use St. Louis, also noted this tendency (Robbins and Murphy, 1967). Nurco, Lukoff and O'Donnell in three widely different studies all independently confirmed this tendency of most heavy opiate users to spontaneously--that is, with or without treatment, stop opiate use as the user grows older (Nurco et al., 1975; Kleinman and Lukoff, 1978; O'Donnell et al., 1976).

Robbins, in her careful, elegant post Viet Nam drug abuse studies, carefully destroyed the myth that opiate addiction need be a lifelong condition (Robbins, 1975). She found the overwhelming majority of heavy and regular Viet Nam heroin users promptly and spontaneously reverted to an opiate free state when returned to the United States. Zinberg (Zinberg et al., 1977) documented the many varieties of heroin users and their change in usage patterns. These studies of the natural history of opiate use and dependence in the United States at this time overwhelmingly point to the condition as being episodic, relapsing, but also self limited for most users (O'Donnell, et al., 1976).

When methadone maintenance therapy was introduced in the 1960's the now current concept of prospective double blind random assignment clinical trials with crossover medication-placebo design wherever possible was not a commonly used method of demonstrating clincial efficacy. Instead a "before-after" study design was used (Campbell and Stanley, 1963). Dole and Nyswander (1965), and also Gearing (1970 and 1974), compared certain social outcome measurements before and during methadone therapy. These included illicit drug (opiate) use, illegal acts, and employment, homemaker or student status while the same patient was stabilized on the legal opiate--methadone, as contrasted with his or her own behavior at a younger age while on the illegal opiate--heroin.

Frequently, the year preceding admission, which may have been a crisis year leading to the admission, was the "before" time period. Patients were highly selected for admission--most patients were not accepted. Those accepted were older, better educated, had no known alcohol problems, no psychiatric problems, non minority patients to a disproportionate extent, and were inpatients for many months before becoming outpatients. In addition, there were few options in those early days--prison was a frequent alternative to methadone maintenance and, therefore, motivation was high in the first methadone patients. Statistical analysis was frequently performed employing as denominator only those patients remaining in treatment rather than all patients entering therapy. Lukoff, in a thorough study of a methadone maintenance treatment program located in a black ghetto and which had markedly relaxed admission standards compared to the original programs, was quite pessimistic with regard to the therapeutic efficacy of methadone in that treatment program (Lukoff and Quatrone, 1973). He felt that careful patient selection, the maturing out with age of many sociopathic traits, increased crimial activities immediately prior to program entry and other methodologic considerations biased the earlier optimistic reports.

By contrast, in the LAAM clinical trials, LAAM was compared with methadone for efficacy. No matched untreated addicts or double blind LAAM--placebo trials were ever envisioned or conducted. To date no studies have compared methadone with untreated or placebo treated randomly assigned persons as controls. This leaves us with methadone maintenance therapy, which is considered the primary chemotherapeutic approach to opiate addiction at this time, with something less than certainty regarding its therapeutic efficacy.

The above review is not meant to belittle the efficacy and importance of methadone therapy but to put narcotic antagonists, the other chemotherapeutic approach to opiate dependence rehabilitation, into perspective. Whatever the reason, we know that some patients become rehabilitated during the course of methadone rehabilitation. We also know that certain patients seem to do quite well while on narcotic antagonist therapy. Last, we know that the natural history of opiate dependence at this time indicates that for most of those addicted, there will be a maturing out with time--if the person can be kept alive. It is important to remember that Sells describes a 1-2% yearly mortality for addicts in treatment in his large sample (Sells et al., 1976).

Last, it seems extremely unrealistic to confuse treatment utilizing a medication with the 'cure' of certain diseases. Insulin does not cure diabetes. Penicillin alone, in the absence of minimal rest and nutritional status would not be expected to cure a streptococcal throat infection. Digitalis may be necessary for the treatment of a patient after a myocarial infraction, but it alone is not sufficient. Duretics and hypertensives alone will not "cure" hypertension; diet, no cigarette smoking, exercise, and a low fat and cholesterol diet are also indicated. The ingestion of tranquilizers or mood elevators alone is usually not considered appropriate treatment for certain of the mental illnesses.

We know that narcotic antagonists effectively block the effects of opiates. Further tests of therapeutic efficacy along a methadone model do not seem indicated. Instead, other directions seem to warrant further study.

The type of programs in which narcotic antagonists will be most useful need to be studied. Programs now vary from behavior modification clinics (Project HALT in Oxnard, California, Haas et al., 1976) to very low intervention clinics which are similar to methadone clinics.

Since it seems that most addicts will spontaneously mature out with time after multiple drug use episodes and several relapses, it seems appropriate to return to clinics utilizing several modalities of treatment at a given site. Only in this way can continuity of care between patients and therapists be assured. The appropriate therapy at a given time for any given patient may be methadone maintenance or a therapeutic community--or a narcotic antagonist--or a day care, or vocational or legal counseling service at different points in time. Only by offering multimodality service at a given clinic can these changing needs be met.

REFERENCES

Campbell, D.T. and Stanley, J.C. 1963. *Experimental and Quasi-experimental Designs for Research.* Rand McNally, Chicago.

Curran, S. and Savage, C. 1976. Patient response to naltrexone: issues of acceptance treatment effects and frequency of administration, in *Narcotic Antagonists*: Naltrexone Progress Report, Research Monograph Series No. 9, Julius, D. and Renault, P. (eds.). National Institute on Drug Abuse, Department of Health, Education and Welfare, (ADM) 76-387, Rockville, Md.

Dole, V.P. and Nyswander, M.A. 1965. A medical treatment of diacetylmorphine (heroin) addiction. *J.A.M.A.*, 193, 646.

Gearing, F.R. 1970. Evaluation of methadone maintenance treatment programs. *Int. J. Addict.* 5(3): 57.

Gearing, F.R. 1974. Methadone maintenance treatment five years later --Where are they now? *J. Am. Public Health Assoc.* Supplement 64, 44.

Goldstein, A. 1975. On the role of chemotherapy in the treatment of heroin addiction. *Am. J. Drug Alcohol Abuse.* 2, 279.

Goldstein, A. 1976. Heroin addiction, sequential treatment employing pharmacologic supports. *Arch. Gen. Psychiatry.* 33, 353.

Haas, N., Ling, W., Holmes, E., Blakis, M., and Litaker, M. 1976. Naltrexone in methadone maintenance patients electing to become "drug free", in *Narcotic Antagonists*: Naltrexone Progress Report, Research Monograph Series No. 9, Julius, D. and Renault, P. (eds.), National Institute on Drug Abuse, Department of Health, Education and Welfare (ADM) 76-387, Rockville, Md.

Hollister, L.E. et al. 1977. Problems of Drug Dependence, Proceedings of the 39th Scientific Meeting, Committee on Problems of Drug Dependence, Clinical Evaluation of Naltrexone Treatment of Opiate Dependent Individuals, Report of the National Research Council Committee on clinical Evaluation of Narcotic Antagonists.

Hurzeler, M., Gerwitz, M.S., and Kleber, H. 1976. Varying clinical contexts for administering naltrexone, in *Narcotic Antagonists*: Naltrexone Progress Report, Research Monograph Series No. 9, Julius, D. and Renault, P. (eds.), National Institute on Drug Abuse, Department of health, Education and Welfare, (ADM) 76-387, Rockville, Md.

Kleinman, P.H. and Lukoff, I.E., Ethnic differences in factors related to drug use. *J. Health Soc. Behav.*

Lewis, D.C. 1975. The clinical usefulness of narcotic antagonists: preliminary findings of the use of naltrexone. *Am. J. Drug Alcohol Abuse.* 2(3,4): 403.

Lukoff, I.F. and Quatrone, D. 1973. A two year follow-up of the Addiction Research and Treatment Corporation, in *Heroin Use and Crime in a Methadone Maintenance Program.* Hayim, G., Lukoff, I.F., and Quatrone, D. (eds.). L.E.A.A., U.S. Dept. of Justice, Washington, D.C.

Martin, W.R., Gorodetzky, C.W., and McClane, T.K. 1965. A proposed method for ambulatory treatment of narcotic addicts using a long-acting, orally effective narcotic antagonist cyclazocine: an experimental study, in *Committee on Problems of Drug Dependence.* National Research Council, Washington, D.C.

Martin, W.R., Gorodetzky, C.W., and McClane, T.K. 1966. An experimental study in the treatment of narcotic addicts with cyclazocine. *Clin. Pharmacol. Ther.* 7, 455.

Martin, W.R. 1968. The basis and possible utility of the use of opioid antagonists in the ambulatory treatment of the addict, in *The Addictive States*, Wikler, A. (ed.). Williams and Wilkens.

Martin, W.R., Jasinski, D.R. and Mansky, P.A. 1971. Characteristics of the Blocking Effects of EN-1639A (N-Cyclopropylmethyl-7, 8-dihydro-14-hydroxnormorphinone HCL), presented to the Committee on Problems of Drug Dependence, National Research Council, Toronto.

Martin, W.R. and Jasinski, D.R. 1973. Characteristics of EN- 1639A, *Clin. Pharmacol. Ther.* 14, 142.

Martin, W.R., Jasinski, D.R., and Mansky, P.A. 1973. Naltrexone, an antagonist for the treatment of heroin dependence effects in man. *Arch. Gen. Psychiatry*. 28, 784.

Nurco, D.N., Bonito, A.J., Learner, N., and Baltes, M.D. 1975. Studying addicts over time: methodology and preliminary findings. *Am. J. Drug Alcohol Abuse*. 2(2): 183.

O'Donnell, J.A., Voss, H.L., Clayton, R.R., Slatin, G.T., and Room, R.G.W. 1976. Young Men and Drugs: A Nationwide Survey, Research Monograph Series No. 5, National Institute on Drug Abuse, Department of Health, Education and Welfare, (ADM) 76-311, National Technical Information Service, Springfield, Va.

Parwatikar, J., Crawford, J., and Unverdi, C. 1975. Antagonist study in St. Louis. *Am. J. Drug Alcohol Abuse*. 2(3,4): 379.

Resnick, R., Volavka, J., and Freedman, A.M. 1974. Short-term effects of naltrexone: a progress report, in *Proc. Committee on Problems of Drug Dependence*. National Academy of Sciences, National Research Council, Washington, D.C.

Resnick, R., Volavka, J., Freedman, A.M., and Thomas, M. 1974. Studies of EN-1639A (naltrexone): a new narcotic antagonist. *Am. J. Psychiatry*. 131, 646.

Robbins, L.N. and Murphy, G.E. 1967. Drug use in a normal population of young Negro men. *Am. J. Public Health*. 57(9): 1580.

Robbins, L.N. 1975. *Veterans' Drug Use Three Years After Vietnam*. Department of Psychiatry, Washington University School of Medicine, (mimeograph)

Schecter, A.J. and Grossman, D.J. 1974. Naltrexone in a clinical setting: preliminary observations. *Proc. Committee on Problems of Drug Dependence*. National Academy of Sciences--National Research Council, Washington, D.C.

Schecter, A.J. and Grossman, B.A. 1974. Experience with Naltrexone: a suggested role in drug abuse treatment programs, in *Developments in the Field of Drug Abuse*, Senay, E., Shorty, V., and Alksne, H. (eds.). Schenkman Publishing, Cambridge, Mass.

Schecter, A., Friedman, J., and Grossman, D. 1974. Clinical use of naltrexone (EN-1639A): efficacy and safety in pilot studies. *Am. J. Drug Alcohol Abuse*. 1(92): 253.

Schecter, A. 1975. Clinical use of naltrexone (EN-1639A). II. Experience with the first 50 patients in a New York City treatment clinic. *Am. J. Drug Alcohol Abuse*. 2: 433.

Sells, S.B., Simpson, D.D., Joe, G.W., Demaree, R.G., Savage, L.J., and Lloyd, M.R. 1976. A national follow-up study to evaluate the effectiveness of drug abuse treatment: a report on Cohort 1 of the DARP five years later. *Am. J. Drug Alcohol Abuse.* 4: 545.

Taintor, Z., Landsberg, R., Wicks, N., Plumb, M., D'Amanda, C. and Greenwood, J. 1975. Experiences with naltrexone in Buffalo. *Am. J. Drug Alcohol Abuse.* 2: 391.

Winick, C. 1962. Maturing out of narcotic addiction. *Bull. Nar.* 14(1): 1.

Zinberg, N.E., Harding, W.M., Stelmack, S.M., and Marblestone, R.A. 1977. Patterns of Heroin Use. *Advances in Clinical Pharmacology of Drug Dependence*. New York Academy of Sciences Conf., Washington, D.C.

A SUMMARY OF THE PHASE II CLINICAL STUDIES OF THE NARCOTIC ANTAGONIST, NALTREXONE

H. Alex Bradford, M.S. and Samuel C. Kaim, M.D.
Biometric Research Institute

Walter Ling, M.D.
Veterans Hospital, Sepulveda, California

INTRODUCTION

In an effort to develop a pharmacologic alternative to methadone in the treatment of opiate dependent individuals, the Special Action Office for Drug Abuse Prevention (SAODAP) contracted, in June of 1973, with the Division of Medical Sciences at the National Academy of Sciences National Research Council (NAS/NRC) to select one of several narcotic antagonists for use in a NAS-administered research study. Among the various narcotic antagonists available at the time were cyclazocine, naloxone, and naltrexone. The Committee on Clinical Evaluation of Narcotic Antagonists (CENA), formed by the NAS/NRC, selected naltrexone as the "best" among the various contenders because of its effectiveness as an antagonist, its extended span of action, its expected margin of safety, and its relative lack of agonist activity (Renault, 1976).

One of the responsibilities of the newly formed National Institute on Drug Abuse (NIDA), which was to supersede SAODAP, involved the funding and monitoring of various research programs evaluating naltrexone from several different perspectives. NIDA in conjunction with the Food and Drug Administration (FDA) determined that the collection of a common set of data from these research projects would be very useful if it were decided that naltrexone had a place in the treatment of opiate addiction. Subsequently, Biometric Research Institute, Inc. (BRI) was contracted to collect and analyze the data from the NAS-administered study (NAS Study) and selected NIDA-funded studies (NIDA Study). The following sections present relevant study characteristics and summarize the results of the analyses and subsequent interpretation of these results.

NAS STUDY

Introduction

In this placebo-controlled double-blind parallel group study, 192 patients from the five participating clinics were randomly assigned to receive weekly dosages of 350 mg. of either naltrexone or matching placebo for a period of up to 40 weeks (Bradford and Kaim, 1977). The objectives of the study, as defined by the CENA Committee, were to 1) assess the feasibility of this type study design in an addict population; 2) evaluate patient acceptance of narcotic antagonist therapy; 3) assess the safety and toxicity of naltrexone; 4) establish methods for the evaluation of naltrexone's efficacy; and 5) provide a preliminary assessment of naltrexone's efficacy in this clinical setting.

To ensure proper perspective when considering study findings, the following study characteristic are relevant.

Addict classification. Three broadly defined "addict" groups were included in the study.

Street addicts - Established users recently detoxified from opiate use (one clinic).

Post-addicts - Former addicts currently drug-free following incarceration or participation in a drug-free therapeutic program.

Methadone-Maintenance Addicts - Former addicts currently enrolled in a methadone maintenance program for a period of at least six months.

Patient eligibility. Participation was limited to opiate-dependent males, 18 years and older. A documented diagnosis of opiate dependency was required. Documentation included 1) a verified history of past or present opiate dependence plus either 2) the presence of opiates in urine, 3) signs or symptoms of opiate withdrawal, or 4) evidence of current participation in a methadone maintenance program for at least six months. Patients having serious and/or chronic physical or psychiatric problems were excluded.

Detoxification and study medication induction schedules. The study consisted of a Baseline Period from two to 70 days in duration, a Study Medication Period which could last up to 40 weeks, and a six-month Study Follow-up Period. The detoxification and induction schedules for the methadone maintenance patients are of particular interest as this group exhibited such low proportions of eligible patients entering the Study Medication Period (26% vs. 50% for all other other patients) and of patients completing the first two weeks of study medication (58% vs. 79% for all other patients). Two detoxification schedules were recommended in the protocol, and all methadone maintenance patients had to be able to achieve a stabilization dose of 50 mg./day or less in order to meet eligibility requirements. The first schedule was suggested for subjects who had been in a methadone maintenance program for at least two years, and called for a gradual reduction over an eight-week period. The second was suggested for subjects who had been in a methadone maintenance program for between six and 24 months and called for a reduction over a four-week period. Conversely, the one-week study medication induction schedule for methadone maintenance patients indicated a much more rapid increase (150 mg. on the fifth day) than either of the other two schedules, which were 70 mg. on the fifth day for street addicts and 50 mg. on the fifth day for post-addicts.

Study Findings

Study feasibility. Although there are several difficulties in carrying out a double-blind placebo-controlled study in an addict population, such studies are feasible. If future studies of this nature are to be undertaken, particular attention must be given to the manner in which the study is introduced to potential participants and the amount of support patients receive throughout the study. Furthermore, the investigator must anticipate less patient interest in study participation and a much greater dropout rate than would be expected in other patient populations. These problems can be lessened to some extent by selecting only suitable patients, providing the patient with special attention and care, allowing for a protracted patient recruitment phase and ensuring that sufficient numbers of patients are started in the study.

Patient acceptance of narcotic antagonist therapy. In reviewing the reasons for failure to start study medication among the 589 eligible patient candidates, 118 patients indicated that they were uninterested in a prolonged treatment program or felt that they could make it on their own, and 13 patients refused to sign the Informed Consent. Using these patient groups, a conservative estimate would then be that 131 patients among the 589 eligible patients (22.2%) were not interested in this type of program. In evaluating patient commitment to the program it is important to note that only 18 of the 192 patients starting (9.4%) successfully completed the program. This ratio is certainly not impressive, but patient completion figures for other long-term programs are not a great deal better (National Research Council Committee on Clinical Evaluation of Narcotic Antagonists, 1978). Of the three addict groups participating in the study, the post-addict group exhibited the best study retention rate. It is possible, however, that the experience of methadone maintenance patients could be improved if the detoxification schedule were more or less rapid than those used in this study, and that the especially high patient dropout rate during the first two weeks of study medication might be lessened if the dosage induction schedules were modified for this group. Additionally, only one of the five clinics had street addict patients participating in this study. Therefore, additional research and review of existing data from other studies is warranted before it is determined that naltrexone therapy is inappropriate for treating street addicts or methadone maintenance patients. In summary, it is apparent that naltrexone therapy for the treatment of opiate-dependence should not be considered as the treatment of choice for the majority of opiate-dependent patients, but that it does have its place and seems to be best suited for patients who are well-motivated and not too heavily addicted, or are in the post-addict phase and anxious not to relapse.

Safety. One of the primary goals of this study was to make a preliminary evaluation of naltrexone's safety. A thorough review of the data and the many statistical analyses of the data revealed that naltrexone appears to have a wide margin of safety. As the safety data, the collection points, and subject retention for both studies were so similar, each study was analyzed separately and these results were then reviewed together. A summary of the results of all three evaluations is included in the NIDA Study safety section.

Efficacy. Of a total of five categories of information related to the efficacy of naltrexone treatment, only two--detected drug use and subject retention--were considered to provide objective measurement. These two, singly and in combination, were selected by the CENA Committee as the primary yardsticks in the evaluation of naltrexone's efficacy as an adjunct in the treatment of opiate dependence. The other three--psychosocial change, global evaluations and heroin craving--were chosen by CENA as secondary sources in the evaluation of naltrexone's efficacy. However, at least tentative findings were made based on all five categories, as follows.

1) Detected drug use and study subject retention. Because of the study's anticipated sample size and the fact that well formulated combinations of these two variables would be more meaningful than either measure alone, it was decided to use them in combination as the primary measure of naltrexone's efficacy.

Combination of these variables using four different approaches afforded consistent indications that subjects in the naltrexone groups, in contrast to placebo medicated subjects, presented less detected use of morphine and/or methadone while remaining in the study for somewhat longer periods. Thus these combination scores all indicate that antagonist treatment can contribute significantly as an adjunct in a comprehensive opiate addiction treatment program.

Analyses of each variable alone revealed that naltrexone subjects used the other illicit drugs for which tests were run--barbiturates and amphetamines--to no greater extent than did the placebo subjects.

It was also shown that, following a first instance of detected opiate use, there was significantly less subsequent detected opiate use among the naltrexone subjects than among the placebo subjects. Study subject retention was also somewhat higher in the naltrexone groups than in the placebo group.

2) Social/psychological change and post-study medication period global evaluation. No meaningful differences between medication groups were observed with respect to these categories. Both groups showed slight improvement with respect to certain items in both categories. These instruments appear to be relatively insensitive to whatever subtle changes they were intended to measure in this study.

3) Changes in craving for heroin. Analyses of the data resulting from the use of a new experimental scale, the Heroin Craving Scale, revealed that naltrexone subjects were reporting significantly greater decreases in their craving for heroin than placebo subjects. (It is interesting to speculate whether these findings are a result of some intrinsic property of naltrexone, or due to some related factor, e.g., is craving merely lessened because the addict realizes that a "fix" would not give him the desired effect? In any case, future studies should certainly employ this scale again and, if the findings of this study are replicated, research on interpreting scale response is warranted.)

NIDA STUDY

Subject Characteristics and Retention

At the termination of data collection a total of 1005 subjects had been started on naltrexone medication with the participation of 903 of these subjects terminated prior to this date (Bradford and Kaim, 1977).

Although study plans called for admission of male subjects only, exceptions were made in the cases of ten females. Of subjects for whom ages were recorded, 775 (79.2%) were age 30 or younger, and 428 (43.7%) were 25 or younger. Some variation among clinics was evident with respect to subjects' race distribution; while 389 (39.6%) of the total study population were black more than three-quarters of the subjects at three clinics were black. Following are the subjects' age and race distribution. (See Table I.)

Most recent Study Medication Period participation of 46 subjects was terminated reportedly for medical reasons (study medication for an additional seven subjects was discontinued for medical reasons during a prior Study Medication Period). The reason given most frequently for the most recent termination (386 subjects) was "prolonged periods of absence." Other terminations included 138 subjects who

Table I.

Subject Frequency Distribution of Population Group

Sex	White	Black	Mexican-American	Puerto Rican	Other	Not recorded	Total
Males	406	383	153	29	1	23	995
Females	4	6	0	0	0	0	10

Subject Frequency Distribution of Age

Sex	≤21	21-25	26-30	31-35	36-40	41-45	>45	Not recorded	Total
Males	49	373	346	117	46	18	20	26	995
Females	0	6	1	1	0	1	1	0	10

"refused to continue study medication," 102 subjects because of the study's completion, and 61 subjects who became "readdicted." "Other" reasons were reported for the most recent termination of study participation of an additional 270 subjects.

Of the 1005 study subjects, 820 subjects had one Study Medication Period, 145 subjects had two, 35 had three, and five had four. As might be expected, the overall duration (counting only time on medication) of the 185 subjects receiving more than one course of naltrexone treatment were, on the average, longer than the durations of the 820 subjects receiving only one course of naltrexone medication. Of interest is the finding that significantly greater percentages of the 185-subject group compared to the 820-subject group with only one course of study medication were active participants at each time point considered over the first four months of their first Study Medication Period.

Table II summarizes the duration of naltrexone medication for all subjects.

Safety

As this study was without a randomly-assigned placebo-control group, it was planned to take advantage of the opportunity to use the results of the double-blind, placebo-controlled NAS study as a basic reference in evaluation of the present data. In order to make the two sets of data commensurable, therefore, the present data forms and collection schedules were patterned after those employed in the NAS trial. The feasibility of between-study comparisons fa-

Table II. Duration of study medication.

Time on Study Medication	Number of Subjects	Percent of Subject Total
At least two weeks	707	70.3
One month or more	584	58.1
Three months or more	287	28.6
Six months or more	118	11.7
Nine months or more	51	5.1
At least 13 months	21	2.1

cilitated a decision to exclude from analysis the data collected in the present study on 39 placebo subjects, too small a group for meaningful comparison. Statistical analysis of the present study data excluded also data collected on the ten female subjects, which were instead reviewed separately, on an individual basis. The following summary presents the findings of both the biostatistical analysis and the medical review of the present safety data, alone and in comparison with the NAS study data.

Medical Terminations

1) NAS Study. The participation of 18 subjects was terminated for medical reasons; 13 were naltrexone subjects and five were placebo subjects. None of the terminations which appeared to be possibly due to actual medical problems were considered to have been drug-related. Gastrointestinal complaints were the most frequent precipitants of medical terminations, and alcohol abuse was an associated factor in many of these cases (Committee on the Clinical Evaluation of Narcotic Antagonists, 1977).

2) NIDA Study. Study participation was terminated for medical reasons in 53 cases. In approximately half of these cases, the primary reasons reported were gastrointestinal problems generally considered unrelated to naltrexone. One subject dropped from the study developed idiopathic thrombocytopenic purpura, which cleared after discontinuation of naltrexone and treatment with prednisone. However, the subject may have been sensitized to naltrexone by three previous periods of its use. The proportion (5.3%) of the present study's subjects whose study participation was terminated for medical reasons was substantially less than that of the NAS naltrexone subject group; otherwise, the types and incidence of problems associated with medical terminations in the two studies were quite similar.

Physical/psychiatric examination. Each subject underwent a physical examination, EKG, psychiatric examination, neurologic examination, and optionally, a slit-lamp examination at baseline and

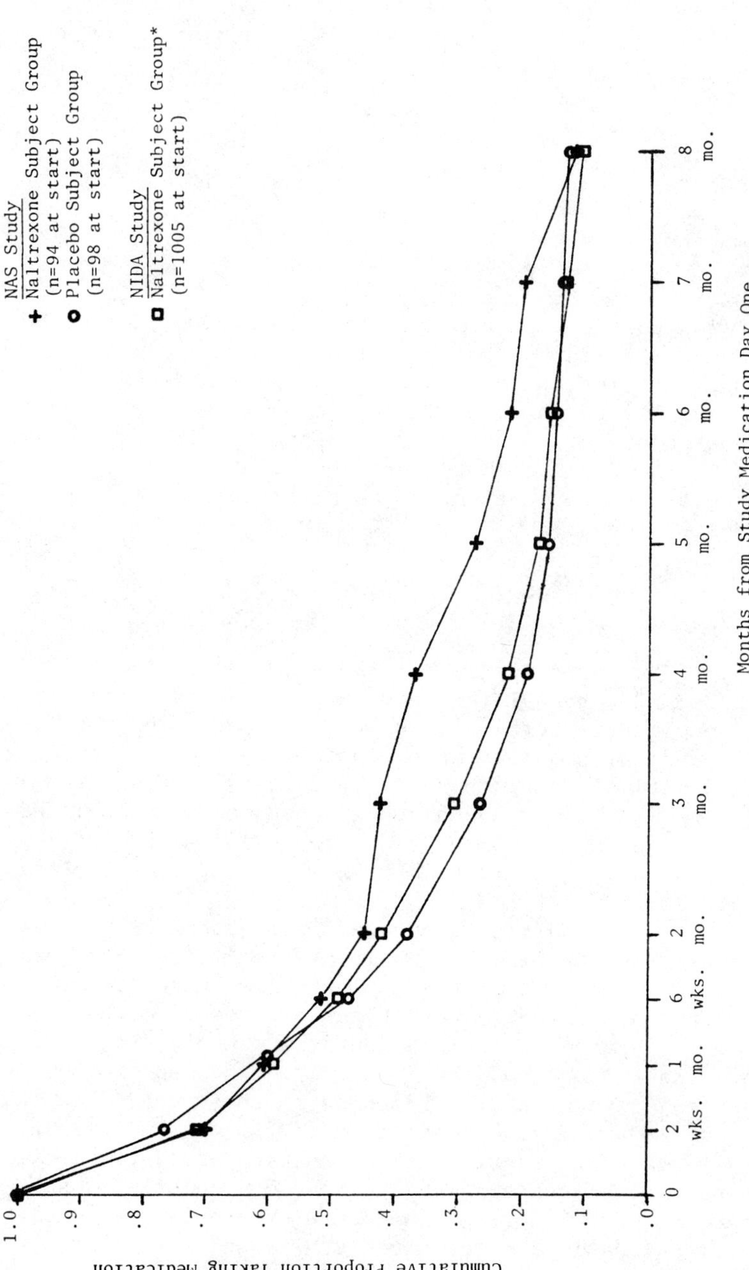

*Cumulative in this group is defined as the expected proportion of subjects active at the end of each interval as there were still 102 subjects actively participating at the time data collection was ended.

Fig 1. Cumulative proportions of subjects participating in the study at the end of the time period indicated.

at monthly intervals during study treatment; a chest x-ray was performed at baseline, at months one, three, six, and twelve, and at termination of each subject's participation. Incidences of abnormalities found in all examinations were slight with the exception of EKG; most EKG abnormalities were clinically insignificant, and the remainder probably not related to naltrexone.

Laboratory data. Twenty-eight required and five optional laboratory tests, including a complete blood count, blood chemistry (SMA/12), and urinalysis were run on each subject at baseline, at two and four weeks, and monthly thereafter during Study Medication Periods. None of the abnormalities found were clinically significant except bilirubin, which shifted slightly toward abnormal in both the present and the NAS naltrexone groups, but not in the NAS placebo groups.

Symptom interview. Symptom/side effect information was collected at baseline and at weekly intervals throughout the course of study medication. Of the twenty-four categories in which symptom/side effects were recorded, three were recorded somewhat more frequently during study medication than at baseline, both in the present study group and in the NAS naltrexone groups, but not in the NAS placebo groups. These were headaches, skin rash, and a miscellaneous category of "other" symptoms.

Blood pressure. Blood pressure readings were taken at baseline, daily for the first two weeks of study medication, and at weekly intervals thereafter. No significant changes were seen. The present study did not reproduce the NAS study findings of a minimal blood pressure increase during the first two weeks of study medication among subjects treated with naltrexone.

Weight data. Body weights were measured at baseline and at weekly intervals during study treatment. Weight changes were slight and clinically insignificant. The tendency to slight weight loss seen in the NAS naltrexone subjects (not statistically significant) was not seen in the present study.

Female subjects. Evaluation of female subjects was virtually the same as evaluation of males. Separate review of the data revealed no clinically important trends.

In summary, with the exception of one case of possibly drug-related idiopathic thrombocytopenic purpura in a subject who may have been sensitized to naltrexone in previous courses of treatment, the information pertaining to medical terminations of subjects' participation revealed no evidence of medical problems related to the drug. This finding applies to the cardiovascular and pulminary systems as well as the parameters evaluated in the psychiatric and neurological examinations. In the context of an opiate addict population, and with the exception of a shift in bilirubin values, the laboratory data were reassuring with respect to renal, hepatic and blood chemistry values. Three symptom/side effect categories deserve further attention: headaches, skin rash, and the miscellaneous category of "other" symptoms. Study medication produced no important changes in subjects' weights or blood pressures. Since the present study sample was considerably larger than that of the NAS study, this would tend to discount the importance of the marginal trends of change seen in these two categories in that study. It is concluded that these studies establish that naltrexone has a wide margin of safety in clinical use.

ACKNOWLEDGEMENTS

Members of the National Research Council Committee on Clinical Evaluation of Narcotic Antagonists include the following: Leo Hollister, M.D., Chairman, Veterans Administration Hospital, Palo Alto, CA; Jacob E. Bearman, Ph.D., University of California, Berkeley; Daniel X. Freedman, M.D., University of Chicago; Donald M. Gallant, M.D., Tulane University, New Orleans; Louis S. Harris, Ph.D., Medical College of Virginia, Richmond; Murray E. Jarvik, M.D., Veterans Administration Hospital, Los Angeles; Donald R. Jasinski, M.D., Addiction Research Center, Lexington, KY; C. James Klett, Ph.D., Veterans Administration Hospital, Perry Point, MD.

Activities of the Committee on Clinical Evaluation of Narcotic Antagonists were supported by the National Institute on Drug Abuse (NIDA) Contract 271-76-3301 supervised by Pierre F. Renault, M.D. and Demetrius Julius, M.D.

The NAS/NRC staff (Samuel C. Kaim, M.D., Joan Rittenhouse, Ph.D., Henry S. Parker, M.D.) cooperated with this study. Biometric Research Institute was supported by NIDA Contract 271-75-3050 and NAS subcontract MS-45-74-93.

The principal investigators in the NAS study include Herbert Kleber, M.D., Marc Hurzeler, M.D., Charles Savage, M.D., Kenneth Schoof, M.D., Walter Ling, M.D., Neil Haas, M.D., and Sadashiv Parwatikar, M.D. The principal investigators in the NIDA Study include, in addition to those mentioned above, Richard Resnick, M.D., Leonard Brahen, M.D., David Lewis, M.D., Ralph Landsberg, D.O., Charles O'Brien, M.D., Roger Meyer, M.D., and Robert Lieberman, M.D.

REFERENCES

Bradford, A. and Kaim, S. 1977. *Final Report: Double-Blind Placebo-Controlled Study, Administered by the National Academy of Sciences to Evaluate the Safety and Efficacy of the Narcotic An-*

tagonist, Naltrexone. Washington, D.C.: Submitted to the National Institute on Drug Abuse (March 30).

__________________ 1977. *Final Report: National Institute on Drug Abuse Studies Evaluating the Safety of the Narcotic Antagonist, Naltrexone*. Washington, D.C.: Submitted to the National Institute on Drug Abuse (May 26).

Committee on the Clinical Evaluation of Narcotic Antagonists (Division of Medical Sciences Assemby of Life Science National Research Council) 1977. *Pilot Studies in the Clinical Evaluation of Narcotic Antagonists.*

National Institute on Drug Abuse Research (Monograph Series 9) 1976. Julius, D. and Renault, P. (eds.). *Narcotic Antagonists: Naltrexone: Progress Report*. Washington, D.C.: U.S. Department of Health, Education and Welfare, pp. 163-171.

National Research Council Committee on Clinical Evaluation of Narcotic Antagonists 1978. "Clinical Evaluation of Naltrexone Treatment of Opiate-Dependet Individuals." *Archives of General Psychiatry*, 35.

Note: The references cited in this summary derive from data developed along parallel lines of investigation. In some instances, therefore, conclusions can be attributed to more than one source.

TECHNIQUE AND SIGNIFICANCE OF RAPID DETOXIFICATION BY ACUPUNCTURE AND ELECTRICAL STIMULATION (AES), NALOXONE AND VALIUM

H.L. Wen, M.D.

Kwong Wah Hospital

In the past, when an individual was addicted to narcotics, he was treated either by an abrupt or a gradual discontinuation of the narcotics. It is now more than a decade since methadone was introduced as a treatment modality for the maintenance and detoxification of drug addicts. Often, the period of detoxification is very long, especially if used in an outpatient setting. Sometimes, patients cannot be detoxified but have to be maintained on methadone. It had been reported that rapid detoxification could be accomplished with the aid of injections of naloxone (Blachly et al., 1975; Resnick et al., 1977). One report advocated that one or two days' technique be used. We had reported using acupuncture and electrical stimulation (AES) to suppress the abstinence syndrome before and during the naloxone injections (Wen, 1977; Wen, in press). More recently, we have modified our own technique by injecting 10 mg. of valium 15 minutes after the first naloxone injection. The paper deals with 41 cases of AES and naloxone technique, compared with 41 cases of AES, naloxone and valium combined.

MATERIAL

Out of these 82 patients, 41 (all male, Group One) received AES plus naloxone and the other 41 (39 male, two female, Group Two) were given AES, naloxone and valium. All were voluntary patients. They had been treated by SARDA (the Society for the Aid and Rehabilitation of Drug Addicts, Hong Kong) but for administrative reasons, they could not be accepted for a rehabilitation program with SARDA until some time had elapsed since their discharge from the Shek Kwu Chau Rehabilitation Centre. The average age of the patients was 29.2 years for Group One and 28.8 years for Group Two.

All the participants volunteered for the treatment and at any stage of the investigation he could withdraw himself from receiving the treatment.

The 82 patients began their addiction by smoking, "shooting the anti-aircraft gun" and "chasing the dragon". Fifteen cases graduated to mainlining, in Group One and 20 members of Group Two became mainliners. The total average addiction period in Group One was 8.28 years and, in Group Two, 7.76 years. The cost of the drugs taken by Group One averaged about HK $54.00 and, in Group Two, HK $63.00 per day.

METHOD

Upon admission, the patients were put on methadone 10 mg. t.i.d. on the first day, 10 mg. b.i.d. on the second day and 5 mg. b.i.d. on the third day. The technique for AES treatment has already been reported (Wen and Cheung, 1973; Wen, 1975; Wen and Teo, 1975). On the fourth day after admission, AES was first started for half an hour and then continued for another three hours, until the treatment ended with increasing doses of naloxone being injected. A total of six injections (a total of 2 cc.=8 mg.) of naloxone was given to Group One and also to Group Two. Fifteen minutes after the first naloxone injection, 10 mg. of valium were given intramuscularly to the patients in Group Two. The whole procedure for both Groups One and Two was concluded in three and a half hours. After that, the patients went back to their rooms and were observed for 48 hours. During the treatment period, vital signs (blood pressure, pulse, respiration and oral temperature, pupillary changes and symptoms and signs of abstinence) were monitored for comparison in Group One and Two. The abstinence symptoms were categorized as very severe (the failure case), severe, moderate and mild.

Following is the explanation for the severity of the abstinence syndrome:

Very severe (failure) - Procedure had to be stopped.

Severe - All the abstinence symptoms and signs were present but the patients did not ask for stoppage of the procedure.

Moderate - The patients had some abstinence symptoms but did not complain about them.

Mild - Yawning, tearing and blocked nose.

The abstinence symptoms are nausea, pain and cramps, nervousness, chills, sweating and shaking. Abstinence signs are yawning, tearing, rhinorrhea, sweating, tremor, agitation, vomiting, anorexia, gooseflesh, sneezing, salivation and diarrhea.

RESULTS

During the three and a half hours of detoxification, the abstinence symptoms and signs of the two groups were compared as seen in Table I.

The vital signs monitored in both groups during the treament period showed changes in the failure and severe cases, even when valium was injected, but not in the moderate and mild cases. The same was true of the pupillary sizes, which showed marked increases in size in the failure and severe cases. In the moderate and mild cases, changes in the pupillary sizes were present but not remarkable.

We have found that during the second and third naloxone injections, the symptoms precipitated are much more severe (which agrees with other reports [Resnick et al., 1977]) and therefore declined to the base-line level. That is why the valium was injected prior to the second injection of naloxone.

It is noted that the severity of the naloxone-precipitated abstinence symptoms and signs depends directly on how long, how often, how much and by which method the patients had been addicted to narcotics. The longer the period of addiction, the more severe the symptoms and signs precipitated by naloxone. Another important factor is whether or not the patient is a mainliner. More symptoms and signs are present in mainliners and the abstinence syndrome is also affected by patients taking drugs immediately prior to treatment (this group of patients was not put on methadone before treatment and are not included in this paper). Better results were found in those patients who inhaled their drugs or who had been addicted for

Table I. Abstinence syndrome.

	Group One (41 cases)	Group Two (41 cases)
Very severe	8	2
Severe	3	5
Moderate	12	11
Mild	18	23

Group One = AES and Naloxone
Group Two = AES, Naloxone and Valium

only a short time. In such cases, the treatment produced very mild symptoms.

Group One (see Table I) revealed more failures than did Group Two. During treatmtent, Group Two patients tended to be drowsy and to sleep more, because of the valium injections, and they also tolerated the procedure much better than did Group One patients. On the day following treatment, it became apparent (from remarks made by the subjects) that patients in Group Two experienced fewer abstinenece symptoms and signs than did members of Group One.

DISCUSSION

AES and naloxone are able to detoxify drug addicts and cleanse them by displacement of the opiates from the receptor sites that give rise to the abstinence syndrome as advocated by Shen and Way (1976). This suggests that the number of opiate-occupied receptor sites is an important factor in determining the severity of the abstinence. The more these receptor sites are cleansed, the less likely it is that severe abstinence symptoms will be present in the patients. This theory of Shen and Way has not been duplicated by Catlin et al. (1977). From our clinical studies, it is our impression that somehow naloxone displaces opiate from the receptor sites. It is known that naloxone, a pure opiate-antagonist, blocks the development of tolerance and physical dependence, if administered concomitantly with morphine (Cochin and Mushlin, 1976). In several patients, two days after AES, naloxone and valium treatment, Narcan tests were done and we found no appreciable abstinence syndrome on the third and fourth post-treatment days (details to be published).

More recently, it has been hypothesized that AES increases the opiate-like peptide (Hughes, 1975; Pasternak et al., 1975; Teschemacher et al., 1975; Terenius and Wahlstrom, 1975; Li and Chung, 1976; Guillemin et al., 1976; Bradbury et al., 1976) in the brain. In experiments with mice treated with AES for 30 minutes, prior to the injection of naloxone, it was found that the animals had a significantly lower score in abstinence behavior ($0.02<p<0.05$, by the T-test). Compared with the untreated group, a reduction in score of close to 50 percent was observed. When the brains of the animals were analyzed for opiate-like activity at the end of the experiment, the AES-treated group showed a significant increase of opiate-like activity in the Peak I fraction ($0.00<p<0.01$). There was no difference in the opiate-like activity between the two groups in the Peak 2 fraction (0.8 p 0.9). We believe that Peak 1 is beta-endorphin and Peak 2 enkephalin. This would suggest that AES works primarily on the endorphin and not on the enkephalin system (Ho et al., to be published).

The significance of this modality of treatment is that it is fast and the suffering produced by the naloxone injections is tolerated by the patients, especially with the supplement of valium. In our experience, certain groups of patients would prefer this technique to others, because of the shorter period of treatment and suffering. It could easily be given to patients who do not want to receive institutionalized treatment. After detoxification, these patients could be put on a long-acting opiate antagonist, such as naltrexone or they could be given AES whenever they felt the urge for the drug. Should a drug addict require emergency surgery, this technique can be used to detoxify the patient before the operation. This obviates the danger of drug overdose in this group of surgical patients and also there is no worry that he or she will experience the abstinence syndrome post-operatively. This technique can be recommended for patients who do not want a long period of detoxification and it can be given prior to psycho-social rehabilitation. The cost of naloxone is not prohibitive and is compensated for by the short stay in hospital.

SUMMARY

The technique of fast detoxification was used to clear 82 patients of drug addiction. Forty-one of the cases were detoxified by AES and naloxone (Group One) and the other 41 were treated in a similar way but also given valium 10 mg. intramuscularly (Group Two). It was found that the abstinence syndrome produced by the increasing dosage of naloxone was less and more to the liking of the patients of Group Two.

ACKNOWLEDGEMENT

I wish to thank Doctors A.D. Mehal and Y.H. Ng for their assistance and also the Society for the Aid and Rehabilitation of Drug Addicts of Hong Kong (SARDA) for referring the patients to us.

REFERENCES

Blachly, P., Casey, D., Marcel, L., and Denny, D. 1975. Rapid Detoxification from Heroin and Methadone using Naloxone: A Model for study of the Treatment of the Opiate Abstinence Syndrome, in Senay, E., Shorty, V., and Alkesne, H. (eds.), *Developments in the Field of Drug Abuse*. Cambridge, Mass.: Shenkman Publishing Co.

Bradbury, A.F., Smyth, D.G., Snell, C.R., Birdsell, N.S.M., and Hume, E.C. 1976. C fragment of Lipotropin has a High Affinity for Brain Opiate Receptors. *Nature*. 260: 793-795.

Catlin, D.H., Schaeffer, J.C., and Liewen, M.B. 1977. 2-Diazo-morphine directed Antiserum: Determination of Morphine in Brain after Naloxone Challenge in Morphine Pellet Implanted Mice. *Life Sci.* 20: 123-132.

Catlin, D.H., Liewen, M.B., and Schaeffer, J.C. 1977. Brain Level of Morphine in Mice Following Removal of a Morphine Pellet and Naloxone Challenge: No evidence fo Displacement. *Life Sci.* 20: 133-140.

Chen, G.S. 1977. Enkephalin, Drug Addiction and Acupuncture. *Am. J. Chinese Med.* 5: 25-30.

Cochin, J. and Mushlin, B.E. 1976. Effect of Agonist-Antagonist Interaction on the Development of Tolerance and Dependence. *Ann. N.Y. Acad. Sci.* 281: 244-251.

Guillemin, R., Ling, N., and Burgus, R. 1976. Endorphines, Peptides, D'origine Hypothalamique et Neuropypophysiare a Activite Morphinomimetique Isolement et Structure Moleculaire de - l endorphine. *C.R. Acad. Sci. Paris.* 282: 783-785.

Ho, W.K.K., Lam. S., Ma, L. and Wen, H.L. Elevation of Brain Level During treatment of Morphine Addicted Mice by Acupuncture. To be published.

Hughes, J. 1975. Isolation of an Endogenous Compound from the Brain with Pharmacological Properties Similar to Morphine. *Brain Res.* 88: 295-308.

Li, C.H. and Chung, D. 1976. Isolation and Structure of an Untriakontapeptide with Opiate Activity from Camel Pituitary Glands. *Proc. Natl. Acad. Sci.* 73: 1145-1148.

Mayer, D.J. 1975. Opiates Receptor Mechanisms. *Neurosci. Res. Prog. Bull.* 13: 98.

Pasternak, G.W., Goodman, R., and Snyder, S.H. 1975. An Endogenous Morphine-Like Factor in Mammalian Brain. *Life Sci.* 16: 1765-1769.

Pomeranz, B. 1975. Brain Opiates at Work in Acupuncture. *New Scientist.* 73: 12-13.

Pomeranz, B. and Chiu, D. 1976. Naloxone Blockage of Acupuncture Analgesia: Endorphin Implicated. *Life Sci.* 1757-1762.

Pomeranz, B., Cheng, R., and Law, P. 1977. Acupuncture Reduce Electro Physiological and Behavioral Responses to Noxious Stimuli: Pituitary is Implicated. *Experimental Neurology.* 54: 172-178.

Resnick, R.B., Kestenbaum, R.S., Washton, A., and Poole, D. 1977. Naloxone--Precipitated Withdrawal: A Method for Rapid Induction onto Naltrexone. *J. Clin. Pharmacol. and Therap.* 21: 409-413.

Shen, J.W. and Way. E.L.Y. 1976. Antagonist Displacement of Brain Morphine during Precipitated Absinence. *Life Sci.* 16: 1829-1830.

Terenius, L. and Wahlstrom, A. 1975. A Search for Endogenous Ligand for the Opiate Receptor. *Acta. Physiol. Scand.* 94: 74-81.

Teschemacher, H., Opheim, K.E., Cox, B.M., and Goldstein, A. 1975. A peptide-like Substance from Pituitary that acts like morphine. *Life Sci.* 16: 1777-1781.

Wen, H.L. 1977. Fast Detoxification of Drug Abuse by Acupuncture and Electrical Stimulation (AES) in Combination with Naloxone. *Mod. Med. Asia.* 13: 13-17.

Wen, H.L. Fast Detoxification of Heroin Addicts by Acupuncture and Electrical Stimulation (AES) in Combination with Naloxone. *Comp. Med., East and West.* In press.

Wen, H.L. Fast Detoxification of Heroin Addicts by Acunpuncture and Electrical Stimulation in Combination with Naloxone - Treatment Technique. Proceeding of the 7th International Institute on the Prevention and Treatment of Drug Dependence. In press.

Wen, H.L. and Cheung, S.Y.C. 1973. Treatment of Drug Addiction by Acupuncture and Electrical Stimulation. *Asian J. Med.* 9: 138-141.

Wen, H.L. 1975. The Role of Acupuncture in Narcotic Withdrawal. *Med. Progress.* May, 15-16.

Wen, H.L. and Teo. S.W. 1975. Experience in the Treatment of Drug Addiction by Electrical Acupuncture. *Mod. Med. Asia.* 11: 23-24.

"NARCAN CHALLENGE" A FOUR YEAR STUDY OF ADDICTION DETERMINATION

Jordan Scher, M.D.; Khalid Baig, M.D.; and Si Bom You, M.D.

Methadone Maintenance Institute, the National Council on Drug Abuse

The determination of the presence or absence of addiction is often a difficult and uncertain task. Dependence upon clearly defined "tracks", other stigmas of heroin addiction, history and verifying third parties can be chancy, if not downright unreliable. This is particularly the case with so-called "snorters". Withdrawal symptoms can be fabricated, particularly before the less experienced examiner. And the mere presence of a "dirty" urine is no proof of addiction. These issues are vital in avoiding the inadvertent admission of a non-addict, a user who has not become fully addicted, or an individual who is attempting to simulate addiction. To minimize or avoid mistakes in determining addiction so that non-addicts are not admitted, a Narcan Challenge test is given all doubtful, questionable or border-line patients applying to the Methadone Maintenance Institute.

This procedure was made standard practice for doubtful cases over four years ago at MMI. In mid-1974, in an effort to weed out and eliminate those potential applicants who may have been improperly motivated or non-addicted, we decided, initially and very tentatively, to give such individuals a Narcan Challenge test. At first we were quite concerned that the administration of narcan to addicts might throw them into intensive and disturbing withdrawal. However, after some experience with the procedure, we found that most patients accepted the procedure willingly, and that even the worst reactions were manageable and completely within our ability to safely induce.

Prior to the introduction of the narcan approach, try as we might to be sure that we were not admitting non-addicts under the guise of addiction, there were quite possibly some who escaped even our sincerest efforts to exclude them. In some of these dubious cases, we would employ a rapid screen thin layer chromatography test which, even if positive, did not indicate whether or not the individual was currently addicted. We would also, in some instances, employ the use of a finger galvanometer to determine whether or not the individual was lying about his asserted addiction. Both of these procedures have been previously described (Scher, et al., 1974).

Despite the use of rapid screen TLC and finger galvanometry, we were not at all satisfied that there was still not the possibility of missing individuals who claimed to be addicts, but were not. Obviously, so-called needle marks were not necessarily conclusive, since an individual who was so inclined could, with the use of a straight pin, simulate fairly adequately the presence of needle marks that might be convincing to those who may not be sufficiently experienced in determining bona fide tracks, even trained professionals such as physicians and nurses. The testimony of presumably reliable witnesses such as family members, those who had known the individual on the street, etc., might not ultimately be telling. This was, of course, the case particularly prior to the newer methadone regulations, which no longer require proof of two year addiction.

As a consequence, in the face of disinterested testimony and the evaluation of the signs of addiction, we felt that we might be far more secure in making a diagnosis of addiction if we were to place the patient in a safe, but significant withdrawal state when he did not come to us in that condition. This is particularly significant in the case of the so-called "snorter", who does not usually show tracks and whose nasal mucosa may not be sufficiently excoriated and/or scarified as to make a presumptive diagnosis.

As a result, in mid-1974 we began testing all snorters and all of those individuals whose tracks, history and/or objective witnesses may have led us to wonder about the reliability of the whole package of proof of addiction.

After some experience, we developed three categories for the determination of results. They are as follows:

I. Negative: no withdrawal symptoms and no evidence of physical dependence on opiates after two successive injections of 1 ml. narcan each time.

II. Moderately Positive: no or mild withdrawal symptoms and signs after first injection, but definitive signs and symptomology after a second injection in the opposite arm.

III. Definitely Positive: definite and severe withdrawal symptoms and signs after first injection.

During the last half of 1974, 54 candidates for admission were initially tested. Of these, 5 were diagnosed as negative, 7 moderate and 42 positive. Of the definitely positive, 22 were mainliners and 18 snorters. Of the moderately positive, 4 were mainliners and 5 snorters. Of the negative, 1 was a mainliner and 4 were snorters.

During 1975, 175 candidates for admission were narcaned out of 675 who applied. Of these, 18 were diagnosed as negative, 35 moderately positive and 122 definitely positive. About half of these were snorters and half were mainliners. In 1976, out of 750 individuals applying for admission to the program, 105 were narcaned. The results proved 19 to be negative, 22 moderately positive and 64 definitely positive.

In 1977, out of 450 new candidates for admission, 87 were given narcan. Of these, 18 were clearly negative, 15 moderately positive and 54 definitely positive. As of January 27, 1978, only 1 negative narcan had been recorded out of 11 done. Of these, 2 were moderate and the remaining 8 were clearly positive. In each of these years the number of so-called snorters and mainliners was approximately 50-50.

The procedure for examining border-line or doubtful candidates for admission is as follows: 1 ml. (0.4 mg.) narcan is injected subcutaneously into the deltoid area. If the result is negative, the injection is repeated intramuscularly into the opposite deltoid. If both of these injections appear to be negative after a delay of 15 minutes to half an hour following each injection, a third injection may be given intravenously. Resuscitation equipment is, of course, in the clinic in case the reaction should prove so overwhelming that such intervention would be necessary.

A response to narcan may take from 5 to 20 minutes or more following any of the above injections. New candidates for whom this procedure is employed are told that they will be given a drug which will determine how much of a narcotic is in their system. By this device, it is our hope that potential narcan patients will not be induced to feign symptoms of withdrawal. Unfortunately, many narcan patients are preliminarily "educated" by other patients as to their concept of what narcan does, and what symptoms we may be looking for.

Among the common symptoms that may indicate a positive effect are the following: clamminess, increased agitation and irritability, abdominal cramping, nausea, vomiting, "gooseflesh", diaphoresis, rhinorrhea, lacrimation, tremors, pupillary dilation and skeletal muscular cramping (upper, lower extremities and lower back).

Body temperature may increase 0.1° to 0.2° F. Pulse rate may increase 10 to 20 beats per minute over the baseline value. Blood pressure may increase 10 to 20 mm. of mercury in either the systolic and/or diastolic readings. In some instances, systolic and/or diastolic readings may drop 10 to 20 mm. of mercury and/or the pulse rate may decline 10 or more beats per minute paradoxically.

Pupil size will usually dilate in 0.2 increments over the baseline to 0.4 or 0.6 mm. of dilatation. In a positive reaction, the skin will inevitably become moist, sweaty, and in many instances, flush bright red over the face, neck and upper cheek and back.

It is our custom to have the patient sign a release form. On the form card (Scher, 1977), the following items are included and evaluated: baseline blood pressure, pulse, skin moisture and turgor, pupil size and other items indicative of withdrawal from the list above. As stated above, the methadone candidate may be given one, two or three injections. In some cases, although rarely, he may be given a fourth injection, perhaps a second intravenous injection. In some cases candidates may receive a first or second injection which is not narcan but a placebo.

In the latter instance, such injections may be given to candidates for admission who, in the opinion of the physician, have possibly been "educated" by other patients. We may then amplify our controls to avoid the possibility of feigning, or simulating, a positive response. It has, indeed, been edifying on occasion to have such particularly suspicious candidates attempt to simulate nausea, vomiting and other symptoms of withdrawal following a placebo injection. Generally speaking, such simulated narcan responses have neither been effective nor productive.

In this instance of the more honest narcan responses, the major criteria for the determination of a positive result have been blood pressure changes, pulse changes, pupil size changes, skin temperature, turgor and moisture and such other minor symptoms which the patient may show. In extreme responses, withdrawal is accompanied by nausea, vomiting, muscular tension, agitation and trembling. It is important to realize that the patient may appear to be negative after the first or second injection, but may exhibit an extreme and very severe withdrawal response after the intravenous injection. Why the first or second injection fails to elicit this response on occasion is unclear, but that the third may, particularly in snorters, evoke an extreme withdrawal reaction is certainly the case.

Although we cannot be sure, it is our impression that once the narcan procedure was introduced in mid-1974, the number of fictitious narcotic addicts who came to the clinic seeking admission to the methadone program was markedly diminished. It is our belief that word got around among the drug-using population that if you were not a bona fide addict, you might as well not enter the Methadone Maintenance Institute. In fact, one can occasionally hear patients in the waiting room advising new candidates for admission that if they are not really addicted they might as well leave, since we have a test which will determine the authenticity of their addictive state.

The Narcan Challenge test has in our experience the unique advantage that in those who are not addicted, despite the fact that they may have taken a narcotic that day or even be pre-addicted, non-tolerant "chippers" of longer or shorter duration, the withdrawal reaction will not occur. Withdrawal as a response to narcan seems to be restricted to those individuals who are clearly addicted, and usually addicted for some considerable duration of time.

Without the administration of narcan, it is in some instances hazardous or even dangerous to assume that some candidates for admission are authentic addicts, despite tracks, history and various forms of substantiating testimony.

In some states such as California, on the other hand, the degree of intensity of addiction verification is so great that admission to a methadone maintenance program may become almost prohibitively difficult. In fact, in at least one instance, a bona fide twenty year addict chose not to undergo two preliminary withdrawal attempts required by California law. He felt it was necessary to come to Chicago in order to be admitted to a methadone program.

There must be control, but where the barriers to admission for bona fide addicts are raised prohibitively, true addicts may be compelled to remain on the street, rather than seek necessary help. Similar irrational restrictions in the assignment of dosages may even compel many addicts to refuse to enter programs altogether. Once entered, arbitrary and restrictive dosage scheduling may compel some patients to continue to use street narcotics as well as methadone. There is no question that we must make every effort to keep non-addicts off of methadone, but many of our rules seriously discourage true addicts from making the effort to find treatment.

Arbitrary decisions even after admission such as theoretical maximum dosages may result in other problems; for example, very high dirty and dropout rates (Scher et al., 1977). Despite the need for validation, excessive admission criteria may not protect the public and non-addicts from entry into programs, but instead may discourage addicts from attempting to receive any kind of help at all. Almost as though to underscore these points, in states like California where admission criteria are based not on the degree and authenticity of addiction, but rather on the presence or absence of two arbitrarily determined efforts at detoxification, several kinds of fallout have occurred.

One of these fallouts is the fact that California is said to currently have the highest rate of heroin over-dose deaths in the country. Another is that California is said to have the highest number of street addicts in the country who are not in treatment, as well as the highest rate of new addicts per 100,000 of anywhere in the country. We do not cite these figures to put down California or anyplace else. We cite them only to point out that the issue in addiction should be determination of a true addicted state, whether that state is the result of a six month intensive run or the result of two to ten years of chronic street addiction.

One of the greatest problems that we have in attempting to employ methadone in the treatment of addiction is the degree of credibility such a program has in the eyes of the street addict. If an addict feels, rightly or wrongly, that he cannot trust the confidentiality of a program, he will not appear for treatment. If an addict, judging from the experience of others, feels that he will not get a sufficient dose of methadone, he may well not appear for methadone treatment. It is meaningless to him that there are experts who say 40 mg. of methadone is enough for anyone. If it is his experience that the maximum dosage a program will allow leaves him feeling uncomfortable, an addict will continue to use street heroin to satisfy his concept of how much narcotic he may need. It is difficult enough to entice addicts at the present time into programs, so that despite the need for caution in admission determination, many other problems must also be overcome if we are to serve the addicted community best.

Addicts may also be discouraged from entering methadone programs because of the fearsome myths about methadone they have heard from their friendly local pusher and/or street peers. Addicts may be discouraged if they fear they will "get two habits" or that methadone is harder to withdraw from than heroin.

One of the biggest problems that initial methadone treatment patients run into is that they must attend a clinic from six to seven days a week in order to get their methadone for the first three months. The attrition, as a consequence, is very high in new patients, unused to discipline and inured to the difficult, but self-adjusting street life. Nonetheless, despite the word-of-mouth admonition about Narcan Challenge that many patient candidates for admission receive, this barrier has not proved an excessively discouraging one for authentic and properly motivated new patients.

One of the major problems facing methadone treatment today is the fact that all too many potential candidates have been touted away from methadone for many of the wrong reasons noted above. If patients are afraid they are going to get two habits, never get enough methadone, etc. they are going to stay away from effective treatment in large numbers.

Recently we have been testing this hypothesis. We have been sending teams of interviewers into the streets in the copping quarters to survey and evaluate attitudes and potential for change of the street addicts. These findings will be reported separately elsewhere. But suffice it to say here, if it is the impression that the current apparent decline in new admission to methadone programs is a reflection of a decline in the number of street addicts available for treatment, our experience does not support this contention, for Chicago at least.

What we have found is that a considerable number of street addicts currently remain on the street with a high level of resistance to methadone treatment for many, or all, of the reasons we have alluded to above. If we cannot lower the artificial barriers to admission, make more methadone accessible on a more realistic basis and educate the street addict better, we will have failed to provide the most effective treatment available.

Narcan Challenge, although a seemingly stressful introduction for a new methadone candidate, seems to have a very high degree of reliability in the determination of a true addicted state. Most entering patients will accept this burden willingly, if it means that a positive result will indicate immediate relief from withdrawal through the administration of methadone. Undoubtedly, all of the currently approved tests for methadone addiction are cumulatively of assistance in determining the validity of addiction. The availability of such a simple, and apparently reliable, procedure as the Narcan Challenge should enable methadone clinics to simplify and enhance the reliability of addiction determination.

Methadone treatment works, whether it be for the purposes of maintenance and/or detoxification. If we are to make this modality available to that far larger number of street addicts who have never seen the inside of a methadone clinic, we must find ways to make the process considerably less painful and more accessible for their entry. The Narcan Challenge is a dependable method of addiction determination. It offers a simple and, in our experience, highly reliable method of determining true addiction in doubtful cases. We therefore strongly recommend the Narcan Challenge as a supplement to, or in the place of, other, perhaps less effective but far more burdensome methods of assuring oneself of addictive authenticity.

If we are to make methadone truly available to that vast army of street addicts who are currently either terrorized by it or grossly misinformed about it, we must enter upon a broad and massive educational program. We must also do our best to eliminate or minimize unnecessary and unsatisfactory rules and practices which only serve to discourage or prohibit entry of bona fide patients into necessary treatment programs. The latest revision of the Federal rules for the operation of methadone maintenance programs seems to take into consideration the overcoming of some of the excessive entry barriers enacted by previous regulations.

Methadone is an effective treatment. If it is to fulfill its promise, it must become available to the population that needs it most. Narcan Challenge very often permits one to make a clear-cut decision about the degree of addiction of a patient with a fair degree of certainty in those cases where clinical judgment alone would lead one to be extremely doubtful. It would be a great disservice to send a true addict back to the street once he had made the enormous effort to attempt to seek help, when we ourselves may be clinically unsure of the validity of his addiction. Anyone can be found fooled by the clinical claim of addiction, but Narcan Challenge gives the opportunity to minimize the instances in which this may occur.

REFERENCES

Scher, J.M. et al. 1974. *Confirmatory Procedures in Addiction Evaluation: Narcan Challenge, Rapid-Screen TLC and Finger Galvonometry.* San Francisco: North American Congress on Alcohol and Drug Problems.

Scher, J.M. et al. 1977. *Methadone Maintenance: Patient Participant Known Dosage Setting, Dosage Changing, and Self-Determined Detoxification Scheduling.* San Francisco: National Drug Abuse Conference.

Scher, J.M., Baig, K., Crown, B., and Roman, R. 1977. *Methadone Maintenance: A National Comparative Study of Dosage, Dirty Urines and Dropouts.* San Francisco: National Drug Abuse Conference.

NALTREXONE IN CLINICAL PRACTICE - SOME LESSONS FROM THE PHASE II EXPERIENCE AS RELATED TO THE PROPOSED PHASE III WORK*

Walter Ling, M.D. and Neil B. Haas, M.D.
Veterans Administration Hospital, Sepulveda, California

H. Alex Bradford, M.S.
Biometrics Research Institute

INTRODUCTION

On the eve of Phase III naltrexone investigation, it seems appropriate to recall that it was almost excatly three and one-half years ago when the first naltrexone investigators' conference was convened in this very city. We were preparing to launch the Phase II study and there was great excitement and considerable anxiety as we gathered to hear Drs. Martin, Resnick and others relate their early experience with narcotic antagonists. The questions then were whether a naltrexone study was clinically feasible, whether it would prove safe, whether it would be acceptable, and whether we could arrive at some preliminary indication of its clinical efficacy. Summarized in the preceding report were the results of a concerted effort in the intervening three years by many investigators and others associated with these projects. We have answered some of the questions and raised new ones. It seems apropos to review this experience as we prepare for the next steps in the clinical development of naltrexone.

ISSUES OF SAFETY AND TOXICITY

Naltrexone has proved to be safe on a relatively short-term basis. Of the some 1200 patients exposed to naltrexone treatment for varying periods, averaging between two to three months, we have seen only infrequent G.I. complaints which, in many instances, were prob-

* This work is supported in part by Grant #DA4PG023, NIDA.

ably residual withdrawal symptoms precipitated by initiation of naltrexone. Headache and skin rash were seen in a few patients. One patient developed idiopathic thrombocytopenic purpura while on naltrexone. The drug was discontinued and the patient successfully treated with prednisone. No causal relationship has been established between this clinical entity and naltrexone, but a possible link cannot be completely ruled out.

No definite conclusion can be drawn with regard to long-term safety because of the small number of patients treated and this would be one area where the Phase III study could provide valuable additional information.

PATIENT SELECTION

Three groups of patients were studied in Phase II: the street addicts, post-addicts, and those detoxified from methadone maintenance. Viewed in isolation, the NAS study seemed to indicate that post-addicts tend to remain in treatment longer than the other two groups. However, this advantage was not apparent in the NIDA clinics, nor when all the clinics are combined. Table I shows the relative retention rate for the three groups. Of the 135 post-addicts, 43 (31.9%) were in treatment longer than one month, 11 (8.1%) longer than three months and four (3%) longer than six months. Of the 753 street addicts, 408 (54.2%) were in treatment longer than one month, 206 (27.4%) were longer than three months and 81 (10.8%) were longer than six months. There were too few methadone maintenance patients to make any meaningful comparison.

No clear-cut outcome predictors emerged from the large amount of psychosocial data, although a few variables did reach a level of statistical significance. For instance, late terminators had more arrests and less heroin use for the 90 days preceding admission, and patients who were employed, had a stable family relationship and good rapport with the staff, tended to remain in treatment longer. However, none of these appeared to be strong enough indicators to be useful for patient preselection. In an earlier study with cyclazo-

Table I. Relative retention rate.

	Post-addicts		M.	M.	Street addicts	
	N	%	N	%	N	%
Baseline	135	100	15	100	753	100
1 month	43	31.9	3	20	408	54.2
3 months	11	8.1	0	0	206	27.4
6 months	4	3.0	0	0	81	10.8

cine, Resnick et al. found that patients who attributed some medicinal value to opiates tended to reject antagonist treatment whereas those whose use of opiates was more related to environmental factors tended to view antagonist therapy more favorably.

At present, it seems wise not to entertain any preconception about attributes which may predict treatment outcome one way or the other, less such predictions become self-fulfilling prophesies in which certain groups of patients are viewed as good or poor candidates and are treated as such consciously or unconsciously, since such attitudes themselves are likely to have significant effects on the treatment results.

INDUCTION

It now appears that the NAS study induction schedule was probably too conservative and it could have accounted for some early dropouts. We know now that a more rapid induction schedule is safe and well-tolerated, provided that patients are completely detoxified. A more rapid induction may be desirable in the coming Phase III protocols. A few patients in the Phase II study did complain of what appeared to be residual withdrawal symptoms during induction. These occured predominantly in the methadone maintenance group.

PATIENT RECRUITMENT AND RETENTION

By far the most difficult problems experienced during the Phase II trials have been patient recruitment and program retention. The NAS report pointed out that low enrollment and retention rates are problems of any long-term treatment programs with drug abusers and are not peculiar to naltrexone. In fact, the retention rate in the NAS study compared rather favorably with a number of other studies reported in the literature in spite of the stringent protocol requirements. On the other hand, there is evidence from these and other studies that favorable outcome at follow-up is related to the length of treatment. The fact that other long-term treatment efforts also show poor retention rates does not really help improve the clinical usefulness of naltrexone. A major effort in planning the Phase III study would therefore be likely to be devoted towards maximizing patient participation. This subject will undoubtedly be treated with greater detail and in more depth in the remainder of this panel.

As alluded to earlier, the NAS study indicated that patients tend to remain in treatment longer where there is good rapport between patients and clinic personnel and where the clinic staff is enthusiastic about and familiar with the research protocol. An educational effort directed toward the research team would seem therefore worthwhile.

More efforts will no doubt be devoted to defining subgroups of patients for whom naltrexone may be especially useful. Several special patient populations have not been adequately explored to date, including adolescents and women.

Perhaps the greatest effort will be directed toward motivating patients to remain in treatment by various methods of reinforcement. Small amounts of money have been offered as incentives with some degree of success by Dr. Meyer and his associates. Traditional psychotherapy and behavior modification techniques have also been used. One of the more promising approaches might be an educational program for the patients. There are some reasons to believe that this may be quite useful since it addresses the patients' cognition directly. Meyers and others have pointed out that when addicts are treated with an antagonist, they stop craving and using heroin as if the drug were not available. This depends to a much greater extent on their conscious recognition and understanding of what naltrexone is doing instead of on the process of extinction. Moreover, as O'Brien et al. have shown, the extinction of secondary physiological addictive phenomena brought about by repeated self-administration of drugs under blocked conditions bears little relationship to subsequent cessation of heroin use. An educational approach therefore seems to us to make good sense at this time.

Eventually, there may emerge an integrated program combining different aspects of a number of approaches including some forms of overt reward, staff and patient education and various forms of psychotherapeutic and behavioral interventions.

In comparing the results of the NAS study to the NIDA study, it appears that the double-blinding and the use of a placebo-control did not adversely affect the retention rates of the NAS study. Moreover, the use of a placebo control group not only allowed for conventional statistical analysis of the NAS study data but permitted meaningful group comparison between the two studies. It would seem desirable to include a double-blind placebo control in at least a portion of the Phase III study.

EFFICACY

It has been suggested that one way to deal with the problem of evaluating efficacy would be to declare that there is no problem since the narcotic blocking property of naltrexone has already been amply demonstrated. One could insist that all the rest is a problem of human motivation. Unfortunately, such reasonings do not help us clinically and we are forced to accept a number of secondary outcome criteria with their shortcomings and imperfections.

In the Phase II studies we chose to use as measurements of efficacy, program retention, use of opiates as reflected by urinalysis, subjective craving for heroin, certain psychosocial adjustments and a global rating as judged by patients and staff. It was pointed out that program retention and urine results need to be interpreted together and changes in craving may not adequately predict post-treatment abstinence. Meyers et al. have shown that patients stop craving heroin when they are treated with an antagonist and realize that heroin is in effect psychologically unavailable. This lack of craving may therefore be related to this recognition rather than to the pharmacological properties of naltrexone. Measurable psychosocial changes are by and large possible only if there is concomitant cessation of opiate use for relatively extended periods.

In the next phase, with permission for treatment re-entry, several other parameters could be examined and may prove useful. Some of these are the number of times patients return to treatment programs, the length of each succeeding treatment episode, and the interval between each. Of even greater importance may be the patients' opiate dependency status at each re-entry point. A patient who has undergone a long treatment period and who has a long inter-treatment interval but returns readdicted is obviously not doing as well as one who returns when readdiction is imminent but has not yet occured regardless of the lengths of the latter's treatment episodes or the inter-treatment intervals.

In summary, it seems that we have profited considerably from the Phase II experience. We have learned that naltrexone is relatively safe, and that it is acceptable to at least some patients. We have discovered some of the difficulties in patient recruitment and retention and have also found some useful measures of efficacy and have been led to explore some new ones which may aid us in the planning of the Phase III work and in the interpretation of its results.

REFERENCES

Bradford, A. and Kaim, S. 1977. Final Report: Double-Blind Placebo-Controlled Study, Administered by the National Academy of Sciences to Evaluate the Safety and Efficacy of the Narcotic Antagonist, Naltrexone. Submitted to National Institute on Drug Abuse. March 30, 1977.

Bradford, A. and Kaim, S. 1977. Final Report: National Institute on Drug Abuse Studies Evaluating the Safety of the Narcotic Antagonist, Naltrexone. Submitted to National Institute on Drug Abuse. May 26, 1977.

Meyer, R.E. et al. 1976. Heroin Self-Administration: The Effects of prior Experience, Environment and Narcotic Blockade. Proceedings, 38th Annual Scientific Meeting, Committee on Problems of Drug Dependence. *Nat. Acad. Sciences*: 272-295.

National Research Council Committee on Clinical Evaluation of Narcotic Antagonists. 1978. Clinical Evaluation of Naltrexone Treatment of Opiate-Dependent Individuals. *Arch. Gen. Psychiatry.* 35: 335-340.

Resnick, R.B. and Washton, A.M. 1978. Clinical Outcome with Naltrexone: Prediction Variables and Follow-up Status in Detoxified Heroin Addicts. *Annals of the N.Y. Acad. of Sciences.* In press.

Resnick, R., Fink, M., and Freedman, A.M. 1970. A Cyclazocine Typology in Opiate Dependence. *Am. J. Psychiatry.* 126: 1256-1260.

CHEMOTHERAPY OF ADDICTION - A PHARMACOLOGICAL PERSPECTIVE

Donald R. Jasinski, M.D.

National Institute on Drug Abuse - Addiction Research Center

Accompanying the greater use of drugs to treat narcotic addicts has been the development of new drugs for such treatment. It is conceivable that in the near future the therapist will choose from an array of drugs to treat addict patients. Such a situation would be analogous to that with the chemotherapy of infectious diseases. In such a situation, knowledge of the comparative pharmacology of the available drugs is one factor necessary for rational therapy. This presentation will review the pharmacology of the narcotic antagonists in relation to each other and in relation to those narcotic analgesics used as maintenance drugs. This presentation will not review the relative efficacies of the various classes of agents used to treat addicts nor the rationales for choosing among the agents.

For comparative purposes, certain pharmacologic characteristics of the maintenance drugs need to be noted. All are narcotic analgesics, and as such, share essential pharmacologic features with morphine (Jasinski, 1977). Methadone differs from morphine in that methadone is more effective orally and suppresses withdrawal for a longer time (Martin et al., 1973a). LAAM (l-acetylmethadol) in turn is similar in pharmacological action but is longer acting even than methadone (Fraser and Isbell, 1953). Thus, all maintenance drugs would have the ability to produce morphine-like euphoria, physical dependence and toxicity (Jasinski, 1977). Therapeutically, these properties would lead to high patient acceptability and cross-tolerance to the effects of other narcotics. Because of these properties, however, it is difficult to terminate treatment; patients must be selected; and the drugs are toxic and may be diverted for abuse.

The group of drugs known as the narcotic antagonists share the common ability to block the effects of morphine-like drugs if administered before the morphine-like drug or to reverse the effects if administered after the morphine-like drug. The mechanism of this antagonism is felt to be based upon the antagonist competing with the morphine-like drug for chemical attachment at the site of action of the morphine-like drug. This is termed competitive antagonism. Some antagonists can also produce pharmacologic effects in their own right. These are termed agonist effects. For therapeutic purposes, antagonists can be classed into three groups based upon the character of their agonist effects (Jasinski, 1977).

One group of antagonists are those whose agonist effects resemble those of cyclazocine, the prototype drug of this class. Cyclazocine is orally effective and long-lasting (Martin et al., 1965). The agonist effects include dysphoric subjective changes. With repeated administration, tolerance develops to the agonist but not to the antagonist effects of cyclazocine. Thus, subsequent administration of heroin or other narcotics is without effect (Martin et al., 1966). Further, cyclazocine is less toxic than morphine-like drugs; treatment is easily terminated; and the drug is unlikely to be diverted for abuse. The chief limitation of cyclazocine is the unpleasant subjective effects that make stabilization on blocking doses difficult and which in turn limit patient acceptability.

A second group of antagonists are the so-called "pure antagonists". They are compounds which antagonize morphine but produce virtually no agonist effects. The prototype drugs are naloxone and naltrexone, with naltrexone having the more extensive clinical trial as a treatment drug. Naltrexone is the least toxic drug in treating addicts and is unlikely to be diverted for abuse (Martin et al. 1973b). Chronic administration of blocking doses of naltrexone prevents a patient from experiencing the effects of narcotics (Martin et al., 1973b).

A third group of antagonists that have potential in treating addicts are the partial agonists of morphine. These drugs can act as antagonists and at the same time produce agonist effects that are morphine-like. However, the maximum degree of morphine-like effects produced by these drugs is limited and is less than that produced by narcotic analgesics such as morphine, methadone or heroin. These partial agonists have not yet undergone therapeutic trials; however, one of these drugs, buprenorphine, has had extensive clinical pharmacologic studies at the Addiction Research Center and shows promise as a treatment agent (Jasinski et al., 1978). These studies indicated that buprenorphine chronically administered in a single daily dose of 8 mg. produces morphine-like subjective effects and euphoria equivalent to that produced by 30 mg. of morphine administered subcutaneously four times daily. By extrapolation, this degree of euphoria and morphine-like subjective effects produced by chronic buprenorphine was equivalent to that produced by a single daily oral dose of methadone in the range of 40 to 60 mg. Thus, it would be predicted that chronic buprenorphine would produce subjective changes that would be acceptable to the addict patient. Buprenorphine produced a low level of physical dependence with a maximum degree of abstinence not occurring until 10-14 days after abrupt withdrawal. Thus, it is concluded that buprenorphine would produce little if any clinically significant physical dependence, indicating that the addicts maintained on buprenorphine can be easily detoxified. These observations also suggest that addicts possibly could be maintained on doses less frequent than once daily. Chronically given, buprenorphine is at least no more and possibly less toxic than other maintenance drugs. During chronic administration, buprenorphine blocked the effects of large doses of morphine, probably through two mechanisms. Because buprenorphine is a partial agonist, subjects can become tolerant and show cross-tolerance to other narcotics. On the other hand, buprenorphine can also act as an antagonist and as a competitive antagonist of narcotics. Other investigators have estimated that in man buprenorphine is equally effective on a milligram for milligram basis to naloxone as an antagonist.

In summary, buprenorphine would produce methadone-like and morphine-like subjective effects that would be acceptable to the addict patient. On the other hand, buprenorphine could not produce significant methadone- or morphine-like physical dependence and toxicity. Thus, it is felt that buprenorphine would have the advantages of agents such as naltrexone and cyclazocine which also do not produce clinically significant physical dependence or serious toxicity. Buprenorphine would also have a low abuse potential.

REFERENCES

Fraser, H.F. and Isbell, H. 1952. Actions and addiction liabilities of alpha-acetylmethadols in man. *J. Pharmacol. Exp. Ther.* 105: 458-465.

Jasinski, D.R. 1977. Assessment of the abuse potentiality of morphine-like drugs. In: *Drug Addiction. I. Morphine, Sedative-Hypnotic and Alcohol Dependence. Handbook of Experimental Pharmacology*. Vol. 45, W.R. Martin (ed.), pp. 197-258. Berlin-Heidelberg-New York: Springer-Verlag.

Jasinski, D.R., Pevnick, J.S., and Griffith, J.D. 1978. Human pharmacology and abuse potential of the analgesic buprenorphine: A potential agent for treating narcotic addiciton. *Arch. Gen. Psychiatry*. 35: 501-516.

Martin, W.R., Fraser, H.F., Gorodetzky, C.W., and Rosenberg, D.E. 1965. Studies of the dependence-producing potential of the narcotic antagonist 2-cyclopropylmethyl-2'-hydroxy-5,9-dimethyl-6,7-benzomorphan (cyclazocine, Win 20-740; ARC II-C-3). *J. Pharmacol. Exp. Ther.* 150: 426-436.

Martin, W.R., Gorodetzky, C.W., and McClane, T.K. 1966. An experimental study in the treatment of narcotic addicts with cyclazocine. *Clin. Pharmacol. Ther.* 7: 455-465.

Martin, W.R., Jasinski, D.R., Haertzen, C.A., Kay, D.C., Jones, B.E., Mansky, P.A., and Carpenter, R.W. 1973a. Methadone--a reevaluation. *Arch. Gen. Psychiatry.* 28: 286-295.

Martin, W.R., Jasinski, D.R., and Mansky, P.A. 1973b. Naltrexone, an antagonist for the treatment of heroin dependence. *Arch. Gen. Psychiatry.* 28: 784-791.

NALOXONE "FLUSH": A NOVEL APPROACH TO COMPLETE HEROIN ABSTINENCE

Alvin J. Cronson, M.D.

Texas Tech University School of Medicine

Naloxone hydrochloride is a specific narcotic antagonist whose main use is the reversal of coma due to opiates. In 1973 the Drug Research Unit of New York Medical College used it to test the patients' freedom from opiates, prior to the administration of naltrexone, which is more potent and has a longer duration. Both patients and staff called this procedure a "flush." If naloxone was administered too soon, generally less than a week after termination of methadone maintenace, abstinence symptoms such as muscle cramps, nausea and hypersecretions would occur. These could be symptomatically treated with diazapam, prochlorpromazine and atropine.

Patients desiring to discontinue methadone consulted the author with complaints of not being able to complete their withdrawal from methadone treatment programs because of the discomfort of residual abstinence symptoms even weeks after discontinuation of the drug. The author developed the hypothesis that naloxone would neutralize the residual opiates and might "flush out" residual withdrawal symptoms. He had surmised that what was called a "flush" was actually a reaction to the deactivation of the opiates. Perhaps the administration of naloxone would truncate and terminate abstinence symptoms that would otherwise cause patients to relapse. To test this hypothesis, the patients were asked to return after a week of symptomatic treatment and if discomfort persisted, they would be given a naloxone "flush."

MATERIAL AND METHODS

The subjects were ambulatory. The initial group of fourteen patients were seen in private practice in New York. This group consisted of nine patients that returned for the naloxone flush and five patients that were members of the Methadone Treatment Unit at Metropolitan Hospital in New York. From 1977 to the present, the study has been continued at the Drug Abuse Program at the Lubbock (Texas) Mental Health and Mental Retardation Center. This second cohort consists of eleven subjects who are still under continuing observation.

The study consisted of twenty-five subjects--sixteen males and nine females ranging in age from 20 to 45 years with a mean age of 29. Nine of the subjects were white, five were black, three were Puerto Rican, and eight were Mexican American. This was a relatively educated group with from nine to fifteen years ($\bar{x}$=12) of education. The length of addiction varied between two and twenty-six years with a mean of 4.5 years and they had been abstinent for between one to four weeks.

All the subjects presented signs and symptoms as described in the Himmelsbach Scale which had persisted at least a week. The Lubbock cohort, consisting of eleven patients, was rated according to a Himmelsbach Scale modified as follows: -2--Calm, relaxed, no craving or interest in opiates, sleepy; -1--Calm; 0--Anxiety, drug craving, insomnia; 1--Yawning, perspiration, lacrimation, rhinnorhea; 2--Midriasis, goose flesh, tremors, hot and cold flashes, aching bones, leg cramps, and anorexia. Vital signs were recorded and taken q 30 minutes x 3 and 0.4 mg. was administered IV or IM. The patients were then rated in five and in thirty minutes. If the first dose was tolerated a second dose IM was given and the patient observed and responses were recorded again.

RESULTS

The original (New York) group of fourteen patients were followed for a minimum of six months. Nine of these patients were abstinent from four months to one year before the author moved to Texas although it is only fair to mention that four of these patients were treated with lithium. On the eleven Lubbock patients rated with the Himmelsbach Scale, there were thirteen trials. All nine subjects reported and demonstrated an improvement of withdrawal symptoms. However, two patients required a repetition of the procedure two weeks after the first one. The eleven subjects were rated to be on a range from 1 to 2 ($\bar{x}$=1.76) at initial evaluation before naloxone "flush." Immediately after the first injection the modified Himmelsbach score improved to a range of -1 to 2 ($\bar{x}$=-.77). Thirty minutes after injection the range was further improved to -2 to 2 ($\bar{x}$=.31). There was a slight further improvement thirty minutes after the second (IM) naloxone .4 mg. injection to $\bar{x}$=-.54.

While the mean pulse rate decreased from 89.9 to 84.7 after thirty minutes and thirty minutes later was further down to 76.3, this is not regarded as significant, nor were there significant changes of blood pressure. Furthermore, of the eleven patients, nine are currently abstinent (seven are on lithium maintenance).

DISCUSSION

Fink et al. (1971), Zaks et al. (1971), Eddy and May (1973), and Jaffee and Martin (1975) describe naloxone as a pure narcotic antagonist of short duration of activity. "Pure" opiate antagonists compete with morphine-like drugs for the stereospecific opiate receptor without enlisting any response themselves. Subcutaneous doses of up to 12 mg. produce no discernible effects and 24 mg. causes only slight drowsiness. Doses of 0.4 and 0.8 mg. of naloxone given IV or IM prevent or promptly reverse the effects of 25 mg. of heroin. Agonistic effects depend on the dosage and route of administration.

Kurland et al. (1967) experimented with administration of naloxone in dosages of 1000 and 2500 mg. orally after his subjects showed evidence of heavy narcotic use. He found that following the precipitation of a severe abstinence syndrome subsequent administration of the drug elicited no further abstinence syndrome. Of fifteen paroled narcotic abusers who received this rapid detoxification, six were reported as having enjoyed lengthy periods of abstinence after treatment. Blackley et al. (1974) reported the use of naloxone in precipitating opiate abstinence and noted increased abstinence in twelve of twenty-nine who received a second series of naloxone injections. Kurland and McCabe (1976) administered parenteral doses of naloxone hydrochloride to twenty-nine parolees who showed evidence of increasing opiate use while participating in an aftercare abstinence program. Thirty-eight percent of the patients showed a "detoxification effect" characterized by a negative reaction to subsequent administrations. However, all twenty-nine subjects are reported to have relapsed.

Observing the patients desiring to detoxify from opiates often are deterred by the fear of being "sick," (as a residual withdrawal syndrome has been reported to last as long as six months after detoxification), the author decided to treat the residual syndrome symptomatically for one week. If the symptoms persisted, then naloxone in doses fo 0.4 mg. was administered, the first IV and the second IM, in an attempt to terminate the symptoms. Only two patients necessitated a repetition of the procedure. The remaining twenty-three felt almost immediate relief of symptoms of abstinence. The two patients that needed second treatments a week or two after the first have remained abstinent.

These results suggest that a naloxone "flush" is an approach to the residual withdrawal symptoms that have been consistently reported as a major cause for recidivism. The results are gratifying although one also should note the number of patients on lithium maintenance. The fact that many of these patients were treated for primary affective disorders may have significantly contributed to the success of the treatment modality.

ACKNOWLEDGEMENTS

The author wishes to express appreciation to Abraham Flemenbaum, M.D., M.S., Associate Professor and Research Psychiatrist with the Texas Tech University School of Medicine for his major critical review of the paper, his support on reporting the methodology, and the editing of the scientific material. The author also expresses his appreciation to Ms. Lupe Castro, LVN, and Ms. Petra Ramos of the Lubbock Mental Health and Mental Retardation Center for their cooperation and assistance in the performance of this project.

REFERENCES

Blackley, P.A., Casey, D., Marcel, L.J., and Denny, D.D. 1974. Titration of the opiate syndrome and naloxone: a mode for the study of the opiate abstinence syndrome. Presented at the First National Drug Abuse Conference, March 1974.

Eddy, M.B. and May, E.L. 1973. The search for a better analgesic. *Science*. 181: 407-414.

Fink, M., Fredman, A.H., Zaks, A.M., and Resnick, R.B. 1971. Narcotic antagonists. Another approach to addiction therapy. *American Journal of Nursing*. 71: 1359-63.

Jaffe, J.H. and Martin, W.R. 1975. Narcotic Analgesics and Antagonists. In L.S. Goodman and A. Gilman (eds.), *The Pharmacological Basis of Therapeutics* (5th ed.). New York: MacMillan Publishing Co.

Kurland, A.A., Krantz, J.C., and Kerman, R. 1967. Naloxone (N Allylnoroxymorphone) a pilot study of brief high dose administration. *Report of the 34th Annual Scientific Meeting of Committee of Problems of Drug Dependence*. Ann Arbor. May, 1967.

Kurland, A.A. and McCabe, L. 1976. Rapid detoxification of the narcotic addict with naloxone hydrochloride. A preliminary report. *Journal of Clinical Pharmacology*. 16: 68-74.

Zaks, A., Jones, T., Fink, M., and Freeman, A. 1971. Naloxone treatment of opiate dependence. *Journal of the American Medical Association*. 215: 2108-2110.

THE EFFECTIVENESS OF SHORT TERM TREATMENT WITH NALTREXONE

David C. Lewis, M.D., Rebecca Black, Ph.D., Joseph Mayer, Ph.D., and Ronald G. Hersch, Ph.D.

Washingtonian Center for Addictions

It was originally hoped that the widespread clinical availability of a narcotic antagonist would provide a vehicle for addicts to achieve long-term abstinence, and that a significant number of addicts would choose to take an antagonist for a long period of time. While naltrexone seemed particularly promising because it was free of the side effects of other opiate antagonists, our experience, and the experience of others, indicates that many patients are reluctant to take naltrexone and when they do so, many continue its use for only short periods of time.

Yet, for many patients, even short-term naltrexone use may have a significant impact on their craving for opiates, the nature of their controls and their therapeutic commitment both during and after they discontinue naltrexone use. Short-term experience with a narcotic antagonist may be useful in helping the addict achieve and preserve abstinence. If so, then the assumption that the major impact of naltrexone is limited to its period of actual use may not be correct. Unlike methadone maintenance treatment, in which many patients remain for several months to severl years, naltrexone treatment, in contrast, may be a treatment in which only a few patients remain for months to years. However, many patients may remain in treatment a short period of time, but may be significantly influenced by this treatment experience. Given this possibility a study has been initiated at The Washingtonian Center for Addictions which compares three randomly assigned groups of addicts. One group receives naltrexone, one group receives placebo medication, and the last group receives no supportive medication. The medication phase of the study is limited to a duration of one month with a six month follow-up evaluation. Since this study is double blind and now in progress we cannot report the results at this time. However, since the rationale for this study developed from our experience with offering long term naltrexone treatment to addicts, we would like to describe the current status of that experience which began in 1974.

SUBJECT SELECTION

During a 29-month period beginning in January of 1974, 738 patients hospitalized at the Washingtonian Center for Addictions for opiate detoxification were eligible to receive naltrexone.

Of these 738 patients, 133 (18%) expressed serious interest in such treatment, 47 (6%) eventually signed informed consents, and 22 (3%) completed the baseline evaluation of physical and psychological tests and received naltrexone.

SUBJECT CHARACTERISTICS

Table I shows the demographic characteristics of the 22 naltrexone subjects.

Table I. Characteristics of 22 Naltrexone Subjects

Characteristics	Subjects (22)
Age	27.7 years
Length of addiction	7.1 years
Age of initial addiction	20.6 years
Previous number of addiction treatments	7.5 years
Education	11.6 years
Legal Status:	
Previous number of arrests	7.4
Previous number of convictions	3.1
Race:	
White	77%
Black	23%
Religion:	
Catholic	64%
Protestant	18%
Jewish	9%
Other	9%
Marital Status:	
Separated/divorced	36%
Married	32%
Single	32%
Employment Status:	
Employed	14%
Unemployed	86%

The mean age of subjects receiving naltrexone was 27.7 years at the onset of treatment, with an average age of 20.6 years at the onset of addiction and an average duration of addictive opioid use of 7.1 years. The average patient completed high school and was of normal intelligence. This mean number of previous treatment attempts was 7.5 and included psychotherapy, detoxification, methadone maintenance, and participation in therapeutic communities. Three of the 22 subjects were employed at the time of naltrexone induction.

RESULTS

Duration of Treatment

The mean duration of receipt of naltrexone was 43.3 days (6.2 weeks); the median duration was 35.0 days (5.0 week). Time on naltrexone ranged from less than 1 week to 21 weeks. Twelve subjects (55%) continued in abstinence-oriented addiction treatment after stopping naltrexone. The mean length of post naltrexone treatment for these 12 subjects was 5.6 weeks.

Subjects were divided into two groups, those who received naltrexone for less than 8 weeks (15 subjects) and those who received naltrexone 8 weeks or longer (6 subjects). Subjects who remained on naltrexone 8 weeks or longer were not substantially different in age, race, education, age at onset of addiction, and employment status from subjects who terminated treatment before 8 weeks.

However, subjects who remained in treatment for 8 weeks or longer had fewer prior arrests (mean 4.6 vs 13.2), and more of them were married than the group of subjects who terminated treatment before 8 weeks (66 vs 13%).

Treatment Outcome: Addiction and Employment

At the time the data were analyzed, 2 of the 22 subjects were still taking naltrexone and were abstinent. Nine other subjects who had terminated naltrexone treatment were interviewed using a structured questionnaire. Additional information on these 9 subjects was available through the staff of the Center, other patients, and the subjects' families, and corroborated the information gathered in the structured interview. Of the remaining 11 subjects, 1 was deceased, 2 were incarcerated, and 8 subjects could not be located. These data were collected at an average of 45.3 weeks (range: 1-97 weeks) following receipt of the last dose of naltrexone and are shown in Table II.

Excluding the 2 subjects currently receiving naltrexone, the subject who died and the 2 incarcerated, 17 subjects remain. Of

Table II. Addiction Status of 22 Naltrexone Subjects

Addiction Status	Number of Subjects
Abstinent (receiving naltrexone)	2
Abstinent (not receiving naltrexone)	9
Deceased	1
Incarcerated	2
Known Readdicted	0
Unknown	8
Total	22

these 17 subjects, 9 were abstinent and the addiction status of 8 was unknown. Readdiction may have contributed to our inability to contact these 8 subjects. Thus at least 9 (53%) of the 17 of the subjects following naltrexone treatment have remained abstinent.

Excluding the 3 subjects whose death or incarceration renders themunable to work, 9 (47%) of the remaining 19 subjects were employed. Only 3 (16%) of these 19 were employed at the onset of naltrexone treatment.

Addiction status and employment were also examined in relation to the duration of treatment on naltrexone in order to determine whether or not those subjects who received naltrexone treatment longest had more positive treatment outcomes. There was no significant difference in addiction status and employment between these two groups.

Clinical Responses

A central psychological issue in the treatment of opioid addicts with antagonists is the issue of control. Many addicts become overwhelmed in the absence of external supports and controls. The naltrexone serves as an external control against the impulse to use heroin, a control which the patients value, but about which all are ambivalent.

Initially, almost all issues of control are focused on the "power of the drug, naltrexone, to counteract the effects of heroin. Some patients want to test the power of this external control against the power of heroin by experimenting early in their treatment with the effectiveness of the antagonist. After only one such test many are convinced that the drug is effective and most discontinue all opioid testing.

Once they are convinced of the drug's effectiveness, many

patients experience a reduction in the desire for heroin and a marked decrease in obsessional thinking about heroin acquisition and the heroin high. For some patients this initial reaction is short-lived and there is a resurgence of obsessive thinking about heroin, or similarly, of phobic thinking about the abstinence state, often accompanied by fears of never again being able to get high. Some patients, disillusioned that the naltrexone did not eliminate their thoughts about heroin, terminate naltrexone treatment at this time. For some patients, fear of abstinence is a continuing issue, although the conflict may diminish in intensity over time.

Within the first 3 or 4 weeks on naltrexone, many patients test the necessity of using naltrexone as an external control against the desire to use heroin. In an attempt to achieve a balance between the external control afforded by the drug and their own internal controls, and to lessen the degree of dependency they feel on the drug and on the clinic, clients start to miss clinic appointments to see if they can abstain from heroin use. For example, they may leave the city for a day or lock themselves in their houses to avoid temptation. Again, it appears that they are attempting to show that they can exercise adequate control over the urge to use heroin without the help of naltrexone.

As patients recognize that they can exercise partial control themselves, they become more conscious of their participation in establishing this balance between external and internal controls. One patient stated, for example, that he now found it necessary to "change everything at once." Naltrexone helped him to release energy for uses other than total control or complete change and the impact of "external circumstances" was reduced, allowing him to introspect and to take time to think through decisions.

The use of naltrexone reduces the fear that external environmental factors will cause a loss of control. After "testing" themselves in a nondrug environment on weekends, patients became aware that they can be on the street, in the places where they normally would take heroin, without taking heroin. This enlargement of the nonthreatening, formerly self-destructive life space seems to allow patients to feel freer.

The naltrexone treatment program has provided a unique experience for the clinical staff as well as the patients. Not only did staff need to learn about the pharmacological effects of naltrexone, but also to accept the use of a drug in a program that was oriented toward abstinence. In contrast to the regular, more predictable methadone and LAAM maintenance schedules with which the staff were familiar is this modality in which patients frequently test their own self-control by skipping doses. The staff had to become sensitive to this testing and also accepting of the relatively short periods of time that patients would use the naltrexone.

As the final phase of the investigational use of naltrexone begins, a larger number of clinical settings will be providing naltrexone and considerably more experience will be gained to evaluate its effectiveness. As with other successful interventions in the addictions, the use of this drug can be viewed as one element of the support that can be used by skilled clinicians in developing a treatment plan tailored to the needs of each client.

ACKNOWLEDGEMENT

This research was supported by Contract No. HSM 43 73-264 from the National Institute on Drug Abuse.

NARCOTIC ANTAGONISTS AND STRATEGIES FOR ENHANCING THEIR EFFECTIVENESS

Albert A. Kurland, M.D.

Maryland Psychiatric Research Center

Abstract: A series of clinical trials employing the narcotic antagonist, naloxone, administered on a maintenance and on a contingent basis, have indicated that the therapeutic promise of these compounds has not been borne out. Similar investigative experiences appear to be in the process of being reduplicated by the more potent narcotic antagonist naltrexone. These have emphasized the need for treatment approaches which might enhance the patient's motivation relative to the treatment endeavor. One such possiblity is the structuring of a dramatic treatment experience prior to the initiation of treatment with a narcotic antagonist. For this purpose the use of psychedelic psychotherapy is suggested.

Ever since Martin, Gorodetsky and McClane reported on the use of Opioid Antagonists in 1966 for the purpose of preventing the relapse of "detoxified" addicts, the hopeful promise that such therapeutic management held out has continued to be an elusive quest. More recently as newer antagonists, naloxone (Blumberg et al., 1961; Fink, 1970) and naltrexone (Blumberg and Dayton, 1972; Martin, Jasinski and Mansky, 1973) were introduced possessing "pure" antagonist properties which could be administered orally and were relatively free of toxic effects at high dose, their therapeutic promise took on a new brightness. However, the brightness has begun to fade as controlled studies have replaced the initial pilot investigations yielding data of increasing uncertainty characterized by high dropout rates and relatively little difference between drug and placebo as reflected in the studies of this writer and his associate (Hanlon et al., 1977).

These studies, designed to shed light on the relevance of such issues as: what specific therapeutic actions do antagonists have when placebo and programmatic elements are controlled?; must antagonists be administered daily or can they be effectively utilized on a "crisis-intervention" basis?; do their side effects affect outcome?; are there specific categories among the addict population that might respond better to antagonist treatment than others?; and is antagonist treatment uniformly effective over diverse outcome criteria? have confronted the clinician with a paradoxical state of affairs in which there appears to be "the failure of an effective approach."

The apparent paradox arises from the fact that there is little reason to doubt that a "pure" antagonist such as naloxone and naltrexone successfully blocks the euphoric and analgesic effects produced by opiates, theoretically reducing an individual's incentive to use narcotics when an antagonist is being administered. Perversely, if anything rather than inhibiting the drug-taking behavior of the street or parolee felon addict the principal effect of these compounds seems to be that of reinforcing the avoidance of medication. For many such individuals there is simply no incentive or reward in the taking the medication that denies them the relief they obtain from narcotics. When viewed on this basis antagonists can be expected to be no more effective in alleviating chronic narcotics use than disulfiram (antabuse) has been in alleviating chronic alcohol use.

Analysis of the lacklustre therapeutic accomplishments of the antagonists outlines two major obstacles which may be attenuating their treatment promise. First, there is no uncertainty that with the administration of the narcotic antagonists, the urge to use narcotics is not reduced. If anything because of its neutralizing effect on the presence of opiates, the responsiveness to the opiate experience may be enhanced with the discontinuation of the antagonists. In essence, whenever an individual wishes to receive his customary reinforcement from narcotics use, he simply avoids antagonist medication, thereby setting up his own reinforcement schedule wherein avoidance of medication becomes the conditioned response.

The complex factors interacting in this process as conceptualized by Wikler (1974) proposed that in man, relapse is due to evocation by drug-related environmental stimuli of fragments of the opioid-abstinence syndrome that had become classically conditioned to such stimuli during previous episodes of addiction. Wikler states: "On the street," the opioid user who is both tolerant and physically dependent frequently undergoes withdrawal phenomena before he is able to obtain and self-administer the next dose. Given certain more or less constant exteroceptive stimuli (physical environment, "pushers" other opioid users) that are temporally contiguous with such episodes, the cycle of opioid abstinence and its termination can become classically conditioned to such stimuli while opioid-seeking behavior is operationally conditioned, both types of conditioning being attributable to the reinforcement provided by the efficacy of opioids in suppressing withdrawal phenomena. In consequence, long after "detoxification," transient withdrawal phenomena, experienced and interpreted as "craving" or narcotic "hunger," together with renewed "hustling" for opioids (relapse) can occur as conditioned responses when the subject returns to his original environment and is exposed to such stimuli. Also, certain of the interoceptive actions of the opioids, not involved in the suppression of abstinence phenomena, can acquire similar conditioned properties inasmuch as in a tolerant and physically dependent individual, they are often followed by withdrawal phenomena before termination of the latter by the next dose. Other interoceptive events can likewise acquire the property of evoked conditioned self-administration of opioids. For example, "anxiety" is frequently associated with the opioid abstinence syndrome, and probably the two phenomena are medicated in part, by the same central nervous system pathways. Hence, the occurrence of "anxiety" for whatever reason long after "detoxification" may result in relapse.

If one assumes the existence of "conditioned abstinence" the question as to how this may thus be attenuated arises from the possibilities suggested by the theories of classical and operant conditioning, a rather simplistic approach suggests that the detoxified or abstinent addict on being returned to his usual environment would go through a deconditioning treatment in which he would be exposed to a series of trials in which the addict would be administered an opiate and immediately afterward a narcotic antagonist. Over a period of time there should occur an extinction of the exteroceptive stimuli. Another major facet of the problem, namely, the linkage or association of anxiety or dysphoric states with the craving experience and the inability to differentiate between the two raises the question whether this might be countered with the use of a psychotropic agent combined with the administration of the narcotic antagonist, a course that requires a careful selection of the psychotropic compound in order to avoid the introduction of an agent which would further enhance the opiate experiences or give rise to a physical dependency of its own.

The second major consideration is focused on the treatment and the steps which might be taken to enhance the individual's motivation. Currently, these rely primarily on a combination of mandatory and psychotherapeutic measures. In the case of the former, the motivation is derived from the possibility of being sent to a correctional institution should the individual fail to comply with the requirements of the program. In the latter, the attempts to utilize group and individual psychotherapy to assist in the rehabilitative process necessitates a sustained commitment to the psychotherapeutic undertaking and for many of the addicted its meaningfulness appears remote to their existential life styles. Consequently, these usual types of therapeutic intervention generally offer little in staying their deviant course.

In the search for more effective approaches for altering the addict's course, a phenomenon which attracted considerable attention is the dramatic impact of a conversion experience in altering the lifestyle which is occasionally observed in the addictive disorders. It would seem, as a consequence of this experience with its intense emotional overtones, there occurs a marked attenuation to the conditioning and operant reactions which contribute toward the maintenance of the matrix of behavior that makes for the continuance of the drug-consuming behavior. Growing interest in replicating this phenomenon with the patient exposed to carefully controlled but powerful experiences of a transcendental nature has suggested an approach which has yielded some rather encouraging observations but which require a great deal of additional study and validation.

For this purpose, there have been some encouraging experiences reported utilizing the LSD experience as an adjunct to psychotherapy (Kurland, 1978). The therapeutic impact brought about by the intense nature of the drug experience has, in the addictive disorders relieved a persisting depression, generalized anxiety and dysphoric discomfort and in the treatment aftermath a renewed sense of motivation. An aspect of this experimental treatment form frequently referred to as psychedelic drug-assisted psychotherapy is its apparent effectiveness in culturally deprived patients who tend to score low on standard tests of intelligence and who ordinarily would be considered poor candidates for conventional psychotherapy. This approach to emotional reeducation represented by the experimental psychedelic treatment form does not, in our experience, seem prejudiced against individuals lacking in verbal skills or sophistication. Many of these patients who also might be considered difficult or inappropriate candidates for conventional psychotherapy of one type or another would be, because of the nature of their personalities, so frequently dominated by a range of disorders of a sociopathic or psychopathic makeup, that it would be difficult for them to enter into the conventional psychotherapeutic relationship which can apparently be engaged to some degree by the psychedelic treatment form into a wholesome productive rehabilitative relationship. As a result of this sense of renewal, compliance with the use of a narcotic antagonist may be much more readily carried out since the addict has had demonstrated to him in a dramatic manner the potency of a treatment that offers him a ready and more healthy recourse for working through the dysphoric states which so prevail and flourish in the vulnerable and traumatized recesses of his personality structure.

However, before concluding these putative comments concerning the mutually supportive elements suggested by the adjunctive use of this experimental paradigm designed to add to the motivational forces brought about by a renewed sense of self-worth, even if it is of relatively brief duration of a few weeks to a few months, this could provide the necessary support for the the addict to accept the protective use of the narcotic antagonist at a time when his vulnerability would ordinarily be at its greatest. Thus by providing an experience which takes the individual "out of himself" which is accompanied by a system of symbols and beliefs and encourages reflection in the experiencer with a desire for their understanding, and guided by the therapeutic maneuvers of a knowledge and experienced clinician, we offer the patient a greater opportunity for initiating the healing of the troubled and perhaps in the healing, the individual may emerge better than he or she was to begin with.

REFERENCES

Blumberg, H., Dayton, H.B., George, M. and Rapaport, D.N. 1961. N-allylnoroxymorphine: a potent narcotic antagonist. *Fed. Proc.* 20:311.

Blumberg, H. and Dayton, H.B. 1972. Narcotic antagonist studies with EN-161639A (N-cyclopropylmethylnoroxymorphone). In: *Fifth International Congress on Pharmacology,* San Francisco. Abstracts of volunteer papers.

Fink, M. 1970. Narcotic antagonists in opiate dependence. *Science* 169: 1005-1006.

Hanlon, T.E. et al. 1977. Narcotic antagonist treatment of addict parolees--the future of an effective approach. *Comp. Psych.* 18: 211-219.

Kurland, A.A. 1978. *Psychiatric Aspects of Drug Dependence.* West Palm Beach, Florida. CRC Press, Inc.

Martin, W.F., Gorodetzky, C.W., and McClane, T.K. 1966. An experimental study in the treatment of narcotic addicts with cyclazocine. *Clin. Pharmacol. Ther.* 7: 455-465.

Martin, W.R., Jasinski, F.P. and Mansky, P.A. 1973. Naltrexone, an antagonist for the treatment of heroin dependence: effects in man. *Arch. Gen. Psych.* 28: 784-791.

Report of the National Research Council Committee on Clinical Evaluation of Narcotic Antagonists, 1978. Clinical evaluation of naltrexone treatment of opiate-dependent individuals. *Arch. Gen. Psych.* 35: 335-340.

Wikler, A. 1974. Opioid Antagonists and Deconditioning in Addiction Treatment. Paper presented at the syposium on drug dependence-treatment and treatment evaluation. Spandia Interna. Symposium. Stockholm, Sweeden.

NALTREXONE: IN OPIATE TREATMENT PROGRAMS AND DETOXIFICATION PROCEDURES

Leonard S. Brahen, Ph.D., M.D., Thomas Capone, Ph.D., Harold E. Adams, C.S.W., and Raymond J. Condren, Ph.D.

County of Nassau, Department of Drug and Alcohol Addiction

Methadone is currently the primary drug treatment for narcotic addiction. It is widely used as a substitute for other opiates in maintenance programs, and is also used for opiate detoxification. Methadone is addictive and it has been criticized for perpetrating an addictive cycle (Coghlan et al., 1974; Brown et al., 1975). Further many deaths have been attributed to methadone. Methadone diversion, into the illicit channels has resulted in the promotion of criminality.

Therefore a means has been sought to treat the more motivated addict with a new chemical approach. More recently narcotic antagonists have been investigated in different treatment settings (Brahen et al., 1973; Schecter, 1975). One such treatment program is being conducted by the Medical Research and Education Unit of the Nassau County Department of Drug and Alcohol Addiction (Brahen et al., 1974; Brahen et al., 1976; Brahen et al., 1977; Brahen et al., 1977). This program operated two separate services. The first, The Narcotic Antagonist Work-Release Program is located in the Work-Release Facility of the Nassau County Correctional Center (Institutional). It is designed to incorporate eligible inmates from the maximum security Correctional Center, into the regular Work-Release Program by immunization against opiates with blocking doses of the non-addictive antagonist, naltrexone. The second, The Antagonist Treatment Clinic is located at the Nassau County Medical Center and is operated by the same staff. The hospital Outpatient Clinic is an extension of the Work-Release Program for post-incarceration treatment. The outpatient program now accepts referrals from probation, parole, community agencies, and "street" friend referrals as well.

NARCOTIC ANTAGONISTS

The mechanism of action of narcotic antagonists has been extensively dealt with in other publications (Brahen et al., 1977; Jaffe, 1970; Goldstein, 1974; Dundee and McCaughey, 1976). However, it may be helpful to review the three main effects associated with narcotic antagonists: (1) The precipitation of acute withdrawal symptomology in opiate dependent persons. If the narcotic antagonist is administered after the opiate, competitive displacement of the opiate occurs from the receptor with abrupt termination of narcotic effects and the precipitation of an acute withdrawal syndrome. (2) The reversal of actions of opiate drugs. Displacement as described above is the pharmacological basis for the use of narcotic antagonists in the treatment of acute opiate overdoses and in establishing opiate dependence. (3) The blockade of opiate effects. Administration of sufficient, established doses of a narcotic antagonist prior to opiate administration prevents opiate linkage to the receptor sites and results in a blockade of opiate effects. Frequent, adequate doses of an antagonist prevent the characteristic opiate response and physical dependence.

Currently, naltrexone appears to be the antagonist of choice as an oral blocking agent. It is relatively free of agonistic effects thus permitting rapid induction and little or no withdrawal symptoms on abrupt discontinuation. It can be effectively administered orally and has a long duration of action (Schecter, 1975). Oral blocking doses of 50 mg., 100 mg. and 150 mg. offer protection for approximately 24 hours, 48 hours and 72 hours, respectively.

THERAPEUTIC USES

Rapid Detoxification

A rapid opiate detoxification procedure has been devised by this unit as an alternative to costly and long term (10 day) methadone detox or the "cold turkey" method (Protocol, 1976). It is designed to detoxify select opiate dependent persons and transitionally induct them onto the narcotic antagonist in one to two days. Management of the withdrawal syndrome is accomplished by the pre-withdrawal administration of symptom suppressing drugs. For example, Compazine and Valium may be administered orally in combination to initially reduce anxiety, the anticipated muscular pains and twitching and autonomic gastrointestinal symptoms such as nausea, vomiting and diarrehea. Naltrexone is administered initially at a low dose level and is incrementally increased over time to a total blocking dose of 50 mg. The resulting induced withdrawal is symptomatically treated by the periodic and sequential administration of Valium and Compazine orally or parenterally and either aspirin or Tylenol as required. This regimen of vigorous drug treatment of induced withdrawal symptoms minimizes the symptoms and patient discomfort. The patient is continuously monitored by frequent recording of vital signs, symptoms, complaints and the required directive procedures. After detoxification, the patient may continue with the Antagonist Treatment Program as a patient or be referred to another treatment modality as indicated.

There are several advantages to this method of detox compared to traditional ten day methadone detox in a psychiatric facility or the drugless "cold turkey" method. (1) Compared to lengthy institutional detoxification, only one or two days are lost from a patient's regular activities, such as, work, school, family supervision, and so forth. (2) The expected and usual discomfort of narcotic withdrawal is minimized by the administration of appropriate symptom suppressing drugs and (3) the cost of detoxification accomplished in one to two days is small compared to expensive institutional detox of approximately 1,200 to 1,400 dollars per patient.

Narcotic Antagonist Work-Release Program and Outpatient Program

The general operational format to be described below for the Antagonist Treatment Clinic is also adhered to for the Work-Release Program. Only minor procedural changes have been made to accomodate patients and maintain efficient liaison personnel relationships within the more rigid institutional setting. However, since the generalized required format is essentially the same for both programs discussion here will be basically limited to the Antagonist Treatment Clinic.

Criteria for admission to the program include a history of documented opiate addiction. However, no time limit is specifically stipulated. Males in the age group 18 to 45 years can be candidates. Females require special consideration regarding the risk to benefit ratio. Pregnancy during antagonist treatment is contra-indicated. Therefore, preventive measures such as an IUD or the "pill" are mandatory. Special approval from F.D.A. must also be obtained.

All subjects must, by examination, be found in reasonably good physical and mental health. Their motivational status should also be at such a level as to give some assurance that program requirements will be adhered to. When the court requires a probationer, or parolee be protected from addiction by chemical means the antagonist can serve as a non-addictive, relatively safe methadone alternative having no "street value."

The admission procedure requires that complete history be obtained, a physical examination be given and laboratory tests be performed. This is done before the patient begins induction: (The starred examinations and laboratory tests are done as required by F.D.A.)

a) Physical examination*
b) Neurological examination*
c) Chest X-ray
d) Slit-lamp eye examination
e) E.K.G.*
f) Urinalysis*
g) Prothrombin time
h) CBC (with differential)*
i) SMA 12/60 and 6/60*
j) Australian Antigen
k) VDRL
l) Reticulocyte*
m) Platelet count*
n) Sickle Cell

A mental status evaluation and psychological testing for baseline data and treatment planning are also conducted on entering the program, and repeated at appropriate time intervals. The Rorschach, MMPI, as well as a Psychosocial History are administered routinely upon admission.

Each patient voluntarily agrees to enter this drug treatment program after it has been fully explained to him (including sufficient information about the possible harmful effects of the investigational antagonist) so that he can make and sign an informed consent. At the time of the interview for work-release the informational brochure is reviewed with the candidate.

In preparation for naltrexone induction the patient is detoxified from opiates in one of three ways: (a) Hospitalized for ten days in a psychiatric facility and detoxified using methadone with daily reduction of dosage and controlled withdrawal. (b) Self-detoxification--"cold turkey"--outpatient. (c) Transitional one-day detoxification-naltrexone induction procedure whereby detoxification and naltrexone induction are accomplished in twelve to twenty-four hours.

In general all patients on the day of induction receive naltrexone as follows: 5 mg. initially, followed by 10 mg., 10 mg. and 25 mg. with 20 - 30 minute intervals between doses. Patients are then maintained on the program with the equivalent of 50 mg. of naltrexone daily. This is administered on a twice a week basis. The patient usually receives 150 mg. of naltrexone on Mondays and 200 mg. on Thursdays. Urines for drugs of abuse and vital signs are monitored at each visit. Patients are also seen at least one evening a week for individual, group and vocational counseling. Every attempt is made, where appropriate and with patients approval to engage families in treatment. Additionally, referrals are often made to other units within our agency to provide further vocational evaluation and assistance, family services, as well as other specialized services. It is our belief that a strong emphasis on such supportive services is necessary for treatment success.

CONCLUSIONS

Our experience suggests that the two times a week dosage regimen is feasible and practical for both the institutionalized work-release and outpatient populations in a strictly treatment oriented setting for highly motivated persons (Brahen et al., 1978). However, with this regimen, some therapeutic intimacy is lost compared to daily dosing so that opportunity for effecting behavioral change is lessened. This is one reason daily contact is still maintained in the Work-Release Facility. In an attempt toward greater motivation for program participation, antagonist treatment is presented as a high status program for select and treatable persons.

Preparation of our incarcerated population for re-entry into the community is an important facet of our treatment package. This is evidenced by the fact that approximately 23 percent of Work-Release Programs patients are taken off the program and sent to maximum security for work-release infractions. Aggressiveness and negativistic behavior emerge as the release date grows nearer, usually from days to a month before release. Patients often express a fear of freedom in the community and going back to family situations. Retention of patients in the controlled incarcerated setting is approximately 70 to 75 percent.

Retention of patients in the outpatient clinic is more difficult ranging from 4 days to 590 days for the first treatment phase with a mean of 62.3 days (N=33, S.D. 114.9). Those who entered the second treatment phase stayed in treatment ranging from 1 day to 91 days with a mean of 20.97 days (N=14, S.D. 25.8). Even with the imposition of rules and regulations and the conditions of probation or parole there is sporadic absence from treatment.

It should be emphasized that naltrexone patients unlike those in methadone programs are not motivated by the availability of an addicting agent. Consequently, the stress of a new narcotic free life may result in a premature termination from the program. Such behavior may represent an attempt to operate more autonomously. Such patients when unable to continue drug-free are permitted program re-entry. Such returns may reflect the need for emotional "refueling" and may represent patient growth toward a greater self-regulatory capacity. Typically at a latter point in antagonist therapy, varying with the patient, but usually within 1 to 1½ years we note more successful program separation efforts. We believe it important to allow and even encourage appropriate attempts by patients to give up the antagonist as a "transitional object", in favor of more independent functioning. While we have not yet quantified the percentages of patients who have achieved a permanent remission, our clinical observations suggest that relapses subsequent to a relatively long period (1 to 2 years) of antagonist treatment are less severe and lengthy than after a short term program exposure. In essence this treatment differs from methadone in that more rapid dynamic changes may occur in the patient. Further, this new chemotherapeutic approach, appears to offer a greater opportunity for ego growth and the acquisition of self regulatory behavioral and emotional mechanisms. We have presented elsewhere a more detailed analysis of the difference in the psychodynamic processes of the antagonist and methadone treatment programs (Capone et al., 1978).

It appears, from our studies, that naltrexone is a relatively safe antagonist. A full blocking dose can be administered to the opiate-free patient without incremental induction doses, without producing any untoward effects. Naltrexone maintenance dose can be abruptly withdrawn without any withdrawal effects (Brahen et al., 1977). In general our follow-up laboratory and other examinations showed a few significant changes from predrug baseline examinations, but all still fell within normal limits (Brahen et al., 1976).

Not all somatic and behavioral symptoms can be attributed to drug effect. Certain individuals who can be identified by psychological examination show a greater tendency to experience untoward placebo and drug effects (Capone et al., 1976). The importance of supportive services cannot be over emphasized. Presently, and in the near future, orally administered naltrexone will be used as a transitional modality from an opiate, such as methadone, to a drug free state. Beyond this, with the emergence of long-acting naltrexone for long term immunization, this antagonist will be an important drug in the control of motivated opiate addicts in other settings, including probation and parole (Brahen et al., 1978).

REFERENCES

Brahen, L.S., Capone, T. and Wiechert, V. 1973. A new antagonist treatment program for narcotic addicts. *J. Drug Educ.* 3:4.

Brahen, L.S., Capone, T. Wiechert, V. and DeJulio, P. 1974. Nassau County pioneers work-release program for addicted inmates. *Amer. J. of Correction*. 16-18.

Brahen, L.S., Weichert, V. and Capone, T. 1976. Narcotic antagonist treatment of the criminal justice patient; institutional vs. outpatient, including a 24 hour detox naltrexone induction regiment with oral medication. Proceedings of the 38th Annual Scientific Meeting of the Committee on Problems of Drug Dependence, Richmond, Va.

Brahen, L.S., Capone, T., Wiechert, V. and Desiderio, D. 1977. Naltrexone and cyclazocine: a controlled treatment study. *Archives of General Psychiatry*. 34: 1181-1184.

Brahen, L.S., Capone, T., and Wiechert, V. 1977. Antagonist Treatment Clinic: Long Term Management Approach to Opiate Dependence. Nassau County Medical Center Proceedings. 4(4).

Brahen, L.S., Capone, T. Heller, R.C., Linden, S., Landy, H. and Lewis, M. 1978. Controlled Clinical Study of Naltrexone Side Effects Comparing First-Day Doses and Maintenance Regimens. *Am. J. of Drug and Alcohol Abuse*. 5(2).

Brahen, L.S., Capone, T., Wiechert, V. et al. 1976. Controlled studies of the narcotic antagonists; an assessment of cyclazocine and naltrexone drug effects and laboratory data. Proceedings of the Third National Drug Abuse Conference. New York.

Brahen, L.S., Capone, T. and Bloom, S. 1978. An alternative to methadone for probationer-addicts: Narcotic Antagonist Treatment. *Contemporary Drug Problems--A Law Quarterly*.

Brown, B.S., Jansen, D.R. and Benn, G.J. 1975. Changes in attitude towards methadone. *Arch. Gen. Psychiat.* 32: 214-218.

Capone, T., Brahen, L.S., Brahen R. 1978. Psychoanalytic and behavioral considerations in antagonist and methadone treatment. *J. Contem. Psychother.* 9(2): 139-150.

Capone, T., Brahen, L.S., and Wiechert, V. 1976. Personality factors and drug effects in a controlled study of cyclazocine. *J. Clin. Psychol.* 32: 489-485.

Coghlan, A.J., Pixley, L. and Zimmerman, R.S. 1974. Community mental health concepts and methadone maintenance: Are they compatible? *Community Mental Health*. 10: 4.

Dundee, J.W. and McCaughey, W. 1976. In Avery, G.S. (ed.), *Drug Treatment: Principles and Practice of Clinical Pharmacology and Therapeutics*. Publishing Services Group, Inc.: Acton, Mass.

Goldstein, A. 1974. Interactions of narcotic antagonists with receptor sites. In Braude, Harris, May, Smith, and Villarreal (eds.), *Advances in Biochemical Psychopharmacology*. Vol 8. Raven Press: New York.

Jaffe, J.H. 1970. Narcotic analgesics. In Goodman, L.S. and Gilman, A. (eds.), *The Pharmacological Basis of Therapeutics* (4th edition). MacMillan: New York.

Protocol 1976. 24- to 48-hour opiate detoxification regimen and a subsequent naltrexone induction treatment regimen. On file with F.D.A.

Schecter, A. 1975. Naltrexone in a clinical setting: observations with 50 patients during 14 months. Proceedings of the Thirty-Seventh Annual Scientific Meeting of the Committee on Problems of Drug Dependence. Washington, D.C.

NALTREXONE: FACTORS CONTRIBUTING TO RETENTION IN TREATMENT

R.A. Greenstein, M.D., J. Grabowski, Ph.D., C.P. O'Brien, M.D., Ph.D., Melody Long, R.N., G. Woody, M.D., and S. Livingston, R.N.

University of Pennsylvania and Veterans Administration Hospital

The narcotic antagonist naltrexone has been evaluated in extensive clinical trials (Bradford, Hurley, Golondzowski, and Dorrier, 1976; Brahen, Capone, Heller, Landy, Linden, and Lewis, in press; Lansberg, Taintor, Plumb, Amico, and Wicks, 1976; Martin, Jasinski, and Mansky, 1973; O'Brien, Greenstein, Mintz, and Woody, 1976). However a variety of factors appear to have contributed to the relatively low acceptance of naltrexone as a treatment form and the short duration of naltrexone maintenance usually observed. Thus at least some of the factors contributing to naltrexone's desirability from a therapeutic viewpoint result in it being a less preferred treatment agent from the viewpoint of many patients. Several reports (e.g., Callahan, Rawson, Glazer, McCleave, and Arias, 1976; Meyer, Randall, Barrington, Mirin, and Greenberg, 1976; Stitzer and Bigelow, 1975; Wikler, 1976) have described or suggested procedures which might facilitate treatment. It appears that in combination with other approaches naltrexone may serve as a useful adjunct in treatment of opiate users (Brahen et al., in press; Callahan et al., 1976) although no optimal combination has yet emerged (Callahan, Rawson, McCleave, Arias, Glazer, and Lieberman, in press).

The present paper describes clinical experience with naltrexone in terms of treatment, use of other drugs and approaches to enhancing treatment effectiveness.

METHOD

Subjects were recruited from the patient population at the University of Pennsylvania/Philadelphia Veterans Administration Hospital Drug Dependence Treatment Service where the investigation was undertaken, and from other drug dependence treatment programs in the Philadelphia area. Individuals engaging in street opiate use or currently in methadone maintenance as well as those who were drug free were considered for treatment. Presence of medical disorders, including diabetes, hepatitis, or gastrointestinal and cardiovascular problems, and in the case of females, being of child bearing age, were criteria for exclusion. Severe behavioral-psychological disturbances other than those directly related to drug use were also considered to warrant exclusion in the selection process. Physical examination and laboratory tests including CBC, SMA 12, serology, Australian antigen, urinalysis, EKG and chest X-ray were done at intake. Psychological evaluation included the Brief Psychiatric Rating Scale, Wonderlic Personnel Test, Gordon Personal Profile, Beck Depression Inventory as well as anxiety and symptom checklists.

During the period from late 1973 to early 1978, 360 individuals were screened for the Naltrexone-Behavioral Research Project. Following opiate detoxification and a 2-5 day drug-free period, the short acting antagonist naloxone (0.4-0.8 mg.) was administered intravenously to assure absence of opiate dependence prior to administration of the longer acting antagonist naltrexone (Blachly, 1973). Appearance of moderate or severe abstinence symptoms following naloxone administration resulted in postponement of naltrexone induction. Absence of symptoms or presence of mild symptoms was followed by induction with orally administered naltrexone. During induction both placebo and active solutions were administered. Dosage was increased over a five day period and naltrexone was then administered in doses of 100 mg. on Monday and Wednesday and 150 mg. on Friday. Subsequently some patients were given the option of a 200 mg., twice per week regimen. Throughout the course of naltrexone treatment, physical examinations and laboratory analysis were repeated monthly and urine samples were obtained regularly.

In addition to naltrexone administration a variety of services including individual as well as group therapy, job placement and educational guidance was available. Staff members providing services included a rehabilitation counselor, nurses, physicians and psychologists. Some patients participated in investigations involving analysis of self-injection behavior and conditioned abstinence (O'Brien and Greenstein, 1976) for which monetary reimbursement was provided. Analysis of the effects on treatment duration of paying patients for clinic attendance/naltrexone ingestion was recently initiated (February 1978). In addition to examining the effects of payment itself, various payment schedules including regular, irregular, visit dependent and visit independent conditions have been evaluated.

RESULTS

Individuals screened for the project were using street opiates (46%), were in methadone maintenance (35%) or were opiate free (19%). Naltrexone was administered at least once to 209 (58%) patients who remained opiate free or detoxified from the original sample of 360. In general, those patients who had been drug free or on low doses of methadone at screening were more likely to proceed to and through induction. In addition it was generally the case that patients who experienced severe withdrawal symptoms either during the naloxone test or in initiation of naltrexone took no more than two doses of naltrexone. Induction was completed with 131 (63%) of the 209 patients who received an initial dose. As experience accrued the results led to an increase in the required opiate free period prior to induction which produced fewer occurrences of withdrawal symptoms and reduced likelihood of patients departing during stabilization on naltrexone.

Following induction 47 (36%) of the 131 patients remained in treatment less than one month. Of the original number completing induction 84 (64%) continued naltrexone treatment for one-two months, 52 (40%) for two-three months, 34 (26%) for three-four months, 13 (9%) for six-seven months and two patients remained for more than 12 months. Termination of treatment followed by later return to naltrexone therapy occurred twice for 43 patients (33%) and three or more times for 9 (4%) of the patients. In cases of multiple treatment episodes patients with longer first treatment periods remained longer during subsequent episodes. Preliminary results involving a limited number of patients under several schedules of payment (up to $40.20/month, $3.25/dose ingested) indicate a substantial effect since 5 (70%) remained in treatment 3-4 months with the naltrexone/payment combination compared to 26% prior to explicit monetary reinforcement of naltrexone ingestion.

Subjects receiving naltrexone differed in some respects from a sample of patients in the Drug Dependence Treatment Service clinic. A slightly higher frequency of "any alcohol use", 88%, but lower frequency of "excessive alcohol use", 8% was evident compared to the reference groups frequencies of 73% and 12% respectively. Opiate, barbiturate, marihuana and amphetamine use was less frequent at intake for the subjects compared to a reference sample.

Several changes in subsequent patterns of drug use appeared. Many patients did use opiates initially to "test" the effectiveness of the naltrexone blockade. Thereafter likelihood of opiate use was low but did increase if a patient missed a scheduled naltrexone dose. Of patients who used alcohol 7% reported an increase in consumption. In general use of any particular drug reflected continuation rather than initiation of use of previously untried drugs.

Additionally 21 subjects (16%) received therapeutic doses of anxiolytics or antidepressants for varying periods during treatment with naltrexone.

DISCUSSION

The results generally paralleled those described by others administering naltrexone in outpatient settings (e.g., Thomas, Kauders, Harris, Cooperstein, Hough, and Resnick, 1976; Lewis, Hersch, Black, and Mayer, 1976; Landsberg et al., 1976). That is, a substantial number of individuals may express interest in a naltrexone regimen but a large proportion leave treatment at some point prior to completion of induction. Additionally a number of self initiated departures occur within the first month of treatment. Several aspects of the results which relate to duration of treatment and use of other drugs warrant further consideration.

Successful treatment with naltrexone is often attributed to general "motivational" factors (e.g., Goldstein, 1976; Greenstein, O'Brien, Woody, Long, Coyle-Perkins, and Grabowski, in preparation; Keegan, Lavenduski, and Schoof, 1976; Landsberg et al., 1976) the nature of which is difficult to assess. However, it is apparent that a variety of acute and immediate conditions related to detoxification and naltrexone induction may be important determinants of patient continuation in treatment. Detoxification itself is a difficult process and patients may experience considerable anxiety (Haas, Ling, Holmes, Blakis, and Litaker, 1976). Typically a portion of detoxification and the period prior to initiation of antagonist treatment is spent as an inpatient. Hospitalization may generate additional discomfort. It also permits a patient to reflect on treatment and to discuss the matter with peers often to his detriment. If the patient continues on an outpatient basis (s)he is continually exposed to those conditions previously correlated with opiate use and this too is likely to prove difficult. The combined factors dictate that more liberal but judicious administration of supplemental drugs, e.g., anxiolytics, may be essential during this period.

The naloxone test may constitute a particularly stressful point in the induction sequence. Therefore, as mentioned previously, efforts should be made, e.g., longer detoxification to test interval, to assure absence of withdrawal symptoms. Since the longer drug free "unprotected" state may itself be more difficult and result in opiate use the need for supplemental pharmacotherapy is evident. Preliminary results of the naltrexone ingestion/payment procedure suggest that enrolling patients during the drug free period, thereby signalling impending monetary reinforcement, may increase the likelihood that a patient will remain to initiate the naltrexone regimen.

When naltrexone induction has been completed several factors which dictate clinical usefulness of the agent also detract from its effectiveness in maintaining patients in treatment. Compliance with drug regimens by medical and psychiatric patients generally is difficult to assure (Swinyard, 1975). Resnick and Schuyten-Resnick (1976) have noted a critical factor in naltrexone treatment is compliance and assuring that patients do not depart prematurely. The problem may be particularly important when an agent has no readily discernible effect and when cessation of treatment has no immediate consequence (Zifferblatt, 1975). Naltrexone, unlike methadone, has no intrinsic reinforcing effects. Unlike both methadone and LAAM the absence of physical dependence dictates that no adversive consequences result from missing scheduled doses. In fact the only point at which the effectiveness of the agent is apparent to the patient is when he "tests" the blockade through self administration of opiates. Diverse efforts to maintain patients in treatment and facilitate adjustment through counseling or job placement are likely to be of importance. However, these efforts typically do not provide *immediate* desirable consequences which might strengthen the behavior of *consistent* clinic attendance and naltrexone ingestion. As was noted in the results section each occurrence of failure to ingest a scheduled naltrexone dose is associated with an increased likelihood of opiate use. In some cases an instance of nonattendance/noningestion may be a clear behavioral statement of intent to use opiates (Resnick and Schuyten-Resnick, 1976). Again results of the payment procedures described here and in a report by Meyer, Randall, Barrington, Mirin and Greenberg (1976) suggest that providing a modest but *immediate* favorable consequence for naltrexone ingestion may enhance attendance. Continued attendance in turn provides opportunities for other reinforcers, e.g., patient-therapist interaction, to be established.

Since opiate use would probably continue in the absence of naltrexone the fact that patients continue to use other drugs during treatment should not be surprising. Concern has existed about increased alcohol use by some patients. Administration of supplemental drugs may serve to alleviate discomfort in a controlled fashion and attenuate alcohol use. Secondarily it should be noted that in some cases alcohol use appears to reflect a return to social environments and behaviors which, if not entirely healthful, are unrelated to opiate use.

Overall it appears that naltrexone can serve a useful function as an adjunct in treatment of opiate use. Continuous blockade may be essential during the early opiate free period. Thereafter the option for return to obtain blockade protection when deemed necessary may be sufficient for many patients. However, this procedure would probably be most effective if regular follow-up contact was maintained to provide "reminders" of availability. Clearly a variety of reinforcers, monetary and otherwise, may provide the necessary immediate favorable consequence which would ensure continued treatment and contact. Finally it is equally clear that there is a need to determine the most effective treatment combinations and the optimal duration of naltrexone treatment.

REFERENCES

Blachly, P.H. 1973. Naloxone for diagnosis in methadone programs. *Journal of the American Medical Assoication.* 224, 334.

Brahen, L.S., Capone, T., Heller, R.C., Landy, H.J., Linden, S.L., and Lewis, M.J. Controlled clinical study of naltrexone maintenance regimens. (In press).

Bradford, H.A., Hurley, F.L., Golondzowski, O., and Dorrier, C. 1976. Interim Report on clinic intake and safety data collected from 17 NIDA funded naltrexone studies. In D. Julius and P. Renault (eds.), *Narcotic Antagonists: Naltrexone.*

Callahan, E., Rawson, R., Glazer, M., McCleave, B., and Arias, R. 1976. Comparison of two naltrexone treatment programs: naltrexone alone versus naltrexone plus behavior therapy. In D. Julius and P. Renault (eds.), *Narcotic Antagonists: Naltrexone.*

Callahan, E., Rawson, R., McCleave, B., Arias, R., Glazer, M., and Lieberman, R.P. The treatment of heroin addiction: Naltrexone alone and with behavior therapy. *Int. Journal of the Addictions.* (In press).

Goldstein, A. 1976. Naltrexone in the management of heroin addiction: critique of the rationale. In D. Julius and P. Renault (eds.), *Narcotic Antagonists: Naltrexone.*

Greenstein, R., O'Brien, C.P., Woody, G., Long, M., Coyle-Perkins, G., and Grabowski, J. Naltrexone: defining its place in multimodality drug treatment. (In preparation).

Hass, N., Ling, W., Holmes, E., Blakis, M., and Litaker, M. 1976. Naltrexone in methadone maintenance patients electing to become "drug free". In D. Julius and P. Renault (eds.), *Narcotic Antagonists: Naltrexone.*

Keegan, J., Lavenduski, C., and Schoof, K. 1976. Comments and findings from a naltrexone double-blind study. In D. Julius and P. Renault (eds.), *Narcotic Antagonists: Naltrexone.*

Landsberg, R., Taintor, Z., Plumb, M., Amico, L., and Wicks, N. 1976. An analysis of naltrexone use - its efficacy, safety and potential. In D. Julius and P. Renault (eds.), *Narcotic Antagonists: Naltrexone.*

Lewis, D., Hersch, R., Black, R., and Mayer, J. 1976. Use of narcotic antagonists (naltrexone) in an addiction treatment program. In D. Julius and P. Renault (eds.), *Narcotic Antagonists: Naltrexone.*

Martin, W.R., Jasinsky, F.P., and Mansky, P.A. 1973. Naltrexone, an antagonist for the treatment of heroin dependence. *Archives of General Psychiatry*. 28: 784-791.

Meyer, R., Randall, M., Barrington, B.A., Mirin, S., and Greenberg, I. 1976. Limitations of an extinction approach to narcotic antagonist treatment. In D. Julius and P. Renault (eds.), *Narcotic Antagonists: Naltrexone.*

O'Brien, C.P. and Greenstein, R. 1976. Naltrexone in a behavioral treatment program. In D. Julius and P. Renault (eds.), *Narcotic Antagonists: Naltrexone.*

O'Brien, C.P., Greenstein, R., Mintz, J., and Woody, G. 1975. Clinical experience with naltrexone. *American Journal of Drug and Alcohol Abuse*. 2: 365-377.

Resnick, R. and Schuyten-Resnick, E. 1976. A point of view concerning treatment antagonists. In D. Julius and P. Renault (eds.), *Narcotic Antagonists: Naltrexone.*

Stitzer, M. and Bigelow, G. 1975. Contingency management in a methadone maintenance program: availability of reinforcers. *International Journal of the Addictions*. 13.

Swinyard, E.A. 1975. Principles of prescription order writing and patient compliance instruction. In L. Goodman and A. Gilman (eds.), *The Pharmacological Basis of Therapeutics*. New York: MacMillan Publishing Co., Inc.

Thomas, M., Kauders, F., Harris, M., Cooperstein, J., Hough, G., and Resnick, R. 1976. Clinical experience with naltrexone in 370 detoxified addicts. In D. Julius and P. Renault (eds.), *Narcotic Antagonists: Naltrexone.*

Wikler, A. 1976. The theoretical basis of narcotic addiction treatment with narcotic antagonists. In D. Julius and P. Renault (eds.), *Narcotic Antagonists: Naltrexone.*

Zifferblatt, S. 1975. Increased patient compliance through the applied analysis of behavior. *Preventive Medicine*. 4: 173-182.

NOW THAT LAAM'S HERE, IS TAKE HOME METHADONE DESIRABLE?

Arnold Schecter, M.D.

New Jersey Medical School

It seems fitting that this introductory paper to the LAAM session should briefly review the status of Methadone--including long acting Methadone-therapy before turning to reports from individual principal investigators.

In 1965, Drs. Vincent Dole and Marie Nyswander published their now classic article on methadone maintenance treatment of heroin addiction (Dole and Nyswander, 1965). Ten years later, after enormous success and also enormous controversy, the same two authors described many problems connected with methadone therapy as it exists in the 1970's (Dole and Nyswander, 1976). Gearing and others have also, in employing the same patients with pre-treatment as the control group, convincingly described social improvement which occurred during methadone treatment. Employing a different methodology, Lukoff has described the outcome of methadone maintenance patients in a different treatment program which, employing less selective criteria for program entry, had a less favorable outcome (Lukoff and Quatrone, 1973).

Baden has frequently described problems with take home methadone which include illicit street use and deaths in non tolerant individuals. Avram Goldstein has, lucidly and repeatedly, advised against the use of take home methadone for opiate addicts (Goldstein, 1975).

The social scientists have provided us with a description of patterns of opiate use in contemporary Americans. Winick, as early as 1962, has described an "aging out" of opiate (and other drug) use with maturity for many, if not the overwhelming majority of opiate users (Winick, 1962 and 1974).

Preble and Casey and Zinberg in separate studies of untreated opiate users in New York and Boston describe a wide variety of patterns of opiate and other drug use with maturing out occurring in many "addicts" or regular users with advancing age, if the users do not die in youth of complications associated with the drug culture. Nurco, in his study of opiate use in Maryland again notes a usually spontaneous, that is, non treatment program influenced, cessation of opiate use with time (Nurco, Bonito, Learner, and Baltes, 1975).

Robbins and Murphy, in their survey of drug use among black males in St. Louis, published in 1967, noted multiple drug and alcohol use and abuse, but also the same maturing out of drug use with increasing age (Robbins and Murphy, 1967). Robbins' Viet Nam studies shattered the hypothesis of an irreversible permanent enzymatic change in opiate addicts which would usually necessitate lifelong opiate, frequently methadone, maintenance (Robbins, 1973 and 1975).

She found that the overwhelming majority of Viet Nam Army heroin addicts, who used to up to 5 grams of heroin daily, rather than 2-10 milligrams which is customary in the United States, stopped opiate use completely after returning to the United States. Usually a very short (under 2 weeks) enforced detoxification in Viet Nam was the only "treatment". It is worthy of note that heavy alcohol use or misuse was frequently seen in most of Robbins' cohorts in reports to date.

Lukoff, in his survey of drug use in Bedford Stuyvesant, Brooklyn, not to be confused with his treatment outcome studies, also noted this same maturing out of drug use with time (Kleinman and Lukoff, 1979). O'Donnell's monumental survey of drug use in the United States also suggests a maturing out of most drug use (O'Donnell et al., 1976).

Indeed, whether we consider opiate, cigarette or alcohol use in our culture or ganja use in Jamaica, as described by Vera Rubin and Lambros Comitas, this maturing out for many, if not most or all users, is so ubiquitous as to seem almost a universal pattern for drugs of abuse (Rubin and Comitas, 1975).

What is the significance of the above to methadone treatment today? At this time methadone treatment is less popular with addicts than it has been previously in many parts of the country--including but not limited to such disparate parts of the country as New York City and Louisville, Kentucky. The methadone patient census is currently markedly lower than several years ago in these areas and nationwide. Community sentiment is overwhelmingly for closing down existing programs and for not permitting new programs to open (Gearing, 1977). There are many reasons for this.

The early methadone patients were an elite, highly selected

group. This was as it should have been in a radical experiment in medical-social-legal care of suffering and frequently deviant human beings. At the present time, in many cities almost all program applicants are placed in methadone treatment programs and all too frequently, few, if any, social demands are imposed on patients. Criminal activity is common and continued drug abuse frequently does not or cannot lead to patients moving from outpatient methadone therapy to a therapeutic community, halfway house or inpatient care facility where there would be more appropriate therapy for many patients. The sale of methadone and other drugs in or near clinic sites by patients at present is very common and readily apparent to most persons regularly in the area. Reasons for permitting this vary from sloppy or poor program administration, an over concern for "patient rights" and "due process", a permissive attitude on the part of some program directors, to a refusal to acknowledge such behavior by a few of the old time methadone clinician administrators who were closer to patient contact at a time when patients were very carefully screened prior to program admission--unlike many of today's patients.

The bad reputation of methadone today is, surprisingly pointed out by Frances Rowe Gearing, the epidemiologist whose pioneering work, scientific competence and personal integrity, did the most to give credibility to this modality of treatment (Gearing, 1977).

Given this state of affairs and with LAAM certainly, one hopes at this late date, within months of general availability to treatment programs, we have a simple positive public health step which suggests itself as a way to continue the usefulness of methadone by helping to assure its survival.

Assuming LAAM will remain a non-take-home medication, and that methadone's chances for survival as the humane opiate maintenance therapy--which I believe it is--partially depends on a markedly decreased amount of "street" or illicit methadone, it seems reasonable to suggest that it is time to consider the following steps:

First, methadone should not be prescribed as a take home medication in its first 3 months use in a given treatment cycle for any individual. LAAM can be used as a weekend medication or methadone can be given 7 days a week in a clinic or pharmacy setting.

Second, LAAM should be encouraged for that majority or minority as the case will prove to be--of methadone patients who will prefer this slightly different methadone preparation to the shorter acting form. Presumably, when methadone is not a common take home medication the temptation to prefer methadone in order to occasionally sell it will not exist. Also, at that time presumably the convenience of three times weekly only clinics or pharmacy visits will appeal to many if not most of the emotionally stable methadone patients.

Third, take home privileges for long time methadone patients should be thoroughly reviewed and decreased wherever possible, certainly for high dose level users, where the possibility of 'skimming' and partial resale is easiest. Because a patient has been ingesting methadone for years in a clinic setting does not *a priori*, mean that the patient is doing well. In some cases, quite the contrary may be the case. Good clinical judgment and close patient contact combined with common sense is essential to tell the difference.

Fourth, if the already overly cumbersome governmental regulations do become still more rigid, it seems reasonable to make changes in order to make our programs more humane and more effective. Let us combine several modalities of treatment at each treatment facility to assist our patients at whatever stage of habituation, dependence or relapse they are in at a given phase of what hopefully will be an ultimate maturing out of drug misuse. Methadone, narcotic antagonist, day care and therapeutic community programs need not be adversaries to one another. Ten years ago Jaffe, Senay, Chappel and others demonstrated this in Chicago--offering all accepted and experimental treatment options at several sites to their patients. Many of these former patients are leading clinicians and administrators today. It seems both unnecessary and unwise to forget this lesson from our recent past. Despite the need for prospective controlled random assignment blind, double blind and even double blind crossover studies with multivariate and regression analyses, which are so much in vogue today--and which were not in vogue during the earlier methadone trial years, let us not become so overly rigid in our love of slow, expensive elequent methodology--that we forget the urgent needs of patients now. It seems reasonable that the Federal government evaluate LAAM at this time in order to ascertain whether the drug is now at the stage where a limited or full NDA approval is indicated.

Last, let us not forget the lesson of the last era of opiate maintenance clinics in the earlier years of this century. They were closed not for lack of proof of efficacy, but rather because of an unsavory public image. If we do not remember history and learn from it, we are bound to repeat its mistakes.

REFERENCES

Dole, V.P. and Nyswander, M.A. 1965. A Medical Treatment of diacetylmorphine (heroin) addiction. *J.A.M.A.*, 193, 646.

Dole, V.P. and Nyswander, M.A. 1976. Methadone Maintenance Treatment. A Ten-Year Perspective. *J.A.M.A.*, 235. 19: 2117-9.

Gearing, F.R. 1970. Evaluation of a methadone maintenance treatment program. *Int. J. Addict.* 5(3): 517.

Gearing, F.R. 1977. Evaluation of Treatment Programs in Schecter, A.: *Rehabilitation Aspects of Drug Dependence.* C.R.C. Press.

Goldstein, A. 1975. On the role of chemotherapy in the treatment of heroin addiction. *Am. J. Drug Alcohol Abuse*, 2(3, 4) 279.

Kleinman, P.H. and Lukoff, I.F. 1979. Ethnic differences in factors related to drug use. *J. Health Soc. Behav.* (in press).

Lukoff, I.F. and Quatrone, D. 1973. A two year follow-up of the Addiction Treatment and Research Corporation, in *Heroin Use and Crime in a Methadone Maintenance Program.* Hayim, G., Lukoff, I.F. and Quatrone, D. (eds.), Law enforcement Assistance Administration, U.S. Dept of Justice, Wash, D.C.

Nurco, D.N., Bonito, A.J., Learner, N., and Baltes, M.B. 1975. Studying addicts over time: methodology and preliminary findings. *Am. J. Drug Alcohol Abuse*, 2(2): 183.

O'Donnell, J.A., Voss, H.L., Clayton, R.R., Slatin, G.T., and Room, R.G.W. 1976. Young Men and Drugs; A Nationwide Survey, Research Monograph Series No. 5, National Institute on Drug Abuse, Department of Health, Education, and Welfare, (ADM) 76-311, National Technical Information Service, Springfield, Va.

Robbins, L.N. and Murphy, G.E. 1967. Drug use in a normal population of young Negro men. *Am. J. Public Health.* 57(9): 1580.

Robbins, L.N. 1973. *The Vietnam Drug User Returns.* Special Action Office for Drug Abuse Prevention, U.S. Government Printing.

Robbins, L.N. 1975. *Veterans' Drug Use Three Years After Vietnam.* Department of Psychiatry, Washington University School of Medicine. (mimeograph).

Rubins, V. and Comitas, L. 1975. Ganja in Jamaica. Vol 26 of *New Babylon Studies in the Social Sciences.* The Hague, Mouton.

Winick, C. 1962. Maturing out of narcotic addiction. *Bull. Narc.*, 14(1): 1.

Winick, C. 1974. Careers of chronic heroin users, in: *Drug Use: Epidemiological and Sociological Approaches.* Josephson, E. and Caroll, E. (eds.). Wiley-Halsted.

LAAM INDUCTION: BEHAVIORAL CORRELATES

Elaine D. Holmes, Ph.D., Walter Ling, M.D., and William E. Carter, P.A.
Veterans Administration Hospital, Sepulveda, California

Bryan S. Finkle, Ph.D. and Thomas A. Jennison, B.S.
University of Utah

INTRODUCTION

There has been great interest for the past few years in the development of 1-alpha-acetylmethadol (LAAM) as an alternative maintenance treatment agent for heroin addicts. This methadone derivative is capable of suppressing abstinence for up to 72 hours and this clinical effect is related to its biotransformation and accummulation of active metabolites, in particular noracetylmethadol (N-LAAM) and dinoracetylmethadol (DN-LAAM). As part of a cooperative pharmacokinetic study of LAAM between us and the University of Utah, we attempted to make certain observations on the patients' cognitive functions and other behavioral parameters after each dosing, both during initial administration and after repeated dosing for several weeks, to see if these observations correlated with the level of active metabolites. This report deals with the correlation between the plasma level of the major metabolite N-LAAM and clinical observations including the results of several specific psychometric tests. It was anticipated that the results of these tests may reflect impairment or the lack thereof, in everyday functioning such as driving, operating equipment, ability to concentrate, memory, etc.

MATERIAL AND METHOD

Subjects were 10 male patients chronically maintained on 60-80 mg. daily methadone, who volunteered to participate in the metabolic study. They ranged from 30-47 years with a mean age of 39 years. Five subjects were of Hispanic origin, three Caucasian and two Black. All had IQ's within the normal range with normal liver profiles and no recent medical illnesses. They were crossed over abruptly to LAAM with a crossover ratio of 1.0-1.5 mg./kg. body weight with an average of 1.16 mg./kg. After crossover, all patients received LAAM on Monday, Wednesday and Friday for a total of 10 doses over three weeks. Thereafter, they were returned to methadone. During their tenure in the study, serial blood samples were obtained for analyses of LAAM and its metabolites according to a schedule which emphasized frequent sampling after doses 1, 2, 3 and 10. After returning to methadone, their LAAM metabolite levels continued to be followed daily for four days and twice weekly thereafter for the remainder of another three weeks.

Baseline evaluations included a general physical examination and routine laboratory tests and a battery of psychometric tests, including the Trails B Test from the Reitan Battery, Digit Span (WAIS), Threshold Measurement of Sensation from an Electric Current, Finger Tapping (Reitan), Reaction Time and a Paired-Association Test. They also completed the symptom-sign record, used in a previous cooperative study (Ling et al., 1976), which elicits feelings of underdosing, overdosing and other somatic complaints. The baseline evaluations were completed on the day prior to the first dose of LAAM. The scheduling of these tests was fixed relative to the time of dosing and selected blood drawings and this relationship was maintained throughout the entire study such that the test batteries were always given at a fixed hour after patients received their doses of LAAM. A detailed blood-drawing and testing schedule is shown in Table I. The LAAM metabolites were analyzed by a Gas Chromatography-Mass Spectrometry method developed by Dr. Bryan Finkle at the Center for Human Toxicology, University of Utah. The blood-drawing schedule was heavily concentrated around the first two hours after dosing and then every two hours up to the sixth hour and at 12 hours, 16 hours, and 24 hours after dosing.

RESULTS

The results will be presented in terms of clinical observations and the patients' performance on the various tests.

Clinical Observations

All the patients completed the three-week study and found no particular problem with the blood-drawing or testing schedules. No patient required supplemental medication during the study. There was no increase in consumption of alcohol. Daily urines taken Monday through Friday showed no abuse of heroin, amphetamines or barbiturates. Of the nearly 800 blood samples required, only two were not obtained on schedule. We did not observe any oversedation or suppression of conscious level after dosing and patients were generally

Table I. Activity schedules for the LAAM pharmacokinetics and metabolism study and its behavioral correlates adjunct study.
X - indicates blood was drawn
* - indicates psychological tests were administered
(At some times, psych. testing=symptom-sign material only)

Time of Day	Baseline	1	2	3	4	5	6	7	8	9	10	11	12	13	14	15	16	17	18	19	20	21
		Study Period--Day Numbers																				
LAAM Dose:		D			D		D		D			D		D		D			D		D	
8:00		X	X	X	X	X	X	X	X	X	X			X	X				X	X		
a. m.	*	*	*	*	*	*	*	*	*	*	*			*	*				*	*		
8:15		X			X		X															
8:30		X			X		X															
9:00		X			X		X		X					X					X			
	*	*			*		*		*					*					*			
9:30		X			X		X															
10:00		X			X		X															
	*	*			*		*															
12:00		X			X		X															
noon	*	*			*		*															
2:00		X			X		X															
p. m.	*	*			*		*															
8:00		X			X		X		X					X					X			
	*	*			*		*		*					*					*			
12:00		X			X		X		X													
mdnt.	*	*			*		*															

Time of Day	Baseline	22	23	24	25	26	27	28	29	30	31	32	33	34	35	36	37	38	39	40	41	42	43
LAAM Dose:		D																					
8:00		X	X	X	X				X			X				X			X				X
a. m.		*	*	*	*	*	*	*	*	*	*	*	*	*	*	*	*	*	*	*	*	*	*
8:15		X																					
8:30		X																					
9:00		X																					
		*																					
9:30		X																					
10:00		X																					
		*																					
12:00		X																					
noon		*																					
2:00		X																					
p. m.		*																					
8:00		X																					
		*																					
12:00		X																					
mdnt.		*																					

able to maintain alertness when engaged in performance tasks. Although some of the patients, when left alone, did tend to doze off, they were easily arousable.

No untoward side effects were observed in any patients. All patients appeared comfortable for the first 48-60 hours. Several showed discomfort on the third day and on Monday mornings they reported feeling "icky." These patients also reported craving for heroin on Sundays. These same patients continued to complain of not feeling well after they were returned to methadone (see Fig. 1). Several patients complained of anxiety, one patient related his anxiety to the prospect of his going to jail at the end of the study. Complaints of constipation and twitching muscles were quite common. One patient complained of sexual impotence, another of lack of sexual interest. Several patients worried about not taking medications daily. One patient reported loss of interest in cigarettes. Several patients reported feeling more "normal" but did not consider this a positive attribute. They felt that too many feelings came to the surface and the medication did not "cover up" as well as methadone. All the patients appeared physically healthier at the end of the study, probably due to a regular diet, exercise and sleep-waking cycle.

Psychometric Tests

The first type of tests included Digit Span and Paired Associates. In the former, patients were required to repeat forward and backward seven random sequences of numbers as in the WAIS. In the latter, patients were required to make an association between com-

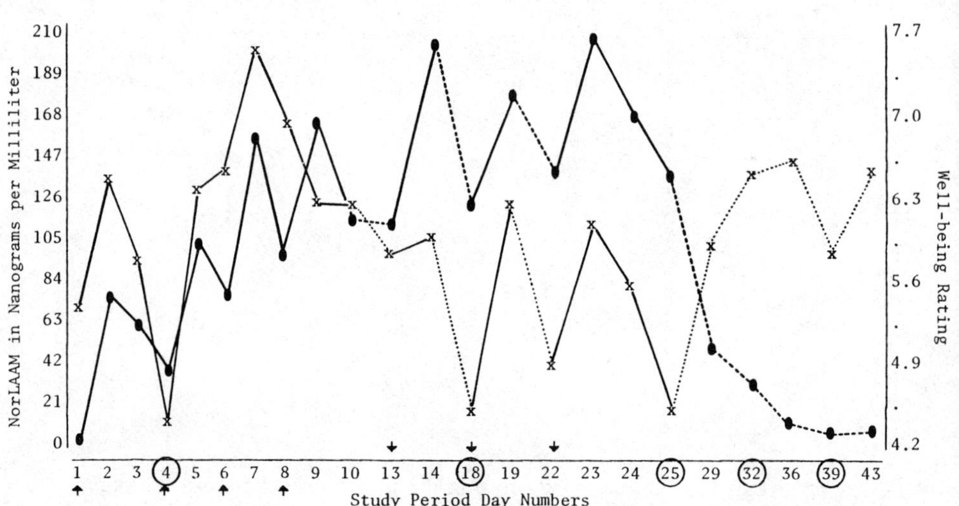

Fig. 1. Mean NorLAAM blood levels (●—●) and mean subjective ratings of feelings of well-being (x—x) from ten subjects, at 8 a.m. on 23 selected days from the 43-day study period. Subjects were asked to rate their current well-being status on an eight-point scale. Arrows (↑,↓) indicate the days on which LAAM was given, shortly after the rating was obtained. Those days circled were Mondays.

mononly used words and trigrams of high association value. The cards were presented for five seconds and the patient was requested to repeat the trigram immediately after the presentation and again one-half hour later. There did not appear to be any significant correlation between performance on these tasks and LAAM metabolite levels, either initially when LAAM was first introduced, or at the end of the study when there was considerable LAAM metabolite accumulation.

Another group of tests required patients to react to a test stimulus with a motor response: the Finger Tapping Test and Trails B from the Reitan Battery, and Reaction Time using a Hewlett-Packard programmable calculator. This latter test required the patient to press a button as soon as he saw a light which appeared at varying intervals. All three tests showed considerable practice effects. Finger tapping and Trailmaking were not significantly affected by the rising metabolite levels after chronic dosing. However, reaction time appeared to be somewhat influenced by rising metabolic levels after dosing. This effect was noticeable on the first day of LAAM administration and appeared to be preserved after chronic administration where there was considerable increase in baseline level of metabolites. However, this difference was not significant statistically ($p > .10$). The practice effects seen in all three tests persisted throughout the entire study and suggest that LAAM and its active metabolites do not interfere with learning.

We also tested the patients' threshold perception of a mild electrical stimulus to their fingertips and here again acute and chronic LAAM administration did not appear to significantly affect the threshholds.

DISCUSSION

The prolonged clinical effect of LAAM has been attributed to the biotransformation of this relatively inactive parent compound to its more active metabolites and the accummulation of these metabolites over time. It seems therefore important to ascertain if such accummulation affects the patients cognitive function and their other behaviors, if patients must be treated with LAAM on a long-term basis. Henderson et al. (1977) have shown that active LAAM metabolites after chronic administration are much higher compared to those after initial administration. Grevert, Masover, and Goldstein (1977) in a cohort of patients maintained on LAAM for one to three months showed that there was no significant memory deficit among these patients. However, no serial testing after dosing, either with acute or chronic LAAM administration, has been previously reported. This would seem crucial to us since if LAAM and its metabolites do in fact interfere with a patient's mental status, it is likely to occur at some time after each dosing in a manner related to the absolute or changing metabolic levels. This may have important clinical and medico-legal implications if a patient should become impaired for some time after receiving his medication. For example, accidents could occur if their reaction is slowed; or there may be interference with job performance.

CONCLUSION

Although the present study involved only a small number of patients, the fact that no significant impairment was seen in reaction time, perceptual-motor performance and learning seems reassuring.

ACKNOWLEDGEMENT

This work is supported by V.A. Medical Research Funds and by Grant #DA01391-01, NIDA.

REFERENCES

Grevert, P., Masover, B., Goldstein, A. 1977. Failure of Methadone and Levothadyl Acetate (Levo-Alpha-Acetylmethadol, LAAM) Maintenance to Affect Memory. *Arch. Gen. Psychiatry.* 34: 849-853.

Henderson, G., Weinberg, A., Hargreaves, W., Lau, D., Tyler, J., and Baker, B. 1977. Accummulation of Levo-Alpha-Acetylmethadol (LAAM) and active Metabolites in Plasma Following Chronic Administration. *J. Analytical Toxicology.* 1: 1-5.

Ling, W., Charuvastra, V.C., Kaim, S.C., and Klett, C.J. 1976. Methadyl Acetate and Methadone as Maintenance Treatment for Heroin Addicts. *Arch. Gen. Psychiatry.* 33: 709-720.

SELF-SELECTION AND PERSONALITY FACTORS IN THE INITIATION OF LAAM MAINTENANCE

John G. Cisco, M.A. and Allen Blasucci, Psy.D.

Community Drug and Alcohol Program of Hudson County

Jersey City, New Jersey

L-alpha-acetyl-methadol (LAAM) has been in the clinical trials phase of pharmacological testing since approximately 1970 when Jaffe et al. (1970) performed a double blind study of the effectiveness of LAAM as a substitute for methadone in the treatment of heroin addicts. What they and virtually all succeeding researchers found was that LAAM was a suitable and desirable alternative to methadone. At present LAAM is being administered by methadone clinics throughout the country, conceivably within the framework of the last pharmacological testing stage before eventual approval for widespread use in treatment. Because of the experimental nature and purpose of this dispensing program, a predominantly volunteer subject pool is being utilized. Primarily as a result of this necessary use of volunteers, several phenomena have occurred which warrant closer examination. First of all, what Jaffe et al. (1970) and virtually all subsequent researchers also found was that the experimental group of clients maintained on LAAM consistently produced higher dropout rates than a comparable group of methadone clients used as a control. In some instances this ratio reached proportions of three experimental dropouts to every control dropout. The second phenomenon which has occurred is that the number of clients volunteering for LAAM maintenance has been surprisingly below expected frequencies. It is within the context of these two phenomena that the existence of a self-selection process may be implicated.

In reviewing the literature on LAAM, and more precisely on LAAM dropouts, several tentative findings have arisen. Resnick et al. (1976) found that dropouts tended to have shorter addiction histories and higher mean MMPI scores on the depression (D) and psychopathic (Pd) scales than did the clients who remained in the study. Savage et al. (1976), in a similar study, found that dropouts had higher paranoid (Pa) scores than non-dropouts. What these and similar studies appear to indicate is that there well may be a distinguishable difference between dropouts and those clients that remain in the LAAM study. In addition, dropouts from the LAAM program may possess characteristics similar to those held by the general non-volunteer client population.

A related area of investigation concerned with the potential differentiation of successful and unsuccessful methadone clients also warrants consideration. Pittel, (1972), in a study with this purpose found that unsuccessful clients scored slightly higher on 8 of 10 clinical scales of the MMPI, with the greatest difference occurring on the scales in the psychotic triad. Stewart and Waddell (1972) essentially replicated these results in finding that street addicts and methadone maintenance dropouts tended to have higher psychotic symptoms than successful clients. Williams and Johnson (1972) found several demographic differences between successful and unsuccessful clients including age, addiction history, criminal history, marital status, and pattern of drug use. These studies are related to the subject of LAAM research in that the distinguishing factors noted may be valuable in the characterization of volunteers and non-volunteers for the LAAM program.

One final area of study must also be considered, and that is the topic of volunteering per se. Rosenthal and Rosnow (1969) have addressed the topic of self-selection, or volunteering, and have compiled a possible hierarchy of characteristics which may distinguish the likely volunteers from the non-volunteers. They conclude that volunteers are generally better educated, have higher occupational status, have a higher need for approval, are more intelligent, less authoritarian, and better adjusted than non-volunteers. They also make the important point that besides these factors, one possible overriding consideration would be the specific task that is required of the volunteers.

The presence of a self-selection process and the variables which define this process may provide valuable information to the study of the LAAM dropout phenomena, as well as to the investigation of the causes for the unexpectedly low number of volunteers for the LAAM study. This research study was designed to examine demograpic, psychological and attitudinal information of volunteers and non-volunteers for evidence of a significant selection factor, and if found, to define the parameters of this factor and its implications for the experimental LAAM program.

METHODS

The setting of this research project consisted of six satellite methadone clinics, operating within the New York-New Jersey metropolitan area, under the auspices of a central community based, parent agency. Within these clinics, whose total population consists of approximately eight hundred methadone maintained clients, male volunteers are being sought to take part in a national experimental program to test the efficacy of LAAM. Each volunteer for this program is either randomly placed into an experimental group consisting of clients who will be switched from methadone to LAAM maintenance, or placed in a control group of volunteers who will continue to be maintained on methadone. The progress of both groups will be monitored, and eventually compared, in a final analysis of LAAM's efficacy.

Demographics

Seventeen categories of demographic information were collected on volunteers and non-volunteers for the LAAM program. The information was collected through the client's respective social worker or other staff member in order to insure that the information was valid and current. The control group consisted of fifty male methadone clients who did not volunteer for the LAAM program, and whose names were randomly drawn from computer files. Fifty-two methadone clients who volunteered for the LAAM program within the 6 months following its initiation comprised the experimental group. This group consisted of all clients who volunteered regardless of whether they were selected for LAAM maintenance, remained on methadone as a control, or elected to discontinue LAAM maintenance at a later date.

Attitudes

A subjective questionnaire consisting of twenty-one Likert type questions was distributed to approximately fifteen percent of the male methadone population. Most questions had five response categories, ranging from an extreme positive to an extreme negative position, and with a distinctly neutral category in the center. The questionnaires were distributed without requiring the respondents' names in order to insure anonymity and response spontaneity. The respondents did report their age, race, marital status, and education for statistical purposes. The topics of the questions were chosen according to their reported saliency as determined in a prior informal survey. All of the questions dealt with the clients' subjective attitudes towards LAAM and methadone, and the perceived hazards or benefits of each. Some of the questions were formulated to allow the comparison of attitudes held towards LAAM and methadone on similar parameters.

Psychological

The psychological measure used in this study was a 166 answer short form MMPI (Faschingbauer, 1974). They were administered to 11 LAAM control subjects and a matched group of 13 male methadone clients. The groups were matched according to age, race, education, marital status, and time spent on methadone maintenance. Only volunteers who were selected as LAAM controls were used in the experimental group in order to prevent possible contamination of results due to the effects of LAAM. MMPIs with a validity T score of eighty or more were excluded from the study.

RESULTS

Demographics

Of the seventeen measures used in the demographic survey, seven showed statistically significant differences ($p<.05$) between the volunteer and the non-volunteer groups. The most pronounced was that of racial composition with the volunteer group showing a significantly higher white to non-white ratio than the non-volunteer group ($p<.005$). In the number of dependent children, the volunteer group tended to have significantly more clients without dependent children than the non-volunteer group ($p<.05$). Educationally, the volunteers averaged 11.64 years of education versus the non-volunteers' mean of 10.2 years, a difference that's significant ($p<.005$). When evaluating the number of previous treatment attempts the volunteer group had significantly more clients with three or more previous attempts than the non-volunteer group ($p<.025$). The volunteer group also was found to have significantly more clients with two or more felony convictions ($p<.025$). The two groups also differed significantly on two measures that could be considered indicative of their functioning while on the methadone program. The non-volunteer group had significantly more clients with three or more take-homes, and fewer with one or no take home doses than did the experimental group have before they volunteered ($p<.01$). The volunteer group also was reported to have significantly more clients with one or more positive urinalysis reports per month over the previous six months than did the non-volunteer group ($p<.05$).

Several other demographic measures did not reach statistical significance though noteworthy trends were evident. In terms of employment status, the volunteer group showed a ratio of 3.8 employed clients to every unemployed client while the non-volunteer group demonstrated a lower ratio of 2.6 to 1. The volunteer group also reported a mean age of 27.6 years versus the non-volunteers' mean of 28.6 years. The remaining 8 demographic measures (birth order, age of addiction, age of first use, time on methadone, psychiatric hospitalizations, psychotropic use, in-patient pill detoxifications, and marital status) did not demonstrate significant differences, nor trends, between groups.

Attitudes

Data drawn from the subjective questionnaires was analyzed from three standpoints: whether the opinions of non-volunteers, taken as a group, differed from an expected normal distribution, whether the distributions of expressed opinions towards LAAM differed from those expressed towards methadone, and whether the distributions of opinions differed between homogenous categories of clients.

Results indicate that non-volunteers, as a group, believed that LAAM would be hard to detoxify off of ($p<.001$) and that it has not been tested enough ($p<.001$). They believed that LAAM would probably affect their sex life ($p<.001$), would probably be harmful to their body ($p<.001$), and would probably not prevent them from having children ($p<.01$). The non-volunteers also believed that LAAM would probably not produce a "rush" like methadone ($p<.001$). They tended to be non-committal as to whether LAAM would be easy to overdose ($p<.001$), whether it would be harmful to their minds ($p<.001$), or whether it would hold them for two ($p<.02$) or three ($p<.01$) days. They also stated that they would not go on LAAM regardless of whether they had no take home doses of methadone ($p<.001$)or four take home doses ($p<.001$). The distribution of their opinions did not differ significantly from normal when asked if the government is using LAAM to control people, nor when asked if they consider LAAM an essentially "good" or "bad" drug.

When comparing the non-volunteers' opinions towards LAAM versus their opinions towards methadone, two differences emerged. They expressed a generally more positive opinion of methadone than LAAM ($p<.001$) and also believed that it would be safer to use cocaine with methadone than LAAM ($p<.05$). They did not differ as to which would be harder to detoxify from, and held the same opinions as to whether LAAM would hold for two or three days.

When comparing different subgroups of clients, it was found that blacks tended to see LAAM as a controlling device of the government more than whites ($p<.05$), and unmarried clients tended to hold this same view more than married ones ($p<.05$). Volunteers for the LAAM program expressed a higher general opinion of LAAM ($p<.01$), did not see it as hard to detoxify from ($p<.02$), and did not believe as strongly that it could prevent them from having children ($p<.02$), as did the sample of non-volunteers. They also did not believe as strongly that it had not been tested enough ($p<.05$). The volunteers also believed more strongly than non-volunteers that LAAM would be able to "hold" them for two ($p<.001$) or three ($p<.01$) days.

Psychological

An analysis of variance performed on the MMPI returns indicated a significant main effect due to the classification of clients into volunteer and non-volunteer groups ($p<.05$). Overall, the volunteer group had lower mean T scores on eight of the ten clinical scales, surpassing the non-volunteer group only on the scale for hypomania (Ma) and social introversion (Si). The largest differences between the mean scores of the two groups occurred on the scales for psychopathic deviance (Pd), paranoia (Pa), psychasthenia (Pt), and schizophrenia (Sc) with the volunteer group having lower mean scores on all four of these scales. Only on the scale for paranoia (Pa) was there a significant difference with the volunteers having a mean T score of 54.4 and the non-volunteer group a mean of 65.7 ($p<.01$).

DISCUSSION

In surveying the demographic information it becomes obvious that a single causative factor is not responsible for the self-selection of clients for the LAAM program. As with most behavioral patterns several factors, resulting in a multi-causal relationship, seem to be evidenced in the data. The single most influential factor appears to be the number of take home doses, with volunteers having had much fewer than non-volunteers. Assuming that having no, or few, take home accounts, for obvious reasons, for the volunteering of these clients, a plausible explanation must be determined why clients with two or three take home doses volunteered in a proportion representative of the normal population. In grouping the volunteers into those who had no or one take home, versus those who had two or three take homes, two somewhat homogeneous categories of volunteers appear to emerge. The first consists primarily of white, higher educated, married, and employed clients with fewer dependents, more felony convictions, and two or three take homes. This group, which could be called the stable volunteers, appears to have volunteered for reasons related, but not exclusive to, the take home factor. The second group of clients, which may be termed the unstable volunteers, consisted of clients who were again predominately white, higher educated, have fewer dependents, more previous treatment attempts and felony convictions than the regular methadone maintained population, but also more positive urinalysis reports and fewer take home doses. In essence, if a continuum of functional stability can be conceived of, the LAAM program has attracted the extreme ends of this continuum. The unstable clients volunteered for reasons of convenience due to the freedom the LAAM dispensing schedule provides, the stable clients also because of this freedom, plus other reasons emanating from a more stable and open lifestyle. Both groups appear to exhibit less of the suspicious, paranoid opposition to LAAM evidenced in the psychological characterization of the non-volunteer group of methadone clients.

The results of the subjective survey generally lends support to these conclusions. The most potent indicator of the populations' opposition to LAAM appear in questions concerning the government's use of LAAM and the relative experimental status of LAAM. These questions seem to indicate a general, non-specific suspicion and opposition to LAAM. This observation is supported by the finding that non-volunteers expressed no opinions different than what they hold towards methadone when asked pointedly direct questions concerning LAAM's potential actions, except that it may be more unsafe with the use of cocaine. This contention of the existence of a predominately non-specific suspicious attitude towards LAAM is further supported by the finding that volunteers tended to have lower paranoid (Pa) scores on their MMPI's, thus accounting for their less suspicious and wary attitude towards LAAM. The level of formal education of volunteers also tends to support this finding in that if education is conceived to help prevent the development of unfounded and misconceived beliefs, the volunteers, who were more educated than non-volunteers, would be less likely to develop an unfounded suspicious attitude towards LAAM. To further support this contention, only on the questions of governmental control, a tacit indicator of suspicion, did blacks or unmarried clients hold significantly more negative opinions towards LAAM than did white or married clients.

In reaching some conclusions from the data, it appears that the high number of drop-outs from the LAAM program may be due to the attraction it has for some less stable clients. Though the majority of clients who volunteer for LAAM may tend to be psychologically stable, a small percentage may be attracted for very tangible reasons (take homes) and may experience more serious psychopathological symptoms, especially of a paranoid nature. This small group may be the clients who eventually drop out of LAAM study as Savage et al. (1976) indicate.

Another topic which warrants consideration is the possible implications of the existence of a self-selection factor to the national research effort now being conducted with LAAM. If the LAAM program does not in fact attract a very specific population of volunteers, which most notably excludes the black methadone maintained client, as this research indicates, will the findings of the national project be generalizable to the large majority of the methadone maintained population? More specifically, will LAAM meet the same type of opposition from clients when it is eventually approved for widespread use? These possibilities certainly warrant further consideration and follow-up research.

A more timely topic is the exploration of possible strategies to attract more clients to the present LAAM research project. One possible tactic would be the implementation of an instructional program centering on some of the topics noted in the subjective questionnaire. A problem arises with this approach in that, as this research indicates, the reluctance of clients to volunteer for LAAM maintenance is based on a rather non-specific suspicion and opposition which would render such an instructional approach of questionable value, at least for a large part of the population. This approach may be effective for the small category of clients who are amenable to volunteering, and who demonstrate a less suspicious personality, but again the sampling would be questionable in representing the general methadone maintained population. Future research may well be directed in determining if such an instructional approach would prove effective in attracting the bulk of the methadone maintained population to LAAM maintenance. Another approach to this problem may be the restructuring of certain policies now employed in methadone maintenance programs, most notably the policies surrounding the dispensing of take home doses of methadone. A problem with this approach would be the negative philosophical and moral ramifications of such a potentially coercive policy change. Whichever approach is implemented, the essential psychological and practical nature of the problem must be accounted for, as well as the requirements of a LAAM research project whose results must be eventually generalized to a large portion of the methadone maintained population.

REFERENCES

Faschingbauer, T.A. 1974. A 166 item short form of the group MMPI: The fam. *Journal of Counseling and Clinical Psychology*. 42: 645-655.

Jaffe, J.H., Schuster, C.R., Smith, B.B. and Blachley, P.H. 1970. Comparison of Acetylmethadol and methadone in the treatment of long-term heroin users. *JAMA*. 211: 1834-1836.

Pittel, S.M. 1972. Three studies of the MMPI as a predictive instrument in methadone maintenance. Proceedings of Fourth National Conference on Methadone Treatment, San Francisco: NAPAN.

Resnick, R.B., Orlin, L., Geyer, G., Resnick, E.S., Kestenbaum, R.S. and Freedman, A.M. 1976. 1-Alpha-Acetylmethadol (LAAM): prognostic considerations. *American Journal of Psychiatry*. 133(7): 814-819.

Rosenthal, R. and Rosnow. 1969. *Artifact and Behavioral Research*. N.Y.: Appleton-Century Crofts.

Savage, C., Karp, E.G., Curran, S.F., Hanlon, T.E. and McCabe, O.L. 1976. Methadone/LAAM maintenance: A comparison study. *Comprehensive Psychiatry*. 17(3): 415-424.

Stewart, G.T. and Waddell, K. Attitudes and behavior of heroin addicts and patients on methadone. Proceedings of Fourth National Conference on Methadone Treatment. San Francisco: NAPAN.

Williams, H.R. and Johnson, W.E. 1972. Factors related to treatment retention in a methadone maintenance program. Proceedings of Fourth National Conference on Methadone Treatment. San Francisco, NAPAN.

A FOLLOW-UP STUDY OF PATIENTS PREVIOUSLY ON LAAM MAINTENANCE

Mara Blakis, M.D., Elaine D. Holmes, Ph.D., and Walter Ling, M.D.

Veterans Administration Hosptial, Sepulveda, Calif.

INTRODUCTION

Many studies in the past decade, including several large-scale cooperative clinical trials, have demonstrated the safety and efficacy of LAAM as an alternative treatment agent to methadone maintenance treatment. About 3000 patients are currently involved in a multi-clinic Phase III study and there is hope that LAAM may soon become available for general clinical use. However, very little follow-up information of patients previously treated with LAAM is presently available. This report describes such a study of two cohorts of patients treated by us in the two Phase II cooperative studies comparing LAAM and methadone. We wanted to see if there were any unforeseen late adverse effects of LAAM treatment in these patients as well as to see how they perceived their treatment experience with LAAM some time after the pressure of being a study subject had ceased. We also took the opportunity to examine several aspects of their lives to see if treatment with LAAM and with methadone affected these areas differently.

MATERIAL AND METHOD

Between 1973 and 1976 we conducted two comparative studies between LAAM and methadone at the Sepulveda V.A. Hospital Drug Dependence Treatment Center. In the first study 90 patients were randomly assigned 30 per group, to methadone 50 mg., methadone 100 mg. and LAAM 80 mg. in a 40-week double-blind trial. In the second study 76 patients were assigned in an open study of LAAM three times per week (25 patients), methadone daily (27 patients), or methadone Monday through Thursday and LAAM Friday (24 patients). Twenty-eight patients participated in both studies. Because of this there were actually 138 patients when the two studies were combined. Of these, 51 were treated with LAAM Monday, Wednesday, and Friday; 24 were on LAAM Friday only and 63 were on methadone. Of the 138, 90 patients were treated double-blind and 48 were in the open study.

A follow-up interview was attempted between August 1976 and May 1977. Of the 138 patients 18 could not be traced, 15 were incarcerated in local, state and federal penal facilities, 14 had moved from the area and could not be interviewed, and four were dead--one from a gunshot wound, one from an auto accident, and two from multiple drug overdose. Two of the deaths were patients from the LAAM group and two were from the methadone group. The remaining 87 patients were interviewed in person (except one who was interviewed by phone) and their responses form the basis of this report; 53/79 LAAM patients were available compared 53/87 methadone patients. Table I shows the distribution of the patients by study groups.

The interviews were in the format of an hour-long conversation guided by a structured protocol outlining the different areas to be explored. These included an inquiry of the current state of health,

Table I. Distribution of patients by study groups.

Groups	Subjects					
	Total	Interviewed	In Jail	Dead	Moved	Untraceable
Double-blind LAAM	30	20	1	1	4	5
Double-blind Methadone	60	30	8	2	6	10
Open LAAM	25	19	2	0	2	2
Friday LAAM[a]	24	14	3	1	2	1
Open Methadone	27	23	2	0	1	1
Totals: Participations	166	106	16	4	15	19
Individuals[b]	(138)	(87)	(15)		(14)	

a. Methadone daily Monday through Thursday, LAAM on Friday for the weekend.

b. Because 28 individuals participated in two studies, the actual number of individuals--indicated by this total--does not always equal the number of participations.

extent of drug use, employment, social adjustments interpersonal relation and involvement with law enforcement agencies. Patients were asked to reflect on their experience with LAAM and to compare it to methadone. They were asked to recall how they were functioning while on the study, more specifically, whether the study drug (LAAM or methadone) affected their memory, concentration, or their mood. Those who terminated early from the study were asked to restate what they felt to be the reason for the dropout and finally all patients were asked if they would avail themselves of LAAM if it was available in a routine clinic or in a research setting.

RESULTS

There were no serious late side effects reported by the patients. None of the subjects interviewed reported being ill and no one attributed any serious symptoms or complaints to their prior experience with LAAM. In general, their health status did not differ from that of other patients on methadone maintenance and while this lack of reported serious aftereffects does not constitute any systematic data, nevertheless, it was reassuring in terms of LAAM's long-term safety profile.

There was no difference between the LAAM and methadone groups in terms of reported drug use, drinking behavior, changes in living arrangments, or time spent in jail. Of the 50 LAAM patients on whom we have current treatment status 24 were on methadone maintenance, two were on LAAM, one was receiving naltrexone, and 21 were not involved in any pharmacologically supported treatment.

Of the 46 previous methadone patients whose current treatment status was available 26 were on methadone maintenance, one was on naltrexone, and 19 were receiving no pharmacological treatment. One patient in the last group candidly reported being addicted to heroin.

More methadone (15/21) than LAAM patients (9/25) recalled having difficulties with memory when the two studies were combined ($p<0.02$). This difference remained significant only when those receiving their study drugs blind were compared. However, no difference was evident when those involved in the open study were compared.

Roughly a fourth of all the patients recalled having difficulties with concentration but no difference was noted between groups.

Slightly more than half of the patients receiving LAAM blind recalled some discomfort on Sundays (third day from their last LAAM dose) whereas nearly 3/4 of those in the open study recalled having problems in spite of the fact that their dosage was adjustable and many of them received a dose increment on Friday. Clearly the open-nature of the study design contributed to this difference.

The overwhelming majority of LAAM early terminators continued to believe "medication not holding" or "feeling unmedicated" to be the reason for their dropout. No methadone patients dropped for these reasons. Generally, early termination from methadone resulted from reasons which were not drug-related. Only one patient from the combined studies reported dropping out of the study because he did not like the study.

Almost two-thirds (63%) of LAAM patients reported that LAAM had fulfilled their expectation as a treatment agent. Those on LAAM alone were more satisfied than those on methadone and LAAM combined ($p<.02$) and those on the blind study were more favorably impressed than those in the open study ($p<.02$). On the other hand, more than half (57%) of the patients would rather stay on methadone than return to LAAM. Only 37% were certain they would prefer LAAM and 6% were undecided. There was no substantial difference in this regard whether LAAM was offered as part of a clinic treatment routine or in the context of a research project. Interestingly, however, patients who were in the double-blind study generally accepted LAAM much more readily whether they were previously treated with LAAM or methadone.

DISCUSSION

Needless to say, this was not a study comparing LAAM and methadone, rather it was a comparison of what the patients recalled as their treatment experience with either drug and this recollection must necessarily be influenced by many other factors impacting favorably or otherwise on their particular treatment. Both studies suffered from problems of protocol designs as we have pointed out elsewhere (Ling et al., 1976 and Ling, Klett, and Gillis, 1978). The first study (VA study) had a very conservative induction schedule and many patients suffered withdrawal symptoms early in the study which undoubtedly had an impact on their later recollection of this experience. On the other hand, the open design of the SAODAP study (the second study) inevitably exposed patients on the new drug (LAAM) to a greater degree of anxiety, and this in turn affected their later feelings about the drug. Therefore, the fact that fewer patients said they would return to LAAM does not mean LAAM is any less acceptable to patients in general. For instance, in a study measuring the acceptability of LAAM and methadone by patients who had been treated by methadone and LAAM for at least three months, Dr. Goldstein's group (Trueblood, Judson, and Goldstein) found that patients preferred LAAM over methadone in a number of specific parameters. Our results indicated that we need to communicate openly with our patients in order that when the patients account for their difficulties in the course of a study they do not blame any and every, real or imagined, symptom on the drug they receive from us. The fact that nobody seemed to have suffered any late ill effects is again reassuring and perhaps this knowledge alone makes such follow-up efforts worthwhile at the present stage of the clinical development of LAAM.

ACKNOWLEDGEMENTS

We wish to thank Ms. Jan Harrell and Dr. Eric McPherson for their assistance in conducting many of the interviews and compiling the data.

This investigation was supported in part by a Veterans Administration Research Grant and by Grant DA 3-AC723 NIDA from the National Institute on Drug Abuse.

REFERENCES

Ling, W., Charuvastra, V.C., Kaim, S.C., and Klett, C.J. 1976. Methadyl Acetate and Methadone as Maintenance Treatments for Heroin Addicts. *Arch. Gen. Psychiatry*. 33: 709-720.

Ling, W., Klett, C.J., and Gillis, R.D. 1978. A Cooperative Clinical Study of Methadyl Acetate. *Arch. Gen. Psychiatry*. 35: 345-353.

Trueblood, B., Judson, B.A., and Goldstein, A. Acceptability of Methadyl Acetate (LAAM) as Compared with Methadone in a Treatment Program for Heroin Addicts. In press.

RESTABILIZATION ON METHADONE AFTER METHADYL ACETATE (LAAM) MAINTENANCE

Walter Ling, M.C., Mara Blakis, M.D., Elaine D. Holmes, Ph.D., and William E. Carter, P.A.
V.A. Hospital, Sepulveda, Ca.

C. James Klett, Ph.D.
V.A. Hospital, Perry Point, Md.

Abstract: Sixty-eight heroin addicts maintained for 40 weeks on LAAM or methadone in a double-blind study were transferred to a uniform dose of 60 mg. methadone daily at the end of their tenure in the study and followed double-blind for the ensuing six weeks during which their daily methadone doses were adjusted according to their clinical needs. Patients were observed for symptoms and signs of discomfort and for the amount of illicit drug use during this period of transition. The results indicate that patients maintained on LAAM can be readily restabilized on methadone and that sudden decrease of methadone dose tends to result in patients' supplementing with illicit heroin and, conversely, increasing methadone doses resulted in a corresponding reduction in illicit drug use. It is suggested that a chronic covert abstinence syndrome may exist in some patients on long-term methadone maintenance and that while it may contribute to their continued illicit drug use, it may have a different pathophysiological basis and require different therapeutic considerations.

INTRODUCTION

Methyadyl acetate (LAAM), a long-acting methadone derivative, has been undergoing clinical investigation for the past decade as an alternative maintenance treatment agent for chronic heroin addiction. Previous studies have indicated that LAAM given three times per week in appropriate doses is effective clinically and that its clinical safety is comparable to methadone when used in this manner (Ling,

Charuvastra, Kaim, and Klett, 1976; Ling, Klett, and Gillis, 1978). LAAM has certain advantages over methadone due to its long duration of action and while it is not anticipated that LAAM will totally replace methadone there is reason to believe that it may eventually become the drug of choice for maintenance treatment (Ling and Blaine, 1978). Currently, several thousand patients are enrolled in a nationwide Phase III testing which is expected to culminate in a new drug application approval by the FDA in the near future.

Although LAAM is similar to methadone in its clinical effects, its metabolism differs sufficiently so that somewhat different considerations are necessary for direct patient induction and for crossover from methadone maintenance. These issues have been examined in a number of earlier clinical studies and are being actively addressed by a consortium of clinics as part of the Phase III trial. On the other hand, there has been virtually no published information on the relative ease or difficulty in restabilizing patients on methadone after LAAM maintenance. However, one of the earlier studies provided an opportunity to examine this problem in a double-blind setting and it is expected that the results will be of some use to the clinicians when LAAM becomes generally available since they will undoubtedly be faced with the need to transfer patients from maintenance treatment with one drug to the other.

MATERIAL AND METHODS

The present study was carried out in the context of a previously described fixed-dose double-blind multihospital trial comparing one dose of LAAM (80 mg. twice weekly) and two doses (50 mg. and 100 mg. daily) of methadone after a period of induction. At the end of 40 weeks all patients were returned to methadone. Instead of breaking the study code, the double-blind was maintained for an additional six weeks during which time patients were readjusted to a dose of methadone judged clinically optimal by the treating physician, and by the patients. All patients initially received 60 mg. methadone daily. They were seen from daily to several times a week during the subsequent weeks and their dosages were adjusted according to how they felt and how alert or sedated they appeared. A symptom-sign checklist used in the main study was obtained at the beginning of each week. Random weekly urine samples were analyzed for illicit drugs but no punitive actions were taken.

The study sample consisted of 68 patients with 24 previously assigned to low dose methadone (M-50); 25 to high dose methadone (M-100) and 19 to LAAM (80 mg.).

RESULTS

The results were examined with respect to three sets of related questions:

1. What dosage adjustments were necessary and how soon?

2. How much discomfort was experienced by patients and when?

3. To what extent was there supplementation with illicit opiate use?

Table I presents the distribution of patients whose dose was increased and when this first occurred. Of the 55 patients whose dose was increased, 40 had their increase during the first two weeks. Nearly half of these were in the M-100 group. The time of the first dosage adjustment was significantly different between the M-50 and M-100 groups ($p<0.05$) and between the M-50 and L-80 groups ($p<.01$) but not between M-100 and L-80 groups (Fisher's exact test, dichotomizing time at two weeks). Both M-100 and L-80 groups required earlier dosage adjustments than the M-50 group. However, the total number of patients whose dose was eventually increased during the entire period of observation was not different among the three groups (Fisher's exact test).

The next question of interest was what the patients' methadone doses were at the end of the six weeks. Fifteen patients were stabilized at 60 mg. or less, seven were in the M-50 mg. group, three in the M-100 group and five in the LAAM 80 mg. group. Seven patients were stabilized between 60-70 mg.: six in M-50 group and one in L-80 group. Of the 46 patients stabilized on doses between 70 and 80 mg. of methadone, 11 were M-50 patients, 22 were M-100 patients and 13 were L-80 patients. No patients required more than 80 mg. per day.

The percentage of patients who stabilized in the 70-80 mg. range was significantly higher in the M-100 group (88%) than the M-50 group

Table I. Number of patients having their first dose increase by week.

Week	M-50	M-100	L-80	Total
1	1	7	2	10
2	7	12	11	30
3	6	3	1	10
4	3	1	0	4
5	0	0	0	0
6	1	0	0	1
TOTAL	18	23	14	55

(46%) by Fisher's exact test ($p<0.05$). The percentage of LAAM patients stabilized on 70-80 mg. range of methadone (68%) was not significantly different from the percentage of either methadone group.

The second major research question was the extent of discomfort experienced by patients in the three groups during restabilization. Only one patient (in the M-100 group) had complaints reaching a severe degree and these symptoms disappeared after dosage adjustment. Furthermore, the number and severity of symptoms-signs reported by the entire sample during the six-week period was not great. When symptoms-signs were reported, they tended to appear in the first week or two and then disappear. The three groups were compared in several ways but there was no evidence of systematic difference among groups in any of the symptoms-signs.

Finally, the opiate urine indexes for the six weeks immediately preceding the crossover to 60 mg. of methadone was compared to those obtained after the crossover. This index was obtained by a method which takes into account the frequency as well as the pattern of illicit opiate use. A higher index indicates more extensive opiate use than a lower one. The general methodology had been discussed in detail in a previous report! Table II shows the urine indexes for the three groups before and during restabilization. The M-100 and L-80 groups showed a slight increase in illicit opiate use, but this change was not statistically significant. The M-50 group, on the other hand, showed a significant decrease in opiate use ($p<.05$).

DISCUSSION

Most patients appeared to undergo restabilization with relatively little discomfort. Except for one patient in the M-100 group, all of the symptoms-signs reported were in the mild-to-moderate range and did not differ significantly among the three groups. Some patients, while not complaining of severe symptoms, were visibly uncomfortable to the clinician and were given dose increases before symptoms became severe. The M-100 and L-80 patients received dosage increments earlier than M-50 patients and eventually they were stabilized at somewhat higher dosage levels. This was not entirely unexpected since for the M-100 the starting dose of 60 mg. represented a substantial reduction from their previous maintenance doses, and the change from

Table II. Comparison of pre-post stabilization urine index.

	Pre	Post	Change	
M-50	15.1	11.0	-4.1	*($p<0.5$)
M-100	11.5	14.3	2.8	N.S.
L-80	5.7	7.0	1.3	N.S.

L-80 mg. twice weekly to 60 mg. of methadone daily probably represents a reduction in dose. Patients in these two groups also tended to use slightly more illicit opiates after crossover compared to the preceding six weeks.

Many M-50 patients also received dosage increases after they were switched to methadone 60 mg. and there was a significant diminution in the amount of illicit opiate use in this group compared to their prestudy performance. Some of the dosage adjustments undoubtedly occurred simply because patients were free to ask for them. However, the fact that patients were blind to their previous maintenance doses and that these dose changes occurred earlier in M-100 and L-80 patients would seem to indicate that something besides chance was contributing to the dosage increments. The corresponding reduction in illicit opiate use among the M-50 patients was quite surprising and was to us a most interesting observation. Since these patients were also blind to their previous doses, there was no reason for them to think that in receiving a starting dose of 60 mg. they were in fact receiving a 10 mg. increase from their maintenance dose. Moreover, this increase, while unknown to them, affected them in such a way that their illicit drug use was significantly reduced. It is even more intriguing when we recall that in the parent study, M-50 patients used significantly more opiates than the other two groups and were also more symptomatic with significantly more complaints of aching bones and joints, insomnia and anxiety. They did not appear to be in acute withdrawal as did some of the early terminators. Nevertheless, they may have been feeling chronically ill in some vague manner that resulted in their continued illicit opiate use despite rather long term maintenance treatment. In retrospect, during their tenure in the main study, they may have been suffering from a type of chronic abstinence syndrome which eventually responded to their dosage adjustment. Were these the kind of complaints that post-addicts have long after their withdrawal from opiates? Clinicians are generally aware that a chronic abstinence syndrome exists among postaddicts for many months after they achieve a drug-free state, and it is generally assumed that these are residual symptoms of the acute abstinence syndrome. But if these symptoms exist, covertly or overtly, in patients on long term maintenance treatment, they may well be related to some other pathophysiological process and perhaps demand different therapeutic considerations, since they are obviously related to the patients' drug-seeking behavior and are responsible for many relapses. The fact that these symptoms seem to respond favorably to dose increases is interesting but does not necessarily mean higher dose maintenance is the answer. Perhaps they were only temporarily suppressed in much the same manner that raising phenothiazine levels temporarily suppresses symptoms of tardive dyskinesia. In that event we will be faced with yet another dilemma in treating patients with chronic maintenance. On the other hand, if a covert chronic abstinence syndrome can be shown to exist in other patients on relatively long term maintenance treatment perhaps this issue should be addressed more thoroughly and other therapeutic approaches considered. There has been recent interest in treating addicts and postaddicts with antidepressants and other drugs because some of their complaints resemble depression. Whether we believe these to be symptoms of depression or whether it is possible they are related to a form of chronic abstinence may be of more than academic interest. It is of some practical importance that they be recognized and that innovative therapeutic measures be explored.

CONCLUSION

1. Patients maintained on LAAM for extended periods can be restabilized on methadone with relative ease provided that appropriate dosage adjustments are made. Eighty milligrams of LAAM given three times per week is probably more potent clinically than 60 mg. of daily methadone.

2. In patients on long term methadone maintenance, there may exist a chronic abstinence syndrome which is related to their continued drug-seeking behavior and which may respond at least temporarily to methadone dosage increments with a corresponding decrease in illicit opiate use by these patients. It is not concluded nor recommended, however, that these patients should be treated by simply raising their doses.

REFERENCES

Ling, W., Charuvastra, V.C., Kaim, S.C, and Klett, C.J. 1976. Methadyl Acetate and Methadone as Maintenance Treatments for Heroin Addicts. *Arch. Gen. Psychiatry*. 33: 709-720.

Ling, W., Klett, C.J., and Gillis, R. 1978. A Cooperative Clinical Study of Methadyl Acetate. *Arch. Gen. Psychiatry*. 35: 345-353.

Ling, W. and Blaine, J. 1978. The Use of LAAM in Treatment. In Press.

PHASE III LAAM STUDY: AN INTERIM REPORT

Miriam Burns, M.A. and Mary Carol Newmann, M.A.
John A. Whysner Associates, Inc., New York City

Gail L. Levine
John A. Whysner Associates, Inc., Washington, D.C.

Jack D. Blaine, M.D.
Division of Research, NIDA, Rockville, Maryland

Levo-alpha-acetylmethadol (LAAM) is a derivative of methadone not inappropriately referred to by patients as "long-acting methadone." While LAAM was originally developed during World War II as an analgesic, it was found to be inappropriate for that purpose due to the slow-acting onset of its psychoactivity. Further testing of the drug was delayed until the early 1950's when Isbell and Fraser (Fraser and Isbell, 1952; Fraser, Nash, Vanhorn, and Isbell, 1954) demonstrated the drug's opiate-like profile. LAAM was found by them to be equally as efficacious as methadone in preventing the occurrence of withdrawal symptoms in morphine dependent subjects. Furthermore, since LAAM has an extremely long duration of action which can persist over 72 hours after a single oral dose it can be clearly advantageous for use in drug-dependent individuals.

Subsequent investigations (Irwin, Kinochi, Cooler, and Bottomly, 1973; Irwin, Blachly, Marks, and Carter, 1973) demonstrated that the subjective effects of LAAM were similar to, but less intense than methadone. Furthermore, LAAM was primarily activating while depressant effects predominated with methadone. LAAM and methadone were found to be quantitatively equipotent in many other physiological effects characteristic of opiates.

The ability of LAAM to provide blockage of intravenous heroin in maintenance subjects was demonstrated in other studies (Levine et al., 1973). At a dosage level of 50 mg. three times weekly, no effects were perceived by the subjects from 25 mg. of heroin given 72 hours after previous LAAM dose.

In the late 1960's the first real clinical trials of LAAM were initiated. In a series of controlled double-blind (Jaffe et al., 1970; Jaffe and Senay, 1971; Jaffe et al., 1972; Senay et al., 1975) studies the researchers investigated the safety and efficacy of LAAM compared to methadone in chronic heroin-dependent persons previously stabilized on methadone and in subjects entering treatment dependent on heroin. There were few differences between LAAM and methadone patients on outcome measures such as use of heroin and other illicit drugs, illegal activity and arrests, employment, education, clinic attendance, patient acceptance or dosage changes. They confirmed earlier findings that LAAM can be administered at 48 to 72-hour intervals without the development of an opiate abstinence syndrome. Few unusual reports of toxicity or side effects were noted. The adverse reactions which did occur were particularly associated with excessive dosage and were not uncommonly reported with methadone and would be considered common to all opiates. Results of hematology, blood chemistry, urinalysis and physical examinations revealed few differences between LAAM and methadone patients.

In the ensuing years several additional teams of investigators conducted clinical trials (Zaks et al., 1972; Savage et al., 1976; Goldstein and Judson, 1974) in which approximately 750 patients were treated with LAAM. These studies confirmed the findings of comparable safety and effectiveness of LAAM given three times weekly and daily methadone (Blaine, 1978).

In 1973, double-blind well controlled fixed-dose Phase II studies were initiated in Veterans Administration Drug Dependence Treatment Centers comparing LAAM maintenance and high and low dose methadone maintenance in street heroin addicts. Four hundred thirty male heroin addicts were treated in this 40 week study. Of these, 142 received LAAM.

Another large cooperative Phase II clinical study of LAAM was initiated in 1974 by SAODAP. Sixteen (16) outpatient drug treatment clinics throughout the country were chosen for participation. Seven hundred sixty seven (767) male patients on methadone maintenance of whom 383 received LAAM, participated in this open study. Both of these larger studies confirmed safety evaluated by frequent clinical and laboratory observations and efficacy evaluated by illicit drug (Blaine and Renault, 1976) use, program retention and global ratings of treatment performance.

The current Phase III IND study is a large scale cooperative clinical trial involving 83 treatment programs nationwide. The aim is to further confirm LAAM's effectiveness and safety in a large number of heroin dependent persons under the precise clinical conditions they would encounter when the drug is finally approved for general use by opiate-addicted persons. The clinic trials are carefully monitored and controlled and are intended to establish the most desirable dosage regimen. Information is obtained on incidence of side effects, adverse reactions, drug-drug interactions and medical contraindications. The study is designed to gain at least 40 week data on 2,000 LAAM patients and a suitable methadone control group.

METHOD

Patients are eligible for the study if they are males, 18 years or older, and meet the criteria for methadone maintenance therapy as defined by the methadone regulations. Subjects are only excluded for serious medical conditions requiring imminent hospitalization or the use of many medications, inability to voluntarily sign an informed consent, and imminent move or imprisonment. Subjects are enlisted from new admissions to methadone treatment as well as those already stabilized on methadone. After the experimental status of the LAAM study is explained, the subject voluntarily signs an informed consent to participate in the study. The criteria for admission are broad enough to provide a cross-sampling of the patient-population including those patients who have presented management problems. It should be noted further that, while the study has to date only involved male patients, approval has been granted by FDA for use by women who are not pregnant at the point of enrollment in the program. Until all of the necessary pre-natal studies are completed these women will be placed on methadone should they become pregnant and will be able to resume LAAM treatment after the birth of their babies. A small scale study will soon begin on proper dosaging for females and, when this is completed, the drug will be made available to women in all programs currently using LAAM.

Clinical Evaluation

Pretreatment evaluation consisting of medical history, general physical exam, hematological profile, SMA 12 blood chemistry screen, urinalysis, chest x-ray or tine test, background data and urine screen for illicit drugs is obtained. During the trial for all study patients, there is regular reporting of drug dosage level, observed drug effects, serious medical illnesses, adjunct medication used, adverse reactions, urine tests weekly for morphine and monthly for amphetamines, barbiturates and methadone. At completion of the trial or termination from LAAM treatment, post-treament evaluation is performed including repeat of the baseline medical history, physical exam and laboratory tests and determination of reasons for termination, evaluation of detoxification experience, and global evaluation of treatment outcome.

Drug Dosage

Three protocols are being utilized in this study. In protocol I, all patients are assigned to LAAM dispensed Monday, Wednesday and Friday or Tuesday, Thursday and Saturday. In protocol II, 60% of the subjects are randomly assigned to LAAM three times weekly and 40% to daily methadone and thereafter treated on an open basis. LAAM dosage is flexible within limits. In these two protocols, the recommended dosage regimen for new admissions to treatment is an initial 20 to 30 mg dose of LAAM with 10 mg dose increments on the following clinic visits as needed to adequately control opiate abstinence symptoms. The suggested nominal induction is given in Table I. For patients already stabilized on methadone, the intial TIW dosage should be 1.2 to 1.3 times the daily methadone dose with an additional 10 mgs given on the third or weekend dose. A target dose of 80 mg TIW is recommended as reasonable but the protocol recommends the stabilizing dose should be individualized for each patient. Under no circumstances is a dosage exceeding 120 mg -120 mg -140 mg or a total of 380 mg weekly permitted.

Because LAAM is taken three days a week, one of the three doses must last 72 hours. Several suggestions are made to physicians to eliminate discomfort on the third day although earlier studies demonstrated that no withdrawal signs appeared when the interval between LAAM doses was over 72 hours, and subsequent double-blind studies confirmed that there were no significant differences in opiate abstinence signs or symptoms experienced by patients treated with methadone daily or LAAM three times weekly (Ling et al., 1976; Jaffe et al., 1970); other studies ascribe the weekend complaints to psychologic rather than pharmacologic factors (Goldstein and Judson, 1974). In patients complaining of such discomfort, the three day dose may be raised to 10-30 mg. higher than other doses; the three day dose can be given late in the day and the subsequent dose early in the morning; clinic visits can be scheduled so that the third day is a week-

Table I. LAAM induction (TIW).

1st dose	20
2nd dose	30
3rd dose	40
4th dose	40
5th dose	50
6th dose	50
7th dose	60
8th dose	60
9th dose	70
10th dose	70
11th dose	80
12th dose	80

day when clinical staff is available and the patient has activities planned, or if necessary, a patient with intolerable symptoms may be supplemented with up to ½ dose of LAAM or methadone on the third day.

Because of the necessity for active LAAM metabolites to build up over a period of one or two weeks, some patients may be uncomfortable at the beginning of LAAM treatment, especially if they are dependent on large doses of heroin or methadone. If assurance that this effect is only temporary does not adequately alleviate problems, supplementation with small, usually less than 20 mg, doses of methadone is permitted during early induction.

The usefulness of supplementary medication is currently being more systematically studied in protocol III, being performed as a sub-study in 5 (five) protocol I clinics.

Ling, Klett and collaborators are also independently investigating the usefulness of supplementary methadone and LAAM during the early phases of induction of randomly assigned street heroin dependent subjects and methadone maintained subjects in a double-blind study, the results of which will be reported elsewhere at this conference.

DRUG ADMINISTRATION

LAAM used in this study is supplied in concentrated liquid form (10 mg/ml) and must be diluted in 15 ml of orange juice, Tang, Kool-Aid or similar fruit flavored drinks. Dilutions made prior to dispensing must be refrigerated and utilized within 72 hours. LAAM may only be dispensed by maintenance programs approved to dispense methadone by the Food and Drug Administration. All medication must be consumed under direct observation of a physician, pharmacist or nurse. Under no circumstances will take-home privileges be granted for LAAM nor will LAAM be allowed to leave the clinic.

MONITORING SAFETY DURING THE STUDY

The first level of responsibility for the safety of patients in this study is assumed by the principal investigator or designated study physician at each participating clinic who is in a position to make direct clinical observations and review all laboratory and other data prior to their submission to the data processing center. Review of clinic performance is conducted by the medical staff of John A. Whysner Associates, Inc., who are regularly provided an updated computer listing of dosages, laboratory values and adverse reactions and concomitant illnesses for each patient in the study. All reported adverse reactions and medical illnesses are thoroughly investigated to determine clinical course and relationship to LAAM. Frequent com-

Table II. Demographic characteristics.

	Protocol I		Protocol II			
			Street		X-Over	
	Street	X-Over	L	M	L	M
% Total sample	21%	79%	5%	4%	22%	15%
Mean Age	29.89	29.52	27.59	28.65	28.48	28.94
Race						
White	46%	53%	32%	38%	51%	47%
Black	23%	33%	41%	36%	27%	30%
Sp./Am.	28%	13%	26%	25%	22%	22%
Other	3%	1%	1%	1%	1%	1%
Education						
Attended college	26%	30%	29%	35%	26%	28%
High School grad.	38%	39%	27%	29%	35%	32%
High School	28%	25%	34%	23%	30%	31%
No High School	8%	6%	10%	13%	9%	8%
Employment						
Full-time	38%	55%	36%	34%	48%	51%
Part-time	7%	9%	7%	5%	10%	11%
Unemployed	55%	35%	57%	61%	42%	37%
Years Daily Use--Mean	8.36	7.49	7.41	8.07	6.86	7.40
Months current maintenance treatment						
mean	15.70				16.07	19.24
1-3	27%				32%	22%
4-6	15%				12%	14%
7-12	21%				18%	18%
13-24	17%				16%	19%
24	21%				22%	27%
Take-home privileges						
Yes	35%				35%	45%
No	65%				65%	55%
Required clinic attendance						
2 days	1				0	4
3 days	11				13	15
4 days	1				1	2
5 days	11				14	18
6 days	17				17	19
7 days	57				54	42
Pre-study methadone dose--mg.					48.95	46.98
Mean	46.80				44	49
≤ 40	49				31	31
41-60	31				18	14
61-80	16				6	5
81	5					

munication with principal investigators is maintained; site visits and investigators' meetings are held when indicated. A medical advisory panel of outside counsultants periodically reviews study data. Whysner Associates and each participating program have Institutional Review Boards which semi-annually review protocols and study progress to assure adequate protection of human subjects.

RESULTS

Characteristics of Sample

The Phase III LAAM study is now well underway. Currently (2/28/78) 83 methadone programs have initiated 2673 patients in the study which began in July, 1976. Approximately 78% of these subjects were crossed-over from methadone and 22% entered LAAM treatment directly.

The demographic characteristics are displayed in Table II. The subjects are approximately 29 years of age with 50% white, 31% black and 19% Spanish/American or other. The sample appears to be well educated with over 50% having graduated from high school or attended college. Over 50% of the crossover subjects are employed at baseline while approximately 55% of the street sample are unemployed at entrance to treatment.

Clinical Progress

Fifty-three (53) clinics are participating in protocol I and have medicated 1552 patients with LAAM. Of these 686 (44%) are currently in treatment and 255 (16%) have completed 40 weeks of treatment. Thus 39% of the subjects have terminated prior to completion of the 40 week treatment period.

Twenty-five (25) clinics are participating in protocol II and have medicated 1121 total patients (LAAM 673, methadone 448). 76% of the subjects were on methadone maintenance prior to randomization. Of these 227 LAAM subjects (34%) and 187 methadone subjects (42%) are currently in treatment and 117 LAAM subjects (17%) and 120 methadone subjects (26%) have completed 40 weeks of treatment. Thus, 49% of the LAAM subjects and 32% of the methadone subjects have terminated early.

The available data indicate that the 40 week retention will be about 46% for protocol I patients. Protocol II 40 week retention will be about 50% for methadone subjects and 43% for LAAM subjects. Protocol III shows a retention rate of 67% at 40 weeks for LAAM subjects. This is a significant increase over protocol I and II retention and is comparable to the methadone retention rate in protocol II.

While conclusive results are not available regarding the different crossover schedules in protocol III preliminary results suggest that LAAM supplementation is more satisfactory than methadone supplementation or gradual crossover.

The analysis thus far reveals that the major reasons for early termination from the study, across all groups are clearly not drug related. Methadone patients may be having some of the same difficulties as LAAM patients but they are not terminated from the study because this would mean total termination from maintenance treatment.

Safety

All of the reported adverse reactions have been investigated thoroughly to determine the clinical course and relationship to LAAM. All are known reactions to opiates. There have been 10 out of 2225 LAAM subjects (0.4%) who have terminated due to adverse reactions to LAAM.

Preliminary review of data from medical histories, physical exams, hematological tests, blood chemistry tests and urinalysis at completion or termination do not indicate any clinically significant changes related to LAAM treatment.

Discussion

Opiate maintenance treatment with methadone has been demonstrated to be an effective and safe treatment modality for many chronic heroin-dependent individuals. However, the many clinical and societal advantages of a longer lasting medication prompted the development of LAAM. It is impossible to overstate the value attendance upon interrupting the cycle of drug diversion which is prevalent in any area permitting a take-home schedule of methadone. Patients have expressed relief at the prospect of ingesting their medication in the clinic and not having to cope with peer-pressure to sell take-home medication. They often have fears of children accidently ingesting toxic doses of medication which they must keep refrigerated and which presents great temptation to curious youngsters.

LAAM retention has improved as more is learned about appropriate dosing in protocol III clinics using predetermined supplementary schedules. Further, retention rate will improve as more clinics and patients have experience with the drug. As long as the drug is classified as "experimental" any open study will favor the more familiar drug. It is imperative that the principal investigator and the clinic staff feel comfortable with the drug since their own anxieties and discomfort with the drug translate into helping the patient confirm his own suspicions that his problems are drug-related. It is not un-

Table III. Reasons for early termination for all LAAM patients.

	Crossover		Street	
	LAAM	Meth	LAAM	Meth
Starters	1728	342	497	106
Terminators	774	99	246	42
Completors	296	92	76	28

Reasons for Early Termination

	Crossover		Street	
	LAAM	Meth	LAAM	Meth
Clearly not drug related	331	85	137	34
Possibly drug related	286	11	81	8
Clearly drug related	162	3	27	0

common for patients to attribute long-standing problems to the use of the new drug. For example, a patient in a protocol I drug clinic advised the investigator that he had to terminate from the program because it was rendering him unable to work and, as a consequence, his rent was two months in arrears. At that time the patient had been in LAAM programs for only two weeks. Counseling helped the patient gain insight into his use of the drug as an excuse for his long-time inability to be productive in an employment situation. He agreed to remain in the study and attack the real problem of finding work.

It should be noted that a symptom checklist in a double blind early study (Ling et al., 1976) showed that there were no important differences between LAAM and methadone for signs and symptoms of underdosing, overdosing or somatic complaints. In the early experience with methadone, similar problems were encountered. Adams et al. (1971) found that 24% of patients attributed methadone treatment failure to side effects. Side effects of LAAM and methadone are very similar and range from constipation, to sweating, nausea and loss of interest in sex.

Motivation for entering a research study is varied. Patient motivation for entering a methadone program is perhaps even more varied. At one end of the spectrum are patients who are in treatment without having any desire to stop opiate use. These patients will possibly find that a drug which has characteristics close to heroin will have the greatest acceptability. These characteristics may include euphoria, economic value from the sale of take-home doses, and the ability to use heroin after blockade has diminished. For these patients a drug like LAAM which has little euphoria, eliminates take-home doses and provides a prolonged blockade of heroin will be less acceptable than methadone.

However, for those patients who look at a maintenance drug for its therapeutic value, LAAM may offer advantages. Trueblood et al. (1977) in a study of patients at the Addiction Research Foundation (ARF) reported that LAAM was more acceptable than methadone. Patients felt more comfortable and in better health. Heroin craving was less and heroin blockade was better on LAAM compared to methadone.

This Phase III study is the final stage in the development of LAAM for marketing. Utilizing a greater sample of clinics within the current study, a significant factor of clinic variability has emerged which has often been suppressed by the presentation of aggregate results. Although as yet unanalyzed, the routine monthly monitoring of individual clinic progress indicates marked variability in subject recruitment and patient retention across clinics. While some clinics average less than one new study client per month, other clinics have been able to recruit more than one-half of their total population. Retention, which appears to correlate positively with recruitment success, ranges from a high of 88% to a low of 15% for the midway point, 20 weeks of participation, in the study. Clinics displaying marginal performance are readily identified, contacted and often visited. The problems usually encountered involve negative staff attitudes towards LAAM due either to a general lack of information or resentment of the introduction of study for economic, political or any other non-clinically motivated reason. Other factors negatively biasing clinic performance include lack of strong medical leadership or inaccessible physicians, ultra-conservative or otherwise inappropriate dosing procedures, or a "bad rap" of LAAM by a few patients who often times have not even tried the drug.

The factors responsible for the successful use of LAAM are perhaps more difficult to isolate. Experience, as attested to by the fact that 23% of the subjects of the Phase III study have been recruited by the six Phase II clinics whose staff are essentially unchanged, must certainly play a role. However, that one new investigator has contributed 35% of the 40 week completors suggests that time alone is not responsible. The success of these investigators lends the most convincing support to the planned widespread use of LAAM. Efforts are currently being directed to identify and alleviate the subjective non-pharmacologic factors in clinic staffs and prospective patients which at times undermine the delivery of a pharmacologically effective and preferable treatment agent to this population.

REFERENCES

Adams, R.G., Bloom, W.A., Capel, W.C., and Stewart, G.T. 1971. Heroin addicts on methadone replacement: A study of dropouts. *Int. J. Addict.* 6(2): 269-277.

Blaine, J. 1978. Early clinical studies of levo-alpha-acetylmethadol (LAAM): An opiate agonist for use in the medical treatment of chronic heroin dependence. In The International Challenge of Drug Abuse. R.C. Peterson (ed.) *NIDA Research Monograph (Ser)*. (19), Winter.

Blaine, J.D. and Renault, P.F. 1976. Introduction. Rx: 3x/week LAAM: Alternative to methadone. *NIDA Research Monograph Ser.* (8): 1-9, July.

Senay, E.C., Renault, P.F., Dimenza, S., Collier, W.E., Daniels, S.J., and Dorus, W. 1975. Three times a week LAAM equals seven times a week methadone. A preliminary report of a control study. *Problems of Drug Dependence*. 543-550.

Fraser, H.F. and Isbell, H. 1952. Actions and addiction liabilities of alpha-acetylmethadols in man. *J. Pharm. Exp. Ther.* 105(4): 458-465.

Fraser, H.F., Nash, T.L., Vanhorn, G.D., and Isbell, H. 1954. Use of miotic effect in evaluating analgesic drugs in man. *Arch. Internationales de Pharmacodynamie et de Therapie*. 98: 443-451.

Goldstein, A. and Judson, B. 1974. Three critical issues in the management of methadone programs: critical issue 3: Can the community be protected against the hazards of take-home methadone. *Addiction*. Peter G. Bourne (ed.). New York: Academic Press.

Irwin, S., Kinochi, R., Cooler, P., and Bottomly, D. 1973. Acute time-dose-response effects of cylazocine, methadone, and methadyl in man. (Unpublished). Prepared under NIMH Contract ND-72-115 at the University of Oregon Medical School, Portland, Oregon.

Irwin, S., Blachly, P., Marks, J., and Carter, C. 1973. Preliminary observations with acute and chronic methadone and LAAM administration in humans. (Unpublished). University of Oregon Medical School, Portland, Oregon.

Jaffe, J.H., Schuster, C.R., Smith, B.B., and Blachly, P.H. 1970. Comparison of acetylmethadol and methadone in the treatment of long-term heroin users: A pilot study. *J.A.M.A.* 211: 1834-1836.

Jaffe, J.H. and Senay, E.C. 1971. Methadone and 1-methadyl acetate. Use in management of narcotic addicts. *J.A.M.A.* 216: 1303-1305.

Jaffe, J.H, Senay, E.C., Schuster, C.R., Renault, P.F., Smith, B., and Dimenza, S. 1972. Methadyl acetate vs. methadone. A double-blind study in heroin users. *J.A.M.A.* 222(4): 437-442.

Levine, R., Zaks, A., Fink, M., and Freedman, A.M. 1973. Levomethadyl acetate. Prolonged duration of opioid effects including cross tolerance to heroin, in man. *J.A.M.A.* 226(3); 316-318.

Ling, W., Charuvastra, V.C., Kaim, S.C., and Klett, C.J. 1976. Summary of Veteran's Administration Phase II Cooperative Study for LAAM and Methadone. In: Rx: 3x/week LAAM: Alternative to Methadone. *NIDA Research Monograph Ser.* (8): 94-102.

Trueblood, B., Judson, B.A., and Goldstein, A. 1977. Acceptability of methadyl acetate (LAAM) as compared with methadone in a treatment program for heroin addicts. *Drug and Alcohol Dependence.* In press.

Savage, C., Karp, E.G., Curran, S.F., Hanlon, T.E., and McCabe, O.L. 1976. Methadone/LAAM maintenance: A comparative study. *Comprehensive Psychiatry.* 17(3), May/June.

Zaks, A., Fink, M., and Freedman, A.M. 1972. Levo-methadyl in maintenance treatment of opiate dependence. *J.A.M.A.* 220(6): 811-813.

CLINICAL EXPERIENCES WITH LAAM IN A RESEARCH ORIENTED CLINIC

Ronald G. Hersch, Ph.D. and Martha S. O'Bryan, R.N.

South Shore Mental Health Center

Levo-alpha acetyl methadol (LAAM) was introduced as an adjunct to treatment in the research oriented outpatient clinic of the Washington Center for Addictions in October of 1976. The focus of this paper is to discuss our experiences with this medication in the context of the general clinical atmosphere of the program. To give a comprehensive picture of our experiences, we will discuss the assumptions of our treatment model, the procedures of the LAAM project, the staff's perception of the impact of LAAM, and the reports of clients after receiving LAAM for 40 weeks.

ASSUMPTIONS

1. The basic assumption of our model is that people who have an opiate dependency require an extended period of retraining during which they will frequently experience an urge to return to excessive drug use. This return of craving and probable increase in drug consumption is a sign that they are currently experiencing some biological, psychological or social stress, for which no suitable alternative has yet been learned. A clinical implication of this assumption is that clients are encouraged to discuss their drug cravings and usage with staff. In our program no negative consequences accrue as a result of illicit drug usage off the program premises.

2. A second assumption of our model is that drug usage, by the time most people ask for treatment, is an habitual behavior which is both a response to numerous internal and external stimuli, and a stimulus for other behaviors. A clinical consequence for controlling drug usage, then, is that uncontrolled usage of opiates can be brought under control through multiple interventions which may occur simul-

taneously or sequentially. Opioid maintenance is a logical component of the total treatment regimen and is indicated when the client is afraid that he can no longer personally control his opioid usage and is unwilling to totally give up opiates.

3. Addicts requesting opioid maintenance treatment (i.e., methadone or LAAM) typically come to treatment under duress, imposed by the courts or by their families. Their drug-taking which was once experienced as a self-determined sign of defiance, escape, secret pleasure, etc., has become a burden, i.e., an additional force by which to be controlled. The unspoken goal of treatment for most addicts is to get the forces off one's back. The addict wants to regain experiential control over his drug usage. He wants the externally experienced force to again become an internally experienced control. The goal then is controlled drug use rather than abstinence.

4. Opiate dependence is not a purely physical phenomenon. High dosages of an opioid cannot totally control the physical signs of withdrawal. If one has used an opiate with any degree of regularity and has developed a physical dependency, the result is that numerous internal and external stimuli can elicit the abstinence syndrome, even in a state of physical satiation. The physical state of need is only one of the antecedents to signs of physical withdrawal and craving. Illicit opiate usage and withdrawal symptomatology, while on maintenance treatment, have no correlation to actual maintenance and medication dosage. A clinical implication of this assumption is that therapeutic interventions other than medication are an essential component of a comprehensive treatment program.

5. When one is being treated with a maintenance opioid such as methadone or LAAM, there will always be the possibility of an actual physically induced high. The pioneering methadone maintenance research of Dole and Nyswander led many of us to believe that if a client was maintained on a sufficiently high level of methadone, termed a "blockade dose", he could not purchase sufficient additional opiates to experience a narcotic high. However, several readily accessible medications such as Valium, Darvon, and Phenergan can potentiate the effects of opioids at whatever dosage, and consequently, even if the narcotic blockade can be recognized on an economic level (i.e., can't afford enough dope), there is no way to avoid the potentiating effects of other drugs.

6. There are no clear indications that there will always be a necessity for a maintenance opioid, a narcotic antagonist, or a totally medication free approach for any one personality type or individual drug abuser. Rather, there seems to be a logic to implementing different approaches at different times. A clinical consequence of this assumption is that individual addicts can be taught to have different goals for treatment at various points in their treatment.

In order to facilitate the changing of treatment goals, we have integrated the treatment of patients who are receiving medications with those who are not, and have developed groups composed of individuals in different phases of treatment. According to the assumption, this approach should lead individuals to consider drug free alternatives more readily.

CLINICAL PROCEDURE

Initial Evaulation Phase

All prospective clients are asked to come to the Center at a specified intake time which is held only once a week. After being registered as clients and paying a mandatory $40 evaluation fee, they are oriented in groups of three to five to the treatment program. This is followed up by a comprehensive individualized interview wherein social, psychological, educational, vocational, legal, and substance use histories are gathered. When this is completed, the individual receives a complete physical examination, and a urine sample is obtained. While waiting to see the physician, clients fill out a Client's Aim of Treatment Scale.

A second day in the initial week, the client sees a counselor for a brief follow-up interview and a second drug screen. On this day he also receives a chest x-ray. This second visit typically takes place during a medication dispensing hour so that we can assess the client's interaction with other clients.

At the end of one week the staff discusses the information that was gathered and a treatment plan is decided upon. The essential components of all plans are: 1) assignment to a mandatory weekly group therapy session, 2) assignment to an individual counselor, 3) assignment of a medication or non-medication regimen, and 4) assignment of a weekly fee for service.

Initial Treatment

Following the discussion and evaluation by staff, the client meets with the Program Director to review the treatment model and to sign a contract which specifies some of the treatment program.

During the first two weeks of treatment, medication is adjusted. An individual session with the group leader is held, and there are two sessions with the individual counselor to review responses on the Client's Aim of Treatment Scale.

Once a month there is posted a list of all clients eligible to participate in Protocol II of the Whysner LAAM Study. The criteria for eligibility are:

1. Males who have completed a minimum of six weeks of outpatient treatment in the current program admission.

2. Are in good standing in the program (i.e., have met the minimal behavioral criteria of the contract which include regular group therapy attendance, payment for treatment, and absence of aggressive behavior, and drug dealing and use on the grounds of the program.)

All eligible clients are approached by their nurse-counselor to discuss their potential participation. At this time, the client is given appropriate literature about the medication, is informed of the study protocol and is encouraged to ask other patients about their experiences with LAAM.

Evaluation for the LAAM Protocol

The standard Whysner Protocol II baseline data is collected over a two week period which starts in a group meeting with the Program Director. At this meeting all questions about the Protocol are answered and the contract is signed. This is followed up by SMA 12/60 Hematology Urinalysis and urine screening and PPD tuberculin testing in the clinic. One week after signing the informed consent, the client has a complete physical and review of laboratory data. Two days prior to starting the study medication clients meet individually with the Program Director to open a sealed envelope which designates whther the client will receive LAAM or methadone.

The client's exact verbal statements prior to and just after the opening of the envelope were recorded.

LAAM Treatment

All LAAM medication begins on Mondays and is dispensed on Monday, Wednesday and Friday. If a client misses a day he is offered partial medication dosage on the off day to carry him till the next regular dosage day. Throughout the study, clients who chose to do so were allowed to come to the dispensing area seven days a week and, for the first four Sundays, a minimal dose of methadone was an option. In all respects other than requiring fulfillment of the Whysner protocol which include completing the Profile of Mood States form five times in the course of the 40 weeks, participants were treated in the same fashion as the non-participating program clients.

Levels of medication dosages were based upon the Whysner recommended conversion formula of approximately 1.2 to 1.3 times the methadone dose for weekdays. The "3 day" week-end dose dispensed on Friday varied 5 to 20 mgs. above the crossover dose.

When clients complained about the LAAM medication they were encouraged by counselors (and frequently by other patients) to stay on it for as long as possible. If they insisted on switching to methadone they were not stopped, but were asked to talk to the Program Director to explain their discomfort. In many instances the client reported discomforts similar to those recorded in their record when they first used methadone. Where this was the case, clients were reminded of their previous experience in an effort to dissuade them from a premature switch to methadone.

At the conclusion of 40 weeks of treatment, all clients were interviewed by the Program Director. Clients were asked 1) what factors they considered when requesting to participate in the project, 2) what concerns they had about LAAM prior to taking it, 3) about initial physical and psychological effects of LAAM, 4) whether there were any long-term maintenance effects, 5) if they perceived changes in relationship to significant others as a function of participating in an experimental study, and 6) reasons for requesting an early termination of LAAM, if this was the case.

RESULTS

To date, 56 clients have participated in the study. This represents about 90% of all male clients who have been treated in the program during the same time of the study. Statistical analysis of the demographic data, dose level distribution, urinalysis results (for the first year of the study) and reasons for termination of treatment are presented in Tables I-IV. Tables I-III are taken from an evaluation of the program done by the Stanford Research Institute. This study was undertaken to assess the impact of LAAM on the total outpatient program. These data then include information about methadone clients who did not participate in the study (primarily females).

Table I shows that the average client is a white male in the age range of 26 to 35. Most are single and have completed high school. About a third of the participants are employed.

Table II shows that most of the LAAM clients were maintained in the dose range of 40-59 mgs. This is quite similar to the median dose range for the methadone clients, suggesting it might be possible to make a direct 1:1 conversion from methadone dose to LAAM dose. However, it would be an incorrect conclusion to state that there is a direct 1:1 conversion formula for an individual subject. We did in fact have to increase clients' doses by a factor of about 1.2 to 1.3. Also, as shown in Table II, no subject was maintained on a LAAM dose of less than 40 mgs.

Table I. Client characteristics.

	Methadone clients			LAAM clients	
	October 1976	April 1977	October 1977	April 1977	October 1977
Race					
Black	13%	9%	8%	12%	6%
White	87%	91%	92%	88%	94%
(N)	(47)	(45)	(37)	(16)	(18)
Sex					
Male	73%	58%	62%	100%	100%
Female	27%	42%	38%	0%	0%
(N)	(48)	(45)	(37)	(16)	(18)
Age					
Under 18	0%	0%	0%	0%	0%
18-20	2%	2%	0%	0%	0%
21-25	30%	20%	41%	19%	22%
26-35	60%	69%	51%	75%	78%
36-45	4%	7%	3%	6%	0%
Over 45	4%	2%	5%	0%	0%
(N)	(47)	(45)	(37)	(16)	(18)
Marital status					
Married	36%	33%	38%	31%	28%
Unmarried	64%	67%	62%	69%	72%
(N)	(48)	(45)	(37)	(16)	(18)
Education					
8th grade or less	4%	2%	2%	0%	0%
9th grade	0%	0%	0%	0%	6%
10th grade	15%	9%	8%	25%	17%
11th grade	9%	9%	8%	12%	6%
12th grade (or GED)	53%	44%	56%	38%	43%
more than 12th grade	19%	36%	26%	25%	28%
(N)	(47)	(45)	(37)	(16)	(18)
Employment status					
Employed (including students, trainees, and homemakers)	52%	36%	36%	44%	33%
Not employed	48%	64%	64%	56%	67%
(N)	(48)	(45)	(37)	(16)	(18)
Time in treatment (time since most recent admission)					
Less than one month	2%	0%	8%	0%	12%
One to six months	22%	38%	30%	25%	18%
Seven to twelve months	18%	24%	22%	25%	18%
13 to 24 months	28%	18%	16%	12%	18%
25 to 48 months	30%	16%	19%	38%	34%
More than 48 months	0%	4%	5%	0%	0%
(N)	(46)	(45)	(37)	(16)	(18)

Table II. Dose level distribution.

Percentage of Clients

Daily dose (mg.)	Methadone October 1976	Methadone April 1977	Methadone October 1977	LAAM* April 1977	LAAM* October 1977
1-19	4%	13%	11%	6%	0%
20-39	25%	18%	5%	0%	0%
40-59	44%	54%	49%	27%	53%
60-79	25%	11%	30%	47%	35%
80-99	2%	4%	5%	20%	12%
	100%	100%	100%	100%	100%
(N)	(48)	(46)	(37)	(15)	(17)

* Monday dose.

Table III shows the percentage of urine samples positive for illict drugs for both groups of subjects. As can be seen, only a small percentage of our clients' samples turn up positive for illicit opiates regardless of the type of study medication. However, many of the samples, for both groups, show a positive for Darvon (Propoxyphene). During the time these samples were collected, Darvon was a readily available drug on the street, and its cost was relatively small. These data do not indicate a considerable difference between the two groups in their illicit drug usage.

Table III. Urinalysis Results.

Percentage of samples

Type of drug	Methadone clients July-September 1976	Methadone clients February-April 1977	Methadone clients August-October 23, 1977	LAAM clients Feb.-April 1977	LAAM clients August-October 23, 1977
Opiates (excluding Darvon)	10%	21%	4%	5%	6%
Sedative-Hypnotics	4	0	5	0	6
Amphetamines	1	0.3	0	0	0
Negative Methadone	3	0	0	N.A.	N.A.
Darvon	9	25	45	25	38
(Number of samples tested)	(630)	(611)	(528)	(81)	(120)

Table IV. Pattern of completion and termination of study medication.

	LAAM	Methadone
Total number of subjects	18	14
Number of subjects completing study	9	2
Number of subjects prematurely terminating study medication	9	12
Mean number of days on study medication	74	89
Reasons for premature termination:		
Completed opioid treatment	-	2
Transferred	-	3
Administrative termination	2	5
Jail	1	2
Side effects	6	-

Table IV shows the pattern of completion and reasons for non-completion of the study for the first 32 subjects, those subjects who at this writing could have completed the 280 day study period. As can be seen, one half of the LAAM subjects finished the study as opposed to one-seventh of the methadone subjects. None of the LAAM subjects voluntarily detoxed from opioid treatment whereas two methadone clients did so. Similarly, three methadone clients transferred to other clinics, whereas none of the LAAM clients transferred. Two methadone clients went to jail while only one LAAM client was incarcerated. These data make some sense when one remembers that clients draw to determine which medication they receive, and only people who want to try LAAM participate in the study. When one does not get what he wants it is highly probable that he will be disappointed and either terminate treatment voluntarily or behave in such a way as to be terminated by the program. Further, within the greater Boston area there was a Protocol I study conducted in three clinic locations where clients could readily receive LAAM (essentially for the asking) and a Protocol II study in still another location.

Table IV also shows that the average number of days on study medication was 15 days shorter for the early LAAM terminators then for the early methadone terminators. This is a function of six LAAM clients terminating due to early experienced "side effects", usually within the first 4 weeks of treatment. The most common "side effects" are: 1) a spacy high feeling, 2) a withdrawal reaction on Sundays, and 3) gastrointestinal difficulties (primarily due to increased Propoxyphene abuse).

CLINICAL IMPRESSIONS OF LAAM

All subjects were interviewed by the Program Director at the end of 40 week study dose medication. These interviews show that most patients participated in the study so that they would have more free time and be less inconvenienced by daily visits to the Center for medication. It was reported in these interviews that prior to receiving LAAM there was a wide variety of concerns ranging from whether the dose would hold them to whether it would get them high, and what the long term side effects might be. However, every subject said that they had come to trust the staff prior to participating in the study and their knowledge of the staff's prior experience with other successfully completed studies was an encouraging factor.

The patients reported three types of initial physical and psychological responses. Some clients reported no adverse effects at all. The second response was a feeling of spaciness by a very few clients, described as similar to the effect of Demerol. Subjects who experienced this effect dropped out of the study very quickly. A third effect of the medication was the feeling of a high on Fridays and a withdrawal reaction on Sundays. We found this effect to be the most frequent upsetting factor for our clients and as a result learned that with a slightly higher Friday dose, 5-10 mgs., in comparison to the Monday and Wednesday dose rather than larger spreads of 10 mgs. and up that this effect was minimized.

Those subjects who remained on LAAM the full 40 weeks experienced no adverse long term side effects, whereas those who terminated earlier were largely people who had difficulties over the 3-day weekends as well as various other inconsistent complaints. We believe that these are the people who abused more pills including Propoxyphene and were also less ready to give up illicit opiates. Subjects reported that they experienced a loss of the intensity of their relationships to other clients due to their less frequent visits. Our data suggest that the LAAM subjects made less use of the optional aspects of our program (e.g., individual counseling, vocational rehabilitation, and parenting groups) than the methadone clients. Most subjects reported a tremendous relief as a function of fewer clinic visits, but the staff's impression is that only subjects who were psychologically ready to use this additional time effectively did so. Less well prepared clients just spent more time in non-functional activities such as watching TV or hustling.

In summary, our experiences with LAAM have been generally quite positive. There were no significant negative physical side effects and LAAM seems to be a helpful adjunct for treating opioid addicts who do not need daily contact with a clinic to give their life some consistent organization and structure.

ALCOHOL AND OPIATE NARCOTIC DEPENDENCIES: POSSIBLE INTERRELATEDNESS VIA CENTRAL ENDORPHIN: OPIATE RECEPTOR SYSTEM

Carl Pinsky, Ph.D., Frank S. LaBella, Ph.D., and Leonid Leybin, C.Sc.

University of Manitoba

INTRODUCTION

It is widely known that opiate narcotics and alcohol (ethanol) share an extreme dependency-inducing liability. Behavioral interactions have been observed between these two classes of drug, especially the effects of opiate agonists and antagonists on voluntary consumption of alcohol in rats and mice (Sinclair, Adkins, and Walker, 1973; Blum et al., 1975; Ho et al., 1975; Ho, Chen, and Morrison, 1977) and on severity of the ethanol withdrawal syndrome in those species (Blum et al., 1975, 1976, 1977; Ho et al., 1975). Kissin (1974) has cited human studies reported by Roizin (1969) to suggest that prolonged exposure to opiate narcotics sensitizes the individual to alcohol. Gelfand and Amit (1976) sought a reciprocal ethanol-morphine interaction, but found no effect of a single injection of ethanol on morphine ingestion in rats. Jones and Spratto (1977), however, found that ethanol, given every 8 hours for 4 days during development of morphine dependence, suppressed the severity of naloxone-precipitated withdrawal in 4-day morphine pellet-implanted rats. In contrast, simultaneous ingestion of methadone and ethanol over a five-month period appeared to enhance naloxone-precipitated opiate withdrawal symptoms provoked at the end of the long-term study (Friedman, Geonjian, and Cummins, 1974).

Blum, Hamilton, and Wallace (1977) have reviewed the neuro-chemical actions and effects of opiates and ethanol that might account for the interactions just described. Prominent among these is the depletion of Ca^{++} ion content of brain by either morphine or ethanol and the ability of the opiate antagonist naloxone to block this effect of either drug (Ross, Medina, and Cardenas, 1974). Another possibility is that acetaldehyde, the initial oxidation product of ingested alcohol, forms condensation products with catecholamines (especially dopamine) to produce tetrahydroisoquinoline (THQ) compounds in the brain (Davis and Walsh, 1970) or in the adrenal medulla (Cohen and Collins, 1970). These compounds have certain opiate-like effects on the central nervous system, on smooth muscle and on neurotransmitter kinetics (Hirst, Hamilton, and Marshall, 1977). Davis and Walsh (1970) suggested that heavy and prolonged consumption of alcohol might lead to the formulation of opioid alkaloids in the body. It is known that the THQ, salsolinol, shares the brain Ca^{++} depleting action of morphine and ethanol and is antagonized by naloxone (Ross, Medina, and Cardenas, 1974). Nevertheless, this possible mechanism for a similarity of action of opiates and ethanol remains highly speculative, since the formulation of salsolinol after alcohol administration in animals or humans appears to take place either to an insignificant degree or only under very special conditions of pretreatment (Sandler et al., 1973; Collins, and Bigdeli, 1975; Collins, 1977; O'Neill and Rahwan, 1977).

There are some similarities in the actions of ethanol and opiates on various central neurotransmitter systems (Blum, Hamilton, and Wallace, 1977), but the role of these effects on development of dependency and expression of withdrawal from the two classes of drugs is far from clear. A simplistic interpretation of results involving the apparent effects of the two different kinds of drugs on central neurotransmitter systems can be misleading because of the complex interactions among such systems (Collier, Francis, and Schneider, 1972). It would appear, therefore, that although opiates and ethanol do exhibit several reproducible behavioral interactions it has not as yet been possible to specify the precise central nervous system locale or locales for such interactions. Nor has an explanation as yet been forthcoming to account for their almost equal primacy as drugs of dependency, addiction and compulsive self-administration. The recent demonstration of endogenous opioid substances (Terenius and Wahlstrom, 1974; Hughes, 1975; Hughes et al., 1975; Goldstein, 1976; Guillemin, Ling, and Burgus, 1976) in brain, by general agreement termed endorphins (Simon, 1975), had led to considerable speculation concerning a physiological (Goldstein, 1974, 1976; Kosterlitz and Hughes, 1975; Jacquet and Marks, 1976) and pathophysiological (Kosterlitz and Hughes, 1975; Bloom et al., 1976; Leybin et al., 1976) role for the central opiate receptor:endorphin system. We shall attempt to demonstrate in this report that this system is a likely candidate for a good part of the opiate-alcohol interactions that have been cited here, and may be the vulnerable central site in the process of alcohol over-consumption in humans.

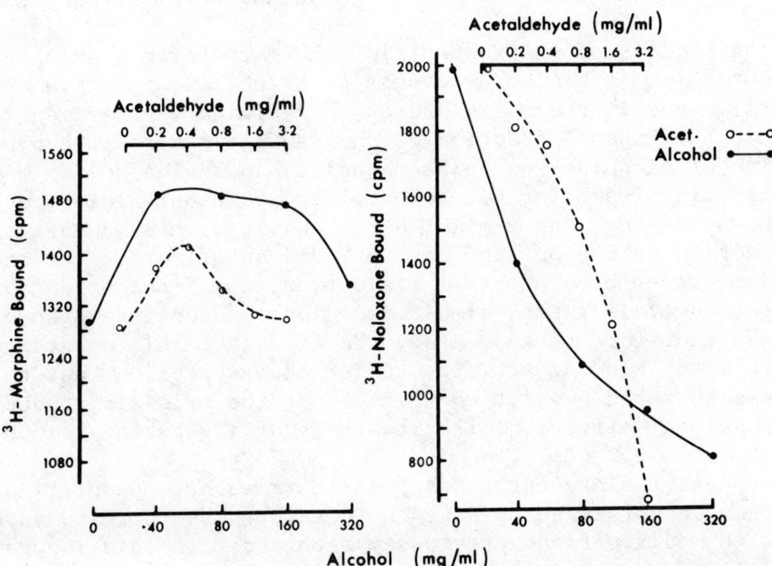

Fig. 1. Effects of ethanol (Alcohol) and acetaldehyde on binding of ^{3}H-morphine (left panel) and ^{3}H-naloxone (right panel) to membranes from rat brain.

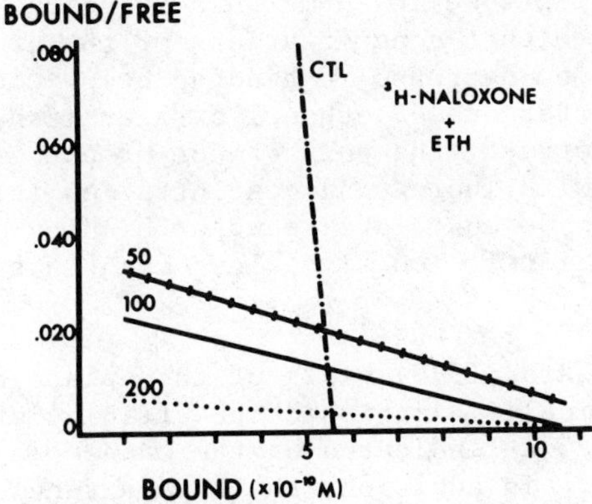

Fig. 2. Scatchard plot of ^{3}H-naloxone binding in presence of varying concentrations of ethanol (ETH); 0, 50, 100, and 200 mg. ml^{-1}. The abscissa represents the concentration of ^{3}H-naloxone bound to membranes equivalent to 15 mg. brain tissue. The X-intercept represents the number (molar concentration) of receptor sites in the tissue aliquot and is seen to increase in the presence of ETH.

EXPERIMENTAL OBSERVATIONS SUGGESTING A MOLECULAR BASIS FOR INTERACTION BETWEEN ALCOHOL AND OPIATE RECEPTORS

During studies on stereospecific opiate receptor binding done in this laboratory, it became necessary to solubilize putative opiate receptor ligands in ethanol mixtures. Preliminary experiments indicated that the ethanol fraction of the vehicle altered the binding affinities of morphine, an opiate agonist, and naloxone, an opiate antagonist, in dissimilar but dose-related fashion. This immediately suggested to us that there might be a behavioral consequence of this ethanol action on the opiate receptor, via an effect on the endorphin:opiate receptor system. Acetaldehyde, the first oxidation product of ethanol, is formed almost immediately upon ingestion of alcohol. We therefore tested, also, the effect of this substance on opiate receptor binding affinity, under the assumption that the in vitro results would predict whether or not the oxidized product might exert behavioral effects similar to those of its parent compound.

The opiate radioreceptor assay was carried out on purified membranes from rat brain homogenates as described by Pert and Snyder (1973) with modifications previously reported from this laboratory (Queen, Pinsky, and LaBella, 1976). Briefly, an aliquot of membranes derived from 15 mg. fresh brain tissue was incubated with 3.0 nM^{3}H-opiate in tris-HCl buffer, pH 7.4, in a total volume of 1 ml., at 0° for 30 minutes. The incubation tubes were centrifuged and the membrane pellet dissolved and counted for radioactivity.

Addition of ethanol or acetaldehyde to the incubating medium caused an initial increase in ^{3}H-morphine binding, followed by decreased binding at higher concentrations of the drugs (see Fig 1). However, there was no decrease in binding below control level at even the highest concentrations of ethanol or acetaldehyde. On ^{3}H-naloxone binding, in contrast, the only effect of either drug was a progressive decrease with increasing concentration (see Fig. 1). The action of acetaldehyde on opiate receptor binding occurred at concentrations approximately 100 times less than that of ethanol (see Fig. 1).

Scatchard (Rodbard, 1973) plots of the opiate binding data show that ethanol drastically diminished the affinity of the opiate receptor for naloxone, as indicated by the change in the slope of the curve (see Fig. 2). In addition, the apparent number of receptor sites was approximately doubled in the presence of all doses of ethanol tested. The X-intercept of the line occurred at approximately 10×10^{-10} M in the presence of ethanol, as compared to 5×10^{-10} M for the control. This apparent doubling in receptor density probably reflects elution of endogenous ligands from the receptors, which would increase the number of binding sites available to exogenous labelled naloxone. Acetaldehyde had effects (see Fig. 3) similar to those of ethanol. The plotted slopes decreased

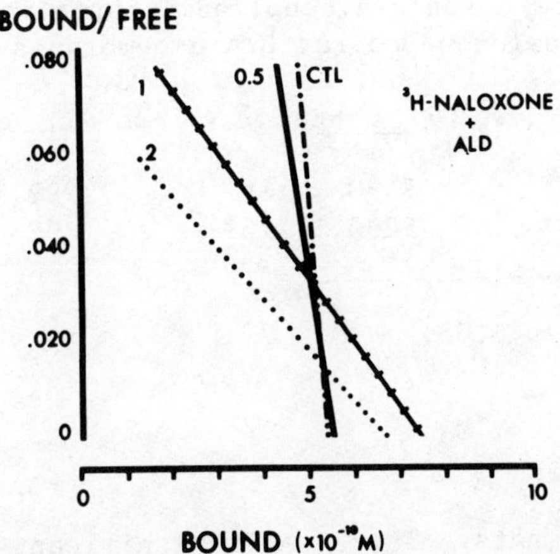

Fig. 3. Scatchard plot of ^{3}H-naloxone binding in presence of varying concentrations of acetaldehyde (ALD); 0, 0.5, 1.0, and 2.0 mg. ml^{-1}.

with increasing concentration of drug, and the X-intercept tended to change in the same direction as with ethanol. Presumably, the addition of higher concentrations of acetaldehyde could have produced still greater increases in the apparent number of receptors, comparable to the doubling with ethanol. We have not yet done a similar analysis on the effects of the two drugs on ^{3}H-morphine binding. Our results suggest, nevertheless, that ethanol and acetaldehyde, in addition to promoting apparent displacement of bound ligand, favor the agonist (morphine) form of the receptor, because binding of ^{3}H-morphine tends to be increased and that of ^{3}H-naloxone decreased.

EFFECTS OF CHRONIC ADMINISTRATION OF ALCOHOL ON ENDORPHIN LEVELS AND OPIATE RECEPTOR OCCUPANCY IN RAT BRAIN: PRELIMINARY OBSERVATIONS

Rats were given ethanol in a nutritionally balanced liquid diet for 15 days. Diets for control and ethanol groups were made equicaloric and given ad lib. Brains were rapidly removed from decapitated animals and homogenized in 10% sucrose. A membrane fraction prepared according to Pasternak, Wilson, and Snyder (1975) was used for measuring ^{3}H-naloxone binding. In cortex area of the brains tested, there was a significant increase in ^{3}H-naloxone binding (see Table I), compatible with the hypothesis that more opiate receptors were available due to the action of ethanol in displacing endogenous ligand.

Total brain endorphin was estimated by the opiate radioreceptor assay on acetic acid extracts of the two brain regions from control

Table I. Effect of chronic alcohol administration on binding of ^{3}H-naloxone to rat brain membranes.

	Cortex (cpm)	Midbrain (cpm)
Control	7490 ± 186	9270 ± 240
Alcohol	8230 ± 276*	9746 ± 361

Mean ± S.E. * $p \lesssim 0.05$.

and ethanol-treated rats. There was a significant decrease in brain endorphin in the midbrain of alcohol-treated rats (a higher count for bound ^{3}H-naloxone represents lesser concentration of competing ligand [endorphin] in the extracts) (see Table II). Endorphin content in brain cortex was also lowered although not to a statistically significant level.

These observations on the effects of alcohol given in vivo correlate well with our in vitro observations (see Figs. 1 and 2). They may be interpreted as indicating that chronic administration of alcohol can deplete endogenous opiate ligands in brain. This depletion is reflected in an apparent increase in the number of opiate receptors (due to displacement of bound endorphin) and in a decreased level of total brain endorphin. An important issue to be clarified by future studies is whether or not the increased binding of ^{3}H-naloxone is, in fact, due to an increased availability rather than an actual increase in number of receptors.

Table II. Endorphin activity in rat brain extracts as determined by displacement of ^{3}H-naloxone.

	Cortex (cpm)	Midbrain (cpm)
Control	9,640 ± 941	8,160 ± 1100
Alcohol	12,000 ± 1351	10,900 ± 695*

Mean ± S.E. * $p \lesssim 0.05$.

EFFECT OF ACUTE INTRAPERITONEAL ADMINISTRATION OF ALCOHOL ON RESPONSES TO SUSTAINED MILD PAIN

Working Hypothesis

The in vitro and in vivo studies just described suggested that administration of ethanol or acetaldehyde might influence the binding of endorphins to their target central opiate receptors in the living animal. Goldstein (1974) suggested that an endogenous antinociceptive ligand for stereospecific opiate receptors might exist as part of an adaptive process to protect the organism from the impairing effects of chronic pain. By an extension of this reasoning we postulated that mild chronic pain would optimally activate such a system.

The Warmplate Test

An experimental model of mild chronic pain was found in a warmplate on whose surface the animal was placed within a vented plastic cylinder (see Fig. 4). Placing a mouse or rat in such an environment provokes behavior that is apparently modulated by central endorphin activity; the method will, therefore, be described in some detail below.

The warmplate test has been utilized in several studies already reported from this laboratory (Pinsky, Koven, and LaBella, 1975;

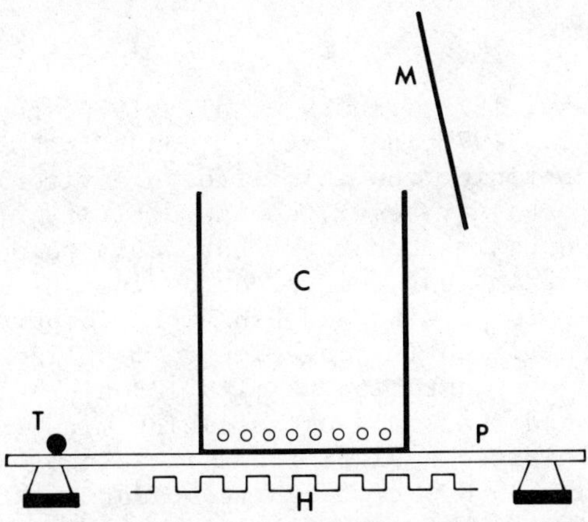

Fig. 4. Warmplate-cylinder test apparatus. P = teflon-coated aluminum plate; H = 112-watt heater (optimal for minimal thermal overshoot); T = thermostat probe (in practice--at edge of cylinder C); C = clear acrylic cylinder, vent holes as shown, 20 cm. dia, open at top; 35 cm. ht.

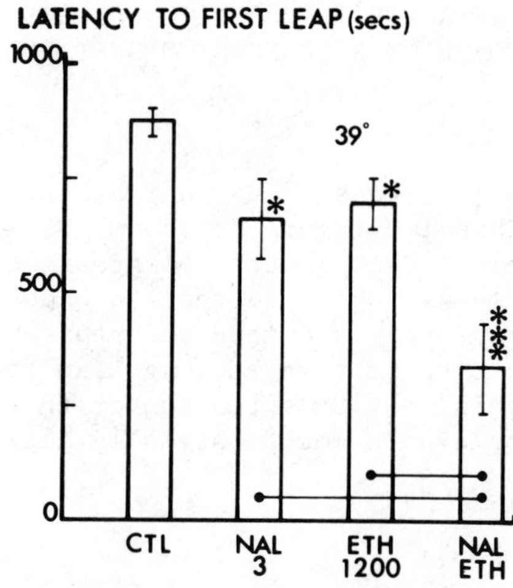

Fig. 5. Effects of naloxone, ethanol and their combination on latency to the first upward leap in mice on the warmplate at 39°C. Drugs in normal saline were given i.p. Mice were placed on warmplate 5 minutes after injection; n=8 per group. Vertical bars show mean values ± standard errors. Asterisks over bars indicate significance of difference from saline control group; one asterisk, $p \leq 0.05$; two, $p \leq 0.01$; three, $p \leq 0.005$. Horizontal lines join groups which were significantly different from each other with $p \leq 0.05$. CTL - saline-injected controls; NAL 3 - naloxone HCl, 3.0 mg. kg^{-1}; ETH 1200 - ethanol, 1.2 g. kg^{-1}; NAL/ETH - combination of the drugs at the just-specified doses (by two separate injections, bilaterally i.p.; spaced 30 sec. apart).

Gigliotti and Pinsky, 1975; Pinsky et al., 1976; Leybin et al., 1976; LaBella et al., 1977; Ho, LaBella, and Pinsky, 1978). The test consists of confining the animal to an environment which provokes nocifensive behavior, e.g., escape activity, exploration, restlessness or some purposive behavior that would reduce the effects of a specific noxious stimulus. We tried to find the lowest surface temperature that would elicit such behavior, to permit exposure of the animal for a sufficiently long time to mobilize the postulated central mechanisms of adaptation to mild chronic pain. Cumulative experiments indicated that male Swiss albino mice (18-22 g., Canadian Breeding Farms) exhibited a significant incidence of upward leaping in the cylinder on a 39°C warmplate but not at a surface temperature of 22°C. The nature of the upward leaping was identical to that seen in the morphine abstinence syndrome in this species (Way, Loh, and Shen, 1969; Marshall and Weinstock, 1971). The frequency of leaping was sensitive to the dimensions of the vented cylinder, and those described in Fig. 4 (caption) were optimal for invoking this response, as determined from our preliminary experiments. In mice

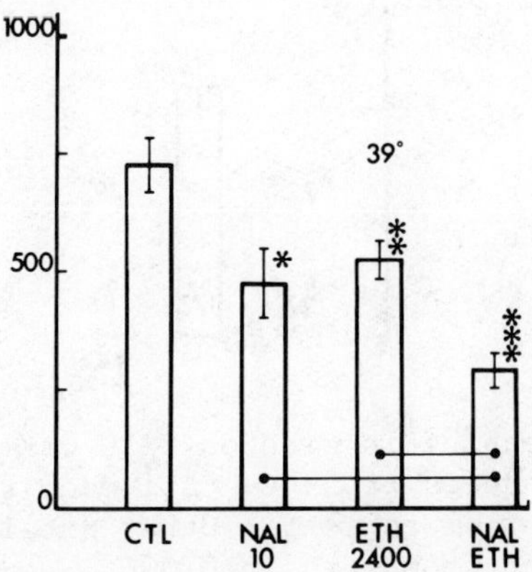

Fig. 6. Effects of naloxone, ethanol and their combination on latency to the first leap on the warmplate at 39°C. Protocol and details as for Fig. 5. NAL 10 - naloxone HCl, 10 mg. kg^{-1}; ETH 2400 - ethanol, 2.4 g. kg^{-1}; NAL/ETH - combination of just-specified doses.

that leaped in this test, the average latency to the first leap was approximately 12 minutes (Figs. 5 and 6), suggesting that the noxious component of the experimental gradually activates a central mechanism responding specifically to this low-intensity thermal stimulus. In contrast, other independent central nociceptive mechanisms might mediate responses to more painful stimuli acting over a much shorter time. Hence the low-temperature warmplate test used in this study is qualititavely different from standard tests for analgesia in that it does not measure reflex responses to cutaneous pain (Ankier, 1974). Instead, the test measures the activity of a much more slowly-responding nocifensor system, sensitive to a noxious stimulus whose intensity is below that needed to activate the more rapidly-acting pain reflex mechanism.

Sprague-Dawley rats tested at warmplate temperatures of up to 44.5°C (Pinsky, Koven, and LaBella, 1975; Leybin et al., 1976). For periods of up to 15 minutes, neither the mouse at 39° to 41°C nor the rat at 44.5°C displayed signs of distress, e.g., vocalization, excitement, aggressive behavior, or fighting behavior when replaced in the colony. Mice continued to groom themselves between leaps, and total grooming activity was not altered by the 39°C stimulus even when leaping was prominent. Rats appeared very calm and some became drowsy, although hindpaw licking was always enhanced. A few rats displayed periods of immobility and apparent sleep over part of the 15 minute test period. Therefore, a routine period of 15 minutes was chosen for these experiments.

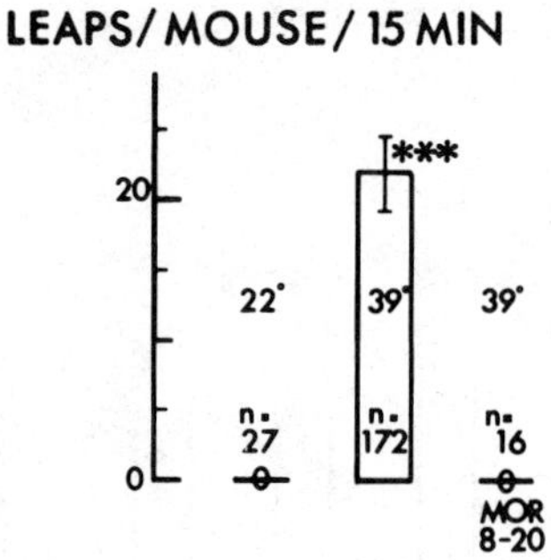

LEAPS/MOUSE/15 MIN

	22°	39°	39°
n =	27	172	16
			MOR 8-20

Fig. 7. Absence of leaping in mice on 22°C warmplate; blockade by morphine of leaping induced at 39°C. All mice received saline injections i.p. 5 minutes prior to test. Other details in text. (* * *) - Statistically different from zero; $p < 0.005$.

We first tested the hypothesis that a central endorphin:opiate receptor system would be activated in the warmplate test. At 39°C mice leaped an average of 22 times in 15 minutes,but not at all at 22°C (Fig. 7). The leaping response was very sensitive to morphine sulphate given by intraperitoneal (i.p.) injection, and doses between 8.0 and 20 mg. kg^{-1} abolished it (Fig. 7). Lower doses of morphine diminished the incidence of leaping in dose-related fashion, with a threshold dose at about 1.0 mg. kg^{-1}. Thus, whatever central pathways may be activated by the warmplate environment, their behavioral expression is sensitive to opiate agonists.

Effects of Naloxone, Alcohol and Acetaldehyde on Responses to the 39°C Warmplate

Batches of naive mice were tested on the 39°C warmplate system after i.p. injection of naloxone, ethanol, acetaldehyde or various combinations of these drugs (see Figs. 5,6, 8-10). Naloxone HCl at 3.0 and 10.0 mg. kg^{-1} increased the frequency of leaping, as compared to saline-injected controls, in a somewhat dose-related manner (see Figs. 8 and 9). Latency to first leap was significantly ($p \leq 0.05$), diminished by both doses of naloxone (see Figs. 5 and 6). Ethanol significantly increased the tendency to leap, at doses of 1.2 and 2.4 g. kg^{-1} (see Figs. 5,6, 8, and 9). These doses of ethanol caused no impairment of righting reflex nor of tiltboard clinging when tested at room temperature in other mice from the same shipments. The effects of naloxone combined with ethanol were greater than additive (see Figs. 5, 8, and 9), except for that on latency at the higher doses (see Fig. 6). Acetaldehyde by itself never provoked leaping on the warmplate, even at doses up to 480 kg^{-1} (see Fig. 10). However, at 240 mg. kg^{-1} it added profoundly to the effect of naloxone HCl at 10 mg. kg^{-1} (see Fig. 10).

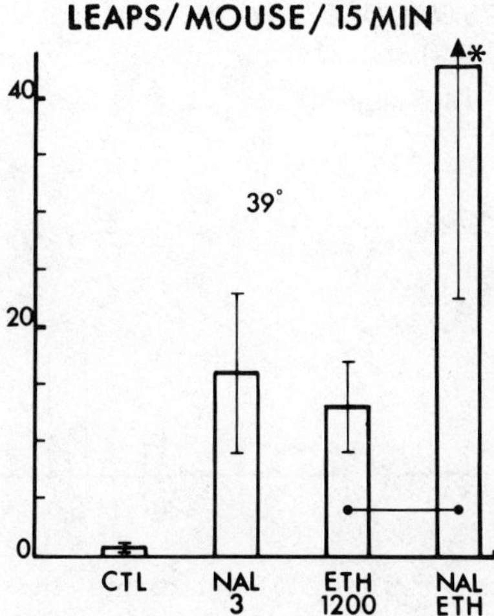

Fig. 8. Effects of naloxone, ethanol and their combination on frequency of leaping in mice on the warmplate at 39°C. Data from same animals as in Fig. 5; all details and procedures as given in legend for Fig. 5.

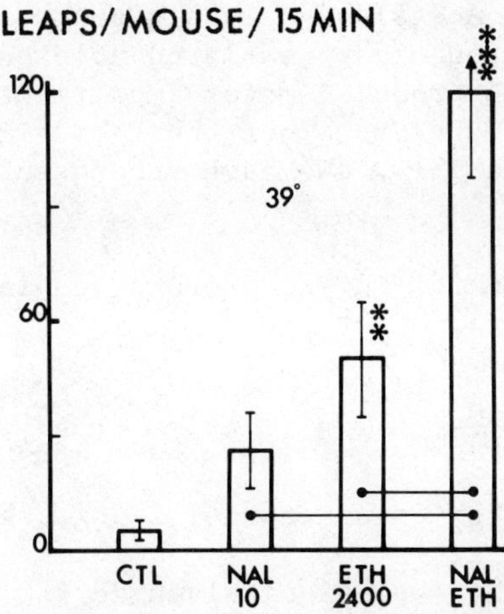

Fig. 9. Effects of naloxone, ethanol and their combination on frequency of leaping in mice on the warmplate at 39°C. Data from same animals as in Fig. 6; all details and procedures as given in legend for Fig. 6.

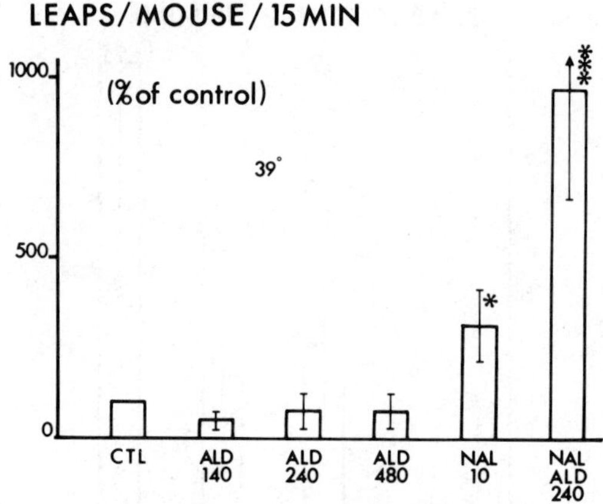

Fig. 10. Effects of acetaldehyde, naloxone and their combination on leaping in mice on 39°C warmplate. Details as for Fig. 5. ALD 140, 240, 480 - acetaldehyde in normal saline, at 140, 240, and 480 mg. kg^{-1} respectively. NAL/ALD 240 - combination of naloxone HCl 10 mg. kg^{-1} acetaldehyde 240 mg. kg^{-1}.

Experiments were done to test the possibility that leaping provoked by naloxone, ethanol and acetaldehyde was due simply to tissue irritation or, for naloxone, to a nonspecific excitant effect. Mice were tested in the cylinder-warmplate system at a surface temperature of 22°C ± 1° (ambient). They were given injections of naloxone and ethanol in doses and in combinations which enhanced leaping at 39°C (see Figs. 6 and 9). Acetaldehyde was given at 480 mg. kg^{-1}, twice the dose that had profoundly potentiated naloxone (see Fig. 10) but still below that which produced motor impairment when tested in pilot experiments. None of the drugs nor their combinations provoked leaping at 22°C (see Table III). We have subsequently tested over one

Table III. Treatments failing to provoke leaping in mice on warmplate at 22°C (ambient).

Saline
Naloxone HCl, 10 mg. kg^{-1}
Ethanol, 2.4 g. kg^{-1}
Naloxone HCl, 10 mg. kg^{-1} plus ethanol 2.4 g. kg^{-1}
Acetaldehyde, 480 mg. kg^{-1}
Naloxone HCl, 10 mg. kg^{-1} plus acetaldehyde 480 mg. kg^{-1}

Each drug treatment was tested on at least 12 and saline treatment on almost 200 mice.

hundred mice with various doses of saline, naloxone and ethanol at 22°C, and have never observed upward leaping. Hence, the enhancement of leaping appears to be a specific interaction of the drugs with the opiate-sensitive mechanisms (see Fig. 7) mobilized by the prolonged mildly noxious stimulus of the 39°C warmplate test.

These results are compatible with the hypothesis that both naloxone and ethanol impair the binding of an antinociceptive substance mobilized by sustained noxious stimulus. The ability of naloxone to enhance leaping activity in mice on a hotplate has already been noted (Jacob, Tremblay, and Colombel, 1974[1]; Jacob and Ramabadran, 1977) and a similar interpretation given. Several other reports have confirmed a pronociceptive effect of naloxone (Gigliotti and Pinsky, 1974, 1975; Pomeranz and Chiu, 1976; Akil, Mayer and Liebeskind, 1976; Grevert and Goldstein, 1977; Frederickson, Burgis, and Edwards, 1977; Buschbaum, Davis, and Bunney, 1977).

EFFECTS OF ALCOHOL AND ACETALDEHYDE GIVEN INTRACEREBROVENTRICULAR INFUSION IN UNRESTRAINED RATS

The inability of acetaldehyde, in contrast to ethanol, to enhance leaping by itself seems anomalous; in vitro experiments had shown it to be even more potent than ethanol in reducing the affinity of radiolabelled opiates for their stereospecific receptor sites (Figs. 1-3). We therefore tested the effects of intracerebroventricular (ICV) infusion of naloxone, acetaldehyde and ethanol on pawlicking in the rat in the cylinder-warmplate system at a surface temperature of 44.5°C. This route of administration minimizes the effects of absorption and metabolism that might contribute to the differences between the relative in vitro and central in vivo effects of the drugs.

Rats (Sprague-Dawley albino strain, 190-230 g.) were implanted with chronic ICV cannulae into the right lateral ventricle, according to the technique of Rezek and Havlicek (1975). Volume of infusion was 10.0 ul, given over 65 sec. for all animals. India ink was infused at the end of these experimental trials, to permit confirmation of proper cannulae placement and success of infusion. At 1.0 min. after the end of infusion the rat was placed on the warmplate at 44.5°C for 10 min. Hindpaw licking was prominent in all rats tested under these conditions and this behavior was selected for examination

---

1 It is interesting to note that Jacob's group provided their 1974 report, and its interpretation in terms of an endogenous ligand for the stereospecific central opiate receptor, before there was any publication, abstract or title describing the demonstration of an endogenous opioid substance.

since it was easily distinguishable from normal grooming activity (Gigliotti and Pinsky, 1975; Pinsky, Koven and LaBella, 1975; LaBella et al., 1977; Ho, LaBella, and Pinsky, 1978). We (Leybin et al., 1976) have shown that morphine, given ICV, significantly reduces hindpaw licking in the 44.5°C warmplate test. In the experiments here ICV infusion of ethanol (710 ug) and of acetaldehyde (80 ug) both promoted significantly ($p \lessapprox 0.01$) more hindpaw licking than did saline-infused rats (see Fig. 11). Mean latency to first hindpaw lick (which was 304 ± 45 sec., mean ± S.E., for the saline-infused controls) was significantly ($p \lessapprox 0.02$) shorter for the drug-treated animals. In a separate group of rats the infusion of naloxone HCl, 2.0 ug in 10 ul over 65 sec., also gave significantly shorter latency to first hindpaw lick. Thus, in the rat, where hindpaw licking appears to be a reproducible nocefensive response, both acetaldehyde and ethanol given ICV mimic a nocifensor-enhancing property of the opiate antagonist naloxone; by this route the acetaldehyde is more potent than ethanol by almost an order of magnitude (see Fig. 11).

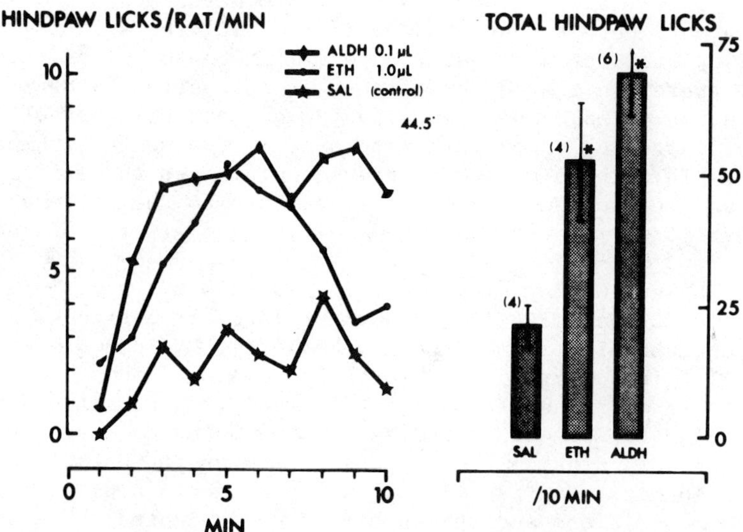

Fig. 11. Effects of intracerebroventricular infusion on hindpaw licking in rats on 44.5°C warmplate. Details in text. Left panel: (diamonds) - infusion of 80 ug. acetaldehyde in 10 ul saline (-1.0% v/v); ETH (dots) -curve obtained with 710 ug ethanol (- 10.0% v/v); SAL (stars) - control saline infusion. Right panel: Summarized data; vertical bars represent mean values ± standard errors. (*) - significantly different from saline controls, $p \lessapprox 0.05$.

SOME INTERPRETATIONS AND IMPLICATIONS OF THE EVIDENCE FOR ALCOHOL-OPIATE RECEPTOR-ENDORPHIN INTERACTIONS

The experiments reported here, although they support the concept of an endorphin:opiate receptor mediation of ethanol- and acetaldehyde-invoked enhancement of nociception, do not rule out the possible involvement of other neuromediator systems in the observed responses. We feel it is significant, however, that there are no reports to indicate that the acute administration of ethanol or acetaldehyde, in the dose ranges used here, alter any neurotransmitter system sufficiently to account for specific behavioral effects (Kalant, 1975). Those studies showing changes in neurotransmitter levels or turnover have used either high doses (Thadani et al., 1976), repeated doses (Orenberg et al., 1976) or prolonged administration (Chopde, Brahmankar, and Shripad, 1977). In studies where the doses and route of administration of ethanol (Frankel et al., 1974; Rawat, 1974) or of acetaldehyde (Rawat, 1974) were comparable to those used here, no effect was seen on 5-hydroxytryptamine (5-HT) turnover (Frankel et al., 1974) nor on cerebral content of noradrenaline, choline and 5-HT (Rawat, 1974). Acetylcholine (ACh) was the only central neurotransmitter whose level in mouse brain was altered by the acute i.p. injection of ethanol 3 g. kg^{-1} or by a single intravenous injection of acetaldehyde 40 mg. kg^{-1} (Rawat, 1974).

The reported lack of effect of acutely administered ethanol or acetaldehyde on neuroamine metabolism is important to our hypothesis, since it diminishes the likelihood of ethanol- or acetaldehyde-invoked impairment of either tryptaminergic or noradrenergic neurotransmission. Impairment of either of these might interfere with the antinociceptive effect of the opiate agonist endorphins (Calcutt et al., 1973) or with the test animal's temperature regulation (Cooper, Pittman, and Veale, 1976). Although the effect on brain ACh might be responsible for some of the pronociceptive responses observed in our study it is difficult to reconcile this possibility with the lack of effect of acetaldehyde by itself at even very large doses. The data of Rawat (1974) showed acetaldehyde to have a much greater effect than ethanol on brain levels of ACh. We refer therefore to our original hypothesis that the pronociceptive effects of the drugs we have tested are due to their impairment of the binding of endorphins to their receptors on target brain cells.

The postulated interaction of ethanol and its metabolite, acetaldehyde on the endorphin:opiate receptor system has implications in the etiology of alcohol overconsumption. If the original proposition of Goldstein (1974) is correct we should expect endorphins to be released at appropriate central nervous system loci whenever the organism encounters sustained, relatively mild, noxious stimuli. In the human this could possibly express itself as an ability to cope with the various stresses of everyday life. Presumably, an unusual degree of noxious stimulus (e.g., pain from injury, chronic irritation or malignancy) would further mobilize antinociceptive endorphins, with a consequent dulling of the affective component, or "appreciation" of the pain. Ingestion of alcohol, therefore, might unmask previously-tolerated chronic pain, with both physical and mental sequelae, by impairment of binding of endorphin to central opiate receptors. Although this appears to contradict the more usual concept of alcohol as being a crude analgesic (e.g., "feeling no pain" after alcohol ingestion), it should be recalled that alcohol is not universally analgesic. There are many instances where ingestion of alcohol is reported to *cause* pain in humans, as in Hodgkin's Disease or certain other neoplastic conditions (Conn, 1957; Brewin, 1966; Perkin, 1973). The description of alcohol-induced pain in Hodgkin's Disease corresponds to that given for the response to alcohol after the administration of disulfiram (Antabuse) (Jacobsen, 1952), a substance known to increase the blood levels of acetaldehyde after alcohol ingestion. "Alcohol pain" has been known to occur also in osteomyelitis (Alexander, 1953), simple fracture (Braun and Schnider, 1953), and in inflammatory conditions and reactions such as pancreatis (Joske, 1955). These conditions might correspond to the "sustained, mildly noxious stimulus" generated by our warmplate tests. Certain individuals respond to alcohol with muscle, joint or "neuritic" pains, in the apparent absence of any lesion, injury or disease (Conn, 1957; Snell, 1966; Perkin, 1973). In such persons there may be sufficient stimulation of nociceptive pathways, even in the absence of specific pain-inducing lesions, to render ethanol (or acetaldehyde) impairment of opiate receptor binding capable of unmasking a pain response referred to a somatic locale. The susceptibility might of course originate in the CNS, with subsequent referral to the somatic locale.

The foregoing observations open the possibility that alcohol ingestion may in some individuals evoke an aversive affective response which is not recognized as pain, but which predisposes the individual to consume more or less of the beverage further to the initial ingestion. At least three clinically-important consequences may potentially result from such a possibility: 1) Initial ingestion of alcohol may result in feelings of psychic discomfort, or even of overt pain, and induce aversion to the consumption of alcohol. An individual responding in this fashion is protected from overconsumption of alcohol, i.e., he will not likely become an alcoholic. 2) Alcohol-induced pain in some individuals might stimulate greater consumption in an attempt to obtund the noxious response in a non-specific manner--not involving the endorphin:opiate receptor system but, rather, by suppressing nociception via a generalized ethanol impairment of cellular excitability within the central nervous system. This individual would be *prone* to repeated self-administration of alcohol, and hence might represent the alcoholism-disposed personality for whom the "first drink" sets off a bout of alcohol drinking and overconsumption. 3) Individuals with a "punishment seeking" alcoholic personality might seek alcohol *because* it causes psychic and/or physical discomfort. The final result, of course, would be overconsumption as in (2).

These concepts represent a departure from conventional views on the etiology of alcohol overconsumption. They were introduced even to our own thinking by way of an unexpected observation. Hence our interpretation of the experimental results must be regarded as speculative, and much further work is required, particularly with human subjects. We welcome the critical examination that would result from future experimentation, in both humans and animals, as additional tests for our hypotheses. We firmly believe that a rational therapy for the disease state that constitutes or accompanies alcoholic overconsumption will be soonest provided by research uncovering the neurobiological origins of the overconsumption itself.

ACKNOWLEDGEMENTS

We thank Mr. J.P. Beaulac, Endo Laboratories, Montreal, for gifts of naloxone hydrochloride. These studies were supported by the Non-Medical Use of Drugs Directorate (NMUD) of Health and Welfare Canada, the Medical Research Council of Canada and the Manitoba Medical Service Foundation, Inc. L. Leybin was supported by the Sellers Foundation (Manitoba).

REFERENCES

Akil, H., Mayer, D.J., and Liebskind, J.C. Antagonism of stimulation produced analgesia by naloxone, a narcotic antagonist. *Science*. 191: 961-962.

Alexander, D.A. 1953. Alcohol-induced pain in Hodgkin's Disease (correspondence). *Brit. Med. J.* 2: 1376.

Ankier, S.I. 1974. New hot plate tests to quantify antinoceptive and narcotic antagonist activities. *European J. Pharmacol.* 27: 1-4.

Bloom, F., Segal, D., Ling, N., and Guillemin, R. 1976. Endorphins: profound behavioral effects in rats suggest new etiological factors in mental illness. *Science*. 194: 630-632.

Blum, K, Futterman, S., Wallace, J.E., and Schwertner, H.A. 1977. Naloxone-induced inhibition of ethanol dependence in mice. *Nature*. 265: 49-51.

Blum, K., Hamilton, M.G., and Wallace, J.E. 1977. Alcohol and opiates: A review of common neurochemical and behavioral mechanisms: Chapt. 12, pp. 203-236. *Alcohol and Opiates*. New York: Academic Press.

Blum, K., Wallace, J.E., Eubanks, J.D., and Schwertner, H.A. 1975. Effects of naloxone on ethanol withdrawal, preference and narcosis. *The Pharmacologist.* 17: 1976 (abstr.).

Blum, K., Wallace, J.E., Schwertner, H.A., and Eubanks, J.D. 1976. Morphine suppression of ethanol withdrawal in mice. *Experientia.* 32: 79-82.

Braun, W.E. and Schnider, B.I. 1958. Alcohol-induced pain in Hodgkin's Disease. *JAMA.*

Brewin, T.B. 1966. Alcohol intolerance in neoplastic disease. *Brit. Med. J.* 2: 437-441.

Buschbaum, M.S., Davis, G.C., and Bunney, W.E., Jr. Naloxone alters pain perception and somatosensory evoked potentials in normal subjects. *Nature.* 270: 620-622.

Calcutt, C.R., Handley, S.L., Sparkes, C.G., and Spencer, P.S.J. 1973. Roles of noradrenaline and 5-hydroxytryptamine in the antinociceptive effects of morphine. pp. 176-191. *Agonist and antagonist actions of narcotic analgesic drugs.* Baltimore: University Park Press.

Chopde, C.T., Brahmankar, D.M., and Shripad, V.N. 1977. Neurochemical aspects of ethanol dependence and withdrawal reactions in mice. *J. Pharmacol. Exp. Ther.* 200: 314-319.

Cohen, G. and Collins, M. 1970. Alkaloids from catecholamines in adrenal tissue: possible role in alcoholism. *Science.* 167: 1749-1751.

Collier, H.O.J., Francis, D.L., and Schnider, C. 1972. Modification of morphine withdrawal by drugs interacting with humoral mechanisms: some contraindications and their interpretation. *Nature.* 237: 220-223.

Collins, M.A. 1977. Identification of isoquinoline alkaloids during alcohol intoxication. Chapt. 9, pp. 155-166. *Alcohol and Opiates.* New York: Acadmic Press.

Collins, M.A. and Bigdeli, M.G. 1975. Tetrahydroisoquinolines in vivo. I. Rat brain formation of salsolinol, a condensation product of dopamine and acetaldehyde, under certain conditions during ethanol intoxication. *Life Sciences.* 16: 585-602.

Conn, H.O. 1957. Alcohol-induced pain as a manifestation of Hodgkin's Disease. *A.M.A. Arch. Intern. Med.* 100: 241-247.

Cooper, K.E., Pittman, Q.J., and Veale, W.L. 1976. The effect of noradrenaline and 5-hydroxytryptomine injected into a lateral cerebral ventricle, on thermoregulation in the new-born lamb. *J. Physiol. (Lond.)*. 261: 223-234.

Davis, V.E. and Walsh, M.J. 1970. Alcohol, amines and alkaloids: A possible basis for alcohol addiction. *Science*. 167: 1005-1007.

Frankel, D., Khanna, J.M., Kalant, H., and LeBlanc, A.E. 1974. Effect of acute and chronic ethanol administration on serotonin turnover in rat brain. *Psychopharmacologia*. 37: 91-100.

Frederickson, R.C.A., Burgis, V., and Edwards, J.D. 1977. Hyperalgesia induced by naloxone follows diurnal rhythm in responsitivity to painful stimulus. *Science*. 198: 756-757.

Friedman, R., Geonjian, A., and Cummins, J.T. 1974. Alcoholism and methadone maintenance. *Proc. West. Pharmacol. Soc.* 17: 132-134.

Gelfand, R. and Amit, Z. 1976. Effects of ethanol injections on morphine consumption in morphine-preferring rats. *Nature*. 259: 415-416.

Gigliotti, O. and Pinsky, C. 1974. Naloxone antagonism of accommodation to unpleasant stimuli. *RODA Summer Scholarships* 1974 Abstracts. Ottawa: Health and Welfare (publ.).

Gigliotti, O and Pinsky, C. 1975. Naloxone effects and a possible endogenous morphinomimetic substance. *RODA Summer Scholarships 1975 Abstracts*. Ottawa: Health and Welfare (publ.).

Goldstein, A. 1974. Opiate receptors and narcotic dependency. *Proceedings Can. Fed. Biol. Soc. 17*, Symposium Address.

Goldstein, A. 1976. Opioid peptides (endorphins) in pituitary and brain. *Science*. 193: 1081-1086.

Grevert, P. and Goldstein, A. 1977. Some effects of naloxone on behavior in the mouse. *Psychopharmacology*. 53: 111-113.

Guillemin, R., Ling, N., and Burgus, R. 1976. Endorphines, peptides, d'origine hypothalamique et neurohypophysaire a activite morphinometique. Isolement et structure moleculaire de l'alpha-endorphine. *C.R. Acad. Sc. (Paris) Serie D*. 282: 783-785.

Hirst, M., Hamilton, G., and Marshall, A.M. 1977. Pharmacology of isoquinoline alkaloids and ethanol interactions. Chapt. 10, pp. 167-187. *Alcohol and Opiates*. New York: Academic Press.

Ho, A.K.S., Tsai, D.S., Chen, R.C.A., and Braude, M.C. 1975. Drug interaction between alcohol and narcotic agents in rats and mice. *The Pharmacologist*. 17: 197 (abstr.).

Ho, T.K., LaBella, F.S., and Pinsky, C. 1978. Opiate properties of SKF-525A. *Can. J. Physiol. Pharmacol.* 56: 550-554.

Hughes, J. 1975. Isolation of an endogenous compound from the brain with pharmacological properties similar to morphine. *Brain Research*. 88: 295-308.

Hughes, J., Smith, T.W., Kosterlitz, H.W., Fothergill, L.A., Morgan, B.A., and Morris, H.R. 1975. Identification of two related pentapeptides from the brain with potent opiate agonist activity. *Nature*. 258: 577-579.

Jacob, J.J.C. and Ramabadran, K. 1977. Opioid antagonists, endogenous ligands and nociception. *European J. Pharmacol.* 46: 393-394.

Jacob, J.J., Tremblay, E.C., and Colombel, M.C. 1974. Facilitation de reactions noceceptives par la naloxone chez la souris et chez le rat. *Psychopharmacologia*. 37: 217-223.

Jacobsen, E. 1952. The metabolism of ethyl aclohol. *Pharmacol. Rev.* 4: 107-135.

Jacquet, Y.F. and Marks, N. 1976. The C-fragment of beta-lipotropin: an endogenous neuroleptic or antipsychotogen? *Science*. 194: 632-635.

Jones, M.A. and Spratto, G.R. 1977. Ethanol suppression of naloxone-induced withdrawal in morphine-dependent rats. *Life Sciences*. 20: 1549-1556.

Joske, R.A. 1955. Aetiological factors in pancreatitis syndrome. *Brit. Med. J.* 2: 1477-1481.

Kalant, H. 1975. Direct effects of ethanol on the nervous system. *Federation Proc.* 34: 1930-1941.

Kissin, B. 1974. Interaction of ethanol with morphine and its derivatives. pp. 126-127. *The Biology of Alcohol*, Vol. 3, Kissin and Begleiter (eds.) New York: Plenum Press.

Kosterlitz, H.W. and Hughes, J. 1975. Some thoughts on the significance of enkephalin, the endogenous ligand. *Life Sciences*. 17: 91-96.

LaBella, F.S., Havlicek, V., Pinsky, C., and Leybin, L. 1977. Opiate-like, naloxone-reversible effects of androsterone sulfate in rats. *Ann. Mtg. Soc. for Neuroscience*. 7: 295.

Leybin, L., Pinsky, C., LaBella, F.S., Havlicek, V., and Rezek, M. 1976. Intraventricular met 5-enkephalin causes unexpected lowering of pain threshold and narcotic withdrawal signs in rats. *Nature*. 264: 458-459.

Marshall, I. and Weinstock, M. 1971. Quantitative method for assessing one symptom of the withdrawal syndrome in mice after chronic morphine administration. *Nature*. 234: 223-224.

O'Neill, P.J. and Rahwan, R.B. 1977. Absence of formation of brain salsolinol in ethanol-dependent mice. *J. Pharmacol. Exp. Ther.* 200: 306-313.

Orenberg, E.K., Zarcone, V.P., Renson, J.F., and Barchas, J.D. 1976. The effects of ethanol ingestion on cyclic AMP, homovanillic acid and 5-hydroxyindoleacetic acid in human cerebrospinal fluid. *Life Sciences*. 19: 1669-1672.

Pasternak, J.W., Wilson, H.A., and Snyder, S.H. 1975. Differential effects of protein modifying reagents on receptor binding of opiate agonists and antagonists. *Molec. Pharmacol*. 11: 340-351.

Perkin, G.D. 1973. Alcohol-induced pain in chordoma. *Brit. Med. J.* 33: 478.

Pert, C.B. and Snyder, S.H. 1973. Opiate receptor: Demonstration in nervous system tissue. *Science*. 179: 1011-1014.

Pinsky, C., Koven, S.J., and LaBella, F.S. 1975. Evidence for role of sex steroids in morphine antinociception. *Life Sciences*. 16: 1785-1786.

Pomeranz, B. and Chiu, D. 1976. Naloxone blockade of acupuncture analgesia: endorphin implicated. *Life Sciences*. 19: 1757-1762.

Queen, G., Pinsky, C., and LaBella, F.S. 1976. Subcellular localization of endorphine activity in bovine pituitary and brain. *Biochem. Biophys. Res. Comm.* 72: 1021-1027.

Rawat, A.K. 1974. Brain levels and turnover rates of presumptive neurotransmitters as influenced by administration and withdrawal of ethanol in mice. *J. Neurochem*. 22: 915-922.

Rezek, M. and Havlicek, V. 1975. A chronic cerebral cannula. *Physiol. Psychol*. 3: 263-264.

Rodbard, D. 1973. Mathematics of hormone-receptor interactions. *Adv. Exp. Med. Biol.* 36: 327-364.

Roizin, L. 1969. Interaction of methadone and ethanol. Presentation at meeting of *Eastern Section of American Psychiatric Assoc.* New York.

Ross, D.H., Medina, M.A., and Cardenas, H.L. 1974. Morphine and ethanol: selective depletion of regional brain calcium. *Science*. 186: 63-65.

Sandler, M., Carter, S.B., Hunter, K.R., and Stern, G.M. 1973. Tetrahydroisoquinoline alkaloids: in vivo metabolites of L-dopa in man. *Nature*. 241: 439-443.

Simon, E.J. 1975. Goldstein (*Life Sciences* 1975 17: 1643-1654; *Science* 1976 193: 1081-1086) credits Simon with having devised the term "endorphin" in 1975.

Sinclair, J.D., Adkins, J., and Walker, S. 1973. Morphine-induced suppression of voluntary alcohol drinking in rats. *Nature*. 246: 425-427.

Snell, W.E. 1966. Pain and alcohol (correspondence). *Brit. Med. J.* 2: 645.

Terenius. L. and Wahlstrom, A. 1974. Inhibitor(s) of narcotic receptor binding in brain extracts and cerebrospinal fluid. *Acta. Pharmacol. Toxicol. (Suppl.) K b h.* 35(1): 55.

Thadani, P, Kulig, B.M., Brown, F.C., and Beard, J.D. 1976. Acute and chronic ethanol-induced alterations in brain norepinephrine metabolites in the rat. *Biochem. Pharmacol.* 25: 93-94.

Way, E. Leong, Loh, H.H., and Shen, F.H. 1969. Simultaneous quantitative assessment of morphine tolerance and physical dependence. *J. Pharmacol. Exp. Ther.* 167: 1-8.

THE CENTRAL ANTICHOLINERGIC SYNDROME - ETIOLOGY, DIAGNOSIS, AND MANAGEMENT

Richard C.W. Hall, M.D., Michael K. Popkin, M.D., Sondra K. Stickney, R.N., and Earl R. Gardner, Ph.D.

Departments of Psychiatry and Medicine
Medical College of Wisconsin, Milwaukee, Wi 53226

The potentiation of the anticholinergic effects of one drug by another is frequently overlooked as a cause of acutely developing psychosis. This is not surprising considering that more than 600 drugs with significant anticholinergic properties are currently commercially available (Alpern and Marriot, 1973). This paper defines the central anticholinergic syndrome (CAS), its forms of presentation, etiology, pharmacology, and management.

Complex clinical presentations are often seen in patients who have ingested a mixture of drugs, either by prescription or inadvertent self-medication (Drachman and Leavitt, 1974). The CAS occurs in patients ingesting what they regard as safe over-the-counter sedative hypnotics, (i.e., Sominex, Triptone). The incidence of intentional self-intoxication with plant alkaloids such as Angel's Trumpet and Jimson Weed, is also increasing (Hall et al., 1977). It is seen in children following the therapeutic use of anticholinergics or secondary to the accidental ingestion of medicinal agents or plants (El-Yousef et al., 1973). Ordinary doses of psychotropic medications may produce it in geriatrics, patients with organic brain disease, and the mentally retarded.

THE SYNDROME AND ITS FORMS

Much of the confusion existing in the literature concerning the CAS centers on which elements are necessary for its diagnosis. This is not surprising since various components of the syndrome may appear in different patients. Classically, it is thought of as a psychosis (characterized by confusion, agitation, auditory and visual hallucinations and short-term memory impairment) accompanied by signs of peripheral muscarinic blockade, (i.e., blurred vision, dilated and poorly reactive pupils, flushed face, dry skin and mucous membranes, tachycardia, urinary retention, abdominal distress, paralytic ileus, constipation, fever and hyper-reflexia). Other symptoms include ataxia, poor motor coordination, myotonic twitching and increased muscular tone followed by profound muscular weakness. Clinical confusion stems from physician expectation of finding a majority of these symptoms in constellation, in a given patient. The dilemma is that any or all of these signs and symptoms may be seen at any moment in time and that characteristically, the clinical picture changes dramatically and rapidly.

Clinical presentation is determined by the nature of the ingested drug, its dose, age of the patient, premorbid personality, presence of concomitant psychiatric illness, previous tolerance to the medication, presence of other medications with atropinic properties, and rate of drug administration. Children typically present with agitation, restlessness, hyperactivity and sudden violent states. Toxic delirium is more common in children while adults present with signs of apprehension, paranoid ideation, hallucinations and overwhelming episodes of fear. As the syndrome progresses, both adults and children appear confused, disoriented and incoherent in their speech. Picking at bed clothing or the gathering or grasping of imaginary objects is an important behavior as it is seen late in the course of the syndrome in at least one-third of these patients. Fever, which is frequently considered a sine qua non of the fully developed syndrome, occurs in only 20% of adult cases and 25% of pediatric cases. Convulsions, rash, vomiting, urinary retention, paralytic ileus, constipation, abdominal distress and leukocytosis occur in less than 10% of cases. The most reliable signs of central anticholinergic syndrome include: dilated and poorly reactive pupils, incoherence, confusion, disorientation, short term and recent memory disturbances, flushed face, dry mucous membranes, auditory and visual hallucinations, tachycardia, restlessness, hyperactivity or agitation, picking or grasping movements, ataxia and motor coordination. The incidence of auditory and visual hallucinations is twice as great in adults as in children. Urinary retention, although rare in the general population, occurs frequently in geriatrics. Somnolence and/or coma, which are often considered characteristic of the syndrome, occur in less than one-third of fully developed cases. When coma does occur, it is twice as likely to be seen in an adult as a child and typically occurs late in the course of the syndrome.

PATHOPHYSIOLOGIC MECHANISMS

It has been proposed that the toxic confusional states caused by various psychotropic drugs (phenothiazines, tricyclic antidepressants, antiparkinsonian agents) represent an atropine psychosis produced by the competitive inhibition of the neurotransmitter acetylcholine at central and peripheral muscarinic sites (Forrer and Milken, 1958). Animal experiments involving correlative EEG monitoring indicate that behavioral effects of anticholinergic agents occur from a primary blockade of central inhibitory cholinergic pathways (Gramacher, 1975). Anticholinergics, particularly scopolamine, disrupt memory acquisition and retrieval by blockage of inhibitory cholinergic receptor systems in the hippocampus and hypothalamus (Gramacher and Baldessarini, 1975; Greenblatt and Shader, 1973; Hall, Popkin, and McHenry, 1977). Mechanisms other than pure cholinergic blockade are also involved (Drachman and Leavitt, 1974).

ANTICHOLINERGIC POTENCY AND DRUG INTERACTION

Psychotropic drugs have significant anticholinergic effects and physicians are often unaware of the relative potencies and interactions which exist between these commonly used drugs. Antiparkinsonian drugs have a strong tendency to produce the CAS, particularly when given in combination with tricyclic antidepressants and/or phenothiazines (Cole, 1972). Even when given alone, these agents will produce central nervous system toxicity in up to 20% of Parkinsonian patients (Alpern and Marriot, 1973). Thirteen percent of all patients treated with tricyclic antidepressants and 35% of patients over the age of 40 treated with phenothiazines and antiparkinsonians, develop toxic confusional states (Burks, et al., 1964). These drugs, when used in combination, summate their cholinergic properties in a geometric rather than arithmetic fashion. Consequently, small dose changes of one drug may produce the CAS in a previously unaffected patient (Forrester, 1975).

The anticholinergic activity of phenothiazines is inversely proportional to their propensity to produce extrapyramidal symptoms. The antipsychotic and extrapyramidal effects of phenothiazines are thought to be due to their ability to block central dopamine receptors (Brimble and Buxton, 1972). Phenothiazines having high muscarinic affinity exert an antagonistic effect on the development of extra-pyramidal symptoms (Brimble and Buxton, 1972) and are least likely to produce extrapyramidal side effects. Clozapine has the greatest anticholinergic effect followed by Thioridazine, Promazine, Chlorpromazine, Perphenazine, Fluphenazine, Trifluoperazine and Haloperidol. The antimuscarinic potency of Thioridazine, for example, is 7 times greater than that of Chlorpromazine and about 320 times greater than that of Haloperidol (Snyder, Greenberg, and Yamamura, 1974). The practical consequence of this information is that when multiple medications must be prescribed, combinations of drugs having high anticholinergic activity should be avoided. The anticholinergic potency of the antiparkinsonian agents is considerably greater than that of the phenothiazines. For example, Benzatropine has an anticholinergic effect 6 times greater than that of Thioridazine.

DIAGNOSIS OF CENTRAL ANTICHOLINERGIC SYNDROME

Diagnosis of the CAS is predicated on a careful history and detailed physical examination. It is often difficult to obtain an accurate history from these patients because of coexisting confusional states. Responsible individuals who may be able to clarify the patient's history and medication use should be contacted and all medications available to the patient brought to the physician for inspection.

Physical examination may reveal signs of elevated blood pressure, mydriatic and unresponsive pupils, dry mucous membranes, foul breath, rapid thready pulse, widened pulse pressure and hot, dry, flushed skin, particularly of the face. Bowel sounds may be hypoactive. Hyperthermia may be present with temperatures as high as 104 degrees and is usually associated with the presence of a desquamative rash over the face, neck and trunk. In severe cases, the patient may be comatose and evidence decreased or absent corneal and ovulovestibular reflexes, hyperactive deep tendon reflexes, clonus and Babinski sign. Clonus usually precedes the onset of seizures, which are not an uncommon sequeli.

In cases of mild to moderate intoxication, mental status reveals diminished concentration, agitation, tangentiality, circumstantiality, auditory and/or visual hallucinations, misperception of sensory stimuli and disorientation to time or place. Ability to perform first and second order mathematics is diminished as is digit recall, with reverse digits being significantly more impaired. Proverbs, similarities and absurdities show concreteness and diminished ability to abstract. In severe cases, signs of profound delirium with disorientation, impaired recent memory, inability to concentrate, hallucinosis and agitation are seen (Janowsky, 1972).

TREATMENT

It is essential that the treatment of CAS proceed immediately upon diagnosis since the majority of signs clear rapidly if the syndrome is of short duration. In cases where the CAS has been present for more than 48 hours, recovery of mental function may be considerably delayed and signs of organic impairment or hallucinosis may persist for up to 10 days.

When CAS is suspected, a test dose of physostigmine (Antilirium) is indicated. The adult patient is given an intramuscular injection of 1 to 2 mgs. of physostigmine salicylate and closely observed for 30 minutes (Anderson and Narasimhan, 1975). Cardiac monitoring during the observation period permits the physician to determine whether expected therapeutic effects are occuring (i.e., slowing of heart rate), and alerts to the cardiac toxicity associated with parasympathetic overstimulation (profound bradycardia). If the diagnosis is correct, improvement will be dramatic; if incorrect, signs of parasympathetic stimulation will occur.

Another diagnostic test consists of the application of 1 to 2 drops of 1% Pilocarpine opthalmic solution to the eye. The muscarinic agonist effect of this drug will constrict a neurologically dilated pupil, but have no effect on the mydriatic pupil of the CAS. Exceptions to this rule are when pupilary dilation is secondary to glaucoma or occular trauma (Atkinson and Ladinsky, 1972).

The methacoline test may also be used. Here, 10 to 30 mgs. of methacoline chloride are given subcutaneously or intramuscularly. The patient is then observed for 30 minutes for signs of parasympathetic stimulation. Failure to elicit bradycardia, rhinorrhea, sialorrhea, diaphoresis or increased GI motility, is a strong indication of the presence of an atropine-like blockade (Pearl et al., 1976).

Physostigmine is the drug of choice for treatment of CAS. It is efficacious in treating CAS produced by pure anticholinergic agents (Ross; McDermott and Grossman, 1975), tricyclic antidepressant overdose (Slovis et al, 1971) or overdose of antiparkinsonian agents (Anderson and Narasimhan, 1975). Its pharmacology makes it particularly suitable, since it is a tertiary amine that rapidly crosses the blood-brain barrier. The drug's rapid action and short half-life (1 hour) endow it with an excellent margin of safety (Brimble and Buxton, 1972).

Another useful treatment agent is the anticholinesterase, tetrahydroaminacridine (THA), which also has a rapid onset, effectively crosses the blood-brain barrier, and has been shown to reverse both the clinical and EEG abnormalities associated with anticholinergic poisoning (Warnes and Aranth, 1971).

Physostigmine is *relatively* contraindicated in patients with diabetes mellitus, gangrene, renal hypertension, coronary artery disease, heart block, significant cardiac arrythmia, hyperthyroidism, hypothroidism, asthma, severe bronchitis, peptic ulcer, colitis, urinary or renal obstruction, glaucoma, pregnancy, myotonia atropia or cognetia and in patients with indications of vagotonia. Clinical evaluation for these considerations should be initiated prior to treatment (Pearl et al., 1976). The most serious acute complications following administration of physostigmine are the precipitation of acute asthmatic episodes and heart block resulting in myocardial infarction. The most frequent toxicities encountered are nausea and vomiting (Atkinson and Ladinsky, 1972).

In the absence of contraindications, an initial dose of 1 to 2 mgs. of physostigmine salicylate, may be given intramuscularly or by slow intravenous injection (i.e., over a 1 to 2 minute period). Rapid intravenous injection may produce seizures. In pediatric cases a dose of 0.5 mgs. physostigmine salicylate I.M. or slow I.V. is indicated. If no signs of cholinergic stimulation appear following the test dose, or if there is no clinical improvement within 15 to 30 minutes, an additional 1 to 2 mgs. of physostigmine may be administered. Should the patient clear rapidly, it is essential to remember the drug's short half-life and subsequent need to readminister approximately q.30 minutes until stable reversal is achieved.

Treatment efficacy is indicated by a lowered pulse rate, a return toward normal pulse pressure, diminished hypertension, improved alertness and diminuation of hallucinations, agitation and confusion. Pupils may remain dilated for up to a week following successful treatment. In true anticholinergic syndromes, dramatic recovery occurs within 15 minutes after physostigmine is administered (Itil and Fink, 1968). If significant hyperthermia is present before treatment, a hypothermic response may be noted for a short period following physostigmine administration (Leff and Bernstein, 1968). In patients who develop anticholinergic syndrome secondary to the administration of an antiparkinsonian drug used to control dystonia, the dystonia may recur within 5 to 10 minutes following physostigmine treatment, but usually abates within the next 10 to 20 minutes (Anderson and Narasimhan, 1975).

Physostigmine is efficatious in reversal of the coma, choreoathetosis, myoclonus and life threatening arrhythmias seen in patients who have overdosed with tricyclic antidepressants (Alpern and Marriot, 1973).

The clinician must be aware of symptoms of physostigmine overdose if he is to successfully titrate its dose. Physostigmine should be discontinued if any of the following signs of cholinergic toxicity appear: myosis, seborrhea, rhinorrhea, dyspenea, hyperhidrosis, sudden bradycardia, fasiculation, tremor, explosive diarrhea or admoninal colic. Toxicity may be reversed with atropine sulphate, 0.5 mgs. for each milligram of physostigmine to be counteracted (Cole, 1972).

MEDICAL AND PSYCHIATRIC MANAGEMENT

Management of the delirious patient with CAS should include a protected environment and careful documentation of vital signs, pupilary size and bowel sounds. Frequent mental status examinations should be conducted. Careful monitoring of electrolytes and state of hydration are essential. In patients who are frankly delirious, diazepam is the drug of choice for sedation. It may be given in doses of 5 to 10 mgs. I.V. or P.O.q.4 hours. It is essential to remember that, because of their anticholinergic properties, phenothiazines are contraindicated; their use may further worsen the CAS, produce cardiovascular collapse through their alpha blocking effects or lower the seizure threshold sufficiently to precipitate convulsions.

Patients who have overdosed with tricylcic antidepressants should be placed on a cardiac monitor. Once the patient's medical condition has stabilized, careful psychiatric evaluation is essential. In situations where the CAS has been present for periods in excess of 48 hours, the patient may remain psychotic for several days (Castellano, 1972).

REFERENCES

Alpern, H.P. and Mariott, J.G. 1973. Short term memory: facilitation and disruption with cholinergic agents. *Physical Behavior*. 2: 571-575.

Anderson, J.A. and Narasimhan, M.J. 1975. Poisoning from food, chemicals, drugs and metals. *Nelson's Textbook of Pediatrics*. Philadelphia: W.B. Saunders, Co. 10th Ed. Edited by Vaughn, V.C. 1666-1667.

Atkinson, J. and Ladinsky, H. 1972. A quantitative study of the anticholinergic action of several TCA on the rate isolated duodenal strip. *Brit. J. Pharmac*. 45: 519-524.

Bartholow, R. 1873. The antagonism between atropia and physostigmine. *The Clinic*. 5: 61-63.

Bell, R.C. and Hall, R.C.W. 1977. The mental status examination: guidelines for administration and interpretation. *American Family Physician*. 16:5: 145-152.

Brimble, C.R.W. and Buxton, D.A. 1972. Behavioral actions of anticholinergic drugs. *Biochemical and Pharmacological Mechanisms Underlying Behavior*. New York: Elsevie Publishing Company. Ed. Bradley, P.B. and Brintlecombe, R.W. 115-126.

Burks, J.S. et al. 1964. TCA poisoning: reversal of coma, choreatherotosis and myoclonus by physostigmine. *JAMA*. 230: 1405-1407.

Castellano, C. 1972. Effects of some anticholinergic drugs on water maze learned behavior in mice. *Psychopharmacologia*. (Berl.) 21: 361-368.

Cole, J. 1972. Atropine-like delirium and anticholinergic substances. *Am. J. Psy*. 128: 898.

Drachman, D.A. and Leavitt, J. 1974. Human memory and the cholinergic system. *Arch. Neurology*. 30: 113-121.

El-Yousef, M.E. et al. 1973. Reversal of antiparkinsonian drug toxicity by physostigmine: a controlled study. *Am. J. Psych.* 130: 2: 141-145.

Forrer, G.R. Milken, J.J. 1958. Atropine coma: a somatic therapy in psychiatry. *Am. J. Psych.* 115: 455-458.

Forrester, P.A. 1975. An anticholinergic effect of general anaesthetics on cerebro-cortical neurones. *Br. J. Pharmacol.* 55: 275-276.

Frasher, T.R. 1863. The characters, actions and therapeutic uses of the ordeal bean of calabar. *Edinburgh Medical J.* 9: 36-55 pssim.

Gramacher, R.P. 1975. Anticholinergic action of Thioridazine. *Am. J. Psych.* 132:2: 302-303.

Gramacher, R.P. and Baldessarini, R.J. 1975. Physostigmine: its use in acute anticholinergic syndrome with antidepressant and antiparkinson drugs. *Arch. Gen. Psy.* 32: 375-380.

Greenblatt, D.J. and Shader, R.I. 1973. Anticholinergics. *N. Engl. J. of Med.* 288:23: 1215-1219.

Hall et al. 1977. Psychosis induced by datura suaveolens: hallucinosis and anticholinergic delirium. *World J. of Psychosynthesis.* 9:3: 19-22.

Hall, R.C.W., Popkin, M.K., and McHenry, L. 1977. Angel's Trumpet psychosis. *Am. J. Psych.* 134:3: 312-315.

Itil, T. and Fink, M. 1968. EEG and behavioral aspects of the interaction of anticholinergic hallucinogens with centrally active compounds. *Progress in Brain Research.* 28: 149-168.

Janowsky, D. et al. 1972. Anticholinergic agents and atropine-like delirium. *Am. J. Psych.* 129:3: 360-361.

Leff, R. and Bernstein, S. 1968. Proprietary hallucinogens. *Dis. Nerv. Syst.* 29: 621-626.

Pearl, J. et al. 1976. Anticholinergic activity of antipsychotic drugs in relation to their extrapyramidal effects. *J. Pharm. & Pharmac.* 28: 302-304.

Ross, J.F., McDermott, L.J., and Grossman, S.P. 1975. Disinhibitory effects of introhippacampal and intrahypothalamic injections of antichoinergic compounds in the rat. *Pharmacology, Biocehmistry and Behavior.* 3: 631-639.

Slovis, T.L. et al. Physostigmine therapy in acute TCA poisoning. *Clinical Toxicology*. 4: 451-454.

Snyder, S., Greenberg, D., and Yamamura, H.I. 1974. Antischizophrenic drugs and brain cholinergic receptors. *Arch. Gen. Psy*. 31: 58-61.

Warnes, H. and Aranth, J.V. 1971. Complications of psychotropic medications in high doses. *Psychiat. Q*. 45: 87-91.

ANTAGONISM OF COCAINE HIGHS BY LITHIUM

Alvin J. Cronson, M.D. and Abraham Flemenbaum, M.D., M.S.

Texas Tech University School of Medicine

INTRODUCTION

According to Jefferson and Greist (1977), the interest on the beneficial effects of lithium for amphetamine abuse was aroused by a report by Flemenbaum (1974) on two patients who spontaneously stopped abusing amphetamines because they no longer experienced the expected high while taking lithium. A third patient receiving lithium for weight reduction did not experience appetite suppression or highs of the subjective effects until lithium was discontinued. Flemenbaum's concept was further supported by a placebo-control double blind study of nine severely depressed patients in whom lithium treatment resulted in a 60% reduction in the activation and euphoria responses to d-amphetamine and almost complete abolition of response to l-amphetamine (Van Kammen and Murphy, 1975).

Directly or indirectly, for amphetamine there is a large body of evidence regarding the beneficial effects of interaction with lithium, although the reports are still controversial (Jefferson and Greist, 1977; Furukawa, Ushizima, and Ono, 1975; Poitou, Boulu, and Bohuon, 1975; Cox, Harrison-Read, Steinberg, et al., 1971; Lal and Sourkes, 1972).

Recently, Flemenbaum (1977) demonstrated an inhibition of hyperactivity and SB in 10 rats serving as their own controls that received doses of cocaine up to 19 mg./kg.

We are now reporting on some human clinical experiences that appear to support the hypothesis of an inhibition of cocaine highs by lithium.

CASE VIGNETTES

J. L., a 28-year-old Hispanic married male, had been detoxified from heroin and was drug-free. Because of affective disorder symptoms he had been stabilized on lithium. He reported he had obtained and used "pure coke" but had noticed no reaction. Having once been on cyclazozine and knowledgeable that it blocked heroin, he asked if lithium was like the former drug in that it blocked cocaine.

R. A., a 25-year-old white male, was on methadone, 90 mg., and was difficult to manage because of drug abuse and behavior that was arrogant, manipulative, and impulsive until placed on lithium. He also reported no response to cocaine despite his companions "getting quite high on it."

P. L., a 28-year-old white male former addict and alcoholic, had been placed on lithium to control his mood swings, violent temper, and impulsiveness. He came across cocaine, tried it, and found to his surprise that he had no reaction. Since he was grossly obese, and felt he could not get a job because of obesity, he was given amphetamine on a careful experimental basis. He reported that while it controlled his appetite, he felt no high from it and did not crash when the effects wore off, even if he took more medication than was prescribed. He did lose 40 pounds and was only 20 pounds overweight when he completed treatment.

M. C., a 23-year-old white single male manic depressive, was placed on lithium with excellent results. He became stable in his interpersonal relations and employment and no longer abused hypnotics and alcohol. However, one Christmas he went to a party where a group of people were indulging themselves on cocaine. Having had some brief experiences in college with cocaine, he looked forward with enthusiasm to getting high on it. To his surprise, he received no response, even though "everyone else thought it was great."

K. J., a 32-year-old white single actress with a history of manic depression (bi-ploar) had been diagnosed and successfully treated at several well known lithium clinics. In spite of her good response, she had had multiple re-hospitalizations and in her last one cocaine was discovered in her urine and her lithium level was negligible. Once stabilized, she confided that she liked the high produced by cocaine but could not obtain one while on lithium. Because of that she had stopped her lithium medication frequently.

DISCUSSION

Even though in their concise and superb book Jefferson and Greist (1977) concluded that lithium has not been shown to be effective in the treatment of drug abuse and still must be considered experimental, they still summarized that "evidence that in man lithium's ability to attenuate the activating and euphorogenic effects of amphetamine seem to be firm, but still conclusions cannot be possible until there are more well designed studies."

Although this one particular report is an anecdotal one, the authors feel that it further illustrates the potential ability of lithium inhibition of drug highs. The authors recognize that lithium administration may work in more ways than one, i.e., by improving a masked effective disorder and thus indirectly decreasing the need for highs obtained from a drug. The authors also recognize potential biases of non-blind evaluations, but conclude that the evidence continues to be supportive of a potential use of lithium in drugs of abuse, and in this particular case of a potential activity of lithium in reducing cocaine "highs".

REFERENCES

Cox, C., Harrison-Read, P.E., Steinberg, H., et al. 1971. Lithium attenuates drug-induced hyperactivity in rats. *Nature*. 232: 336-338.

Flemenbaum, A. 1974. Dose lithium block the effects of amphetamine? A report of three cases. *Am. J. Psychiatry*. 131: 820-821.

Flemenbaum, A. 1977. Antagonism of behavioral effects of cocaine by lithium. *Pharmacol. Biochem. Behav*. 7: 83-85.

Furukawa, T., Ushizima, I, and Ono, N. 1975. Modifications by lithium of behavioral responses to methamphetamine and tetrabenazine. *Psychopharmacologia* (Berl.) 42: 243-248.

Jefferson, J.W. and Greist, J.H. 1977. *Primer of Lithium Therapy*. Baltimore: Williams and Wilkins Co.

Lal, S. and Sourkes, T.L. 1972. Potentiation and inhibition of the amphetamine sterotypy in rats by neuroleptics and other agents. *Arch. Int. Pharmacodyn. Ther*. 199: 289-301.

Poitou, P., Boulu, R., and Bohuon, C. 1975. Effect of lithium and other drugs on the amphetamine chlordizaepoxide hyperactivity in mice. *Experientia*. 31: 99-101.

Van Kammen, D.P. and Murphy, D.L. 1975. Attenuation of the euphoriant and activating effects of d- and l-amphetamine by lithium carbonate treatment. *Psychopharmacologia*. 44: 215-224.

DRUG ABUSE AND CONVULSIVE STATES

Kenneth F. Lampe, Ph.D.

University of Miami School of Medicine

With the exception of alcohol, the influence of psychoactive drugs on epileptic patients does not seem to have undergone systematic examination. Yet this information is of considerable importance to the drug counselor since drugs, particularly in combination with other physical factors which increase seizure frequency, can interfere with the subject's ability to hold a job or, in some cases, even to be able to maintain normal social relationships. This, in turn, may result in more extensive drug abuse and ultimate socio-psychological deterioration.

A single drug or physiological influence by itself may not elicit a seizure in a predisposed individual, whereas a combination of such influences may be sufficient to do so. It is essential to analyze an increased seizure frequency from such a standpoint. A second significant consideration is that the most important reason for failure of seizure control is inadequate dosage of anti-epileptic medication. In general, this may be translated as non-compliance by the subject. The causes may include attempts at concealment of the disorder, an inadequate understanding of the importance of regular medication, an attempt to conquer the condition by will power, irregular habits, and a host of other reasons.

The principle non-drug causes of increased seizure activity have been discussed by Sibley (1974). He was particularly impressed by the frequency with which seizures could be correlated with lack of sleep; patients receiving only three or four hours of sleep at night were especially vulnerable on the following day. This had been noted earlier by Janz (1962), who considered it the single most important precipitating factor in patients with a history of epileptic seizures predominantly occurring characteristically during a period of two hours after awakening. This group of patients also showed increased seizure activity if they were awakened prematurely.

In female patients, menses is often accompanied by increased seizures, but the relationship to pregnancy or the use of oral contraceptives is quite variable and unpredictable. It has been presumed that these endocrine influences may be related to water retention and its consequent electrolyte factor.

Stress and emotional shocks are a possible but infrequent cause of seizures. Overexertion, however, may be of significance in some patients, particularly if related to loss of sleep time.

The effect of alcohol consumption on the epileptic patient has been reported frequently in the literature but there are few carefully documented studies. It is apparent that a few epileptics cannot tolerate alcohol in even small amounts and will have a seizure on the day of ingestion. Others, on the other hand, have decreased seizures if their medication is supplemented by a daily drink and some are even tolerant of heavy alcohol abuse. However, the combination of sleep deprivation, moderate to heavy drinking over a weekend or holiday, and early morning wakening on the following back-to-work Monday can be considered as specifically provocative in patients with awakening epilepsies.

Oversedation with hypnotic drugs, e.g., the short acting barbiturates, methaqualone, and gluthetimide, may elicit seizures in patients with a history of epileptic attacks appearing predominantly during sleep.

The antihistamine drugs, e.g., diphenhydramine and tripelennamine, are encountered only occasionally as drugs of abuse. In toxic overdosage these drugs may induce convulsions in patients without a history of seizure disorder. In epileptic patients they are well recognized as capable of eliciting seizures in even sub-clinical doses (Churchill and Gammon, 1949; Diaz-Guerrero et al., 1956).

A number of drugs may produce convulsions in patients who have neither epilepsy nor other diseases affecting the central nervous system. Some of these physician induced drug seizures recorded in in-patients by the Boston Collaborative Drug Surveillance Program (1972) include penicillin, anti-diabetic agents, local anesthetics, prednisone, and isoniazid. The use of aminophylline in patients with status asthmaticus might also be included. Previously controlled epilepsy can be exacerbated by the phenothiazines, and this group of drugs can also produce seizures in non-epileptic patients. A similar observation for both epileptic and non-epileptic patients has been made for the tricyclic antidepressants, such as imipramine (Kiloh et al., 1961; Dallos and Heathfield, 1969). Scopolamine may also activate focal actively and is occasionally associated with convulsions following anesthetic premedication in patients who have discontinued their anti-epileptic drugs.

Epileptic patients do not exhibit a particular sensitivity to central nervous system stimulants. Indeed, amphetamines and methylphenidate have been employed to counteract the sedative influence of some types of anticonvulsant medication. It was noted in the early days of drug-induced convulsant therapy in psychiatry that non-epileptic patients responded with a single generalized convulsive seizure whereas epileptic patients tended to show some of the characteristics of their spontaneous seizures (Laidlaw and Richens, 1976). The prolonged self-administration of amphetamines or an acute overdosage of cocaine will induce convulsions in any subject.

Chronic alcoholics, that is those who consume an equivalent of one pint of whiskey each day, may exhibit seizures about 24 hours after an abrupt cessation of drinking. Such seizure activity is seen more commonly in patients who are physically dependent on short acting barbiturates and other sedative hypnotics. This was first reported by Dunning (1940) and has been investigated extensively with particular regard to dose-treatment time relationships to evoke this effect by Fraser (1958), Wulff (1960) and by Ronse and Vervarcke (1972). Concealed sedative-hypnotic drug abuse with withdrawal convulsions may be difficult to diagnose and may subject the non-epileptic patient to needless nuerologic examination and a course of anticonvulsant therapy. The epileptic patient is probably more sensitive to this withdrawal phenomenon, but this does not seem to have been studied. For drug counselors, it may be of some significance to note that new born infants of sedative-hypnotic dependent mothers may also suffer a severe withdrawal syndrome.

Marijuana has not been found to increase seizure frequency in the epileptic patient. A few individual case reports have appeared citing convulsive activity in such patients subsequent to marijuana use (Keeler and Reifler, 1967), but there is insufficient clinical evidence in the facts of extremely widespread usage, including high dose chronic abuse, to support an adverse effect on epilepsy control. One neurologist found that 29% of his epileptic patients in the 15-30 age group smoked marijuana (Feeley et al., 1976). The major psychoactive component of marijuana, Δ^9-tetrahydrocannabinol, has been shown to have a dose-response anticonvulsant effect in laboratory animals (Consroe et al., 1976). In high dosage it produces seizure in rabbits (Consroe et al., 1977) and in naturally epileptic beagles (Consroe et al., 1976; Weiss et al., 1976). A similar dose dependent anti-convulsive effect can be demonstrated with certain currently employed anti-epileptic drugs.

Martin and Consroe (1976) were not able to produce convulsions in Δ^9-tetrahydrocannabinol sensitive rabbits with a number of hallucinogens including lysergic acid diethylamide (LSD), mescaline, phencyclidine, or psilocybin. It should be noted, however, that psilocybin-containing mushrooms have produced convulsions in small, non-epileptic children as have the muscimol-containing mushrooms, *Amanita muscaria* and *A. pantherina* (Lampe, 1978).

SUMMARY

Convulsions subsequent to acute overdose of a drug or during abrupt withdrawal of a sedative-hypnotic drug from a dependent individual do not constitute epilepsy and do not warrant chronic anticonvulsant medication. An increased frequency of seizures in an epileptic patient, however, deserves careful attention and counseling. An attempt has been made to identify those agents most likely to be abused that may be responsible for such an exacerbation. It should be remembered that drug abuse may be only a part of an altered life style. Thus, non-drug factors such as lack of sleep and poor anticonvulsant drug compliance may be the predominant determinants of the increased seizure frequency.

REFERENCES

Churchill, J.A. and Gammon, G.D. 1949. The effect of antihistaminic drugs on convulsive seizures. *JAMA*. 141: 18-21.

Consroe, P., Jones, B., Laird III, H., and Reinking, J. 1976. Anticonvulsant effects of delta-9-tetrahydrocannabinol. In Cohen, S. and Stillman, R.C. (eds.), *The Therapeutic Potential of Marihuana*. New York: Plenum.

Consroe, P., Martin, P., and Eisenstein, D. 1977. Anticonvulsant drug antagonism of delta-9-tetrahydrocannabinol-induced seizures in rabbits. *Res. Comm. Chem. Path. Pharmacol*. 16: 1-13.

Dallos, V. and Heathfield, K. 1969. Iatrogenic epilepsy due to antidepressant drugs. *Brit. Med. J*. 4: 80-82.

Diaz-Guerrero, R., Feinstein, R., and Gottlieb, J.S. 1956. EEG findings following intravenous injection of diphenhydramine hydrochloride (Benadryl). *EEG Clin. Neurophysiol*. 8: 229-306.

______________. 1972. Drug-induced convulsions. Report from Boston Collaborative Drug Surveillance Program. *Lancet*. ii: 677-679.

Dunning, H.S. 1940. Convulsions following withdrawal of sedative medication. *Int. Clinics*. 3: 254-264.

Feeley, D.M., Spiker, M., and Weiss, G.K. 1976. Marihuana and epilepsy: Activation of symptoms by Delta-9-THC. In Cohen, S. and Stillman, R.C. (eds.), *The Therapeutic Potential of Marihuana*. New York: Plenum.

Fraser, H.F., Wikler, A., Essig, C.F., and Isbell, H. 1958. Degree of physical dependence induced by secobarbital or pentobarbital. *JAMA*. 166: 126-129.

Janz, D. 1962. The grand mal epilepsies and the sleeping-waking cycle. *Epilepsia* 3: 69-109.

Keeler, M.H. and Reifler, C.B. 1967. Grand mal convulsions subsequent to marijuana use. *Dis. Nerv. System*. 28: 474-475.

Kiloh, L.G., Davison, K. and Osselton, J.W. 1961. An electroencephalographic study of the analeptic effects of imipramine. *EEG Clin. Neurophysiol*. 13: 216-223.

Laidlaw, J. and Richens, A. 1976. *A Textbook of Epilepsy*. London: Churchill Livingstone.

Lampe, K.F. 1978. Pharmacology and therapy of mushroom intoxications. In Rumack, B.H., and Salzman, E., (eds.), *Mushroom Poisoning: Diagnosis and Treatment*. Cleveland: CRC Press.

Martin, P. and Consroe, P. 1976. Cannabinoid induced behavioral convulsions in rabbits. *Science* 194: 965-967.

Ronse, H., and Vervarcke, J. 1972. *Acta psychiat. belg*. 72: 241-255.

Sibley, W.A. 1974. Diagnosis and treatment of epilepsy: overview and general principles. *Pediatrics* 53: 531-535.

Weiss, G.K., Feeney, D.M. and Spiker, M. 1976. Effect of delta-9-THC on physical symptoms and EEG in the totally naturally epileptic beagle. *Fed. Proc*. 35: 644 abs. 2418.

Wulff, M.H. 1960. The barbiturate withdrawal syndrome. *EEG Clin. Neurophysiol. Suppl*. 14.

THE DEVELOPMENT OF A TRAINING EVALUATION SYSTEM FOR MONITORING AND EVALUATING TRAINING OUTCOMES*

George N. Sousa

Menlo Park, California

INTRODUCTION

With support from the National Institute on Drug Abuse (NIDA), state and local drug abuse intervention agencies have committed substantial resources to establish training delivery systems for drug abuse workers. As drug abuse treatment programs become more fully recognized under the law and by the larger social system, the need for training drug abuse workers increases.

The success of federal, state and local training delivery systems is dependent upon several factors, including accurate assessments of training needs, availability of appropriate training resources, effective management methods and timely mechanisms for tracking and evaluating the outcomes of training. One of the most important factors for continued life of a training system are the mechanisms set up for tracking the outcomes of training, and for comparing outcomes against training standards or performance criteria. Without these mechanisms, any attempts to implement a training system would be like trying to sail a boat without a rudder.

The process by which information is gathered in the determination of the success of a training system (or parts of it) is called training evaluation. Specifically, training evaluation is a process of defining and selecting important from trivial information, devising ways of obtaining minimally obtrusive information, analyzing data without grossly violating statistical assumptions that must be met, and providing evaluation results to decision makers in time for them to make "better informed" decisions. The purpose of training evaluation, then, is to provide a valid basis for improving the quality of training system outcomes.

The roles that training evaluation can serve is put into clear perspective when the training needs of the drug abuse intervention system are considered.

The NIDA *Manpower and Training Strategy 1977*, estimates that at the beginning of Fiscal Year 1977:

1. An estimated 173,000 positions make up the drug abuse field, with about 72,000 in treatment, 96,000 in school and community-based prevention, 4,000 in programmatic evaluation and information gathering, a little over 1,000 in research. Only an estimated 140,000 trained workers are available to fill those positions. A shortfall of over 33,000 trained workers exists.

2. Ninety-four percent of this shortfall is in the area of drug abuse education and community-based prevention.

3. About half the program-level evaluation workers (about 2,000) lack adequate training.

4. Combining this shortfall with annual turnover, nearly 56,000 persons require preservice or inservice training in drug abuse intervention.

5. BUT the National Training System is only budgeted to provide training to 14,000 in FY 1977.

6. Available research manpower...is extremely limited.

(From NIDA Manpower and Training Strategy 1977, NIDA-ADAMHA, 1977, p. 1.)

Although these estimates may not be very accurate at this point in time, they serve to point out the continuing need to provide training to drug abuse personnel not only in 1977, but in the foreseeable future as well, and the critical need for accurate training needs assessment data and other evaluative data on the quality of training services that are being developed at the federal, state and local levels.

The NIDA strategy is to provide training and technical assistance to drug abuse treatment providers at federal, state and community levels through the development and implementation of the National Manpower and Training System (NMTS). The NMTS includes four major components:

1. The National Drug Abuse Center (NDAC) is charged with developing information resources and course material, and to effect technology transfer at the national level.

2. Five Regional Support Centers (RSCs) function to extend and complete NDAC's role by conducting direct training activities.

3. The State Training Support Program (STSP) functions to provide direct financial support to states to develop their own training delivery systems.

4. The Career Development Center (CDC) functions to raise the standards of drug treatment by providing nontraditional educational opportunities.

5. The Physician Education Program (PEP) provides support to medical school faculty for research and teaching on drug treatment issues.

6. The NIDA International Training Support Program (ITSP) provides assistance to other nations on drug abuse prevention, treatment and rehabilitation efforts through the U.S. Department of State.

It will be interesting to see how well the NMTS has responded to the challenge of 1977, not only in terms of numbers of drug abuse treatment persons "served" by training, but also in terms of the quality of training and cost-effectiveness. Current evaluation technology at the federal level may permit an examination of the question of how many persons received various types of training, but unfortunately will not allow an examination of the question of the quality of training.

TRAINING EVALUATION ISSUES

The question of the quality of training revolves around two basic issues in evaluation theory and practice. The first issue is the formative role that training evaluation serves by providing immediate feedback on the short-term outcomes of training. Types of feedback that may be useful includes demographic descriptions of the participants (e.g., for sex-ethnic affirmative action criteria), participant reactions to the training environment and to the training staff, and assessments of changes in performance of certain skills learned in the training situation that may be later observed to be useful in the drug abuse intervention system.

The second evaluation issue is the summative role that training evaluation serves in providing information for end-of-cycle decisions to continue, modify or terminate support for training on the basis of feedback on short-term and/or long-term effects of training (Scriven, 1963; Worthen and Sanders, 1975). Summative training evaluation also examines the question of the extent to which material learned in training is of practical use to participants when they have returned to the realities of their work place.

It is clear that an information gathering system is needed to record and monitor the flow of drug abuse personnel who received training, and to provide feedback on the effects of training. What is specifically needed is an evaluation system that can perform *two* roles: a formative evaluation function which records and monitors immediate training outcomes so that refinements and improvements can be made in the training curricula or resources, and a summative evaluation function which compares performance against training standards, job-performance criteria, cost-benefit factors or other drug abuse intervention strategies.

A prototype Training Evaluation System (TES) is being developed by the evaluation staff of the Western Regional Support Center (WRSC) to perform these two roles.

THE WRSC TRAINING EVALUATION SYSTEM

The Western Regional Support Center (WRSC) was funded by NIDA as one of five Regional Support Centers (RSCs) to provide training and technical assistance to the eleven states that comprise the Western Region. WRSC is specifically charged with providing over 55 days of training and technical assistance to each state. The major purposes for NIDA assistance to the RSCs and the states (through the Single State Agencies and the STSP) are to improve professional skills, and to provide resources so that the states can develop and maintain their own training delivery systems.

A major vehicle for providing direct training is through the delivery of NIDA-funded courses, developed by NDAC using RSC training staff resources and NDAC course packages. A number of courses have been developed by NDAC, and each has been developed as part of an extensive, planned development process; most of the courses have been reviewed, field tested and revised prior to publication and dissemination by NDAC. In cases where field test data are available, all courses are reported to show that participants exhibited considerable learning gains and had responded favorably to the course and its delivery under field test conditions.

One of the contract obligations of the WRSC evaluation staff is to evaluate the immediate outcomes of training as it is delivered to drug abuse treatment workers in the Western Region. The remainder of this paper describes the prototype TES under development.

ASSUMPTIONS

In this age of computers large and small, there is little doubt that access to a reliable computer can simplify the job of recording and processing information needed to monitor the development of a training system. Without the aid of a computer we in fact cannot manage the flow of necessary information required to operate a training system. We assume, then, that a local, state or federal training evaluation system can be developed using computer technology, and that there will be staff and other resources available for maintaining a timely data flow of information input, processing and output in the form of evaluative feedback. We can also assume that there are fiscal resources available for initial development costs and for relatively low maintenance and operational costs.

SOFTWARE PROCEDURES FOR TRACKING IMMEDIATE TRAINING OUTCOMES

Procedures for operating the WRSC Training Evaluation System (WRSC-TES) include defining needed information; specifying the flow of information; development instruments required for collecting needed information; scoring and coding collected data; analyzing data in appropriate ways; and most important, providing feedback about training outcomes to the original participants in training.

The core of the WRSC-TES is a computer program that was developed to meet the needs of the WRSC training staff, NIDA (as part of our reporting and evaluation requirements) and the STSPs in the Western Region. TESPROG3 is a FORTRAN computer program that analyzes data collected from trainees and from training staff about their training experience. The TESPROG3 program produces a two-page printout of results of data analyses on participant demographic data, pretest to posttest measures of learning gains, and participants' ratings of training staff performance (see Figures 1 and 2 for a xerox-reduced copy of the TES printout). Group averages for all variables are not only printed out in the report, but they are also saved in a separate data set for second generation, summative evaluation analyses. Thus, TESPROG3 performs both a tracking function for each training event for which data were collected, and a summative evaluation function for end-of-cycle analyses, including the generation of a Quarterly Report that aggregates and summarizes the outcomes of training during the preceding fiscal quarter.

SOFTWARE PROCEDURES FOR EVALUATING LONGTERM EFFECTS

Tracking the outcomes of each training event permits the development of a data base which contains potentially useful information for outcome evaluation purposes.

When training outcome data are aggregated for each state over a period of time, for example, it may be possible to make comparisons between states, if one can first demonstrate some validity in the methods of measurement (i.e., collecting comparable data).

WESTERN REGIONAL SUPPORT CENTER TRAINING EVALUATION SYSTEM DATE 03-15-78 PAGE 1

TRAINING EVENT NO. M43.FACTS ABOUT DRUG ABUSE HELENA,MT FEB 6-9,1978

****

THERE WERE 11 PARTICIPANTS IN THIS TRAINING EVENT 6 MALES 5 FEMALES AVERAGE AGE: 33.9 YEARS

ETHNIC BREAKDOWN	N	%
AMERICAN INDIAN	6	54.5
ASIAN AMERICAN	0	0.0
BLACK	0	0.0
WHITE	5	45.5
SPANISH BACKGROUND	0	0.0
MEXICAN	0	0.0
PUERTO RICAN	0	0.0
CUBAN	0	0.0
OTHER SPANISH	0	0.0
OTHER	0	0.0
MISSING DATA	0	0.0

EDUCATIONAL BACKGROUND	N	%
UNDER 12 YEARS	2	18.2
HIGH SCHOOL GRAD	1	9.1
SOME CCLLEGE	6	54.5
BA-BS	1	9.1
MA-MS	1	9.1
DOCTORATE (PHD OR MD)	0	0.0
TECH SCHOOL OR NURSING	0	0.0
MISSING DATA	0	0.0

GEOGRAPHIC AREAS REPRESENTED BY TRAINEES	N	%
RURAL	6	54.5
SUBURBAN	0	0.0
URBAN	1	9.1
ENTIRE STATE	1	9.1
MISSING DATA	3	27.3

TREATMENT PROGRAM FUNDING SOURCES	N	%
NIDA FUNDED PROGRAMS	1	9.1
NON NIDA FUNDED		
FEDERAL	2	18.2
STATE	5	45.5
COUNTY	3	27.3
OTHER	4	36.4

OCCUPATIONAL TITLES	N	%
COMMUNITY ORGANIZATION SPECIALIST	0	0.0
SOCIAL WORKER	0	0.0
RELIGIOUS WORKER-CLERGY	0	0.0
THERAPEUTIC COUNSELOR	0	0.0
REHABILITATION COUNSELOR	3	27.3
DRUG COUNSELOR	2	18.2
SHORT-TERM CLIENT CRISIS COUNSELOR	1	9.1
OTHER COUNSELOR	3	27.3
CLINICAL DIRECTOR	1	9.1
ADMINISTRATIVE DIRECTOR	0	0.0
PROGRAM DIRECTOR	0	0.0
MANAGEMENT ASSISTANT	1	9.1
PHYSICIAN	0	0.0
NURSE	0	0.0
OTHER MEDICAL OR PARAMEDICAL JOB	0	0.0
POLICE OFFICER	0	0.0
ATTORNEY	0	0.0
INVESTIGATOR-DETECTIVE	0	0.0
PROBATION-PAROLE OFFICER	0	0.0
OTHER LEGAL OR PROTECTIVE SERVICE	0	0.0
ELEMENTARY-KINDERGARTEN TEACHER	0	0.0
HIGHSCHOOL-VOCATIONAL SCHOOL TEACHER	0	0.0
COLLEGE-UNIVERSITY TEACHER	0	0.0
TRAINER-FACILITATOR	0	0.0
SOCIAL OR BEHAVIORAL SCIENTIST	0	0.0
STUDENT	0	0.0
SECRETARY	0	0.0
RESEARCH AND EVALUATION SPECIALIST	0	0.0
OTHER-NOT LISTED	0	0.0
UNEMPLOYED	0	0.0
MISSING DATA	0	0.0

AVERAGE LENGTH OF TIME IN THE DRUG ABUSE FIELD IS 33.6 MONTHS
6 PARTICIPANTS WORK DIRECTLY WITH CLIENTS
AVERAGE MONTHLY CLIENT WORKLOAD IS 28.0 CLIENTS PER MONTH
AVERAGE PERCENTAGE OF MINORITY GROUP CLIENTS IS 39.5 PERCENT
AVERAGE PERCENTAGE OF WOMEN CLIENTS IS ***** PERCENT

****

PARTICIPANT GAIN SCORES ON PRE- AND POSTTEST ACHIEVEMENT MEASURES OF COURSE CONTENT

PRETEST MEAN SCORES	POSTTEST MEAN SCORES	AVERAGE GAIN SCORE	T-TEST AND SIGNIFICANCE
9.9	16.4	6.5	1.0 P= 0.340

****

Fig. 1. Printout of the Training Evaluation System

TRAINING EVENT NO. M43.FACTS ABOUT DRUG ABUSE HELENA.MT FEB 6-9.1978 PAGE 2

PARTICIPANTS' FINAL EVALUATION QUESTIONNAIRE-RESULTS

TRAINEE EVALUATIONS OF TRAINING STAFF EFFECTIVENESS - AVERAGE RATINGS- ON A SCALE OF SIX POINTS

	ITEM	TRAINER WILLIAM CRAWFORD	TRAINER	TRAINER
1.	MADE APPROPRIATE USE OF HUMOR	5.2	***	***
2.	MADE AN EFFORT TO HELP ME FEEL SAFE	5.5	***	***
3.	MANAGED THE EQUIPMENT WELL (I.E., FLIPCHARTS, ETC)	5.6	***	***
4.	DEFINED TERMS AND CONCEPTS CLEARLY	5.2	***	***
5.	RELATED EACH MODULE TO PREVIOUS MODULES	5.1	***	***
6.	ADEQUATELY EXPLAINED THE OBJECTIVES FOR EACH MODULE	5.1	***	***
7.	GAVE CLEAR INSTRUCTIONS FOR EACH EXERCISE	5.4	***	***
8.	PROVIDED STRUCTURE FOR PROCESSING EACH EXERCISE	5.3	***	***
9.	COORDINATED GROUP DISCUSSION AND FEEDBACK	5.5	***	***
10.	ENCOURAGED SELF-DISCOVERY	5.3	***	***
11.	MODELED THE SKILLS BEING TRAINED	5.4	***	***
12.	SHOWED PATIENCE	5.9	***	***
13.	GAVE NON-JUDGMENTAL FEEDBACK TO INDIVIDUALS IN GROUP	5.9	***	***
14.	ALLOWED ADEQUATE TIME FOR PRACTICING	5.0	***	***
15.	SET A TIME SCHEDULE AND STUCK TO IT	5.3	***	***
	OVERALL AVERAGE RATING	5.4	***	***

OVERALL AVERAGE RATINGS FOR WRSC STAFF IN THE SECOND QUARTER WAS FIVE AND FOUR-TENTHS POINTS ON A SCALE OF SIX POINTS

Fig. 2. TES Printout Continued

One of the critical issues in summative training evaluation (and other outcome or impact evaluation) studies is the degree of *control* the evaluation has over the environment in which the evaluation takes place. Control here means not only having access to reliable and valid data sources, or at least being able to develop a data flow, but also being able to account for the sources of variation that may threaten the validity of measurement.

The classic true experimental design exemplifies a research paradigm in which there is a high degree of control over the environment under study. The hallmark of the design is randomization of participants, e.g., the random selection of persons for training, and the use of control groups with participants randomly selected and assigned to either a training group or to a no-training, delayed training, or alternative treatment (like reading the training materials only) group. The true experimental design has historically worked best under laboratory conditions, but is extremely difficult to effect in most social action programs in the "real world."

Alternatives to the experimental design offer the possibility of effecting *statistical* control through the use of covariance designs and multiple regression analyses. When it is not possible to use a control group, or when randomization is not possible, the use of natural groups may be useful. Given a number of training programs or courses delivered in a number of the states in the Western Region to various groups of participants one may obtain a representative sample of training programs, states and groups of participants having certain characteristics. In such a representative design, statistical control can be used over the natural variations occurring across kinds of programs and kinds of inputs in kinds of situations (Brunswik, 1956; Snow, 1974).

In using either the experimental, quasi-experimental or representative designs, one is interested in obtaining the most useful information that relates to the outcome, or dependent, variable. A procedure for deciding more useful from less useful information is to analyze sets of variables to determine those which have first order correlations with the dependent variable, and to test for hypothesized specific interaction effects. An analysis of variance can be used if certain assumptions can be met. One would then enter the selected variables into a multiple regression analysis to explore those variables that make the greatest contribution to the criterion variable.

The procedures being explored and developed by the WRSC evaluation staff are presented in Figure 3. They include determining the sources of variation among the important factors of the study (i.e., determine the "variables" from the "constants"), and the degree of control one can maintain; using natural groups of drug abuse and alcoholism treatment personnel; obtaining a probability sample of the actual training population to determine the appropriate sample size needed for analyses of each course (or training program) in each state under study; negotiating with participants and their superiors (and yours) for the collection of data and then collecting data at several points in time (a minimum of two times); computing the variables which have first order correlations or specific interaction effects (from an analysis of variance (ANOVA) design); computing the relative contribution of each selected variable to the criterion, dependent variable; and preparing and reporting the evaluation findings.

SUMMARY

As we have endeavored to successfully implement these procedures, we have discovered several pitfalls and limitations to be aware of as we have groped and muddled through our explorations. So far they include:

1. The reliability of most of the NDAC-developed tests for the courses have not been adequately demonstrated; our preliminary estimates of the reliability of measurement of the pretest and posttest instruments have been very low, and suggest that the tests need to be examined.

2. The validity of the skills that are being transmitted by the courses have not been shown to be related to effective drug abuse treatment performance; this is endemic to a field that cannot define what an effective drug abuse counselor "does" (just as educational research has yet, in over fifty years of research, been able to define what an effective teacher is or does).

3. Computers can compute anything! But remember if you put garbage in you'll get garbage out, and it could very well be expensive garbage at that.

4. The greatest degree of success in conducting an evaluation occurs in an environment that allows and encourages flexibility and adaptability in a relatively open climate; fear, distrust and reluctance to collaborate in an evaluation effort is in itself a sign of a system that does not effectively utilize its feedback mechanisms for improvement.

The future of training evaluation is bright, and is an area filled with opportunities for further work. The WRSC Training Evaluation System for tracking and monitoring training outcomes is continually being updated and improved as we learn from our experience. As a prototype it has potential usefulness to NIDA as a model upon which to develop its proposed National Participant Registry that will

1. Determine the degree of control over sources of variance.

2. Estimate the sample size needed for a representative sampling over courses, states (or regions) and persons in different categories of interest.

3. Negotiate the most valid sampling design available.

4. Collect data at several points in time, depending on available resources for collecting and processing data.

5. Test for violations of statistical assumptions of additivity and linearity.

6. Analyze data using a strategy that determines the nature of variation among relevant training variables, rather than focusing on differences between training and non-training groups.

7. Interpret statistical findings only within the constraints of the study design.

Figure 3. Procedures for Developing a Summative Training Evaluation Design

register and certificate persons who have successfully completed NIDA-sponsored training programs, or who wish American Council on Education (ACE) accreditation for having successfully completed a course and for credentialling purposes). It is also being utilized by WRSC to provide an evaluation service for the eleven states in the region, for monitoring and evaluating the outcomes of their training delivery systems.

REFERENCES

Brunswik, E. Perception and the Representative Design of Psychological Experiments. Berkeley, California: University of California Press, 1976.

NIDA-ADAMHA. Manpower and Training Strategy 1977. National Institute on Drug Abuse, 1977.

Scriven, M. The Countenance of Educational Evaluation, American Educational Research Journal, Monograph Series No. 1, 1963.

Snow, R.E. Representative and Quasi-Representative Designs on Teaching, Review of Educational Research, 1974, 44(3), 265-291.

Worthen, B.R. and Sanders, J. (eds.) Educational Evaluation: Theory and Practice. Worthenton, Ohio: Charles A. Jones, 1973.

SOCIAL CORRELATES OF IN-TREATMENT "SUCCESS": AN ANALYSIS OF LINKED CLIENT ADMISSION/DISCHARGE RECORDS*

Thomas Vale Rush, Ph.D.

Social Security Administration

Baltimore, Maryland

This paper focuses on social correlates associated with success at the time of discharge from treatment. The population studied was composed of alcohol and drug abuse clients who were receiving substance abuse treatment in publicly supported facilities within a large Eastern state. Using three different criteria of success and controlling for important variables, it assesses the relative importance of several demographic characteristics in predicting treatment success.

INTRODUCTION

It is not an easy task to measure the effect of treatment on clients, especially when the behavior to be treated is the result of illicit actions and/or is considered to be immoral. Regardless, any attempt to measure the effect of treatment should contain both a control and experimental group, along with information about pre-treatment and post-treatment conditions. This may be the minimum state of affairs for testing the effectiveness of treatment under laboratory conditions, but to this point in time it has been found nowhere in the literature (Sirotnik and Bentler, 1975).

---

* This publication was prepared independent of the professional status of the author as a former employee of the Governor's Council on Drug and Alcohol Abuse, Commonwealth of Pennsylvania, and the conclusions and interpretations contained herein are solely those of the author.

Probably the crucial reason for the lack of even the minimal standards of an experimental design is the elusive control group. A true control group would match the subjects in treatment to other persons with substance use problems but who were not in treatment for the duration of the experiment. As is obvious, the selection of such a control group is nearly impossible. This has meant that researchers must be satisfied with a quasi-control group composed of persons who have signed up for treatment but have withdrawn quickly afterward (Sells and Simpson, 1974; Macro Systems Inc., 1975; Burt and Pines, 1976). However, one serious shortcoming with this procedure is that many of the control group also enter treatment, consequently contaminating themselves as a control on treatment.

The direction of this paper is not to measure the effectiveness of treatment, rather it is to consider the social correlates of success at discharge. Three variables are individually considered as criteria of success, and they are productivity, felony arrests, and completion of treatment. These criteria have general acceptance in the field (Mandell, Goldschmidt, and Jillson, 1975; Norris, 1975; Collins and Kelly, 1975; and Keil, Rush and Dickman, 1975a,b). Most of the predictor variables adopted had been found to be related to success in previous studies (Eichberg and Bentler, 1975; Keil, Rush, and Dickman, 1975; Cutter, et al., 1975; and Sirotnik and Roffe, 1977).

There have been some studies which have addressed similar questions. The most complete recent study on treatment effectiveness has been the major analysis of persons in the DARP system as reported by Sells and Simpson (1974). Of a much more limited scope has been the work of Keil, Rush, and Dickman (1975). Those studies followed the work of Keil, Dickman, and Tower (1975), and used a data system similar to the one being used in this paper. The former two papers controlled for primary substance of abuse, looking at opiate abusers and non-opiate abusers who were in treatment. It was found that the variables associated with in-treatment changes varied according to the success criterion analyzed and to a lesser degree, the class of the primary substance of abuse. In general, it appears that those who were productive or active in the eyes of society were more likely to be successful at the end of treatment, than were those who were not productive.

SAMPLE

The analysis is based on treatment experiences rather than individuals, which means that if a particular person were in treatment two times, he/she accounted for two experiences. The time period covers all experiences in treatment from July 1, 1975 through December 31, 1976. All clients who had entered treatment in that period and were discharged from either inpatient therapeutic community or outpatient counseling and/or psychotherapy modalities.

Inpatient non-hospital/therapeutic communities and outpatient counseling and/or psychotherapy were chosen because they represented contrasting styles of treatment and they were both long term treatment facilities. An inpatient therapeutic community is an environment which provides services twenty-four hours a day and is used as a residence (it has a work-in/live-in milieu), which seeks to modify the client's attitude and behavior toward substance abuse, life style, social responsibility, and interpersonal relationships. This contrasts with outpatient counseling and/or psychotherapy; treatment services are of short duration (less than five hours per day) and are limited to the provision basically of advise or guidance as a means of dealing with the client's maladjustments and make-up.

Analysis was performed on information derived from two points in time, at admission and at discharge. The data was self report, with counselors completing the forms at those two times. A certain percentage of the records could not be linked, but this number was small (less than 20 percent) and there is no reason to believe they substantially differed from those records which were linked.

A comparison between the different groups (by age and by environment/approach) revealed some interesting points as seen on Table I. The juveniles differed from the adult groups on most demographic characteristics, as well as on substance use dimensions. Most (64 percent) of the juveniles were in treatment for non-opiate drug problems (of which the most used substance was marijuana), a pattern similar to that of young adults but clearly much more accentuated in the case of juveniles. Much of their drug use took place in conjuction with other drugs; indeed most of the juveniles receiving treatment in these two environment/approaches were multiple drug abusers.

Considerable differences were found among the groups for persons in counseling/psychotherapy as contrasted to therapeutic community. Those in counseling/psychotherapy were relatively split on the sex characteristic and seemed to have less involvement with the drug subculture. Most were currently in an education or training program when they entered treatment and they had never been convicted of a crime. They had started drug use at a later age. Few (17.1%) were in treatment because of legal pressure, and even fewer (14.9%) had been in treatment before. Less than one-half of the persons were multiple drug users. In contrast, the juveniles in therapeutic communities appeared to have been more involved with the drug and deviant subculture. A higher percentage had been convicted of crimes, and nearly one-half were in treatment because of legal pressure. More than three of every four juvenile clients were a multiple drug user. A higher percentage of the therapeutic community clients had experienced drug treatment before.

Table I. Comparison of the two populations on selected items (first set of figures is for clients in outpatient counseling and/or psychotherapy, second set is for clients in inpatient therapeutic communities).

Variables	Juveniles	Young Adults	Adults
Percentage of males	55.4% 69.8%	70.0% 79.3%	74.6% 83.6%
Percentage of whites	86.8% 87.7%	80.6% 79.7%	67.7% 66.3%
Percentage of students in an education program	70.6% 37.8%	12.4% 17.2%	3.0% 8.2%
Percentage of students in a training program	5.3% 3.3%	4.1% 2.3%	1.6% 3.6%
Percentage of clients, no employment in last two years	60.8% 40.8%	24.6% 23.8%	31.5% 25.8%
Percentage in treatment, legal pressure	17.1% 44.3%	29.3% 44.3%	19.6% 23.1%
Percentage with no prior convictions	85.5% 61.4%	72.8% 55.2%	75.4% 64.9%
Median age of first use	14.3 13.1	14.9 14.3	16.7 16.3
Percentage with no previous treatment experience	85.1% 63.6%	72.8% 55.2%	53.7% 38.4%
Median time in treatment	123 days 36 days	100 days 34 days	91 days 29 days
Percentage of opiate abusers	2.6% 8.0%	14.3% 17.0%	14.4% 21.6%
Percentage of non-opiate drug abusers	64.2% 74.6%	50.3% 45.2%	11.0% 8.5%
Percentage of multiple drug abusers	47.2% 76.3%	50.9% 59.2%	17.1% 26.7%
N =	2417 503	1360 458	21789 6969

METHOD

Multivariate analysis was used in this paper, with the specific technique being the stepwise regression procedure as found in the Statistical Package for the Social Sciences (Nie, et al., 1975). Each of the predictors was entered into the equation individually to explain the variation around the criterion. The remaining variables are entered into the equation on the basis of their ability to explain the remaining residual variation unaccounted for by the previous variables entered into the equation, not necessarily on the basis of their own correlation with the criterion. The standard regression method of decomposition was used. The alpha acceptance level was established at the 0.05 level. In some instances multicollinearity was present and it was handled by dropping one of the two variables from the analysis.

The procedure was used largely in a descriptive sense. It was used to describe the relationship between the various predictor variables and each selected criterion. There is no attempt to establish causality, nor should any of the findings be considered to suggest causal connections.

In this paper there were three sets of variables used: independent (predictors), dependent (criteria), and control variables. The two control variables were age and treatment modality. The independent variables were patient demographics or behaviors prior to treatment admission and the time spent in treatment, while the dependent variables were behavioral conditions at the time of discharge.

The operational definitions of the adopted variables were as follows. Sex, race, and substance abused were dummy variables, with sex indicating either the presence or absence of male, race indicating the presence or absence of whites, and substance of abuse pulling out opiate and non-opiate abusers (the reference category here was mostly alcohol abusers). A patient was considered employed if he/she was working at least part-time on a job for which taxable income was derived, and then the number of months worked in the past two years was established. If the person at admission to treatment was at least part-time enrolled in either an education or a training program, then he/she was considered to be a current student.

Items pertaining to the individual's involvement with the criminal justice system included total arrests for felonies, total convictions, and commitment to treatment under legal pressure. Over a period of two years preceding admission to treatment, the client was to indicate the number of times he/she had been arrested for felonies, and the total number of conviction (for any type of offense) for the past two years. If entry to treatment was the result of a court order or other direct official order, then the person entered under legal pressure.

Several variables related to type of abuse and treatment experience. A multiple drug user was one who indicated at admission that more than one drug was a problem. Information was also gathered on the age of first use and the number of years of continued use. If a patient has previously been in publicly supported treatment, then it was so counted. The time spent in treatment was figured on a daily basis from admission to discharge. The client was either discharged with his/her goals successfully completed or he/she was considered to have been discharged for some other reason (the latter being the reference category).

RESULTS

Outpatient Counseling and/or Psychotherapy

Productivity Criterion. Three general criteria (employment at discharge, student in an education program, and student in a training program) were combined in such a manner as to have an overall productivity measure and several factors were found to be useful predictors of productivity at time of discharge. For the juvenile subsample, the strongest predictor of productivity was whether the individual was a student in an education program when he/she was admitted to treatment. This was followed in magnitude by employment status; those who were employed at admission were more likely to be productive at discharge. The less the involvement with drugs and delinquency, the more likely that a person would be productive at discharge. This is reflected in the following variables: the more felony arrests, the longer habit (in years of continuous use), and the earlier the initial use of drugs, the less likely that the juvenile client would be productive at discharge. Youths abusing non-opiate drugs were more likely to be productive at discharge, as were non-multiple drug abusing persons with no previous treatment experience, and persons who had formerly been students in training programs. Time in treatment accounted for very little of the variability, and it had a negative effect; the less time the person was in treatment, the better the chance that he/she was productive at discharge. Those factors accounted for a fairly substantial amount (R^2=.28) of the variation in the productivity variable.

The young adult groups differed from the juvenile group in the ordering of the variables and the magnitude of their explanatory ability. Because of their age involved, the best predictor of productivity at discharge among juveniles was enrollment in an education program, following by the employment status prior to treatment. The reverse was true for the young adults and adults. Moreover, the next variables with explanatory ability were time in treatment and race. Only one of these, time in treatment, was significant for the juveniles and for the juveniles it was the opposite direction. Young adults and adults in treatment for abusing opiates were significantly less likely to be productive and this was also true for persons with prior treatment experiences.

The most important difference on the productivity criterion was not in the predictor variables, rather it was in the magnitude of variability explained. The variables picked in the adult sample were able to explain 58% of the variation, this figure dropped to 39% for young adults and 28% for juveniles. Thus the variables selected for the adult sample were much better in explaining the variation around the criterion than were those for either the young adults or the juveniles.

Arrest Criterion. Among persons under 18 years of age, the best predictor of arrests for felonies while in treatment was the number of felony arrests prior to treatment (see Table II). After the arrest variable had been entered, two more law-related variables were placed in the equation: total convictions and treatment entry under legal pressure. The only other variable found to be significant was race, with whites being less likely to be arrested while in treatment.

Once the main predictor of the felony arrest criterion had been entered (total number of felony arrests), the three groups differed substantially. In all three cases the total number of convictions was a significant factor, yet the direction of its relationship differed. For both juveniles and adults, the direction was positive as expected. But for young adults it was negatively related. For young adults this inconsistency conceivably is a statistical artifact of the technique used, for the simple correlation between the total number of convictions for young adults and felony arrests at discharge was positive (r=+.12).

Treatment Completion Criterion. The other criterion to be considered was completion of treatment according to the goals of the program. The time spent in treatment was the best predictor of completion of treatment for juveniles. What that exactly means is uncertain: does it mean that the longer a person is in treatment the more his/her mind is changed, or does it mean that the longer a person stays around, the more likely that the program will say he/she has completed the goals just to get rid of the person. Other significant factors positively related to the completion of treatment were: being enrolled in an eduction program, being a non-opiate abuser, being white, and being older when the drug of abuse was first tried. Those juveniles who were opiate abusers and multiple drug abusers were less likely to complete treatment. The more felony arrests a youth had, the less likely he/she was to complete treatment.

As was true with the juvenile group, the time spent in treatment was the first variable entered in the equation for young adults--the longer the person was in treatment the more probable that he/she

Table II. Predicting treatment outcomes, outpatient counseling and/or psychotherapy.

		DEPENDENT CRITERIA																	
		PRODUCTIVITY						FELONY ARRESTS						TREATMENT COMPLETION					
		Juveniles		Young Adult		Adult		Juveniles		Young Adult		Adult		Juveniles		Young Adult		Adult	
	Independent Predictors	Beta Weight	R^2	Beta Weight	R^2	Beta Weight	R^2	Beta Weight	R^2	Beta Weight	R^2	Beta Weight	R^2	Beta Weight	R^2	Beta Weight	R^2	Beta Weight	R^2
Demographics	Males					.03	.001					.01	.000					-.03	.001
	Whites			.11	.000	.05	.003	-.04	.002			.01	.000	.04	.002	.07	.004	.03	.001
	School status	.43	.233	.23	.061	.13	.021			-.04	.001			.13	.025	.12	.017	.04	.002
	Training status	.06	.033			.04	.002												
	Employment	.14	.017	.51	.293	.71	.544					-.03	.000			.12	.019	.15	.029
Treatment history	Length time in treatment	-.05	.002	.10	.012	.07	.006							.24	.065	.29	.097	.24	.068
	Prior treatment experience	-.05	.002	-.04	.001	-.03	.001					.01	.000					-.07	.006
Drug use history	Opiate abuser	.04		-.06	.005	.02	.000					.04	.002	-.06	.007				
	Non-opiate abuser	-.05	.002	.04	.001									.05	.002			.03	.000
	Multiple drug abuser	-.11	.003			.02	.000					.03	.001	-.08	.006	-.05	.002	-.02	.000
	Age first use primary drug	-.11	.009									.03	.000	.04	.001			.06	.003
	Years continued use	-.12	.004			.02	.000					.02	.000					.06	.002
Criminal Justice Use	Legal referral							.05	.003			.03	.001						
	Prior convictions							.05	.002	-.09	.005	.05	.002						
	Felony arrests	.07	.008			.03	.001	.30	.132	.34	.092	.20	.068	-.04	.002	-.06	.005	-.05	.005
	Total N	1178		1167		16781		1810		1186		17592		1810		1186		17582	
	Total R	.53		.63		.76		.38		.32		.27		.33		.38		.34	
	Total R^2	.28		.39		.58		.14		.10		.07		.11		.14		.12	

would complete the treatment successfully. However, the second variable entered for both groups was prior employment status, a variable which was not even significant for juvenile clients. With adults, those who had previously been in treatment were significantly less likely to complete treatment, a factor not important for the two other groups. Being a student in an education program, not being a multiple drug abuser, not having previous felony arrests, and being white were significant for all groups as far as successfully completing treatment.[1]

Inpatient Therapeutic Community

Productivity Criterion. The most important variables correlated with being productive at discharge for juveniles were being enrolled in either education or training and the time spent in treatment (this was the best predictor). These were the only significant factors, and they clearly were associated with the age of the clients. The juveniles were much different from the young adults in this regard; for the latter, time in treatment was closely followed by employment at admission. Other significant and important variables among young adults were total arrests for felonies at admission, enrollment in an education program, not being white, and having a long continuous use pattern.

In comparison to the strength of the regression equation for adults (R^2=.42) the equations for preceding two groups lacked explanatory strength. Only two of the ten significant variables need be mentioned. If the adult client was employed upon admission to treatment, this by far would be the best predictor of productivity at discharge. A distant second was the time spent in treatment. All other significant variables were of only minor importance in explaining productivity at discharge.

Arrest Criterion. The equation for arrests for felonies was not particularly effective for juveniles, explaining only four percent of the variation. Four variables were significant, and they were lumped relatively closely in their explanatory ability. Total arrests for felonies at the time of admission was the best predictor,

---

1 Regressions were run for all three age groups using the criterion of completing treatment or leaving under favorable circumstances (transferred or referred to another facility) as success. For all three groups, this cut the explained variation in half. For juveniles it dropped the number of significant variables and did result in a reordering of their rank of importance after the first two variables. A similar pattern was noted for young adults and adults.

with a positive relationship between the two variables being present (see Table III). Persons who had been in treatment previously were significantly more likely to be arrested for felonies while in treatment, as were persons who had started using the abused drug later in life.

The regression equation for young adults explained ten percent of the variation, but instead of doing it with one main predictor, it had several which were nearly as good as the best predictor (which was the number of felony arrests before entering treatment). As was found with juveniles, the later the person started using the drug of abuse, the more likely he/she was of being arrested for felonious offenses after admission, and this was both statistically significant and strong in explanatory value. There was an inverse relationship between the number of convictions prior to admission and the number of arrests following admission. With adults, as with juveniles and young adults, the best predictor was the number of felony arrests prior to admission. After that, all other significant variables were negatively correlated with the criterion.

Treatment Completion Criterion. If a juvenile was a non-opiate drug abuser in treatment, that was the best predictor of whether he/she would complete treatment or not. They were unlikely to complete treatment, as were opiate abusers, indicating that juveniles with alochol abuse problems in this environment/approach probably did very well in completing treatment since they constituted most of the reference category. The time a juvenile spent in treatment was also strongly related to whether the individual successfully completed treatment or not as the longer he/she was in treatment the higher the success rate. Juveniles who were students in an education program were significantly more likely to complete treatment.

Both young adults and adults had as their best predictor opiate abuse--those in treatment for opiate abuse were significantly less likely to complete treatment according to the regimen established by the program, and they were closely followed by persons in treatment for non-opiate abuse problems. Among young adults, only three other variables were statistically significant with regard to this criterion: not having previous treatment experience, being white, and spending a longer time in treatment. For adults several (ten) additional varables were significant, the more important ones were: being employed when admitted to treatment, not having previous felony arrests, being white, and having a longer time of continued use.[2]

---

[2] As was the case for clients in outpatient counseling and/or psychotherapy, a separate analysis was executed using completed treatment or left under favorable circumstances as the success criterion. The regression equation explained half again more variance (R^2=.18

Table III. Predicting Treatment Outcomes, Inpatient Therapeutic Communities

	DEPENDENT CRITERIA																	
	PRODUCTIVITY						FELONY ARRESTS						TREATMENT COMPLETION					
	Juveniles		Young Adult		Adult		Juveniles		Young Adult		Adult		Juveniles		Young Adult		Adult	
INDEPENDENT PREDICTORS	Beta Weight	R^2	Beta Weight	R^2	Beta Weight	R^2	Beta Weight	R^2	Beta Weight	R^2	Beta Weight	R^2	Beta Weight	R^2	Beta Weight	R^2	Beta Weight	R^2
Males					.03	.001					.02	.001					-.05	.003
Whites			-.10	.007	-.06	.003			.09	.011			.09	.008	.10	.009	.08	.007
School Status	.22	.054	.12	.014	.05	.003							-.09	.009			-.03	.001
Training Status	.10	.012			.03	.001											-.02	.001
Employment			.18	.030	.64	.374											.13	.020
Length Time in Treatment	.23	.049	.19	.055	.17	.032							.22	.047	.12	.012	.03	.001
Prior Treatment Experience					.02	.000	.10	.009					-.06	.004	-.13	.016		
Opiate Abuser					-.02	.000			-.20	.023	-.05	.003	-.16	.019	-.23	.040	-.20	.081
Non-Opiate Abuser							-.09	.009	-.21	.011	-.03	.002	-.26	.032	-.26	.025	-.16	.047
Multiple Drug Abuser					.03	.001					-.03	.001					-.02	.000
Age 1st Use Primary Drug							.07	.004	.21	.012							.05	.002
Years Continued Use			.08	.007	-.05	.003			.11	.007							.06	.003
Legal Referral											-.04	.001						
Prior Convictions									-.14	.009								
Felony Arrests			.12	.013			.12	.019	.25	.019	.20	.022					-.09	.011
Total N	468		415		5470		487		442		5632		487		442		5632	
Total R	.35		.36		.65		.21		.31		.17		.35		.32		.42	
Total R^2	.12		.13		.42		.04		.10		.03		.12		.10		.18	

SUMMARY

This paper has been concerned with various correlates of success at the time of discharge, with success defined in three different ways (being productive, being free of felony arrests, and completing the treatment program). It was found that the two control variables did indeed have an effect upon the results. The results for clients who were in inpatient therapeutic communities were substantially different from those results for clients in outpatient counseling and/or psychotherapy treatment. Likewise, the same is true for the three age groups studied, with juveniles differing from young adults and adults, and vice versa.

For persons in outpatient counseling and/or psychotherapy treatment, the pretreatment productive status was of primary importance in predicting outcome status. For all three age groups, two strong-

---

vs. R^2=.12), and there was a reordering of the predictors for juveniles. When just completed treatment was considered, the time spent in treatment had the second highest standardized regression coefficient, but as the criterion was expanded, the time spent in treatment dropped to a relatively minor role (it still was statistically significant). The most important predictors remained being in treatment for opiate and for non-opiate abuse (they were negatively related), but there were a flock of new significant predictors. The strongest of these new predictors were pretreatment employment status (standard regression coefficient = +.13), arrests for felony offenses (standard regression coefficient = -.09), number of years continued use (standard regression coefficient = +.06), sex status (standard regression coefficient = +.05). It would appear that those who were referred or transferred had their use patterns for a longer period of time, but they were able to cope with drug use--as is evident with their employment status and the lack of felony arrests.

Surprisingly, the variable time in treatment had the opposite effect for young adults when compared to juveniles. Its standard regression coefficient increased dramatically with the expanded criterion (up from +.12 to +.21). A new significant variable was added, that being the number of years of continued use (standard regression coefficient = .07). There was a slight increase in the explanatory ability of the regression equation (R^2 =.12 vs. .10).

The pattern for adults was different from the preceding, in that the regression equation for the expanded criterion was less than for the completed treatment version (R^2=.14 vs. 18.). The variable which changed the most was time in treatment, increasing as the criterion was expanded (standard regression coefficient = +.03 to +.08).

est variables were pretreatment employment and education program status. In contrast, the time spent in treatment was a strong predictor variable of productivity for clients in inpatient therapeutic communities. As for the age of the clients, the most significant finding was the importance and strength of pretreatment employment to productivity for adults. Clearly those who were employed at admission were most likely to be productive at discharge, regardless of the treatment environment/approach. Moreover, it was both statistically and substantively significant.

If arrests for felonies were the criterion, a stark contrast to the productivity criterion emerged. For all age groups and both environment/approaches, the same variable was the best predictor, that being the number of previous arrests for felonies. The data reveal that those who had been arrested for felonies prior to treatment were significantly more likely to be arrested while in treatment.

The best predictor of completing treatment for clients in outpatient counseling and/or psychotherapy treatment was the time spent in treatment, regardless of age. This was followed by a productivity related variable (employment or student status). Thus, in this environment/approach the longer an individual remained in treatment and the more productive he/she was before entering treatment, the more likely that he/she would complete treatment. Contrasting with this pattern was the one found among clients in inpatient therapeutic communities. There it was not time in treatment, (although this was very important for juvenile clients) but the substances abused which were the best predictors. Persons who abused either opiates or nonopiates were less likely to complete treatment which is reflective of the fact that clients in this environment/approach have a high treatment split rate.

REFERENCES

Burt, Marvin R. and Sharon Pines 1976. Evaluation of the District of Columbia's Narcotic Treatment Administration programs 1970-1973. Draft manuscript, Burt Associates, Inc.

Collins, William P. and William P. Kelly, Jr. 1975. Methadone treatment and crime reduction--differential impact: an analysis and a case study. In Edward Senay et al. (eds.), *Development in the field of drug abuse*. Cambridge: Schenkman Publishing Co.

Cutter, Henry S.G., Albert Samaraweera, Barton Price, David Haskell, and Clement Schaeffer 1977. Prediction of treatment effectiveness in a drug-free therapeutic community. *Internat. J. Add.* 12(2-3): 301-321.

Eichberg, Robert H. and Bentler, Peter M. 1975. Current issues in the epidemiology of drug abuse as related to psychosocial studies of adolescent drug use. In Dan J. Lettieri (ed.), *Predicting adolescent drug abuse: a review of issues, methods and correlates.*

Governor's Council on Drug and Alcohol Abuse, Commonwealth of Pennsylvania 1975. Harrisburg, Pennsylvania: The Uniform Data Collection System.

Keil, Thomas J., Dickman, Frances B., and Tower, Barbara S. 1975. Social correlates of in-treatment client success. Paper presented at National Drug Abuse Conference, New York.

Keil, Thomas J., Rush, Thomas V., and Dickman, Frances B. 1975. Client demographics and therapeutic approaches as predictive factors in the in-treatment outcome of opiate users. Paper presented at Alcohol and Drug Problems Association of North America Conference, Chicago.

Keil, Thomas J., Rush, Thomas V., and Dickman, Frances B. 1975. Pre-treatment client roles and therapeutic environment as correlates of in-treatment client success: the case of the non-opiate users. Paper presented at American Society of Criminology Conference, Toronto.

Marco Systems, Inc. 1975. Three year follow-up study of clients enrolled in treatment programs in New York City, phase III final report. Report submitted to National Institute of Drug Abuse.

Mandell, Wallace, Goldschmidt, Peter, and Jillson, Irene 1975. Evaluation of treatment programs for drug abusers. In Edward Senay, et al. (eds.), *Developments in the Field of Drug Abuse.* Cambridge: Schenkman Publishing Co.

Nie, Norman H., Hull, C.H., Jenkins, J.G. Steinbrunner, K. and Bent, D.H. 1975. *SPSS: Statistical Package for the Social Sciences.* Second Edition. New York: McGraw-Hill.

Norris, Thelma L. 1975. The role of program evaluation in therapeutic communities. In Edward Senay, et al. (eds.), *Development in the Field of Drug Abuse.* Cambridge: Schenkman Publishing Co.

Sells, Saul B. and Simpson, Dwayne D. 1974. *The Effectiveness of Drug Abuse Treatment, Vol I-V.* Cambridge: Ballinger Pub. Co.

Sirotnik, Kenneth A. and Roffe, Michael W. 1977. An investigation of the feasibility of predicting outcome indices in the treatment of heroin addiction. *Internat. J. Add.* 12(6): 755-775.

DRUG TREATMENT OUTCOME AMONG MINORITIES*

John G. Phin, Ed.D. and Paul Phillips, M.A.

Creative Socio-Medics Corporation

INTRODUCTION

"Drug Treatment Outcome Among Minorities" is part of a larger study describing the nature and extent of drug abuse problems and effective treatment for minorities. The minorities study was conducted over a 12-month period and explored the issue of minority treatment both in the literature and in the field. Five cities (Los Angeles, Seattle, Denver, Minneapolis/St. Paul, and New York) with large concentrations of Asian Americans, Native Americans, and Puerto Ricans were visited by minority representatives who culled over 1,452 client records from 41 programs to collect needed data. In addition, program staff and a sample of minority and majority clients were interviewed to obtain current information about the status of minority treatment.

The purpose of the minorities study was to learn:

- The nature and extent of drug abuse among minority groups and how they differ, if at all, from drug abuse among the white majority population.

- What types of drugs are used among these groups and how their choices differ from those of the majority population.

---

* This study was supported by National Institute on Drug Abuse Contract No. ADM-271-76-4415. The project officer was Barry S. Brown, Ph.D., Chief, Services Research Branch, Division of Resource Development.

- What drug abuse patterns occur among these groups and how they differ from those found among whites.
- How minority user characteristics compare to those of the majority.

In this report, quantifiable drug abuse treatment outcomes for selected minority clients--namely, Asian Americans, Native Americans, and Puerto Ricans--are described. The objective of this study was to determine the degree to which outcomes for each of these groups differed, if at all, from the observed outcomes achieved by (white) majority clients.

METHODOLOGY

For the outcome study, three separate but related procedures were used to gather treatment outcome data. First, record searches were conducted in treatment programs selected on the basis of their minority populations. Change from admission to discharge (in those clients who were discharged) defined treatment outcome in the program record data. Second, project staff returned to a subset of those programs for one day and asked all minority and white clients present to complete a self-administered questionnaire (client self-report). The third and final procedure involved returning to a smaller subset of programs and interviewing a sample of minority and white clients using a structured interview form.

In all, 41 programs contributed subjects to the program record data, 27 to the client self-report data, and 19 to the client interview data. Data were collected during November and December, 1976 and were limited to admissions in the preceding 18 months. Minority fieldworkers familiar with the selected programs and their geographical areas were used to collect data. All data were subjected to rigorous manual and automated editing.[1]

ANALYSIS OF DATA

A three-step process was required for full utilization of the

---

[1] Marginal frequencies from the program record search and client self-report data were compared to the Client Oriented Data Acquisition Process (CODAP) to verify correspondence on key demographic and drug use variables.

outcome data. First, the characteristics of discharged clients were summarized and minority and white clients were compared. (Puerto Rican clients were excluded from this procedure because too few discharge records were found for them.) Second, the admission and discharge characteristics of discharged patients were crosstabulated. Any changes found constituted the first level of outcome measures. Third, the relatedness of the data sets was evaluated, and the outcome data from the client self-report and interviews were summarized.[2]

OUTCOME MEASURES

The guidelines for choosing specific outcome measures were suggested by the conceptual framework for this research design. In general, variables which could be quantitatively measured were selected, i.e., variables more expressive of behaviors than attitudes.

On the program record search the variables for examination were:

1) client status at discharge
2) length of time in treatment
3) employment status at discharge
4) drug use

In addition, although they are not strictly "outcome" criteria, the following variables also were analyzed:

5) education status at discharge
6) participation in a skill development program
7) participation in an education program

Further, the following functional areas were covered by the intensive client interview and, to a lesser extent, by the self-administered questionnaire:

1) health, physical and emotional
2) employment
3) education
4) criminal involvement
5) socio-family functioning
6) drug use

---

2 The reader should understand that the three data sets may contain portions of the same treatment population and that the white majority group clients were taken from the same program bases as were the minority clients.

The samples drawn for this study were not random samples, therefore, generalization must be cautious and conservative. Since the minorities of interest are not distributed randomly in the population, though, and the admission characteristics of the subjects in this study compare favorably to the admission characteristics for these minorities in CODAP, the outcome measures obtained do provide a base of preliminary information which can be used for future, follow-up studies.

LITERATURE REVIEW

In general, there is a paucity of studies that specifically examine drug treatment outcomes among minorities. To derive an interpretive context for discussion of this study's findings, it was necessary to review:

- Drug abuse treatment outcome studies *per se*.
- Descriptions of indigenous treatment and cultural issues found in mental and physical health related studies.

The contribution of ethnic reference to treatment outcome is disputed by researchers. In general, employment, marriage, older age, and limited criminal justice problems have been cited as positive characteristics. Beginning drug abuse later, using for a shorter period, and less involvement in the addict subculture also have been associated with success in treatment.

Other studies report that treatment modality rather than clients' ethnicity are more relevant to treatment outcome. Similarly, clients' use of secondary drugs, clients' attitudes, and program orientation have been reported as significant influences on treatment outcome.

In contrast, other authors have found evidence to support the association of specific character disorders and subsequent prescriptions for treatment with race/ethnicity and socioeconomic class.

In a study of 900 patients (65 percent black, 19 percent Spanish, 16 percent white) from Phoenix House in New York, Wexler (N.D.) found evidence to support his hypothesis that in the therapeutic community the white, non-ghetto person is expressing emotional disturbance in his/her drug abuse, while the black is expressing social and behavioral deficits. He found that Hispanics demonstrated both psychological disturbance and social deficit, suggesting stress from their bi-cultural status.

In his studies of Phoenix House, DeLeon (N.D.), supports the same hypothesis. He suggests that black and Hispanic ghetto residents in therapeutic community treatment are people in need of habilitation or the development of social skills, while whites in treatment with psychological disturbances are better treated by rehabilitation, or the restructuring of values and goals.

Clients' socioeconomic, cultural, political, and psychological characteristics have been related to treatment outcomes, but the relationships are unclear. Fortunately, socioeconomic characteristics are more amenable to statistical control. This approach was used by Joe, et al. (N.D.) in the recent analysis of outcome data in the Drug Abuse Reporting Program (DARP). They considered the interaction between the black and white racial composition of treatment programs and "community structural variables" (unemployment, number of youths, minority density, poverty, and crime rates from census data). They found that race was related to outcome only in combination with these other variables. This study is the most recent of a succession of outcome studies based on the DARP and each successive study has relied more heavily on the statistical control of variables.

Cultural, political, and psychological variables are less amenable to control but have received attention, particularly in the mental health literature. Anthropologists have frequently noted that groups which are forced into situations of culture conflict and of partial, unorganized acculturation seem prone to a higher frequency of neurotic and personality trait disorders (e.g., chronic anxiety and tension, psychosomatic complaints, alcoholism, narcotic addiction, delinquency and crime, regressive or stunted personality development). Likewise, sociologists have reported that among migrant groups not only the incidence of these milder disorders, but also, the incidence of psychosis are measurably higher (Wallace, 1970).

It is critical to recognize the relation of culture to mental health statistics. If cultural beliefs about mental illness and/or economic position prevent or inhibit a person from obtaining help at an early stage of mental stress, the risk of the individual eventually becoming a mental health statistic is appreciably enhanced. If problems of mental stress increase severely, seeking help in outpatient clinics or confinement in a mental health hospital may become inevitable. Thus, mental health statistics among minorities may reflect a lack of the preventive and intermediary mental health care measures used by the white majority.

Further, studies have shown that when a minority member does come into contact with the mental health system, she/he has a greater chance of receiving a more severe psychiatric diagnosis than a white person who is equally disturbed (Chestler, 1972). Often mental health personnel are ignorant of the minority individual's culture, the patient's behavior is seen in a more bizarre light, and the psychiatric label ascribed is far more severe than is warranted.

In conclusion, regardless of their positions on the influence of minority status *per se* on outcome, most authors agree that the success or failure of drug abuse treatment is linked inextricably to change in the larger socioeconomic, cultural, political, and psychological context in which the client lives.

TREATMENT OUTCOME DATA

Data for each minority group studied--namely, Asian Americans, Native Americans, and Puerto Ricans--are presented in separate sections. The reader will recall that the objective of this study is to compare outcomes attained by minority clients with outcomes attained by majority clients, not to compare minorities.

Asian Americans

Findings

- Asian Americans tended to stay in treatment longer than whites (7.2 mean months versus 3.4 mean months), but were no more likely to complete treatment. In fact, 20 percent of Asian Americans were terminated for noncompliance compared to 6 percent of whites.

- Forty-two percent of Asian Americans were employed at discharge versus 20 percent of whites. However, with the exception of Asian American clients receiving methadone, the employment status of the majority of both groups remained unchanged as a result of treatment (33 percent on methadone employed at admission compared to 46 percent employed at discharge for Asian Americans; 29 percent and 36 percent for whites).

- Asians stayed in treatment longer than whites, and while retention decreased with age in whites, the oldest group of Asian clients stayed in treatment a relatively long time. This difference suggests that Asian Americans and whites may not mature out of drug abuse in the same way or to the same extent.

- The majority of white and Asian clients decreased drug use during treatment.

- Smaller proportions of both groups were arrested during treatment than in the year preceding entry (whites: 76 percent before versus 26 percent during; Asians: 85 percent before versus 19 percent during.

. White clients treated previously were retained longer than first admissions, but previous treatment did not differentiate Asian American clients.

. More Asian American clients reported an improvement in living conditions (66 percent) and/or health (61 percent) as a result of treatment than did whites (42 percent and 51 percent, respectively).

Conclusions

In general, Asian Americans paralleled whites in most outcome measures, with the exception of three areas. The first is the high percentage (20 percent) of Asian American clients in the terminated category; the second is the relatively large number of clients over 30 retained in treatment for long periods; and the third is long retention in treatment in treatment for pill users.

The high noncompliance rate cannot be explained, but some reasonable hypotheses can be offered. The primary program contributing clients to this sample operates a therapeutic community and termination for noncompliance is a commonly-accepted technique within this environment. Similarly, in detoxification programs, use of contraindicated substance usually means that a client must be terminated for safety reasons.

As mentioned earlier, the number of clients over 30 suggests that among the Asian Americans sampled, "maturing out" of drug abuse may not be a part of the usage pattern. The data show that Asian Americans tend to use drugs longer than whites before seeking treatment and are more likely to be employed. The lag between onset and treatment, coupled with the employment statistic, could indicate that some Asian American clients manage their drug taking more acceptably (using legitimate income) and for longer periods of time than whites. Whatever the cause(s) for continued drug use, its occurence suggests that treatment programs should provide services responsive to a middle-aged population, as well as to the young adults traditionally found in majority group programs.

Similarly, among the Asian American clients studied, pill users were retained in treatment for a relatively long time (8.0 months for Asians compared to 4.5 months for whites). If outcome data were monitored for this group, it would be possible to detect patterns of use and develop some notions about appropriate duration of treatment, relapse, and so on.

Native Americans

Findings

. Most Native Americans (84 percent) were discharged, 19 percent completed treatment, and the mean duration of treatment was 1.4 months. In contrast, 48 percent of whites were discharged, 31 percent completed treatment, and the mean duration of treatment was 3.4 months.

. The majority of both client groups decreased drug use during treatment.

. Heroin users were retained the longest in both groups, while alcohol users were retained for the shortest period of time. In all substance categories, whites were retained longer, with the most notable difference between the two in the retention time of heroin users (10.3 mean months for whites compared to 3.9 mean months for Native Americans).

. Smaller proportions of both groups were arrested during treatment than in the year preceding entry (whites: 76 percent before versus 26 percent during; Native Americans: 77 percent before versus 20 percent during).

. The employment status of the majority of whites (76 percent) and Native Americans (74 percent) remained unchanged as a result of treatment. (Five percent of Native Americans were employed at discharge as compared to 20 percent of whites.)

. More Native Americans reported an improvement in living conditions (62 percent) and/or health (69 percent) than did whites (42 percent and 51 percent, respectively).

Conclusions

The outcome results for Native Americans compared with white clients are equivocal. Few Native Americans stay in treatment, but that may be a function of a receiving center system rather than a true representation of treatment efficacy. For example, if the program is designed primarily to offer detoxification and the client is interested in a different approach to treatment, then a high dropout rate could be anticipated. This could occur even if the program is operated effectively.

Although Native Americans sampled reported improvements in living conditions and health, it seems unlikely that such improvements could have taken place given their short treatment epoch. For these clients, it may be that reports of improvement are as much related to the relative disadvantage of the minority group as they are to the effectiveness of treatment programs.

Puerto Ricans

Findings

- Ninety-two percent of Puerto Ricans reported decreased drug use as compared to 81 percent of whites.

- Both white (10.3 mean months) and Puerto Rican (11.0 mean months) heroin users had been in treatment longer than users of other drugs, and heroin users treated with methadone reported the longest retention times (whites: 12.8 mean months and Puerto Ricans: 30.1 mean months).

- Arrests while in treatment were less likely for both Puerto Ricans (28 percent) and whites (26 percent) compared with arrests in the year before admission (Puerto Ricans: 86 percent; whites: 76 percent).

- Recidivist Puerto Rican clients reported shorter retention times (8.1 mean months) than did white clients with previous treatment experience (11.2 mean months). On the other hand, Puerto Rican clients with no prior experience in treatment were retained longer (11.1 mean months) than whites (7.8 mean months) during their initial admission. This trend suggests that habilitation is the treatment of choice for Puerto Rican clients during the initial admission, while practical, shorter term goals are addressed in successive admissions. For white clients, the reverse appears true.

- Fifty-eight percent of Puerto Rican clients reported improvements in living conditions and 73 percent reported improvements in health.

Conclusion

Overall, Puerto Ricans fared better than whites in decreasing heroin use and improving both living conditions and health while in treatment. Further, the Puerto Rican sample reduced the number of arrests during treatment as compared to pre-treatment at approximately the same rate as the white group (3.3 percent arrests before versus .6 mean arrests during treatment for Puerto Ricans; 2.2 mean arrests before versus .4 arrests during treatment for whites). Despite these positive outcomes, however, the influence of treatment on employment is as ambiguous for Puerto Rican clients as for the other groups studied.

One of the most interesting findings in the study concerns the issue of retention as a function of previous treatment experiences. Typically, time in treatment tends to increase with prior treatment episodes; with the Puerto Rican clients studied, however, the opposite trend was observed.

One theory proffered to explain this pattern was the Puerto Rican drug user's need for habilitation. Since study restrictions did not permit the kind of intensive examination required to adequately explore this possibility, no definitive conclusions can be drawn. However, when this trend is linked to the short time in treatment experienced by Native Americans, it suggests that retention rates may reflect program orientation as well as client needs.

REFERENCES

Chestler, Phyllis. 1972. *Women and Madness*. New York: Avon Press.

DeLeon, George. Therapeutic Community: Correlates of Opiate, Alcohol and Polydrug Abusers. Phoenix House Foundation, Inc., N.D.

Joe, G.W.; Singh, B.K.; Finklea, D.; Hudiburg, R.; and Sells, S.B. The Effects of Racial and Ethnic Compositions of Treatment Programs on Treatment Outcomes. IBR Report, N.D. (typed).

Wallace, Anthony F.C. 1970. Culture and mental illness. In: Wallace, A.F.C., *Culture and Personality*. New York: Random House.

Wexler, Harry K. Psychological Differences among Black, White and Hispanic Drug Abusers in the Therapeutic Community. Phoenix House Foundation, Inc., New York, N.D.

DETERMINANTS OF RELAPSE: IMPLICATIONS FOR TREATMENT

G. Alan Marlatt and Lisa F. Friedman

University of Washington

Few would doubt that the techniques of behavior therapy or behavior modification are capable of bringing about meaningful changes in behavior. In the past two decades, we have developed a plethora of intervention procedures which have proven effective in modifying a wide variety of target behaviors ranging from snake phobias to sexual dysfunctions. Despite our successes, in being able to initiate changes in behavior, we are still grappling with the difficulties involved in *maintaining* behavioral change over time and across situations.

The study of addictive behaviors provides ample opportunity to examine the maintenance problem, for recidivism rates are notoriously high across the spectrum of such behaviors. In fact, the use of the word maintenance is ubiquitous in the area of addiction treatment programs (i.e, methadone maintenance, heroin maintenance) and this usage stands as further evidence of the highly recalcitrant nature of addiction. The commonality of relapse rates across these problem behaviors is striking: about two-thirds of all relapses occur within the first ninety days following treatment (Hunt, Barnett, and Branch, 1971). These data strongly suggest the possibility of common elements underlying the mechanism of relapse itself. While the importance of individual properties of various substances cannot be ignored, particularly as they affect the development of abuse patterns within user groups, the approach we have adopted seeks to understand the determinants and reactions to the relapse episode itself, viewed as a discrete behavioral entity.

Our approach applies specifically to the study of relapse in human subjects, where cognitive factors play a paramount role as they interact with other behavioral and physiological factors. For this purpose, we have defined relapse as any discrete violation of an imposed rule or set of rules governing the rate or pattern of consummatory behaviors. The criterion of abstinence, the most stringent and absolute rule one can adopt in this regard, is violated by a single occurrence of the target behavior.

The study involves the comparative analysis of a large number of relapse episodes, drawn from individuals who have received treatment of problems associated with the use of alcohol, tobacco, and heroin. Rather than focusing on the internal, physiological factors associated with the particular addictive properties of these three substances, we have based our analysis on the situational or environmental determinants of relapse, and on the individual's cognitive interpretation of the relapse episode. Because we have been concerned with the connection between the initial relapse episode and subsequent use of the substance, our theoretical model has been designed to predict reactions to the relapse based on cognitive-behavioral theoretical principles.

The development of this model began with a study evaluating the effectivenss of aversive conditioning procedures with chronic alcoholics (Marlatt, 1973), the results of which revealed the presence of powerful interpersonal forces as determinants of relapse. Over fifty percent of all relapse situations fell into one of the following categories: (a) situations in which the patient was frustrated or angered, usually in an interpersonal or social situation; and (b) situations in which the patient was confronted by social pressures to resume drinking, usually from a drinking partner or family member. These results indicated the need for an expanded classification system of relapse situations and a new study which would evaluate its effectiveness over a larger sample of subjects. Accordingly, we obtained detailed accounts of relapse episodes from a total of 137 individuals, all of whom were involved in treatment programs for either alcoholism, smoking or heroin addiction.

The alcoholism sample consisted of seventy subjects, all male chronic alcoholics (average age in the mid-forties range) drawn from two inpatient treatment programs (forty-eight subjects from a state hospital inpatient program and twenty-two from an inpatient Veterans Administration hospital program). All alcoholics particpated in programs with abstinence as the primary goals of treatment. Data on smoking relapses were obtained from thirty-five subjects, college-age males and females, who participated in a smoking cessation program conducted at the University of Washington. Descriptions of relapse were also obtained from thirty-two heroin addicts (both male and female young adults), who were assessed following termination of treatment as part of a statewide evaluation of services. A relapse was defined as the first use of illegal heroin or other opiates for this group of subjects. For the alcoholics and the smokers, the first drink or the first use of tobacco constituted the relapse episode.

All subjects were contacted within ninety days of the termination of the subject's treatment program for follow-up assessment. The alcoholics and heroin addicts were interviewed in person at this time, whereas the smoking subjects filled out a detailed questionnaire designed to obtain the same information. Most of the questions asked were patterned after the follow-up version of the Drinking Profile (Marlatt, 1976). As in the previous study, each subject was asked to give the date and time of the relapse episode and to describe the setting in which the episode occurred (place, and presence/absence of other individuals). In addition, the following open-ended questions were asked, to provide the descriptive basis for the subsequent content-analysis ratings: (a) "When I took my first drink (or cigarette, or fix of heroin), the situation was as follows (briefly describe the important features of the situation which led you to take the first drink),"; (b) "What would you say was the *main reason* for taking that first drink?"; (c) "Describe any inner thoughts or emotional feelings (things *within* you as a person) that triggered off your need or desire to take the first drink at that time," and "Describe any particular circumstances or set of events, things which happened to you in the outside world, which triggered off your need or desire to take the first drink." Other questions were asked concerning the subject's reactions to the relapse episode, including whether or not the first "slip" was followed by subsequent use of the substance on the same or on following occasions.

The next step involved the development of a classification system to enable the assignment of each relapse episode to an independent category. The first category, *intrapersonal environmental determinants*, used whenever the relapse episode involved a response to primarily psychological or physical events (e.g., coping with intrapersonal emotional states, giving in to "internal" urges, etc.), or a response to a nonpersonal environmental event (e.g., misfortune, accident, financial loss, etc.). Here, the emphasis is on precipitating events in which another person or group of individuals is *not* a significant factor. The second major category, *interpersonal determinants*, applied whenever the relapse episode involved the significant influence of other individuals (e.g., coping with interpersonal conflict, social pressure, etc.).

After the classification system was derived, two students were trained to assign category scores for the relapses described in the questionnaires mentioned above. Training continued until a high degree of agreement between raters was obtained: in an independent test of reliability, the inter-rater agreement was eighty-eight percent for category assignment. Each rater was then asked to score approximately half of the 137 relapse episodes given by subjects in the alcoholism, smoking and heroin addiction treatment groups. The results of this classification system are presented in Table I.

Table I. Analysis of relapse situations with alcoholics, smokers, and heroin addicts.

Situation	Alcoholics (N=70)	Smokers (N=35)	Addicts (N=32)	All Subjects (N=137)
Intrapersonal Determinants	61%	57%	53%	58%
Negative Emotional States	38%	43%	28%	37%
Negative Physical States	3%	---	9%	4%
Positive Emotional States	---	8%	16%	6%
Testing Personal Control	9%	---	---	4%
Urges and Temptations	11%	6%	---	7%
Interpersonal Determinants	39%	43%	47%	42%
Interpersonal Conflict	18%	12%	13%	15%
Social Pressure	18%	25%	34%	24%
Positive Emotional Stress	3%	6%	---	3%

The data in Table I indicate that over three-quarters (76%) of all relapse episodes fall into just three categories: coping with negative emotional states (37%), social pressure (24%), and coping with interpersonal conflict (15%). For the remaining 24% of all relapses, the distribution was as follows: giving into temptations or urges (7%), enhancement of interpersonal positive emotional states (39%), negative physical states (4%), testing personal control (4%), and enhancement of intrapersonal positive emotional states (6%).

The majority (58%) of all relapses involved intrapersonal/environmental determinants with 42% involving primarily interpersonal determinants. In the interpersonal settings, almost all (82%) of the relapse episodes involved coping with frustration or anger. In the intrapersonal settings however the results were reverse, with 85% of relapses triggered by emotional states other than frustration and anger. These results seem to indicate that frustration and anger associated with relapse stem primarily from interpersonal sources (arguments with others etc.), whereas other negative emotions (fear, anxiety, etc.), seem to predominate as determinants of relapse when significant other individuals were not involved. These findings were consistent for each of the three subject groups in our sample.

Coping with negative physical states accounted for only 4% of all relapses and mostly involved the heroin addict group. None of the alcoholics or smokers cited physical withdrawal as a determinant of relapse, a finding which casts doubt on theories which posit withdrawal symptoms as the primary precipitating factor in relapse (c.f., Marlatt, 1978). Temptations or urges may be related to the subjective experience of craving, sometimes posited as a conditioned response to clues associated with prior withdrawal (Ludwig and Wikler, 1974). Yet this category accounted for only 7% of all responses, most of which occurred in the alcoholic group, with relatively few smokers and no heroin addicts citing this as a determinant. When giving in to urges or temptations was mentioned, however, most of these feelings were experienced in the absence of substance cues. In addition, the use of a substance to enhance positive emotional states, whether in an intrapersonal or interpersonal situation, accounted for relatively few relapses (9%). These findings suggest that coping with stress, interpersonal conflicts, or a response to social pressure etc.) is a much stronger determinant of relapse than the desire to "feel good," to enhance already existing positive emotional states, or to cope with negative physical states.

Only among the alcoholic sample was testing personal control, given as a determinant of relapse. For a small subgroup of the alcoholics (9%), the desire to either test one's ability to have a drink or two and then stop, or the intention to test the effects of the abstinence oriented treatment program was given as the precipitating event. As none of the smokers or heroin addicts cited tests of personal control as the primary determinant, it seems likely that some alcoholics are particularly susceptible to this temptation.

Approximately one-fourth of all relapses were classified under the social pressure category. Whether the influence of social pressure was direct or indirect in nature interacted in an important way with the substance or drug used. For alcoholic and heroin addicts, the predominant relapse situation involved direct social pressure, or actual contact between users (e.g., meeting an old drinking buddy who put pressure on the person to begin drinking again; or the addict running across someone who offers some of his/her stash). For the smokers, on the other hand, the pattern was reversed. Smokers showed a greater tendency to succumb in situations where others were smoking (observation of models), even though the other smokers put no direct pressure on the observers to join in the indulgence. This finding may reflect the fact that smoking is a far more public behavior (exhibited across a wide variety of situations) than is drinking, or the illicit use of narcotics.

Our three groups relapsed within a relatively short time of completing their respective treatment programs. For our sample, the average number of days between beginning abstinence and the subsequent date of relapse was seventeen days for smokers, thirty days for the alcoholics, and thirty-two days for the heroin addicts. These figures must be interpreted with caution, however, because of differences among the treatment programs involved.

To account for the similarity of the relapse process across different consummatory behaviors, we have constructed a theoretical model based on a cognitive-behavioral orientation. Underlying this model of relapse is a cognitive process which we have labeled the Abstinence Violation Effect (AVE). The AVE is postulated to occur under the following conditions: (a) The individual is personally committed to an extended or indefinite period of abstinence from engaging in a specific behavior; (b) The behavior occurs during this period of voluntary abstinence.

We hypothesize that the AVE itself is characterized by two key cognitive elements:

1. A cognitive dissonance effect (Festinger, 1957, 1964), wherein the occurrence of the previously restricted behavior is dissonant with the cognitive definition of oneself as abstinent. Cognitive dissonace is experienced as a conflict state, and underlies what most people would define as guilt for having "given in to temptation".

2. A personal attribution effect (cf., Jones, Kanouse, Kelley, Nisbett, Valins, and Weiner, 1972), wherein the individual attributes the occurrence of the taboo behavior to internal weakness or personal failure, rather than to external situational or environmental factors. The additive effects of both reactions will greatly increase the probability of repeating the restricted behavior and engaging in a full blown relapse. From this perspective, relapse is determined in large part by the alcoholic's perception of having "lost control" when the first slip occurred.

The conditions under which the AVE would come into play are mediated by the individual's overall sense of personal self-efficacy and the dynamics of the relapse situation itself. While the factors that determine relapse are interwoven in a network of complex interactions, the basic sequence of cognitive events appears consistent across substances. Thus, our model for the conditions of a full blown relapse is summarized as follows:

(a) The abstinent individual feels "in control" until he encounters a high-risk situation which challenges his perception of control;

(b) The individual lacks an appropriate method of coping with the high risk situation, or fails to engage in a coping response;

(c) He has positve expectancies about the effects of the substance or behavior he is abstaining from;

(d) He engages in that behavior or consumes that substance;

(e) He or she experiences one or both components of the Abstinence Violation Effect;

(f) The probability of continued drinking markedly increases.

Most traditional treatment programs for addictive behaviors, including those for alcoholism, smoking control, and heroin addiction, tend to ignore the relapse issue altogether in their intervention procedures. Yet, teaching skills that may help an individual to cope successfully with a relapse, would seem to be a matter of common sense. Skills offer more help to the individual than relying on vague constructs such as "will power" or trying to adhere to the advice implied by various slogans. With this in mind, the following is a brief description of intervention strategies for coping with a potential relapse situation. (See Figure 1.)

High Risk Situation

The first step to take in the prevention of relapse is to train

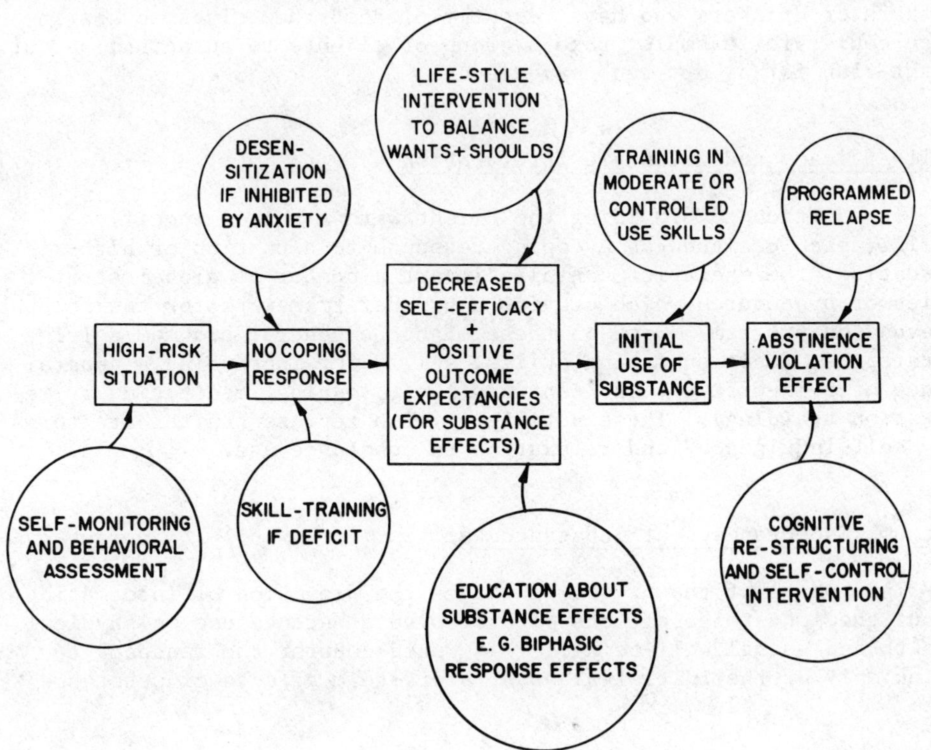

Fig. 1. Points of intervention for prevention of relapse.

the client to recognize those high-risk situations which are likely to increase the probability of relapse. The goal would be to teach the client to recognize the discriminative stimuli that are associated with "entering" a high-risk situation and to use these warning signals as cues to implement an alternative sequence of behavior. *Self-monitoring procedures* (McFall, 1977) provide an effective means of identifying potential high-risk situations as do the *Drinking Profile* (Marlatt, 1977) and the *Situational Competency Test* (Chaney, 1976).

Coping Responses

If the client has never learned the appropriate coping response to high-risk situations, the emphasis should be on teaching the requisite skills involved. On the other hand, if the response is available in the client's repertory but is blocked by an inhibiting influence, the therapist must first deal with the client's reactions so that the response can be performed with minimal anxiety. One suggestion that may increase the generalization of newly acquired coping skills is to require the client to practice the adaptive behavior in the actual high-risk situation. The therapist who is working with smokers or drinkers who have recently pledged themselves to abstinence could, for example, take a group of clients to an actual bar or night-club for a "dry run" experience.

Self-Efficacy and Lifestyle Intervention

In addition to providing the client with a set of specific skills, each designed as a coping response to a particular high-risk situation, the therapist can also impart a number of global strategies or procedures which provide a broader framework for the relapse prevention program. Some of these might be general problem-solving strategies, decision-making skills, skills for increasing a general sense of self-efficacy and control (i.e., jogging, meditation or relaxation training). These activities also represent alternate forms of "self-indulgence" and relaxation to substance use.

Outcome Expectancies of Substance Use

This phase of the program involves the provision of information about the long-range effects of excessive substance use on physical health and social well-being. This would counter the tendency to think only of the initial pleasant short-term effects of substance use.

Initial Use of Substance: What To Do If A Slip Occurs

Here, the client can be prepared in advance to cope with this possible outcome, and to try to apply some "brakes" so that the slip does not escalate into a full-blown relapse. A combination of specific skills and cognitive intervention strategies would seem to offer the greatest advantage for relapse prevention.

REFERENCES

Chaney, E.F. 1976. *Skill Training with Alcoholics*. Unpublished doctoral dissertation, University of Washington.

Festinger, L. 1964. *Conflict, decision and dissonance*. Stanford: Stanford University Press.

Hunt, W.A., Barnett, L.W., and Branch, L.G. 1971. Relapse rates in addiction programs. *Journal of Clinical Psychology*. 27: 455-456.

Jones, E.E., Kanhouse, D.E., Kelley, H.H., Nisbett, R.E., Valins, S., and Weiner, B. (eds.) 1972. *Attribution: Perceiving the causes of behaviors*. Morristown, N.J.: General Learning Press.

Ludwig, A.M. and Wikler, A. 1974. "Craving" and relapse to drink. *Quarterly Journal of Studies on Alcohol*. 35: 108-130.

Marlatt, G.A. 1973. *A comparison of aversive conditioning procedures in the treatment of alcoholism*. Paper presented at the meeting of the Western Psychological Association, Anaheim, Ca.

Marlatt, G.A. 1976. The Drinking Profile: A questionnaire for the behavioral assessment of alcoholism. In E.J. Mash and L.G. Terdal (Eds.). *Behavior therapy assessment: Diagnosis, design, and evaluation*. New York: Springer.

Marlatt, G.A. 1978. Craving for alcohol, loss of control, and relapse: A cognitive-behavioral analysis. In P.E. Nathan, G.A. Marlatt, and T. Loberg (eds.), *Alcoholism: New directions in behavioral research and treatment*. New York: Plenum.

McFall, R.M. 1977. Parameters of self-monitoring. In R.B. Stuart (ed.), *Behavioral self management: Strategies, techniques, and outcomes*. New York: Brunner/Mazel.

THE INTEGRATION OF TREATMENT AND EVALUATION*

Demosthenes Lorandos, Ph.D.

Central Michigan University

What constitutes "success" or "failure" in the treatment situation? Is the therapist's assessment of the client's adjustment or change resultant of treatment the criterion? Do we concern ourselves with the client's perceptions of their well being? Can we assess their functioning in society as a criterion measure?

Each of us has at one time or another been faced with making a decision at discharge as to whether the treatment has been successful. How do we arrive at those judgements? What variables do we consider? To illustrate this dilemma, I offer these hypothetical examples:

A thirty-eight year old man began treatment at the outpatient portion of our substance abuse clinic six months ago. According to our policy, a problem oriented record was started immediately upon intake. His presenting complaints were: Episodic excessive drinking with "benders" following homosexual encounters; difficulty in holding onto a job because of his "nerves"; money problems that came as a partial consequence of his prolonged unemployment; problems at home where he lived with his elderly mother (a practicing alcoholic); and an admittedly low self concept.

During his treatment at the clinic, he attended evening groups twice weekly and individual counseling sessions once each week. His

---

* The author wishes to acknowledge the aid of Paul H. Miller, Ph.D. in developing these concepts.

attendance in both was satisfactory until he was discovered to have manipulated a local psychiatrist into the prescription of antidepressants. When confronted with the evidence he pleaded an acute anxiety attack following the discovery by his local Alcoholics Anonymous group that he was gay. Although he had stopped drinking upon entering the evening group, the episode with the pills and subsequent confrontation preceded a decline in his attendance. At a case load review meeting some weeks later his group leader and his individual counselor both agreed that he should be taken off active client status and that he was not likely to return until another personal crisis compelled him to do so.

The experience based professionals (read: paraprofessionals and/or ex-addicts) noted that he was no longer actively abusing alcohol and in fact, had been dry for five months. They voted for "successful" to be entered in his discharge notes. The academically based professionals (read: psychologists, psychiatrists, social workers) pointed to what they considered underlying ego deficits and personality structure problems. They voted for "unsuccessful" to be affixed to his discharge papers. No one spoke to his behavior in the larger community.

In another example: A twenty-three-year-old man was referred by the courts for treatment in lieu of jail. When he was evaluated by the intake worker he presented problems of chronic unemployment; severe depression with disorganized thinking; a history of PCP and polydrug abuse; low self concept; and petty criminality. He was living with "friends" and had no relatives in our state. During the nine months he stayed in treatment with our therapeutic community, he began to gain insight into his drug taking behavior. He participated in all phases of treatment and adjusted well to the TC's milieu. Early on in his stay with us staff noticed the characteristics of PCP induced personality deterioration. He had attention difficulties, repeated bouts with depression, and at times, earned the nickname "Space Cadet".

Staff was divided as to whether this was due to PCP use or indications of previous personality and/or thought disorder. When he chose to leave us to accept a job as a carpenter's helper (which he hoped would lead to a union apprenticeship), staff was again divided as to whether treatment had been successful or a failure. While all agreed that he would have benefited from a longer stay, several staff noted his good adjustment in the TC, his acceptance of personal responsibility and the fact that he had actually landed a job for himself. The hardliners were pessimistic about his continued abstinence from drugs once out of the TC and were unable to term his treatment successful. Interestingly enough, the academically trained (read: psychologists, psychiatrists, social workers), felt that he had adjusted well, had a good chance of reintegrating into the larger community and saw this as a result of the TC experience, therefore successful was their vote.

In each case, staff was divided on the decision of classifying treatment success or failure. In each case, varying perspectives gave voice to differing views. I am sure, if we polled this room, we would obtain many differing views on the relative success or failure of treatment in the two examples given.

In my capacity as consultant and treatment supervisor for these and other clinics, I was frustrated over having to hash out the same pros and cons each time it came to a determination of treatment outcome. Considering myself an experience based professional as well as an academically trained psychologist, I was pulled in both directions during each discussion. One side of me wanted to demand abstinence as the criterion. Another leaned toward an assessment of change based upon objectively defined criteria, test data and so on. Finding myself in the role of Soloman all too often, I plunged into the literature of evaluation, seeking a way out of the problem.

I am unhappy to report, after many hours with *Psychological Abstracts*, I was unable to come up with any tactic which felt satisfactory. With the possible exception of some work done in goal attainment scaling (Kiresuk and Sherman, 1968; Romney, 1977), there is little of value! Indeed, in an incredible two volume work on program evaluation (Guttentag and Struening, 1975; Struening and Guttentag, 1975), which seems to go on forever, no one spoke to the issue of the determination of success in treatment. It seems the researchers leave it up to the folks on the firing line and then step in to "follow-up" on those who were a success, comparing them with those determined to have been a failure.

In two delightful articles on a three part approach to treatment outcome evaluation, Strupp and Hadley point to overwhelming methodological problems in evaluation research which seem to be carried forward into each new study! "The fact that many of the deficiencies identified over a decade earlier persist in current studies documents regrettable neglect on the part of researchers to heed important lessons." (Strupp and Hadley, 1977)

Many researchers have identified problem areas (Kiesler, 1966; Gross and Miller, 1975), but Strupp and Hadley point out that evaluators everywhere seem to be charging off in their own divergent direactions: ". . .We contend that profound discrepancies in evaluative criteria are continuing to confound our best efforts at evaluating the outcomes of psychotherapy" (Strupp and Hadley, 1977). Indeed, my research led me to the same unhappy conclusion. While there is extensive literature on what constitutes mental health (Jahoda, 1958) and mountains on evaluation of outcome criteria in psychotherapy (Bergin and Garfield, 1971; Strupp and Bergin, 1969) no one is reaching any consensus.

To underline the point, just last year, psychologist Theodore Blau was calling for a " . . . consortium of human service practitioners and evaluation specialists. Those who are in constant contact with people struggling to achieve quality of life . . . must somehow be brought together with researchers of various kinds, but in particular, with evaluation specialists" (Blau, 1977). Being unwilling to wait for an "evaluation specialist" to walk through my door, I began to work on a second area of concern.

In the examples given above, we noted that different perspectives spoke to differing areas of concern. The academically trained folks spoke to issues of personality structure and underlying psychodynamic problems. Our experience based people spoke to the role of the client's self concept as being pivotal in the recovery process. As a community psychologist, I kept on wondering about how the clients were actually "doing" in the world. Were they paying their share of the rent? Were they continuing to use drugs on the sly? Were they taking out their share of the garbage?

Again and again I watched as these differing viewpoints confounded our ability to reach consensus on a determination of outcome. We were all painfully aware that the clients' own opinions as to treatment outcome found their way into an already complicated discussion all too rarely. With all of this percolating inside me I went back to the literature to seek some help with the issue of differing viewpoints and their influence upon outcome decisions. "Perhaps some sort of variably weighted regression equation was in order?" I wondered.

This time I found something. Kelman and Parloff (1957) and Parloff, Kelman and Frank (1954) contributed some valuable insights when they discussed intercorrelations among the three criteria of comfort, effectiveness, and self awareness as a means to gauge improvements in psychotherapy. The work cited above by Strupp and Hadley (1977) offers another three part model for assessment of outcome which focuses upon society's judgment, the view of the mental health professional and opinions of the client as well. They spoke nobly to our problem: ". . .If one is interested in a *comprehensive* picture of the individual, evaluations based on a single vantage point are inadequate and fail to give necessary consideration to the *totality* of an individual's functioning" (Strupp and Hadley, 1977).

Although Strupp and Hadley present these ideas in two important journals within the same year, they give no hint as to how to put this three part orientation into the service of practical decision making. Alas, a second comprehensive search of the current literature delivered only provocative food for thought.

With all this incubating inside me, I was reminded again and again of the importance of consulting the clients about their own treatment. Judgments made without their input seem destined to be made with a strong overlay of values that are not their own. We are aware of the literature concerning middle class values and treatment (Szasz, 1970 and 1974). How then, could a system be created which would allow for all viewpoints concerned to have their perspectives factored into the process of decision making?

In the spring of last year I developed the plan that I wish to share with you today. In April I discussed it in detail with the chief research people at NIAA and the next day, with their counterparts at NIDA. All were supportive and urged me to move ahead with a research grant request. While I was flattered and a bit heady with all of this, I could not forget what Lewin used to say about research. He demanded that it be socially useful as well as theoretically meaningful. It is with this adage in mind that I wish to present these ideas to you, the ultimate judges of their value, before applying for research funds. I have prepared a short questionnaire in which I would very much appreciate your feedback.

It is my feeling that the client, therapist and community all must be heard from in any comprehensive evaluation of person change. For, ultimately, this micro-evaluation will serve as part of each macro-evaluation of effectiveness. I have been continually turned off by the after-the-fact boys, who ride in to review your "statistics" as if they were bank auditors. Person change is the heart's blood of our business and it cannot be treated so without doing it an incredible disservice.

This model provides for a system of in-process evaluation that becomes a meaningful part of the on-going counseling milieu. The data generated can be used as grist for the after-the-fact boys *and* may be used to establish a common understanding of treatment goals and progress for therapist and client.

We are here concerned with three dimensions of person change over time. The first dimension speaks to the concern of the therapist. It is a dimension of personality structure. It can be assessed to a greater or lesser degree by such classical measurement techniques as the MMPI, the CPI, checklists, sentence completion blanks, ink blots, etc.

The second dimension speaks to the concern of the client. Am I feeling better? Do I like myself more? Am I OK now? It is a dimension most closely represented by the construct of self concept. It can be assessed simply and easily by Q Sort, Tennessee Self Concept test, and so on.

The third dimension of concern to a comprehensive statement of evaluation is the social or community sphere. Although we are all too often too busy to aquire them, the statements of collaterals are usually of great value in a client's treatment. Research in social and community psychology is now providing us with short, easily administered inventories of community integration and behavior in the "real" world. outside of the clinic.

It can be seen at once that these dimensions are highly intercorrelated. Therein lies the beauty of this particular three point approach. We will create a vehicle to display intercorrelations of these three perspectives and thereby integrate an immediately comprehensive picture of "person" into our counseling sessions.

I propose that these three perspectives be presented to the client *visually* and in three dimensions. The measure of personality structure, the "intrinsic" variable, shall be *depth*. The estimate of the client's self concept, a variable molded significantly by his internal state as well as the world around him, the "intrinsic/extrinsic" variable shall be *width*. The perspective of society, the "extrinsic" variable, the perception of his behavior outside the clinic, shall be *height*.

A table of standard scores will factor the raw scores on each measure into an appropriate regression equation and read out a single number score. This single number score on each variable will be placed on a three dimensional grid (I propose a simple set of blocks and graph paper grid). As each dimension is completed the client and therapist can compare onset of therapy scores and their corresponding three dimensional display to a three month, six month or post-treatment display. In this way, client and therapist can evaluate progress in each dimension.

It is my particular prejudice that symmetry of display dimensions is the goal. An asymmetrically long first dimension display in correlation with fairly symmetrical second and third dimensions might denote a client in one of our behavioristic therapeutic communities. He has learned how to act and how to talk, but down there somewhere, something is wrong. It is precisely this sort of simple explanation that the three dimensional model will facilitate. The client need not take the word of the therapist, the model display faces him directly with his score!

In advanced stages, the three dimensions could be assessed with a short battery of strongly intercorrelated (and criterion correlated) tests whose raw scores would be fed to a computer. With an appropriate topological program, the computer would deliver a three dimensional grid display through a high speed printer. When we develop easily administered parallel forms of the measuring devices we will be able to assess change more frequently during the process of treatment.

All data generated have as a by-product to their main teaching and consensus functions, a compellingly simple effect. While they are depicting in physical form constructs that were often misunderstood or not assessed at all, they create easily stored information on part processes of therapy. Summed, these pieces lend themselves to as complicated or as simple a macro-assessment as third parties might want. Taken in varying ratio, they will yield significant information on single variables or on many variables in multi-dimensional array.

REFERENCES

Bergin, A. 1971. The evaluation of therapeutic outcomes. In Bergin and Garfield (eds.), *Handbook of psychotherapy and behavior change*. New York: Wiley.

Blau, T. 1977. Quality of life, social indicators and criteria of change. *PP*. 8: 464-473.

Gross, S. and Miller, J. 1975. A research strategy for evaluating the effectiveness of psychotherapy. *PP*. 37: 1011-1021.

Guttentag, M. and Struening, E. 1975. *Handbook of evaluation research*. (vol. 2) Beverly Hills: Sage.

Jahoda, M. 1958. *Current concepts of positive mental health*. New York: Basic Books.

Kelman, H. and Parloff, M. 1957. Intercorrelations among three criteria of improvement in group therapy: Comfort, effectiveness, and self-awareness. *JASP*. 54: 281-288.

Kiesler, D. 1966. Some myths of psychotherapy research and the search for a paradigm. *PB*. 65: 110-136.

Kiresuk, T. and Sherman, R. 1968. Goal attainment scaling: A general method for evaluating comprehensive community mental health programs. *CMHJ*. 4: 443-453.

Parloff, M.; Kelman, H. and Frank, J. 1954. Comfort effectiveness and self-awareness as criteria of improvement in psychotherapy. *AJP*. 3: 343-351.

Romney, D. 1977. Treatment progress by objectives: Kiresuk's and Sherman's approach simplified. *CMHJ*. 13: 286-290.

Struening, E. and Guttentag, S. 1975. *Handbook of evaluation research* (vol. 1). Beverly Hills: Sage.

Strupp, H. and Bergin, A. 1969. Some empirical and conceptual bases for coordinated research in psychotherapy. *IJP*. 7: 18-90.

Strupp, H. and Hadley, S. 1977. A tripartite model of mental health and therapeutic outcomes. *AP*. 33: 187-196.

Strupp, H. and Hadley, S. 1977. Evaluations of treatment in psychotherapy: Naiveté or necessity? *PP*. 8: 478-490.

Szasz, T. 1974. *The age of madness*. New York: Aronson.

Szasz, T. 1970. *The manufacture of madness*. New York: Harper and Row.

DRUG DEPENDENCE TREATMENT BENEFIT STUDY PROJECT OVERVIEW

Arthur Leyland

Research and Development Division
Blue Cross Association
840 North Lake Shore Drive
Chicago, Illinois 60611

In January, 1978, the Blue Cross Assocation's Research and Development Division, Product Development Department, began a one-year preliminary study to assess the feasibility and desirability of offering a comprehensive drug dependence treatment benefit, nationwide, through the Blue Cross organization. The Association will conduct the investigation under contract and in cooperation with the National Institute on Drug Abuse.

This overview provides background information on the project, and describes its goals and principal activities.

BACKGROUND

Drug abuse is increasingly recognized as a major social/health problem in the United States. Although once primarily associated with the counter-culture of the 1960's or low socioeconomic status, the use of illicit drugs and misuse of controlled prescription drugs has become a significant problem among all age groups and socioeconomic strata. A wide variety of drugs has been found to be subject to abuse, ranging from such legal addictive drugs as caffeine and nicotine to illicit drugs, such as LSD or cocaine.

Most communities now report drug abuse problems in one form or another. In a report to the President published in 1975, the Domestic Council Drug Abuse Task Force estimated that 1.2 million people were abusing or addicted to either amphetamines, barbiturates or heroin. Additional data from federally funded treatment programs indicated that approximately 1/3 of those admitted to treatment were employed and approximately 11% had some college education.

In addition to the personal impact of the drug dependent individual and his family, the cost of the problem on a national scale is great. In 1975 the federal government was estimated to have spent over 700 million dollars to reduce the supply of and demand for illicit drugs. During the same period, the cost of drug-related crime, reduced productivity and treatment and prevention programs has been estimated at between 10 and 17 billion dollars.

Steps are being taken to curb drug abuse through law enforcement and preventative education, but an equally important approach to the drug problem is medical and psychological therapy for the drug dependent person. However, in order for this approach to be successful, three requirements must be satisfied. First, the patient must recognize his problem and be motivated to seek help. Second, treatment programs must be of high quality and readily accessible. Third, financial resources must be available to pay for the treatment.

The first two requirements are increasingly being met. According to statistics from federally funded treatment programs, 83% of admissions were voluntary during the first quarter of 1976 and initial inspection suggests that qualified providers are becoming increasingly available. But notable deficiencies exist in the way in which drug abuse treatment has been financed. At government monitored treatment programs, the majority of funding comes from federal sources, with less than 10 percent from private sources, and only .5 percent from private health insurance. This stands in marked contrast to the total expenditure for hospital care, for example, for which private insurance contributes most of the hospital industry's funding.

The heavy reliance by drug dependence treatment facilities on government funding reduces their financial stability and their ability to adequately plan program development and expansion. In addition, drug dependent individuals, who are typically slow to recognize their own drug related problems, may be even more reluctant to seek treatment if it is not covered by private insurance. Thus, the lack of private, third party payment may greatly limit the ability of treatment programs to reach important segments of the population and to provide a full range of services.

Since its inception, the Blue Cross organization has demonstrated its commitment to eliminate financial barriers to health care for its members. Consistent with this commitment and in the interest of developing greater accessibility and availability of treatment services for drug dependent persons, the Blue Cross Association and the National Institute on Drug Abuse are cooperating to study the problems incurred when a health service benefit for drug dependence treatment is offered and to propose solutions to these problems. The following section provides a general outline of the project and explains what will be accomplished during the one year, preliminary study.

THE PROJECT

The ultimate project goal is to develop tools and techniques needed by the Blue Cross Association to support a nationwide comprehensive drug dependence treatment benefit. Because such a program is virtually unprecedented, its development will proceed in an incremental manner to promote careful analysis of problems and the development of appropriate solutions. Thus, the current project constitutes Phase I of an anticipated three-phase program. Phase I is the planning period during which preliminary research will be done, prototype materials developed and the determination made whether a Phase II demonstration is warranted. Phase II would be a field test in which the prototype benefit materials would be tested, evaluated and refined. If the results of Phase II demonstrate that benefit delivery is both feasible and desirable, Phase III would be initiated to integrate the drug dependence treatment benefit into existing health benefit packages offered by Blue Cross Plans on a nationwide basis.

PHASE I

During Phase I of the drug dependence benefit study, project staff will analyze the issues and problems in offering a comprehensive drug dependence treatment benefit. The project will include both research and development aspects aimed specifically at determining the benefit feasibility/desirability and the best methods to deliver a drug dependence treatment benefit. Among the questions to be answered which related to the feasibility of the proposed benefit are:

* What is the state of the art of drug dependence treatment services?
* Can effective benefit, marketing and administrative packages be developed?
* Do Blue Cross Plans have the authorization, under existing state regulations, to offer a drug dependence benefit?

Among the questions which relate to the desirability of the proposed benefit are:

* Is there sufficient need among Blue Cross Plan members to support the benefit?
* Is there sufficient demand for the drug dependence benefit among benefit purchasers and subscribers?
* Can the benefit be provided at a cost acceptable to the marketplace?

The answers to those questions, in addition to information from Blue Cross Plans and subjective evaluations by the Blue Cross Association and the National Institute on Drug Abuse, will determine whether the proposed drug dependence treatment benefit is feasible and desirable and whether a Phase II study is needed or warranted to gather additional information.

The research aspect of Phase I will draw from 3 major sources:

1) Literature review: with the assistance of the NIDA, current relevant information will be reviewed and integrated.

2) Field interface: project staff will solicit information from providers and other professionals with direct experience in drug dependence treatment.

3) DABSAC: project staff will establish a Drug Abuse Benefit Study Advisory Committee comprised of experts on drug dependence treatment and third party payment and representatives of both labor and management. Throughout the course of Phase I, the DABSAC will assist in the development of a clinically valuable and technically feasible benefit.

Prototype materials to be researched and developed during Phase I can be divided into 3 categories: a model drug dependence treatment benefit, a model marketing package and a model administrative package.

MODEL DRUG DEPENDENCE TREATMENT BENEFIT

As part of Phase I activities, a prototype comprehensive drug dependence treatment benefit will be developed. The specific form this benefit will take will depend on the solutions developed during Phase I to clinical and technical issues such as:

Clinical Issues

* For which drugs of abuse have successful treatments been developed?

* Which treatment settings and services are most effective?

* Which patient populations show the greatest response to treatment?

Technical Issues

* What factors determine the cost of each type of treatment?
* What controls can be established to minimize misuse of the benefit?
* What should be the coverage amounts and durations?

MODEL MARKETING PACKAGE

To effectively meet the needs of subscribers and to help local Plans institute the drug dependence benefit, a model marketing package will be developed. This package will include an equation to assist in determining the rate or premium of the model benefit and educational materials to inform potential purchasers and subscribers about drug dependence treatment and the Blue Cross drug dependence benefit.

Because national awareness of the scope of the drug dependence treament problem is relatively new and recognition of drug dependence as an illness has been slow to develop, two separate but coordinated marketing themes will be developed. The first will focus on achieving desired levels of enrollment under the benefit and will be directed toward both employers and subscribers. The second theme, aimed at employers, subscribers and providers, will be designed to promote utilization of the benefit through educational materials communicating the concept of drug dependence as a treatable illness.

MODEL ADMINISTRATIVE PACKAGE

Administrative guidelines and materials will be designed so that the drug dependence benefit can be administered efficiently. The administrative package will deal primarily with 3 issues:

1) Plan/Provider Relationships: Recommended criteria for identifying providers who meet quality of care standards, and model Blue Cross Plan/Provider Contracts will be developed.

2) Plan/Member Relationships: A subscriber brochure to explain the benefit and how it should be used, a manual to assist Plans in answering subscriber questions, a model subscriber contract and benefit forms will be designed.

3) Administrative Cost: The issues involved in determining the cost of initiating and administering the drug dependence treatment benefit will be assessed.

CONCLUSION

In summary, the primary anticipated activities of Phase I of this study will be to:

1) Analyze the problem of drug dependence in terms of epidemiology and symptomology for various drugs of abuse; analyze treatment methods and sites as potential resources for drug dependence benefit coverage; and design a prototype benefit which is both clinically and technically appropriate.

2) Develop recommended treatment standards for provider participation in the drug dependence benefit.

3) Develop model administrative and marketing packages.

4) Analyze the legal ramifications of a drug dependence treatment benefit and develop a plan for overcoming any legal barriers.

5) Decide, on the basis of the information gathered in Phase I, whether the drug dependence treatment benefit proposed is feasible and desirable and determine the need for a benefit demonstration as proposed for Phase II.

RESEARCH AND DEVELOPMENT DIVISION, BLUE CROSS ASSOCIATION

This study is being conducted by the Product Development Department of the Blue Cross Association's Research and Development Division.

The Research and Development Division provides Blue Cross Plans with relevant information and products to meet the demands and opportunities of the future. It is the specific function of the Product Development Department to identify, develop, test and help implement new health care benefits, procedures to control health care costs and utilization and methods to increase the efficiency and capacity of Blue Cross Plan operations.

PRETREATMENT CRIMINAL JUSTICE STATUS, RETENTION, AND POSTTREATMENT OUTCOMES

Robert J. Harford, Ph.D.

Yale University School of Medicine

The majority of applicants for drug abuse treatment have been involved in criminal activities during the time period immediately preceding admission. Coercion by the criminal justice system (CJS), either explicit or anticipated, is the principal reason for seeking treatment. Perhaps 20% are forced by legal authorities to choose between treatment and incarceration (Burt, 1977), and an additional 40-50%, while not conditionally stipulated to treatment programs, are on probation, parole or have pending litigation when they apply. Since legal pressure increases the probability that applicants will accept treatment (Kleinhans and Harford, 1976), the percentage of nonvolunteers (persons who have an active CJS status) in intact treatment populations may range as high as 80%.

Many treatment programs solicit referrals from the CJS and offer preferential admissions procedures to applicants with legal pressure in the belief that rehabilitation of drug abusers requires the stimulus of strong external pressure. According to this view, legal pressure facilitates treatment to the extent that it provides a motivation to achieve abstinence from drugs. Consequently, addicts without legal pressure are viewed as inferior prospects and often receive closer motivational scrutiny. Their admission to programs characteristically is more difficult than that of nonvolunteers. Not coincidentally, the diminution of criminal activities has become the major goal of most treatment facilities.

Empirical investigations of the relationships between CJS status and treatment efficacy have not yielded unequivocal support for these policies. Most investigators have been concerned with documenting differences in program performance and retention between volunteers and nonvolunteers. Since retention is assumed to be positively correlated with prosocial behavioral changes, positive relationships between CJS status and retention are cited as indications of the beneficial effects of legal pressure. Recent followup studies have contradicted the assumption that long-term client behavioral changes are correlated with the amount of time spent in treatment (Burt, 1977). Consequently, evidence concerning the therapeutic consequences of legal pressure should be based upon its relationships with behavioral outcomes as well as retention. Some investigators have discovered negative relationships between CJS status and retention. Harford, Ungerer and Kinsella (1976), for example, showed that older persons who accepted methadone maintenance as their initial treatment modality while on probation were less likely to be retained than either younger persons on probation or those not on probation. All other CJS statuses were unrelated to retention. They attributed this negative relationship to the inhibiting effects of external motivators like legal pressure on the internal permanent behavioral improvements. They concluded that legal pressure might diminish rather than enhance the probability of successful rehabilitation.

The suggestion that legal pressure may be dysfunctional for treatment prognosis has received indirect support from investigations of program performance. Perpich, Dupont and Brown (1973), among others, reported higher rates of arrest and illicit drug use by CJS referrals. Since the predominance of involuntary clients may adversely affect the treatment received by volunteers, clarification of motivations and past history of criminal activities of the volunteers is an important concern.

This research investigated differences in treatment retention and outcomes in relation to pretreatment CJS status and prior criminal history. The purpose of these analyses was to assess the validity and usefulness of the assumptions that underlie the social control functions of drug abuse treatment programs with respect to criminal behavior. A secondary purpose was to examine the past criminal history and posttreatment illegal activities of the volunteer minority in order to distinguish between internal motivational factors and other sources of external pressure for accepting treatment.

METHOD

1974-75 Treatment Group

The 1974-75 treatment group consisted of 72 persons who entered the methadone maintenance program and 115 who accepted treatment in one of four drug-free modalities of the Drug Dependence Unit (DDU) of the Connecticut Mental Health Center in New Haven. Six measures of pretreatment CJS involvement included probation, parole, and whether the applicant had litigation actively pending in the courts.

Also included were whether the applicant had been referred for evaluation by the CJS, whether acceptance of treatment was stipulated as a condition of probation or parole, and a composite measure of pretreatment CJS status that indicated whether the applicant had at least one of the five other types of CJS involvement. This group was part of a larger investigation of the effects of several classes of client and clinic variables on attrition during referral and treatment that has been reported in detail by Harford and Kleinhans (1977). Seventy-four percent of this group were male, 59% were white, 72% abused opiates primarily, and 64% were positive on the composite measure of CJS status. Their mean age was 25.7 years.

Since outcomes of drug abuse treatment may require a minimum of three years after admission to stabilize, no data concerning treatment outcomes have been collected for this group. Consequently, only the relationships between initial CJS status and retention in treatment were available for analysis.

1968-72 Treatment Group

The 1968-72 treatment group was comprised of a random sample of 321 persons who applied for treatment at the DDU during the period from June 1968 through December 1972 and who subsequently accepted treatment either at the DDU or another facility. The outcome data reported here were collected as part of a long-term followup investigation that was conducted by Gould, Forrest and Kleber (1975). Seventy-nine percent of this group were male, 48.3% were white, 87.2% abused opiates primarily, and 76.6% were positive on the composite measure of CJS status. Their mean age was 22.9 years. The first DDU treatment modality most frequently was one of the three drug-free therapeutic communities (28.3%) followed by methadone maintenance (27.1%) and drug-free outpatient (23.7%). Sixty-seven applicants (20.9%) chose to receive their only treatment during this period in a non-DDU facility.

RESULTS

Retention

The relationships between retention and the six measures of CJS status in the 1974-75 treatment group were assessed by a stagewise multiple regression procedure that is described by Harford and Kleinhans (1977). Since methadone patients differ in several important respects from those in drug-free modalities (Ungerer, Harford, and Kleinhans, 1976) and since mean length of retention was substantially greater in methadone than in the drug-free modalities (311 days vs. 164 days), two separate regression analyses were conducted. Probation and the composite CJS measure yielded statistically significant Beta weights for retention in methadone but not in the drug-free modalities. None of the other CJS measures was significantly related to retention. The negative zero-order correlations for probation (-.32) and the composite (-.14) indicate that volunteers were retained longer than nonvolunteers in methadone maintenance. This result replicated the principal finding of Harford, Ungerer and Kinsella (1976) that was derived from an independent DDU sample. Both of these studies defined retention as the number of consecutive weeks of active participation in the initial treatment modality.

Data concerning retention in the 1968-72 group were available only for the 254 persons who accepted DDU treatment. For these analyses retention was defined alternatively as either the total number of months each person was enrolled in all DDU treatment programs combined or the percentage of time at risk (the period of time intervening between application and followup, excluding time spent incarcerated) that each person was enrolled in the DDU. The relationships among the two measures of retention and the four measures of pretreatment CJS status were assessed by a series of four three-way analyses of variance (ANOVAs) that were performed on both measures of retention. The three independent variables were modality (3 levels), age (2 levels--above and below the median within each modality) and CJS status (2 levels). The latter variable was alternatively probation status, parole status, pending litigation status or the composite index of CJS status. The results for the unadjusted measure of retention showed the expected large modality main effect indicating longer retention in methadone than in the other two modalities which did not differ, and an age main effect such that older clients were retained longer than younger clients, but there were no significant main effects or interactions involving any of the four measures of legal pressure. The results for the risk-adjusted retention measure were similar with two exceptions. (1) The difference in retention between therapeutic community clients and those in the drug-free modality was statistically significant. Therapeutic community clients spent 38% of their time at risk enrolled in treatment while drug-free clients were receiving treatment 27% of the time. (2) Clients who had pending litigation spent a higher percentage of their time at risk in treatment than did those with no pending litigation (47% vs. 33%); $F_{(1,240)} = 6.77$; $p = .010$. This result contradicts Harford, Ungerer and Kinsella (1976).

Initial Difference Related to CJS Status

Of the 321 members of the 1968-72 group, 124 (39.0%) were on probation when they initially applied for treatment, 33 (10.4%) were on parole, and 148 (46.4%) had litigation pending in the courts. The composite measure of legal pressure indicated that the 246 (76.6%) nonvolunteers had at least one of these three types of involvement with the CJS. Comparisons of individuals with and without active

CJS status according to their standing on the composite measure indicated that those lacking legal pressure at the time of admission were significantly more likely to be (1) female (32.0% vs. 17.1%; $x^2_{(1)}$= 6.95; p = .008), (2) employed (38.0% vs. 18.4%; $x^2_{(1)}$= 10.85; p=.001), and (3) not living with parents (59.5% vs. 42.9%; $x^2_{(1)}$=5.57; p=.018). In addition, the volunteers were likely to be (4) earning more money ($30.85 vs. $18.19; $t_{(305)}$=2.04, p=.042), (5) using greater amounts of heroin (2.60 vs. 2.15; $t_{(153)}$=4.42. $p < .001$), and (6) older when first arrested (18.9 years vs. 16.6 years; $t_{(319)}$=4.31; $p < .001$). The 24 pretreatment measures on which the two groups did not differ significantly included age (mean=23.0 years), ethnicity (51.7% non-white), years of schooling (mean= 11.0 years), current school status (10.1% enrolled) and 20 measures of drug use (amount currently used, amount of heaviest use, and years of use for heroin, cocaine, marihuana, barbiturates, amphetamines, psychedelics and other opiates).

Illegal Activities of Volunteers

Arrests prior to admission for treatment were quite prevelant among the volunteers. Although they had no current CJS involvement, 50 of the 75 volunteers had been arrested previously; 12 were first arrested after they had applied for treatment, and the remaining 13 had never been arrested. The 13 included 7 white females, 3 white males and 3 black males.

At the time of the followup interview, nearly half of the volunteers were actively involved with the CJS. Seven (9.3%) were on probation, 4 (5.3%) were on parole; 3 (4.0%) were currently incarcerated, 4(5.3%) had pending litigation; and 2 (2.7%) had outstanding warrants for their arrest. The majority (62.7%) had been arrested at least once after their admission for treatment (mean = 1.4) and 31 (41.3%) had been convicted of at least one postadmission offense (mean = .7). Somewhat fewer, 28 (37.3%), were arrested after being discharged from treatment (mean = .9) and 19 (25.3%) were convicted of postdischarge offenses (mean = .4). The lower rates of arrest and conviction of postdischarge offenses as compared with postadmission rates may be attributable to the differences in time of risk. Thirty-four (45.3%) of the volunteers had been incarcerated at least once after being admitted for treatment (mean = 4 months). The composite measure of posttreatment CJS status indicates that 17 (22.7%) were on probation, on parole or had pending litigation, the equivalent legal pressure category that had defined them as volunteers according to preadmission CJS status. The change in CJS status was statistically significant; $x^2_{(1)}$= 3.85; p = .05. The composite measure of postadmission illegal activities showed that a total of 36 (48.0%) volunteers either had been convicted of a postadmission offense or admitted recently engaging in illegal activities. These results demonstrate that persons who have no apparent legal motivations to seek treatment for addiction and are normally categorized as volunteers frequently are involved in criminal activities after treatment has been initiated.

CJS Status and Treatment Outcomes

Statistically significant posttreatment differences between volunteers and nonvolunteers (as defined by pretreatment CJS status) were evident in employment and several measures of criminal activity but not in use of opiates, nonopiates or marihuana. Volunteers, who were more likely to be employed prior to treatment, also had higher employment rates than nonvolunteers after treatment (73.3% vs. 59.8%; $x^2_{(1)} = 3.98$; $p = .046$). Similarly, the rates of posttreatment criminal activity favored volunteers; probation (9.3% vs. 24.4%; $x^2_{(1)} = 4.36$; $p = .008$), pending litigation (5.3% vs. 19.1%; $x^2_{(1)} = 7.16$; $p = .007$), currently incarcerated (4.0% vs. 13.0%; $x^2_{(1)} = 3.92$; $p = .048$), postadmission incarceration (45.3% vs. 61.0%; $x^2_{(1)} = 5.13$; $p=.024$), postdischarge incarceration (25.3% vs. 39.4%; $x^2_{(1)} = 4.36$; $p = .037$), CJS status and composite criminal activities (48.0% vs. 77.6%; $x^2_{(1)} = 22.98$; $p < .001$). The lower rates of criminal activities by the volunteers occurred despite the fact that their mean time at risk (3.7 years) was significantly greater than the 3.3 years for the nonvolunteers ($t_{(157)} = 2.57$; $p = .011$). The decrease in positive CJS status by the volunteers from 100% to 54.9% was statistically significant, $x^2_{(1)} = 72.99$; $p < .001$. The two groups did not differ with respect to heroin use, marihuana use, nonopiate use, income, longest period of unaided abstinence, retention in either methadone maintenance or therapeutic communities or parole status. These results indicate that applicants for treatment of drug abuse who were involved with the CJS were more likely than volunteers to have difficulties related to continued illegal activities after treatment even though the groups did not differ with respect to posttreatment drug use.

Substantial pretreatment to posttreatment reductions in rates of CJS involvement occurred with respect to probation (39.0% vs 20.9%; $x^2_{(1)} = 5.45$; $p = .015$), pending litigation (46.4% vs. 16.0%; $x^2_{(1)} = 30.27$; $p < .001$), and the composite measure (76.6% vs. 47.4%; $x^2_{(1)} = 23.02$; $p < .001$), but not for parole (10.4% vs. 10.3%). Although the results suggest diminished CJS involvement, the posttreatment rates are not strictly comparable to the pretreatment rates since drug abusers seek admission to programs precisely at the time of maximum involvement with CJS. Results from a comparison group of 36 applicants who did not accept treatment showed a statistically significant decline in the percentage with pending litigation (38.9% vs. 13.9%; $x^2_{(1)} = 5.82$; $p = .013$). Reductions in the percentage on probation (38.9% vs. 22.2%) and with composite CJS involvement (66.7% vs. 50.0%) were not statistically significant. These results from the untreated comparison group indicate that the pretreatment to posttreatment reductions in CJS activity cannot unequivocally be attributed to the effects of treatment for addiction but rather could be the result of unrelated factors such as maturation, statistical regression, criterion contamination, or the addicts' becoming more skillful in avoiding detection by CJS agents.

Since volunteers differed from nonvolunteers on six pretreatment measures, differences in posttreatment outcomes might be attributable to preexisting differences in these variables. A stepwise discriminant analysis was performed according to a hierarchical model that statistically equated the volunteers and nonvolunteers for these pretreatment differences and assessed the importance of remaining differences in the eight significant outcome variables. The results of this analysis showed that three pretreatment variables (sex, current heroin use and age when first arrested) and three outcome variables (postadmission incarceration, composite CJS status, and composite illegal activities) comprised the final discriminant function which accounted for 18.1% of the criterion variance. The absolute values of the respective standardized discriminant weights (.01, .43, .41, -.09, -.32, -.40) indicate that although sex and postadmission incarceration made significant contributions to the discriminant function, they were not as important as the other four variables in distinguishing volunteers from nonvolunteers. This analysis demonstrates that higher posttreatment rates of involvement with the CJS and criminal activity can be expected from nonvolunteers even when preexisting differences between the groups are minimized by a conservative statistical procedure.

To assess the relationship between retention and posttreatment CJS status eight t-tests were performed, one for each combination of two measures of retention and the four CJS statuses. The only significant relationship showed that persons who were positive on the composite had significantly longer retentions adjusted for risk, $t_{(254)} = 2.32$; $p = .021$. A similar set of eight analyses that statistically adjusted for differences in pretreatment CJS status using a stepwise discriminant procedure showed only one significant relationship. Persons who were positive on the composite were retained for shorter periods of time according to the unadjusted measure of retention, $F_{(1,251)} = 3.89$; $p = .041$.

To assess the pretreatment correlates and predictiveness of the posttreatment illegal activities index, a stepwise multiple regression analysis was performed that initially included all pretreatment variables. Four variables (in order of entry: sex, ethnicity, years of heroin addiction and pretreatment composite CJS status) yielded a multiple correlation of .41 accounting for 16.8% of the criterion variance. Lower rates of posttreatment criminal activities were associated with being female, white, fewer years of heroin addiction and volunteer status. The correlation between pretreatment CJS status and the posttreatment illegal activities composite was .28 ($p < .001$).

DISCUSSION

Relation of CJS Status and Retention

Does legal pressure facilitate retention in drug abuse programs? The answers to this question require a set of complex qualifications and, moreover, may be totally irrelevant to the more pertinent question concerning the effects of legal pressure on treatment efficacy. If retention is defined as the duration of program participation during the initial treatment episode, CJS status is either uncorrelated, or in the cases of probation in methadone programs, negatively correlated with retention. This finding is consistent with IBR report (Simpson et al., 1977) of nonsignificant correlations between a 4-point composite criminal history index and retention in methadone maintenance, TC, and detoxification only modalities. Simpson et al. also found a negative correlation (-.13) between pretreatment criminality and retention in drug-free modalities but did not report its significance. In this study when retention was defined as the total amount of time spent in treatment aggregated across all treatment episodes, correlations with CJS status again were not significant. When retention was adjusted for time at risk, however, the relationship with pending litigation was positive. Since drug abusers who initially enter treatment with pending litigation are more likely to be imprisoned, a relatively greater proportion of their time at risk is spent in treatment.

How is retention related to posttreatment CJS status? Depending upon whether statistical adjustments were made for time at risk and pretreatment CJS status, retention was either positively or negatively related to composite posttreatment CJS status and unrelated to probation, parole or pending litigation. Simpson et al. also reported negative correlations between unadjusted retention and posttreatment incarceration status in therapeutic communities (-.37) and outpatient drug-free modalities (-.42). Retention measures must be interpreted extremely cautiously since they are particularly susceptible to a source of bias known as criterion contamination. If drug abusers continue their drug use or illegal behavior they are likely to be discharged by treatment authorities or incarcerated by CJS agents. In either case, the negative relationships between retention and the outcome criteria are artificially enhanced so that those who stay longer, on the average, have better outcomes. These relationships may have nothing to do with any benefits conferred by treatment. Evidence for this phenomenon was found in this study. The expected negative relationship between CJS status and retention in treatment was obtained, but the direction was reversed when retention scores were adjusted for time incarcerated. Since CJS and treatment authorities partially control how long clients remain in treatment, the potential for criterion contamination is so great that use of retention as an index in an evaluative context has almost invariably produced misleading and erroneous conclusions concerning the efficacy of treatment.

Relation of CJS Status to Outcomes

How is pretreatment CJS status related to outcomes? There were substantial pretreatment to posttreatment improvements in employment, drug use, and CJS status. Although volunteers were using more heroin before treatment, the posttreatment difference between the two groups became negligible, and volunteers maintained their pretreatment advantage in employment. Despite their significant increase in CJS involvement and the significant decrease in the nonvolunteers, only 22.7% of the volunteers were on parole, probation or had pending litigation as compared with 54.9% of the nonvolunteers. These results are consistent with the IBR summary: "the most consistent result was that low preDARP criminality . . . . was related to more favorable postDARP outcomes" (Simpson et al., 1977).

There was no evidence that volunteers increased their criminal activity as a result of treatment, nor is treatment to be credited with causing decreases in criminal activity of nonvolunteers. Drug abusers are either self-selected or selected by the CJS for treatment primarily on the basis of legal pressure and only secondarily because of drug use. When CJS status or any measure of illegal activity which is correlated with the selection criterion is also used as an outcome variable, statistical regression effects that would cause decreased criminality of nonvolunteers and increased criminality of volunteers are to be expected. Since the nonvolunteers predominate in treatment populations of drug abusers, the overall effect is an inevitable pretreament to posttreatment decrease in all measures correlated with criminal behavior. Maturation effects might also account for some of the decreases in undesirable outcomes. The fact that changes of similar magnitudes and directions occurred in the untreated comparison group illustrates the crucial fact that statistical regression and maturation occur equally in treated and untreated groups. For these reasons, general decreases in undesirable behavior by members of treatment groups cannot be attributed to treatment without adequate demonstrations that changes of similar magnitude do not occur in untreated comparison groups. It is precisely on this point that most evaluations of drug abuse treatment programs have failed to demonstrate unequivocal treatment benefits.

Volunteers had substantially better outcomes with respect to criminal behavior than nonvolunteers even after pretreatment differences between the two groups were removed by statistical procedures. However, even the most advanced statistical technology is an inadequate substitute for random assignment in assuring the pretreatment equivalence of groups. The volunteers and nonvolunteers in this study may well have differed on important but unmeasured variables that would have accounted for their differences in posttreament performance. The causal reationships between CJS status and prosocial behavior could be assessed more validly either by assigning stipulations at random to qualified candidates or by randomly removing the stipulations once the clients have accepted treatment. The fact that volunteers and nonvolunteers differed significantly with respect to criminal behavior but not drug use suggests that traditional explanations for criminal behavior of drug abusers may be quite inaccurate. The criminal activities index was 60% higher for nonvolunteers (77.6%) than for volunteers (48.0%). Since posttreatment drug use levels of the two groups were indistinguishable, the difference in their criminal activities provides a rough estimate of the proportion of criminal behavior of drug abusers that is not attributable to the need for drugs.

This research documents many of the problems involved in reaching definitive conclusions concerning outcomes of drug abuse treatment. Adequate comparison groups, preferably achieved by random assignment wherever possible, are essential for attributing causal relationships between treatment and outcomes. Even within these limitations, this study strongly suggests that volunteers make more significant improvements in drug abuse treatment than nonvolunteers. Their advantages might be even greater if clinical programs were structured specifically to meet the needs of the volunteers rather than the demands of the majority of nonvolunteers. The results showed that 16% of the volunteers were arrested for the first time after accepting treatment. It is conceivable that the predominant influence of nonvolunteers in delimiting the treatment processes contributed to the increased CJS involvement of the volunteers. Separate programs for the two groups that clinically account for their differences in motivation might enhance treatment prognosis for both. If criminal behavior and drug abuse are more independent than the simple causal model implies, then treatment in lieu of incarceration for nondrug-related offenses may be ill-advised. One way of screening nonvolunteers for treatment motivation would be to punish drug abusers in the appropriate CJS agency for their criminal behavior and curtail or eliminate CJS stipulations to treatment as an alternative to incarceration, encouraging them to volunteer for treatment after serving their sentences.

This research was supported in part by grants DA 01077 and DA 16356 from the National Institute on Drug Abuse, Herbert D. Kleber, M.D., Principal Ivestigator.

REFERENCES

Burt Associates, Inc., 1977. Drug treatment in New York City and Washington, D.C.: Followup studies. DHEW Publication No. (ADM)

77-506. Services Research Monograph Series, Rockville, MD: National Clearinghouse on Drug Abuse Information.

Gould, L.C., Forrest,C.K. and Kleber, H.D. 1975. Five-year followup of methadone maintenance patients. Unpublished manuscript. Yale Medical School.

Harford, R.J. and Kleinhans, B. 1977. Correlates of attrition during referral and treatment for drug dependence. Unpublished manuscript. Yale Medical School.

Harford, R.J., Ungerer, J.C. and Kinsella, J.K. 1976. Effects of legal pressure on prognosis for treatment of drug dependence. *Am. J. Psychiat.* 133(12): 1399-1404.

Kleinhans, B. and Harford, R.J. 1976. Predicting pretreatment attrition in drug abuse programs. In J.H. Lowinson (ed.): *Proceedings of the Third National Drug Abuse Conference.* New York: Marcel Dekker (in press).

Perpich, J., Dupont, R.L. and Brown, B.S. 1973. Criminal justice and voluntary patients in treatment for heroin addiction. In *Proceedings, Fifth National Conference on Methadone Treatment.* New York: NAPAN.

Simpson, D.D., Savage, L.J., Lloyd, M.R. and Sells, S.B. 1977. Evaluation of drug abuse treatments based on the first year after DARP. (IBR Report 77-14) Fort Worth: Texas Christian University, Institute of Behavioral Research.

Ungerer, J.C., Harford, R.J. and Kleinhans, B. 1976. Interpersonal trust and heroin abuse. In J.H. Lowinson (ed.): *Proceedings of the Third National Drug Abuse Conference.* New York: Marcel Dekker (in press).

Wieland, W.F. and Novack, J.L. 1973. A comparison of criminal justice and non-criminal justice related patients in a methadone treatment program. In *Proceedings, Fifth National Conference on Methadone Treatment.* New York: NAPAN.

SOME LIKELY ALTERNATIVE EXPLANATIONS FOR IMPROVEMENTS DURING DRUG ABUSE TREATMENT: MATURATION, STATISTICAL REGRESSION, HISTORY, CRITERION CONTAMINATION

Robert J. Harford, Ph.D.

Yale University School of Medicine

ASSESSING THE EFFECTS OF DRUG ABUSE TREATMENT

The intensive funding of drug abuse treatment programs is predicated on the assumption that treatment processes are effective in reducing criminal behavior, drug use and other asocial behaviors of drug abusers. Evaluations of the extent to which these goals are accomplished usually proceed by way of comparing well measured indices of pretreatment to post-treatment changes in levels of drug usage, criminal activities, and employment. Somewhat less frequently changes in other variables such as psychosocial adjustment, personality characteristics, and utilization of institutions of social support also are investigated. These data are then analyzed by a variety of methods ranging from straightforward univariate comparisons of pretreatment and posttreatment levels to technologically more advanced multivariate techniques such as analysis of covariance and stagewise multiple regression and correlation, which are used primarily for comparing effectiveness of different modalities of treatment. These latter multivariate techniques statistically equate members of diverse groups for significant preexisting differences in treatment modalities as rival explanations for the observed changes in the outcome measures. One undesirable consequence of the ease with which such multivariate analyses can be performed is that investigators of treatment outcomes have eschewed the use of untreated control groups in favor of statistical procedures for assessing improvements that use intact treatment groups as their own controls. The resulting absence of untreated comparison groups in the evaluation designs severely limits the basis for crediting treatment with a causal influence on these improvements.

The results of drug abuse treatment evaluations have sustained

Federal officials and treatment authorities in their conviction that these efforts have been beneficial and perhaps even cost-effective (Rufener, Rachal and Cruze, 1976). In general, these studies have credited treatment with improving employment and effecting profound reductions in levels of drug use and criminal behavior. The fact that addicted treatment populations sustain improved prosocial functioning beyond the immediate postdischarge period is now beyond dispute. What is still unknown is the extent to which these gains can be attributed to the influence of treatment processes per se. The skeptic's alternative hypothesis (which is seldom examined) holds that the variables responsible for these improvements are extraneous to the treatment process. The actual relative contributions of treatment and incidental factors to improvements in the outcome measure undoubtedly lie somewhere between these two extremes. The most important task now facing those responsible for evaluating treatment effectiveness and formulating social policy is to determine the magnitude of treatment effects when the potential biasing effects of extraneous influences are eliminated or controlled as completely as possible. Although arcane and relatively minor reservations can be raised concerning the findings of previous outcome studies with respect to such issues as reliability and validity of measures, the use of unverified program or self-report data, and inappropriateness of statistical procedures, the main causes of unjustified conclusions concerning the effects of treatment lie in the failure to account for errors that are inherent in the pretest-posttest research design. The main purposes of this paper are to (1) describe several probable sources of variation in the outcome measures that erroneously have been attributed to treatment effects, (2) iterate some of the reasons treatment effects cannot adequately account for all of the improved prosocial functioning of drug addicts on the basis of internal evidence contained in previous outcome research, and (3) derive a crude estimate of the relative magnitude of some of these extraneous factors based upon previously unpublished results of a followup study that included an untreated comparison group.

The plausible rival hypotheses of Campbell and Stanley (1963) are nowhere more prevalent than in the area of treatment evaluation. Rightly or wrongly therapists have resisted on ethical grounds random assignment of clients to no-treatment control groups. Even when ethical objections are surmounted, clients can be expected to vote with their feet and leave no-treatment groups producing unacceptably high attrition rates that jeopardize the validity of the findings (Bale et al., 1976). Without properly constituted control groups several major sources of error identified and defined by Campbell and Stanley are indistinguishable from treatment effects. The most potent of these errors are maturation, statistical regression, and history.

Maturation technically refers to the tendency of subjects to change systematically with the passage of time between pretreatment and posttreatment measurements. Such changes are invariably confounded with the effects of treatment, and since treatment of drug abuse requires such long periods of time during a phase of life when prosocial changes occur most rapidly, the possibilities for maturation effects are quite real. Most drug abusers reach the attention of treatment authorities at the peak of their asocial acitivities. Criminality and unemployment in normal populations attain their highest levels during late adolescence and decline quickly to near zero levels by the mid-twenties. Similar trends have been observed for drug abuse, although the use of opiates may peak a few years later than unemployment and criminality curves. Maturation in this context refers to the totality of biological and social processes that combine to diminish deviant behavior during young adulthood. It is to be distinguished from the maturing out process experienced by heroin addicts (Winick, 1962) in that it generally occurs at a much earlier age, requires less time and in principal is unrelated to weariness caused by adversities associated with long term heroin addiction. Instead maturation in this population has many similarities to the colloquial connotation of the term--growing up. For some clients certain aspects of the process may be delayed, and for a few it may never happen, but for the majority of contemporary drug abusers, maturation is an inevitable process that is characterized by an acceleration of prosocial changes that occur regardless of whether or not they are enrolled in drug treatment facilities.

If there is a random tide that flows through human events, then drug abusers arrive for treatment at the lowest ebb. Perhaps 80% have had recent encounters with the criminal justice system (CJS) and many have lost their jobs in the process. Their drug use is likely to be at its highest levels, and family pressures may be impelling them to seek changes. *Statistical regression*, unlike maturation, does not hypothesize the sort of permanent improvements in drug use, employment and criminal behavior associated with spontaneous remission. Technically it refers to results of measurement error that is compounded by the assignment of individuals to groups on the basis of their standing on criteria which are correlated with outcome variables. If the outcome measures are unreliable (attributable to chance), changes in status away from the original state are inevitable. The majority of clients seek treatment because of exceptional legal pressure, temporary unemployment and excessive drug use. To the extent that self-reports, urine tests and CJS status are unreliable measures of the underlying outcome variables, statistical regression effects will occur that produce significant changes in status. This principle implies that, when outcome measures are administered, many of the applicants who were referred for treatment because of legal pressure with a related problem of unemployment will show decreases in these variables regardless of the true level of their posttreatment criminal behavior. Those who were not referred because of CJS involvement will show increases in these variables for the same reasons. Since clients with legal pressure outnumber those without it by four to one, the overall trend is an apparent decrease in CJS involvement and unemployment that occurs equally in treated and untreated samples. According to the laws of probability, bad luck gets better and good luck gets worse. To the extent that luck determines involvement with the CJS, drug abusers, as a group, will evince apparent improvements with regard to CJS status, even if there are no actual changes in their criminal behavior.

History is Campbell and Stanley's term for events that intervene between pretreatment and followup in addition to those associated with treatment. For example, decreases in quality and availability of heroin may cause reduced heroin use independent of treatment effects. The potential sources of history effects in drug abuse treatment are so numerous as to preclude identifying and measuring their separate effects.

Since maturation, statistical regression and history effects would be equally likely to occur in randomly constituted no-treatment control groups as they would in treatment populations, only improvements in treatment outcomes over and above those found in the control group can be attributed to the effects of treatment. In this scene, the control group eliminates the effects of these and several other less likely sources of error. The use of control groups cannot, however, eliminate the possibility of *criterion contamination*, which technically refers to systematic attrition from treatment groups on the basis of outcome criteria or variables correlated with outcome criteria. Widespread clinical practices commonly result in several types of criterion contamination. Some programs expel persons who use drugs excessively and then point to diminished mean rates of drug use among the remaining members as an indication of treatment efficacy. Employment or other prosocial activities routinely are required for advancements in incarceration which improves the average legal standing of the remaining members. Since criterion contamination refers to systematic attrition in the treatment group, it can be avoided by including all persons who received pretests in statistical comparisons with the members of control groups and refraining from statistical adjustments for during-treatment variables such as retention that are particularly susceptible to criterion contamination.

A study by MACRO Systems, Inc. (1975) reported two findings that implicate extraneous factors as causes of improvements: (1) retention in treatment was uncorrelated with changes in drug use, criminal activity, and employment and (2) outcomes did not differ in the three treatment modalities (i.e., methadone maintenance (MM), therapeutic communities (TC), and ambulatory drug-free). Although the report notes in passing "the healing effects of time," it concludes that *availability* of treatment has a positive impact on outcomes irrespective of the type of modality or amount of treatment received.

Since the two findings in conjunction imply that duration and specific activities in treatment are irrelevant to outcomes, the conclusion asserting the importance of treatment availability, in the absence of results from untreated drug abusers, is at best, unsupported by the data.

Like the MACRO study, a study by Burt Associates (1976) found that improvements were unrelated to the type and amount of treatment received or the number of treatment episodes. In a comparative analysis of the Washington and New York data Burt also concluded that successful outcomes are unrelated to demographic and background factors. The authors speculated that perhaps the client is the best judge of the most appropriate modality and how much treatment is enough. A more likely possibility not considered by these studies is that improvements were entirely unrelated to differences between clients or treatment interventions. Burt nominated changes in community attitudes and availability, purity and price of heroin as possible extraneous causes for the improvements. These history effects undoubtedly contribute to the undifferentiated improvements shown, but other such extraneous forces including maturation and statistical regression, whose influence cannot be assessed without an untreated comparison group, also may be important.

In the most thorough outcome evaluation yet undertaken, the Institute of Behavioral Research (IBR) has included a group of clients who completed only the intake and admission phase for use as a no-DARP treatment comparison group. The representativeness of these individuals for untreated DARP drug abusers is questionable since they comprised only 7% of the followup sample (Simpson et al., 1977) as compared with 14% of the total DARP population (Simpson and Joe, 1976), and the criteria used to include them in the DARP varied from agency to agency. Smart (1977) has argued that persons in this group may be less motivated and therefore more difficult to treat than those who received treatment. Since 53% of the intake-only (IO) group had received treatment before entering DARP and 41% were subsequently treated during the first year after DARP but before the followup interview, it cannot be considered as an untreated group. In fact, it appears as if the IO group spent more time in treatment prior to the followup interview than did members of all other modalities except methadone maintenance, yet it had the poorest outcomes.

The IBR study contradicts the MACRO and Burt studies in reporting small but statistically significant preDARP to postDARP improvements in several outcome measures that were related to the type of modality. In deriving this finding, however, the univariate analyses failed to make appropriate statistical adjustments in the change scores for pretreatment differences among the different modalities with respect to client characteristics. Similarly, the analyses of covariance (ANCOVAs) which compared treatment differences in outcome measures adjusted for pretreatment differences in client characteristics but failed to enter the pretreatment levels of the outcome variables as covariates. In addition, the multiple regression analyses were conducted separately within each modality and therefore provide no basis for comparing the modalities with respect to their differing effects on outcomes. The magnitude of treatment differences in the covariance analyses was not reported, but the F values in six out of eight measures were less than those for the covariates which were described as "not large . . . accounting for only 2-9% of the variance of the individual criteria" (Simpson et al., 1977). However, most of the differences among the five treatment groups resulted from the poorer outcomes of members of the 10 and outpatient detox categories, groups that are not examined in the Burt and MACRO studies. These groups are often default categories that were used at the discretion of the programs to classify prospective clients who refused treatment in other modalities. Consequently, they tend to overrepresent recalcitrant clients and those less motivated toward prosocial behaviors.

The IBR report also contradicted the previous followup studies with respect to the relationship between retention and treatment outcomes. Within each of the four treatment classifications DARP retention was positively related to improvements in 16 out of the 17 significant partial correlations with the ten outcome measures. Positive correlations were obtained in 6 of the 12 possible instances involving retention and postDARP opiate use, employment and jail status. In past evaluations such correlations have been interpreted as de facto indications of treatment benefits, and retention often has been considered as an outcome measure. In actuality, the relationship between retention and outcomes may be too confounded to permit valid interpretation. Such correlations are necessary but not sufficient evidence that improvements in outcomes (within certain limits) are attributable to the effects of treatment. Since such correlations can be produced artifactually by the process of criterion contamination, it is necessary to demonstrate that the higher retention rates of clients with good outcomes is not the result of selective elimination of high-risk clients by treatment agencies or the CJS on the basis of their behavior as indicated by the outcome measures. When retention is the main variable of interest, persons who terminate treatment because of incarceration or decision by the program should have their retention scores adjusted by an appropriate risk factor. Since this procedure was not followed by the IBR, we must reserve judgement concerning the implications of the positive correlations as evidence for the benefits of treatment.

This report reanalyzes outcome data collected by Gould, Forrest and Kleber (1975) as part of a long-term followup investigation. It compares the amount of improvement with respect to CJS status, opiate use and employment that was experienced by 323 members of four treatment groups and 40 members of a comparison group who, at the time of the followup interview, had received no treatment for drug abuse whatsoever. The use of the control group allowed a demonstration of the magnitude of improvements to be expected in the absence of treatment. The analyses also illustrate the phenomenon of criterion contamination.

METHOD

The group of study patients was comprised of a random sample of 363 persons who applied for treatment at the Drug Dependence Unit (DDU) of the Connecticut Mental Health Center in New Haven during the period from June, 1968 through December, 1972. The majority accepted treatment in one of three DDU treatment types: MM 88 (24.2%), TC 92 (25.3%), and outpatient (OP) 76 (20.9%) while 67 others (18.5%) did not enter the DDU but subsequently accepted treatment in a non-DDU modality. Forty applicants (11.0%) refused treatment and had received no drug abuse treatment at the time of the followup interviews which were conducted during the 12-month period ending June, 1975. Males comprised 79.6% of the population, 47.7% were black and 1.9% Hispanic, 86% abused opiates primarily, 71.5% were using heroin when they applied, 75.6% were positive on a composite measure of CJS status, and 77% were unemployed. Their mean age at referral was 22.9 years.

RESULTS

Univariate analyses of variance (ANOVAs) and chi-square tests were conducted on 32 pretreatment variables (including 21 measures of current and past drug use) to identify the variables on which the five treatment groups differed significantly. The resulting set of eight variables was reduced to six by a discriminant analysis which deleted variables that did not make independent contributions to the discriminative criterion. A plot of the group centroids indicated that those who entered methadone differed from those who entered all other groups primarily with respect to age, but the other four groups did not differ significantly among themselves. The reduced set of variables including age, total years of heroin addiction, length of time between application and followup, vocational status and family status were then entered simultaneously as in a series of 3-way ANCOVAs that were performed on each of the outcome measures (CJS status, heroin use, employment). The three independent variables were sex, ethnicity (white and nonwhite) and modality (5 levels). For each ANCOVA the pretreatment level of the particular outcome variable and all other pretreatement variables that were significantly correlated with the outcome variable also were included among the set of covariates. The ANCOVA model was hierarchical in the sense that it removed differences in the criterion associated with all of the covariates before assessing the degree of relationship of the independent variables with the remaining criterion variance. Only the covariates that were significantly associated with the criterion were retained in each of the final analyses.

With respect to posttreatment heroin use, none of the independent variables was significantly related to the criterion and only one covariate, years of heroin addiction (r=.14), was a significant source of variation. With respect to employment, there were no main effects and only pretreatment employment status emerged as a significant covariate (r=.24). When posttreatment CJS status was analyzed, a significant sex main effect emerged which indicated that females were less likely than males $F_{(1,327)}=6.20$, $p=.013$ to engage in illegal activities. Significant covariates included years of heroin addiction (r=.15) and the composite measure of pretreatment CJS status (r=.24). These results indicate that when the sexes were statistically equated for differences in years of heroin addiction and pretreatment CJS involvement, females were less likely than their male counterparts to be on probation, parole or have pending litigation.

Since no program main effects on the three outcome measures were found, a second series of ANCOVAs was performed that compared the no-treatment group with the combined members of the four treatment groups. The principal result was that most of the covariates were deleted because methadone clients were no longer distinguishable as a separate group. No significant differences between treated and untreated groups were found.

Pretreatment to posttreatment changes in heroin use, employment and CJS status were examined separately in the combined and the no-treatment groups. For clients who received treatment (1) unemployment levels increased from 63.3% to 28.8%; $x^2_{(1)}=50.24$, $p<.001$, but the increases did not depend upon initial employment status, (2) heroin use decreased from 71.4% to 13.2%, $x^2_{(1)}=106.47$, $p<.001$, and posttreatment heroin use was equally likely for pretreatment users and nonusers and (3) CJS involvement decreased from 76.6% to 47.4%; $x^2_{(1)}=26.94$, $p<.001$ and pretreatment CJS involvment was associated with higher rates of posttreatment CJS involvement (54.9% vs. 22.7% for those with no initial CJS status). For clients not receiving treatment the direction of these changes was the same. Unemployment levels decreased from 60.0% to 20.0%; $x^2_{(1)}=5.63$, $p<.05$; heroin use decreased from 67.5% to 20.0%; $x^2_{(1)} = 8.01$, $p<.01$, and CJS involvement decreased nonsignificantly from 67.7% to 50.0%. For each of these outcome variables the posttreatment levels were unrelated to pretreatment levels in the no-treatment group.

The improvements in unemployment and CJS status provided evidence of statistical regression effects in both the treated and untreated groups. The percentage of clients with pretreatment CJS status who were clear at posttreatment was 45.1% and 45.8% respectively. The percentage of clients without pretreatment CJS status who subsequently became involved was 22.7% and 41.7% in the treated and untreated groups. The percentage of initially unemployed clients who subsequently were employed was 56.3% and 62.1% in the untreated and treated groups while the percentage of those who became unemployed were 25.0% and 32.4%, respectively. Since the rates of gains and decrements in these outcome measures were not significantly different in the treated and untreated clients, treatment is not to be implicated as a causal factor. The size of the discrepancy between gains and decrements can be considered as an inverse indication of the importance of regression effects in determining results. For heroin use the discrepancies were large, on the order of 15% vs. 85%, suggesting that regression contributed more to apparent improvements in unemployment and CJS activity than in heroin use. This analysis suggests that (1) of the 29% decrement in CJS involvement perhaps as much as 23% could be attributed to statistical regression and (2) of the 35% gain in employment as much as 32% might result from regression. This conclusion depends on the possibly unwarranted assumption that treatment does no harm since it uses the percentage of persons who apparently got worse as the estimate for the upper limit of the amount of regression. More accurate estimates are possible when the reliability of the outcome measures are known.

To assess the relationships between pretreatment status, retention and outcomes, six discriminant analyses were performed for the combination of the two measures of retention and CJS status, unemployment and heroin use. Retention was defined as either (1) the total number of months of treatment received at the DDU between the date of application and followup or (2) the proportion of time spent in treatment less the proportion of time incarcerated. These analyses were restricted to DDU patents. The results showed that, after adjustments for differences in initial status on the outcome variables, both measures of retention were significantly related to post-treatment CJS status and unemployment, but neither was related to heroin use. However, the relationships were positive for the raw retention measure and negative for the adjusted measure (for raw retention--positive CJS mean=12.2 months, no-CJS mean=15.3 months; $F_{(1,253)}=4.17$; $p=.041$; employed mean=14.8 months, unemployed mean=12.2 months; $F_{(1,244)}=4.65$; $p=.031$ and for adjusted retention--positive CJS mean= 54.9%; no-CJS mean=30.3%; $F_{(1,253)}=3.25$; $p=0.62$; employed mean=35.6%, unemployed mean=46.4%; $F_{(1,253)}=3.30$; $p=.068$). These results show that when adjustments were made for retention of persons removed from risk of treatment by the CJS, the amount of treatment received was negatively rather than positively related with outcome. This finding suggests that the positive correlations between retention and outcomes are more likely to be indications of criterion contamination rather than benefits of treatment.

DISCUSSION

The results of this study confirmed previous followup investigations in showing sharply reduced heroin use, CJS involvement and unemployment during the interval between application for treatment and followup. As in the MACRO and Burt studies, differences among modalities in outcomes were not significant. The results also showed that these improvements were shared equally by drug abusers who were untreated. Preexisting differences between the treated and untreated groups were minimal and uncorrelated with outcomes. Thus, there was no evidence that untreated drug abusers had equally good outcomes because they were less severely addicted. Instead, it seems probable that most of the major causes for improvements are independent of any benefits derived from treatment. Some hypotheses have been advanced concerning the reasons for the general improvement which invoke the concepts of maturation, history and statistical regression, and an internal analysis showed that regression could account for most of the improvement in employment and CJS involvment. No-treatment comparison groups are indispensable for controlling maturation and history while random assignment to treatment conditions is the surest remedy for regression.

The analyses of the relationships between adjusted and raw retention scores with the outcome criteria indicated that criterion contamination is likely to bias estimates of treatment effectiveness based on retention. For this reason, use of raw retention scores either as indications of efficacy or as predictors of outcomes should be avoided. The negative relationships between outcomes and adjusted retention scores are a further indication that treatment received in general is not a value-added process in that the amount of treatment may be unrelated to improvements.

The results indicate that the apparently large benefits of treatment are illusory when compared against gains made by untreated drug abusers. This conclusion must be tempered by the knowledge that selection to the no-treatment group was not random, and the groups may have differed in a crucial unmeasured variable that accounts for gains of the untreated group. This unlikely possibility should be investigated in a large-scale evaluation including treatment assignments on a random basis. The ethical question has now become whether drug abusers should be obliged to submit to demanding regimens which are of questionable value. The results also raise doubts concerning the benefit side of the cost-effectiveness equation. If improvements shown by untreated addicts approach the gains made by those in treatment, the marginal benefits of treatment necessarily are lower than have been estimated heretofore. For these reasons research that compares the importance of treatment and extraneous factors using representative groups of untreated addicts should be given highest priority.

ACKNOWLEDGEMENT

This research was supported in part by grants DA 01077 and DA 16356 from the National Institute on Drug Abuse, Herbert D. Kleber, M.D., Principal Investigator.

REFERENCES

Bale, R.N., VanStone, W.W., Kuldau. J.M., Engelsing, T.M.J. and Zarcone. V.P. 1976. Therapeutic communities versus methadone treatment: Further preliminary results from a randomized comparison. In J.H. Lowinson (ed.): *Proceedings of the Third National Drug Abuse Conference*. New York: Marcel Dekker (in press).

Burt, M.R. 1976. A followup study of former clients at the District of Columbia's Narcotics Treatment Administration. NIDA report, Bethesda, MD: Burt Associates, Inc.

Burt, M.R. 1977. Drug treatment in New York City and Washington, D.C.: Followup studies. DHEW Publication No. ADM (77-506). Services Research Monograph Series, Rockville, MD. National Clearinghouse on Drug Abuse Information.

Campbell, D.T. and Stanley, J.C. 1963. *Experimental and Quasi-experimental Designs for Research.* Chicago: Rand McNally.

Gould, L.C., Forrest, C.K. and Kleber, H.D. 1975. Five-year follow-up of methadone maintenance patients. Unpublished manuscript, Yale University School of Medicine.

MACRO Systems, Inc. 1975. Three year follow-up study of clients enrolled in treatment programs in New York City. NIDA report. Silver Springs, MD: MACRO Systems, Inc.

Rufener, B.L., Rachal, J.V. and Cruze, A.M. 1976. A management effectiveness measure for NIDA drug abuse treatment programs. Vol. 1: Cost benefit analysis. NIDA report. Research Triangle Park, NC: Research Triangle Institute.

Simpson, D,D, and Joe, G.W. 1976. Research design and sample selection for the followup study of 1969-71 DARP admissions. Special IBR Report. Fort Worth: Texas Christian University, Institute of Behavioral Research.

Simpson, D.D., Savage, L.J., Lloyd, M.R. and Sells, S.B. 1977. Evaluation of drug abuse treatments based on the first year after DARP. IBR report 77-14. Fort Worth: Texas Christian University, Institute of Behavioral Research.

Smart, R.G. Comment on Sells' paper. In J.D. Rittenhouse (ed.): *The Epidemiology of Heroin and Other Narcotics.* Research Monograph 16, Rockville, MD: NIDA

Winick, C. 1962. Maturing out of narcotic addiction. *U.N. Bulletin on Narcotics*. 15(L): 1-7.

THE RELATIONSHIP BETWEEN PROGRAM-RELATED CHARACTERISTICS AND PRETREATMENT AND POSTADMISSION ATTRITION IN METHADONE MAINTENANCE AND DRUG-FREE MODALITIES*

Robert J. Harford, Ph.D.

Yale Universiy School of Medicine

Two chronic problems confronting programs for rehabilitating drug dependent individuals have been failure of applicants to accept treatment after referral and premature termination of treatment. Upwards of 55% of referrals "drop out" before initiating treatment. Of those who actually enter programs, the postadmission attrition rates during the first three months range as high as 88% for drug-free modalities and 46% in methadone maintenance. Our own experience involving more than 4,000 applicants during the past nine years has shown that fewer than 25% of referrals are enrolled in programs three months after their initial application.

Understanding the causes of attrition affords at least three potential benefits in rehabilitating drug abusers. (a) Correlates of attrition could be used in multivariate prediction equations during the screening process to identify applicants who are least likely to be retained. When sufficient resources are available, these high risk cases could receive more immediate and intensive clinical attention. During periods of scarce resources, well-validated predictors of attrition could be useful in selecting the applicants most likely to benefit from treatment. (b) Attrition may result from mismatching patient typology with type of treatment. Investigating the correlates of attrition as a joint function of patient and program variables could improve referrals of applicants

* This research was supported in part by grants DA 01077 and DA 16356 from the National Institute on Drug Abuse, Herbert D. Kleber, M.D., Principal Investigator.

to the most appropriate type of program. (c) Attrition might be caused by inappropriate clinical ideologies and practices. Understanding attrition, therefore, could lead either to improvements in existing therapies or to the development of more effective modalities.

In their recent review of the attrition literature, Baekeland and Lundwall (1975) observed that investigations of the causes of attrition have examined differences in treatment in entry rates and retention in programs primarily as a function of client characteristics. For example, Smart (1977) has characterized the typical dropout as "male, younger, single, living alone, poorly educated, previously unemployed, and has a history of juvenile delinquency and many arrests. He tends to be residentially mobile, to have no previous treatment, and to deny his addiction." This client-centered emphasis in attrition studies has resulted in the neglect of program-related variables and factors external to the treatment environment. The goal of finding the best type of treatment for individual patients requires more detailed taxonomic investigations of treatment characteristics than those which have been completed to date (Cole and Watterson, 1976).

One of the objectives of this study was to examine the relationship between attrition and two sets of three program-related variables while controlling statistically for pretreatment differences among patients. The first set of variables (i.e., previous application, previous admission, methadone recommendation) are presumed to reflect joint characteristics of patients and programs. Whether or not a patient previously has applied or been admitted to a specific treatment facility or receives a referral for methadone maintenance are indications of the patient's attitudes and perceptions of the program and the program's evaluation of the patient based upon past experience. These variables measure potential causes of attrition at the level of the patient-program interaction. The second set of variables (i.e., number of current applicants for admission, number of patients awaiting admission, current clinic census) are program characteristics that are relatively independent of individual patient characteristics. They reflect such factors as staff/client ratios, need and efficiency in processing application, the local reputation of the program, and community levels of demand for treatment. The study was undertaken with the knowledge that these indices of program-related characteristics are primitive at best, but at least they have the advantages of being virtually free of measurement error and are easily obtained by any treatment facility.

Most investigators have been concerned with attrition during treatment and only a few studies have examined attrition during referral. Treatment entry requires a decision to seek treatment and a subsequent decision to accept the treatment which is offered. The decision not to accept treatment is a form of self-selection which may produce systematic biases in subsequent retention and treatment outcomes. For example, if applicants must surmount a number of obstacles to receive a referral, those who persist in gaining admission probably will be more highly motivated to perform well. By not considering the process of pretreatment attrition, researchers focusing on program retention and outcomes may be missing important sources of variance in these criteria. The second aim of this study was to investigate attrition sequentially during both stages and to assess the effects of attrition during the referral process on retention in programs.

METHOD

Subjects

Two groups of applicants for treatment at the Drug Dependence Unit (DDU) of the Connecticut Mental Health Center in New Haven provided data for this study. The pretreatment attrition group consisted of 168 persons who were referred for treatment by the evaluation staff during the 15-month period ending October, 1975. The postadmission attrition group consisted of 101 members of the pretreatment group who accepted treatment and 94 other persons who entered treatment directly before being interviewed for the pretreatment attrition study during this same period. Seventy-three percent of the pretreatment group were opiate abusers, 73% were male, 68% were white and 54% were either on probation, parole or had pending litigation. Their mean age was 23.3 years. Seventy-two percent of the 94 additional members of the postadmission group were opiate abusers, 74% were male, 57% were white and 62% had criminal justice system involvement. Their mean age was 25.8 years. These groups were part of a comprehensive examination of the effects of client characteristics (demographic, personal history, psychological characteristics), clinic variables and their interactions on attrition during referral and treatment that has been reported in detail by Harford and Kleinhans (1977).

Predictor Measures

The six measures of program and program-related characteristics included whether or not the applicant had (1) previously applied for admission to the DDU, (2) been admitted previously to the DDU, and (3) received a referral for methadone maintenance. (4) "Applicants for admission" was the number of persons who applied during the three-day interval prior to, during and subsequent to the day each applicant was first evaluated. This variable was introduced to reflect the effects of the case load of the evaluation staff on the referral's probability of entering treatment. (5) "Patients awaiting admission" was the total number of active referrals who had been referred for treatment but had not yet entered a program during the same three-day period. This variable reflected activity in the clinical program as well as in the evaluation unit. (6) Clinic census size indicated the number of patients in the initial phase of treatment on the day of application. For opiate abusers the measure included all patients in the premaintenance stage of the methadone program. For patients whose primary drug of abuse was not opiates the measure included all those in the initial treatment stage in clinics other than methadone. Since only half of the opiate abusers received a referral to methadone maintenance, current clinic census was less than an ideal measure of case loads in programs for opiate abusers. Data concerning prior DDU treatment history, clinic census and retention in treatment were extracted from clinic records. Referrals were counted as admissions if they entered a program within 30 days after referral. Retention during the 18 months following admission was defined on the basis of continuous clinic attendance including absences of no more than two consecutive weeks during the first treatment episode. For 17 admissions who remained in treatment for more than 18 months, retention was projected on the basis of survival curves generated from an independent treatment population.

RESULTS

Pretreatment Attrition

Of the 108 referrals, 101 (60.1%) accepted treatment. A stagewise discrimination function procedure, described more fully by Harford and Kleinhans (1977), was used to identify the program-related characteristics that distinguish dropouts from admissions after the two categories were equated statistically for client-related differences. Four of the six variables were significantly related to pretreatment attrition. Among opiate abusers, blacks were more likely to accept treatment if they previously had applied for admission, $F_{(1,50)}=6.56$; $p<.05$. Whites were more likely to accept treatment if they received a methadone referral, $F_{(1,56)}=7.47$; $p<.01$ or if there were fewer patients awaiting admission, $F_{(1,56)}=4.94$; $p<.05$. The size of the clinic census was the most important of the predictors for whites even though the univariate difference was not statistically significant. Dropouts were more likely to have applied when clinic census was high even though there were no waiting lists during the period of study. The multiple correlation of the three-variable discriminant function for black opiate abusers, in which previous application was the most important predictor, was .51, and it accounted for 25.8% of the variance. For the five-variable function describing whites, which included program-related variables as the three most important predictors, the multiple correlation was .58, and it accounted for 33.3% of the variance. When these indices were adjusted for estimated shrinkage according to procedure recommended by Schmitt, Coyle and Rauschenberger (1977), the proportions of explained variance were revised downwards to 14.5% and 18.3% respectively.

Postadmission Attrition

The criterion for postadmission attrition was the length of retention during the initial treatment episode. A stagewise multiple regression procedure was used to equate the clients for individual differences. Five of the six program-related predictors received significant Beta weights. Clients were likely to be retained longer in methadone maintenance if they (1) had not previously applied for admission, (2) sought admission when the number of other applicants was low and (3) sought admission when the clinic census was high. Clients in drug-free modalities were less likely to be retained longer if they (1) sought admission during times when relatively large numbers of applicants were awaiting admission and (2) previously had received DDU treatment. The multiple regression equation for methadone retention contained a total of eight predictors which produced a multiple correlation of .56, accounting for 31.0% of the retention variance. The seven-variable equation for drug-free retention produced a multiple correlation of .44 accounting for 19.1% of the retention variance. When shrinkage corrections were applied the proportions of explained variance were reduced to 11.0% and 7.4%, respectively.

DISCUSSION

Pretreatment Attrition

Sixty percent of the applicants accepted treatment. Dropouts among the opiate abusers were more likely when the number of persons awaiting admission and the number enrolled in methadone maintenance were above their mean levels for the period of study, despite the fact that vacant treatment slots were always available and there were no waiting lists. One possible explanation for this result is that favorable staff/client ratios in screening and the treatment programs facilitated acceptance of treatment. When caseloads are high, screening and program personnel may be more selective and concentrate their energies on applicants who intuitively seem to be the risks with the result that many applicants may feel discouraged from pursuing their treatment. A second possibility is that sudden influxes of applicants in the referral process and in treatment may indicate unusual external conditions, such as temporary heroin scarcity, that stimulate marginal applicants to seek treatment. If these applicants are less motivated than those who apply during periods of normal heroin supplies, higher pretreatment attrition rates would be expected. Dropouts also were likely to be applying for the first time. Again there are any number of possible explanations.

Subsequent applications might indicate higher levels of motivation. First-time applicants may examined especially carefully by evaluation personnel because they are more likely to be unknown. Persons who have applied more than once may have higher acceptance rates because they are more familiar with the admission procedures. Although there are an indefinite number of possible explanations of these results, the fact that the applicants were equated statistically for differences in personal histories and psychological characteristics suggests that program or external factors are more likely reasons for these relationships than client characteristics.

Postadmission Attrition

The number of previous applicants was positively related to admission but negatively related to retention. Patients who had not previously applied to the DDU were retained longer in the methadone program than those who had previous applications. Therefore, having made a previous application for admission *increased* their probability of admission but *decreased* their probability of retention. This result demonstrates that pretreatment attrition can produce biases in variables that are related to retention in treatment. The result also illuminates the processes underlying pretreatment attrition. If reapplicants were more highly motivated than first applicants, they would also be expected to be retained longer, which was not the case. However, if first applicants and reapplicants were equal in pretreatment motivation and the screening procedures were more selective for first applicants, then higher motivations and higher retentions would be expected among those who were admitted. Thus, it seems likely that screening practices tended to eliminate a higher portion of first applicants which resulted in a selection bias favoring retention of the first applicants who were admitted.

Methadone patients were retained longer if they had applied during times when the methadone census was high. This variable was *negatively* related to pretreatment attrition and *positively* related to postadmission retention and, therefore, is another indication of the biasing effects of selection on retention. Again, self-selection on the basis of motivation is a likely explanation. Encouragement of staff to enter treatment varies inversely with the size of clinic census so those who are admitted when the census is high may be more highly motivated and, therefore, stay longer.

The correlations between retention and patients awaiting admission were positive for methadone (.16) and drug-free modalities (.19), but the variable received a significant Beta weight only in the drug-free equation. Like the census, this variable was negatively related to pretreatment attrition and positively realted to postadmission retention.

Why should these three program-related variables seem to have opposite effects on pretreatment as opposed to postadmission retention? Admissions to treatment are not unaffected by organizational needs, demands for patients and clinic workloads. Screening staffs act as gatekeepers for treatment programs by regulating the flow of prospective patients. One of the consequences of this process is that patients who are admitted seem to covary systematically with respect to their motivation. As the admission process becomes more restrictive, the admitted patients are more highly motivated and retention increases. This phenomenon implies that the goals of increased admissions and increased retention rates are mutually inclusive. Barriers to admission that challenge the applicants' motivation will reduce admission rates and improve retention. Conversely, changes in screening policies during times of low census that are intended to increase admissions run the risk of causing decreased retention and vitiating the rehabilitative effectiveness of treatment by inflating the number of unmotivated patients in the treatment population. Motivations of peers in treatment seems to be an important but generally unrecognized influence on prospects for success. When unmotivated patients predominate, the therapeutic climate is not likely to be conducive to effective rehabilitation even of highly motivated individuals. Treatment of a small number of marginal cases might be attempted in an otherwise supportive environment, but admission of large numbers of unmotivated patients may adversely affect the prognosis of those who otherwise would have been successful. In order to maximize retention the inherent conflict between admission rates and retention rates should be resolved by adopting uniform admission standards that are independent of the number of patient vacancies. Mandatory waiting periods and a nominal application fee charged to all candidates for admission might be useful nondiscriminatory policies to minimize unmotivated applications.

The development of a valid instrument for measuring personal motivations for abstinence directly would be a valuable contribution to the treatment of drug abusers. In the absence of such an instrument, motivations can only be inferred tentatively from patterns of correlations among other well-measured variables such as legal pressure. Under these circumstances multivariate prediction equations constitute the most valid method of identifying high-risk individuals at the time of their application.

The chief advantage of the statistical prediction approach to analyzing attrition is that direct measures of personal motivations for abstinence are not essential for identifying patients who are unlikely to be retained. Although the addition of valid measures of motivation probably would improve prediction, variables that are merely correlated with motivation can be sufficient for predictive purposes. The statistical techniques have the additional advantage of objectivity so that the basis for referral decisions can remain invariant despite differences in diagnostic skills and job turnover among screening personnel. The cut-off points for admission can be varied systematically in response to changes in programmatic needs, and periodic corrections of the indices through cross-validation would assure that the indices reflect relevant changes in applicant populations or clinical practices.

How adequate were the indices obtained in this investigation for the purpose of identifying high risk candidates for treatment? Predictive utility is assessed by the proportion of criterion variance that is accounted for by the set of predictors. In this study, R^2 ranged from .191 to .333. When these figures were adjusted for the amount of shrinkage in R^2 to be expected in replications, however, the results were less impressive. The predictability of whether an individual applicant would decline a treatment referral was acceptable (mean adjusted R^2=.177), but, for the clinically more important problem of identifying high risk cases, the corrected R^2's were not large enough to justify their use for prediction of individual retention. Although the multivariate analyses clarified the descriptive and inferential processes, the predictor variables did not measure enough relevant characteristics to be used in selecting patients for differing levels of treatment intervention.

What can be done to improve the predictive indices? First unmeasured variables that are potentially relevant to treatment outcomes can be identified and included in subsequent studies. The desirability of measuring motivations for abstinence has been discussed. Other variables such as severity of addiction, individual psychopathology, expectations for positive treatment outcomes of individual patients, levels of illicit drug use by peers in treatment during the preadmission process, patient turnover rates, staff expectations and morale of staff and patients all would seem to exert important influences on attrition. Second, the effects of these variables on treatment outcomes, rather than attrition, also should be investigated. It may be clinically important to identify different sets of predictors for each dimension of treatment outcome such as cessation of drug use, termination of criminal activities, and employment among other possible criteria of rehabilitative success. Understanding the processes associated with these various aspects of treatment success is a complex problem that requires multivariate methods of analysis similar to those used in this investigation.

The results have shown that even the rudimentary indices of organizational processes used in this research had moderately strong and independent relationships with the criterion of attrition. The findings suggest that systematic attempts to investigate the effects of programmatic variables may be more productive of practical treatment benefits than a continuation of intensive research on the effects of client-related variables. To the extent that treatment has assumed the role of a social control agency, retention and outcome have less to do with differences among clients and are more strongly affected by variables related to treatment policies and procedures.

I have argued elsewhere in these Proceedings (Harford 1978a, 1978b) that retention as a criterion variable is so greatly affected by systematic biases of treatment organizations and the criminal justice system that it is not a useful indication of treatment efficacy. In the absence of criterion contamination, the gains attributed to treatment effects should be at least moderately correlated with the amount of treatment received. If the goal of evaluation research is to demonstrate the effects of variables on outcomes rather than the amount of treatment, then future investigations should be concerned with assessing the relationshops between program-related variables and treatment outcome measures rather than retention.

REFERENCES

Baekeland, F. and Lundwall, L. 1975. Dropping out of treatment: A critical review. *Psychol. Bull.* 82: 738-751.

Cole, S.G. and Watterson, O. 1976. A treatment typology for drug abuse in the DARP: 1971-1972 admissions. In S.B. Sells and D.D. Simpson (eds.); *The Effectiveness of Drug Abuse Treatment* (vol. 3) *Further Studies of Drug Users, Treatment Typologies, and Assessments of Outcomes During Treatment in the DARP.* Cambridge, Mass.: Ballinger.

Harford, R.J. 1978. Some likely alternative explanations for improvement during drug abuse treatment: Maturation, statistical regression, history, criterion contamination. In: *Proceedings of the Fifth National Drug Abuse Conference.*

Harford, R.J. 1978. Pretreatment criminal justice status, retention, and posttreatment outcomes. In: *Proceedings of the Fifth National Drug Abuse Conference.*

Harford, R.J. and Kleinhans, B. 1977. Correlates of attrition during referral and treatment for drug dependence. *Am. J. Psychiat.* 133(12): 1399-1404.

Schmitt, N., Coyle, B.W. and Rauchenberger, J. 1977. A Monte Carlo evaluation of three formula estimates of cross-validated multiple correlation. *Psychol. Bull.* 84: 751-758.

Smart, R.G. 1977. Comment on Sell's paper. In J.D. Rittenhause (ed.): *The Epidemiology of Heroin and Other Narcotics.* Research Monograph 16, Rockville, MD: NIDA.

AN INVESTIGATION OF OUTCOMES OF TRADITIONAL DRUG TREATMENT/SERVICE PROGRAMS*

Marvin R. Burt, D.P.A. and Barbara Sowder, Ph.D.

Burt Associates, Incorporated

This paper reports on one of eight major investigations conducted by Burt Associates, Incorporated, on the subject of drug abusing women. This research was funded by the National Institute on Drug Abuse (Contract No. 271-76-4401).

This particular investigation focused on answering the following research questions:

* What are the treatment outcomes of females compared to males (controlling for relevant variables)?
* What are the relationships between the characteristics, backgrounds, etc., of female and male drug abusers and their treatment outcomes?
* What are the implications of the characteristics of male and female abusers with respect to service provisions in terms of:
    * Evidence regarding the types of treatment that are most effective for clients with certain characteristics?
    * The state of knowledge with respect to the effectiveness of treatment for women? For female clients with certain characteristics? What knowledge gaps exist?

* What percent of women with drug problems come into contact with various systems (including hospital emergency rooms and crisis centers)?

* How do rates of entry into treatment relate to the prevalence of drug abuse for women compared to men?

TREATMENT OUTCOMES FOR MALES VS. FEMALES

Treatment outcomes were analyzed in two parts: 1) progress of clients while in treatment and/or his or her status upon leaving treatment and, 2) the client's behavioral status at some point in time *after* leaving treatment.

Status After Leaving Treatment

Studies on the status of clients sometime after leaving treatment appear infrequently in the literature and few of these provide data by sex of subject. Two of these studies dealt with very small samples of women, in each study only 12 (Zahn and Ball, 1974; Ross et al., 1974). In view of the small samples and lack of comparisons with males, these two studies will not be discussed further.

Table I presents summaries of pertinent data on each of the four followup studies for which there were published results reported separately for males and females and/or additional data could be obtained from the authors which separated outcome by sex. Three of the studies reported no statistically significant differences in behavioral outcomes between men and women; one reported that females experienced greater behavioral change (in a desirable direction) than males. Behavioral outcomes were measured in terms of drug use, arrests, and employment for all four studies, although there were differences in the specific measures employed. Only one of the studies considered other prosocial activities such as keeping house as equivalent to posttreatment employment (Burt, 1976).

Status While In Treatment

The most extensive data sources on the status of clients while in treatment is the CODAP system. This source was tapped and an extensive literature review was conducted.

The best presentation, in our view, would be the *change* in behavior realized by clients while in treatment. Unfortunately, the na-

Table I. Follow-up studies of male and female clients.

Study	Initial Sample	Final Sample (i.e., no. & % interviewed)	% Females	Age	Race/ Ethnicity	Pre-Treatment Drug Use	Modality	Significant Differences in Outcome Between Males & Females
Chambers and Incardi (1975)	209	92 (44%)	32	13-17 = 21% 18 20 = 35% 21-24 = 26% 25-34 = 14% 35-49 = 4%	Whites = 37% Blacks = 43% Latins = 20%	Not Given	Residential Therapeutic Community	No 2)
Burt (1976)	360	291 (81%)	18	<18 = 0% 18-20 = 26% 21-25 = 42% 26-30 = 20% 31-36 = 5% >36 = 7%	Black = 86% White = 14%	Heroin = 97% Other = 3%	Methadone Maintenance & Abstinence	No 2)
Burt & Glynn (1976) 1	782	457 (58%)	22	<18 = 3% 18-20 = 17% 21-25 = 40% 26-30 = 22% 31-36 = 10% >36 = 8%	Black = 50% White = 44% Other = 6%	Heroin = 81% Other = 19%	Methadone Maintenance, Residential Therapeutic Community & Ambulatory	No 2)
Gould, Forrest, and Kleber (1975)	513	337 (73%)	20	<18 = 10% 18-20 = 29% 21-25 = 37% 26-30 = 13% 31-36 = 7% >36 = 4%	Black = 50% White = 47% Other = 3%	Heroin = 65% Other = 35%	Methadone Maintenance	Yes. Females experienced greater behavioral changes.

1) Data collected by Macro Systems (1975)
2) Statistical tests showed no significant difference
3) No statistical tests of significance were conducted. However, many of the differences were large.

Table II. Client performance in treatment.

Study/ Data Set	Sample	Modality	Significant Differences in Outcomes Between Males and Females ?
CODAP (This report)	All clients discharged in 1975-1st 6 months of 1976 from Federally funded treatment programs	Detoxification Methadone Maintenance Drug Free Other	Males more likely to be employed upon discharge, but females more likely to be in education program. Differences in drug use generally slight.
Demaree, et al. (1975)	33,611 clients of 45 agencies located throughout the U.S.	Methadone Maintenance Therapeutic community Drug Free Outpatient Detoxification	No; except that females performed slightly better than males in drug-free out patient programs
Sells (1974)	11,384 clients of agencies located throughout the U.S.	Methadone Maintenance Therapeutic community Drug Free Outpatient Detoxification	No; except that males had substantially higher death rates than females.
Rosenthal, et al. (1976)	13,268 CODAP clients throughout the U.S.	All Federally funded modalities	No; other than slight differences.
Cohen and Woerner (1976)	284 clients in New York City	Youth centers and ambulatory treatment programs	No
Richman, et al. (1971)	500 clients in New York City	Detoxification and methadone maintenance	Males were more likely than females to enter a methadone maintenance program subsequent to detoxification.
Gearing (1971)	718 clients in New York City	Methadone Maintenance	Women more likely to be not employed and/or on welfare than men upon entry and discharge; but cannot determine whether men or women realized greater improvement.
Aron and Daily (1976)	286 clients in California	Residential Therapeutic community	Cannot be determined.
Bell (1976)	392 clients in Maryland and Virginia (suburban Washington D.C.)	Residential Therapeutic community	No

Table II (cont'd.). Client performance in treatment.

Study/ Data Set	Sample	Modality	Significant Differences in Outcomes Between Males and Females?
DeLeon (1974)	208 clients in New York City	Residential Therapeutic community	Females slightly more likely than males to leave treatment against clinical advice.
Newman, et al. (1973)	330 clients in New York City	Methadone Maintenance	Females realized greater decline in arrest rates while in treatment than males.
Sechrest and Crim (1977)	473 clients in New York City and 271 in Santa Clara County, California	Methadone Maintenance	No consistent differences in changes in arrest rates among males and females.
Edwards and Goldner (1975)	100 clients in Detroit	Not specified	Female's arrest and conviction rates declined more than male's.
Edwards and Jackson (1975)	294 clients in Detroit	Not specified	Females referred for job placement were less likely than males to obtain jobs.

ture of the data and lack of comparability between intake and discharge data often makes such comparisons difficult at best and usually impossible. Therefore, this section is focused almost entirely on clients' behavior upon leaving a program or while in treatment.

Analysis of the CODAP data reveals that females were somewhat less likely than males to leave treatment during the first four weeks, regardless of the type of treatment modality. Females were slightly: 1) more likely to complete treatment than males, 2) less likely to be discharged for non-compliance, and 3) less likely to be incarcerated. However, these and all other male/female differences with respect to outcome were small. The general conclusion is that male and female outcomes are not substantially different.

Thirteen published studies were analyzed; the results are presented in Table II. Few of these reported significant differences among male and female outcomes during treatment.

On the basis of the data considered, it is concluded that the preponderance of the evidence points to few differences in female vs. male treatment outcomes, and those differences that exist are generally slight. This is not to say that *no* differences could appear if other measures were considered. However, in terms of the traditional generally accepted measures considered in this report, differences are indeed few and those that exist are not large.

EXPLANATORY FACTORS

A large number of multiple regression or stepwise multiple regression analyses were conducted on the Washington, D.C., New York City, DARP, and CODAP data sets. The purpose of these analyses was to determine to what extent sex, in combination with other variables, could explain treatment outcomes. None of the independent variables, used in various linear combinations, explained very much of the dependent (i.e., outcome) variables' variance. Sex explained no more than 4 percent of the variance in the dependent variables for any regression analysis.

PREVALENCE COMPARED TO ENTRY INTO TRADITIONAL DRUG TREATMENT PROGRAMS

An analysis was conducted, encompassing the 15 most populous SMSA's, of the relationship between male/female prevalence of drug problems, as indicated by contacts with DAWN emergency rooms and crisis centers, and entries into Federally funded traditional drug treatment programs.

Table III. Comparison of prevalence of drug problems with entries into federally funded drug treatment programs.

Female:Male Ratios

Drug	Ratios
Heroin	1:1.5
Cocaine	1:2
Barbiturates, Tranquilizers, Sedatives	1:2.3
Amphetamines	1:2
Hallucinogens	1:1.2
Inhalants	1:1
Marijuana	1:1.6

Conceptually, we sought to determine the relationships between prevalence and entries into traditional drug treatment programs for women and compare these to men. Two data sources are employed in this analysis:

* Prevalence--DAWN

* Entries into treatment--CODAP

Table III depicts the results in terms of the relationship between prevalence of drug problems and entry into traditional treatment for females and males with each specific drug as the primary drug of abuse. Males experienced 1.5 (heroin) to 2.3 (barbiturates, tranquilizers, sedatives) times as many entries per drug problem (mentions) as females.

This analysis suggests, despite the limitations, that females with drug problems are underrepresented in Federally funded drug treatment programs. However, females less than 18 years old are generally slightly overrepresented in treatment compared to their male counterparts. Both male and female blacks are generally overrepresented in treatment compared to their white counterparts, but underrepresented compared to their counterparts of "other" races.

PREVALENCE COMPARED TO RATES OF ENTRY INTO TRADITIONAL TREATMENT

This section addressed the following research question:

How much prevalence is required to produce one entry into Federally funded treatment for women compared to men during specified time periods?

Table IV. Comparison of prevalence of drug use with rate of entry into Federally funded drug treatment programs, female:male ratios.

Drug	Age	
	≤18	18+
Heroin	Prevalence too low	Prevalence too low
Other opiates	10:1	3.3:1
Cocaine	2.5:1	3.8:1
Marijuana	1.3:1	4.5:1
Hallucinogens	3:1	2.5:1
Inhalants	4:1	(prevalence too low)

Special tabulations of data from a national household survey (conducted by The George Washington University and Response Analysis Corporation) were used for prevalence of drug use; CODAP data were used to provide the number of male and female clients entering Federally funded drug treatment.

Table IV depicts the results in terms of the extent to which males using specific drugs are more likely than females to enter a Federally funded drug treatment program. Prevalence of heroin use was too low to permit conducting this analysis for that drug. For other drugs examined (other opiates, cocaine, hallucinogens, and marijuana) adult males abusing these drugs were 2.5 to 4.5 times more likely than females to enter a Federally funded treatment program.

IMPLICATIONS FOR TREATMENT

Careful inspection of the data concerning various outcomes for males and females leads to the conclusion that there are no clear implications for treatment based upon the findings of this study. Further, the lack of differences between males and females in outcomes for different treatment modalities suggests no clear implications with respect to choice of modalities.

There are few differences in retention rates for males and females and the CODAP data show that females are, in fact, slightly *less* likely than males to drop out of treamtment during the first four weeks after admission. Thus, there is no evidence that females, once they enter treatment, are more likely than males to become discouraged and leave. Therefore, there are no implications with respect to changes in practices that could eliminate any alleged, but unproven disparities in retention for males and females.

There is substantial evidence suggesting that female drug abusers are less likely to enter traditional treatment than males using those drugs. This disparity in entry rates among male and female drug abusers could be due to several factors. Among these may be:

* Female drug users tend to be less heavily involved and less in need of treatment than males;

* Females are less likely to be referred to treatment by the criminal justice system and, therefore, feel less coercion to enter treatment;

* Females are turned-off from seeking treatment because of:

    * Sexist practices in the male orientation of traditional treatment

    * Lack of what they consider to be viable treatment

    * Inconvenient treatment (e.g., lack of day care for children)

    * Lack of treatment applicable to their particular drug problem (e.g., use of psychotropics)

    * Dependency relationships with a male, etc.

This disparity in entry rates requires further discussion; yet such discussion must, at this point, fall into the realm of speculation. In order that this speculation be converted to answers, certain basic questions will have to be answered:

* Are female drug abusers who do not enter treatment different from males who do enter treatment?

* Are females remaining in treatment different from females who drop out?

Certain implications currently exist which may help answer these questions. Females may use drugs less heavily, in terms of dosage, a phenomenon that would not be reflected in the prevalence figures used. Males, who may be heavier users, may therefore be more likely to enter treatment. However, the analysis showed that males were overrepresented in Federally funded treatment programs relative to prevalence of drug problems irrespective of drug type. And another analysis, while limited by rather small samples, indicates that men and women abusing a drug are equally likely to contact a hospital emergency room or crisis center for treatment of a problem with that drug (Burt et al., 1977).

Non-Federally funded programs, not included in this analysis, may be serving other types of abusers. Therefore, the exclusion of the two thirds of the treatment population not served by Federally funded programs could act to bias the picture presented in this report.

Another possible explanation is that males are more likely than females to abuse the types of drugs for which the criminal justice system might refer users to treatment. Males are more likely to be referred by the criminal justice system than females, not only due to their heavier use of different drugs, but also because they are more likely to be arrested and convicted. Any of the above could help to explain the differences between male and female entry rates, especially in view of the small number of cases in question.

In order to determine with any degree of confidence what the implications for treatment might be, further research must be conducted which goes beyond the realm of speculation. The most vital point here is that research should be conducted with *nontreatment* populations as well as those currently in treatment.

REFERENCES

Aron, W.S. and Daily, D.W. 1976. Graduates and splitees from therapeutic community drug treatment programs: A comparison. *Int. J. of Addictions*. 11(1): 1-18.

Bell, M.D. 1976. *Therapeutic communities in the treatment of drug abuse*. (Doctoral Dissertation, The George Washington University) Washington, D.C.

Burt, M. 1976. *A follow-up study of former clients of the District of Columbia's Narcotics Treatment Administration: Volume I.* Bethesda, Maryland: Burt Associates, Incorporated.

Burt, M. and Glynn, T. 1976b. *A follow-up study of former clients of drug treatment programs in Washington, D.C. and New York City.* Bethesda, Maryland: Burt Associates, Incorporated.

Burt, M., Glynn, T., Sowder, B., and Gilden, A. 1977. *An investigation of the characteristics of drug-abusing women: Phase II of the investigation of the characteristics of drug-abusing women, the treatment process as it affects women and men differently and treatment outcomes for women.* Bethesda, Maryland: Burt Associates, Incorporated.

Chambers, D.C. and Incardi, J.A. 1975. Three years after the split. In: E. Senay, V. Shorty, and H. Alksne (eds.), *Developments in the field of drug abuse. Proceedings 1974 of the National Association for the Prevention of Addiction to Narcotics.* Cambridge, Massachusetts: Schenkman Publishing Co. 124-131.

Cohen, M. and Woerner, M. 1976. Variables related to length of stay in day programs for drug abusers. *Am. J. of Drug and Alcohol Abuse.* 3(2): 303-313.

DeLeon, G. 1974. Phoenix House: Psychopathological signs among male and female drug-free residents. *Addictive Diseases: An International Journal.* 1(2): 135-152.

Demaree, R.G. et al. 1975. *Effectiveness measures of treatment programs: DARP admissions, 1969-73.* Institute for Behavioral Research.

Edwards, E.D. and Goldner, N.S. 1975. Criminality and addiction: Decline of client criminality in a methadone treatment program. In: E. Senay, V. Shorty, and N. Alksne (eds.), *Developments in the field of drug abuse. Proceedings 1974 of the National Association for the Prevention of Addiction to Narcotics.* Cambridge, Massachusetts: Schenkman Publishing Co. 878-885.

Edwards, E.D. and Jackson, J. 1975. Rehabilitative services provided women versus men in a substance abuse treatment program. In: E. Senay, V. Shorty, and H. Alksne (eds.), *Developments in the field of drug abuse. Proceedings 1974 of the National Association for the Prevention of Addiction to Narcotics.* Cambridge, Mass.: Schenkman Publishing Co. 505-508.

Gearing, F.R. 1971. Successes and failures in methadone maintenance treatment of heroin addiction in New York City. *Proceedings Third National Conference on Methadone Treatment November 14-16, 1970.* Washington, D.C.: U.S. Government Printing Office. 2-16.

Gould, L., DeForrest, C., and Kleber, H. 1975. Five-year follow-up of methadone maintenance patients. New Haven Connecticut Mental Health Center.

Newman, R.G., Bashkow, S., and Cates, M. 1973. Arrest histories before and after admission to a methadone maintenance program. *Contemporary Drug Problems.* 2(3): 417-430.

Richman, A. et al. 1971. *Follow-up of 500 heroin users by means of a case register. Report of the Thirty-third Annual Meeting of the Scientific Committee on Problems of Drug Dependence.* National Academy of Sciences, Toronto. 1: 853-890.

Rosenthal, B.J. et al. 1976. Drug treatment outcomes: Is sex a factor? (Paper prepared for the National Drug Abuse Conference. March 28, 1976).

Ross, W.F., McReynolds, W.T., and Berzins, J.I. 1974. Effectiveness of marathon group psychotherapy with hospitalized female narcotics addicts. *Psychological Reports*. 34(2): 611-661.

Sechrest, D. and Crim, D. 1977. Methadone programs and crime reduction: A comparison of New York and California addicts. (Paper presented at the National Conference on Criminal Justice Evaluation, Washington, D.C.).

Sells, S.B. 1974. *The effectiveness of drug abuse treatment*. (2 volumes). Cambridge, Mass.: Ballinger Publishing Co.

Zahn, M.A. and Ball, J.C. 1974. Patterns and causes of drug addiction among Puerto Rican females. *Addictive Diseases: An International Journal*. 1(2): 203-214.

EVALUATION OF THE LOS ANGELES COUNTY SYSTEM OF DRUG ABUSE SERVICES

Lucille Burlew-Lawler, M.P.H.
Department of Public Health, San Francisco, Ca.

Herman L. DeBose, M.S.W.
Los Angeles Country Drug Abuse Program Director's Office

INTRODUCTION

An operational research model and such basic research principles as the ethics of confidentiality of data, the completeness, accuracy and objectivity of that data and the relevance to desired outcomes of the evaluation criteria adopted, form the basis for the evaluation procedures and instruments developed by the Los Angeles County Drug Office.

Positive results have accrued from the application of these principles. The following paper describes how the principles were implemented into a functioning system of evaluation and offers examples of those results.

EVALUATION RESEARCH

The field of evaluation research is still comparatively new. Dr. E.A. Suchman wrote the first definitive text in 1967. He defined evaluation as "the determination of the results . . . attained by some activity . . . designed to accomplish some goal or objective."

In his discussion of these four elements of evaluation, determination, results, activity, and objective, Suchman defines each of them in such a way as to permit the application of the scientific method and its research techniques to the entire evaluation process. *Determination* is the dimension of *process* (the evaluation process, not process in the sense of program process); *results* is the dimension of *criteria*; *activity* is the dimension of *stimulus*; and objective is the dimension of *value*. Insofar as possible, evaluation should be a process for determining the relationship of the *stimulus* to the *value* in terms of measurable *criteria*.

In order to conceptualize and expand the Suchman definition, a comprehensive model based on a similar model of R.E. Stake (1965) was developed. This model (Figure 1) is a structured way of looking at the components of a total evaluative process and suggests their dynamic interactions. His model also includes the "judgment matrix" which, through feedback to programs of information on the elements and outcomes of those programs, offers positive recommendations for change.

A description of the components of the model are included as an appendix to this paper.

LEVELS AND STRUCTURE OF PROGRAM EVALUATION

The three levels or phases of evaluation which are described by the model, process evaluation and short and long-term outcome evaluation, have been implemented during the four-and-one-half years since the beginning of the Los Angeles County Drug Abuse Program Director's Office in 1973. The structure of evaluation suggested by the "judgment matrix" of Stake has also been implemented. It can be termed an operational research model, in that findings and recommendations from all levels or phases of evaluation are fed back to program providers to implement these recommendations. Program operations are thus changed and affect program outcomes. The process of evaluation, feedback and training is repeated periodically, leading to continued improvements in program efficiency and effectiveness.

PRINCIPLES OF EVALUATION AND THE RESULTS OF THEIR APPLICATION

Basic principles of evaluation suggested by the model and elsewhere in the literature (Scriven, 1967; Borgatta, 1966; Weiss, 1972) were followed in the development of the instruments and procedures of the evaluation system. The most important of these and some results of their application are discussed as follows.

1. Information gathered must be confidential. Protection of the rights of confidentiality for both clients and programs has been a high priority since the beginning of the evaluation process. An advisory group recommended the use of a unique client identifying number for the automated client information system. Census tract numbers were used to locate client residence. There have been no known instances of clients having been traced by law enforcement agents using the client identifier number or the census tract number.

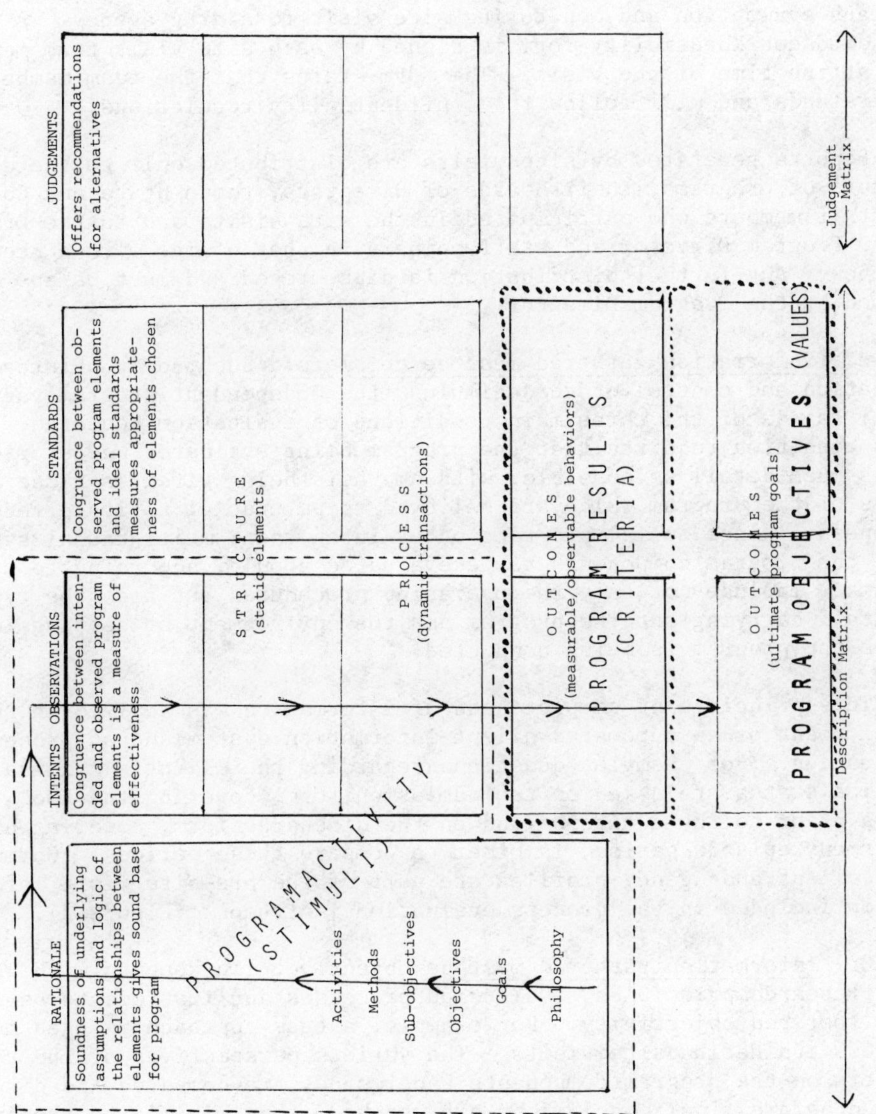

Fig. 1. Program evaluation model.

The staff of drug programs, of county departments and clients themselves have been kept informed about confidentiality regulations protecting drug clients.

Confidentiality of records is an issue which has been addressed on each evaluation and monitoring site visit to a drug agency. A standard confidentiality form is signed by each site visit team member at the time of the visit. The form states that the team member understands and will follow the confidentiality regulations.

Reports generated by site visits are distributed only to the directors of programs, their Boards of Directors, those State and County staff members who participated in the site visits and to the Drug Abuse Program Director and staff members in that office with a need to know. Any further distribution is discouraged and must be approved by the Program Director.

2. Information gathered must be complete. Suchman stipulates isolation and control of the stimulus (the independent program variable) as one of the three main conditions of evaluation research. This condition requires that the program being evaluated be described in as much detail as possible, with emphasis being given to those aspects of the program which are believed to be crucial to its effectiveness. Since social programs occur in an open, multi-causal system, the program components to be evaluated should, according to Suchman, include not only the operating procedures but also the type of staff carrying out the program and the environment or setting in which the program is being conducted.

This principle of completeness of information was applied in the development of an automated client information system and of process evaluation. For example, questions regarding the client's criminal justice status are asked on the admission form, covering the two years prior to the admission and on the discharge form, covering his treatment episode period, in order to compare these periods. Community, client and agency profiles are part of the pre-site visit information included in the process evaluation instrument (Figure 2).

3. Information gathered must be objective. In keeping with general research principles, evaluation processes and instruments used have fostered objectivity. For example, a team approach is used in process evaluation site visits. The various perspectives of the team members on the program components under study are compared and discussed before final conclusions are reached, thus minimizing the bias of a single perspective. In the present site visit instrument, criteria are operationalized. Specific questions concerning the same criterion are asked of the administrator, as many staff members as

Substantiation of units, hours billed by documentation review:
Number of charts, documents reviewed______
Time period covered______________________
Number of units, hours billed____________
Units, hours substantiated vs number billed

Mode of Service	Units, hours substantiated	Units, hours billed

Percent of units substantiated %__________
Percent of units NOT substantiated %__________

Criterion #2
The program will serve the geographic area for which it is being funded. (Contract)

PRE-SITE VISIT INFORMATION

a. Area:___________________. (Contract, Exhibit, 2.)

ON-SITE VERIFYING INFORMATION

Documentation (Case Records):

b. Substantiation of area served:
Of a 10% random sample of active case records, how many document that the client lives in the above area? #__________

Criterion #3
The program will serve the target population which it purports to serve and for which it is being funded. (Contract, DAO)

PRE-SITE VISIT INFORMATION

a. Contract definition of persons to be served (Contract Exhibit, 2.):
____

b. Program's definition of its target population (Prior Reports):
____

____

c. Statistical Profile of Client Population (DAO Data)

ON-SITE VERIFYING INFORMATION

Administrator Interview:

d. Has the target population changed since the previous year? If so, how has it changed?

Fig. 2.

possible, clients and referral agencies to determine the extent to which that criterion is met.

Validation through a review of specific documents is also required (see Figure 2).

Confidence on the part of program providers that the recommendations made by evaluators are based on information which is as unbiased as possible is a result of the effort to obtain objective data.

4. Information gathered must be accurate. Training of provider staff in the completion of client information forms was initiated in 1973 and has continued on a regular and as-needed basis. Questions on those forms which have gathered information that was proved to be consistently inaccurate have been dropped from those forms.

Information gathered on site visits must be verified by actual documentation, whenever possible. Documented accuracy of the information has been especially helpful in backing up unfavorable findings and ratings which have been challenged by providers.

5. Criteria used in evaluation must be relevant to the desired outcomes. The necessity of a valid relationship between objectives and ultimate goals has been discussed.

To assure this relationship, expert opinion has been consulted in the development of evaluation criteria. It was acknowledged that valid drug treatment criteria were not "state of the art" four years ago. For that reason, in 1974, the drug abuse office sponsored a week-long workshop during which drug abuse program directors and staff and others with expertise in this field developed together an instrument and criteria used to measure drug abuse program process.

Criteria have been changed and increased in number to reflect the growing knowledge of valid administrative, fiscal and treatment criteria. In October of 1977 the site visit instrument was revised to include a total of 114 criteria, many of which had been developed by the respected State Substance Abuse Research and Evaluation Department under Dr. Roberta Marlowe.

The relevance of the criteria used has resulted in a general acceptance of those criteria by the providers and a resulting willingness to make internal changes in their programs in order to meet those criteria. Overall, the effectiveness and efficiency of the drug programs evaluated have improved during the four years since the evaluation process was initiated. Programs and their staff have benefited by being prepared for feature accreditation, and clients have benefited by receiving higher quality care.

OTHER PROCESSES AND THEIR RELATIONSHIPS TO EVALUATION

Evaluation and Planning

As James (1963) points out, the evaluation of available resources, both their quantity and quality, is a prerequisite to adequate planning. The evaluation of program process, including as it does a detailed specification of all program components, helps planners to determine whether specific programs are a good "fit" to the population at risk and to the prevailing patterns of drug use in the area served. Both Weiss (1972) and Suchman (1967) suggest that an analysis of program process must also include the specification of the recipients of the program and the situational context within which it operates, in order for rational decisions to be made regarding the allocation of resources to meet local needs.

A major goal for the 1977-78 Short/Doyle Drug Plan was "the development of a uniform assessment procedure to be used Countywide . . . to determine the following: 1) are programs providing effective and efficient services, and, 2) are these services the most needed services in the community at the present time." The ratings and recommendations from each site visit report are summarized and included in the current Short/Doyle Drug Plan.

Evaluation and Training and Technical Assistance

The use of program process evaluation for gathering the information needed to improve program practices and procedures through training and technical assistance is most timely and relevant when programs are new and unproven. Scriven's (1967) term "formative evaluation" is that phase of evaluation which is fed back to programs in order to improve them. Many drug programs are still in that stage in Los Angeles County, and the major stated purpose of the present evaluation site visits is to help programs improve deficiencies.

The analysis of program process is necessary to identify the major training needs, so that training monies can be spent appropriately. Other factors identified by process evaluation which affect planning for training are those which describe clients and staff. To be relevant, training must be offered to staff at appropriate levels of sophistication and must have content which addresses the primary problems of the target populations.

Overviews summarizing by modality common findings and recommendations based on the site visit evaluations of drug programs have been provided to the NIDA 409 funded training contractors since that money was made available in 1977. The criteria on which programs are being evaluated were also provided to the training contractors, so that their training content and technical assistance efforts would be compatible with the County standards.

Evaluation and Contract Monitoring

All phases of contract management, from negotiations through ongoing monitoring of contract compliance, to decisions on termination or renewal, benefit from a detailed knowledge of the program processes of each contracted program.

The Los Angeles County Health Department Contracts and Grants Management Division has worked closely with the Drug Abuse Office in all phases of the contracting process. Important program criteria, such as the presence within each drug program of an in-service training component, are included in the contract boiler plate and exhibits for drug abuse program contracts.

Contract monitors are based in the five Health Services Regions and are responsible for quarterly visits to all contracted programs in their Regions. Contract monitoring and program evaluation act as complementary and interrelated processes. Contract monitors from each Region are team members on each site visit for program evaluation. Correspondingly, the contract monitoring reports are shared with the drug abuse evaluation section, for inclusion on the pre-site visit information portions of the evaluation instrument.

Activities like these, which promote a continuing communication among evaluator, monitor, and directors and staff of programs foster a dynamic evaluation/monitoring/feedback process which promotes program effectiveness and efficiency, both in contract compliance and the delivery of services.

Evaluation and Program Management

As Suchman says, "Evaluative research is the basic ingredient of 'scientific' program management." To the extent that program directors wish to increase the effectiveness of their operations and not simply to perpetuate their own existence, they will use some form of self-assessment, as well as outside evaluation to:

Point out specific strong and weak points of program operation and suggest changes and modifications of procedures and objectives.

Increase effectiveness by maximizing strengths and minimizing weaknesses.

Weiss (1972) makes the point that evaluation is most often called on to help with decisions about improving programs rather than to decide whether they should be discontinued. She states that this purpose, that of producing information on the effectiveness of comparative strategies within programs and thus offering program managers help in deciding what specific changes to make in program process, is seldom accomplished. Evaluation of outcome of "overall impact" is more frequently attempted, according to Weiss, than an approach which will link specific elements of program activity to specific results.

It has been the primary purpose of program evaluation in the Los Angeles County Drug Abuse Office to work toward program improvement, rather than toward developing evidence that programs should be discontinued. Recommendations are made in very specific terms. Strengths as well as weaknesses are reported, with the effect that agencies are encouraged to build upon those strengths, as well as to correct their weaknesses.

Evaluation and Coordination

Coordination among the various departments concerned with and responsible for program evaluation has been an ongoing concern of the Drug Abuse Office. Coordination is an official function of the Drug Abuse Program Director's Office. For the Evaluation Section, the emphasis on coordination has minimized the impact on drug programs of multiple visits from a number of different kinds of agencies.

Since 1975, the visits of the Drug Office have coincided with the visits of the Health Services Regions' contract monitors and, to the greatest extent possible, with the visits of the locally based State Substance Abuse Research and Evaluation Office team. The reports made as a result of these visits have been shared. In 1978, the sharing of State and County visits and reports to Drug agencies was futher extended and officially sanctioned. The Los Angeles County fiscal auditor's visits have also been made to coincide with the annual evaluation site visit.

Although differences in requirements by the various agencies still prohibit achieving the ideal of only one visit a year per agency for process evaluation and other kinds of evaluation or monitoring, efforts are continuing to minimize the impact on programs of repeated vistis to drug treatment agencies.

REFERENCES

Borgatta, E.F. 1966. *Research Problems in Evaluation of Health Service Demonstrations*, Health Services Research Study Section of the U.S. Public Health Services, Milbank Memorial Fund Quarterly, V. XLIV, No. 4, Part 2: 195. New York.

James, G. 1963. The Present Status and Future Development of Community Health Research - A Critique from the Viewpoint of Community Health Agencies. *Annals of the New York Academy of Sciences*. 107: 761.

Scriven, M. 1967. The Methodology of Evaluation. In *Perspectives of Curriculum Evaluation*, American Educational Research Association Monograph Series on Curriculum Evaluation. Chicago: Rand McNally.

Stake, R.E. 1965. *The Countenance of Educational Evaluation*. Chicago: University of Illinois.

Suchman, E.A. 1967. *Evaluative Research*. New York: Russell Sage Foundation.

Weiss, C.H. 1972. *Evaluation Research, Methods of Assessing Program Effectiveness*. New Jersey: Prentice Hall Methods of Social Science Series.

APPENDIX

A MODEL FOR PROGRAM EVALUATION

Definitions: Program Monitoring and Program Evaluation

The model for program *evaluation* shown here is simply a structured way of looking at the components of a total evaluative process.

Description of the Model

Rationale: This is a hierarchy which begins with program philosophy and extends to goals, and specific measurable results and the methods chosen to implement these results. Results are the criteria by which the achievement of the valued program objectives is measured. These elements should be examined in regard to their internal relationships. Are they logically and functionally contingent upon each other sequentially? Are the assumptions underlying these relationships reasonable and logical? The soundness of the program's rationale is one predictor of program effectiveness. From it should flow the structures and processes which are chosen to lead to the desired outcomes.

Description matrix: This part of the model includes all required program elements. The congruence between the stated or intended elements and those observed to actually be present is a measure of program effectiveness.

Outcomes (results/criteria) are the measurable and observable client behaviors which are the result of the whole treatment process. What is the client supposed to be like at the completion of treatment? Is he to be drug-free? For how long? Employed? In other words, outcomes are the criteria for "successful" treatment, the psycho-physio-social indicators which have been chosen by the program for their clients. They also include such measures of program activity as numbers of clients treated over time and population served.

Outcomes (program objectives/values) are the broad, ultimate goals of the program, the achievement of which is dependent on accomplishing the results/criteria.

The *process* portion includes those dynamic transactions of a program such as case-finding, screening and enrollment, treatment, and community interfacing.

The *structure* of the program includes such static elements as staff functions and ratios to clients; standard procedures; resources available; setting and equipment.

There should be logical relationships among the three parts of the program: outcomes, process and structure.

The Description Matrix is linked to the Judgment Matrix by the next part of the model, *Standards*. This portion measures the appropriateness of the program elements chosen by the provider. We assume there are certain ideal elements of structure, process and outcome, which are common to all drug programs and additional program-specific elements which should be present in each type of program, i.e., recovery house, in-patient detox, etc. These elements can be identified from the literature and/or from other "expert opinion." When these standards are established, components of existing programs can be compared to them.

Judgment Matrix: This is the component which feeds back recommendations of alternatives to existing program elements for the improvement of program efficiency and effectiveness.

EVALUATION PROCEDURES FOR FIELD INTERVIEWS WITH FORMER CLIENTS OF DRUG TREATMENT PROGRAMS

Chris Balisky, B.A., Eugene Lofton, Ray Margarella, B.A., Leslie Rubicam, B.A., and Janet Scanga, B.A.

Center for Addiction Services

Locating and interviewing former clients of drug treatment programs and a follow-up evaluation study calls for a field staff with specialized skills, knowledge and tools plus a stronger, fuller support system than is usually required in other social science surveys. The respondents in this sample population are suspicious of someone who is looking for them, transient and sensitive about the confidentiality of the information they give. The field interviewer is subject to serious personal risks, many frustrations and situations demanding all their skill in order to gather valid data. Several recent follow-up and evaluation studies of drug treatment agency clients, using various methods and procedures, have come up with some of the same assessments of the field problems (Nurco, 1977; Boudouris, 1976; Robins, 1966; O'Donnell, 1968).

The Center for Addiction Services and Affiliate Agencies Evaluation Project is still in the process of gathering field data at this writing (March 1978). Field work will continue through October 1978. We feel that the methods and procedures we are using contribute to successful completion rates and can be useful in other evaluation projects.

The Evaluation Project, funded by a grant from NIDA, was designed at CAS by Karl Fischer. The field portion was developed by the University of Washington Center for Social Welfare Research Design Lab, under a contract with the Research and Planning Department of the Center for Addiction Services. The Design Lab's role was phased out at the end of the 02 grant year. The project calls for a baseline interview done at the time of a client's intake into CAS or one of seventeen affiliate drug treatment agencies--six therapeutic communities, eight mental health centers, two methadone maintenance facili-

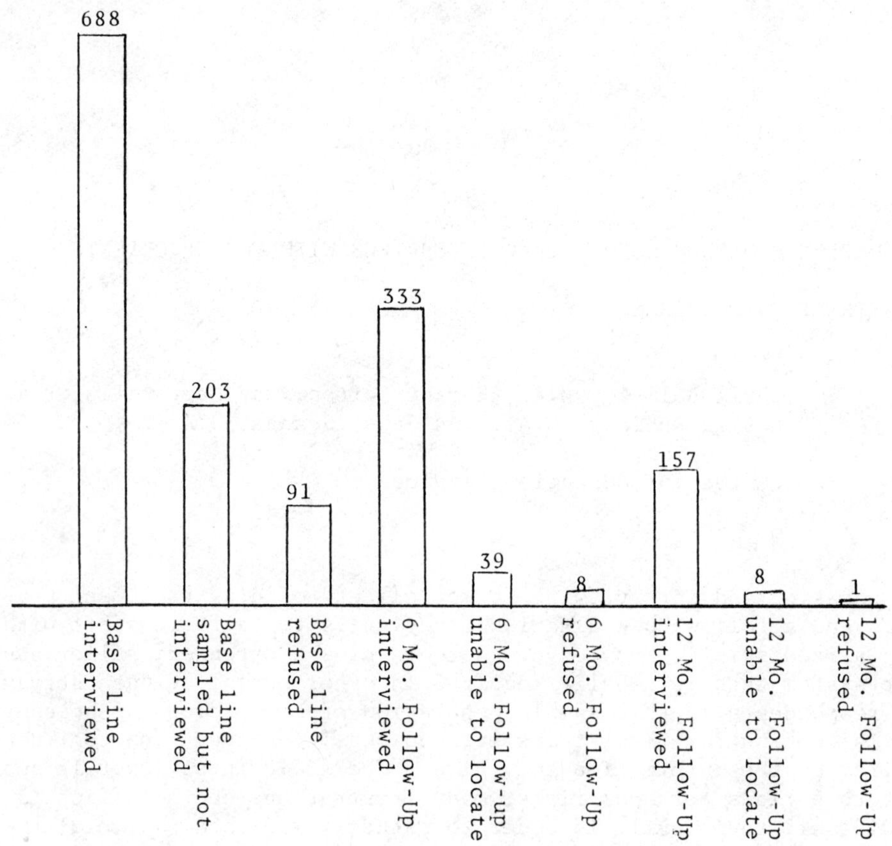

Fig. 1. Interviewing progress. # cases completed through March 31, 1978.

ties and a detoxification unit. Follow-up interviews are scheduled for six months and twelve months after a client leaves the treatment agency where the baseline was done. Figure 1 shows our progress to date.

An understanding of the sample population is important so we will begin with the profile of the respondents in our sample. Quantitatively, there is demographic data from the baseline questionniare along with information about homelife, family stability, drug use patterns, illegal activities, arrest history and history of employment. This data is incorporated in Tables I and II.

Against this family background, 47% of the respondents reported that they first used drugs or alcohol on a regular basis before they were 15 years old. Another 29% said their first regular use started between the ages of 16 and 18 (see Figure 2). Though heroin was the

Table I. Demographic data to respondents at baseline.

	Number	%
Male	428	62.2
Female	260	37.8
White	515	74.9
Black	144	20.9
Asian	7	1.0
Nat. Amer.	10	1.5
Other	12	1.7
Age		
14-17	8	1.2
18-25	306	44.0
26-30	191	27.7
31-40	145	21.0
41+	38	5.5
Married	228	33
Unmarried	460	67
Education		
No H.S. or G.E.D.	218	31.7
H.S.	336	48.8
Post H.S.	26	3.8
Never employed	26	3.8
Employed during "clear period"*	253	36.8

* "Clear period" is defined as a 2 month period of time nearest to date of intake when client was not incarcerated, hospitalized or in a situation different from normal routine.

drug that caused the principal problem for 59% of the respondents at the time they entered treatment, only 25% reported it was the first drug they had used on a regular basis. 69% of the respondents had prior treatment experiences.

In addition to involvement with drugs, respondents indicated involvement with the legal system. Figure 3 shows reported illegal activities, arrests and court pressures to enter drug treatment.

To these figures we need to add some qualitative observations from the field. The interviewers' knowledge about the sample popula-

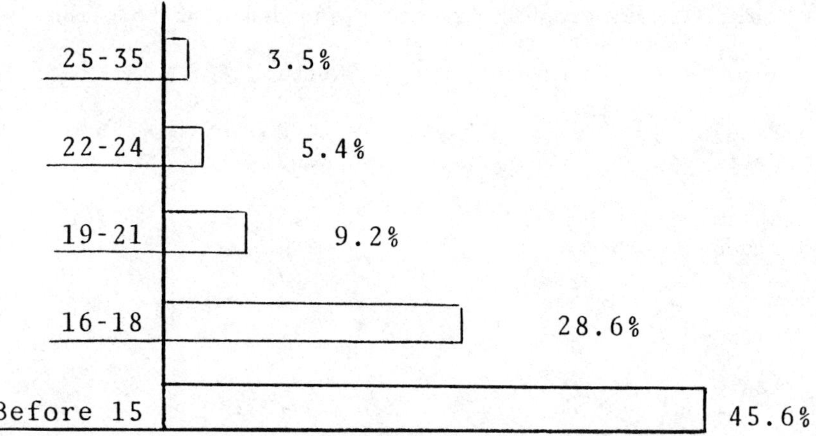

Fig. 2. Age of first drug use involvement on regular basis (greater than three times per week).

Table II. Early childhood, family background up until age 13.

	Number	&
Location of household		
Urban	591	86
Rural	92	13.3
Stability		
Stable: moved up to 2 times	566	82.2
Moved 3 times or more	69	10.1
Unstable: not more than 2 years in the same place	53	7.7
Lived with		
Both parents	454	66.0
One parent only	189	27.4
Family non-medical drug use		
Yes	89	12.9
No	591	85.9

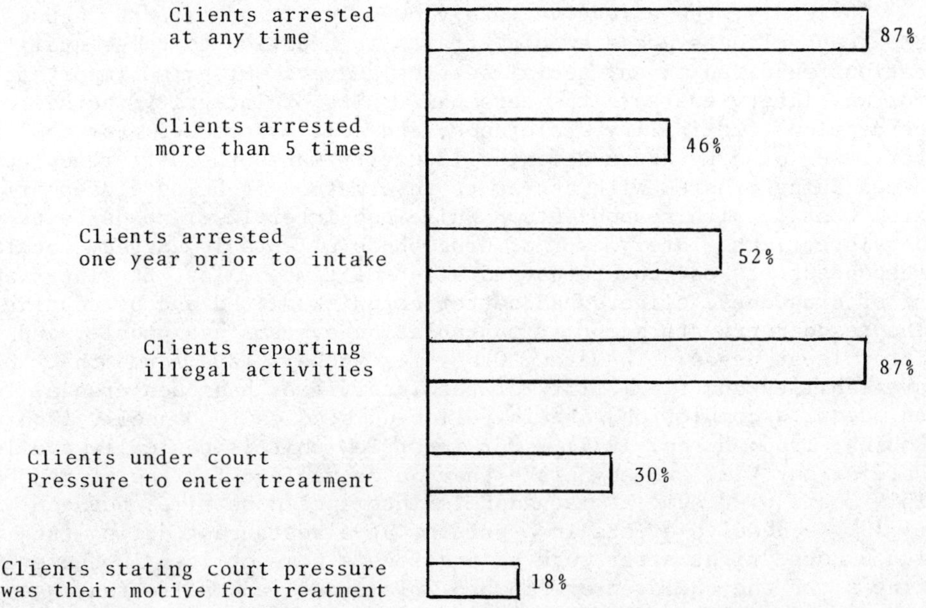

Fig. 3. Client arrest history at time of intake.

tion is important, but of equal importance are the interviewers' perceptions of the respondents and the respondents' perceptions of the interviewers. Being a drug addict puts a person through many trying times. A condescending attitude must be avoided. The interviewers must be accepted and trusted by the respondent and the drug sub-culture, which is one of their first tasks. As can be seen in Figure 1, since participation is voluntary, a small percentage of the sample does refuse to be interviewed. However, once respondents agree, they are cooperative and give each question a thoughtful answer. They like to talk and let us know the details of their lives.

Transience is not unusual. It is common for a respondent to be at a different address each time an interview is to be completed. Sometimes they go underground as a result of past or impending legal difficulties and sometimes they are located easily because they have returned to the treatment system.

Most striking to an interviewer from the onset is the lack of stereotype. Respondents in our sample came from nearly every stratum of society. Some had college degrees and some were nearly illiterate. There were professionals who were employed and could afford their habit and many who had little job training, were chronically unemployed and could not afford their habit. Though we find many in sleazy hotels or incarcerated at follow-up, we also find many at home with their families in suburbia.

Obviously, the effective interviewer in a study of drug abuse treatment clients needs special traits in addition to those qualifications required in any social service survey. Of prime importance for any interviewer are the personal traits of integrity, persistence perception, flexibility, tolerance, and self-assurance. For the interviewer of drug abuse clients, integrity makes the difference between being trusted with access to information or being closed off from contact with respondents. While any interviewer needs to be persistent, the interviewer of drug abuse clients is dogged. Locating respondents takes many turns and a lot of time. Also, the interviewer of drug abuse clients had better be quick-witted and perceptive. The respondents are adept at manipulating systems and people, and the interviewer needs to beware. Other researchers have found that the availability and flexibility of the interviewer has been crucial for an adequate completion rate in follow-up studies (O'Donnell, 1966; Robins, 1966; Nurco, 1977). Our interview unit is basically available at any time and any place that is convenient for the respondent. This means 6:00 a.m. at methadone maintenance dose time, between rigidly scheduled events in a prison, at a restaurant during the lunch hour, right after work at a client's home or late in the evening after the children are in bed. Tolerance and self-assurance make it possible for the interviewer of drug abuse clients to accept what they find, and move around in a sometimes unfamiliar, sometimes hostile environment.

In addition to those traits, there is a need for openness to learning. An interviewer must be receptive to new ideas. The interviewer must learn the techniques of neutral probing, reading questions exactly as written, and maintaining an objective attitude. When trying to get a full and accurate answer to a question, these techniques are important. Asking the respondent "What do you mean?" or "Can you give me an example?" or repeating part of the question (*not* rephrasing) elicits data that is less biased. Reading questions exactly as worded insures that each respondent hears the same questions no matter which interviewer asks them. Being objective involves the entire appearance and attitude of the interviewer. A mere change in intonation, facial expression or physical movement may affect the response. There is also the need to learn to control the interview situation in such a way that the respondent does not go off on tangents. Celia Homans (1977) found that "The best interviewers are people who feel comfortable in relating to others, who are not dependent on the opinions of others to maintain their personal image and who find a non-response or hostility in others a challenge rather than a threat."

All these demands on the interviewers lead to the real possibility of staff "burn-out". Our project has been fortunate in keeping this at a minimum. We feel a primary reason for this is the support system set up for the interviewers. Our system has five elements that we found important to successful completion of the field work:

1. Salaried staff positions for interviewers - the concentration and commitment called for in a study of the drug addiction population is more likely to be achieved when the interviewers are considered employees rather than self-employed part-time contractors.

2. Adequate office space and equipment - the field staff should have space enough to provide each interviewer with a place to work. There should be desks or tables, locking storage space for materials, telephones and extra chairs for respondents. The value of this arrangement cannot be over-emphasized. It increases the productivity of the interviewing unit greatly because it makes the unit easier to manage, centralizes all information about field resources and improves the morale of the interviewers by making them feel part of the project instead of dislocated outsiders.

3. Relevant files and records - resources for locating clients should be kept on file and accessible to the interviewers. We gathered and kept tracking sheets for each client interviewed while they were still in treatment. These are kept on file for use during the follow-up interviewing. We also keep card files that identify clients in our sample with records of their identification number, their original treatment program, the first interviewer, when they entered and left treatment and when follow-ups are due and completed. These resources are stored in cabinets that are locked when the office is closed.

4. Close supervision - the management of an interviewing unit entails much more than making assignments and checking in completed work. The supervisor is the organizational center of the unit, maintaining the files such that information is current and accessible; reporting progress on a regular basis not only to superiors but to the interviewers enabling them to plan their work; providing a sounding board and source of encouragement; and monitoring the quality and validity of the completed interviews.

5. Routine progress appraisals - the interviewing unit should know where it is going and how it is doing at all times. Weekly and monthly records and reports of progress in all areas of the field work should be done by the supervisor. Goals and procedures to meet these goals should be set up from the beginning. A constant analysis of various field data such as the number of interviews completed, the time it takes for completion, problem areas and countless bits of information monitor the status of the project. This makes it possible to change procedures and adjust priorities to meet the goals.

All five of these elements contribute to a team approach. Interviewer stress is increased and production suffers unless positive measures are taken to minimize field problems. The interviewers need support from the supervisor and one another--a feeling that they are respected.

A follow-up study of a drug treatment network presents the interviewer with an especially sensitive sample in regard to locator methods. At the conclusion of the initial interview, we asked the respondent for addresses of parents, brothers, sisters, other relatives, and friends. Then each respondent is sent a reminder letter four to five months in advance of the second interview. A self-addressed post paid post card is enclosed requesting their current address and telephone number. Returned cards are filed with the locator sheets. We call all possible contacts, leave written and/or verbal messages at the client's home or in care of a possible contact, write another letter to the respondent at home, in care of an employer, parole or probation officer, attorney, friend or relative. We also check the booking office of local city and county jails and state and federal prisons. We include in the messages that the respondent will be paid for the interview and that it can be scheduled at the most convenient time and place for the respondent. Someone is available to answer our phones during working hours and can set appointments for interviewers. Of course, information conveyed while tracking a respondent must be worded without reference to drugs, drug use or abuse, or study of drug abusers.

Reviewing previous records of locator steps for a particular respondent or talking with the other interviewers is often helpful in determining likely avenues to pursue. If there is an alias, this is often discovered via sharing and/or reviewing records and greatly affects the interviewers' success in tracking the respondent. Being open and friendly with the client's family without divulging confidential information helps establish trust between the interviewer and the family member or friend so that good locator information can be obtained. Also an interviewer needs to gain acceptance with the drug user's community. Each person that appears on a regular basis where clients of drug treatment programs are to be found is going to be identified and placed. Is that person another drug addict, a drug dealer, a narcotics agent, a social worker or an interviewer for a survey? It is important for an interviewer to be identified as someone whose interest is objective. The interviewer's commitment to protect the confidentiality of information should be made clear so that neither the respondent nor the interviewer is put in jeopardy. Once acceptance and trust is established, leads on respondents can be obtained. Previously interviewed respondents are sometimes our first contact. When we go to treatment agencies, talking with workers and clients is valuable as an ice-breaker in new surroundings. To quote Homans (1977) again, "People who can learn to be skillful in finding hard-to-locate respondents are usually the same people who can conduct interviews."

The training of field interviewers is pretty well standardized by social science research organizations that maintain a large field staff and have several studies on the field over a year's time. The process is usually divided into basic training and then briefing for the specific study. Any basic training process that effectively trains interviewers in locating and contracting, establishing rapport, interviewing technique, and accuracy in recording and editing data is directly applicable for training interviewers of drug treatment clients. The skills required to read questions exactly as written, record verbatim responses to open end questions, protect confidentiality, and follow complex directions are just as necessary here as in any other social science research survey. Our own procedures were designed to provide intensive training in all these skills. We went over the interview schedules explaining the purposes for each question. We used practice interviews with staff and paid respondents who knew it was a training session. We used a video tape machine to show interviewers how they were doing and each interviewer was observed as they conducted their first one or two interviews. At each point in this process critiques were written by the trainers and discussed with the trainee. A newly hired interviewer is on probation for three months. In our opinion, training must be ongoing--staff training meetings should be routine for at least three months after field work begins. If there are relays of hiring and training as the study gains momentum, then experienced interviewers can help new ones. But when all the interviewers are new, allow time and provide support as they get established with the client environment.

We have already discussed the special attributes of the field interviewer but it needs to be added here that every affordable incentive used to keep competent trained interviewers on the staff and keep staff turnover at a minimum is well worth it. Selective hiring, thorough training, full support, and adequate compensation of interviewers saves time, money and even the evaluation study in the long run.

In order to gather valid data, the interviewers must be successful at solving the difficult cases. The respondents are hard to locate, or they are unreliable about keeping appointments, or they are behind layers of bureaucratic red tape. As can be seen in Figure 1 there are a certain number of cases that are closed out as "unable to locate". The effort spent to keep this number low takes its toll in time and frustration. For instance, respondents are not home even though an appointment is made; addresses are often unreliable, i.e., vacant lots or no such number; warrants, legal problems, or dangers and threats cause the respondent to go underground; mothers and other well-meaning contacts run interference for the respondent fearing that the interviewer may do harm to the respondent or might be a drug dealer.

Some of the difficulties are generated by the respondents after we find them. The respondent is sick or does not have time; the respondent looks to the interviewer as a way to get back into the treatment system, or just does it for the $10.00 fee; or the respondent may exaggerate or extend the life story to impress the interviewer. Other difficulties are imposed by treatment programs that try to shield the respondent or bureaucratic procedures at jails, prisons and other government agencies. Given these constraints the interviewer has to pace the work, meet deadlines, maintain quality, protect confidentiality and learn to drop (i.e., close out) a long pending case in which a lot of unrewarding effort has been expended.

In spite of all these difficulties and frustrations there are rewards in the field work ranging from the tangible paycheck to an often elusive sense of accomplishment. We focus the competitive challenge between the "team" and the field rather than between staff members and depend heavily on intra-staff support systems. Not only is the variety in our sample population the spice of life, but noting improvements and building up a rapport over time gives us a personal sense of commitment. This commitment to the evaluation project is enhanced with each completed interview. We also have the additional opportunity, unusual for most field interviewers, to review the data as it is compiled and to take part in some of the preliminary analysis tasks.

In summary, we feel that any evaluation project can improve its success record by giving close attention to the field problems and the field staff. Even the most elegant research design and the most brilliant analysis cannot be supported without accurate data, collected within a given time frame, by motivated field staff.

ACKNOWLEDGEMENTS

This project was supported by Grant number H81-DA-01781-03 from the National Institute of Drug Abuse. Special thanks to R.M.J. Haglund, Principal Investigator, for his assistance in preparing this report.

REFERENCES

Boudouris, J. 1976. A follow-up study of dropouts from a methadone maintenance program. *Inter. J. of Addiction*. II(5): 807-818.

Homans, C. 1977. Interviewing respondents. *Conducting Follow-up Research on Drug Treatment Programs*. 85-97.

Nurco, D.N. and Lerner, M. The feasibility of locating addicts in the community. *J. of the Addictions*. 6(1): 51-62.

O'Donnell, J.A. 1966. Research problems in follow-up studies of addicts. *Rehabilitating the Narcotic Addict*. Feb: 16-18.

Robins, L.N. 1966. *Deviant Children Grown Up*. Baltimore: Williams and Wilkins.

DIFFERENTIAL USE OF SELF WITH ADDICTED VERSUS PSYCHIATRIC PATIENTS*

Maryanne Vandervelde, Ph.D.

President
Pioneer Management, Inc.
Seattle, Washington

We need to learn more as therapists about how we use ourselves differently with addicts versus other psychiatric patients. During the four year period 1971 to 1975, five social workers at Hall-Brooke Hospital in Westport, Connecticut, had the rather unique experience of working on both the addiction unit and one or more psychiatric units. This is an unusual situation because addiction programs tend to attract a particular type of professional, and social workers who get into that specialized area tend to remain there. Similarly, social workers experienced in psychiatric settings often express distaste for working with alcoholics and drug addicts. By contrast, the "musical chairs" philosophy of this institution allows all of its treatment staff to move back and forth, usually at one year intervals, between various kinds of treatment units, and to work in team relationships with a variety of other staff members. The accumulated experience of these five well-trained, highly professional social workers points to some rather interesting differences in therapeutic work with addicted persons versus other psychiatric patients. Of course, there is a basic core of knowledge and a basic sense of professionalism which remains extant, but the different ways these people used themselves as change agents with the two groups seemed striking and significant. Although some of the conclusions may be peculiar to these social workers in this particular setting, most of them can be generalized.

---

* Special appreciation is due to Carol Greenberg, Mary Luminoso, Suzanne Metcalf, and Miriam Singer for their contributions.

Drug and alcohol treatment programs have often had a bad press within the professions. Is this situation warranted? Do we really understand how a social worker uses himself/herself differently in these two settings? Can we be more clear about what kind of qualities and personal resources are called for in each situation? What are the gratifications in working with addicted personalities? And what kinds of lessons learned from working with addicts can change one's life as a therapist?

Differential use of self has been a part of mental health literature and education for many years. Even the most orthodox Freudian analyst has some experience with the need to change his/her attitude or posture in certain situations, such as the need to establish a therapeutic alliance with a very anxious patient, the need to temporarily withhold all interpretations from an angry, furious, borderline patient, or the need to talk intensively and "be with" a suicidal patient (Burger, 1974). The more recent emphasis on creativity in a therapist suggests the need "to be able to utilize different therapeutic approaches selectively, to be a true eclectic, to know how and when to be a Gestaltist or to use free association, when to interact, when to analyze a dream, when and how to confront, or when to use behavior therapy or whatever" (Quaytman, 1974, p. 172). There is still a fascinating dialogue going on among psychotherapists as to how much one's own feelings, thoughts and behavior should be disclosed.

Essentially, there seems to be little consensus yet about just how one works differently with addicts. And there seems to be some kind of antagonism and resentment between those who treat "straight" psychiatric patients and those who are on the firing line, so to speak, with addicts. There seems to have been precious little sharing about what is alike and what is different in the way one uses oneself with these patient populations, and very little acknowledging of how lessons from treating one group can be carried over to the other. Experience from this sample of professionals, thoughtfully analyzed, may shed some light on the subject and provoke further analyses.

Ten specific comparisons can be made about working with these two types of patients:

1. In psychiatric units the therapist's tendency or inclination is to believe until the facts prove otherwise; with addicted patients, the therapist develops a tendency to <u>dis</u>believe until the facts prove otherwise. To borrow from legal jargon, the psychiatric patient is innocent until proven guilty; the addict is guilty until proven innocent. Social workers do not find this switch easy, but they quickly learn that if they are trusting and believing

with addicted patients and do not question intensively, they are seen as some kind of patsy - one of those middle-class people who can be easily fooled. Alcoholics and street-wise drug addicts are slightly different in their styles: Alcoholics tend to be older and more middle class, and their games of manipulation and deception may be more subtle and sophisticated. Drug addicts are more obviously tough and confronting and antagonistic to any "establishment" standards; their intricate systems of deception are more angry and anti-social, and of course involve many illegal acts. But both habits are basically built on a system of lies, and any therapist who starts out as a believer soon finds himself/herself in big trouble.

As one worker said, it took her a while on the addiction unit to grasp that, in order to be effective and get somewhere, she had to be pretty mean. Most therapists are trained to see themselves as kind and understanding, and that's the way they want to operate. Then they get "taken to the cleaners" a few times, and suddenly they have to re-evaluate that. They have trouble thinking of themselves as mean, direct, obnoxious, but that approach seems to work. Eventually, they develop a way of using both - i.e., at times using the disbelief, cut-the-crap, confronting attitude, while at other times really talking to the person and being deeply interested. Most of the time they need to be quite guarded in what they believe and accept, but they come to see this as an important way of using themselves. There may be trouble at first knowing how to handle the blatant, almost constant lieing and exaggerations; most of it seems so obvious and ridiculous that workers are reluctant to "get in there more", especially with people who seem to have so little commitment to treatment. It seems easier to just ignore it, rather than confront every issue, every lie. But invariably, when the therapist just accepts things, more lies come. The important issues seem to be: How is the therapist seen? Can the therapist be easily conned or tricked? Is the therapist sharp enough to dig behind the patient's facade? Is the therapist sturdy enough to stick with the patient, and interested enough to get down to his core?

Having learned this skill, however, there is a feeling of growth and accomplishment. At times with all patients, one must be very direct and confronting, and not worry about whether the patient is feeling helped. Having gotten comfortable with confronting, having gotten to the point where you don't need to hear how much you are helping, you are better able to get under the surface level with patients, and able to do this part of the work much faster.

2. The move from a psychiatric unit to an addiction unit makes most workers poignantly aware of how dependent they have been on positive feedback from patients. People who go into the helping professions usually do so at least partly because they like the image of themselves as helpers, and most of our egos thrive on

reinforcement of this fact. Consequently, we tend to put various kinds of pressures on our patients, often unconsciously, to meet our own needs, and are reluctant to antagonize or differ with a patient even if that might be more helpful.

On a regular psychiatric unit, at least in this institution, the social worker is a very powerful, important person, and has a distinct role. On the addiction unit, you as an individual are not very important. With the possible exception of the ex-addict counselor or alcoholic counselor, team members on that unit are seen as necessary evils. There you are; you come along with the program; and the program is something patients need to stay out of jail or appease their families, employers, or whatever. So in most cases you cannot establish the same kind of relationship that you can with other patients who feel a need for you, want to talk to you, seek you out.

Psychiatric social workers are really "spoiled" because most of the time people want what they have to give. They are not very accustomed to selling themselves or selling their services, as is de rigeur in many other parts of the social work profession. And to find themselves constantly rejected! - that is very draining, and it can have a very negating effect. It often seems that whatever the addict wants, the social worker isn't going to be able to give it. For example, if they want methadone, the social worker can't give it; if they want absolution from their families, the social worker can't order it. Whatever the worker <u>can</u> provide, they don't seem to want. They don't want the insight that you've used to help other people because their only problem is the alcohol or drugs. They don't want the communication skills that you could teach them because they're much better at it than you are. So there's a process of negation that you have to expect. But this forces you to deal with what your own strength is rather than depending so much on external gratification.

Addicted patients use the hospital just the way they use the world. When a patient on that unit comes to you and seeks you out, you had better understand that he wants something, but it won't be in the form of help you want to give. If the patient feels comfortable, he might ask you directly for it; if not, he'll try to con you and ask indirectly. It will usually be something like, "Talk to my probation officer and tell him this," or "Go to court with me and see if you can get me welfare," or "See if you can convince my boss that I won't drink anymore." It almost never is, "I would really like to talk; I feel terrible." If he ever comes to you and says that, watch what's coming next, because he really wants something else. Most addicts just don't operate by baring their souls and getting relief that way.

As a result of this situation, something very personal happens to the therapist. He/she is somehow strengthened and is really less burdened by the need to be seen in a certain way. It is a kind of toughening experience, but it is also reality. It is a pleasant feeling to be valuable to the patient as a person, but also pleasant to know that you are not dependent on this.

3. There is something about working with addicts that forces the therapist to get comfortable with himself/herself. Unless you know where you stand in every area and what your honest feelings are, addicts will pick it up and zero in on that area. If what you want from your patients is gratitude, acceptance, thanks, improvement, you're going to be hurt over and over again. There is a great deal of testing that goes on, and you are better off to start out with the idea that alliances are unlikely. Somehow, once the therapist is found to be sturdy and not very connable, there often evolves a certain kind of respect, and liking, and even trust. But any kind of mother or pal or God-like feelings on the part of the therapist spell doom. And if the therapist does not come across as real, he/she will be clobbered.

One social worker was left by the psychiatrist to handle a particularly difficult privilege meeting alone. She was furious with him, and felt quite unwilling to make certain decisions. The patients picked this up and put intense pressure on her with all kinds of conning and pleading. The meeting was chaotic until she leveled about her feelings. After that, some help came from patients and other staff, and decisions eventually got made.

Workers also learn on the addiction unit to deal with defensiveness they have about lacks in the hospital. These patients are anxious to attack the staff for any little mistake the hospital makes. It's almost as if their antennae are out for whatever the staff is feeling a little guilty about. The worker must learn more rapidly on this unit to deal with his/her own feelings about the hospital and what things can and cannot be controlled. The worker must learn to separate out inadequacies on the part of the hospital from guilt feelings toward the patients. You must learn quickly how to explain and understand hospital actions without getting defensive about them. By contrast, psychiatric patients tend to be much more trusting of the institution and much less confronting about bureaucratic "mistakes."

4. The therapist has to be guarded in different ways with psychiatric patients and with addicts. There are various kinds of fragilities in psychiatric patients, and the therapist, especially in a group, is constantly screening for appropriateness and timing. This does not mean that all psychiatric patients are fragile, but with certain patients, for example, you might withhold a negative response, store it in the back of your head and perhaps use it later. Some psychiatric patients are easily frightened; the therapist would

be careful to screen out things that might set off paranoid fantasies In most psychiatric groups, the ages, symptoms and backgrounds are so diverse that the therapist feels intense pressure to meet all needs in different ways.

At first glance, then, an addict group may seem like a relief. The symptoms and backgrounds are similar, the preferred topics of discussion are alcohol and drugs, and the therapist may think he or she can relax. One worker even described feeling like a peer at first - so many of the patients were her age, and they seemed to demand so little of her. But, as this worker quickly learned, there must be a different kind of guardedness with this group. You must be careful here to screen out things that you can later be hit over the head with. You don't have to curb negative response with this group, but you had better be sure you know what you're doing.

The same worker described her first experiences with an outpatient addict group. She attended two sessions with the previous therapist, and thought the patients were benign and easy going. Consequently she relaxed, opened up about herself, and answered their questions rather loosely. On the first night she took over alone, they let her have it. Who did she think she was, with her middle class values, making judgments about them, coming in when she knew nothing about them, etc. She was properly stunned, and her guardedness increased.

5. Both treatment situations require the ability to be an authority figure, but on the addiction unit, you must almost be a negative authority; you must be comfortable with being fought against, being challenged. The addict's usual approach to authority is negative, and authorities in his experience have often been inconsistent and unfair.

To the surprise of many workers, there is an excitement in being challenged and confronted. The patients look you over and size you up, and you find yourself doing the same with them. It's almost a question of what kind of game you two will be playing. Who's going to get the first move, and what is it going to be? It's almost a strategy session, and it's amazingly exhilarating. Being the target can be very exciting and can get you quite charged up. This almost never happens with psychiatric patients. One favorite tactic of addicts is to attack a staff member in a large meeting. To some staff members this may look like a horrible, frightening experience. But if the therapist knows he/she is right, it can be experienced as a marvelously exciting challenge. It's a bit like fighting a battle, with the opposition using all kinds of sneaky, manipulative tactics; and winning.

6. Work with addicted personalities helps the therapist to confront his or her moral judgments. Self awareness is, of course, an

important part of work with psychiatric patients, too. But an addiction unit puts you in touch with people who lead sub-terrainian kinds of life styles, people you never would have been in touch with in any other way. These may be people who are in and out of jail, constantly robbing, beating people up. The therapist finds that he/she gets more accepting - not of the acts, but of the person despite all his acts. The effect is to become a little less up-tight, a little less moralistic.

This is not an easy process. Two therapists, for example, had had their houses robbed; these had been terrible experiences for them, and hearing patients talk about robberies they had done reawakened dreadful feelings. Crimes of violence are even worse to hear about. One worker was horrified to learn of a patient who made his living by beating up people on assignment; that was how he got his money for drugs. It is often hard to know how to handle one's distaste for these acts. Expressing displeasure may simply be a repeat of the parents they've heard a million times, but the verbalization of distaste is occasionally helpful. Certainly anti-social behavior should be treated as a symptom, and most of the patients know when their behavior is offensive or wrong anyway. The guy who made his living beating up people might be told that that was a horrible, astonishing thing to hear. He might also be told that it is a rotten way to earn a living because it is very dangerous. But he knew all of this already, and himself said that it was repugnant. He was so caught by that need for drugs that he saw it as a survival tactic.

Some female therapists have a harder time accepting the prostitution of female addicts than the male type of acting out. Even though many addict-prostitutes feel they are using the man too, they usually end up with feelings of humiliation, degradation, and worthlessness. The usual "male" crimes of breaking and entering or assault do not demean the person and are not self-destructive things from the patient's perspective.

On the other end of the spectrum, the therapist may at times find his/her own values more liberal than those of the law or the institution. A particular worker might agree with Garbin that "any individual should have the legal right to use or abuse any drug he chooses" (1972, p. 333). Garbin feels that a worker can indeed hold the belief that the use of drugs should be a matter of individual choice. The most usual controversy comes over the use of marijuana; patients can ask some hard questions about your feelings and your own use of drugs, and many middle-class people these days have tried pot. Professional tensions can result, and you must be clear about judgmental attitudes in either direction.

If the therapist can put aside moral judgments and get beneath the guardedness of the patient, it is possible to get a very rich picture of a complicated life style. Usually you can get a good impression of a psychiatric patient's life, or how he spent his time, rather easily. With the drug addict, the quality or flavor of life style is more elusive at first. But the conning and planning is an art in itself. All of life is drug related. The relationships in the sub-culture are fascinating. All the addicts in a particular area know each other; they have relationships which are based on need. None of them really trust each other, but they will all use one another, and there are contracts made between them. They know who is helping whom; what man is living off what girl; who did which robbery. If there was a B and E, they know who did it because of they way it was done. They talk about having to establish themselves physically, especially in jail, in order to avoid being inundated by other addicts or prisoners. (That is the big fear about going to jail and the reason most choose treatment instead of jail.) The older men worry about the young, attractive boys, especially the small, thin ones; they try to explain to young male patients that they would be used in jail and would lose their manhood. There is a detailed, complicated life pattern under the empty facade.

So the therapist really has to be in touch with middle class values and be aware that they might not apply to certain patients. Even middle class alcoholics sometimes have rich, deceptive, subterrainian lives that can be shocking indeed. But the more you are aware of the realities of life, and the more confronting you can be of your own judgmental attitudes, the more effective you can be as a therapist.

7. The importance of team work is especially clear when working with addicts. Most workers feel that good teamwork is equally important on a psychiatric unit, but it often doesn't happen because the patients don't make the staff so conscious of it. On an addiction unit, the patients will zap you with any weakness in the team structure, and will manipulate one team member against another. The addiction team might frequently find themselves together in an office just before going in to group, almost to present a united front and "get themselves together" for the barrage from patients. Obviously, then, the team members need to be comfortable with one another and work out their differences, because it's almost like clashing armies at times. The team needs to be strong; there must not be any breaks in the ranks. If there is one dissenting team member, if one member doesn't like what's going on with a particular patient, the group finds the weak spot. If the staff isn't on the same wave length, or if staff problems develop, the patients move into the breach quickly.

8. The kind of group and the use of group process is very different in the two places. The people in the addict group are seen as the most helpful treatment tools, and the role of the therapist is to get them working on each other. One must learn how to get

them to pick up the conning in each other and to use their sociopathy on each other. The usual material in this group is day-to-day activities and relationships within the group. They confront each other about styles of using people: "You two have a contract not to talk about certain things, and you're not letting the group in." It's usually the interaction of the group that does the work if any work gets done. The team really has to provide the framework and be very clear. Of course, these patients can also reinforce each other's negative impulses, and the team has to help them to see when they're doing that. The team has to watch for collusion in a group that keeps it from getting any real work done; and the team has to assess when a group is capable of doing good work because not all groups are. When they are at their best, groups of addicts can be given most of the responsibility for privileges. When the group doesn't do its work for some reason, a particular privilege might be taken away from the entire group. Patients can exert more pressure on each other than staff can. This is the kind of group where people talk. There is something buzzing all the time, always some verbal exchange, and their effect on each other is really where it's at.

By contrast, a psychiatric group almost always has some depressed, withdrawn patients who don't talk; they don't even relate that much to other patients and are terribly self-involved. In a psychiatric group, the staff often has to work hard to get things going. While the addicts are looking for confrontation, most psychiatric patients are trying to avoid it. However, there is usually more focus in a psychiatric group on underlying causes of symptoms, and in a good group, the discussion is often more meaningful than the usual addict confrontation.

Because of the differences in group process, the team reviews of the groups tend to be quite different. The psychiatric teams are usually more concerned with individuals or how the individual operated in the group that day. Because the symptoms are so different, these patients seem to get more individual attention, even in groups. On the other hand, addicts have much more common personalities, and much of the review is spent assessing where the group is at any particular time, what the sub-groups mean, how to deal with pairing, etc.

9. The families of addicts are more disturbed and harder to work with, while the families of psychiatric patients tend to be more treatable. For example, the parents of addicted adolescents often do not get along, and give conflicting messages; consequently the problems of these kids are easier to trace but harder to treat. Another complication is the fact that parents of addicted adolescents have usually gone through so much by the time they get to a hospital; after courts and lawyers and police and a great outlay of money, they have often become turned off and skeptical, and have

removed themselves from the treatment process. In most cases, parents of an acutely ill adolescent on a psychiatric unit are very distressed and want to be involved: The kind of parental investment in these two situations is quite different.

There have been several theories about the type of pathology found in addicts' families, such as over-protective mothers and ineffectual fathers; but no consistent pattern or categories of pathology have been scientifically demonstrated. There simply seems to be less strength in these families, less willingness to change patterns of communication, and the goals with addicts' families must of necessity be more limited.

There also seems to be validity to the theory that families of addicts often have an investment in keeping the addiction going. There was the daughter of a policeman who had been on heroin for ten years; during that time he had pulled every string imaginable to keep her out of jail. There was the mother of another addict who covered for her son's stealing by paying back the money he stole; when confronted, she got very angry and wanted to know how many children the therapist had. There are the straight wives of addicts who work and quietly pay for their husbands' habits for years. Part of the therapy with these families is to get them to stop denying. Sometimes the kid has done everything but put heroin on the table, but it is still hard to get to just what the family investment is. Some of these parents seem to get a charge out of their kids acting out or doing sneaky things, but the therapist can seldom get directly into that because the families are so inaccessible. At times almost a behavioral approach is indicated. For example, you might teach a mother who had discovered $50 missing from her purse that it is useless to ask, "Did you take it?" The kid would say no, and a crazy hassle would develop. Rather, you would deal with her on the surface level, such as, "If you know the answer to your question, why do you ask it?" This is not terribly different from the A.A. or Al-Anon approach, which encourages the family to stop rewarding or encouraging or covering up the drinking, and rather to decathect from it.

With a somewhat sharper focus on psychiatric units, workers find themselves getting more into what is going on with each family member and what the relationship patterns are. The classical kinds of family therapy can be done much more frequently on psychiatric units than in an addiction program. And yet, just this kind of expertise is necessary to determine how to use one's skills with families differentially in the two settings.

10. Work with addicted people helps therapists to become more realistic about recidivism and "cures" in general. You come to quickly learn on an addiction unit that patients will repeat. You dismiss the grandiose idea that a short in-patient stay will cure

them of alcohol or drugs for life; you must think in terms of a long out-patient program and possible re-admissions. Even if the staff doesn't think this way, the patients do. This somewhat controversial view, in our experience, is valid: "Limited goals seem to be the only feasible ones. It is important to view addiction as a chronic illness in which periodic abstinence, even for a few weeks or months, is considered a boon to a patient, family and community. For most patients, total abstinence for extended periods may be a goal beyond realization" (Moffet et al., 1974, pp. 394-5).

The trick, of course, is that you must never give the patient permission to fail. Many therapists, although they do not expect immediate success, see an element of fatality in alcohol and drugs, and they honestly warn patients about that. The physical complications of alcoholism are often crippling or fatal; the statistical chances of death from an overdose of drugs may be slight, but there are also the dangers of death from a cop's bullet during a robbery or the emotional death of spending most of one's life in jail. In a general way, both addictions are extremely destructive to one's life style and to any worthwhile relationships.

Thus, the discharge planning is somewhat different. To a psychiatric patient anxious about leaving, you can say, "Just relax; don't get so upset if you start to feel a little confused or whatever." But you can never really give addicted patients permission to feel this relaxation or operate this way. When addicts talk of their anxieties about leaving, your whole response must be different. You have to help convince them that they don't have to hit bottom again, even though you know that they possibly will.

Becoming more realistic about discharge planning and goals invariably makes you a better treatment person because it removes some of the pressure and expectations. It's much more comfortable for both patient and therapist if you know you're not going to accomplish everything. It takes the pressure off the patient to meet your needs and desires. It gives him/her more freedom to level. As one worker put it: "The way I look at it, I have very little control over their success; it is primarily up to them! That's where it's at. That really has helped me accept my tiny little role in helping someone get better."

SUMMARY

There is much evidence that the successful therapist works differently with addicts versus other psychiatric patients. Experience in treating addicts is often difficult and painful, but it is also extremely beneficial for the therapist. There certainly aren't a lot of successes, and the patients don't give you much gratification, but it is undoubtedly a growth experience and a strengthening process. It is hard to quantify it, but one worker even said she "had a ball". Perhaps it is the sense of excitement you get from knowing that you can tangle with these rather tough people and come through intact - maybe even stronger for it. Another positive aspect is the team work; the team is challenged to be straight with each other and really make close professional relationships. And occasionally there are the dramatic successes. It is probably those dramatic changes that keep you going in psychiatric treatment too; but to see those kinds of changes in an addict who is on the road to self-destruction, if not fatality, can be especially gratifying. To sense one's strength in a new way, to learn to stretch and use oneself even more differentially, is to really grow as a therapist and as a person.

NOTES AND REFERENCES

Burger, M. - Piaget. 1974. "Changes of Psychoanalytic Technique," *Dynamic Psychiatry,* Vol. 7, No. 5.

Garbin, J.P. 1974. "Professional Values vs. Personal Beliefs in Drug Abuse," *Social Work,* Vol. 19, No. 3, May.

Moffet, A.D., Bruce, J.D. and D. Horvitz. 1974. "New Ways of Treating Addicts, *Social Work,* Vol. 19, No. 4, July.

Quaytman, W. 1974. "What makes A Creative Therapist?" *Journal of Contemporary Psychotherapy,* Vol. 6, No. 2, Summer.

PSYCHIATRIC ILLNESS WITH CONCOMITANT DRUG ABUSE

Sondra K. Stickney, R.N.C., Richard C.W. Hall, M.D., and Earl R. Gardner, Ph.D.

INTRODUCTION

The correlation between drug abuse and intrapsychic problems is not a new issue as can be substantiated by the number of specialty units designated for the treatment of drug and alcohol abuse. The psychiatric literature emphasizes specialization theories regarding drug abuse as a specific form of mental illness. A review of current literature reveals a preponderance of articles dealing with specific psychiatric symptomatology caused by the use of particular substances. Few articles examine the prevalance of substance abuse in psychiatric patients carrying unrelated diagnoses. Those studies which have been done indicate a wide range of prevalence figures for drug abuse, 31%-78%, in selected psychiatric populations (Butzer and Cline, 1974; Fisher et al, 1975; Westermeyer and Walzer, 1975; Blumbert et al, 1971; Hall et al, in press; Cohen and Cline, 1970; Whitlock et al, 1967). The majority of these studies have focused on the young, ages 15-25, rather than considering the psychiatric population as a whole.

This prospective study was undertaken to define the prevalence and patterns of substance abuse in a general psychiatric population and their effects on hospital course and diagnosis, as well as to investigate the question of whether drug abuse precipitated the psychiatric illness or the psychiatric illness had precipitated the drug abuse.

METHOD

Fifty-seven consecutive admissions to a university affiliated, inpatient psychiatric research facility were systematically evaluated for demographic data, psycho-social history, history of substance abuse, type of substance abused, duration of abuse, patient rationale for abuse, previous psychiatric admissions, current hospital course, AMA discharge rate and diagnostic confusion. All admissions to this unit were extensively prescreened to meet specific admission criteria. A pre-admission psychiatric interview was conducted in an attempt to eliminate those patients with primary diagnoses involving drug or alcohol abuse, characterological disorder (particularly sociopathy) or significant preexisting mental disease. This information was obtained on the basis of patient self report, history provided by family members and available records. No minors were accepted for admission to this unit and a sexual and racial balance was maintained whenever possible. Emphasis was placed on admitting patients with a variety of diagnoses because of both the research and teaching components of the facility.

All diagnoses were evaluated on the basis of mental status examination and ability to meet Endicott Spitzer Research Diagnostic Criteria (1975). Patients were evaluated for the following classifications of drug abuse: hallucinogens, amphetamines, barbiturates, opiates, marijuana, alcohol, physician prescriptions and over-the-counter drugs.

Patients were evaluated for total number of psychiatric hospitalizations, length of stay during current hospitalization, and proclivity for AMA discharge.

Data was analyzed using the test for significance of difference between two proportions (Bruteline and Kinty, 1968). Differences were defined as significant at the $P \leq 01$ level.

DATA

Thirty-three of the fifty-seven consecutive admissions evaluated (58%) had a history of drug abuse. Demographic, social and historical data are shown in Tables I-III. Data concerning past and present psychiatric hospitalizations are shown in Table IV. Primary diagnosis at time of discharge is shown in Table V. Categories and extent of abuse are depicted in Table VI and Figure 1.

The average age of onset for abuse was 19.2 years with the average duration being 7.5 years. Mean age of the male abuser was 24.5 years while that of the female abuser was 27.2 years. Twenty-seven percent of the previous hospitalizations for the drug abuse population were directly related to their substance abuse.

Only one patient was found to abuse alcohol alone, seventeen abused alcohol in addition to other drugs.

Table I. Demographic Characteristics

	TOTAL (N=57)	KNOWN ABUSE (N=33)	NON-ABUSERS (N=24)
SEX			
Male	49%	61%	33%
Female	51%	39%	67%
AGE			
25 and Under	51%	67%	29%
Over 25	49%	33%	71%
Mean Age	27.4 yr.	26.8 yr.	30.1 yr.
Median Age	35 yr.	31 yr.	35 yr.
Age Range	18-52 yr.	18-44 yr.	18-52 yr.
MARITAL STATUS			
Single	42%	55%	25%
Married	25%	15%	37.5%
Widowed, Separated, Divorced	33%	30%	37.5%
RACE			
White	61%	70%	50%
Non-White	39%	30%	50%
EMPLOYED	25%	24%	25%

Table II. Demographic Characteristics

	TOTAL (N=57)	KNOWN ABUSE (N=33)	NON-ABUSERS (N=24)
EDUCATION			
Less than high school completion	39%	39%	38%
High school graduate	27%	36%	12%
Some college	34%	25%	50%
Mean education	12.1 yr.	11.5 yr.	12.9 yr.
Learning disability	18%	21%	13%

Table III. Differences Between Abusers and Non-Abusers in Relationship to Home Environment

	TOTAL (N=57)	ABUSERS (N=33)	NON-ABUSERS (N=24)
Alcoholic Parent	26%	40%	8% Significant at P < .01 level
Unstable Home In Childhood	33%	45%	17% Significant at P < .01 level

Table IV. Past and Present Psychiatric Hospitalizations

	TOTAL (N=57)	ABUSERS (N=33)	NON-ABUSERS (N=24)
TOTAL HOSPITAL ADMISSIONS	113	67	44
HOSPITALIZATION RATE PER PATIENT	2.0	2.03	1.8
AVERAGE HOSPITAL STAY (current hospitalization)	58.3 days	60.9 days	54.8 days
AMA DISCHARGE (current hospitalization only)	16%	21%	8%

Table V. Primary Diagnostic Classification at Time of Discharge

	ABUSERS	NON-ABUSERS
Acute Psychotic Episode	6	0
Hysterical Personality	6	1
Psychotic Organic Brain Syndrome	6	3
Schizophrenia	7	3
Paranoid Type	5	0
Hebephrenic Type	1	0

Table V. Primary Diagnostic Classification at Time of Discharge (Continued)

	ABUSERS	NON-ABUSERS
Chronic Undifferentiated Type	1	3
Borderline Personality	2	0
Schizo-Affective Disorder	2	2
Sociopathic Personality	1	0
Antisocial Personality	1	1
Depression	1	7
Adult Adjustment Reaction	1	1
Manic/Depressive Disease	0	2
Anxiety	0	3
Dissociative Reaction	0	1

Table VI. Categories of Abuse (N=33)

Hallucinogens	45%
Amphetamines	45%
Barbiturates	45%
Opiates	21%
Marijuana	76%
Alcohol	50%
Doctor's Prescription	30%
Over-the-Counter Drugs	12%

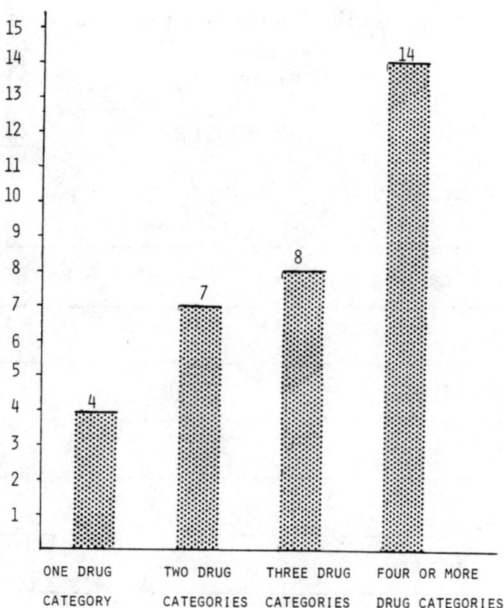

Figure 1. Extent of Abuse

Of the demographic variables analyzed, two proved significant at the $P < .01$ level:

1. the presence of an alcoholic parent
2. an unstable home during childhood (i.e., parents separated, divorced or physically abused their children).

Analysis for motivation of abuse conducted by 3 researchers utilizing a forced choice system, showed 8 patients to have abused for relief of anxiety, 9 for the reduction of internal psychic chaos and ego disruption, 4 to gain peer acceptance and 12 in association with an initially undiagnosed characterological disorder.

Ten of the 57 patients (18%) were found to have a learning disability, based on their placement in special education classes while in school. Prevelance of learning disability in the abuse population was 21% as compared to 13% in the non-abuse group.

Only 12% of the abuse group abused a single category of drug. The most common pattern (43%) was the abuse of 4 or more categories.

The incidence of acute psychotic episodes was more than twice as high (45%) in the abuse group as in the non-abusers.

DISCUSSION

It is well known that historical data furnished by psychiatric patients is often inaccurate and inadequate at best. Therefore, it is not surprising that substance abuse is frequently missed on initial psychiatric contact. Fifty-eight percent of our patients were found to have a significant abuse history even though they had been extensively prescreened to eliminate the admission of patients with this diagnosis.

Patterns of abuse revealed that no user of hallucinogens, amphetamines or opiates abused only that class of drug. Forty-two percent of the abuse population used 4 or more classes of drug while 12% abused only one class.

Thirty percent of the population abused prescription drugs, the most common being Valium, Demerol, "diet pills", "pain pills", "sleeping pills" and Marax. Twelve percent regularly abused over-the-counter drugs, the most frequently abused being Excedrin PM and Sominex, which contain antihistimines and/or belladonna alkaloids.

We believe that the patient's expectation of drug effect defined both the quantity and nature of their drug abuse. Patients who developed overwhelming anxiety symptoms, which they attempted to self-medicate through drug abuse, used only those agents that produced relief of anxiety and strenuously avoided taking any drugs that would worsen their condition. Patients who developed severe psychiatric symptoms, related to ego dysfunction and psychic turmoil, tended to abuse large quantities of any medication available to them, in an attempt to alter their state of internalized dysphoria. Those individuals who began their abuse of drugs as the result of peer pressure chose drugs that were abused by the peer group and, in general, their frequency of abuse was low. Patients with characterological disorders on the other hand, who abused drugs for their euphoric effects, tended to be highly selective in the agents that they abused.

Forty-eight percent of the patients in this study, who abused drugs, did so <u>following</u> the development of significant psychiatric symptoms. Their abuse was an effect rather than a cause of psychiatric dysfunction. Such abuse lengthened their hospital stay by 10%, made them twice as likely to sign out of the hospital AMA and made diagnosis of their initial psychiatric problems more difficult. The average patient experienced this first drug abuse at age 19 and had continued to abuse drugs for the following 7-1/2 years.

Ten patients in the total population were found to have learning disabilities. Of these 10, 9 had dropped out prior to the completion of high school and had, within the following three year period, manifested psychiatric symptomatology. It was noteworthy that the incidence of learning disabilities in the drug abusing population was 2 to 4 times that expected in the general population (Berlin, 1975). Careful inquiry revealed, with the exception of one patient with mild mental retardation, an absence of physical factors normally associated with learning disabilities such as: fetal anoxia, prematurity, sensory defects, chronic otitis media, lead poisoning and malnutrition. We believe, therefore, that the learning disabilities present in these patients were of a psychogenic nature. Analysis of these patient's family support systems during childhood substantiates this conclusion. There was a statistically significant difference between the rate of parental alcoholism or severe disruption in the home, as manifested by divorce, separation of the parents or clear cut physical abuse from one or both parents, between the abusing and non-abusing groups. In general, these patients gave a history of apathy, social withdrawal, academic underachievement, and the absence of childhood friends and normal patterns of socialization.

If psychiatric intervention is to be effective in severely psychotic patients who abuse drugs, the clinician must understand the patient's motivation for abuse. Twenty-four percent of our patients had specifically begun to abuse drugs to relieve severe anxiety and had had no history of drug abuse prior to the clear cut onset of psychiatric symptoms. In 27%, initial drug abuse was temporally associated with the development of symptoms of psychotic disorganization. In these cases drugs were either used to avert full blown psychotic manifestations or to prevent their further development. In 12% of the drug abusing patients, we felt that abuse was directly related to a need for peer acceptance. In these cases the drugs of abuse were those of the peer group and the patient gave a specific history of feeling socially inadequate and having a desire to "belong." The remaining 37% had some sort of character pathology and abused drugs for the pleasure they produced rather than the alleviation of social or intraphysic tension. These findings suggest that, at least in this selected inpatient population, the majority of severely impaired patients who abused drugs, did so as a <u>result</u> of their psychiatric illness. These findings, if confirmed by others, would suggest that clinicians need to carefully reevaluate their concept that drug abuse, in the main, produces psychosis in previously stable individuals or tips marginally compensated patients "over the edge."

CONCLUSIONS

On the basis of our data it seems reasonable to draw the following conclusions:

1. Substance abuse is a significant factor in a majority of the patients admitted to general psychiatric hospitals.

2. A history of an unstable childhood home and/or parental alcoholism predisposes to abuse.

3. The incidence of learning disability among psychiatric patients who abuse drugs is 2 to 4 times higher than that expected in the general population.

4. If abuse of one class of drug is illicited on intake, it is wise to evaluate for abuse of other classes as well.

5. AMA discharges are seen nearly three times as often in patients with a history of substance abuse.

6. Motivations for substance abuse differ and must be defined if treatment is to be effective. Such abuse may be directly related to the patient's attempt to modify internal turmoil, relieve anxiety, or enhance peer status as well as being the result of a characterological disorder.

7. In patients admitted to psychiatric hospital, substance abuse is at least as likely to follow the onset of a psychiatric disorder as it is to preceed it.

REFERENCES

Berlin, I.N. 1975. Psychiatry in the School. *Comprehensive Textbook of Psychiatry*. Baltimore, Md.: Williams and Wilkins. V.2., 2nd Ed., pp. 2251-2256.

Blumbert, A.G. et al. 1971. Covert Drug Abuse Among Voluntary Hospitalized Psychiatric Patients. *JAMA*. *217:* 1959-1961.

Bruteline, J.L. and Kinty, B.L. 1968. *Computational Handbook of Statistics*. Chicago, Il.: Scott, Foresman and Company. Section 5.2.

Butzer, S.C. and Cline, D.W. December 1974. Covert Drug Abuse Among Young Hospitalized Psychiatric Patients. *Minn. Med. 57:* 961-962.

Cohen, M. and Klein, D. 1970. Drug Abuse in a Young Psychiatric Population. *Amer. J. Orthopsychiat. 40:* 448-455.

Fischer, D.E. et al. 1975. Frequency and Patterns of Drug Abuse in Psychiatric Patients. *Dis. Nerv. Syst. 36:* 550-553.

Hall, R.C.W. et al. 1978 in press. Covert Outpatient Drug Abuse: Incidence, Therapist Recognition and Management. *J. of Nerv. and Ment. Dis.*

Spitzer, R.L., Endicott, J.M. and Robins, E. November 1975. *Research Diagnostic Criteria (R.D.C.) for a Selected Group of Functional Disorders*. New York: Biometrics Research, N.Y. State Psychiatric Institute, 2nd Ed., pp. 1-34.

Whitlock, F.A. et al. 1967. Drug Dependence in Psychiatric Patients. *Med. J. Aust. 1:* 1157-1166.

Westermeyer, J. and Walzer, V. 1975. Sociopathy and Drug Use in a Young Psychiatric Population. *Dis. Nerv. Syst. 36:* 673-677.

INCIDENCE OF CLINICAL DEPRESSION IN HEROIN ADDICTS*

Robert A. Steer, Ed.D., Maria Kovacs, Ph.D., and Aaron T. Beck, M.D.

West Philadelphia Community Mental Health Consortium, Inc.

Trying to estimate the incidence of clinical depression in heroin addicts seeking treatment has been hampered by different definitions as to what constitutes specific levels of depression. For example, Weissman et al. (1976) in studying 106 young methadone-maintenance patients at a Connecticut program found that approximately one third were moderately to severely depressed as assessed by the Raskin Depression Scale (Raskin) (Raskin et al., 1970) and the Hamilton Psychiatric Rating Scale for Depression (Hamilton) (Hamilton, 1960). Dorus and Senay (1977) had reported that approximately 25% of the 266 heroin addicts admitted for methadone maintenance treatment at an Illinois program were suffering with moderate to severe depressions; they used the Hamilton and the Beck Depression Inventory (Beck) (Beck et al., 1961). A number of other psychiatric and self-report instruments have been used to assess the level of clinical depression in heroin addicts and give varying estimates for the incidence of clinical depression (DeLeon et al., 1973; Fisch et al., 1975; Steer, 1977). Unfortunately, most of these studies do not employ the same scales and have drawn samples from different regions and socioeconomic strata.

The purpose of the present study was to assess the incidence of clinical depression in daily heroin users requesting methadone maintenance by using the Beck Depression Inventory (Beck), the Raskin Depression Scale (Raskin), and the Hamilton Psychiatric Rating Scale for Depression (Hamilton). These three instruments had been employed by Weissman et al. (1976) and Dorus and Senay (1977) and allowed for

---

* Supported by NIDA Grant No. 5-R01-DA01319-01.

ready comparison to their clinical populations. The study also sought to investigate whether or not the levels of depression which were identified could differentiate among the psychosocial characteristics of the heroin addicts.

METHOD

Subjects

The subjects were 206 daily heroin users who volunteered for this particular depression project after applying to the Drug Abuse Rehabilitation Program of the West Philadelphia Community Mental Health Consortium, Inc., for methadone maintenance therapy. The subjects were limited to daily heroin users, and none was displaying manifest signs of withdrawal at the time of admission.

There were 152 (73.8%) men and 54 (26.2%) women. The mean age was 27.61 (*SD* = 5.81) years, and the mean educational attainment was 11.00 (*SD* = 1.64) years. Employment was only reported by 13.1%, and 44.2% were single at the time of admission. The usual occupational level of the patients was low; 54.9% were classified as unskilled workers, whereas 45.1% could be classified as being semi-skilled or above. At admission, 15.5% of the addicts were living with other addicts, and 12.8% admitted to at least one past suicide attempt. The majority of the patients reported no legal involvement at the time of admission (70.4%), but the mean number of arrests was 5.92 (*SD* = 9.62).

The age at which daily heroin use had commenced was 20.49 (*SD* = 4.36) years, and 77.7% reported that they had been previously treated for drug abuse problems. Surprisingly, less than seven percent of these daily heroin users admitted to using other illicit drugs. The modal additional illicit drug being used was marijuana, but less than four percent reported that they were currently smoking it on a daily basis.

Instruments

As previously mentioned, the Beck, the Hamilton, and the Raskin were chosen as the three instruments for ascertaining the levels of depression; these instruments were chosen because they had been employed by previous researchers and allowed ready comparison with the results reported by these researchers for other heroin addict samples.

The Beck is a 21 item self-report inventory which covers a wide range of symptoms associated with depression; these symptoms cover affective, cognitive, motivational, and vegetative signs of depression. It had been used previously with heroin addict populations by DeLeon, Skodol, and Rosenthal (1973) and Dorus and Senay (1977).

The Hamilton is a 17-item scale developed to measure the severity of depression, and this scale has been used extensively in clinical studies of depression, particularly psychopharmacological trials. Both Weissman et al. (1976) and Dorus and Senay (1977) had used this scale for the assessment of depression in heroin addicts.

The second clinical rating scale for the assessment of depression was the Raskin. The Raskin is a five-item scale which measures the patient's self-report, behavior, and secondary symptoms of depression. It too has been used in a number of psychopharmacological trials and was used by Weissman et al. (1976) for the assessment of depression in heroin addicts.

The Raskin and Hamilton ratings were made by a trained clinician after a standard psychiatric interview which focused on various aspects of depressive symptomatology. A detailed set of background information was also gathered at this time, and the Beck was administered at the end of the clinical interview. The clinicians completing the Raskin and the Hamilton had established high levels of reliability in assessing depressive disorders as part of previous research in the mood clinic at the University of Pennsylvania where a wide variety of affective disorders were assessed daily.

RESULTS

Rather than relying on either the self-report or clinical ratings of depression alone for ascertaining the levels of depression in the heroin addicts, it was decided to use the standard cutoff scores for all three instruments to decide what the incidence of depression was within the heroin addicts. Three levels of depression were chosen - none (asymptomatic), mild, and moderate to severe depression. The persons with Beck scores $\geq$ 16, Raskin scores $\geq$ 7, and Hamilton scores $\geq$ 14 were considered to be moderately to severely depressed. Addicts with scores below all three of these respective instruments' cutoff scores were considered as not depressed; and persons with one or two scores either below or above these cutoff scores were considered to be mildly depressed. The present diagnostic scheme indicated that 10.7% were moderately to severely depressed, 27.7% were mildly depressed, and 61.6% were not depressed.

Since the Raskin, Beck, and Hamilton had been used previously with other heroin addicts, it was important to compare the current sample's scores with those described by Weissman et al. (1976) and Dorus and Senay (1977). The Kolmogorov-Smirnov two sample test (Siegel, 1956) was used to compare the present study's Raskin scores with those reported by Weissman et al. (1976), and $D = .16$ ($p < .10$).

The mean Raskin score for the present sample was 4.99 (SD = 2.14); the sample was mildly depressed.

The Beck had also been used by Dorus and Senay (1977), and their mean score was 14.84 (SD = 8.27) for 266 heroin addicts. The present sample's mean Beck score was 13.18 (SD = 9.35) for 206 heroin addicts; again, the mean Beck score indicated that the present sample was only mildly depressed. The difference between these two mean Beck scores was significant beyond the .05 level, two-tailed test ($t(475) = 2.05$). Dorus and Senay's (1977) sample was significantly more depressed than the present one according to the Beck.

The Hamilton had been used by both Dorus and Senay (1977) and Weissman et al. (1976). The means for both of these studies were 11.27 (SD = 6.82) and 17.47 (SD = 5.73), respectively. The mean Hamilton score for the present sample was 8.46 (SD = 5.11). The mean comparisons among all three samples indicated that Weissman et al.'s (1976) mean Hamilton score was significantly higher than Dorus and Senay's (1977) mean Hamilton score ($t(370) = 8.73$), $p < .001$); and Dorus and Senay's (1977) mean Hamilton score was, in turn, higher than the present sample's mean score ($t(470) = 5.20$, $p < .001$).

To ascertain whether or not the three levels of depression which had been derived by using simultaneous cutoff scores on the Raskin, Hamilton, and Beck would differentiate among the addicts' background characteristics, chi-square tests and oneway analyses of variance were used. Table 1 presents the classification of selected background characteristics by the three levels of depression - none, mild, and moderate to severe. Only those variables for which significant associations were found are reported. As Table 1 shows, a higher proportion of the moderately to severely depressed addicts lived with other addicts, were unemployed, and had made suicide attempts than the mildly and nondepressed addicts. They were also more likely than the other two groups to have made suicide attempts within the last three years, threatened suicide in advance, suffered a real or imagined personal loss within the six months prior to seeking treatment, and had been permanently separated from their mothers during childhood.

DISCUSSION

The present findings indicated that the levels of depression reported for this sample's heroin addicts were less than those reported by Dorus and Senay (1977) for Illinois addicts and by Weissman et al. (1976) for Connecticut addicts; these latter researchers had indicated that perhaps 20 to 35% of heroin addicts suffered from severe levels of depression, and the present results would suggest that only 10.7% of the addicts suffered from moderate to severe levels of depression. It may be assumed that geographical factors

Table I. Selected Background Characteristics by Levels of Depression

Characteristic	Level of Depression: None $\underline{n}$	Mild $\underline{n}$	Moderate/Severe $\underline{n}$	Total	$\chi^2(2)$
Living with Other Addict?					
Yes	14	7	11	32	
No	113	50	11	174	
Total	127	57	22	206	22.35***
Employment					
Unemployed	104	54	21	179	
Employed	23	3	1	27	
Total	127	57	22	206	7.29*
Previous Suicide Attempt?					
Yes	8	8	10	26	
No	117	48	12	177	
Total	125	56	22	203	25.70***
Time Since Last Attempt					
$\leq$ 3 years	3	3	10	16	
> 3 years	5	4	1	10	
Total	8	7	11	26	6.86*
Previous Suicide Threat					
Yes	11	13	14	38	
No	94	30	8	132	
Total	105	43	22	170	31.86***

Table I. Continued

Characteristic	Level of Depression None n	Mild n	Moderate/Severe n	Total	$\chi^2(2)$
Patient Suffered Loss in Last 6 Months?					
Yes	35	23	16	74	
No	78	28	5	111	
Total	113	51	21	185	15.85***
Separation From Mother During Childhood?					
Yes	97	41	13	151	
No	16	9	9	34	
Total	113	52	22	185	9.08**

*$p < .05$
**$p < .01$
***$p < .001$

could be associated with the incidence of depression for a particular addict population. However, the present study employed a stringent definition of depression based on three cutoff scores, and the incidence reported here may be overly conservative.

The validity of the present diagnostic scheme for differentiating among the addicts was shown by its associations with several important background characteristics. The moderate to severely depressed addicts were more likely to be living with other addicts, were more likely to be unemployed, and had attempted suicide in the past; these addicts attempted suicide more recently than the mildly and nondepressed addicts. The moderately to severely depressed patients had also threatened suicide more frequently in advance and had suffered a real or imagined personal loss more often within the last six months than the mildly and nondepressed addicts. The moderately to severely depressed addicts described themselves as having been more frequently separated from their mothers during childhood than the mildly and nondepressed addicts.

Since the more depressed addicts described previous suicide attempts and past attempts have been shown to be related to subsequent attempts (Beck et al., 1974), the moderately to severely depressed addicts should be closely watched for signs of suicide potential. The finding that the moderately to severely depressed addicts were unemployed and that they were living with other addicts suggested that such addicts had few societal supports which might dissuade them against suicide. It is important to note that these moderately to severely depressed persons had also experienced recent losses as well as having suffered from maternal separation during childhood. The moderate to severely depressed addict certainly needs psychiatric help, and strenuous efforts should be made to ascertain what the nature of the immediate loss had been and whether or not any cognitive distortions might be involved.

REFERENCES

Beck, A.T., Resnick, H.L.P. and Lettieri (Eds.). 1974. *The Prediction of Suicide*. Bowie, Maryland: Charles Press.

Beck, A.T., Ward, C.H. and Mendelson, M. 1961. An Inventory for Measuring Depression. *Archives of General Psychiatry, 4:* 561-571.

DeLeon, G., Skodol, A., and Rosenthal, M.S. 1973. Phoenix House: Changes in Psychopathological Signs of Resident Addicts. *Archives of General Psychiatry, 28:* 131-135.

Dorus, W., and Senay, E.C. May, 1977. Severity and Course of Depression in Opiate-dependent Patients. Paper presented to the Fourth National Drug Abuse Conference, San Francisco.

Fisch, A., Patch, V.D., Greenfield, A., Raynes, A.E., McKenna, G., and Levine, M. 1973. Depression and Self-concept as Variables in the Differential Response to Methadone Maintenance Combined With Therapy. Proceedings of the Fifth National Conference on Methadone Maintenance Treatment. New York: *NAPAN*, 440-445.

Hamilton, M. 1960. A Rating Scale for Depression. *Journal of Neurology and Neurosurgical Psychiatry, 23:* 56-62.

Raskin, A., Schulterbrandt, J.G., Reatig, N., McKeon, J.J. 1970. Differential Response to Chlorpromazine, Imipramine, and Placebo: A Study of Subgroups of Hospitalized Depressed Patients. *Archives of General Psychiatry, 23:* 164-173.

Siegel, S. 1956. *Nonparametric Statistics for the Behavioral Sciences.* New York: McGraw-Hill.

Steer, R.A. 1977. Affect Dimensions of Heroin Addicts and Methadone Patients. *JSAS Catalog of Selected Documents in Psychology, 7:* 78 (Ms. No. 1536).

Weissman, M.M., Slobetz, F., Prusoff, B., Mezritz, M., and Howard, P. 1976. Clinical Depression Among Narcotic Addicts Maintained on Methadone in the Community. *American Journal of Psychiatry, 133:* 1434-1438.

EVALUATION OF LABORATORY ABNORMALITIES IN AN ADDICT POPULATION: A PATIENT PROFILE

Bruce D. Miller, M.D. and Ahmed Nakah, M.D.

Denver, Colorado

INTRODUCTION

Most of the studies on the medical and laboratory abnormalities found in heroin addicts have focused on specific, predictably abnormal parameters and attempted to quantitate the degree and significance of such laboratory abnormalities. The majority of such investigations have involved relatively small numbers of patients who are most often taken from selected populations, i.e., hospital clinics, in-patient settings, city morgues, etc.[1-8] The usefulness of these studies in predicting or describing general trends in heroin addicts is somewhat limited by their selectivity.

The drug treatment program in Newark, N.J., which is under the purview of the Dept. of Preventive Medicine and Community Health of the New Jersey Medical School, offers a unique setting for the study of a relatively unselected heroin addict population. The Central Intake and Referral Unit (CIRU), which is the hub of the entire city drug treatment program, provides a comprehensive intake and referral process for any drug addict voluntarily seeking or otherwise referred for drug treatment throughout the greater Newark area. Included as a routine part of the intake is a complete medical screening for all new patients entering the program. This includes a thorough medical history and physical examination, and the following laboratory examinations: SMA-18; serum electrolytes; complete blood count; sickle cell preparation; venereal disease serology; australia antigen; urine analysis for opiates, amphetamines and barbiturates; chest x-ray and electrocardiogram (EKG). The results of these tests are recorded on the patient's permanent record and are used in conjunction with clinical information in making appropriate medical referrals whenever indicated.

In the present retrospective study we have analyzed the laboratory data (excluding the chest x-rays and EKG's) of all patients who had entered into treatment from the inception of the program in July, 1974, through August, 1975, in order to determine any significant abnormal trends.

PATIENTS AND METHODS

Although no attempt was made to group patients demographically, clinical observation enables us to make certain generalizations concerning our patient population. By far the greatest percentage of the patients in this study were blacks coming from the heart of the Newark urban ghetto. The next most prevalent group were Hispanic and the remainder consisted primarily of whites of various ethnic backgrounds. In almost all cases, these patients were abusing heroin parenterally at the time of entry into the program. The source of heroin was primarily from New York City. Many of the patients reported traveling daily to New York City in order to obtain a supply of their drug.

Patients often used illicit methadone in lieu of heroin or in conjunction with heroin and/or pills in order to augment the high and prolong the effect of the drugs. Diazepam, chlordiazepoxide, amitriptyline and various other sedatives and tranquilizers were used liberally. It is difficult to determine the incidence and degree of alcoholism in the study group, but heavy alcohol use was rarely the major problem for these patients. Marijuana was used regularly among our patients, but potent psychedelic agents, such as LSD, were used very infrequently. Amphetamine use was also infrequent among this patient group.

As previously mentioned, this study was conducted as a review of laboratory records of all patients who entered treatment for drug abuse. The laboratory records consisted of a computerized print-out sheet with each laboratory function reported numerically followed by the control or normal range for that specific test. For each patient record, 43 individual tests were reviewed and any test falling outside the indicated normal range was scored qualitatively as high (↑), low (↓), or positive (+) accordingly. Results were then translated into the percentage of patients having abnormalities for each of the 43 laboratory functions studied. The only patient variables considered were age and sex, and for each sex the following age categories were used: ≤ 20 years; 21-30 years; 31-40 years; ≥ 40 years.

All laboratory tests were performed by a well-recognized, fully licensed clinical laboratory. The normal values were obtained by sampling 1,000 clinically well patients undergoing routine physical examination and laboratory evaluation and then taking the centralized 95% of reported values as the normal or control range. We found these values to be in agreement with generally accepted control ranges and they were used as the reference guides from which the qualitative analyses were made.

RESULTS

The results in Table I show the percentage of patients having abnormally high or low values in each of the 23 chemistries studied. Those laboratory functions for which more than 5% of the total number of patients had abnormally high or low values are included with an arrow indicating the direction of the abnormality. In 9 out of the 23 blood chemistries studied, the results showed that greater than 5% of all patients studied had values above or below the normal range as follows: glucose = ↑5.2%; total protein = ↑17.5%; albumin/globulin ratio = ↓9.5%; direct bilirubin = ↑6.1%; serum glutamic oxalacetic transaminase (SGOT) = ↑34.0%; serum glutamic pyruvic transaminase (SGPT) = ↑36.5%; alkaline phosphatase = ↑9.4%; beta lipids = ↓9.8%; gamma glutamyl transpeptidase (GGPT) = ↑27.6%. Of these 9 chemical constituents of blood found to be frequently altered in addicts, 5 are direct indicators of liver function (Bilirubin, SGOT, SGPT, Alkaline Phosphatase and GGTP). Of the remaining 14 chemical constituents measured in our patients, 7 (blood urea nitrogen, uric acid, total bilirubin, lactic dehydrogenase, cholesterol, sodium and potassium) showed greater than 5% abnormalities in one or more patient age/sex grouping, although the total population showed overall less than 5% abnormalities.

Table II summarizes the abnormalities found in the hematologic profile among heroin addicts. In 7.8% of the patients, white blood counts were lower than the accepted 4.0 x 1,000/CuMM; 10.9% of the patients had decreased percentage of polymorphonuclears and 5.5% had an increased percentage of lymphocytes in the differential count. In 5.2% of male patients > 40 years of age, there were abnormally high eosinophil counts, although the overall patient population showed only 1.7% (statistically insignificant) abnormally high eosinophils. The hemoglobin was abnormally low in 9% of female patients, and only 0.5% (statistically insignificant) of males. Hematocrit was abnormally low in 22.5% of females and only 1.4% (statistically insignificant) of males. Red blood counts were decreased in 17.4% of female patients and in 0.7% (statistically insignificant) of males.

Table III shows the patient abnormalities on routine urinalysis. 10% of patients had greater than trace ketones in the urine and 13% of patients had greater than trace albuminurea.

Table IV shows the results of tests for sickle hemoglobin, venereal disease serology and australia antigen. 6.8% of patients had positive serologies, and in 3% australia antigen was positive. The sickle cell trait appeared in 7% of patients.

Table I. Blood Chemistry Determinations in Addicts
(measured as percentage of patients with abnormal lab findings)

TEST (NORMAL RANGE)	Men Age (Years)				Women Age (Years				TOTAL		
	$\leq$20	21-30	31-40	>40	$\leq$20	21-30	31-40	40	M	F	M/F
Number of Cases	97	461	108	19	35	107	12	1	685	155	840
Calcium (8.5-11.0 MG/DL)									*	*	*
Phosphorus (2.0-4.8 MG/DL)									*	*	*
Blood Urea Nitrogen (5.0-25.0 MG/DL)							↓8.3		*	*	*
Uric Acid (2.0-8.0 MG/DL)				↑10.5					*	*	*
Glucose (Less than 120 MG/DL)		↑5.0	↑6.5	↑15.8	↑8.6		↑8.3		↑5.1	↑5.8	↑5.2
Total Protein (6.0-8.2 MG/DL)	↑17.5	↑17.7	↑17.5	↑36.8	↑5.7	↑17.7	↑8.3	↑100.0	↑18.2	↑14.8	↑17.5
Albumin (3.4-6.0 GM/DL)							↓8.3		*	*	*
Albumin: Globulin (More than 1.0)	↓6.1	↓7.8	↓13.8	↓5.2	↓8.5	↓15.8	↓8.3	↓100.0	↓8.4	↓14.1	↓9.5

TEST (NORMAL RANGE)	Men Age (Years) ≤20	21–30	31–40	>40	Women Age (Years) ≤20	21–30	31–40	>40	TOTAL M	F	M/F
Total Bilirubin (Less than 1.6 MG/DL)				↑5.2					*	*	*
Direct Bilirubin (Less than .3 MG/DL)	↑6.1	↑6.2	↑9.2	↑5.2					↑6.7	*	↑6.1
SGOT (Less than 50 I.U./L	↑37.1	↑35.5	↑37.9	↑26.3	↑25.7	↑27.1	↑8.3	↑100.0	↑35.9	↑25.8	↑34.0
SGPT (Less than 50 I.U./L)	↑43.2	↑39.0	↑35.1	↑21.0	↑42.8	↑23.3	↑16.6	↑100.0	↑38.5	↑27.7	↑36.5
Alkaline phosphatase (Less than 40 I.U./L)	↑18.5	↑7.8	↑11.1	↑5.2	↑10.2				↑9.7	↑8.3	↑9.4
LDH/(Less than 225 units)	↑6.1			↑5.2					*	*	*
Amylase (less than 180 units)									*	*	*
Cholesterol (120–290 MG/DL)	↓14.4								↓5.5	*	*
Beta Lipids (Less than 15 units)	↓8.2	↓10.6	↓6.5	↓5.3	↓8.6	↓10.3	↓25.0		↓9.5	↓11.0	↓9.8

Table I (Cont.). Blood Chemistry Determination in Addicts
(measured as percentage of patients with abnormal lab findings)

TEST(NORMAL RANGE)	Men Age (Years				Women Age (Years)				TOTAL		
	≤20	21-30	31-40	>40	≤20	21-30	31-40	>40	M	F	M/F
Total Lipids (0.3-1.0 GM/DL)									*	*	*
Sodium (134-145 MEQ/L)			↑6.5	↑10.5	↑5.7	↑5.6	↑8.3		*	↑7.1	*
Potassium (3.2-5.1 MEQ/L)				↓5.3	↓5.7	↓5.6			*	↓5.2	*
Chloride (90-110 MEQ/L)									*	*	*
GGPT (Less than 40 U/L)	↑20.6	↑29.0	↑43.5	↑36.8	↑17.1	↑14.9	↑16.6	↑100.0	↑30.3	↑16.1	↑27.6
Creatinine (0.1-1.5 MG/DL)				↑5.3					*	*	*

* Note: All blank spaces indicate values less than 5% of patient outside normal range.

Table II. Hematologic Determinations in Addicts by Age and Sex (measured as percentage of patients with abnormal lab findings)

TEST (NORMAL)	Men Age (Years) ≤20	21-30	31-40	>40	Women Age (Years) ≤ 20	21-30	31-40	>40	TOTAL M	F	M/F
Number of Cases	97	461	108	19	35	107	12	1	685	155	840
Red Blood Cells (4.4-6.2x1,000,000/cuMM)					↓17.1	↓16.8	↑8.3 / ↓25.0		*	↓17.4	*
Hemoglobin (14-18 GM%)				↓5.2	↓11.4	↓8.4	↓8.3		*	↓9.0	*
Hematocrit (42-52%)				↓5.2	↓25.7	↓23.3	↓8.3		*	↓22.5	*
Mean Corpuscular Hemoglobin concentration (31-36%)		↓13.0		↓5.2		↓6.5	↓8.3		*	↓5.8	*
White Blood Count 4.0-12.0x1,000/cuMM)	↓13.4	↑5.2	↓5.5	↓5.2	↓8.5	↓11.2	↓8.3		↓6.2	↓10.3	↓7.8
Polys (55-80%)	↓10.3	↓10.8	↓12.0	↑5.2 / ↓10.5		↓14.9			↓10.9	↓10.9	↓10.9
Stabs (0-5%)									*	*	*
Lymphs (20-45%)	↑5.1		↑5.5	↑15.7		↑8.4			↑5.4	↑5.8	↑5.5
Monos (0-8%)									*	*	*
Eosinophils (0-5%)				↑5.2					*	*	*
Basophils (0-1%)									*	*	*

*Note: All blank spaces indicate values less than 5% of patients outside of the normal range.

Table III. Urinalysis Findings in Addicts by Age and Sex
(measured as percentage of patients with abnormal lab findings)

TEST (NORMAL RANGE)	Men Age (Years)				Women Age (Years)				TOTAL		
	≤20	21-30	31-40	>40	≤20	21-30	31-40	>40	M	F	M/F
Number of Cases	94	457	114	23	35	143	22	0	688	200	888
Acetone (Neg or Trace)	+7.4	+11.3	+7.8	+8.6	+11.4	+9.7			+10.3	+9.5	+10.0
Albumin (Neg or Trace)	+6.3	+14.8	+12.2	+13.0	+11.4	+11.8	+18.0		+13.2	+12.5	+13.0
Bilirubin (Neg or Trace)*									*	*	*
Blood (Neg or Trace)*									*	*	*
Glucose (Neg or Trace)*									*	*	*

* Note: All blank spaces indicate values less than 5% of patients outside of normal range.

Table IV. Miscellaneous Lab Findings in Addicts by Age and Sex (measured as percentage of patients with abnormal lab findings)

TEST (NORMAL RANGE)	Men Age (Years)				Women Age (Years)				TOTAL		
	≤20	21-30	31-40	>40	≤20	21-30	31-40	>40	M	F	M/F
Number of Cases	97	461	108	19	35	107	12	1	685	155	840
Hemoglobin (neg)	+8.2	+7.5	+7.4		+2.8	+7.4			+7.4	+5.8	+7.0
Serology/Non-Reac	+5.1	+5.6	+12.0	+15.7	+5.7	+7.4			+6.8	+6.4	+6.8
Australian Antigen (Neg)	+3.0	+3.4	+2.7		+2.8	+0.9			+3.2	+1.2	+3.0

DISCUSSION

It is well-known that laboratory analysis in asymptomatic drug addicts often reveals abnormalities in liver function. There has been much controversy over the mechanisms and etiology of this hepatic dysfunction as well as the frequency with which it occurs. Marks and Chapple (1966) reported elevations of SGPT in 69% and SGOT in 61% of patients studied. Cushman and Grieco (1973) reported SGOT elevations in 65% of untreated addicts screened for liver function abnormalities. Both groups found alkaline phosphatase to be elevated in 28% of addicts. In contrast to these results, a report by Cherubin et al. (1972) indicates that SGOT values were elevated in 25% and SGPT values in 41% of addicts entering treatment. Our results showed a lower incidence of enzyme elevations than those reported by Marks and Chapple or Cushman and Grieco. Our overall averages were 24%, 36.5% and 9.4% for SGOT, SGPT and alkaline phosphatase, respectively. Even when considering our results according to age and sex sub-groups, the highest frequency of elevation was for males less than 20 years of age and the percentages were still considerably lower than those reported by either Marks and Chapple or Cushman and Grieco (see Table I). Our results are more concordant with those of Cherubin, et al. (1972).

In addition to using the transaminases and alkaline phosphatase to assess liver function, we used gamma-glutamyl transpeptidase (GGTP) as an indicator of liver dysfunction. This test has been useful not only as an indicator of generalized liver disease, but also in differentiating between liver disease due to biliary obstruction and that due to viral hepatitis, GGPT being disproportionately more active in relation to the transaminases in cases of intra- or extra-hepatic biliary obstruction and less active in cases of viral hepatitis (Lum and Gambino, 1972). Keeping in mind this distinction, it is interesting to note that GGPT was less frequently elevated relative to transaminases in patients of both sexes under 31 years of age. These findings suggest that the type of hepatic dysfunction found in younger addicts (with shorter duration of drug use) may be quite different from that found in older, more chronic users. These differences may be due to duration of drug use or to complicating factors such as chronic infection, excess alcohol use, nutritional deficiencies or combinations of these factors. Although the etiologic mechanism for liver dysfunction in heroin addicts is not fully understood, and there are indeed discrepancies concerning the percentage of patients with liver abnormalities according to which study is being read, it is clearly evident that heroin addicts are at risk and must be carefully looked at long term for the development of cirrhosis.

The decreased Beta-lipoprotein (low density lipoprotein) fraction of the serum is most likely related to the eating habits in this patient group. Clinical observation revealed that most of our patients were lean if not outright malnourished. Their diets often consisted of foods rich in carbohydrates (sugar, candy and soda), and these foods provided the most significant component of their caloric intake. We have not seen mention of this tendency toward lower Beta-lipoprotein levels in drug addicts and we feel that this finding merits further investigation.

Our results showed a high percentage of patients with increased total protein in the serum, a decrease in the albumin/globulin ratio, and no increase in albumin. These findings suggest an increase in the globulin component. This has been carefully documented by Cushman and Grieco (1973) in their study of hyperimmunoglobulinemia associated with narcotic addiction. Although they speculated on the causes for these immunoglobulin changes in addicts, they were unable to define the specific etiology. This is another area in which further investigation is warranted.

In reviewing the hematologic findings, some interesting observations can be made. The lack of anemia, as measured by hemoglobin and hematocrit, in males is stricking, especially in view of the poor nutritional status generally found in these patients. This is in contrast to a variable amount of anemia found in female patients. Schoenfeld and Samula (1966) have investigated the hematologic findings in greater detail and have suggested there is a tendency to erythrocytosis in heroin addicts, and postulated chronic hypoxia secondary to respiratory depression as one possible mechanism to account for this finding. They related the variable anemia found in female drug addicts to different menstrual patterns in these patients in combination with nutritional deficiencies.

It is interesting to note that the WBC is below the normal range in 6.2% of men and 10.3% of women. This is accompanied by an increased number of patients having a higher percentage of lymphocytes and a lower percentage of polymorphonuclears. Although no conclusive results can be derived from these data, it can be speculated that chronic viral infection (sub-clinical) plus antigenic impurities introduced into the blood stream during I-V injection of drugs may explain both the leukopenia and the reversal in the differential pattern of lymphocytes and polymorphonuclear cells.

The positive venereal disease serology found in approximately 6.8% of patients in this study may result from syphilitic infection or may represent the biologic false positives often found in the addict population (Cherubin and Millan, 1968). In those with positive venereal disease tests, the most sensible approach is to obtain a confirmatory fluorescent antibody test and proceed with treatment accordingly.

Australia antigen was present in 3% of patients tested in this study, a figure slightly higher than that found by Cherubin et al.

(1970), who reported 2.2% antigenemia in addicts without clinical evidence of hepatitis. These figures can be contrasted with blood donors who show 0.1% to 1.5% antigenemia. The wide discrepancy in these control figures is probably due to the control samples, i.e., volunteer donors on the one hand versus commercial donors on the other. It is obvious from this study and other studies that addicts do have increased frequency of australia antigenemia.

Consideration of the urinalyses in this study shows two interesting findings. There was a ketonuria in approximately 10% of patients and albuminuria in 13%. Given the dietary habits of these patients, nutritional deprivation probably accounts for the ketonuria without concomitant glycosuria. There is no simple explanation for the albuminuria. Treser, et al. (1974), studied renal lesions and their correlation with clinical and laboratory evidence of renal disease in drug addicts. They too noted proteinuria among the addicts studied. However, they were unable to find a relationship between drug use and the development of specific clinical disease or histopathologic renal lesions. They also found several addicts who manifested the nephrotic syndrome, an infrequent occurrence apparently due to an immunologic reaction. Our own findings, however would not indicate a greater frequency of nephrotic syndrome in addicts based on laboratory findings in this group.

In summary, the usefulness of the results presented herein are as follows: They can serve as a baseline or reference point for the types and frequencies of laboratory abnormalities which may be expected in a general addict population, and can serve as an aid in the clinical assessment and treatment of these patients. Furthermore, these data supplement and add to the fund of knowledge on which both clinical management and investigative endeavors are founded. And finally, this study will hopefully encourage further research into areas not yet adequately explored or understood in connection with the medical complications and treatment of patients with drug addictions.

REFERENCES

Treser, G., Cherubin, C., et al. 1974. Renal Lesions in Narcotic Addicts. *Amer. J. Med., 57:* 687.

Cushman, P., Jr., Grieco, M.H. 1973. Hyperimmunoglobulinemia Associated with Narcotic Addiction. *Amer. J. Med., 54:* 320.

Gorodetzky, C.W., Sapira, J.D., et al. 1968. Liver Disease in Narcotic Addicts. *Clin. Pharmacol. and Therapeut., 9(6):* 720.

Cherubin, C.E., et al. 1972. Chronic Liver Disease in Asymptomatic Narcotic Addicts. *Ann. Intern. Med., 76:* (3)391.

Cherubin, C.E., et al. 1972. Persistence of Transaminase Abnormalities in Former Drug Addicts. *Ann. Intern. Med., 76(3):* 385.

Cherubin, C.E., Hargrove, R.L. and Prince, A.M. 1970. The Serum Hepatitis Related Antigen (SH) in Illicit Drug Users. *Amer. J. Epidemiol., 91(5):* 510.

Hunter, J., et al. 1971. The Australia (Hepatitis Associated) Antigen Amongst Heroin Addicts Attend-ng a London Addiction Clinic. *J. Hyg., Camb. 61:* 565.

Marks, V. and Chapple, P.A.L. 1966. Heptic Dysfunction in Heroin and Cocaine Users. *Brit. J. Addict., 62:* 189.

Schoenfeld, M.R. and Samala, R. 1966. Absence of Anemia in Hepatitis Due to Heroin Addiction: Evidence for Tendency to Erythrocytosis Among Heroin Addicts. *J. New Drugs, 6:* 149.

Degroote, J., et al. 1968. A Classification of Chronic Hepatitis. *Lancet,* 626-629.

Sapira, J.D., Jasinski, D.R. and Gorodetzky, C.W. 1968. Liver Disease in Narcotic Addicts. II. The Role of the Needle. *Clin. Pharmacol. and Therapeut., 9(6):* 725.

Lum, G. and Gambino, S.R. 1972. Serum Gamma Glutamyl Transpeptidase Activity as an Indicator of Disease of Liver, Pancreas or Bone. *Clin. Chem., 18(4):* 358.

Louria, D.B., Hensle, T. and Rose, J. 1967. The Major Medical Complications of Heroin Addiction. *Ann. Intern. Med., 67(1):*

Cherubin, C.E., Millan, S.J. 1968. Serological Investigations in Narcotic Addicts. *Ann. Intern. Med., 64:* 15.

INDEX

Randall Library – UNCW
RC564 .N37 1978 v.1 NXWW
National Dru / Drug dependence and alcoholism
3049002599918